Lecture Notes in Artificial Intelligence **13014**

Subseries of Lecture Notes in Computer Science

More information about this subseries at http://www.springer.com/series/1244

Xin-Jun Liu · Zhenguo Nie · Jingjun Yu ·
Fugui Xie · Rui Song (Eds.)

Intelligent Robotics and Applications

14th International Conference, ICIRA 2021
Yantai, China, October 22–25, 2021
Proceedings, Part II

Editors
Xin-Jun Liu
Tsinghua University
Beijing, China

Zhenguo Nie
Tsinghua University
Beijing, China

Jingjun Yu
Beihang University
Beijing, China

Fugui Xie
Tsinghua University
Beijing, China

Rui Song
Shandong University
Shandong, China

ISSN 0302-9743 ISSN 1611-3349 (electronic)
Lecture Notes in Artificial Intelligence
ISBN 978-3-030-89097-1 ISBN 978-3-030-89098-8 (eBook)
https://doi.org/10.1007/978-3-030-89098-8

LNCS Sublibrary: SL7 – Artificial Intelligence

This Springer imprint is published by the registered company Springer Nature Switzerland AG
The registered company address is: Gewerbestrasse 11, 6330 Cham, Switzerland

Preface

With the theme "Make Robots Infinitely Possible", the 14th International Conference on Intelligent Robotics and Applications (ICIRA 2021) was held in Yantai, China, during October 22–25, 2021, and designed to encourage advancement in the field of robotics, automation, mechatronics, and applications. The ICIRA series aims to promote top-level research and globalize quality research in general, making discussions and presentations more internationally competitive and focusing on the latest outstanding achievements, future trends, and demands.

ICIRA 2021 was organized by Tsinghua University, co-organized by Beihang University, Shandong University, YEDA, Yantai University, and IFToMM China-Beijing, undertaken by the Tsingke+ Research Institute, and technically co-sponsored by Springer. On this occasion, three distinguished plenary speakers and 10 keynote speakers delivered their outstanding research works in various fields of robotics. Participants gave a total of 186 oral presentations and 115 poster presentations, enjoying this excellent opportunity to share their latest research findings.

The ICIRA 2021 proceedings cover over 17 research topics, with a total of 299 papers selected for publication in four volumes of Springer's Lecture Note in Artificial Intelligence. Here we would like to express our sincere appreciation to all the authors, participants, and distinguished plenary and keynote speakers. Special thanks are also extended to all members of the Organizing Committee, all reviewers for peer review, all staff of the conference affairs group, and all volunteers for their diligent work.

October 2021

Xin-Jun Liu
Zhenguo Nie
Jingjun Yu
Fugui Xie
Rui Song

Organization

Honorary Chair

Youlun Xiong	Huazhong University of Science and Technology, China

General Chair

Xin-Jun Liu	Tsinghua University, China

General Co-chairs

Rui Song	Shandong University, China
Zengguang Hou	Institute of Automation, CAS, China
Qinchuan Li	Zhejiang Sci-Tech University, China
Qinning Wang	Peking University, China
Huichan Zhao	Tsinghua University, China
Jangmyung Lee	Pusan National University, South Korea

Program Chair

Jingjun Yu	Beihang University, China

Program Co-chairs

Xin Ma	Shandong University, China
Fugui Xie	Tsinghua University, China
Wenguang Yang	Yantai University, China
Bo Tao	Huazhong University of Science and Technology, China
Xuguang Lan	Xi'an Jiatong University, China
Naoyuki Kubota	Tokyo Metropolitan University, Japan
Ling Zhao	Yantai YEDA, China

Publication Chair

Zhenguo Nie	Tsinghua University, China

Award Chair

Limin Zhu	Shanghai Jiao Tong University, China

Advisory Committee

Contents – Part II

Human-Robot Collaboration

Hybrid System Modeling and Human-Machine Interface

Generation of Collision-Free Five-Axis Sweep Scanning Path Using Graphics Engine

Yijun Shen, Li Hua, Wenze Zhang, Yang Zhang(✉), and Limin Zhu

School of Mechanical Engineering, Shanghai Jiao Tong University, No. 800 Dongchuan Road, Minhang District, Shanghai 200240, China
meyzhang@sjtu.edu.cn

Abstract. Five-axis sweep scanning is an emerging technology for freeform surface inspection. This new inspection manner greatly improves measurement efficiency and reduces the dynamic errors. However, the existing five-axis scanning path planning methods ignore the obstacle avoidance issue during sweep scanning, which may cause damage to the probe stylus if any collision occurs. In this paper, the collision detection is performed by a graphics engine HOOPS. An incremental algorithm is proposed to reduce the computation of the collision detection. A collision-free sweep scanning path for a freeform surface patch can be calculated within 30 s. In real experiments, the tangential velocity and acceleration in the proposed five-axis sweep scanning path are only 39.3% and 20.1% of those in the traditional three-axis scanning, which manifests that the sweep scanning path helps to reduce the kinematic loads and the dynamic error of the inspection machine.

Keywords: Five-axis sweep scanning · Collision detection · Path planning · Graphics engine · CMM

1 Introduction

Coordinate measuring machines (CMM) are the most commonly used instrumentations for precise dimensional measurement of industrial products. With the rising demand for a shorter time of the measurement cycles, continuous scanning rather than point-by-point probing is more preferred in the CMM inspection. However, when the scanning speed gets higher, the dynamic error of the CMM will become more serious [1]. When the CMM works at high speed, the heavy CMM gantry causes considerable inertia load and geometric distortion because of its large inertia and low stiffness structures [2]. Currently, several dynamic error compensation methods have been proposed [3–5], while this problem has not been thoroughly solved. Thus, in practical three-axis CMM scanning inspection, the scanning speed is usually limited, i.e. less than 25 mm/s, which greatly reduces the inspection efficiency.

Five-axis inspection is an emerging technology to perform accurate and efficient inspection on parts with free-form surfaces. A typical five-axis inspection machine is composed of a traditional three-axis CMM and a two-axis rotary probe head, as shown

X.-J. Liu et al. (Eds.): ICIRA 2021, LNAI 13014, pp. 3–13, 2021.
https://doi.org/10.1007/978-3-030-89098-8_1

in Fig. 1. The stylus is so light that it can swing quickly around the pivot point, i.e. the intersection of the two rotary axes, to accomplish the majority of the scanning work. This kind of inspection manner is named as "sweep scanning". Different from the traditional three-axis scanning, five-axis sweep scanning maximizes the usage of the stylus, while keeping the heavy CMM gantry move smoothly and slowly. In this way, the dynamic error of the CMM gantry is reduced and the measurement efficiency is improved.

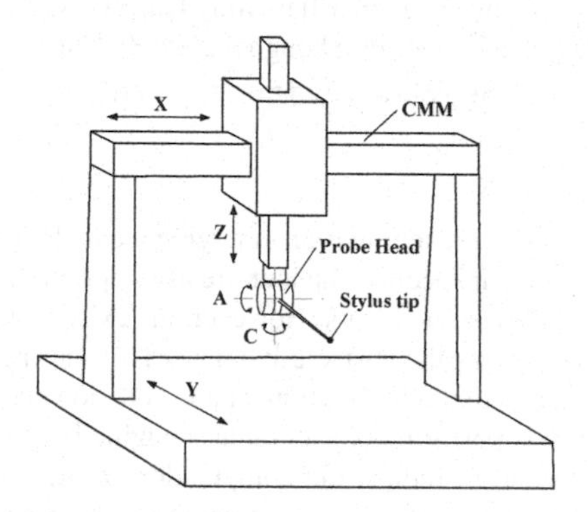

Fig. 1. Typical structure of a five-axis inspection machine.

The novel inspection manner improves the efficiency but heightens the difficulty of path planning. The most suitable sweep scanning path should be a smooth probe head trajectory and an oscillating probe tip trajectory, which caters to the kinematic performance of the five-axis inspection machine. As an emerging technology, the existing researches on path planning for five-axis sweep scanning is really scarce. Hu et al. [6] proposed a semi-automatic method to generate a five-axis sweep scanning path on a relatively flat surface. Zhang et al. [7] proposed a method to subdivide the freeform surface into elementary shapes and generate a sweep scanning path on each of them. Chen et al. [8] proposed a new strategy to partition the surface into a single spiral strip. A continuous sweep scanning curve is generated within the strip area. The existing researches main focus on surface partitioning and planning strategy. However, they ignored the collision-free condition during the sweep scanning. As collision detection is considered to be time-consuming in the five-axis scanning path planning, they assumed that there is no obstacle existing in the measurement environment. While in the practical inspection, the fixtures or the measured object itself may become obstacles for the sweep scanning.

This paper focuses on the collision avoidance issue in the five-axis sweep scanning and proposes an efficient method to calculate the collision-free area using a graphics engine. First, the collision-free condition at a specific stylus orientation is checked with the collision detection function of the graphics engine. Then the whole collision-free area is generated efficiently with an incremental algorithm. Compared to the existing five-axis sweep scanning path planning methods, the proposed method has the following advantages:

1. The graphics engine is applied to perform accurate and efficient collision detection for five-axis sweep scanning paths.
2. An incremental algorithm is proposed to reduce the computation of collision detection and improves the efficiency of path planning.

The remainder of this paper is organized as follows. Section 2 introduces some basic concepts in five-axis scanning path planning. In Sect. 3, the generation of the collision-free area with a graphics engine is explained in detail. In Sect. 4, the performance of the proposed path planning method is verified through real experiments. In Sect. 5, the superiority and limitations of the proposed method are discussed. Finally, conclusions are drawn in Sect. 5.

2 Preliminary

In this section, the common procedures of generating a continuous five-axis sweep scanning path are briefly introduced. For a given surface S_p, a five-axis sweep scanning path is uniquely defined by the trajectory curve (TC) of the probe head (driven by the CMM gantry) and the sweeping curve (SC) of the probe tip (driven by the stylus).

A diagram of the five-axis sweeping path planning is shown in Fig. 2. First, a guiding curve $G(t)$ is extracted on S_p, which is used to approximate the topology of the surface. Usually, the medial axis [9] or the iso-parametric curve [10] is used as the guiding curve. In practice, the guiding curve is discretized into n_p points, i.e. $G(t) = G_i|_{i=1,2,\ldots,n_p}$. By establishing a local coordinate system (LCS) at G_i, the stylus orientation in the LCS can be represented by a tilt angle ω and a yaw angle φ as

$$\mathbf{v} = \begin{bmatrix} \cos\varphi \cdot \cos\omega \\ \cos\varphi \cdot \sin\omega \\ \sin\varphi \end{bmatrix} \tag{1}$$

Then, the corresponding probe head positions C_i in the part coordinate system (PCS) can be calculated as

$$C_i = \mathbf{R}_i^{3\times3} \cdot l_{stylus} \cdot \mathbf{v}_i^{3\times1} + \mathbf{t}_i^{3\times1} \tag{2}$$

where $\mathbf{R}_i^{3\times3}$ and $\mathbf{t}_i^{3\times1}$ represent the rotation and translation matrix from LCS to PCS respectively. The probe head positions C_i form the trajectory curve TC.

After the TC is obtained, the reference curves are calculated as auxiliary curves for generating SC. The reference curve R_i is defined as the intersection between the sphere O_i and the surface S_p, where O_i is centered at C_i with its radius equal to the stylus length l_{stylus}. If there are several segments of the intersection curve, only the segment, which contains the guiding curve point G_i is regarded as the reference curve R_i. Assume that the length of R_i is $\left|\widehat{r_i^s r_i^e}\right|$. The sample point W_i on the i-th reference curve can be sampled as

$$\left|\widehat{r_i^s W_i}\right| = \left|\widehat{r_i^s r_i^e}\right| * \frac{1 + \lambda\sin(2\pi n_s t_i)}{2} \tag{3}$$

where n_s is the total number of the oscillation period, t_i represents the i-th isometric parameter of the trajectory curve $C(t)$ and λ is a scaling factor. With W_i connecting from beginning to end, SC is obtained as a sinusoidal-shaped curve.

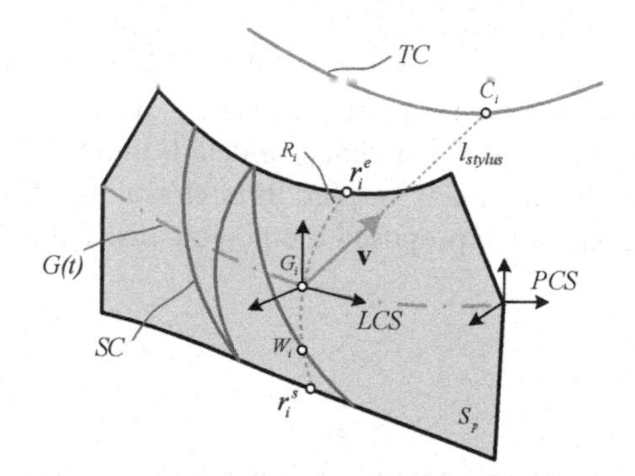

Fig. 2. A schematic diagram of five-axis sweep scanning path planning.

3 Collision-Free Scanning Path Planning

In order to plan the five-axis sweep scanning path, the collision-free area of the stylus should be calculated first. In this section, the collision-free condition is introduced. Then the collision detection method with a graphics engine is described in detail. Finally, an incremental algorithm is proposed to calculate the global collision-free area, which has a better performance than the existing exhaustive algorithm does.

3.1 Image-Based Representation of Collision-Free Conditions

Before we perform the collision detection, the collision-free condition for five-axis sweep scanning should be defined first. As described in Sect. 2, the final TC is sinusoidal and the path points are sampled from the reference curves. Thus, it should be noted that the collision-free condition should be checked on the whole reference curve rather than the single guiding curve point. For a given surface S_p and the environment E, the collection set V_{cf} of all the collision-free orientations at guiding curve point G_i is defined as

$$V_{cf} = \left\{ \begin{array}{l} \mathbf{v} : \overline{R_i^j T_i} \cap S_p = R_i^j, \overline{R_i^j T_i} \cap E = \emptyset, \\ T_i = G_i + l_{stylus} \cdot \mathbf{v}, j = 1, 2, \ldots, n_r \end{array} \right\} \quad (4)$$

where $\overline{R_i^j T_i}$ represents the probe head system defined by the probe head position T_i and probe tip position R_i^j. In practice, the orientation is defined in spherical coordinates and discretized into finite angles. According to Eq. (1), the stylus orientation can be uniquely defined by the tilt angle ω and yaw angle φ. Since the "drag" mode is preferred in the five-axis sweep scanning [11], the range of ω should be limited to $[-\pi/2, \pi/2]$. Furthermore, the stylus should always lie above the tangent plane of the guiding curve point G_i. In another word, the range of φ should be limited to $\left[0, \pi/2\right]$. By setting an angle resolution ϕ (i.e. $\pi/30$ used in this paper), the rectangular $\omega - \varphi$ domain can be discretized into $\left(\pi/\phi + 1\right) \times \left(\pi/(2\phi) + 1\right)$ pixels. Each pixel (u, v) represents a specific orientation with ω and φ equals

$$\begin{cases} \omega = -\frac{\pi}{2} + v \cdot \phi \\ \varphi = \frac{\pi}{2} - u \cdot \phi \end{cases} \tag{5}$$

If the orientation meets the collision-free condition (4), the pixel is set to be white. Otherwise, it is set to be black. The white pixels form the collision-free area, namely the admissible area, while the black pixels form the inadmissible area. Thus, the overall collision-free condition of a given guiding curve point can be represented as a binary image as shown in Fig. 3, which is referred to as the admissible map.

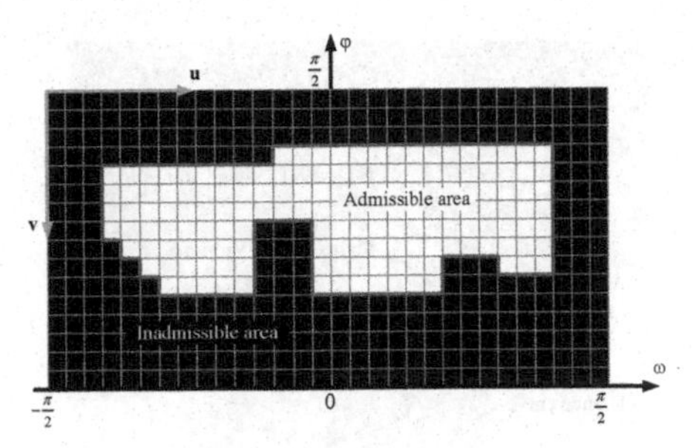

Fig. 3. The image-based representation of collision-free conditions

3.2 Collision Detection by Graphics Engine

In order to perform efficient collision detection between the probe head system and the measuring environment, the commercial graphics engine HOOPS is applied. HOOPS is a C++ graphics toolkit, which provides the creation, querying and editing of the graphical information in an application. HOOPS provides the function of detecting the collision between triangular meshes and an object. In order to simplify the calculation,

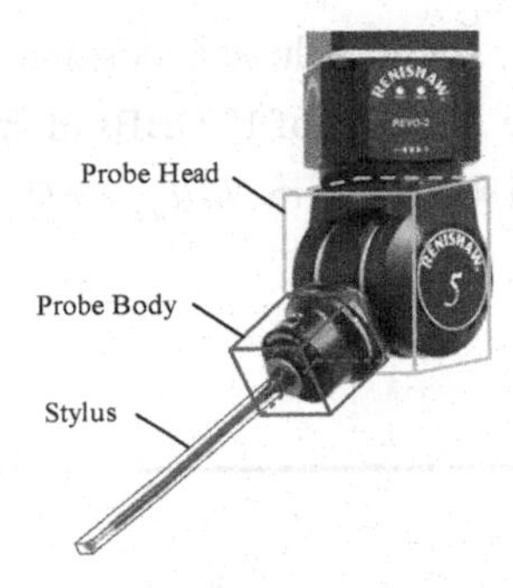

Fig. 4. The oriented bounding box of the probe head system.

the oriented bounding box (OBB) is used to approximate the real probe head system. As shown in Fig. 4, the five-axis probe head system consists of a probe head, a probe body and a slender stylus. According to its designing dimensions, three bounding boxes can be constructed to represent the maximum collider of the probe head system.

For a given pixel in the admissible map, the stylus orientation is determined, so is the probe head position. Thus the corresponding reference curve can be calculated as explained in Sect. 2. When the stylus sweeps along the reference curve from one end to the other, the OBBs of the stylus and probe body form a fan-shaped swept volume, as shown in Fig. 5. In order to check the collision-free condition along the whole reference curve, several sample positions on the reference curve are selected and the swept volume can be generated. Then the collision detection between the object and the swept volume can be conducted by HOOPS functions. The pseudocode of the proposed collision detection algorithm is shown in Algorithm 1.

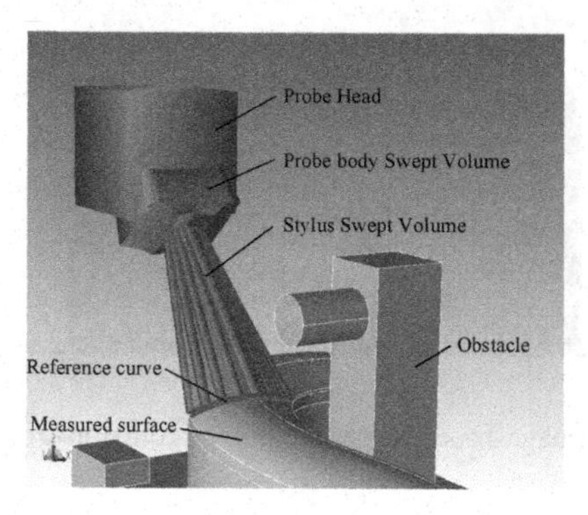

Fig. 5. The swept volume of the stylus at a reference curve.

Algorithm 1: Collision detection algorithm for reference curve.

Input: Reference curve $R(t)$, probe head position T and measured parts $Shell_{meas}$.

Output: A true/false result *hascollision* indicting whether there exists collision.

1: Discretize $R(t)$ into 10 points, i.e. $R_i \quad (i = 1, 2, K, 10)$.

2: Construct the OBB $Shell_{head}$ for probe head at position T using HOOPS function.

3: Construct the swept volume $Shell_{stylus}$ of 10 OBBs of the stylus at positions $R_i T$.

4: **if** ($Shell_{head} \cap Shell_{meas} \neq \varnothing \mid\mid Shell_{head} \cap Shell_{stylus} \neq \varnothing$) **then**

5: $\quad hascollision = true$;

6: **else**

7: $\quad hascollision = false$;

8: **return** *hascollision*;

3.3 Incremental Algorithm for Generating Admissible Areas

Although the continuous workspace has been discretized into finite pixels, checking the collision-free condition of all the pixels is still quite time-consuming. Considering the fact that the collision-free area of the two adjacent guiding curve points should be similar, we propose an incremental algorithm to iteratively check those pixels, which are likely to change.

The boundary pixels will certainly change if the collision-free area changes. We iteratively check the pixels from the boundary of the admissible area whether their collision-free condition is changed. A checking queue $\boldsymbol{Q}$ is constructed to store the candidate pixels. Each time the top element is popped out for checking. Also, a status map $\boldsymbol{M}$, which is used to record the checked pixels, is necessary to avoid the repeated checking. In the beginning, the white boundary pixels are put into $\boldsymbol{Q}$ for checking. There are totally four cases when performing the collision detection. A simplified admissible map is shown in Fig. 6 to introduce the proposed algorithm.

- ***Case 1***: If the current pixel is white (i.e. pixel ***a***) and it remains admissible, the adjacent black pixels (i.e. pixel ***d*** and ***e***) should be checked.
- ***Case 2***: If the current pixel is white (i.e. pixel ***a***) and it turns inadmissible, the adjacent white pixels (i.e. pixel ***b*** and ***c***) should be checked.
- ***Case 3***: If the current pixel is black (i.e. pixel ***e***) and it turns admissible, the adjacent black pixels (i.e. pixel ***f*** and ***g***) should be checked.
- ***Case 4***: If the current pixel is black and it remains inadmissible, no more pixels are needed to be checked. The adjacent white pixels must have been checked.

The collision detection is iteratively performed until $\boldsymbol{Q}$ is empty. The pseudocode of the whole algorithm is shown in Algorithm 2.

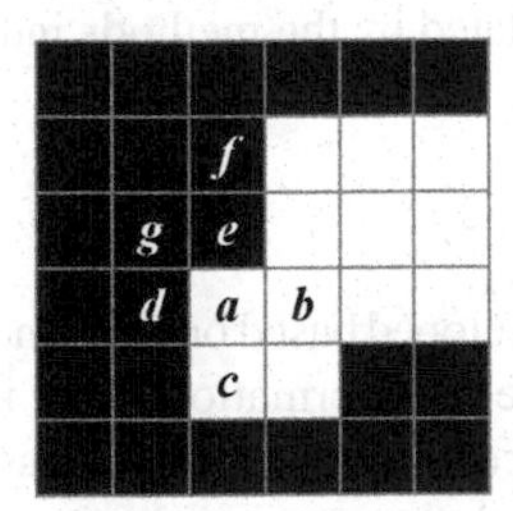

Fig. 6. Different cases in collision detection.

Algorithm 2: Algorithm of calculating admissible map.

Input: last admissible map I_{last}.
Output: next admissible map I_{next}.
1: Initialize $\boldsymbol{Q} = \boldsymbol{null}$, $\boldsymbol{M} = [0]$, $I_{next} = I_{last}$.
2: Extract boundary pixels in I_{next} and push them into $\boldsymbol{Q}$.
3: **while** ($\boldsymbol{Q}$ is not empty)
4: pop out pixel $\boldsymbol{p}$ from the top of $\boldsymbol{Q}$ and set $M(\boldsymbol{p})$=1;
5: perform collision detection of $\boldsymbol{p}$ with **Algorithm 1** and get result $hascollision$;
6: **switch** ($\boldsymbol{p}$, $hascollision$)
7: ***case 1***: push adjacent unchecked black pixels into $\boldsymbol{Q}$;
8: ***case 2***: $I_{next}(\boldsymbol{p}) = 0$, push adjacent unchecked white pixels into $\boldsymbol{Q}$;
9: ***case 3***: $I_{next}(\boldsymbol{p}) = 1$, push adjacent unchecked black pixels into $\boldsymbol{Q}$;
10: ***case 4***: **continue**;
11: **end switch**
12: **end while**
13: **return** I_{next};

The proposed incremental algorithm only checks those pixels, which are possible to change. Once there are no candidate pixels for checking, the collision detection is terminated. Thus, compared to the traditional exhaustive algorithm which checks all the pixels, the proposed method is more timesaving. In real practice, when calculating admissible area for 100 guiding curve points, the calculation time is below 30 s. After the admissible area of each guiding curve point G_i is calculated, the center of the admissible area is extracted by Hu moments [12] and the corresponding probe head position C_i is calculated by Eq. (2). After all the probe head positions C_i are obtained, the probe head trajectory curve can be obtained by fitting C_i into a cubic B-spline [13]. Then, the probe tip sweeping curve can be calculated by the methods introduced in Sect. 2.

4 Experiment

The path planning platform is developed based on C++ and QT framework. A 3D modeler ACIS is used to obtain the geometric information of the input model. A graphics engine HOOPS is used to render the scene and perform the user interaction. An object with free-form surface is measured with three-axis and five-axis scanning paths respectively. The velocity and the acceleration of the CMM are calculated in order to compare the kinematic loads in two different scanning modes.

The inspection machine is a five-axis inspection machine, which is composed of a Renishaw Agility CMM and a Revo probe head. The stylus length is 250 mm and the tip radius is 3 mm. The Renishaw CMM controller supports the Inspection-plusplus Dimensional Measurement Equipment Interface (I++ DME). The planned five-axis scanning path can be sent to the controller through a local area network. The experiment setup is shown in Fig. 7.

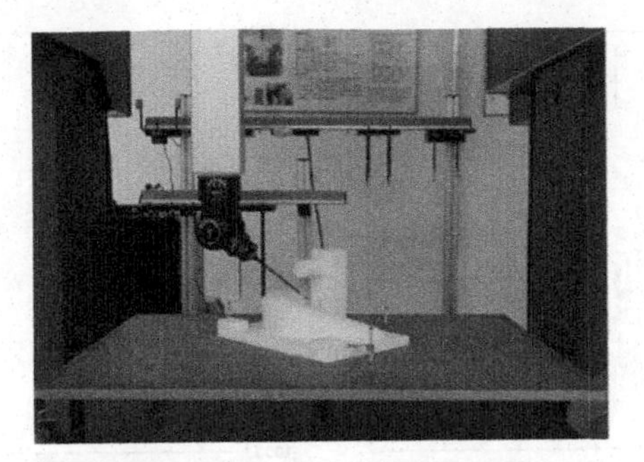

Fig. 7. The experiment setup.

The proposed five-axis scanning path is compared with a traditional three-axis zigzag scanning path [14]. Figure 8 shows the scanning paths, including the probe head trajectory curve and the probe tip sweeping curve. The reference velocity and the acceleration of the CMM is set to be 40 mm/s and 400 mm/s^2. The tangent velocity and acceleration can be respectively calculated by the first and second-order difference of the position. The total length of the scanning path, the velocity and the acceleration of the CMM are listed in Table 1. The velocity and acceleration of the CMM under two different scanning modes are shown in Fig. 9.

It can be seen from Table 1 that the length of the probe head trajectory in sweep scanning mode is merely 15.8% of that in zigzag mode. While the length of the probe tip trajectory is larger in sweep scanning mode, which indicates that the coverage of sweep scanning paths is better. The scanning time of sweep scanning and zigzag scanning is 54.6 s and 123.8 s respectively. Thus, the efficiency of five-axis sweep scanning is much better than that of the three-axis zigzag scanning.

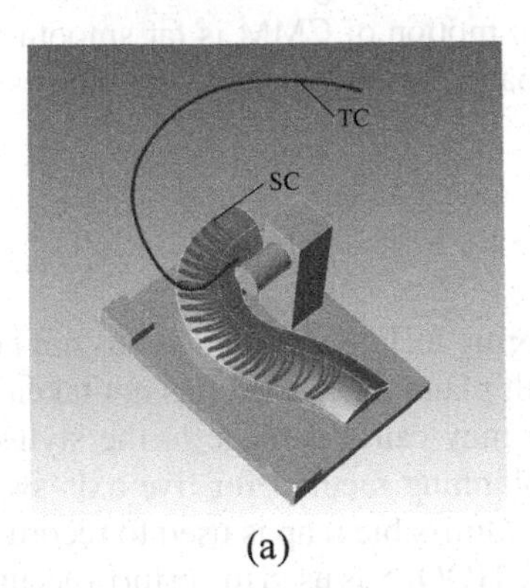

(a)

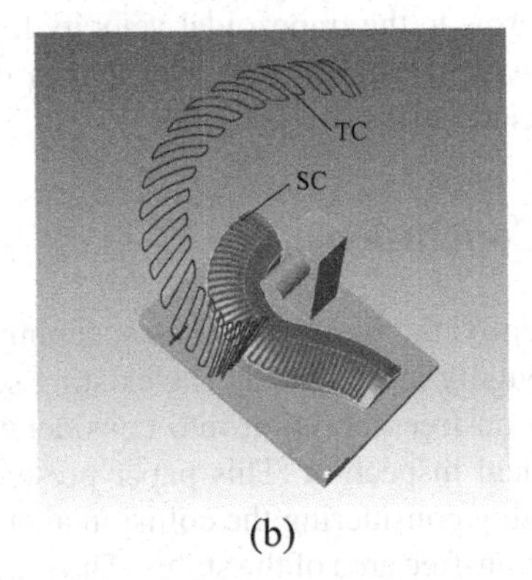

(b)

Fig. 8. Scanning paths with different scanning modes. (a) Five-axis sweep scanning path. (b) Three-axis zigzag path.

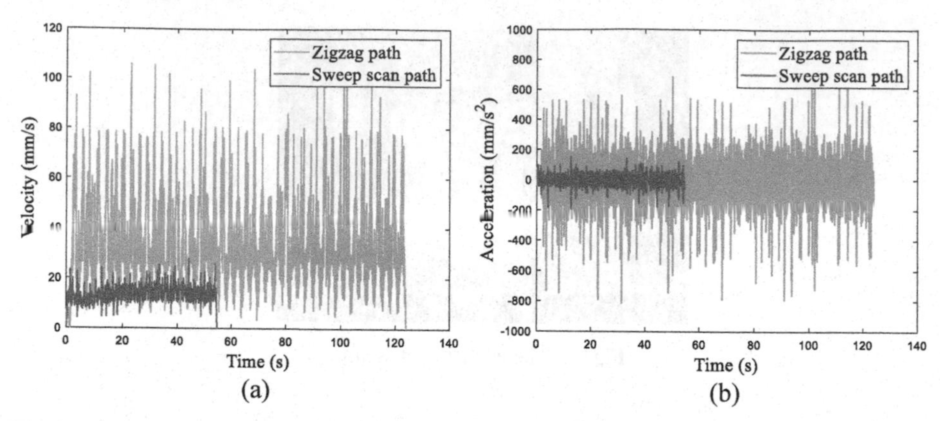

Fig. 9. Velocity and acceleration of different scanning modes. (a) Velocity of the CMM. (b) Acceleration of the CMM.

Table 1. Analytical results of two scanning paths.

	Total length (mm)		Velocity (mm/s)		Acceleration (mm/s^2)	
	TC	SC	max	mean	max	mean
Zigzag path	4139.7	3961.1	105.7	35.4	861.9	165.8
Sweep scan path	655.0	5217.8	27.9	13.9	152.4	33.3

In view of the velocity and acceleration, the mean velocity and acceleration of the sweep scan path are respectively 39.3% and 20.1% of that of zigzag path. It can be seen from Fig. 9(a) that the velocity of the CMM is stable along scanning and basically conforms to the trapezoidal velocity law. Thus, the motion of CMM is far smoother in the sweep scanning path than that in the zigzag path, which significantly reduces the kinematic loads of the CMM.

5 Conclusion

Currently, the five-axis sweep scanning is an emerging technology, which has not been thoroughly investigated. The existing scanning path planning method has not taken the collision-free condition into consideration, which may cause damage to the stylus in practical inspection. This paper presents a path planning method for five-axis sweep scanning considering the collision avoidance. The admissible map is used to record the collision-free area of the stylus. The graphics engine HOOPS is used to perform accurate and efficient collision detection. Moreover, an incremental algorithm is proposed to reduce the computation of collision detection. Compared to the existing five-axis path planning methods, the proposed method not only guarantees the collision-free condition during the whole scanning process, but also simplifies the time-consuming collision detection operation to realize efficient five-axis path planning. In the experiment, the velocity of the CMM in the planned sweep scanning path maintains at a low level, while the scanning efficiency is much better than that of the traditional zigzag scanning path. The planned sweep scanning path is verified to be collision-free and the kinematic loads of the machine are greatly reduced.

Acknowledgments. This research was supported by the National Natural Science Foundation of China (Grant No. 51905346 and No. 91948301) and the Science & Technology Commission of Shanghai Municipality (Grant No. 19511106000).

References

1. Echerfaoui, Y., El Ouafi, A., Chebak, A.: Experimental investigation of dynamic errors in coordinate measuring machines for high speed measurement. Int. J. Precis. Eng. Manuf. **19**, 1115–1124 (2018)
2. Sousa, A.R.: Metrological evaluation of a Coordinate Measuring Machine with 5-axis measurement technology. In: 15th CIRP Conference on Computer Aided Tolerancing, vol. 75, pp. 367–372 (2018)
3. Wu, X., Zhang, M., Li, G., et al.: Analysis of CMM dynamic measurement error based on decision regression tree. J. Phys.: Conf. Ser. 1605, 012103 (2020)
4. Zhang, M., Fei, Y.-T.: Hybrid modeling of CMM dynamic error based on improved partial least squares. Nanotechnol. Precis. Eng. **10**, 525–530 (2012)
5. Vermeulen, M., Rosielle, P., Schellekens, P.H.J., et al.: Design of a high-precision 3D-Coordinate Measuring Machine (1998)
6. Hu, P., Zhang, R., Tang, K.: Automatic generation of five-axis continuous inspection paths for free-form surfaces. IEEE Trans. Autom. Sci. Eng. **14**, 83–97 (2017)
7. Zhang, Y., Zhou, Z., Tang, K.: Sweep scan path planning for five-axis inspection of free-form surfaces. Robot. Comput.-Integr. Manuf. **49**, 335–348 (2018)
8. Chen, L.F., Zhang, R., Tang, K., et al.: A spiral-based inspection path generation algorithm for efficient five-axis sweep scanning of freeform surfaces. Comput.-Aided Des. **124**, 17 (2020)
9. Quan, L., Zhang, Y., Tang, K.: Curved reflection symmetric axes on free-form surfaces and their extraction. IEEE Trans. Autom. Sci. Eng. **15**, 111–126 (2018)
10. Hu, P., Zhou, H., Chen, J., et al.: Automatic generation of efficient and interference-free five-axis scanning path for free-form surface inspection. Comput.-Aided Des. **98**, 24–38 (2018)
11. Zhou, Z., Zhang, Y., Tang, K.: Sweep scan path planning for efficient freeform surface inspection on five-axis CMM. Comput.-Aided Des. **77**, 1–17 (2016)
12. Ming-Kuei, H.: Visual pattern recognition by moment invariants. IEEE Trans. Inf. Theory **8**, 179–187 (1962)
13. Piegl, L.A., Tiller, W.: The NURBS Book. Springer, Heidelberg (1997). https://doi.org/10.1007/978-3-642-59223-2
14. ElKott, D.F., Veldhuis, S.C.: Isoparametric line sampling for the inspection planning of sculptured surfaces. Comput.-Aided Des. **37**, 189–200 (2005)

Multi-robot Cooperative System Modeling and Control Software Development

Tianhong Cheng, Zhiwei Wu, and Wenfu Xu(✉)

School of Mechanical Engineering and Automation, Harbin Institute of Technology Shenzhen, Shenzhen 518055, China
chengtianhong@stu.hit.edu.cn, wfxu@hit.edu.cn

Abstract. In order to solve the problem that it is difficult to achieve better multi-robot coordination operation using the original control system in multi-robot cooperative systems, this paper conducts system modeling, control framework design and software development and verification experiments for multi-robot cooperative systems. We firstly model the system composed of multiple robots and operated objects. The coordination between any two robots can be divided into two categories: loose coordination and tight coordination according to the constraint relationship between the robots and the objects, based on which the kinematics and dynamics analysis of the multi-robot system is carried out. A multi-robot cooperative control framework is designed, which contains modules for task planning, force and position assignment, sensor information acquisition, position/force control and synchronization clock. On the basis of the multi-robot cooperative control system, the robot cooperative control software is designed and developed, and the software system contains, with excellent characteristics. Finally, an experimental environment containing three large-load EFORT industrial robots was established and experiments were conducted using the developed multi-robot cooperative control software to verify the usability and reliability of the software system.

Keywords: Multi-robots cooperative system · Control software development · Robot control

1 Introduction

Collaborative multi-robot operation is the focus of current robotics applications, especially in industrial robots, where multi-robot collaboration has the advantages of large loads, high efficiency, and higher flexibility and organization to better meet the demands of increasingly complex applications [1, 2].

However, in current industrial robot applications, control is usually performed through the control system accompanying the robot, where each robot corresponds to a dedicated set of control devices. And the independent motion of each robot is achieved by operating and programming each controller separately. For systems with multiple robots, which often do not truly achieve the characteristics of synergy in time and space [3]. This control method neither ensures the synchronization of robot movements nor

X.-J. Liu et al. (Eds.): ICIRA 2021, LNAI 13014, pp. 14–24, 2021.
https://doi.org/10.1007/978-3-030-89098-8_2

enables fast information exchange between robots, which makes it impossible for multiple robots to perform the same task simultaneously in an overlapping operation space, and at the same time. Each robot has to be operated or programmed separately for each robot, which greatly increases the difficulty and complexity of implementing a multi-robot cooperative operating system.

The common controller architectures are "PC + motion control card" [4] and "PC + real-time operating system + industrial bus" [5, 6]. The EtherCAT bus is becoming a hot spot for controller research [7, 8].

Traditional controllers only support position control, which has large limitations in compliance [9]. Commonly used compliance control methods include impedance control, force/position hybrid control, etc. [10]. And coordinated compliance control is a concern in multi-robot coordinated motion [11–13].

The main objective of this paper is to propose a general multi-robot cooperative control framework and design a scalable, easy-to-use, and high-performing multi-robot cooperative control software based on this.

2 Multi-robot Collaborative System Modeling

2.1 Coordinated Operation Model

Modeling and kinematic and dynamical analysis of multi-arm robots are the basis for studying multi-robot cooperative control systems. Generally, multi-robot systems consist of a base, robots and operated objects, and for the basic two-arm cooperative system, there are two cases of tight coordination and loose coordination between two robots according to the kinematic and dynamical constraint relationships between robots.

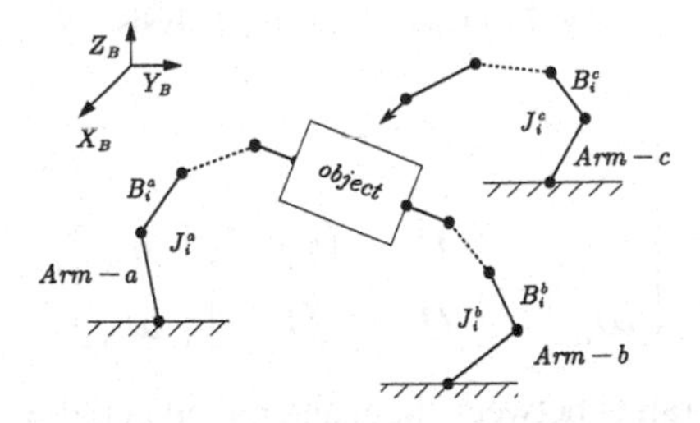

Fig. 1. Multi-arm collaborative system configuration

And for the case of more robots with the same operated operation, a tight coordination relationship is presented between robots with a fixed constraint with the same operated object, and a loose coordination relationship is presented between robots without a constraint relationship. Consider a multi-robot cooperative system consisting of three robots and an operated object, three robots with the same object (see Fig. 1), assuming that robot A, robot B have constraint relationship with the object and robot C has no constraint relationship with the object, then robot A has a tight coordination relationship with robot B and has a loose coordination relationship with robot C.

2.2 Kinematic Modeling

For a single robot, the positional relationship between its position and the joint vector can be expressed by Eq. 1.

$$ {}^{0i}\boldsymbol{T}_{ei} = {}^{0}\boldsymbol{T}_{e}^{i} = \text{fkine}(\boldsymbol{q}_i) \tag{1} $$

The relationship between end velocity and joint angular velocity can be expressed by the Eq. 2.

$$ \begin{bmatrix} \boldsymbol{v}_e^i \\ \boldsymbol{\omega}_e^i \end{bmatrix} = \boldsymbol{J}_i(\boldsymbol{q}_i)\dot{\boldsymbol{q}}_i \tag{2} $$

For the case where there are constraints between the robot and the object can be expressed by the Eq. 3.

$$ {}^{0i}\boldsymbol{T}_{ei} = \left({}^{B}\boldsymbol{T}_{0i}\right)^{-1} \cdot {}^{B}\boldsymbol{T}_{L} \cdot \left({}^{ei}\boldsymbol{T}_{L}\right)^{-1} \tag{3} $$

The velocity of the object is analyzed as shown in the Fig. 2 and satisfies the Eq. 4.

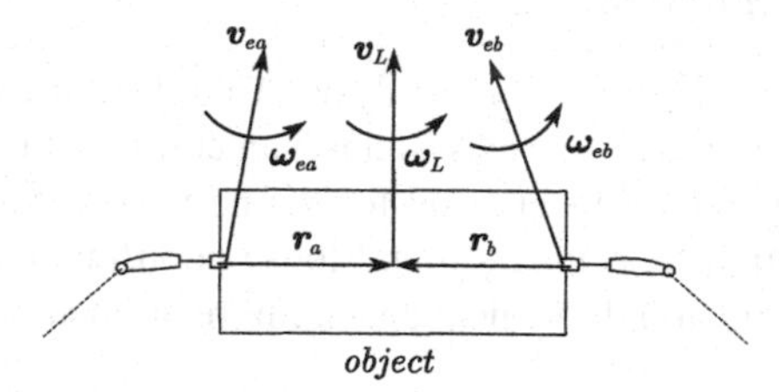

Fig. 2. Object velocity analysis

$$ \begin{bmatrix} \boldsymbol{v}_L \\ \boldsymbol{\omega}_L \end{bmatrix} = \begin{bmatrix} \boldsymbol{E} & -(\boldsymbol{r}_i)^{\times} \\ \boldsymbol{0} & \boldsymbol{E} \end{bmatrix} \begin{bmatrix} \boldsymbol{v}_{ei} \\ \boldsymbol{\omega}_{ei} \end{bmatrix} \tag{4} $$

For robots with constraints between them and the object, the end velocity is the same as the object.

$$ \dot{\boldsymbol{q}}_i = \boldsymbol{J}_i^{\mathrm{T}}(\boldsymbol{q}_i) \begin{bmatrix} \boldsymbol{E} & -(\boldsymbol{r}_i)^{\times} \\ \boldsymbol{0} & \boldsymbol{E} \end{bmatrix}^{-1} \begin{bmatrix} \boldsymbol{v}_L \\ \boldsymbol{\omega}_L \end{bmatrix} \tag{5} $$

2.3 Dynamics Modeling

For each robots.

$$ \boldsymbol{M}_i\ddot{\boldsymbol{q}}_i + \boldsymbol{C}_i(\boldsymbol{q}_i, \dot{\boldsymbol{q}}_i)\dot{\boldsymbol{q}}_i + \boldsymbol{G}_i(\boldsymbol{q}) = \boldsymbol{\tau}_{li} + \boldsymbol{\tau}_{exti} \tag{6} $$

The joint torque and the robot end-action force are expressed by Eq. 7.

$$\boldsymbol{\tau}_{exti} = \boldsymbol{J}_i^T \boldsymbol{F}_{ei} = \boldsymbol{J}_i^T \begin{bmatrix} \boldsymbol{f}_{ei} \\ \boldsymbol{\tau}_{ei} \end{bmatrix} \tag{7}$$

The object is subjected to the robot end-action and environmental forces that are bound to its existence, which can be expressed as Eq. 8.

$$\begin{gathered} \begin{bmatrix} -\boldsymbol{G}_L \\ \boldsymbol{0} \end{bmatrix} + \begin{bmatrix} \mathrm{m}_L \dot{\boldsymbol{v}}_L \\ \boldsymbol{I}_L \dot{\boldsymbol{\omega}}_L + \boldsymbol{\omega}_L \times (\boldsymbol{I}_L \boldsymbol{\omega}_L) \end{bmatrix} \\ = \sum \begin{bmatrix} \boldsymbol{E} & \boldsymbol{0} \\ \boldsymbol{r}_i^{\times} & \boldsymbol{E} \end{bmatrix} \begin{bmatrix} -\boldsymbol{f}_{ei} \\ -\boldsymbol{\tau}_{ei} \end{bmatrix} - \begin{bmatrix} \boldsymbol{E} & \boldsymbol{0} \\ \boldsymbol{r}_L^{\times} & \boldsymbol{E} \end{bmatrix} \begin{bmatrix} -\boldsymbol{f}_L \\ -\boldsymbol{\tau}_L \end{bmatrix} \end{gathered} \tag{8}$$

3 Collaborative Control System Frame and Methods

3.1 Control Frame Design

The controller system mainly consists of task planning, machine force/position controller, and sensor information acquisition.

Task planning acquires the position and force planning of the robot, performs position and force assignment. And sends it to the robot's force controller or position controller, which outputs the desired joint trajectory, sends it to the robot and executes it (Fig. 3).

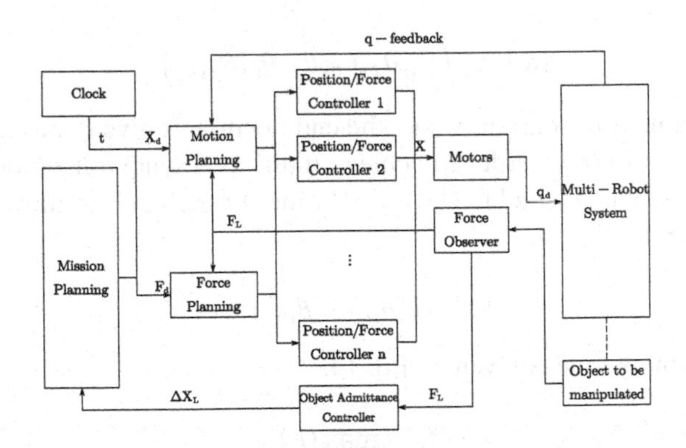

Fig. 3. Overall diagram of the coordinated control algorithms.

3.2 Position and Force Distribution

The case where there are two robots exerting forces on the object can be expressed by the Eq. 9.

$$\boldsymbol{F}_{bed} = \boldsymbol{\Gamma}_{eb}^{-1} \left(-\boldsymbol{\Gamma}_{ea} \boldsymbol{F}_{ea} + \boldsymbol{\Gamma}_L \boldsymbol{F}_L + \begin{bmatrix} \boldsymbol{G}_L \\ \boldsymbol{0} \end{bmatrix} - \begin{bmatrix} \dot{\boldsymbol{P}}_L \\ \dot{\boldsymbol{L}}_L \end{bmatrix} \right) \tag{9}$$

With the task trajectory, the robot planning trajectory with constraint relationship with the object can be found.

$$^{0i}\boldsymbol{T}_{ei} = {}^{0i}\boldsymbol{T}_{B} \cdot {}^{B}\boldsymbol{T}_{L} \cdot {}^{L}\boldsymbol{T}_{ei} \tag{10}$$

3.3 Motor Control and Motion Planning

The multi-robot cooperative control system designed in this paper is mainly for industrial robots with large mass and inertia, and a primary issue is to ensure smooth and continuous joint motion in position, velocity, and acceleration. The servo motors used in general robots can be controlled relatively easily in terms of position or velocity. In order to reduce the vibration caused by velocity changes, the control of a single motor axis is achieved by applying PD control on the motor velocity loop.

In order to ensure the synchronization of multi-axis and multi-robot position motion, taking into account the smoothness of motion and efficiency requirements, a five times spline method is used for planning each motor axis.

$$\begin{cases} \theta(t) = a_0 + a_1 t + a_2 t^2 + a_3 t^3 + a_4 t^4 + a_5 t^5 \\ \dot{\theta}(t) = a_1 + 2a_2 t + 3a_3 t^2 + 4a_4 t^3 + 5a_5 t^4 \\ \ddot{\theta}(t) = a_2 + 6a_3 t + 12a_4 t^2 + 20a_5 t^3 \end{cases} \tag{11}$$

The vector of coefficients can be obtained by the method of coefficients to be determined.

$$[a_i] = f\left(\theta_0, \theta_f, \dot{\theta}_0, \dot{\theta}_f, \ddot{\theta}_0, \ddot{\theta}_f, t_f\right) \tag{12}$$

The initial value is the current value, the end position is given, the end velocity and acceleration are set to 0. In order to achieve multi-axis synchronization, a maximum average velocity of v is set, and the estimated time for each axis to move to the current command is given by Eq. 13.

$$[t_i] = ([\theta_{if}] - [\theta_{i0}])/v \tag{13}$$

Then the planning time is given by Eq. 14.

$$t_f = \max(t_i) \tag{14}$$

3.4 Multi-arm Compliance Control Method

In the actual application of robot, the pure position control cannot meet the requirement of flexibility, and it is easy to cause damage to robot or object. For the robot without constraint, the traditional impedance control is used, and for the robot with object with constraint, the main concern is the compliance control between object and environment. A dual closed-loop impedance controller is used to perform force control between robot/object and object/environment at the same time (Fig. 4).

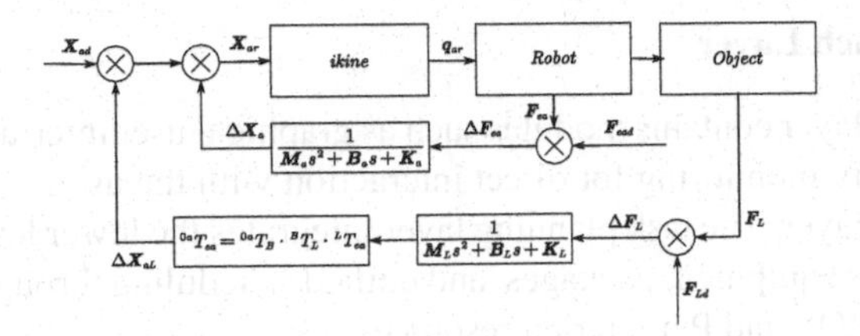

Fig. 4. Object-based impedance controller

4 Control System Software

4.1 Demand Analysis

The control system software has three main requirements.

Scalability. Heterogeneous collaboration of multiple robots is possible, compatible with existing industrial solutions, and allows for rapid expansion of multi-robot systems.

Ease of use. Provides good interfaces at different levels to facilitate the expansion of operating environments, algorithm development and verification.

High performance. Meet real-time requirements, provide multiple computing unit expansion interfaces, and fully mobilize the resources of self-developed hardware.

4.2 Overall Design

The entire software system is divided into five main layers, namely user interface layer, task planning layer, controller layer, device management layer, and hardware interface layer (Fig. 5).

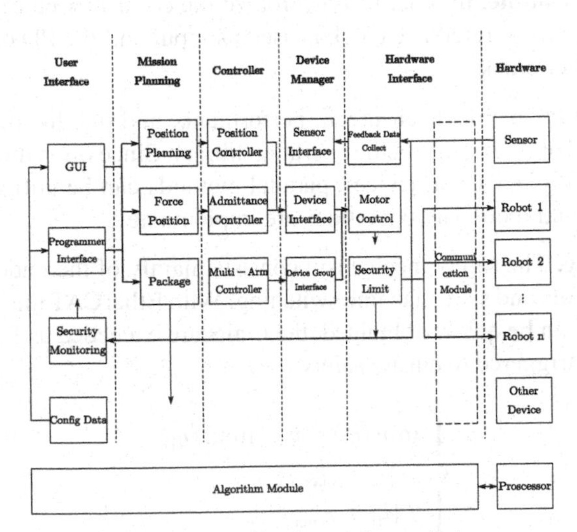

Fig. 5. Overall software frame design

4.3 Design of Each Layer

The user interface layer contains modules such as graphical user interface, programming interface, and safety monitoring for direct interaction with the user.

Task planning layer. The task planning layer integrates the lower level position/force controllers, process equipment packages, and unified scheduling through Python scripts, and can integrate ROS and Python rich resources.

The controller layer implements advanced control algorithms, including impedance control, Cartesian space trajectory control, and multi-robot cooperative control algorithms, accepts feedback data from the lower layer and sends commands to the device or device group, facilitating the development and application of advanced algorithms.

Device management layer. The device interface layer integrates the basic control units of the lower layer into independent devices or groups of devices from multiple devices for dynamic management of execution and sensing devices, facilitating the application of complex coordination algorithms for multiple robots to a specific set of devices for easy development and use.

The hardware interface layer communicates directly with the underlying hardware, performs motor axis motion planning through multi-axis synchronization algorithms, and accepts feedback from encoders and other sensors, and is compatible with a variety of communication modules, enabling the connection of different kinds of devices in industrial solutions.

The hardware environment can include robots, sensors, process equipment, etc.

4.4 Main Modules

Some important modules of the software system are described.

Communication Module. In order to synchronize the controller clock with the EtherCAT reference clock, the reference clock is used as input and the PD control is used to correct the controller clock.

Calculation Module. In order to ensure the performance and stability of the controller, a coprocessor is used for the design of the operation library. Since often the same algorithm is operated for multiple axes or devices, parallel methods can be utilized and dynamic load optimization and redundant operations are implemented.

Safety Monitoring. The safety mechanism consists mainly of the underlying position, speed and force limits and state machine switching. With EtherCAT the current position, speed and current can be quickly obtained, the limits are exceeded as Eq. 15 and a state machine switch is triggered to ensure safety.

$$\begin{cases} \min(q_i) < q < \max(q_i) \\ |v_i| < v_{i\max} \\ T|i_i| < \tau_{i\max} \end{cases} \tag{15}$$

4.5 Software Interface

The graphical user interface contains the main functions of the force control system. It includes communication module, robot control, package usage, script management and running (Fig. 6).

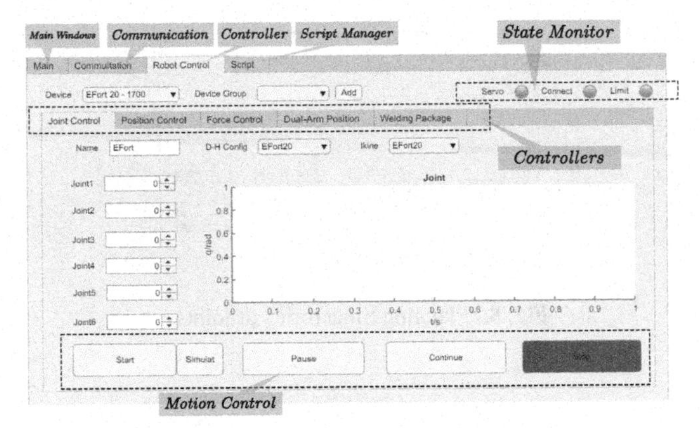

Fig. 6. Graphical user interface

5 Experiment Verification

5.1 Experimental System

The experimental environment was set up with three Efort ER20-1700 industrial robots, all connected to the control system software using EtherCAT.

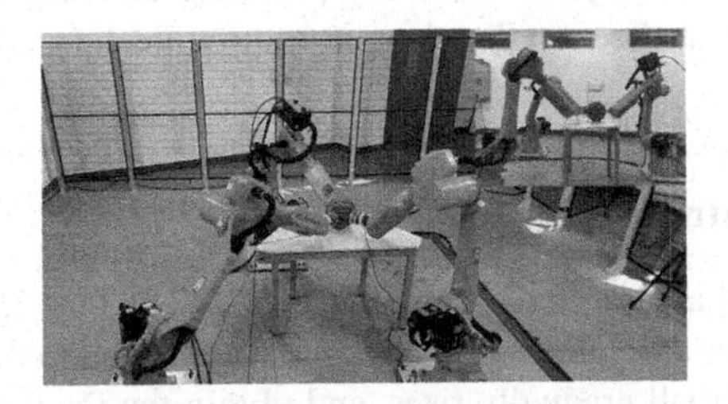

Fig. 7. Experimental System

5.2 Motion Planning Experiment

The three robots are kept in a constant relative position to each other and move linearly along the x-axis in space to obtain the actual robot end x-axis direction position and compare it with the planned position (Figs. 8 and 9).

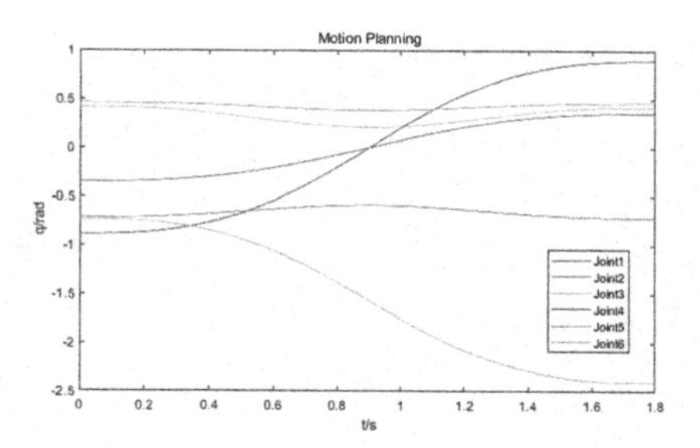

Fig. 8. Planning trajectories of joints

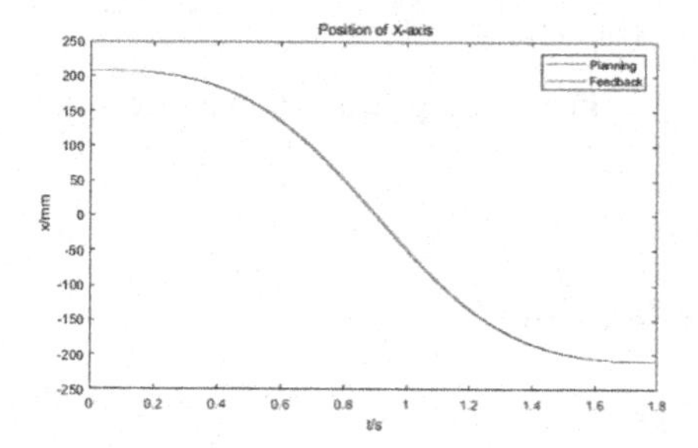

Fig. 9. Actual end x-axis position of robot A

The figure represents the joint trajectory planning of robot A and the actual end x-axis trajectory, and it can be seen that the minimum trajectory error is 0.31 mm, which has sufficient accuracy.

5.3 Compliance Control Experiment

Make three robots hold a basketball at an angle of 120° as Fig. 7, one robot is position controlled, and the other two robots are impedance controlled in the direction of the end z-axis. Make the basketball gradually rises, and obtain the feedback value of the robot end sensor to prove that this multi-robot motion control system is capable of complex force control algorithms (Fig. 10).

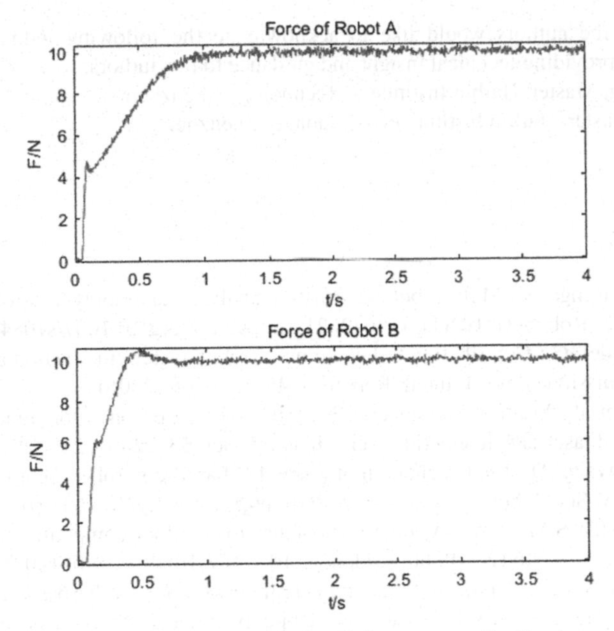

Fig. 10. Robot A and Robot B force sensor feedback

The figure shows the end force sensor feedback values for Robot A and Robot B using force control with a desired force of 10 N. It can be seen that the force control accuracy of the impedance controller is less than 1%.

6 Conclusion

In this paper, we concern the problem of multi-robot cooperative control. Firstly, a basic multi-robot operating system model is established, and on the basis of the model, kinematics and dynamics are analyzed for the two cases of constrained and unconstrained between robots and objects, respectively. Then the multi-robot cooperative motion control system framework is designed, the force and position distribution methods are given, the single-axis motor control and multi-axis synchronous motion planning methods are designed. And the multi-robot cooperative soft-smooth control algorithm based on object impedance is given. Based on the control framework, the multi-robot cooperative control system software is designed and developed. The software system has hardware interface layer, device management layer, control layer, task layer and user layer, and contains communication module, scalable computing module, safety monitoring and fault handling module, with strong scalability, ease of use and usability. Finally, an experimental environment containing three industrial robots was built, and experiments on multi-robot trajectory motion and multi-robot coordinated impedance control were conducted with good results to verify the usability of the control system software.

Acknowledgement. This work was supported by the Key-area Research and Development Program of Guangdong Province [grant numbers 2019B090915001].

In addition, the authors would like to acknowledge the following individuals for their contributions by providing technical insight and guidance to the authors.

Bowen Wang, Master. Harbin Institute of Technology Shenzhen.

Tao Chen, Master. Harbin Institute of Technology Shenzhen.

References

1. Verma, J.K., Ranga, V.: Multi-robot coordination analysis, taxonomy, challenges and future scope. J. Intell. Rob. Syst. **102**(1), 1–36 (2021). https://doi.org/10.1007/s10846-021-01378-2
2. Saïdi, F., Pradel, G.: Contribution to human multi-robot system interaction application to a multi-robot mission editor. J. Intell. Rob. Syst. **45**, 343–368 (2006)
3. Zhuang, Y., et al.: Multi-robot cooperative localization based on autonomous motion state estimation and laser data interaction. Sci. China Inf. Sci. **53**, 2240–2250 (2010)
4. Farooq, M., Wang, D.: Implementation of a new PC based controller for a PUMA robot. J. Zhejiang Univ. Sci. A **8**(12), 1962–1970 (2007). https://doi.org/10.1631/jzus.2007.A1962
5. de Oliveira, R.W.S.M., et al.: A robot architecture for outdoor competitions. J. Intell. Rob. Syst. **99**(3–4), 629–646 (2020). https://doi.org/10.1007/s10846-019-01140-9
6. Han, S.J., Oh, S.Y.: An optimized modular neural network controller based on environment classification and selective sensor usage for mobile robot reactive navigation. Neural Comput. Appl. **17**, 161–173 (2008). https://doi.org/10.1007/s00521-006-0079-1
7. Aristova, N.I.: Ethernet in industrial automation: Overcoming obstacles. Autom. Remote. Control. **77**(5), 881–894 (2016). https://doi.org/10.1134/S0005117916050118
8. Huang, R.-Y., Chen, Y.-J., Chen, Y.-X., Cheng, C.-W., Tsai, M.-C., Lee, A.-C.: Advanced application of centralized control for a scanning mirror system based on etherCAT fieldbus. Int. J. Control Autom. Syst. **19**(3), 1205–1214 (2021). https://doi.org/10.1007/s12555-019-0754-5
9. Xu, G., Song, A., Li, H.: Adaptive impedance control for upper-limb rehabilitation robot using evolutionary dynamic recurrent fuzzy neural network. J. Intell. Rob. Syst. **62**, 501–525 (2011)
10. Furtado, G.P., Americano, P.P., Forner-Cordero, A.: Impedance control as an optimal control problem: a novel formulation of impedance controllers as a subcase of optimal control. J. Braz. Soc. Mech. Sci. Eng. **42**(10), 1–20 (2020). https://doi.org/10.1007/s40430-020-02586-x
11. Vukobratović, M.K., Rodić, A.G., Ekalo, Y.: Impedance control as a particular case of the unified approach to the control of robots interacting with a dynamic known environment. J. Intell. Rob. Syst. **18**, 191–204 (1997). https://doi.org/10.1023/A:1007915307723
12. Dubowsky, S., Sunada, C., Mavroidis, C.: Coordinated motion and force control of multi-limbed robotic systems. Auton. Rob. **6**, 7–20 (1999). https://doi.org/10.1023/A:1008816424504
13. Yan, L., Xu, W., Hu, Z., Liang, B.: Virtual-base modeling and coordinated control of a dual-arm space robot for target capturing and manipulation. Multibody Sys. Dyn. **45**(4), 431–455 (2018). https://doi.org/10.1007/s11044-018-09647-z

Gesture Recognition and Conductivity Reconstruction Parameters Analysis with an Electrical-Impedance-Tomography (EIT) Based Interface: Preliminary Results

Xiaodong Liu and Enhao Zheng(✉)

The State Key Laboratory of Management and Control for Complex Systems, Institute of Automation, Chinese Academy of Sciences, Beijing 100190, China
enhao.zheng@ia.ac.cn

Abstract. With the development of Human-machine interface (HMI), the requirements of perceiving the human intention are much higher. Electrical Impedance Tomography (EIT) is a promising alternative to existing HMIs because of its portability, non-invasiveness and inexpensiveness. In this study, we designed an EIT-based gesture recognition method achieving the recognition of 9 forearm motion patterns. We analysed the parameters, including current level and contact impedance, which are relevant for practical applications in robotic control. The gesture recognition method produced an average accuracy of 99.845% over nine gestures with PCA and QDA model on one subject. The preliminary results of parameter analysis suggested that the resolution increased with the current amplitude less than a threshold (5.5 mA) but decreased when the current amplitude was over 5.5 mA. The mean value of Region of Interest (ROI) nodes didn't change obviously when the contact impedance increased. In future works, extensive studies will be conducted on the priori information of forearm and biological-model-based methods to further improve recognition performances in more complicated tasks.

Keywords: Human-machine interface (HMI) · Electrical Impedance Tomography (EIT) · Gesture recognition · Quadratic Discriminant Analysis (QDA)

1 Introduction

With the rapid development of robotics, human-robot collaboration has become an important direction of development in the field of robotics. Human-machine interface (HMI) is the bridge between human and machine, and has recently received considerable attention in the academic community, in labs, and in technology companies [1–4]. On primary task in human robot collaboration is that

X.-J. Liu et al. (Eds.): ICIRA 2021, LNAI 13014, pp. 25–35, 2021.
https://doi.org/10.1007/978-3-030-89098-8_3

robots need to perceive or understand the human intention. There are many ways for robots to obtain the human intention, such as visual signal [2] and audio signal [3]. The biological signal may be more suitable in the area of wearable robots, because the sensor is more portable and wearable. On the other side, the above-mentioned sensing approaches cannot extract human joint dynamic information, such as joint torques and joint stiffness [5]. It hinders the applications in the scenarios with external dynamic uncertainties. Therefore, the interfaces with biological signals are necessary to facilitate human-robot interaction with physical couplings.

Human-Robot Interface with biological signals has been widely researched for years. Surface Electromyography (sEMG) is one of most widely used biological signals. Many researchers designed machine-learning based methods to classify the discrete motion patterns. For instances, in [6], researchers used Electromyography (EMG) and force myography (FMG) to recognize different gestures with Linear Discriminant Analysis (LDA). Some other researchers designed biological-model-based method to map the EMG-features to corresponding motion patterns. For instances, in [7], researchers integrated the forward dynamics of human joint movement into the Hill-based Muscle Model (HMM), EMG features are developed to construct measurement equations for the extended HMM to form a state-space model. However, the information of surface muscles is limited and not enough to analyze complex motion. sEMG couldn't get the information of deep muscles [8]. Capacitive sensing technology, which can measure medium changes that is conductive or has a dielectric different from air, has been used to detect the shape change information of the muscles. According to [9], capacitive sensing method allowed the metal electrodes not being in contact with human skin. The drawback of our previously proposed capacitive sensing is that its spatial resolution is limited for forearm motion recognition tasks. Many researchers devoted themselves to solving this issue [10].

The Electrical Impedance Tomography (EIT) is a technology for making images, which contain the information of impedance throughout the measured domain. EIT is an alternative to sEMG and capacitive sensing technologies because it is portable, non-invasive and inexpensive. Researchers present a wearable system based on EIT to recover the interior impedance of a user's arm [11]. This system can monitor and classify gestures in real-time. In [12], both 2D and 3D EIT electrode arrangements were tested for hand gesture recognition. In [13], an interface based on EIT is applied to the Human-Robot collaboration task, sawing, and the experiment shows that the EIT signals can be used to estimate the continuous variables, such as force.

The contributions of this study were listed as follows. Firstly, we proposed a gesture recognition method based on our previously designed EIT-based human-machine interface. We evaluated the method with experimental evaluations. Secondly, we analyzed the relevant parameters in the EIT construction algorithms, including current level and contact impedance.

2 The Hardware and Algorithm of Measurement System

2.1 Hardware

The measurement system was constructed with sensing front-end and the control circuits. The sensing front-end was made with a soft elastic fabric band, in which 16 electrodes were embedded. The control circuits contain a micro control unit (MCU), a voltage-controlled current source (VCCS), an analog-to digital converter (ADC), an oscillator circuit, a multiplexer module, and the pre-amplifier circuit. More details can be found out in our previous works [13].

2.2 Construction Algorithms

EIT problems can be divided into two parts: the forward problem and the inverse problem. The forward problem is simulating the voltages everywhere on the domain given a set of input currents and the admittance distribution. The inverse problem is discovering the admittance distribution over the domain according to the measured voltage on the boundary and the input currents. The admittance γ can be expressed as:

$$\gamma = \sigma + i\omega\epsilon \tag{1}$$

where σ is the conductivity on the domain, ω is the angular frequency, and ϵ is the electric permittivity. However, the real part of γ is more important than others [14]. Therefore, we only take the conductivity σ into consideration to simplify the problem.

Forward Problem
According to the Maxwell's equations for electromagnetics, charge conservation, Ohm's law and the quasi-static assumption, a Laplace equation can be derived:

$$\nabla[\sigma(x)\nabla\phi(x)] = 0, x \in \Omega \tag{2}$$

where $\phi(x)$ is the electrical potential,Ω is the domain to be imaged.

In order to solve the partial differential equation, Finite Element Model (FEM) is applied. As for the boundary condition to this situation, a Complete Electrode Model (CEM) is more realistic, which can be implemented as Neumann boundary condition. The Neumann boundary condition is

$$\sigma(x)\frac{\partial\phi(x)}{\partial\hat{n}} = \begin{cases} J & \text{electrodes} \\ 0 & \text{otherwise} \end{cases}, x \in \Omega \tag{3}$$

where $\hat{n}$ is unit vector, outward normal from the boundary $\partial\Omega$. J is current density, $\int JdS = I$.

Through FEM and weak form transformation [15], we can deduce the governing equation for EIT to linear equations.

$$\boldsymbol{Y}(\boldsymbol{\sigma})\boldsymbol{\phi} = \boldsymbol{I} \tag{4}$$

where $Y(\sigma)$ is FEM admittance matrix, I is the boundary condition, ϕ is the voltage on the nodes. This is the resolution of the forward problem. It can be expressed as

$$\phi = F(\sigma) = Y(\sigma)^{-1}I \tag{5}$$

which means we can get the voltage on the nodes by the forward model. And then we can obtain the measured voltage V on the boundary through an extraction matrix T. It is

$$V = T(\phi) = T(F(\sigma)) \tag{6}$$

Inverse Problem

As for the solution of the inverse problem, we used time-difference (td-EIT) to reconstruct the conductivity difference $\Delta\sigma$, because it remains stable and is less sensitive to disturbance.

According to the study [16], the measured signal z is called dynamic signal.

$$z = \frac{2(V_m - V_0)}{V_m + V_0} \tag{7}$$

V_m is the measured voltage at real time, and V_0 is the predefined baseline.

The linearized model of EIT can be stated as: Estimate the conductivity parameter x given a measured signal z.

$$z = Hx + n \tag{8}$$

where $x = \sigma_m - \sigma_0$. H is the sensitivity matrix which can be derived from $Y(\sigma)$. n is the random noise.

Least squares estimate is a common method of parameter estimation. However, owing to the ill-posed and ill-conditioned of EIT inverse problem, it does not work out well. In this paper, we used Maximum a Posteriori [16] to estimate x,$\hat{x}$.

We model the distribution of x as Gauss Distribution, $P(x)$, which means priori. The likelihood of (8) is $P(z|x)$. According to the Bayes formula, the posteriori is

$$P(x|z) = \frac{P(z|x)P(x)}{P(z)} \tag{9}$$

so the estimated $\hat{x}$ is

$$\hat{x} = \arg\min_{x} P(x|z) \tag{10}$$

The result is

$$\hat{x} = (H^T W H + \mu Q)^{-1} H^T W z \tag{11}$$

where $W = R_n^{-1}$ represents the inverse covariance of noise, $Q = R_x^{-1}$ represents the inverse covariance of image(or conductivity).$\mu = \sigma_x^2/\sigma_n^2$ is hyper-parameter. The algorithms were implemented with MATLAB2016a with the aid of EIDORS toolbox [17].

3 Methods

3.1 Gesture Recognition Based on EIT

Experimental Protocol

In this study, we carried out an experiment, which was about discrete gesture recognition, to evaluate the interface's performance. All the participants provided written and informed consent. The experiments were approved by the Ethical Review Board of Institute of Automation, Chinese Academy of Sciences. In this experiment, we employed one subject with an age of 26, a height of 175 cm, and a weight of 65 kg. During the experiment, the subject wore the sensing front-end with the electrodes align to the most prominent part of the forearm, and then performed nine gestures in a sequence of Middle Position (MP), Wrist Ulna Deviation (WUD), Wrist Radius Deviation (WRD), Wrist Flexion (WF), Wrist Extension (WE), Wrist Pronation (WP), Wrist Supination (WS), Clench Fist (CF), Stretch Hand (SH). Each gesture was held for 90 s. In order to reduce the influence of gesture switch, the data between 3–5 s before and after the gesture switch were removed.

Data Preprocessing

Using the algorithms mentioned in chapter 2, we can obtain the reconstructed conductivity image which has 5868 nodes. The down sampling operation is applied in order to lessen the burden of training. In short, a sample (208 measured voltages) maps to a conductivity image (382 nodes).

Feature Extraction: Principal Component Analysis

The down sampled conductivity image reflects the activities of forearm Internal muscles. Some of these features may be correlative. The reason is that some muscles are co-operative when some gestures and motions are performed. Therefore, dimensionality reduction is necessary. Principal Component Analysis (PCA) is one of the common methods.

Amusing that there are m samples in the dataset, and each sample has n dimensions. The dataset can be identified as a matrix D of m rows by n columns. PCA was carried out based on the following steps [18] (Fig. 1).

(1) The data should be centered for PCA. The centered dataset is denoted by D_0.
(2) Calculate the covariance matrix C of D_0.
(3) Calculate the eigenvalues and eigenvectors of C.
(4) Obtain the projection matrix, P, by selecting the k most important eigenvectors.
(5) The new dataset is $D_{train} = D_0P$, where D_{train} is a m-by-k matrix.

In our dataset, the max eigenvalue was around 4.75×10^6. When $k > 69$, the eigenvalue was less than 1. Taking the computation efficiency, we determined $k = 40$.

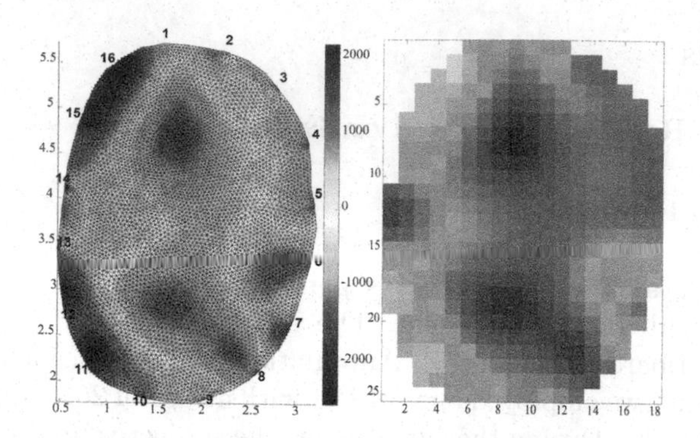

Fig. 1. The original conductivity image (left) and the down sampled conductivity image (right)

Classifier: Quadratic Discriminant Analysis (QDA)
In Quadratic Discriminant Analysis (QDA), the covariance matrices of each class are not necessarily equal. One way to obtain the equation of boundary of classes is equating the posterior probabilities. [19] Take multiple classes into consideration, the discriminant equation is

$$f_k(x)\pi_k = \frac{1}{\sqrt{(2\pi)^d|\Sigma_k|}} \exp\left(-\frac{(x-\mu_k)^T\Sigma_k^{-1}(x-\mu_k)}{2}\right)\pi_k \tag{12}$$

where $f_k(x) = P(X = x|x \in C_k)P(x \in C_k)$ is *likelihood* and π_k is *prior probability.* After performing natural logarithm and dropping the constant term, the equation above becomes

$$\delta_k(\mathbf{x}) = -\frac{1}{2}\ln(|\Sigma_k|) - \frac{1}{2}(x-\mu_k)^T\Sigma_k^{-1}(x-\mu_k) + \ln(\pi_k) \tag{13}$$

The class of a sample $\mathbf{x}$ is estimated as:

$$\hat{C} = \arg\max_k \delta_k(\mathbf{x}) \tag{14}$$

In our problem, x is one row in the D_{train}. C is the set of different kinds of gestures. k represents one kind among nine kinds.

3.2 Conductivity Reconstruction Parameters Analysis

Experimental Protocol
We collected the EIT data of the motion from Wrist Flexion (WF) to Wrist Extension (WE). In this experiment, we employed one subject with an age of 24, a height of 165 cm, and a weight of 67 kg. The perimeter of their measured forearms (maximum part) was 25 cm. The length of their forearms was 23.5 cm. During the experiment, every subject wore the sensing front-end as the same

as that of the experiment in 3.1, and IMU was placed on subject's hand. They carried out the motion from WF to WE for 15–20 cycles, and each cycle was around 3–5 s. The simulation of current level and contact impedance was carried out through EIDORS toolbox.

Current Level
The maximum current, which can be applied to the electrodes within the safety limits, is 5 mA [20]. We reconstructed images at the current level from 0.5 mA to 15.0 mA and the step size between each iteration was 0.5. We tried to see its effect to the resolution, which is defined as the number of pixels utilized in the construction of a digital image [21]. In our study, we defined the resolution as the area ratio between different muscles. In order to simplify the representation of the spatial resolution, we focused on the change of the most distinguishable area, where we chosen based on our observation of the simulation result. The resolution in our preliminary study is defined as:

$$R = \frac{S_d}{S_f} \tag{15}$$

where R is the resolution, S_d is the area of the most distinguishable area, and S_f is the area of the forearm section.

Contact Impedance
The complete electrode model (CEM) allows a complex impedance for each electrode, and the contact impedance could affect the model. But the point electrode model (PEM) doesn't consider the contact impedance of an electrode. In EIDORS, the parameter *z_contact* means the impedance of the material out of which the electrode is constructed [22].

In [23], the researchers found out that the mean electrode contact impedance increases did not result in artefacts in the reconstructed image. An experiment about the effect of contact impedance was carried out based on our data. We collected the EIT data of the motion from WF to WE. Then we down sampled the reconstructed image and calculated the R^2 between each node and the movement, which is represented by the angle detected by IMU. We selected the region with high R^2 as Region of Interest (ROI, shown in Fig. 2(a)) and simulate the relation between the mean value of ROI nodes and contact impedance. The contact impedance ranged from 0.1 to 0.6 and the step size between each iteration was 0.01.

4 Results

4.1 Recognition Accuracy

In our dataset, there are about 10,606 samples, each of which is 208 measured voltages. We separate them into training set and testing set as a ratio of 1:1. After Cross Validation, the average accuracy is 99.845%, and the confusion matrix is shown in Fig. 2(b). The x-axis means the real labels, and the y-axis means the prediction of QDA.

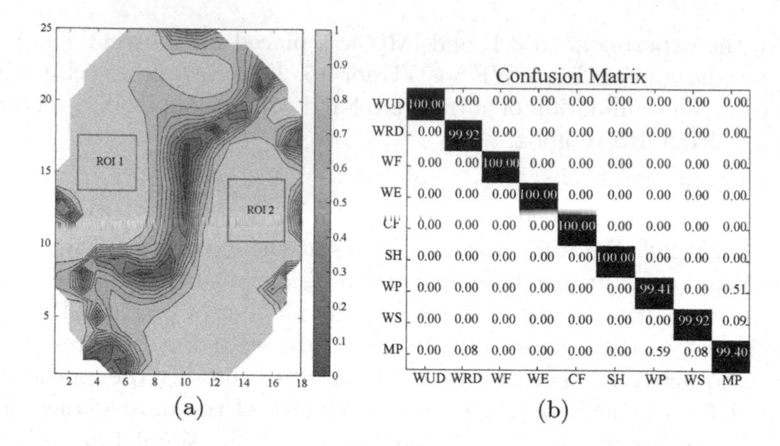

Fig. 2. (a) The Region of Interest of the motion from WF to WE. The regions in red boxes were selected manually based on R^2. (b) The Confusion Matrix of result (Color figure online)

4.2 The Effect of EIT Parameters

The variation of resolution with the increase of current level is shown in the Fig. 3(a).

The relation between the mean value of ROI nodes and contact impedance is shown in Fig. 3(b). The mean values of ROI decreases with the increase of contact impedance from 0.1 to 0.6. However, the value change is not obvious (From 311.7 to 311.35 in ROI1 and from 144.4 to 143.8 in ROI2).

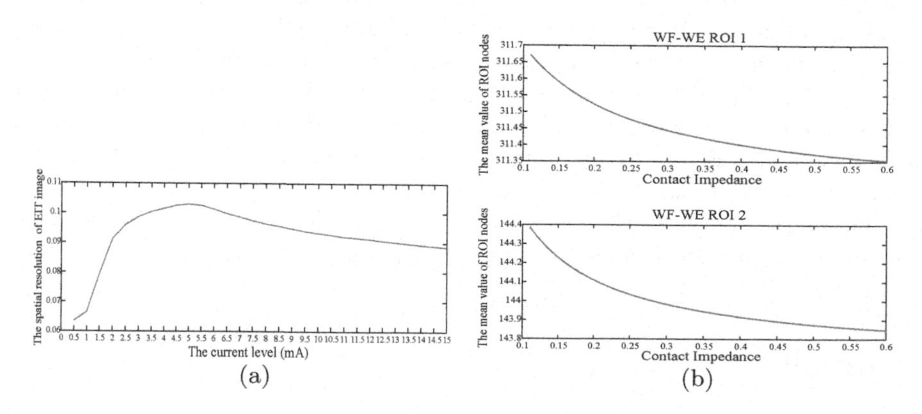

Fig. 3. (a) The resolution changes with the increase of current level from 0.5 mA to 15.0 mA. (b) The relation between the mean value of ROI nodes and contact impedance

5 Discussions and Conclusions

In this pilot study, we performed PCA and QDA to recognize the gestures based on EIT signals, and the accuracy can reach 99.845%. In [6], Hand gesture classification accuracy was 81.5 ± 10.3% for EMG only, 80.6 ± 8.5% FMG only, and 91.6 ± 3.5% for co located EMG-FMG. In [11], the accuracy can reach 97% on gross hand gestures and 87% on thumb-to-finger pinches. Reference to [12], eight hand gestures can be distinguished with accuracy of 97.9% for 2D EIT and 99.5% for 3D EIT. It can be concluded that the EIT signals could detect more slight muscle movement compared with sEMG signals, and that our EIT signals could produce comparable recognition results with other EIT-based strategies for up-limb motion recognition. Wrist Pronation (WP) and Middle Position (MP) are easily confused according to our result.

We discussed the effect of conductivity reconstruction parameters. In our simulation, the current level is from 0.5 mA to 15 mA. When the level is lower than 5.5 mA, the higher current level is, the better the resolution is. It means that the contrast between different muscles become more obvious with the increase of current level. But when the level is higher than 5.5 mA, the higher current level is, the smaller the distinguishable area is. And the similar changes also appeared in [24]. In this article, images had been generated by different currents from 0.1 mA to 5 mA. The image using 0.1 mA was not detailed in components. The best image reconstruction was the one with 1 mA. With higher current level near to 5 mA the reconstructed image performs a low resolution, because only high intensity points are distinguishable. In future studies, we will conduct more extensive experiment on forward model and stimulation current. As for the contact impedance, the mean value of ROI nodes didn't show obvious change when the contact impedance increases. Taking the factors of safety and performance into consideration, the current level can be set at 5 mA and the contact impedance can be set at 0.01 in our simulation.

EIT is a technology which is portable, non-invasive and inexpensive. Besides, compared with sEMG or other biological signals, EIT can obtain the information from deep muscles. The results proved that the feasibility of human discrete gesture recognition based on our proposed EIT-based HMI. However, the resolution is not high enough to obtain every muscle's activity precisely. In order to address this issue, some priori information of forearm should be introduced to the FEM model. The musculoskeletal model can be taken into consideration to raise recognition accuracy and improve the performance of Human-machine interface applied to robotic control. Future works should be done to obtain more exciting and extensive outcomes.

Acknowledgements. This work was supported by the National Natural Science Foundation of China (NO. 62073318).

References

1. Berg, J., Lu, S.: Review of interfaces for industrial human-robot interaction. Curr. Robot. Rep. **1**(2), 27–34 (2020). https://doi.org/10.1007/s43154-020-00005-6
2. Zhang, X.: Human-robot collaboration focusing on image processing (2020)
3. Badr, A.A., Abdul-Hassan, A.K.: A review on voice-based interface for human-robot interaction. Iraqi J. Electr. Electron. Eng. **16**(2), 91–102 (2020)
4. Li, K., Zhang, J., Wang, L., Zhang, M., Li, J., Bao, S.: A review of the key technologies for sEMG-based human-robot interaction systems. Biomed. Sig. Process. Control **62**, 102074 (2020)
5. Ajoudani, A.: Transferring Human Impedance Regulation Skills to Robots. Springer, Cham (2016). https://doi.org/10.1007/978-3-319-24205-7
6. Jiang, S., Gao, Q., Liu, H., Shull, P.B.: A novel, co-located EMG-FMG-sensing wearable armband for hand gesture recognition. Sens. Actuators A: Phys. **301**, 111738 (2020)
7. Han, J., Ding, Q., Xiong, A., Zhao, X.: A state-space EMG model for the estimation of continuous joint movements. IEEE Trans. Ind. Electron. **62**(7), 4267–4275 (2015)
8. Sikdar, S., et al.: Novel method for predicting dexterous individual finger movements by imaging muscle activity using a wearable ultrasonic system. IEEE Trans. Neural Syst. Rehabil. Eng. **22**(1), 69–76 (2013)
9. Zheng, E., Mai, J., Liu, Y., Wang, Q.: Forearm motion recognition with noncontact capacitive sensing. Front. Neurorobot. **12**, 47 (2018)
10. Zheng, E., Zeng, J., Xu, D., Wang, Q., Qiao, H.: Non-periodic lower-limb motion recognition with noncontact capacitive sensing. In: 2020 IEEE/ASME International Conference on Advanced Intelligent Mechatronics (AIM), pp. 1816–1821 (2020)
11. Zhang, Y., Harrison, C.: Tomo: wearable, low-cost electrical impedance tomography for hand gesture recognition. In: Proceedings of the 28th Annual ACM Symposium on User Interface Software & Technology, pp. 167–173 (2015)
12. Jiang, D., Wu, Y., Demosthenous, A.: Hand gesture recognition using three-dimensional electrical impedance tomography. IEEE Trans. Circuits Syst. II: Express Briefs **67**(9), 1554–1558 (2020)
13. Zheng, E., Li, Y., Zhao, Z., Wang, Q., Qiao, H.: An electrical-impedance-tomography-based interface for human-robot collaboration. IEEE/ASME Trans. Mechatron. (2020)
14. Zong, Z., Wang, Y., Wei, Z.: A review of algorithms and hardware implementations in electrical impedance tomography. Progr. Electromagn. Res. **169**, 59–71 (2020)
15. Boyle, A.J.S.: The effect of boundary shape deformation on two-dimensional electrical impedance tomography. Ph.D. thesis, Carleton University (2010)
16. Adler, A., Guardo, R.: Electrical impedance tomography: regularized imaging and contrast detection. IEEE Trans. Med. Imaging **15**(2), 170–179 (1996)
17. Adler, A., Lionheart, W.R.B.: Uses and abuses of EIDORS: an extensible software base for EIT. Physiol. Meas. **27**(5), S25 (2006)
18. Ghojogh, B., Crowley, M.: Unsupervised and supervised principal component analysis: tutorial. arXiv preprint arXiv:1906.03148 (2019)
19. Ghojogh, B., Crowley, M.: Linear and quadratic discriminant analysis: tutorial. arXiv preprint arXiv:1906.02590 (2019)
20. Lionheart, W.R.B., Kaipio, J., McLeod, C.N.: Generalized optimal current patterns and electrical safety in EIT. Physiol. Meas. **22**(1), 85–90 (2001)

21. Chitturi, V., Farrukh, N.: Spatial resolution in electrical impedance tomography: a topical review. J. Electr. Bioimpedance **8**, 66–78 (2017)
22. Aadler. The effect of contact impedance. [EB/OL]. http://eidors3d.sourceforge.net/tutorial/EIDORS_basics/contact_impedance.shtml Accessed 28 Feb 2017
23. Boyle, A., Adler, A.: The impact of electrode area, contact impedance and boundary shape on EIT images. Physiol. Meas. **32**(7), 745 (2011)
24. Canales-Vásquez, D.: Electrical impedance tomography (EIT) image reconstruction for the human forearm (2016)

Robot Manipulation Skills Learning

Depth from Shading Based on Online Illumination Estimation Under RAMIS Environment

Jiacheng Fan, Yuan Feng, Jinqiu Mo, Shigang Wang, and Qinghua Liang(✉)

School of Mechanical Engineering, Shanghai Jiao Tong University,
800 Dongchuan Road, Shanghai, China
qhliang@sjtu.edu.cn

Abstract. Shape From Shading (SFS) algorithm has attracted many attentions in the Robot-Assisted Minimally Invasive Surgery (RAMIS) environment due to the superior texture-free characteristic. But this algorithm is limited to shape reconstruction rather than depth reconstruction. Unlike natural illumination environment, the main reason of this limitation in RAMIS environment is that the illumination condition does not calibrated. In this paper, the imaging principle, incorporating surface reflectance and camera parameter, is fully modeled. Based on the model and the practical environment of RAMIS, we present an online illumination estimation method. Without requiring additional equipment, our method only uses images captured from two positions to perform. The illumination estimation is accomplished based on a Gradient Descent formulation. The result and accuracy of the method is shown through synthetic image experiment.

Keywords: Shape From Shading · Robot-assisted minimally invasive surgery · Depth reconstruction

1 Introduction

Robot-Assisted Minimally Invasive Surgery (RAMIS) systems have drawn attentions extensively since they have the advantages of small trauma and precise operation. There are many researches dedicated to improve the intelligence of the systems. The inability to perceive the geometric information [7] of the surgical environment is one of the major obstacles to help them build intelligence. There is one significant characteristic of the environment: the visual area is illuminated by a near-field point light source. How to reconstruct the 3D information (depth) under this environment is still an open case. Shape From Shading (SFS) [2] is a unique monocular 3D reconstruction algorithm. The basic principle of this algorithm is to reversely infer the shape of the image based on the bidirectional reflectance distribution function (BRDF) reflectance model. That means the complex illumination condition is not the obstacle but the basis of reconstruction. The solution approach makes SFS a very suitable algorithm for the applications under the RAMIS environment.

X.-J. Liu et al. (Eds.): ICIRA 2021, LNAI 13014, pp. 39–47, 2021.
https://doi.org/10.1007/978-3-030-89098-8_4

Before SFS can be implemented, the reflectance parameters of the BRDF reflectance model should be calibrated. There are two types of reflectance parameters [5]: illumination related and material related. Material related reflectance parameters affect the computation of shape (normal field). Usually illumination related parameters are neglected since the illumination condition under natural environment is uniform everywhere [12]. That is why SFS algorithm only focuses on shape reconstruction rather than depth reconstruction. However, under RAMIS environment, the illumination is near-field point light without any ambient light. This characteristic means that the illumination related parameters directly affects the image intensity. As a result, depth computation can be accomplished by calibrating illumination related parameters.

Typically, the calibration of the parameters require additional equipment [13]. But the equipment is too sophisticated and large to be utilized under RAMIS environment. In order to bypass the equipment, many researches have been studied on the calibration of reflectance parameters only rely on multi-view images. We summarize them as multi-view approaches. K. Hara et al. [4] estimate light source position and reflectance from a single image. Yet the impractical assumptions limit the universality of this method. T. Yu et al. [16] derive a View Independent Reflectance Map from BRDF model. They convert the estimation of many view dependent reflectance maps into the estimation of one view independent reflectance map. C., Yu et al. [15] present a global minimization method based on branch-and-bound algorithm along with convex minimization. K.J., Yoon et al. [14] propose a method based on variational framework. This framework can refine surface shape and estimate reflectance parameters simultaneously. G. Oxholm et al. [12] demonstrate a reflectance estimation method based on Bayesian formulation. D. Maurer et al. [10] also propose a variational method that combines data terms from multi-view stereo and SFS based on a separate parametrisation for depth and reflectance parameters. Compared with [14] and [12], this method does not require initial shape as input. With respect to the reflectance calibration under RAMIS environment, there are also some studies. They are regarded as RAMIS approaches. A. J., Chung et al. [1] are the first to calibrate reflectance parameters under RAMIS environment. Abed Malti and Adrien Bartoli [9] further extend the method presented in [1] to a more realistic reflectance model which contains the modeling of specular component. Most recently, A.L., Nunes et al. [11] introduced a non-retroreflective method to efficiently measure the reflectance parameters in vivo.

The purpose of reflectance estimation in multi-view approaches is to provide essential information for SFS, such that SFS can recover surface details. The depth calculation of the surface is usually accomplished by stereo-type methods. Besides, the aim of RAMIS approaches is to help with the realism of synthetic simulations in RAMIS, for examples, virtual and augmented reality techniques. To sum up, existing methods do not resolve the calibration of illumination related parameters. SFS algorithm is still restricted to the shape reconstruction.

To the best of our knowledge, there is no research aiming for enabling SFS with the ability of reconstructing depth instead of only shape. To this end, we need to calibrate illumination related parameters, which is quite different from existing methods. As for the material related parameters, they can be chosen by building the material database same as [8]. In this paper, we introduce an online illumination parameters estimation method. This method also requires no additional equipment. It can be performed by one commonly used monocular endoscope. By moving the endoscope vertically on two positions, we can calibrate the illumination parameters. Through synthetic image experiment, we show the performance of our method.

2 Shape from Shading

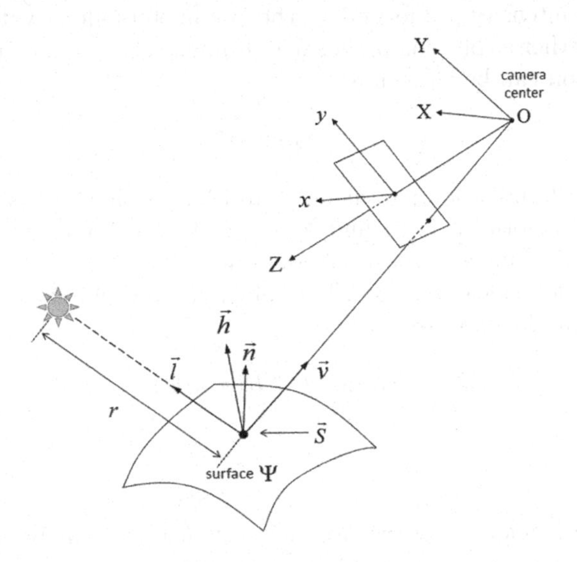

Fig. 1. SFS model

SFS model is the fundamental of all subsequent derivation. In order to better represent the characteristics of RAMIS environment, we follow the model of our former research [3] in this paper. The hybrid surface assumption makes the model more suitable for this environment than other models. As shown in Fig. 1, $\boldsymbol{l}$ is normalized light direction vector pointing from surface point $\boldsymbol{S}$ to light source. $\boldsymbol{v}$ is the normalized viewing direction vector. It points from surface point $\boldsymbol{S}$ to the camera center O. $\boldsymbol{h}$ is the bisector vector of $\boldsymbol{l}$ and $\boldsymbol{v}$. $\boldsymbol{n}$ is the surface normal of the surface point $\boldsymbol{S}$. The reflectance equation of the model can be written as:

$$R = I_0\{k\rho\frac{\boldsymbol{ln}}{r^2} + (1-k)\frac{[F_0 + (1-F_0)(1-(\boldsymbol{lh}))^5](\boldsymbol{ln})}{4\pi m^2(\boldsymbol{nh})^4(\boldsymbol{lh})^2 r^2} e^{(\frac{(\boldsymbol{nh})^2-1}{m^2(\boldsymbol{nh})^2})}\} \quad (1)$$

where I_0 is the illumination intensity of the light source. It is one of the illumination related parameters. r is the distance from light source to the surface point S. k, ρ, m, $F0$ are material related parameters of the surface.

In order to express the model more concise, we rewrite Eq. (1) into a simplified version:

$$R = R(I_0, \theta, \Psi) \tag{2}$$

where θ represents the collection of material related parameters. Ψ represents the surface geometry, i.e. a collection of $\boldsymbol{r}$, $\boldsymbol{l}$, $\boldsymbol{v}$, $\boldsymbol{n}$ and $\boldsymbol{h}$. Note that Ψ can also be regarded as point cloud of the surface. I_0 has a linear relationship with R. θ and Ψ have a complex non-linear relationship with R.

However, Eqs. (1) and (2) are not fully equivalent to the intensity (gray value) of the image. According to the basic principle of image formation [6], there are many other parameters of the camera further contribute to the image intensity such as exposure time, gain and so on. Since they have linear relationship with the image intensity, we formulate the parameters along with illumination intensity I_0 together as one comprehensive illumination parameter σ. The image intensity equation can be written as:

$$I = R(\sigma, \theta, \Psi) \tag{3}$$

With known illumination parameter σ and known material parameter θ, the solution of SFS algorithm is to find a proper surface geometry Ψ which can make the theoretical image intensity (right side of Eq. (3)) coincide with the real image intensity (left side of Eq. (3)). The solution is a sophisticated optimization scheme, which can be expressed as:

$$\Psi = argmin([I - R(\sigma, \theta, \Psi)]^2) \tag{4}$$

3 Method

As we illustrated before, material related parameter θ can be determined by selecting the parameters in the material database which are consistent with the target soft tissue in the visual field. Therefore, that leaves the comprehensive illumination parameter σ and the surface geometry Ψ to be computed. As one image can only provide one constraint, solving this problem requires at least two images. Besides, the complex non-linear relationship between Ψ and I makes the solution beyond simple linear computation.

As shown in Fig. 2, we move the endoscope vertically in two positions 1 and 2. Image I_1 and image I_2 on positions 1 and 2 are captured respectively. The motion is controlled by surgical robot. Therefore, the transformation matrix M between these two positions can be obtained easily. The surface geometries of the overlapped area under O_1 and O_2 coordinate systems are denoted as Ψ_1 and Ψ_2. If the surface geometries Ψ_1, Ψ_2 and comprehensive illumination parameter σ are correct, there should exist following equation:

$$\begin{aligned} \Psi_2 &= M\Psi_1 \\ I_1 &= R(\sigma, \theta, \Psi_1) \\ I_2 &= R(\sigma, \theta, \Psi_2) \end{aligned} \tag{5}$$

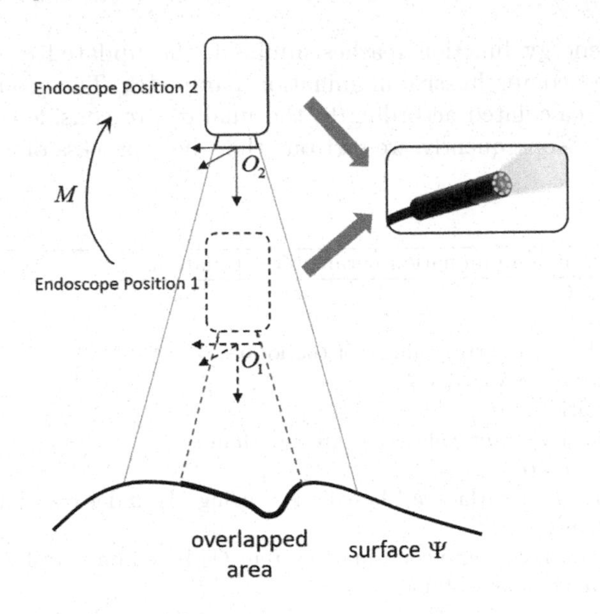

Fig. 2. Method setup

Based on above equation, we can derive our solution accordingly. We take lower position 1 as the primal position. By giving arbitrary comprehensive illumination parameter σ', SFS algorithm can reconstruct the surface $\Psi_1{}'$. By transforming $\Psi_1{}'$ onto coordinate system O_2 using transformation matrix M, we obtain $\Psi_2{}'$ of the overlapped area. Based on this geometry $\Psi_2{}'$, we can back render a image $R(\sigma', \theta, \Psi_2{}')$ according to Eq. (3). $R(\sigma', \theta, \Psi_2{}')$ should be equal to the overlapped area of image I_2. If not, the comprehensive illumination parameter σ' can be updated to minimize the error between the two images. We denote the overlapped area of image I_2 as $I_2{}'$, then the error between $R(\sigma', \theta, \Psi_2{}')$ and I_2 is modeled as an energy function:

$$e = [R(\sigma', \theta, \Psi_2{}') - I_2{}']^2 \tag{6}$$

Apparently, the error can be minimized by Gradient Descent method. Based on the energy function (6), the iterative calculation of σ' is:

$$\sigma'^{n+1} = \sigma'^{n} + 2\lambda[R(\sigma'^{n}, \theta, \Psi_2{}'^{n}) - I_2{}']\frac{\partial R(\sigma'^{n}, \theta, \Psi_2{}'^{n})}{\partial \sigma'^{n}} \tag{7}$$

where λ is the coefficient of the iteration step. $\frac{\partial R(\sigma'^{n}, \theta, \Psi_2{}'^{n})}{\partial \sigma'^{n}}$ is the partial differential term of R w.r.t. σ'. The value of σ'^0 should be chosen according to the approximate $\Psi_2{}'$. If approximate $\Psi_2{}'$ is about 50 mm–100 mm, we can choose the depth as 75 mm to build a $\Psi_2{}'^0$ and substitute it into Eq. (3) to calculate σ'^0.

When the energy function reaches minimal, the updated σ' is considered as the estimated comprehensive illumination parameter. The geometries (point clouds) $\Psi_1{}', \Psi_2{}'$ calculated according to the final σ' are considered as the final true geometries. Consequently, we sort out the whole process of our method as shown below.

Method: Online illumination parameter estimation

Input: I_1, I_2, M

Initialize:

Set threshold error value ϵ of the loop
Set iterative variable λ
Set initial σ'
Set initial e (any value that greater than ϵ)

While $e < \epsilon$ **do**

Reconstruct surface $\Psi_1{}'$ based on σ' using (4) under coordinate sy--stem O_1
Transform $\Psi_1{}'$ onto coordinate system O_2 based on transformation matrix M to obtain $\Psi_2{}'$
Re-project surface $\Psi_2{}'$ back onto image plane coordinate and mark out image $I_2{}'$ of the overlapped area from I_2
Render theoretical image $R(\sigma', \theta, \Psi_2{}')$
Update σ' according to (7)
Re-compute error value e according to (6)

End While

Output: Final $\sigma', \Psi_1{}', \Psi_2{}'$

---

4 Experiment and Result

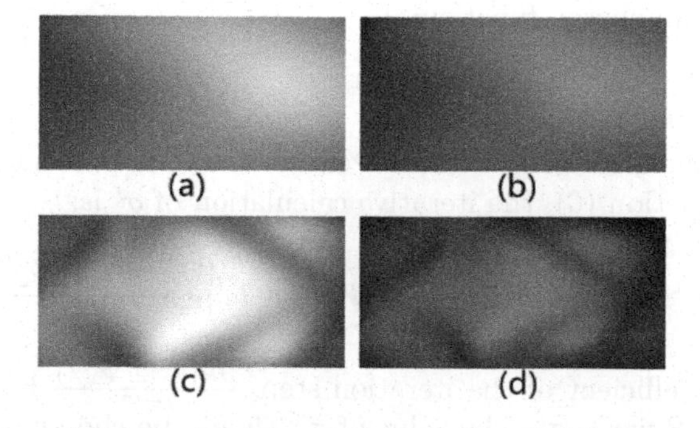

Fig. 3. Experimental images. (a), (c) are the image captured from position 1. (b), (d) are the image captured from position 2.

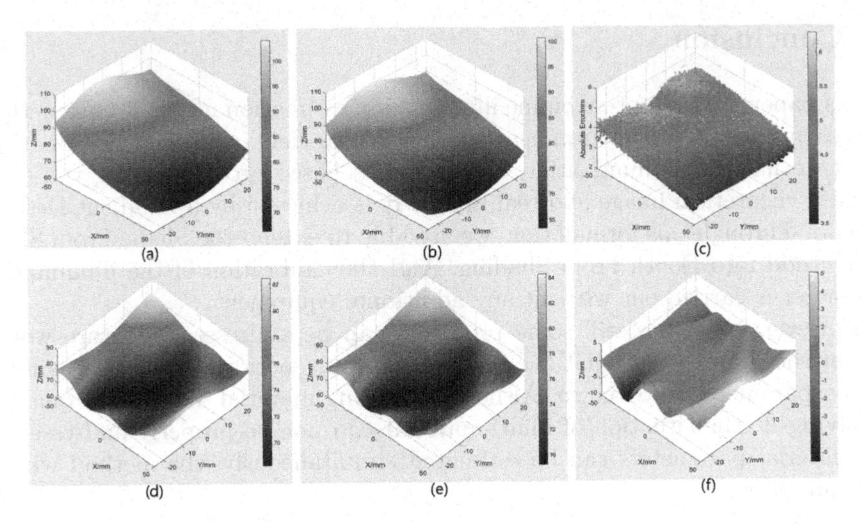

Fig. 4. Reconstruction result. (a), (d) are the depth result of our method. (b), (e) are the ground-truth result. (c), (f) are absolute error map w.r.t depth result of our method and ground-truth result.

We conduct our experiment based on two sets of synthetic images. The surfaces are smooth surfaces, which are common in RAMIS environment. To mimic the endoscope, the light sources of the experiments are located at the camera center. The rendering parameters of the experiment are $I_0 = 1.8e6, \rho = 0.8, k = 0.7, m = 0.7, F_0 = 0.7$. The rendered images are shown in Fig. 3. Figure 3 (a), (c) are the image captured from position 1. Figure 3 (b), (d) are the image captured from position 2. The transformation matrix of Fig. 3 (a), (b) is $M = \begin{bmatrix} 1 & 0 & 0 & 0 \\ 0 & 1 & 0 & 0 \\ 0 & 0 & 1 & 10 \end{bmatrix}$. The transformation matrix of Fig.3 (c),(d) is $M = \begin{bmatrix} 1 & 0 & 0 & 0 \\ 0 & 1 & 0 & 0 \\ 0 & 0 & 1 & 20 \end{bmatrix}$.

We employ our method on the images. The estimated illumination parameter σ' of Fig. 3 (a), (b) is $1.744e6$. The estimated illumination parameter σ' of Fig. 3 (c), (d) is $1.769e6$. Compared with ground-truth illumination parameters, the relative errors of the estimation are 3.11% and 1.72% respectively. After estimating the illumination parameters, geometries are further computed based on parameter. The depth results of the primal position 1 are shown in Fig. 4. Figure 4 (a),(d) are the depth results of our method. Figure 4 (b), (e) are the ground-truth results. Figure 4 (c), (f) are absolute error maps w.r.t depth result of our method and ground-truth result. Compared with the ground-truth depth, our reconstruction result has good accuracy.

The results of our experiment shows that our method can estimate the illumination parameter well and provide the geometry with promising accuracy.

5 Conclusion

In this paper, we present an online illumination estimation method. Our method focuses on the RAMIS environment which has near-field point light illumination. The true nature of illumination estimation is based on minimizing the error between theoretical image and real image. It is achieved by a Gradient Descent method. Through this formulation, we are able to extend the Shape From Shading method into Depth From Shading. And the calibration of the illumination parameter is carried out without any additional equipment.

However, there are still some problems can be addressed in future works. One of the limitation is that the method can only perform when the material related parameters are known. Although building material database is one way of solving it, the situation of multi-material can not be properly addressed. If the material parameters can be estimated simultaneously, the method will be more adaptive.

Funding. National Key R&D Program of China (No. 2017YFB1302901).

References

1. Chung, A.J., Deligianni, F., Shah, P., Wells, A., Yang, G.Z.: Patient-specific bronchoscopy visualization through BRDF estimation and disocclusion correction. IEEE Trans. Med. Imaging **25**(4), 503–513 (2006)
2. Distante, Arcangelo, Distante, Cosimo: Shape from shading. In: Handbook of Image Processing and Computer Vision, pp. 413–478. Springer, Cham (2020). https://doi.org/10.1007/978-3-030-42378-0_5
3. Fan, J., Chen, M., Mo, J., Wang, S., Liang, Q.: Variational formulation of a hybrid perspective shape from shading model. Vis. Comput. 1–14 (2021). https://doi.org/10.1007/s00371-021-02081-x
4. Hara, K., et al.: Light source position and reflectance estimation from a single view without the distant illumination assumption. IEEE Trans. Pattern Anal. Mach. Intell. **27**(4), 493–505 (2005)
5. Hoffman, N.: Background: physics and math of shading. Phys. Based Shading Theor. Pract. **24**(3), 211–223 (2013)
6. Wietzke, L., Sommer, G., Schmaltz, C., Weickert, J.: Differential geometry of monogenic signal representations. In: Sommer, G., Klette, R. (eds.) RobVis 2008. LNCS, vol. 4931, pp. 454–465. Springer, Heidelberg (2008). https://doi.org/10.1007/978-3-540-78157-8_35
7. Li, M., et al.: Intra-operative tumour localisation in robot-assisted minimally invasive surgery: a review. Proc. Inst. Mech. Eng. **228**(5), 509–522 (2014)
8. Lim, M., Yoon, S.: A method for estimating reflectance map and material using deep learning with synthetic dataset. arXiv preprint arXiv:2001.05372 (2020)
9. Malti, A., Bartoli, A.: Estimating the cook-torrance brdf parameters in-vivo from laparoscopic images. In: Workshop on Augmented Environment in Medical Image Computing and Computer Assisted Intervention (MICCAI), Nice, France, pp. 768–774 (2012)

10. Maurer, D., Ju, Y.C., Breuß, M., Bruhn, A.: Combining shape from shading and stereo: a joint variational method for estimating depth, illumination and albedo. Int. J. Comput. Vis. **126**(12), 1342–1366 (2018). https://doi.org/10.1007/s11263-018-1079-1
11. Nunes, A.L., Maciel, A., Cavazzola, L., Walter, M.: A laparoscopy-based method for BRDF estimation from in vivo human liver. Med. Image Anal. **35**, 620–632 (2017). https://doi.org/10.1016/j.media.2016.09.005
12. Oxholm, G., Nishino, K.: Multiview shape and reflectance from natural illumination. In: Proceedings of the IEEE Conference on Computer Vision and Pattern Recognition, pp. 2155–2162 (2014)
13. Wu, C., Narasimhan, S.G., Jaramaz, B.: A multi-image shape-from-shading framework for near-lighting perspective endoscopes. Int. J. Comput. Vis. **86**(2), 211–228 (2010)
14. Yoon, K.J., Prados, E., Sturm, P.: Joint estimation of shape and reflectance using multiple images with known illumination conditions. Int. J. Comput. Vis. **86**(2), 192–210 (2010)
15. Yu, C., Seo, Y., Lee, S.W.: Global optimization for estimating a BRDF with multiple specular lobes. In: 2010 IEEE Computer Society Conference on Computer Vision and Pattern Recognition, pp. 319–326. IEEE (2010)
16. Yu, T., Xu, N., Ahuja, N.: Shape and view independent reflectance map from multiple views. Int. J. Comput. Vision **73**(2), 123–138 (2007)

Self-learning Visual Servoing for Robot Manipulation in Unstructured Environments

Xungao Zhong[1](✉), Chaoquan Shi[1], Jun Lin[1], Jing Zhao[1], and Xunyu Zhong[2]

[1] School of Electrical Engineering and Automation,
Xiamen University of Technology, Xiamen 361024, China
[2] School of Aerospace Engineering, Xiamen University, Xiamen 361005, China
zhongxunyu@xmu.edu.cn

Abstract. Current visual servoing methods used in robot manipulation require system modeling and parameters, only working in structured environments. This paper presents a self-learning visual servoing for a robot manipulator operated in unstructured environments. A Gaussian-mapping likelihood process is used in Bayesian stochastic state estimation (SSE) for Robotic coordination control, in which the Monte Carlo sequential importance sampling (MCSIS) algorithm is created for robotic visual-motor mapping estimation. The Bayesian learning strategy described takes advantage of restraining the particles deterioration to maintain the robot robust performance. Additionally, the servoing controller is deduced for robotic coordination directly by visual observation. The proposed visual servoing framework is applied to a manipulator with eye-in-hand configuration no system parameters. Finally, the simulation and experimental results demonstrate consistently that the proposed algorithm outperforms traditional visual servoing approaches.

Keywords: Bayesian filtering · Gaussian-mapping estimation · Visual servoing

1 Introduction

The current visual feedback schemes are mainly deployed in structured environments and requires a calibration process based on known target model, and the robotic autonomy is limited due to predefined coding with certain kinematics model. Especially, the rigid-soft integrated robotic with varied stiffness would bound to make those modeling methods invalid. Thus, it has brought about new challenges for the robot perceptual coordination in unstructured environment, and it is also a cutting-edge problem in field of robotics visual servoing (VS) control [1–4].

The VS is a promising solution to control the robots interacting with their working environments [5–7]. Among various kinds of VS methods, the position-based visual servoing (PBVS), and the image-based visual servoing (IBVS) are popular ones and widely deployed in many manufacturing applications [8]. PBVS retrieves 3D pose information in Cartesian space based on known camera projection parameters. But it was sensitive to calibration errors [9], and the 2D image features beyond the controlling may disappear

X.-J. Liu et al. (Eds.): ICIRA 2021, LNAI 13014, pp. 48–57, 2021.
https://doi.org/10.1007/978-3-030-89098-8_5

from camera's field-of-view (FOV) [10]. In contrast, IBVS estimates the target position through the 2D image features, However, IBVS cannot keep the 3D Cartesian movement of robot insider its workspace, particularly when a large displacement of posture coordination is required.

It is clear that the mentioned VS methods face the robust camera calibration or depth information problems. And the visual feedback schemes are mainly deployed in specific environments requires definitive target modeling. These may not be possible for many real-world applications that have unknown system parameters and contain uncertain and dynamic changes [11–14].

Thus, this work tries to develop a visual servoing which the crucial problem is online learning mapping between robotic visual-motor spaces, we consider it as a dynamic state self-learning problem without the hand-eye calibration parameters, the depth of target and the robot kinematics modeling. More specifically, our ideas behind self-learning visual servoing with visual-motor global mapping is treated as SSE in the Markov sense. The Bayesian-based MCSIS algorithm is then presented to estimate the system state by sampling particles with their corresponding weights.

Differing from the traditional PBVS and IBVS [8], our new VS framework is very flexible for hand-eye feedback system without parameters. The method can be expanding for other filed robotic tasks, including future stiffness-varied robotics. The paper has made the following contributions, 1) The global mapping was defined for robotic 2D visual and 3D motor spaces, and the mapping is treated as a SSE in the Markov sense. The Bayesian-based MCSIS prediction and updating recursive process is utilized to solve the mapping on-line learning problem. 2) The servoing controller is developed by using the MCSIS. This new visual servoing framework is implemented on a 6-DOFs robot with eye-in-hand configuration, which allows the robot to adapt its motion to image feature changes without system calibration and kinematics modeling, also the depth information.

2 Descriptions on Visual Servoing

Figure 1 shows the facilities, in which the manipulator is mounted with an on-board camera and the target is assumed to be stationary with respect to the base frame of robot. The objectives of VS is described to derive the end-effector moving by using a set of observed image features to minimize the error, that is defined as:

$$\mathbf{e}_s(t) = \mathbf{S}(t) - \mathbf{S}^d = \left(s_1(t) - {s_1}^d, s_2(t) - {s_2}^d, \ldots, s_n(t) - {s_n}^d\right)^{\mathbf{T}} \tag{1}$$

where $\mathbf{S}(t) = (s_1(t), \ldots, s_n(t))^{\mathbf{T}} \in \Re^{2n\times 1}$ and $\mathbf{S}^d = \left(s_1^d, \ldots, s_n^d\right)^{\mathbf{T}} \in \Re^{2n\times 1}$, the elements $s_i(t) = (u_i(t), v_i(t))^{\mathbf{T}}$, $s_i^d = \left(u_i^d, v_i^d\right)^{\mathbf{T}}$ are the image position of the ith feature corresponding to the current robot pose and the desired ones, respectively.

Due to the camera motion, the time variation of the image feature $\mathbf{S}(t)$ can be written by [15]:

$$\dot{\mathbf{S}}(t) = \mathbf{L}_t(\mathbf{S}(t), z(t))\mathbf{U}_c(t) \tag{2}$$

where $\mathbf{U}_c(t) \in \Re^{6\times1}$ is the camera's instantaneous velocity screw in the camera frame, $\mathbf{L}_t(\mathbf{S}(t), z(t))$ is the interaction matrix, which depending on camera-object projection configuration, one has:

$$\mathbf{L}_t(\mathbf{S}(t), z(t)) = \left(L_1(s_1(t), z_1(t))\big|_{S=S_1} \cdots L_n(s_n, z_n(t))\big|_{S=S_n}\right)^{\mathbf{T}} \in \Re^{2n\times6} \quad (3)$$

where $L_i(s_i(t), z_i(t))$ is the ith feature point's interaction matrix, it takes the same forms as in the IBVS, we have:

$$L_i(s_i(t), z_i(t)) = \begin{pmatrix} \frac{\ell}{z_i(t)} & 0 & -\frac{u_i(t)}{z_i(t)} & -\frac{u_i(t)v_i(t)}{\eta} & \ell+\frac{u_i^2(t)}{\ell} & -v_i(t) \\ 0 & \frac{\ell}{z_i(t)} & -\frac{v_i(t)}{z_i(t)} & -\ell-\frac{v_i^2(t)}{\ell} & -\frac{u_i(t)v_i(t)}{\ell} & u_i(t) \end{pmatrix} \quad (4)$$

where ℓ is the focal length of the camera, $z_i(t)$ is depth of the target point i.

A good calculation of the interaction matrix in (2) is very crucial to apply VS. The production of that matrices at each time instant must require: 1) an accurate depth of the target point; 2) the camera parameters; 3) the camera calibration related to the robot frame. These are complicated and inevitably causes a large amount of transformation calculation and the problem of system local stability, also the robot applications are limited.

3 Robust MCSIS Learning Estimator

To overcome the difficulties regarding the interaction matrix calculation, we propose a nonparametric VS which is treated as a SSE problem with the MCSIS. As shown in Fig. 1, the mapping between the image space and the robotic moving space, is defined as:

$$\mathbf{L}(t) = \frac{\delta\mathbf{S}(t)}{\delta\mathbf{U}_c(t)} \in \Re^{n\times6} \quad (5)$$

Let the stochastic state vector be formed by concatenations of the row and the column elements of $\mathbf{L}(t)$, *i.e.*

$$\mathbf{L}(t) \to \mathbf{X}(t)\text{: } \Re^{n\times6} \mapsto \Re^{6n\times1}$$

Yet the general, we have the state transition and state measurement formulated by a Gaussian-likelihood process with hidden state variables, according to:

$$\mathbf{X}(t) = f_t(\mathbf{X}(t-1), u(t), \vartheta)+w(t)\ , \quad w \sim N(0, Q(t)) \quad (6)$$

$$\mathbf{Z}(t) = h_t(\mathbf{X}(t), \vartheta)+v(t)\ , \quad v \sim N(0, R(t)) \quad (7)$$

where the nonlinear function f_t is described by a prior probability density function (PDF) $f_t \propto p(\mathbf{x}(t)|\mathbf{x}(t-1))$, and h_t is described by a normal likelihood PDF $h_t \propto p(\mathbf{x}_{(t)}|\mathbf{z}_{(t)})$, $\mathbf{Z}(t) \in \Re^{r\times1}$ is observed random variable, the measurements value can be represented by a finite set $\mathbf{Z}_{(1:t)} = ([Z]_{k,1}, \ldots, [Z]_{k,t})$, which is characterized by a posterior PDF

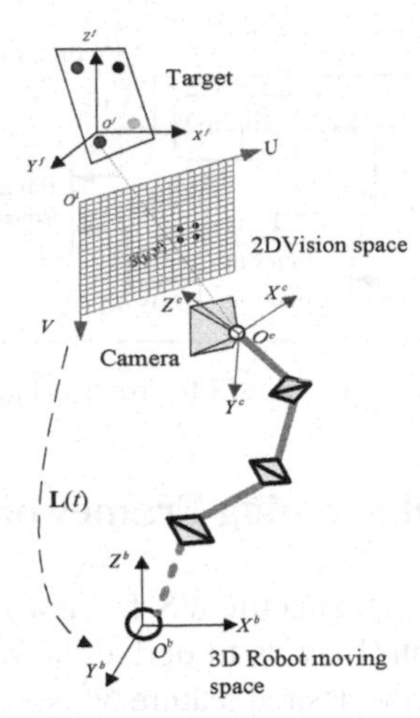

Fig. 1. The flow of vision based feedback control without parameters. The $\boldsymbol{O}^C$ is camera frame and $\boldsymbol{O}^b$ is the robot base frame, $\mathbf{L}(t)$ is the global mapping between 3D movement space of robotic and 2D image space.

$\mathbf{z}_{(1:t)} \propto p\left(\mathbf{x}_{(t)} \middle| \mathbf{z}_{(1t)}\right)$, $u(t)$ is the control forcing data, ϑ is the model parameter, $w(t) \in \Re^{6n\times 1}$ and $v(t) \in \Re^{r\times 1}$ are the process and measurement noise terms, respectively.

According to the basic Bayesian-based method, the accurate estimation of the posterior PDF at each time step will constitute a complete solution for minimum-variance state estimation equation [16]. Thus, the MCSIS is to enlarge the sampling variance to cover the possible state distribution so that the posterior density can be represented nonparametrically by N particles through a generalized PDF, we have:

i) The updating equation of Bayesian given by [17]:

$$p(\mathbf{X}(t)|\mathbf{Z}(1:t)) = \frac{h_{(t)}p(\mathbf{X}(t)|\mathbf{Z}(1:t-1))}{p(\mathbf{Z}(t)|\mathbf{Z}(1:t-1))} \approx \sum_{i=1}^{N} \tilde{\mathbf{W}}^i(t)\delta(\mathbf{X}(t) - \tilde{\mathbf{X}}^i(t)) \tag{8}$$

where $\delta(\cdot)$ is a Dirac delta function.

The sampling particles-weights pair $\left\{\left(\tilde{\mathbf{X}}^i(t), \tilde{\mathbf{W}}^i(t)\right)\right\}_{i=1}^{N}$ by MCSIS can then be used to compute an estimate of the system state $\hat{\mathbf{X}}(t)$. Hence, the state estimation equation in Bayesian is replaced by the sum of products of the particles-weights pair.

ii) The minimum mean squared state estimate is defined as [17]:

$$\hat{\mathbf{X}}(t) = \int_{\mathbf{Z}(1:t)} \mathbf{X}(t)p(\mathbf{X}(t)|\mathbf{Z}(1:t))d\mathbf{X}(t) \approx \sum_{i=1}^{N} \tilde{\mathbf{W}}^i(t)\tilde{\mathbf{X}}^i(t) \tag{9}$$

Then the global mapping can be recovered from state estimation value, *i.e.* $\hat{\mathbf{X}}(t) \rightarrow \mathbf{L}(t)$: $\Re^{6n\times 1} \mapsto \Re^{n\times 6}$. The MCSIS scheme for robust state estimation is shown in the Fig. 2

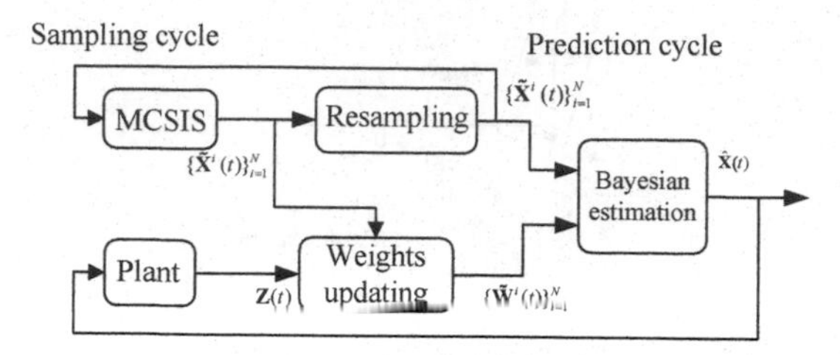

Fig. 2. The scheme of MCSIS for robust state estimation.

4 Nonparametric Visual Servoing Framework

In this section we propose a nonparametric VS framework, which based on learning MCSIS scheme, where the control law is to derive the robot motion $\mathbf{U}_e(t)$ by image feedback $\mathbf{S}(t)$. Considering that the desired feature S^d is a constant parameter due to the fixed goal pose, the differential coefficient of image error $\mathbf{e}_s(t)$ in (1) is

$$\dot{\mathbf{e}}_s(t) = \frac{d}{dt}\left(\mathbf{S}(t) - \mathbf{S}^d\right) = \dot{\mathbf{S}}(t) \tag{10}$$

And substituting (2) into (10), we have

$$\dot{\mathbf{e}}_s(t) = \mathbf{L}(t)\mathbf{U}_c(t) \tag{11}$$

Selecting a non-zero constant λ to make the following equation establish

$$\dot{\mathbf{e}}_s(t) = -\lambda\mathbf{e}_s(t) \tag{12}$$

Then substituting (12) into (11), we obtained

$$-\lambda\mathbf{e}_s(t) = \mathbf{L}(t)\mathbf{U}_c(t) \tag{13}$$

According to (13), the robot moving can be derived to:

$$\mathbf{U}_c(t) = -\lambda\hat{\mathbf{L}}^{\dagger}(t)\mathbf{e}_s(t) \tag{14}$$

where λ is the control rate, and $\hat{\mathbf{L}}^{\dagger}(t) \in \Re^{6\times n}$ is the inverse mapping matrix, given by:

$$\hat{\mathbf{L}}^{\dagger}(t)(t) = \mathbf{L}(t)^T\left(\mathbf{L}(t)\mathbf{L}(t)^T\right)^{-1} \tag{15}$$

Thus, by using the Euler forward discretization rule, the control law (14) can be to approximate as:

$$\mathbf{U}_c(t_{k+1}) = \mathbf{U}_c(t_k) - \lambda\hat{\mathbf{L}}^{\dagger}(t_k)\mathbf{e}_s(t_k)\Delta t_k \tag{21}$$

where Δt_k denotes the sampling period at time t_k.

5 Results and Discussions

The image features have four-feature-points and are used for testing the robot. The feature vector $\mathbf{S}(t)$ is obtained at each time instant through:

$$\mathbf{S}(t) = (u_1, v_1, u_2, v_2, u_3, v_3, u_4, v_4)^{\mathbf{T}} \in \Re^{8\times 1} \tag{16}$$

The desired features vector $\mathbf{S}^d$ does not change over time, and therefore can be calculated before the main control loop of the experiment.

The robot moving control is $\mathbf{U}_c(t) = \left(v_c(t)\, w_c(t)\right)^{\mathbf{T}} \in \Re^{6\times 1}$, $v_c(t) = \left(v_x(t), v_y(t), v_y(t)\right)$ and $w_c(t) = \left(w_x(t), w_y(t), w_y(t)\right)$ denote the instantaneous linear and angular velocities respectively.

The size of the mapping matrix is 8×6 and initialized by introducing the robot moving at the neighborhood of its initial position n times $\partial \mathbf{U}_c(t)^{i-n-1}$, ..., $\partial \mathbf{U}_c(t)^i$. The corresponding features displacements are $\partial \mathbf{S}(t)^{i-n-1}$, ..., $\partial \mathbf{S}(t)^i$. The mapping $\mathbf{L}(0)$ then could be obtained by:

$$\mathbf{L}(0) = \left(\partial \mathbf{S}(t)^{i-n-1}, \ldots, \partial \mathbf{S}(t)^i\right)\left(\partial \mathbf{U}_c(t)^{i-n-1}, \ldots, \partial \mathbf{U}_c(t)^i\right)^T \tag{17}$$

The size of the system state is 48×1 which was initialized according to $\mathbf{L}(0) \rightarrow \mathbf{X}(0)$, $\Re^{8\times 6} \mapsto \Re^{48\times 1}$. The control rate λ is 0.25.

This test case is to compare the performances among classic PBVS, IBVS [8], and our method. In this case, the task requires the initial and desired locations of image features are close to edge of image plane. The results are illustrated in Fig. 3 and Fig. 4.

The Comparison Between the Performance of Image Feature Trajectories: Figure 3(a) shows that the image feature trajectories are constrained by the IBVS controller to move on straight lines from the initial toward desired feature points. Figure 3(b) shows the image feature easily leaves the FOV when the PBVS method for this task. Figure 3(c) shows the feature trajectories obtained by our method were kept on the FOV for the same task, and the feature points did not leave the image plane. Hence, in terms of feature trajectories, the PBVS is not perfect compared with IBVS and our method.

The Comparison Between the Performance of Robot Moving Trajectories: Figure 4(a) shows that the robot motion in the 3D workspace obtained by IBVS often becomes slight odd, i.e., the robot undergoes an abrupt movement at the initial period. For the same task, Fig. 4(b) shows that PBVS method enables the robot motion to be constrained by the controller on a straight line from the initial pose toward the desired pose. Figure 4(c) shows the result obtained by our method, in which the robot moves trajectory seemingly straight line from the initial pose positioning to the desired pose with smooth stability. Hence, in terms of robot trajectory, the IBVS is not perfect compared with PBVS and our method.

Note that the PBVS and IBVS methods need the system calibration and target depth to calculate the mapping matrix, those traditional servo control method, there is sensitive to the modeling error, while our method is on-line learning identified techniques without requiring any calibration parameters and the depth information, the method is independent from the robot's kinematic model. The results by our VS method, shown that, the proposed approach takes the perfect advantages in both trajectories of the image feature on image plane and robot movement in the Cartesian space, since the mapping estimator is learning to simulate IBVS and PBVS, the controller performed the robot robust movement in workspace as PBVS did and keep the features in FOV as IBVS did.

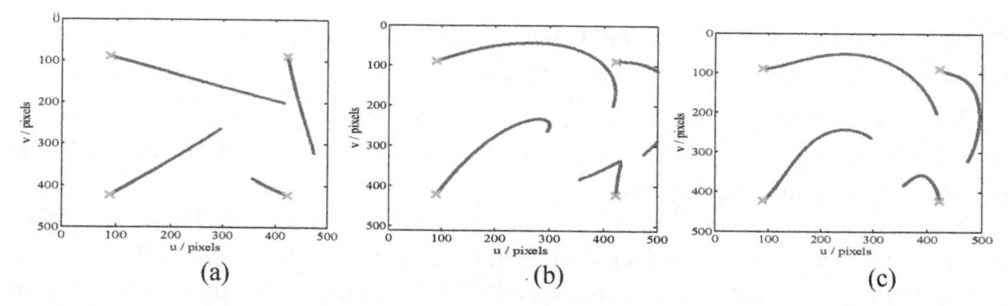

Fig. 3. Results of feature trajectories, (a) the IBVS method, (b) the PBVS method, (c) the proposed method.

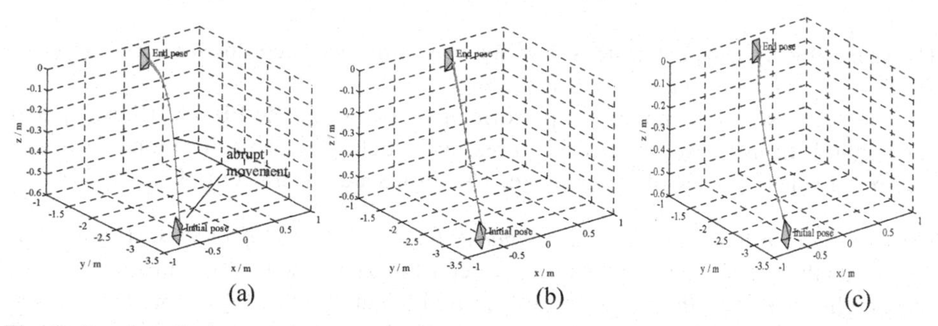

Fig. 4. Results of robot moving trajectory, (a) the IBVS method, (b) the PBVS method, (c) the proposed method.

The real experiments have been carried out by using an eye-in-hand robot in our lab. Figure 5 shows the VS system, including a DENSO RC7M-VSG6BA controller and a computer with an Intel Corei5 1.8-GHz CPU, 4 GBs RAM for image processing and servo control algorithm. The robotic controller and computer can communicate through RS232C serial interface. A DENSO VS-6556GM 6-DoF robotic arm has a USB camera randomly mounted at its end-effector. The object is an board with the printed small circular disks on it. The images are captured at 30 Hz, with a resolution of 640 × 480. The center points of small circular disks are used as image feature.

The experimental results with different representative tests are given to show the performance of the nonparametric VS with learning MCSIS method, as follows.

In Test 1, the initial and desired locations of the image features are chosen in a way that the robot needs to move in a combination of long distance translational and small scale rotational. Figure 6 shows that our method did well with the rotation and the feature trajectories did not leave the image plane although the image points were very close to

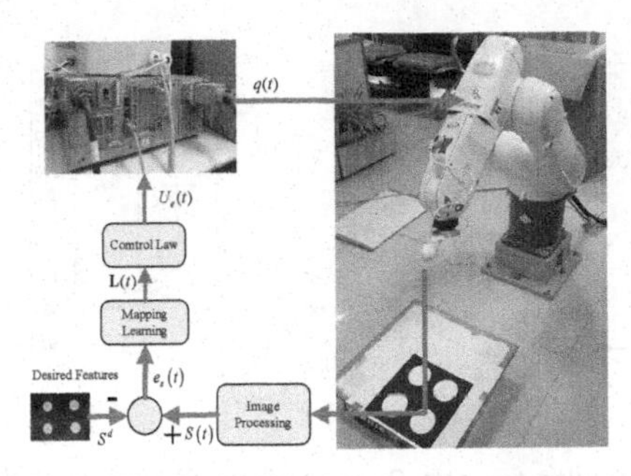

Fig. 5. The eye-in-hand system, the robot without kinematic modling, and the camera randomly mounted at its end-effector.

the FOV edges (see Fig. 6(a)), and the robot moving trajectory was a straight line from the initial pose positioning to the desired pose smoothly and stably without retreat (see Fig. 6(b)). The image errors were reduced to zero when the task was completed, which can be clearly seen in Fig. 6(c).

In Test 2, the servoing task is prepared which requires the robot combination of long distance translational and large scale rotational movement. The results of this test are shown in Fig. 7, where although the robot with a large scale of rotation, the performances of the feature trajectories are within image plane (see Fig. 7(a)). The results of the robot moving in the Cartesian space show that the trajectory reached the desired pose after a long distance translation and rotation (see Fig. 7(b)). At the end pose, the steady-state errors in the image plane are close to zero (see Fig. 7(c)), hence the servo tasks were completed with robust stability.

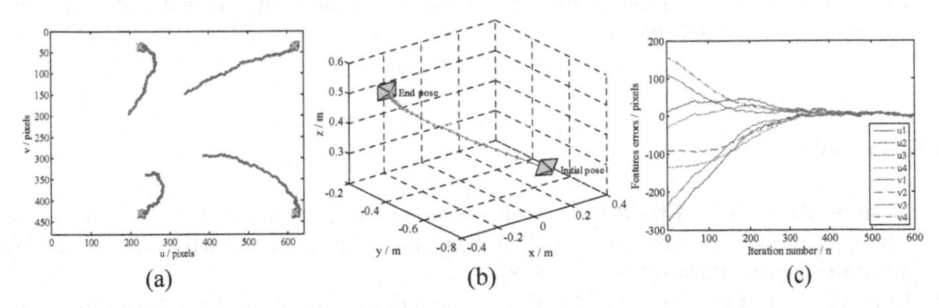

Fig. 6. The experimental results for test 1, (a) image feature trajectories, (b) robot moving trajectory, (c) image errors.

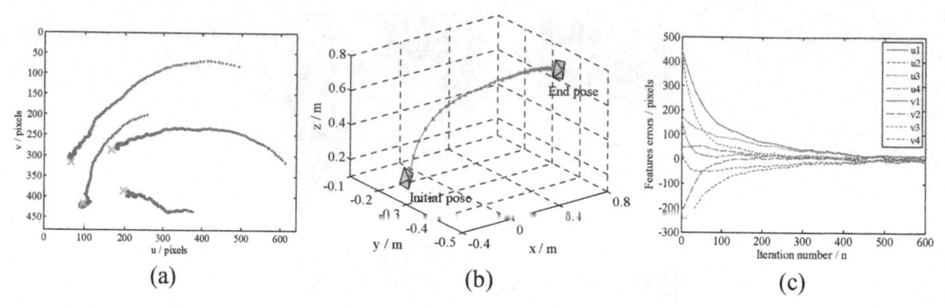

Fig.7. The experimental results for test 2, (a) image feature trajectories, (b) robot moving trajectory, (c) image errors.

6 Conclusion

In this paper, based on current visual servoing methods, a sequential importance sampling Bayesian-based algorithm is proposed for visual-motor spaces mapping and identification of robotic manipulators. The proposed approach does not require the calibration of system parameters and the depth knowledge of the target. It is a more flexible nonparametric VS framework and has been applied to an eye-in-hand robotic manipulator with six degrees-of-freedom (DOFs) no need considering kinematic. Several simulation and experiments were conducted to demonstrate the feasibility and good performances of the proposed approach in comparison to some existing VS methods.

Acknowledgments. This work was supported in part by the National Natural Science Foundation of China under Grant (No. 61703356), in part by the Natural Science Foundation of Fujian Province under Grant (No. 2018J05114 and 2020J01285), in part by the Innovation Foundation of Xiamen under Grant (No. 3502Z20206071).

References

1. Khansari-Zadeh, S.M., Khatib, O.: Learning potential functions from human demonstrations with encapsulated dynamic and compliant behaviors. Auton. Robots **41**(1), 45–69 (2016). https://doi.org/10.1007/s10514-015-9528-y
2. Fang, G., et al.: Vision-based online learning kinematic control for soft robots using local Gaussian process regression. IEEE Robot. Autom. Lett. **4**(2), 1194–1201 (2019)
3. He, W., Dong, Y.: Adaptive fuzzy neural network control for a constrained robot using impedance learning. IEEE Trans. Neural Netw. Learn. Syst. **29**(4), 1174–1186 (2018)
4. Kong, L., He, W., Yang, C., Li, Z., Sun, C.: Adaptive fuzzy control for coordinated multiple robots with constraint using impedance learning. IEEE Trans. Cybern. **49**(8), 3052–3063 (2019)
5. Calli, B., Dollar, A.M.: Robust precision manipulation with simple process models using visual servoing techniques with disturbance rejection. IEEE Trans. Autom. Sci. Eng. **16**(1), 406–419 (2019)
6. Huang, B., Ye, M., Hu, Y., Vandini, A., Lee, S., Yang, G.: A multirobot cooperation framework for sewing personalized stent grafts. IEEE Trans. Ind. Inf. **14**(4), 1776–1785 (2018)

7. He, W., Xue, C., Yu, X., Li, Z., Yang, C.: Admittance-based controller design for physical human-robot interaction in the constrained task space. IEEE Trans. Autom. Sci. Eng. (2020). https://doi.org/10.1109/TASE.2020.2983225
8. Chaumette, F., Hutchinson, S.: Visual servo control. part I: basic approaches. IEEE Robot. Autom. Mag. **13**(4), 82–90 (2006)
9. Do-Hwan, P., Jeong-Hoon, K., In-Joong, H.: Novel position-based visual servoing approach to robust global stability under field-of-view constraint. IEEE Trans. Ind. Electron. **59**(10), 4735–4752 (2012)
10. Farrokh, J.S., Deng, L.F., Wilson, W.J.: Comparison of basic visual servoing methods. IEEE/ASME Trans. Mechatron. **16**(5), 967–983 (2011)
11. Chang, W.: Robotic assembly of smartphone back shells with eye-in-hand visual servoing. Robot. Comput. Integr. Manuf. **50**, 102–113 (2018)
12. Zhao, Y., Lin, Y., Xi, F., Guo, S.: Calibration-based iterative learning control for path tracking of industrial robots. IEEE Trans. Ind. Electron. **62**(5), 2921–2929 (2015)
13. Kudryavtsev, A.V., et al.: Eye-in-hand visual servoing of concentric tube robots. IEEE Robot. Autom. Lett. **3**(3), 2315–2321 (2018)
14. Xu, F., Wang, H., Wang, J., Samuel Au, K.W., Chen, W.: Underwater dynamic visual servoing for a soft robot arm with online distortion correction. IEEE/ASME Trans. Mechatron. **24**(3), 979–989 (2019)
15. Farrokh, J.S., Deng, L., Wilson, W.J.: Comparison of basic visual servoing methods. IEEE/ASME Trans. Mechatron. **16**(5), 967–983 (2011)
16. Atchad, Y., Rosenthal, J.S.: On adaptive Markov chain Monte Carlo algorithms. Bernoulli **11**(5), 815–828 (2005)
17. Isard, M., Blake, A.: Condensation-conditional density propagation for visual tracking. Int. J. Comput. Vis. **29**(6), 5–28 (1998)
18. Liu, C., Wen, G., Zhao, Z., Sedaghati, R.: Neural network-based sliding mode control of an uncertain robot using dynamic model approximated switching gain. IEEE Trans. Cybern. (2020). https://doi.org/10.1109/TCYB.2020.2978003

Obstacle Avoidance Methods Based on Geometric Information Under the DMPs Framework

Ming Shi, Yongsheng Gao(✉), Boyang Ti, and Jie Zhao(✉)

State Key Laboratory of Robotics and Systems, Harbin Institute of Technology, Harbin 150001, China
{gaoys,jzhao}@hit.edu.cn

Abstract. Dynamical Movement Primitives (DMPs) can equip the robot with humanoid characteristics and make it efficiently complete the task. However, research on obstacle avoidance strategies combined with DMPs is still being explored nowadays. In this paper, we proposed new obstacle avoidance methods based on geometric information under the DMPs framework, in order that the robot can still complete the task in an unstructured environment. We first generalized new trajectories by DMPs after demonstration. Then according to the interference area of a static or a moving obstacle after extracting the geometric information of the obstacle, we quantitatively adjusted the generalized trajectories by adding proper offset at a certain direction. Finally we used proportional-derivative (PD) control to ensure that the modified trajectories converge to the original one and the goal. The methods make the robot successfully avoid the interference of a static or a moving obstacle, and besides, they maintain all the advantages of the DMPs framework, such as fast goal convergence, simple mathematical principles and high imitation similarity. We verified the effectiveness of the methods in two dimensional and three dimensional simulation, and set up the real experiments using a static or a moving obstacle respectively, and finally proved the feasibility and validity of the avoidance methods.

Keywords: Dynamical movement primitives · Obstacle avoidance · Geometric information

1 Introduction

In order to complete different complex tasks, robots are usually required to have high adaptability and reliability in motion planning. Traditional motion planning methods like rapidly exploring random tree and artificial potential field have some disadvantages, such as slow convergence speed, easy to fall into local minimum state, etc. [1, 2]. In contrast, Learning from Demonstration (LfD) can intuitively and conveniently let the robot imitate and reproduce the trajectory obtained from demonstration. At the same time, it can generate new trajectories which can adapt to a new situation such as the change of the environment structure, the goal or initial state [3]. In recent years, intelligent algorithms based on LfD like Gaussian mixture model [4, 5] have been emerging endlessly.

X.-J. Liu et al. (Eds.): ICIRA 2021, LNAI 13014, pp. 58–69, 2021.
https://doi.org/10.1007/978-3-030-89098-8_6

However, how to make the robot attain the ability of autonomous obstacle avoidance is also an important research direction in motion planning. Although it is a common topic, the research on obstacle avoidance based on LfD is very limited. To improve the performance of robot avoiding obstacles, LfD methods may provide new ways. In recent years, more and more researchers have been carrying out obstacle avoidance experiments based on Dynamical Movement Primitives (DMPs) [6, 7], one of the most typical algorithms in LfD. For example, Park [8] used the DMPs combined with artificial potential field to achieve obstacle avoidance. Hoffmann [9] modified the dynamic equation of DMPs combined with the characteristics of human obstacle avoidance behavior, adding the coupling term that can avoid the obstacles in the plane. Rai [10] used neural network to train and learn the parameters of the coupling term, effectively improving the ability of different obstacle avoidance, but it was easy to fail in the face of large obstacles. In [11] the improved coupling term was used to describe the dynamically changing spatial obstacles, which made the adaptability of the algorithm more extensive.

In this paper, new obstacle avoidance methods based on geometric information of obstacles under the DMPs framework are proposed. First, the DMPs learns the characteristic parameters corresponding to the trajectory through demonstration. Then, change the environmental conditions such as the position of the initial point and the goal to make DMPs generate trajectories that adapt to the new environment. Add unified description of obstacles in the new environment, and perform the generalization process according to their static or moving states. Perform real-time collision detection, and employ corresponding offset strategies to generate new trajectories avoiding obstacles, until the trajectory is completely away from the interference area.

We prove that the methods inherit all the advantages of DMPs, greatly maintain the shape features of demonstration and most importantly, avoid static or moving obstacles successfully.

The rest of this paper is arranged as follows: the second part introduces the basic principle of DMPs, the obstacle avoidance methods based on ellipsoidal offset and dangerous zone offset; the third part performs simulation in the static and dynamical environment and take the effect in static one to analyze its inheritance of DMPs and high retention of the demonstration; the fourth part is real experiments, which verify the methods in different LfD tasks, and use the real manipulator to prove the feasibility of the methods.

2 Obstacle Avoidance Based on Geometric Information

In LfD tasks, DMPs is generally used to imitate the behavior from demonstration, which can generate discrete or periodic trajectories according to different forms and characteristics of the movement [6]. To retain the advantages of DMPs, the methods in this paper are under DMPs framework and employ a principle of offset to avoid a static or moving obstacle.

2.1 The Fundamental Principle of DMPs Framework

The core of DMPs is the weakly nonlinear second-order attractor system derived from the spring damping system, which can be divided into three parts, a transformation system, a canonical system and a nonlinear force term, respectively. Among them, the transformation system is a second-order linear differential equation written in the form of two first-order differential equations, such as formula (1),

$$\begin{aligned} \tau\dot{v} &= \alpha_v(\beta_v(g-y)-v) \\ \tau\dot{y} &= v \end{aligned} \tag{1}$$

The variables y, v and $\dot{v}$ represent the position, velocity and acceleration of the system. g represents the goal point, τ represents the time constant, α_v and β_v are normal numbers to guarantee that y achieves fast convergence to g.

In addition, DMPs use the canonical system to transform the time independence t into a vector s, which is shown in Eq. (2).

$$\tau\dot{s} = -\alpha_s s \tag{2}$$

For the imitation of demonstration, DMPs mainly rely on the nonlinear forcing term to learn the shape characteristics of teaching trajectory and its specific form is shown in Eq. (3),

$$f(s) = \frac{\sum_{i=1}^{N} \psi_i(s) w_i}{\sum_{i=1}^{N} \psi_i(s)} s \tag{3}$$

$$\psi_i(s) = \exp\left(-h_i(s-c_i)^2\right) \tag{4}$$

where $\psi_i(s)$ is a radial basis function, usually in the form of Gaussian kernel function shown in Eq. (4). N represents the number of kernel functions while h and c represent the width and center of the kernel function respectively.

In conclusion, the weak nonlinear dynamic system of DMPs is as shown in (5),

$$\tau\dot{v} = \alpha_v(\beta_v(g-y)-v) + f(s) \tag{5}$$

The system can fit different forms of the movement. When the s weakens from 1 to 0, the effect of nonlinear forcing term weakens to 0 until the system finally approaches to the spring damping system, and namely, the convergence of the goal is achieved as $(v, y) = (0, g)$.

2.2 Obstacle Avoidance Based on Ellipsoidal Offset of Static Obstacles

In static environment the obstacles are static or move very slowly. Through the depth camera, radar or other external sensors, the real shape and pose of the obstacle can be obtained. Therefore, in order to make the shape of obstacles easier to express, this paper employs spatial quadric surface ellipsoid to describe volumetric obstacles and ellipse to describe planar obstacles [12], which can completely envelope the obstacle and are as small as possible so as not to reduce the space of the robot's reachable workspace.

The obstacle avoidance based on ellipsoidal offset of static obstacles mainly takes the following steps:**1 Collision detection**. After demonstration, the trajectory is generated in real time by DMPs while constantly calculating the distance d between the current position of the trajectory and the ellipse o by formula (6).

$$d = ||y - o||_2 = \sqrt{(y_x - o_x)^2 + (y_y - o_y)^2} \tag{6}$$

2 Obstacle avoidance module. When d is less than the safe distance S, it indicates that the current trajectory is within the interference area of the obstacle. Meanwhile, the starting and the end point of avoidance can be connected to form a straight line, from which the reference points can be taken at a certain interval, in order to generate new points at a certain distance Δ in the centrifugal direction by formula (7), as shown in Fig. 1 (a).

$$\begin{aligned} y'_x &= (a + \Delta) \cdot \cos\theta \\ y'_y &= (b + \Delta) \cdot \sin\theta \end{aligned} \tag{7}$$

3 End obstacle avoidance. When the trajectory has arrived beyond the interference area, the original one generated by DMPs will be used again.

The final avoidance trajectory is shown in Fig. 1 (a). The solid ellipse and the dotted ellipse represent the unified description of the obstacle and its interference area considering the safe distance, respectively. The green line represents the original generalized trajectory while the red line represents the obstacle avoidance trajectory by this method. Similarly, the illustration of obstacle avoidance in 3D space is shown in Fig. 1 (b).

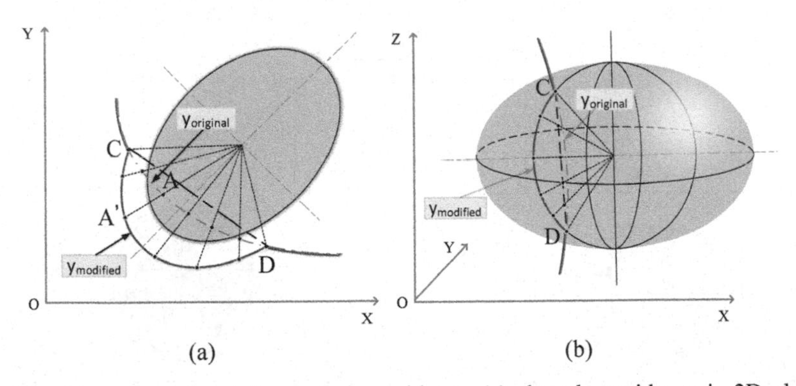

Fig. 1. The illustration of a static obstacle avoidance. (a) obstacle avoidance in 2D plane. (b) obstacle avoidance in 3D space

2.3 Obstacle Avoidance Based on Dangerous Zone Offset of a Moving Obstacle

In the real unstructured environment, it is significant for the robot to avoid moving obstacles in a fast reaction. In this section, based on the dangerous zone offset of a moving obstacle, a new avoidance method for dynamic environment is proposed.

The interference area can be represented by a dangerous zone, as shown in Fig. 2 (a) with its length and width are shown in the Eq. (8),

$$
\begin{aligned}
l_{\parallel} &= 2\alpha v \cdot \mathrm{S} \\
l_{\perp} &= 2\beta \cdot \mathrm{S}
\end{aligned} \tag{8}
$$

where α and β are adjustable constants and meet the requirements of $\alpha, \beta \in (0, 1)$, $v = f \cdot \Delta x$ is the moving speed and S is the artificially-set safe distance which, in this paper, is made to be the maximum between the long and the short half axis of the obstacle ellipse.

1 Collision detection. During the movement of obstacles, it can be described by a certain dangerous zone in every moment. Therefore, the collision between the robot and the moving obstacle can be judged by detecting whether the current robot position and the dangerous zone collides.

2 Obstacle avoidance module. When the trajectory generalized by DMPs intersects with the dangerous zone, the partial trajectory in the zone can be added by the offset along the direction of the obstacle, as shown in Fig. 2 (a), where AB is the original trajectory $y_{original}$, and AB' is the avoidance trajectory $y_{modified}$. We make the offset value *length* increase exponentially, and the variable $d_{\perp}$ is the longitudinal distance between the current trajectory and the center of the zone, as shown in Eq. (9), where constant $\mathrm{l_m}$, b are the peak value and the time constant for adjusting the speed of offset changing respectively. Equation (10) transforms the independent variable $d_{\perp}$ into the intermediate variable x.

$$
length = \mathrm{l_m}(1 - e^{-\mathrm{b}x}) \tag{9}
$$

$$
x = -\frac{8}{b}\left(\frac{d_{\perp}}{\mathrm{S}} - 1\right) \tag{10}
$$

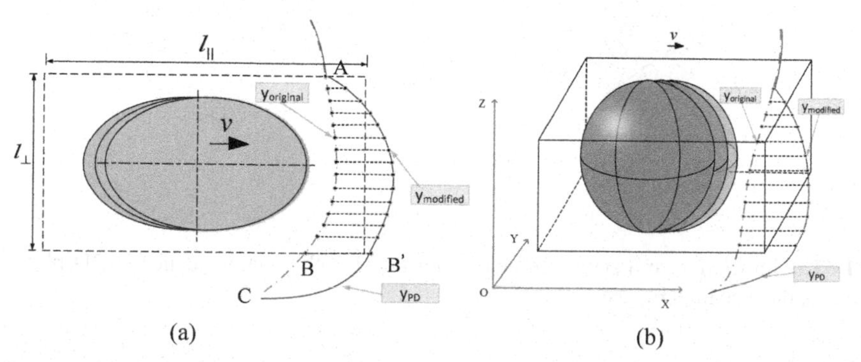

Fig. 2. The illustration of moving obstacle avoidance. (a) The dangerous zone and spatial avoidance of a moving obstacle. (b) The avoidance of a moving obstacle in spatial dangerous zone.

3 End obstacle avoidance. When the current trajectory is away from the zone, the avoidance process ends. However, since there will be a certain deviation between the current and the original position, we add a proportional-derivative (PD) control, as Eq. (11), to guarantee that the trajectory can still converge to the goal.

$$u_k = K_P\left[e_k + T_D\frac{e_k - e_{k-1}}{T}\right] \tag{11}$$

Through multiple iterations, appropriate points are generated to reduce the deviation until the current trajectory approaches the goal generated by DMPs. Obviously, this method still converges to the goal anyway. As is shown in Fig. 2 (a), curve BC is the original trajectory, and B'C segment is the modified trajectory after PD control. Similarly, the illustration of dangerous zone and its obstacle avoidance in 3D space is shown in Fig. 2 (b).

3 Simulation and Analysis

In order to verify the feasibility and validity of the methods, we carried out simulation and analyzed the effect of the obstacle avoidance. For demonstration, we use the letter G in the dataset of [13] as the teaching data to input into the algorithm for training, and then single or multiple obstacles are randomly assigned into the environment.

3.1 Simulation

1 Static environment. As shown in Fig. 3 (a) and Fig. 3 (b), the effect of single obstacle and multiple obstacles avoidance in static environment are shown respectively. The black and the dotted ellipse represent the obstacle and the corresponding interference area considered a certain distance Δ, respectively. The red curve is the original trajectory by DMPs which is the letter G, and the blue curve is the obstacle avoidance trajectory obtained by our method. We can find that our method can generate a new obstacle avoidance trajectory along the periphery of the interference area.

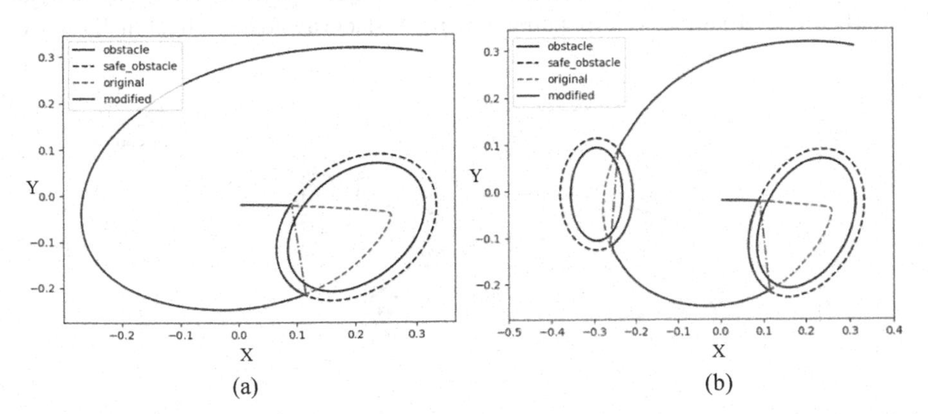

Fig. 3. Obstacle avoidance of letter G in 2D plane. (a) With single static obstacle. (b) With multiple static obstacles.

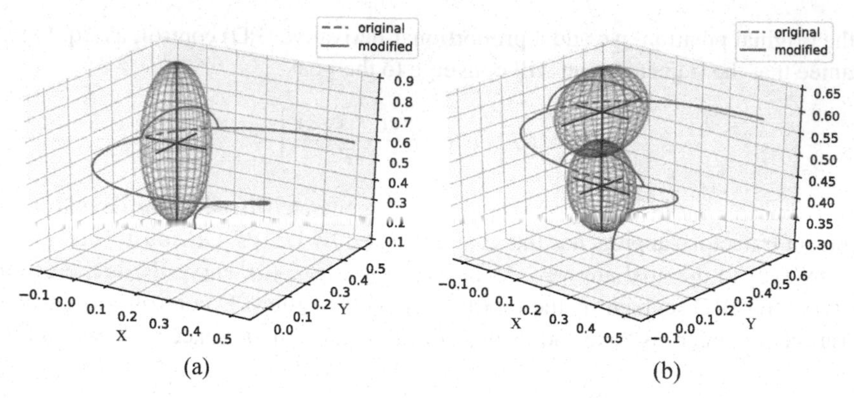

Fig. 4. Obstacle avoidance of spatial G in 3D space. (a) With single static obstacle. (b) With multiple static obstacles.

For the demonstration in 3D space, we make the data in the Z axis monotonously decrease and then we form a teaching data like a spatial letter G. After using our method, we can get the result as shown in Fig. 4, in the case of single obstacle and multiple obstacles in 3D space respectively. The blue ellipsoid is the interference area of the obstacle considered safe distance Δ. The green line is the original trajectory and the red line is the obstacle avoidance trajectory obtained by this method.

2 Dynamic environment. In the dynamic environment, we make the obstacle move along a straight line AB. In order to imitate the effect of external sensors collecting the pose of the obstacle at a certain frequency, we make the obstacle refresh its pose at a certain time interval.

As shown in Fig. 5 (a), it is the result of the moving obstacle avoidance in 2D plane. Among them, the gray ellipse is the unified description of the obstacle, of which the movement from A to B forms overlapping trajectory. The red curve is the original trajectory and the blue one is the obstacle avoidance trajectory generated by our method. Similarly, Fig. 5 (b) is the corresponding moving obstacle avoidance effect in 3D space.

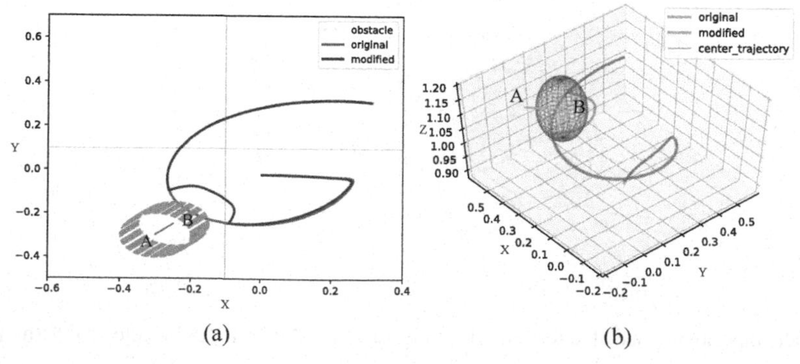

Fig. 5. Moving obstacle avoidance of letter G. (a) The moving obstacle avoidance in 2D plane. (b) The moving obstacle avoidance in 3D space.

3.2 Trajectory Evaluation in Static Obstacle Avoidance

To verify the effect of avoidance trajectories obtained by ellipsoidal offset of static obstacles, we verified that the obstacle avoidance trajectories preserve the shape features of the original trajectory and the generalization features of DMPs framework.

1 Shape feature. In order to prove the high retention of the shape feature of our method based on ellipsoidal offset of static obstacles, the DMPs-based obstacle avoidance method proposed by Ginesi [12] is specially selected for comparison in static environment. As shown in Fig. 6, differences of the shape between these trajectories and the original one are greatly obvious.

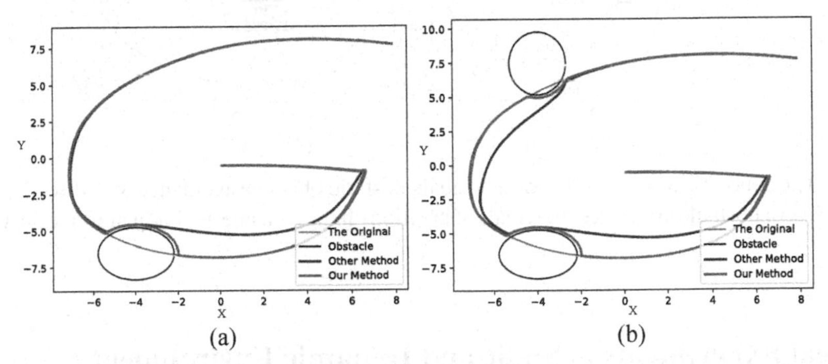

Fig. 6. The difference of two methods in static obstacle avoidance. (a) With single obstacle. (b) With multiple obstacles.

Table 1. The Frechet distance of our method and Ginesi's method [12].

Method	Obstacle information	Frechet distance
Our method	Single obstacle	3.378738109324879
	Multiple obstacles	3.5525857563925287
Ginesi's method	Single obstacle	11.016560138942637
	Multiple obstacles	11.02512320525695

We used the revised Frechet distance [14] to evaluate the similarity between the avoidance trajectories and the original trajectories. The results are shown in the Table 1, indicating that the trajectories generated by our method largely maintained the shape feature of the original trajectory.

2 Generalization feature of DMPs framework. Since DMPs is the basic framework of our obstacle avoidance methods, the inherent characteristics of DMPs are still maintained while avoiding obstacles. To test this, the generalization of variable goal points is carried out in simulation.

As shown in the Fig. 7 (a) and Fig. 7 (b), the results are respectively the tracking effect when the goal point changes continuously during generalization process and the convergence effect after the position of the goal changes suddenly. Through observation, we can find that this method not only achieves the generalization of different goal, but also successfully avoids obstacles.

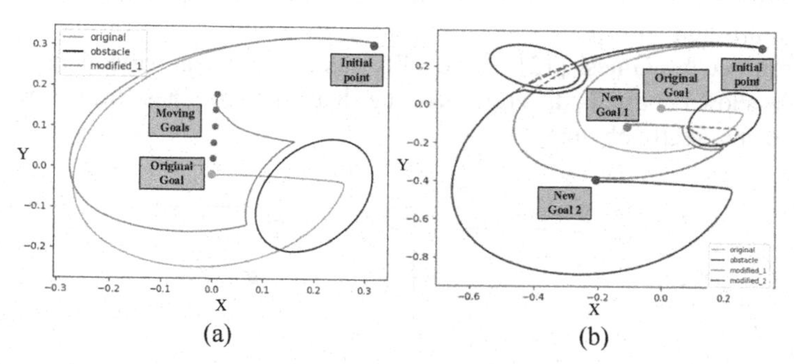

Fig. 7. The generalization effect of variable goals in static obstacle avoidance. (a) Moving goal in the process of obstacle avoidance. (b) Goals of position changed while avoiding multiple obstacles.

4 Real Experiments in Static and Dynamic Environment

We verify the obstacle avoidance effect of the methods on UR10 robot. In the aspect of obstacle recognition, the depth camera Intel RealSense D435 is used to obtain the point cloud data of the obstacle surface, and the minimum ellipsoid of the envelope is obtained according to the least square fitting. The hand eye calibration of the system adopts the method in [15].

For our method tested in static environment, we used one and two bottles as obstacles to avoid when performing the trajectory of pouring task, as shown in Fig. 8 (a) and Fig. 8 (b).

Through observation, we can clearly find that the trajectories modified by our method can avoid the influence of obstacles on the basis of the original generalized trajectory, and still converge to the original goal.

In the dynamic environment, we made the bottle move along the desktop at a certain speed and direction. The original task and obstacle avoidance effect is shown in the Fig. 9 (a) and Fig. 9 (b). Through observation, we can find that our method can generate a trajectory to avoid the moving obstacle. Additionally, we test the writing of letter G on board while making the moving obstacle interrupt during the process. The effect of moving obstacle avoidance is shown in Fig. 9 (d).

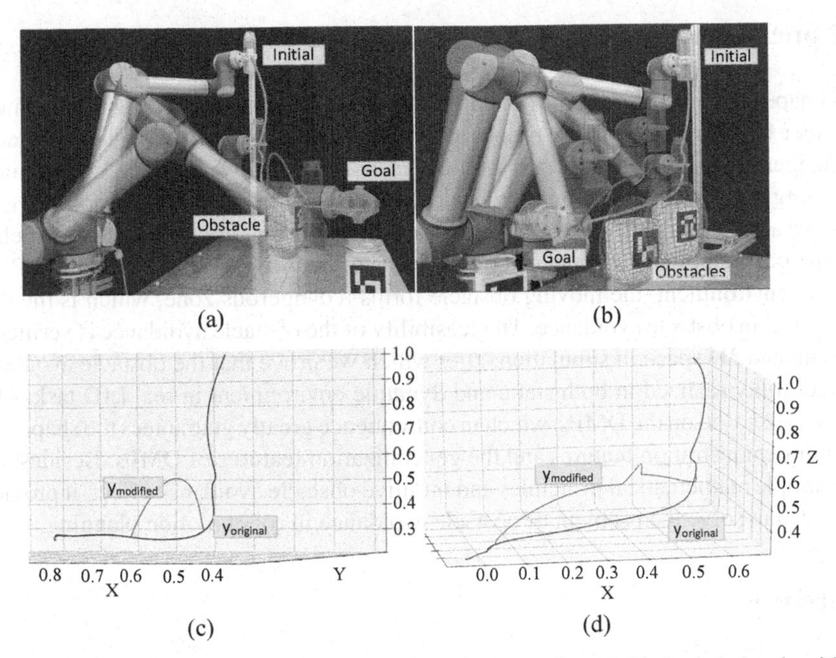

Fig. 8. The effect of static obstacle avoidance in pouring tasks. (a) The pouring task with one obstacle. (b) The pouring task with two obstacle. (c) The original and modified trajectory in pouring task. (d) The generalized and modified trajectory in pouring task.

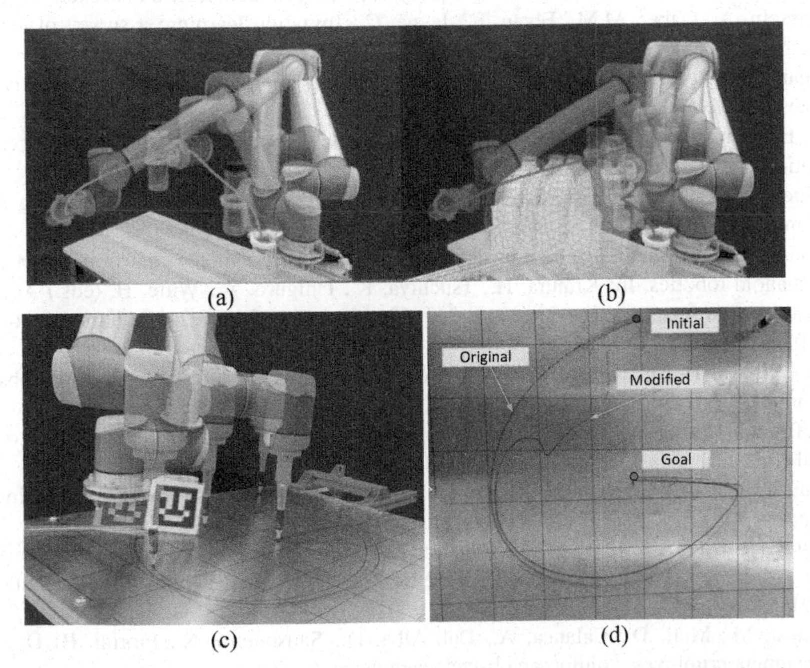

Fig. 9. The effect of moving obstacle avoidance in two LfD tasks. (a) The original trajectory of a pouring task. (b) The modified trajectory after avoiding an obstacle moving from far to near. (c) The setup of planar writing. (d) The writing effect after introducing a moving obstacle.

5 Conclusion

In this paper, according to the movement of the obstacles, new obstacle avoidance methods under DMPs framework are proposed based on ellipsoidal offset of static obstacles and dangerous zone offset of moving obstacles, respectively. In static environment, employing the principle of minimum enclosed volume, the obstacle is simplified to an ellipse or an ellipsoid and then base on the geometric information of the ellipse or ellipsoid, the position offset is calculated on the basis of the original trajectory of DMPs. In dynamic environment, the moving obstacle forms a dangerous zone, which is the base of the offset in obstacle avoidance. The feasibility of the obstacle avoidance is verified in 2D plane and 3D space in simulation, after which we prove that the obstacle avoidance is successfully realized in both static and dynamic environment in real LfD tasks. The methods are based on the DMPs, which in consequence greatly guarantees the shape similarity of demonstration features and the generalization features of DMPs. Besides, due to the simple mathematical principles and intuitive obstacle avoidance effect, it provides a new idea to solve the problem of obstacle avoidance in robot motion planning.

References

1. Zhang, T., Zhu, Y., Song, J.: Real-time motion planning for mobile robots by means of artificial potential field method in unknown environment. Ind. Robot. **37**, 19 (2010)
2. Xu, J., Duindam, V., Alterovitz, R., Goldberg, K.: Motion planning for steerable needles in 3D environments with obstacles using rapidly-exploring random trees and backchaining, p. 6
3. Hussein, A., Gaber, M.M., Elyan, E., Jayne, C.: Imitation learning: a survey of learning methods. ACM Comput. Surv. **50**, 35 (2017)
4. Khansari-Zadeh, S.M., Billard, A.: Learning stable nonlinear dynamical systems with Gaussian mixture models. IEEE Trans. Robot. **27**, 15 (2011)
5. Ti, B.: Human intention understanding from multiple demonstrations and behavior generalization in dynamic movement primitives framework. IEEE Access **4**, 9 (2016)
6. Ijspeert, A.J., Nakanishi, J., Schaal, S.: Learning attractor landscapes for learning motor primitives, p. 8
7. Schaal, S.: Dynamic movement primitives -a framework for motor control in humans and humanoid robotics. In: Kimura, H., Tsuchiya, K., Ishiguro, A., Witte, H. (eds.) Adaptive Motion of Animals and Machines, pp. 261–280. Springer, Tokyo (2006). https://doi.org/10.1007/4-431-31381-8_23
8. Park, D.-H., Hoffmann, H., Pastor, P., Schaal, S.: Movement reproduction and obstacle avoidance with dynamic movement primitives and potential fields, p. 8
9. Hoffmann, H.: Biologically-inspired dynamical systems for movement generation: automatic real-time goal adaptation and obstacle avoidance, p. 6
10. Rai, A., Sutanto, G., Schaal, S., Meier, F.: Learning feedback terms for reactive planning and control, p. 8
11. Pairet, È., Ardón, P., Mistry, M., Petillot, Y.: Learning generalizable coupling terms for obstacle avoidance via low-dimensional geometric descriptors. IEEE Robot. Autom. Lett. **4**, 8 (2019)
12. Ginesi, M., Meli, D., Calanca, A., Dall'Alba, D., Sansonetto, N., Fiorini, P.: Dynamic movement primitives: volumetric obstacle avoidance, 6
13. pbdlib-matlab. https://gitlab.idiap.ch/rli/pbdlib-matlab/

14. Har-Peled, S., Raichel, B.: The Fréchet distance revisited and extended. ACM Trans. Algorithms **10**, 22 (2014)
15. Garrido-Jurado, S.: Automatic generation and detection of highly reliable fiducial markers under occlusion. Pattern Recognit. **13** (2014)

Integrated Classical Planning and Motion Planning for Complex Robot Tasks

Jinzhong Li[1(✉)], Ming Cong[1], Dong Liu[1], and Yu Du[2]

[1] Dalian University of Technology, Dalian 116024, China
lijz_1994@mail.dlut.edu.cn
[2] Dalian Dahuazhongtian Technology Co., Ltd., Dalian 116023, China

Abstract. In order to execute complex tasks in complex environment successfully, robot need to make detailed planning for high-level task and low-level motion according to the requirements of task scenes. In this paper, a method based on the combination of classical planning and motion planning is discussed. In terms of high-level tasks, the robot decomposes a complex task into a collection of simple task, organizes simple task by logical reasoning and finally achieves the goal required to complete the task. In order to improve the adaptability of robot operation skills, the dynamic motion primitives were used to model the robot motion by imitating the interaction characteristics between human and objects. We set up a home service robot task scene to verify the feasibility of the proposed method.

Keywords: Task planning · Classical planning · Dynamic motion primitives

1 Introduction

At present, in structured environment such as industrial scenes, robot has made great research progress and can be competent for accurate and repetitive work. Today, robot are leaving the industrial environment and gradually becoming a part of our lives. We hope that robot can replace human to complete a variety of complex and boring tasks (such as making tea, pouring water, etc.). However, the performance of robot tasks in a home service scenes is quite different from that in industrial scenes, robot need to overcome problems such as uncertainty, high dimension and long sequences. Robot need to adjust its movements as the task scene changes, therefore, humans hope robot can have ability of intelligent decision to cope with complex task challenges.

Automatic planning, also known as artificial intelligence planning, is a field of artificial intelligence that studies the deliberative process of selecting and organizing actions to achieve goal [1]. Human can predicte the result of their behavior automatically, even if they are not fully aware of it. Automatic planning attempt to reproduce human reasoning and behavior and can be used to model the skills and strategies of robot, so as to robot don't need expensive manual code when operating in a different environment. The focus is developing task planners composed of efficient search algorithms to generate solutions to problems. Given a domain that describes the state and behavior of the world, and a problem that describes the initial state and goal state, the task planner

X.-J. Liu et al. (Eds.): ICIRA 2021, LNAI 13014, pp. 70–78, 2021.
https://doi.org/10.1007/978-3-030-89098-8_7

generates a set of actions that guarantee the transition from the initial state to the goal state. In order to achieve the correct transition between different world states, behaviors are defined as precondition and effects, representing the state of the world before and after the behavior is executed respectively. Planning can be thought of as the process of selecting appropriate actions to bring the state of the world into the goal state.

In terms of low-level motion planning, robot need to consider both the function of the end-effector and the operation requirements of the object. For example, the movement of picking up a cup with water is completely different from picking up a cup without water. Robot need to add constraints in the trajectory to meet the execution requirements of the task. Traditional model-based motion planning and model-based control strategy search both need a accurate robot model, and difficult to meet the needs of complex robot motion in home service scenes. Human arm can manipulate the object in various flexible ways according to the task requirements. Imitating the motion mechanism of human arm, and establishing the motion mode of robot for the specific task requirements, which can improve the task execution ability of robot. Dynamic motion primitives (DMP) is a framework for learning from demonstration trajectory, which can represent steady goal-directed or periodic movements [2]. learning from observation or demonstration, DMP depends on the proper function approximation, if these approximators are flexible enough, it can represent arbitrarily complex motions and are guaranteed to converge in point-to-point motions [3]. In this paper, DMP is used to encode trajectory to learn complex robot motion.

2 Literture Review

There are generally two types of models (Behavior-Tree and Finite State Machines) that can be used to model task planning problems. Behavior-Tree [4, 5] (BTs) is a mathematical model which can be used for computer science, robotics, control systems, and video games. BTs represents a complex task as a switch between simpler tasks based on the states and conditions of other tasks. With the development of robot technology, it has been used in robot task scheduling to describe finite task set in a modular way. Marzinotto designed a behavior tree manually to operate on the NAO robot [6]. Hu described the application of behavior tree in semi-autonomous simulated surgery [7]. Ligot explored the possibility of using behavior tree as the architecture of robot swarm control software [8]. Finite State Machines (FSMs) are mathematical models that represent a finite number of states and behaviors such as transitions or actions between them, the next state and output are determined by the input and the current state [9]. Allgeuer applied the FSMs as the control architecture of humanoid soccer robot [10]. Li applied the probabilistic FSMs to solve the multi-target search problem of swarm robots [11]. In addition, NASA's Autonomous Systems and Operations (ASO) program makes extensive use of AI technology to help monitor, plan, execute, and manage failures in future human space missions, the ASO program has successfully demonstrated autonomous capabilities in the three areas: occupant autonomy, vehicle systems management, and autonomous robotics [12].

Motion planning can be generalized into two categories: generative motion planning and generative control strategy. Generate motion planning includes potential field method, heuristic search technology, dynamic programming and sample-based algorithm. Generative motion planning take into account the robot's ability to execute in the real world, uncertainties in the execution (such as slipping), inaccuracies in the execution, and dynamic changes in the environment (such as changes in targets or obstacles), change and uncertainty often lead to inaccurate operations and replanning. Generative control strategy establish a mechanism to adapt to the uncertainty and dynamic change of the environment, and incorporates them into the control strategy. The control strategies based on motion primitives are usually used to represent and learn the basic movements of robot. Motion primitives formula is the compact parameterization of robot control strategy. Adjusting parameters allows for imitation and reinforcement learning. The control strategy based on motion primitives mainly includes: principal component analysis method [13], probability motion primitives method [14], and dynamic motion primitives [15]. Dynamic motion primitives can record the action data containing the context to build the action library, semantic information can be used to select a particular DMP to perform a particular task, this allows to encode object-oriented operations.

3 Methods

3.1 High-Level Task Planning

Classical planning is used to model the task planning. Classical planning can be represented as a state transition system:

$$\Sigma = (S, A, \gamma) \tag{1}$$

$S = \{s_1, s_2, \ldots\}$ is a finite set of states;
$A = \{a_1, a_2, \ldots\}$ is a set of actions;
$\gamma : S \times A \to S$ is the state transition function;

The state transition system Σ is represented as a directed graph, where S is the node of the state and the arc is the action. Applying action a to state s will produce a new state $s' = \gamma(s, a)$. From s to s' is called state transition, In classical planning, it is deterministic for all state S and action A and the transition function $\gamma(s, a)$.

Logical representation is one of the most commonly used representations of classical planning. World state is represented by a set of logical proposition p, that represent the true facts of the world. If p is not in the S state, it is considered false. Operators alter the state of the world by modifying the truth values of propositions, expressed in terms of preconditions and effects.

operator o is a tuple $o =$ (name(o), precond(o), effect(o)), where:
name (o) is the name of the operator;
precond (o) is a set of literal that must be true to apply the operator o;
effect (o)− is a set of literal that are false when the operator o is applied;
effect (o)+ is a set of literal that are true when the operator o is applied;
The following constraint relationship exists:

$$\text{effect}(o) = \text{effect}(o)^{-} \cup \text{effect}(o)^{+} \ \text{effect}\ (o)^{-} \cap \text{effect}^{+}(o) = \emptyset \tag{2}$$

An action is any basic instance of an operator. We think an operator as a generalized action. If a is an action and s is a state, precond (a) is true in s, applying action a to state s will causes a new state s', such that:

$$s' = \gamma(s, a) = \left(s - \text{ effects }^{-}(a)\right) \cup \text{ effects }^{+}(a) \tag{3}$$

The planning problem can be defined as a a triple (Σ, s_0, g), where:
Σ is the planning domain;
$s_0 \subseteq S$ is the initial state of the world;
g is a set of goal propositions that describe the goal to be achieved.
a planning is a set of actions $\pi = \langle a_1, \cdots, a_k \rangle$, where k ≥ 0.

The state produced by applying π to the state S is the state obtained by applying each action of π in turn. If a plan π exists, then state s_n can be reached from state s_0. When a complete specification (that is, the planning domain and problem) is provided to the planner, it uses the available actions to generate a plan to achieve the specified goal. Thus, planning π is the solution to the planning if it guarantees a transition from the initial state s_0 to the target state g.

Most Planners are search programs, whose difference lies in the search space: state space planning, planning space planning and planning diagram. The simplest planners are state space planners, which rely on search algorithms, where the search space is a subset of the state space. Each node corresponds to a world state, each arc represents a state transition, that is, an action, and a plan is a path in the search space from the initial state to the target state. The problem with state-space search is that it promises to plan sequence of steps, which means considering many different sequences of the same operation, even if none of them leads to a solution. The main open source planning toolkit is the lightweight Fast Forward based automation planning toolkit [16]. PDDL4J is a schematic domain description library for Java cross-platform developers.

3.2 Low-Level Motion Planning

The theory of dynamic motion primitives (DMP) has been established [2]. DMP is a motion learning method to obtain trajectories in joint space or task space from human demonstrations. Demonstration learning is mainly about the control strategy and the attractor space of nonlinear dynamic equation behind the learning trajectory. The core of the model is a point attractor system modulated by nonlinear functions to generate complex motions. DMP can be applied to the goal-directed or periodic motion. This paper only studies the goal-directed motion. The application of DMP to the single-point attraction system is the spring damping system:

$$\tau \ddot{y} = \alpha_z(\beta_z(g - y) - \dot{y}) \tag{4}$$

Where, τ is the duration of motion, α_z and β_z are constants to ensure that the system becomes critically damped, g is the target state, and y, $\dot{y}$, and $\ddot{y}$ are the desired position, velocity, and acceleration respectively. Equation 4 can be written as follows:

$$\begin{aligned} \tau \dot{z} &= \alpha_z(\beta_z(g - y) - z) \\ \tau \dot{y} &= z \end{aligned} \tag{5}$$

The intuition of using spring-damping systems is that, after excitation, the spring always converges to the target state in a finite amount of time. However, these generated trajectories are trivial as they always form similar shapes. More complex patterns are needed to control complex movements. Equation (6) is realized by using the nonlinear function f modulation, and it is obtained

$$\begin{aligned} \tau\dot{z} &= \alpha_z(\beta_z(g-y)-z)+f \\ \tau\dot{y} &= z \end{aligned} \tag{6}$$

the nonlinear function f is the forced function, which are generally defined as linear combinations of basis functions:

$$f(t) = \frac{\sum_{i=1}^{N} \Psi_i(t) w_i}{\sum_{i=1}^{N} \Psi_i(t)} \tag{7}$$

Ψ represents basis function and w_i represents the corresponding weight. The number of basis functions, whose purpose is to reproduce motion smoothly, is defined by trial in advance. Compared with the basis function, the w_i needs to be learned and cannot be set in advance.

4 Experimental Results

The experimental scene is shown in the Fig. 1, which can represents most robot service scenes. The service scene includes a robot, two cups (cup_blue and cup_glass), a tea, a coca, and two trays (plate_1 plate_2). The robot should put the coca and cup_glass into plate_1 and put the tea and cup_blue into plate_2. We define three types: wp (the initial position (wp1, wp2, wp3, wp4), and the target position plattle_1, plattle_2), robot(robot) and oject (tea, coca, cup_blue, cup_glass), four properties represents spatial relationships in the scene as shown in Fig. 2. Three operators: *grasp*, *move*, *place*. Figure 3 shows the detail of the operators.

Fig. 1. Robot perform the service scene task, left is the initial state, right is the target state

PDDL is a standard coding language for classical planning. The components of the PDDL planning task include five parts: Objects (the objects in the environment that are relevant to the task), Predicates (the properties of the objects in the world), Initial state (the state of the world at the start of the task), Goal state (the state of the world that is expected to be achieved), Operators (the methods that change the state of the world). The planning tasks specified in PDDL are divided into two files: a domain file for predicates and operators, a problem file for objects, initial state and goal state.

```
(:predicates
(robot_at_wp ?robot - robot ?wp - wp) robot in location wp
(object_at_wp ?object - object ?wp - wp) oblect in location wp
(object_on_robot ?object - object ?robot - robot) robot carry object
(robot_does_not_have_object ?robot - robot)) robot not carry object
```

Fig. 2. Predicates list of the service scene

```
(:durative-action move
    :parameters( ?robot - robot  ?from - wp  ?to - wp)
    :duration( = ?duration (move_cost) )
    :condition(and (at start (robot_at_wp ?robot ?from)))
    :effect(and(at start (not(robot_at_wp ?robot ?from)))
              (at end (robot_at_wp ?robot ?to))))

(:durative-action grasp
    :parameters(?robot - robot ?object - object ?wp - wp)
    :duration(= ?duration (manipulate_object_cost))
    :condition(and (at start(object_at_wp ?object ?wp))
          (over all(robot_at_wp ?robot ?wp))
          (at start(robot_does_not_have_object ?robot)))
    :effect(and(at start(not(object_at_wp ?object ?wp)))
              (at end(object_on_robot ?object ?robot))
              (at start(not(robot_does_not_have_object ?robot)))))

(:durative-action place
    :parameters(?robot - robot ?object - object ?wp - wp)
    :duration(=?duration (manipulate_object_cost))
    :condition(and(at start(object_on_robot ?object ?robot))
              (over all (robot_at_wp ?robot ?wp)))
    :effect(and(at end(not(object_on_robot ?object ?robot)))
              (at end (object_at_wp ?object ?wp))
              (at end (robot_does_not_have_object ?robot)))))
```

Fig. 3. The details of operators

Rosplan integrates classical programming into ROS. The high layer schedules the robot tasks through PDDL language, and controls the execution of the bottom motion of the robot through the form of server-client, the framework of Rosplan is shown in Fig. 4 [17]. The system execution process needs a domain file and a problem file. It is generally compiled manually. Rosplan includes two ROS nodes, namely the knowledge base and the planning system. A knowledge base is used to store the latest environment models. The planning system acts as a wrapper for planners and also dispatchs plans. Details of the Rosplan framework can be found in Reference [18]. The planning result is shown in Fig. 5.

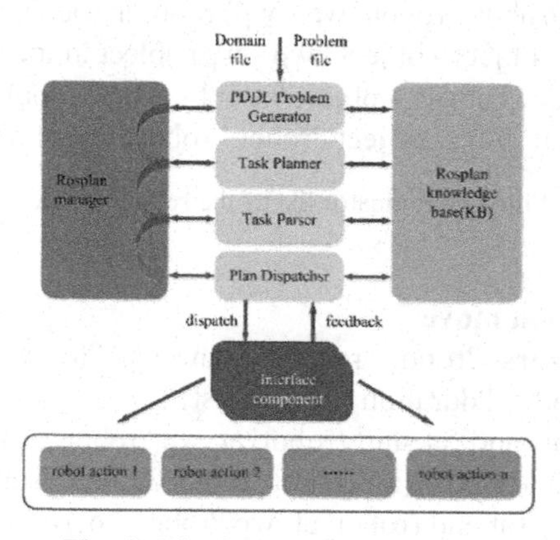

Fig. 4. The framework of Rosplan

0.000: (grasp robot tea wp1) [5.000]
5.000: (move robot wp1 plate_2) [10.000]
15.000: (place robot tea plate_2) [5.000]
20.000: (move robot plate_2 wp2) [10.000]
30.000: (grasp robot coca wp2) [5.000]
35.000: (move robot wp2 plate_1) [10.000]
45.000: (place robot coca plate_1) [5.000]
50.000: (move robot plate_1 wp3) [10.000]
60.000: (grasp robot cup_blue wp3) [5.000]
65.000: (move robot wp3 plate_2) [10.000]
75.000: (place robot cup_blue plate_2) [5.000]
80.000: (move robot plate_2 plate_1) [10.000]
90.001: (move robot plate_1 wp4) [10.000]
100.001: (grasp robot cup_glass wp4) [5.000]
105.001: (move robot wp4 plate_1) [10.000]
115.001: (place robot cup_glass plate_1) [5.000]

Fig. 5. Planning result of robot service task

Robot motion planning includes Cartesian coordinate space or joint coordinate space. No matter joint coordinate space or Cartesian coordinate space, DMP models each dimension of the motion trajectory, and finally achieves tight synchronization through a coupling term. Cartesian coordinate space is of more practical significance than joint coordinate space. However, in Cartesian space, robot movement trajectory not only needs to consider the landscape of trajectory, orientation, as a special dimension, has an important constraint on the trajectory of DMP. The nonlinear characteristics of DMP will produce oscillating behavior for orientation modeling in Cartesian space, quaternion can solve the nonlinear problem in orientation modeling. We tend to model the robot end-effector trajectory in the Cartesian coordinate space to learn the complex motion learning. In the scene we set, the action *move* means the robot move the object from the initial position to the target position. We learn about this trajectory so that it can be generalized to different objects. Figure 6 shows the robot motion trajectory learning using DMP, the purple line represents the original trajectory. We conducted three kinds of experiments, the first is to regenerate without changing the initial position and target position (red line), the second modified the initial position (yellow line), and the third modified the initial position and the target position (blue line).

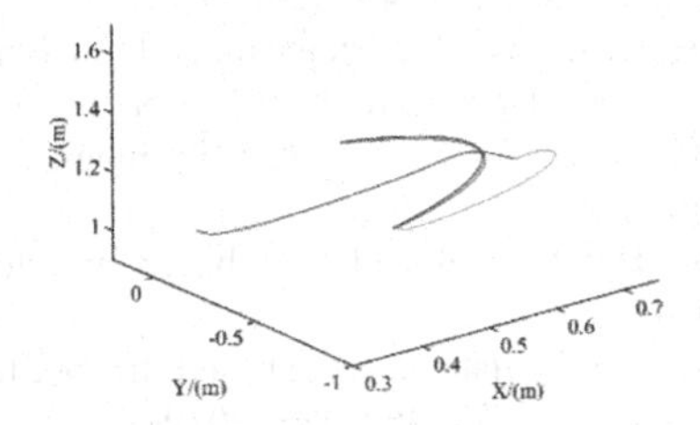

Fig. 6. Learning and generating robot trajectories. (Color figure online)

5 Conclusion

We discuss how robot can better perform tasks in home service. Robot should have deliberative ability and model this ability using classical planning methods. PDDL was used to describe the task and use Rosplan framework to support the implementation of robot deliberative ability. Although in the actual scene, the robot should deal with uncertainty and other problems, this paper assumes that the execution of the robot task is deterministic, models the robot task execution as a classical planning problem, and focuses on verifying the feasibility of the method. In terms of motion planning, we hope to add more constraints in the process of motion planning, so that robot can better adapt to the home service scene. The motion planning method based on Moveit usually focuses on the target, with little consideration of the constraints on the trajectory landscape. DMP was used to model the trajectory of the robot. This will be the main direction of future service robot motion planning.

Acknowledgments. This research was supported by the National Natural Science Foundation of China (Grant No. 61873045).

References

1. Ghallab, M., Nau, D., Traverso, P.: Automated planning: theory & practice. In: Handbook of Knowledge Representation (2004)
2. Schaal, S.: Is imitation learning the route to humanoid robots? Trends Cogn. Sci. **3**(6), 233–242 (1999)
3. Fanger, Y., Umlauft, J., Hirche, S.: Gaussian processes for dynamic movement primitives with application in knowledge-based cooperation. In: IEEE/RSJ International Conference on Intelligent Robots & Systems. IEEE (2016)
4. French, K., Wu, S., Pan, T., et al.: Learning behavior trees from demonstration. In: 2019 International Conference on Robotics and Automation (2019)
5. Colledanchise, M.: Behavior Trees in Robotics (2017)
6. Marzinotto, A., Colledanchise, M., Smith, C., et al.: Towards a unified behavior trees framework for robot control. In: 2014 IEEE International Conference on Robotics and Automation (ICRA). IEEE (2014)
7. Hu, D., Gong, Y., Hannaford, B., et al.: Semi-autonomous simulated brain tumor ablation with RAVENII Surgical Robot using behavior tree. In: Proceedings IEEE International Conference on Robotics & Automation, pp. 3868–3875 (2015)
8. Ligot, A., et al.: Automatic modular design of robot swarms using behavior trees as a control architecture. PeerJ Comput. Sci. **6**(9), e314 (2020)
9. Foukarakis, M., Leonidis, A., Antona, M., Stephanidis, C.: Combining finite state machine and decision-making tools for adaptable robot behavior. In: Stephanidis, C., Antona, M. (eds.) UAHCI 2014. LNCS, vol. 8515, pp. 625–635. Springer, Cham (2014). https://doi.org/10.1007/978-3-319-07446-7_60
10. Allgeuer, P., Behnke, S.: Hierarchical and state-based architectures for robot behavior planning and control (2018)
11. Li, J., Tan, Y.: A probabilistic finite state machine based strategy for multi-target search using swarm robotics. Appl. Soft Comput. **77**, 467–483 (2019)
12. Frank, J.D.: Artificial intelligence: powering human exploration of the moon and mars (2019)
13. Lim, B., Ra, S., Park, F.C.: Movement primitives, principal component analysis, and the efficient generation of natural motions. In: IEEE International Conference on Robotics & Automation. IEEE (2006)
14. Paraschos, A., Daniel, C., Peters, J., et al.: Probabilistic movement primitives (2013)
15. Kober, J., Peters, J.: Policy search for motor primitives in robotics. In: NIPS, pp. 849–856 (2009)
16. Ramirez, M., Lipovetzky, N., Muise, C.: Lightweight automated planning ToolKiT. http://lapkt.org/ (2015). Accessed 12 Mar 2019
17. Azimi, S., Zemler, E., Morris, R.: Autonomous robotics manipulation for in-space intra-vehicle activity. In: Proceedings of the ICAPS Workshop on Planning and Robotics (2019)
18. Cashmore, M., Fox, M., Long, D., et al. ROSplan: planning in the robot operating system (2015)

Deep Reinforcement Learning for an Anthropomorphic Robotic Arm Under Sparse Reward Tasks

Hao Cheng, Feng Duan(✉), and Haosi Zheng

Department of Artificial Intelligence, Nankai University, Tianjin 300350, China
duanf@nankai.edu.cn

Abstract. Training operation skills of a robotic arm using Reinforcement Learning algorithms requires a process of reward shaping, which needs considerable time to adjust. Instead, the setting of sparse rewards makes the tasks clear and easy to modify when the task changes. However, it is a challenge for the agent to learn directly from sparse reward signals because of the lack of reward guidance. To solve this problem, we propose an algorithm based on the DDPG algorithm and add three techniques: Hindsight Experience Replay for improving sample efficiency, Expert Data Initialization for accelerating learning speed in the early stage, Action Clip Scaling for acting stable. For validating our algorithm, we built a simulation environment of an anthropomorphic robotic arm based on the Pybullet module and set up the training interface. There are two tasks trained under sparse reward signals: *Push* task, *Pick and Place* task. The experimental results show that the agent can quickly improve the operation skill level in the early stage. It also has a good convergence effect in the later stage, which effectively solving the sparse rewards problem.

Keywords: Reinforcement learning · Sparse rewards · Anthropomorphic arm

1 Introduction

Reinforcement Learning (RL) combined with Deep Neural Network is currently a research highlight in recent years. Deep Reinforcement Learning (DRL) has achieved many breakthroughs in video games, robotics, and other fields due to its strong versatility. For example, the proposal of the Deep Q Network (DQN) algorithm in 2013 made the agent comparable to human-level operation in video games for the first time [1]. High-dimensional continuous control methods like PPO, TRPO have been successfully applied in the field of robotics [2,3]. Robots are trained to complete operation skills [4–6], and humanoid robots can also be controlled by the RL algorithm [7,8].

For robots with RL algorithms, solving different tasks requires different reward shaping, which means that once the task targets change, the rewards need to be reset and adjusted. It will be a tedious process. In addition, a simple

X.-J. Liu et al. (Eds.): ICIRA 2021, LNAI 13014, pp. 79–89, 2021.
https://doi.org/10.1007/978-3-030-89098-8_8

linear transformation of the set rewards will produce a significant effect on the training result and cause unacceptable fluctuations. When the rewards is inappropriate, it may cause the agent to converge to a local optimum. If the reward function is too complicated, the agent's behavior after training may deviate from the experimenter's intention. As a result, reward shaping engineering seriously damages the versatility of RL, making it hard to transplanted fast.

As an alternative, sparse rewards can overcome the above problems. The sparse rewards we set are binary signals. i.e., 1 when completing the task, otherwise 0. The advantage of sparse rewards is that the task goals are clear and unambiguous. Moreover, when the task goal changes, only a tiny amount of the code needs to be modified. Nevertheless, the ensuing problem is that it is challenging for the agent to gain successful experience without dense rewards guiding, which requires us to design an effective algorithm to ensure the agent can obtain good training results.

In this paper, we adopted the Deep Deterministic Policy Gradient (DDPG) algorithm to deal with the robot physical problem of the high-dimensional continuous action space. Based on DDPG, we added three techniques: Hindsight Experience Replay (HER), Expert Data Initialization, and Action Clip Scaling. We built an anthropomorphic robotic arm simulation environment based on the Pybullet module and set up the training interface for RL. There are two tasks set under the sparse rewards: *Push* task, *Pick and Place* task. The experimental results show that the agent can learn successful experience in the early stage due to the initialization of the expert data. With the utilize of the HER algorithm, the sample efficiency is greatly improved. In the later stage, due to the Action Clip technique, the agent's actions change to be stable, and finally it have a good convergence effect.

2 Related Work

In order to solve the challenge of sparse rewards, there are mainly three types of methods in the literature: Curiosity Module, Curriculum Learning, and Hierarchical RL. These methods will be introduced respectively below.

2.1 Curiosity Module

Curiosity is a characteristic of human agents. When facing sparse rewards in the real world, humans can still learn a skill because of their intrinsic curiosity motivation. Therefore, we can apply the idea of curiosity to the RL algorithm. The pivotal idea is to treat the rewards from the environment as external rewards and then set additional intrinsic rewards to encourage the agent to explore. The sum of external rewards and internal rewards is the training goal of the agent.

The difference among curiosity methods in the literature mainly lies in how to set intrinsic rewards. In [9], Intrinsic Curiosity Module is proposed to generate intrinsic rewards. It is mainly divided into two sub-modules: forward and inverse dynamics models. The loss of the inverse dynamics model is defined as the difference between the real action and the predicted action, and the forward model is the difference between the prediction of the next state and the real state. Combine the loss of the two, the self-supervised method is used to update the network and prediction error is treated as the intrinsic rewards. Similar to the above method, in [10], a method named Random Network Distillation (RND) is proposed. The RND input the observed state to a randomly initialized fixed target network as a prediction task $f(x)$, and then send the predicted value after the state is input from the predictor network to $\hat{f(x)}$, regard the prediction error as an inherent reward. Finally, it uses MSE to update the network.

2.2 Curriculum Learning

Curriculum Learning (CL) is a methodology of optimizing the order in which the agent acquires experience so as to accelerate training convergence on the final target [11]. A curriculum can be thought of as an ordering over experience samples. CL can first train on one or more source tasks and then transfer the acquired knowledge to solve the target task. The agent can learn prior knowledge from a simple, intermediate task and gradually transit to the difficulty of the final target task.

2.3 Hierarchy Reinforcement Learning

Hierarchical reinforcement learning (HRL) is a methodology of multiple layers of policy training [12]. In HRL, goals are generated in a top-down fashion [13]. The upper-level agent conducts long-term planning and decision-making over a more extended period (lower time resolution) to form a high-level reasoning policy. By setting intrinsic rewards, It converts the goals of the high-level planning policy into a low-level action policy under the higher time resolution. In a nutshell, the manager (high-level agent) is responsible for planning, and the work (low-level agent) is responsible for making the actions.

3 Background

3.1 Reinforcement Learning

Markov decision process (MDP) is a universal framework to model the problem in RL. The agent continuously interacts with the environment to obtain data, i.e., The agent makes actions, and the environment responds to these actions and presenting new situations to the agent [14].

MDP can be presented as a tuple (S, A, P, R, γ), which consists of a set of states S, a set of actions A, a transition function $P(s'|s, a)$, a reward function R and a discount factor γ. The goal of RL is to find an optimal policy $\pi(a|s)$, and maximize the discounted return G_t. It can be described as (1),

$$G_t = \sum_{k=0}^{\infty} \gamma^k R_{t+k+1}, 0 < \gamma < 1 \tag{1}$$

3.2 Policy Gradient Optimization

As mentioned above, The goal of RL is to maximize the discounted return. In RL, there are two types of methods to solve this problem, i.e., the value-based method and the policy-based method. However, the policy obtained from the value-based method is usually a finite set of mappings from state space to action space. As a comparison, a policy-based method generates a paramotorizod policy target function described as (2),

$$U(\theta) = \sum_{\tau} P(\tau|\theta)R(\tau) \tag{2}$$

where θ is the parameters of the policy, τ is a sample trajectory. The parameterized policy makes the problem easier to better convergence and is suitable for solving problems with a large or continuous action space.

The policy-based method optimizes the parameters through the gradient ascent algorithm (also called the policy gradient) and obtains the final parameters through iterative calculations.

$$\theta_{new} = \theta_{old} + \alpha \nabla_{\theta} U(\theta) \tag{3}$$

By computing the derivative of $U(\theta)$ with respect to θ, we can get the policy gradient (4),

$$\nabla_{\theta} U(\theta) \approx \hat{g} = \frac{1}{m} \sum_{i=1}^{m} \nabla_{\theta} log P(\tau;\theta) R(\tau) \tag{4}$$

where m is the number of the trajectories. However, It still a problem of complex computing, which needs further simplification and estimation.

4 Method

We use Deep Deterministic Policy Gradient (DDPG) algorithm [15] of the Actor-Critic (AC) framework as the basis and add three techniques to it: Hindsight Experience Replay (HER), Expert Data Initialization, and Action Clip Scaling. The schematic view of the algorithm is shown in Fig. 1.

First, we use expert data to initialize the Replay Buffer and then start training. We collect k episodes of data and put it into the replay buffer. When updating the network, we take out N mini-batches of data from the Replay Buffer and replace goals through the HER module. The replaced data are used to update the Actor and the Critic network. Finally, the updated Actor output the action in the following sample episodes with the utilization of the Action Clip module to limit the range of the actions.

4.1 Base Algorithm: Deep Deterministic Policy Gradient

DDPG is a model-free RL algorithm that can effectively deal with continuous control problems. It is also an Actor-Critic framework based on the Deterministic

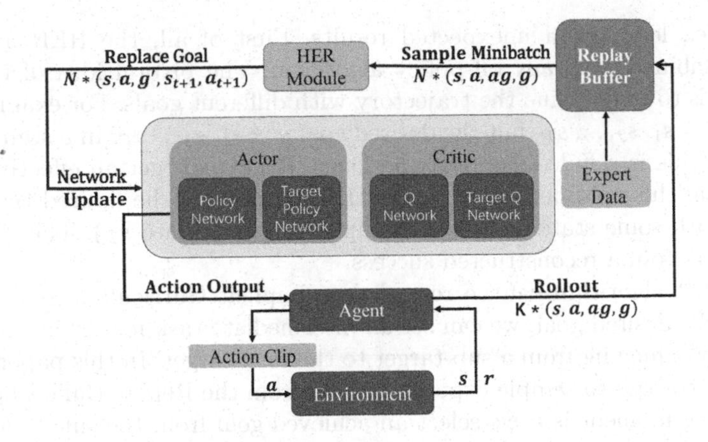

Fig. 1. Schematic view of the algorithm.

Policy Gradient (DPG) algorithm [16]. Different from the output of a probability distribution of actions in a stochastic policy algorithm, in DDPG, the output is a deterministic action in the state s, which has better stability in performance. The algorithm maintain two networks: the Actor Network $\mu(s|\theta^{\mu}) : S \rightarrow A$ and the Critic Network $Q(s, a|\theta^{Q}) : S \times A \rightarrow R$. The actor as the target policy outputs the action a in the state s, and the critic evaluates the actions. The actions will be added an exploration noise, the expression as shown in (5),

$$a_t = \mu(s_t|\theta^{\mu}) + N_t \tag{5}$$

where N_t is a noise distribution. The Critic updates the network by minimizing the loss $L = \frac{1}{N}\sum_i (y_i - Q(s_i, a_i|\theta^{Q}))^2$, and the Actor updates the network through computing the gradient of the expected return on the trajectories:

$$\nabla_{\theta^{\mu}} J \approx \frac{1}{N} \sum_i \nabla_a Q(s, a|\theta^{Q})|_{s=s_i, a=\mu(s_i)} \nabla_{\theta^{\mu}} \mu(s|\theta^{\mu})|_{si} \tag{6}$$

In addition, DDPG incorporated the successful experience of DQN and introduces two techniques: Replay Buffer and Target Network. The Replay Buffer is used to store the sample data and execute random uniform sampling from it when the network updates, so as to break the correlation between the sample data and improve the sample efficiency. The target network is an independently set up network used to estimate the TD error when the network updates, which can effectively avoid the fluctuation caused by the action-value update to be brought into the next round of iterative calculations.

4.2 Techniques for Sparse Rewards

Hindsight Experience Replay. Hindsight Experience Replay (HER) [17] is motivated by the experience of human agents. Human agents can learn from failures, i.e., learn from unexpected results. First of all, the HER algorithm can be combined with any off-policy algorithm. The pivotal idea of the HER algorithm is to re-examine the trajectory with different goals. For example, in a episode, $S = s_1, s_2, ..., s_T$, but the desired goal $g \neq s_1, s_2, ...s_T$. In a standard RL algorithm, this is a failed sample experience. RL cannot get an effective policy update from this episode. By using the HER algorithm, the desired target G is replaced with some state in S (for example, the final state s_T). The agent can finally learn from a reconstructed success.

The HER algorithm can be regarded as implicit Curriculum Learning. By replacing the desired goal, we can set an intermediate task for the agent to help it gradually achieving from a sub-target to the final target. In this paper, we use the *future* strategy to sample experience data from the Replay Buffer. i.e., when the sampling moment is i, we select an achieved goal from the subsequent time step as the sampling goal. The expression is as (7):

$$g_i \leftarrow ag_i, j \in [i+1, T] \tag{7}$$

Through *future* sampling, we can construct more successful transitions than the *final* (S_t as the sampling goal) or *random* (sample from all transition) sampling goal strategy.

The adoption of the HER algorithm greatly improves the sampling efficiency and provides a huge effect for solving the sparse rewards problems.

Expert Data Initialization. In the early stage of the training, it takes considerable time for the agent to have a fundamental breakthrough before the HER algorithm can highlight its role. For example, if the agent has not touched the object in the *Push* task or the object has not been picked up by hand in the *pick and place* task, no successful experience will be obtained or reconstructed.

In order to accelerate the learning speed of the early stage and make the agent can directly learn successful experience, we use the program to generate trajectories that try to complete the task and record the transitions of the episodes when the result is a success. At the beginning of the training, the collected expert data are poured into the Replay Buffer as the initialization process, effectively solving the problem of large fluctuation and slow learning efficiency in the early training stage.

Action Clip Scaling. The actions output by the agent needs to be restricted. On the one hand, unlimited action space will affect the training effect. On the other hand, a motion of large range is also prohibited in real robots for damaging the motors. We can limit the range of the actions by clipping it.

$$\begin{gathered} action \leftarrow clip(action, lim_l, lim_h), \\ clip(x, x_1, x_2) = \begin{cases} if\ x < x_1 & x \leftarrow x_1 \\ if\ x_1 < x < x_2 & x \leftarrow x \\ if\ x_2 < x & x \leftarrow x_2 \end{cases} \end{gathered} \tag{8}$$

At the early period of training, the agent is allowed to take larger actions to explore. As the training progresses, the agent has learned a part of the experience. We narrow down the range of the output actions (Scale down lim_l and lim_h). There is no requirement for force output in the task. Moreover, in MDP, interactive data satisfies the Markov property, i.e., the next state of the system is only related to the current state. As a result, such a change will not excessively affect the training effect and improve the accuracy and precision of motion control.

5 Experiments and Results

5.1 Environments

The RL environment built in this paper is based on the Pybullet module [18]. We built a 7 DOF anthropomorphic arm [19,20] simulation environment to test our algorithm.

First, we consider two different tasks:

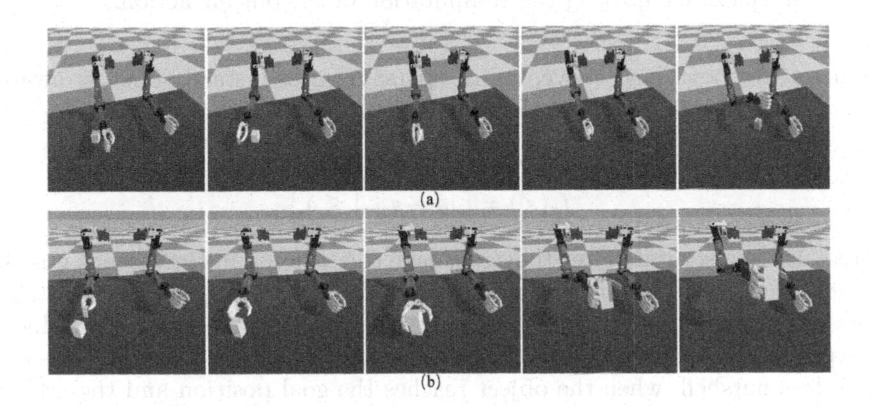

Fig. 2. (a) *Push* task (b) *Pick and Place* task

Push. The agent pushes the object placed on the table to the target position. The end-effector of the arm is blocked.

Pick and Place. The agent picks up the object on the table and places it at the target position in the air.

The details of the RL interface settings are as follows:

Observations. Observation is the state vector returned after each time the agent interacts with the environment. It will be input into the policy for gradient optimization. It includes the position of the end-effector (3), the orientation of the end-effector (3), the linear velocity of the end-effector(3), the angular velocity of the end-effector (3), the position of the object (3), the orientation of the object (3), the linear velocity of the object (3), the angular velocity of the object (3), the position of the target (3), the relative position of the object and the end-effector (3). The angle-related elements are represented by Euler angles. The total dimension of the observation vector is 30.

Actions. The actions is represented by a four-dimensional vector $[d_x, d_y, d_z, d_{grip}]$. In the *Push* task, the hand is blocked and the value of the d_{grip} is 0. The position of the hand performs the output action based on the current position. i.e., in the Cartesian coordinate, the next position reached by the hand is:

$$P_{next} = [\ p_{curx} + d_x,\ p_{cury} + d_y,\ p_{curz} + d_z,\ p_{curgrip} + d_{grip}\ ] \tag{9}$$

We then use inverse kinematics to convert the hand position into the joint angles in the joint space, completing the computation of the output action.

Rewards. As mentioned above, we use sparse rewards instead of dense rewards to guide the agent to complete training. The sparse rewards is defined as:

$$\begin{aligned} r(s,a,g) &= -[\ f_g(s') = 0\ ], \\ f_g(s') &= [\ |g - s_{obj}| \le \delta\ ] \end{aligned} \tag{10}$$

where s' is the next state after the agent take the action a in the state s, g is the goal position, s_{obj} is the object position in the state s, δ is the tolerance distance between the object and the goal ($\delta = 5$ cm is set in the two tasks). The value of the expression in brackets means that if the expression is true, it is 1, otherwise it is 0. In a nutshell, when the object reaches the goal position and the error is below δ, the reward is 0, otherwise is -1.

Initial States. The object is randomly set to a position where the arm can reach, and the initial position and posture of the robotic arm is fixed.

Goal. The goal position is set at a position reachable by the end of the arm. In the *Push* task, the goal position is set on the table, while in the task of *Pick and Place*, the goal position is set in the air.

5.2 Experimental Details and Results

Before the training, we first collected 1000 episodes of expert data for *Push* task and *Pick and Place* task separately by using program. Before the training, we put the expert data into the Replay Buffer.

There are 100 episodes of samples in an epoch, and the agent is allowed to perform actions for 150 timesteps in each episode. After the agent performs an action, it needs to wait for 20 sub-steps (1/240 s per sub-step), i.e., the agent's action frequency 12 Hz. We update the network every two episodes sampling. After 100 epochs, we change action clip limits from $(-0.5\,\mathrm{m}, 0.5\,\mathrm{m})$ to $(-0.1\,\mathrm{m}, 0.1\,\mathrm{m})$ in *Push* task, $(-0.5\,\mathrm{m}, 0.5\,\mathrm{m})$ to $(-0.15\,\mathrm{m}, 0.15\,\mathrm{m})$ in *Pick and Place* task. The experiment was conducted on 3 random seeds. After each epoch training, we evaluate the each epoch trained model for 50 episodes, and finally the success rate curves was generated. As a comparison, we add a set of experiments without expert data.

The result is shown as Fig. 3. Due to the role of Expert Data Initialization, the agent can improve quickly in the early stage. Note that the training effect of the *pick and place* task is worse than that of the *push* task. This is because the *pick and place* task requires the agent to use the hand to pick up the object and put it in the air, which is much more difficult than the *push* task. This leads to the fact that in the *Push* task, the agent achieved a success rate of about 90% in 75 epochs, and the fluctuation is comparable small. While in the *Pick and Place* task, the success rate improve relatively slower and exist fluctuation. But it converges in the later period at a success rate of 90%. However, in the comparison group without expert data, it takes a long time for the agent to learn successful experience in the push task, and there are large fluctuations in three random seed experiments. In the Pick and Place task, the agent lacking expert data cannot learn successful experience at all.

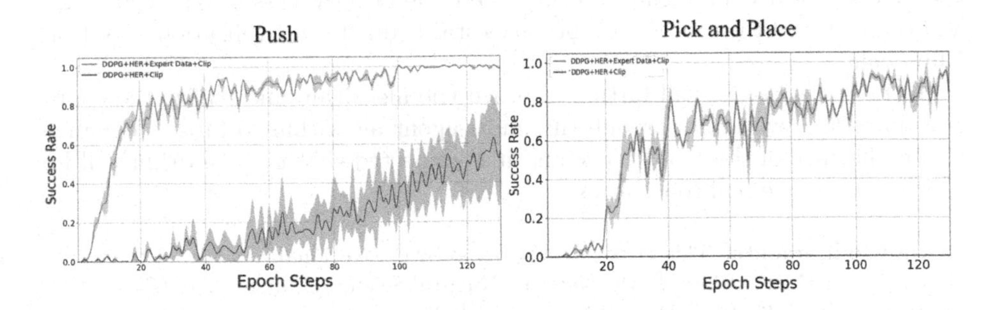

Fig. 3. Success rate

After 100 epoch, we change action clip limits. We can observe that the success rate after 100 epoch obviously fluctuates less, which means the agent's actions more stable. We drew the action curves using models of 99 epoch and 101 epoch in the *Push* task, as shown in Fig. 4. In the two curves, the initial positions, orientation of the object and the target are the same. The blue curve is when the agent's action is limited to (−0.5, 0.5), while the light blue is (−0.1, 0.1). The two curves record the hand positions of the agent from the beginning to the completion of the task. Although the blue curve takes less time to complete the task, the acceleration during the movement is very large, and there are some

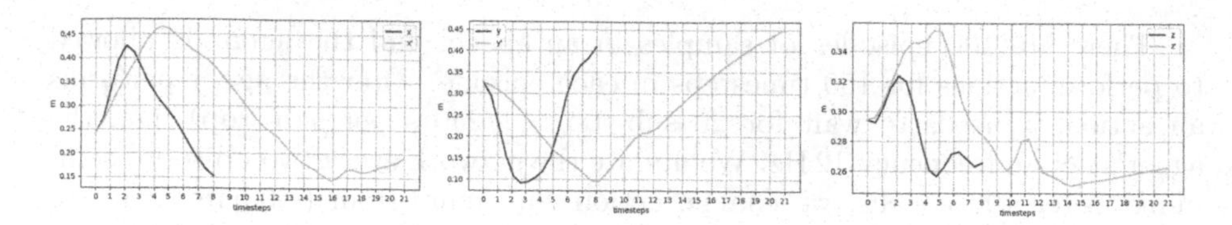

Fig. 4. The use effect of Action Clip

peaks, which are not allowed on a real robot because it will damage the motor. As a comparison, in the light blue curve, the action to complete the task is more stable and gentle.

6 Conclusion

Aiming at the sparse rewards problem, we propose an algorithm based on DDPG and add three techniques: HER, Expert Data Initialization, Action Clip Scaling. We built a simulation environment for anthropomorphic robotic arms based on the Pybullet module and set two tasks under sparse rewards: *Push* task, *Pick and Place* task. The experiments show that the agent has a better convergence effect in the *Push* task. For the *Pick and Place* task has more complex actions, it is more challenging to learn. The learning process slow and fluctuating in the early stage, but in the end, the agent learns good results. Finally, we solved the two sparse rewards problems, which proved the effectiveness of our algorithm, and the action of the robotic arm becomes stable due to the utilization of Action Clip Scaling.

In future work, we will further improve the algorithm and realize the cooperation of the two arms through the multi-agent algorithm to further enhance the intelligence of the robot arms, and it is expected that the algorithm will be transferred to the real robot arms.

Acknowledgements. This work was supported by the National Key R&D Program of China (No. 2017YFE0129700), the National Natural Science Foundation of China (Key Program) (No. 11932013), the Tianjin Natural Science Foundation for Distinguished Young Scholars (No. 18JCJQJC46100), and the Tianjin Science and Technology Plan Project (No. 18ZXJMTG00260).

References

1. Mnih, V., Kavukcuoglu, K., Silver, D., Rusu, A.A., Veness, J., Bellemare, M.G., et al.: Human-level control through deep reinforcement learning. Nature **518**(7540), 529–533 (2015)
2. Schulman, J., Levine, S., Abbeel, P., Jordan, M., Moritz, P.: Trust region policy optimization. In: International Conference on Machine Learning, pp. 1889–1897. PMLR (2015)

3. Schulman, J., Wolski, F., Dhariwal, P., Radford, A., Klimov, O.: Proximal policy optimization algorithms. arXiv preprint arXiv:1707.06347 (2017)
4. Niroui, F., Zhang, K., Kashino, Z., Nejat, G.: Deep reinforcement learning robot for search and rescue applications: exploration in unknown cluttered environments. IEEE Robot. Autom. Lett. **4**(2), 610–617 (2019)
5. Nguyen, H., La, H.: Review of deep reinforcement learning for robot manipulation. In: 2019 Third IEEE International Conference on Robotic Computing (IRC), pp. 590–595. IEEE (2019)
6. Li, F., Jiang, Q., Zhang, S., Wei, M., Song, R.: Robot skill acquisition in assembly process using deep reinforcement learning. Neurocomputing **345**, 92–102 (2019)
7. Garcia, J., Shafie, D.: Teaching a humanoid robot to walk faster through Safe Reinforcement Learning. Eng. Appl. Artif. Intell. **88**, 103360 (2020)
8. Abreu, M., Lau, N., Sousa, A., Reis, L.P.: Learning low level skills from scratch for humanoid robot soccer using deep reinforcement learning. In: 2019 IEEE International Conference on Autonomous Robot Systems and Competitions (ICARSC), pp. 1–8. IEEE (2019)
9. Pathak, D., Agrawal, P., Efros, A.A., Darrell, T.: Curiosity-driven exploration by self-supervised prediction. In: International Conference on Machine Learning, pp. 2778–2787. PMLR (2017)
10. Burda, Y., Edwards, H., Storkey, A., Klimov, O.: Exploration by random network distillation. arXiv preprint arXiv:1810.12894 (2018)
11. Narvekar, S., Peng, B., Leonetti, M., Sinapov, J., Taylor, M.E., Stone, P.: Curriculum learning for reinforcement learning domains: a framework and survey. J. Mach. Learn. Res. **21**(181), 1–50 (2020)
12. Vezhnevets, A.S., et al.: Feudal networks for hierarchical reinforcement learning. In: International Conference on Machine Learning, pp. 3540–3549. PMLR (2017)
13. Nachum, O., Gu, S., Lee, H., Levine, S.: Data-efficient hierarchical reinforcement learning. arXiv preprint arXiv:1805.08296 (2018)
14. Sutton, R.S., Barto, A.G.: Reinforcement Learning: An Introduction. MIT Press, Cambridge (2018)
15. Lillicrap, T.P., Hunt, J.J., Pritzel, A., Heess, N., Erez, T., Tassa, Y., et al.: Continuous control with deep reinforcement learning. arXiv preprint arXiv:1509.02971 (2015)
16. Silver, D., Lever, G., Heess, N., Degris, T., Wierstra, D., Riedmiller, M.: Deterministic policy gradient algorithms. In: International Conference on Machine Learning, pp. 387–395. PMLR (2014)
17. Andrychowicz, M., Wolski, F., Ray, A., Schneider, J., Fong, R., Welinder, P., et al.: Hindsight experience replay. arXiv preprint arXiv:1707.01495 (2017)
18. Pybullet. https://pybullet.org/wordpress/. Accessed 29 April 2021
19. Li, W., et al.: Design of a 2 motor 2 degrees-of-freedom coupled tendon-driven joint module. In: 2018 IEEE/RSJ International Conference on Intelligent Robots and Systems (IROS), pp. 943–948. IEEE (2018)
20. Li, W., et al.: Modularization of 2-and 3-DoF coupled tendon-driven joints. IEEE Trans. Robot. (2020)

Design and Experiment of Actuator for Semi-automatic Human-Machine Collaboration Percutaneous Surgery

Leiteng Zhang(✉), Changle Li, Yilun Fan, Gangfeng Liu, Xuehe Zhang, and Jie Zhao

State Key Laboratory of Robotics and Systems, Harbin Institute of Technology, Harbin 150001, China
18b908094@stu.hit.edu.cn, {lichangle,liugangfeng,zhangxuehe, jzhao}@hit.edu.cn

Abstract. This paper designs a manual percutaneous puncture actuator, and proposes a semi-automatic human-machine collaboration workflow that fully integrates the high precision of the robot and the flexibility of the doctor. This paper verifies the second-order polynomial relationship between the friction force of the needle in the tissue and the deflection through experiments. In the framework of the workflow proposed in this paper, combined with the damage-minimizing puncture method proposed in this paper, manual high-precision puncture is realized, and the accuracy of manual puncture is improved by 86% when the penetration depth is 120 mm.

Keywords: Human-machine collaboration · Percutaneous · Needle deflection

1 Introduction

Needle-based percutaneous minimally invasive surgery is commonly used in surgical, which is often used for biopsy, radiofrequency ablation, drainage and brachytherapy. Whether the tip of the needle can reach the lesion in the body directly determines the effect of the operation or the result of the diagnosis. Some operations can improve the accuracy of needle tip targeting through real-time ultrasound guidance, such as prostate seed implantation. More lesions require CT or MRI positioning, which cannot be guided in real time.

Robot-assisted percutaneous surgery can solve this problem well when the robot frame is registered to the CT or MRI frame. Generally, such a system will include a tracking and navigation device, such as the commonly used NDI, which facilitates the registration of these frames and can also track the motion state of the robotic. Currently existing systems, include commercial laboratories, can be divided into three categories: teleoperation, automatic and completely passive. Teleoperation is the most widely used and mature operation mode of medical robots. The Da Vinci system is a classic of teleoperation medical robots system. Hiraki et al. experimented with the first human percutaneous trial based on the teleoperation guided by CT to evaluate the feasibility

X.-J. Liu et al. (Eds.): ICIRA 2021, LNAI 13014, pp. 90–99, 2021.
https://doi.org/10.1007/978-3-030-89098-8_9

and safety [1]. Teleoperation will greatly increase the time cost and economic cost of doctor training due to changes in the telepresence. The automatic mode is obviously able to maximize the advantages of the robot's high precision. Most relevant research can achieve this goal [2–5]. But from an ethical point of view, automated medical robot systems are limited to the laboratory research stage. Most of the current commercial systems for percutaneous surgery are completely passive, for example. Dou et al. designed a four degree-of-freedom robot to assist the template positioning for lung brachytherapy treatment, where this template contains a lot of guide holes to guide the doctor to insert in the correct position [6]. Remebot is a passive percutaneous puncture system currently promoted in the market [7]. During surgery, the robot acts as a flexible stent that can be accurately positioned. The robot end tool is usually a simple guide to constrain the motion of the needle in the direction of the axis. A large number of studies have shown that the needle with bevel tip will deflect obviously after penetrating the skin [8–10]. Obviously proper rotation can improve the accuracy of puncture, but this is not easy for doctors to control.

We think that robot-assisted surgery should fully integrate the high-precision and repeatable characteristics of robots and the advanced decision-making capabilities of humans. Robots should better assist doctors rather than replace them. Therefore, robot-assisted surgery needs to model surgical tasks well and assign appropriate tasks to humans or robots to achieve better collaboration [11]. In this paper, we propose a semi-automatic human-machine collaboration workflow for high-precision percutaneous surgery. We also designed a special robot end actuator to adapt to the proposed workflow, which is manual. Under the framework of this workflow, the actuator designed by us is used to achieve a high-precision percutaneous method with minimal damage in the gelatin phantom.

2 Actuator Design

2.1 Actuator Structure

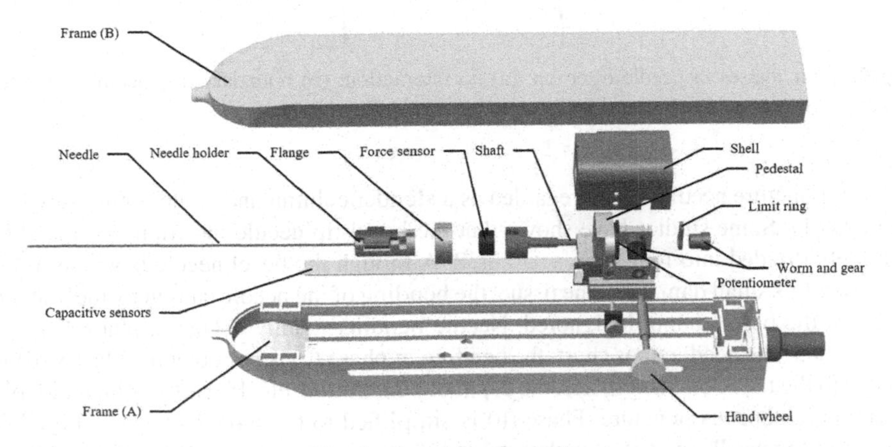

Fig. 1. Structure of actuator

The structure of the actuator is shown in Fig. 1. The needle is inserted into a flip-top needle holder. The needle holder fixes the needle through the attraction of a magnet, and this structure can easily release the needle from the actuator. This is crucial for post-operative verification. The needle holder is flanged to the force sensor. This connection method allows the needle holder and needle to accept a higher level of disinfection without considering the waterproof level of the actuator. The force sensor is fixed on the front end of the spindle. The spindle passes through the pedestal to connect the potentiometer, limit ring and the gear in turn. Capacitive sensors are used to measure the position of the needle, they are fixed on the bottom of the frame (A). The structure on both sides of the pedestal allows it to slide along the groove on the side of the frame (A). The hand wheel is inserted through the slot on the side of the frame (A) to the pedestal to connect to the worm wheel. The translation of the hand wheel along the slot controls the movement of the needle along the axis. The rotation of the hand wheel is controlled by the worm gear mechanism to control the rotation of the needle around the axis. Frame (A) and frame (B) are combined together by magnets. The front ends of frame (A) and frame (B) are designed with a semi-cylindrical structure. When the two frames are combined, the guide hole of the needle is formed, which is important to improve the puncture accuracy.

2.2 Guide Hole

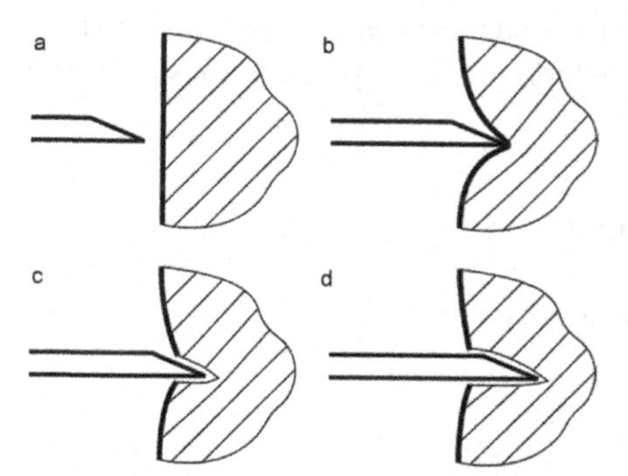

Fig. 2. Four phases in needle insertion: (a) no interaction; (b) boundary displacement; (c) tip insertion; (d) tip and shaft insertion

The puncture needle can be regarded as a slender column and its flexibility satisfies formula (1). Some studies have shown that the bevel-tip needle insertion into the skin is mainly divided into four phases as Fig. 2. Although the bevel needle is widely used because of its small damage to the tissue, the bending of the needle caused by the uneven force on the bevel cannot be ignored. Needle bending mainly occurs in phase (b) and (d). Because the needle tip is short, the bending at phase (c) is not obvious. Phases (d) is related to the tissue characteristics, the depth of penetration, etc. Phase (b) is unavoidable in any percutaneous puncture. Phase (b) is simplified to the model shown in Fig. 3-I. The tip of the needle in contact with the skin is simplified as a hinge. When the needle is in phase (b), the axial force of the needle reaches its peak. When the axial force reaches Euler load as formula (2), the original straight needle will become unstable, resulting in a large bending. The needle will be pierced in a curved state and seriously deviated from the predetermined trajectory. As described in Sect. 2.1, there is the guide hole at the front end of the actuator. The simplified model of phase (b) is transformed into Fig. 3-II. This means that the Euler load has increased by nearly twice under the same conditions. Therefore, the design of the guide hole is of great significance to improve the accuracy of puncture. The guide hole ensures that the needle can be inserted straight into the tissue instead of being bent in advance.

$$\lambda = \frac{\mu l}{i} \geq \pi \sqrt{\frac{E}{\sigma_P}} \tag{1}$$

$$F_{cr} = \frac{\pi^2 EI}{(\mu l)^2} \tag{2}$$

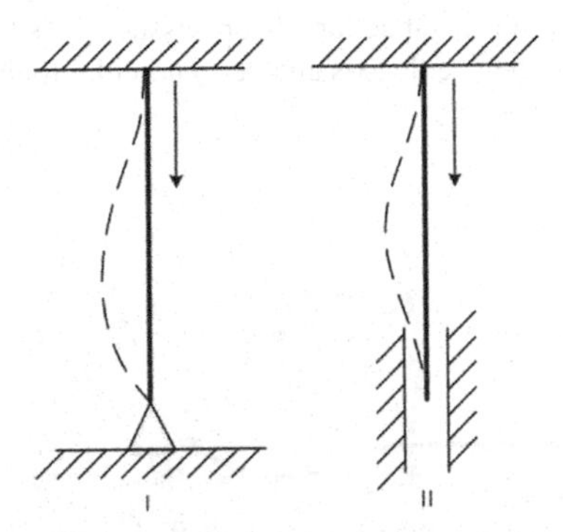

Fig. 3. Simplified model of phase (b) (I. One end is fixed and the other end is hinged: μ = 0.5, II. One end is hinged and the other end slides: μ = 0.7)

3 Minimize Damage Insertion

Although there are many force interaction models between the needle and the tissue, the research is still going on because some theories cannot be verified. Many studies have fixed the base of the needle on a six-dimensional force/torque sensor, and pierced the needle into the phantom through a mobile platform to study the force interaction between the needle and the tissue. But obviously this method cannot directly measure the force of the tissue loaded onto the needle. Various models have been proposed to reconstruct the load distribution through external force measurements [12–14]. The analysis in Sect. 2.2 shows that this method is useful for experimental research, but actual use will increase the deflection of the needle compared to the case with a pilot hole. When using the actuator described in the second section, a classic model of the force after a needle penetrates the tissue is shown in Fig. 4. F_q is the main cause of needle deflection. Therefore, the rotation of the needle can change the direction of F_q and reduce the deflection of the tip. Compared with the case of using six-dimensional force/torque sensor, we obtain less force information. So it is impossible for us to reconstruct the load distribution. Fortunately, a large number of experiments have shown that the axial force is the most informative and the largest in the process of needle penetration into the tissue, which makes it possible to infer the deflection of the needle only from the axial force.

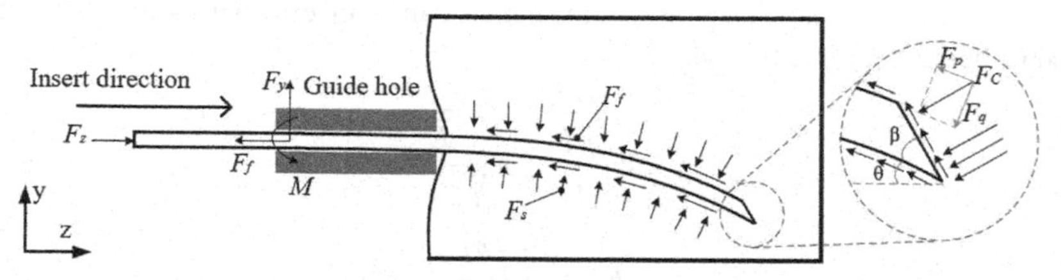

Fig. 4. Force acting on the needle after insertion into tissue with guide hole in 2-D plane (F_z is the axial force of needle base, F_y is the lateral force on the needle provided by the guide hole, M is the moment from the guide hole, F_f is the friction force, which is distributed in multiple positions of the needle, F_s is the resistance force from tissue, F_c is the cutting force, F_p and F_q are the projections of F_c along the needle axis and perpendicular to the needle axis, respectively)

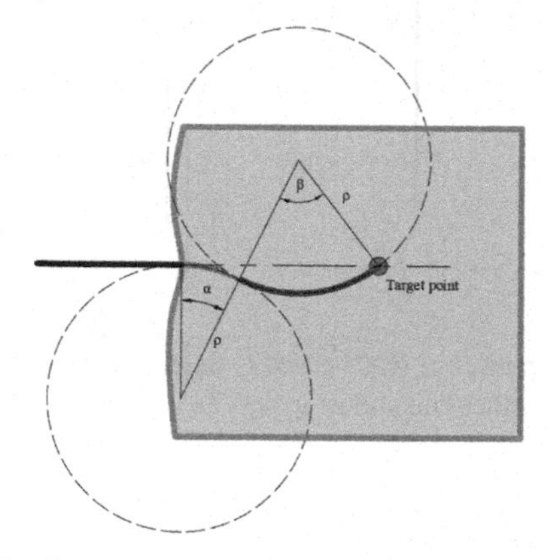

Fig. 5. Accurate puncture path in one rotation

When studying the kinematics of needle penetration into the tissue, the unicycle model is classic and effective assuming that the tissue is isotropic [15]. In this model, the trajectory of the needle with bevel tip after inserting the tissue is a circular arc, that is, the curvature ρ is constant. This gives us inspiration to explore the method of minimizing damage that can accurately reach the lesion with only one rotation with axial of needle as Fig. 5. Rotate the needle 180° around the axis at the appropriate position to change the direction of the needle bend. When the insertion depth is known, if the curvature can be identified, the position of the appropriate point is very easy to determine.

Usually the organization is modeled as a complex viscoelastic model. It is difficult to identify viscosity and elasticity at the same time. We propose a step-by-step puncture strategy, keeping a certain time between each step to reduce the effect of viscosity and ignore it. When the needle penetrates the tissue, the needle continues to penetrate for a while, as the blue segment shown in Fig. 5. During the penetration process, the force F_z will increase due to the friction F_f, the pressure F_s from the tissue. In the process of piercing the tissue without considering piercing the epidermis

$$F_z = F_f + F_1(F_s) + F_2(F_c) \tag{3}$$

Where $F_1(F_s)$ and $F_2(F_c)$ represent functions of F_s and F_c. If we keep it for a while and do not continue to penetrate in each step, the friction F_f will gradually disappear due to the lack of relative movement. This means that the difference between the peak value and the valley value of F_z in each step of the step-by-step puncture strategy is the friction force, where the cutting force F_c is ignored due to the small area of the needle tip. Obviously the friction F_f is positively related to the deflection of the needle tip y. Here we combine the experiment without proof and give the hypothesis that the relationship between friction F_f and deflection y is a second-order polynomial, as in formula (4).

$$y = aF_f^2 + bF_f + c \tag{4}$$

Since the deflection of the needle is small deformation, the position of the needle tip in the x direction can be approximated by the length of the needle, that is, the arc length. The radius is calculated by formula (5).

$$(\rho - y)^2 = \rho^2 - L^2 \tag{5}$$

We can estimate the curvature through the measurement of the axial force to achieve precise puncture in one rotation if we know the parameter a, b and c.

4 Human-Machine Collaboration Workflow

The operating room will not allow robots to move autonomously on a large scale. Our proposed human-machine collaboration workflow as Fig. 6 using the actuator designed in Sect. 2 avoids this problem. First, the doctor manually drags the end of the robot to roughly reach the working position under the guidance of navigation system. The robotic arm may be fixed or on a mobile platform. Then the base of the robot is fixed and automatically adjusted to the correct working position under the guidance of the navigation system. Then we use the actuator designed in Sect. 2 to manually push the needle into the tissue. Then continue to pierce ΔL and keep still until F_z stable to estimate K_1, K_2 and ρ. This process may be repeated several times to estimate the correct ρ. Then calculate the appropriate rotation position, and let the needle rotate 180° at this position. Finally, continue to push the needle to reach the target point.

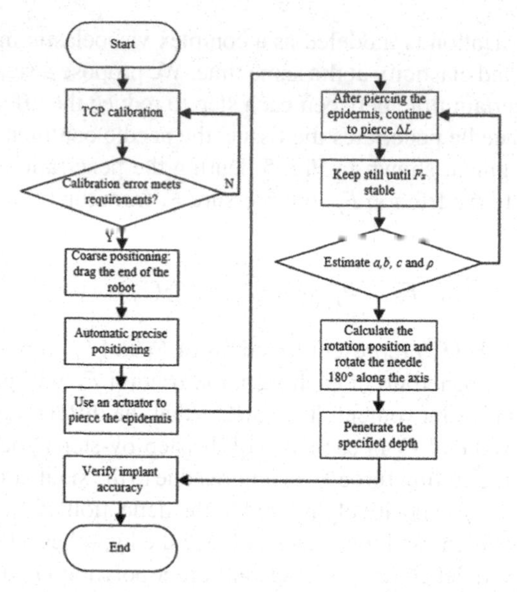

Fig. 6. Human machine collaboration workflow

5 Experiment and Results

The experimental system shown in Fig. 7 includes the robot (UR5, UNIVERSAL ROBOTS Co., Ltd.), the actuator designed in Sect. 2, the force/torque sensor (Gama, ATI Industrial Automation, Inc.), the phantom made of gelatin (mass score of 18%) and the vision system (MER-1810-21U3C, DAHENG IMAGING Co., Ltd.).

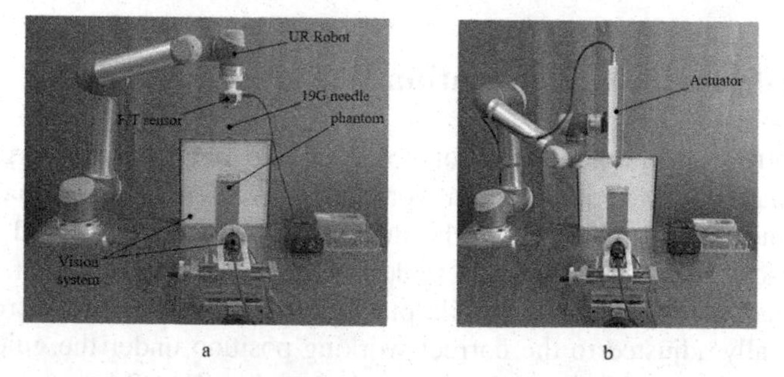

Fig. 7. Experiments system

The first step is to fix the puncture needle (19G) on the measuring surface of the force/torque sensor. The puncture needle is directly driven by the robot to penetrate 10mm each step and then hold for 10s. Take pictures and record every step. We got the force response curve in the whole process as shown in Fig. 8. These pictures are processed to extract the deflection of the needle tip in each step. Fit friction F_f and deflection y using the method proposed in Sect. 3 as shown in Fig. 9. We can see that the second-order polynomial can fit the relationship between F_f and y well. The fitting result is $y = 14.322x^2 - 0.819x + 0.197$, $R^2 = 0.9987$. The ρ of the unicycle model is 650 mm. But it should be noted that these parameters are related to the characteristics of needles and gelatin and are not universal.

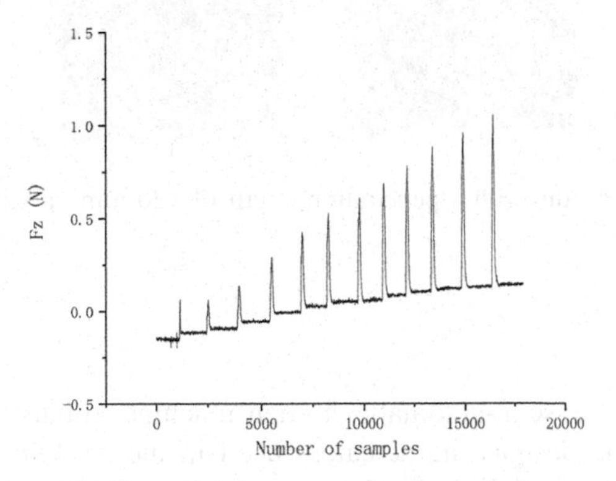

Fig. 8. Axial force response in each step

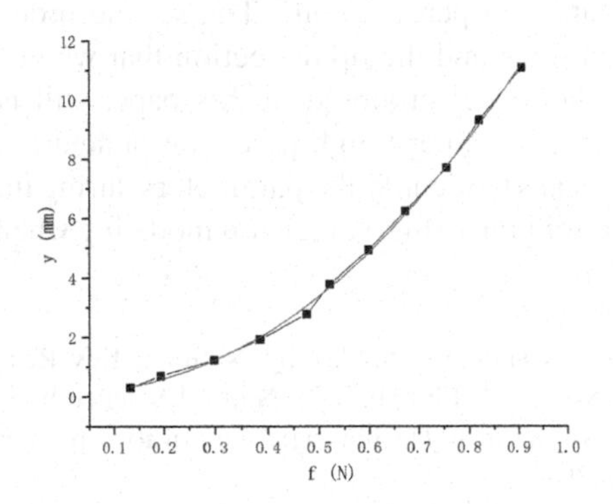

Fig. 9. The relationship between friction and deflection

The second step is to manually pierce the needle into the gelatin phantom using the actuator designed in Sect. 2 using the method proposed in Sect. 3. Penetration depth is 120 mm. The experimental results are shown in Fig. 10. The experimental results clearly show that using our proposed method can significantly improve the accuracy when the penetration depth is deep.

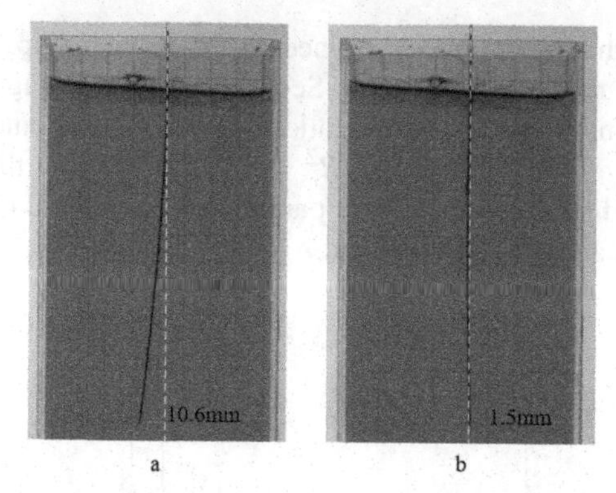

Fig. 10. Image of the needle with a penetration depth of 120 mm (a. without rotation; b. one rotation)

6 Discussion

This paper proposes a semi-automatic human-machine collaborative percutaneous surgery workflow and designs the actuators used in the workflow. Using the actuator designed in this paper, a high-precision puncture method that minimizes damage is realized. The results show that the one-rotation method proposed in this paper can greatly improve the accuracy of percutaneous. The second-order polynomial relationship between the friction force and the tip deflection that we guessed is confirmed by experiments. However, the model proposed in this paper still needs to identify relevant parameters in advance to achieve high-precision puncture. Although we can use B-ultrasound and other means to identify the parameters during the first penetration into the tissue, there is still no solution to high-precision modeling when the deflection cannot be directly observed.

Acknowledgements. This research was funded by National Key Research and Development Program of China (Grant No. 2019YFB1311303), Natural Science Foundation of China (Grant No. U1713202) and Major scientific and technological innovation projects in Shandong Province (Grant No. 2019JZZY010430).

References

1. Hiraki, T., et al.: Robotic needle insertion during computed tomography fluoroscopy–guided biopsy: prospective first-in-human feasibility trial. Eur. Radiol. **30**(2), 927–933 (2019). https://doi.org/10.1007/s00330-019-06409-z
2. Kwoh, Y.S., Hou, J., Jonckheere, E.A., Hayati, S.: A robot with improved absolute positioning accuracy for CT guided stereotactic brain surgery. IEEE Trans. Biomed. Eng. **35**, 153–160 (1988). https://doi.org/10.1109/10.1354
3. Haidegger, T.: Autonomy for surgical robots: concepts and paradigms. IEEE Trans. Med. Robot. Bionics **1**, 65–76 (2019). https://doi.org/10.1109/TMRB.2019.2913282
4. Abedin-Nasab, M.H.: Handbook of Robotic and Image-Guided Surgery. Elsevier, Amsterdam (2019)

5. Podder, T.K., et al.: AAPM and GEC-ESTRO guidelines for image-guided robotic brachytherapy: report of task group 192. Med. Phys. **41** (2014)
6. Dou, H., Jiang, S., Yang, Z., Sun, L., Ma, X., Huo, B.: Design and validation of a CT-guided robotic system for lung cancer brachytherapy. Med. Phys. **44**, 4828–4837 (2017). https://doi.org/10.1002/mp.12435
7. Remebot. https://remebot.com.cn/index.php/site/index
8. Yi, L.: Research on key technology of radioactive seed implantation robot for prostate (2017)
9. Abolhassani, N., Patel, R.V., Ayazi, F.: Minimization of needle deflection in robot-assisted percutaneous therapy. Int. J. Med. Robot. Comput. Assist. Surg. **3**, 140–148 (2007)
10. Li, P., Yang, Z., Jiang, S.: Needle-tissue interactive mechanism and steering control in image-guided robot-assisted minimally invasive surgery: a review. Med. Biol. Eng. Comput. **56**(6), 931–949 (2018). https://doi.org/10.1007/s11517-018-1825-0
11. Roberti, A., Piccinelli, N., Meli, D., Muradore, R., Fiorini, P.: Improving rigid 3-D calibration for robotic surgery. IEEE Trans. Med. Robot. Bionics **2**, 569–573 (2020). https://doi.org/10.1109/TMRB.2020.3033670
12. Okamura, A.M., Simone, C., O'Leary, M.D.: Force modeling for needle insertion into soft tissue. IEEE Trans. Biomed. Eng. **51**, 1707–1716 (2004). https://doi.org/10.1109/TBME.2004.831542
13. Yang, C.J., Xie, Y., Liu, S., Sun, D.: Force modeling, identification, and feedback control of robot-assisted needle insertion: a survey of the literature. Sensors **18**, 38 (2018). https://doi.org/10.3390/s18020561
14. Lehmann, T., Rossa, C., Usmani, N., Sloboda, R., Tavakoli, M.: A real-time estimator for needle deflection during insertion into soft tissue based on adaptive modeling of needle-tissue interactions. IEEE/ASME Trans. Mechatron., 1–1 (2016)
15. Webster, R.J., Cowan, N.J., Chirikjian, G., Okamura, A.M.: Nonholonomic modeling of needle steering. Int. J. Robot. Res. **25**, 509–525 (2006)

A Robot Learning from Demonstration Platform Based on Optical Motion Capture

Hengyuan Yan[1], Haiping Zhou[2,3], Haopeng Hu[1], and Yunjiang Lou[1](✉)

[1] Harbin Institute of Technology Shenzhen, Shenzhen 518000, China
louyj@hit.edu.cn
[2] Beijing Institute of Precision Mechatronics and Controls, Beijing, China
[3] Laboratory of Aerospace Servo Actuation and Transmission, Beijing, China

Abstract. Motion capture (MoCap) is the technology of capturing the movement of a target through sensors such as optical equipment or inertial measurement units. It is widely used in industrial fields. In this work, a robot learning from demonstration platform is established including motion capturing, data pre-processing, policy learning, and a robot controller. It takes the optical MoCap system as the sensor to acquire the motion data of the target. Since the data obtained through the MoCap system always suffer from problems such as noise and data loss, a data processing strategy that can be divided into data pre-processing and policy learning is proposed to obtain a smooth robot motion trajectory. Then the robot trajectory will be transmitted to the designed robot controller to drive a real robot. The proposed robot learning from demonstration platform, which is designed for the rapid deployment of robots in industrial production lines, is convenient for secondary development and enables non-robotics professionals to operate the robots easily.

Keywords: Learning from demonstration · Motion capture · Robot learning

1 Introduction

Motion capture (MoCap) refers to the process of tracking and recording the movements of operators or objects in space, which can be used in many aspects, such as clinical medicine, animation production, etc. In the field of industrial manufacturing, it also has very broad application prospects [7], such as remote control of robots, worker safety inspection [12], teleworking, and so on. At present, various MoCap systems have been developed for motion capture, including optical motion capture systems as shown in Fig. 1, motion capture systems based on inertial sensors, electromagnetic motion capture systems, and so on.

This work was supported partially by the NSFC-Shenzhen Robotics Basic Research Center Program (No. U1713202) and partially by the Shenzhen Science and Technology Program (No. JSGG20191129114035610).

X.-J. Liu et al. (Eds.): ICIRA 2021, LNAI 13014, pp. 100–110, 2021.
https://doi.org/10.1007/978-3-030-89098-8_10

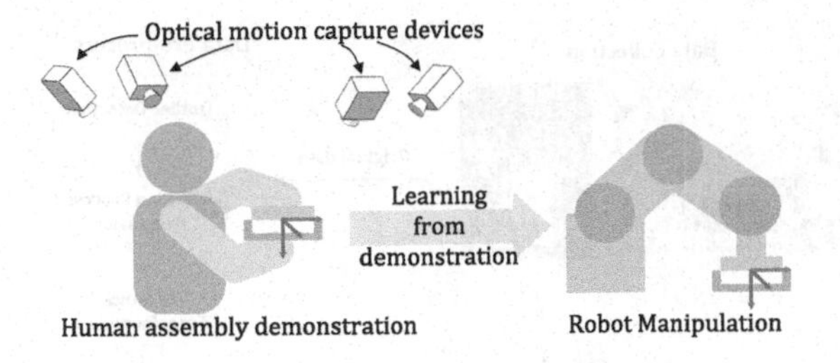

Fig. 1. Robot learning from demonstration with the optical MoCap system.

Among these different systems, the optical MoCap system is based on the optical infrared camera to obtain the position reflective markers on the target object and so on to track the movement of the object [7]. Compared with the MoCap system based on the inertial measurement unit, the optical MoCap system has more advantages. First, the optical MoCap system has fewer restrictions in many aspects, through which we can track more diverse trajectories. In addition, the optical motion system is more convenient in obtaining data and the data drift error is smaller. Moreover, the optical MoCap system has less impact on the operator's natural motion, resulting in a more natural trajectory. Therefore, we use the optical MoCap system to obtain motion data to adapt to various application environments. When programming the movements of robots with complex behaviors, a large amount of expertise in different fields is often required, therefore only trained professionals can complete such operations. The script-based traditional robot programming technology is not only expensive in labor costs, but also takes a long time [8]. However, through the MoCap system, we can conveniently obtain the desired trajectory. In this case, non-professionals can also achieve their target needs well through simple operations [5].

In the article by Tian Y et al., they used the Kincect sensor to collect human body motion data and then recognize related instructions to control the motion of the robot [7]. However, they cannot make the robot complete complex movements. In another article, Du G used Leap Motion to capture and recognize the gestures of the operator to control the movements of the industrial robot in real time [3]. Compared with Tian's study, Du G can control the manipulator to complete more diversified tasks. However, the gestures of the operator are not intuitive and complex, which is not convenient for non-professionals to operate. There are also some other problems, such as the control accuracy is mediocre as well as safety risks. R. Xiong et al. designed a perception system which can transfer knowledge from human demonstrations to robots to a certain extent [13]. Different from us, they focused on understanding the semantics of the parts and skills in the skill videos they collected. With this regard, we build a robot learning from demonstration platform based on an optical MoCap system. In the platform, to make the system convenient and respond quickly with high stability and extensibility, it's necessary to design a special trajectory acquisition device, solve the communication problem between various parts of the software and hardware, process the data collected by the optical MoCap system, design the

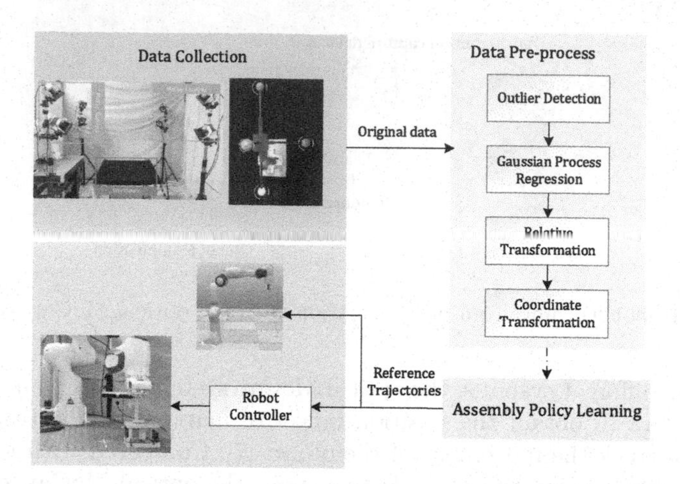

Fig. 2. Work flow of the proposed robot skill learning system.

application program that is convenient for non-robot professionals to operate, and design the robot controller.

The rest of the article is organized as follows. The adopted MoCap system and a novel human-machine interface are introduced in Sect. 2. In Sect. 3, we introduce a data pre-process strategy and a policy learning method. In Sect. 4, a robot assembly experiment is carried out to verify the effectiveness and practicability of the designed system. Finally, Sect. 5 summaries this article.

2 System Setup

Work flow of the proposed robot skill learning system is exhibited in Fig. 2. Through an exclusively designed demonstration platform, the original data of some demonstrated skills are collected. After being processed with a series of data *pre-process* operations, the demonstration data are utilized to learn a *policy*. In this work, a robot assembly policy is taken as a case study. The learned policy serves to generate the motion trajectory of the robot given the goal pose.

2.1 The Optical MoCap System

The OptiTrack MoCap system [5] is selected in this work to obtain the object motion trajectory. The camera used is Prime X41 of which specific parameters are listed in Table 1.

The layout of the optical MoCap system is shown in Fig. 3. There are a total of 8 cameras placed on two floors in a limited real space, with four cameras on each floor. Placing multiple lenses at different heights helps to obtain a diversified perspective and a full range of object motion trajectories in the workspace, so to a certain extent, the problem of point loss caused by occlusion and the influence of external interference factors on data collection are reduced. This space-saving

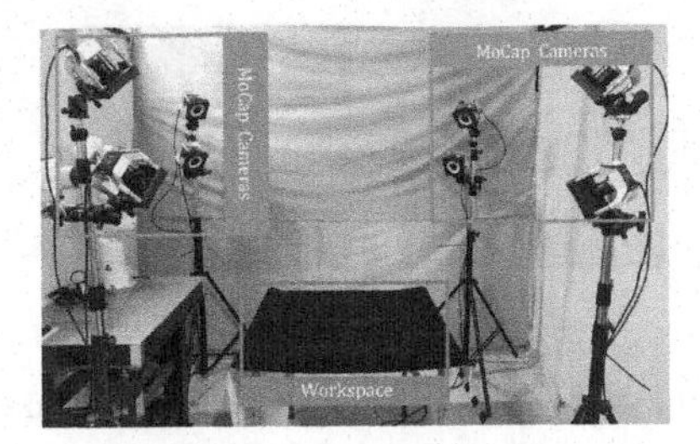

Fig. 3. The spatial layout of the MoCap system.

Table 1. Camera parameters of Prime X41.

Dimensions (mm^3)	Resolution	FOV(°)	Latency (ms)	Frame rate (Hz)
$126 \times 126 \times 126$	2048×2048	51×51	5.5	180

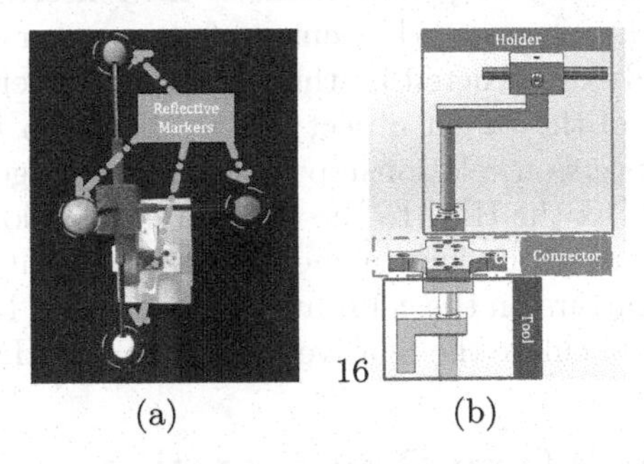

Fig. 4. The human-machine interface designed for the platform.

deployment method has application value in the industrial field. At the same time, orientations of the cameras are carefully adjusted so that all cameras can cover the entire workspace, and make the center of the camera's field of view coincide with the center of the workspace.

2.2 Human-Machine Interface

The optical MoCap system obtains the motion information of the object through reflective markers. Reflective markers can be directly pasted on the operator's hands or target object [4,11]. As for us, a special bracket, which is also called the human-machine interface (HMI), is used to connect the target object and the reflective markers, as shown in Fig. 4.

This novel HMI can be divided into three parts. The lower part is a tool used for the demonstrated tasks, which is connected to a markers' holder through a connector. When the tool varies, it is only necessary to redesign the connector to make it firmly connected to the holder. The proposed HMI also enjoys other advantages. For instance, 4 reflective markers with constant relative translation

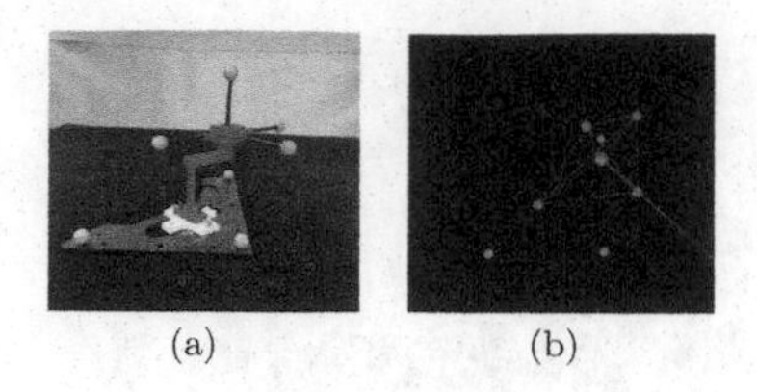

(a) (b)

Fig. 5. The calibration device (left) and captured markers (right).

can be used to create a rigid body with six degrees of freedom. Since the demonstrated tasks may involve complex movements, the position and orientation of the target should be recorded at the same time. The designed HMI allows six degrees of freedom data to be fully recorded. Moreover, the HMI can also play the role of an amplifier. With the holder, the movement of the target will be transformed into a relatively large movement of the reflective markers.

Owing to the machining error, the center of the markers isn't fixed and the center of the rigid body constructed by the reflective markers does not coincide with the body center of the target object, so a calibration is needed to obtain the corresponding conversion relationship between the target object and HMI. As shown in Fig. 5, we fix the HMI to the designated position of the calibration triangle of which parameters are known and the error is small enough. Obtain the position data of the markers on the calibration triangle and HMI at rest through the Mocap system, to calculate the relative position of the HMI's bottom center.

3 Robot Learning from Demonstration

As shown in Fig. 2, the captured data (motion trajectories of two parts) are used to learn an assembly policy. In this work, the policy is in the form of a pre-structured stochastic model. However, the original data cannot be exploited directly due to some problems such as the inevitable missing points and the *correspondence issue* [1]. Therefore, a data pre-process strategy is applied in advance to the policy learning process.

3.1 Data Pre-process

Generally, data obtained by the optical MoCap system suffer from the data loss problem which is caused by occlusion, noise, and some others. When performing complex or dexterous operations, the reflective markers will inevitably be covered by objects or human bodies, resulting in occlusion. Subsequently, occlusion will result in data loss points and blank trajectories. Unexpected reflective objects in the field of view can introduce noise, which can be reduced by the spatial layout of the MoCap system in Fig. 3. However, there are also systematic noise issues that can influence trajectory data. When the number of lost points is small, the accuracy and efficiency of the collected trajectory generally do not be affected much. However, if the number of lost points is large, it could lead to the captured trajectory less effective or even unusable. This will affect the final experimental results and may have potential safety hazards when it is applied to a real machine. Therefore, the data needs to be pre-processed to obtain a more complete and safe trajectory. In this section, we focus on how to deal with data loss issues based on Gaussian process regression which is appropriate to retrieve the data when most of the motion characteristics are still retained.

Gaussian process regression (GPR) is a method in machine learning that is derived from Bayesian theory and statistical principles [10]. GPR uses the kernel function to define the covariance of the prior distribution on the objective function and uses the observed training data to define the likelihood function. Based on Bayes' theorem, we define the posterior distribution of the objective function. It can realize multi-degree-of-freedom trajectory data filling.

Treat the output data or any subset of the output data as a sample from the multivariate Gaussian distribution:

$$y \sim \mathcal{N}(0, \Sigma) \tag{1}$$

$$k(x, x^*) = G_f^2 exp(-\frac{1}{2}(x, x^*)W^{-1}(x, x^*)) \tag{2}$$

where Σ is the covariance matrix; G_f^2 is the signal variance; W^{-1} is the Gaussian kernel covariance matrix. Given the kernel function $k(\cdot)$ and the training data set, the data to be predicted obey the following Gaussian distribution:

$$y \sim \mathcal{N}(0, k(x, x)) \tag{3}$$

Then predict x_q as a brand new input:

$$\begin{aligned} y_q|x, y, x_q \sim \mathcal{N}(&k(x_q, x)k(x, x)^{-1}y, \\ &k(x_q, x_q)k(x, x)^{-1}k(x_q, x)^T) \end{aligned} \tag{4}$$

Therefore,

$$\begin{bmatrix} y \\ y_q \end{bmatrix} = \mathcal{N}\left(0, \begin{bmatrix} k(x, x) & k(x_q, x)^T \\ k(x_q, x) & k(x_q, x_q) \end{bmatrix}\right) \tag{5}$$

$$k(x_q, x) = [k(x_q, x) \ldots k(x_q, x_N)] \tag{6}$$

$$\bar{y} = k(x_q, x)k(x, x)^{-1} = k(x_q, x)w^* = \Sigma_{n=1}^{N} w_n^* k(x_q, x) \tag{7}$$

In this work, we treat sampling time as the independent variable and the trajectory as the output. For the position trajectory, we can directly use the corresponding three-degree-of-freedom coordinate value as the training data and then estimate the movement trajectory of the object at the time of missing sampling. For orientation data, we can use the method proposed in [6] to map the original orientation data (unit quaternion) to its tangent space under the selected auxiliary variable through logarithmic transformation for processing.

Performance of GPR is shown in Fig. 6 in which the blue trajectories are the original data with missing points and the red trajectories are data estimated by GPR. It can be seen that the trajectory after GPR compensation is relatively smooth and so close to the original trajectory that the effect of GPR for data loss is good. In view of this, the Gaussian process regression method is suitable for data pre-processing of the optical MoCap system.

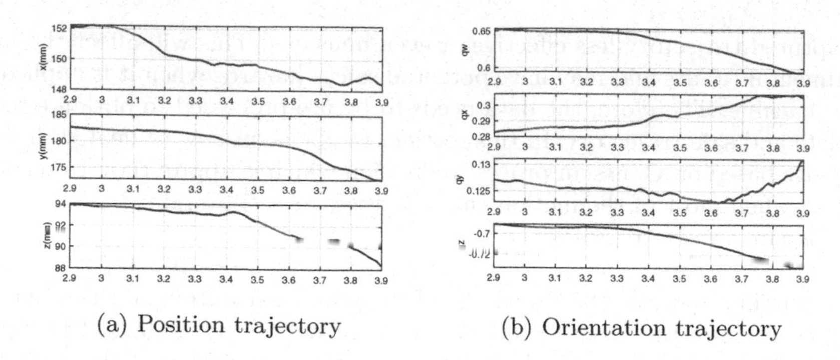

(a) Position trajectory

(b) Orientation trajectory

Fig. 6. The original motion trajectory (blue) and the compensated trajectory (red). (Color figure online)

3.2 Policy Learning

As discussed in [1], a policy can be defined as a mapping that takes the *state* s as input and the *action* a as output:

$$a = \pi(s, \theta), \quad s \in \mathcal{S}, a \in \mathcal{A}, \theta \in \Theta \tag{8}$$

Policies given by (8) are referred to as the *pre-structured* policies whose parameters are estimated in the policy learning process. In the literature, many deterministic or stochastic models, such as the dynamic movement primitive (DMP) [14] and Gaussian mixture model [2] (GMM), can be used to pre-structure the policy. In this work, the *hidden semi-Markov model* (HSMM) [15] is selected as the policy model. Compared with DMP, it is a stochastic model that is capable of encapsulating the variance of motion. Moreover, with HSMM it is unnecessary to apply the temporal alignment operation to the demonstration data, which is always inevitable in LfD studies based on GMM [2].

However, the typical HSMM is far too complicated. With regards to the assembly skill learning task, its parameters are simplified into

$$\theta \triangleq \{a_{ij}, p_j(d), b_j, \rho_j\}_{i,j=1}^{K} \tag{9}$$

where a_{ij} is the *transition probability* from latent state i to j, $p_j(d)$ denotes the *duration* probability, b_j represent the observation/emission probability and ρ_j is the initial state probability. It can be seen here that the duration d_j depends only on the latent state j itself, which is a popular simplification and makes the usage of HSMM much more convenient [15]. In this work, both the duration d_j and the observation probability b_j are defined as multi-variate Gaussian distributions with center μ and covariance Σ, i.e.

$$p_j(d) = \mathcal{N}\left(\mu_j^{(d)}, \Sigma_j^{(d)}\right), \quad b_j = \mathcal{N}\left(\mu_j^{(b)}, \Sigma_j^{(b)}\right)$$

Given the demonstration data, parameters of the policy (9) can be estimated by expectation-maximization algorithm [15].

As exhibited in Fig. 2, once a policy is learned it can generate a reference trajectory for the robot to track. For the HSMM based policy, to generate the motion trajectory we calculate the *forward variable* first. The forward variable of HSMM is defined as

$$\begin{aligned}\alpha(j,t,d) &\triangleq P\left[s_{[t-d+1:t]} = j, o_{1:t} \middle| \theta\right] \\ &= \sum_i \sum_d \alpha(i,t-d,d')a_{(i,d')(j,d)}b_{j,d}(o_{t-d+1:t})\end{aligned}$$

where $b_{j,d}(o_{t-d+1:t})$ represents the observation probability of latent state j whose duration is d. In view of the simplified parameters (9), the forward variable can also be simplified into

$$\alpha(j,t) = \sum_i \sum_d \alpha(i,t-d)a_{ij}p_j(d) \tag{10}$$

Note that in (10) we drop the observation probability $b_{j,d}(\cdot)$ as the forward variable here serves to generate motion of the robot rather than evaluate a given observed motion. Given the forward variables of each latent state j and time step t, a state sequence $S \triangleq \{s_t\}_{t=1}^{T}$ is calculated by

$$s_t = \max_j \alpha(j,t) \tag{11}$$

It is a discrete state sequence instead of a motion trajectory. As done in [9], the least quadratic tracking (LQT) controller can play the role of interpolating the state sequence (11) into a smooth motion trajectory.

4 Robot Assembly Experiment

In order to verify the effectiveness and practicability of the designed demonstration platform for robot manipulation skills based on the optical MoCap system, a robot assembly task is taken as a case study.

4.1 Experiment Settings

Figure 7 shows the components used in the experiment together with the HMI devices. Part A is the assembly base and Part B is the target to be captured. Part C is the designed HMI whose lower part is connected with Part B in Fig. 7, and the upper part is connected with four markers. The entire assembly movement of Part A and Part B can be simplified into three actions, namely downward insertion, horizontal movement, and vertical downward movement in sequence. In this work, orientation variance during the assembly process is not taken into account. We also set a LED light on the assembly parts to assist in verifying whether the assembly is accomplished.

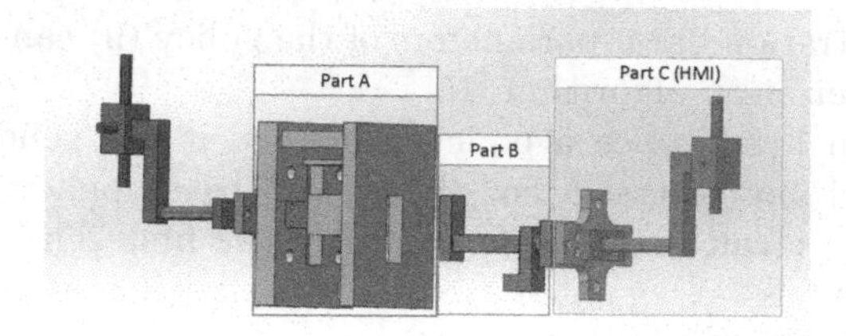

Fig. 7. Design of the parts to be assembled and the HMI devices

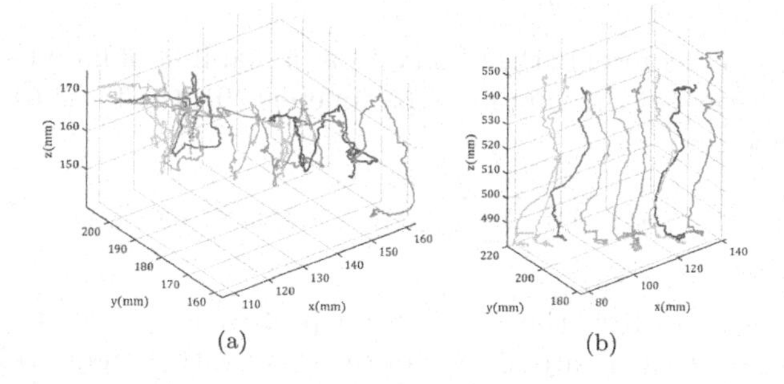

Fig. 8. The original trajectories of two parts captured through the MoCap system.

9 sets of assembly demonstration data are collected and then processed by designed data pre-process and learning policy strategy. The generated trajectory is transmitted to the controller of a 7-degree-of-feedom Franka Panda robot which is used in the experiment to perform the demonstrated assembly task.

4.2 Experimental Results and Discussion

The original data captured through the optical MoCap system are shown in Fig. 8. Since the human operator can never perform the task twice from the same initial position and the robot in use cannot perform bi-manual tasks, it is necessary to transform the motion trajectories of one part to the frame defined by the other part. Then through the proposed data pre-processing strategy, motion trajectories used for policy learning is gotten as shown in Fig. 9 (a).

Figure 9 (b) exhibits the learned HSMM in which centers of the ellipsoids denote the centers of the observation probabilities $\mu_j^{(b)}$ and radius of them denotes the covariance $\Sigma_j^{(b)}$. Figure 9 (c) shows the motion trajectory generated by the policy and LQT controller. It can be seen the shape features of the demonstrated trajectories are encapsulated by the generated motion trajectory. Figure 10 exhibits the probabilities of the state at time t belonging to each latent state.

The Franka robot is controlled to reproduce the assembly operation by following the trajectory shown in Fig. 9 (c), and its movement is shown in Fig. 11. The led light is on finally, which means the assembly experiment is accomplished.

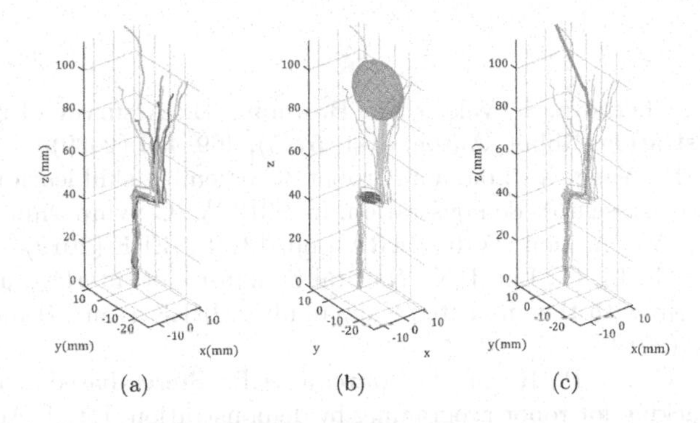

Fig. 9. The pre-processed demonstration data (a), the learned HSMM (b), and the robot motion trajectory generated by the policy and LQT controller (c).

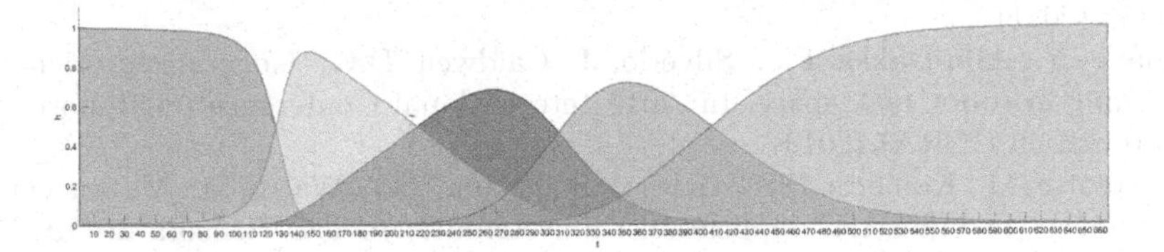

Fig. 10. The probabilities of the state at time t belonging to each latent state.

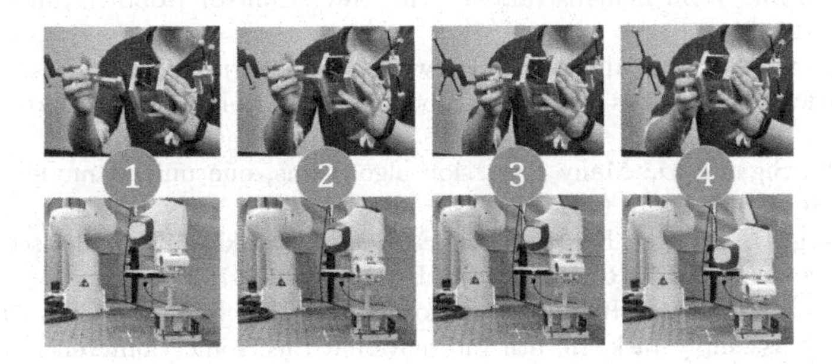

Fig. 11. The human and robot assembly process.

5 Conclusion

A robot learning form demonstration platform based on the optical MoCap system is proposed in this work. In this platform, we designed a data pre-processing strategy to solve problems such as data loss. We also propose a robot learning policy based on HSMM. After data processing, we perform the simulation to verify the safety and effectiveness of the trajectory, which improves the robustness of the system. There is a robot controller designed, which can effectively control the seven-degree-of-freedom robot arm to complete the desired trajectory. The platform we established helps industrial production lines to quickly deploy robots to complete frequently updated assembly tasks and is convenient for non-professionals to operate.

References

1. Argall, B.D., Chernova, S., Veloso, M., Browning, B.: A survey of robot learning from demonstration. Robot. Auton. Syst. **57**(5), 469–483 (2009)
2. Cao, Z., Hu, H., Yang, X., Lou, Y.: A robot 3C assembly skill learning method by intuitive human assembly demonstration. In: 2019 WRC Symposium on Advanced Robotics and Automation (WRC SARA), pp. 13–18. IEEE (2019)
3. Du, G., Yao, G., Li, C., Liu, P.X.: Natural human-robot interface using adaptive tracking system with the unscented Kalman filter. IEEE Trans. Hum. Mach. Syst. **50**(1), 42–54 (2019)
4. Ferreira, M., Costa, P., Rocha, L., Moreira, A.P.: Stereo-based real-time 6-DoF work tool tracking for robot programing by demonstration. Int. J. Adv. Manufact. Technol. **85**(1–4), 57–69 (2014)
5. Hu, H., Cao, Z., Yang, X., Xiong, H., Lou, Y.: Performance evaluation of optical motion capture sensors for assembly motion capturing. IEEE Access **9**, 61444–61454 (2021)
6. Huang, Y., Abu-Dakka, F.J., Silvério, J., Caldwell, D.G.: Generalized orientation learning in robot task space. In: 2019 International Conference on Robotics and Automation (ICRA) (2019)
7. Menolotto, M., Komaris, D.S., Tedesco, S., Flynn, B.O., Walsh, M.: Motion capture technology in industrial applications: a systematic review. Sensors **20**(19), 5687 (2020)
8. Ravichandar, H., Polydoros, A.S., Chernova, S., Billard, A.: Recent advances in robot learning from demonstration. Ann. Rev. Control Robot. Auton. Syst. **3**, 297–330 (2020)
9. Rozo, L., Silverio, J., Calinon, S., Caldwell, D.G.: Learning controllers for reactive and proactive behaviors in human-robot collaboration. Front. Robot. AI **3**, 30 (2016)
10. Stulp, F., Sigaud, O.: Many regression algorithms, one unified model: a review. Neural Netw. **69**, 60–79 (2015)
11. Tanneberg, D., Ploeger, K., Rueckert, E., Peters, J.: Skid raw: skill discovery from raw trajectories. IEEE Robot. Autom. Lett. **6**, 4696–4703 (2021)
12. Tuli, T.B., Manns, M.: Real-time motion tracking for human and robot in a collaborative assembly task. In: 6th International Electronic Conference on Sensors and Applications (2019)
13. Wang, Y., Xiong, R., Yu, H., Zhang, J., Liu, Y.: Perception of demonstration for automatic programing of robotic assembly: framework, algorithm, and validation. IEEE/ASME Trans. Mechatron. **23**(3), 1059–1070 (2018)
14. Xie, Z., Zhang, Q., Jiang, Z., Liu, H.: Robot learning from demonstration for path planning: a review. Sci. China Technol. Sci. **63**, 1325–1334 (2020)
15. Yu, S.Z.: Hidden semi-Markov models. Artif. Intell. **174**(2), 215–243 (2010)

Real-Time Collision Avoidance in a Dynamic Environment for an Industrial Robotic Arm

Uchenna Emeoha Ogenyi[1,2](✉), Qing Gao[3], Dalin Zhou[1], Charles Phiri[1], Zhaojie Ju[1], and Honghai Liu[1]

[1] University of Portsmouth, Portsmouth, UK
{uchenna.ogenyi,dalin.zhou,charles.phiri,zhaojie.ju, honghai.liu}@port.ac.uk

[2] Institute of Robotics and Intelligent Manufacturing and School of Science and Engineering, The Chinese University of Hong Kong, Shenzhen, Shenzhen 518172, China

[3] Shenzhen Institute of Artificial Intelligence and Robotics for Society, Shenzhen 518129, China
gaoqing@cuhk.edu.cn

Abstract. This paper proposed learning from demonstration control policy that permits an industrial robotic arm to generalise in an unstructured and dynamic environment. The approach combines a probabilistic model with a reactive approach that learns the cost function of an unknown state of the environmental constraints. The approach redefines the robot's behaviour towards generating a trajectory that satisfies the task and scene constraints. In the end, experiments on a real industrial robotic arm were presented to show how the proposed approach works and facilitates enhanced human-robot coexistence.

Keywords: Robot learning from demonstration · Dynamic obstacle · Probabilistic modelling

1 Introduction

With the recent increase in the demand for collaborative robotic manipulators in the industries, the need to develop control policies to improve the robot's ability to work in non-static environments while effectively collaborating with human users is rapidly growing [1]. Despite this increase in demand, the state-of-the-art robot LfD approaches (e.g. [2,3]) have only strived to enable a robot to learn control policies to avoid stationary objects. Consequently, how a robot should respond to new incidents at the reproduction state is of concern. For example, a sudden appearance of an obstacle category not considered at the demonstration stage or even a case that causes a target object to change its position at the reproduction stage is still a challenging problem yet to be addressed.

X.-J. Liu et al. (Eds.): ICIRA 2021, LNAI 13014, pp. 111–121, 2021.
https://doi.org/10.1007/978-3-030-89098-8_11

An approach to deal with this situation is using a corrective method, which allows the human teacher to provide additional task examples to the robot at the reproduction stage [4,5]. This approach is fit for the purpose; however, it is time-consuming and increases the model complexity. Another approach is to explicitly consider all possible scenarios during the training session [6,7]. This approach, as employed in [7] for motion encoding framework to extracts essential features of motion trajectories to enable a robot to autonomously repeat a task and adapt to task variations such as changes in target objects starting and goal positions. This approach can extract important patterns to a given task and generalise them to different scenarios to achieve dynamic obstacle avoidance. Nevertheless, the obstacle must be considered at the model learning stage. Thus, the unavailability of the training data limits the applicability of the approach in different scenarios.

An alternative to the probabilistic method is the Dynamic Movement Primitives (DMPs), which can be used to compute the motion planning in a complex environment [8]. The DMPs allow robot trajectory demonstrations to be represented as a set of damped linear string disturbed by an external force. The DMPs make it possible for a learned skill to be generalised to a new goal state, but it is sensitive to temporal perturbation because it is implicitly time-dependent. This technique was used in [9] to create a planner which applies non-linear differential equations to create a control function that permits a trajectory generalisation to a new goal point. This approach is only suitable when the start and goal positions of the reaching points are defined with a definite obstacle control policy which is difficult for non-experts to design.

Recently, there has been a growing tendency towards increasing the generalisation capability of robot trajectory learning approaches. Many researchers have strived to merge more than one learning approach to exploit their unique learning strengths and provide enhanced learning approaches suitable for handling complex tasks. For example, [10] used the Gaussian Mixture Model (GMM) to encode a set of demonstrations for a virtual spring damper and employed the Gaussian Mixture Regression (GMR) to generate the desired motion direction for the manipulator to reach the initial goal position. Then the DMP was employed in finding a control function that pulls the robot manipulator to the position of the new goal point at the reproduction state. Similarly, in [11], the authors explored GMM/GMR's strength to learn the optimal control for a demonstrated trajectory motion and employed inverse reinforcement learning in reshaping the motion trajectory to satisfy both task and environment constraints.

Although the LfD models and obstacle avoidance has widely been studied in the literature, the existing methods barely considered situations where the robot must generalise a task in a dynamic environment, such as in an industrial setting with moving obstacles. Thus, this paper proposes a framework that combines a probabilistic method and underlying real-time cost functions of the environmental constraint to provide a control policy that satisfies the task and environmental constraints. The probability model provides the task-related attributes, while the cost function guarantees the robot corrects its motion within a restricted task environment and to cope with real-life dynamic obstacles while reproducing the learned trajectory.

2 Methods

This section presents the proposed approach for learning robot trajectory and generalising it in a perturbed environment. Firstly, it is presumed that the dataset is provided using Kinethestic teaching, as demonstrated in the experimental result in our previous work [12]. Once the demonstration has been recorded, the task parameter is learned using the GMM/GMR and DMP. The Cartesian space is chosen because most manipulation tasks can be defined in Cartesian space. In the sweeping task, the model is simplified by using some prior knowledge of the experimental setup, e.g. the brush must be in contact with the surface of the table, i.e. z = 0.23[m] value of the Cartesian coordinate is set to offset the distance between the end-effector and the table. More details of the steps are presented in the following subsections.

2.1 Learning Dynamic Movement Primitives

Based on the proposed approach, we first used the GMM/GMR to compute a generalised trajectory across a set of demonstrated trajectories $x = (x_1, x_2, ..., x_N)^T$ with K-components of a D-dimension. For simplicity sake, we sampled demonstrated trajectories in time-space T = 100 [s]. The parameters of the model with K-components are defined by π_m (μ_m, Σ_m)— m = 1, . . ., K, where π_m represents the mixing coefficients, and (μ_m, Σ_m) are the model mean and covariance matrix of the Gaussian respectively. The GMM model allows us to extract the relevant value of multi-demonstrations across slightly different environments with similar starting and goal positions. By encoding the demonstrated trajectories, the noise in the demonstration is removed. This operation is followed by the GMR, which generates the nominal trajectory.

From the GMM, we recovered the expected position x of the robot's end-effector at each timestamp by using the GMR algorithm. The recovered trajectory is used as input for the DMPs model. The yellow line depicts the DMPs trajectory denoted by ξ^{dm} which is good enough to accomplish the task of sweeping successfully, should the scene remain the same throughout the operation as shown in Fig. 1(a). We empirically chose the damping factor c = 28, spring constant k = 139.26 and τ = 1 to ensure that the Eq. 1 forms a critically damped system. The outcome is a learned model that is used to generate a smooth path from the start position $P_s(-5.2, -0.6)$ to the goal position $P_g(5.2, 0.6)$ in the absence of obstacles Fig. 1(b).

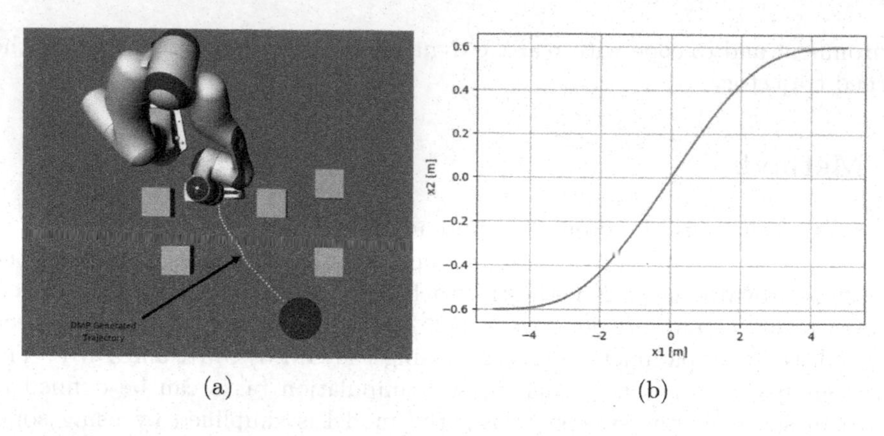

Fig. 1. (a) A sample of the DMP trajectory generated from the performed sweeping task. (b) The optimal path generated from the trained DMP. Ordinarily, this trajectory is enough to enable the robot to move its end-effector from the start point to the goal point, assuming there is no obstacle in its path.

2.2 Cost Function Formulation

The trajectory optimisation in this context consists of all the possible cost of the transition from the initial state to the desired position. In general, the cost of the robot's transition from the current state to the goal state is the sum of the cost of the task constraint plus the scene cost. The previous work [12] proposed approach is a semi-open loop strategy in which the positions of the obstacles are known before generating a collision-free trajectory. In contrast, we hypothesise that a skill learned in a static scene can be generalised in a dynamic environment. Thus, the learning is carried out in a stationary scene, while the reproduction is done in a dynamic scene with moving obstacles.

Let us denote the smooth trajectory generated by the DMPs as shown in Fig. 2 as ξ^{dm} and J^{dm} as the cost of allowing the robot to optimally perform a task by following the path ξ^{dm}. The cost function satisfies the task constraint, which entails the ability to move the robot end-effector from the start position to the goal position, but it does not satisfy the scene constraint caused by the presence of a moving obstacle. To compute the optimal solution, we postulate that a utility function J^G can generate a trajectory that satisfies both the task and the scene constraints if a Gaussian cost is included in the task cost function J^{dm}.

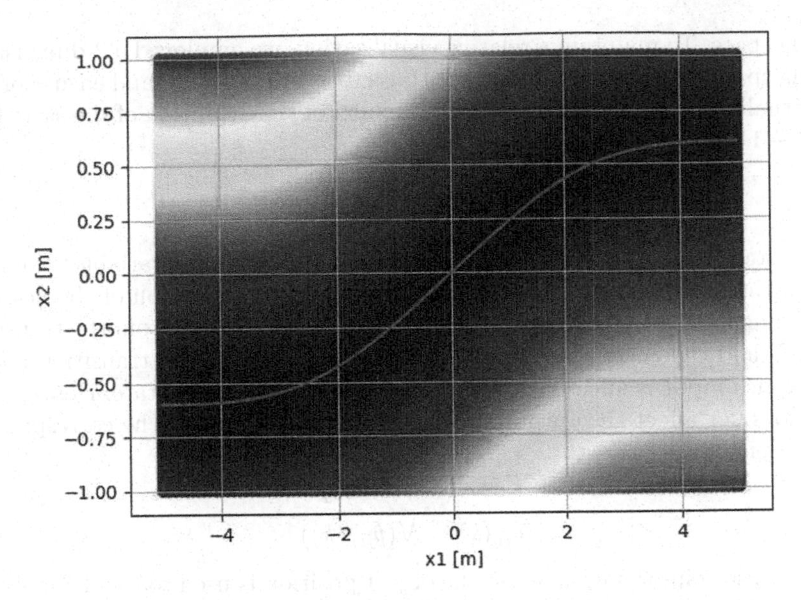

Fig. 2. The dark blue region represents the utility with low-cost and it increases towards the thick red region which has the highest cost function. (Color figure online)

The DMPs control generates the trajectory in the thick blue line, which runs from P(−0.9, −0.63) through P(5.6, 0.63). The cost function formed around the optimal path is illustrated with colour regions. The dark blue region represents the utility with low-cost, and it increases towards the thick red region, which has the highest cost function.

$$J^{obst}(\chi_k : \xi^{dm}, J^{obst}) = \xi^{dm} + \sum_{i=1}^{N} \alpha^{obst} \, e^{(\wp^2)\theta} - \iota^{obst} \tag{1}$$

where α^{obst} ι^{obst} are constant values defined by the height and width of the ith obstacle respectively. Also, $\wp = (\chi_k - \rho_k)$ where χ_k is a subset of ξ^{dm} and ρ_o denotes obstacle position detected by the vision sensor. When an obstacle is detected within the robot vision sensor's range, the exponential function will have a value, otherwise, the value is zero.The variable θ is the covariance matrix of the utility function which must be learned from the demonstration.

2.3 Obstacle Position Estimation

Knowing that the proposed trajectory algorithm goal is to move the end-effector from the starting position along the path with the minimum cost to arrive at the target position while avoiding collision with unknown moving obstacles in the environment. It becomes pertinent that the future position of the moving obstacle is included in the control policy. Also, since we are dealing with a continuous and dynamic process, the process would require a multidimensional working space. To maintain a relative visualisation, we employed a 2-dimensional working space that attracts working with vectors and matrices and ensure proper conceptualisation of the system. Hence a model of the evolution of the state from time $k-1$ to time k is defined as:

$$\hat{x}_k = Fx_{k1} + Bu_k\hat{a}1 + w_{k-1} \tag{2}$$

The state transition describes the transition of the state variables from one point in time to the next. The function performs the estimation of the new systems dynamic matrix by using the state (position and speed) from the transition matrix **F** and the current mean vector x. Each element of the transition matrix shows which input is affected and how they affect the output state. Hence, if the input does not affect the output state, there will be a zero in the corresponding matrix element.

$$KF = \hat{y}_{i1}(k), ..., \hat{y}_{itp}(k) \tag{3}$$

$$\tilde{y}_{itp}(k) = \mathcal{N}(\hat{y}_{itp}(k)) \tag{4}$$

The sensory information about the object position is used to build a real-time feedback control for the robot to perform object manipulation in the dynamic environment safely. We employed the Kalman Filter to filter the noisy information provided by the object tracker, which gives more accurate information about the position of the obstacles at every time step. The predicted state of the obstacle, which is a Gaussian, is incorporated into the control policy.

$$J^G(\chi_k + 1 : \xi^{dm}, J^{obst*}) = \xi^{dm} + \sum_{i=1}^{N} \frac{1}{n} \sum_{s=1}^{n} \alpha^{obst} e^{(\wp^2)\theta} - \iota^{obst} \tag{5}$$

From the Eq. 5, ρ_{obst} denotes the obstacle position detected by the robotâs sensor and incorporated as shown here y = χ_k - ρ_{obst}

2.4 Experimental Results

A PANDA collaborative robot arm is used in this experiment. It is a 7-DOF robotics arm with a maximum reaching length of 855mm and supports integration with ROS, C++. Configuring Panda is straightforward, and it has excellent features for interfacing external devices. It comes with moderate Cartesian velocity and torque sensors in each of the joints to minimise any collision impact. The SDK is readily available for adequate use of the research version for research projects in the universities.

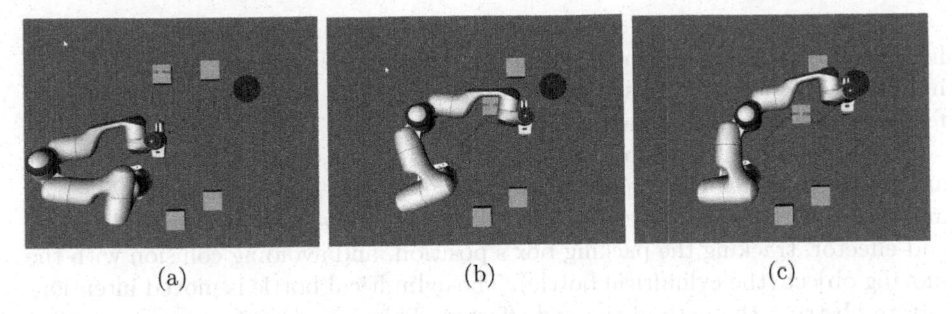

Fig. 3. The screenshots demonstrate that the robot is capable of avoiding collision with the moving cube obstacle which is moved deliberately to cause a collision the robot.

The experiment's goal is for the robot to move its end-effector from a defined starting position to a goal position to imitate the action of sweeping a crumb of rubbish on a table into a dustpan while avoiding a moving obstacle in its operational space. At the experiment's reproduction stage, we used a vision sensor to track the goal object and obstacles and extract their real-time positions.

After the DMPs is learned, we conducted a test to verify the effectiveness of the proposed approach in generalising to a moving obstacle avoidance situation. Figure 3 (a)-(c) show the screenshots of the robot perform trajectory motions to prevent the possible collision of the end-effector with the moving obstacle. The test consists of a robot avoiding moving obstacle where the square-cube is moved intentionally to cause a collision with the robot. The robot starts position is the robotic arm's home configuration as shown in Fig. 3 (a) the blue dotted lines represent the DMPs which satisfies the task constraints. Figure 3 (b) shows the robot following the DMPs path until it encountered a moving obstacle. In Fig. 3 (c), the robot then switched to avoiding obstacle mode while still moving towards the goal position. The position of the stationary and moving obstacle can change from one execution to another. The experiment verified the effectiveness of our task and environmental-oriented robot control strategy for avoiding moving obstacle.

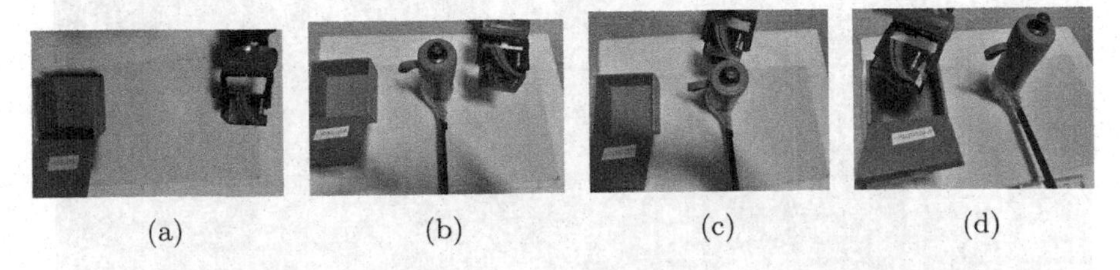

Fig. 4. (a) and (b) show the robot picking and tracking the position of the box package. (c) is the snapshot of the robot avoiding collision with the moving obstacle (the water bottle) which is moved intentionally to cause a collision. (d) shows the robot reaching the placing point of the location.

The second experiment consists of a robot avoiding moving object in a real-life scenario. This experiment's goal is for the robot to pick and track the box to imitate the action of a person picking and packaging items in a manufacturing facility while avoiding obstacles.

With the help of a vision sensor, we tracked the positions of the target object and obstacles in real-time as the robot operates in such a dynamic scene. The pick and place task consists of three actions: picking the cube box object with the robot end-effector, tracking the packing box's position, and avoiding collision with the moving object (the cylindrical bottle). The cylindrical bottle is moved intentionally to obstruct the path of the end-effector. This experiment was repeated five times while a person intentionally tries to move the object into the robot workspace in order to cause collision between the robot end-effector and the obstacle object. Nevertheless, the robot kept moving away from the obstacle anytime it is brought closer to the robot. This experiment illustrates the ability of the proposed approach to cope with moving obstacles of different velocity profiles in a real-life scene involving a shared workspace for human-robot collaboration.

Kalman-filter was used in all the experiments to track the position of the moving object. As the cube object moves, the vision sensor tracks its position, and the noisy measurement is fed into the Kalman-filter to estimate the location and the speed of the object. The state vector of the object is a random variable with a prior belief corresponding to the initial values of the mean and the covariance. In our case, the prior belief is initialised as the first detected position of the object within the workspace, then each time a new measurement is received, the previously estimated mean and the covariance are used as the prior belief. Since the likelihood and prior are both Gaussian distributed, we compute the product of the distributions and estimate the posterior belief as a weighted average of the prior belief mean and covariance. Consequently, the probability density function with regards to the standard deviation becomes smaller, resulting in estimating a more precise value of the object position. Shown in Fig. 5 is the iteration values of the prediction and correction recordings from the Kalman-filter computation.

```
Shell
[INFO] [1607371839.126651, 348.415000]: correct_step: [9.61579117182, 0.0493901531919]
[INFO] [1607371839.760967, 348.908000]: predict step: [9.27078167962, 4.04939015319]
[INFO] [1607371839.762196, 348.908000]: correct_step: [9.84756956138, 0.0493901531919]
[INFO] [1607371840.599909, 349.450000]: predict step: [-4.68176043296, 4.04939015319]
[INFO] [1607371840.601430, 349.454000]: correct_step: [9.95496265151, 0.0493901531919]
[INFO] [1607371841.226730, 349.911000]: predict step: [11.05256589, 4.04939015319]
[INFO] [1607371841.231351, 349.914000]: correct_step: [0.101716399006, 0.0493901531919]
[INFO] [1607371841.980220, 350.410000]: predict step: [0.232267735265, 4.04939015319]
[INFO] [1607371841.980705, 350.410000]: correct_step: [0.763582917128, 0.0493901531919]
[INFO] [1607371842.654744, 350.907000]: predict step: [0.202199467704, 4.04939015319]
[INFO] [1607371842.655187, 350.907000]: correct_step: [0.595614915828, 0.0493901531919]
[INFO] [1607371843.354882, 351.422000]: predict step: [-0.88933437066, 4.04939015319]
[INFO] [1607371843.355755, 351.423000]: correct_step: [0.778170415157, 0.0493901531919]
[INFO] [1607371844.131001, 351.926000]: predict step: [-0.602838390106, 4.04939015319]
[INFO] [1607371844.131861, 351.926000]: correct_step: [1.73630776847, 0.0493901531919]
[INFO] [1607371845.024167, 352.422000]: predict step: [2.62174133487, 4.04939015319]
[INFO] [1607371845.024977, 352.422000]: correct_step: [1.60437535317, 0.0493901531919]
[INFO] [1607371845.686011, 352.913000]: predict step: [4.05619630966, 4.04939015319]
[INFO] [1607371845.687066, 352.913000]: correct_step: [2.09135687479, 0.0493901531919]
```

Fig. 5. The updated and corrected position of the object tracked using Kalman filter

To visualise the Kalman filter's effect in estimating the object's position, we plotted a graph Fig. 6(a) to compare the raw sensor data and the filtered signal. From the graph, the y-axis indicates the object position as published by the position state topic from the ROS. The value starts from 0 and moves to 10 as the obstacle moves in a specified direction. The pink zig-zag line shows the sensor reading from the camera. The line's fluctuation was due to the sensor noise, which we used the Kalman filter to minimise. The graph shows that the Kalman filter, represented with the green line, is more stable than the pink line (camera reading). This sensor signal's stability is because the Kalman filter estimates and outputs only the signal with the lowest uncertainty percentage. Furthermore, we plotted a graph of the Kalman filter and the camera readings and then compared the two with the ground truth. As shown in Fig. 6(b), it could be seen that the filtered position is much closer to the ground truth than the camera readings is from the ground truth.

Figure 6(c) represents the end-effector trajectory in correspondence with the obstacle object's position. The black lines represent the correspondence between the robot's end-effector and the obstacle object when it approaches the robot's

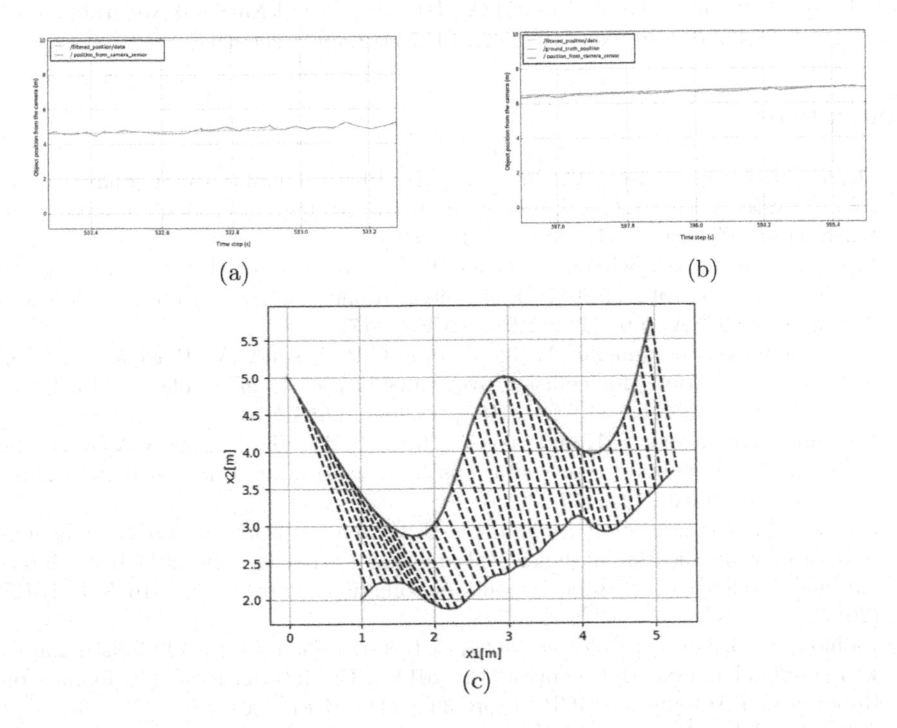

Fig. 6. (a) A snapshot of the filtered position and the position from a camera sensor. (b) The filtered position and the sensor data compared with the ground truth. (c) The figure depicts the corresponding trajectory of the end-effector (represented with a blue line) and that of a moving obstacle which is represented by a red line in the x1, x2 plane of the robot coordinates. (Color figure online)

end-effector. The figure illustrates the robot's responses with an allowable range of obstacle avoidance motion. This result indicates that the model enhanced the robot's ability to adapt to the obstacle's varying positions and speed without colliding with it.

2.5 Conclusion

This paper presents an extended framework of obstacle avoidance introduced in the previous paper [12] to allows a robot to perform object manipulation in a dynamic scene. Unlike various other approaches, this method is based on a probabilistic model and the concept of a cost-function, to enable the industrial robotic arm to quickly and smoothly achieve obstacle avoidance. The presented experimental results show the robustness of the proposed approach in obstacle avoidance cases. The approach is appropriate for enabling a robot to avoid moving obstacles of different shapes and sizes and still fulfil the task constraints.

Acknowledgment. The authors would like to acknowledge the support from the AiBle project co-financed by the European Regional Development Fund, National Key R&D Program of China (Grant No. 2018YFB1304600), and National Natural Science Foundation of China (grant No. 52075530, 51575412, and 62006204).

References

1. Ogenyi, U.E., Liu, J., Yang, C., Ju, Z., Liu, H.: Physical human-robot collaboration: robotic systems, learning methods, collaborative strategies, sensors, and actuators. IEEE Trans. Cybernet. **51**, 1888– 1901 (2019)
2. Rai, A., Sutanto, G., Schaal, S., Meier, F.: Learning feedback terms for reactive planning and control. In: 2017 IEEE International Conference on Robotics and Automation (ICRA), pp. 2184–2191. IEEE (2017)
3. Sangiovanni, B., Rendiniello, A., Incremona, G.P., Ferrara, A., Piastra, M.: Deep reinforcement learning for collision avoidance of robotic manipulators. In: European Control Conference (ECC) 2018, pp. 2063–2068 (2018)
4. Niekum, S., Osentoski, S., Konidaris, G., Chitta, S., Marthi, B., Barto, A.G.: Learning grounded finite-state representations from unstructured demonstrations. Int. J. Robot. Res. **34**(2), 131–157 (2015)
5. Karlsson, M., Robertsson, A., Johansson, R.: Autonomous interpretation of demonstrations for modification of dynamical movement primitives. In: 2017 IEEE International Conference on Robotics and Automation (ICRA), pp. 316–321. IEEE (2017)
6. Calinon, S., Bruno, D., Caldwell, D.G.: A task-parameterized probabilistic model with minimal intervention control. In: 2014 IEEE International Conference on Robotics and Automation (ICRA), pp. 339–3344. IEEE (2014)
7. Tanwani, A.K., Calinon, S.: Learning robot manipulation tasks with task-parameterized semitied hidden semi-Markov model. IEEE Robot. Autom. Lett. **1**(1), 235–242 (2016)
8. Ijspeert, A.J., Nakanishi, J., Hoffmann, H., Pastor, P., Schaal, S.: Dynamical movement primitives: learning attractor models for motor behaviors. Neural Comput. **25**(2), 328–373 (2013)

9. Kober, J., Gienger, M., Steil, J.J.: Learning movement primitives for force interaction tasks. In: 2015 IEEE International Conference on Robotics and Automation (ICRA), pp. 3192–3199. IEEE (2015)
10. Calinon, S., Li, Z., Alizadeh, T., Tsagarakis, N.G., Caldwell, D.G.: Statistical dynamical systems for skills acquisition in humanoids. In: 2012 12th IEEE-RAS International Conference on Humanoid Robots (Humanoids 2012), pp. 323–329. IEEE (2012)
11. Ghalamzan, A.M., Paxton, E.C., Hager, G.D., Bascetta, L.: An incremental approach to learning generalizable robot tasks from human demonstration. In: 2015 IEEE International Conference on Robotics and Automation (ICRA), pp. 5616–5621 (2015)
12. Ogenyi, U.E., Zhou, D., Ju, Z., Liu, H.: Adaptive collision-free reaching skill learning from demonstration. In: Qian, J., Liu, H., Cao, J., Zhou, D. (eds.) International Conference on Robotics and Rehabilitation Intelligence. ICRRI 2020, pp. 105–118. Springer, Singapore (2010). https://doi.org/10.1007/978-981-33-4932-2_8

DOREP 2.0: An Upgraded Version of Robot Control Teaching Experimental Platform with Reinforcement Learning and Visual Analysis

Shuai Wang[3], Lujun Li[1,2,4], Zhen Zhang[5], and Hualiang Zhang[1,2](✉)

[1] Key Laboratory of Industrial Control Network and System, Shenyang Institute of Automation, Chinese Academy of Sciences, Shenyang, China

[2] Institutes for Robotics and Intelligent Manufacturing, Chinese Academy of Sciences, Shenyang, China

Zhanghualiang@sia.cn

[3] Faculty of Robot Science and Engineering, Northeastern University, Shenyang, China

[4] School of Information Science and Technology, University of Science and Technology of China, Hefei, China

[5] Faculty of Mechanical Engineering and Automation, Northeastern University, Shenyang, China

Abstract. The Deep Open Robot Experiment Platform (DOREP) is an experimental system for general robot control. It includes a robot toolbox, a Linux based real-time controller and corresponding environment deployment tools. It is compatible with ROKAE robots, Universal robots, ABB robots, AUBO robots and other 6-DOF small low load general robots. It aims to provide users with a direct, high-level, more open and comprehensive programming interface. The toolbox of the original version of DOREP system contains more than 30 functions, including forward and inverse kinematics calculation, point-to-point joint and Cartesian control, trajectory generation, graphic display, 3D animation and diagnosis. On the basis of the original version, DOREP 2.0 system adds some new functional modules such as reinforcement learning and visual analysis, which further improves the performance of DOREP system. Taking the newly added module as an example, this paper expounds the functions of reinforcement learning module and visual analysis module, and applies them to simulation and experiment successfully.

Keywords: Robot control · Reinforcement learning · Vision · Experiment platform · MATLAB/Simulink.

1 Introduction

Since the 20th century, the factory began to carry out the reform of mechanization, the mechanical arm gradually replaced the manual into the factory.

Supported by National Key Research and Development Program of China under Grant (NO.2018YFE0205803).

X.-J. Liu et al. (Eds.): ICIRA 2021, LNAI 13014, pp. 122–132, 2021.
https://doi.org/10.1007/978-3-030-89098-8_12

As a new type of production tool, manipulator plays an important role in some repetitive production scenarios [9]. At present, the typical working scene of the manipulator is to grasp and place objects and to work on a given trajectory [16], but the manipulator can not complete the task of dynamic control target stably, because this kind of task needs to change the control strategy of the manipulator according to the environmental state [1]. However, this kind of task exists widely in real life, such as polishing the surface of the object, balance control of inverted pendulum, etc. [11], which will become a big obstacle to the mechanization of the factory.

At present, the main direction of manipulator control is the motion control of manipulator. The commonly used methods are inverse kinematics control [7] and linear kinematics control [10]. This kind of method mainly solves how to calculate the desired joint angle of the manipulator at the target point. It usually needs to know the position of the target point in advance. It is very suitable for the control task of fixed scene, but it can not deal with the time-varying complex system. Therefore, in order to solve this problem, we need to propose a more intelligent and flexible control method, which is to realize the autonomous learning of robot through the "trial and error" mechanism of reinforcement learning algorithm.

Soft Actor-Critic (SAC) is a reinforcement learning algorithm based on maximum entropy. It tries to maximize both the expected return and the strategy entropy. The purpose of adding the maximum entropy term to the reward is to encourage the exploration of the environment, so that the learned strategies can be randomized as much as possible while optimizing the objectives, while maintaining the possibility in each promising direction, rather than quickly converging to a local optimum [4].

The maximum entropy optimization strategy is more robust. If the strategy can allow highly random actions in the training process, it is more likely to deal with unexpected disturbances in the test. Google evaluated SAC algorithm in two tasks: 1) Walking of Minitaur quadruped robot in ghost robotics environment; 2) Using a three finger dynamixel claw to turn the valve. Google observed that the strategy learned through SAC can cope with the disturbance in the test without any additional learning [5].

In addition, we introduce Jacobian matrix [3] to learn only the end velocity control of the robot, not the joint angles of the robot's six joints. In this way, the inverted pendulum only moves back and forth on the y-axis, so as to improve the learning speed, avoid unexpected mechanical wear or impact, and maximize the safety.

2 Overview of the DOREP 2.0

2.1 DOREP 2.0 Components and Operational Procedures

In this section the constitutes of DOREP 2.0 will be briefly described.

The left part of Fig. 1 illustrates the communication scheme between DOREP and manipulator. It consists of four parts:

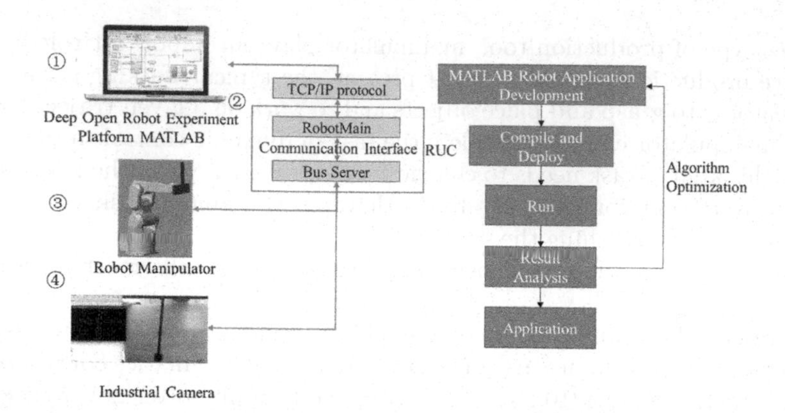

Fig. 1. The communication scheme and operational procedures between DOREP and the robot manipulator.

① A remote computer running DOREP in MATLAB,

② The part circled by the red line in Fig. 1 is the Robot Universal Controller (RUC) based on real-time Linux operating system.

③ The Robot Manipulator

④ Industrial vision camera

In order to establish the connection between the remote computer and the robot controller, DOREP provides robotmain, a C multi thread server running on the RUC. Robotmain communicates with the bus server through the communication interface. The bus server supports a variety of industrial bus implementations, including EtherCAT, CANopen and UDP running on the RUC, and manages the information exchange with the robot manipulator in a one millisecond soft real-time cycle. Robot main and bus server constitute RUC (robot universal controller). The part circled by the red line in Fig. 1 is RUC.

The operation procedure of DOREP is illustrated in the right part of Fig. 1. The first step is the development of MATLAB robot program, and then write the program and deploy it to RUC. It can run after the program is deployed. The algorithm is optimized according to the output results, and can be applied as long as the results are correct.

2.2 Synchronization Between MATLAB and Bus Server

Using MATLAB as the development tool, the communication cycle between the bus server and the manipulator is 1ms. Due to the problem of control cycle delay in the control process, our research team proposed a solution in reference [12] to ensure the synchronization of operation and communication. This paper continues to use this method, and will not repeat here.

2.3 SIA Robot Toolbox

SIA Robot Toolbox is a part of DOREP. It is a robot toolbox based on MATLAB developed by Shenyang Institute of automation. It has been introduced in detail in reference [12]. As an upgrade to the original DOREP, the SIA robot toolbox of DOREP 2.0 adds a visual analysis module and a reinforcement learning controller module.

For more information about SIA Robot Toolbox module, see reference [12]. Here, the visual analysis module and reinforcement learning module are introduced as examples.

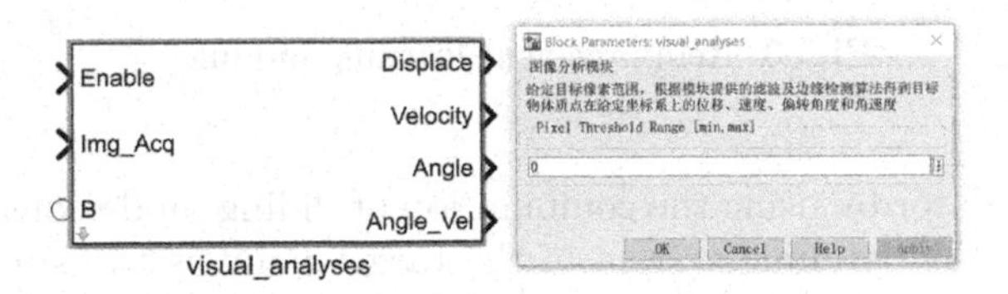

Fig. 2. Visual analysis module

The visual analysis module is shown in Fig. 2, and its input is the three-dimensional array of RGB images collected by the binocular industrial camera. The output of the module is the displacement, velocity, deflection angle and angular velocity of the target. The visual toolbox of Matlab can only be used in Windows system, while the robot controller adopts Linux system. The team compiles the library in MATLAB with C language, then encapsulates the library into the visual module, and deploys the module to the robot controller, so as to realize the robot control.

Although MATLAB has developed a set of reinforcement learning toolbox, it is only suitable for simulation of virtual environment and verification of algorithm. DOREP 1.0 platform system is an experimental system for general robot control. Inspired by the robot chemists developed by the University of Liverpool team [2], our development team hopes to develop a mature intelligent learning and control system which can autonomously learn for a variety of control tasks and realize real-time learning and control. DOREP 2.0 is created by the development team based on such a vision.

Reinforcement learning module showed in Fig. 3 can not only train in simulation environment, but also learn autonomously in the actual control process. Taking the actual control of the robot inverted pendulum as an example, we have initially built a robot autonomous learning control platform.

3 Visual Robot Inverted Pendulum Control Based on Reinforcement Learning

Inverted pendulum control has always been a hot topic in the field of robotics. In 2010, Wang Hongrui et al. [8] realized the tracking experiment of two degree of

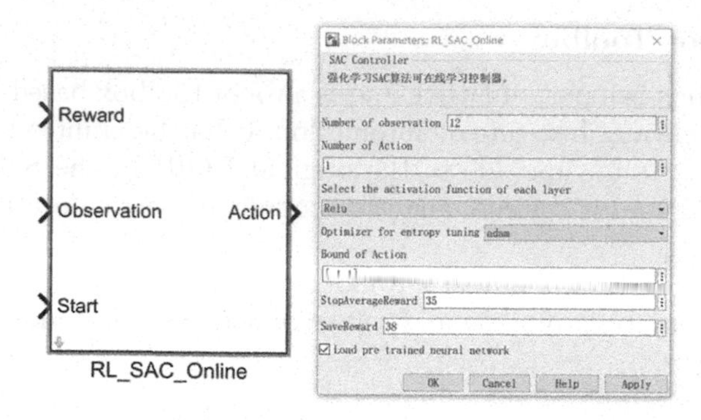

Fig. 3. Reinforcement learning module

freedom manipulator by using the combination of sliding mode control and fuzzy logic. However, the dynamic system of the target point is too simple. In 2017, Imran Mohsin introduced [13] a force controlled flexible robot polishing system, which uses a specially designed active end-effector for multi-step path planning to complete the polishing task. However, the method has strict restrictions on the shape of the experimental object, which makes its generalization ability weak.

Reinforcement learning algorithm [14] can make up for the above shortcomings. The traditional algorithm needs to simulate the physical model of the environment, while reinforcement learning has the characteristics of autonomous learning [6]. Its control strategy is learned from the actual task, so that it can adapt to any complex and learnable system. With the application of images, the controller can carry out strategy learning [15] without information of the controlled object, which greatly enhances the generalization of the algorithm.

3.1 Introduction of Reinforcement Learning Algorithm

Reinforcement learning learns by trials and errors. Its purpose is to guide the behavior through the reward obtained by interacting with the environment, so that the agent can get the maximum reward. In reinforcement learning, the reward signal provided by the environment is a kind of evaluation (usually scalar signal) of the action. Through this evaluation method, the system obtains knowledge in the Actor-Critic environment and improves the action plan to adapt to the environment.

As shown in Fig. 4, the agent receives the state value s_t and reward value r_t from the environment at time t, then it updates the strategy according to the reward value r_t, and gives the action a_t according to the updated strategy. After the environment receives the action, it gives the following reward value $r_t + 1$ and state value $s_t + 1$ at time $t + 1$. In this way, we continue to cycle and iterate until we train excellent strategies.

In this paper, SAC (Soft Actor-Critic) is used as reinforcement learning algorithm.

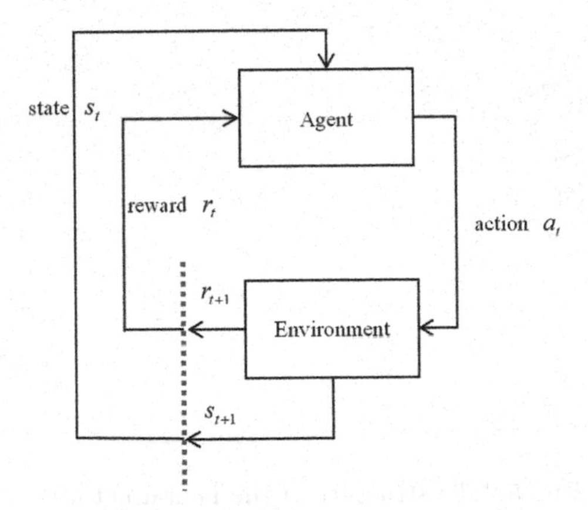

Fig. 4. Schematic diagram of reinforcement learning

The general reinforcement learning method learns a deterministic strategy, that is, only one optimal action is considered.

$$\pi^* = \arg\max_{\pi} \mathbb{E}_{(s_t,a_t)\sim\rho_\pi}\left[\sum_t R(s_t, a_t)\right] \tag{1}$$

where s_t is the state value that the environment feeds back to the agent at time t, a_t is the action of the agent according to the current strategy, and π^* is the best action strategy which can maximize the expected value of cumulative reward.

However, most of the time, the best action is not only one. Reinforcement learning algorithm needs to be able to give a random policy, which can output the probability of each action in each state, so that when there are multiple optimal actions, one can be randomly selected with the same probability.

For this reason, SAC algorithm takes entropy into account. Let the probability density function of the random variable x be P, then the entropy $H(x)$ of x can be expressed as:

$$H(x) = \underset{x\sim P}{\mathrm{E}}[-\log P(x)] \tag{2}$$

Then, the policy expression of SAC algorithm is as follows:

$$\pi^* = \arg\max_{\pi} \underset{\tau\sim\pi}{\mathrm{E}}\left[\sum_{t=0}^{\infty} \gamma^t \left(R\left(s_t, a_t, s_{t+1}\right) + \alpha H\left(\pi\left(\cdot \mid s_t\right)\right)\right)\right] \tag{3}$$

α represents the importance of entropy. π is the actual strategy of agent training. τ is a set of action sequences given by the agent according to the current strategy π. SAC algorithm does not leave any useful action, it has stronger exploration ability and generalization ability. So the trained policy is more robust.

The left part of Fig. 5 depicts the structure of the actor network that generates random policy $u(s)$: The actor takes observation s and returns the action

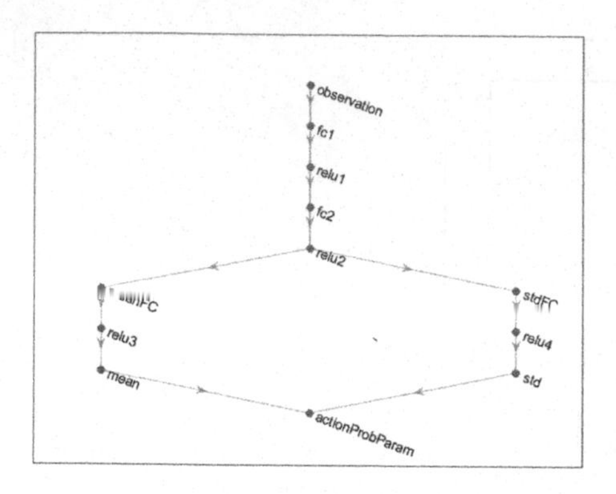

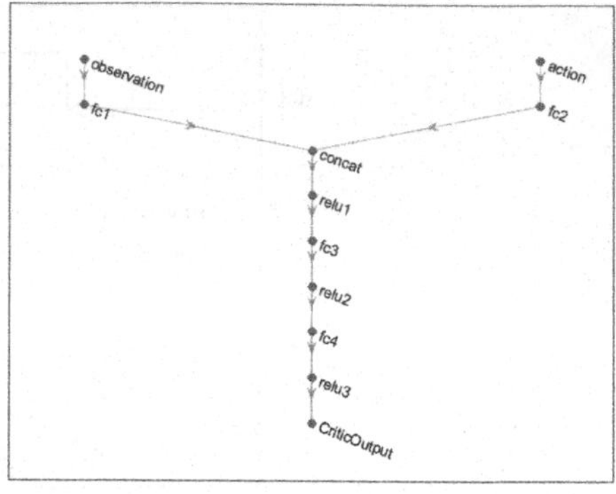

Fig. 5. The structure of the neural network

probability density function. The agent randomly selects actions based on this density function.

The right part of Fig. 5 depicts the structure of the critic networks. The critic network includes Q-value critics and Target critics.

Q-value critics $Q_1(s, a)$ and $Q_2(s, a)$: The critics take observation s and action a as inputs and return the corresponding expectation of the value function, which includes both the long-term reward and entropy. The existence of $Q_1(s, a)$ and $Q_2(s, a)$ is to select the minimum value to weaken the overestimation of Q-value, so they have the same structure but different initial parameters.

Target critics $V_1(s, a)$ and $V_2(s, a)$: To improve the stability of the optimization, the agent periodically updates the target critic $V_k(s, a)$ based on the latest parameter values of the critic $Q_k(s, a)$.

3.2 Introduction of Visual Analysis

The vision system used in this experiment is composed of two Daheng Mercury monocular cameras. In the movement process of the placementpiece, the two monocular cameras respectively collect the information of tilt Angle and centroid position. And then the actual displacement of the placementpiece can be calculated through the principle of binocular imaging. The displacement information is transmitted to the robot for further processing.

The steps of image contour recognition are as follows: (1) Grayscale and Gaussian filtering; (2) OTSU threshold segmentation; (3) Canny operator extraction of edge contour; (4) search for contour points of binary image; (5) output Angle and centroid information.

As the left part of Fig. 6 shows, the position of the swing bar is calculated by scanning each frame image in two lines. Select two rows of pixels with fixed position and 50 rows apart from each other from the image, find out the feature points belonging to the swing rod in the corresponding rows, record their positions r_1 and r_2, and then use the arc tangent function $\theta = \arctan((r_1 - r_2)/50)$ to calculate the current angle of the pendulum. Because it is necessary to swing the

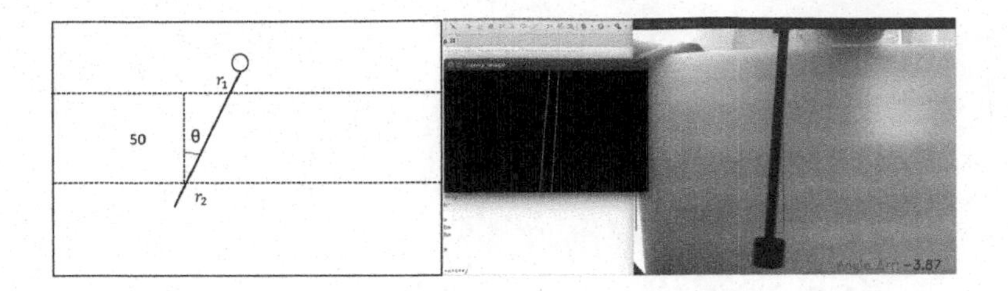

Fig. 6. Schematic diagram of swing angle calculation

pendulum bar from the vertical downward angle to the vertical upward angle, two pairs of four rows of pixels with fixed positions and 50 rows apart are set in the collected image to calculate the angle of the pendulum bar in the lower half plane and the upper half plane respectively.

The right part of Fig. 6 shows the real-time acquisition of inverted pendulum angle by camera.

3.3 Results of Experiment

In order to achieve better training effect, we define the reward functions of reinforcement learning.

$$r_\theta = \exp(-0.1 * \theta^2) \tag{4}$$

θ is the angle of inverted pendulum, r_θ is the reward function to evaluate the angle of inverted pendulum. The smaller θ^2 is, the more stable the inverted pendulum is, so the greater r_θ is.

$$r_x = -0.3 * x^2 \tag{5}$$

x is the displacement of end effector, r_x is the reward function to evaluate the deviation of the end effector from the ideal position. The smaller x^2 is, the greater r_x is.

$$r_v = -(0.02 * v)^2 \tag{6}$$

v is the command value of robot, r_v is the reward function to evaluate the robot effort.

$$R = r_\theta + r_x + r_v \tag{7}$$

R is the sum of the above reward functions to give the final reward of the manipulator.

As shown in Fig. 7, the blue line represents the reward received by the agent in each episode, and the red line represents the average reward. After 600 episodes training, the robot finally realized the successful swing up and stabilized the inverted pendulum in the balance position in the simulation.

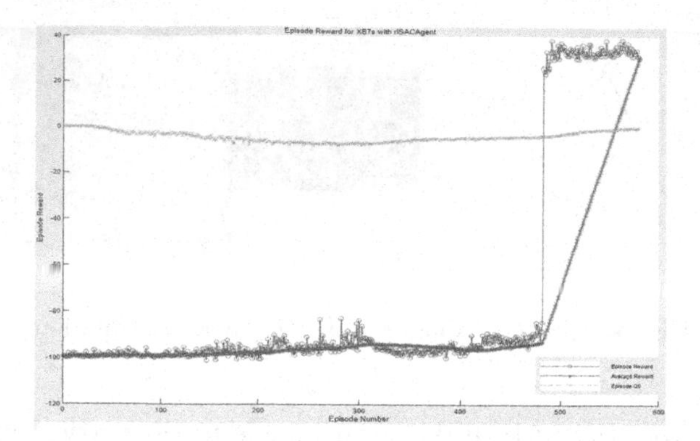

Fig. 7. The reward of reinforcement learning algorithm in training

Figure 8 shows that the robot can swing up and stabilize the inverted pendulum at any initial angle in the simulation environment.

However, it is not easy to apply the simulation results directly to the actual control. Because the error between the simulation environment and the real experimental platform is very large. In order to realize the migration from the simulation environment to the real environment, we add noise to the output action and the movement of robot joints, and randomize the parameters of inverted pendulum. Of course, these are not enough, so we enable the online learning function of reinforcement learning module, allowing the robot to continue training in the real environment after loading the trained parameters.

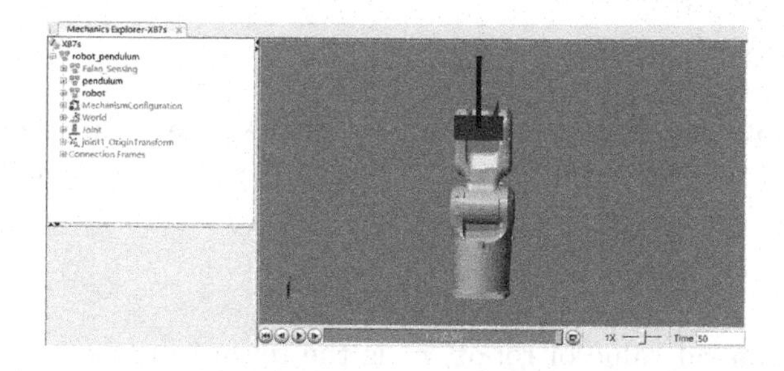

Fig. 8. Animation of robot swing up and stabilization in simulation environment

Figure 9 is the project diagram of real-time training and control of robot inverted pendulum.

As can be seen from Fig. 10, the effect of real-time control on the swing up and stability of the robot inverted pendulum is very stable.

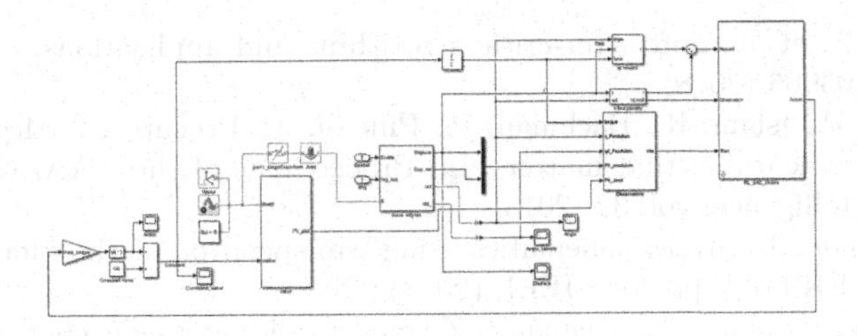

Fig. 9. Project diagram of robot inverted pendulum

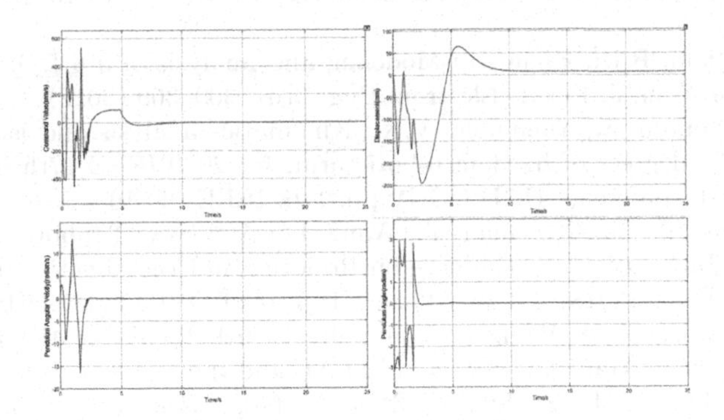

Fig. 10. Effect picture of robot inverted pendulum control in real environment

4 Conclusion

DOREP experimental platform attaches great importance to the development needs of intelligent and flexible robot. It adds reinforcement learning and visual analysis functions on the original basis, combines robot control with artificial intelligence, and proposes an upgraded version of DOREP 2.0 system. Taking the experiment of robot vision inverted pendulum based on reinforcement learning as an example, it proves that DOREP 2.0 is powerful in experiment and control. In the future, the DOREP system will continue to upgrade and make new progress in deep learning and reinforcement learning to meet the growing needs of researchers.

References

1. Abdelaziz, O.T., Maged, S.A., Awad, M.I.: Towards dynamic task/posture control of a 4dof humanoid robotic arm. Int. J. Mech. Eng. Rob. Res. **9**(1), 99–105 (2020)
2. Burger, B., et al.: A mobile robotic chemist. Nature **583**(7815), 237–241 (2020)
3. Craig, J.J.: Introduction to Robotics: Mechanics and Control, 3/E. Pearson Education India (2009)
4. Haarnoja, T., Zhou, A., Abbeel, P., Levine, S.: Soft actor-critic: off-policy maximum entropy deep reinforcement learning with a stochastic actor. In: International Conference on Machine Learning, pp. 1861–1870. PMLR (2018)

5. Haarnoja, T., et al.: Soft actor-critic algorithms and applications. arXiv preprint arXiv:1812.05905 (2018)
6. Henderson, P., Islam, R., Bachman, P., Pineau, J., Precup, D., Meger, D.: Deep reinforcement learning that matters. In: Proceedings of the AAAI Conference on Artificial Intelligence, vol. 32 (2018)
7. Hock, O., Sedo, J.: Inverse kinematics using transposition method for robotic arm. In: 2018 ELEKTRO, pp. 1–5. IEEE (2018)
8. Hongrui, W., Congna, L., Qiguang, Z.: Fuzzy sliding mode control for robotic manipulators based on passivity theory. In: 2010 International Conference on Logistics Systems and Intelligent Management (ICLSIM), vol. 1, pp. 535–538. IEEE (2010)
9. Iqbal, J., Islam, R.U., Khan, H.: Modeling and analysis of a 6 DOF robotic arm manipulator. Can. J. Electr. Electron. Eng. **3**(6), 300–306 (2012)
10. Kaur, M., Sondhi, S., Yanumula, V.K.: Kinematics analysis and jacobian calculation for six degrees of freedom robotic arm. In: 2020 IEEE 17th India Council International Conference (INDICON), pp. 1–6. IEEE (2020)
11. Li, J., Zhang, T., Liu, X., Guan, Y., Wang, D.: A survey of robotic polishing. In: 2018 IEEE International Conference on Robotics and Biomimetics (ROBIO), pp. 2125–2132. IEEE (2018). https://doi.org/10.1109/ROBIO.2018.8664890
12. Liu, G., Han, B., Li, Q., Wang, S., Zhang, H.: DOREP: an educational experiment platform for robot control based on MATLAB and the real-time controller. In: Yu, H., Liu, J., Liu, L., Ju, Z., Liu, Y., Zhou, D. (eds.) ICIRA 2019. LNCS (LNAI), vol. 11745, pp. 555–565. Springer, Cham (2019). https://doi.org/10.1007/978-3-030-27529-7_47
13. Mohsin, I., He, K., Cai, J., Chen, H., Du, R.: Robotic polishing with force controlled end effector and multi-step path planning. In: 2017 IEEE International Conference on Information and Automation (ICIA), pp. 344–348. IEEE (2017)
14. Sutton, R.S., Barto, A.G.: Reinforcement Learning: An Introduction. MIT press, Cambridge (2018)
15. Zhang, F., Leitner, J., Milford, M., Upcroft, B., Corke, P.: Towards vision-based deep reinforcement learning for robotic motion control. arXiv preprint arXiv:1511.03791 (2015)
16. Zheng, Z., Ma, Y., Zheng, H., Gu, Y., Lin, M.: Industrial part localization and grasping using a robotic arm guided by 2d monocular vision. Ind. Rob. Int. J. **45**(6), 794–804 (2018)

An Improved TCN-Based Network Combining RLS for Bearings-Only Target Tracking

Yingqi Zhao[1], Jie Liu[2](✉), Kaibo Zhou[1], and Qi Xu[1]

[1] School of Artificial Intelligence and Automation, Huazhong University of Science and Technology, Wuhan 430074, China

[2] School of Civil and Hydraulic Engineering, Huazhong University of Science and Technology, Wuhan 430074, China

jie_liu@hust.edu.cn

Abstract. Traditional approaches like extended Kalman filter may suffer performance degradation in bearings-only target tracking, because of the non-linear relationship between the measured angle and target position. In this paper, a temporal convolutional network-based target position estimation method is proposed, which has strong capability of fitting any linear or non-linear mapping. The recursive least squares is combined to further improve the accuracy and ability of target tracking in an iterative way, as well as estimate the velocity of target. Simulation results show that the proposed method achieves a higher accuracy compared to traditional approaches during the whole tracking process.

Keywords: Bearings-only target tracking · Temporal convolutional network · Recursive least squares · Nonlinear filtering

1 Introduction

Underwater vehicles, such as submarine, torpedoes and exploration ship, often need to locate and track targets during missions. Due to mission requirements, they usually work in passive mode that the observer can remain hidden. Only target angles are measured, which is called bearings-only target tracking [1]. In this case, the mapping between the measured angle and the target position is non-linear.

Several methods have been proposed for bearings-only target tracking, in which non-linear filtering is the majority. For example, the extended Kalman filter (EKF) linearize the measurement equation using Taylor series expansion. It was mainly utilized in early researches on target tracking [2, 3], having the advantage of simplicity. However, the large estimation error of the first-order Taylor series approximation can make EKF suffer performance degradation [4]. Gordon et al. first proposed the use of particle filter algorithm in target tracking [5]. The particle filter algorithm has the advantage of not being affected by the nonlinearity in the model, but its random sampling leads to excessive calculations and problems such as particle degradation, limiting its practical application [6]. In contrast, deterministic sampling, which helps reduce the computational burden, has been used for unscented Kalman filter (UKF), cubature Kalman filter (CKF), central

X.-J. Liu et al. (Eds.): ICIRA 2021, LNAI 13014, pp. 133–141, 2021.
https://doi.org/10.1007/978-3-030-89098-8_13

difference Kalman filter (CDKF), etc. [7–9]. And UKF is widely used in target tracking recently. Compared to EKF, the accuracy of UKF is significantly higher under the same conditions, reaching a second-order Taylor series approximation at least. However, if the dimension of the state vector is higher than three, its sample scheme may need some adjustments to overcome the propagation of a non-positive definite covariance matrix. It leads to an increase in estimation error [10].

Recent advances in deep learning have demonstrated its ability to extract the high-level features from raw sensory data. Temporal convolutional network (TCN) is a convolutional architecture designed for sequence signals [11]. It performs better than recurrent architectures like long-short term memory (LSTM) on many sequence model tasks such as Seq. MNIST and Music Nottingham [12]. Moreover, as the universal approximation theorem states, TCN is capable of approximating any continuous function to a reasonable accuracy. It means that the mapping between angle measurement and target dynamics can be fitting by TCN, although its non-linearity is existing.

Despite the above advantages, TCN is not suitable to be used directly in target tracking. Underwater targets maintain a uniform linear motion most of the time, which indicates the relationship between the target position and speed, as well as the relationship between the target positions at different times. But for a TCN, the estimation of the current target state only depends on the input observations, having nothing to do with other factors. Therefore, it is a natural idea to establish a state equation describing the target motion to help TCN utilize the relationship mentioned above, where the key is the identification of the equation parameters. Least squares method (LS) is widely used in parameter identification. It has good adaptability in linear and non-linear systems. However, a large amount of data is required to use it, which brings great computational burden. The recursive least squares method (RLS) proposed later overcomes the above-mentioned shortcomings in an iterative manner while maintaining accuracy.

In this paper, a TCN-based network combining RLS named TCN-RLS is proposed to improve the accuracy and stability of bearings-only target tracking. Main contributions are summarized as follows. 1) A TCN-based network is proposed to fit the mapping between angle measurement and target position. It is capable of estimating the target position with a higher accuracy. 2) The RLS is combined to further improve the accuracy and stability of target tracking, as well as estimate the target speed.

The rest of this paper is organized as follows. The dynamic model of bearings-only target tracking is review in Sect. 2. The TCN-RLS we proposed is illustrated in Sect. 3. In Sect. 4, simulation results for the performance of the TCN-RLS is shown and analyzed, followed by conclusions in Sect. 5.

2 Dynamic Model of Bearings-Only Target Tracking

Underwater targets maintain a constant speed most of the time, and only the case of vehicles moving on a two-dimensional plane is discussed in this paper. Thus the state of the target at time k can be expressed as

$$X_k = \left[x_k \; v_x \; y_k \; v_y \right] \tag{1}$$

where x_k, y_k represent target position, and v_x, v_y represent target velocity. The state equation of the system is defined by

$$X(k+1) = \boldsymbol{\Phi}X(k) + \boldsymbol{\omega}(k) \tag{2}$$

where $\boldsymbol{\Phi}$ represents the state transition matrix, $\boldsymbol{\omega}(k)$ represents the system noise with mean 0 and variance $\boldsymbol{Q}$,

$$\boldsymbol{\Phi} = \begin{bmatrix} 1 & \Delta t & 0 & 0 \\ 0 & 1 & 0 & 0 \\ 0 & 0 & 1 & \Delta t \\ 0 & 0 & 0 & 1 \end{bmatrix} \tag{3}$$

$$\boldsymbol{Q} = \delta_q^2 \begin{bmatrix} \Delta t^3/3 & \Delta t^3/2 & 0 & 0 \\ \Delta t^3/2 & \Delta t & 0 & 0 \\ 0 & 0 & \Delta t^3/3 & \Delta t^3/2 \\ 0 & 0 & \Delta t^3/2 & \Delta t \end{bmatrix} \tag{4}$$

where δ_q^2 represents the process noise intensity of the system, and Δt represents the interval of sampling.

The measurement equation of the system can be expressed as

$$\boldsymbol{Z}_k = f(\boldsymbol{X}_k) + \boldsymbol{v}_k \tag{5}$$

where $\boldsymbol{v}_k$ represents the measurement noise with mean $\boldsymbol{0}$ and variance $\boldsymbol{R}$. And the observation function can be defined by

$$f(\boldsymbol{X_k}) = \arctan[(x_k - x_k^{(o)})/(y_k - y_k^{(o)})] \tag{6}$$

where $x_k^{(o)}$ and $y_k^{(o)}$ are the observer position at time k.

3 TCN-Based Network Combining RLS

3.1 TCN-Based Network

A TCN-based network is presented to fit the mapping between the observation sequence and target position. As shown in Fig. 1, the sequence composed of the last 20 target angles measured by the observer is the input, represented as $\boldsymbol{Z}_{k\text{-}19}, \ldots, \boldsymbol{Z}_{k\text{-}1}, \boldsymbol{Z}_k$, and the target position at the current moment $\left[\hat{x}_k \ \hat{y}_k\right]$ is output accordingly.

Unlike traditional convolutional neural networks such as one-dimensional convolution, causal convolutions are used in TCN. It means the output at time k has nothing to do with elements later than time k in the previous layer.

In addition, dilated convolutions that achieve an exponentially large receptive are employed in TCN. For the proposed network, the dilated convolution operation $F(\cdot)$ on elements s is expressed as

$$F(s) = (\boldsymbol{X} *_d f)(s) = \sum_{i=0}^{k-1} f(i) \cdot \boldsymbol{X}_{s-d \cdot i} \tag{7}$$

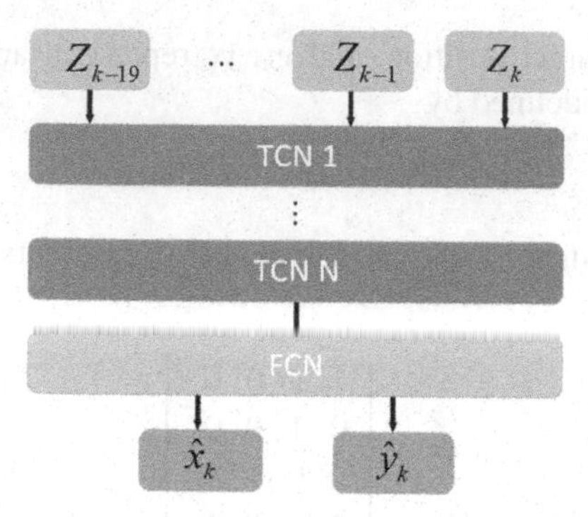

Fig. 1. Proposed TCN-based network.

where $\boldsymbol{X}$ represents the input of TCN, f is a filter, $*_d$ presents the dilated convolution operation, d is the dilation factor. k represents the size of filter, and $s\text{-}d{\cdot}i$ indicates the direction of the past.

3.2 Recursive Least Squares

A state equation describing the target motion based on TCN is represented as

$$\boldsymbol{Y}(k) = \boldsymbol{\Phi}(k)\boldsymbol{\theta} + \boldsymbol{e}(k) \tag{8}$$

$$\boldsymbol{Y}(k) = \begin{bmatrix} \hat{x}_k \; \hat{y}_k \end{bmatrix} \tag{9}$$

$$\boldsymbol{\Phi}(k) = \begin{bmatrix} k \; 1 \end{bmatrix} \tag{10}$$

$$\boldsymbol{\theta} = \begin{bmatrix} \hat{v}_x \; \hat{v}_y \\ \hat{x}_0 \; \hat{y}_0 \end{bmatrix} \tag{11}$$

where $\hat{x}_k$ and $\hat{y}_k$ in $\boldsymbol{\Phi}(k)$ are estimated by TCN, $\boldsymbol{\theta}$ is the estimated parameter, and $\boldsymbol{e}(k)$ is the error vector. The RLS is used to update parameter in an iterative manner as shown below, then target velocity at time k can be estimated.

$$\boldsymbol{K}(k) = \boldsymbol{P}(k)\boldsymbol{\Phi}(k) \tag{12}$$

$$\boldsymbol{P}(k) = \boldsymbol{P}(k-1) - \frac{\boldsymbol{P}(k-1)\boldsymbol{\Phi}(k)\boldsymbol{\Phi}^{\boldsymbol{T}}(k)\boldsymbol{P}(k-1)}{1 + \boldsymbol{\Phi}^{\boldsymbol{T}}(k)\boldsymbol{P}(k-1)\boldsymbol{\Phi}(k)} \tag{13}$$

$$\hat{\boldsymbol{\theta}}(k) = \hat{\boldsymbol{\theta}}(k-1) + \boldsymbol{K}(k)[[\hat{x}_k \; \hat{y}_k] - \boldsymbol{\Phi}(k)\hat{\boldsymbol{\theta}}(k-1)] \tag{14}$$

where $\hat{\boldsymbol{\theta}}(0)$ is selected based on experience, $\boldsymbol{P}(0)$ can be taken as $\boldsymbol{E}$, the identity matrix.

The target position at time k is finally estimated as

$$\begin{bmatrix} \hat{x}_k \; \hat{y}_k \end{bmatrix} = \begin{bmatrix} k \; 1 \end{bmatrix}\hat{\boldsymbol{\theta}}(k) \tag{15}$$

The pseudocode of the TCN-RLS is given hereinafter.

Algorithm 1 TCN-RLS

Input: target angles sequence $\boldsymbol{Z}_1, \ldots, \boldsymbol{Z}_2, \ldots, \boldsymbol{Z}_k$; a trained TCN-based network

Output: target position estimations $\hat{\boldsymbol{X}}_{20}, \hat{\boldsymbol{X}}_{21}, \ldots, \hat{\boldsymbol{X}}_k$, $\hat{\boldsymbol{X}}_k = \left[\hat{x}_k \quad \hat{v}_x \quad \hat{y}_k \quad \hat{v}_y\right]$

Initialization: $\hat{\boldsymbol{\theta}}(0) = \boldsymbol{0}$, $\boldsymbol{P}(0) = \boldsymbol{E}$

1: **for** $i = 20, k$ **do**
2: Get position estimation $[\hat{x}_i \quad \hat{y}_i]$ by TCN
3: Update $\boldsymbol{K}(i)$ based on Equation (12)
4: Update $\boldsymbol{P}(i)$ based on Equation (13)
5: Update $\hat{\boldsymbol{\theta}}(i)$ based on Equation (14), including $[\hat{v}_x \quad \hat{v}_y]$
6: Update $[\hat{x}_i \quad \hat{y}_i]$ based on Equation (15)
7: $\hat{\boldsymbol{X}}_k = \left[\hat{x}_k \quad \hat{v}_x \quad \hat{y}_k \quad \hat{v}_y\right]$
8: **end for**

4 Simulation

4.1 Initial Conditions

The observer takes a zigzag movement, with initial state of [0 m 0 m/s 100 m 10 m/s]. The initial state of the target is set randomly in a certain range to verify the generalization of the algorithm, with $x_0, y_0 \in$ [800 m 1200 m], and $v_x, v_y \in$ [5 m/s 10 m/s] (see Fig. 2).

Target angle β is measured by observer, which contains white noise with a mean value of 0 and a standard deviation of 1°. A total of 80 iterations of observation is taken, with a sampling interval of 10 s.

4.2 Simulation Results

10000 samples that meet the initial conditions are generated to train the TCN, with the ratio of the training set to the test set being 9:1. 50 samples randomly selected test set are used to compare the tracking performance of TCN-RLS with EKF and UKF. The initial estimation of the traditional algorithm is set to be [1000 m 7.5 m/s 1000 m 7.5 m/s], which is the expected value of the initial target state. Since TCN does not provide its estimation until 20th iteration, only the 20th to 80th iterations of simulation results are counted. The root mean square error (RMSE) is utilized to represent the tracking error, which can be expressed as

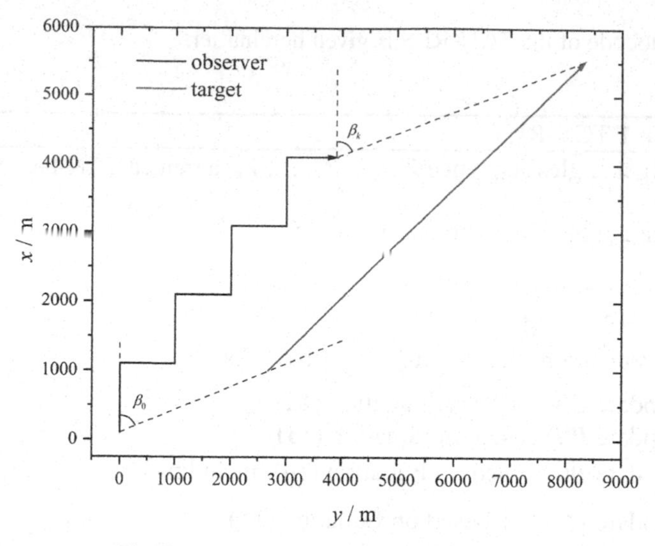

Fig. 2. An instance of the tracking process.

$$RMSE_{position} = \sqrt{\frac{1}{k}\sum_{i=1}^{k}[(x_i - \hat{x}_i)^2 + (y_i - \hat{y}_i)^2]} \tag{16}$$

$$RMSE_{velocity} = \sqrt{\frac{1}{k}\sum_{i=1}^{k}[(v_x - \hat{v}_{xi})^2 + (v_y - \hat{v}_{yi})^2]} \tag{17}$$

The estimation errors for an instance of tracking with TCN-RLS, EKF, and UKF are shown in Figs. 3 and 4. EKF has the largest estimation error due to the excessive linearization error, while UKF maintains a relatively small estimation error. Compared to these traditional methods, the TCN-RLS has the smallest estimation error during the entire tracking process. After 50 Monte Carlo experiments, the average $RMSE_{position}$ of EKF, UKF, and TCN-RLS are calculated to be 654.5, 360.1, and 217.4, respectively, and $RMSE_{velocity}$ are 1.8, 1.3, and 1.2, respectively, which supports the analysis mentioned above.

The estimation of the current target state provided by TCN only depends on the input observations, which may cause large fluctuations in the estimation sequence. And RLS is utilized to establish a state equation describing the relationship between the target position and speed, as well as the relationship between the target positions at different times.

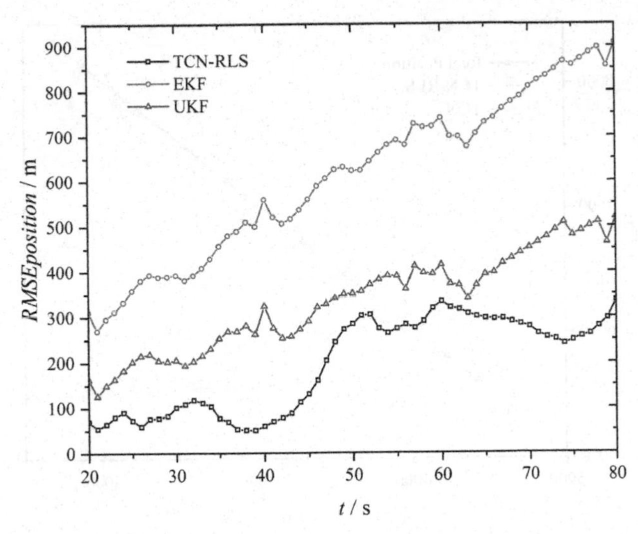

Fig. 3. Comparison of the $RMSE_{position}$ for TCN-RLS, EKF and UKF.

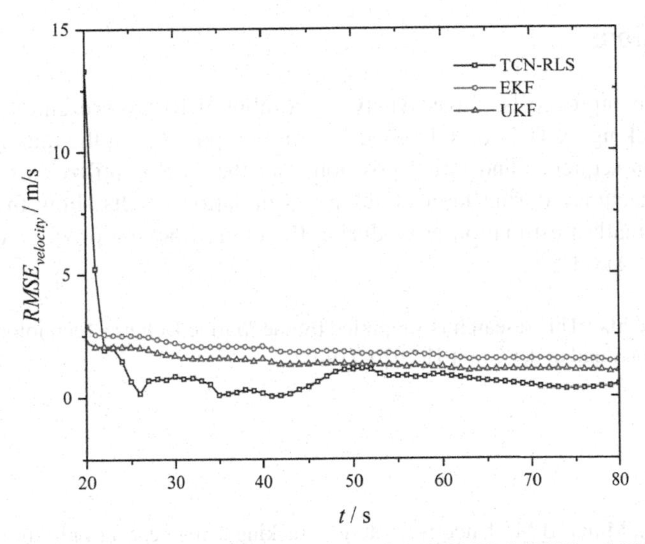

Fig. 4. Comparison of the $RMSE_{velocity}$ for TCN-RLS, EKF and UKF.

As shown in Fig. 5, the target position sequence estimated by TCN-RLS forms almost a straight line,with points remains basically the same distance in it. While the target position sequence estimated by TCN fluctuates greatly as predicted. Compared with the single TCN, the target position sequence estimated by TCN-RLS is more stable and close to the true value.

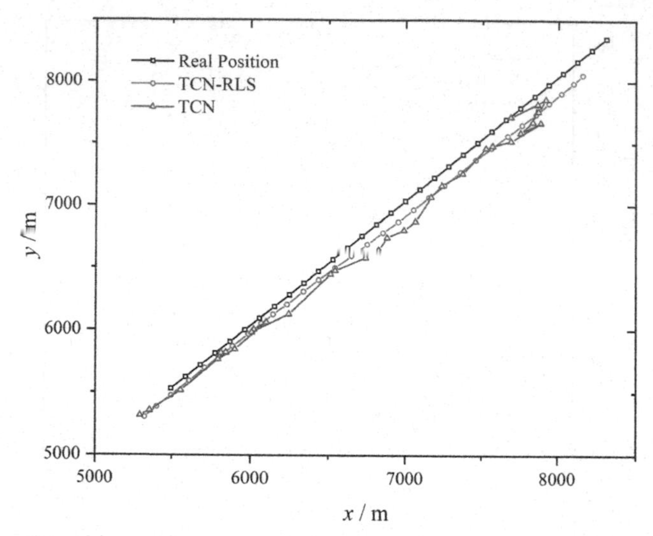

Fig. 5. Estimated target position of the algorithm before and after adding RLS in one instance.

5 Conclusions

In this paper, an improved TCN-based network combining RLS is presented for bearings-only target tracking. A TCN-based network is first proposed to fit the mapping between the observation sequence and target position, and the RLS is utilized to improve the stability and accuracy during target tracking. Simulation results show that the TCN-RLS has the smallest estimation error during the entire tracking process, compared to traditional EKF and UKF.

Acknowledgements. This research is supported by the Marine Defense Technology Innovation Center Innovation Fund under Grant JJ-2020-719-02.

References

1. Miller, A.B., Miller, B.M.: Underwater target tracking using bearing-only measurements. J. Commun. Technol. Electron. **63**(6), 643–649 (2018)
2. Xin, G., Xiao, Y., You, H.: Bearings-only underwater track fusion solutions with feedback information. In: Proceedings 7th International Conference on Signal Processing, Beijing, pp. 2449–2452 (2004)
3. Modalavalasa, N., Rao, G.-S., Prasad, K.S.: A new method of target tracking by EKF using bearing and elevation measurements for underwater environment. Robot. Auton. Syst. **74**, 221–228 (2015)
4. Konatowski, S., Kaniewski, P., Matuszewski, J.: Comparison of estimation accuracy of EKF UKF and PF filters. Ann. Navigat. **23**(1), 69–87 (2016)
5. Gordon, N.J., Salmond, D.J., Smith, A.: Novel approach to nonlinear/non-Gaussian Bayesian state estimation. IEE Proc. F. Radar. Signal Process. **140**(2), 107–113 (2002)

6. Zhang, H., Xie, W.: Constrained auxiliary particle filtering for bearings-only maneuvering target tracking. J. Syst. Eng. Electron. **30**(4), 684–695 (2019)
7. Julier, S.J., Uhlmann, J.K.: Unscented filtering and nonlinear estimation. Proc. IEEE **92**(3), 401–422 (2004)
8. Wan, M., Li, P., Li, T.: Tracking maneuvering target with angle-only measurements using IMM algorithm based on CKF. In: 2010 International Conference on Communications and Mobile Computing, pp. 92–96. IEEE, Shenzhen (2010)
9. Hu, Z.-T., Jin, Z., Fu, C.: Maneuvering target tracking algorithm based on CDKF in observation bootstrapping strategy. High Technol. Lett. **2017**(02), 35–41 (2017)
10. Arasaratnam, I., Haykin, S., Elliott, R.J.: Discrete-time nonlinear filtering algorithms using Gauss-Hermite quadrature. Proc. IEEE **95**(5), 953–977 (2007)
11. Oord, A., Dieleman, S., Zen, H.: WaveNet: a generative model for raw audio (2016). https://arxiv.org/abs/1609.03499
12. Bai, S., Kolter, J.Z., Koltun, V.: An empirical evaluation of generic convolutional and recurrent networks for sequence modeling (2018). https://arxiv.org/abs/1803.01271

Learning Robot Grasping from a Random Pile with Deep Q-Learning

Bin Chen[1,2], Jianhua Su[1,2(✉)], Lili Wang[1,2], and Qipeng Gu[1,2]

[1] University of Chinese Academy of Sciences,
Beijing 100190, People's Republic of China

[2] State Key Lab. of Management and Control for Complex Systems, Institute of Automation, Chinese Academy of Science,
Beijing 100190, People's Republic of China

jianhua.su@ia.ac.cn

Abstract. Grasping from a random pile is a great challenging application for robots. Most deep reinforcement learning-based methods focus on grasping of a single object. This paper proposes a novel structure for robot grasping from a pile with deep Q-learning, where each robot action is determined by the result of its current step and the next n steps. In the learning structure, a convolution neural network is employed to extract the target position, and a full connection network is applied to calculate the Q value of the grasping action. The former network is a pre-trained network and the latter one is a critical network structure. Moreover, we deal with the "reality gap" from the deep Q-learning policy learned in simulated environments to the real-world by large-scale simulation data and small-scale real data.

Keywords: Grasping · Deep Q-learning · Simulation to real

1 Introduction

The ability of robots to reliably grasp a wide range of novel objects can benefit applications in logistics, manufacturing, and housing robot. Traditional methods try to obtain the full knowledge of the model and pose of an object to be grasped, and then select the grasping configuration on the surface of the object. Contrary to the analytic approaches using the geometric information, learning-based methods pay more attention on the feature extraction, the pose estimation, and the object classification, etc. The resulting data is then used to retrieve grasps from some knowledge base or samples and rank them by comparison to existing grasp experience. Those works started to be popular in recent ten years.

Some researchers imitate human grasps to deal with the grasping problem. Based on the statistical modelling, statistical learning technology such as Gauss

This work was supported in part by NSFC under Grant number 91848109, supported by Beijing Natural Science Foundation under Grant number L201019, and major scientific and technological innovation projects in Shandong Province Grant number 2019JZZY010430.

X.-J. Liu et al. (Eds.): ICIRA 2021, LNAI 13014, pp. 142–152, 2021.
https://doi.org/10.1007/978-3-030-89098-8_14

process regression model (GPR) [1], Hidden Markov Models (HMMs) [2], Support Vector Machine (SVM) [3], neural network [4], have been employed to deal with imitation-based grasping problem. A comprehensive review work on this area can be referred to[5].

Some researchers regard a grasp as a problem from "RGB image" to "grasping configuration", thus the end-to-end deep learning approaches have been developed for grasping planning. Pinto [6] trained a Convolutional Neural Network (CNN) for the task of predicting grasp locations without severe overfitting. They recast the regression problem to an 18-way binary classification over image patches. Mahler et al. [7] trained Grasp Quality Convolutional Neural Networks (GQ-CNNs) using synthetic depth images, and then combined them to plan grasps for objects in the point cloud.

The deep learning-based methods inevitably require a considerable initial training data, which is tediously labelled by a human supervisor. However, it is reasonable to expect the robot can perform a new task more simply. The method is very challenging in practice. Some works show deep reinforcement learning (Deep RL) has achieved greet success in domains ranging from games [8,9] to robotic control [10], and perform well in the field of robotic manipulation, such as pushing [11], grasping [12] etc. Popov et al. [13] discussed the object grasping and precise stacking based on Deep RL. They provided a simulation study that investigated the possibility of learning manipulation end-to-end with Deep Deterministic Policy Gradient (DDPG) [14]. But their works are limited to single objects, and only give the simulation result. Some other works also have presented autonomous data collections through trial and error [2] or in simulated scenes [15], however, the policies trained on synthetic data may have reduced performance on a physical robot due to inherent differences between models and real-world systems. This simulation-to-reality transfer problem is a long-standing challenge in robot learning [15]. In our previous work [16], we employ the DDPG for grasping of an object. The DDPG-based grasping strategy will perform plenty of actions adjustment in training, and has many super parameters need to be adjusted. Avoiding the tedious parameters adjustment, this work considers a grasping action as an end-to-end problem. We pay more attentions on the learning of target not the trajectory for grasping. The Deep Q-learning network (DQN) is then employed to deal with the grasping problem. The input of the model is RGB image, while the output is a robot grasping action. Compared with our previous work [16], there are less parameters need to be adjusted using DQN. And only one decision should be made in a grasping task.

The contributions of this work are summarized as follows:

We learn the robot grasping from a random pile with a Deep Q-network (DQN). To improve the grasping successful rate, each robot grasping action is determined by the result of its current step and the next n steps.

We deal with the "reality gap" from a deep Q-learning policy learned in simulated environments to the real-world by large number of simulation data and small number of real data.

More closely related to our work is that of Zeng et al. [17], which learned synergies between pushing and grasping from experience through Deep Q-learning. They show good grasping performance on the pile of objects by sequence actions that combine the pushing and grasping actions. However, their intrinsic reward of DQN does not explicitly consider whether a push enables future grasps. In contrast, our grasping reward is considered by the current grasp and the next n steps. That is, we also can grasp from a pile of object only with grasping action.

The rest of this paper is arranged as follows: In Sect. 2, some concepts related to RL and DQN are introduced. In Sect. 3, the construction of the DQN-based grasping strategy is presented. In Sect. 4, the performances of the algorithms are given and simulation. In Sect. 5, the method of simulation to real is tested by the experiment on picking a target from a pile of objects.

2 Background

In this section, we introduce several concepts and definitions related to the reinforcement learning and Deep Q-learning network (DQN).

2.1 Reinforcement Learning

Reinforcement learning deals with learning in sequential decision-making problems, where the only feedback consists of a scalar reward signal [18]. In this work, we formulate the robot grasping problem as a Markov Decision Process (MDP) framework like most RL-based grasping learning works.

We denote continuous space Markov Decision Process by the tuple $(S, A, P, r, \gamma, S_{-})$, where S is a set of continuous sates of robot, A is a set of continuous actions of robot, P is a state transition probability function, r is a reward function, γ is a discount factor determining the agent's horizon and S_{-} is a set of initial states. The goal of reinforcement learning is to learn a policy $a_t = \pi(s_t)$ to maximize the expected return from the state s_t, where the return can be denoted by $R_t = \sum_{i=t}^{T} \gamma^{i-t} r_i$ in which T is the horizon that the agent optimizes.

The learning of the policy involves constructing an estimate of the expected return from a given state after taking the action a_t:

$$Q_\pi(s_t, a_t) = E_{r_i, s_i, a_i}(\sum_{i=t}^{T} \gamma^{i-t} r_i | s_t, a_t) \tag{1}$$

where $Q_\pi(s_t, a_t)$ is the action-value function, which is known as Q-function. A recurse vision of Eq. 1 is given as Eq. 2, which is known as the Bellman equation.

$$Q_\pi(s_t, a_t) = E_{r_t, s_{t+1}, a_i} E[r_t + \gamma E_{a_{t+1}}](Q_\pi(s_{t+1}, a_{t+1})) \tag{2}$$

A more comprehensive introduction of the equation can be found in [18].

2.2 Deep Q-learning Network

Deep Q-learning network (DQN) is a value approximation method which combines reinforcement learning with artificial neural network to deal with complex sequential decision-making problems [8].

The update of the Q-network is similar to a typical supervised learning method, and the update at iteration t uses the following loss function:

$$L(\theta^Q) = \frac{1}{N} \sum_t |y_t - Q(s_t, a_t|\theta^Q)|_2 \tag{3}$$

where $|\cdot|_2$ denotes 2-norm, θ^Q is the parameters of the function approximators, and y_t can be seen as a 'label':

$$y_t = r_t + \gamma Q(S_{t+1}, \mu(s_{t+1})|\theta^Q) \tag{4}$$

where r_t is immediate reward assigned to each action, γ is the discount factor, s_{t+1} is the next state, $\mu = \underset{a}{\operatorname{argmax}}\, Q(s,t)$.

3 Design of Grasping Strategy with DQN

The problem of grasping an object from a random pile is shown in Fig. 1, where the objects are laid closely side by side, and the robot is required to pick every object. This work proposes the grasping strategy based on DQN, where a CNN is employed to estimate the pose of the objects, and a fully connected network is used to approximate the optimal action-value function.

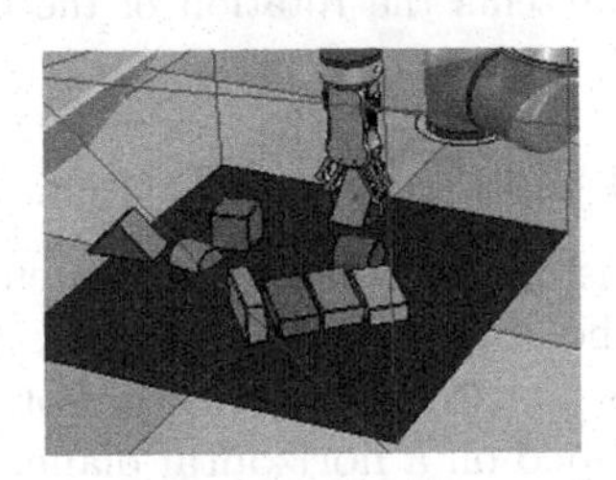

Fig. 1. A target should be picked from a random pile.

3.1 Representations of the Target State

Firstly, a high-dimensional RGB-D image of the target object is used to represent its state. The RGB-D image includes color image and depth image, which are with different views and sizes. As shown in Fig. 2(a), the size of the RGB image is 1920×1080 and the size of depth image is 520×414. An image alignment method is initially employed to match the RGB image with the depth image. The aligned images are shown Fig. 2(b), which make the object position description be consistent in the pixel spatial coordinate frame. It should be noted that the alignment of the images will improve the efficiency of the actions training.

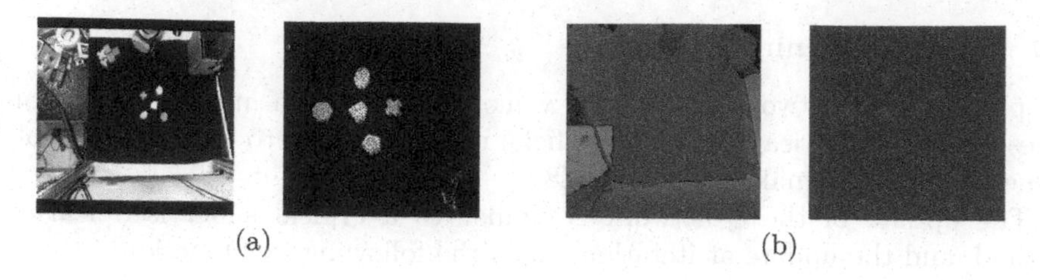

Fig. 2. The image has been projected onto a 3D point cloud, and orthographically back-project upwards in the gravity direction to construct a height map image representation with both color (RGB) and height-from-bottom (D) channels. (a) shows the RGB image and its height map image, and (b) shows the depth image and its height image.

The rotation angles of the target are separated into 18 orientations, where the interval between two adjacent angles is 20°. The input height map represents the rotation of the input image at 18 different angles, as shown in Fig. 3.

Fig. 3. Input height map represents the rotation of the input image at 18 different angles.

3.2 Representations of the Grasping Action

We represent the robot grasping action by a 4-dimensional vector (x, y, z, r_z), where x, y and z describe the object position in the Cartesian coordinates, r_z is the angle around z axis that is used to control the rotation of the gripper. Note that since the object is grabbed on a horizontal plane, the posture of the object will only change along the z axis, so we fixed the grasping angle r_x and r_y along the x and y axis, and only learned the grasping angle r_z along the z axis.

3.3 Design of the Rewards

Grasping of a target from a pile of objects suffers from the narrow space between objects. To solve the problem, each grasping action is determined by the result of its current step and the next n steps, that is, the current reward is decided by its immediate and the later rewards. It should be noted that if only one object lefts on the table, the immediate reward is determined by the result of the current grasping. Therefore, the immediate reward r_t is expressed as follows:

$$r_t = \sum_{i=0}^{n} \gamma^i r_{t+1} \tag{5}$$

where $\gamma \in [0, 1]$ is a discount factor.

In the training, an RGB-D image is taken as the input state, and the output is a 4-dimensional state-action value function, which describes the target pose of the gripper. The selection of the best action a_t is defined by:

$$a_t = \underset{a}{\operatorname{argmax}}\, Q(s_t, a_t) \tag{6}$$

where the action-value function Q relates to the robot action.

3.4 Design of the Q-networks

This work extends the vanilla deep Q-network by enhancing its feature extraction capability as shown in Fig. 4, where a convolutional neural network is in front of the full connect layers, which is used to extract the features of the input image.

The output $Q(s_t, a_t)$ represents the desired cumulative reward of (x, y, z, r_z) for the action a_t in the current state s_t. Thus, the corresponding action is executed with the maximum Q value. In the network, a resnet-v2 network [19] is employed as feature extraction module, which is pre-trained on ImageNet [20]. The parameters of the shallow network are fixed and only the parameters of the last few layers are tuned in the training.

3.5 Training of the Grasping DQN

Like reference [17], this work employs DQN algorithm to train the grasping action. The grasping features learned by the deep network are visualized as Fig. 4, which represents the optimal grasping point.

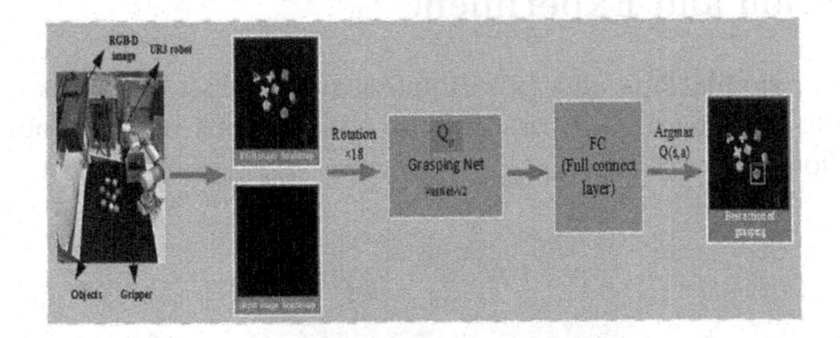

Fig. 4. The Q-networks proposed to train the grasping action.

When grasping from a pile of objects, we hope that the current grasp will be successful and also make the next grasp easily. That is, the current grasp picks up a target object meanwhile change the distribution of objects to make the objects scatter. Thus, the current reward of the grasp is design by the result of the current round and the results of the next episode; hence, each grasping action not only considers whether the current episode is successful, but also considers the results of the following rounds. The current reward (5) is re-written as:

$$R(t) = \sum_{i=0}^{n} \gamma^i r^{t+i} \tag{7}$$

where r is the sparse reward for the grasping action at t , that is, if the grasping action is successful, the reward function r is 1, otherwise it is 0.

$$r = \begin{cases} 1, & if\ successed \\ 0, & otherwise \end{cases} \tag{8}$$

The overview of the whole strategy is given as Algorithm 1.

Algorithm 1. Training of the single grasping action

Initialize parameters of the grasping net Q_g.
for index=1 to N: **do**
  Initialize the simulation environment in random state.
  Capture the RGB-image and Deep-image.
  Process the images and then input them to the network.
  Get the pixel position (p_x, p_y) of each image and obtain the grasping action through coordinate transformation.
  if $t < T_{prop}\%$ where T_{prop} is the exploration probability **then**
    Set a_t=random action as final action.
  else
    Set $a_t = \text{argmax}_a Q_g(s_t, a_t)$ as final action.
    if $a_t == \text{argmax}_a Q_g(s_t, a_t)$ **then**
      Update parameter of Q_g by $\min_{\theta_g}(y - Q_{\theta_g}(s, a))^2$.
    end if
  end if
end for

4 Simulation and Experiment

In this section, simulations and experiments are conduct to illustrate the proposed grasping learning algorithm, and the comparisons between our work and others are also given.

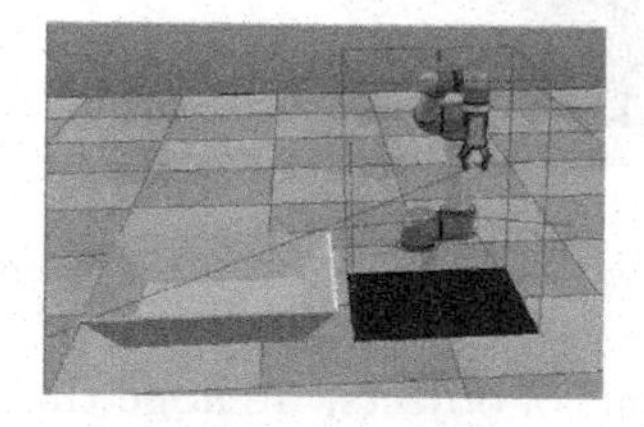

Fig. 5. The simulation of the pushing and grasping actions on V-REP.

4.1 Simulations

We firstly evaluate the proposed algorithm with simulation on V-REP, as shown in Fig. 5. The simulation environment includes a UR5 robot, a RG2 gripper, a Kinect2.0, and a pile of objects with different shapes.

We focus on the performances of the convergence speed, the total number of the grasping, and the success rate of grasping at the same condition. We assume that 10 objects should be picked from the table. The convergence rates of the training is about 75%, and the number of training episodes is about 1000.

Figure 6(a) shows the training times of grasping objects. We test on the grasping of 3, 5, 8 and 10 objects, respectively. If grasping times more than the pre-set threshold, the grasp is still unfinished, then the threshold represents the result.

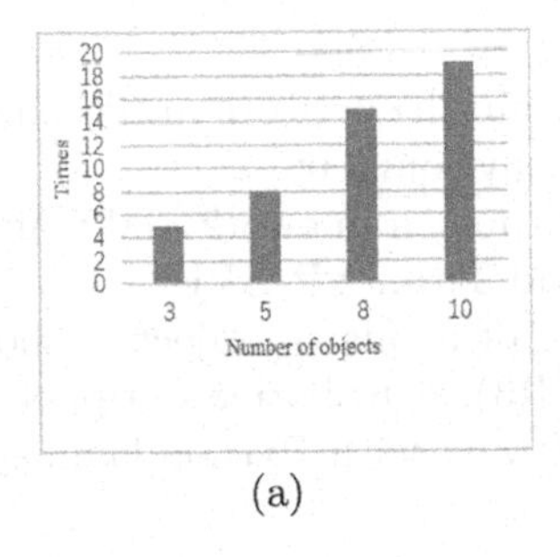

(a)

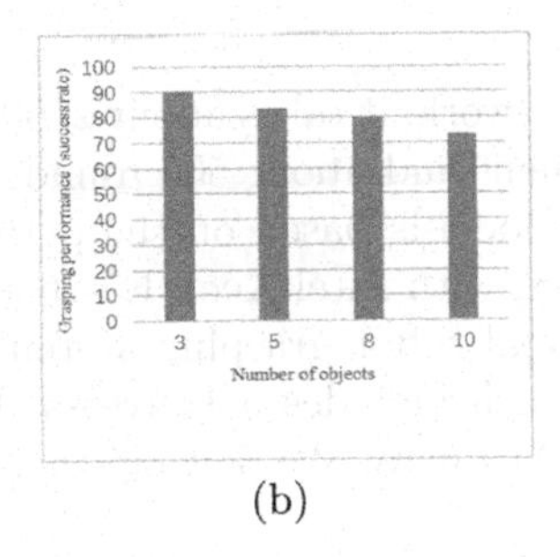

(b)

Fig. 6. (a) the times for training the grasps; (b) the success rate in grasping of different number of objects.

The success rate of grasping is defined the number of objects successfully grasped divided by the total number of objects. We use the average value of 10 grasping as the final success rate of grasping, and test the success rate in grasping of 3, 5, 8, and 10 objects as shown in Fig. 6(b).

We compare the proposed method to the two baseline methods in a simulation where 10 objects are randomly dropped onto a table. Our grasp success is 72%. As reported in reference [17], their grasp success of 30 objects is 67.7%. It is interesting that the success rate of the proposed grasping with long term rewards is similar to that of visual pushing and grasping (VPG).

4.2 Strategy Transferring from Simulation to Real

It is notoriously costly and dangerous to train a real grasp when the deep Q-learning networks have initialized weights values as random. One way to deal with the problem is to leverage the power of simulation to produce the parameters of the deep Q-network. That is, training grasping model on simulation and then transferring them to the real robot. Using domain adaptation methods to cross this "reality gap" requires a large amount of unlabelled real-world data, while domain randomization could waste modelling power [21].

In this work, we deal with the "reality gap" by "large-scale simulation data & small-scale real data". We firstly train the deep Q-network in the simulated environment in which the grasped objects are shown in Fig. 7(a). And then, the deep Q-network is transferred to the real robot for appropriate training in which the grasped objects are shown in Fig. 7(b). The proposed method would avoid the inconsistency between the simulation environment and the real environment, such as the camera installation position, the light information, etc.

(a) (b)

Fig. 7. The objects used in training, where (a) shows the objects used in the simulated environment, and (b) shows the objects used in the real robot.

In this work, it takes about 2000 rounds to train the network in the simulated environment and about 500 rounds in the real environment.

Our model is based on the python deep learning framework and trained on our server, with Intel Xeon E5-2630V3 CPU and Titan XP GPU.

The real robot grasping system for picking of a pile of object is shown in Fig. 8, which includes a Universal Robot 3 (UR3), a Robotiq85 gripper and a Kinect2.0 camera. We develop the grasping software with Python language.

Fig. 8. The real grasping system.

The RGB image and depth image captured by Kinect are with different views and image sizes; hence, we perform an image alignment operation for the RGB image, and map it to the same perspective as the depth image. This operation ensures the consistency of object position description in the pixel spatial coordinate system for RGB image and depth image.

The captured image contains many objects, which will affect the grasping decision of the agent. Thus, we project the image onto a 3D point cloud, and orthographically back-project upwards in the gravity direction to construct a height map image that represents both color (RGB) and height-from-bottom (D) channels. In the real environment, we test the grasping success rate and the generalization ability of the DQN-based grasping model.

In the following, we will describe how to use the algorithm we proposed to control the robot to complete a grasping process in the case of mixed distribution of 8 objects shown in Fig. 9.

Figure 9 shows a complete process of the robot's grasping object. The robot system first initializes, then collects the image from the camera, calculates the position of the object on the plane, and then moves directly to the top of the

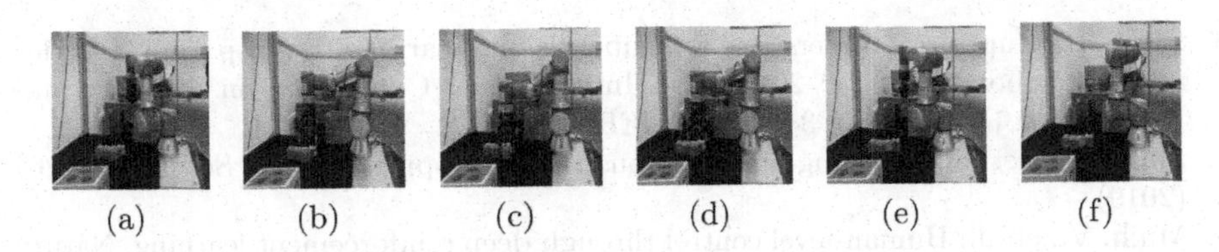

Fig. 9. A pile of objects on the table are picked up by the robot one by one. (a) Robot system initialization. (b) Robot moves directly above the object position. (c) Robot adjusts the gesture of the gripper. (d) Robot is grasping the object. (e) Robot determines if the object is grasped successful. (f) Robot puts the object into the tray.

object. Then the robot began to calculate the position on the z axis and adjust the jaw to grab the object along the z axis. After grasping, the robot determines whether to successfully capture the object through the force of the hand. If the failure is refetched, the captured object is placed in the tray.

5 Conclusion

This paper proposed a robot grasping method based on the Deep Q-network. The full convolutional network is used to extract the input image features, and the full connection layer outputs the action value. Unlike other similar methods, our method determines each robot grasping action through the result of its current step and the next n steps. In the process of grasping, the distribution of objects on the table is constantly changing. Previously obscured objects appear in the robot's field of view. The proposed method still has the ability to generalize to previously unseen objects. To eliminating "reality gap" between simulation and real-world, we use large number of simulation data and small number of real data in the real-world training.

References

1. Goins, A.K., Carpenter, R., Wong, W.K., Balasubramanian, R.: Evaluating the efficacy of grasp metrics for utilization in a gaussian process-based grasp predictor. In: 2014 IEEE/RSJ International Conference on Intelligent Robots and Systems, pp. 3353–3360 (2014). https://doi.org/10.1109/IROS.2014.6943029
2. Boularias, A., Kroemer, O., Peters, J.: Learning robot grasping from 3-d images with markov random fields. In: 2011 IEEE/RSJ International Conference on Intelligent Robots and Systems, pp. 1548–1553 (2011). https://doi.org/10.1109/IROS.2011.6094888
3. Balaguer, B., Carpin, S.: Learning end-effector orientations for novel object grasping tasks. In: 2010 10th IEEE-RAS International Conference on Humanoid Robots, pp. 302–307 (2010). https://doi.org/10.1109/ICHR.2010.5686826
4. Saxena, A., Driemeyer, J., Ng, A.Y.: Robotic grasping of novel objects using vision. Int. J. Rob. Res. **27**(2), 157–173 (2008)
5. Bohg, J., Morales, A., Asfour, T., Kragic, D.: Data-driven grasp synthesis-a survey. IEEE Trans. Rob. **30**(2), 289–309 (2013)

6. Pinto, L., Gupta, A.: Supersizing self-supervision: Learning to grasp from 50k tries and 700 robot hours. In: 2016 IEEE International Conference on Robotics and Automation (ICRA), pp. 3406–3413. IEEE (2016)
7. Mahler, J., et al.: Learning ambidextrous robot grasping policies. Sci. Rob. **4**(26) (2019)
8. Mnih, V., et al.: Human-level control through deep reinforcement learning. Nature **518**(7540), 529–533 (2015)
9. Silver, D., et al.: Mastering the game of go with deep neural networks and tree search. Nature **529**(7587), 484–489 (2016)
10. Gu, S., Holly, E., Lillicrap, T., Levine, S.: Deep reinforcement learning for robotic manipulation with asynchronous off-policy updates. In: 2017 IEEE International Conference on Robotics and Automation (ICRA), pp. 3389–3396. IEEE (2017)
11. Nair, A., Pong, V., Dalal, M., Bahl, S., Lin, S., Levine, S.: Visual reinforcement learning with imagined goals. arXiv preprint arXiv:1807.04742 (2018)
12. Pinto, L., Andrychowicz, M., Welinder, P., Zaremba, W., Abbeel, P.: Asymmetric actor critic for image-based robot learning. arXiv preprint arXiv:1710.06542 (2017)
13. Popov, I., Heess, N., Lillicrap, T., Hafner, R., Barth-Maron, G., Vecerik, M., Lampe, T., Tassa, Y., Erez, T., Riedmiller, M.: Data-efficient deep reinforcement learning for dexterous manipulation. arXiv preprint arXiv:1704.03073 (2017)
14. Levine, S., Pastor, P., Krizhevsky, A., Ibarz, J., Quillen, D.: Learning hand-eye coordination for robotic grasping with deep learning and large-scale data collection. The International Journal of Robotics Research **37**(4–5), 421–436 (2018)
15. Sadeghi, F., Toshev, A., Jang, E., Levine, S.: Sim2real view invariant visual servoing by recurrent control. arXiv preprint arXiv:1712.07642 (2017)
16. Chen, B., Su, J.: Addressing reward engineering for deep reinforcement learning on multi-stage task. In: Gedeon, T., Wong, K.W., Lee, M. (eds.) ICONIP 2019. CCIS, vol. 1143, pp. 309–317. Springer, Cham (2019). https://doi.org/10.1007/978-3-030-36802-9_33
17. Zeng, A., Song, S., Welker, S., Lee, J., Rodriguez, A., Funkhouser, T.: Learning synergies between pushing and grasping with self-supervised deep reinforcement learning. In: 2018 IEEE/RSJ International Conference on Intelligent Robots and Systems (IROS), pp. 4238–4245. IEEE (2018)
18. Wiering, M., Van Otterlo, M.: Reinforcement learning. Adapt. Learn. Optim. **12**(3) (2012)
19. He, K., Zhang, X., Ren, S., Sun, J.: Deep residual learning for image recognition. In: Proceedings of the IEEE Conference on Computer Vision and Pattern Recognition, pp. 770–778 (2016)
20. Deng, J., Dong, W., Socher, R., Li, L.J., Li, K., Fei-Fei, L.: Imagenet: a large-scale hierarchical image database. In: 2009 IEEE Conference on Computer Vision and Pattern Recognition, pp. 248–255. IEEE (2009)
21. James, S., et al.: Sim-to-real via sim-to-sim: Data-efficient robotic grasping via randomized-to-canonical adaptation networks. In: Proceedings of the IEEE/CVF Conference on Computer Vision and Pattern Recognition, pp. 12627–12637 (2019)

Surplus Force Control Strategy of an Active Body-Weight Support Training System

Chao Wei[1], Tao Qin[1,2(✉)], Xin Meng[1], Jinxing Qiu[1], Yikun Wang[1], and Bo Li[1,2]

[1] School of Mechanical Engineering, Hubei University of Arts and Sciences, Xiangyang 441053, China

[2] Technical Center, Xiangyang Institute of Advanced Manufacturing Engineering Research of Huazhong University of Science and Technology, Xiangyang 441053, China

Abstract. In order to improve the problem of inadequate control of the body's center of gravity (COG), an active body-weight support training system (BWSTS) with double-shoulder suspension based on cable-driven was designed. Firstly, the mathematical model of the system driven unit was established by the mechanism analysis method; secondly, the dynamic characteristics of the system forward channel transfer function was analyzed, and the active loading compound control strategy (ALCCS) of the active BWSTS was designed. The simulation results demonstrated that the ALCCS can meet the requirements of system loading accuracy; Finally, the frequency characteristics of the surplus force transfer function was analyzed, and the surplus force compensation control strategy for the active BWSTS was designed. The simulation results demonstrated that the designed surplus force compensation control strategy can reduce effectively the surplus force and improve the loading accuracy of the active BWSTS.

Keywords: Body-weight support training system (BWSTS) · Rehabilitation training · Active loading compound control strategy (ALCCS) · Surplus force

1 Introduction

Stroke is the most important cause of damage to nervous system and impairment of lower limb motor function, which has the characteristics of high incidence and disability [1]. Therefore, improving the walking ability of patients' lower limbs has become one of the urgent problems for post-stroke rehabilitation [2]. Modern medical research has proved that continuous repetitive training can help improve the body's motor function and the plasticity of nerve function [3]. At present, BWSTS is often used for gait rehabilitation training for stroke and spinal cord injury patients [4], which was first proposed by Finch and Barbeau in 1986. Some quantitative studies have shown that BWSTS is more effective than traditional training for correcting gait [5], improving balance [6], link muscle spasm [7] and reducing cardiopulmonary load [8].

The current research has found that the cable has the advantages of high load/weight ratio, high velocity/accuracy ratio, large working space, and adjustable stiffness [9]. This makes the suspended type BWSTS based on cable-driven widely used in rehabilitation

X.-J. Liu et al. (Eds.): ICIRA 2021, LNAI 13014, pp. 153–162, 2021.
https://doi.org/10.1007/978-3-030-89098-8_15

robots, astronaut training and other fields [10]. For example, the single-cable suspension system Zero G [11], the multi-cable suspension system FLOAT [12], and the exoskeleton lower limb rehabilitation robot system LOPES [13] are all using a cable suspension type BWSTS.

Based on the above analysis, in order to improve the problem of insufficient movement control of the COG of the human body for the existing BWSTS, an active BWSTS with double-shoulder suspension based on cable-driven was designed to control the up and down movement of the body's COG during gait training, and effectively improve the patient's exercise balance ability.

2 Mechanical Structure of BWSTS

The three-dimensional structure of the BWSTS with double-shoulder suspension is shown in Fig. 1. The system is mainly composed of the weight-loss frame, treadmill, human-machine interface (HMI) system, suspension vest, two identical cable-driven units and force measuring armrests. The BWSTS can individually control the up and down, left and right movements of the patient's COG to ensure the effectiveness and coordination of training.

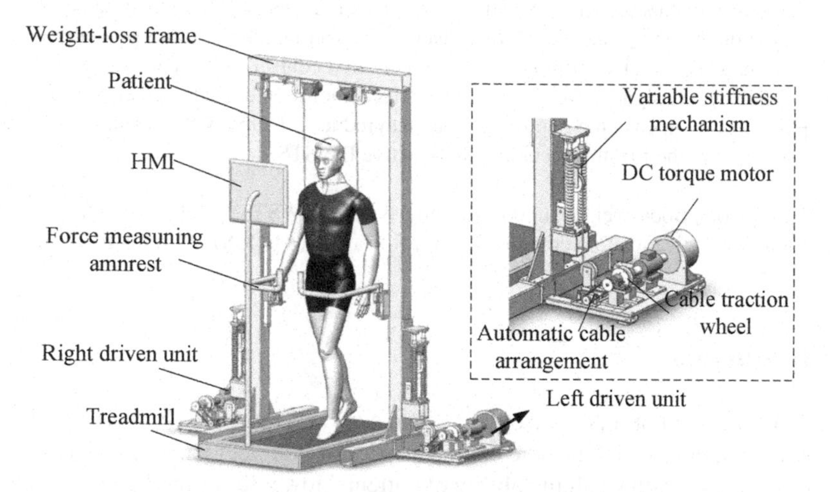

Fig. 1. The three-dimensional structure of BWSTS

The unilateral cable-driven unit of the BWSTS is simplified into a mass-spring model by using the mechanism analysis method. The mechanism model for the cable-driven unit of the BWSTS is established as shown in Fig. 2. y is the displacement of the cable end (the vertical displacement of the patient's COG), F is the internal tension of the cable (the weight-loss of the patient), M is the mass of the patient, K is the stiffness coefficient of the cable, B is the damping coefficient of the cable.

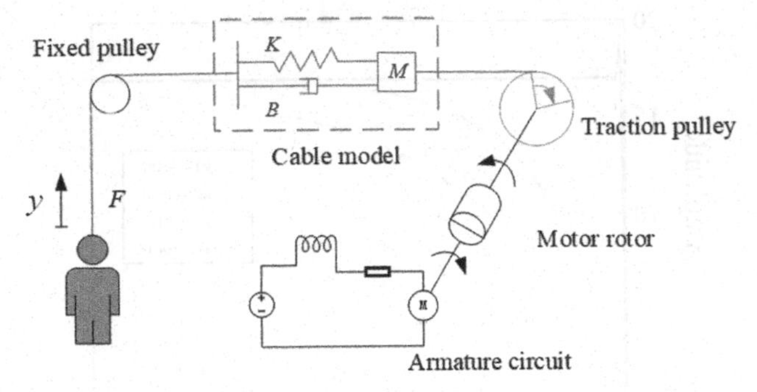

Fig. 2. The mechanism model for the cable-driven unit of BWSTS

3 Active Loading Compound Control Strategy

According to the cable-driven dynamics model in Fig. 2, and taking the influence of patient's movement during rehabilitation training into consideration, the block diagram of the open-loop model of the cable-driven unit loading system is shown in Fig. 3.

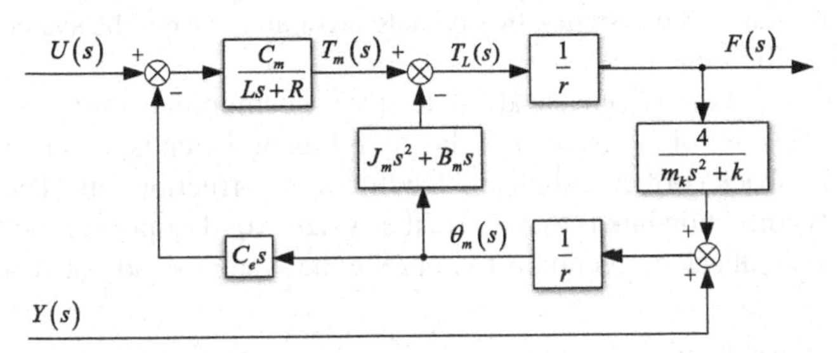

Fig. 3. Block diagram of the cable-driven unit loading system

The simulation model of the cable-driven unit loading system is established in Matlab/Simulink environment. The unit step response curves under different spring stiffness coefficient k are obtained as shown in Fig. 4.

As show in Fig. 4, the cable-driven unit loading system is with a fast response speed and no overshoot when the spring stiffness coefficient $k = 5000$. Therefore, the cable-driven unit loading system model with spring stiffness coefficient $k = 5000$ is selected as the main research object, and the system's forward channel transfer function is:

$$M_1(s) = \frac{13.6s^2 + 68000}{s^3 + 625s^2 + 5814.5s + 4000} \tag{1}$$

According to the final value theorem of Laplace transform, the system steady-state error under the action of a unit step input signal is calculated as:

$$e_{ss} = \lim_{s \to 0} s \frac{1}{1 + M_1(s)} \frac{1}{s} = \frac{1}{1 + \lim_{s \to 0} M_1(s)} \approx 0.056 \tag{2}$$

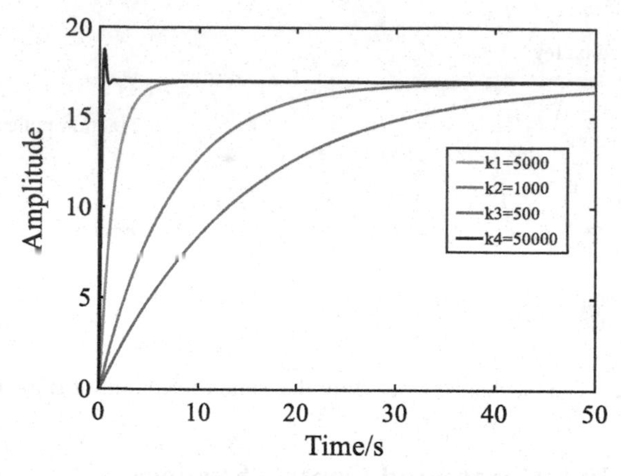

Fig. 4. Unit step response curves of the cable-driven unit loading system

As show in Eq. (2), the system is a typical *0 - type* system. Although it can track the unit step signal, it has a certain steady-state error and cannot meet the high-order and no static error requirements of the system loading. Therefore, corresponding measures must be taken to reduce the system's steady-state error and correct the system's dynamic characteristics at the same time.

In order to reduce the system steady-state error, a compound control strategy based on traditional PID control is designed as shown in Fig. 5. which is mainly composed of PID negative feedback correction link and feedforward correction link. The function of the feedforward correction link is to improve the system rapid response, and the function of PID negative feedback correction is to improve the system steady-state accuracy.

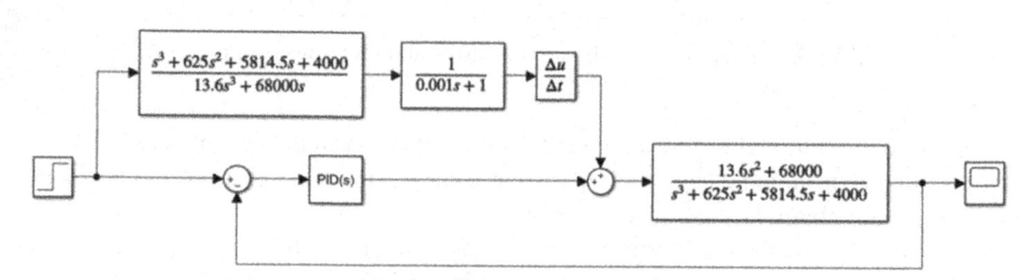

Fig. 5. Simulation model of ALCCS

The PID negative feedback control system model and the ALCCS simulation model are built in Simulink, and the results are shown in Fig. 6.

As shown in Fig. 6, the designed ALCCS is better than the traditional PID control strategy, which meets the system's requirements for force command loading accuracy and speed, and improves the dynamic performance of system loading.

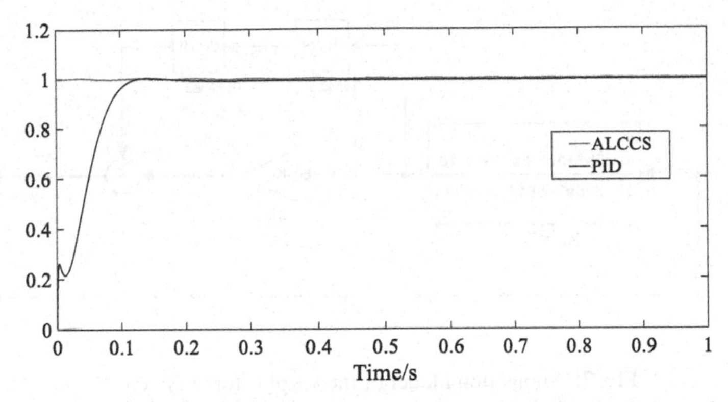

Fig. 6. Unit step response curves of ALCCS and PID feedback control

4 Surplus Force Compensation Control Strategy

When the patient is actively moving, the position disturbance introduced by the patient's movement will generate surplus force, and the generated surplus force will affect the weight reduction of the system. Therefore, it is also necessary to study the dynamics characteristics for the surplus force of the cable-driven unit of the active BWSTS, and design a corresponding compensation controller to reduce surplus force and improve the loading accuracy of the cable-driven unit of the active BWSTS.

According to Fig. 2, the surplus force transfer function of the cable-driven unit of the active BWSTS can be written as Eq. (3). After bringing in the actual system parameters, it can be written as Eq. (4).

$$M_2(s) = \frac{\left(m_k s^2 + k\right)\left[J_m L s^2 + (B_m L + J_m R)s + B_m R + C_m C_e\right]}{(m_k r^2 + 4J_m)Ls^3 + (m_k r^2 R + 4J_m R + 4B_m L)s^2 + \left(4C_e C_m + 4B_m R + k^2 L\right)s + k^2 R} \tag{3}$$

$$M_2(s) = \frac{0.25\left(s^2 + 5000\right)(s + 613)(s + 9.48)}{(s + 616)(s + 8.69)(s + 0.748)} \tag{4}$$

Affected by the second-order differential link and the first-order inertia link (the turning frequency is $\omega_{sd3} = 0.748$ rad/s), the amplitude of the system's surplus force in the low frequency band decreases slightly with the increase in the frequency of the moving speed of the carrying object, and the phase is slightly lagging. With the increase of the moving frequency of the load-bearing object, the amplitude of the system's surplus force will decay faster, and the phase lag will be larger but not more than 90°. When the moving frequency of the load-bearing object is equal to 70.7 rad/s, the amplitude of the system's surplus force will be attenuated to 0 and the phase will be 90° under the influence of the second-order differential link.

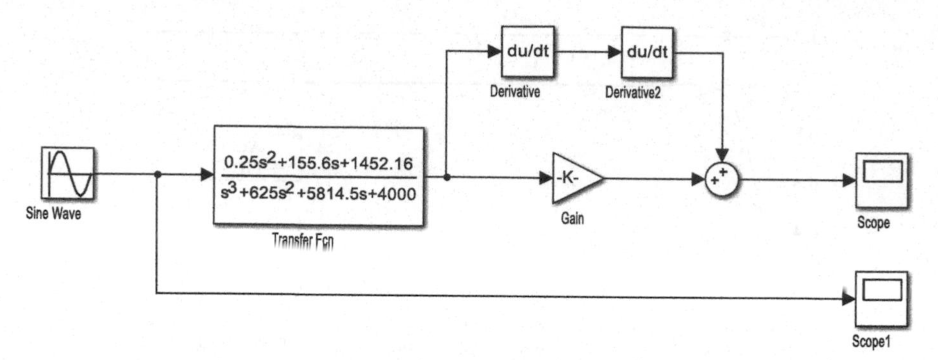

Fig. 7. Simulation model of the surplus force system

The surplus force system simulation model is established by Matlab/Simulink as shown in Fig. 7. Considering the patient's exercise frequency during normal walking is generally about 1Hz, and the exercise displacement is about $\pm$ 0.1m. Select a sine curve with an amplitude of 0.1m/s and a frequency of 1Hz as the input speed curve. The output surplus force curve is obtained by simulation as shown in Fig. 8.

As shown in Fig. 8, due to the patient's movement disturbance, a surplus force about 21N will be generated, which will seriously affect the weight-loss force output by the cable-driven unit, thereby affecting the system loading accuracy.

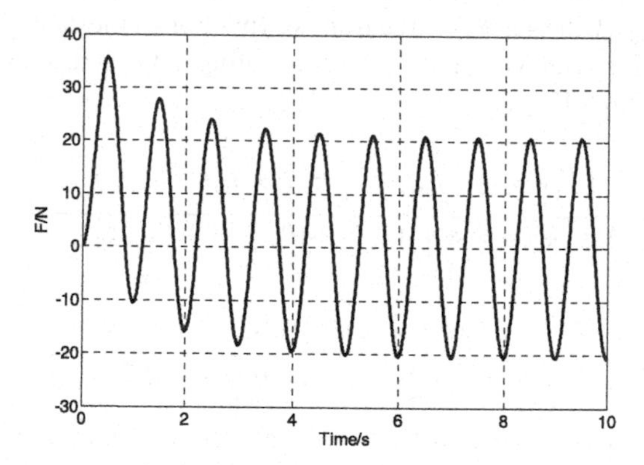

Fig. 8. Surplus force output curve

Since the system needs to have a strong disturbance suppression capability during operation, the disturbance feedforward compensation control (DFCC) shown in Fig. 9 is adopted to improve the system's anti-interference ability.

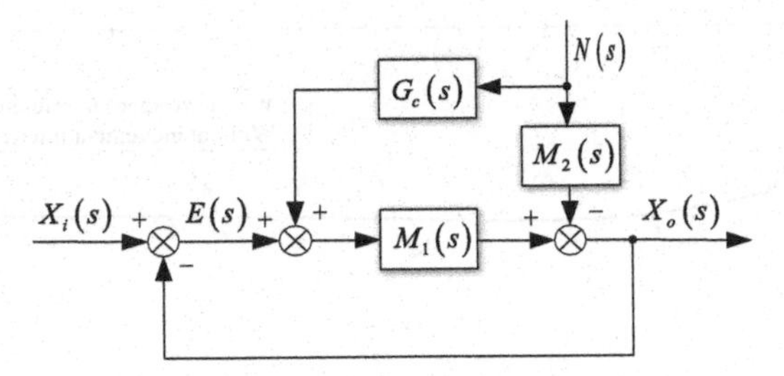

Fig. 9. Schematic diagram of DFCC

According to Fig. 9, the system's output under disturbance action is derived as:

$$X_O(s) = \frac{G_c(s)M_1(s) - M_2(s)}{1 + M_1(s)} N(s) \tag{5}$$

To eliminate the system error caused by the disturbance action, it is required that $X_O(s) = 0$. The transfer function of the compensation controller is designed as:

$$G_c(s) = \frac{M_2(s)}{M_1(s)} \tag{6}$$

Due to modeling errors and nonlinear factors, there will always be errors between the established theoretical model and the actual system. The compensation controller designed according to the theoretical model cannot completely eliminate the surplus force caused by the disturbance. Therefore, using a composite control method combining with DFCC and closed-loop negative feedback control, the closed-loop negative feedback main controller can reduce the impact of the uncompensated part of the feed-forward control, reduce the design requirements for the compensation link, and improve the compensation effect.

The simulation model of the surplus force compensation control system of the active BWSTS is built as shown in Fig. 10, and the given weight reduction force is selected as 200N. The response curves are obtained as show in Fig. 11.

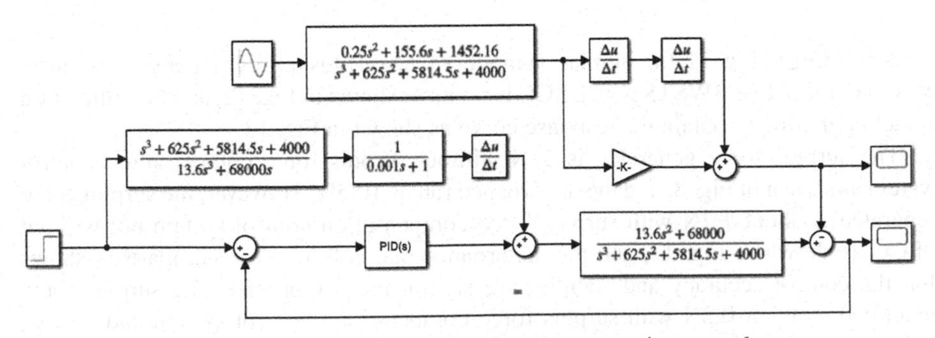

Fig. 10. Simulation model of surplus force compensation control system

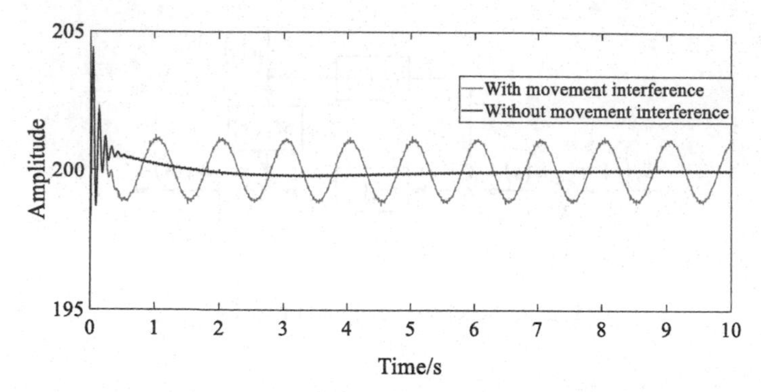

Fig. 11. The response curves of the surplus force compensation control system

As show in Fig. 11, the negative feedback active controller has limited adjustment capabilities, and it is difficult to effectively suppress the patient's movement interference, which seriously affects the system's loading accuracy.

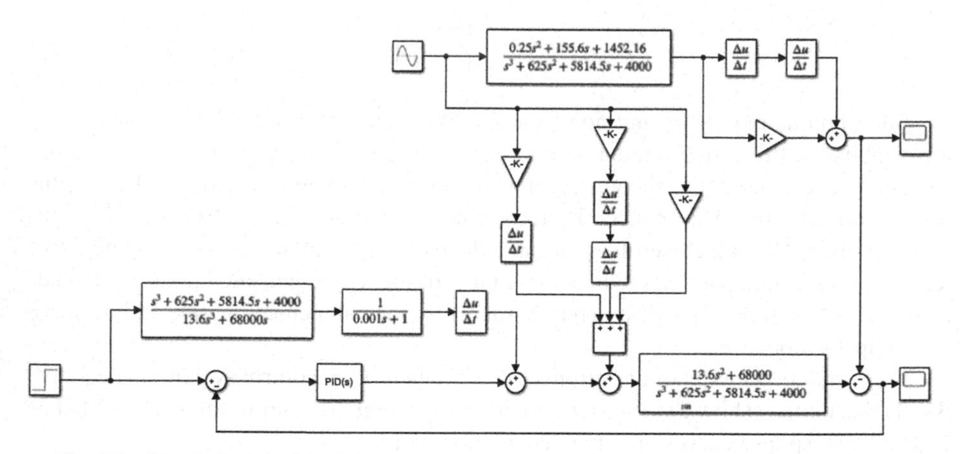

Fig. 12. Simulation model of the surplus force compensation control system with DFCC

According to Fig. 11, the simulation model of the surplus force compensation control system of the active BWSTS with DFCC is set up as shown in Fig. 12, and the simulation model is running to obtain the response curve as shown in Fig. 13.

The surplus force generated is 21N without surplus force compensation control system as shown in Fig. 8, and the disturbance rate is 10.5%. However, the surplus force generated is about 1.2N with surplus force compensation control system but without DFCC as shown in Fig. 13, and the disturbance rate is 0.6%. The comparison shows that the control accuracy and stability are significantly improved. The surplus force generated is about 0.2N with surplus force compensation control system and DFCC, and the disturbance rate is only 0.1%. It can be seen that the anti-interference ability of the system is improved after adding the DFCC, and the position disturbance caused by

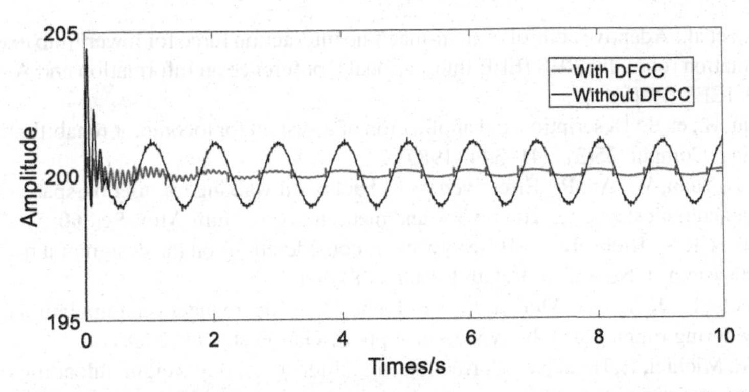

Fig. 13. Response curve of the surplus force compensation control system with and without DFCC

the patient's movement can be effectively suppressed, thereby improving the system's loading accuracy.

5 Conclusion

An active BWSTS with double-shoulder suspension based on cable-driven was designed. The mathematical model of the cable-driven unit of BWSTS was established by using the mechanism analysis method. An ALCCS based on the traditional PID control strategy was designed. The simulation results demonstrated that the designed ALCCS can meet the requirements of system loading accuracy. The frequency characteristics of the surplus force transfer function was analyzed, and the active BWSTS surplus force compensation control strategy was designed. The simulation results demonstrated that the surplus force generated by the designed surplus force compensation control strategy is 0.2N and the disturbance rate is 0.1%, so it can effectively reduce the surplus force and improve the system's loading accuracy.

Acknowledgments. This work supported by Hubei Provincial Natural Science Foundation under Grant 2018CFB313, Xiangyang Science and Technology Plan Project under Grant 2017YL12 and 2019ZD03, Hubei Superior and Distinctive Discipline Group of "Mechatronics and Automobiles" under Grant XKQ2021041, Hubei Provincial Higher Institutions Excellent Young and Middle-aged Technological Innovation Team Project under Grant T201919, National Natural Science Foundation of China under Grant 11902112.

References

1. Liang, X., et al.: Position based impedance control strategy for a lower limb rehabilitation robot. In: 2019 41st Annual International Conference of the IEEE. Engineering Medicine & Biology Society (EMBC). IEEE (2019)
2. Keynote talk 2: Passive and active control for rehabilitation robots. In: 2016 10th International Conference on Software, Knowledge, Information Management & Applications (SKIMA) (2017)

3. Zhu, A., et al.: Adaptive control of man-machine interaction force for lower limb exoskeleton rehabilitation robot. In: 2018 IEEE International Conference on Information and Automation (ICIA). IEEE (2018)
4. Barbeau, H., et al.: Description and application of a system for locomotor rehabilitation. Med. Biol. Eng. Comput. **25**(3), 341–344 (1987)
5. Wang, J., Yuan, W., An, R.: Effectiveness of backward walking training on spatial-temporal gait characteristics: a systematic review and meta-analysis. Hum. Mov. Sci. **60**, 57–71 (2018)
6. Gonçalves, R.S., Krebs, H.I.: MIT Skywalker: considerations on the design of a body weight support system. J. Neuroeng. Rehabil. **14**(1), 88 (2017)
7. Masumoto, K., Joerger, J., Mercer, J.A.: Influence of stride frequency manipulation on muscle activity during running with body weight support. Gait Post. 473 (2018),
8. Salil, A., Michiel, P., Heike, V.: Correction to: influence of body weight unloading on human gait characteristics: a systematic review. J. Neuroeng. Rehabil. **15**(1), 73–76 (2018)
9. Roche-Seruendo, L.E., et al.: Effects of different percentages of body weight support on spatiotemporal step characteristics during running. J. Sports Sci. **36**(13), 1–6 (2017)
10. Zhang, L., et al.: Force control strategy and bench press experimental research of a cable-driven astronaut rehabilitative training robot. IEEE Access **5**(99), 9981–9989 (2017)
11. Hidler, J., et al.: ZeroG: overground gait and balance training system. J. Rehabil. Res. Dev. **48**(4), 287–298 (2011)
12. Vallery, H., Lutz, P., Zitzewitz, J.V., et al.: Multidirectional transparent support for overground gait training. In: IEEE International Conference on Rehabilitation Robotics. IEEE (2013)
13. Veneman, J.F., et al.: Design and evaluation of the LOPES exoskeleton robot for interactive gait rehabilitation. IEEE Trans. Neural Syst. Rehabil. Eng. **15**(3), 379–386 (2017)

Process Learning of Robot Fabric Manipulation Based on Composite Reward Functions

Tianyu Fu[1,2], Fengming Li[1,2], Yukun Zheng[1,2], and Rui Song[1,2(✉)]

[1] Center for Robotics, School of Control Science and Engineering, Shandong University, No. 17923, Jingshi Road, Jinan 250061, China
[2] Engineering Research Center of Ministry of Education for Intelligent Unmanned Systems, Jinan 250100, China
rsong@sdu.edu.cn

Abstract. The robot's weak ability to manipulate deformable objects makes robots rarely used in the garment manufacturing industry. Designing a robot skill acquisition frame that can learn to manipulate fabrics helps to improve the intelligence of the garment manufacturing industry. This paper proposes a process learning framework for robot fabric stacking based on a composite reward function. A limited task flow model describes the overall process, and robot skills represent a single task. Based on the robot's acquisition of operational skills, a priori knowledge of the technological process is embedded into the reward function to form a composite reward function, and then it takes to guide the robot to use the acquired skills to complete the overall process task. Experiments are conducted on the UR5e robot to prove the effectiveness of this method, and results show that the robot guided by the composite reward function can complete the process task of fabric manipulation before garment sewing.

Keywords: Robot manipulate skill · Process learning · Composite reward.

1 Introduction

Garment manufacturing is a labor-intensive process [1], which can be divided into three processes before, during, and after sewing. The process before sewing includes fabric quality inspection, cloth cutting, cloth printing, etc. All of these processes involve the handling and stacking of soft materials. Besides, Completing the sewing of a garment requires multiple pieces of fabric to be spliced. Therefore, sorting, picking and, stacking the cut fabric in advance according to the sewing process is the basis for completing of the garment sewing. However, at present, a

Supported by Shandong Major Science and Technology Innovation Project(No.2019JZ ZY010430), Shandong Major Science and Technology Innovation Project(No.2019JZ ZY010429), Shandong Provincial Key Research and Development Program(Grand No.2019TSLH0302).

X.-J. Liu et al. (Eds.): ICIRA 2021, LNAI 13014, pp. 163–174, 2021.
https://doi.org/10.1007/978-3-030-89098-8_16

large part of the fabric handling and stacking work is still done manually. According to statistics, about 80% of the labor cost of a finished product in the garment manufacturing process comes from the handling of fabrics [2]. Today, when labor costs are gradually increasing, using robots to replace humans to complete the fabric handling and stacking during the sewing process can improve the production efficiency of the garment manufacturing process and save labor costs [4].

In automobile manufacturing and other industries, robots have been widely used in handling and stacking rigid objects [1]. However, due to the characteristics of fabric with high flexibility, low friction, and material anisotropy [12], wrinkles and deformation will occur during manipulation, while the poor ability of robots to manipulate deformable objects, making it still less used in the garment manufacturing industry. Many scholars have done related research for this. Firstly, by studying the trajectory planning of the end-effector of the robot, the manipulation of the deformed object is realized. Mark et al. studied a computational method for path planning of deformed objects subject to manipulation constraints [10]. A PID-based robust control strategy for deformable objects is proposed by T. WaDa et al. [13]. Furthermore, more scholars tend to manipulate deformed objects based on sensor feedback. Navarro [11] based on visual serving to realize robot manipulation of deformable objects. In [5], force control was used to design a deformation manipulation strategy that maintains the tension of the cloth, while Angel et al. apply haptic servo to the manipulation of deformed objects [3]. However, these methods are mostly used in home service backgrounds and are rarely used in industrial manufacturing. In addition, due to the large shape space of deformable objects, solutions using modeling methods require a lot of engineering work. To avoid that, learning skills through an end-to-end manner may be a better way [9].

The data of fabric manipulation can be generated by interaction with the environment through reinforcement learning. Reinforcement learning has been extensively studied in the manipulation of rigid objects in manufacturing industries, such as assembly [6] and grinding [15], and a few of them have applied reinforcement learning to the manipulation of deformable objects. Matas first applied deep reinforcement learning to robotic fabric dragging and hanging, and studied the migration of the algorithm from simulation to the real environment [9], Wu et al. completed the manipulation of the robot to unfold the towel through model-free visual reinforcement learning. For robot stacking manipulation skills, most of the work is to apply reinforcement learning to stacking rigid objects, such as regular-shaped Lego blocks [14], irregularly-shaped building dry stacking [8]. It does not involve the stacking of soft materials in garment manufacturing, and the manipulation skills learning mentioned above only involves the learning of individual discrete skills, and lacks continuous learning of the overall process.

In this paper, a learning framework for robot soft material stacking process based on compound reward function is proposed. The main contributions of this paper are as follows:

- Proposed the Composite Reward Deep Deterministic Policy Gradient (CR-DDPG) method based on limited task flow description, which has the ability of single learning and process learning.

- A procedural compound reward mechanism suitable for stacking soft materials is proposed.
- The method was verified on a real UR5E manipulator, and the soft material stacking task was successfully completed by learning.

This remainder of this paper is organized as follows. Section 2 introduces the model of limited task of robot and formulates the problem to be solved. Section 3 contains the description of the proposed method. Experiments were performed to validate the proposed method, and the experimental results are presented and discussed in Sect. 4. Finally, in Sect. 5, we summarize the results of the current work and discuss future directions.

2 Problem Description

The robot's task to realize fabric handling and stacking among each functional unit of the garment factory can be divided into the manipulation behaviors of "robot arrives at the pre-grabbing point of the fabric, robot grabs material, robot moves to the pre-stacking position and robot stacks material." Each step can be further abstracted into basic manipulation skills such as the "arrival, grasp, move, release" of the robot. If the robot can learn the sequence of basic manipulation skills according to the expected process sequence, then it can finally complete the procedural task of robot handling and stacking.

Therefore, we propose a finite task flow behavior description model to solve the above problem. As shown in Fig. 1, First, define the robot action process to complete the task as robot behavior, denoted as $\mathcal{A}$. The behavior includes a task sequence $\{\phi_1(m_1, f_{g_1}), \phi_2(m_2, f_{g_2}), ..., \phi_n(m_n, f_{g_n})\}$ consisting of n behavior element ϕ. Where m_i is the robot manipulate skill to complete the i behavior, and f_{g_i} is the finish flag of the behavior element ϕ. When the mark is True, proceed to the next behavior element learning.

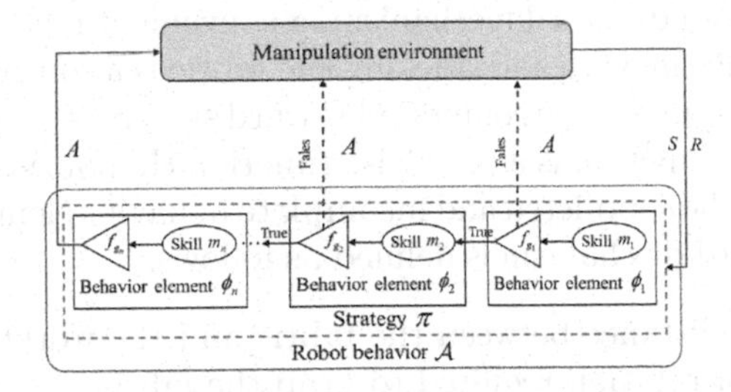

Fig. 1. The finite task flow behavior description model

The manipulate skill m is realized by a series of robot states $S = \{s_1, s_2, ..., s_t\}$ and robot actions $A = \{a_1, a_2, ..., a_t\}$. s_t represents the observed value of the robot state at time t, a_t represents the action of the robot at time t, and each manipulate skill m can be represented by a state-action chain $\{s_0, a_0, s_1, a_1,..., a_{t-1}, s_t\}$. For the finish flag f_{g_i}, set the target set $G = \{g_1, g_2, ..., g_n\}$, where g_i is the completed state flag of the robot behavior element, that is, the target set G maps the state set S. During the interaction between the robot and the environment, the state s_t at the next moment is saved as the completion goal g_t at the previous moment. When $g_i = f_{g_i}$, the behavior element learning ends.

Our method is to use deep reinforcement learning to observe the state S of robots and fabrics in the fabric operating environment, combined with delayed rewards R, and learn a strategy π to non-linearly couple each robot's behavior primitives (m_i, f_{g_i}). In this way, the robot actions A are output, and the individual actions in the task process are completed successively, and finally, the robot handling and stacking fabric process behavior $\mathcal{A}$, As shown in Eq. 1.

$$\mathcal{A} = \pi\{(m_1, f_{g_1}), (m_2, f_{g_2}), ..., (m_n, f_{g_n})\} \tag{1}$$

3 Proposed Method

This section describes a method for the robot to acquire process Manipulation skills in the process task of moving and stacking fabrics. The proposed method is based on deep reinforcement learning, and the robot can learn process skills under the guidance of the compound reward function, which can realize the process task of distributing the stacking process in sequence. It provides a solution for the generalization of robot skills.

3.1 Compound Reward Function

For the whole fabric handling and stacking process of the robot, the sparse reward mechanisms may complete the individual primitive tasks in the behavioral process, but it is very little help to complete the whole behavioral task chain. Without the guidance of further reward mechanisms, it is difficult for the robot to realize the value-adding of process-based behavioral ability. Therefore, this paper proposes a process-based compound reward mechanism, a convenient method for embedding prior knowledge. Since moving and stacking the fabric is a continuous process, the reward value obtained under the compound reward system is a cumulative process. The complex reward function is a piecewise function that varies in different state space regions under the completed and incomplete behavior elements. The content of the complex reward mechanism is defined as follows.

- m_z : the vertical distance between the fabric and the worktable.
- m_θ : The angle of rotation required to grab the fabric.
- $P^{start}_{x,y,z}$: The initial position of the fabric.
- $P^{target}_{x,y,z}$: Fabric target stacking position.
- $P^{m}_{x,y,z}$: The current spatial position of the fabric.

Based on the above, we can define the relevant behavior primitives completion criteria.

Reach the fabric pre-grabbing point.

$$reach = (\left|P_x^m - P_x^{start}\right| < \Delta_{x_{\max}}^{reach}) \wedge (\left|P_y^m - P_y^{start}\right| < \Delta_{y_{\max}}^{reach}) \\ \wedge (\left|P_z^m - P_z^{start}\right| < \Delta_{z_{\max}}^{reach})$$

where $\Delta_{(x_{\max},y_{\max},z_{\max})}^{reach}$ is the maximum error of the position that allows the robot to complete the arrive task.

Fabric grab.

$$grasp = (m_\theta \approx \theta) \wedge (m_z > Z)$$

where θ is the optimal grasping angle of the robot end and Z is the minimum height for the robot to complete the grasping task.

Fabric stack.

$$stack = (\left|P_x^m - P_x^{t\arg et}\right| < \Delta_{x_{\max}}^{stack}) \wedge (\left|P_y^m - P_y^{target}\right| < \Delta_{y_{\max}}^{stack}) \\ \wedge (\left|P_z^m - P_z^{t\arg et}\right| < \Delta_{z_{\max}}^{stack})$$

where $\Delta_{(x_{\max},y_{\max},z_{\max})}^{stack}$ is the maximum error of the position that allows the robot to complete the stacking task.

Define the distance-based reward and construct sparse reward function, as shown in Eq. 2.

$$r_{d_{start}} = \sqrt{(|P_x^m - P_x^{start}|^2 + \left|P_y^m - P_y^{start}\right|^2 + |P_z^m - P_z^{start}|^2)}/d_{start_{\max}}$$

$$r_{d_{target}} = \sqrt{(\left|P_x^m - P_x^{target}\right|^2 + \left|P_y^m - P_y^{target}\right|^2 + \left|P_z^m - P_z^{target}\right|^2)}/d_{target_{\max}}$$

$$r = \begin{cases} 1 & if\ stack\ or\ arrive \\ 0 - r_{start}/0 - r_{d_{target}} & otherwise \end{cases} \tag{2}$$

Then a compound reward function related to the process of arrival, grab, and stacking is designed by combining the definition of element complete, as shown in Eq. 3.

$$r = \begin{cases} 1 & if\ stack(P_{x,y,z}^{start}, P_{x,y,z}^{targett}, \Delta_{(x_{\max},y_{\max},z_{\max})}^{stack}) \\ 0.25 - r_{d_{target}}(P_{x,y,z}^m, P_{x,y,z}^{target}) & if\ \neg stack(P_{x,y,z}^{target}, \Delta_{(x_{\max},y_{\max},z_{\max})}^{stack}) \\ & \wedge grasp(\theta, Z) \\ 0.25 & if\ \neg stack(P_{x,y,z}^{target}, \Delta_{(x_{\max},y_{\max},z_{\max})}^{stack}) \\ & \wedge \neg grasp(\theta, Z) \\ & \wedge reach(P_{x,y,z}^{start}, \Delta_{(x_{\max},y_{\max},z_{\max})}^{reach}) \\ 0 - r_{d_{start}}(P_{x,y,z}^m, P_{x,y,z}^{start}) & otherwise \end{cases} \tag{3}$$

3.2 Network Structure

The essential thing in the robot limited task process behavior model is to solve an optimal combination strategy π^* in the model, which maximizes the long-term reward. For the action set of each skill of the robot, it is a continuous value, and the spatial dimension is huge, so the deterministic strategy is adopted to simplify the problem. Based on the Deep deterministic policy gradient (DDPG) algorithm [7], in the actor-critic framework, combined with the compound reward function to realize the learning of the robot fabric operation process.

The policy network mainly selects the current action a_t according to the current state s_t of the robot, and used in the fabric stack environment to interactively generate s_{t+1}, r_t. Ornstein-Uhlenbeck noise δ is added to the actions selected by the network in order to increase the ability of strategy exploration. $a_t = \mu(s_t|\theta^\mu) + \delta$. According to the gradient of objective function $J(\theta^\mu)$, combining with Behrman equation, the policy network adopts gradient ascent algorithm to continuously update network parameter θ^μ iteratively.

$$J(\theta^\mu) = -\frac{1}{N}\sum_{i=1}^{N} Q(s_i, a_i|\theta^\mu)$$

$$\nabla_{\theta^\mu} J(\theta^\mu) = \frac{1}{N}\sum_{i=1}^{N} \nabla_{\theta^\mu}\mu(s_i)\nabla_a Q(s_i, a_i)\,|_{a_i=\mu(s_i)}$$

The target-policy network sampled the next state s_{t+1} based on the experience buffer pool, select the best next action a_{t+1}, Adopt a soft update strategy, regularly use θ^μ to update network parameters $\theta^{\mu'}$.where inertial renewal rate $\tau \in (0, 1)$.

$$\theta^{\mu'} \leftarrow \tau\theta^\mu + (1-\tau)\theta^{\mu'}$$

The evaluation network is used to calculate the current Q value $Q(s_i, a_i|\theta^Q)$, update the network parameter θ^Q based on the state value function to minimize the timing difference error. The target-evaluation network is to calculate the $Q'(s_i', a_i'|\theta^{Q'})$ part of the target Q value y_i.

$$y_i = r_{i+1} + \gamma Q'(s_{i+1}, \mu_\theta'(s_{i+1}|\theta^{\mu'})\,\Big|\theta^{Q'})$$

The mean square error is used to define the loss function, and the calculate the gradient

$$L(\theta^Q) = \frac{1}{N}\sum_{i=1}^{N} [y_i - Q(s_i, a_i\,|\theta^Q\,)]^2$$

$$\nabla_{\theta^Q} L(\theta^Q) = \frac{1}{N}\sum_{i=1}^{N} (y_i - Q(s_i, a_i))\nabla_{\theta^Q} Q(s_i, a_i)$$

$$\theta_{t+1}^Q = \theta_t^Q + \alpha_Q \nabla_{\theta^Q} L(\theta^Q)$$

Also adopt a soft update strategy, regularly use θ^Q to update network parameters $\theta^{Q'}$

$$\theta^{Q'} \leftarrow \tau\theta^Q + (1-\tau)\theta^{Q'}$$

The overall process of the algorithm is shown in the Fig. 2.

4 Experiments

In this section, our purpose is to verify whether the CR-DDPG algorithm allows the robot to learn fabric manipulation skills and realize the task of fabric handling and stacking before sewing. Therefore, we first PROVED that the CR-DDPG framework has the learning ability of a single sub-skill on the experimental platform of robotic fabric stacking and handling. Then verify that the CR-DDPG algorithm can learn process tasks and compare it with the DDPG and Normalized Advantage Function (NAF) algorithms. Finally, the composite reward function is applied to NAF and compared with the HER-DDPG algorithm to prove that the composite reward function plays a better a priori guiding role in procedural tasks and is suitable for different continuous action algorithm frameworks.

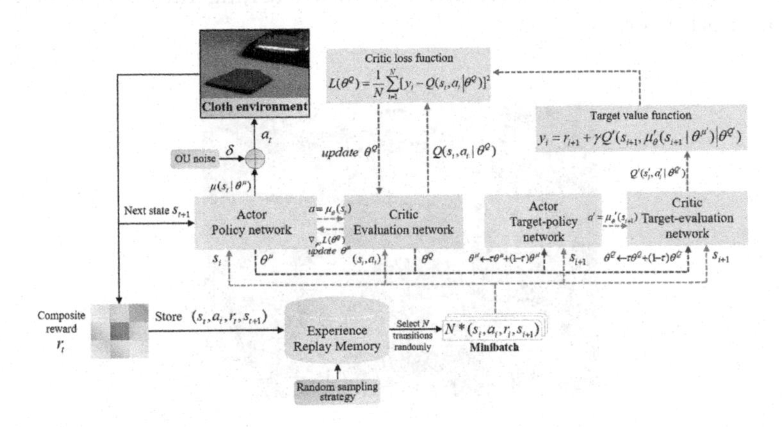

Fig. 2. The overall algorithm structure

The white sweatshirt pocket with cotton 60% polyester fiber 40% was selected as the manipulation object. The experimental platform is shown in Fig. 3, which includes the UR5e collaborative robot, pneumatic soft beak gripper, air pump, and algorithm development computer. The flexible beak gripper is composed of several hollow rubber gripper, open under positive air pressure, closed under negative air pressure, gripper control cabinet through a serial port connected to the manipulator control cabinet I/O module, through a digital signal control. The robot control cabinet is connected to the computer through a network cable. The algorithm environment is configured in Windwos10, I7-9700K, NVIDIA GeForce RTX2070. Using Python language, the algorithm network construction and training are completed under the framework of TensorFlow, and the motion control of the manipulator is realized through the URX function package.

We use robot pose to define the fabric environment. The maximum distance between the initial pose of the robot and the pose of the pre-grabbing point is 0.15 m. When the distance error is less than 0.005 m, the robot is considered to have arrived at the pre-grabbing point. The robot grasping point pose and the pre-grabbing point pose differ by a horizontal rotation angle of 1 rad. When the rotation error is less than 0.01 rad, it is considered that the robot completes the grasping task. The maximum distance between the robot's grasping point pose and the stacking pose is 0.1 m. When the distance error is less than 0.01, the robot is considered to complete the stacking task.

4.1 Single Task

When the robot completes every single task, the pose is shown in Fig. 4 and the terminal data as shown in Table 1. Set the total episodes is 1000, and the maximum step of attempts for each episode is 10. The replay buffer size is 4000, and the minibatch size is 128. The discount factor is 0.95. The initial value of $\varepsilon - greedy$ is 1, the attenuation value is 0.995, and the minimum is 0.01. Actor network learning rate is 0.0001, Critic network learning rate is 0.01 and adopt the Adam optimizer.

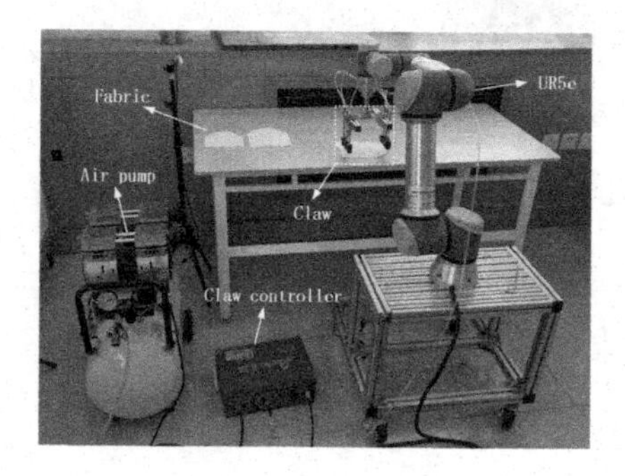

Fig. 3. The fabric manipulation experimental platform

Table 1. Robot single task pose

Task	Terminal position	Terminal attitude
Arrive	(−0.1154 m, 0.3916 m, 0.412 m)	(2.1233 rad, 2.2914 rad, −0.0507 rad)
Grab	(−0.1154 m, 0.3916 m, 0.412 m)	(1.1233 rad, 2.2914 rad, −0.0507 rad)
Stack	(−0.1154 m, 0.4916 m, 0.412 m)	(1.1233 rad, 2.2914 rad, −0.0507 rad)

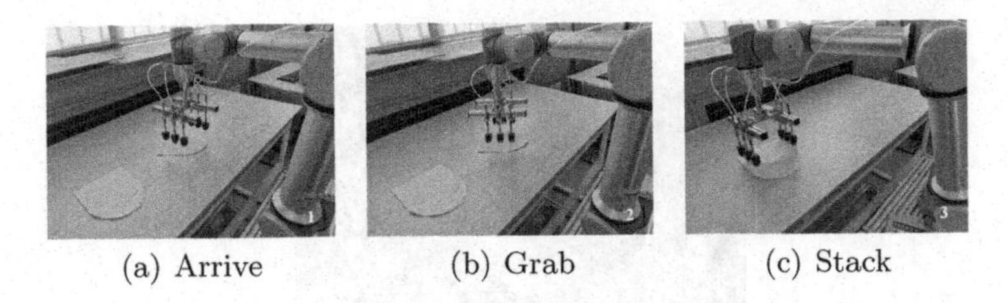

(a) Arrive (b) Grab (c) Stack

Fig. 4. The single task

The experimental results are shown in Fig. 5. Around 300 episodes, the robot learns all the single tasks required for fabric stacking, and it tends to be done in one step.

4.2 Process Task

Behavior element judgment is added based on setting the same network parameters as the single task skill learning. It should be noted that to meet the repeatability of the environment in reinforcement learning, we default that the robot can pick up the fabric after reaching the allowable error range. Using the complete composite reward function in Eq. 3, the robot is tested to learn the whole process tasks of arrival and stacking as shown in Fig. 6. At the same time, DDPG and NAF are used to learn the same task, and the network parameters are the same. Both algorithms use sparse reward functions as shown in Eq. 2. The experimental results are compared as shown in the Fig. 7. It can be seen that our algorithm learned the process task around 200 episodes. In comparison, the

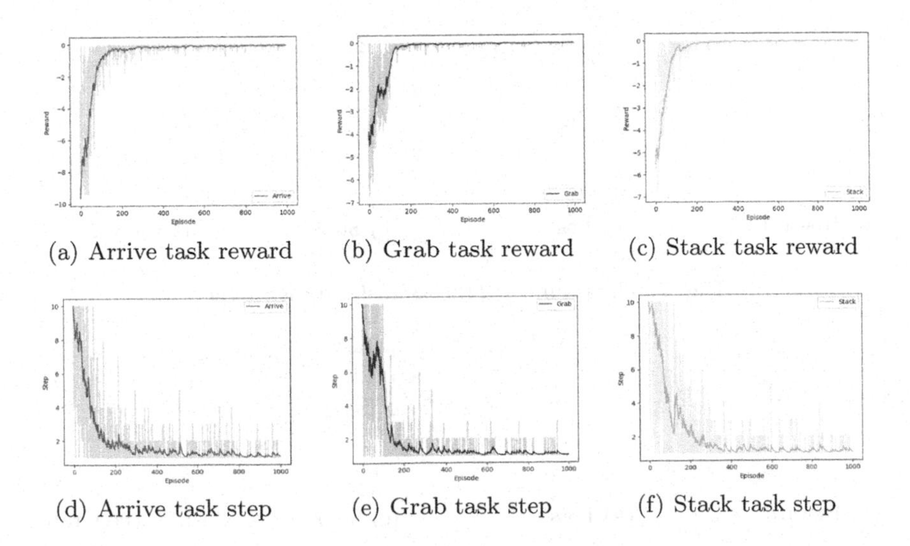

(a) Arrive task reward (b) Grab task reward (c) Stack task reward

(d) Arrive task step (e) Grab task step (f) Stack task step

Fig. 5. The results of single task skill learning

Fig. 6. The process task

DDPG algorithm learned the procedural task after 420 episodes, while the NAF algorithm kept oscillating without convergence, and the reward function eventually showed a downward trend. The success rate in the learning process is defined as $success = \frac{\sum_j \varphi_j}{\sum_i \omega_j}$. Where $\sum_j \varphi_j$ represents the number of successful assemblies and $\sum_i \omega_j$ represents the current training episode

4.3 Compound Reward Function

To prove that the compound reward function plays a better a priori guiding role in procedural tasks, it is applied to the NAF algorithm and compared with the HER-DDPG algorithm. The experimental results are shown in Fig. 8. It can be seen that under the guidance of the compound reward function, the CR-NAF algorithm can complete the procedural task better than the NAF algorithm, and the algorithm guided by her does not complete the procedural task.

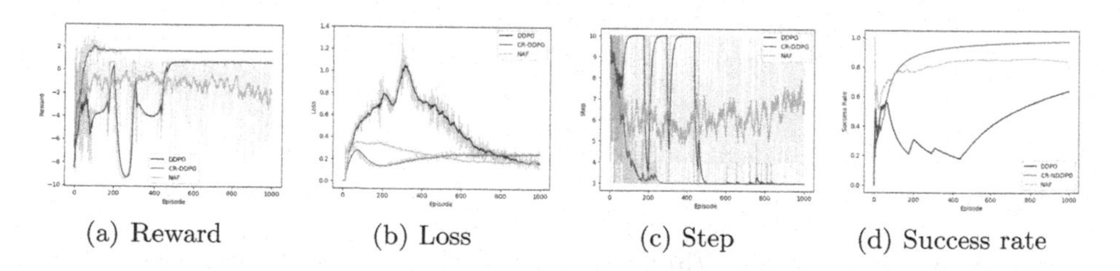

(a) Reward (b) Loss (c) Step (d) Success rate

Fig. 7. The results of process task skill learning

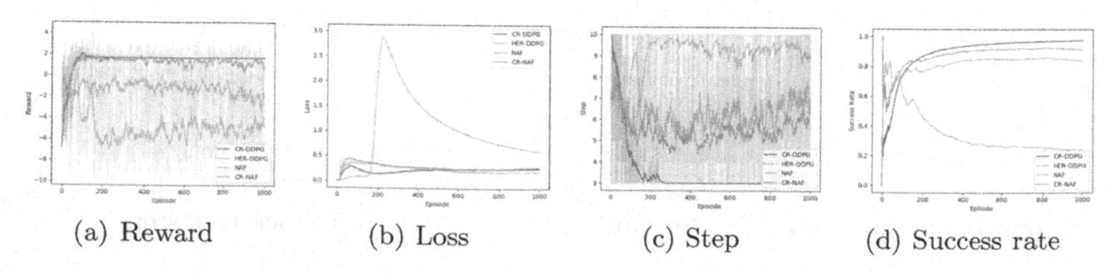

(a) Reward (b) Loss (c) Step (d) Success rate

Fig. 8. The learning result guided by compound reward function

5 Conclusions

This paper proposes a method for acquiring robot fabric manipulation skills based on deep reinforcement learning by designing a composite reward function, guiding the prior knowledge into the robot learning process, guiding the robot to complete procedural task learning. It can be used to improve industrial robots'application and flexible operation capabilities in the garment sewing industry. Finally, through experiments on the fabric stacking process before seam, the experimental results verify the algorithm's effectiveness.

In this work, the design of the flexible gripper prevents large deformation during the operation of the fabric, and the universality is still lacking. Therefore, one aspect of future work is exploring the acquisition of more complex deformation soft material operation skills in the algorithm. In addition, fabric handling is used as preparatory work for garment sewing. On this basis, we will study how to apply robots to the garment sewing process.

References

1. Nayak, R., Padhye, R.: Introduction to automation in garment manufacturing. In: Nayak, R., Padhye, R. (eds.) Automation in Garment Manufacturing. The Textile Institute Book Series, pp. 1–27. Woodhead Publishing, Sawston (2018). https://doi.org/10.1016/B978-0-08-101211-6.00001-X
2. Nayak, R., Padhye, R.: Automation in material handling. In: Nayak, R., Padhye, R. (eds.) Automation in Garment Manufacturing. The Textile Institute Book Series, pp. 165–177. Woodhead Publishing, Sawston (2018). https://doi.org/10.1016/B978-0-08-101211-6.00007-0
3. Delgado, A., Jara, C.A., Torres, F.: Adaptive tactile control for in-hand manipulation tasks of deformable objects. Int. J. Adv. Manuf. Technol. **91**(9), 4127–4140 (2017). https://doi.org/10.1007/s00170-017-0046-2
4. Gale, C.: The robot seamstress. Adv. Res. Text. Eng. **1**(1), 1–2 (2016)
5. Kruse, D., Radke, R.J., Wen, J.T.: Collaborative human-robot manipulation of highly deformable materials. In: 2015 IEEE International Conference on Robotics and Automation (ICRA), pp. 3782–3787 (2015). https://doi.org/10.1109/ICRA.2015.7139725
6. Li, F., Jiang, Q., Quan, W., Cai, S., Song, R., Li, Y.: Manipulation skill acquisition for robotic assembly based on multi-modal information description. IEEE Access **8**, 6282–6294 (2020). https://doi.org/10.1109/ACCESS.2019.2934174
7. Lillicrap, T.P., et al.: Continuous control with deep reinforcement learning (2019)
8. Liu, Y., Shamsi, S.M., Fang, L., Chen, C., Napp, N.: Deep Q-learning for dry stacking irregular objects. In: 2018 IEEE/RSJ International Conference on Intelligent Robots and Systems (IROS), pp. 1569–1576 (2018). https://doi.org/10.1109/IROS.2018.8593619
9. Matas, J., James, S., Davison, A.J.: Sim-to-real reinforcement learning for deformable object manipulation. In: Billard, A., Dragan, A., Peters, J., Morimoto, J. (eds.) Proceedings of The 2nd Conference on Robot Learning. Proceedings of Machine Learning Research PMLR, vol. 87, pp. 734–743, 29–31 October 2018
10. Moll, M., Kavraki, L.E.: Path planning for deformable linear objects. IEEE Trans. Rob. **22**(4), 625–636 (2006). https://doi.org/10.1109/TRO.2006.878933

11. Navarro-Alarcon, D., et al.: Automatic 3-d manipulation of soft objects by robotic arms with an adaptive deformation model. IEEE Trans. Rob. **32**(2), 429–441 (2016). https://doi.org/10.1109/TRO.2016.2533639
12. Parker, J.K., Dubey, R., Paul, F.W., Becker, R.J.: Robotic fabric handling for automating garment manufacturing. J. Eng. Ind. **105**(1), 21–26 (1983). https://doi.org/10.1115/1.3185859
13. Wada, T., Hirai, S., Kawamura, S., Kamiji, N.: Robust manipulation of deformable objects by a simple pid feedback. In: Proceedings 2001 ICRA. IEEE International Conference on Robotics and Automation (Cat. No.01CH37164), vol. 1, pp. 85–90 (2001). https://doi.org/10.1109/ROBOT.2001.932534
14. Zhang, J., Zhang, W., Song, R., Ma, L., Li, Y.: Grasp for stacking via deep reinforcement learning. In: 2020 IEEE International Conference on Robotics and Automation (ICRA), pp. 2543–2549 (2020). https://doi.org/10.1109/ICRA40945.2020.9197508
15. Zhang, T., Xiao, M., Zou, Y., Xiao, J.: Robotic constant-force grinding control with a press-and-release model and model-based reinforcement learning. Int. J. Adv. Manuf. Technol. **106**(1), 589–602 (2020). https://doi.org/10.1007/s00170-019-04614-0

Surface Crack Detection of Rubber Insulator Based on Machine Vision

JingWen Wang[1], Chen Li[2], Xu Zhang[1(✉)], and YuJie Jiang[1]

[1] School of Mechatronic Engineering and Automation, Shanghai University, Shanghai, China
xuzhang@shu.edu.cn

[2] School of Mechanical Science and Engineering, Huazhong University of Science and Technology, Wu Han, China

Abstract. The surface crack detection of rubber insulator is an essential part of its quality inspection. Aiming at the problem of low efficiency and complicated operation of manual inspection, an automatic surface crack detection algorithm of rubber insulator based on local threshold algorithm is presented in this paper. Firstly, the source image is filtered by a Dimension-increased Bilateral Filter to weaken the effects of noise and the inherent texture of the rubber surface. Then, the filtered image is segmented by Sauvola Local Threshold to separate the cracks from the background. Subsequently, an algorithm combined morphological processing with Seed Filling algorithm is applied to connect the discontinuous cracks. Finally, the real cracks are located by measuring the connected domain and using the distance threshold. The experimental results show that the proposed method can effectively remove the background interference and accurately locate the cracks, with an accuracy of 94.3%.

Keywords: Rubber insulator · Crack detection · Bilateral filter · Sauvola local threshold segmentation · Seed filling algorithm

1 Introduction

Rubber insulator is a charring ablative material and is usually used as inner insulation of solid rocket engine combustion chamber [1]. During the production process of rubber insulator, its surface quality is one of the most critical evaluation indexes of product quality. Among which, the surface cracks will significantly affect the thermal insulation performance of the material.

Traditional surface defect detection relies on manual visual inspection. This detection method has many shortcomings, such as high cost, strong subjectivity, low detection efficiency, and so on. With the significant advances in optics and image processing techniques, the defect detection method based on machine vision has become the mainstream of surface quality detection. And it has been widely used in practical production.

At present, there are few pieces of research on the surface crack detection algorithm of rubber insulator with complex texture interference. The studies are

X.-J. Liu et al. (Eds.): ICIRA 2021, LNAI 13014, pp. 175–185, 2021.
https://doi.org/10.1007/978-3-030-89098-8_17

mostly within the fields of metal, ceramic, leather, road, and so on. According to the characteristics of rubber ring surface defects, Wei Haixia [2]proposed a non-contact defect depth detection algorithm based on edge detection and coordinate transformation, which effectively conducted detection under noise interference. Zhang Xuewu [3]used Gaussian pyramid decomposition and Gabor filter to extract features of defect images and synthesized features saliency map to detect copper strip surface defects based on visual bionic mechanism to achieve the correct classification of defects. Li Junfeng [4]proposed a detection method of micro surface defects in magnetic tile based on machine vision, reaching the detection accuracy of 93.5%

Figure 1 is a detailed photograph of the rubber insulator. It can be noticed that the surface texture of the rubber insulator is complex and uneven in depth. Significantly, the micro-cracks are easily confused with the texture areas, which causes much interference to the defect detection. Aiming at these problems, the algorithm proposed in this paper can effectively extract crack information and improve the accuracy and efficiency of detection.

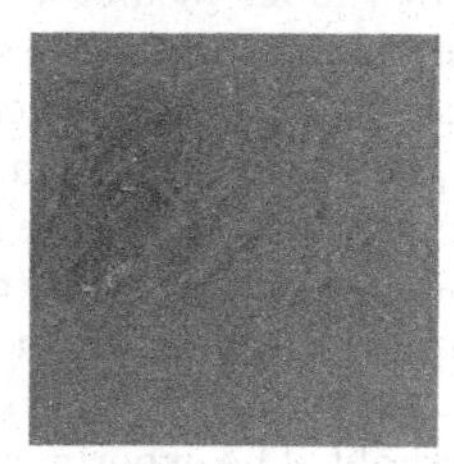

Fig. 1. Detail photograph of the rubber insulator

2 Surface Crack Detection System of Rubber Insulator

As shown in Fig. 2, the rubber insulator surface defect detection system consists of three parts: the central control part, the visual detection part, and the execution part. The central control part includes a human-computer interaction interface, motion control operation, PLC control, and data reading&writing communication; The visual detection part includes image acquisition and image processing algorithm. The execution part includes a defect marking machine and so on. The specific workflow of the system is as follows:

1) Image acquisition: when the conveyor belt starts to move, the encoder keeps sending signals to the camera, and the visual imaging platform starts to collect high-quality images of the rubber insulator and transmit them to the computer.
2) Image processing: the position information of the defect area in the image is obtained through the image processing algorithm in the computer, and then the information will be sent to the defect marking machine and recorded in the database.
3) Defect marking: after receiving the signal, the marking machine moves to the crack and marks according to the obtained crack position information. Afterward, the following image detection cycle starts.

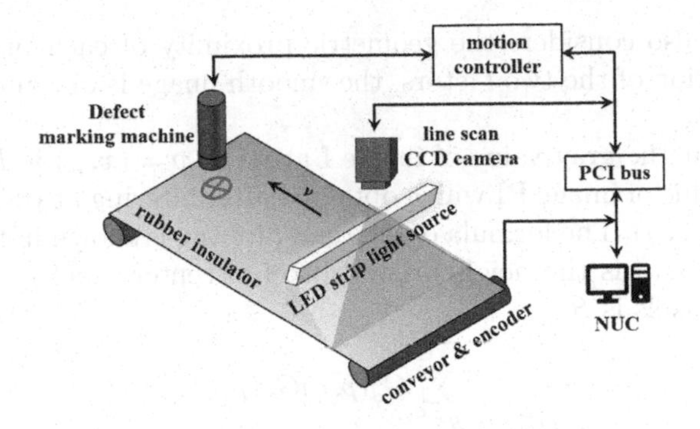

Fig. 2. System structure diagram

3 Surface Crack Detection Algorithm of Rubber Insulator

In order to accurately detect the cracks that exist in the image of the rubber insulator, an algorithm for the detection of cracks on the rubber insulator surface is designed based on machine vision technology in this paper. The algorithm process is shown in Fig. 3.

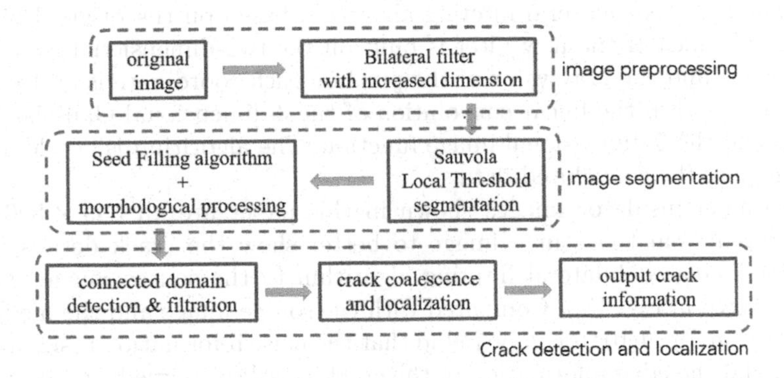

Fig. 3. Flow chart of image processing algorithm

3.1 Dimension-Increased Bilateral Filtering

Because the working environment of the imaging system is relatively complex, image noise is inevitable when imaging. Moreover, the rubber insulator itself has obvious texture, leading to a large amount of useless background information on the image. Therefore, it is necessary to preprocess the original image.

Bilateral Filter has the effect of removing noise and retaining edge information at the same time. It is a filtering function with two Gaussian bases. Therefore, it not only considers the similarity of the brightness of each pixel in the image but also considers the geometric proximity of each pixel. By a non-linear combination of the two factors, the smooth image is obtained by adaptive filtering.

Suppose that the gray value of image I at point $p = (x, y)$ is $I(p)$ or $I(x, y)$, and the gray value of image BI which obtained after filtering at point P is defined as $BI(p)$ or $BI(x, y)$. The formula of bilateral filtering is shown in the formula(1–3), where $q = (u, v)$ is the neighborhood pixel of center pixel p, and the set of neighborhood pixels is S.

$$BI(\mathrm{p}) = \frac{\sum_{q \in S} G_s(p, q) G_r(p, q) I(q)}{\sum_{q \in S} G_s(p, q) G_r(p, q)} \tag{1}$$

$$G_s(p, q) = e^{-\frac{(x-u)^2+(y-v)^2}{2{\sigma_s}^2}} \tag{2}$$

$$G_r(p, q) = e^{-\frac{[I(p)-I(q)]^2}{2{\sigma_r}^2}} \tag{3}$$

The original bilateral filtering algorithm cannot operate linear image convolution, resulting in a large computation and a slow computation speed when processing the image. Therefore, in 2006, Paris and Durand [5] proposed a dimension-increased bilateral filtering algorithm based on the original bilateral filtering. The idea of the new filter is built on the two-dimensional coordinates of the image, and the gray value of the pixel on each coordinate is added as the third space to form the linear convolution of the 3-dimensional Gaussian kernel function and the 3-dimensional image function. The algorithm is accelerated by downsampling the convoluted data.

The rubber insulator images shown in this paper are all 500 × 500-pixels images taken from the source image to better show the crack details. Apply dimension-increased bilateral filtering algorithm to the source image, and the result is shown in Fig. 4(c). Compared with the source image and the median filtered image in Fig. 4(b), it can be found that the noise information is significantly reduced, and the edge information of the crack is well preserved simultaneously.

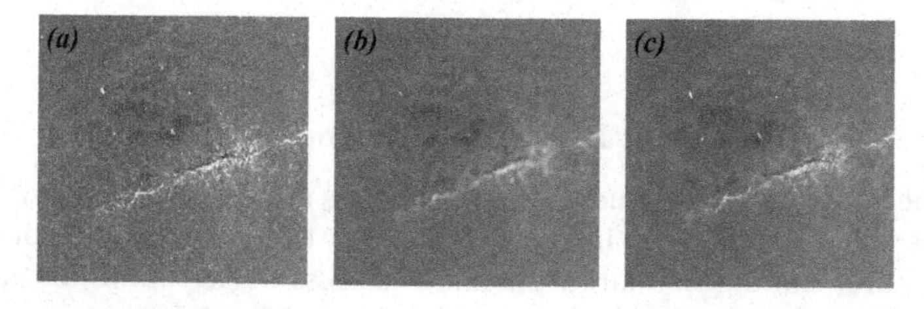

Fig. 4. Results of image filtering (a) source image (b) median filtering (c) dimension-increased bilateral filtering

3.2 Image Segmentation

Sauvola Local Threshold Segmentation. After the above processing, the rubber insulator image needs to be divided into background and foreground. The threshold segmentation method is based on the gray value difference between the object and the background. Therefore, choosing the optimal threshold is the key to the success of threshold segmentation. The global threshold segmentation methods process the whole image by a fixed threshold, so it is not suitable for uneven illumination or complex texture occasions.

Variable local threshold segmentation methods can calculate the local threshold according to the brightness distribution of different regions.Sauvola Local Threshold Segmentation algorithm is an efficient local threshold method. The algorithm's input is a grayscale image, which takes the current pixel as the center and dynamically calculates the threshold of the pixel according to the grayscale mean value and standard variance in the neighborhood of the current pixel. Assuming that the coordinate of the current pixel is (x, y), the neighborhood centered on the pixel is $r \times r$, $g(x, y)$ represents the gray value at (x, y), the steps of Sauvola algorithm are as follows:

Step1: Calculate the gray mean $m(x, y)$ and the standard variance $s(x, y)$ in $r \times r$ neighborhood of pixel (x, y).

$$m(x, y) = \frac{1}{r^2} \sum_{i=x-\frac{r}{2}}^{x+\frac{r}{2}} \sum_{j=y-\frac{r}{2}}^{y+\frac{r}{2}} g(i, j) \tag{4}$$

$$s(x, y) = \sqrt{\frac{1}{r^2} \sum_{i=x-\frac{r}{2}}^{x+\frac{r}{2}} \sum_{j=y-\frac{r}{2}}^{y+\frac{r}{2}} (g(i, j) - m(x, y))^2} \tag{5}$$

Step2: Calculate the threshold value T of pixel (x,y).

$$T(x, y) = m(x, y) \cdot [1 + k \cdot (\frac{s(x, y)}{R} - 1)] \tag{6}$$

where R is the dynamic range of standard variance. If the current input image is an 8-bit grayscale image, then $R = 128$. k is a self-defined correction parameter, the value of K has no significant influence on the algorithm result. Generally speaking, $0 < k < 1$ [6].

It can be seen in Fig. 5 that the potential crack areas are extracted, but the crack part contains a large number of cavities, loses part of the edges, with some noise and false defects simultaneously. Therefore, it is necessary to connect the discontinuous crack edges and filter the small noise through morphological processing.

Morphological Processing Combined with Seed Filling Algorithm. Morphological operations are a series of operations based on shapes, which are mainly used in the following aspects: image preprocessing; enhancing object structure;

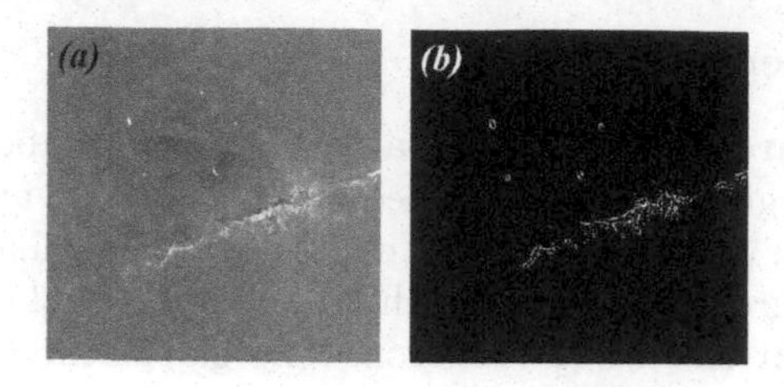

Fig. 5. Sauvola Local Threshold Segmentation (a) Preprocessed image (b) Sauvola local threshold segmentation

separating objects from the background; quantified description of objects. The basic operations of morphological operation include: erode, dilate, open operation, closed operation, morphological gradient, top cap, black cap, etc. [7].

Among the above operations, the closed operation (dilate +erode) can bridge the narrow discontinuous and eliminate the holes, but it requires a relatively high requirement of selecting the kernels. As shown in Fig. 6, when the kernel is selected as a circle with a radius of 6 and 8, the holes are not completely filled. When the radius of the circle increases to 12, all the holes are filled, but the area of cracks is also significantly enlarged, which is inconsistent with the actual size of cracks.

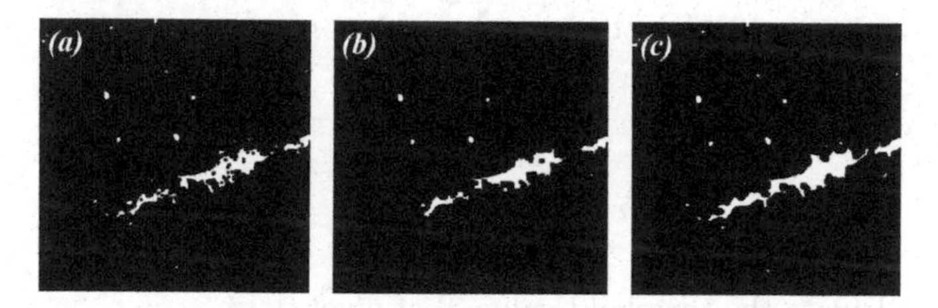

Fig. 6. The closed operations with different kernel sizes (a) Closed operation with kernel 6 (b) Closed operation with kernel 8 (c) Closed operations with kernel 12

In order to tackle the problem above, this paper proposes a method that combines morphological processing with Seed Filling algorithm. Morphological processing eliminates the holes by growing or coarsening the objects in the image, while Seed Filling algorithm eliminates the holes by filling all the pixels in the contour. The combined algorithm can not only connect the discontinuous edges but also guarantee the size of the contours. The steps are as follows:

Step1: connecting edges. As shown in Fig. 7(a), the dilate operation is applied to the segmented image, with the kernel of a circle(radius-6). As a result, the discontinuous edges are connected to form a closed contour but still remain some holes inside the contours.

Step2: filling holes. Seed Filling algorithm is used to fill the holes in closed contours. As shown in Fig. 7(c), all the holes in the connected domain are filled with a value of 255.

Seed Filling algorithm is a commonly used region filling algorithm. It selects one pixel inside the binary contour as the seed point. Then other points in the neighborhood are searched one by one to observe whether the pixel value is the same as the edge pixel. If different, fill in pixel value; if the same, it is considered as a boundary. Scan-Line Seed Filling algorithm fills the pixel segments along the horizontal scan line, and processes the neighboring points one by one, reaching a higher processing efficiency [8].

Step3: Thining. To restore the size of the connected domain, as shown in Fig. 7(d), erode operation is applied to the image with the same kernel in Step1.

Eventually, the discontinuous crack edges after segmentation are connected. The segmentation of the crack information from the background is completed preliminary.

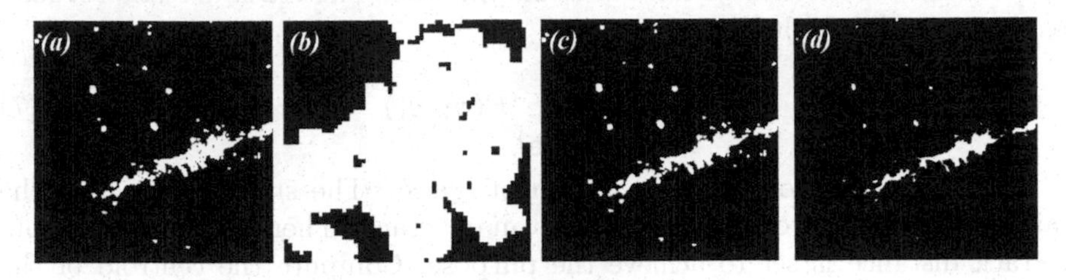

Fig. 7. Results of the proposed algorithm (a) Dilate with kernel 6 (b) Enlargement of details (c) Seed Filling algorithm (d) Erode with kernel 6

3.3 Crack Detection and Localization

After connecting discontinuous edges, some false defects still remain in the image. In this case, traditional denoising algorithms, such as median filtering, mean filtering, or Gaussian filtering, are not applicable [9]. Therefore, the algorithm based on connected domains is used to remove the false defects in this paper.

The specific idea is to use the algorithm of labeling connected domains to identify each connected domain in the image, calculate the area of each connected domain, and filter those connected domains with size under the given threshold to achieve the purpose of removing the false defects.

Connected Domain Detection. Since a connected domain is a set of pixels composed of adjacent pixels with the same pixel value. It can be determined whether the two pixels belong to the same connected domain by the two conditions: whether the two pixels have the same value; whether they are adjacent. Each connected domain found is given a unique identity to distinguish it from other connected domains. The steps of connected domain labeling algorithm steps are as follows:

Step1: Traverse the binary image from left to right and from top to bottom. If the pixel value is 255 (white) and is not marked, mark the point and scan its neighborhood with this point as the center point.

Step2: If a white and unlabeled point is found in the neighborhood, mark the point and scan its neighborhood.

Step3: Return to Step2, until there is no satisfying situation, then the search of this connected block ends.

Step4: Return to Step1, until the binary image traversal is completed and the search of all connected blocks is ended. The detection and labeling results of the connected domain are shown in Fig. 8(a).

Crack Determination and Localization. The determination of the authenticity and localization of cracks is as follows:

Step1: Filter false defects. Compute the size of all the connected domains obtained above: compute the number of pixels in each connected domain to be the size of it. Formula (7) is the statistical formula for the size. When its size is larger than the given threshold, the connected domain can be determined as a crack defect and marked with a red bounding box. The crack detection results are shown in Fig. 8(b)(c).

$$D = \sum_{(x_1, y_1) \in I} P(x_1, y_1) \tag{7}$$

Step2: Combination and localization of cracks. The small cracks near the main crack are regarded as part of the main crack. Therefore, the threshold of crack distance is set to achieve the purpose. Compute the centroid of the bounding boxes in step 1, the two cracks with a distance less than the threshold will be determined as the same crack. Then, combine the bounding boxes of these two connected domains, and push the new bounding box centroid into the original centroid array for the next iterative calculation until the centroid distance between all the bounding boxes is greater than the distance threshold.

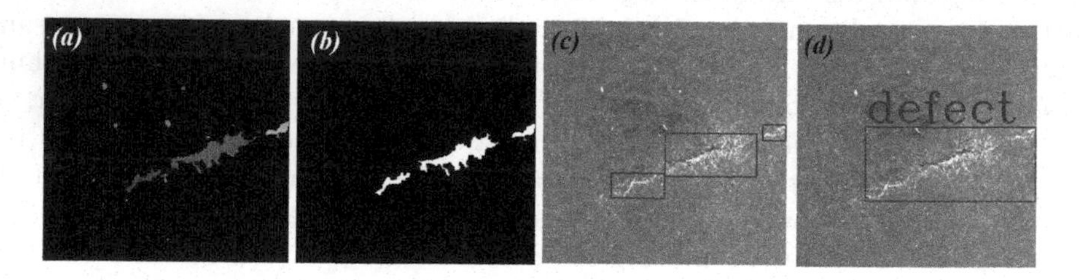

Fig. 8. Crack detection and localization (a) Connected domain detection and labeling (b) Crack determination (c) Bounding boxes (d) Crack localization

Eventually, the connected domains that remained in the image are the final cracks, and the centroids of them are output as the location of the final cracks. The results of crack determination and localization are shown in Fig. 8.

4 Experiment and Analysis

Figure 9 is a photograph of the rubber insulator surface defect detection equipment. The system uses a Line-Scan CCD camera (Dalsa-GigE) to collect the surface image of the rubber insulator. When the conveyor belt starts to move, the encoder (OMRON) installed on the conveyor's shaft starts to send signals (1000 signals/revolution)to the camera. The camera in hard-triggered mode takes one image for each signal it receives from the encoder and composes the continuous images automatically. Then the NUC host (I7-10710U, 64-bit Windows10 operating system) is used to process the images received from the camera. If a crack is detected, through the output information of NUC, the marking machine moves to the crack and executes an action.

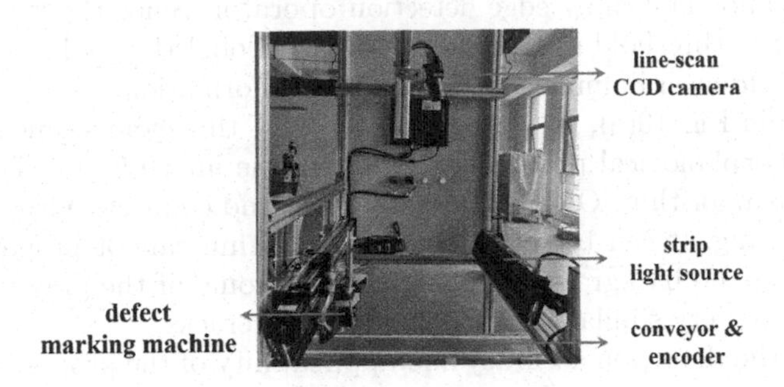

Fig. 9. Photograph of the equipment

In order to verify the effectiveness of the proposed algorithm, three existing algorithms are compared in this paper: Sobel algorithm [10], Canny algorithm based on iterative threshold segmentation [11], OUST segmentation combined with morphological processing algorithm [12]. The detection results is shown in Fig. 10.

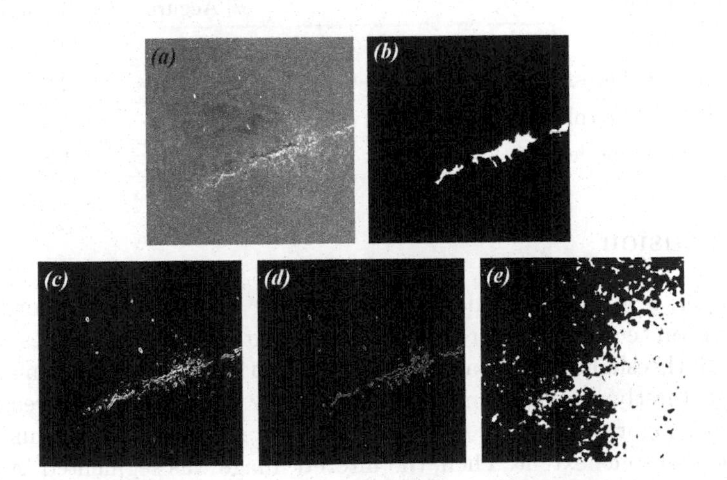

Fig. 10. Comparison of detection effects by different algorithms (a) The original image (b) Algorithm proposed in this paper (c) Sobel (d) Canny+Iterative Threshold segmentation (e) OTSU+ morphological processing

As shown in Fig. 10(c), the Sobel algorithm firstly preprocesses the original image with median filtering and then detects the crack contour with the Sobel edge detection operator. It is found that the Sobel operator needs to find the gradient graph in X and Y directions respectively, so the algorithm has a high cost of time. And the algorithm is sensitive to noise.

As shown in Fig. 10(d), the Canny algorithm based on iterative threshold segmentation firstly enhances the input image, then uses the threshold iteration method and the Canny edge detection operator. Since the threshold value obtained by the threshold iteration method is a global threshold, the algorithm is sensitive to uneven illumination and texture information.

As shown in Fig. 10(e), the algorithm of OUST threshold segmentation combined with morphological processing, segments the image by maximizing interclass variance algorithm (OUST), removes noise and connects edges by morphological processing. It can be seen that this algorithm cannot properly separate the crack from the background. This is because some of the gray values in the background are very similar to the gray values of cracks.

To verify the detection accuracy and applicability of the proposed algorithm, a sample size of 100 images (images from discontinuous rubber insulator parts to obtain more defects) is tested by the algorithms. As shown in Table 1, the average accuracy of this algorithm is about 94.3%, and the detection time for a single 4096 × 1000-pixels image is about 1.8 s. Experimental results show that the cracks on the rubber insulator surface can be detected through the proposed algorithm effectively.

Table 1. Accuracy and time consumption of different algorithms

Algorithm	Accuracy/%	Time/s
sobel [10]	79.6	2.3
Iterative Threshold Segmentation+Canny [11]	88.4	2.8
OUST + morphological processing [12]	56.7	1.4
Proposed method	94.3	1.8

5 Conclusion

In this paper, a surface crack detection algorithm for rubber insulator based on machine vision technology is proposed. The algorithm mainly includes image filtering, local threshold segmentation, crack connection, crack detection and localization, etc. Firstly, the source image was filtered by the dimension-increased bilateral filtering algorithm, which smoothed the noise and the background texture information to some extent. Then, the filtered image was segmented by Sauvola Local Threshold segmentation to extract cracks from the background. Afterward, the discontinuous cracks were connected by the method of dilating + Seed Filling algorithm + erode. After that, the connected domain in the processed image was detected, and the false cracks are filtered by size threshold. Finally, to locate the cracks, micro-cracks are merged with the distance threshold.

The experimental results show that the method presented in this paper can remove the interference of uneven illumination and the surface texture of rubber. The cracks can be extracted with an accuracy of 94.3%, which provides an effective method for the detection of rubber insulator surface cracks. Of course, this algorithm also has some shortcomings that need to be improved. Because the surface texture of the rubber insulator is extremely uneven in-depth, so the deeper non-defect texture area may be detected by mistake, and the micro shallow cracks may also be missed. These problems will be studied in future work.

Acknowledgement. This research is partially supported by the key research project of the Ministry of Science and Technology (Grant No. 2018YFB1306802), the National Natural Science Foundation of China (Grant No. 51975344) and China Postdoctoral Science Foundation (Grant No. 2019M662591).

References

1. Carlos, J., Amado, Q., Ross, P.G., et al.: Evaluation of elastomeric heat shielding materials as insulators for solid propellant rocket motors: a short review. Open Chem. **18**(1), 1452–1467 (2020)
2. Wei, H., Jiang, D.: Research on the depth inspection method of non-contact rubber surface defects. Chin. Sci. Technol. J. Database (Full-text Ed.) Eng. Technol. 00308–00310. (in Chinese)
3. Xuewu, Z., Yanqiong, D., et al.: Surface defects inspection of copper strips based on vision bionics. J. Image Graph. **16**(04), 593–599 (2011). (in Chinese)
4. Li, J., Hu, H., Shen, J.: Research on micro defect detection method of small magnetic tile surface based on machine vision. Mech. Electr. Eng. **184**(002), 117–123 (2019). (in Chinese)
5. Paris, S., Durand, F.A.: Fast approximation of the bilateral filter using a signal processing approach. Int. J. Comput. Vis. **81**(1), 24–52 (2009)
6. Lazzara, G., Geraud, T.: Efficient multiscale sauvola's binarization. Int. J. Doc. Anal. Recogn. **17**(2), 105–123 (2014)
7. Sonka, M., Hlavac, V., Boyle, R.: Image Processing, Analysis, and Machine Vision. Posts and Telecom Press, Thomson Brooks/Cole, Beijing (2002)
8. Lasheng, Yu., Deyao, S.: A refinement of scanning line seed filling algorithm. Comput. Eng. **10**, 70–72 (2003). (in Chinese)
9. He, L.F., Chao, Y.Y., Suzuki, K., et al.: Fast connected-component labeling. Pattern Recogn. J. Pattern Recogn. Soc. **42**(9), 1977–1987 (2009)
10. Chunjian, H., Xuemei, X., Ying, C.: Feature extraction of workpiece circular arc contour based on sobel operator. J. Laser Optoelectron. Prog. **55**(002), 233–240 (2018). (in Chinese)
11. Xiao, M., Li, J., Peng, Y.: A detection algorithm for the inner wall crack of ceramic bottles. Chin. Ceram. Ind. **27**(04), 43–48 (2020). (in Chinese)
12. Jiawei, Z., Xing, F., He, C., et al.: Detection of crack in drying wood based on image processing. Heilongjiang Sci. Technol. Inf. (016), 14–16 (2019). (in Chinese)

A Hierarchical Force Guided Robot Assembly Method Using Contact State Model

Ziyuo Wang[1], Tianyu Fu[2,3], Xuting Yang[2,3], and Yanghong Wang[2,3](✉)

[1] Guilin University of Technology, Guilin 541006, China
[2] Shandong University, Jinan 250100, China
wangyanhong@sdu.edu.cn
[3] Engineering Research Center of Ministry of Education for Intelligent Unmanned Systems, Jinan 250100, China

Abstract. In the industrial robotic assembly process, it is difficult to establish a precise physical model. Moreover the assembly process is generally described by contact state. A robot phased guided assembly method based on the contact state model is proposed. An assembly contact state model with the force, torque, position, and posture in each stage of the assembly is built, combined with the mapping relationship between each stage in the assembly process and the robot's execution actions. The algorithm verification is carried out on the built KUKA robot assembly platform. The results showed that the recognition accuracy rate of each assembly stage through the established assembly process model reached 100%, and the assembly success rate under the phased guidance strategy reached 99%.

Keywords: Robot assembly · Assembly modeling · Contact state · Stage guidance.

1 Introduction

In the manufacturing of products in all walks of life, assembly is a typical process widely used. It is the last process of the entire product molding, accounting for about 50% of the entire manufacturing process, such as the assembly of air-conditioning compressor crankshafts and hot sleeve rotors in the household appliance industry, battery valve body assembly [1], large reducer assembly [14] in automobile manufacturing, etc. The assembly of turbine shaft holes of engines in the aviation industry, the assembly of tony parts of 3C electronic products, and the assembly of a sleeve in the printing industry [8] and so on.

Supported by Guangdong Key Research and Development Program(No.2020B090925001), Shandong Major Science and Technology Innovation Project(No.2019JZZY010429), Shandong Provincial Key Research and Development Program(Grand No.2019TSLH0302).

X.-J. Liu et al. (Eds.): ICIRA 2021, LNAI 13014, pp. 186–197, 2021.
https://doi.org/10.1007/978-3-030-89098-8_18

In recent years, more and more manufacturing companies have introduced robots for assembly and manufacturing operations to reduce labor costs in the assembly process and improve assembly efficiency. The assembly of robots is mainly based on various sensors such as vision sensors and force sensors. The collected information is preprocessed and combined with algorithms to complete the assembly. Luo in [10] matched and recognized the image collected by the robot vision with the known template, used inverse kinematics calculation to obtain the coordinate offset and other information of the workpiece, and transmitted it to the robot, and finally adjusted the position and posture of the end effector to achieve the purpose of obtaining the workpiece accurately. Lu in [9] aimed at improving the product quality of the precision assembly of the production line and achieving the goal of industrial upgrading, developed a machine vision positioning guidance system, which improved the detection speed and positioning accuracy. Wang in [13] proposed an intelligent recognition and assembly guidance method for an important application in the field of industrial machinery-part recognition, which realized the offline training of aerospace connector images, online recognition, and intelligent guidance of assembly information, reducing the time-consuming of the assembly of precision products in small space is time-consuming. Wu in [15] introduced correction pose algorithms and error elimination strategies to make the positioning more accurate and provide a guarantee for the assembly work under visual guidance. Wu in [16] established a physical model based on kinematics equations to provide support for the robot's target positioning. In the application of assembly robots, a single-purpose vision positioning algorithm is used to complete the positioning. SONG in [11] used a multi-module calibration method to transform the pose relationship between the robot, the conveyor belt, and the two cameras in the robot assembly system. For workpieces with incomplete motion information, a minimum matrix method is proposed for position detection.

However, the above vision-based positioning process will be affected by light stability, visual occlusion, light source, and viewing angle, resulting in inaccurate positioning and affecting the final assembly success rate.

Therefore, some scholars introduced force information to complete the assembly task. Yin in [18] used a six-dimensional force sensor to sense the force signal in the hole search process to adjust the posture of the shaft to eliminate the posture deviation between the accessories quickly, and through high-precision assembly experiments, it was verified, and the accuracy reached 50 microns. Wu in [17] proposed an automatic assembly positioning method guided by the resolution force and applied it in the virtual assembly prototype system. Li in [7] combined with virtual assembly technology, adjusted the position of the parts to be assembled with the help of space control equipment and completed the precise positioning of the parts under the action of the simulated assembly force. Literature [3] proposes a force-guided assembly positioning method, which completes the assembly by calculating the assembly force in each state. Zhang in [19] carried out gravity compensation on the robot's end posture, which eliminated the dynamic influence of gravity in different postures of the robot and combined with admittance control to achieve high precision requirements. Due to the complexity of the assembly process, it is difficult to establish an accurate physical model.

Therefore, a method based on the contact model was proposed. The method of statistical learning is used to identify the contact state. The recognition of the state is defined as a classification problem. The commonly used methods are Fuzzy Classifier (FC), Neural Network (NN), Support Vector Machine (SVM), Gaussian Mixed Model (GMM), and Hidden Markov Model (HMM). JASIM in [5] proposed a Gaussian Mixture Model based on the measure of distributed similarity, which is used to model the contact state of the robot assembly process of force-directed flexible rubber parts. JAKOVLJEVIC in [4] proposed a method for generating Fuzzy Inference Mechanism (FIM) based on flexible active motion. Support Vector Machine method is used to identify the contact state of robot parts in the process of matching. JASIM in [6] judges the corresponding contact state based on the force perception information. After the contact state recognizes the result, the control instruction is derived, and the robot executes the assembly action. Sun in [12] proposed an improved impedance control method for multi-joint glass installation robots. The simulation of the PUMA560 robot proved that the fuzzy adaptive controller has a good control effect on the end contact force of the robot. Qiu in [2] added the decoupling control algorithm of the end coordinate system to the position control module and verified through simulation that the force-position hybrid control method can effectively improve the response speed and steady-state accuracy of the control.

The above methods are independent research on the contact state's recognition rate and the improvement of the control strategy. There are few studies on the relationship between the recognition of the contact state and the control strategy and how to combine these two stages. This paper combines contact state recognition and control decision-making and proposes a robot step-by-step guided assembly method based on process modeling. Based on the assembly process analysis, the assembly stage is described and identified, and the relationship between the assembly stage and the robot action is established. The mapping relationship, and finally complete the flexible assembly.

2 The Description of Robot Assembly

Taking the assembly process shown in Fig. 1 as an example, suppose there are n consecutive stages in the entire assembly process, respectively are $P = \{s_1, s_2, s_3, ..., s_n\}$. Each stage has i states, and each stage is described by 12 characteristic attributes:

$$s_m = \{f_x, f_y, f_z, \tau_x, \tau_y, \tau_z, X, Y, Z, \phi_x, \phi_y, \phi_z\},\ m\ is\ positive\ integer \quad (1)$$

Among them, f_x, f_y, f_z and τ_x, τ_y, τ_z represent Cartesian force and moment, X, Y, Z and ϕ_x, ϕ_y, ϕ_z represent the end pose of the manipulator, and the state information is obtained by the force sensor at the end of the robot and the end pose.

It is known that s_1 is the shaft in the free state. The free refers to the process of the shaft moving from the initial pose state to the contact with the target-hole,

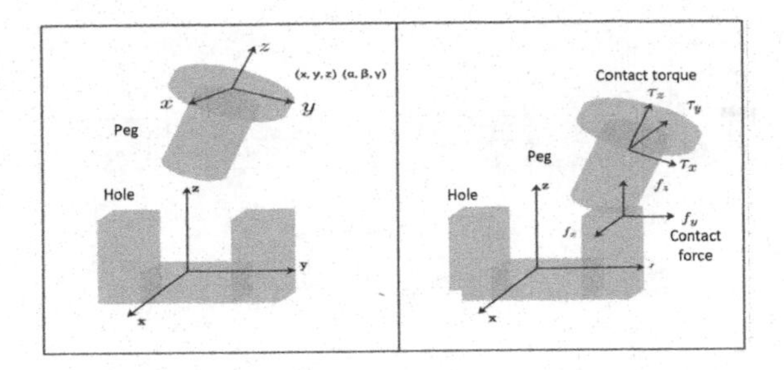

Fig. 1. Schematic diagram of assembly force and position description

and s_n represents the characteristic properties of the shaft in the final assembled state.

Assuming the action of the A robotic arm, it is expressed as:

$$A = \{a_1, a_2, a_3, ..., a_m\}, \ m \ is \ positive \ integer \tag{2}$$

Input state s , traversal state P, exists:

$$\begin{cases} y = 1, if \ s \in P \\ y = 0, otherwise \end{cases} \tag{3}$$

If $y = 1$, there is a non-linear mapping of f:

$$f : s \rightarrow A \tag{4}$$

How to identify the current assembly stage of the robot and find the mapping relationship of the action of the manipulator is the problem to be solved in this article.

3 Hierarchical Force Guided Assembly Strategy

Figure 2 is a block diagram of the robot's phased guided assembly. First, the robot is started for the data collection stage. The data collected each time is a complete process, and the beginning of each process is in the first stage-the space roaming stage. Input the collected data into the neural network for training and learning, export the network model with the highest success rate after multiple pieces of training, use the trained network to predict the test set and output the classification results at the same time, and the classification accuracy rate, and perform multiple experiments and take the average value.

3.1 Assembly Process Modeling

The SVR maps the contact state data into the robot cation space via a nonlinear mapping. Assume the $f(x)$ takes the linear form:

$$f(x) = w^T \cdot \phi(x) + b \tag{5}$$

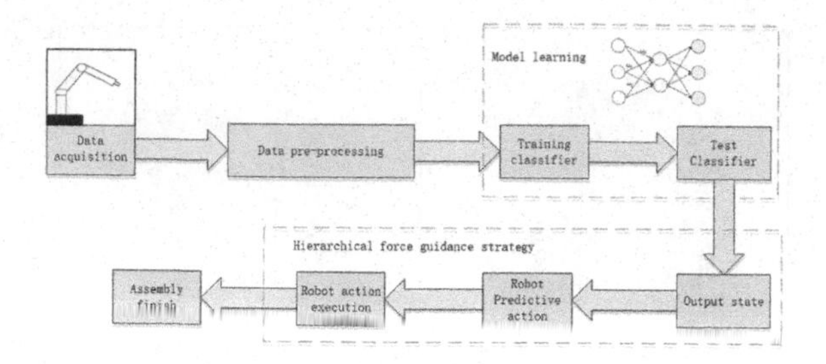

Fig. 2. Phased force guided robot assembly frame diagram

Where $\phi(x)$ is called the assembly contact state features; b is the bias term and w is the weight vector in the primal weight space. The $\varepsilon - insensitive$ loss function is defined as:

$$|f(x_i) - y_i|_\varepsilon = \begin{cases} |f(x_i) - y_i| - \varepsilon \; for |f(x_i) - y_i| \geq \varepsilon \\ 0 \qquad\qquad\qquad\qquad otherwise \end{cases} \tag{6}$$

which means that the loss function exists that approximates all pairs (x_i, y_i) with ε precision, and ε is a prescribed parameter controlling the tolerance to error. Two positive slack variables ξ_i and ξ_i^* are introduced for representing the distance from actual values to corresponding boundary values of the ε-tube and coped with the following minimization problem:

$$\min_{\omega, \xi_i, \xi_i^*} \frac{1}{2} \omega^T \omega + C \sum_{i=1}^{l} (\xi_i + \xi_i^*) \tag{7}$$

subject to

$$\begin{aligned} & y_i - (\omega^T \cdot \phi(x_i) + b) \leq \varepsilon + \xi_i \\ & (\omega^T \cdot \phi(x_i) + b) - y_i \leq \varepsilon + \xi_i^* \\ & \xi_i \cdot \xi_i^* \geq 0, \; i = 1, 2, \cdots l \end{aligned} \tag{8}$$

parameter C controls the trade-off between the complexities of the model and frequency of error. C is normally optimized using cross validation. This constrained optimization problem is solved by the following primal Lagrange function:

$$\begin{aligned} & L(\omega, b, \xi_i, \xi_i^*, \alpha_i, \alpha_i^*, \eta_i, \eta_i^*) \\ & = \frac{1}{2} \omega^T \omega + C \sum_{i=1}^{l} (\xi_i + \xi_i^*) \\ & - \sum_{i=1}^{l} \alpha_i (\varepsilon + \xi_i - y_i + \omega^T \cdot \phi x_i + b) \\ & - \sum_{i=1}^{l} \alpha_i^* (\varepsilon + \xi_i^* + y_i - \omega^T \cdot \phi x_i - +b) \\ & - \sum_{i=1}^{l} \eta_i \xi_i + \eta_i^* \xi_i * i \end{aligned} \tag{9}$$

The dual variables $\eta_i, \eta_i^*, \alpha_i, \alpha_i^*$ are Lagrange multipliers satisfying positivity constraints, i.e. $\alpha_i \cdot \alpha_i{}^* \geq 0$, $\eta_i \cdot \eta_i^* \geq 0$. The dual form is given by maxmize:

$$\begin{aligned} &\sum_{i=1}^{l} \frac{1}{2}(\alpha_i - \alpha_i^*)(\alpha_j - \alpha_j^*)k(x_i, x_j) \\ &+ \varepsilon \sum_{i=1}^{l}(\alpha_i + \alpha_i^*) - \sum_{i=1}^{l}(\alpha_j - \alpha_j^*) \end{aligned} \tag{10}$$

subject to

$$\begin{aligned} &\sum_{i=1}^{l}(\alpha_i - \alpha_i^*) = 0 \\ &\alpha_i \cdot \alpha_i^* \geq 0 \end{aligned} \tag{11}$$

here $k(x_i, x_j) = \phi(x_i) \cdot \phi(x_j)$ is the kernel function, after calculating these by solving Eq. (10) subject to constraints given by Eq. (11), final form of Nonlinear SVR function is:

$$f(x) = \sum_{i=1}^{l}(\alpha_i - \alpha_i^*)k(x_i, x_j) + b \tag{12}$$

for variable b, it can be computed using Karush-Kuhn Tucker (KKT) condition.

3.2 Assembly Process

According to the manual process of shaft hole assembly, the entire assembly process is divided into five stages-the space roaming stage, the non-vertical hole finding stage, the hole finding stage, the posture adjustment stage, and the hole entering completion.

As shown in Fig. 3, during the entire assembly process, the first stage is a free state, and there is no contact between the accessories in this stage. The second stage is the hole finding stage in a non-vertical state when the axis falls skewed and finally falls on the periphery of the hole without success. The state when entering the hole, when the force sensor detects that the two objects are in contact. The purpose of setting the second stage is to provide different contact states and generalize the assembly process to complete the assembly under different conditions. The third stage is the hole finding stage. When the hole entry fails in the second stage, the robot will perform the hole finding operation again. At this time, the shaft overcomes the friction with the workpiece hole and moves to the target-the hole. In the process of left or right translation, the influence of friction is ignored, and the force information f_x, f_y, f_z and τ_x, τ_y, τ_z charge. The fourth stage is the posture adjustment stage. Since the shaft in the non-vertical state cannot successfully enter the hole, the shaft needs to be twisted slightly to adjust the posture so that it is perpendicular to the hole and no longer receives a force in the horizontal direction. The final stage is the stage of entering the hole to the completion state. At this stage, the shaft falls in the vertical state until the six-dimensional force sensor feedbacks force information, indicating that the shaft is in contact with the accessory and the assembly is complete. During the whole assembly process, the ideal state is that the force information in the vertical direction is zero and the force in the horizontal direction is zero. At this time, the accessory shaft can enter and exit the hole smoothly.

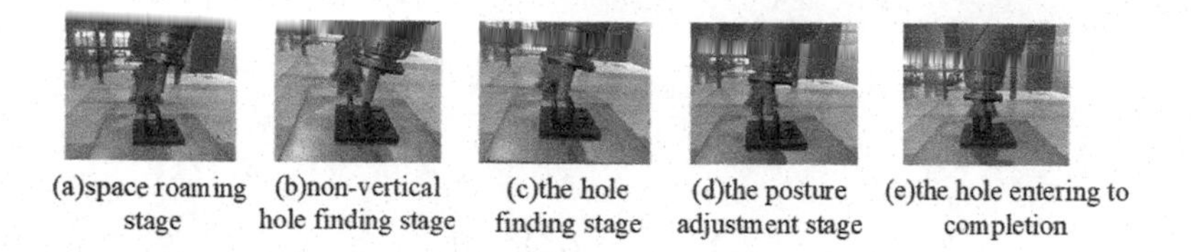

(a)space roaming stage
(b)non-vertical hole finding stage
(c)the hole finding stage
(d)the posture adjustment stage
(e)the hole entering to completion

Fig. 3. Schematic diagram of assembly stage

3.3 Phased Guidance Stage

In the whole assembly, the assembly process is complete according to five stages: space roaming stage, non-vertical contact hole stage, hole finding stage, posture adjustment stage, and vertical drop stage. In the assembly process, the initial position is often random, which can be roughly divided into four situations: pose offset, position offset, attitude offset and vertical, corresponding to the four stages of assembly. The contact state of the assembly is identified through the SVM network. According to the corresponding relationship between the assembly contact state and the action of the robot arm, the corresponding stage of the assembly corresponding to the current initial position can be judged, and then the guidance strategy of the subsequent stages can be executed in sequence, which can be stable and smooth to complete the assembly action to avoid damage to the work piece to be assembled.

The state containing force and pose information is input to the trained SVM neural network, and the network outputs stage labels. According to the mapping relationship between stage labels and robot actions, phased guided actions are realized, which are completed by the position control of the robot arm. Points a to f in Fig. 4 represent the initial position of the center of gravity of the five stages in the simulation assembly process. The rotation angle of the square simulates the deflection angle of the accessory hole. Table 1 shows the corresponding guidance strategies for different stages. When the network determines which stage the current stage belongs to, the execution action corresponds to the starting point of the next stage.

4 Experiments

4.1 Experimental Platform

In order to verify the effectiveness of the algorithm, a physical experiment platform consisting of the robot body, assembly table, server, and assembly model

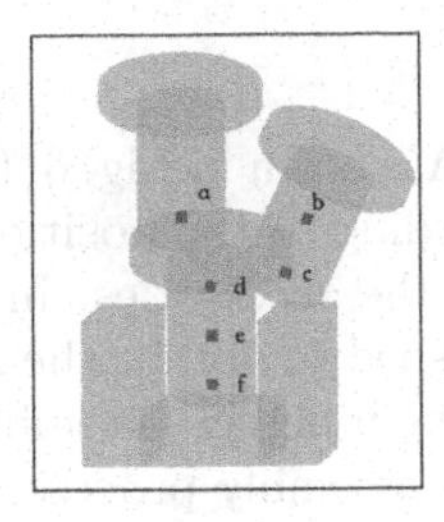

Fig. 4. Schematic diagram of assembly pose during phased guidance

Table 1. Tiered guidance strategy table

Assembly phase	Guidance strategy	Corresponding stage
Space roaming stage	$(X_{a\to b}, Y_{a\to b}, Z_{a\to b})$ $(\phi_{xa\to xb}, \phi_{ya\to yb}, \phi_{za\to zb})$	$a \to b$ move to the target work piece and adjust its pose
Non-vertical hole finding stage	$(X_{b\to c}, Y_{b\to c}, Z_{b\to c})$ $(\phi_{xb\to xc}, \phi_{yb\to yc}, \phi_{zb\to zc})$	$b \to c$ failed to enter the hole in the skewed state, contact between the two work piece
Hole finding stage	$(X_{b\to c}, Y_{b\to c}, Z_{b\to c})$ $(\phi_{xc\to xd}, \phi_{yc\to yd}, \phi_{zc\to zd})$	$c \to d$ work piece translation to find holes
Posture adjustment stage	$(X_{d\to e}, Y_{d\to e}, Z_{d\to e})$ $(\phi_{xd\to xe}, \phi_{yd\to ye}, \phi_{zd\to ze})$	$d \to e$ adjust its posture so that the horizontal direction is not force
Vertical drop stage	$(X_{e\to f}, Y_{e\to f}, Z_{e\to f})$ $(\phi_{xe\to xf}, \phi_{ye\to yf}, \phi_{ze\to zf})$	$e \to f$ vertical drop to complete flexible assembly

is built, as shown in Fig. 5. The shaft of the assembly model is composed of a steel shaft with a Yang's modulus $E_{stl} = 2.1 \times 10^{11}$ Pa and a plastic sleeve with a Yang's modulus $E_{stl} = 2.8 \times 10^{6}$ Pa. The diameter of the shaft is 20.0 mm, and the width of the hole is 20.1 mm. In this article, a communication process is established between the assembly robot and the server through TCP/IP. The robot is used as the Client side. Data transmission and command issuance are carried out in the form of Socket.

Fig. 5. Assembly platform diagram

4.2 Results

Assembly Stage Analysis. As shown in Fig. 6, the phased guidance is divided into corresponding stages according to the position relative to the environment and the contact state between the accessories. In the training model, the data of different stages are distinguished to obtain the ideal classification results and then feedback. There will also be a higher assembly completion rate in the execution of the robot. During the assembly process, the force information did not change significantly in the first two stages and changed at the end of the second stage and the beginning of the third stage. The reason is that shaft, and the hole is in contact, and the force sensor will feedback the force and torque. The contact force f_x, f_y, f_z changes when the accessory is in contact. In the third stage of the hole-finding process, the change of pose is due to the adjustment of the angle of the axis and the change of the relative position with the environment.

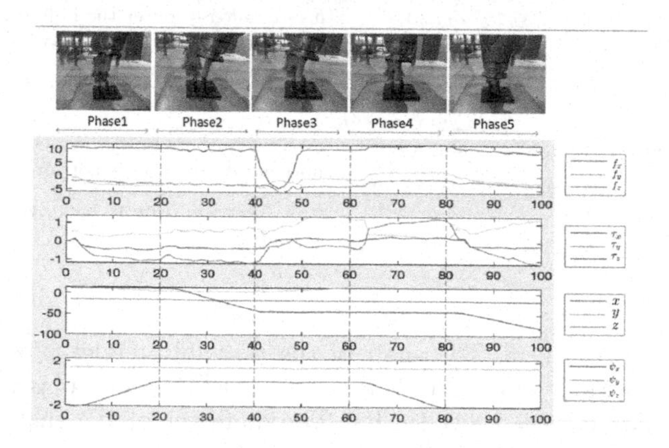

Fig. 6. Trend diagram of assembly force position transformation

Analysis of Experimental Results. In the experiment, 50 complete processes were collected, a total of 5000 samples, each stage of the sample is 1000, of which the training set and the test set are randomly divided according to the ratio of 8:2, 80% of data is used of training, and the remaining 20% of data is used of testing. The input layer to the hidden layer of the SVM neural network model is the Sigmoid function. The phase recognition rate is 100%. Figure 7 shows the learning curve and ROC curve. The initial position in each stage is randomly selected within the range set before the experiment, and 100 experiments are performed in each stage. It can be seen in Table 2 that the assembly success rate of stage 1 and stage 2 are both 100%. In the first stage, the Y vector in the position and the Z vector of the posture is changed, and the posture is shifted. The central axis gradually comes into contact with the hole during the vertical drop, and the force information changes significantly at the end of this stage. The vector of posture Z is randomly selected in ± 0.04 rad, so the distinction in the stage is obvious, and it is not easy to be confused. In the third stage, the movable distance range of the shaft is set to 10 *mm*. At this time, the shaft is the same as the posture after the first stage descends. The judgment result is confused with the first stage, and the success rate is 97%. In the fourth stage, the posture of the axis needs to be adjusted to be perpendicular to the hole. The range of the radian value is ± 0.04 *rad*. The posture adjustment in the stage is obvious, easy to judge, and the success rate is 100%. The final stage of action is to drop vertically to complete the assembly. The position change at this moment is 40 *mm*. The judgment is accurate, and the success rate is 100%.

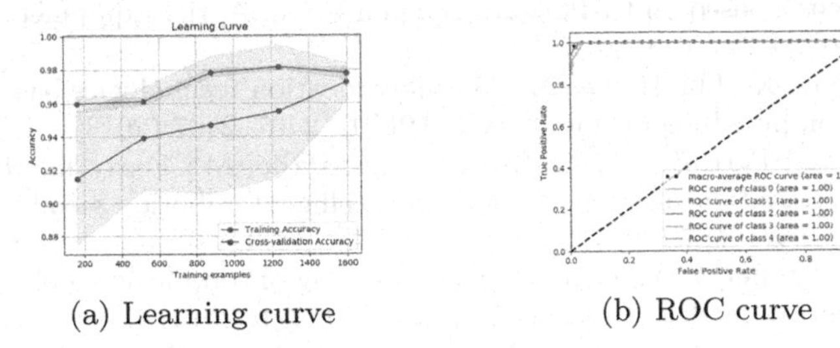

(a) Learning curve (b) ROC curve

Fig. 7. The model results

Table 2. The assembly success rate of different initial postures

Phase	Test times	Number of successful	Test success rate	Average assembly success rate
Phase 1	100 times	100 times	100%	99.2%
Phase 2	100 times	100 times	100%	
Phase 3	100 times	97 times	97%	
Phase 4	100 times	100 times	100%	
Phase 5	100 times	99 times	99%	

5 Conclusion

Aiming at the difficulty of establishing an accurate physical model of the robot assembly process, which leads to poor robot assembly flexibility, this paper proposes a hierarchical force guidance method based on the contact state model. Traditional assembly problems need to be taught first and then executed. This method makes it difficult to meet the assembly accuracy requirements under different assembly conditions, and the execution cycle is long. In this paper, the robot end contact force, torque, and pose describe the assembly state. Based on the assembly process, five stages state samples are collected for learning and recognition. After the contact state is judged, the assembly state is combined with the mapping relationship between the assembly state and the robot action to complete the assembly and improve. The assembly success rate and accuracy are improved.

References

1. Chen, H., Xu, J., Zhang, B., Fuhlbrigge, T.: Improved parameter optimization method for complex assembly process in robotic manufacturing. Ind. Rob. Int. J. (2017)
2. Chou, P., Fang, P.: Research on mixed force and position control system of a 3R manipulator based on feedforward compensation. Agric. Equipment Veh. Eng. **56**(323), 39–42+47 (2018)
3. Gao, W., Shao, X., Liu, H., Ge, X.: Assembly location technologies based on force-guidance. Comput. Integr. Manuf. Syst. **19**(10), 2407–2407 (2013)
4. Jakovljevic, Z., Petrovic, P.B., Mikovic, V.D., Pajic, M.: Fuzzy inference mechanism for recognition of contact states in intelligent robotic assembly. J. Intell. Manuf. **25**(3), 571–587 (2014)
5. Jasim, I.F., Plapper, P.W.: Contact-state modeling of robotic assembly tasks using gaussian mixture models. Procedia Cirp **23**, 229–234 (2014)
6. Jasim, I.F., Plapper, P.W., Voos, H.: Contact-state modelling in force-controlled robotic peg-in-hole assembly processes of flexible objects using optimised gaussian mixtures. Proc. Inst. Mech. Eng. Part B J. Eng. Manuf. **231**(8), 1448–1463 (2017)
7. Li, K.: Assembly-force guided virtual assembly methodology. Ph.D. thesis, Xidian University (2006)
8. Li, X., Cao, S., Xiang, H., Jin, Z., Wang, M.: Modeling of the robot of sleeve installation for printing machine. In: Zhao, P., Ouyang, Y., Xu, M., Yang, L., Ren, Y. (eds.) Applied Sciences in Graphic Communication and Packaging. LNEE, vol. 477, pp. 553–561. Springer, Singapore (2018). https://doi.org/10.1007/978-981-10-7629-9_68
9. Lu, J., Sun, S., Song, Y.: Visual positioning guidance system for automatic precision assembly. Modular Mach. Tool Autom. Manuf. Tech. **2**, 111–114 (2020)
10. Luo, K.: Application of vision positioning assembly for industrial robots. Woodworking Mach. **000**(002), 20–22 (2019)
11. Song, R., Li, F., Fu, T., Zhao, J.: A robotic automatic assembly system based on vision. Appl. Sci. **10**(3), 1157 (2020)
12. Sun, X., Yang, S., Wang, X., Xia, Y., Liu, F.: Study on fuzzy adaptive impedance control of multi joint robot. Process Autom. Instrum. **39**(01), 47–50 (2018)
13. Wang, J., Wang, L., Fang, X., Yin, X.: Vision based intelligent recognition and assembly guidance of aerospace electrical connectors. Comput. Integr. Manuf. Syst. **11**, 2423–2430 (2017)
14. Wu, B., Qu, D., Xu, F.: Improving efficiency with orthogonal exploration for online robotic assembly parameter optimization. In: 2015 IEEE International Conference on Robotics and Biomimetics (ROBIO), pp. 958–963. IEEE (2015)
15. Wu, G., Zhang, W., Yang, B.: Research on accurate positioning method in robot visual assembly system. Mach. Electron. **038**(003), 75–80 (2020)
16. Wu, H.: The research of assembly robot based on visual shaft hole identification and positioning technology. Ph.D. thesis, Henan Polytechnic University
17. Wu, W., Shao, X., Liu, H.: Automatic assembly location technologies based on phased force-guidance. Comput. Integr. Manuf. Syst. **24**(10), 106–119 (2018)

18. Yin, L.: Research on robot flexible position control system based on six dimensional force
19. Zhang, X., Han, J., Li, X., Xiu, K.: Design of force control system for peg-in-hole assembly robot. Mech. Des. Manuf. **70**(012), 63–66 (2019)

A BAS-Based Strategy for Redundant Manipulator Tracking Continuous Moving Object in 3D Relative Velocity Coordinate

Mengde Luo[1,2], Dongyang Bie[1,2], Yanding Qin[1,2], and Jianda Han[1,2](✉)

[1] College of Artificial Intelligence, Nankai University, Tianjin, China
hanjianda@nankai.edu.cn
[2] Tianjin Key Laboratory of Intelligent Robotics, Nankai University, Tianjin, China

Abstract. The ability of real-time tracking of manipulator end-effect is critical for intelligent manipulators. This paper presents a control strategy for tracking the continuous moving object. Our research aims at converting the problem of tracking moving object to a planning problem of relative velocity vector by establishing Relative Velocity Coordinate (RVC) on manipulator end-effect. The key novelty of the proposed strategy about constructing the optimization problem lies in considering both relative velocity and relative position simultaneously. The Beetle Antennae Search (BAS) algorithm is introduced to tackle the formulated optimization function. Theory analysis and different simulation experiments are provided to illustrate the feasibility and efficiency of the proposed control strategy.

Keywords: Manipulator end-effect · Track moving object · Optimization problem · Relative velocity coordinate · Beetle antennae search

1 Introduction

In recent years, intelligent robot have been widely used in various industry and daily life. With the industrialization development and enrichment of information [16], intelligent robots play significant role in intelligent industrial production [6], intelligent transportation system [12], medical health and intelligent robot services [16].

Among those intelligent robots, the industrial manipulators are most widely used in industry product lines for repeating tasks [1]. One common task for manipulator is tracking predefined trajectory in repeat works. Tracking control for manipulator refers to move manipulator from one point to another

This work was supported in part by the National Key Research and Development Program of China (2018AAA0103003) and National Natural Science Foundation of China (NSFC 52005270).

X.-J. Liu et al. (Eds.): ICIRA 2021, LNAI 13014, pp. 198–210, 2021.
https://doi.org/10.1007/978-3-030-89098-8_19

point following predefined trajectory in cartesian space. And the tracking control have increased much attention for industry manipulator. The tracking control of manipulator can be addressed using dynamic or kinematic. Kinematic tracking control for manipulators has long been the basic and typical problem in robotic literature [7,9]. Kinematic models are usually provided to users for designing more higher level operating systems for specific operation [10]. Controllers, using dynamic models, can get excellent performance but highly rely on designers' experience [17] and accuracy of dynamic model [14].

However, the increasing autonomous and complex mission [5] require manipulators to plan trajectory online instead of tracking predefined trajectory [4]. Typically, the mission of tracking moving object, is a critical requirement for manipulator in dynamic unstructured environment. There are numbers of literatures about trajectory planning online for manipulator. Virtual controller-based method [2] established virtual position error between manipulator end-effect and the moving object. But this method can't track object moving fast. Fixed proportion method [11] with fixed gain was direct and simple but it cannot cover both closer and farther away object. Law of conservation of energy method [17] used two kinds of energy to move manipulator instead of establishing constraints or models, but this method was only used in redundant planner manipulator.

In this paper, a novel strategy is proposed to tackle the problem of tracking moving object in 3D cartesian space. Firstly, establish the Relative Velocity Coordinate (RVC) on manipulator end-effect. Then relative velocity and relative position between manipulator end-effect and object were both considered to construct the total optimization function based RVC. Finally, the Beetle Antennae Search (BAS) algorithm [15] was used to addressed the optimization problem. BAS algorithm is computationally efficient [8] and the convergence of BAS algorithm has been proven [18]. The main contributions of this paper are presented as follows:

- The model of RVC is used on manipulator end-effect to formulate the tracking problem.
- A new optimization function is build for tracking problem by using both the relative velocity and relative position.
- A heuristic method based on BAS algorithm is built to tackle the optimization function. The joint space trajectorys of manipulator are always smooth because of two mapping function for BAS algorithm.

The remainder of this article is organized as follows. Section 2 present the problem of manipulator tracking moving object and constructs a novel optimization function based RVC. Section 3 introduces the BAS algorithm to tackle the formulated optimization function. Section 4 provide the simulation methodology and results. Section 5 concludes this article.

2 Problem Formulation

2.1 Tracking Moving Object

Considering the problem of tracking moving objects, manipulators can only get the real-time position of the target object by sensor information. And no 3D cartesian tarjectory is available to manipulators. Considering this tracking problem directly, the distance of manipulator end-effect and the object should be as small as possible. This tracking control can be formulated to the following optimization problem:

$$\min : D(\boldsymbol{P}_E, \boldsymbol{P}_{Ok}) = \|\boldsymbol{P}_E - \boldsymbol{P}_{Ok}\|_2 \tag{1}$$

where $\|\cdot\|_2$ is 2-norm, $\boldsymbol{P}_E$ is the end-effect positon of manipulator and $\boldsymbol{P}_{Ok}$ is the observed object positon at k time step.

For a given manipulator, the end-effect pose is determined by its joint angle and mechanical structure. For example, Franka Panda [13] manipulator has 7 DOF, the manipulator forward kinematic [3] can be calculated as follows:

$$\boldsymbol{x} = f(\boldsymbol{\theta}) \tag{2}$$

where $\boldsymbol{x} \in \mathbb{R}^6$ and $\boldsymbol{\theta} \in \mathbb{R}^7$ represent cartesian space and joint space, respectively. The end-effect positon is $\boldsymbol{P}_E = \boldsymbol{x}(1:3)$.

2.2 3D Relative Velocity Coordinate

For the complex nonlinearity of manipulator forward kinematic, it is difficult to solve optimization problem Eq. 1 directly. In this paper, the RVC method is used to formulated the problem of tracking moving object. Figure 1 shows the relation of RVC and manipulator end-effect.

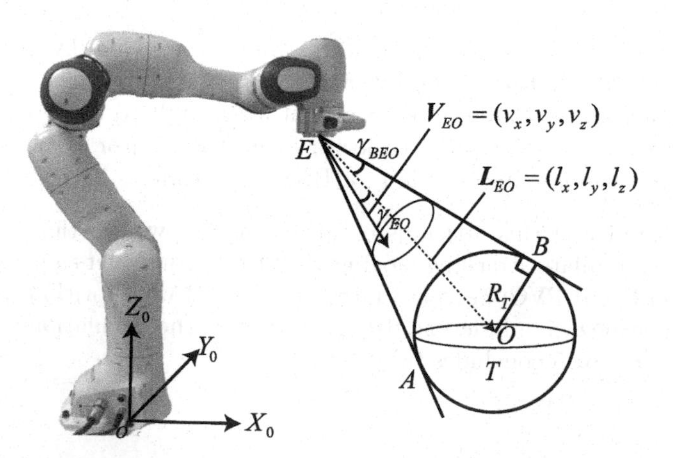

Fig. 1. Relative Velocity Coordinate on manipulator end-effect.

The moving object is wrapped by a ball. The ball denotes as T, and ball center indicates O. The ball radius denotes R_T. Connecting the position of end-effect (refers E) with the position of moving object (refers O), we can get the relative position vector:

$$\boldsymbol{L}_{EO} = \boldsymbol{P}_O - \boldsymbol{P}_E = (l_x, l_y, l_z) \tag{3}$$

where $\boldsymbol{P}_E$ and $\boldsymbol{P}_O$ are expressed in manipulator base coordinate.

Similarly, the relative velocity vector is obtained as shown in Eq. 4:

$$\boldsymbol{V}_{EO} = \boldsymbol{V}_E - \boldsymbol{V}_O = (v_x, v_y, v_z) \tag{4}$$

where $\boldsymbol{V}_E$ and $\boldsymbol{V}_O$ are cartesian translational velocity of manipulator end-effect and object, respectively. $\boldsymbol{V}_O$ can be obtained by numerical differentiation, that is $\boldsymbol{V}_O = (\boldsymbol{P}_{O,k} - \boldsymbol{P}_{O,k-1})/\Delta t$. Δt refers to the control period. The relationship between manipulator cartesian velocity and joint velocity is as follows:

$$\dot{\boldsymbol{x}} = \boldsymbol{J} \cdot \dot{\boldsymbol{\theta}} \tag{5}$$

where $\boldsymbol{J}$ is Jacobian matrix which is determined by manipulator configuration [3] and $\dot{\boldsymbol{\theta}}$ is manipulator joint velocity. The end-effect translational velocity is $\boldsymbol{V}_E = \dot{\boldsymbol{x}}(1:3)$. $\boldsymbol{V}_E$ and $\boldsymbol{V}_O$ are also expressed in manipulator base coordinate.

The object can be wrapped in a minima cone by the rays emitted from end-effect. Take a section of the spherical cone through ball center O, the intersection point on sphere are A, B. The angle γ_{BEO} is defined as:

$$\gamma_{BEO} = \arcsin(R_T/\|\boldsymbol{L}_{EO}\|_2) \tag{6}$$

We difine tracking angle γ_{EO} as follows:

$$\gamma_{EO} = \arccos(h) \quad h = \cos(\gamma_{EO}) = \frac{\boldsymbol{V}_{EO} \cdot \boldsymbol{L}_{EO}}{\|\boldsymbol{V}_{EO}\|_2 \cdot \|\boldsymbol{L}_{EO}\|_2} \tag{7}$$

The tracking angle γ_{EO} is included angle between relative velocity vector $\boldsymbol{V}_{EO}$ and relative position vector $\boldsymbol{L}_{EO}$.

It is obvious that if the direction of relative velocity vector $\boldsymbol{V}_{EO}$ is inside the cone, the manipulator end-effect will be closer to the object. That is to say, if relative velocity vectory $\boldsymbol{V}_{EO}$ satisfies the Eq. 8, the manipulator moves in a correct direction.

$$\gamma_{EO} \leq \gamma_{BEO} \tag{8}$$

2.3 The Principle for Tracking Moving Object

Now, the problem of tracking moving object is transformed to optimizing the relative velocity vector $\boldsymbol{V}_{EO}$. With the object moving randomly, we can only control the $\boldsymbol{V}_E$ to reach the positon of target. As shown in Fig. 2, the relative velocity vector $\boldsymbol{V}_{EO}$ is decomposed into two velocity component $\boldsymbol{V}_A$ and $\boldsymbol{V}_C$.

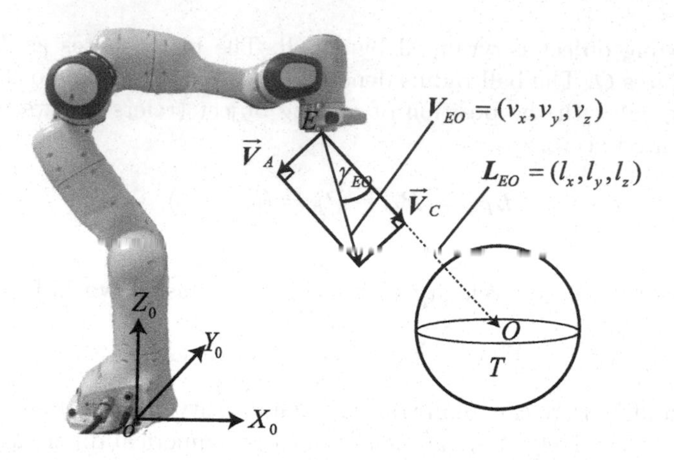

Fig. 2. The principle of tracking moving object.

The velocity component $\boldsymbol{V}_A$ lets manipulator end-effect far away from the moving object. And the velocity component $\boldsymbol{V}_C$ brings the end-effect closing to object. So we can get two principles for tracking mission as shown in Eq. 9 and Eq. 10. Increasing V_C and decreasing V_A can both lead manipulator end-effect closing to the moving target.

$$\max: V_C = \|\boldsymbol{V}_C\|_2 = \|\boldsymbol{V}_{EO}\|_2 \cdot \cos\gamma_{EO} \tag{9}$$

$$\min: V_A = \|\boldsymbol{V}_A\|_2 = \|\boldsymbol{V}_{EO}\|_2 \cdot \sin\gamma_{EO} \tag{10}$$

Besides the moving direction, we also consider the relative distance between the manipulator end-effect and object. This is also one main contribution of this article. For rapid and stable tracking, the relative distance should also be an optimization item. As shown in Eq. 11, we construct the total optimization function. Then, the tracking mission is translated to the optimization of minimum J. Through keeping the moving direction and decreasing relative distance simultaneously, the optimization equation Eq. 11 leads the manipulator end-effect closing to the moving object rapidly and stably.

$$\min: J = |l_x| + |l_y| + |l_z| + V_A - V_C \tag{11}$$

3 Control System Design

After the problem of tracking moving object was formulated in Sect. 2.3, this tracking problem ultimately was converted to total optimization equation Eq. 11. To tackle this optimization function, we take a heuristic algorithm, called BAS algorithm, and the control process is shown as Algorithm 1:

At time k step, the franka manipulator joint space state were $\boldsymbol{\theta}_k$. The positon and translational velocity of end-effect were $\boldsymbol{P}_E$ and $\boldsymbol{V}_E$. Meanwhile, we can get the positon and translational velocity of moving object as $\boldsymbol{P}_O$ and $\boldsymbol{V}_O$.

The BAS algorithm generated a random normalization vector, inspired by beetle antennae explored the environment, which described as Eq. 12.

$$\boldsymbol{b} = \frac{\text{rand(m, 1)}}{\|\text{rand(m, 1)}\|_2} \tag{12}$$

Consider the manipulator arm to a bettle, the manipulator ("virtual bettle") joint space states can be calculated as Eq. 13:

Algorithm 1: BAS Algorithm-Manipulator Tracking Moving Object

---

Input: Initial joint state $\boldsymbol{\theta}_0$, manipulator kinematic $f(\cdot)$, jacobian matrix $\boldsymbol{J}$,
Output: A continuous joint position trajectory
while ($k < k_{max}$) or (stop criterion) **do**
  Read joint positon vector $\boldsymbol{\theta}_k$, get $\boldsymbol{P}_E$, $\boldsymbol{V}_E$, $\boldsymbol{P}_O$, $\boldsymbol{V}_O$
  while (Until current control period is over) **do**
    Generate normalized random vector $\boldsymbol{b}$ according to (12)
    Get $\boldsymbol{\theta_{kL}}$ and $\boldsymbol{\theta_{kR}}$ as stated in (13)
    Calculate g_{kL} and g_{kR} in keeping with (14–17)
    Get the updated joint positon vector $\hat{\boldsymbol{\theta}}_{k+1}$ according to (18)
    Calculate the re-evaluate optimization value g_{k+1} through (19)
    Save the smallest g_{k+1} and the corresponding $\hat{\boldsymbol{\theta}}_{k+1}$
  end
  $\boldsymbol{\theta}_{k+1} \leftarrow \hat{\boldsymbol{\theta}}_{k+1}$
  Control manipulator move to $\boldsymbol{\theta}_{k+1}$
  $k \leftarrow k+1$
end

$$\begin{aligned} \boldsymbol{\theta}_{kL} &= \boldsymbol{\theta}_k + \lambda \boldsymbol{b}, \\ \boldsymbol{\theta}_{kR} &= \boldsymbol{\theta}_k - \lambda \boldsymbol{b}. \end{aligned} \tag{13}$$

where $\boldsymbol{\theta}_{kL}$ and $\boldsymbol{\theta}_{kR}$ are end position of left and right antennate, and λ represents the length of antennae.

Subtracting manipulator joint states from the left and right antennate end positon, the changes of manipulator joint space states can be calculated as Eq. 14.

$$\boldsymbol{e}_X = \boldsymbol{\theta}_{kX} - \boldsymbol{\theta}_k \tag{14}$$

To make sure the length of antennate meets the joint velocity limit, the vector $\boldsymbol{e}_X$ needs to be modified through Eq. 15.

$$P_1(\boldsymbol{\theta}_{kX}, \boldsymbol{\theta}_k) = \boldsymbol{\theta}_k + \max\{-\varDelta, \min\{\boldsymbol{e}_X, \varDelta\}\} \tag{15}$$

where $\varDelta$ denotes the maximum change value of $\boldsymbol{\theta}_k$ in one control period. That is to say, $\varDelta = \boldsymbol{v}_{max} \cdot \varDelta t$. $\boldsymbol{v}_{max}$ is the maximum of joint velocity and $\varDelta t$ is actual manipulator control period.

However, the results of P_1 still might not satisfy the constraints of manipulator joint position limit. Therefore, the results of P_1 need to be restricted by directly projection function as follows:

$$P_{\Omega}(\cdot) = \max\{\boldsymbol{\theta}^-, \min\{P_1(\cdot)\}, \boldsymbol{\theta}^+\} \tag{16}$$

where $\boldsymbol{\theta}^-$ and $\boldsymbol{\theta}^+$ means the lower and upper position limits on manipulator joint space.

After projection function Eq. 16, $\boldsymbol{\theta}_{kL}$ and $\boldsymbol{\theta}_{kR}$ meet the manipulator joint positon limit and joint velocity limit. According the optimization function Eq. 11, $\boldsymbol{P}_{EX}$, $\boldsymbol{V}_{EX}$, $\boldsymbol{P}_O$ and $\boldsymbol{V}_O$ are then used to evaluate the optimization function.

$$g_{kX} = g(\boldsymbol{P}_{EX}, \boldsymbol{V}_{EX}, \boldsymbol{P}_O, \boldsymbol{V}_O), X \in \{L, R\} \tag{17}$$

where $\boldsymbol{P}_{EX}$ and $\boldsymbol{V}_{EX}$ are position and translational velocity of end-effect at the joint states $\boldsymbol{\theta}_{kX}, X \in \{L, R\}$.

Then the optimization values g_{kL} and g_{kR} can be used to update manipulator joint position. In manipulator joint space, the direction and distance of movement at next step can be update as follows:

$$\hat{\boldsymbol{\theta}}_{k+1} = P_{\Omega}(\boldsymbol{\theta}_k - \delta(\lambda)\mathrm{sign}(g_{kL} - g_{kR})\boldsymbol{b}) \tag{18}$$

where $\delta(\lambda)$ denotes the step size of search distance. Namely, $\delta(\lambda)$ determine the distance between $\hat{\boldsymbol{\theta}}_{k+1}$ and $\boldsymbol{\theta}_k$. $\mathrm{sign}(g_{kL} - g_{kR})$ term control the next update direction along the smaller optimization value. So we can get position and translational velocity of end-effect at the joint states $\hat{\boldsymbol{\theta}}_{k+1}$. Similarly, $\hat{\boldsymbol{P}}_E$, $\hat{\boldsymbol{V}}_E$, $\boldsymbol{P}_O$ and $\boldsymbol{V}_O$ are used to evaluate the updated optimization value as Eq. 19:

$$\hat{g}_{k+1} = g(\hat{\boldsymbol{P}}_E, \hat{\boldsymbol{V}}_E, \boldsymbol{P}_O, \boldsymbol{V}_O) \tag{19}$$

During one control period, the procedure from Eq. 12 to Eq. 19 can repeat many times. At each iteration, we always save the smaller $\hat{g}_{k+1}$ and the corresponding joint state $\hat{\boldsymbol{\theta}}_{k+1}$. Until the end of control period, the best joint state $\hat{\boldsymbol{\theta}}_{k+1}$, corresponding to the smallest $\hat{g}_{k+1}$, is used to update actual manipulator.

The iterative procedure of BAS Algorithm for solving optimization function Eq. 11, are given in Algorithm 1. And the algorithm complexity of BAS algorithm is $O(m)$, where m denotes the number of manipulator DOF. The calculation of algorithm complexity can be seen in [8].

4 Simulation Results

4.1 Simulation Methodology

To illustrate the efficiency of our algorithm, simulation experiments are provided. As shown in Fig. 3, experiments are did in Gazebo environment by using Franka Panda manipulator [13]. Franka panda manipulator has 7 revolute DOF and is widely used in research field. This model shows representative potential in real-word implementation. In simulation experiments, the object moves and manipulator controller works independently. The manipulator can get the position of object only by the real-time sensing information. To illustrate the efficiency of proposed method, two different trajectories are designed, as shown in Fig. 4.

Fig. 3. Simulation scenes using Franka Panda model

The first trajectory is a part of sinusoidal wave as shown is Fig. 4(a). In this trajectory, the velocity of moving target keeps changing continuously. Another

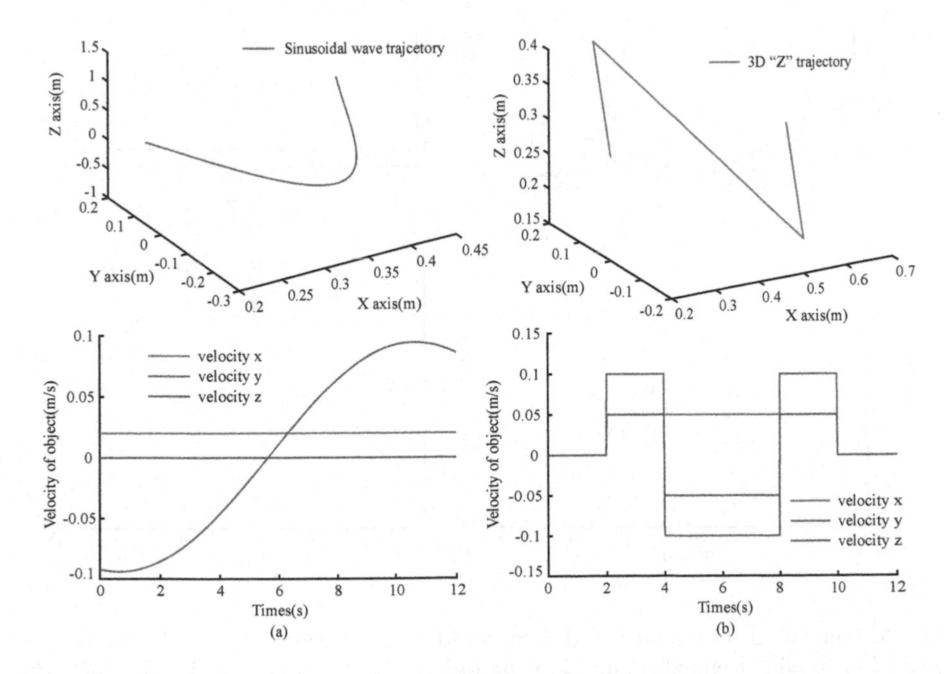

Fig. 4. Two different trajectory for simulation. (a) Sinusoidal wave trajectory and velocity of object in x, y, z axis. (b) 3D "Z" trajectory and velocity of object in x, y, z axis.

trajectory, as shown in Fig. 4(b), is specially designed to verify the algorithm's adaption to trajectories with discontinuous velocity. The velocity of target is discontinuous at the corner of "Z", and the acceleration are also infinity. And at the initial and last 2 s, the object keeps static. This is specially designed to verify the performance on switching between static states and moving states. Without the loss of generality, as long as the tracking points are reachable, the trajectory of object can be arbitrary shape in cartesian space.

As for the moving ability of robot itself, the position limits and velocity limits in joint space are obtained through manuscripts [13]. In equation Eq. 13 and Eq. 18, the hyperparameter parameter $\lambda = \Delta t \cdot 2 \cdot \sqrt{|g_k|}$ and $\delta(\lambda) = 2 \cdot \lambda$ are determined according to similar rules as illustrated in [8]. Δt is a control period in simulation denoting 0.02 s (50 Hz).

4.2 The Results of Tracking Moving Target

These two simulations were used to vertify the strategy of manipulator tracking moving object, described in Sect. 3. The first simulation results for continuous

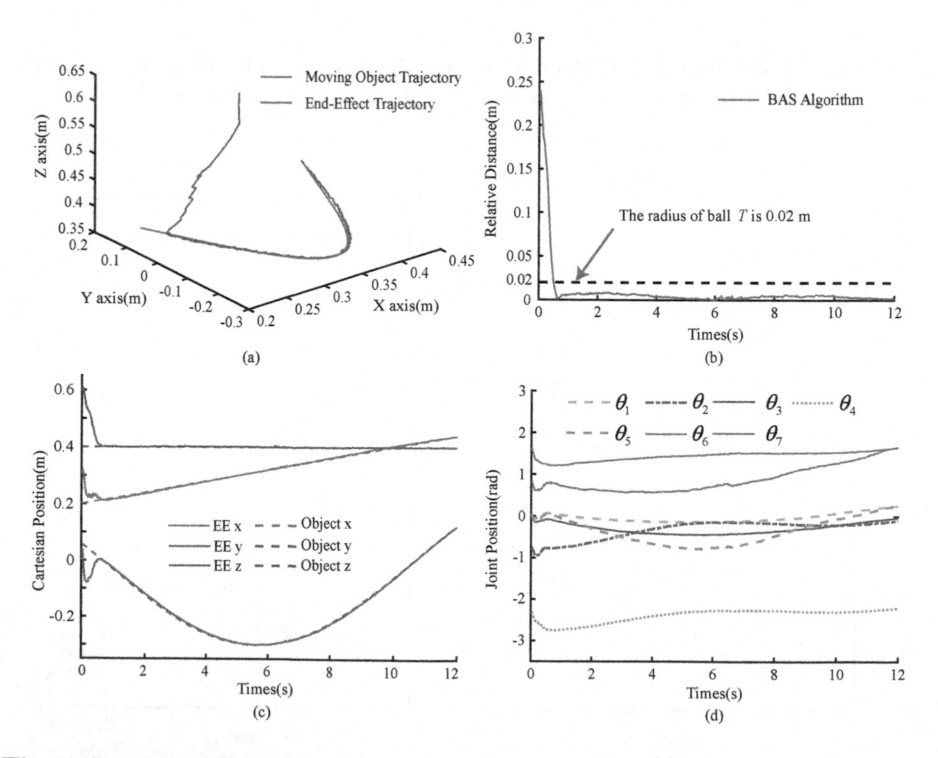

Fig. 5. Simulation results for tracking Sinusoidal wave trajectory. (a) The 3D trajectory of manipulator end-effect and Moving object. (b) The total relative distance. (c) Cartesian positon of end-effect and object presented in x, y, z axis respectively. (d) The joint position trajectory in manipulator joint space.

sinusoidal wave tracking control are shown in Fig. 5. Figure 5(a) shows the 3D trajectorys of manipulator end-effect and object. Figure 5(b) shows the relative distance between manipulator end-effect and object. As shown in Fig. 5(b), the manipulator takes 0.64 s reaching the position of object and then keeps tracking error under 0.02 m in the following process. Figure 5(c) shows x, y, z positions of the end-effect and moving object, presented in manipulator base coordinate. And Fig. 5(c) could clearly shows the tracking process of end-effect. The joint space trajectorys, as shown in Fig. 5(d), are smooth during all simulation process.

Results of another simulation of "Z" trajectory are shown in Fig. 6. Figure 6(a) shows the tracking results of 3D trajectory in cartesian space. Figure 6(b) shows the relative distance between manipulator end-effect and object. In Fig. 6(b), manipulator takes 0.8 s reaching the position of object and keeps tracking error under 0.02 m in the following process. Figure 6(c) shows the cartesian position including end-effect and object in x, y, z axis. At the initial and last 2 s, the object is immobile which can be observed in Fig. 6(b) and Fig. 6(c). During these wo periods, the relative distance is smaller than other tracking process in Fig. 6(b) and the cartesian position keeps the same value

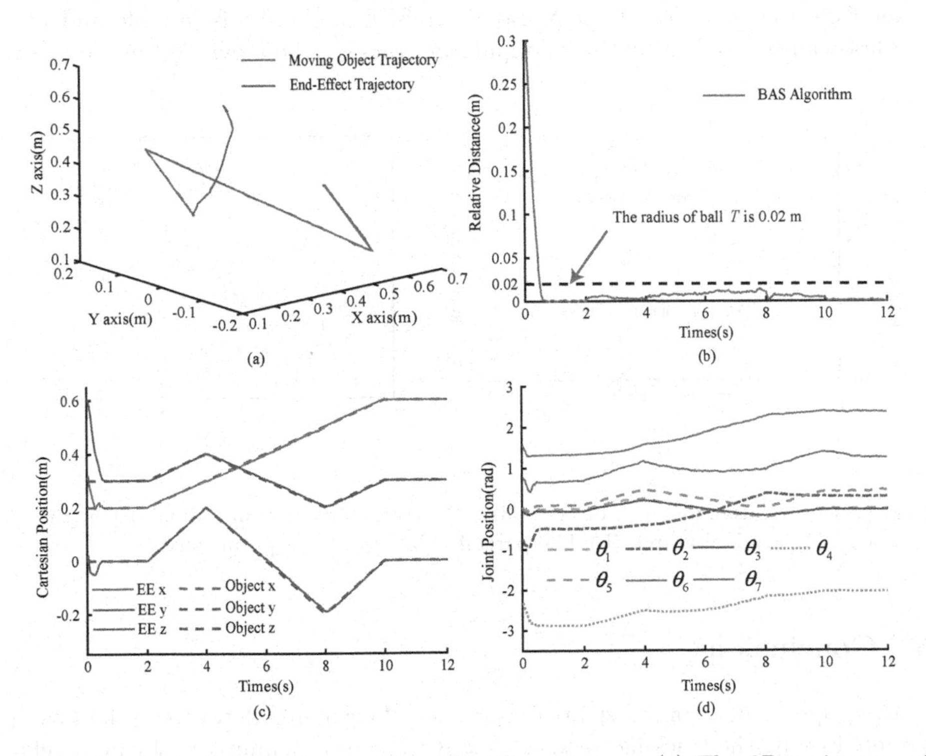

Fig. 6. Simulation results for tracking 3D "Z" trajectory. (a) The 3D trajectory of manipulator end-effect and Moving object. (b) The total relative distance. (c) Cartesian positon of end-effect and object presented in x, y, z axis respectively. (d) The joint position trajectory in manipulator joint space.

in Fig. 6(c). And Fig. 6(d) shows joint position trajectorys in manipulator joint space. Results show that the joint positon trajectory is smooth during all simulation processes even the trajectory of moving object is unsmooth in turning points of "Z". We design two mapping functions, Eq. 15 and Eq. 16, to provent exceeding in joint velocity limit and joint position limit. The joint velocity limit is a key strategy to keep smooth in joint space. And joint position limit keeps the manipulator joint does not exceed the physical positon limit.

The comparison results of relative distance between sinusoidal trajectory 3D "Z" trajectory are shown in Fig. 7. Figure 7(a) shows both trajectory tracking error keep under 0.02 m after reaching the position of object. With the target keeps static in 0 s–2 s and 10 s–12 s in "Z" trajectory, we only compare tracking results during 2 s–10 s, as shown in Fig. 7(b). The relative distance of sinusoidal trajectory is relatively lower than relative distance of "Z" trajectory. During 2 s–10 s, the variance of relative distance for sinusoidal trajectory and "Z" trajectory are 3.77e-06 and 7.96e-06, respectively. The statistical data shows the tracking performance of sinusoidal trajectory is better than "Z" trajectory. This is consistent with our expectations because the velocity of "Z" trajectory is noncontinuous. Besides, the performance of tracking "Z" trajectory is still good. Therefore, the proposed strategy based on RVC and BAS is feasible and this method can be used to control manipulator tracking the moving object rapidly.

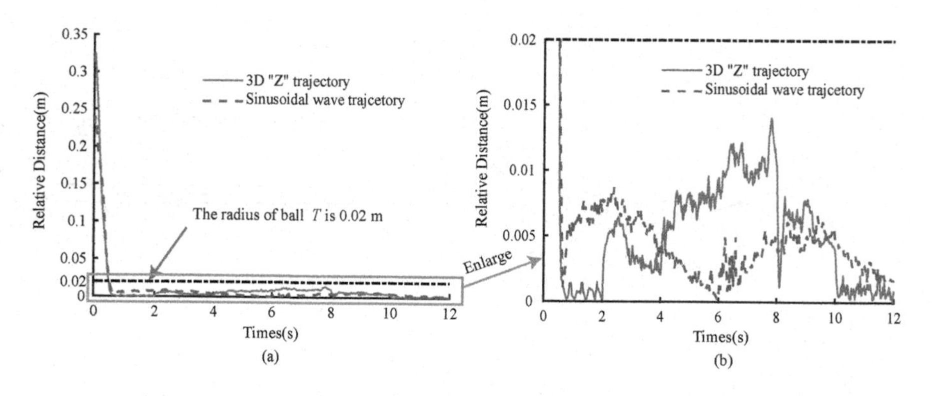

Fig. 7. The Relative distance between 3D "Z" trajectory and Sinusoidal wave trajectory. (a) The original graph. (b) The partial enlargement graph in y-axis.

5 Conclusions

In this paper, a new control strategy is proposed for manipulator to track moving target. Firstly, the tracking mission is converted to a planning problem of relative velocity by establishing RVC on manipulator end-effect. And then a novel optimization function is designed by considering both relative velocity and relative position between manipulator end-effect and the moving target. Finally, a heuristic algorithm, called BAS, was used to tackle the formulated optimization problem. Simulation experiments show that the manipulator end-effect could rapidly reach the position of object and maintain tracking error under 0.02 m. A key feature of proposed strategy is that it can ensure the smoothness of the joint space trajectory even when the velocity of target is discontinuous. A potential application of the proposed strategy includes the manipulator picking and placing package in moving conveyor belt. This article mainly explores how to apply both relative position and relative velocity on optimization function for tracking moving target, and makes some elementary attempt on this problem. In the future, more work will be done on the optimization function with different weight coefficient group and physical experiments.

References

1. Bravo-Palacios, G., Del Prete, A., Wensing, P.M.: One robot for many tasks: versatile co-design through stochastic programming. IEEE Robot. Autom. Lett. **5**(2), 1680–1687 (2020)
2. Buss, S.R., Kim, J.S.: Selectively damped least squares for inverse kinematics. J. Graph. Tools **10**(3), 37–49 (2005)
3. Craig, J.J.: Introduction to Robotics: Mechanics and Control, 3rd edn. Pearson Education India, Bengaluru (2009)
4. Faroni, M., Beschi, M., Visioli, A., Pedrocchi, N.: A real-time trajectory planning method for enhanced path-tracking performance of serial manipulators. Mech. Mach. Theory **156**, 104152 (2021)
5. Hellström, T., Bensch, S.: Understandable robots-what, why, and how. Paladyn J. Behav. Robot. **9**(1), 110–123 (2018)
6. Hu, L., Miao, Y., Wu, G., Hassan, M.M., Humar, I.: iRobot-factory: an intelligent robot factory based on cognitive manufacturing and edge computing. Futur. Gener. Comput. Syst. **90**, 569–577 (2019)
7. Jin, L., Li, S., Luo, X., Li, Y., Qin, B.: Neural dynamics for cooperative control of redundant robot manipulators. IEEE Trans. Industr. Inf. **14**(9), 3812–3821 (2018)
8. Khan, A.H., Li, S., Luo, X.: Obstacle avoidance and tracking control of redundant robotic manipulator: an RNN-based metaheuristic approach. IEEE Trans. Industr. Inf. **16**(7), 4670–4680 (2019)
9. Li, D.P., Li, D.J.: Adaptive neural tracking control for an uncertain state constrained robotic manipulator with unknown time-varying delays. IEEE Trans. Syst. Man Cybern. Syst. **48**(12), 2219–2228 (2017)
10. Lundell, J., Verdoja, F., Kyrki, V.: Robust grasp planning over uncertain shape completions. arXiv preprint arXiv:1903.00645 (2019)
11. Petrič, T., Gams, A., Likar, N., Žlajpah, L.: Obstacle avoidance with industrial robots. In: Carbone, G., Gomez-Bravo, F. (eds.) Motion and Operation Planning of Robotic Systems. MMS, vol. 29, pp. 113–145. Springer, Cham (2015). https://doi.org/10.1007/978-3-319-14705-5_5
12. Schmidtke, H.R.: A survey on verification strategies for intelligent transportation systems. J. Reliab. Intell. Environ. **4**(4), 211–224 (2018)
13. Simon, G.: Franka panda panipulator and interface specifications. https://frankaemika.github.io/docs/control_parameters.html. Accessed 26 Feb 2021
14. Urrea, C., Pascal, J.: Design and validation of a dynamic parameter identification model for industrial manipulator robots. Arch. Appl. Mech. **91**(5), 1981–2007 (2021)

15. Wang, J., Chen, H.: BSAS: beetle swarm antennae search algorithm for optimization problems. arXiv preprint arXiv:1807.10470 (2018)
16. Wang, T.M., Tao, Y., Liu, H.: Current researches and future development trend of intelligent robot: a review. Int. J. Autom. Comput. **15**(5), 525–546 (2018)
17. Zhang, H., Jin, H., Liu, Z., Liu, Y., Zhu, Y., Zhao, J.: Real-time kinematic control for redundant manipulators in a time-varying environment: multiple-dynamic obstacle avoidance and fast tracking of a moving object. IEEE Trans. Industr. Inf. **16**(1), 28–41 (2019)
18. Zhang, Y., Li, S., Xu, B.: Convergence analysis of beetle antennae search algorithm and its applications. arXiv preprint arXiv:1904.02397 (2019)

Leveraging Expert Demonstrations in Robot Cooperation with Multi-Agent Reinforcement Learning

Zhaolong Zhang[1], Yihui Li[1], Juan Rojas[2], and Yisheng Guan[1(✉)]

[1] Biomimetic and Intelligent Robotics Lab (BIRL),
Guangdong University of Technology, Guangzhou 510006, China
ysguan@gdut.edu.cn

[2] Department of Mechanical and Automation Engineering,
Chinese University of Hong Kong, Hong Kong, China

Abstract. While deep reinforcement learning (DRL) enhances the flexibility and intelligence of a single robot, it has proven challenging to solve the cooperatively of even basic tasks. And robotic manipulation is cumbersome and can easily yield getting trapped in local optima with reward shaping. As such sparse rewards are an attractive alternative. In this paper, we demonstrate how teams of robots are able to solve cooperative tasks. Additionally, we provide insights on how to facilitate exploration and faster learning in collaborative systems. First, we increased the amount of effective data samples in the replay buffer by leveraging virtual targets. Secondly, we introduce a small number of expert demonstrations to guide the robot during training via an additional loss that forces the policy network to learn the expert data faster. Finally, to improve the quality of behavior cloning, we propose a Judge mechanism that updates the strategy by selecting optimal action while training. Furthermore, our algorithms were tested in simulation using both dual arms and teams of two robots with single arms.

Keywords: Reinforcement learning · Imitation learning · Robot manipulation · Robot learning

1 Introduction

In the last five years, DRL has achieved impressive results in decision making. DRL has succeeded across many domains including Go [16] and StarCraft [22]. DRL has also made significant in-roads in complex, high-dimensional robot systems as well [6]. Most of the work in robotic DRL has focused on single agent,

This work is partially supported by the Key R&D Programmes of Guangdong Province (Grant No. 2019B090915001), the Frontier and Key Technology Innovation Special Funds of Guangdong Province (Grant No. 2017B050506008), and NSFC (Grant No. 51975126, 51905105).

X.-J. Liu et al. (Eds.): ICIRA 2021, LNAI 13014, pp. 211–222, 2021.
https://doi.org/10.1007/978-3-030-89098-8_20

however, single robot are limited in their ability to solve more complex tasks that are better suited when done in cooperation with one or more agents.

Applying DRL methods to multi-agent system is challenging, which perform excellently in many fields, are inappropriate to multi-agent scenarios directly. Initially, researchers used independent agents across robots in non-contact scenarios to enact their multi-agent systems [19,20]. However, as soon as agents interacted with each other, the policy destabilized and diverged.

Particularly, Multi-Agent Reinforcement Learning (MARL) settings are challenging due to the existence of multiple (Nash) equilibria and non-stationary environments comparing with single-agent DRL settings. Lowe *et al.* improved interaction by leveraging a distributed Actor-Critic framework where one agent learns whilst considering the other in continuous action space [10].

In general, our primary contributions are three fold: (i) based on the Centralized Training and Decentralized Execution (CTDE) framework, we transform the endpoint of interacting data as the virtual target, and increase the number of efficient data in multiple agents. (ii) With a few expert demo sequences, the robot not only learn a skill more quickly (reducing sample complexity) but also learn cooperation more efficiently as well. (iii) A new method is proposed to efficiently update the policy instead of direct behavior cloning, which avoids agents only learns imperfect data.

2 Related Work

2.1 Multi-Agent Reinforcement Learning

Multi-agent issues present specific challenges that render MARL more challenging than single-agent scenarios [5], but researchers have made progress in many aspects of MARL. In this paper, we focus on training the cooperation skill of robot teams.

Deep learning is conducive to the high-dimensional estimation of MARL. MADDPG [10], which is an extended and multi-agent version of DDPG [8], learns a centralized critic based on all agents, enabling robots to generate deterministic policies. COMA [3] utilizes a counterfactual baseline to address the credit assignment problem between cooperative agents. Using a monotonic constraint on the mixing network, QMIX [14] extends additive action-value factorization of VDN [18] to represent the rich joint action-value function, which admits tractable decompositions into a value function of each agent. Currently, QTRAN [17] transforms appropriately from QMIX and factorizes the joint action-value function into an individual one. According to GCN [7] which make a connection graph-structured data, PIC [9] estimate a consistent value by introducing permutation invariant critics.

Most multi-agent works have been implemented in game environments with discrete action spaces but rarely extended to real situations. In this work, our robot agents use continuous space and solve the sparse reward in robotic manipulation. Then we use demonstrations to improve agents learning efficiency and have more benefits on applications. Therefore, our approach allows models to understand it's cooperative agent state needed for rich interactions in cooperative tasks.

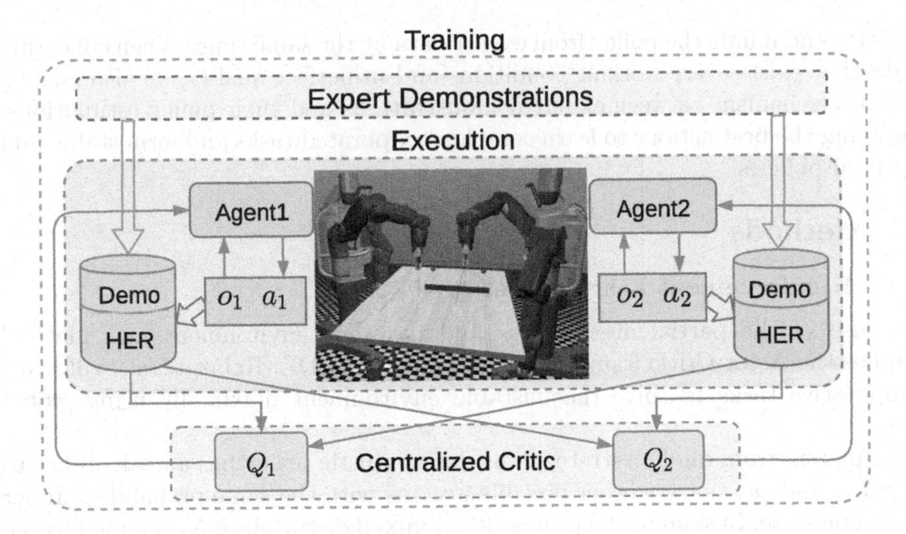

Fig. 1. Framework visualization. Robot teams use MARL and expert demonstrations as prior guidance. The dual-armed manipulators trained with our algorithms are validated to cooperate to carry and transport objects.

2.2 Imitation Learning

In the field of imitation learning, agents are used to generating policy according to demonstrations that provided successful sequences by the experts. A common method is Behavior Cloning (BC) [2,15], which refer to state-action pairs of experts through supervised learning.

Pure imitating demonstrations are limited by the ability of experts, but agents that exceed demo and possess certain generalization ability are essential. For this purpose, imitation learning combines with DRL shows favorable results that promote training policy. DDPGfD [21] overcomes exploration by inserting demonstrations as a prioritized replay buffer along with normal transitions. While demoddpg [1,11] also exploits transition from both buffers and selects sequences by introducing BC loss in actor-network. Comparing with the manipulator, dexterous manipulation augmenting RL with demonstrations [13] corrects the unnatural movement learned in training. Concurrent works put demonstrations into reward functions and guide agents based on state distribution [12]. Different from focusing on DRL, there also have research on solving imperfect demonstrations [4,23]. From above, Learning from Demonstrations (LfD) to bootstrap DRL methods has been studied for a long time. So putting demonstrations into multi-agent scenarios is more helpful in industrial applications.

Possessing expert trajectories can reduce unnecessary learning processes and get natural motions. It is a breakthrough point to help robots quickly master skills of completing cooperative tasks through expert guidance. Therefore, we put forward an imitation method in which agents learn models by their own policies and mimic the policy from expert data at the same time. When collecting data that exist severe working conditions and imperfect quality, we also make a Judge mechanism between expert demonstrations and their policy estimations, applying the best actions to learn complex cooperated tasks and form stable and optimal policies.

3 Methods

3.1 Reinforcement Learning Setup

Because of the partial observations and dynamic environment, our method applies the Actor-Critic framework based on the CTDE. Robot teams will learn cooperative tasks to solve the unstable environment maximally using global critic.

Different from random strategy, the deterministic policy has a weak ability to explore, and can not try other possible actions with the same probability under the same state. Instead of OU process [8], a mixed disturbance $\mathcal{N}_t$ is added to the deterministic output from local observation o_i of each agent during our training to obtain the surrounding exploration value:

$$a_i = a_i^p \oplus \mathcal{N}_t \tag{1}$$

Where predicted action $a_i^p = \mu_{\theta_i}(o_i)$, and $\oplus$ means adding noises. In the random noise $\mathcal{N}_t$ part, Gaussian noise $\mathcal{N}(\mu, \sigma^2)$ is given to the predicted action firstly, and the maximum value of the forward and reverse direction is cut off to prevent the predicted action from being too large, which causes the robot to move too much and unsafe. Secondly, the uniform distribution $U(a, b)$ is adopted to enable agents to explore all states without differences. Updating a_i^p with:

$$a_i^p \leftarrow \text{clip}[a_i^p + \mathcal{N}(\mu, \sigma^2), -a_{\max}, a_{\max}] + U(-a_{\max}, a_{\max}) \tag{2}$$

Noise is added to the policy to enhance the exploration ability of the deterministic algorithm and restrict its exploration. $a_{\max}$ donates maximum action.

Our basic framework shares similarities with MADDPG [10], each agent has an actor network that uses local observation for deterministic action, and a corresponding critic network that learns from the joint states $\mathbf{x}$, next joint states $\mathbf{x}'$ and joint actions $\mathbf{u} = a_1, \cdots, a_N$. The target Q-value is calculated from the next Q estimated by the target critic network and the reward obtained from the mini-batch samples:

$$y = r_i + \gamma Q_i^{\mu'}(\mathbf{x}', \mathbf{u}')|_{a_k' = \mu_k'(o_k)} \tag{3}$$

With one-step lookahead TD-error, we update our centralized critic:

$$\mathcal{L}(\theta_i) = \mathbb{E}_{\mathbf{x},a,r,\mathbf{x}'}[(y - Q_i^{\mu}(\mathbf{x}, \mathbf{u}))^2] \tag{4}$$

Where μ donates the policy of the actor. As the critic learns the joint Q-value function over the update of global information, it sends appropriate Q-value approximations to help the actor training.

$$\nabla_{\theta_i} J = \mathbb{E}_{\mathbf{x},a\sim\mathcal{D}}[\nabla_{\theta_i}\mu_i(o_i)\nabla_{a_i} Q_i^{\mu}(\mathbf{x}, \mathbf{u})|_{a_i=\mu_i(o_i)}] \tag{5}$$

Here the replay buffer $\mathcal{D}$ collected by all agents contains the tuples $(\mathbf{x}, \mathbf{x}', \mathbf{u}, r_1, \cdots, r_N)$. Finally, the target network of each agent i are soft updated.

3.2 Multiple Hindsight Experience Replay

Off-policy MADDPG maintains a replay buffer in the framework, where stores the data of different policies interacting with the environment. When a certain amount of data is reached, mini-batches of data are randomly extracted from the buffer to avoid the correlation between the data before and after. What is more, sampling sequences are used to calculate reward $r_i^t = r(x_i^t, a_i^t, g)$ according actual goal g, then store R into buffer $\mathcal{D}$:

$$R = (x_i^t||g, a_i^t, r_t, x_i^{t+1}||g) \tag{6}$$

To synthesize the problems of invalid data and complex sparse reward, we extend the single-agent algorithm to multi-agent teams and propose Multiple Hindsight Experience Replay (M-HER) as shown in Fig. 1. Due to sparse reward, the rewards of most data, no completion of the task, stored in $\mathcal{D}$ are the maximum step size. Therefore, agents cannot select the optimal values to update the strategies. In the process of random sampling, it is useless for updating when the endpoint of invalid data is not the actual target. We set the endpoint as virtual target $G := \mathcal{S}(\text{current episode})$ to transform invalid data into useful information, then compute the rewards $g' \in G$ of these data and save hindsight sequences R_h into buffer $\mathcal{D}$.

$$R_h = (x_i^t||g', a_i^t, r_t', x_i^{t+1}||g') \tag{7}$$

Where virtual reward $r_t' = r(x_i^t, a_i^t, g')$ is computed with virtual goal to replace the actual value.

3.3 Robot Teams Learning from Demonstrations

The M-HER can increase the number of effective data in the experience buffer, and improve the learning efficiency to a certain extent. The agents training from scratch, however, have to explore plenty of trajectories to accumulate experience. The initial networks learning slowly and perform badly. We need more effective data as prior knowledge to guide how to accomplish the task. In cooperative tasks, with the increase of the number of agents, the dimension of parameters is increasing, which leads to the extension of a training cycle.

Based on behavior cloning, we propose a new method, Robot Team Learning from Demonstrations (RTLfD), which only needs a small number of expert samples, as prior knowledge, and gradually guides the manipulator to update strategy in the process of training.

Different from the original experience replay buffer $\mathcal{D}$, we build another buffer $\mathcal{D}_E$ to store expert demonstrations. Secondly, unlike BC, which only learns expert demonstrations in supervised learning mode, RTLfD introduces a new loss function. As shown in Fig. 1, our method guides agents to learn expert demonstrations while maintaining their interaction with the environment and updating policy functions. When updating strategies, each decentralized actor sample extracts the mini-batch of data from the two buffers and then uses them to update the loss of imitation learning and the policy network respectively.

Under the influence of the imitation part, the agents are forced to learn the expert demonstrations in the way of supervised learning for the output of the action by each agent policy function:

$$L_{IL}(i) = \sum_{t=1}^{N_D} ||\pi(s_t|\theta_\pi) - a_t||^2 \tag{8}$$

Where N_D is the number of expert demonstrations, $\pi(s_t|\theta_\pi)$ is the policy function of current agent. $L_{IL}(i)$ donates the imitation learning loss of i-th agent. Because each agent has a corresponding trajectory sequence in the expert demonstrations, it does not consider the actions of other agents during learning from demonstrations. The loss function of imitation learning only considers its own state, which reduces the unnecessary amount of calculation in the process of calculation. Adding an auxiliary loss function, greatly improves the learning efficiency of DRL.

Finally, L2 regularization L_{reg} added to the actor loss function prevents the trained model more complex, which stabilizes the performance of learning. Every decentralized actor of the agent i is calculated as follows:

$$\lambda_1 \nabla_{\theta_\pi} J(i) - \lambda_2 \nabla_{\theta_\pi} L_{IL}(i) + \nabla_{\theta_\pi} L_{\text{reg}}(i) \tag{9}$$

Where λ_1 and λ_2 are scalar that weights the importance of policy gradient loss against the imitation learning loss.

3.4 Imperfect Demonstrations

The collection of expert data is complicated and it is easy to get imperfect demonstrations, which result in data errors and local optima. In the methods of imitation learning, the training strategy of agents is limited by the unstable expert data, so that the local optimum will be obtained. Based on RTLfD, we introduce imperfect data to influence the training process of agents and propose a Judge mechanism.

After sampling sequence from expert experience buffer, the data (s_i, a_i) will be sent to the value network of the global critic as input and estimate the Q-value of the current state-action pair. At the same time, the predicted action $a_i' = \pi(s_i)$ is computed according to the current state by the policy network π. And then the selected state-action pair from the expert demonstrations are also used to estimate the Q-value. The larger the Q-value is, the more favorable the sampled data is for the agent to update the strategy, and vice versa.

Finally, we choose the largest one instead of blindly following the expert. During the training period, adding an additional estimation of the Q-value can select the correct sequence on its own buffer and avoid imperfect demonstrations. If the expert is better, it will be chosen to exploit imitation learning. Otherwise, the imitation learning will be closed to prevent the agent learning only from identifying the expert data as best, and even using supervised learning to forcibly fit the characteristics of expert data. After adding the Judge, modify the Eq. 8 to calculate the loss function of the imitation learning part:

$$L_{IL}(i) = \sum_{i=1}^{N_D} ||\pi(s_i|a) - a_i||^2, if \quad Q(s_i, a_i) > Q(s_i, \pi(s_i)) \tag{10}$$

The calculation of the critic network determines whether the actions taken by the expert demonstrations are better than the actions selected by the policy of agents. Moreover, the selective loss function uses imitation learning to avoid misleading results, and only updates the strategy in the optimal gradient.

4 Experiments

4.1 Experimental Setup

Before algorithms comparing, two simulated scenarios are built to verify our performance. Lifting objects and relocation are the primary problems in industrial manipulation tasks, where an object is lifted, and moved to a target position. The principal challenge from our opinion here is how to grasp an object, but a DRL perspective is an exploration since, to finish a task successfully, the robot has to open its grippers and reach the object, grasp it and bring it to the target position. Without the guidance of prior knowledge, it would be a hard feat to learn strategy from scratch and accomplish the task.

(i) Lift1: symmetric arms of one robot (Fig. 2(a)). The two arms of the Baxter robot are symmetrical and controlled by two agents to realize the cooperative task. The long strip-shaped object is placed on the table in parallel, and the symmetrical arms move the same distance forward, so they need to move left/right and adjust to adapt to team cooperation when carrying the objects.

(ii) Lift2: staggered arms of two robots (Fig. 2(b)). Two robots provide a manipulation arm respectively and adopt staggered mode. The object is placed vertically on the table. The staggered manipulator arm is not synchronous, and the cooperation difficulty is relatively symmetrical, and the arms need to carry the objects in collaboration.

The Rethink Baxter robot consists of two 7-DoF arms along with a two-fingered parallel gripper. Each arm of Baxter controlled by one agent has a 24D observation including a 3D goal position p^{goal}. The observation space of i-th agent in our environments:

$$\mathbf{x_i} = [p_i^{\text{grip}}, p_i^{\text{obj}}, t_i^{\text{obj}}, d_i^t, v_i^{\text{grip}}, v_i^{\text{obj}}, v_i^t, p^{\text{goal}}] \tag{11}$$

Where position p and rotation t are orientation information of grippers and objects. And d_i^t is the distance of gripper tips. The kinematics information is velocity v. It also has a 4D action space which consists of the 3D end-effector goal pose (d_i^x, d_i^y, d_i^z) and a 1D flag d_i^t that indicates whether the grippers are open or closed. The action frequency of agents is $f = 25\,\text{Hz}$.

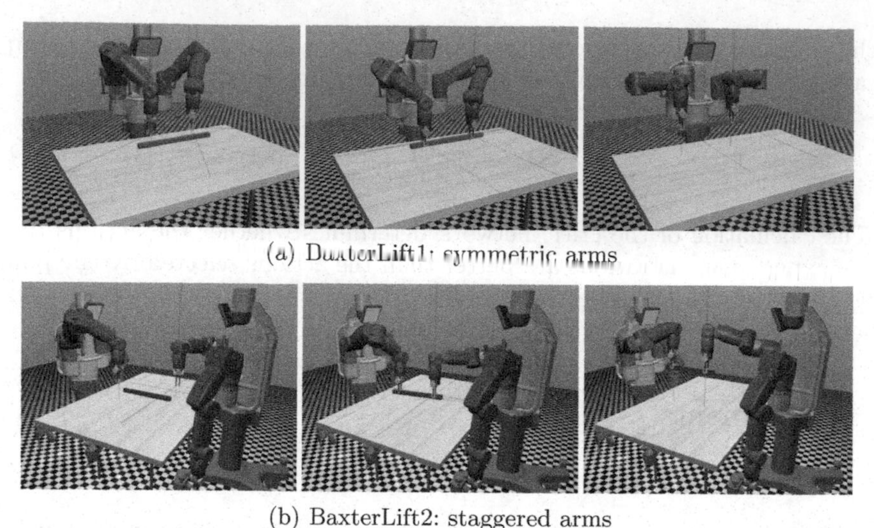

(a) BaxterLift1: symmetric arms

(b) BaxterLift2: staggered arms

Fig. 2. Two simulated environments in MuJoCo. Task: dual-manipulators cooperate to carry and move object to the target position.

Both lifting tasks are sparse rewards. At each time step, each agent will get a reward $R = -1$ if not achieving the goal, else $R = 0$. The overall reward is the sum of all rewards and every agent will use the total reward to predict the Q-value. To prevent one agent from learning quicker than the other, each agent receives an additional reward $R_a = -1$ when it does not reach the block at every step. We limit the range for agents to reach the block as below (unit: mm):

$$\begin{cases} \Delta d_x \in [-40, 40], \\ (\Delta d_y)^2 + (\Delta d_z)^2 <= 20 \end{cases} \tag{12}$$

To bootstrap the policy, lifting demonstrations are provided for the two robots via simulation until 30 and 100 episodes of data were collected. 70% of the data targets are in the air, whilst the rest are on the table.

4.2 Robot Training

Because robot manipulation tasks belong to multi-goal scenarios, the goal and object will appear randomly in the working space around the robots. And the goal is set as a 50% probability on the desktop which is used to guide early learning, 50% in the air. Figure 3 is the result that our methods testing on the two Lifting environments. In the initial stage of training, the robot will learn to push the object using grippers to reach the target. The process, however, is not synchronous and has no cooperation ability, so there is a rapid rise during the first 50 epochs and achieves a 50% success rate in a short time. Moreover, the speed of different methods is similar.

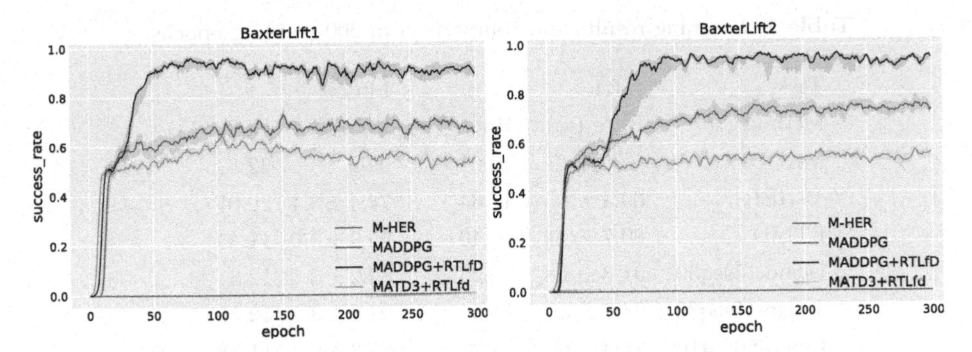

Fig. 3. Training results of robot teams – Incorporating expert demonstrations help teams overcome the sparse reward in continuous environments, and improve the performances than others (Median test success rate line with interquartile range for all robot environments over five random seeds).

a) MADDPG. As DDPG shows in HER [1] paper, only MADDPG can not quickly grasp the robot manipulation in high-dimensional continuous space during 300 epochs. Due to the interference of sparse reward, the whole exploration process is aimless, and it is unable to select suitable data from a large number of invalid data for updating. As a result, the robot cannot know how to grasp objects on the desktop and move to the air.

b) M-HER. When the robot interacts with the environment, the virtual target is used to replace the random exploration estimated target, providing a large amount of reference data to guide the robot how to reach the target position, so that it can finally learn how to grab the object and take it to the target position. However, the trajectory randomly explored by the robot team is not synergistic, resulting in the data with the virtual target can only help the robot to achieve a certain degree of task. Therefore, after 130 epochs, the teams can only achieve a success rate of about 70%. Since the task of Lift2 is interlaced coordination, the sampling range is higher than Lift1, and its performance effect is slightly higher than Lift1.

c) RTLfD. This method adding extra imitation learning forces the robot to interact with the environment with its own strategy and learn expert demonstrations at the same time. Compared with the former two methods, it can sample effective and collaborative data at the beginning of training, and accelerate the learning speed. In the whole process, it will sample data from the expert, which is equivalent to equipped with prior guidance.

On the basis of RTLfD, we extend TD3 which is an upgraded algorithm of DDPG to Multi-Agent TD3 with demonstrations and compare the results. The update of loss function L_{IL}, however, is consistent with the policy of the actor. Delayed updates cause the teams to learn expert data more slowly, so good results cannot be obtained with demo in our experiments.

Table 1. Learning results and comparison in 200 training epochs.

Task	Lift1		Lift2	
Method	Rate (%)	Hours (h)	Rate (%)	Hours (h)
MADDPG	0.00	20.91	0.00	22.32
M-HER	69.4 ± 2.5	17.84	74.4 ± 3.1	20.07
RTLfD	89.7 ± 3.4	10.40	95.9 ± 1.0	11.45
Demo-15seq30	91.3 ± 8.8	10.25	82.2 ± 3.5	12.31
Judge-15seq30	94.7 ± 1.2	11.64	93.8 ± 3.2	12.56
Demo-50seq100	81.3 ± 11.9	10.22	85.3 ± 9.7	11.78
Judge-50seq100	92.7 ± 4.2	11.57	94.1 ± 3.9	12.82

Due to the synchronicity in one direction, The performance of Lift1 is faster than lift2 and has gained a success rate of 90% in nearly 50 epochs. RTLfD quickly reached 93% which is 42% higher than M-HER. And reducing learning time by 74% in 200 epochs. The results indicate that MARL methods with augmenting demonstrations as guidance are beneficial for the robot to explore and achieve tasks in continuous manipulation.

4.3 Imperfect Demonstrations

Our experiments have two sets of imperfect data, each include 30 and 100 data. Half the data in them are mismatched to cooperative teams. Moreover, all the expert data is just to complete the task, not the optimal solution to the task.

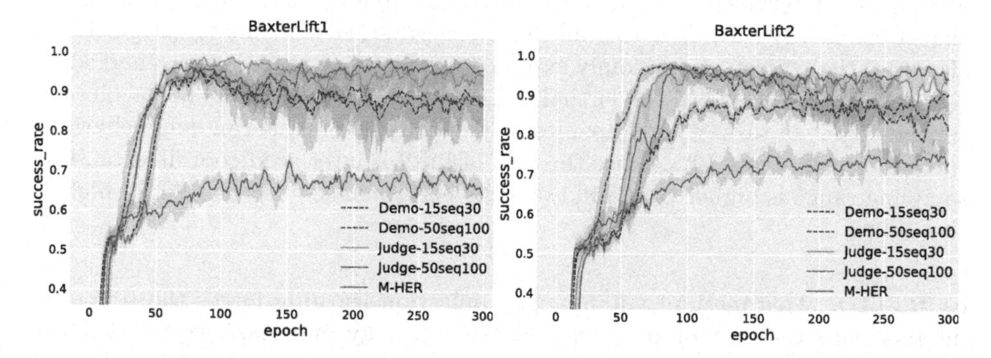

Fig. 4. Comparing the impact of different amounts of imperfect expert data on Demo (RTLfD) and Judge. We use 100 expert demonstrations, including 50 imperfect data (50seq100) and 30 including 15 (15seq30).

While collecting data from the interaction with the environment, the RTLfD algorithm also makes good use of the expert data and completes the task quickly with prior guidance. As shown in Fig. 4 - BaxterLift1, it reaches about 95% success rate within 75 epochs. From Table 1, rate (%) is the success rate of the training, hours (h) represent the robot learning hours to reach the success rate. With the agents using imperfect data, their training situation becomes unstable. Only when the accurate expert data is sampled, the training results are good in this update period. And the success rate of the shaded part fluctuates as 8.8%. With better results reaching more than 95%, and the worse around 80%. In comparison, the learning curve of our Judge mechanism is very stable, and the success rate of the algorithm converges to nearly 98%.

Since our algorithm is used to constrain agents to imitate actions in the process of DRL training, it mainly relies on the agent's own interaction and training the model. Therefore, the introduction of imperfect data can only interfere with the training process, and cannot interrupt learning. When there are only 30 imperfect data, RTLfD cannot achieve the optimal strategy (Fig. 4 - BaxterLift2). In the case of 100 data, it can learn more beneficial information from them. But as the training continues, it will be disturbed by imperfect data, which leads to a decline in learning accuracy. On the contrary, the Judge is to maintain a stable learning efficiency in two situations.

Experiments show that Judge can use the Q-value to evaluate the quality between expert data and their own policy estimates to stabilize performance. From Table 1, the Judge takes one more hour than RTLfD in learning due to the additional calculation, but the performance is best overall other methods and gets stable performance.

5 Conclusions

In this paper, we propose M-HER, which sets the endpoint of random trajectories as virtual targets to provide more valid data. For high-dimensional continuous space robot teams, then collaborative strategies based on imitation learning are proposed and called RTLfD. With DRL drives robots to explore their own strategies, they learn the features of expert data and efficiently solve the deficiencies of multi-agent learning from scratch. Thirdly, given the possible errors and imperfect sequences in the actual collection of experts, a Judge mechanism is proposed to distinguish the pros and cons of expert data and obtain stable strategies. After that, we build robot collaborative environments and verifies the effectiveness of all the above. The result indicates that RTLfD and Judge mechanism can help the robot to form a stable and efficient strategy.

References

1. Andrychowicz, M., Wolski, F., et al.: Hindsight experience replay. In: Advances in Neural Information Processing Systems, pp. 5048–5058 (2017)
2. Bojarski, M., Del Testa, D., et al.: End to end learning for self-driving cars. arXiv preprint arXiv:1604.07316 (2016)
3. Foerster, J.N., Farquhar, G., Whiteson, S., et al.: Counterfactual multi-agent policy gradients. In: Thirty-Second AAAI Conference on Artificial Intelligence (2018)
4. Gao, Y., Xu, H., et al.: Reinforcement learning from imperfect demonstrations. arXiv preprint arXiv:1802.05313 (2018)

5. Hernandez-Leal, P., Kartal, B., Taylor, M.E.: A survey and critique of multiagent deep reinforcement learning. Auton. Agent. Multi-Agent Syst. **33**(6), 750–797 (2019)
6. Hwangbo, J., Lee, J., et al.: Learning agile and dynamic motor skills for legged robots. Sci. Robot. **4**(26), eaau5872 (2019)
7. Kipf, T.N., Welling, M.: Semi-supervised classification with graph convolutional networks. arXiv preprint arXiv:1609.02907 (2016)
8. Lillicrap, T.P., Hunt, J.J., et al.: Continuous control with deep reinforcement learning. arXiv preprint arXiv:1509.02971 (2015)
9. Liu, I.J., Yeh, R.A., Schwing, A.G.: PIC: permutation invariant critic for multi-agent deep reinforcement learning. In: Conference on Robot Learning, pp. 590–602. PMLR (2020)
10. Lowe, R., Wu, Y.I., Tamar, A., et al.: Multi-agent actor-critic for mixed cooperative-competitive environments. In: Advances in Neural Information Processing Systems, pp. 6379–6390 (2017)
11. Nair, A., McGrew, B., Abbeel, P., et al.: Overcoming exploration in reinforcement learning with demonstrations. In: 2018 IEEE International Conference on Robotics and Automation (ICRA), pp. 6292–6299. IEEE (2018)
12. Peng, X.B., Abbeel, P., et al.: Deepmimic: example-guided deep reinforcement learning of physics-based character skills. ACM Trans. Graph. (TOG) **37**(4), 143 (2018)
13. Rajeswaran, A., Kumar, V., et al.: Learning complex dexterous manipulation with deep reinforcement learning and demonstrations. arXiv preprint arXiv:1709.10087 (2017)
14. Rashid, T., Samvelyan, M., et al.: QMIX: monotonic value function factorisation for deep multi-agent reinforcement learning. In: International Conference on Machine Learning, pp. 4295–4304. PMLR (2018)
15. Schaal, S., Ijspeert, A., Billard, A.: Computational approaches to motor learning by imitation. Philos. Trans. R. Soc. Lond. Ser. B: Biol. Sci. **358**(1431), 537–547 (2003)
16. Silver, D., Schrittwieser, J., et al.: Mastering the game of go without human knowledge. Nature **550**(7676), 354–359 (2017)
17. Son, K., Kim, D., Kang, W.J., et al.: QTRAN: learning to factorize with transformation for cooperative multi-agent reinforcement learning. In: International Conference on Machine Learning, pp. 5887–5896 (2019)
18. Sunehag, P., Lever, G., et al.: Value-decomposition networks for cooperative multi-agent learning based on team reward. In: AAMAS, pp. 2085–2087 (2018)
19. Tampuu, A., Matiisen, T., et al.: Multiagent cooperation and competition with deep reinforcement learning. PLoS ONE **12**(4), e0172395 (2017)
20. Tan, M.: Multi-agent reinforcement learning: independent vs. cooperative agents. In: Proceedings of the Tenth International Conference on Machine Learning, pp. 330–337 (1993)
21. Vecerik, M., Hester, T., et al.: Leveraging demonstrations for deep reinforcement learning on robotics problems with sparse rewards. arXiv preprint arXiv:1707.08817 (2017)
22. Vinyals, O., Babuschkin, I., et al.: Grandmaster level in starcraft ii using multi-agent reinforcement learning. Nature **575**(7782), 350–354 (2019)
23. Wu, Y.H., Charoenphakdee, N., et al.: Imitation learning from imperfect demonstration. In: International Conference on Machine Learning, pp. 6818–6827. PMLR (2019)

Fault Diagnosis Method Based on Control Performance of Single-Leg Robot

Xinling Zhuang[1], Yueyang Li[1(✉)], Hui Chai[2], and Qin Zhang[1]

[1] School of Electrical Engineering, University of Jinan, Jinan 250022, China
cse_liyy@ujn.edu.cn

[2] School of Control Science and Engineering, Shandong University, Jinan 250061, China

Abstract. This paper proposes a fault diagnosis method based on the control performance of a single-leg robot. Firstly, the kinematics and dynamics of the single-leg robot are analyzed, and the dynamic model is established. Secondly, the nonlinear system is linearized in the nominal state, and the LQ controller is designed. Thirdly, the control performance index is directly used as the evaluation function, and the function value in the nominal state is used as the threshold. Finally, by comparing the measured performance values and the threshold, a fault diagnosis logic is determined. To illustrate the effectiveness of the proposed approach, the proposed algorithm is directly applied to an actual single-leg robot system.

Keywords: Fault diagnosis · Linearization · Single-leg robotic

1 Introduction

At present, with the rapid development of the robotics industry, the fault diagnosis and reliability of robots have attracted the attention of many people, especially the technology of leg-foot robots. The bionic structure [1] of the leg-foot robot enables it to walk, run, and jump in different environments with the help of flexible legs like a human. However, the performance of the legged robot is susceptible to various factors, and more and more scholars have paid attention to make the legged robot operate with no faults and high efficiency [15]. Generally, quickly detecting faults, and reducing the damage to the robot body, the surrounding environment and people, are the basic requirements for the research of fault diagnosis.

The legged robot is a system with strong coupling, high degree of nonlinearity and susceptible to interference. With the continuous expansion of the

Supported by National Natural Science Foundation of China, Grant/Award Number: 61973135, 91948201, 61773242 and 62073191; Shandong Provincial Key Research and Development Program (Major Scientific and Technological Innovation Project), Grant/Award Numbers: 2019JZZY010441, 2019JZZY020317.

X.-J. Liu et al. (Eds.): ICIRA 2021, LNAI 13014, pp. 223–234, 2021.
https://doi.org/10.1007/978-3-030-89098-8_21

system scale and increasing complexity, researchers need more advanced methods to realize fault diagnosis. Since the professor Chow put forward the idea of using analytical redundancy to replace hardware redundancy [3]. Fault diagnosis technology [2,4] has been constantly innovating in the past decades. Fault diagnosis methods based on models can be roughly divided into three categories: observer-based, parity space method and method based on parameter identification. The parity space method [14,17] is mainly to design residuals. The fault signal of the system is calculated through the measured value of the sensor and the input value of the system. Usually one residual corresponds to a fault signal. The observer-based method [5,18] is mainly to model the fault as the change of the state variable, and then use the state observer to detect the fault. The method of parameter identification [20] takes the deviation of the model parameters from the normal value as the system's fault symptom. In addition, some scholars have taken a different approach and achieved the purpose of fault diagnosis by analyzing the energy balance of the system [6]. In [7], a detection method based on performance degradation is proposed, and the Bellman equation is introduced to obtain performance index. Through performance monitoring, the purpose of fault detection is achieved. However, research on robot failures [13] is in full swing, especially legged robots. To this end, this study has done some research that is different from traditional methods.

Aiming at the specific purpose of detecting the leg-foot robots, a new idea is proposed. By feedback linearization of the model, the controller is designed. And then a new performance index is proposed to fault diagnosis according to the system state, and the effectiveness of the scheme is verified in this study.

2 Robot Model Description

2.1 Physical Model

The dynamic model plays an important role in the control and fault detection of the robot [16]. Suppose the coordinate system of the body is rectangular coordinate system, the x-axis points to the forward direction of the robot, the y-axis is perpendicular to the ground and points upward, and the z-axis is determined by the right-hand rule. Figure 1 is a schematic diagram of the body structure.

The Lagrange method [8] is used for modeling, and the general form of the dynamic model is

$$\tau = D(q)\ddot{q} + H(q,\dot{q}) + G(q) \tag{1}$$

where

$$\tau = \begin{pmatrix} \tau_1 \\ \tau_2 \end{pmatrix}; q = \begin{pmatrix} \theta_1 \\ \theta_2 \end{pmatrix}; \dot{q} = \begin{pmatrix} \dot{\theta}_1 \\ \dot{\theta}_2 \end{pmatrix}; \ddot{q} = \begin{pmatrix} \ddot{\theta}_1 \\ \ddot{\theta}_2 \end{pmatrix}$$

$$D(q) = \begin{pmatrix} m_1 p_1^2 + m_2\left(l_1^2 + p_2^2 + 2l_1 p_2 \cos\theta_2\right) & m_2\left(p_2^2 + l_1 p_2 \cos\theta_2\right) \\ m_2\left(p_2^2 + l_1 p_2 \cos\theta_2\right) & m_2 p_2^2 \end{pmatrix}$$

$$H(q,\dot{q}) = \begin{pmatrix} -2m_2 l_1 p_2 \dot{\theta}_1 \dot{\theta}_2 \sin\theta_2 - m_2 l_1 p_2 \theta_2^2 \sin\theta_2 \\ m_2 l_1 p_2 \dot{\theta}_1^2 \sin\theta_2 \end{pmatrix}$$

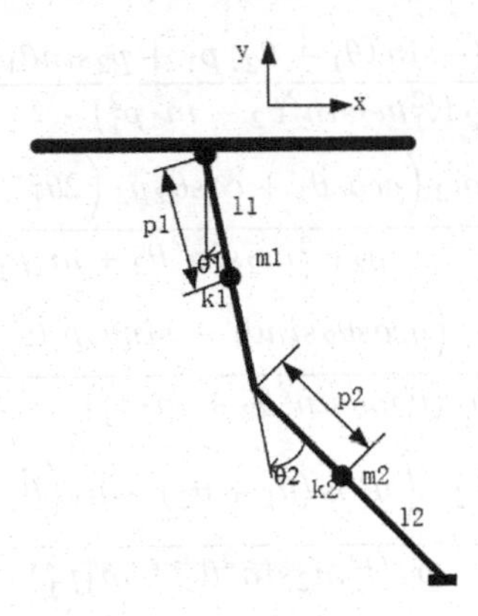

Fig. 1. Schematic diagram of leg and foot structure.

$$G(q) = \begin{pmatrix} (m_1p_1 + m_2l_1)\, g \sin\theta_1 + m_2p_2g \sin(\theta_1 + \theta_2) \\ m_2p_2g \sin(\theta_1 + \theta_2) \end{pmatrix}$$

Here, θ_1 and θ_2 mean the rotation angle variables of connecting rod 1 and connecting rod 2 respectively; m_1 and m_2 symbolize the masses of connecting rod 1 and connecting rod 2 respectively; l_1 and l_2 describe the lengths of connecting rod 1 and connecting rod 2 respectively; k_1 and k_2 betoken the centroids respectively; p_1 and p_2 stand for the distances from the center of the joint respectively; τ_1 and τ_2 represent the torque of joint 1 and joint 2, respectively.

The dynamic equation above shows that the system is a multiple-input multiple-output (MIMO) system, and there is a strong coupling between the joints. Select the joint angle as the system output variable to yield the equation

$$y = (\theta_1 \ \ \theta_2)^T \tag{2}$$

and choose the state variable

$$x = \left(\theta_1 \ \ \theta_2 \ \ \dot{\theta}_1 \ \ \dot{\theta}_2\right)^T$$

then the non-linear system can be expressed as the following state space form [10],

$$\begin{cases} \dot{x} = f_i(x) + g_i(x)u_i \\ y = \begin{pmatrix} h_1(x) \\ h_2(x) \end{pmatrix} = \begin{pmatrix} \theta_1 \\ \theta_2 \end{pmatrix} \end{cases} \tag{3}$$

where

$$u_1 = \tau_1, u_2 = \tau_2, f_1(x) = \dot{\theta}_1, f_2(x) = \dot{\theta}_2$$

$$f_3(x) = \frac{-gm_1p_1sin\theta_1 + cos\theta_2sin\theta_2l_1^2m_2\dot{\theta}_1^2}{l_1^2m_2sin^2\theta_2 + m_1p_1^2} + \frac{sin\theta_2l_1m_2\left(gcos(\theta_1+\theta_2) + p_2\left(\dot{\theta}_1 + \dot{\theta}_2\right)^2\right)}{l_1^2m_2sin^2\theta_2 + m_1p_1^2}$$

$$
\begin{aligned}
f_4(x) = & \frac{gm_1p_1(-sin(\theta_1+\theta_2)p_1+p_2sin\theta_1)}{p_2(l_1^2m_2sin^2\theta_2+m_1p_1^2)} \\
& -\frac{sin\theta_2l_1^2m_2\left(gcos\theta_1+cos\theta_2p_2\left(2\dot{\theta}_1^2+2\dot{\theta}_1\dot{\theta}_2+\dot{\theta}_2^2\right)\right)}{p_2(l_1^2m_2sin^2\theta_2+m_1p_1^2)} \\
& +\frac{l_1\left(m_1p_1\left(gcos\theta_2sin\theta_1-sin\theta_2p_1\dot{\theta}_1^2\right)\right)}{p_2(l_1^2m_2sin^2\theta_2+m_1p_1^2)} \\
& -\frac{sin\theta_2m_2p_2\left(gcos(\theta_1+\theta_2)+p_2\left(\dot{\theta}_1+\dot{\theta}_2\right)^2\right)}{p_2(l_1^2m_2sin^2\theta_2+m_1p_1^2)} \\
& -\frac{sin\theta_2l_1^3m_2\dot{\theta}_1^2}{p_2(l_1^2m_2sin^2\theta_2+m_1p_1^2)}
\end{aligned}
$$

$$
g_1(x)=\begin{pmatrix} 0 \\ 0 \\ \frac{1}{sin^2\theta_2l_1^2m_2+m_1p_1^2} \\ -\frac{l_1cos\theta_2+p_2}{sin^2\theta_2l_1^2m_2p_2+m_1p_1^2p_2} \end{pmatrix}, g_2(x)=\begin{pmatrix} 0 \\ 0 \\ -\frac{l_1cos\theta_2+p_2}{sin^2\theta_2l_1^2m_2p_2+m_1p_1^2p_2} \\ \frac{l_1^2m_2+m_1p_1^2+2l_1m_2p_2cos\theta_2+m_2p_2^2}{m_2\left(sin^2\theta_2l_1^2m_2+m_1p_1^2\right)p_2^2} \end{pmatrix}
$$

2.2 Feedback Linearization of Nonlinear Dynamic Model

In order to achieve control and fault diagnosis, the above-mentioned nonlinear system is linearized. According to [10], the following definitions are preliminarily given.

Definition 1. *Given a smooth scalar function $j(z)$ and a vector field $D(z)$, the derivative of the scalar function along the vector field can be defined as the Lie derivative, which is called the Lie derivative of j to D, which is denoted as L_Dj.*

Let $j(z): R^n \to R$ be a smooth scalar function, $D(z): R^n \to R^n$ is a smooth vector on R^n. $E(z): R^n \to R^n$ is another smooth vector on R^n.

It turns out

$$
L_Dj=\frac{\partial j(z)}{\partial z}\cdot D(z)=\left(\frac{\partial j(z)}{\partial z_1},\frac{\partial j(z)}{\partial z_2},\ldots,\frac{\partial j(z)}{\partial z_n}\right)\cdot D(z)\sum_{w=1}^{n}\frac{\partial j(z)}{\partial z_n}\cdot D_w(z)
$$

Moreover, multiple Lie derivatives can be recursively defined as

$$
\begin{aligned}
L_D^kj(z) &= L_D\left(L_D^{k-1}j(z)\right)=\frac{\partial L_D^{k-1}j(z)}{\partial z}\cdot D(z) \\
L_EL_Dj(z) &= L_E(L_Dj(z))=\frac{\partial L_Dj(z)}{\partial z}\cdot E(z) \\
L_EL_D^kj(z) &= L_E\left(L_D^kj(z)\right)=\frac{\partial L_D^kj(z)}{\partial z}\cdot E(z)
\end{aligned}
$$

here, w and k are non-negative integers.

Definition 2. *For MIMO system, let m be the dimension of the system. If the point $x^0 \in U$ satisfies the following two conditions, the relative order of the system is called $r_1, \ldots, r_m$.*

(1) For all x^0 in the neighborhood of x and all $1 \leq j \leq m, 1 \leq i \leq m, 0 \leq k < r_i - 1$, satisfying

$$L_{gj} L_f^k h_i(x) = 0$$

(2) At a certain point $x = x^0$, the matrix $\alpha(x^0)$ is non-singular.

$$\alpha(x^0) = \begin{pmatrix} L_{g1} L_f^{r_1 - 1} h_1(x^0) & \cdots & L_{gj} L_f^{r_1 - 1} h_1(x^0) \\ \vdots & \ddots & \vdots \\ L_{g1} L_f^{r_i - 1} h_i(x^0) & \cdots & L_{gj} L_f^{r_i - 1} h_i(x^0) \end{pmatrix}$$

Here, the subscript i of r_i and h_i, that is, the row number, refers to the output channel number; r_i is an integer, and the subscript j of g_j, that is, the column number, refers to the input channel number.

Based on the definitions above, the following proposition is given:

Proposition 1. *For model (3), the relative order of the system is $r_1 = r_2 = 2$ and $r_1 + r_2 = 4 = m$, such that the requirements for feedback linearization are satisfied.*

According to proposition 1, the following equation (4) is established

$$\begin{pmatrix} y_1^{(r_1)} \\ y_1^{(r_2)} \end{pmatrix} = b(x) + \alpha(x) \begin{pmatrix} u_1 \\ u_2 \end{pmatrix} \tag{4}$$

where

$$b(x) = \begin{pmatrix} b_1(x) \\ b_2(x) \end{pmatrix} = \begin{pmatrix} L_f^2 h_1(x) \\ L_f^2 h_2(x) \end{pmatrix}$$

$$\alpha(x) = \begin{pmatrix} L_{g1} L_f h_1(x) & L_{g2} L_f h_1(x) \\ L_{g1} L_f h_2(x) & L_{g2} L_f h_2(x) \end{pmatrix}$$

$$= \begin{pmatrix} \frac{1}{sin^2\theta_2 l_1^2 m_2 + m_1 p_1^2} & -\frac{l_1 cos\theta_2 + p_2}{sin^2\theta_2 l_1^2 m_2 p_2 + m_1 p_1^2 p_2} \\ -\frac{l_1 cos\theta_2 + p_2}{sin^2\theta_2 l_1^2 m_2 p_2 + m_1 p_1^2 p_2} & \frac{l_1^2 m_2 + m_1 p_1^2 + 2 l_1 m_2 p_2 cos\theta_2 + m_2 p_2^2}{m_2 \left(sin^2\theta_2 l_1^2 m_2 + m_1 p_1^2\right) p_2^2} \end{pmatrix}$$

In order to achieve linearization, the following form of feedback control is applied to the system

$$u = \alpha^{-1}(x)(-b(x) + \nu) \tag{5}$$

here, $\nu = (\nu_1 \ \ \nu_2)^T$ is defined as a new input and we have

$$\begin{pmatrix} y_1^{(r_1)} \\ y_1^{(r_2)} \end{pmatrix} = \begin{pmatrix} v_1 \\ v_2 \end{pmatrix} \tag{6}$$

Let the new state variable be

$$X = (\gamma \ \ \dot{\gamma} \ \ \varphi \ \ \dot{\varphi})^T$$

and then the transformed system is

$$\begin{cases} \dot{X} = AX + Bv \\ y = CX \end{cases} \tag{7}$$

where

$$A = \begin{pmatrix} 0\,1\,0\,0 \\ 0\,0\,0\,0 \\ 0\,0\,0\,1 \\ 0\,0\,0\,0 \end{pmatrix}, B = \begin{pmatrix} 0\,0 \\ 1\,0 \\ 0\,0 \\ 0\,1 \end{pmatrix}, C = \begin{pmatrix} 1\,0\,0\,0 \\ 0\,1\,0\,0 \\ 0\,0\,1\,0 \\ 0\,0\,0\,1 \end{pmatrix}$$

So far, the original system is transformed into an equivalent linear system [11].

3 LQ Controller Design

For the equivalent linear system (7), many method can be used to design its controller [9,19]. In order to achieve fault diagnosis and achieve the purpose of control, this study chooses to design LQ controller to stabilize the system (7) and minimize the following performance index

$$J(v(\cdot)) = \int_0^\infty \left(X^T QX + v^T Rv\right) dt \tag{8}$$

Here, Q and R are the weighting matrices with appropriate dimensions and $Q^T = Q > 0, R^T = R > 0$.

According to [11] and [12], the following propositions are given.

Proposition 2. *$\{A, B\}$ is completely controllable.*

Proposition 3. *The state feedback law of LQ controller can be designed as follows.*

$$v(t) = -KX(t) \tag{9}$$

where

$$K = R^{-1} B^T P$$

and P is calculated by solving the corresponding matrix Riccati equation

$$-\dot{P}(t) = P(t) A + A^T P(t) + Q - P(t) BR^{-1} B^T P(t)$$

$$P(t_f) = 0, t \in (0, t_f), \ t_f \to \infty$$

4 Fault Detection Based on LQ Control Performance

In this section, a method of fault diagnosis is proposed according to the changes in the performance of the system during actual operation.

As shown in Fig. 1, the structure of the robot is mainly composed of two rigid body components connected by joints. In the nominal (no disturbance and fault) state, under the action of the controller (9), the system repeatedly calculates the spatial coordinates of the center position of foot end, knee joint center, hip joint center, the joint rotation angle θ_1 θ_2, and joint torque τ_1 τ_2 according to the planned route. The above process can be accurately completed in the physical system to achieve the function of minimizing the performance index $J(v(\cdot))$ and obtain the optimal control effect. However, during the operation of the robot, it will inevitably be affected by disturbance or various failures. The physical phenomena caused by the fault signal are mainly induced by joint idling, sole stepping, and component position movement. In addition, changeable terrain, sudden attack, etc. will deteriorate the performance of the robot system.

Specifically, when the system (7) is in a non-nominal scenario (real applications), the system model can be expressed as

$$\begin{cases} \dot{X} = AX_p + Bv + E_d d + E_f f \\ y \ = CX_p \end{cases} \tag{10}$$

Here, E_d and E_f are known matrices, X_p is the state variable under real-time operational conditions, d and f represent the interference signal and the fault signal, respectively.

It can be seen that the performance index $J(v(\cdot))$ of LQ controller is composed of two parts: the terminal cost function $\ell_1 = X^T QX$ and the control input constraint function $\ell_2 = v^T Rv$. For one thing, if the state variable X has a deviation, ℓ_1 is the cost function to measure the size of the deviation. When the deviation is caused by a fault, it will cause ℓ_1 to increase. For another thing, ℓ_2 is obviously on the constraint of v in the dynamic process. The performance index $J(v(\cdot))$ which is sensitive to deviation is the integration of the quadratic function with X and v. Therefore, the above changes can be exploited for fault diagnosis of this system.

Based on the analysis above, performance index $J(v(\cdot))$ can be used as the evaluation function $\Gamma(t)$

$$\Gamma(t) = J(v_p(\cdot)) = \int_0^{\infty} \left(X_p^T QX_p + v_p^T Rv_p\right) dt \tag{11}$$

Here, the state variable X_p and the control input v_p are the measured values under actual operational conditions. Meanwhile, calculate the theoretical performance index $J(v(\cdot))$ of the LQ controller according to Eq. (8) in the nominal system (7), and use it as the threshold $\Gamma_{th}(t)$

$$\Gamma_{th}(t) = J(v_s(\cdot)) = \int_0^{\infty} \left(X_s^T QX_s + v_s^T Rv_s\right) dt \tag{12}$$

Here, the state variable X_s and the control input v_s are values under ideal conditions.

Through the following logic, fault diagnosis can be realized

$$\begin{cases} \Gamma(t) \leq \Gamma_{th}(t), no\ fault\ occurs \\ \Gamma(t) > \Gamma_{th}(t), fault\ occurs \end{cases} \tag{13}$$

Furthermore, it can be seen that the evaluation function $\Gamma(t)$ is a function of time t. In order to further describe the impact of faults on system performance, a new evaluation function $\Gamma_1(t)$ and threshold $\Gamma_{th1}(t)$ can be defined as the derivatives of $\Gamma(t)$ and threshold $\Gamma_{th}(t)$, respectively.

$$\begin{cases} \Gamma_1(t) = \dot{\Gamma}(t) \\ \Gamma_{th1}(t) = \dot{\Gamma_{th}}(t) \end{cases} \tag{14}$$

Hence, the following logic can be used to fault diagnosis,

$$\begin{cases} \Gamma_1(t) \leq \Gamma_{th1}(t), no\ fault\ occurs \\ \Gamma_1(t) > \Gamma_{th1}(t), fault\ occurs \end{cases} \tag{15}$$

It can be seen from (11) and (12), for fault signals with different strength f_λ,

$$||f_\lambda||_2 = \left(\int_0^\infty f_\lambda^T(t) f_\lambda(t)dt\right)^{1/2}, ||f_\lambda||_2 > ||f_{\lambda+1}||_2, \lambda = 0, 1, 2, 3\ldots$$

When $f_\lambda > f_{\lambda+1}$, the evaluation function $\Gamma_{f_\lambda} > \Gamma_{f_{\lambda+1}}$ is derived. Therefore, the proposed method can realize the classification of the fault.

Remark 1. Compared with other fault diagnosis methods, $H_\infty - filter$ is relatively difficult to solve the observer, and the LMI technique for solving the observer is very conservative. This method is generally not universal to different systems. In the traditional case, H_-/H_∞ algorithm requires that the distribution matrix from fault to measurement output should satisfy the condition of full rank. The sensitivity of the system to fault and the robustness to disturbance need also to be balanced. Therefore, in the actual design, the algorithm is very sensitive, which is not convenient to engineering application. The method proposed in this paper, based on the controller, can realize fault diagnosis directly without designing observer. The method is effective and reliable, and can be directly applied to the actual system.

5 Verified Result

The physical model built in this experiment is shown in Fig. 2. The movement basis of a legged robot mainly refers to the movement of a single leg. The physical prototype of a single leg robot fixes the fuselage on a vertical slide rail through a slider. It cannot perform horizontal direction and can only move in the vertical direction. It is mainly composed of rigid connecting rods, hydraulic servo drives and displacement sensors, etc. The rated operational pressure of the hydraulic drive is 21 MPa, and its maximum stroke is 70 mm. The telescopic distance of the displacement sensor is 90 mm. The specific parameters of the physical prototype are shown in Table 1, and the proposed algorithm is verified on this physical prototype. The results are as follows.

To illustrate the result achieved in this study, consider the following situations.

Fig. 2. System physical model.

Table 1. System parameter table.

Parameter	Value
m_1	1.5 kg
m_2	1.2 kg
l_1	355 mm
l_2	310 mm
Sample frequency	1000 Hz

Example 1. Set

$$d = 0.2sint, f_1 = 1.2sin\left(t + 0.1\right), f_2 = 1.4sin\left(t + 0.4\right), f_3 = 1.5sin\left(t + 0.8\right)$$

Step 1: Calculate the state space Eq. (3) according to the parameters in Table 1;

Step 2: According to Eq. (5) and (6), calculate the equivalent linearization system (7);

Step 3: Solve the LQ feedback controller K;

$$K = \begin{pmatrix} 1 & 1.732 & 0 & 0 \\ 0 & 0 & 1 & 1.732 \end{pmatrix}$$

Step 4: Calculate the evaluation function $\Gamma(t)$ of the system in the nominal system and use it as the threshold $\Gamma_{th}(t)$;

$$\Gamma(t) = \frac{1}{2}\int_{t_0}^{t_f} \left(2x_1^2 + 4.00017041x_2^2 + 2x_3^2 \right.$$
$$\left. +4.00017041x_4^2 + 3.4642x_1x_2 + 3.4642x_3x_4\right) dt$$

Step 5: Diagnose the fault phenomenon according to the logical (13).

The results are shown in Fig. 3 and Fig. 4. The line 1, line 2 and line 3 represent the evaluation function of f_1, f_2 and f_3 respectively in Fig. 4. It can be seen from Fig. 3 and Fig. 4 that, the proposed scheme (13) can detect faults with different amplitudes.

Example 2. Set

$$d = 0.2sint, f_1 = 1.2sin(t + 0.1), f_2 = 1.4sin(t + 0.4), f_3 = 1.5sin(t + 0.8)$$

According to (3), (5), (7), (11), (12), (14) and (15), the results are shown in Fig. 5 and Fig. 6. The line 1, line 2 and line 3 represent the evaluation function of f_1, f_2 and f_3 respectively in Fig. 6. It can be seen from Fig. 5 and Fig. 6 that, the proposed scheme (15) can detect faults with different amplitudes.

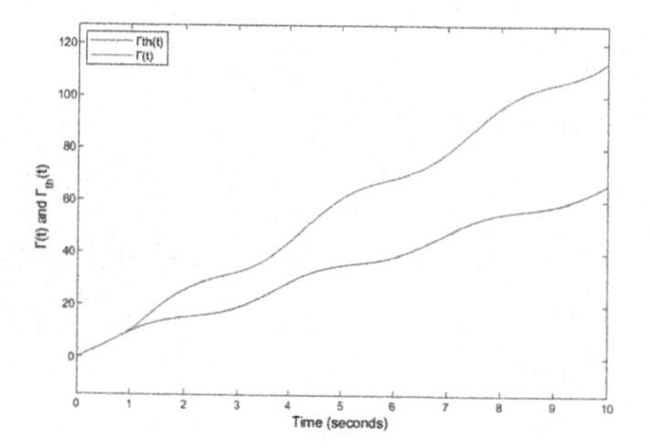

Fig. 3. Fault occurs by (13).

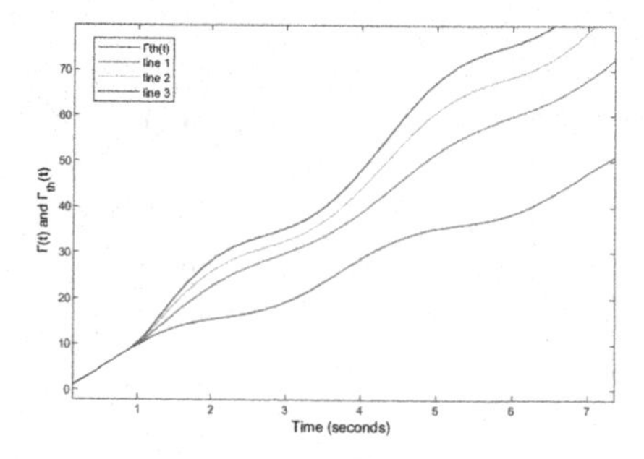

Fig. 4. The grade of the fault by (13).

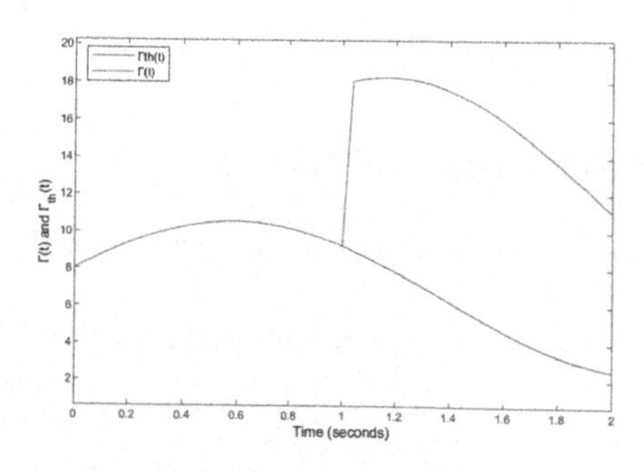

Fig. 5. Fault occurs by (15).

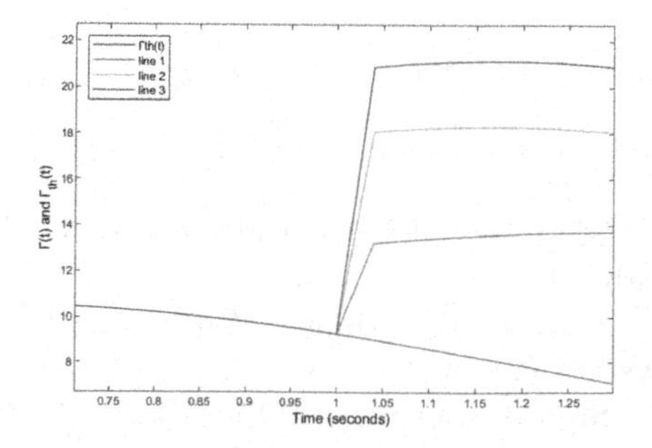

Fig. 6. The grade of the fault by (15).

6 Conclusion

A new fault detection strategy for the motion unit of the leg-foot robot has been proposed, and verified the proposed algorithm through a physical prototype. The algorithm have been directly used in the physical system and realized the function of detecting system failures, which proves the reliability and validity of the algorithm.

References

1. Gao, Z., Shi, Q., Fukuda, T., Li, C., Huang, Q.: An overview of biomimetic robots with animal behaviors. Neurocomputing **332**, 339–350 (2019)
2. Martinez-Guerra, R., Mata-Machuca, J.L.: Fault Detection and Diagnosis in Nonlinear Systems. Springer, Cham (2016). https://doi.org/10.1007/978-3-319-03047-0
3. Chow, E., Willsky, A.: Analytical redundancy and the design of robust failure detection systems. IEEE Trans. Autom. Control **29**(7), 603–614 (1984)
4. Ding, S.X.: Model-Based Fault Diagnosis Techniques: Design Schemes, Algorithms, and Tools. Springer, Heidelberg (2008). https://doi.org/10.1007/978-3-540-76304-8
5. Ma, H.-J., Yang, G.-H.: Simultaneous fault diagnosis for robot manipulators with actuator and sensor faults. Inf. Sci. **336**, 12–30 (2016)
6. Theilliol, D., Noura, H.: Sensor fault diagnosis based on energy balance evaluation: application to a metal processing. Appl. Metal Process. **45**(4), 603–610 (2006)
7. Li, L., Ding, S.X.: Performance supervised fault detection schemes for industrial feedback control systems and their data-driven implementation. IEEE Trans. Industr. Inform. **16**(4), 2849–2858 (2019)
8. Kajita, S., Hirukawa, H., et al.: Introduction to Humanoid Robotics. Springer, Heidelberg (2014). https://doi.org/10.1007/978-3-642-54536-8
9. Akhtar, A., Nielsen, C., Waslander, S.L.: Path following using dynamic transverse feedback linearization for car-like robots. IEEE Trans. Robot. **31**(2), 269–279 (2015)
10. Isidori, A.: Nonlinear Control Systems. Springer, Heidelberg (2013). https://doi.org/10.1007/978-1-84628-615-5
11. Callier, F.M., Desoer, C.A.: Linear System Theory. Springer, Heidelberg (2012). https://doi.org/10.1007/978-1-4612-0957-7
12. Roveda, L., Piga, D.: Robust state dependent Riccati equation variable impedance control for robotic force-tracking tasks. Int. J. Intell. Robot. Appl. **4**(4), 507–519 (2020). https://doi.org/10.1007/s41315-020-00153-0
13. Khalastchi, E., Kalech, M.: On fault detection and diagnosis in robotic systems. ACM Comput. Surv. (CSUR) **51**(1), 1–24 (2018)
14. Zhong, M., Song, Y., Ding, S.X.: Parity space-based fault detection for linear discrete time-varying systems with unknown input. Automatica **59**, 120–126 (2015)
15. Villa, N.A., Englsberger, J., Wieber, P.-B.: Sensitivity of legged balance control to uncertainties and sampling period. IEEE Robot. Autom. Lett. **4**(4), 3665–3670 (2019)
16. Owaki, D., Ishiguro, A.: A quadruped robot exhibiting spontaneous gait transitions from walking to trotting to galloping. Sci. Rep. **7**(1), 1–10 (2017)
17. Halder, B., Sarkar, N.: Analysis of order of redundancy relation for robust actuator fault detection. Control. Eng. Pract. **17**(8), 966–973 (2009)

18. Omali, K.O., Kabbaj, M.N., Benbrahim, M.: New Developments and Advances in Robot Control. Springer, Cham (2019)
19. Zabczyk, J.: Mathematical Control Theory. Springer, Cham (2020). https://doi.org/10.1007/978-3-030-44778-6
20. Poon, J., Jain, P., et al.: Fault prognosis for power electronics systems using adaptive parameter identification. IEEE Trans. Ind. Appl. **53**(3), 2862–2870 (2017)

Robot Bolt Skill Learning Based on GMM-GMR

Zhao Man[1,2], Li Fengming[1,2], Quan Wei[1,2], Li Yibin[1,2], and Song Rui[1,2](✉)

[1] Center for Robotics, School of Control Science and Engineering, Shandong University, No. 17923, Jingshi Road, Jinan 250061, China
rsong@sdu.edu.cn

[2] Engineering Research Center of Ministry of Education for Intelligent Unmanned Systems, Jinan 250100, China

Abstract. Bolt tightening is one of the typical robot screwing operations. In industry, various types of bolts, high repeatability, and frequent task switching have brought challenges for robots to screw bolts. This paper is based on the human-machine skill transfer method to realize the robot bolt screwing. First, the teaching data was aligned with the trajectory through Dynamic Time Warping (DTW), and then the Gaussian Mixture Model, Gaussian Mixture Regression (GMM-GMR) was used for feature extraction and trajectory information. The screwing trajectory was learned and fitted. Finally, method verification was carried out on the built platform. The results show that the robot based on GMM-GMR has acquired the skills of screwing operation.

Keywords: Robot bolt screwing · GMM-GMR · Skill learning and reproduction

1 Introduction

In industry, screw operation is mainly achieved through manual screwing or special machine operation. The screw operation has the characteristics of high repeatability and heavy task, which consumes a lot of labor. One machine is only suitable for special scenes and bolts. When the screw task changes, a lot of time and energy will be spent calibrating. Therefore, it is very important to improve industrial production efficiency and product quality for the robot-based learning method.

At present, the research on the screwing of robot bolts is mainly based on the detection of the screwing state for process control. K. Dharmara et al. [1] used differential edge detection and point cloud matching to detect the real-time position between the stud and the base, and adjust the screwing process in real-time through the detection information. The Harbin Institute of Technology team [2]

Supported by the Joint Funds of the National Natural Science Foundation of China (Grant No. U2013204), Shandong Major Science and Technology Innovation Project (Grand No. 2019JZZY010429), Shandong Provincial Key Research and Development Program (Grand No. 2019TSLH0302).

X.-J. Liu et al. (Eds.): ICIRA 2021, LNAI 13014, pp. 235–245, 2021.
https://doi.org/10.1007/978-3-030-89098-8_22

identified the contact state of the robot end-effector with the environment through logistic regression classification, and then performed the screwing through the force/position hybrid control. The paper [3] proposed a Fuzzy Logic Controller (FLC) with expert knowledge of the tightening process and error detection capabilities. The bolt tightening process was divided into four stages: initial alignment, partial engagement, full engagement, and tightening. Different phases performed different actions. S. Pitipong et al. [4] proposed bolt edge detection combined with Canny and Hough transform, and used visual servoing and corresponding systems which was based on fuzzy control to correct the bolt tightening process of the robot. Another team [5] used multiple cameras to identify the state of the bolt, and used impedance control algorithms to control the screwing process.

The method based on LFD enables the robot to get away from the repetitive simple predetermined behavior in the constrained environment [6] and perform the optimal action in the unstructured environment. This paper models the process of screwing skills from learning to reproduction, and builds a robot screwing system under the LFD framework. The authors dragged and taught multiple times to get multiple screwing trajectories. The trajectory data was preprocessed by DTW, and GMM-GMR was used to extract features of trajectory information to fit a relatively smooth trajectory. The screwing skills can be reproduced, the new screwing start and end-points can be input, and the robot arm can automatically plan the screwing path, realize the learning and generalization of screwing skills, and complete the entire bolt screwing task.

The second chapter introduces the related work of GMM and GMR, the third chapter introduces specific methods, the fourth chapter introduces the experimental platform and results, and the fifth chapter summarizes the full text.

2 Related Work

According to the different interaction modes of robot skill teaching, C. Zeng et al. [7] from South China University of Technology divided LFD into three categories: vision, teleoperation, and human-computer physical interaction. There are different methods for learning and reproducing learned tasks, but they all have the same three steps: behavior acquisition, behavior representation, and behavior representation [15]. According to the use of training data, the paper [6] divided robot operation skill learning methods into reinforcement learning method, teaching-learning method and small data learning method. N.J. Liu et al. [8] roughly divided the demonstration and learning methods into two categories: Behavior Cloning and Inverse Reinforcement Learning. Behavioral cloning is based on the given sequence of multiple teaching tracks to collect the sample set of state-action pairs and directly learn the mapping relationship between the state and action. The commonly used methods include Dynamic Motion Primitive (DMP), Hidden Markov Model (HMM), Gaussian Mixture Model (GMM), etc. The main difference between these models lies in their different perspectives. In DMP, skill characteristics are regarded as motion primitives, and a sequence of motion primitives can be obtained by fitting the DMP model with teaching data. HMM and GMM use probability calculation method to teach and learn skills. In other words, the relevant characteristics of skills are extracted and the corresponding skill feature information is obtained according to the learned state information. The GMM model is only used to represent skill data, and it should be combined with the Gaussian Mixture Regression model (GMR) [9] to realize skill learning.

Calinon et al. [10] proposed a demonstration learning framework for motor skills based on GMM, which can handle the task constraints of joint space and Cartesian space at the same time, and allows the robot to reuse the learned skills to handle new tasks and situation. Muhlig et al. [11] introduced the GMR model into the imitation learning framework of humanoid robots to learn the movement information of the target object, and dynamically adjust the corresponding actions according to the movement information of the target object.

After that, many researchers applied this method to robot skill learning. The paper [12,13] proposed a Gaussian mixture model and Gaussian mixture regression (GMM-GMR) method to analyze the data from human demonstration for the learning of performance skills under the framework of DMP. Y.Q. Wang et al. [14] developed an optimization algorithm based on Simulated Annealing Reinforcement Learning (SA-RL) to refine the number of Gaussian clusters and improved GMM-GMR so that LFD-enabled collaborative robots could effectively perform various complex manufacturing tasks. According to the Bayesian nonparametric model, the characteristic of the appropriate component number can be automatically determined [15], and the Dirichlet process Gaussian mixture model was used to carry out the initial fitting of the demonstration, which solved the problem that stability and accuracy were difficult to be considered in robot skill learning methods.

3 Method

The authors modeled the process from learning to reproduction of the screwing skill and the track information of multiple screws was got from demonstrations. The track data was preprocessed by DTW and was extracted by GMM-GMR. A relatively smooth track was obtained by learning and fitting. By inputting a new starting point, the robot automatically planed the path of the screw and realized the learning and generalization of the screw skill. The framework of this method is shown in Fig. 1:

3.1 Dynamic Time Warping (DTW)

The robot was dragged to screw many times to get the screwing trajectory$\left[x_t\ y_t\ z_t\ \alpha_t\ \beta_t\ \gamma_t\right]$, $t = \{1, 2, \cdots, T\}$, which is represented by the pose of the end of the robot arm relative to the base system, where $\left[x_t\ y_t\ z_t\ \alpha_t\ \beta_t\ \gamma_t\right]$ are the coordinate of the end of the robot arm relative to the base system $o - x_0 y_0 z_0$ and the rotation angle around each coordinate axis, and T is the number of data points contained in single teaching.

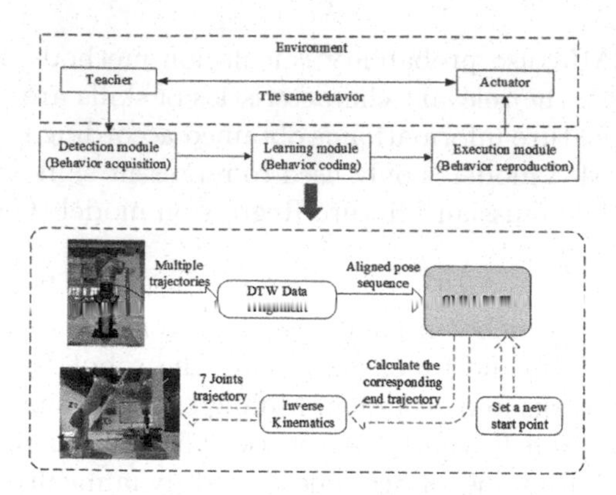

Fig. 1. The block diagram of robot bolt screwing skills using GMM-GMR method.

The saved multiple teaching trajectories will deviate when they are reproduced in a certain time sequence, and the 6 pose sequences are optimized through the Dynamic Time Warping (DTW) method [16]. DTW can characterize the degree of similarity between two sequences and get a method to align the two sequences.

Two time series are Q and C. The lengths are n and m respectively: $Q = q_1, q_2, ...q_i, ..q_n$, $C = c_1, c_2, ...c_j, ..c_m$, If $n = m$, through calculating the distance between the two sequences directly, the authors can characterize the similarity of the two sequences. When n and m are not equal, it is required to align them with DTW. In order to pair two sequences, a matrix grid of $n \times m$ needs to be constructed. The matrix elements (i, j) represent the distance $d(q_i, c_j)$ between the two points q_i and c_j. Euclidean distance $d(q_i, c_j) = \sqrt{(q_i - c_j)^2}$ is used to express the similarity between each point of the sequence Q and each point of the sequence C. The smaller the distance, the higher the similarity. The Dynamic Programming (DP) method is used to find a path through several grid points in this grid, and the grid points through which the path passes are the aligned points for the calculation of the two sequences.

This path is defined as a Warping path, and it is represented by W. The kth element of W is defined as $w_k = (i, j)_k$, which defines the mapping of sequence Q and C. Therefore,

$$W = w_1, w_2, ..., w_k, \max(m, n) \leq K < m + n - 1 \tag{1}$$

This path is not chosen arbitrarily, and several constraints need to be met:

- Boundary conditions: The time interval between $w_1 = (1, 1)$ and $w_k = (m, n)$ teaching has changed, but the order of its parts cannot be changed. Therefore, the selected path must start from the lower left corner and end at the upper right corner.
- Continuity: If $w_{k-1} = (a', b')$, then the next point's $w_k = (a, b)$ of the path needs to meet $(a - a') \leq 1$ and $(b - b') \leq 1$. That is to say, it cannot cross

a certain point to match, only to align with the adjacent point. This ensures that every coordinate in Q and C appears in W.

- Monotonicity: If $w_{k-1} = (a', b')$, then the next point's $w_k = (a, b)$ of the path needs to satisfy $0 \leq (a - a')$ and $0 \leq (b - b')$. This restricts the above points of w must be monotonous over time.

Combined with continuity and monotony constraints, each grid point (i, j) has only three paths. For example, if the path has already passed through the grid point, then the next grid to pass through is one of three conditions: $(i+1, j)$, $(i, j+1)$ or $(i+1, j+1)$.

The final result is the sum of the coordinates (i, j) of all points along the path and the distance between the points of the two time series Q_i and C_i, and the element in the upper right corner of the cost matrix $W(i, j)$ is obtained:

$$W(i,j) = d(i,j) + \min[W(i-1,j), W(i,j-1), W(i-1,j-1)] \tag{2}$$

According to the idea of dynamic programming to minimize the accumulated distance along the path, the mean sequences of x, y, z, α, β and γ of the teaching trajectory were selected as reference sequences respectively to compare the similarity between each pose and its mean sequences. The paths of several points in the grid were found by dynamic programming, and the alignment between the sequences was obtained according to the grid points through which the paths passed, and the corresponding similar points were aligned.

3.2 GMM-GMR

GMM-GMR [13,17] is performed on the end pose data after DTW alignment. The Gaussian Mixture Model (GMM) is used to preprocess the parameters of each teaching point of the robot, and code to extract the behavior characteristics of the robot. The Gaussian Mixture Regression (GMR) method is used to regress and generalize, select the action primitive point closest to the actual screwing task in the stored teaching information, and fit a relatively smooth trajectory curve.

Suppose the first teaching data point is $\xi_j = \{\xi_{s,j}, \xi_{t,j}\}$, $j = \{1, 2, \cdots, P\}$, where P is the number of data points included in a single teaching, $\xi_{s,j}$ is the spatial coordinate value or node angle, $\xi_{t,j}$ is the time value. Assume that each data point obeys the following probability distribution:

$$p(\xi_{t,j}) = \sum_{k=1}^{K} p(k)p(\xi_j|k) \tag{3}$$

Among them, $p(k)$ is the prior probability, and $p(\xi_j|k)$ is the conditional probability distribution, which obeys the Gaussian distribution. Therefore, the entire teaching data set can be represented by the Gaussian mixture model, and K is the number of Gaussian distributions that make up the GMM.

$$p(k) = \pi_k \tag{4}$$

$$p(\xi_j|k) = N(\xi_j;\mu_k,\sum\nolimits_k) = \frac{1}{\sqrt{(2\pi)^D|\sum_k|}} e^{-\frac{1}{2}((\xi_j-\mu_k)^T\sum_k{}^{-1}(\xi_j-\mu_k))} \tag{5}$$

Among them, D is the dimension of the GMM encoding the teaching data; therefore the parameters $\{K,\pi_k,\mu_k,\sum_k\}$ to be determined in the GMM are the prior probability, expectation and variance of the number of components of GMM and the k-th component respectively. Bayesian information criterion is used to estimate parameter K, and model selection is carried out to achieve the tradeoff between model complexity and optimal data fitting performance. EM algorithm is used to estimate parameters $\{\pi_k,\mu_k,\sum_k\}$, and finally model construction is realized.

The teaching data of GMM is reconstructed based on GMR, and the generalized output is obtained. The teaching data ξ_t is used as the query point, and the corresponding spatial value $\xi_s^{'}$ is estimated by GMR. It is known $p(\xi_j|k)$ satisfies the Gaussian distribution, that is $\begin{pmatrix}\xi_{s,k}\\ \xi_{t,k}\end{pmatrix} \sim N(\mu_k,\sum_k)$, where $\mu_k = \{\mu_{s,k},\mu_{t,k}\}$, $\sum_k = \begin{Bmatrix}\sum_{s,k} & \sum_{st,k}\\ \sum_{ts,k} & \sum_{t,k}\end{Bmatrix}$, then under a given condition $\xi_{t,k}$, the conditional probability of $\xi_{s,k}$ also satisfies the Gaussian distribution, that is $\xi_{s,k}|\xi_{t,k} \sim N(\mu_{s,k}^{'},\Sigma_{s,k}^{'})$, $\mu_{s,k}^{'} = \mu_{s,k} + \Sigma_{st,k}(\Sigma_{t,k})^{-1}(\xi_{t,k}-\mu_{s,k})$, $\Sigma_{t,k}^{'} = \Sigma_{s,k} + \Sigma_{st,k}(\Sigma_{t,k})^{-1}\Sigma_{ts,k}$.

The mean value $\mu_s^{'}$ and variance $\Sigma_s^{'}$ of the GMM with K Gaussian components can be obtained.

$$\eta_k = \frac{p(\xi_i|k)}{\sum\limits_{i=1}^{K} p(\xi_i|i)}, \mu_s^{'} = \sum_{k=1}^{K}\eta_k\mu_{s,k}^{'}, \Sigma_s^{'} = \sum_{k=1}^{K}\eta_k^2\Sigma_{s,k}^{'} \tag{6}$$

Suppose the regression function of the GMM is $m(\xi_t)$, as in (7)

$$m(\xi_t) = E(\xi_s|\xi_t) = \sum_{k=1}^{K}\eta_k\mu_{s,k}^{'} = \mu_s^{'} \tag{7}$$

Then the mean value $\mu_s^{'}$ is the required reconstructed value of the teaching data, that is $\xi_s^{'} = \mu_s^{'}$. The generalized data points $\{\xi_s^{'}\xi_t\}$ and the covariance matrix $\Sigma_s^{'}$ were used to extract the task constraints are finally obtained. The generalized data $\xi_s^{'}$ is not included in the teaching data, and the screw trajectory is generated under the relevant constraints $\Sigma_s^{'}$.

4 Experimernts Results

In this part, the platform settings and experimental results are introduced. Average the taught trajectory data and transfer it to the method mentioned above. It is verified that this method can complete the reproduction and generalization of screwing skills.

4.1 The Setting of the Experimental Environment

The KUKA iiwa 7 R800 robot is selected, which has seven degrees of freedom. ROBOTIQ 2-FINGER 85 is selected as the end-effector, which is installed at the end of KUKA iiwa manipulator. The nylon bolt of M30*3.5 is selected and fixed on the bolt screw holder. The screw platform is shown in Fig. 2.

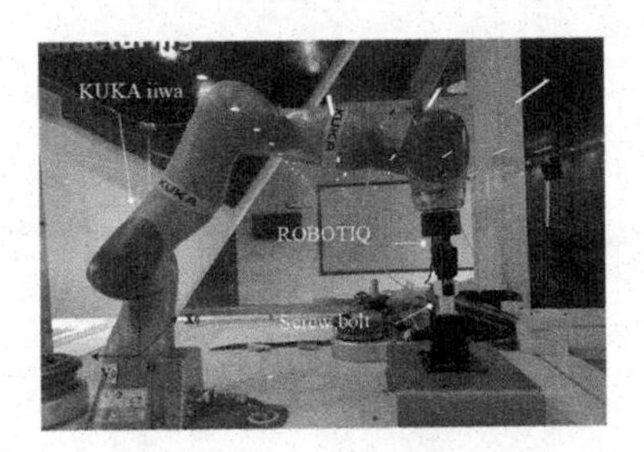

Fig. 2. KUKA iiwa robot arm screwing experiment platform.

The block diagram of the bolt screw system is shown in Fig. 3. The control device was used for processing the position and posture information of the robot arm and making action decisions. The robot arm can adjust the corresponding rotating posture, and the end gripper can grab and screw the bolt. The KUKA Sunrise.OS toolbox is a tool that can be used to implement applications quickly and successfully. With the built-in JAVA database, boxes and object templates, the Sunrise.OS toolbox provides ready-to-use compatible process knowledge. First of all, Sunrise software can be used to control the robot to drag and teach and adjust the appropriate flexibility of the robot arm to make it more smooth. Then, the communication between the gripper and the robot arm was established in the controller, and the gripper was clamped at the top of the bolt at the initial position, and the gripper was opened at the end of the screw. The communication between the robot and the server was established by socket, and the pose sequence of the end joint was recorded while dragging and teaching.

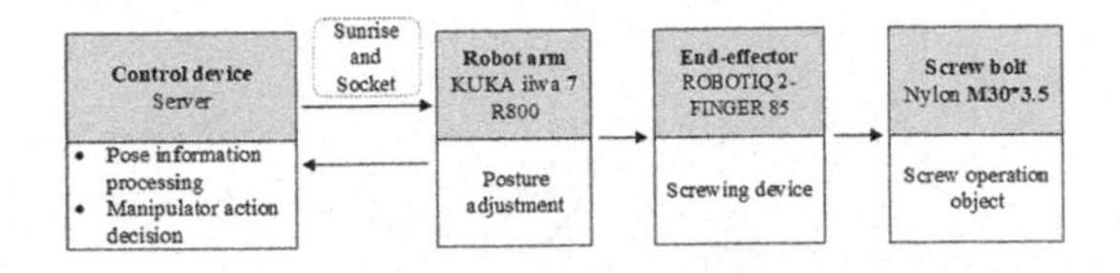

Fig. 3. Block diagram of screw platform.

In the Cartesian coordinate system, the starting screw position of each end of the robot was set as follows: $\left[x_0\ y_0\ z_0\right] = \left[-625.737645\ 8.546893\ 438.795750\right]$ in mm, $\left[\alpha\ \beta\ \gamma\right] = \left[-0.531382\ 0.004931\ 3.097473\right]$ in rad. The Angle range of each rotation of the mechanical arm is 300°. The authors dragged for demonstrations 10 times, and collected the pose of the end of the robot arm every 0.1 s. The data of the ten teaching sessions is shown in Fig. 4. From (a)–(f), they correspond to the terminal positions x, y, z, α, β, and γ. The ten curves in each figure correspond to the pose sequences of the ten demonstrations. Data lengths range from 75 to 93 depending on the speed of each drag. Screwing skills were learned from pose data.

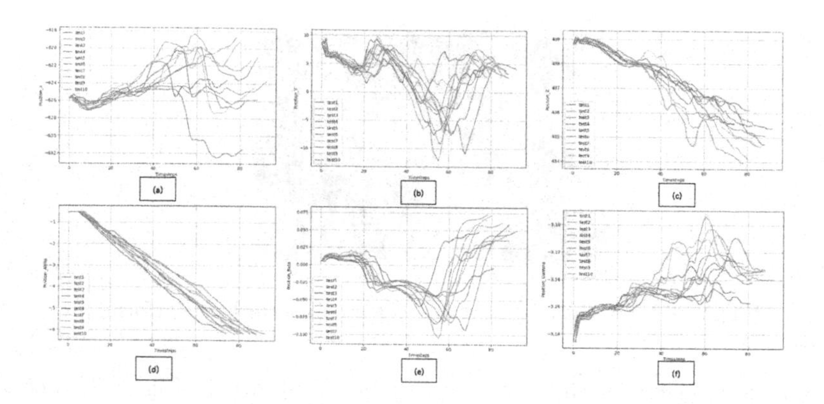

Fig. 4. Ten demonstrations of the end joint position

4.2 The Results of the Experiment

Data Alignment with DTW. Figure 4 shows 10 demonstrations. Many data points were deleted in the long data sequence with the same time interval deletion method, and the data length was modified to 75. Taking the mean series of the data after length alignment as the reference trajectory, the DTW method was used for data alignment. The alignment path sequence is shown in Fig. 5. (a)–(f) correspond to the end pose x, y, z, α, β, and γ. The ten curves in each figure correspond to the pose sequences of the ten experiments. The thick blue line in the figure represents the mean value of each corresponding pose.

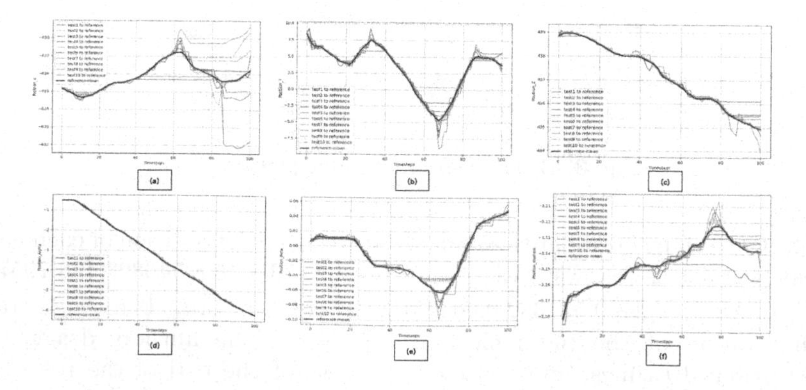

Fig. 5. Trajectories after DTW alignment (Color figure online)

GMM-GMR vs Average Method GMM-GMR was used to cluster and regress the trajectories after DTW adjustment. The parameters to be determined by GMM are $\{K, \pi_k, \mu_k, \sum_k\}$: given the collected ten times of teaching data, the EM algorithm was used to estimate the parameters $\{\pi_k, \mu_k, \sum_k\}$, and K=5 was taken. The GMM model with certain parameters was used to cluster the data after DTW alignment. Then GMR was used to calculate the covariance and mean value, and a trajectory was fitted. The characteristic coding track of the screw is shown in Fig. 6. (a)–(f) correspond to the end pose x, y, z, α, β, and γ. The green dots in each image represent the data points input to the model, and the green curves represent the end pose encoded by GMM. The blue curve coded by GMR consists of two parts: the covariance (light blue) and the mean (dark blue). After GMM-GMR processing, the reconstructed probability trajectory extracted the mean and variance characteristics of several demonstrations.

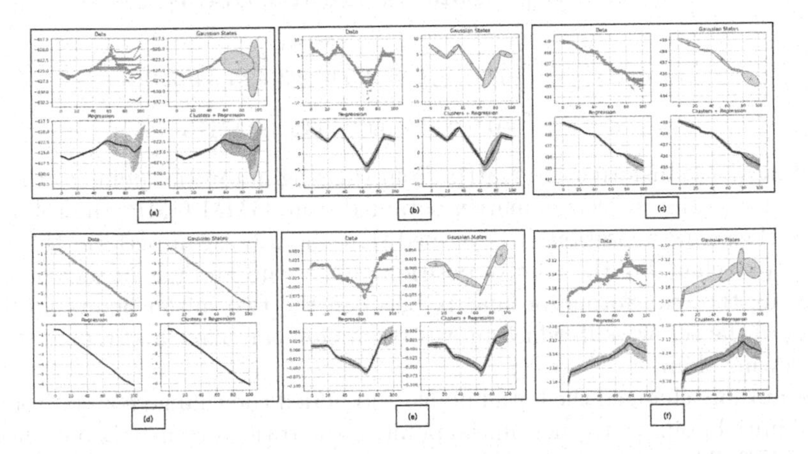

Fig. 6. Trajectories encoded by GMM-GMR. (Color figure online)

Figure 7 shows the variance of the data obtained by the two methods. The blue square bar represents the variance obtained by taking the average of the teaching data, and the orange square bar indicates the variance of the data obtained by the GMM-GMR method. The comparison shows that the data obtained by the two methods are very close at position α, and the variance of the data obtained by GMM-GMR in other positions is smaller, indicating that the obtained trajectory in Fig. 6 is smoother.

Reproduction and Generalization. The trajectory obtained by the mean value and the trajectory calculated by GMM-GMR was respectively sent to the robot so that it can reproduce the screwing process. It is verified that the trajectories obtained by GMM-GMR are more smooth when screwing.

The new screw starting points were set as:

$$\begin{bmatrix} x_1 \; y_1 \; z_1 \end{bmatrix} = \begin{bmatrix} -625.737645 \; 8.546893 \; 446.245450 \end{bmatrix}$$

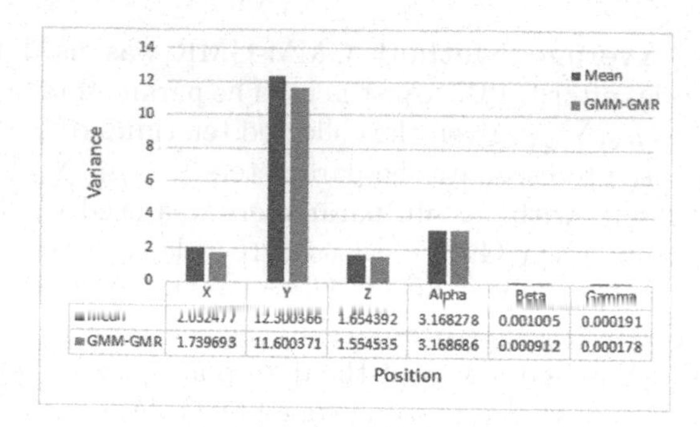

Fig. 7. Variance comparison chart of pose data. (Color figure online)

$$\begin{bmatrix}\alpha\ \beta\ \gamma\end{bmatrix} = \begin{bmatrix}-0.531382\ 0.004931\ 3.097473\end{bmatrix}$$

and

$$\begin{bmatrix}x_2\ y_2\ z_2\end{bmatrix} = \begin{bmatrix}-625.737645\ 8.546893\ 425.346791\end{bmatrix}$$

$$\begin{bmatrix}\alpha\ \beta\ \gamma\end{bmatrix} = \begin{bmatrix}-0.531382\ 0.004931\ 3.097473\end{bmatrix}$$

By adding the transformation matrix from the original initial point to the new initial point to the trajectory matrix obtained from GMM-GMR, the robot can perform the screw operation. It is verified that the robot can perform the screw operation by changing the relative position of the stud and nut in the condition of knowing the transformation matrix between the two initial points after the learning of the screw skill, which proves the model can learn and reproduce the bolt screwing skill.

The authors changed the initial point, and added the change matrix from the original initial point to the new initial point to the trajectory matrix obtained by GMM-GMR. The robot arm can be screwed, and then the teaching and learning of screwing skills can be realized, and the whole bolt can be screwed. When the bolt position changes, the screwing operation can also be performed, as long as the relative position between the two bolts is known.

5 Conclusion

In this paper, the process from learning to reproduction of the twisting skill was modeled by referring to the method of human-machine skill transfer, and the screwing system of the robot under the framework of LFD was established. DTW was used to preprocess the screw track data, and GMM-GMR was used to extract the track information features and fit a relatively smooth track. By inputting a new starting point of the screw, the robot automatically planed the path of the screw, realized the learning and generalization of the screw skill, and completed the screw task of the bolt. In the future, we will study the transfer of screw skills of different types of bolts. Therefore, the robot can adapt to more screw tasks.

References

1. Dharmara, K., Monfared, R.P., Ogun, P.S., Jackson, M.R.: Robotic assembly of threaded fasteners in a non-structured environment. Int. J. Adv. Manuf. Technol. **98**, 2093–2107 (2018). https://doi.org/10.1007/s00170-018-2363-5
2. Zhang, Q., et al.: A novel approach for flexible manipulator conducting screwing task based on robot-environment contact classification. Proc. Inst. Mech. Eng. Part C J. Mech. Eng. Sci. **235**(8), 1357–1367 (2019)
3. Deters, C., et al.: Accurate bolt tightening using model-free fuzzy control for wind turbine hub bearing assembly. IEEE Trans. Control Syst. Technol. **23**(1), 1–12 (2014)
4. Pitipong, S., Pornjit, P., Watcharin, P.: An automated four-DOF robot screw fastening using visual servo. In: 2010 IEEE/SICE International Symposium on System Integration (2010). https://doi.org/10.1109/SII.2010.5708355
5. Shauri, R.L.A., et al.: Sensor integration and fusion for autonomous screwing task by dual-manipulator hand robot. Procedia Eng. **41**, 1412–1420 (2012)
6. Ravichandar, H., Polydoros, A.S., Chernova, S., Billard, A.: Recent advances in robot learning from demonstration. Annu. Rev. Control Robot. Auton. Syst. **3**, 297–330 (2020)
7. Zeng, C., Yang, C.-G., Li, Q., Dai, S.-L.: Research progress on human-robot skill transfer. Acta Automatica Sinica **45**(10), 1813–1828 (2019). https://doi.org/10.16383/j.aas.c180397
8. Liu, N.-J., Lu, T., Cai, Y.-H., Wang, S.: A review of robot manipulation skills learning methods. Acta Automatica Sinica **45**(3), 458–470 (2019). https://doi.org/10.16383/j.aas.c180076
9. Calinon, S., Guenter, F., Billard, A.: On learning, representing, and generalizing a task in a humanoid robot. IEEE Trans. Syst. Man Cybern. Part B (Cybern.) **37**(2), 286–298 (2007)
10. Calinon, S., Billard, A.: Statistical learning by imitation of competing constraints in joint space and task space. Adv. Robot. **23**(15), 2059–2076 (2009)
11. Muhlig, M., Gienger, M., Hellbach, S., Steil, J.J., Goerick, C.: Task-level imitation learning using variance-based movement optimization. In: 2009 IEEE International Conference on Robotics and Automation, pp. 1177–1184 (2009)
12. Ti, B., Gao, Y., Li, Q., et al.: Dynamic movement primitives for movement generation using GMM-GMR analytical method. In: 2019 IEEE 2nd International Conference on Information and Computer Technologies (ICICT), pp. 250–254 (2019)
13. Ti, B., Gao, Y., Li, Q., Zhao, J.: Human intention understanding from multiple demonstrations and behavior generalization in dynamic movement primitives framework. IEEE Access **7**, 36186–36194 (2019)
14. Wang, Y.Q., Hu, Y.D., El Zaatari, S., Li, W.D., Zhou, Y.: Optimised learning from demonstrations for collaborative robots. Robot. Comput. Integr. Manuf. **71**, 102169 (2021)
15. Jin, C.-C., Liu, A.-D., Steven, L., Zhang, W.-A.: A robot skill learning method based on improved stable estimator of dynamical systems. Acta Automatica Sinica **46**(x), 1–11 (2020). https://doi.org/10.16383/j.aas.c200341
16. Müller, M.: Dynamic time warping. Information retrieval for music and motion, pp. 69–84 (2007)
17. Stulp, F., Sigaud, O.: Many regression algorithms, one unified model: a review. Neural Netw. **69**, 60–79 (2015)

Traversability Analysis for Quadruped Robots Navigation in Outdoor Environment

Yingying Kong[1,2](✉), Yingdong Fu[2], and Rui Song[2]

[1] Department of Engineering Information, Shandong University of Science and Technology, Jinan Campus, No.17, Shenglizhuang Road, Jinan 250031, China

[2] Center for Robotics, School of Control Science and Engineering, Shandong University, No. 17923, Jingshi Road, Jinan 250061, China

Abstract. In rough and unstructured environment, robots need accuracy environments understanding around them. Especially for legged robots, they need to decide which patch of the terrain to place their legs. This paper proposes a new traversability analysis algorithm for quadruped robots in rough and unstructured terrain. In order to increase accuracy and efficiency of scene understanding, geometric and appearance information of the terrain are all considered in this algorithm. Geometric features (slope, height, roughness) are obtained from point clouds based on Principle Component Analysis (PCA). Visual-based traversability analysis problem is formulated into an image classification task, a Convolutional Neural Network is trained to classify different terrain patches. increase the accuracy and efficiency. Finally, a fused traversability map based on visual features and typical geometric features are built. Simulation experiments show that this algorithm increases the accuracy and efficiency of traversability analysis.

Keywords: Traversability analysis · Convolutional Neural Network · Quadruped robot

1 Instruction

In rough and unstructured terrain, legged robots have shown more potential than wheeled and tracked robots. Legged robots have a primary advantage that they just need discrete footholds and don't need sequential trajectories. So they could overcome large obstacles which are larger than their bodies.

For safe and efficient navigating, robots need knowledge of their environments. They should know which batches are free or which have obstacles, and especially for legged robots, they should know where to place their feet. Precise perception and terrain traversability analysis is one of the most important part of intelligent robots. Terrain traversability analysis not only concerns with the terrain's properties but also with the characteristics of the robots. For example, the robots' size, gait, locomotion or pose might

Supported by National Key Research and Development Program (No. 2019YFB1312103), National Natural Science Foundation of China (No. U29A20201), Hebei Province Key Research and Development Program (No. 20311803D).

X.-J. Liu et al. (Eds.): ICIRA 2021, LNAI 13014, pp. 246–254, 2021.
https://doi.org/10.1007/978-3-030-89098-8_23

have an influence on the terrain traversability [1]. Now terrain traversability analysis is still a challenge problem for legged robots.

There are two main directions of terrain traversability assessment: approaches based on scene geometry and approaches based on scene appearance. The geometry-based approaches rely on the 3D reconstruction of the environment. They use data from sensors (Lidar or stereo camera) to build a model for the terrain and derive a set of features (slopes, roughness, height). This approach shows good performance in rigid ground and will not be affected by smoke, lighting, or bad weather, but there is a problem when the ground is covered by long grass or snow, the robot could not perceive the true support surface and will make misjudge.

The appearance-based approaches project terrain classification into image processing and classification problems. In this perspective, the approach abstracts a set of features of the terrain, the material of the terrain surface is classified into a discrete set of terrain classes, rather than regress a traversability score. It is more advantage to estimate the support surface, but it might be incorrect to assess the terrain traversability just only rely on the terrain material. For example, flat grassland and slope grassland have the same features but they have different traversability. A robot could cross a flat grassland safely but it might be very difficult to traverse a grassland that with some slope. According to the analysis above, estimating whether a terrain surface will bear a robot or not relies both on geometric and appearance properties of the scene.

This paper contributes with a traversability analysis algorithm based on Convolutional Neural Network. Geometry features are calculated from point clouds obtained from an onboard sensor. A Deep Convolutional Neural Network is trained to classify the terrain batch classes. A comprehensive traversability index is obtained by fusion of geometric and visual features. The remainder of this paper is organized as follows. Section 2 presents some related works. Section 3 describes the geometric and visual processing pipelines and their fusion into a final traversability index. Results and conclusions are presented in Sects. 4 and 5.

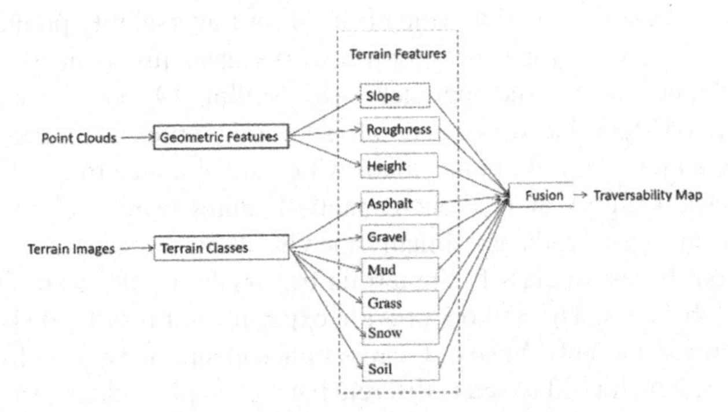

Fig. 1. Traversability analysis algorithm.

2 Related Works

Papadakis [2] presents a comprehensive survey of terrain traversability analysis methods and gives its definition: Terrain traversability analysis has been used as a means for navigating a robot within environments of varying complexity, on the one hand ensuring safety in terms of collisions or reaching unrecoverable states and on the other hand achieving goals in an optimal mode operation. This paper focuses on inferring whether the terrain patch around the robot is safe or dangerous.

The core idea for geometry-based traversability analysis method was mainly proposed by Gennery. In [3], terrain traversability was captured by a cost function that aggregated the elevation, slope, roughness, and data point accuracy, integrated with a path planning algorithm that took into account the distance as well as the probability of traversability.

Chilian [4] presents an algorithm which is based on stereo images and is suitable for wheeled and legged robots. The stereo images are used to model the terrain and estimate the terrain traversability. The authors tested their algorithm on a small six-legged robot: DLR-Crowler. Chavez-Garcia, R.O. [5] addressed traversability estimation as a heightmap classification problem. Given an image representing the heightmap of a terrain patch, the author trains a CNN to predict whether the robot could traverse this patch. There are two problems with this approach: 1. The classifier is trained on simulation data, the real world might be more dynamic and complex. 2. They did not consider the scene appearance and assume the terrain's 3D shape is the only factor which influence its traversability.

Filitchkin and Byl [6] present a feature-based terrain classification approach for Little Dog. Terrain is classified using a bag of visual words created from speeded up robust features with a Support Vector Machine (SVM) classifier. This approach is suitable for small legged robots. The SVM classifier is used to select right gaits to traverse terrain with difficulties. Yan Guo [7] regressed a terrain traversability based on vision features rather than performing terrain classification. In the paper color and texture features are used to train a self-learning function which is used for traversability prediction.

To increase the overall robustness, some works have fused the geometry-based approaches and appearance-based approaches [8]. Shilling [9] presents a muti-sensory terrain classification algorithm with a generalized terrain representation using semantic and geometric features. They use point clouds which are obtained from a laser sensor to compute geometric features, and extract semantic features from a fully Convolutional Network trained on Cityspace dataset. Finally, the visual and geometric features are fused based on Random Forest to classify the terrain traversability, there are three indexes: safe, risky, and obstacles. The authors provide experimental results to show that their algorithm performs accurately. Most of the researches mentioned above do not consider the legged robots, which need to select discrete patches to place their feet.

Wermelinger [10] proposed a framework that planning safe and efficient path for a quadruped robot in rough and unstructured terrain. They computer typical terrain characteristics, such as slop, roughness, and steps to build a terrain traversability map. But they did not consider appearance features.

3 Terrain Traversability Analysis

This paper presents a new traversability analysis algorithm, as shown in Fig. 1. The input of the system is RGB images and point clouds which are obtained from onboard sensors. The problem will be analyzed in three parts: geometric features extraction, visual features from CNN, the fusion of geometric and visual features.

3.1 Geometric Features Extraction

When navigating in rough terrain, the robot needs to know all three dimensions of the surroundings. The most popular approach is to build an elevation map of the environment. For simplicity, elevation map is stored as grid map [11]. For legged robots, several pieces of information are desired at each cell of the grid map, such as slope, roughness and step height. In order to determine whether a legged robot could traverse a grid cell and how much time and energy it will expend in doing so, we need to know the magnitude and direction of the slope of the surface. The roughness represents how much deviation from the smoothed surface exists on a smaller scale. For legged robotics higher roughness will prevent the robot from slipping. The height is also a factor for individual leg placement and the stability of legged robot.

The terrain information is shown in Fig. 2. $h_0 \sim h_8$ is the average height of each cell, L is the cell length, S_{we} and S_{sn} are slope of east-west and north-south respectively. We calculate these values as Function (1–5) [12].

$$S_{we} = \frac{(h_8 + 2h_1 + h_5) - (h_7 - 2h_3 + h_6)}{8L} \tag{1}$$

$$S_{we} = \frac{(h_7 + 2h_4 + h_8) - (h_6 - 2h_2 + h_5)}{8L} \tag{2}$$

$$Slope = \arctan\sqrt{S_{we}^2 + S_{sn}^2} \tag{3}$$

$$r = \sqrt{\frac{1}{8}\sum_{i=1}^{8}(h_i - h_{islopeheight})} \tag{4}$$

$$h = \max(h_i - h_0) \tag{5}$$

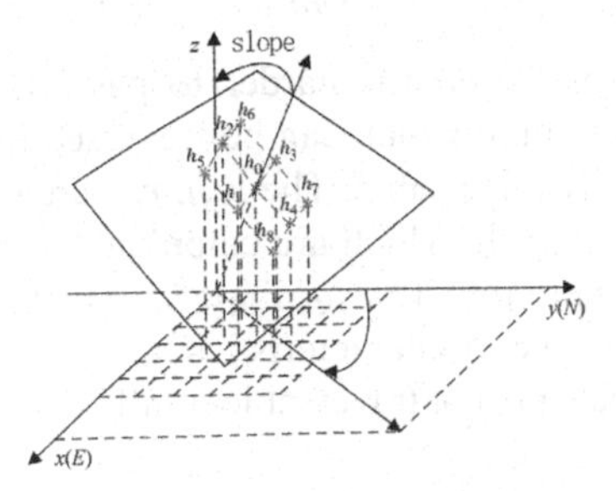

Fig. 2. Terrain information

3.2 Visual Features from CNN

Convolutional neural networks have become the dominant machine learning approach for image classification. VGGnet [14] is one of the most popular Convolutional neural networks. It uses smaller convolution kernel and deeper network to improve parameter efficiency. Here we choose VGG-16 to classify the terrain classes. Our data set is Sucro-Terrain, which is collected around the Shandong University campus. The data set has six different terrain classes: asphalt, gravel, mud, grass, snow and sand, which are typical rough terrains in outdoor environment. The examples of the terrain are shown in Fig. 3.

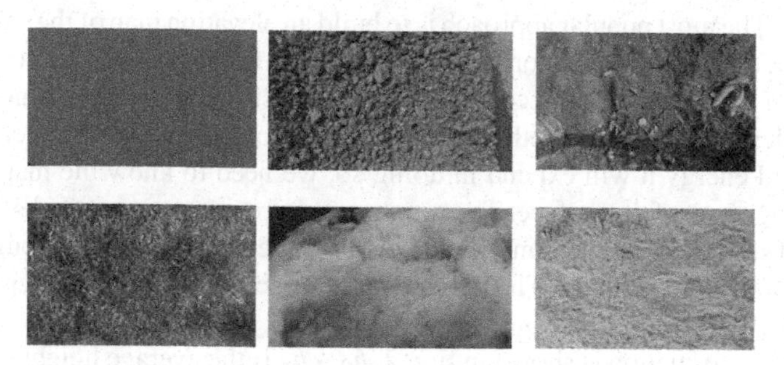

Fig. 3. Examples of dataset

3.3 Fusing of Geometric and Visual Features

The terrain traversability assessment methods presented above could be run stand-alone. Considering the advantages and limitations of each method, we adopt the fusion method based on geometric and appearance features.

The geometric features (slope, roughness, height) are computed from point clouds, as described in Sect. 3.1. At the same time, we apply the softmax operator to the unscaled per-pixel logits of the fully convolutional network to obtain the probability distribution of 6 classes. A local traversability index combines the values computed for slope, roughness, steps and terrain classes of a cell as follows:

$$d = w_1 \frac{s}{s_{crit}} + w_2 \frac{r}{r_{crit}} + w_3 \frac{h}{h_{crit}} + w_4 \frac{c}{c_{crit}} \quad d \in \{[0, 1], \infty\} \tag{6}$$

High danger value means the area is harder to pass. The value of $d = 1$ stands for completely flat, smooth terrain, which can be traversed by the robots easily. $d = 1$ means the terrain is just barely traversable. The s, h, r, c are the values of slope, height, roughness and classes for each cell, which are calculated in Sect. 3.1. w_1, w_2, w_3, and w_4 are the weights which sum up to 1, s_{crit}, r_{crit}, h_{crit} and s_{crit} are the robot-specific maximum allowed values. If one of characteristics s, h, r, c exceeds its critical value, $d = \infty$ means this terrain patch is not traversable at all.

4 Experiment Rusults

In this part, we present the experiment results on terrain classification performance and terrain traversability analysis based on geometric features, appearance features and our fusion method.

4.1 CNN Training Results

In order to train and evaluate this method, we have collected our own dataset, which is called Sucro-terrain dataset. This dataset contains six classes for terrain estimation, they are asphalt, gravel, mud, grass, snow and sand, examples are shown in Fig. 3. The training loss and accuracy is shown in Fig. 4.

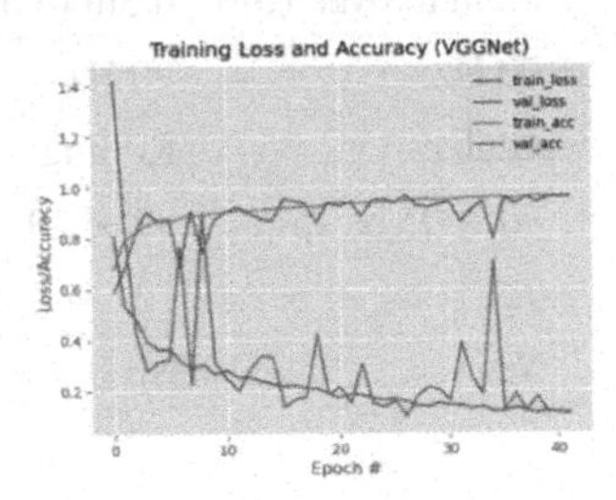

Fig. 4. Training loss and accuracy of VGG 16 (learning rate = 0.01)

We compared accuracy between DCNN and neural networks to predict the terrain classifications. The result is shown in Table 1. From the result, we could see that when the learning rate is 0.01, the accuracy is the highest.

Table 1. Comparison of different classify approach

	Asphalt	Gravel	Mud	Grass	Snow	Sand
DCNN (0.01)	89%	98%	96%	97%	98%	96%
DCNN (0.001)	83%	98%	84%	96%	99%	99%
NN	68%	74%	77%	82%	82%	89%

4.2 Traversability Analysis

The experiment results are collected by a Realsense D435i depth camera. It is an active stereo depth camera with an Inertial Measurement Unit (IMU). Compared to Kinect v2, which can get depth at a resolution of 512 × 424 at up to 30 frames per second, the D435i can get up to 1280 × 720 at up to 90 frames per second. It is smaller and weight less than Kinect v2, Specifications of different depth sensors are shown in Table 2. The small size and ease of integration of the camera provide it integrators flexibility to design into a wide range of products [15]. The camera is calibrated before use, the errors with measurement distance are shown in Fig. 5.

Table 2. Specifications of different depth sensors

	Real sense435di	Real sense 415D	Kinect V2
Active stereo depth resolution(px)	Up to 1280x720	Up to 1280x720	512x424
Depth field of view(h × v)	85.2° × 58°	63.4° × 40.4°	70.6° × 60°
RGB resolution	Up to 1920x1080	Up to 1920x1080	1920x1080
RGB field of view (h × v)	69.4° × 42.5°	69.4° × 42.5°	84.1° × 53.8°
Operating range	0.2m to over 10m	0.3m to over 10m	0.5m to over 4.5m
Frame rate	90 Hz	90 Hz	30 Hz
Shutter type	Global	Rolling	Global
Dimensions(w × d × h) [mm]	90 × 25 × 25	99 × 20 × 23	249 × 66 × 67
Weight(g)	72	72	90
IMU included	Yes	No	No

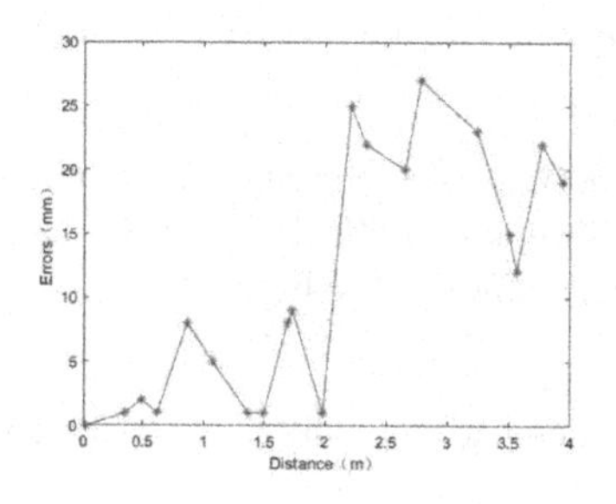

Fig. 5. Measurement errors

The terrain selected for the experiment is shown in Fig. 6. To obtain geometric information of the terrain, the dense point clouds p^C measured by the RGB-D camera in transformed into the robot coordinate system R, that is p^R. The height, slope, roughness of the terrain is shown in Fig. 6(b–d).

In order to compare the accuracy of different approaches, we choose two different traversability methods. One is used geometric features only, and the other is our new algorithm. The results are shown in Fig. 7. Figure 7(a) is the traversability index calculated from geometric features [4] and Figure (b) is obtained from the new algorithm. The terrain patch has mud and gravels, from the practical experiment, it is difficult for the legged to traverse. From Fig. 7(a), almost all the cells have a normal danger index, it means this terrain is safe for the robot. This does not accord with the facts. Figure 7(b) has higher danger index than (a). Take point 1 for example, we could see its slope is almost 70°, it is difficult for the legged robot to traverse. In Fig. 7(a) the danger index is less than 0.2, but in Fig. 7(b) this index is more than 0.8.

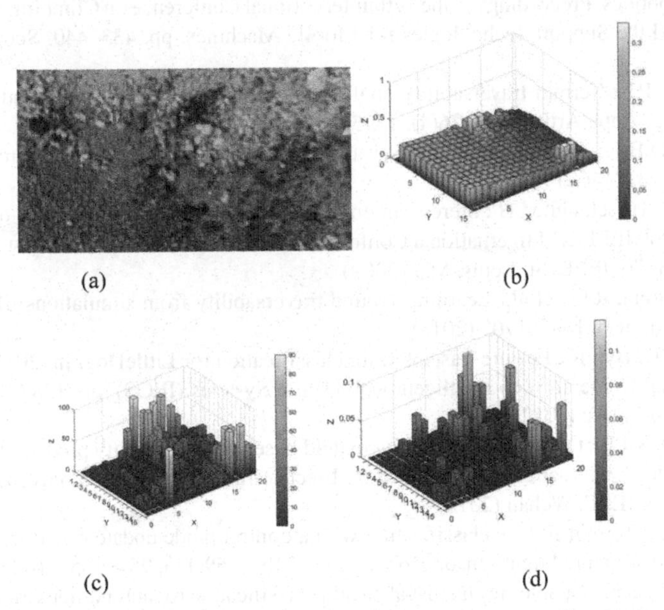

Fig. 6. Terrain image (a) and its geometric features (b) height, (c) slope, (d) roughness.

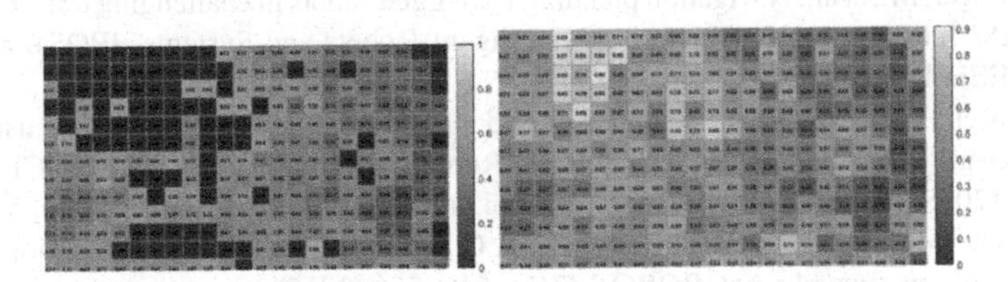

Fig. 7. Danger index of different methods. (a) geometric features only; (b) fusion method

5 Conclusions

In this paper, we introduced an efficient and robust traversability map for off-road navigation. To train the CNN, we have collected a new dataset of off-road imagery. Based on geometric and visual information, we could obtain a precise and robust traversability index. The simulation results show that this approach can improve the accuracy of traversability analysis in complex terrain.

In the future, the dynamic environment will be included, the method of using multiple sensor fusion will be used to obtain more environment information. The network will be trained on a larger dataset to obtain more robust and efficient classification results.

References

1. Fankhauser, P., et al.: Robot-centric elevation mapping with uncertainty estimates. In: Mobile Service Robotics: Proceedings of the 17th International Conference on Climbing and Walking Robots and the Support Technologies for Mobile Machines, pp. 433–440. Scopus, Poznan (2014)
2. Papadakis, P.F.: Terrain traversability analysis methods for unmanned ground vehicles: a survey. Eng. Appl. Artif. Intell. **26**(4), 1373–1385 (2013)
3. Gennery, D.B.F.: Traversability analysis and path planning for a planetary rover. Auton. Robots. **6**, 131–146 (1999)
4. Chilian, A., Hirschmuller, H.: Stereo camera based navigation of mobile robots on rough terrain. In: 2009 IEEE/RSJ International Conference on Intelligent Robots and Systems (IROS), pp. 4571–4576. IEEE, St. Louis, MO (2009)
5. Chavez-Garcia, R.O., et al.: Learning ground traversability from simulations. IEEE Robot. Autom. Lett. **3**(3), 1695–1702 (2018)
6. Filitchkin, P., Byl, K.: Feature-based terrain classification for LittleDog. In: 2012 IEEE/RSJ International Conference on Intelligent Robots and Systems (IROS), pp. 1387–1392. IEEE, Vilamoura-Algarve (2012)
7. Yan Guo, A.S.J.B.: Optimal path planning in field based on taversability prediction for mobile robot. In: 2011 International Conference on Electric Information and Control Engineering, pp. 563–566. IEEE, Wuhan (2011)
8. Lu, L., et al.: Terrain surface classification with a control mode update rule using a 2D laser stripe-based structured light sensor. Robot. Auton. Syst. **59**(11), 954–965 (2011)
9. Schilling, F., et al.: Geometric and visual terrain classification for autonomous mobile navigation. In: 2017 IEEE/RSJ International Conference on Intelligent Robots and Systems (IROS), pp. 2678–2684. IEEE, Vancouver, BC (2017)
10. Wermelinger, M., et al.: Navigation planning for legged robots in challenging terrain. In: 2016 IEEE/RSJ International Conference on Intelligent Robots and Systems (IROS), pp. 1184–1189. IEEE, Daejeon (2016)
11. Fankhauser, Péter., Hutter, M.: A universal grid map library: implementation and use case for rough terrain navigation. In: Koubaa, A. (ed.) Robot Operating System (ROS). SCI, vol. 625, pp. 99–120. Springer, Cham (2016). https://doi.org/10.1007/978-3-319-26054-9_5
12. Zhang, H., Rong, X.W., Li, Y.B., Li, B., Ding, C., Zhang, J.W.: Terrain recognition and path planning for quadruped robot. ROBOT **37**(5), 546–556 (2015)
13. Benet, B., Lenain, R.: Multi-sensor fusion method for crop row tracking and traversability operations (2018)
14. Simonyan, K., Zisserman, A.: Very deep convolutional networks for large-scale image recognition. Computer Science (2014)
15. Ahn, M.S., Chae, H., Noh, D., Nam, H., Hong, D.: Analysis and noise modeling of the intel RealSense D435 for mobile robots. In: 2019 16th International Conference on Ubiquitous Robots (UR), pp. 707–711. IEEE, Jeju (2019)

Micro_Nano Materials, Devices, and Systems for Biomedical Applications

A Calibration Method of Compliant Planar Parallel Manipulator

Jian-feng Lin[1], Chen-kun Qi[1](✉), Yu-ze Wu[2], and Feng Gao[1]

[1] State Key Laboratory of Mechanical System and Vibration, School of Mechanical Engineering, Shanghai Jiao Tong University, Shanghai 200240, China
chenkqi@sjtu.edu

[2] Paris Tech Elite Institute of Technology, Shanghai Jiao Tong University, Shanghai 200240, China

Abstract. Flexure hinge is usually used in micro-nano positioning parallel manipulator. High precision kinematics model is the basis of high performance control. In this paper, the stiffness modeling and calibration methods of 3-RPR micro-nano positioning parallel manipulator are studied. The kinematics model of the 3-RPR parallel manipulator is established and the stiffness model of the compliant mechanism is established based on the virtual work principle. Given the driving force vector of the branch chain, the output pose of the end-effector of the manipulator is obtained by finite element analysis. The parameters of the stiffness model are identified by the least square method. Compared with the theoretical results of the finite element model, the validity of the stiffness modeling and calibration methods is verified, which provides a theoretical basis for the experimental verification.

Keywords: Micro-nano positioning manipulator · Flexure hinge · Stiffness modeling · Stiffness calibration · Finite element analysis

1 Introduction

Due to its high motion accuracy, micro-nano positioning manipulator is widely used in the assembly of tiny parts, surgical robot [1], optical fiber docking, precision machining and other fields. Existing micro-nano positioning manipulators are mostly single degree of freedom (DOF) [2] or two-DOF [3, 4], which are limited in practical application due to the small number of DOF. To address this defect, many researchers have successively proposed multi-degree-of-freedom micro-motion manipulators, such as three-degree-of-freedom [5] and six-degree-of-freedom [6]. In terms of pure movement, Li et al. [7] designed a novel 3-DOF translational flexible mechanism based on flexure hinges. The mechanism is completely decoupled in the XY direction, but due to the asymmetry in the Z direction, there is less coupling motion in the Z direction when the XY-axis is driven. In terms of rotational degrees of freedom, Cai et al. [8] introduced a 3-DOF worktable with T-shaped flexure hinge mechanism, which can be used in precision measuring equipment and micro-nano control system. The motion travel of the positioning manipulator is 6.9 μm along the X-axis, 8.5 μm along the Y-axis and 289 μrad around the Z-axis.

X.-J. Liu et al. (Eds.): ICIRA 2021, LNAI 13014, pp. 257–266, 2021.
https://doi.org/10.1007/978-3-030-89098-8_24

Park et al. [9] proposed a planar 3-DOF micro-nano positioning manipulator driven by flexible hinge and piezoelectric actuators (PZT), with a motion range of 300 μm in X and Y directions and a rotation angle of ±3.9 mrad around the Z direction.

The micro-nano positioning parallel manipulator is generally driven by piezoelectric actuator and characterized by flexible hinge. The main factors affecting the motion accuracy of the manipulator are the hysteresis nonlinear of piezoelectric actuator [10], the size error of the mechanism and the deformation error of the flexible hinge in the non-functional direction [11]. Piezoelectric actuator produces hysteresis between input voltage and output displacement and some scholars have conducted modeling and control for hysteresis nonlinearity of piezoelectric actuators [12].

Existing research on parallel manipulator mainly focuses on mechanism design, kinematics calibration and modal analysis. Wu G-H et al. [13] established statics and dynamics models with pseudo rigid body method for the parallel 3-DOF micro-manipulator and studied the displacement increment ratio and the first-order resonance frequency of the micro-manipulator through the combination of finite element analysis and experiment. On the premise of error modeling for planar flexible 3-RRR parallel mechanism, Shao Z-F et al. [14] calibrated the cable encoder through the instrument and achieved that the terminal position accuracy of micro-nano positioning parallel manipulator was improved to less than 0.5 mm, meeting the accuracy requirements of this kind of flexible parallel test device. Fan R et al. [15] established the error Jacobian matrix of 6-PUS parallel mechanism, adopted the least square method for parameter identification and proposed 54 structural parameters calibration in the form of the whole machine. Experiments proved the validity of the calibration method of the whole machine.

The stiffness modeling of the micro-nano positioning parallel manipulator can combine the dimensional parameter error of the mechanism with the deformation error of the flexure hinge in the non-functional direction. This method not only avoids the error modeling, error measurement, error identification and error compensation in the traditional geometric parameter calibration process, but also can consider the deformation of the flexure hinge, so as to establish a more consistent static model of the mechanism. Dong Y et al. [16, 17] established the relationship between the input displacement of the piezoelectric actuator and the output pose of the terminal of the micro-nano positioning manipulator, identified the relative stiffness matrix of the manipulator through experiments and carried out the trajectory tracking experiment of the manipulator using the calibration-control two-step method. Yue Y et al. [18–20] established the model for the 3-DOF and 6-DOF micro-nano positioning parallel manipulator and verified the accuracy of the model through finite element analysis and experiment.

In order to improve the accuracy of micro-nano positioning parallel manipulator, this paper carries on the kinematics modeling of the 3-RPR micro-nano positioning manipulator and establishes the stiffness model based on the virtual work principle. Finite element analysis is used to solve the terminal output pose and the corresponding stiffness matrix is identified by the least square method. Finally, the accuracy of the identified stiffness model is verified.

2 Kinematics Model of Micro-nano Parallel Manipulator

2.1 3-RPR Micro-nano Parallel Manipulator

The 3-RPR micro-nano positioning parallel manipulator is composed of lower stage, piezoelectric actuator (PZT), branch chain and upper stage. Its physical prototype is shown in Fig. 1 (a) and the three-dimensional diagram is shown in Fig. 1 (b). Three groups of branch chains are distributed in an orthogonality and each group of branch chains consists of one moving pair and two rotating pairs. Three piezoelectric actuators are installed in three groups of moving branches respectively and the upper stage is driven to realize three degrees of freedom movement, which are along X direction, Y direction and around Z direction respectively.

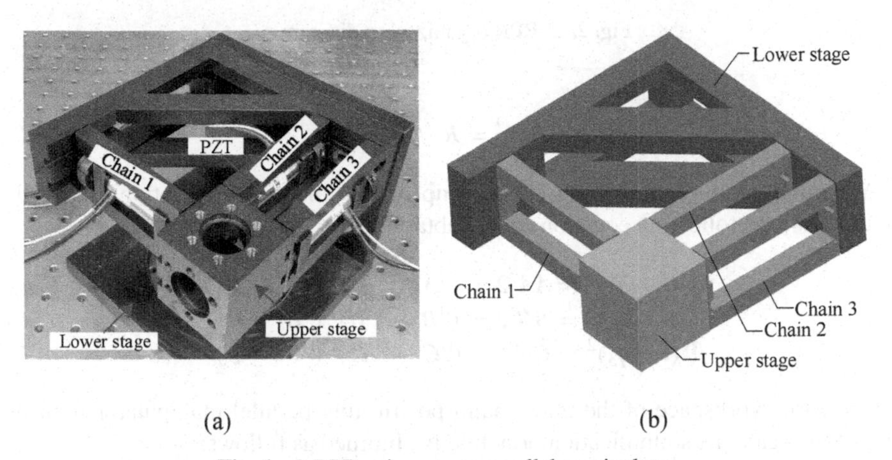

Fig. 1. 3-RPR micro-nano parallel manipulator

2.2 Kinematic Modeling

As shown in Fig. 2, two spatial Cartesian coordinate systems are established on the 3-RPR micro-nano parallel mechanism. CS0 (o-xyz) is a fixed coordinate system whose origin is located at the center of the upper stage. CSL ($o' - xyz$) is the conjoined coordinate system of the upper stage, which is fixedly connected to the upper stage and moves accordingly.

At the initial position, the two frames coincide. Use vector: $\boldsymbol{X} = \left[x\ y\ \gamma\right]^T$ to represent the end pose of the mechanism. Where, A, B and C are respectively the central position of the flexible rotating pair close to the lower stage; ${}^0A'$ and ${}^0A$ are respectively the generalized coordinates of A' and A in CS0; ${}^lA'$ and ${}^lA$ are respectively the generalized coordinates of A' and A in CSL; the rest are similar.

The homogeneous coordinate transformation matrix from CSL to CS0 is:

$$\boldsymbol{R} = \begin{bmatrix} \cos\gamma & -\sin\gamma & x \\ \sin\gamma & \cos\gamma & y \\ 0 & 0 & 1 \end{bmatrix} \tag{1}$$

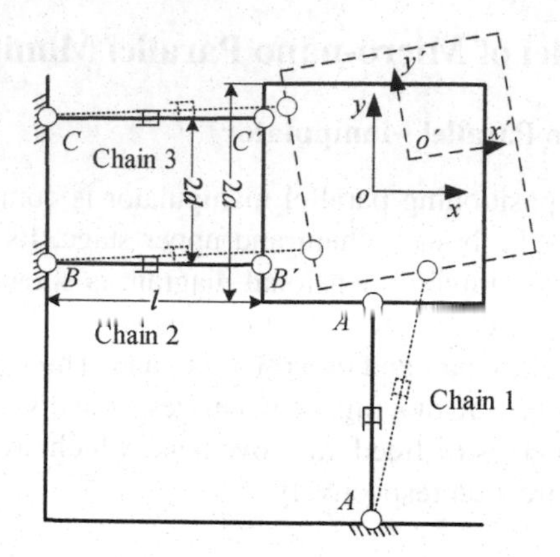

Fig. 2. 3-RPR organization diagram

$$ {}^{0}\boldsymbol{A}' = \boldsymbol{R} \cdot {}^{l}\boldsymbol{A}' \tag{2} $$

According to the input-output relationship of the micro-nano positioning parallel manipulator, the following equation can be abtained:

$$ \begin{cases} (l+q_1)^2 = \boldsymbol{AA}'2 = ({}^{0}\boldsymbol{A}' - {}^{0}\boldsymbol{A})^T({}^{0}\boldsymbol{A}' - \boldsymbol{A}) \\ (l+q_2)^2 = \boldsymbol{BB}'2 = ({}^{0}\boldsymbol{B}' - {}^{0}\boldsymbol{B})^T({}^{0}\boldsymbol{B}' - \boldsymbol{B}) \\ (l+q_3)^2 = \boldsymbol{CC}'^2 = ({}^{0}\boldsymbol{C}' - {}^{0}\boldsymbol{C})^T({}^{0}\boldsymbol{C}' - \boldsymbol{C}) \end{cases} \tag{3} $$

Since the workspace of the micro-nano positioning parallel manipulator is in the micro-nano scale, its simplification principle is obtained as follows:

$$ \begin{cases} \sin\theta = \theta \\ \cos\theta = 1 \\ \sin\theta \cdot \sin\psi = 0 \end{cases} \tag{4} $$

The simplified kinematics equation at micro-nano scale is as follows:

$$ \begin{cases} q_1 = \sqrt{(x+a\gamma)^2 + (y+l)^2} - l \\ q_2 = \sqrt{(x+d\gamma+l)^2 + (y-a\gamma)^2} - l \\ q_3 = \sqrt{(x-d\gamma+l)^2 + (y-a\gamma)^2} - l \end{cases} \tag{5} $$

By differentiating both sides of Eq. (5), the inverse solution model of the 3-RPR parallel manipulator can be obtained:

$$ \Delta\boldsymbol{Q} = \boldsymbol{J}^{-1}\Delta\boldsymbol{X} \tag{6} $$

Where: $\Delta \boldsymbol{Q}$ and $\Delta \boldsymbol{X}$ are the input displacement and the output pose respectively. The terminal output of the parallel manipulator at the initial position is zero and the Jacobian matrix of the parallel mechanism at the initial position can be obtained as follows:

$$\boldsymbol{J} = \begin{bmatrix} 0 & \frac{1}{2} & \frac{1}{2} \\ 1 & 0 & 0 \\ 0 & \frac{1}{2d} & -\frac{1}{2d} \end{bmatrix} \tag{7}$$

Because the working space of micro-nano positioning manipulator is in micro-nano scale. Therefore, the Jacobian matrix at the initial position can be regarded as the Jacobian matrix of the mechanism in the workspace.

3 Stiffness Modeling of Micro-nano Manipulator Based on Virtual Work Principle

3.1 Kinematics Modeling of Branch Chains

The relationship between the flexure hinge movement of each branch chain and the pose of the upper stage is established. Taking branch chain 1 as an example, q_1 is the output displacement of the moving pair, α_1 and β_1 are respectively the output rotation angles of the two rotating pairs close to the lower stage and far away from the lower stage and the relationship can be obtained as follows:

$$\begin{cases} x = -a \sin \gamma - (q_1 + l) \sin \alpha_1 \\ y = a \cos \gamma + (q_1 + l) \cos \alpha_1 - l - a \\ \gamma = \alpha_1 + \beta_1 \end{cases} \tag{8}$$

By differentiating both ends of Eq. (8), the kinematics model of branch chain 1 can be obtained. The branch chain Jacobian matrix is:

$$J_1 = \begin{bmatrix} -l-a & 0 & -a \\ 0 & 1 & 0 \\ 1 & 0 & 1 \end{bmatrix} \tag{9}$$

Similarly, the kinematics model and Jacobian matrix of branch chains 2 and 3 can be obtained.

$$J_2 = \begin{bmatrix} -d & 1 & -d \\ l+a & 0 & a \\ 1 & 0 & 1 \end{bmatrix} \tag{10}$$

$$J_3 = \begin{bmatrix} d & 1 & d \\ l+a & 0 & a \\ 1 & 0 & 1 \end{bmatrix} \tag{11}$$

3.2 Stiffness Modeling of Parallel Manipulator Based on Virtual Work Principle

Stiffness modeling of 3-RPR micro-nano positioning parallel manipulator is carried out by using virtual work principle and the expression can be obtained as follows:

$$\sum_{i=1}^{3} \boldsymbol{P} \cdot \delta\boldsymbol{\lambda}_i \cdot \Delta\boldsymbol{\lambda}_i^T = \boldsymbol{F}^T \cdot \delta\boldsymbol{Q} \tag{12}$$

Where: i represents the serial number of the branch chain; $\boldsymbol{F}$ is the driving force vector; $\boldsymbol{P}$ is the stiffness matrix formed by moving pair and rotating pair. The relation between the virtual displacement $\delta\boldsymbol{X}$ of the upper stage and the virtual displacement $\delta\boldsymbol{\lambda}_i$ of the flexible motion pair is:

$$\delta\boldsymbol{X} = \boldsymbol{J}_i \cdot \delta\boldsymbol{\lambda}_i \tag{13}$$

$$\delta\boldsymbol{\lambda}_i = \Delta\boldsymbol{\lambda}_i = (\alpha_i \; q_i \; \beta_i)^T \tag{14}$$

The relation between the output pose of the upper stage and the input vector of the flexible kinematic pair is:

$$\Delta\boldsymbol{X} = \boldsymbol{J}_i \cdot \Delta\boldsymbol{\lambda}_i \tag{15}$$

Combined with the above formulas, the relation expression between the driving force of the 3-RPR micro-nano positioning parallel manipulator and the pose of the upper stage of the manipulator can be obtained. Here, the stiffness matrix of the manipulator is defined as $\boldsymbol{K}$, then:

$$\boldsymbol{F} = \boldsymbol{K} \cdot \Delta\boldsymbol{X} \tag{16}$$

The stiffness model of parallel manipulator based on virtual work principle is obtained.

4 Finite Element Analysis

The 3-RPR micro-nano positioning parallel manipulator is analyzed by finite element-method. 12 groups of different input force vectors are applied to the moving pair in the finite element model (the force direction is shown in Fig. 3 (a)) and the corresponding end output pose is obtained, which are the movement along the X and Y directions and rotation around the Z direction respectively. The finite element calculation results are shown in Table 1. There is error between the simulation results and the theoretical calculation, in which the maximum displacement error along the X direction is 9.9% and that along the Y direction is 11.62%. The established finite element analysis model is shown in Fig. 3 (b).

The stiffness matrix in the finite element model is identified by the least square method.

Table 1. Finite element analysis results.

Input force vector (unit:N)	Output pose (unit: μm, 10^{-4}rad)
(10,20,30)	10.8732, 4.1985, –0.9875
(10,20,50)	15.2583, 4.0152, –2.9535
(10,60,30)	19.4650, 4.5561, 2.8596
(10,60,50)	23.8505, 4.3728, 0.9015
(40,20,30)	10.8700, 17.0800, 0.8388
(40,20,50)	15.2551, 16.8967, –2.8048
(40,60,30)	19.4620, 17.4374, 3.0016
(40,60,50)	23.8470, 17.2540, 1.0136
(70,20,30)	10.8666, 29.9615, 0.6902
(70,20,50)	15.2519, 29.7780, –2.6561
(70,60,30)	19.4590, 30.3185, 3.1440
(70,60,50)	23.8445, 30.1360, 1.1858

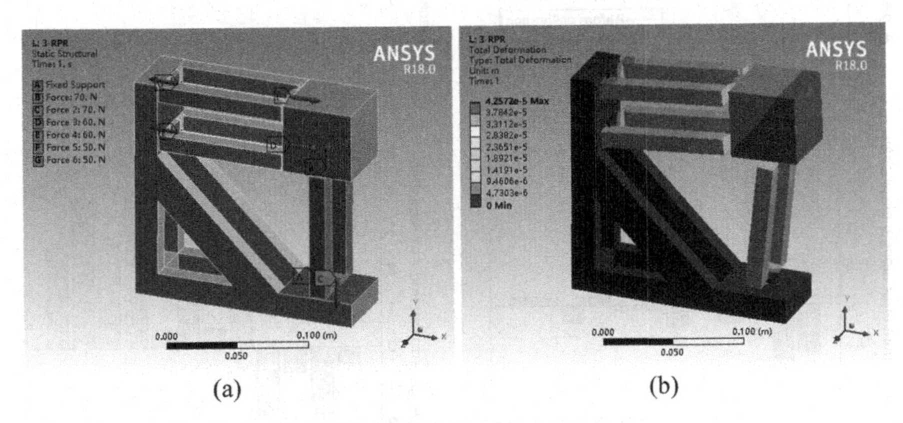

Fig. 3. 3-RPR finite element analysis

5 Simulation Verification

The stiffness model after identification is verified by finite element method. By giving an additional 12 input force vectors, the corresponding terminal output pose can be solved. Comparison of end pose errors before and after calibration is shown in Fig. 4 (a)-(c). The pose error of the terminal output after calibration is shown in Fig. 4 (d)-(f).

By comparing the pose errors of the upper stage end calculated by the stiffness matrices before and after calibration, it can be seen that the accuracy of the stiffness model before calibration has certain errors. The possible reasons are the manufacturing error of the 3-RPR parallel mechanism and the simplification principle used in the modeling process.

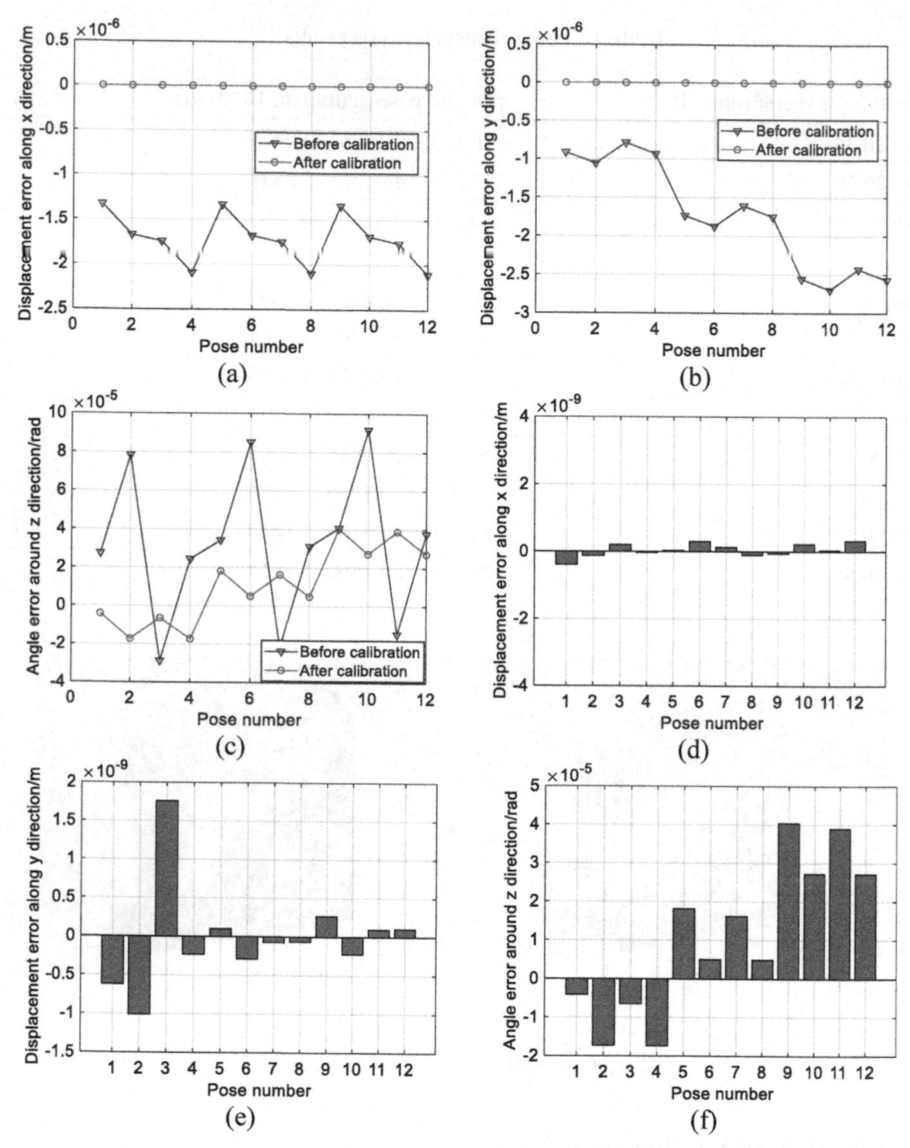

Fig. 4. Comparison of pose errors before and after calibration (a) along X direction; (b) along Y direction; (c) around Z direction. Pose errors after calibration (d) along X direction; (e) along Y direction; (f) around Z direction.

Therefore, it can be concluded that the precision of the stiffness model after calibration has been greatly improved, which lays a theoretical foundation for the subsequent research on the control of micro-nano positioning manipulator.

6 Conclusion

In this paper, the stiffness modeling and calibration of 3-RPR micro-nano positioning parallel manipulator are studied. On the basis of the introduction of the 3-RPR parallel manipulator, the simplified inverse kinematics equation is established. The kinematics models of each branch chain and the output pose of the upper stage are established according to the motion deformation of the flexure hinge and the stiffness model of the whole parallel mechanism is established according to the virtual work principle. Twelve groups of different output pose vectors are obtained by finite element analysis and the stiffness matrix is identified by least square method. Finally, the validity of the calibration method is verified by comparing the output pose errors of the upper stage before and after calibration.

Funding. The work is partially supported by a grant Shanghai Natural Science Foundation (Grant No. 19ZR1425500).

References

1. Dohi, T.: Robot technology in medicine. Adv. Robot. **7**(2), 179–187 (2010)
2. Fleming, A.J.: Nanopositioning system with force feedback for highperformance tracking and vibration control. IEEE/ASME Trans. Mechatron. **15**(3), 433–447 (2010)
3. Fleming, A.J., Yong, Y.K.: An ultra-thin monolithic XY nanopositioning stage constructed from a single sheet of piezoelectric material. IEEE/ASME Trans. Mechatron. 1–1 (2017)
4. Zhang, J., et al.: Kinematic calibration of a 2-DOF translational parallel manipulator. Adv. Robot. **28**(10), 1–8 (2014)
5. Zhu, W.L., et al.: Redundantly piezo-actuated XYθz compliant mechanism for nano-positioning featuring simple kinematics, bi-directional motion and enlarged workspace. Smart. Mater. Struct. **25**(12), 125002 (2016)
6. Yun, Y., Li, Y.: Design and analysis of a novel 6-DOF redundant actuated parallel robot with compliant hinges for high precision positioning. Nonlinear Dyn. **61**(4), 829–845 (2010)
7. Li, Y., Wu, Z.: Design, analysis and simulation of a novel 3-DOF translational micromanipulator based on the PRB model. Mech. Mach. Theory **100**, 235–258 (2016)
8. Cai, K., et al.: Development of a piezo-driven 3-DOF stage with T-shape flexible hinge mechanism. Robot. Comput. Integr. Manuf. **37**(FEB), 125–138 (2016)
9. Park, J., et al.: Note: development of a compact aperture-type XYθz positioning stage. Rev. Sci. Instrum. **87**(3), 036112 (2016)
10. Qi, C., Gao, F., Li, H.-X., Li, S., Zhao, X., Dong, Y.: An incremental Hammerstein-like modeling approach for the decoupled creep, vibration and hysteresis dynamics of piezoelectric actuator. Nonlinear Dyn. **82**(4), 2097–2118 (2015). https://doi.org/10.1007/s11071-015-2302-z
11. Dong, Y., Gao, F., Yue, Y.: Stiffness and workspace analysis of 3-prr micromanipulator. Mech. Des. Res. **27**(02), 15-18+21 (2011)
12. Rui, X., Zhou, M.: A self-adaption compensation control for hysteresis nonlinearity in Piezo-actuated stages based on Pi-sigma fuzzy neural network. Smart Mater. Struct. **27**(4), 045002 (2018). https://doi.org/10.1088/1361-665X/aaae28
13. Wu, G.H., Yang, Y.L., Li, G.P., Lou, J.Q., Wei, Y.D.: A flexible parallel x-y manipulator with high displacement increment. Robot **42**(01), 1–9 (2020)

14. Shao, Z.F., Tang, X.Q., Wang, L.P., Peng, H.: Self-calibration method of planar flexible 3-RRR parallel mechanism. J. Mech. Eng. **45**(03), 150–155 (2009)
15. Fan, R., Li, Q., Wang, D.: Calibration of kinematic complete machine of 6PUS parallel mechanism. J. Beijing Univ. Aeronaut. Astronaut. **42**(05), 871–877 (2016)
16. Dong, Y., Gao, F., Yue, Y.: Modeling and prototype experiment of a six-DOF parallel micro-manipulator with nano-scale accuracy. Proc. Inst. Mech. Eng. C J. Mech. Eng. Sci. **229**(14), 2611–2625 (2015)
17. Dong, Y., et al.: Modeling and experimental study of a novel 3 RPR parallel micro-manipulator. Robot. Comput. Integr. Manuf. **37**(C), 115–124 (2016)
18. Yue, Y., et al.: Relationship among input-force, payload, stiffness and displacement of a 6-DOF perpendicular parallel micromanipulator. J. Mech. Robot. (2010)
19. Yue, Y., et al.: Relationship among input-force, payload, stiffness and displacement of a 3-DOF perpendicular parallel micro-manipulator. Mech. Mach. Theory **45**(5), 756–771 (2010)
20. Yue, Y., Gao, F., Jin, Z., Zhao, X.: Modeling and experiment of a planar 3-DOF parallel micromanipulator. Robotica **30**(2), 171–184 (2012). https://doi.org/10.1017/S0263574711000361

A Jumping Soft Robot Driven by Magnetic Field

Tianliang Zhong and Fanan Wei(✉)

School of Mechanical Engineering and Automation, Fuzhou University, Fuzhou 350108, China
weifanan@fzu.edu.cn

Abstract. In this paper, we designed a magnetically controlled miniature soft robot with the ability to jump, which consisted of a magnetic actuator, main body and an elastic leg. In order to improve the mechanical properties and the jumping ability of the soft robot, we employed polyethylene terephthalate (PET) as the raw material of the main body and an elastic leg. The maximum size of the jumping robot is 10 mm, and the weight is 0.09 g. Driven by an oscillating magnetic field, the jumping robot can constantly jump forward and move at an average speed of 1.1 body length per second. This kind of miniature jumping robot possesses many advantages such as good robustness, small structure and wireless actuation, so that it has great application potential in the fields of environmental investigation, search and rescue, etc.

Keywords: Magnetically controlled · Miniature soft robot · Jump · Wireless actuation

1 Introduction

In nature, because of the ability to jump, many small animals can quickly go through obstacles, complex terrain, which greatly helps them to avoid danger [1–3]. Based on these characteristic advantages, it is very meaningful to design a jumping robot that can carry out environmental detection, post-disaster rescue and military reconnaissance. Inspired by this, researchers began to put their efforts into designing and building jumping robots.

With the development of robot manufacturing technology and bionics theory, researchers have designed many jumping robots with various shapes and driving modes successfully [4–7]. Under the excitation of different actuators, jumping robots with different structures realize intermittent or continuous jumping movement. According to different materials and principles, the actuators of jumping robots can be simply divided into four categories.

The first is elastic actuator, which consists of materials with excellent elastic properties, can store elastic potential energy under compression and further transformed into the kinetic energy of the upward jump in releasing process [8–10]. For example, using elastic strip and motor, Atsushi Yamada *et al.* designed an asymmetric jumping robot that could jump 700 mm far and 200 mm high in a single intermittent jump [11]. Benjamin Pawlowski and Jianguo Zhao employed two symmetrically distributed Nitinol (NiTi) wires as elastic actuators to stimulate the miniature robot to jump [5]. Although

X.-J. Liu et al. (Eds.): ICIRA 2021, LNAI 13014, pp. 267–274, 2021.
https://doi.org/10.1007/978-3-030-89098-8_25

the jumping robot using elastic actuator has excellent jumping ability, the introduction of rigid components such as motor and gear makes this kind of robot too large and the robustness is insufficient.

The second is shape memory alloy (SMA) actuator, which can deform at high temperature and returns to its original state at low temperature [12, 13]. Based on the shape memory alloy's ability to respond quickly to temperature changes, researchers have developed robots that can jump on water. For example, Je-Sung Koh *et al.* utilized shape memory alloy and surface tension to produce a water strider leaping robot [14].

Another type of actuator is pneumatic actuator, which has strict requirements on the air tightness of the structure [6]. For example, Yichao Tang *et al.* proposed to use mechanical instability to improve the speed of the soft jumping robot [15]. By imitating the movement of a cheetah, they designed a soft jumping robot that was driven by gas. This robot had the ability to store and release energy quickly within tens of milliseconds, and could move at speeds of up to 2.68 body lengths per second (bps).

The final type of actuator is piezoelectric actuator. Due to the advantages of short response time and fast motion speed, jumping robot with piezoelectric materials has become a hot spot in the research of jumping robot. For example, M Duduta *et al.* used dieletric elastomer materials as actuator to obtain a jumping robot without feet [16]. When the robot was powered, it could produce bending deformation quickly on the platform, so as to store elastic potential energy; when the opposite voltage was applied, the robot deformed in reverse, released the previous elastic potential energy and thus bounced up quickly. Different from M Duduta *et al.*, Yichuan Wu *et al.* designed a one-legged jumping robot, which was driven by Polyvinylidene difluoride (PVDF) [17]. This jumping robot was the fastest among published reports of insect-scale soft ground robots, whose relative speed could reach at 20 bps. Not content with the one-legged form of tiny soft jumping, Tongil Park and Youngsu Cha carried out the research on bipedal robots [18]. They designed a soft-legged mobile robot with bimorph piezoelectric main body and pre-curved piezoelectric legs, which could move at 0.7 bps.

Even if each of these actuators has its merits and can be used to design jumping robots, they all have their drawbacks. The jumping robot with elastic actuator has poor robustness and large overall structure; the robot using shape memory alloys could only jump intermittent; pneumatic robots need strict air tightness conditions; piezoelectric jumping robots have the ability of fast movement, but they need to be towed by wires. Even if some robots use the way of carrying electricity by themselves [19], the problem of endurance is also a huge challenge.

In order to solve the above problems, we proposed and designed a magnetically controlled miniature jumping robot with one leg. Different from the legless magnetic soft jumping robot [20], this robot has the advantages of simple manufacture, fast response time, and the maximum transverse velocity can reach 13.33 mm/s (1.1 bps).

2 Fabrication Procedure

The magnetic soft robot is composed of three parts: a magnetic actuator, the main body and one elastic leg. In order to obtain good resilience of structure, we used polyethylene terephthalate (PET) film, which had good mechanical properties, high impact strength and good folding resistance, to manufacture the main body part and elastic leg part. As shown in Fig. 1(a), we used 100 μm PET film for laser cutting, so as to get two specifications of small sheets. Among them, the length and width of the main body part and elastic leg part were 10 × 8 mm and 8 × 6 mm, respectively. Then, we chose the smaller PET sheet and bent it to get the elastic leg, as displayed in Fig. 1(b). In order to make the magnetic robot move forward, we controlled the bending angle θ within 90°. The magnetic actuator, which was used as a key part of the response to the external magnetic field, did not have strict requirements for structure and elasticity, so it could be obtained by cutting round magnetic disk. As displayed in Fig. 1(b), the length, width and thickness of the magnetic actuator were 2 mm, 3 mm and 1 mm, respectively. The three parts were bonded by an epoxy resin adhesive, in which the magnetic actuator was pasted above the main body, 4 mm away from the head of the main body; the elastic leg was attached to the bottom of the main film, 2 mm away from the head of the main body, deviating from the center of gravity, as shown in Fig. 1(c). This design can make the structure asymmetrical, which can produce a smooth forward movement trend.

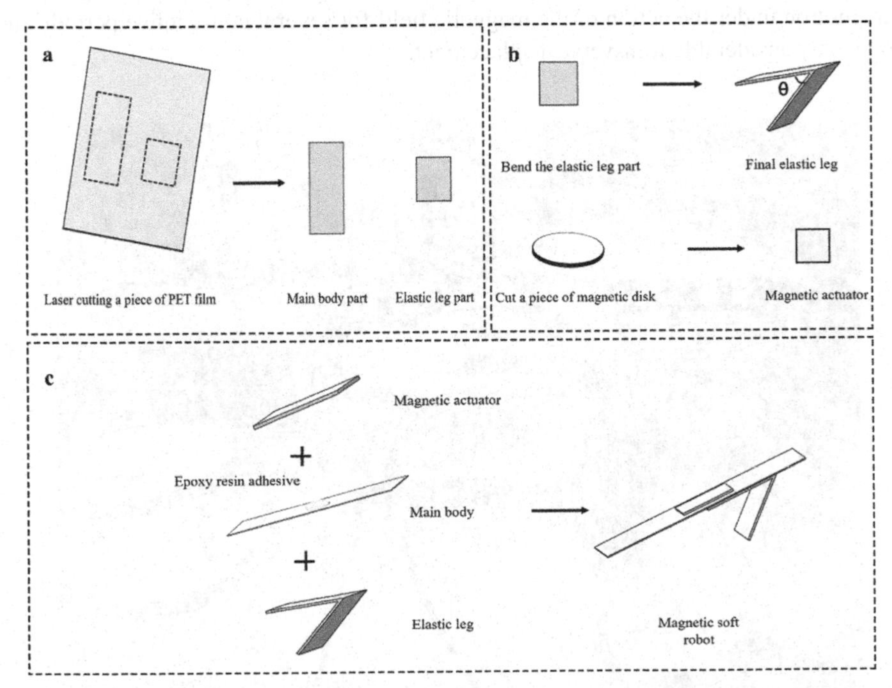

Fig. 1. The manufacturing process of the magnetically controlled soft jumping robot.

3 The Kinematic Performance Experiment of Robot

In the experiment, we set up the experimental platform through signal generator, power amplifier, DC voltage source and one-dimensional magnetic field generator, as shown in Fig. 2. The DC voltage source supplied power to the power amplifier plate; the smaller sinusoidal input became a larger sinusoidal output through the power amplifier, which was then fed into a one-dimensional magnetic field generator to produce an oscillating magnetic field. Under oscillating magnetic field, the magnetic actuator will produce two kinds of responses: 1) When the magnetic force generated by the external magnetic field attracted the magnetic actuator, the robot would press down under the combined action of the platform reaction force and magnetic force, and the elastic leg would deform, thus accumulating the elastic potential energy; 2) When the magnetic force generated by the external magnetic field repulsed the magnetic actuator, the elastic leg released the previously stored elastic potential energy. Under the joint action of the elastic force and the electromagnetic force, the jumping robot had the tendency to move forward and upward. In the preliminary experiment, through the analysis of the movement data of the jumping robot, we found that there were four main movement postures of the magnetically controlled jumping robot under one jump cycle: both-touching, front-touching, back-touching, both-touching, as illustrated in Fig. 3. By controlling the magnetic field frequency, the magnetic soft robot can make the corresponding continuous motion under the action of the magnetic field for several consecutive periods and produce a considerable transverse displacement.

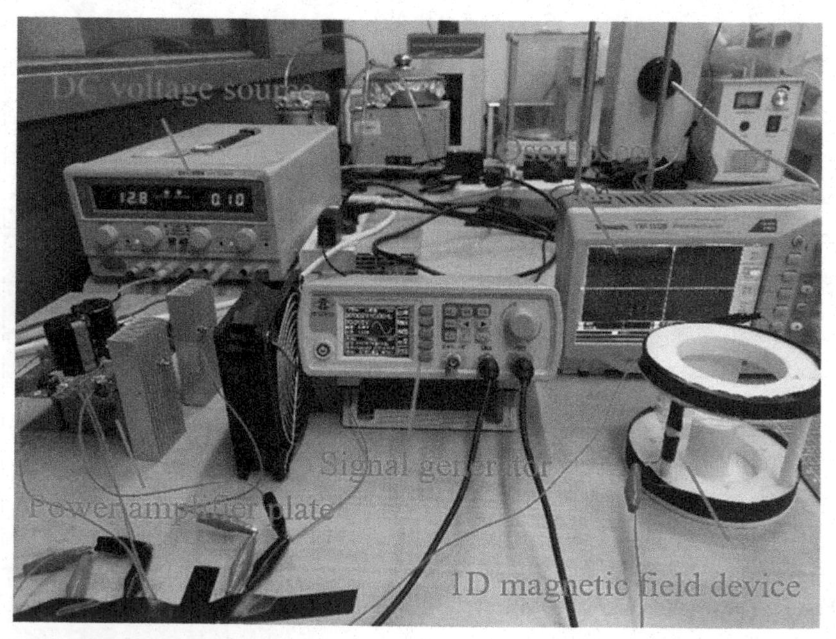

Fig. 2. The schematic diagram of experimental devices for a magnetically controlled jumping robot

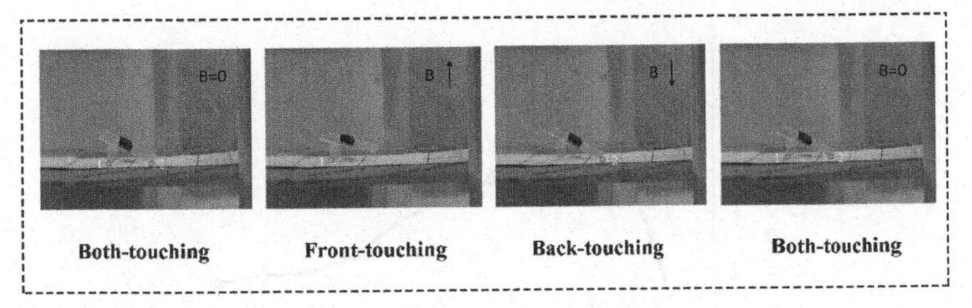

Fig. 3. The main motion postures of the miniature jumping robot in one cycle.

In order to study the range of the magnetic field frequency response of the soft robot, we observed the motion of the magnetic soft robot in the frequency range of 1–200 Hz through the frequency scanning function of the signal generator in the experiment. The results shown that the magnetic soft robot had the ability to move forward in the frequency range of 5–130 Hz, and the lateral motion speed reached the maximum between 20–50 Hz. In order to further determine the optimal driving frequency of the magnetic soft robot and study the relationship between magnetic field frequency and transverse velocity, we conducted motion test experiments in the frequency range of 10–110 Hz. Because the soft robot was small in size, it cannot produce appreciable lateral displacement in a short time, we adopted the control variable method for research, that is, measuring the time taken by the robot to move the same distance under the condition of constant sinusoidal input voltage and changing frequency. We used a narrow lane of 40 mm long and 10 mm wide as the track (to prevent the deflection of the magnetically controlled soft robot in motion), and employed a time meter for timing. To make the results more accurate, we timed three times in the same experiment and took the average value. The experimental results shown that, under the condition of input voltage of 0.7 V, the magnetic soft robot had the maximum lateral motion speed at the frequency of 40 Hz, which could reach 13.33 mm/s, that is, the robot could move 1.1 bps, as shown in Fig. 4.

In addition to studying the relationship between frequency and motion speed, we also studied the experiment of input signal waveform and climbing performance. By changing the upper and lower bias of the input signal, we demonstrated that the greater the downward electromagnetic force and the longer the action time of the jumping robot, the faster the robot moves. It is easy to understand that when the downward electromagnetic force is larger or the action time is longer, the robot will accumulate more elastic deformation and the elastic potential energy will be larger. In the stage of releasing potential energy, the robot could jump further. In the climbing ability test, we increased the slope by increasing the glass shim. Under the sinusoidal input signal of 0.5 V and 40 Hz, the final results showed that the jumping robot could climb the slope at a maximum of 8.4°, as presented in Fig. 5.

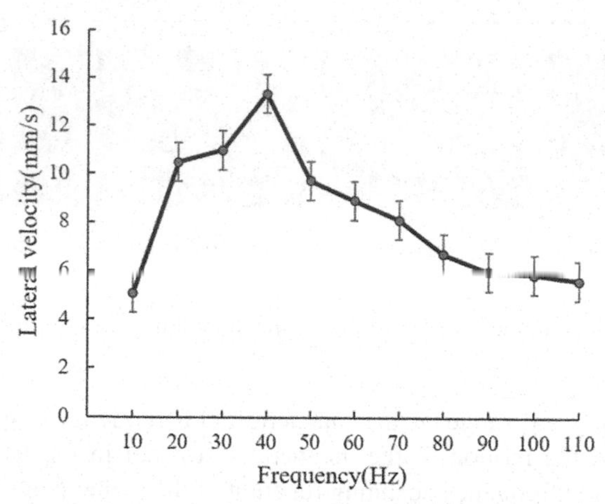

Fig. 4. The relationship between the frequency and the lateral motion rate at an input voltage of 0.7 V.

Fig. 5. Driven by a sinusoidal input signal of 0.5 V and 40 Hz, the jumping robot climbs up an 8.4° slope

4 Discussion

By comparing the different actuators of the jumping robots, it is not difficult to find that there are two main trends in researchers' research on jumping robots: 1) Softening and miniaturizing. Different from rigid jumping robots, the soft jumping robots can greatly reduce the overall size and weight. More importantly, the soft jumping robots have high robustness and more adaptability to the complex terrain environment. 2) Wireless actuator. Pneumatic actuator, shape memory alloy actuator and piezoelectric actuator can quickly drive the jumping robot to complete the jumping movement, but they need air pipes or wires to connect the external functions, which greatly limits the free movement of the jumping robot, especially in the occasions with many curves and obstacles.

In this paper, we proposed an idea to design a single-legged magnetic-controlled soft jumping robot. This jumping robot made up of flexible material instead of rigid structure, which greatly improved its robustness and effectively alleviated the impact of the jumping robot when it lands. Compared with other driving methods, the magnetically controlled jumping robot was not restricted by cables and can move freely on the platform, which provides a reference scheme for the future research of the miniature jumping robot with wireless actuator.

Acknowledgement. The authors want to thank the financial support from National Science Foundation of China (No. 61803088), the Open Project Programs from both the State Key Laboratory of Robotics, Shenyang Institute of Automation, Chinese Academy of Sciences (Grant No: 2017-O02) and Fuzhou University Testing Fund of precious apparatus (No. 2020T016).

References

1. Mo, X., et al.: Jumping locomotion strategies: from animals to bioinspired robots. Appl. Sci. **10**(23), 8607 (2020)
2. Sayyad, A., Seth, B., Seshu, P.: Single-legged hopping robotics research—a review. Robotica **25**(5), 587–613 (2007)
3. Chang, B., Myeong, J., Virot, E., Clanet, C., Kim, H.Y., Jung, S.: Jumping dynamics of aquatic animals. J. R. Soc. Interface **16**(152), 20190014 (2019)
4. Zufferey, R., et al.: Consecutive aquatic jump-gliding with water-reactive fuel. Sci. Robot. **4**(34), eaax7330 (2019)
5. Pawlowski, B., Zhao, J.: 2017 IEEE 7th Annual International Conference on Cyber Technology in Automation, Control, and Intelligent Systems, pp. 460–465. IEEE Press, Honolulu (2017)
6. Hosoda, K., Sakaguchi, Y., Takayama, H., Takuma, T.: Pneumatic-driven jumping robot with anthropomorphic muscular skeleton structure. Auton. Robot. **28**(3), 307–316 (2009)
7. Kovač, M., Schlegel, M., Zufferey, J.-C., Floreano, D.: Steerable miniature jumping robot. Auton. Robot. **28**(3), 295–306 (2009)
8. Zhao, J., Xi, N., Cintron, F.J., Mutka, M.W., Xiao, L.: 2012 IEEE/RSJ International Conference on Intelligent Robots and Systems, pp. 4274–4275. Vilamoura-Algarve, Portugal (2012)
9. Reis, M., Iida, F.: An energy-efficient hopping robot based on free vibration of a curved beam. IEEE/ASME Trans. Mechatron. **19**(1), 300–311 (2014)
10. Li, F., et al.: Jumping like an insect: design and dynamic optimization of a jumping mini robot based on bio-mimetic inspiration. Mechatronics **22**(2), 167–176 (2012)
11. Yamada, A., Watari, M., Mochiyama, H., Fujimoto, H.: 2008 IEEE International Conference on Robotics and Automation, vols. 1–9, pp. 232–2374. IEEE Press, Pasadena (2008)
12. Ho, T., Lee, S.: In 2010 IEEE/RSJ International Conference on Intelligent Robots and Systems, pp. 3530–3535. IEEE, Taipei (2010)
13. Yang, K., Liu, G., Yan, J., Wang, T., Zhang, X., Zhao, J.: A water-walking robot mimicking the jumping abilities of water striders. Bioinspir. Biomim. **11**(6), 066002 (2016)
14. Koh, J.S., et al.: Jumping on water: surface tension–dominated jumping ofwater striders and robotic insects. Science **349**(6247), 517 (2015)
15. Tang, Y., et al.: Leveraging elastic instabilities for amplified performance: Spine-inspired high-speed and high-force soft robots. Science **6**(19), eaaz6912 (2020)

16. Duduta, M., Berlinger, F.C.J., Nagpal, R., Clarke, D.R., Wood, R.J., Temel, F.Z.: Electrically-latched compliant jumping mechanism based on a dielectric elastomer actuator. Smart Mater. Struct. **28**(9), 09LT01 (2019)
17. Wu, Y., et al.: Insect-scale fast moving and ultrarobust soft robot. Sci. Robot. **4**(32), eaax1594 (2019)
18. Park, T., Cha, Y.: Soft mobile robot inspired by animal-like running motion. Sci. Rep. **9**(1), 14700 (2019)
19. Goldberg, B., et al.: Power and control autonomy for high-speed locomotion with an insect-scale legged robot. IEEE Robot. Autom. Lett. **3**(2), 987–993 (2018)
20. Hu, W., Lum, G.Z., Mastrangeli, M., Sitti, M.: Small-scale soft-bodied robot with multimodal locomotion. Nature **554**, 81–85 (2018)

Error Modeling and Analysis of 6-UPS Micro Parallel Mechanism

Duanling Li[1,2], Xingyu Xue[1], Kaijie Dong[1(✉)], Chang Xu[1], Haowei Wang[3], Fangyu Gao[4], and Yuwen Xie[4]

[1] School of Automation, Beijing University of Posts and Telecommunications, Beijing 100876, China

[2] College of Mechanical and Electrical Engineering, Shaanxi University of Science and Technology, Xi'an 712000, China

[3] Beijing Institute of Spacecraft System Engineering, China Academy of Space Technology, Beijing 100094, China

[4] Beijing Guowang Fuda Science and Technology Development, Beijing 10070, China

Abstract. Parallel Mechanisms are utilized widely in the field of high speed pick-and-place manipulation. Error modeling for 6-UPS robots generally simplifies the parallelogram structures included by the robots as a link. As the established error model fails to reflect the error feature of the parallelogram structures, the error model of Parallel Mechanism is established, and the relationship between structural Parameter error and output Pose error is analyze. According to the established error model, MATLAB software is used to draw the change curve of the output Pose error when the structure and Pose Parameters change. The influence of structure and Pose Parameters on the output Pose error of Parallel robot is analyzed.

Keywords: Error modeling · Parameters analysis · Micro parallel mechanism

1 Introduction

With the development of science and technology and the continuous expansion of the application field of parallel mechanisms, the accuracy of parallel mechanisms is required higher and higher in modern astrophysics missions, space exploration and medical research. Generally, the main factors affecting the accuracy of parallel mechanisms are: geometric error, thermal effect error, structural deformation error caused by gravity, dynamic error, sensor error [1–3]. Accuracy analysis is the basis of the accuracy research of parallel mechanisms. By establishing the error model of parallel mechanisms, the influence of various error factors on the pose error of parallel mechanisms is analyzed.

The error modeling is being studied persistently. The accuracy analysis and calculation methods of parallel mechanisms mainly include D-H parameter method, influence coefficient method, significance analysis method [4–6], differential equation method [7, 8] and the dynamics analysis of mechanisms [9]. Early in the 21st century, Ting [10]

X.-J. Liu et al. (Eds.): ICIRA 2021, LNAI 13014, pp. 275–282, 2021.
https://doi.org/10.1007/978-3-030-89098-8_26

studied the effects of multiple clearances in a single loop to obtain the maximum angular error. Based on the principle of virtual work, Venanzi [11] proposed a theory to measure the influence of mechanism kinematics on accuracy.

In this paper, we present a digital error model based on the relationship between the structural parameter error and the output pose error of the parallel mechanisms, which is established by using the differential method of the vector equation of the link, to analyze the influence of structural parameter error on output pose error of parallel mechanisms.

The rest of this paper is organized as follows. In the next section, the proposed error modeling method is introduced in detail. The relationship between the link deformation and length deviation is modeled with a smooth function. The analysis of the proposed method is clearly presented using the comparison with the existing example in the Sect. 3. Moreover, the Influence of structure parameters on output pose accuracy is also analyzed. Finally, we conclude the paper in Sect. 4.

2 Error Modeling

A 6-UPS parallel mechanism (PM) consists of a manipulator and 6 links evenly distributed around the manipulator. Generally, as shown in Fig. 1 the prismatic joint of the UPS link is the actuator. The coordinate and parameter setting of the parallel mechanisms are shown in Fig. 2. The equivalent spherical joints on the moving platform are divided into three groups and evenly distributed on the circumference with radius R, the angle corresponding to the two joint points in each group is α; The Hooke joints on the base are divided into three groups, which are evenly distributed on the circle with radius R_O, and the angle corresponding to the two joint points in each group is β.

a. Model of Mechanism

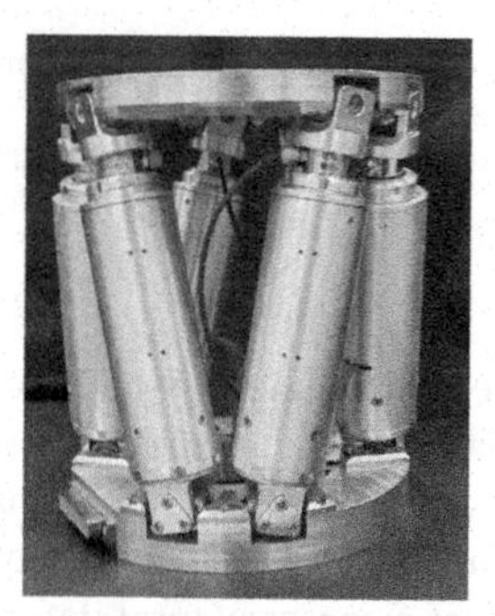

b. 6-UPS mechanism

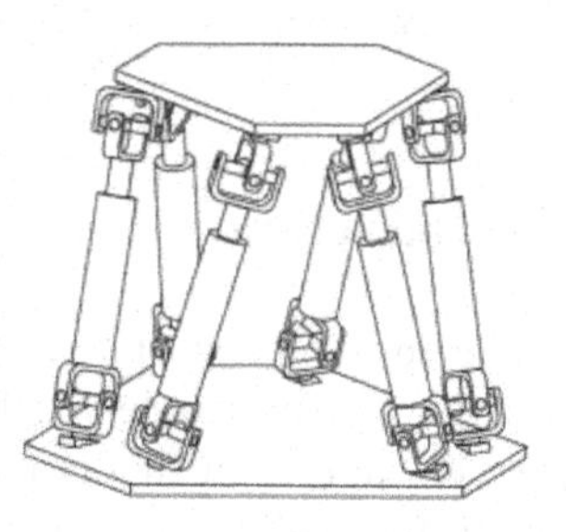

c. Diagram of mechanism

Fig. 1. 6-UPS micro parallel mechanism

The homogeneous transformation matrix $\mathbf{T}$ from the moving platform coordinate system O-XYZ to the body coordinate system O_o-$X_oY_oZ_o$ can be denoted as:

$$\mathbf{T} = \begin{bmatrix} \mathbf{R} & \mathbf{P} \\ 0 & 1 \end{bmatrix} \tag{1}$$

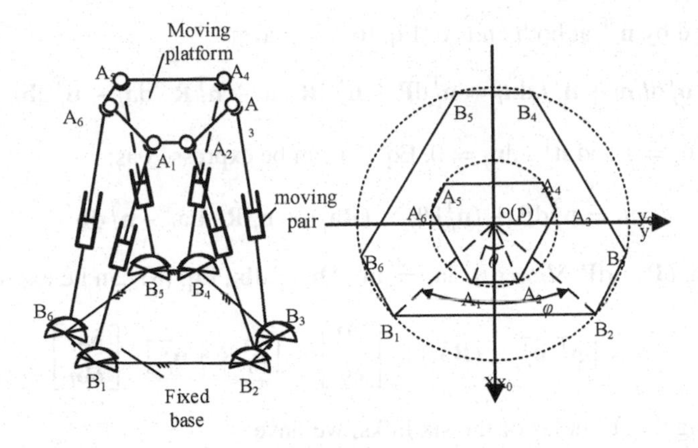

Fig. 2. Coordinate and parameter setting of parallel mechanisms

Where the 3 × 1 matrix $\mathbf{P} = [X_p Y_p Z_p]^{\mathrm{T}}$ denotes 3-dimentional translation. Respectively, the 3 × 3 matrix $\mathbf{R}$, which is shown in Eq. (2), denotes 3-dimentional rotation, each column of which is the direction cosine of X, Y, Z in the coordinate system $O\text{-}XYZ$.

$$\mathbf{R} = \begin{bmatrix} l_1\ m_1\ n_1 \\ l_2\ m_2\ n_2 \\ l_3\ m_3\ n_3 \end{bmatrix} \tag{2}$$

According to the vector expression of each link of the mechanism, there are:

$$l_i \mathbf{n}_i = \mathbf{P} + \mathbf{R} \cdot \mathbf{a}_i - \mathbf{b}_i \ (i = 1 - 6) \tag{3}$$

Where l_i is the length of the link i. Where $\mathbf{n}_i$ is the unit vector of the link $a_i b_i$, the direction of it is from the joint point $\mathbf{a}_i$ to the joint point $\mathbf{b}_i$.

By differential Eq. (3), we have:

$$\mathrm{d}l_i \mathbf{n}_i + l_i \mathrm{d}\mathbf{n}_i = \mathrm{d}\mathbf{P} + \mathrm{d}\mathbf{R} \cdot \mathbf{a}_i + \mathbf{R} \cdot \mathrm{d}\mathbf{a}_i - \mathrm{d}\mathbf{b}_i \tag{4}$$

Since the robot differential relationship, we can get:

$$\mathrm{d}\mathbf{n}_i = \begin{bmatrix} 0 & -\delta n_{iz} & \delta n_{iy} \\ \delta n_{iz} & 0 & -\delta n_{ix} \\ -\delta n_{iy} & \delta n_{ix} & 0 \end{bmatrix} \tag{5}$$

Likewise, from the perturbation matrix [12] of the rotation transformation matrix there are,

$$\mathrm{d}\mathbf{R} = \begin{bmatrix} 0 & -\delta\theta_z & \delta\theta_y \\ \delta\theta_z & 0 & -\delta\theta_x \\ -\delta\theta_y & \delta\theta_x & 0 \end{bmatrix} \cdot \mathbf{R} = \Delta\mathbf{R} \cdot \mathbf{R} \tag{6}$$

Where $\delta\theta_x$, $\delta\theta_y$, $\delta\theta_z$ is the pose error vector of the moving platform, the error components of $\mathbf{R}$ are denoted as $\delta\boldsymbol{\theta} = \left[\delta\theta_x\ \delta\theta_y\ \delta\theta_z\right]^{\mathrm{T}}$.

Multiplied by $\mathbf{n}_i^{\mathrm{T}}$ at both ends of Eq. (6), we have:

$$\mathbf{n}_i^{\mathrm{T}} \mathrm{d}l_i \mathbf{n}_i + \mathbf{n}_i^{\mathrm{T}} l_i \mathrm{d}\mathbf{n}_i = \mathbf{n}_i^{\mathrm{T}} \mathrm{d}\mathbf{P} + \mathbf{n}_i^{\mathrm{T}} \mathrm{d}\mathbf{R} \cdot \mathbf{a}_i + \mathbf{n}_i^{\mathrm{T}} \mathbf{R} \cdot \mathrm{d}\mathbf{a}_i - \mathbf{n}_i^{\mathrm{T}} \mathrm{d}\mathbf{b}_i \tag{7}$$

Since $\mathbf{n}_i^{\mathrm{T}} \mathbf{n}_i = 1$ and $\mathbf{n}_i^{\mathrm{T}} \cdot \mathrm{d}\mathbf{n}_i = 0$, Eq. (7) can be expressed as:

$$\mathrm{d}l_i = \mathbf{n}_i^{\mathrm{T}} \mathrm{d}\mathbf{P} + (\mathbf{n}_i^{\mathrm{T}} \delta\boldsymbol{\theta}) \times (\mathbf{R}\mathbf{a}_i) + \mathbf{n}_i^{\mathrm{T}} \mathbf{R} \cdot \mathrm{d}\mathbf{a}_i - \mathbf{n}_i^{\mathrm{T}} \mathrm{d}\mathbf{b}_i \tag{8}$$

Assuming $\delta\mathbf{P} = \mathrm{d}\mathbf{P}$, $\delta l_i = \mathrm{d}l_i$, $\delta\mathbf{a}_i = \mathrm{d}\mathbf{a}_i$, $\delta\mathbf{b}_i = \mathrm{d}\mathbf{b}_i$, Eq. (8) can be expressed as:

$$\delta l_i = \left[\mathbf{n}_i^{\mathrm{T}} \ \mathbf{n}_i^{\mathrm{T}} \times (\mathbf{R}\mathbf{a}_i) \right] \cdot \begin{bmatrix} \delta\mathbf{P} \\ \delta\boldsymbol{\theta} \end{bmatrix} + \left[\mathbf{n}_i^{\mathrm{T}} \mathbf{R} \ -\mathbf{n}_i^{\mathrm{T}} \right] \cdot \begin{bmatrix} \delta\mathbf{a}_i \\ \delta\mathbf{b}_i \end{bmatrix} \tag{9}$$

Integrating the formulas of the six links, we have:

$$\delta\mathbf{l} = \mathbf{J}_1 \cdot \delta\mathbf{M} + \mathbf{J}_2 \cdot \delta\mathbf{K} \tag{10}$$

Among Eq. (10), where

$$\delta\mathbf{l} = \left[\delta l_1 \ \delta l_2 \ \delta l_3 \ \delta l_4 \ \delta l_5 \ \delta l_6 \right]^{\mathrm{T}} \tag{11}$$

$$\mathbf{J}_1 = \begin{bmatrix} \mathbf{n}_1^{\mathrm{T}} & \mathbf{n}_1^{\mathrm{T}} \times (\mathbf{R}\mathbf{a}_1) \\ \vdots & \vdots \\ \mathbf{n}_6^{\mathrm{T}} & \mathbf{n}_6^{\mathrm{T}} \times (\mathbf{R}\mathbf{a}_6) \end{bmatrix} \tag{12}$$

$$\delta\mathbf{M} = \begin{bmatrix} \delta\mathbf{P} \\ \delta\boldsymbol{\theta} \end{bmatrix} \in R^{6\times 1} \tag{13}$$

$$\mathbf{J}_2 = \begin{bmatrix} \mathbf{n}_1^{\mathrm{T}}\mathbf{R} & -\mathbf{n}_1^{\mathrm{T}} & \cdots & 0 & 0 \\ \vdots & \vdots & \ddots & \vdots & \vdots \\ 0 & 0 & \cdots & \mathbf{n}_6^{\mathrm{T}}\mathbf{R} & -\mathbf{n}_6^{\mathrm{T}} \end{bmatrix} \tag{14}$$

$$\delta\mathbf{K} = \begin{bmatrix} \delta\mathbf{a}_i \\ \delta\mathbf{b}_i \end{bmatrix} \in R^{36\times 1} \tag{15}$$

Since the inverse matrix of $\mathbf{J}_1$ can be obtained, Eq. (10) can be expressed as:

$$\delta\mathbf{M} = \mathbf{J}_1^{-1}\delta\mathbf{l} - \mathbf{J}_1^{-1}\mathbf{J}_2 \cdot \delta\mathbf{K} \tag{16}$$

Through Eq. (16), the influence of joint error and link input error on the output pose error of the moving platform of the parallel mechanisms can be obtained.

3 Case Study and Error Parameter Analysis

3.1 Method Verification

An example of the 6-UPS PM is used to analyze the proposed accuracy analysis method. The parameter settings of parallel mechanisms are presented, as shown in Table 1.

Table 1. Parameter settings

Number	1	2	3	4	5	6
a_{xi}	−13.7819	13.7819	48.5148	34.7329	−34.7329	−48.5148
a_{yi}	−48.0631	−48.0631	12.0961	35.9670	35.9670	12.0961
b_{xi}	−34.7329	34.7329	48.5148	13.7819	−13.7819	−48.5148
b_{yi}	−35.9670	−35.9670	−12.0961	48.0631	48.0631	−12.0961

Assuming the pose parameters are presented as follows:

$$\mathbf{P} = \left[5\ 5\ 120\right]^{\mathrm{T}} \tag{17}$$

$$\mathbf{R} = \begin{bmatrix} 0.9948 & -0.0493 & 0.0888 \\ 0.0521 & 0.9982 & -0.0303 \\ -0.0872 & 0.0348 & 0.9956 \end{bmatrix} \tag{18}$$

Structural error is shown in Table 2.

Table 2. Output pose error of parallel mechanisms.

Number	δa_{ix}	δa_{iy}	δa_{iz}	δb_{ix}	δb_{iy}	δb_{iz}	δl_i
1	−0.02	0.02	−0.01	−0.01	−0.01	0.02	0.01
2	0.01	0.02	0.02	−0.01	−0.01	0.02	0.02
3	−0.01	−0.01	0.01	−0.03	0.01	0.02	0.02
4	0.02	0.02	0.02	−0.01	−0.01	0.02	−0.01
5	−0.01	0.02	−0.02	−0.01	0.01	0.02	−0.01
6	0.01	0.03	0.01	−0.01	−0.01	0.02	0.02

According to the accuracy analysis, the pose error of the parallel mechanisms is shown in Table 3.

Table 3. Output pose error of parallel mechanisms

δP_x (mm)	δP_y (mm)	δP_z (mm)	$\delta\varphi$ (°)	$\delta\theta$ (°)	$\delta\gamma$ (°)
0.0184	0.0354	0.0346	0.0267	−0.0135	−0.0722

3.2 Influence of Structural and Pose Parameters on Output Pose Accuracy

By analyzing the error model, it can be obtained that the values of the structure parameters of the parallel mechanisms affect the accuracy of the output pose of the parallel mechanisms.

According to the coordinate and parameter setting of parallel mechanisms, the structure parameters of parallel mechanisms are related to radius R, circle angle α, radius R_o and circle angle β. And in former analysis we have $\alpha = \beta$ and $R = R_o$. Assuming the output pose of parallel mechanism $\mathbf{P} = [5\ 5\ 120]^{\mathrm{T}}$ and $\omega = [3\ 2\ 2]^{\mathrm{T}}$, the influence of structural R and α on output pose accuracy is analyzed, as shown in Figs. 3 and 4.

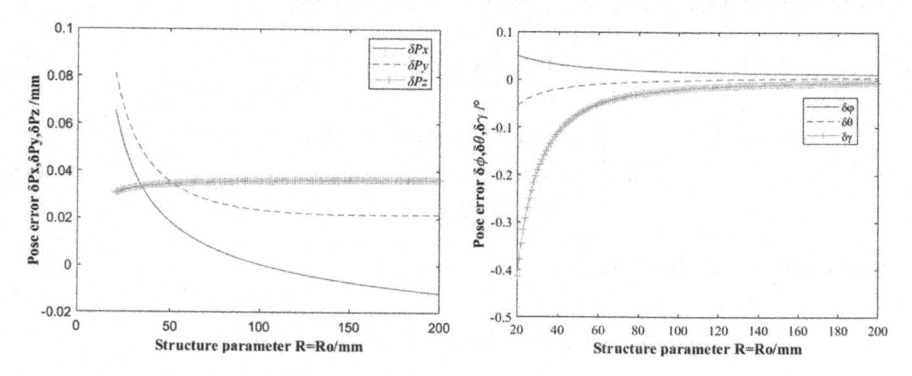

Fig. 3. The influence of radius on output pose accuracy

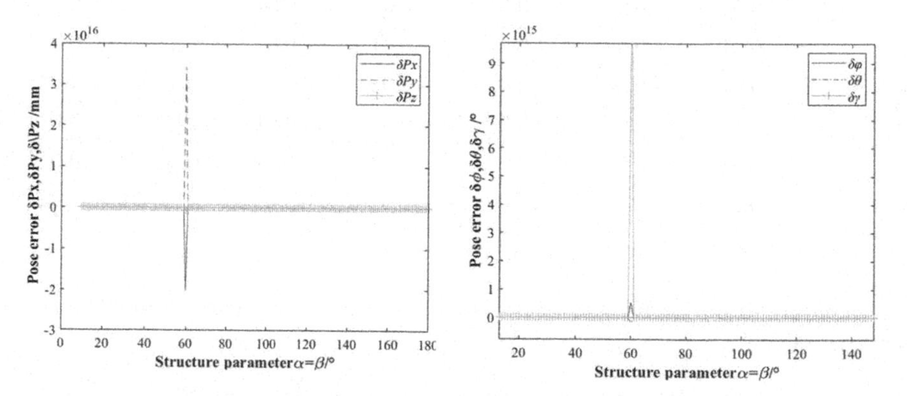

Fig. 4. The influence of α and β setting on output pose accuracy

Through the analysis of the output pose error curves in Figs. 3 and 4, it can be observed that the changes of the structural parameters of the parallel mechanism have the following effects on the output pose accuracy:

Compared with the structural parameters α and β, the structural parameters R and R_o have greater influence on the output pose accuracy. Except for a special position, the structural parameters α and β have little influence on the output pose accuracy; It can be

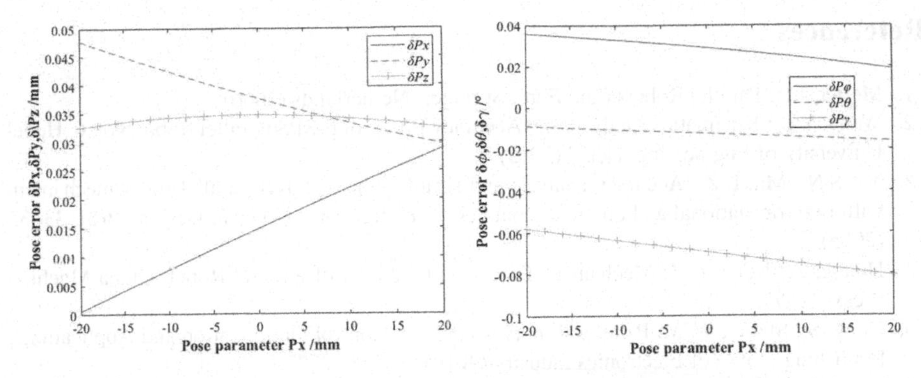

Fig. 5. The influence of P_x and φ on output pose accuracy

found from Fig. 4 that the influence of structural parameters R and R_o on the accuracy of output pose tends to be flat with the increase of R and R_o.

It can be observed from Fig. 5 that when the pose parameter P_x changes individually, the accuracy of the output pose changes as follows:

The error dP_x in X-axis direction increases, as the pose parameter P_x increases from -20 mm to 20 mm; The errors dP_y in Y-axis and the error $d\varphi$ of pose angle φ decrease with the increase of pose parameters; The error dP_z in Z-axis increases at early stage and then decreases with the increase of parameters P_x, and reaches the maximum at $P_x = 0$. The error $d\theta$ and $d\gamma$ increases with the increase of P_x, and the absolute value of them increases constantly.

4 Conclusion

(1) This paper presents the error model of the parallel mechanisms by the differential method of the kinematics equation of the parallel mechanisms. By analyzing a single case, the output pose error of the moving platform is analyzed when the structural parameter error is given. The accuracy performance of a 6-UPS mechanism underload is analyzed to validate the proposed method.
(2) According to the established error model, the output pose error curve is drawn by using MATLAB software when the structure parameters change, and the influence of the structure and pose parameters change on the output pose error of parallel mechanisms is analyzed.
(3) The error modeling method is developed based on the structure of 6-UPS parallel mechanisms. However, the method can be further applied to the accuracy analysis of other parallel mechanisms.

Acknowledgments. This study was co-supported by the National Natural Science Foundation of China (Grant No. 51775052), Beijing Natural Science Foundation (Grant No. 21C10109), Jiangsu Key Laboratory of Advanced Food Manufacturing Equipment & Technology (FMZ202022) and Beijing Municipal Key Laboratory of Space-ground Interconnection and Convergence of China.

References

1. Merlet, J.P.: Parallel Robots (2nd Edn). Springer, Netherlands (2006)
2. Wang, Y.D.: Kinematics Analysis and Accuracy Study of 6-SPS Parallel Robot, vol. 6. Hebei University of Engineering, Hebei (2013)
3. Yu, S.N., Ma, L.Z.: Accuracy analysis and simulations of novel parallel robot mechanism with two translational and one rotational degree of freedom. J. Mach. Design **22**(6), 33–35 (2005)
4. Huang, Z., Kong, L.F.: Mechanism Theory and Control of Parallel Robot. China Machine Press (1997)
5. Cong, S., Shang, W.W.: Parallel Robots-Modeling, Control Optimization and Applications. Publishing House of Electronics Industry (2010)
6. Liu, B.C., Chen, J.J.: Accuracy analysis and simulation of 3-RRR parallel robot. Indust. Control Comput. **24**(12), 30–34 (2011)
7. Jiang, J.X.: 6-PUS/UPS the Structure Design and Theory Analysis of a 6-PUS/UPS Parallel Manipulator, vol. 6. Yanshan University (2010)
8. Zhao, Y.J., Zhao, X.H.: Effects of pose changes on accuracy of 6-SPS parallel robot. J. Mach. Design **20**(9), 42–44 (2003)
9. Tian, Q., Flores, P., Lankarani, H.M.: A comprehensive survey of the analytical, numerical and experimental methodologies for dynamics of multibody mechanical systems with clearance or imperfect joints. Mech. Mach. Theory **122**, 1–57 (2018)
10. Ting, K.L., Zhu, J., Watkins, D.: The effects of joint clearance on position and orientation deviation of linkages and manipulators. Mech. Mach. Theory **35**(3), 391–401 (2000)
11. Venanzi, S., Parenti-Castelli, V.: A 0new technique for clearance influence analysis in spatial mechanisms. J. Mech. Design Trans. ASME **127**(3), 446–455 (2005)
12. Wang, B., Peng, B.B.: Overview on the structure of parallel tools. J. North China Inst. Aerosp. Eng. **16**(5), 8–11 (2006)

Micro-nano Scale Longitudinal Displacement Measurement of Microspheres Based on Digital Holography

Si Tang[1,2,3], Jialin Shi[1,2], Huiyao Shi[1,2,3], Peng Yu[1,2], Chanmin Su[1,2], and Lianqing Liu[1,2](✉)

[1] State Key Laboratory of Robotics, Shenyang Institute of Automation, Chinese Academy of Sciences, Shenyang, China
lqliu@sia.cn

[2] Institutes for Robotics and Intelligent Manufacturing, Chinese Academy of Sciences, Shenyang 110169, China

[3] University of Chinese Academy of Sciences, Beijing, China

Abstract. Most of the existing measurements of microspheres are based on transmission, which use the light field through the microspheres and the substrate to form interference fringes, and then analyzes the movement and variation of the microspheres. Therefore, only transparent samples and transparent substrates can be measured, which greatly limits the measurement range and the application scenarios. In this paper, we present a near-field diffraction measurement for reflective microspheres. This method has no restrictions on the material of the microspheres themselves, and the substrate is no longer required to be the same material as the microspheres. Therefore, the range of measurement using microspheres will be greatly increased. In the case of microspheres with different materials and substrates, microspheres can not only be applied to simple cell measurement, but also are expected to be combined with other measurement methods in the future to further improve the measurement accuracy and measurement range of the existing measurement methods.

Keywords: Measurement · Three-dimensional position of microspheres · Digital holographic microscopy

1 Introduction

Biological tissues, cells, proteins, and so on are all in micro and nano motion, because of metabolism or their movement and response. Traditional measurement methods include optical measurement based on fluorescence labeling, atomic force microscopy based on scanning probe, etc. However, these methods have their limitations, such as requiring fluorescence labeling or a small measurement range. In recent years, some scholars have proposed a marker-free wide range and high precision measurement method based on microspheres. However, it is difficult to measure and calibrate the microscopic morphology of biological or abiotic surfaces under the conventional optical microscope. Microspheres are usually used as biological and abiotic markers for detection and calibration

X.-J. Liu et al. (Eds.): ICIRA 2021, LNAI 13014, pp. 283–291, 2021.
https://doi.org/10.1007/978-3-030-89098-8_27

[1–6]. However, it is very accurate and simple to reflect the three-dimensional morphology of the sample surface through the changes of microspheres. Three-dimensional position measurement of microspheres [7] has gradually become an important single-particle tracking technology (SPT) [8–12] since the beginning of this century, and this emerging technology has also gradually become an important detection method in biomolecular dynamics. At present, common methods for microsphere measurement include cross-correlation method [13], digital holographic reconstruction [14], scattering matching method [15], back focal plane method [16] and defocus imaging method [17]. Although these methods are widely used and have their advantages, they still have their limitations. The cross-correlation method is often used in magnetic tweezers technology, but for magnetic tweezers, the operating object must be magnetic beads; the traditional digital holographic reconstruction method needs a lot of data operation and speed is not suitable for rapid measurement. The scattering law requires accurate measurement of particle diameter, refractive index, environmental refractive index, system magnification, etc. The above measurement methods are all described based on transmitted particles. In the case of transmission, many of the above methods can achieve very high accuracy. If combined with sub-pixel imaging, the resolution of 0.01 pixel can even be reached [18, 19].

However, reflective microsphere imaging is still relatively blank at present. In this paper, a microsphere motion measurement method based on near field diffraction is proposed. By changing the traditional digital holographic imaging using Fresnel diffraction law, when the microspheres are further shrinking, the more accurate Rayleigh-Sommerfeld function is used to reconstruct the measured microspheres in 3D, and the image processing method is used to analyze the obtained microsphere image, to realize the precise measurement of the microsphere position.

2 Methodology

2.1 The Experimental System

The diffraction light path proposed in this paper is shown in Fig. 1. The red light of 632.8 nm is emitted from the laser light source, and the beam emitted by the light source is attenuated through PBS. The attenuated laser beam converges through a plane-convex lens, and an ND filter is placed behind the lens for secondary attenuation of the beam in the whole system. Reduce spot because of the large basement reflection diffraction speckle interference for microspheres, when after a lens to focus the beam together, make the back focal plane of the objective with this point, under the circumstances, through the objective lens into a parallel light, parallel light will irradiate on microspheres on the surface of the silicon substrate, and generating interference. At last, interference image after lens amplification and eventually received by CCD imaging.

Due to the noise of spatial sampling and its system, the digital holographic imaging method will produce errors. The sampling frequency of the image is theoretically due to the transfer function of the microscopic objective imaging system itself and the size of the microsphere used. According to the existing experimental equipment and the diameter of the Airy spot:

$$d_a = 1.22\lambda/NA \approx 183.8\ \text{nm} \quad (1)$$

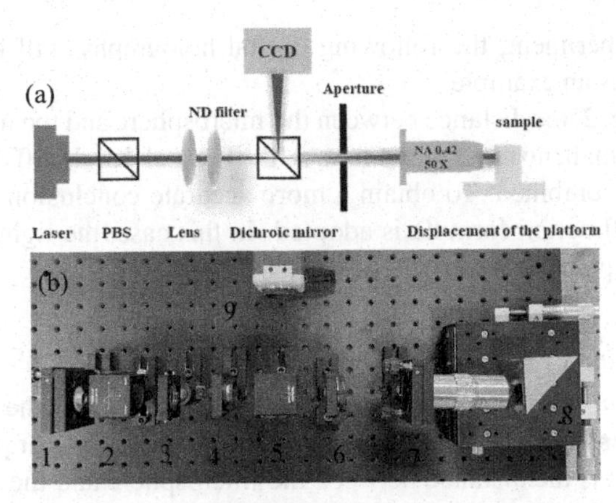

Fig. 1. (a) Schematic of the experimental setup. (b) Diagram of an actual optical measuring system. 1-Laser; 2-PBS; 3-Lens; 4-ND filter; 5-Dichroic mirror; 6-Aperture; 7-Microscope objective; 8-Sample; 9-CCD.

it is recommended that the maximum sampling width for the point spread function (PSF) should not exceed 183.8 nm, and the sampling width used in the experiment is 183 nm.

2.2 Measuring Principle

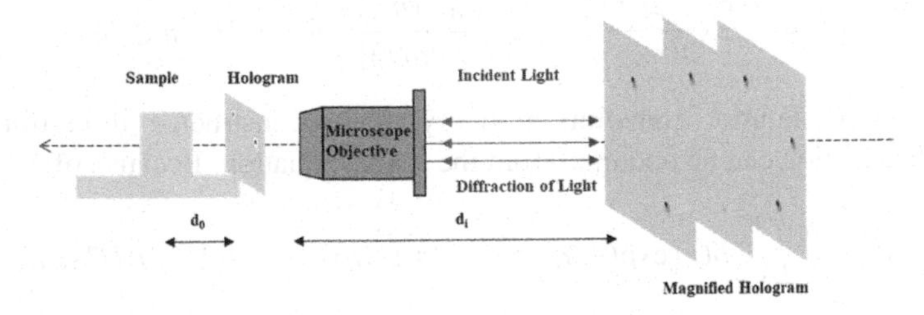

Fig. 2. Coaxial digital holographic schematic with lens.

The principle of digital holography is the same as that of traditional optical holography measurement, that is, it includes two processes of hologram generation, recording, and reproduction. Recording generation is to use the principle of optical interference to record on CCD the interference fringes generated by the interference between the object light reflected by the object and the reference beam that does not pass through the object, to replace the generation of a holographic image recorded by the traditional optical medium. The reconstruction process mainly relies on computer simulation of the diffraction process of the reference light beam to the recorded hologram in the whole recording process, to reproduce the original object light wave. Coaxial digital holography includes lensless technology and lens technology. The former is suitable for microspheres with a diameter of 80–500 μm, while the latter is suitable for microspheres with a diameter of 0.5–20 μm. Since the SiO_2 microspheres with a diameter of 10 μm are used in this experiment, the following digital holography will be described with lenses technology as an example.

As shown in Fig. 2, the distance between the microsphere and the microscopic objective lens is usually small, no more than 1 mm, so the Fresnel-Kirchhoff diffraction integral formula cannot be combined. To obtain a more accurate conclusion, Rayleigh - Sommerfeld integral diffraction formula is adopted. In this case, the light field distribution on the holographic plane is:

$$I(r) = |a(r)|^2 + 2Re\{a^*(u \otimes h_z)\} + |u \otimes h_z|^2 \tag{2}$$

In the formula, $a(r)$ is the parallel reference light, $r = (x, y)$ is the coordinate on the focal plane, $u(r, z)$ is the diffraction light field after the parallel light passes through the microspheres, and z is the distance between the microsphere and the focal plane of the objective lens. h_z is the integral kernel of Rayleigh-Sommerfeld diffraction integral:

$$h_z(r) = -\frac{1}{2\pi}\frac{\partial}{\partial z}\frac{e^{ikR}}{R} \tag{3}$$

In this formula, $R^2 = r^2 + z^2$ and $k = 2\pi/\lambda$.

In Eq. 2. The first term can be eliminated by recording the reference light image when there is no object under measurement, while the intensity of the object light diffracted by the microsphere is far less than the intensity of the reference light. The third term in Eq. 2, which is a high-order term, can be omitted, so:

$$b(r) = \frac{I(r) - |a(r)|^2}{|a(r)|} \approx 2\frac{Re\{a^*(u \otimes h_z)\}}{|a(r)|} \approx 2Re\{u \otimes h_z\} \tag{4}$$

According to Fourier convolution theory, the reconstructed three-dimensional intensity distribution can be obtained from the above equation. Because of:

$$B(q) = \int_{-\infty}^{\infty} b(r)\exp(-iq \cdot r)d^2r \approx U(q)H_z(q) + U^*(q)H_z^*(q) \tag{5}$$

In this case, $U(q)$ is the Fourier transform of $u(r, 0)$, and:

$$H_z(q) = exp\left\{ikz\left[1 - \left(\frac{\lambda q}{2\pi}\right)^2\right]^{\frac{1}{2}}\right\} \tag{6}$$

Equation 5 is the Fourier transform of the integral kernel in the Rayleigh-Sommerfeld diffraction integral.

The Fourier variation of the light field at the distance from the focal plane z' can be expressed as:

$$B(q)H_{-z'}(q) \approx U(q)H_{z-z'}(q) + U^*(q)H_{-z-z'}(q) \tag{7}$$

In the above equation, the first term is the reconstructed light field, and it is focused when $z' = z$. The second term will result in the specular image. In the microsphere measurement, only the intensity distribution of the light field needs to be considered, so the mirror image can be ignored.

Through the above analysis, the reconstructed light field can be obtained as follows:

$$E(r, z) \approx \frac{e^{-ikz}}{4\pi^2} \int B(q)H_{-z}(q)e^{iq\cdot r}d^2q \tag{8}$$

$$I(r, z) = |E(r, z)|^2 \tag{9}$$

The above equation is the holographic reconstruction intensity distribution of the microsphere at the distance from the focal plane z.

2.3 Computer Simulation of Microspheres Position

Since the microspheres used are standard microspheres, their amplitude transmittance is:

$$t(x, y) = 1 - o(x, y) \tag{10}$$

$$o(x, y) = circ\left[\frac{\sqrt{x^2 + y^2}}{D/2}\right] = \begin{cases} 1, \sqrt{x^2 + y^2} \le D/2 \\ 0, else \end{cases} \tag{11}$$

Where, D is the diameter of the microspheres Substitute the above equation into Eq. 9 to simulate the microsphere.

Firstly, the feasibility of coaxial digital holographic reconstruction measurement method was simulated by 2D microspheres. The simulated parameter wavelength is 640 nm, the number of CCD pixels is M × N = 1024 × 1024 pixel, and the pixel size is $\Delta x = \Delta y = 7.4$ μm. The standard recording distance can be calculated by using Eq. 12:

$$d = \frac{\Delta\xi^2 \times N}{\lambda} \tag{12}$$

In the formula, $\Delta\xi$ represents the pixel size of CCD, and substituting the data into Eq. 12, the simulated recording distance is 4.28 cm. Figure 3(a) is a 2D particle field simulated image, Fig. 3(b) is a simulated hologram, and Fig. 3(c) is the reconstructed image obtained according to Eq. 12 and the reconstructed distance d′ = 4.28 cm.

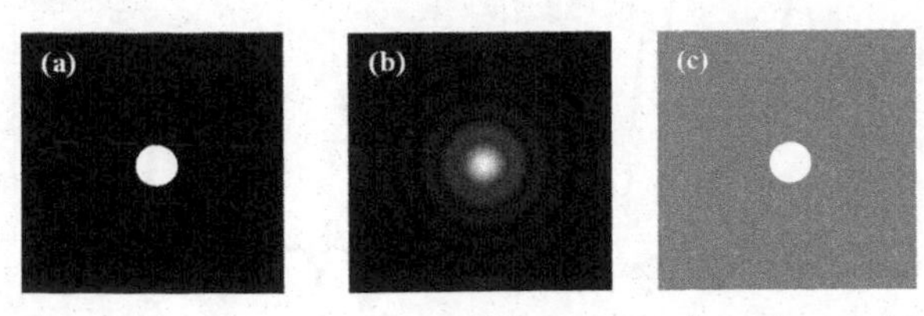

Fig. 3. Digital holographic measurement simulation of 2D microspheres. (a) 2D analog image (b) digital hologram (c) digital reconstruction.

The reconstructed focal plane of the microsphere is the reconstructed plane when the reconstructed distance is equal to the recorded distance. In this case, it is the peak or valley value of the longitudinal intensity curve of the center of the microspheres in the three-dimensional reconstructed intensity diagram. When the background is dark and the microsphere is bright, the maximum value of microsphere center strength in the whole longitudinal range is the peak value. When the background is bright and dark, the minimum value of the central strength is the valley value. However, the holographic recording distance will affect the digital holographic measurement, so the recording distance is discussed.

The diameter of the microspheres was one pixel, and the digital holographic reconstruction was performed for different recording distances, and the step distance of the reconstruction distance was $\delta_p = 1$ nm. A total of 1000 planes were reconstructed, that is, the reconstructed distance space was a longitudinal depth of 1000 nm. The recording distances d' were d'/d = 0.5, 0.8, 1.0, 1.2, 1.5, 2.0, respectively. The intensity distribution curves of the microspheres in three-dimensional space were obtained respectively. Figure 4 shows the intensity distribution curve of the microsphere center in the longitudinal direction.

Through the simulation result shows that under the condition of the microsphere diameter of 10 μm, the overall intensity distribution curve is not obvious, in the light intensity curve of different distance to do bad can see through the results after an operation, in the range 1–5 nm, light intensity difference is 10^{-7}, but when the distance into 6 nm, light intensity difference produced obvious increase, jumped by 3 orders of magnitude. This kind of light intensity difference until close to 1000 nm and promoted an order of magnitude, therefore, we have reason to believe that, for the existing system, weighing measuring accuracy and measuring difficulty, z to the minimum resolution

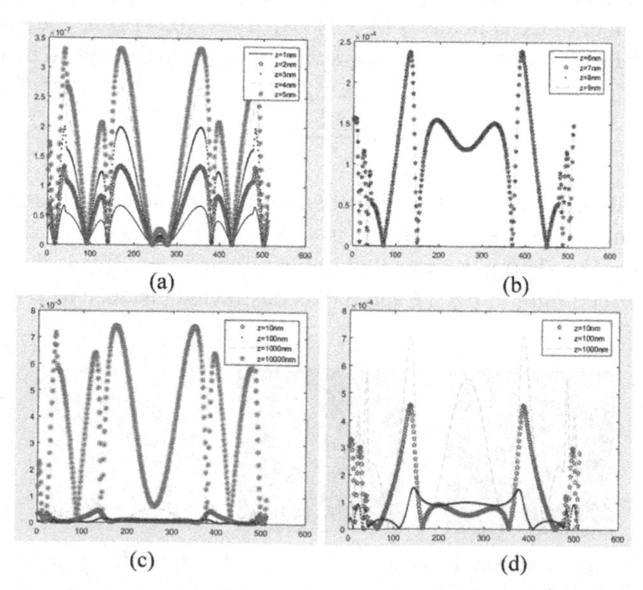

Fig. 4. The strength distribution curve of microsphere center along the longitudinal direction

for 6 nm is the most appropriate, but is limited by the CCD resolution and sampling frequency, the actual experiment accuracy should be more than 6 nm.

3 Experimental Results

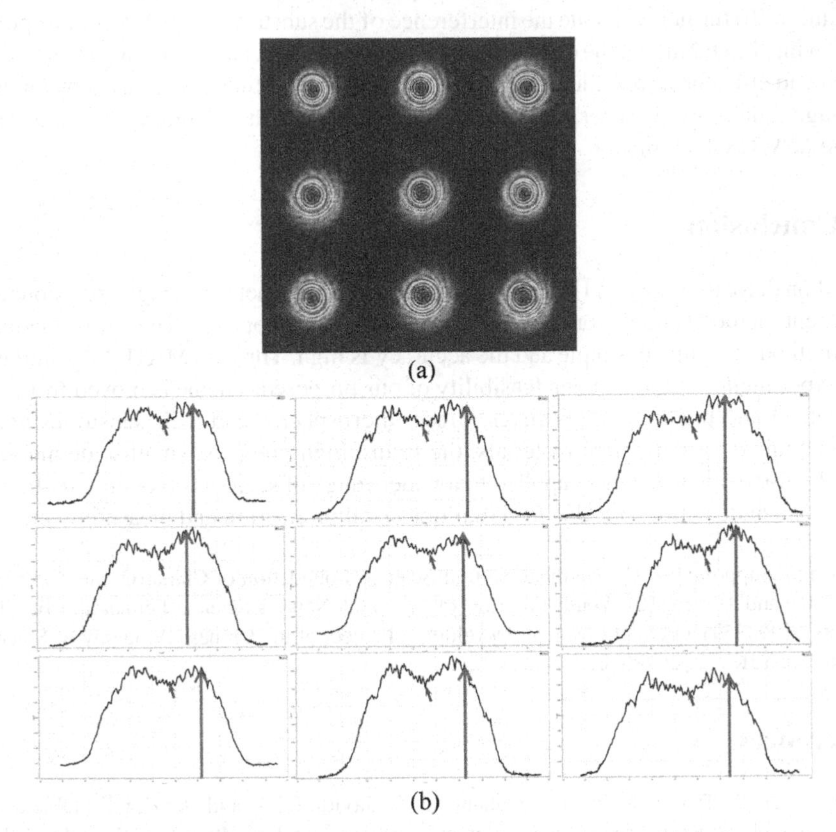

Fig. 5. Diffraction spot and intensity curve of microsphere. (a) Actual diffraction pattern, from top left to bottom right, the distance from the focal point is −1 μm, −0.8 μm, −0.5 μm, −0.1 μm, 0, 0.1 μm, 0.5 μm, 0.8 μm and 1 μm. (b) The corresponding microsphere center strength distribution curve along the longitudinal direction.

Figure 5(a) is the actual system to measure the light diffraction pattern under the different distance of microspheres, the experiments chosen is 10 μm in diameter of silica microspheres, silica microspheres adhere to the surface of a silicon substrate, by microspheres and basal surface of the optical path difference to form the diffraction pattern, the picture shows the actual through a CCD can see the pattern, the slow drift of itself because of the microspheres, therefore, the center of all light spots is not in the same place. At the same time, because the reflective effect of the silicon substrate itself is better, the reflected light beam generated by it will also interfere with the diffraction pattern of the microspheres. Figure 5(b) shows the light intensity distribution curve of the center of the microsphere corresponding to Fig. 5(a). Although in the pattern, the change is not obvious when the distance is changed, the light intensity distribution curve clearly shows the change of light spots when the microsphere is at different positions. Under the existing experimental conditions, the minimum movement range we collected is 100 nm, and when the moving distance is 100 nm, the light intensity curve changes by 3 pixels. In other words, with our current experimental system, we can achieve 100 nm resolution. To further eliminate the interference of the substrate to the diffraction spot, by improving the stability of the displacement mobile platform, and improving the sampling speed and efficiency, we believe that the resolution will be further increased by an order of magnitude or even higher, and finally, we can achieve the accuracy of 6nm as same as the MATLAB simulation.

4 Conclusion

Based on defocus theory and digital holographic imaging method, a high precision measurement method suitable for reflective microspheres is proposed. The whole device of this method is relatively simple and the accuracy is high. Through MATLAB simulation and experiment comparison, the feasibility of our proposed scheme is proved from both theoretical and practical perspectives. Since microspheres and their substrates are no longer limited to transparent materials, the future high-precision small-scale measurement by microspheres will be applied to a wider range of scenarios, not only in the field of biological detection, but also in actual industrial production and measurement.

Funding. Supported by the National Natural Science Foundation of China (Grants 61925307, 61927805 and U1813210), Youth Program of National Natural Science Foundation of China (Grants 61903359) and the Instrument Developing Project of the Chinese Academy of Sciences (Grant No. YJKYYQ20180027).

References

1. Mahajan, S., Trivedi, V., Vora, P., Chhaniwal, V., Javidi, B., Anand, A.: Highly stable digital holographic microscope using sagnac interferometer. Opt. Lett. **40**(16), 3743–3746 (2015)
2. Picazo-Bueno, J.A., Cojoc, D., Iseppon, F., Torre, V., Micó, V.: Single-shot, dual-mode, waterimmersion microscopy platform for biological applications. Appl. Opt. **57**(1), A242–A249 (2018)
3. Verpillat, F., Joud, F., Desbiolles, P., Gross, M.: Dark-field digital holographic microscopy for 3D tracking of gold nanoparticles. Opt. Express. **19**(27), 26044–26055 (2011)
4. Katz, J., Sheng, J.: Applications of holography in fluid mechanics and particle dynamics. Annu. Rev. Fluid Mech. **42**, 531–555 (2010)
5. Neutsch, K., Göring, L., Tranelis, M.J., et al.: Three-dimensional particle localization with common-path digital holographic microscopy. In: Practical Holography XXXIII: Displays, Materials, and Applications, vol. 10944, p. 109440J. International Society for Optics and Photonics (2019)
6. Mann, C.J., Yu, C.J., Lo, C.-M., Kim, M.K.: High-resolution quantitative phase-contrast microscopy by digital holography. Opt. Express **13**(22), 8693–8698 (2005)
7. Wu, M., Roberts, J.W., Buckley, M.: Three-dimensional fluorescent particle tracking at micron-scale using a single camera. Exp. Fluids **38**(4), 461–465 (2005)

8. Cheezum, M.K., Walker, W.F., Guilford, W.H.: Quantitative comparison of algorithms for tracking single fluorescent particles. Biophys. J . **81**(4), 2378–2388 (2001)
9. Thar, R., Blackburn, N., Kuhl, M.: A new system for three-dimensional tracking of motile microorganisms. Appl. Environ. Microbiol. **66**(5), 2238 (2000)
10. Qian, H., Sheetz, M.P., Elson, E.L.: Single particle tracking. Analysis of diffusion and flow in two-dimensional systems. Biophysical Journal **60**(4), 910–921 (1991)
11. Saxton, M.J.: Single-particle tracking: Connecting the dots. Nat. Methods **5**(8), 671–672 (2008)
12. Luo, R., Yang, X.Y., Peng, X.F., et al.: Three-dimensional tracking of fluorescent particles applied to micro-fluidic measurements. J. Micromech. Microeng. **16**(8), 1689 (2006)
13. Charonko, J.J., Vlachos, P.P.: Estimation of uncertainty bounds for individual particle image velocimetry measurements from cross-correlation peak ratio. Measur. Sci. Technol **24**(6), 065301 (2013)
14. Zeng, Y., Lu, J., Hu, X., et al.: Axial displacement measurement with high resolution of particle movement based on compound digital holographic microscopy. Opt. Commun. **475**, 126300 (2020)
15. Ploumpis, S., Amanatiadis, A., Gasteratos, A.: A stereo matching approach based on particle filters and scattered control landmarks. Image Vis. Comput. **38**, 13–23 (2015)
16. Upadhya, A., Zheng, Y., Li, L., et al.: Structured back focal plane interferometry (S-BFPI). Biophoton. Austral. (2019)
17. Leister, R., Fuchs, T., Mattern, P., Kriegseis, J.: Flow-structure identification in a radially grooved open wet clutch by means of defocusing particle tracking velocimetry. Exp. Fluids **62**(2), 1–14 (2021)
18. Chenouard, N., Smal, I., de Chaumont, F., et al.: Objective comparison of particle tracking methods. Nat. Methods **11**(3), 281–290 (2014)
19. Wu, Y., Wu, X., Cen, K.: Development of digital holography in particle field measurement. Chinese J. Laser. **41**(6), 0601001 (2014)

Study on Dual-Deformation of Sheath-Like Carbon Nanotube Composite Yarns

Zeng-Hui Zhao, Su-Feng Zhu, Xu-Feng Dong(✉), and Min Qi

School of Materials Science and Engineering, Dalian University of Technology, Dalian 116024, China
dongxf@dlut.edu.cn

Abstract. Carbon Nanotube Yarns can contract in length when stimulated by electricity, heating, etc. How to endow it with the ability of elongation deformation and realize the bidirectional controllable deformation is an urgent problem to be solved. In this paper, Carbon Nanotube Yarns was used as the core and polydimethylsiloxane (PDMS) as the sheath. The electrostriction characteristics of PDMS were combined with the electrostriction effect of carbon nanofibers to construct a Sheath-like Carbon Nanotube Composite Yarns, and its dual-deformation behavior under different current and voltage was studied. The experimental results show that when the current of 40 mA is applied to the Carbon Nanotube Yarn, the Sheath-like Carbon Nanotube Composite Yarn shows significant shrinkage deformation (1.6%), and the shrinkage stress is twice that of the pure Carbon Nanotube Yarns; when the PDMS was applied with 60 v/μm electric field, the elongation deformation of the Sheath-like Carbon Nanotube Composite Yarn was 0.75%. It lays a foundation for the application of Carbon Nanotube Yarns in the field of flexible robot.

Keywords: Artificial muscle yarns · Carbon nanotube yarns · Sheath-like structure

1 Introduction

Robots play an important role in military, industrial, medical, disaster relief and other fields. However, rigid robots driven by traditional motors and hydraulic devices have some problems, such as slow response, large component volume and complex structure, which are difficult to adapt to complex environment [1]. The new generation of intelligent flexible robot has the advantages of fast response, simple structure, flexibility and so on [2]. Carbon Nanotube Yarns is an intelligent driving material that can untwist [3] and produce shrinkage and rotation deformation under the stimulation of electric, thermal, solvent, electrochemical ion implantation and other external conditions. It has a broad application prospect in the field of the next generation of intelligent flexible robot.

Carbon Nanotube (CNT) Yarns is a kind of visible fiber which is composed of millions of consecutive Carbon Nanotubes arranged in nearly parallel. At present, it is mainly prepared by wet spinning [4], array spinning [5] and floating chemical vapor

X.-J. Liu et al. (Eds.): ICIRA 2021, LNAI 13014, pp. 292–298, 2021.
https://doi.org/10.1007/978-3-030-89098-8_28

deposition [6]. In 2011, Professor Ray H. baughman [7], of Texas State University, Dallas, found that CNT Yarns could rotated at speed of 590 r/min in electrolyte solution. It was proposed for the first time to use it as micro motor and artificial muscle. Then, in order to improve the driving stress and strain of CNT Yarns, the researchers explored the structure and prepared the fiber with Archimedean spiral [8], double Archimedean spiral [9], layered twist [10] and other different structural forms. However, the driving performance of pure fiber is still unsatisfactory. Scholars have compounded the fiber with paraffin [11], epoxy resin [12], PEO-SO_3 [13] and other guest materials, making a new breakthrough in the driving performance and driving mode of the fiber. However, the Carbon Nanotube Yarns studied previously can only achieve unidirectional contraction drive, which will undoubtedly limit its application in future flexible robots. In order to solve this major problem, Leng Jinsong, Ding Jianning team's [14], together with Ray H. Baughman and other teams, used ion-exchange polymers to change the Zero Potential of Carbon Nanotube Yarns, broke through the bottleneck of unidirectional shrinkage, realized bidirectional driving, and greatly expanded the application range of carbon nanotube fibers. However, the fiber can only be driven in the matching liquid or solid electrolyte environment, which has the disadvantages of harsh service conditions and easy failure.

Dielectric elastomer [15] is a kind of intelligent driving material which can stretch along the direction perpendicular to the electric field under the stimulation of electric field. In view of the existing research problems, this paper proposes to construct a sheath like composite fiber with bi-directional deformation by using carbon nanotube fiber as the core and dielectric elastomer PDMS as the sheath. Compared with the previous bi-directional deformation scheme, this sheath like composite carbon nanotube fiber only needs electrical signal driving and does not depend on specific liquid or solid electrolyte, which can adapt to a variety of complex environments, it lays a foundation for the application of carbon nanotube fiber in the field of flexible robot.

2 Experimental Methods

2.1 Materials

CNT yarns, Hebei Tanyuan nanotechnology Co., Ltd; PDMS, Mn = 3000, gelest, USA; 2,6-pyridydicarbonyl chloride ($C_7H_3Cl_2NO_2$, 96%), Shanghai Aladdin Biochemical Co., Ltd; Zinc trifluoromethanesulfonate (Zn $(OTf)_2$, 98%), Shanghai Aladdin Biochemical Co., Ltd; Triethylamine (TEA, AR), Tianjin Fuyu Fine Chemical Co., Ltd; Dichloromethane (CH_2Cl_2, AR), Tianjin Fuyu Fine Chemical Co., Ltd; Anhydrous methanol (CH_3OH, AR), Tianjin Damao Chemical Reagent Factory; Anhydrous ethanol (C_2H_5OH, AR), Xilong Chemical Co., Ltd; Flexible carbon paste, MG chemical (#846).

2.2 Preparation of Uncured PDMS Solution

Using nitrogen as protective gas, 15 g of PDMS capped by aminopropyl group was dissolved in 20 ml dichloromethane, 3 ml triethylamine was added at 0 °C and stirred for 2 h. 1.02 g of 2,6-pyridinedicarbonyl chloride was dissolved in 10 ml dichloromethane and added to the above solution for 2 h after stirring at room temperature for 2 days, 40 ml methanol was added to terminate the reaction. The product was placed in a vacuum drying oven to remove the solvent.

2.3 Fabrication of Sheath-Like Carbon Nanotube Composite Yarns

Take 5 cm CNT yarns, hang it vertically, fix a heavy object at the lower end, connect a stepper motor at the upper end to twist the CNT yarn, shorten the yarn length by 20%, take uncured PDMS solution with a dropper, extrude it from top to bottom at a uniform speed, and evenly coat the PDMS solution on the yarn surface, 1 cm uncoated part should be reserved at both ends of Carbon Nanotube Yarns to connect copper wire. After waiting for PDMS curing at room temperature, the flexible carbon paste should be evenly coated on the surface of PDMS layer to obtain Sheath-like CNT Composite Yarns structure, as shown in Fig. 1.

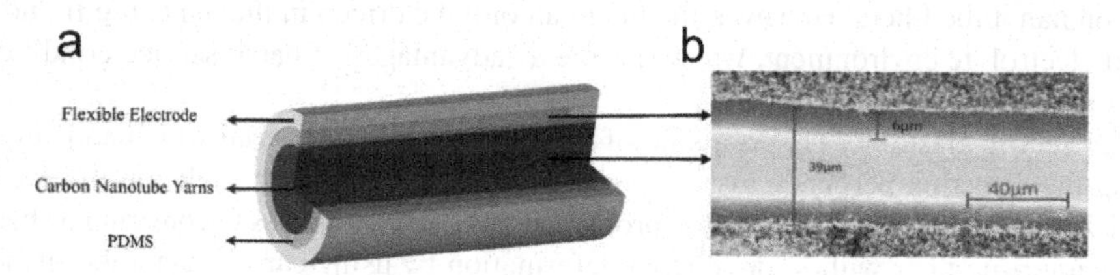

Fig. 1. Structure of Sheath-like CNT Composite Yarns. a) Confocal laser scanning microscope images of CNT Composite Yarns without flexible electrode. b)

2.4 Characterization of Carbon Nanotube Composite Yarns

Shrink Drive Test. Take 5 cm × 3 cm square paper, cut out 2 cm × 1 cm rectangular hole in it. The Sheath-like CNT Composite Yarn were fixed at the upper and lower ends of the hole, and then the copper wires were connected to the two ends of CNT Yarns separately. The uncoating part was reserved at both ends of the CNT yarn and fixed with conductive silver glue.The driving stress was tested by Shanghai Hengyi HY-0580 (single column) microcomputer controlled electronic universal material testing machine and the supporting American Futek LSB 200-1N sensor. The square paper was fixed at both ends of the drawing machine clamp, the paper frame around the rectangular hole was cut off, and the load was loaded at the rate of 0.01 mm/min. The copper wire was connected to the DC power supply, and the current was applied to test the change of force value.

Extension Drive Test. Take 5 cm × 3 cm square paper, cut out 2 cm × 1 cm rectangular hole in it. The sheath like composite carbon nanotube fibers were fixed at the upper and lower ends of the holes, and the copper wires were connected to the uncoated part of the carbon nanotube fibers and the flexible electrode layer on the sheath like composite fiber surface respectively. The driving strain was measured by ZY-H500C high-definition electron microscope, the square paper was placed horizontally, the paper frame around the rectangular hole was cut off, the copper wire was connected to the high-voltage power supply, the voltage was applied between the CNT yarn and the flexible electrode. The Sheath-like CNT Composite Yarns length was calibrated frame-by-frame by professional software, and recorded every 200 ms.

3 Results and Discussion

Figure 2 shows the comparison of driving stress between Sheath-like CNT yarns and pure CNT yarns. It can be seen from Fig. 2 that the driving stress (8.2 MPa) of Carbon Nanotube Composite Yarns is twice as great as that of the pure CNT Yarns (3.3 Mpa) under the current of 40 mA. The internal mechanism may be that a large amount of Joule Heat [16] is generated by the CNT yarns in the core after the current is applied, causing the sheath-like PDMS layer to expand along the radial direction and contract along the axial direction, In this way, the deformation of PDMS is consistent with the deformation of carbon nanotube fiber in the core, which improves the overall shrinkage driving stress.

For driving rate, it takes 1.8 s to reach the maximum driving stress from 0 MPa after the current is applied to Sheath-like CNT Composite Yarns, which is slightly slower than 0.8 s of the pure CNT yarns. This is because the Joule Heat transfer from the core CNT Yarns to the PDMS layer and then to the PDMS layer is heated and the CNT Yarns cooperative deformation takes a certain time. In response time, both of them respond instantaneously after the current is switch on, and the response speed is very fast.

Figure 3 is the bidirectional driving strain diagram of Sheath-like CNT Composite Yarns. Taking its original length as zero, the shrinkage driving strain is positive and the elongation driving strain is negative. When the current of 40 mA is applied, the Sheath-like CNT Yarns shrinks and reaches the maximum shrinkage driving strain value (1.6%) in 1.8 s. After the current is turned off, the original length can be quickly restored within 200 ms. When a 60 V/μm electric field is applied between the core CNT yarns and the flexible electrode, the Sheath-like CNT Composite Yarn shrinks. This is because the sheath-like PDMS layer will produce axial elongation deformation and radial shrinkage deformation due to Maxwell effect [17]. At the same time, the PDMS layer transfers the stress to CNT yarns through the interface to achieve the synergistic elongation deformation. The maximum elongation driving strain (0.75%) is achieved in 1.4 s. The sheath-like CNT Composite yarns recovers its original length in 600 ms after closing the voltage.

Figure 4 shows the stability test of driving stress cycle of Sheath-like CNT Composite Yarns. It can be seen from the figure that under the pulse current of 0–40 mA every 4 s, the driving stress is more than 800 cycles, which is 46% greater than the initial driving stress, similar to the phenomenon of excessive recovery of biological muscle, and the output driving stress is stable without fluctuation. This phenomenon may be related to the release of residual stress during cyclic thermal deformation of PDMS sheath, which is not conducive to the overall shrinkage and deformation of Sheath-like CNT Composite Yarns. It can be proved that the Sheath-like CNT Composite Yarns not only has cycle stability, but also has the characteristics of bionic over recovery.

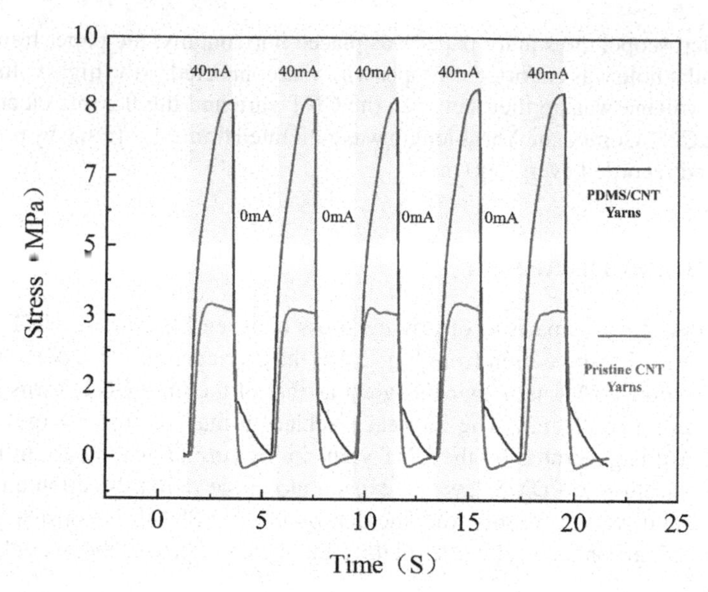

Fig. 2. The difference of driving stress between Carbon Nanotube Composite Yarns and Pristine Carbon Nanotube Yarns

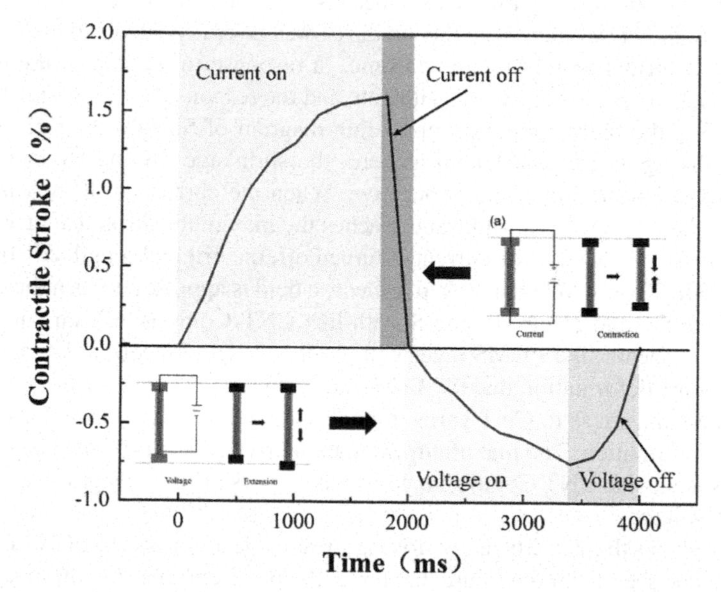

Fig. 3. The dual-deformation stroke of Carbon Nanotube Composite Yarns

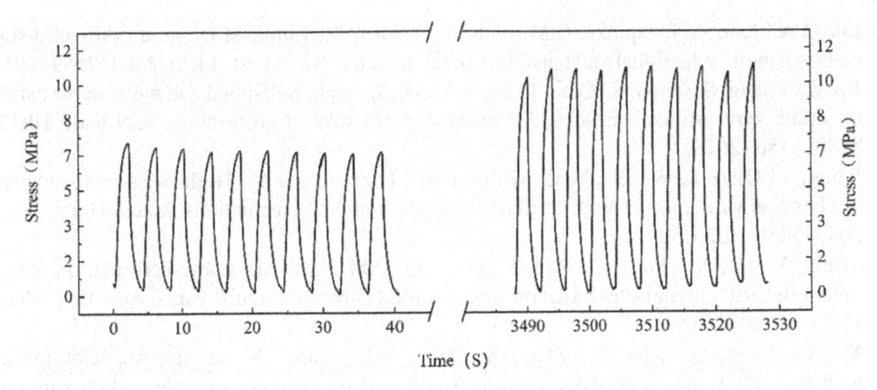

Fig. 4. The circular driving stress of Carbon Nanotube Composite Yarns

4 Conclusion

The Sheath-like CNT Composite Yarns was successfully prepared by using Carbon Nanotube Yarns as core and PDMS and flexible electrode as sheath. The composite yarns can achieve 1.6% shrinkage driving deformation at 40 mA current. Under the electric field strength of 60 V/μm, 0.75% elongation driving deformation can be achieved.

The driving stress of composite yarns was 1.5 times greater than pure CNT yarns.

The composite yarns has excellent cycle driving stability. The driving stress increases by 46% after 800 cycles at a pulse current of 0–40 mA every 4 s, and the driving stress remains stable.

Acknowledgments. This research was financial funded by the National Key R&D Program of China under the Grant No. 2018YFC0705603.

References

1. Yang, G.-Z., Bellingham, J., Dupont, P.E., Fischer, P., Floridi, L., Full, R., et al.: The grand challenges of science robotics. Sci. Robot. **3**(14), eaar7650 (2018)
2. Laschi, C., Mazzolai, B., Cianchetti, M.: Soft robotics: technologies and systems pushing the boundaries of robot abilities. Sci. Robot. **1**(1), eaah3690 (2016)
3. Mirvakili, S.M., Hunter, I.W.: Artificial muscles: mechanisms, applications, and challenges. Adv. Mater. **30**(6), 1704407 (2018)
4. Wang, G., Kim, S.K., Wang, M.C., Zhai, T., Munukutla, S., Girolami, G.S., et al.: Enhanced electrical and mechanical properties of chemically cross-linked carbon-nanotube-based fibers and their application in high-performance supercapacitors. ACS Nano **14**(1), 632–639 (2020)
5. Qiao, J., Di, J., Zhou, S., Jin, K., Zeng, S., Li, N., et al.: Large-stroke electrochemical carbon nanotube/graphene hybrid yarn muscles. Small **14**(38), e1801883 (2018)
6. Janas, D., Liszka, B.: Copper matrix nanocomposites based on carbon nanotubes or graphene. Mater. Chem. Front. **2**(1), 22–35 (2018)
7. Foroughi, J., Spinks, G.M., Wallace, G.G., Oh, J., Kozlov, M.E., Fang, S., et al.: Torsional carbon nanotube artificial muscles. Science **334**(6055), 494–497 (2011)

8. Lee, J.A., Kim, Y.T., Spinks, G.M., Suh, D., Lepro, X., Lima, M.D., et al.: All-solid-state carbon nanotube torsional and tensile artificial muscles. Nano Lett. **14**(5), 2664–2669 (2014)
9. Jin, K., Zhang, S., Zhou, S., Qiao, J., Song, Y., Di, J., et al.: Self-plied and twist-stable carbon nanotube yarn artificial muscles driven by organic solvent adsorption. Nanoscale **10**(17), 8180–8186 (2018)
10. Wang, Y., Qiao, J., Wu, K., Yang, W., Ren, M., Dong, L., et al.: High-twist-pervaded electrochemical yarn muscles with ultralarge and fast contractile actuations. Mater. Horiz. **7**(11), 3043–3050 (2020)
11. Shang, Y., He, X., Wang, C., Zhu, L., Peng, Q., Shi, E., et al.: Large-deformation, multifunctional artificial muscles based on single-walled carbon nanotube yarns. Adv. Eng. Mater. **17**(1), 14–20 (2015)
12. Xu, L., Peng, Q., Zhu, Y., Zhao, X., Yang, M., Wang, S., et al.: Artificial muscle with reversible and controllable deformation based on stiffness-variable carbon nanotube spring-like nanocomposite yarn. Nanoscale **11**(17), 8124–8132 (2019)
13. Mu, J., Jung de Andrade, M., Fang, S., Wang, X., Gao, E., Li, N., et al.: Sheath-run artificial muscles. Science **365**(6449), 150–155 (2019)
14. Chu, H., Hu, X., Wang, Z., Mu, J., Li, N., Zhou, X., et al.: Unipolar stroke, electroosmotic pump carbon nanotube yarn muscles. Science **371**(6528), 494–498 (2021)
15. Qiu, Y., Zhang, E., Plamthottam, R., Pei, Q.: Dielectric elastomer artificial muscle: materials innovations and device explorations. Acc. Chem. Res. **52**(2), 316–325 (2019)
16. Meng, F., Zhang, X., Li, R., Zhao, J., Xuan, X., Wang, X., et al.: Electro-induced mechanical and thermal responses of carbon nanotube fibers. Adv. Mater. **26**(16), 2480–2485 (2014)
17. Klug, F., Solano-Arana, S., Hoffmann, N.J., Schlaak, H.F.: Multilayer dielectric elastomer tubular transducers for soft robotic applications. Smart Mater. Struct. **28**(10), 104004 (2019)

Actuating, Sensing, Control, and Instrumentation for Ultra-precision Engineering

A Compensation Strategy for Positioning Errors in Two-Axis Servo System

Pingfa Yang(✉), Yuanlou Gao, and Chengfu Jiang

Beihang University, Xueyuan Road No. 37, Haidian District, Beijing, China

Abstract. Positioning accuracy is crucial to servo system, therefore, study error identification and compensation methods is significant. Composed of a linear and a rotary axis, the CNC servo positioning platform studied, and kinematics modeling carried out on the DH (Denavit-Hartenberg homogeneous coordinate model) method. Supposing that modeling errors change with driving instructions, in order to obtain the relationship among them, identifying the actual axis equations of the linear and rotating shaft, then the state matrix of the actual coordinated system of each link was re-established. Based on this, the relationship curves between errors of each matrix and driving constructions fitted, the ideal homogeneous coordinate transformation matrix compensated to realize the positioning of the CNC servo platform under different driving instructions. Further, the simulation for compensation strategy carried out, which verify that the results are consistent with the theoretical assumptions and positioning error can be compensated by the compensation method.

Keywords: Servo positioning system · Positioning error · Driving instructions

1 Introduction

Multi-axis servo positioning system is widely used in manufacturing, aircraft and other fields. Meanwhile, it is essential to ensure high accuracy of positioning to complete missions in some cases. Because of errors between kinematic modelling and reality, a servo system cannot be controlled precisely without knowledges, features about those errors.

Roth et al. divided the robot error into three levels [1]. WU proposed a method of identifying kinematic errors, which established a mapping model from all error parameters to the end error of the robot [2]. Using structured laser modules, fixed cameras and other hardware, Park et al. proposed a new technique to estimate the entire kinematic parameter errors of the robot manipulator [3].

This digest proposes a compensation strategy based on relationship between driving instructions and modelling errors. With three-coordinate measuring instrument, actual axis equations of linear axis and rotating axis were identified, and state matrix of each link was re-established. Curve between modelling errors and driving commands fitted, and compensate for the ideal homogeneous coordinate transformation matrix.

X.-J. Liu et al. (Eds.): ICIRA 2021, LNAI 13014, pp. 301–312, 2021.
https://doi.org/10.1007/978-3-030-89098-8_29

2 DH Modeling for Two-Axis Positioning Servo Platform

2.1 Kinematics Modeling with DH Method

Standard DH method (SDH), called Denavit-Hartenberg homogeneous coordinate model, was proposed by Denavit and Hartenberg [4]. Further, Khalil proposed an improved DH model (KDH) [5]. The driving shaft with the SDH located at the far end of link i, while that with KDH located at the close end of link i differently, as shown in Fig. 1.

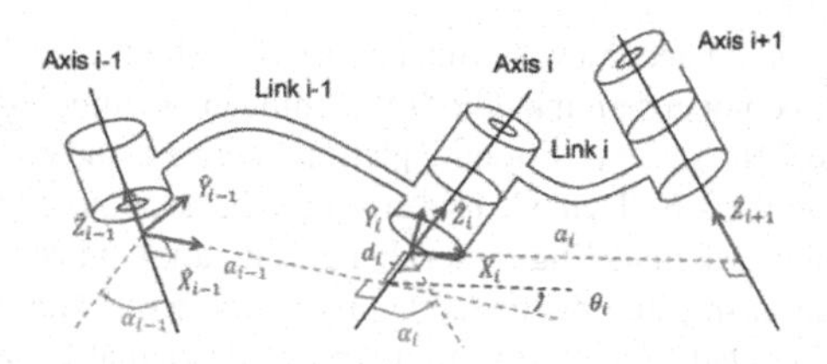

Fig. 1. Improved DH method proposed by Khalil.

The coordinate system {i − 1} established at the front end of link i, and at the rear end establishing {i}. After translation and rotation expressed as ${}^{W}_{B}T$, {i} deduced from {i − 1}. There're four steps from {i − 1} to {i}, rotate Z_{i-1} around X_{i-1} by angle α_{i-1} at first, then move a_{i-1} along X_{i-1}, next, rotate by angle θ_i around Z_{i-1}, finally move d_i along Z_{i-1}, as shown in ${}^{i-1}_{i}T$ in Eqs. (1) and (2).

$$ {}^{i-1}_{i}T = T_{RX_{i-1}}(\alpha_{i-1})T_{PX_{i-1}}(a_{i-1})T_{RZ_i}(\theta_i)T_{PZ_i}(d_i) \tag{1} $$

Where $T_{RX_{i-1}}$ is the transformation matrix around X_{i-1} rotation; $T_{PX_{i-1}}$ is along X_{i-1}; rotation T_{RZ_i} around Z_i; T_{PZ_i} along Z_i.

$$ {}^{i-1}_{i}T = \begin{bmatrix} \cos\theta_i & -\sin\theta_i & 0 & a_{i-1} \\ \sin\theta_i\cos\alpha_{i-1} & \cos\theta_i\cos\alpha_{i-1} & -\sin\alpha_{i-1} & -\sin\alpha_{i-1}d_i \\ \sin\theta_i\sin\alpha_{i-1} & \cos\theta_i\sin\alpha_{i-1} & \cos\alpha_{i-1} & \cos\alpha_{i-1}d_i \\ 0 & 0 & 0 & 1 \end{bmatrix} \tag{2} $$

2.2 Kinematics Modeling for Positioning Servo Platform

As in Fig. 2, composed of two parts, 1 is a linear motion module, and a hollow rotating module named 2. For the platform, Link 1 is 1 plus the stator part of 2, while Link 2 is the rotor part. For the convenience of measurement and calibration, the ground coordinate system {0} does not coincide with {1}, but set to coincide with {W} (reference coordinate system). Origin of {2} coincides with that of {1}, at the center of the rotary reducer, and X_2 coincides with X_1, perpendicular to Z_1 and Z_2 (Table 1).

Translation component represented by ${}^{0}_{1}A$, and that of rotating component is ${}^{1}_{2}A$, hence, {0} to {W} is ${}^{W}_{0}A$. The homogeneous coordinate transformation matrices ${}^{1}_{2}T$, ${}^{0}_{1}T$, and ${}^{W}_{0}T$ corresponding to each coordinate system defined.

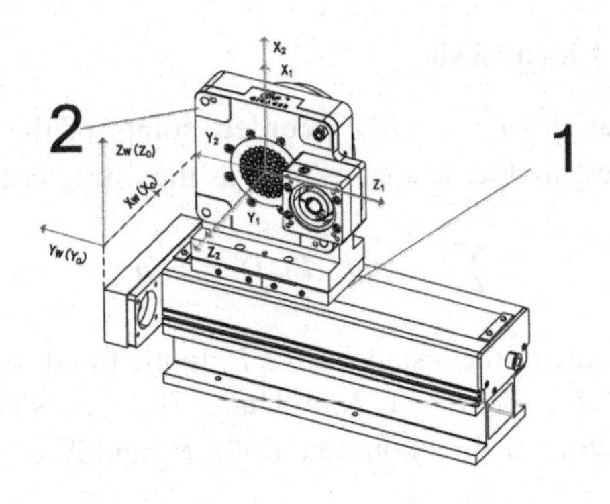

Fig. 2. The establishment of the coordinate system.

Table 1. Nominal DH parameters' table of positioning servo platform.

i	α_{i-1}(rad)	a_{i-1}(mm)	d_i(mm)	θ_i(rad)
1	pi/2	107	d_1	pi/2
2	−pi/2	0	0	θ_1

3 Error Identification Strategy of Axis Method

Assuming that driving instructions related to errors. Set state matrix is A^n ideally, and transformation matrix called T^n; the real state matrix is A^a, hence transformation matrix called T^a, shown in formula (3) and formula (4).

$$^{i-1}_{i}T^a = T_{RX_{i-1}}(\alpha_{i-1} + \Delta\alpha_{i-1})T_{PX_{i-1}}(a_{i-1} + \Delta a_{i-1})T_{RZ_i}(\theta_i + \Delta\theta_i)T_{PZ_i}(d_i + \Delta d_i) \tag{3}$$

$\Delta\alpha_{i-1}$, Δa_{i-1}, $\Delta\theta_i$, Δd_i, errors between nominal parameters and actual parameters.

$$^{i-1}_{i}A^a = \begin{bmatrix} ^{i-1}_{i}R^a_{3x3} & ^{i-1}P^a_{i\,\mathrm{org}} \\ 0 & 1 \end{bmatrix} \tag{4}$$

$^{i-1}_{i}R^a_{3\times3}$, $^{i-1}P^a_{i\,\mathrm{org}}$, the projection of actual {i} to actual {i − 1}.
$^{i-1}_{i}T$, $^{i-1}_{i}A$ satisfies Eq. (5), and $^{i-1}_{i}T = {}^{i-1}_{i}A$ from above, formula (6) proved.

$$^{W}_{i}T = {}^{W}_{0}T^{0}_{1}T^{i-2}_{i-1}T \ldots {}^{i-1}_{i}T \tag{5}$$

$$^{W}_{i}A = {}^{W}_{0}A^{0}_{1}Ai - 2_{i-1} \ldots A^{i-1}_{i}T \tag{6}$$

$$({}^{W}_{0}A^{0}_{1}A \ldots {}^{i-2}_{i-1}A)^{-1}\,{}^{W}_{i}A = {}^{i-1}_{i}T = {}^{i-1}_{i}A \tag{7}$$

Bennett identifies axis from the end [6], while Harb from the base [7]. {0} coincides with {W} completely, and KDH method starts from base coordinate system. {1} substituted into Eq. (7) firstly, and then {2} identified. Collected by coordinate measuring machine, trajectory is to determine direction of Z in {1} and {2}.

3.1 Identification of Linear Axis

Set drive parameters as $d_i^1, d_i^2, \ldots, d_i^n$, recorded points of the platform's motion as $P_i = \left(P_{ix}\, P_{iy}\, P_{iz}\right)$. Weighted least squares used as the fitting criterion, in (8).

$$\sum\nolimits_{i=1}^{n} p_i d^2(P_i, l) = min \tag{8}$$

P_i, weight, $d^2(P_i,l)$, square of the distance from P_i to the fitted straight-line l. Collected points are ${}^W P_i = (P_{ix}, P_{iy}, P_{iz})i = 1, 2...n$. Under (8), l pass center of gravity $P_g = [P_{gx} P_{gy} P_{gz}]^T$, eigenvector $\hat{\nu}$ is direction vector of l. P_g and A are shown in Eqs. (9) and (10).

$$P_g = \left[\sum_i^n p_i P_{ix} / \sum_i^n p_i,\ \sum_i^n p_i P_{ix} / \sum_i^n p_i,\ \sum_i^n p_i P_{ix} / \sum_i^n p_i\right]^T \tag{9}$$

$$A = X_G^T P X_G \tag{10}$$

Among them, $X_G = \begin{bmatrix} P_{1x} - P_{gx} & P_{1y} - P_{gy} & P_{1z} - P_{gz} \\ \cdot & \cdot & \cdot \\ P_{nx} - P_{gx} & P_{ny} - P_{gy} & P_{nz} - P_{gz} \end{bmatrix}, P = \begin{bmatrix} p_1 & & \\ & p_i & \\ & & p_n \end{bmatrix}$.

3.2 Identification of Rotary Axis

To begin with fitting plane with $D \neq 0$, space plane equation transformed into (11).

$$\frac{A}{D}x + \frac{B}{D}y + \frac{C}{D}z = 1 \tag{11}$$

Let $a = \frac{A}{D}$, $b = \frac{A}{D}$, c = $\frac{A}{D}$, collected points are ${}^W P_i = \left(P_{ix}, P_{iy}, P_{iz}\right)$, $i = 1, 2...n$. Then the direction vector of the plane $n = (a, b, c)$ as shown in (12).

$$n = \left(N^T N\right)^{-1} N^T b \tag{12}$$

Among then, $N = \begin{bmatrix} x_1 & y_1 & z_1 \\ \cdot & \cdot & \cdot \\ x_n & y_n & z_n \end{bmatrix}$, and $b = [1, \ldots 1]$. Next, fit the space circle equation, with the center ${}^W C = \left(C_{ix}, C_{iy}, C_{iz}\right)$. ${}^W P_i$ and ${}^W P_j$, with midpoint ${}^W M_{ij} = \left[\frac{P_{jx}+P_{ix}}{2}\ \frac{P_{jy}+P_{iy}}{2}\ \frac{P_{jz}+P_{iz}}{2}\right]$. $\vec{P_i P_j} = \left[P_{jx} - P_{ix}\ P_{jy} - P_{iy}\ P_{jz} - P_{iz}\right]$, then $\vec{CM_{ij}} = \left[M_{ijx} - C_{ix}\, M_{ijz} - C_{iy}\ M_{ijz} - C_{iz}\right]$. Since $\vec{P_i P_j} \perp \vec{CP_{Mij}}$, formula (14) holds

$$\Delta x_{ij} C_x + \Delta y_{ij} C_y + \Delta z_{ij} C_z - l = 0 \tag{13}$$

Where, $l_{ij} = \left[\left(P_{jx}^2 - P_{ix}^2\right) + \left(P_{jy}^2 - P_{iy}^2\right) + \left(P_{jz}^2 - P_{iz}^2\right)\right]/2$, $\Delta x_{ij} = P_{jx} - P_{ix}$, $\Delta y_{ij} = P_{jy} - P_{iy}$, $\Delta z_{ij} = P_{jz} - P_{iz}$.

Center of space circle solved in formula (14).

$$\hat{C} = \left(W^T W\right)^{-1} W^T L \tag{14}$$

Where $W = \begin{bmatrix} \Delta x_{12} & \Delta y_{12} & \Delta z_{12} \\ \Delta x_{1j} & \Delta y_{1j} & \Delta z_{1j} \\ \cdot & \cdot & \cdot \\ \Delta x_{n-1,n} & \Delta y_{n-1,n} & \Delta z_{n-1,n} \end{bmatrix}, L = \left[l_{12}\ l_{1j}\ ...\ l_{n-1,n} \right]$.

After $d_1^1, d_1^2, ...d_1^n$ drove, rotary axis fitted. Then linear axis locked after moved in place, $\theta_2^{11}, \theta_2^{12}, ...\theta_2^{1n}; \theta_2^{21}, ...\theta_2^{2n}; ...\theta_2^{n1}, ...\theta_2^{nn}$ driving meanwhile. Of fitted circle, suppose unit direction vector is $\widehat{Z_2^1}, \widehat{Z_2^2}, ...\widehat{Z_2^i}, ...\widehat{Z_2^n}$, and center is $C_1, C_2, ...C_n$.

3.3 Rebuild the Coordinate System

Vector $\widehat{X_1^i} = \widehat{Z_1^i} \times \widehat{Z_2^i}$, $\widehat{Y_1^i} = \widehat{Z_1^i} \times \widehat{X_1^i}$. Plane π_1 can be determined by $\widehat{X_2^i}$, $\widehat{Z_2^i}$, C_i.Origin of $\{1\}$ is contained in both π_1 and linear axis (Fig. 3).

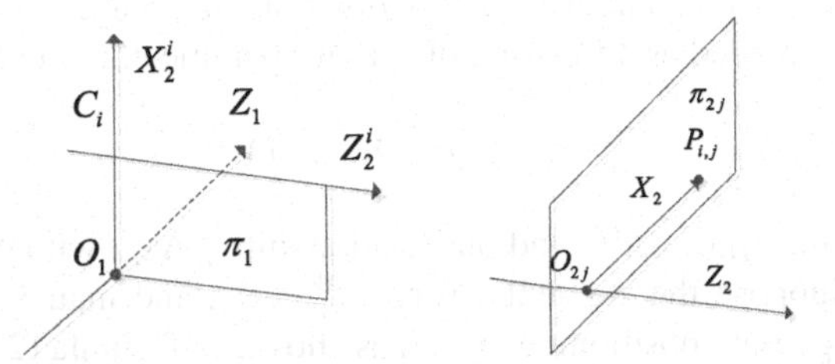

Fig. 3. Coordinate system $\{1\}$ and $\{2\}$ reconstruction to determine the origin.

Plane π_{2j} can be deduced by rotation axis and a collected point. Intersection of rotation axis and π_{2j} expressed as O_2. X_2 pointed from O_2 to $P_{i,j}$. $Y_2 = Z_2 \times X_2$.

After d_1^i drove in linear axis and driving instruction of the rotary axis is 0, the n state matrices ${}_1^W A^a$ of $\{1\}$ and $\{2\}$ n state matrices ${}_2^W A^a$ work out.

3.4 Cubic Spline Curve Fitting Matrix Element Errors

$({}_0^W A0_1...A_{i-1}^{i-2}A)^{-1}{}_i^W A^a$ shown in formula (15), (16) holds. Assuming T^n from the measurement to the end coordinate system as in (17), error matrix shown in (18).

$${}_i^W A = \begin{bmatrix} n_{ix} & o_{ix} & k_{ix} & P_{ix} \\ n_{iy} & o_{iy} & k_{iy} & P_{iy} \\ n_{iz} & o_{iz} & k_{ix} & P_{iz} \\ 0 & 0 & 0 & 1 \end{bmatrix} \tag{15}$$

$${}_i^W T^a = {}_i^W A^a = \begin{bmatrix} n_{ix}^a & o_{ix}^a & k_{ix}^a & P_{ix}^a \\ n_{iy}^a & o_{iy}^a & k_{iy}^a & P_{iy}^a \\ n_{iz}^a & o_{iz}^a & k_{iz}^a & P_{iz}^a \\ 0 & 0 & 0 & 1 \end{bmatrix} \tag{16}$$

$$
{}_{i}^{W}T^{n} = {}_{i}^{W}A^{n} = \begin{bmatrix} n_{ix}^{n} & o_{ix}^{n} & k_{ix}^{n} & P_{ix}^{n} \\ n_{iy}^{n} & o_{iy}^{n} & k_{iy}^{n} & P_{iy}^{n} \\ n_{iz}^{n} & o_{iz}^{n} & k_{iz}^{n} & P_{iz}^{n} \\ 0 & 0 & 0 & 1 \end{bmatrix} \tag{17}
$$

$$
{}_{i}^{W}\Delta = \begin{bmatrix} n_{ix}^{a} - n_{ix}^{n} & o_{ix}^{a} - o_{ix}^{n} & k_{ix}^{a} - k_{ix}^{n} & P_{ix}^{a} - P_{ix}^{n} \\ n_{iy}^{a} - n_{iy}^{n} & o_{iy}^{a} - o_{iy}^{n} & k_{iy}^{a} - k_{iy}^{n} & P_{iy}^{a} - P_{iy}^{n} \\ n_{iz}^{a} - n_{iz}^{n} & o_{iz}^{a} - o_{iz}^{n} & k_{iz}^{a} - k_{iz}^{n} & P_{iz}^{a} - P_{iz}^{n} \\ 0 & 0 & 0 & 1 \end{bmatrix} \tag{18}
$$

As a function of driving, modeling parameters are shown in Eqs. (19) and (20).

$$
\Delta x_{i,j} = f_{c}(\alpha_{0}, \ldots, \alpha_{n-1}, a_{1}, \ldots a_{n-1}, d_{1}, \ldots d_{n}, \theta_{1}, \ldots \theta_{n}) \tag{19}
$$

Where, ${}_{n}^{W}\Delta x_{i,j}$, element error of transformation matrix;

$$
x_{i} = f_{d}(D_{1}, \ldots, D_{n}) \tag{20}
$$

x_i, modeling parameters, D_i, drive instructions of each link.
${}_{n}^{W}\Delta x_{i,j}$ can be regarded as a function of driving parameters of each link, in (21).

$$
{}_{n}^{W}\Delta x_{i,j} = f_{D}(D_{1}, \ldots, D_{n}) \tag{21}
$$

Instead of solving f_D directly, find out relationship ${}_{n}^{W}\Delta x_{i,j}$ with driving parameters by curve fitting. Suppose the element errors is ${}_{n}^{W}\Delta x_{i,j}^{c}$, and matrix ${}_{n}^{W}\Delta^{c}$ to the drive parameters to compensate positioning errors, as shown in formula (22).

$$
{}_{n}^{W}T^{c} = {}_{n}^{W}T^{n} + {}_{n}^{W}\Delta^{c} \tag{22}
$$

4 Simulation

Assuming relationship between modeling parameter errors and driving instructions as in Figs. 4 and 5. After driving with commands and ignoring the errors caused by measuring, actual axis can be fitted.

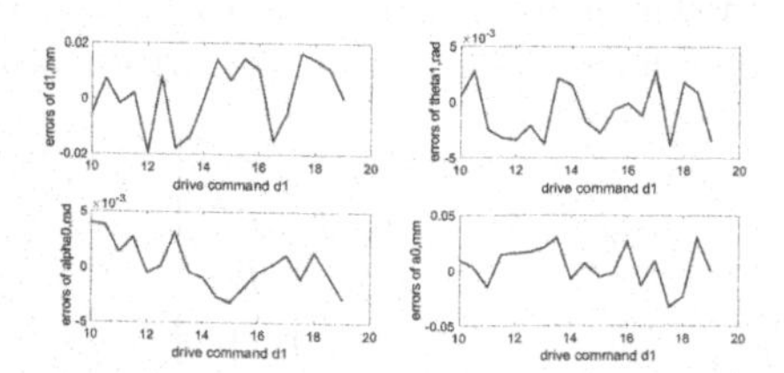

Fig. 4. Modeling error distribution of link 1 in a given interval.

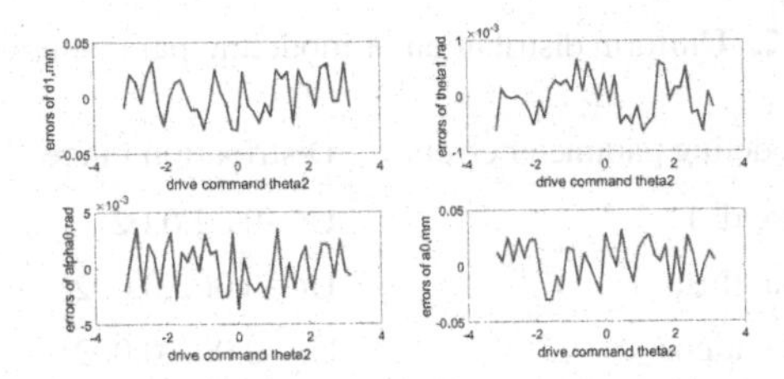

Fig. 5. Modeling error distribution of link 2 in a given interval.

Next, set test parameter: d_1= [16:0.05:17], $\theta_2 = [0{:}2\pi/80{:}\pi/2]$ (Figs. 6 and 7).

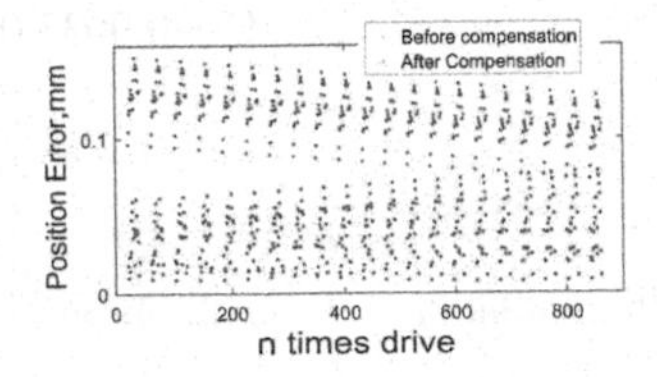

Fig. 6. Comparison of positioning errors before and after compensation.

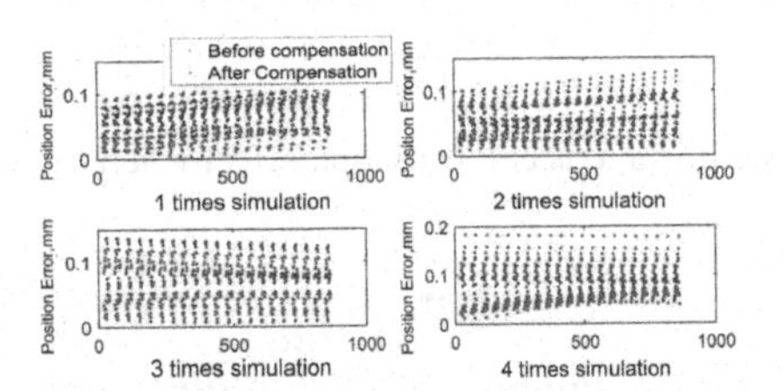

Fig. 7. Results of four times' simulation.

Positioning errors in four simulations are below 0.1 mm, therefore cubic spline curve fitting method can stabilize error under a certain range.

4.1 Effect of Linear Axis Fitting Accuracy on Positioning Error Forward Compensation

Set $\theta_2 = [0{:}2\pi/40{:}2\pi]$, given $d_1 = [10{:}10{:}50]$, $d_1 = [10{:}2{:}50]$ respectively, execute four times for each group of parameters, as shown in Figs. 8 and 9. While drive parameters are $d_1 = [16{:}0.05{:}17]$, $\theta_2 = [0{:}2\pi/80{:}\pi/2]$, assuming errors distribution is in Table 2.

Compare with Fig. 8, in Fig. 9 when $d_1 = [10{:}2{:}50]$, positioning errors after compensation are below 0.1 mm. When posihtioning error caused by the modeling exceeds 0.1, effectiveness of algorithm is less sensitive to error distribution. The algorithm must ensure high accuracy in the linear axis, otherwise will lead to instability of the compensation effect or even increase the positioning error.

Table 2. Uniform distribution of modeling parameter errors.

Modeling parameter errors	Distribution range
deta_d_1	U(−0.02,0.02)
deta_theta_1	U(−0.002,0.002)
deta_alpha_0	U(−0.002,0.002)
deta_a_0	U(−0.0033,0.0033)
deta_d_2	U(−0.0033,0.0033)
deta_theta_2	U(−0.0007,0.0007)
deta_alpha_1	U(−0.002,0.0002)
deta_a_1	U(−0.0033,0.0033)

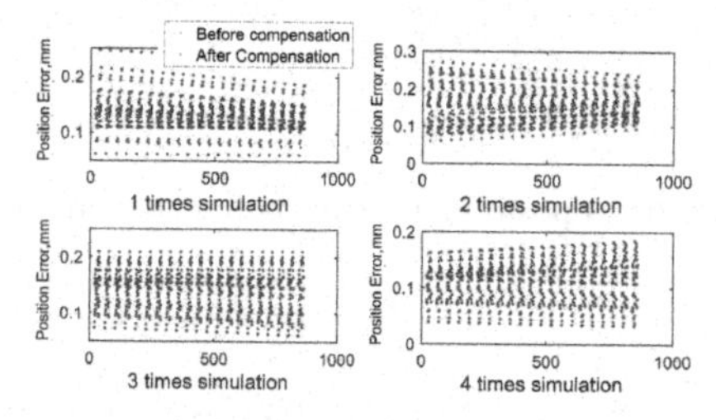

Fig. 8. Fitting linear axis compensation effect when $d_1 = [10:10:50]$.

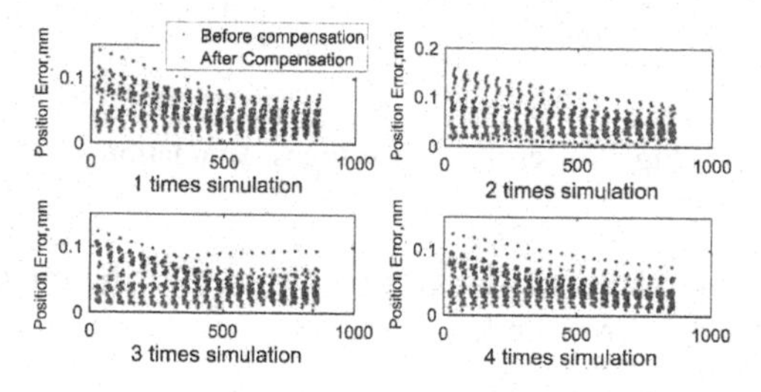

Fig. 9. Fitting linear axis compensation effect when $d_1 = [10:2:50]$.

4.2 The Effect of Fitting Accuracy of Rotating Axis on Compensation

Accuracy of rotation axis fitting can be determined by interval and range of θ_2. To explore the influence of the fitting accuracy of rotating axis on the compensation effect, $d_1 = [10:0.5:19]$ holds, $\theta_2 = [-\pi : 2\pi/10 : \pi]$, $\theta_2 = [-\pi : 2\pi/20 : \pi]$ respectively (Figs. 10 and 11).

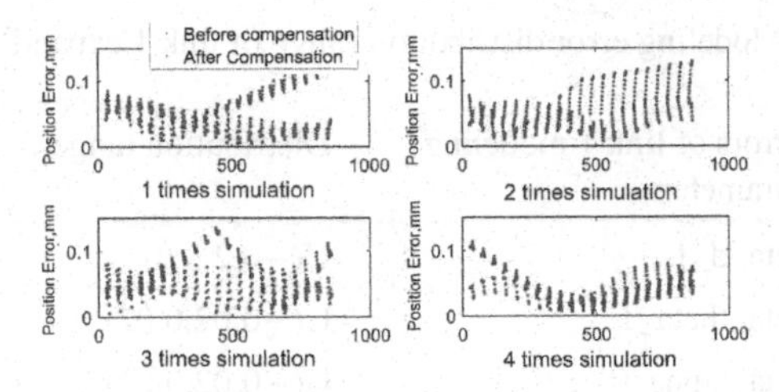

Fig. 10. Fitting rotation axis compensation effect when $\theta_2 = [-\pi : 2\pi/10 : \pi]$.

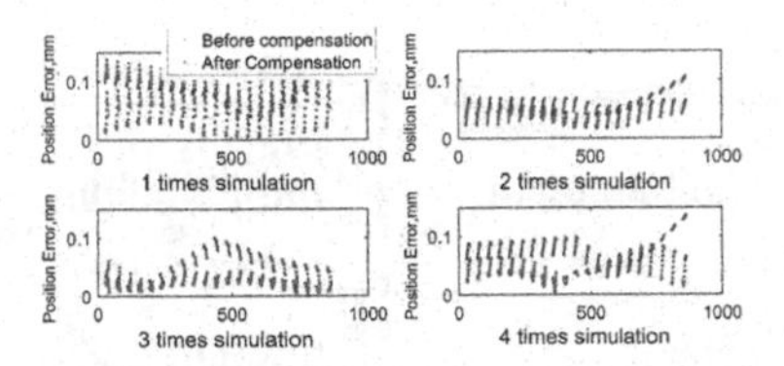

Fig. 11. Fitting rotation axis compensation effect when $\theta_2 = [-\pi : 2\pi/20 : \pi]$.

To explore whether the algorithm is sensitive to the number of fitting points of rotation axis, set $\theta_2 = [-\pi\text{:}2\pi/10\text{:}\pi]$, given again when $d_1 = [16\text{:}0.05\text{:}17]$, $\theta_2 = [-\pi/2\text{:}2\pi/80\text{:}\pi/2]$. Increase interval from $[0,\pi/2]$ to $[-\pi/2,\pi/2]$ in Fig. 12.

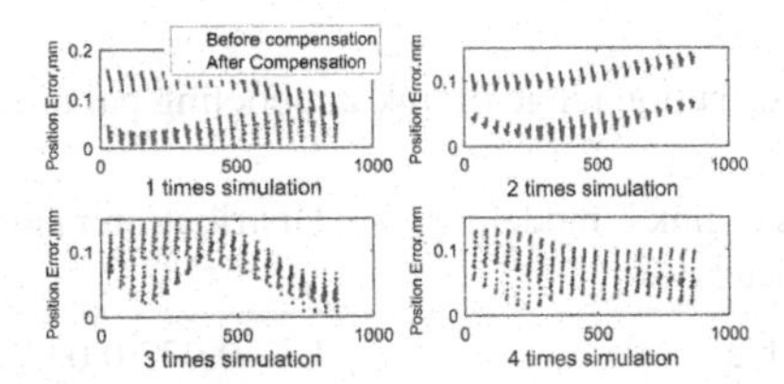

Fig. 12. Results after increasing the driving parameter range of the test process.

Compare Fig. 12 with Fig. 9, when number of sampling points decreases, sensitivity to modeling errors increases. After compensation, the range will not be significantly smaller than 0.1 mm when fitting accuracy is high. Therefore, increasing the sampling points reduce the sensitivity to the error distribution, and make distribution range of position error after compensation more reliable.

4.3 Influence of Modeling Parameter Errors' Distribution on Compensation

After increasing model error of link 1 by 10 times, the algorithm is still effective, but position error increased slightly after compensation. It concludes that the algorithm is not sensitive to the range of the modeling error distribution of link 1 (Fig. 13 and Table 3).

Table 3. Modeling error distribution range of link 1 expand 10 times.

Errors of link 1 modeling parameters	Distribution range
deta_d_1	U(−0.2,0.2)
deta_theta_1	U(−0.02,0.02)
deta_alpha_0	U(−0.02,0.02)
deta_a_0	U(−0.033,0.033)

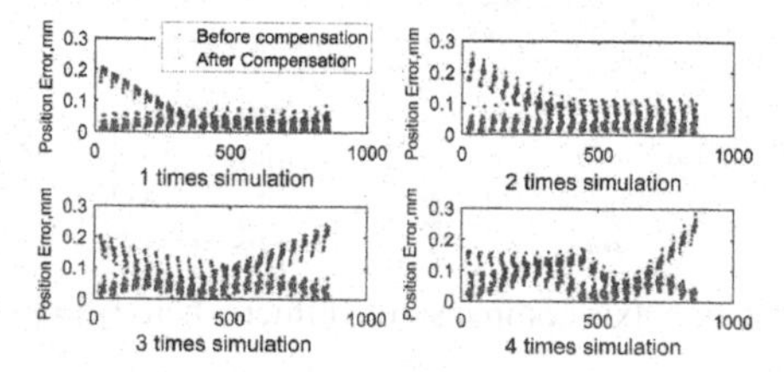

Fig. 13. Compensation after error distribution range of link 1 have been expanded by 10 times.

Algorithm is effective after increasing the modeling error range of link 2 by 10 times (Table 4).

Table 4. The error distribution range of link 2 modeling parameters expand 10 times.

Errors of link 1 modeling parameters	Distribution range
deta_d_2	U(−0.033,0.033)
deta_theta_2	U(−0.007,0.007)
deta_alpha_1	U(−0.02,0.02)
deta_a_1	U(−0.033,0.033)

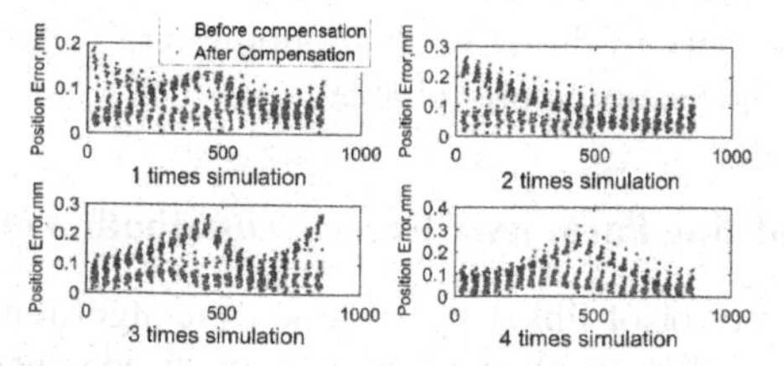

Fig. 14. Compensation after error distribution range of link 2 have been expanded by 10 times.

When modeling error range of link 2 enlarged, compensation effect decreases in Fig. 14. Moreover, modeling error expands to range in Table 5, the algorithm lost effectiveness. Therefore, it is crucial to strictly control machining and installation accuracy of link 2.

Simulation above verified effectiveness of compensation strategy proposed, when applying: 1) accuracy of linear axis needs to be guaranteed; 2) increasing number of sampling points can increase the error distribution reliability after compensation; 3) Need to control machining and installation accuracy of link 2 strictly.

Table 5. The algorithm valid range of link 2's modeling errors distribution.

Errors of link 2 modeling parameters	Distribution range
deta_d_2	U(−0.0825,0.0825)
deta_theta_2	U(−0.0175,0.0175)
deta_alpha_1	U(−0.005,0.005)
deta_a_1	U(−0. 0825,0. 0825)

4.4 The Influence of Modeling Parameter Errors' Distribution on Compensation Effects

Make a comparative table about value θ_2^n, d_1^n in theory and value θ_2^a, d_1^a in practice. Suppose d_1^n and θ_2^n ranges are $[a, b]$, $[-\pi, \pi]$ respectively. Determine transformation matrix after compensation at appropriate intervals as $T_a^i = f\left(d_1^n, \theta_2^n\right)$. When servo platform at initial position, set initial coordinate of center points to P_O^i, $i = 1, 2, ..., 69$, and target position to P_T^n. From $T_a^i = f\left(d_1^n, \theta_2^n\right)$ and P_O^i can calculate positions P_T^a. Look for the closest to P_T^a, and record d_1^a, θ_2^a. It forms a comparative table with the theoretical ${}^{Poi}d_1^n$, ${}^{Poi}\theta_2^n$ that drives initial point P_O^i to target position P_T^n, in Table 6.

Table 6. Control system compensation comparison table.

Orders of target position	1	2	...	69
Command before compensation	$\left({}^1d_1^n, {}^1\theta_2^n\right)$	$\left({}^2d_1^n, {}^2\theta_2^n\right)$	...	$\left({}^{69}d_1^n, {}^{69}\theta_2^n\right)$
Command after compensation	$\left({}^1d_1^a, {}^1\theta_2^a\right)$	$\left({}^2d_1^a, {}^2\theta_2^a\right)$	...	$\left({}^{69}d_1^a, {}^{69}\theta_2^a\right)$

5 Conclusions

Modeled with improved DH, mechanical structure of the two-axis servo platform identified actual motion with axis method. Cubic spline curve is analyzed, and used to complete the modeling error compensation. The influence of accuracy and parameter error range on positioning error compensation studied, in addition, a positioning error compensation strategy is proposed. The innovation is to propose an error compensation strategy, which finds out the relationship between drive command and parameter error of translation matrix, by fitting the actual axis of motion, and compensates based on target position. Meanwhile, influence of fitting accuracy and error range explored, which provides guidance for applying this method in the future.

References

1. Roth, Z., Mooring, B., Ravani, B.: An overview of robot calibration. IEEE J. Robot. Autom. **3**(5), 377–385 (1987)
2. Wu, C.-H.: A kinematic CAD tool for the design and control of a robot manipulator. Int. J. Robot. Res. **3**(1), 58–67 (1984)
3. Park, I., Lee, B., Cho, S., et al.: Laser-based kinematic calibration of robot manipulator using differential kinematics. IEEE-ASME Trans. Mechatron. **17**(6), 1059–1067 (2012)
4. Abdel-Malek, K., Othman, S.: Multiple sweeping using the Denavit-Hartenberg representation method. Comput. Aided Des. **31**(9), 567–583 (1999)
5. Khalil, W., Kleinfinger, J.: A new geometric notation for open and closed-loop robots. In: Proceedings of the 1986 IEEE International Conference on Robotics and Automation, pp. 1174–1179 (1986). doi: https://doi.org/10.1109/ROBOT.1986.1087552
6. Bennett, D.J., Hollerbach, J.M., Henri, P.D.: Kinematic calibration by direct estimation of the Jacobian matrix. In: Proceedings of the International Conference on Robotics and Automation (1992)
7. Harb, S.M., Burdekin, M.: A systematic approach to identify the error motion of an N-degree of freedom manipulator. Int. J. Adv. Manuf. Technol. **9**(2), 126–133 (1994)

A Structured Light Small Field Three-Dimensional Measurement System Based on Three-Frequency Heterodyne Method

ChengCheng Li[1], Chen Li[2], and Xu Zhang[1](✉)

[1] School of Mechatronic Engineering and Automation, Shanghai University, Shanghai, China
xuzhang@shu.edu.cn

[2] School of Mechanical Science and Engineering, Huazhong University of Science and Technology, Wuhan, China

Abstract. As the demand for industrial inspection continues to increase, structured light 3D reconstruction technology is increasingly developing towards miniaturization and high precision in the field of industrial inspection. Aiming at the disadvantages of the small field of view and close-range detection systems, such as high cost, large size and heavyweight, a set of low-cost, high-precision, and high-resolution 3D reconstruction system suitable for measuring small objects with structured light is designed. In this paper, a microscope structured light system which composed of a USB camera and a MEMS programmable projection module with advantages that low-cost, high-accuracy, and high-resolution was proposed. A four-step phase-shifting algorithm and multi-frequency heterodyne algorithm are applied to decode the absolute phase value. Finally to calculate the 3D coordinate by using the correspondence between the absolute phase value and the actual 3D object point (The correspondence can be acquired by system calibration). Experiments show the accuracy of the system can achieve 4μm, and the measuring height range is 1.5 mm. Summarily, this system could greatly satisfy the requirement for the high precision 3D reconstruction of the micro parts.

Keywords: Three-frequency heterodyne · 3D shape measurement · Phase shifting · Temporal phase unwrapping

1 Introduction

In recent years, 3D profile measurement of the tiny object's surface has a broad application prospect in many fields, e.g. defect detection, machine vision, medical mechanical engineering due to its unique engineering value [1]. Generally, 3D profilometers can be concluded into the following ways: moire strip, flight time, triangulation and structured light, etc. Among these solutions, structured

X.-J. Liu et al. (Eds.): ICIRA 2021, LNAI 13014, pp. 313–323, 2021.
https://doi.org/10.1007/978-3-030-89098-8_30

light has developed rapidly, and become one of the most popular non-contact 3D measurement solutions in the industrial produce since high efficiency, high accuracy, and accessibility advantages [2]. It is composed of two key aspects: optical system calculation which is based on triangulation and fringe image processing method based on the phase-shifting fringe.

A lot of researches have been proposed to reconstruct the tiny object's surface. In 2012, Ma Chao et al. [3,4] combined with the advantages of fiber image beams with large freedom and long optical paths based on surface structured light digital fringe projection technology and phase-shift measurement technology. Aiming at the three-dimensional appearance characteristics of small objects such as teeth, a set of three-dimensional micro-measuring system for teeth was designed for direct measurement inside the oral cavity. In 2016, Ai Jia et al. [5] built a set of small field of view three-dimensional. The measurement system successfully reconstructed some characters on the surface of the coin with a field of view of 8 mm $\times$ 6 mm and a depth range of 100 μm, with a system accuracy of 10 μm. In 2019, Shi Yaoqun et al. [6] adopted a reliable path tracking phase unwrapping algorithm, and used a stereo microscope and telecentric lens camera realize a three-dimensional shape measurement system for tiny objects.

It is not difficult to find that the current high-precision and high-resolution 3D reconstruction systems for objects in a small field of view mostly use telecentric lenses or microscope lenses. The entire set of equipment is expensive, bulky and heavy, which is not conducive to lightness and portability. Therefore, based on the above mentioned research, this article has carried out research and designed a set of monocular structured light macro 3D measurement system based on the three-frequency heterodyne method. The system can achieve the goals of small shooting field of view, high image resolution and high reconstruction accuracy without supporting a telecentric lens or microscope lens. The whole set of equipment is lower cost and lightweight which can achieve the goal of 3D measurement and reconstruction of static micro object surface profile.

2 Principle of Fringe Projection

2.1 Fringe Projection Three-Dimensional Measurement Principle

The surface structured light technique (SSL) [7] is one of the most widely applied non-contact 3D measurement solutions in the industrial domains, e.g., 3D measurement, computer vision, and medical mechanical engineering. The principle of SSL-based methods is fringe projection technique, which also can be called phase measuring profilometry (PMP) [8]. Concretely, the principle is that to project encoded stripes into the object surface at first, then to capture the distorted stripes that reflected off the surface using a CCD camera, next according to the stripes to acquire phase distribution by phase decoding and unwrapping algorithm, finally to calculate the 3d surface profile by the relationship between the phase and the altitude of an object point (This relationship can be acquired by system parameters calibration).

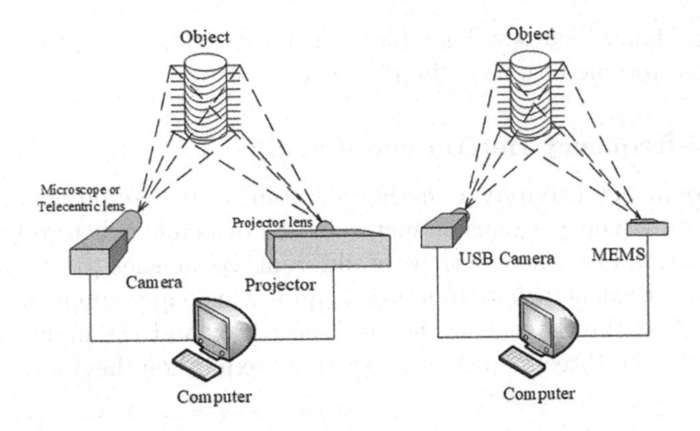

Fig. 1. Schematic diagram of a fringe-projection 3D measurement system

The left of Fig. 1 shows the schematic diagram of a fringe-projection 3D measurement system is mainly composed of a projector, and a camera. The right of Fig. 1 shows a schematic diagram of the 3D reconstruction system built in this article. The system is mainly composed of a MEMS programmable projector and a USB camera.

After collecting the phase-shifted deformed fringe image by a CCD camera, the N-step phase shift method is usually used to demodulate the phase value of the deformed fringe. The standard four-step phase shift method [9] is used to demodulate the phase principal value of deformed fringe image in this paper. Concretely, the original light intensity projected from the projector can be expressed as follows:

$$\begin{cases} I_1(x,y) = A(x,y) + B(x,y)\cos(2\pi f \cdot g(x,y)) \\ I_2(x,y) = A(x,y) + B(x,y)\cos(2\pi f \cdot g(x,y) + \pi/2) \\ I_3(x,y) = A(x,y) + B(x,y)\cos(2\pi f \cdot g(x,y) + \pi) \\ I_4(x,y) = A(x,y) + B(x,y)\cos(2\pi f \cdot g(x,y) + 3 \cdot \pi/2) \end{cases} \tag{1}$$

In this formula, f is the frequency of the cosine periodic function; $A(x,y)$ is the light intensity of the background fringe; $B(x,y)$ is the modulation intensity of the cosine function; $g(x,y)$ is the desired wrapped phase.

Through simultaneous formulas (1), the phase main value can be calculated by following formula:

$$g(x,y) = \arctan\left[\frac{I_4(x,y) - I_2(x,y)}{I_1(x,y) - I_3(x,y)}\right] \tag{2}$$

Since the phase information is calculated by the arctangent function, the phase main value $g(x,y)$ obtained by the formula (2) is distributed in a zigzag shape in the $(0, 2\pi)$ interval. This value is not unique in the entire measurement space so we should unfold the main phase value of the space point which is called phase unwrapping. Compared with the commonly used gray code and phase shift method to unfold the phase, the multi-frequency heterodyne method has better stability and higher accuracy. Therefore, this paper adopts the principle of multi-frequency heterodyne to unfold the phase [10].

2.2 Multi-frequency Heterodyne Principle

The three-frequency heterodyne method combines the heterodyne principle [11] and the three-frequency expansion method. The principle of heterodyne refers to the superposition of phase functions of different frequencies $\varphi_1(x)$ and $\varphi_2(x)$ to obtain a lower frequency phase function $\varphi_{12}(x)$. This paper adopts a compromise method which is three-frequency heterodyne to expand the main phase value. Figure 2 shows the three-frequency heterodyne expansion diagram:

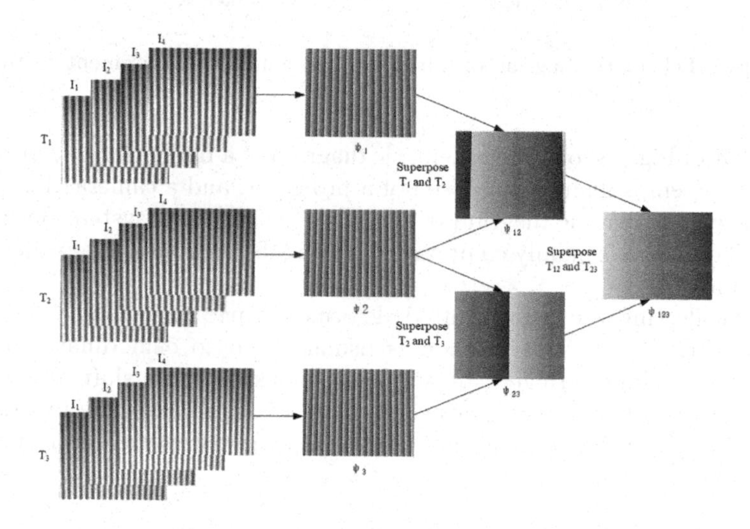

Fig. 2. The three-frequency heterodyne expansion diagram

The phase unwrapping steps are as follows:

Step1: Projecting three sets of four-step phase-shifting which period are T_1, T_2 and T_3 respectively. The four-step phase shift method is used to solve the wrapped phase of each group of fringe $g_1(x, y)$, $g_2(x, y)$ and $g_3(x, y)$. Projecting the decoded phase onto the coded structured light and corrected according to Gaussian filtering.

Step2: According to the heterodyne principle formula $T_{ij} = -(T_i \cdot T_j)/(T_i - T_j)$ to calculate T_{12} , T_{23} and T_{123} , T_{ij} means that the phase function of different frequencies T_i and T_j is superimposed to obtain a phase function $\varphi_{ij}(x)$ with a frequency of T_{ij}, φ_i represents the expansion result of the phase principal value with the number of cycles T_i. It can be seen from the formula deduced from the literature [12]:

$$\varphi_{12} = \varphi_1 + 2\pi \cdot round\left\{\frac{\frac{[(\varphi_1-\varphi_2)-\text{floor}(\frac{\varphi_1-\varphi_2}{2\pi})\cdot 2\pi]\cdot T_{12}}{T_1} - \varphi_1}{2\pi}\right\} \quad (3)$$

Similarly, we can solve φ_{23}. After normalizing φ_{12} and φ_{23}, make them redistribute between $(0, 2\pi)$.

In the same way, unfolding the package phase which frequency is T_{12}:

$$\varphi = \varphi_{12} + 2\pi \cdot round\left\{ \frac{\frac{[(\varphi_{12}-\varphi_{23})-\text{floor}(\frac{\varphi_{12}-\varphi_{23}}{2\pi})\cdot 2\pi]\cdot T_{123}}{T_{12}} - \varphi_{12}}{2\pi} \right\} \tag{4}$$

That means φ is the final unfolding phase.

2.3 The Principle of Calibration of the Relationship Between Phase and Height

Figure 3 shows the geometric model of the fringe projection measurement system in this paper. The meaning of each variable in the model is as follows:

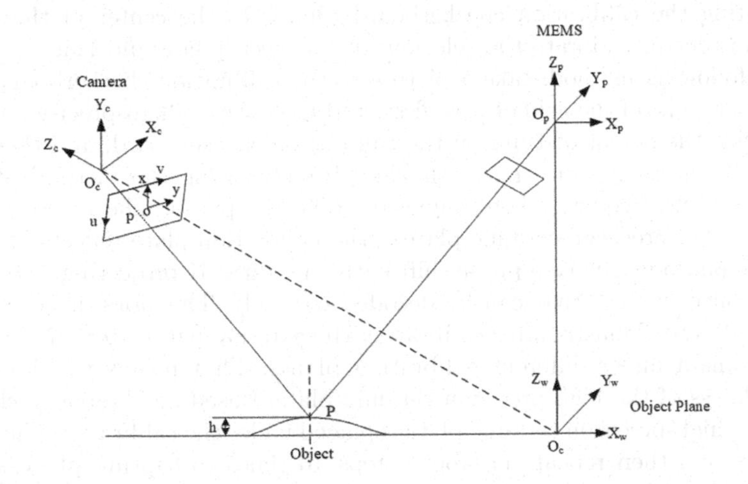

Fig. 3. The geometric model of the fringe projection measurement system

$O_c - X_cY_cZ_c$: Area camera coordinate system.

$O_w - X_wY_wZ_w$: Object plane coordinate system.

$O_p - X_pY_pZ_p$: MEMS laser coordinate system.

$o - uv$: The pixel coordinate system of the area camera which is located on the CCD imaging surface, perpendicular to the axis.

$o - xy$: Image coordinate system.

h: The height of the point on the surface of object surface to be measured.

According to the relationship of each coordinate system, it is easy to know that the K matrix is invertible, $K = \begin{bmatrix} \frac{f}{dx} & 0 & u_0 \\ 0 & \frac{f}{dy} & v_0 \\ 0 & 0 & 1 \end{bmatrix}$ is the internal parameter matrix. The relationship between the camera coordinates (X_w, Y_w, Z_w) and the pixel coordinates (u, v) is $\begin{bmatrix} X_c \\ Y_c \\ Z_c \end{bmatrix} = \mathrm{s} \cdot \mathrm{K}^{-1} \cdot \begin{bmatrix} u \\ v \\ 1 \end{bmatrix}$, s is the scale factor. In the camera coordinate system, the plane equation (point method) that passes the point translation and is perpendicular to can be expressed as $\begin{bmatrix} X_c - t_1 \\ Y_c - t_2 \\ Z_c - t_3 \end{bmatrix}^T \cdot$ $\mathrm{R_z} = \left(\mathrm{s} \cdot \mathrm{K}^{-1} \cdot \begin{bmatrix} u \\ v \\ 1 \end{bmatrix} - \begin{bmatrix} t_1 \\ t_2 \\ t_3 \end{bmatrix} \right)^T \cdot \mathrm{R_z} = 0$, $R_z(3 \times 1)$ is the z-direction component of the rotation matrix. After transformation, the scale factor from the camera coordinate system to the image coordinate system is $s = \frac{\begin{bmatrix} t_1 \; t_2 \; t_3 \end{bmatrix} \cdot R_z}{\begin{bmatrix} u \; v \; 1 \end{bmatrix} \cdot (K^{-1})^T \cdot R_z}$.

As in Fig. 4, the calibration shooting process is to take dozens of photos at different positions and angles of the camera's field of view, and then focus on shooting the calibration checkerboard photos in the center of the field of view. In order to calibrate the relationship between phase and height, here we use the following method: take a picture of the calibration checkerboard in the relatively middle of the field of view first, and then place a high-precision ceramic board with the height of 1mm on the calibration checkerboard, and then shoot three different frequencies phase-shifting. Finally, using the method of phase shift plus multi-frequency heterodyne to solve the phase. The reason why we placing a high-precision ceramic plate is the calibration plate reflects light, and when the photo-mechanical phase shift pattern is directly projecting, the camera cannot collect images that can be decoded correctly. Therefore, it is necessary to calibrate the T matrix after calibrating the external parameter (R, T) matrix of the camera under different calibration planes. That means we should add the thickness of the high-precision ceramic plate. Based on the above shooting process, a high-precision ceramic plate is placed under the calibration plate every 0.15 mm, and then repeat the above steps 10 times to capture phase-shifting images with uniformly spaced height changes.

After the image is processed for distortion correction, calculate the phase of the calibration plane at ten heights φ_i and the phase of the object to be measured and reconstructed $\varphi_{\_object}$.

Solve the s of each point of the calibration plane separately from the equation $s = \frac{\begin{bmatrix} t_1 \; t_2 \; t_3 \end{bmatrix} \cdot R_z}{\begin{bmatrix} u \; v \; 1 \end{bmatrix} \cdot (K^{-1})^T \cdot R_z}$. From the Fig. 5 below, it can be seen that the scale factor s and the phase φ_i are approximately linearly related by observation.

As shown in Fig. 6, the cubic curve fitting effect is the best by drawing residual graphs of fitting linear, quadratic and cubic curves which may be caused by the nonlinearity of the MEMS optical machine. Therefore, it is best to use the cubic curves fitting method to express the relationship between the scale factor s and the phase φ_i. The reason why we don't use high-order fitting is that the accuracy of high-order fitting is almost the same as that of cubic curves fitting,

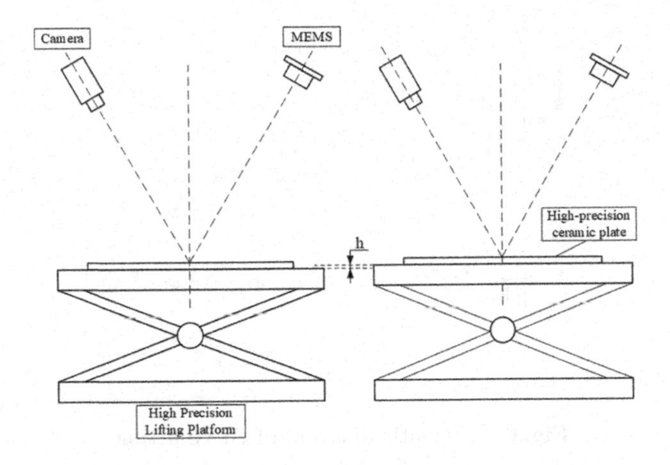

Fig. 4. Schematic diagram of calibration process

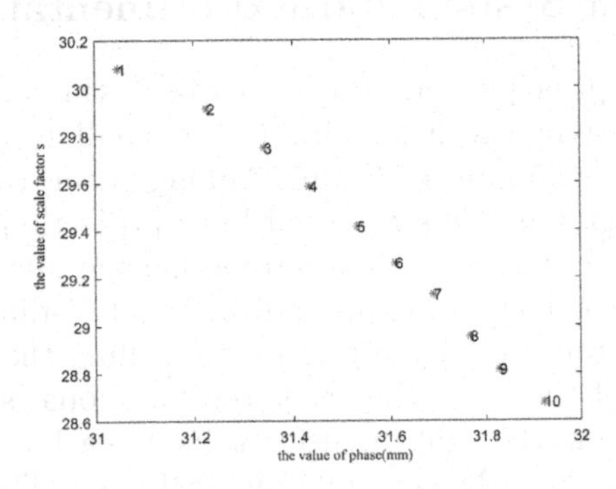

Fig. 5. Schematic diagram of calibration plane fitting

and the time of high-order fitting is longer. The relationship is as follows:

$$s = x_1 \cdot {\varphi_i}^3 + x_2 \cdot {\varphi_i}^2 + x_3 \cdot \varphi_i + x_4 \tag{5}$$

By bringing in the s corresponding to each point of each calibration plane and its corresponding phase to solve the four coefficients x_1, x_2, x_3 and x_4 of s and phase φ_i of each point by using least square method. Finally, bring the phase of the object and four parameters into the equation $s_{new} = x_1 \cdot {\varphi_{\_object}}^3 + x_2 \cdot {\varphi_{\_object}}^2 + x_3 \cdot \varphi_{\_object} + x_4$ to solve the of the object to be measured inversely. Finally, bring in the equation $\begin{bmatrix} X_{c\_object} \\ Y_{c\_object} \\ Z_{c\_object} \end{bmatrix} = s_{new} \cdot K^{-1} \cdot \begin{bmatrix} u \\ v \\ 1 \end{bmatrix}$ to solve the true three-dimensional information of the object.

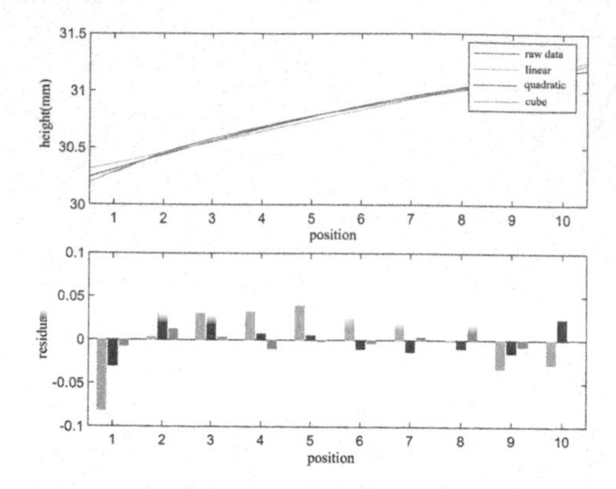

Fig. 6. Schematic diagram of curve fitting

3 Measurement System and Experimental Results

Based on the above principles, as shown in Fig. 7, this article builds a set of 3D meas-urement system which is suitable for small field and small objects and the field of view is 35 mm × 25 mm. Among them, the projection system adopts a MEMS programmable structured light projection module with a resolution of 1280 pixel * 1024 pixel. The projection module generates a parallel beam through a semiconductor laser and reflect by a two-dimensional high-speed vibration MEMS micromirror. By precisely controlling the switching time and input current value of the laser, the set two-dimensional structured light pattern can be realized, and the light intensity value of each spatial position of the structured light pattern can be controlled in real time. The CCD camera uses a USB camera module (resolution 1920 * 1080). At the same time, a precision lifting platform is equipped to calibrate the relationship between system height and phase.

After the system is built, calibrating the system based on Zhang Zhengyou calibration method and MATLAB camera calibration (monocular) toolbox to obtain the camera's internal parameter K, distortion coefficient and every calibration plane's external parameters ($R_{3\times3}$, $T_{3\times1}$).

Projecting the raster image with the period of 28, 26, 24 to the surface of the object with a small field of view (25 mm * 45 mm), and then calibrate the distortion correction and gamma correction of the taken picture. The gamma value of the calibrated system is 3.9. The wrapped phase is obtained by the phase shift method, and the multi-frequency heterodyne method is used Expand the package phase, and further obtain the point cloud data of the three-dimensional object according to the relationship between the calibrated phase and the height.

To verify the measurement accuracy of the measurement system built in this paper, a high-precision ceramic flat plate (with a flatness deviation of 3 μm) was selected as the measurement object. After reconstructing the ceramic flat plate, process the obtained point cloud data into the Geomagic Studio software, using the best fitting method to fit the plane, the error is about 4 μm, as the

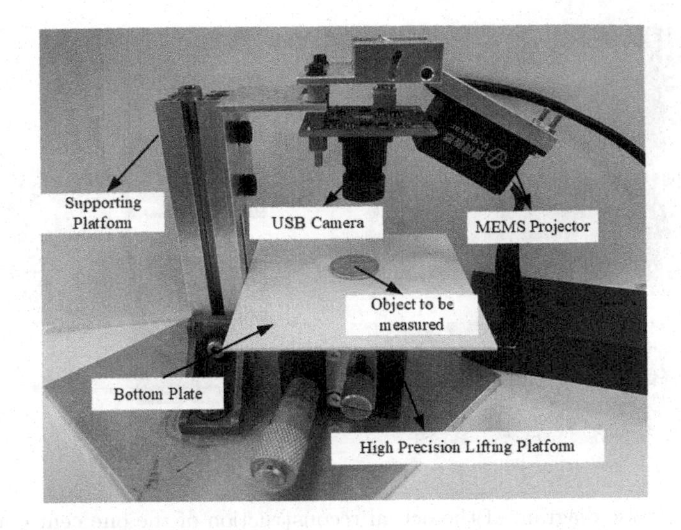

Fig. 7. Schematic diagram of experimental equipment

measurement result is shown in Fig. 8. From the calibrated phase and height relationship, it can be seen that the current measurement range of this measurement system is 1.5 mm. The preliminary evaluation result of the system measurement accuracy by using standard ceramic surfaces shows that the system accuracy can reach 4 microns. If we increase the number of calibration planes (reduce the calibration plane interval) and increase the number of phase shift fringes and optimize the hardware structure, the measurement accuracy of the system will be further improved.

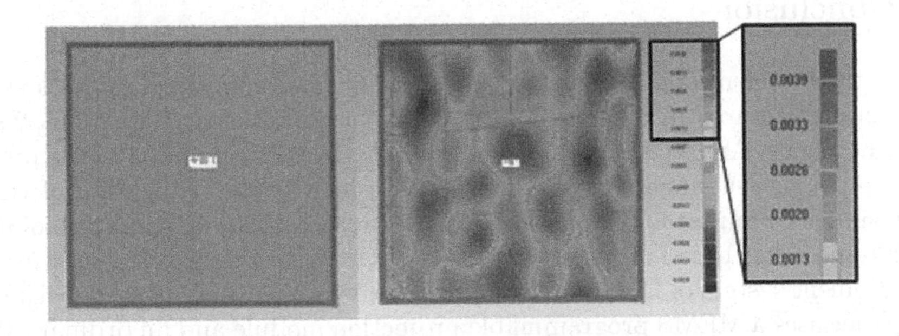

Fig. 8. Image of plane fitting error

After verifying the absolute measurement accuracy of the system, we took physical measurements of a common one cent coin. The schematic diagram of the actual reconstruction of the one cent coin process is shown in the Fig. 9, and the actual three-dimensional shape of the one cent coin is obtained from the relationship between the calibrated phase and height as shown in the Fig. 10.

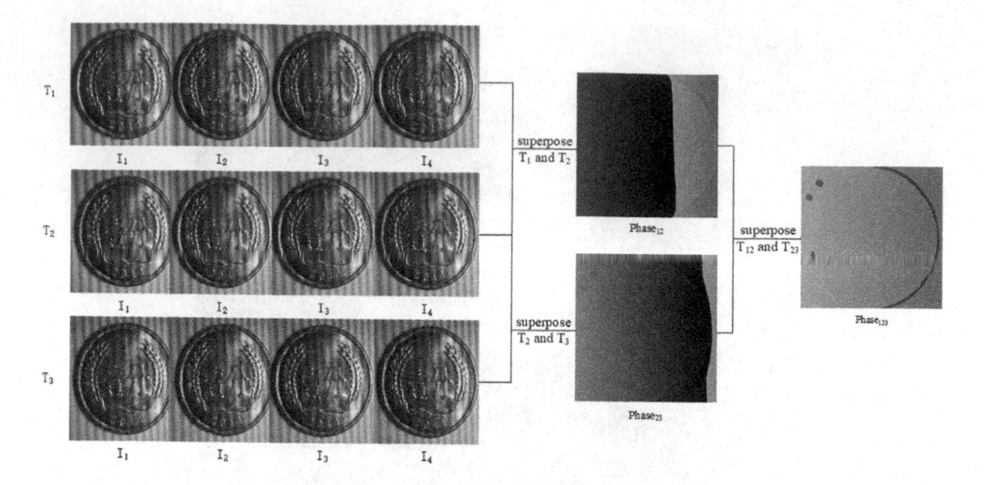

Fig. 9. Schematic diagram of the actual reconstruction of the one cent coin process

Fig. 10. Reconstruction renderings

4 Conclusion

As the three-dimensional surface measurement of small objects plays an increasingly important role in industrial inspection, medical inspection, robot vision and other fields. And its application fields will be further expanded in the future, non-contact measurement based on structured light is the realization of three-dimensional an important method of surface high-precision and high-resolution measurement has broad application prospects. In order to solve the existing 3D measurement system of high cost, large size, heavy weight and other deficiencies, this paper uses a MEMS programmable projection module and an ordinary USB camera to build a small object 3D reconstruction system based on the principle of phase shifting and multi-frequency heterodyne. Under the requirements of high-precision and high-resolution reconstruction, it has the advantages of lower cost and lighter weight. The experimental results show that the measurement depth range of the system built in this paper is 1.5 mm, the system accuracy is 4 μm, and it has a higher measurement accuracy and a larger measurement range.

Acknowledgement. This research is partially supported by the key research project of the Ministry of Science and Technology (Grant No.2018YFB1306802), the National Natural Science Foundation of China (Grant No. 51975344) and China Postdoctoral Science Foundation (Grant No. 2019M662591).

References

1. Ou, P., Wang, T., Li, R., et al.: A Three-Dimensional Teeth Measurement System Based on Structured Light. School of Instrumentation Science and Opto-Electronics Engineering, Beihang University, Beijing 100191, China. (in Chinese)
2. Da, F.: Three Dimensional Precise Measurement of Grating Projection. The Science Publishing Company, Beijing (2011). (in Chinese)
3. Ma, C.: Research on 3D Measurement Technology and System of Teeth. Huazhong University of Science and Technology (2012). (in Chinese)
4. Ma, C., Li, Z.W., et al.: Three dimensional measurement technology and system of micro teeth. Equip. Manuf. Technol. **02**, 30–34 (2012). (in Chinese)
5. Ai, J., Liu, S., Liu, Y., Zhang, Q., et al.: Three-dimensional Small-field Measurement System Based on Tri-frequency Heterodyne Fringe Analysis. Opto-electronics Department, College of Electronic Information, Sichuan University, Chengdu. (in Chinese)
6. Yaoqun, S.H.I., et al.: Micro-object measurement system based on structured light fringe projection. Appl. Opt. **40**(06), 1120–1125 (2019)
7. Feng, X.: Research on object size measurement technology based on structured light vision. Shandong University of Science and Technology (2008)
8. Li, J.: Research on precision measurement method of three dimensional shape of object based on projection fringe. University of Defense Science and Technology (2004)
9. Zhongyi, S., et al.: Three dimensional topography measurement system based on four step phase shifting grating projection. Appl. Opt. **36**(04), 584–589 (2015)
10. Songlin, C., et al.: Improvement of phase unwrapping method based on multi frequency heterodyne principle. Acta Optica Sinica **36**(04), 155–165 (2016)
11. Li, Z.: Research on structured light 3D measurement technology and system based on digital grating projection. Huazhong University of Science and Technology (2009)
12. Zhang, C., Zhao, H., Gu, F., et al.: Phase unwrapping algorithm based on multi-frequency fringe projection and fringe background for fringe projection profilometry. Measurement Science and Technology (2015)

Design and Control of a Normal-Stressed Electromagnetic Actuated Nano-positioning Stage

Xiangyuan Wang[1], Linlin Li[1,2], Wei-Wei Huang[1], and LiMin Zhu[1,3](✉)

[1] State Key Laboratory of Mechanical System and Vibration, School of Mechanical Engineering, Shanghai Jiao Tong University, Shanghai 200240, People's Republic of China
zhulm@sjtu.edu.cn
[2] State Key Laboratory of Fluid Power and Mechatronic Systems, Zhejiang University, Hangzhou 310027, China
[3] The Shanghai Key Laboratory of Networked Manufacturing and Enterprise Information, Shanghai 200240, China

Abstract. This paper presents the design and control of a normal-stressed electromagnetic actuated nano-positioning stage. The principle of the stage is discussed in detail. Induction and assembling relationships related to the cube armature are elaborately designed. As a result, through simple mechanism structure and only one actuator, the designed stage can realize linear motion without parasitic motion. To predict the performance of the stage, both the normal-stressed electromagnetic actuator (NSEA) and the mechanical structure are analyzed via finite element method. A prototype with a 12.60 μm working stroke and a first resonant frequency 4154 Hz is fabricated. To enhance the tracking accuracy, a control scheme with a PI controller and a notch filter is designed, and a −3 dB control bandwidth 2327 Hz is achieved. Triangular wave trajectory tracking tests are carried out. The results show the closed-loop system achieves a rms error of 1.34% when tracking a 10 μm P-V amplitude, 200 Hz triangular wave, which is much lower than the open-loop error of 5.01%, verifying the effectiveness of the nano-positioning stage and the designed control method.

Keywords: Nano-positioning stage · Electromagnetic actuator · Precision motion control · Tracking control

1 Introduction

Nano-positioning has been the supporting technology of a broad spectrum of areas such as ultra-precision manufacturing [1], scanning probe microscopy [2],

This work was partially supported by the National Natural Science Foundation of China under Grant No. U2013211 and Grant No. 51975375, the Open Foundation of the State Key Laboratory of Fluid Power and Mechatronic Systems, China under Grant No. GZKF-202003, and the China Postdoctoral Science Foundation (No. 2021M692065).

X.-J. Liu et al. (Eds.): ICIRA 2021, LNAI 13014, pp. 324–334, 2021.
https://doi.org/10.1007/978-3-030-89098-8_31

and precision manipulating [3]. Due to the requirement of high-precision operations, the design and control of nano-positioning stages have been extensively studied in the past few decades [4,5].

Normally, nano-positioning stages are composed of actuators and flexure hinge mechanisms. According to actuation technologies, nano-positioning stages can be divided into three types [6]: piezoelectric actuated [5], voice coil motor actuated [7], and normal-stressed electromagnetic actuated [8]. Due to the high-stiffness of piezoelectric materials, piezoelectric actuated nano-positioning stages are generally designed with high natural frequency, but short working stroke [4,9]. To improve the stroke of piezo-actuated stages, researchers developed various amplification mechanisms [10,11]. However, due to the contradictory relationship between natural frequency and working stroke, the improvement is not qualitative. In recent years, stepping piezo-actuated stages with large working stroke have been studied [12]. Whereas, it was found that the output characteristics of them are significantly sensitive to the environment and assembling [13]. Moreover, the nonlinearity behaviors of the piezoelectric actuators such as hysteresis and creep, also lead to significant positioning error. Voice coil motors, which use Lorenz force for actuation, can realize approximately linear actuating force–input current relationship [14]. Different from piezoelectric actuators, the stiffness of a voice coil motor actuated nano-positioning stage totally depends on the stiffness of the flexure hinge mechanisms. Considering that the voice coil motor's moving mass indeed limits it's working frequency, the flexure hinge mechanisms of voice coil motor actuated stages are generally designed to be low stiffness. In this sense, voice coil motor actuated stages can usually achieve millimeter or centimeter level working stroke, but very low natural frequency [7,14]. NSEAs, as the name suggests, are driven by Maxwell normal stress electromagnetic force produced by the current excitation and the permanent magnets. In the early years, this kind of actuators were designed in solenoid-type [15], and they could achieve relatively high force density. However, the quadratic relationship between the excitation current and the actuating force brings unavoidable computation challenges to the mechatronic system. In this regard, Lu et al. firstly proposed the configuration of linearized NSEA by using a permanent magnet to provide bias flux, then the excitation current and actuating force are theoretically in proportional relationship [16]. Based on this idea, several actuator configurations were developed to meet different requirements in the latest literature [8,17]. Till now, the existing NESA actuated nano-positioning stages need both complex structure and multi actuators [16]. Otherwise, the undesired parasitic motion will be generated [18]. Therefore, the structural configuration of normal-stressed actuated stages need to be further studied.

In this work, a NESA actuated nano-positioning stage gives attention to both positioning accuracy and structural simplicity is designed, analyzed, and controlled. By elaborately design the induction and assembling relationships related to the cube armature, the designed stage can realize linear motion without parasitic motion through simple mechanism structure and only one actuator. A prototype is fabricated, and a 12.60 μm working stroke and a first resonant frequency 4154 Hz is determined through open-loop tests. To compensate the tracking error as well as damping the lightly-damped resonant mode, control method with a PI controller and a notch filter is implemented in the closed-loop system, and its −3 dB bandwidth is tested 2327 Hz. Triangular wave trajectory tracking results show that, comparing with open-loop tracking results, the rms error for tracking a 10 µm P-V amplitude, 200 Hz triangular wave can be reduced from 5.01% to 1.34%, demonstrating the effectiveness of the control method.

The remainder of this paper is organized as follows. The design and analysis of the stage are presented in Section 2. Basic performances of the stage were tested in Section 3. In Section 4, a control scheme with a PI controller and a notch filter is designed, and trajectory tracking tests are conducted. Section 5 concludes the paper.

2 Design and Analysis of the Nano-Positioning Stage

2.1 Principle and Analysis of the Normal-Stressed Electromagnetic Actuator

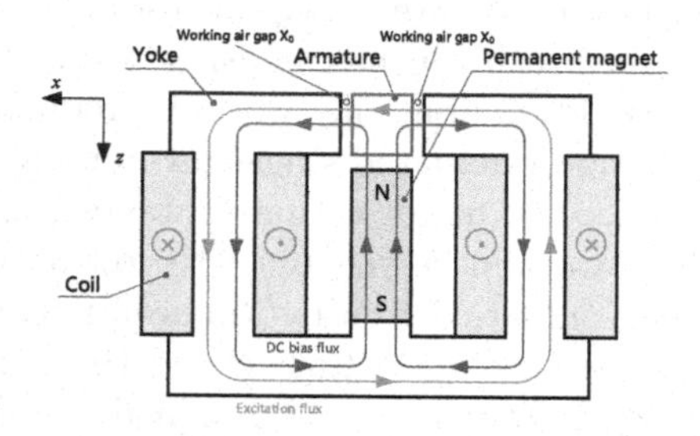

Fig. 1. Working principle of the NSEA.

NSEAs have the ability to perform large force density in a relatively long range [8,16]. In this paper, a NSEA is designed and implemented in the nano-positioning stage. The working principle of the NESA is shown in Fig. 1. A 6 mm × 6 mm × 6 mm armature is the end-effector, and is made of Fe-based amorphous alloy with a saturation flux of 1.56 T. The yoke is also made of Fe-based amorphous alloy for magnetic conduction. A total of 100 turns of Litz coil are wound around the yoke, which will produce the excitation flux of the actuator. 6 mm × 6 mm × 19 mm Nd-Fe-B permanent magnets with a remanence of 1.22 T is chosen to provide the DC bias flux. By superpositioning the excitation flux and DC bias flux, the magnet flux density of the two working air gaps are different. Therefore, the actuating force F_{act} will be generated along x axis. Consider both the assembly tolerance and energy efficiency, the working air gap X_0 on both sides of the armature is determined as 150 µm. It should be noted that, along z axis, the permanent magnet also generates a constant attractive force

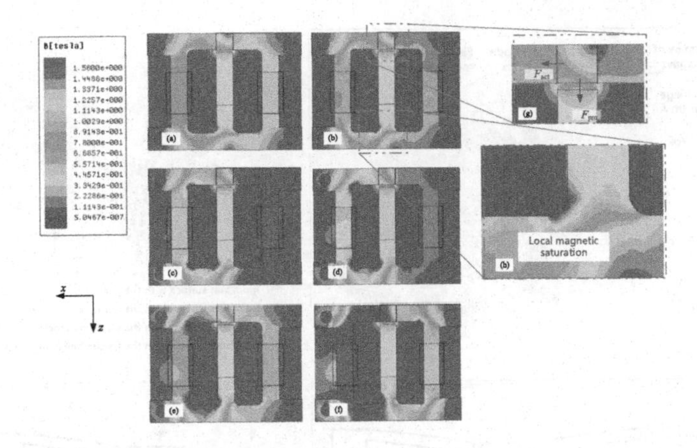

Fig. 2. Electromagnetic simulation results: (a, b, c, d, e, and f) under the excitation of 40 AT, 80 AT, 120 AT, 160 AT, 200 AT, and 360 AT, respectively; (g) force diagram of the armature; (h) enlarged view to show the local magnetic saturation.

F_{pm} on the armature, but it does not affect the actuating process dynamically (i.e., it is decoupled from the actuation force).

To predict the performance of the designed NSEA, finite element simulation is conducted through the commercial software ANSYS Maxwell. The simulated flux distribution under excitation amplitude varying from 40 AT to 360 AT is illustrated in Fig. 2(a-f). The flux difference between the two working air gaps is obtained, agreeing with the previous description. In practical electromagnetic systems, local magnetic saturation effect will appear under large excitation, and the iron losses grow sharply [19]. This will result in a nonlinear relationship between excitation and actuating force. Figure 2(h) gives the enlarged view to show the local magnetic saturation. It can be seen in Fig. 2(a-f), the saturation area correlates with the excitation amplitude. The force diagram of the armature is shown in Fig. 2(g). Simulation results indicate that the attractive force F_{pm} is 7.52 N, and the actuating force F_{act} is 3.33 N under the excitation of 40 AT.

2.2 Design and Analysis of the Mechanical Structure

The assembly model of the proposed nano-positioning stage is shown in Fig. 3(a). For the convenience of assembly process, two independent aluminum alloy bases, i.e. base A and base B are designed and fixed on the experimental platform. The yoke, coil, permanent magnet, and armature constitute the NSEA, which provides the actuating force. The flexure hinge mechanisms (A and B) share the same structure, and are screwed on base A and base B, respectively. Finally, a triangular-shaped plate electrode block is stacked onto the armature for implementing the capacitance displacement sensor during experiments.

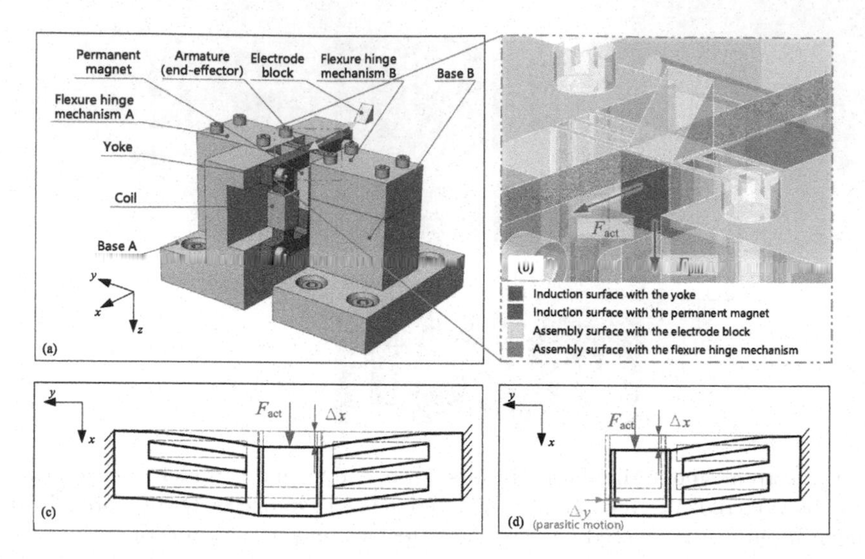

Fig. 3. The mechanical structure of the designed nano-positioning stage: (a) the assembly model; (b) assembly and induction relationship between the armature and other parts; (c) deformation diagram of the designed symmetric flexure hinge mechanisms; (d) deformation diagram of asymmetric flexure hinge mechanisms.

As shown in Fig. 3(b), the induction and assembling relationships related to the cube armature are elaborately designed. In order to restrain the armature's movement, the two xz faces are glued with the flexure hinge mechanisms with sufficient adhesion strength and reliability. To generate the actuating force F_{act}, as analyzed in Sect. 2.1, two yz faces are inductive to the yoke. The lower xy face bears the attractive force F_{pm} of the permanent magnet, and the upper xy face is stacked with the electrode block. Based on this design, the nano-positioning stage can realize linear motion without parasitic motion by implementing simple flexure hinge mechanisms. Meanwhile, only one NSEA is needed. Figure 3(c) shows the deformation diagram of the designed symmetric flexure hinge mechanisms, which indicates that under actuating force F_{act}, the mechanisms only deforms along the actuation axis, i.e. x axis.

Note 1. In some previous works [18], the flexure hinge mechanisms were designed asymmetrically, and the motion is essentially a rotary motion. Therefore, a parasitic motion Δy will be generated together with the major motion Δx as shown in Fig. 3(d). The parasitic motion will lead to nonnegligible positioning error in large working stroke.

The static and dynamic performance of the mechanical structure are simulated through ANSYS/Workbench. In the static analysis, actuating force F_{act} of 3.3 N along x axis and attractive force F_{pm} of 7.5 N along z axis are applied to the armature. As shown in Fig. 4(a), the maximum deformation along x axis is obtained as 1.88 μm under the actuating force F_{act}. Therefore, the simulated

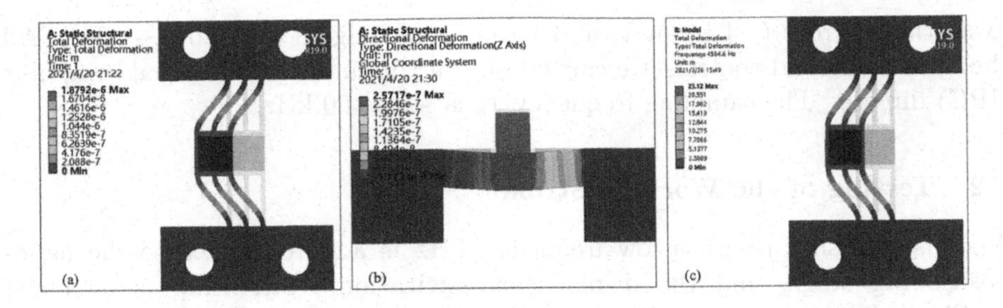

Fig. 4. Simulation results of the mechanical structure: (a) static structural analysis along x axis; (b) static structural analysis along z axis; (c) modal analysis.

stroke should be 3.76 μm under the excitation of ±40 AT. Figure 4(b) shows that, under the attractive force F_{pm}, the deformation along z axis is 0.26 μm. It should be noted that, the attractive force F_{pm} is a constant force, and is totally decoupled from the driving force F_{act}, so it is unnecessary to measure the z axis motion in experiments. In Fig. 4(c), modal analysis results show the simulated first resonant frequency of the mechanical structure is 4565 Hz.

3 Experimental Setup and Performance Evaluation

3.1 Experimental Setup

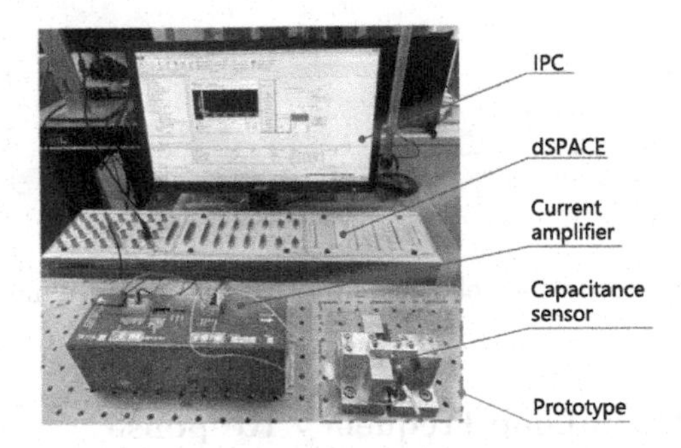

Fig. 5. Experimental setup for the proposed nano-positioning stage.

The experimental setup for the proposed nano-positioning stage is shown in Fig. 5. A high resolution capacitance sensor with measurement range of ±25 μm (MicroSense 8810 with the probe 2823) is gripped on the prototype for measuring the armature's motion. A dSPACE-DS1103 board equipped with 16-bit analog to digital converters (ADCs) and 16-bit digital to analog converters (DACs) is utilized to output the control command for the current amplifier. A current amplifier (Trust Automation Inc., TA115) is adopted to provide the excitation current (−4 A–+4 A) under the control signal of (−10 V–+10 V). All the operations and controls are carried out on the industrial personal computer (IPC) directly. The sampling frequency T_s is set as 50 kHz.

3.2 Testing of the Working Stroke

A sinusoidal signal with a low frequency 1 Hz is adopted to excite the nano-positioning stage, and the displacement-excitation relationship is arranged in Fig. 6. The tested traveling stroke under small excitation of ±40 AT is 3.19 μm, which is 84.8% of the simulation results (3.76 μm). Under large excitation amplitude, the nonlinearity caused by local magnetic saturation, as discussed and simulated in Sect. 2.1, can be clearly read. And the maximum open-loop traveling stroke of the proposed positioning stage is about 12.60 μm under the excitation of ±360 A-turns.

It can also be observed that the designed normal-stressed electromagnetic actuated nano-positioning stage performs lower hysteresis nonlinearity comparing with piezo-actuated stages [20]. Using Fe-based amorphous alloy for magnetic conduction, the hysteresis caused error is also lower than that of some earlier presented normal-stressed electromagnetic actuated stages [8].

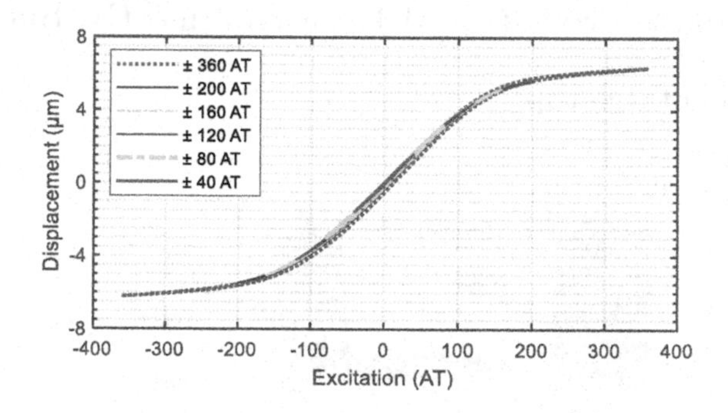

Fig. 6. Displacement-excitation relationship of the proposed nano-positioning stage.

3.3 Testing of Open-Loop Frequency Response

Band-limited noise command with small amplitude is conducted to test the frequency response, which is shown in Fig. 7. DC pre-excitations of 0 AT, 80 AT, and 160 AT are implemented to change the measurement point. Under the condition of zero pre-excitation, the first resonant frequency of the stage is measured 4154 Hz, which is 91% of the simulation results. The slight deviation between experimental and simulation results might be caused by the manufacturing errors.

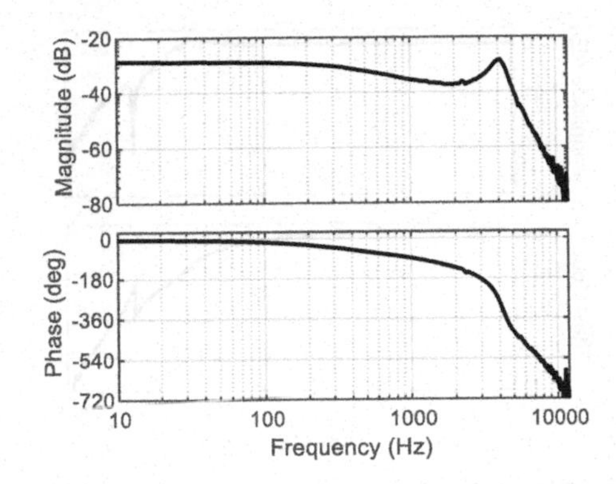

Fig. 7. Frequency responses of the nano-positioning stage.

4 Controller Design and Experimental Verification

4.1 Controller Design

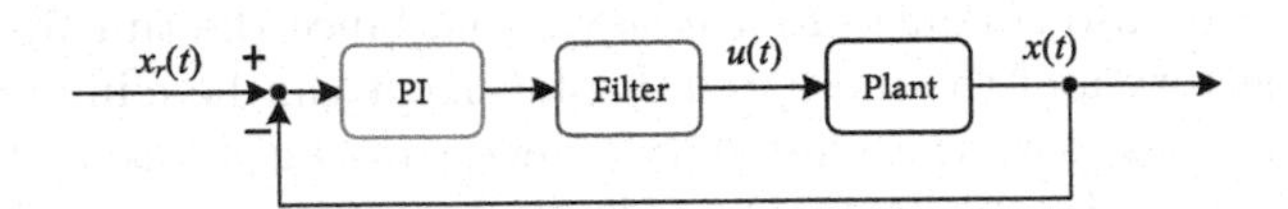

Fig. 8. Block diagram of the control scheme.

To compensate the disturbance as well as damping the lightly-damped resonance, feedback controller incorporating a PI controller and a notch filter is designed and implemented as its simplicity [20]. The block diagram of the control scheme is shown in Fig. 8. Where $u(t)$ denotes the control signal, $x_r(t)$ and $x(t)$ represent the reference and measured displacement, respectively. The transfer function of a second-order notch filter is

$$F_N(s) = \frac{s^2 + 2\zeta_1\omega_n s + \omega_n^2}{s^2 + 2\zeta_2\omega_n s + \omega_n^2} \tag{1}$$

where

$$\omega_n = 2\pi f_n \tag{2}$$

denotes the notch filter's center frequency, and $f_n = 4154\,\text{Hz}$ is the resonant peak frequency of the open-loop system. ζ_1 and ζ_2 are the damping factors, in this work, they are determined as $\zeta_1 = 0$ and $\zeta_2 = 0.707$, respectively. Finally, the proportional gain K_p and integral gain K_i of the PI controller are tuned by trial-and-error method.

To evaluate the performance of the closed-loop system, frequency response of the nano-positioning stage is measured, which is plotted in Fig. 9. The results indicate that, the resonant peak has been successfully damped, and the -3 dB bandwidth of the closed-loop system 2327 Hz.

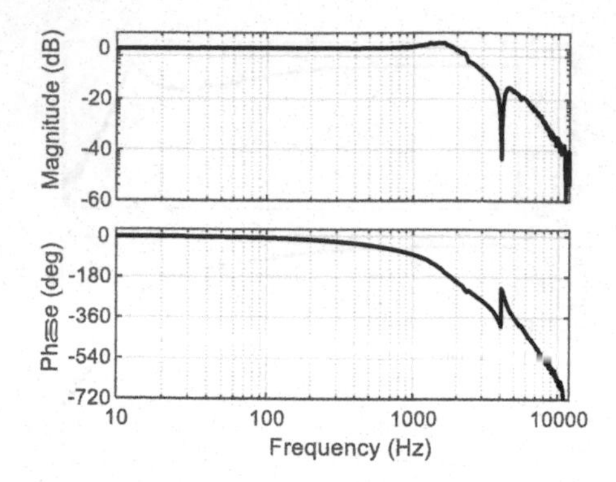

Fig. 9. Closed-loop frequency response of the nano-positioning stage.

4.2 Trajectory Tracking Tests

To investigate the tracking accuracy of the closed-loop system, triangular wave tracking tests are carried out. Triangular waves with 10 μm P-V amplitude are tracked, and the fundamental frequencies are selected 50 Hz 200 Hz. Open-loop tracking tests are also conducted for comparison. Since the time delay between reference trajectory and the measured displacement can be eliminated in some tracking operations [21], the shifted references are introduced. The tracking results are given in Fig. 10. It can be seen that, comparing with the open-loop tracking results, the closed-loop results show better tracking accuracy. It can also be seen that, the tracking accuracy at positive and negative positions are slightly different, and it might due to the assembly errors. In order to quan-

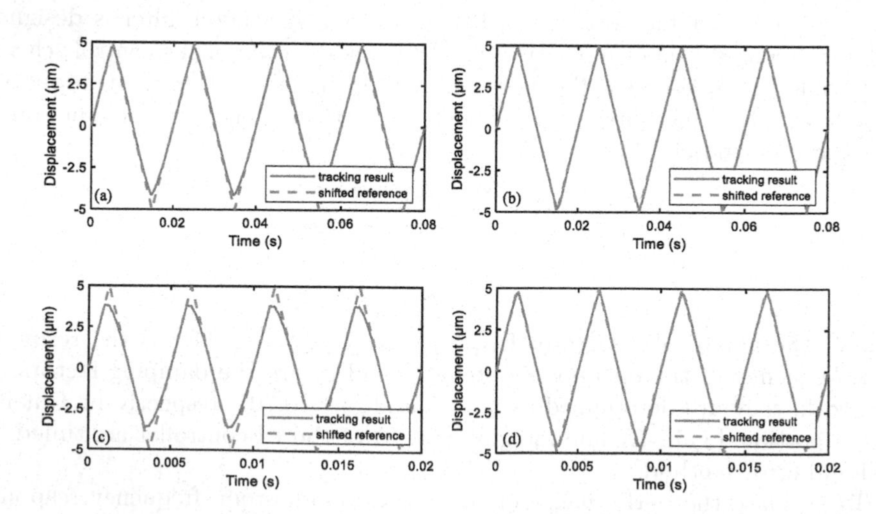

Fig. 10. Triangular wave tracking results: (a) 50 Hz, open-loop; (b) 50 Hz, closed-loop; (c) 200 Hz, open-loop; (d) 200 Hz, closed-loop.

tify the tracking accuracy, the tracking errors of both open-loop and closed-loop tests are computed and listed in Table 1, and two indexes, i.e., root mean square (rms) error e_{rms} and maximum error e_{max} are employed [21]. The error statistics shows that, comparing with open-loop results, the rms tracking error of the closed-loop system 200 Hz is reduced from 5.01% to 1.34%, and the maximum error 200 Hz is reduced from 26.73% to 7.02%. Therefore, the effectiveness of the control method is verified.

Table 1. Error statistics of tracking results.

Method	Fundamental Frequency (Hz)	e_{rms}(%)	e_{max}(%)
Open-loop	50	2.51	12.39
	200	5.01	2.63
Closed-loop	50	0.63	4.04
	200	1.34	7.02

5 Conclusions

In this paper, a novel normal-stressed electromagnetic actuated nano-positioning stage is designed, analyzed, and controlled. It is demonstrated that, through simple mechanism structure and only one actuator, the stage can realize linear motion with no parasitic motion. Open-loop testing results indicate that the working stroke of the proposed stage is 12.60 μm, and the first resonant frequency 4154 Hz. A control scheme with a PI controller and a notch filter is established. The sharp resonant peak is successfully damped and its −3 dB bandwidth is tested 2327 Hz. Trajectory tracking results indicate that, the designed control scheme significantly enhances the tracking accuracy. Therefore, the effectiveness of the nano-positioning stage and the control method is demonstrated. Our future work would include high bandwidth control of normal-stressed electromagnetic nano-positioning stages for better performance.

References

1. Zhu, Z., et al.: Design and control of a piezoelectrically actuated fast tool servo for diamond turning of microstructured surfaces. IEEE Trans. Ind. Electron. **67**(8), 6688–6697 (2020)
2. Li, L., et al.: A smoothed raster scanning trajectory based on acceleration-continuous b-spline transition for high-speed atomic force microscopy. IEEE/ASME Trans. Mech. **26**(1), 24–32 (2021)
3. Su, Q., et al.: A 3-DOF sandwich piezoelectric manipulator with low hysteresis effect: design, modeling and experimental evaluation. Mech. Syst. Sign. Process. **158**, 107768 (2021)
4. Yong, Y.K., et al.: Invited review article: high-speed flexure-guided nanopositioning: mechanical design and control issues. Rev. Sci. Instrum. **83**(12), 121101 (2012)

5. Li, X., et al.: A compact 2-DOF piezo-driven positioning stage designed by using the parasitic motion of flexure hinge mechanism. Smart Mater. Struct. **29**(1), 015022 (2020)
6. Csencsics, E., Schitter, G.: Exploring the pareto fronts of actuation technologies for high performance mechatronic systems. IEEE/ASME Trans. Mechatron. **26**(2), 1053–1063 (2020)
7. Xu, Q.: Design and development of a compact flexure-based XY precision positioning system with centimeter range. IEEE Trans. Ind. Electron. **61**(2), 893–903 (2014)
8. Zhu, Z.H., et al.: Tri-axial fast tool servo using hybrid electromagnetic-piezoelectric actuation for diamond turning. IEEE Transactions on Industrial Electronics (2021)
9. Li, C.-X., et al.: Design, analysis and testing of a parallel-kinematic high-bandwidth XY nanopositioning stage. Rev. Sci. Instrum. **84**(12), 125111 (2013)
10. Watanabe, S., Ando, T.: High-speed XYZ-nanopositioner for scanning ion conductance microscopy. Appl. Phys. Lett. **111**(11), 113106 (2017)
11. Ling, M., et al.: Kinetostatic and dynamic modeling of flexure-based compliant mechanisms: a survey. Appl. Mech. Rev. **72**(3), 030802 (2020)
12. Li, J., Huang, H., Morita, T.: Stepping piezoelectric actuators with large working stroke for nano-positioning systems: a review. Sens. Act. A: Phys. **292**, 39–51 (2019)
13. Wang, X., Zhu, L.M., Huang, H.: A dynamic model of stick-slip piezoelectric actuators considering the deformation of overall system. IEEE Transactions on Industrial Electronics (2020)
14. Xu, Q.: New flexure parallel-kinematic micropositioning system with large workspace. IEEE Trans. Robot. **28**(2), 478–491 (2012)
15. Gutierrez, H.M., Ro, P.I.: Magnetic servo levitation by sliding-mode control of nonaffine systems with algebraic input invertibility. IEEE Trans. Ind. Electron. **52**(5), 1449–1455 (2005)
16. Lu, X.D., Trumper, D.L.: Ultrafast tool servos for diamond turning. Cirp Ann.-Manuf. Technol. **54**(1), 383–388 (2005)
17. Csencsics, E., Schlarp, J., Schitter, G.: High-performance hybrid-reluctance-force-based tip/tilt system: design, control, and evaluation. IEEE/ASME Trans. Mechatron. **23**(5), 2494–2502 (2018)
18. Ito, S., et al.: Long-range fast nanopositioner using nonlinearities of hybrid reluctance actuator for energy efficiency. IEEE Trans. Ind. Electron. **66**(4), 3051–3059 (2019)
19. Zhu, Z.Q., et al.: Influence of local magnetic saturation on iron losses in interior permanent magnet machines. In: 2016 XXII International Conference on Electrical Machines, pp. 1822–1827 (2016). https://doi.org/10.1109/ICELMACH.2016.7732771
20. Gu, G.Y., Zhu, L.M.: Motion control of piezoceramic actuators with creep, hysteresis and vibration compensation. Sens. Act. A: Phys. **197**, 76–87 (2013)
21. Li, C., et al.: Damping control of piezo-actuated nanopositioning stages with recursive delayed position feedback. IEEE/ASME Trans. Mechatrn. **22**(2), 855–864 (2017)

A Two-DOF Ultrasonic Motor Based on Longitudinal and Bending-Bending Vibrations for Manipulator

Dongmei Xu[1](✉), Liangliang Zhao[2], Simiao Yu[3](✉), Bingjie Zhang[1], Yao Xiong[1], Pengcheng Shen[1], and Xuhui Zhang[1]

[1] Xi'an University of Science and Technology, Xi'an 710054, China
dongmeixu@xust.edu.cn
[2] Xi'an Aerospace Engine Co., Ltd., Xi'an 710054, China
[3] Xi'an University of Architecture and Technology, Xi'an 710054, China
simiaoyu@xauat.edu.cn

Abstract. A novel two-DOF ultrasonic motor based on longitudinal and bending-bending vibrations was introduced in this paper. This proposed two-DOF ultrasonic motor is potential in driving the hyper-redundant manipulator, which would significantly decrease the number of the driving motor, thus reduces the cost of the manipulator. The structure and the working principle of the novel ultrasonic motor was illustrated, the driving mechanism of the proposed ultrasonic motor to drive the hyper-redundant manipulator was proposed. The longitudinal vibration was used to drive the proposed motor runs along the manipulator, and bending-bending vibrations were utilized to rotate the joint of the manipulator. Then, the finite element method was used to design and optimize the configuration of the proposed ultrasonic motor. The longitudinal and orthogonal bending-bending vibrations were obtained. This project provides a new idea for optimizing the hyper-redundant manipulator from the perspective of motor, and provides basis for the two DOF ultrasonic motor to drive the hyper-redundant manipulator.

Keywords: Two-DOF · Ultrasonic motor · Longitudinal and bending-bending vibrations

1 Introduction

Hyper redundant manipulator has the advantages of high mobility, obstacle avoidance, singularity overcoming and large motion space [1–3]. It can perform tasks in complex environment or narrow, limited and unstructured environment, and has great application prospects in nuclear power plant pipeline inspection, space station maintenance, aircraft assembly and robot surgery. At present, the most common driving mode of hyper redundant manipulator is electromagnetic motor drive, and the motor and transmission mechanism are usually placed in the joint arm [4–8]. Tappe et al. proposed a concept of hyper-redundant manipulator based on binary electromagnetic tilt actuator and applied

X.-J. Liu et al. (Eds.): ICIRA 2021, LNAI 13014, pp. 335–341, 2021.
https://doi.org/10.1007/978-3-030-89098-8_32

it to endoscope; binary electromagnetic tilt actuator was set in each joint of the manipulator [9]. A redundantly actuated planar 2-DOF parallel manipulator was presented by Liang et al. [10], two set of servo motor and deceleration system were utilized to actuate the manipulator.

There are other driving modes, like cable or pneumatic driven. A cable-driven hyper-redundant manipulator was proposed by Xu et al. [11], in this driven system, $3N$ (N indicate the universal joint's number) cables were needed to drive the $2N$-DOFs manipulator. Karnjanaparichat et al. studied the adaptive position tracking problem of multi-link robot driven by pneumatic muscle group [12].

While introducing multiple degrees of freedom, the hyper redundant manipulator introduces multiple driving motors and complex deceleration system, and so on. Basically, for each additional degree of freedom, it needs to introduce a set of driving motor and deceleration system, which leads to the problems of complex structure, large volume, increased mass, and poor robustness of the manipulator.

Ultrasonic motor has the advantages of fast response speed, high precision, no need of transmission mechanism, large torque mass ratio and so on, which is suitable for the driving of manipulator. In this paper, an ultrasonic motor based on longitudinal and bending-bending vibrations was introduced to drive the hyper redundant manipulator. The proposed ultrasonic motor has two-DOF, which can run along the manipulator and drives the joint to rotate. Thus, the hyper redundant manipulator can be driven by only one two-DOF ultrasonic motor. This paper focuses on the two DOF ultrasonic motor mentioned above. The structure and driving mechanism were illustrated in Part 2, Finite Element Method was used to analyze and verify the working principle in Part 3, a conclusion was proposed in Part 4.

2 Structure and Driving Mechanism

2.1 The Structure of the Two-DOF Ultrasonic Motor

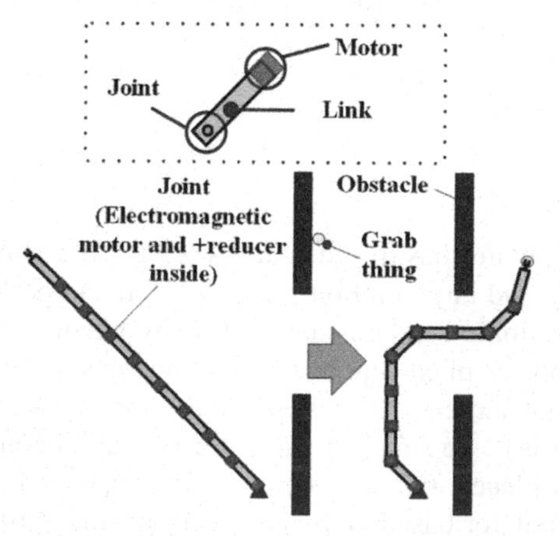

Fig. 1. The proposed two-DOF ultrasonic motor driving mechanism versus the traditional electromagnetic motor driving mechanism.

The proposed two-DOF ultrasonic motor could drive the hyper redundant manipulator, the driving mechanism is shown in Fig. 1. Under the oblique line trajectory, the two-DOF ultrasonic motor could travel along the link of the manipulator; and with the effect of the elliptical trajectory of the driving foot, the ultrasonic motor is able to rotate the joint of the manipulator. Comparing with the traditional electromagnetic motor driven, the proposed two-DOF ultrasonic motor driving mechanism could drive the hyper redundant manipulator with only one ultrasonic motor, which makes the driving device simple, as shown in Fig. 1.

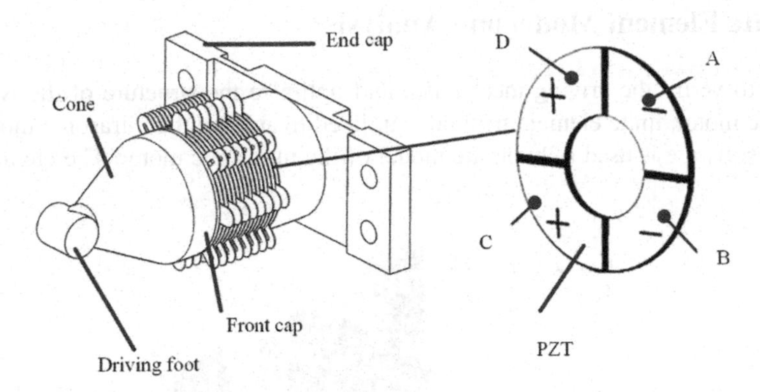

Fig. 2. Structure of the proposed ultrasonic motor and polarization of the PZT elements.

The structure of the proposed ultrasonic motor is shown in Fig. 2, which is composed of the front cap, the end cap and the PZT. The cone is used to magnify the amplitude of the vibration. The PZT elements are clamped between the end cap and the front cap. The PZT has four divisions, the polarization direction of which is illustrated in Fig. 2. The four-division structure of the PZT elements make the ultrasonic motor be able to have two-DOF.

2.2 The Driving Mechanism of the Two-DOF Ultrasonic Motor

When sine signal applying to division-A and C and cosine signal applying to division-B and D simultaneously, the orthogonal bending vibrations with 90° phase shift will be generated, thus the elliptical trajectory is formed, as shown in Fig. 3. The longitudinal vibration will be generated if the two sine signals with 180° phase shift are applied to division-A B and division-C D, respectively. Thus, if the driving foot keeps a certain angle with the rail, the oblique line trajectory will be formed, as shown in Fig. 3.

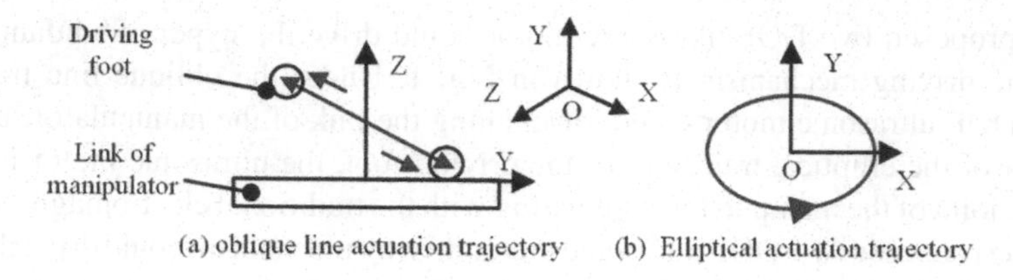

Fig. 3. Two working modes of the proposed ultrasonic motor to obtain two-DOF.

3 Finite Element Model and Analysis

In order to verify the driving mechanism and optimize the structure of the two-DOF ultrasonic motor, finite element method is utilized to analyze the ultrasonic motor. The ANSYS software is used to build the model of the ultrasonic motor. The element type

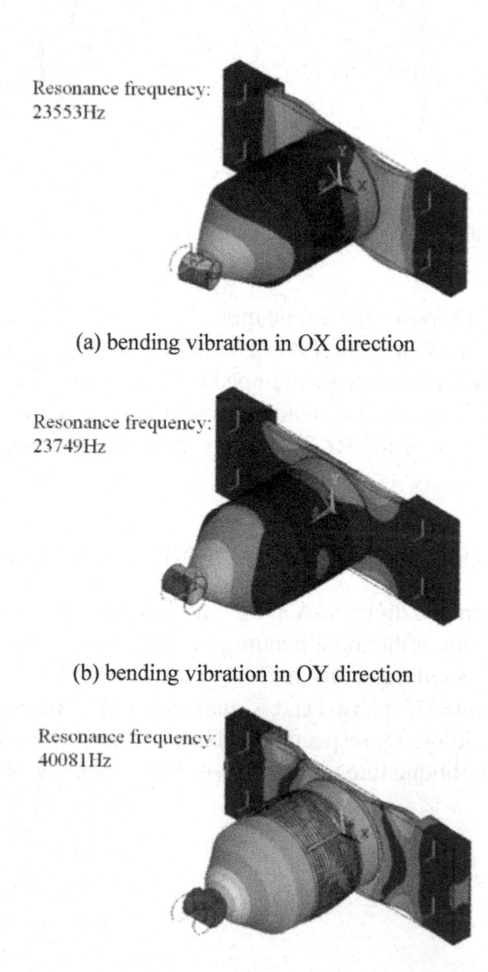

(a) bending vibration in OX direction

(b) bending vibration in OY direction

(c) longitudinal vibration in OZ direction

Fig. 4. Bending and longitudinal vibrations of the proposed two-DOF ultrasonic motor.

of the finite element model of the proposed two-DOF ultrasonic motor is SOLID227. The materials of the ultrasonic motor are duralumin alloy, 45 steel and PZT4. And the front cap, cone and the driving foot are made of duralumin alloy; the end cap is made of 45 steel. The calculated vibration modes are shown in Fig. 4. The frequencies of the bending vibration modes in OX direction and OY direction are 23553 Hz and 23749 Hz, respectively. The frequency of longitudinal vibration mode is 40081 Hz. Two orthogonal bending vibration modes are obtained to generate the elliptical trajectory. The longitudinal vibration is achieved to form the longitudinal trajectory.

Parameter sensitivity analysis of the proposed ultrasonic motor is performed, the related results are shown in Fig. 5. The frequencies of the bending vibration modes in OX and OY can be equal by tuning the main parameter of the ultrasonic motor, thus the elliptical trajectory will be formed to rotate the joint of the manipulator.

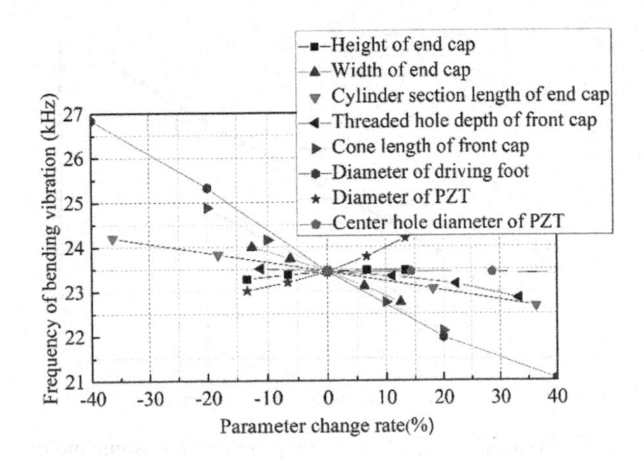

Fig. 5. Main parameter sensitivity of the proposed ultrasonic motor.

Simulation driving trajectories of the proposed ultrasonic motor under two driving modes are shown in Fig. 6. When changing the excitation signal, the ultrasonic motor could work between the oblique line trajectory and the elliptical trajectory, thus, the two-DOF ultrasonic motor could travel along the link of the manipulator or rotate the special joint of the manipulator.

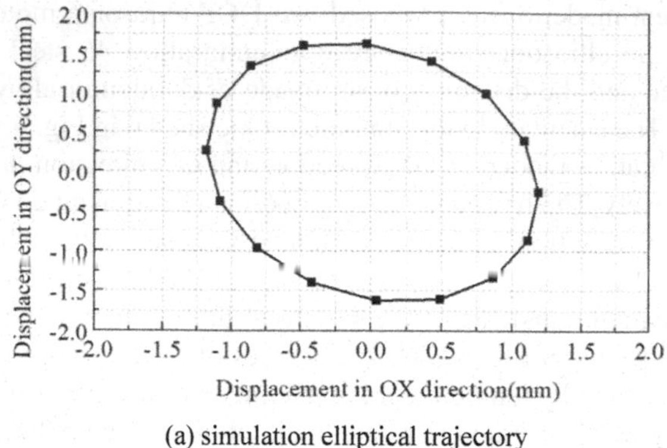

(a) simulation elliptical trajectory

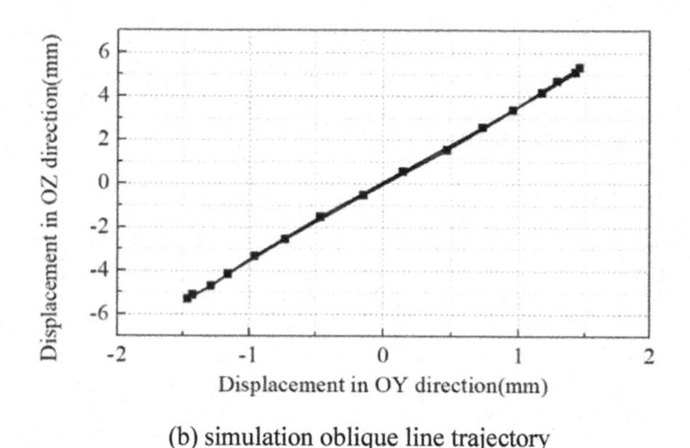

(b) simulation oblique line trajectory

Fig. 6. Simulation trajectory of the proposed ultrasonic motor.

4 Conclusion

A two-DOF ultrasonic motor was presented to drive the hyper redundant manipulator, and only single proposed ultrasonic motor was needed. On account of this proposed ultrasonic motor, the number of the driving motor is significantly decreased, thus the structure of manipulator is simplified and the cost of the driving device is decreased. The driving mechanism of this two-DOF ultrasonic motor was illustrated. The proposed ultrasonic motor can work under two working modes to enable the ultrasonic motor travel along the link of the manipulator and rotate special joint of the manipulator. The ultrasonic motor utilized the longitudinal and bending-bending vibrations to obtain the two-DOF motion. The longitudinal vibration was used to drive the proposed motor runs along the manipulator, and bending-bending vibrations were utilized to rotate the joint of the manipulator. The finite element method was used to design and optimize the configuration of the proposed ultrasonic motor, desired driving trajectories were achieved. The proposed concept provides a new thought for optimizing the hyper-redundant manipulator and provides a basis for the design of two-DOF ultrasonic motor.

Acknowledgments. This work was supported in part by the National Natural Science Foundation of China (No. 52005398), in part by China Postdoctoral Science Foundation (No. 2019M663776), in part by Shaanxi Natural Science Basic Research Program (No. 2019JQ-805), in part by Shaanxi Education Department General Special Scientific Research Plan (No. 20JK0774).

References

1. Liljebäck, P., Pettersen, K.Y., Stavdahl, Ø.: Snake Robots: Modelling, Mechatronics, and Control. Springer, London (2012). https://doi.org/10.1007/978-1-4471-2996-7
2. Da Graça Marcos, M., Machado, J.A.T., Azevedo-Perdicoúlis, T.P.: An evolutionary approach for the motion planning of redundant and hyper-redundant manipulators. Nonlinear Dyn. **60**(1–2), 115–129 (2010)
3. Shukla, A., Singla, E., Wahi, P.: A direct variational method for planning monotonically optimal paths for redundant manipulators in constrained workspaces. Robot. Auton. Syst. **61**(2), 209–220 (2013)
4. Gallardo, J., Orozco, H., Rico, J.M., González-Galván, E.J.: A new spatial hyper-redundant manipulator. Robot. Comput. Integr. Manuf. **25**(4–5), 703–708 (2009)
5. Mu, Z., Yuan, H., Xu, W., Liu, T., Liang, B.: A segmented geometry method for kinematics and configuration planning of spatial hyper-redundant manipulators. IEEE Trans. Syst. Man Cybernet. Syst. **50**(5), 1746–1756 (2018)
6. Xidias, E.K.: Time-optimal trajectory planning for hyper-redundant manipulators in 3D workspaces. Robot. Comput. Integr. Manuf. **50**, 286–298 (2018)
7. Machado, J.T., Lopes, A.M.: A fractional perspective on the trajectory control of redundant and hyper-redundant robot manipulators. Appl. Math. Model. **46**, 716–726 (2017)
8. Xu, W., Mu, Z., Liu, T., Liang, B.: A modified modal method for solving the mission-oriented inverse kinematics of hyper-redundant space manipulators for on-orbit servicing. Acta Astronaut. **139**, 54–66 (2017)
9. Tappe, S., Dörbaum, M., Kotlarski, J.: Kinematics and dynamics identification of a hyper-redundant, electromagnetically actuated manipulator. In: IEEE International Conference on Advanced Intelligent Mechatronics, pp. 601–607 (2016)
10. Liang, D., Song, Y., Sun, T., Jin, X.: Rigid-flexible coupling dynamic modeling and investigation of a redundantly actuated parallel manipulator with multiple actuation modes. J. Sound Vib. **403**, 129–151 (2017)
11. Xu, W., Liu, T., Li, Y.: Kinematics, dynamics, and control of a cable-driven hyper-redundant manipulator. IEEE/ASME Trans. Mechatron. **23**(4), 1693–1704 (2018)
12. Karnjanaparichat, T., Pongvuthithum, R.: Adaptive tracking control of multi-link robots actuated by pneumatic muscles with additive disturbances. Robotica **35**(11), 2139–2156 (2017)

Non-metal Piezoelectric Motor Utilizing Langevin-Type Alumina/PZT Transducer Working in Orthogonal Bending Modes

Jiang Wu[1], Jianye Niu[2], Yixiang Liu[1], Xuewen Rong[1], Rui Song[1](✉), and Yibin Li[1]

[1] School of Control Science and Technology, Shandong University, Jinan 250061, China
rsong@sdu.edu.cn

[2] School of Mechanical Engineering, Yanshan University, Qinhuangdao 066004, China

Abstract. To apply ultrasonic motors (USMs) to the chemical industry, in this study, alumina is employed as vibrating bodies of transducers as it offers not only high chemical resistance but also a possibility to generate high output torques and power of motors due to the high Young's modulus. First, a Langevin-type transducer was constructed by clamping several annular lead-zirconate-titanate (PZT) disks between two rod-shaped alumina vibrating bodies, and a traveling wave was excited to drive the rotor by a superposition of two orthogonal standing waves in bending modes. Subsequently, the performance of the alumina/PZT motor was assessed and compared to those of the metal/PZT motors with identical structures when the same voltage was applied. As predicted, the alumina/PZT motor provides higher performance than the metal/PZT ones. Since the alumina/PZT transducer has relatively high strain in its PZT disks owing to the high Young's modulus of alumina, it exhibits a relatively high force factor, leading to a high output torque of the alumina/PZT motor. In the meantime, the transducer stores higher vibration energy due to the larger force factor; this enables the alumina/PZT motor to exhibit a higher output power. Besides, a higher rotation speed is obtained with this motor because the alumina/PZT transducer has a larger vibration velocity on the end surface. These results indicate the high applicability of alumina to high-power USMs as their vibrating bodies.

Keywords: Ultrasonic motor · Langevin-type bending transducer · Non-metal · Alumina · High Young's modulus

1 Introduction

Exhibiting high torques at low rotation speeds, good transient responses, and simple structures [1, 2], ultrasonic motors (USMs) have been extensively investigated in the past three decades and practically applied to some special fields [1]. Recently, there has existed a demand on USMs as key actuators for robotic application in the chemical industry [3, 4]. However, current USMs are inapplicable as most of their transducers, which are core components of USMs [1, 2], are made of metals lacking chemical resistance, i.e.

X.-J. Liu et al. (Eds.): ICIRA 2021, LNAI 13014, pp. 342–352, 2021.
https://doi.org/10.1007/978-3-030-89098-8_33

stainless steel and aluminum alloy [5, 6]. Rigid ceramics may be promising candidate materials to substitute the metal parts of transducers because they have strong chemical resistance as well as high rigidity, which probably results in high output torques and power of motors [7].

In some previous reports [1, 8, 9], thin ceramic layers were coated (or bonded) on metal/lead-zirconate-titanate (PZT) transducers to improve frictional and/or abrasional properties of motors. However, their vibration characteristics have been rarely discussed as they are assumed to be unaffected by the coating considering that the ceramic layers are very thin [1]. To date, only a few studies [10–12] have employed rigid ceramics as vibrating bodies of transducers. Aoyagi et al. attempted to deposit a thin PZT layer on a ceramic plate to form a multi-degrees-of-freedom motor [10]. Since it was difficult to obtain a sufficiently thick PZT layer, the vibration velocity achievable with the transducer was too small to drive the rotor. Koyama et al. designed a traveling-wave linear motor with twin ceramic ridges [11]. It provided high motion accuracy at small preloads but insufficient thrust because of the large undesirable vibration components, which were unavoidably generated on the twin ridges. In our recent report [12], several ring-shape transducers (this structure has been heavily studied in conventional metal/PZT USMs) were constructed by bonding PZT disks to rigid ceramic vibrating bodies with high-strength epoxy resin. They easily fractured even at low voltage possibly because the stresses on the boundaries between vibrating bodies and PZT disks exceeded the bond strength of the adhesive. Observably, the limited vibration properties of the transducers cause the unsatisfactory performance of the aforementioned motors. It is probably worthwhile to test rigid-ceramic/PZT motors with other types of transducers.

In this study, we first build a Langevin-type alumina/PZT transducer operated in orthogonal bending modes and investigate its basic vibration characteristics. Here, the Langevin-type bending transducer, originally developed by Kurosawa et al. [13], is chosen as it not only achieves firm connection between PZT disks and vibrating bodies by utilizing a bolt instead of epoxy resin but also works in powerful vibration modes for USMs [1, 2, 13]. Besides, alumina is employed for its relatively low ultrasonic attenuation among the commonly-used rigid ceramics [14, 15]. Then, we measure the load characteristics of the alumina/PZT motor and make comparison with the stainless-steel/ and aluminum/PZT motors. Note that, as the first report regarding the Langevin-type alumina/PZT motor, we mainly focus on its load characteristics as they provide fundamental information for discussing performance of USMs [1]. Other characteristics, such as motion accuracy, will be explored in our further studies.

2 Alumina/PZT Transducer

The configuration of the tested transducer is detailedly illustrated in Fig. 1. It incorporates two rod-shaped vibrating bodies made of alumina (A9951, NTK Ceratec, Sendai, Japan) and six annular PZT disks (C213, Fuji Ceramics, Fujinomiya, Japan). The vibrating body has a 35 mm outer diameter and a 48 mm length. A stepped thorough hole is created in the vibrating body. Its first and second steps have inner diameters of 18 and 10 mm, respectively, but the same lengths of 24 mm. The PZT disks 4 mm in thickness, and 35 and 18 mm in outer and inner diameters, respectively, are clamped between two vibrating bodies with a stainless-steel bolt. The PZT disk is divided into two parts polarized in opposite directions. A polar coordinate (r, θ, and z axes) is established on the middle surface of the transducer. The axes corresponding to $\theta = 0$ and 90° are defined as the x and y axes, respectively. The rotor comprises a flywheel and a contact rim, both of which are made of stainless steel. The flywheel has a thickness of 18 mm, and outer and inner diameters of 80 and 35 mm, respectively. The contact rim 2 mm in thickness, and 35 and 33 mm respectively in outer and inner diameters is connected to the flywheel with a screw. The rotor is in contact with the outer edge of the end surface of the transducer.

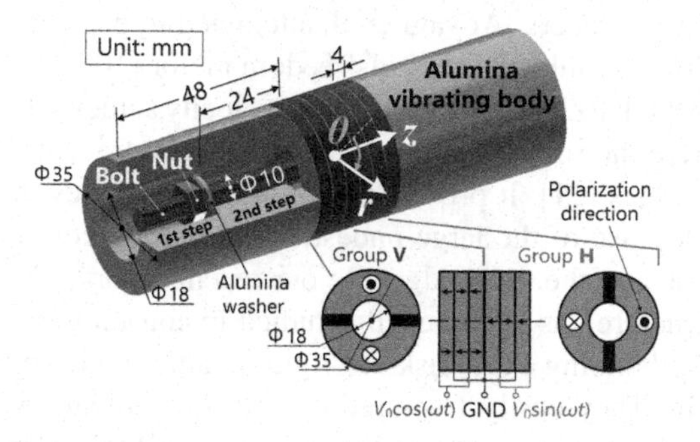

Fig. 1. Configurations of the alumina/PZT transducer.

3 Vibration Characteristics of Transducers

3.1 Vibration Velocity Distribution

First, the vibration velocity distribution of the alumina/PZT transducer was investigated. As shown in Fig. 2, the θ- and z-axis vibration velocities were measured with an in-plane vibrometer (IPV 100, Polytec, Waldbronn, Germany). Their amplitudes and phases were recorded with a lock-in voltmeter (5560, NF Electronic Instruments). The r-axis vibration velocities were measured with an out-of-plane vibrometer (CLV1000, Polytec). The current flowing into the transducer was measured with a current probe (P6021, Tektronix).

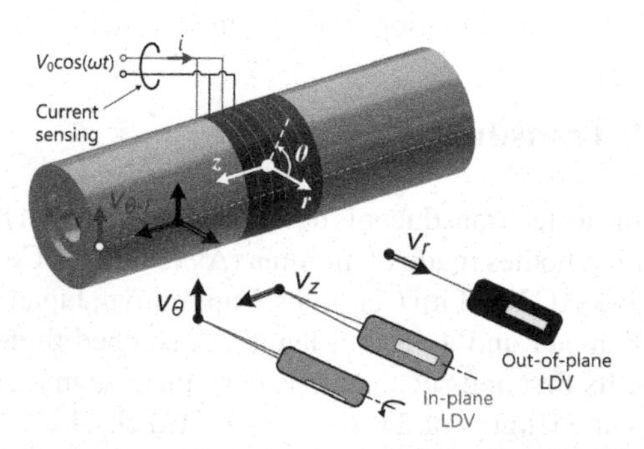

Fig. 2. Experimental setup for measuring vibration velocities in different directions and currents.

Figures 3(a), (b), and (c) respectively illustrate the circumferential distributions of the z-, θ-, and r-axis vibration velocities on the end surface. The z-axis vibration velocities had small variations in amplitude. Whereas their phases linearly decreased in the θ_0 ranges from $-180°$ to $-90°$ and from $-90°$ to $180°$. Clearly, a traveling wave was successfully excited along the circumferential direction. The average vibration velocities in the θ- and r-axes were 425 and 443 mm/s, both of which exceeded twice the value in the z axis. The maximal and minimal θ-axis vibration velocities were 453 and 401 mm/s, respectively. Their ratio, referred to as the standing wave ratio [18], is 1.1, implying that the standing wave component is satisfactorily small [1].

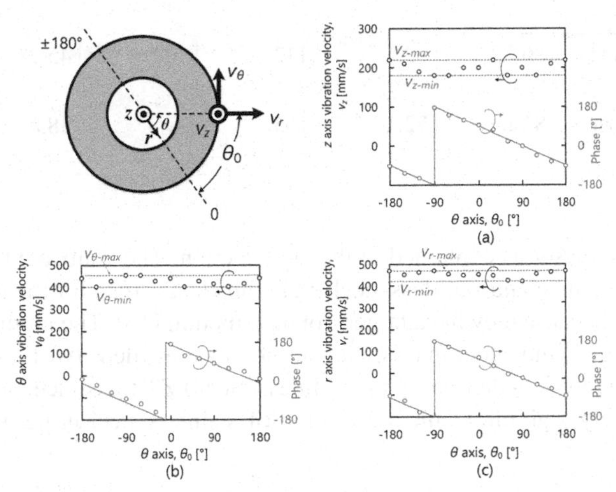

Fig. 3. Circumferential distributions of (a) z-, (b) θ-, and (c) r-axis vibration velocities of the alumina/PZT transducer. Here, $v_{i\text{-}max}$ and $v_{i\text{-}min}$ ($i = z$, θ, and r) represent the maximal and minimal i-axis vibration velocities, respectively. The curves and dots are simulated and measured results, respectively.

3.2 Equivalent Circuit Parameters

Subsequently, the vibration characteristics of the transducers were discussed on the basis of the equivalent circuit model [3, 19, 20]. It consists of an electrical and a mechanical arm coupled with a force factor A, defined as the ratio of the output force generated by PZT disks to the applied voltage, or equivalently the ratio of the current i to the θ-axis vibration velocity on the end surface of the transducer $v_{\theta\text{-}l}$. Here, we focus on the vibration velocity along the θ axis as it is parallel to the rotation direction. The electrical arm has only one element, i.e. the clamped capacitance C_d. Whereas the mechanical arm contains three serially connected elements, i.e. the equivalent stiffness k_m, mass m_m, and damper γ_m.

Table 1 lists the equivalent circuit parameters of the transducers, which are calculated from their admittance characteristics measured with an impedance analyzer (4294A, Agilent). The alumina/PZT transducer exhibits a relatively high force factor among the

Table 1. Equivalent circuit parameters of tested transducers

Indicator [unit]	Ceramic		Metal			
	Alumina		Stainless steel		Aluminum	
	Group V	Group H	Group V	Group H	Group V	Group H
Clamped capacitance C_d [nF]	6.96	7.01	6.94	6.91	6.89	6.97
Force factor A [N/V]	0.451	0.438	0.351	0.336	0.234	0.242
Equivalent stiffness k_m [$\times 10^6$ N/m]	172	163	170	169	98.8	102
Equivalent mass m_m [$\times 10^{-3}$ kg]	56.2	52.5	112	108	45.2	47.4
Equivalent damper γ_m [Ns/m]	82.1	72.1	105	99.4	38.8	39.6

tested ones. The reason is explained in the next section. The equivalent stiffnesses are higher for the alumina/ and stainless-steel/PZT transducers than for the aluminum/PZT one dominantly because they have larger Young's moduli [13]. The alumina/PZT transducer has moderate equivalent mass owing to the moderate density of alumina [7]. The relatively high equivalent damper of the stainless-steel/PZT transducer should originate from the relatively high ultrasonic attenuation of stainless steel in the high-amplitude region [15, 21].

3.3 Resonance Frequency, Elliptical Motion Shape, and Force Factor

Finally, the resonance frequencies corresponding to the 1st bending mode, the elliptical motion shapes, and the force factors of the transducers were explored via finite element analysis. As Fig. 4(a) plots, the resonance frequencies monotonically decrease with increasing length. The alumina/PZT transducer with the 48-mm-long vibrating bodies yields a resonance frequency of 8.7 kHz, relatively high compared to those of the stainless-steel/ and aluminum/PZT ones owing to its relatively high equivalent stiffness

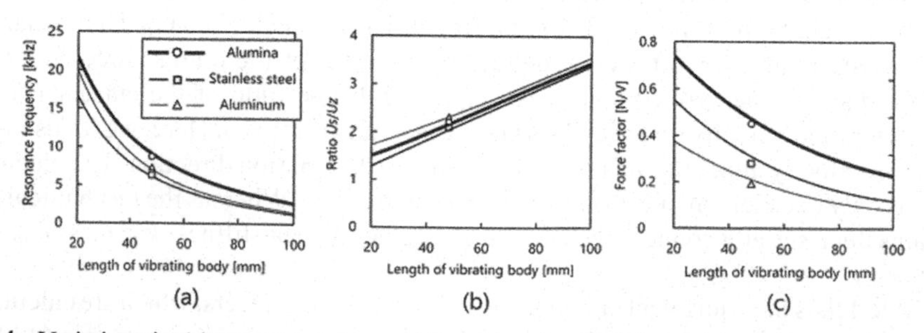

Fig. 4. Variations in (a) resonance frequencies, (b) ratios U_S/U_z, and (c) force factors against lengths of vibrating bodies of alumina/, stainless-steel/, and aluminum/PZT transducers.

and moderate equivalent mass [22]. Figure 4(b) shows that the ratios U_S/U_z linearly increase as the transducers become longer. In addition, there exists slight difference in elliptical motion shapes among these transducers. The force factor of the Langevin-type bending transducer can be estimated with the following equation [23]:

$$A = e_{33} \cdot \frac{\left(D_{out-p}^3 - D_{in-p}^3\right)}{3t_p} \cdot \left(\left.\frac{\partial \xi}{\partial z}\right|_{z=-t_p} - \left.\frac{\partial \xi}{\partial z}\right|_{z=0} \right), \tag{1}$$

where e_{33} is a piezoelectric constant; $D_{in\text{-}p}$, and $D_{out\text{-}p}$ represent inner and outer diameters of PZT disks, respectively; t_p is the entire thickness of one group of PZT disks (=12 mm); and ξ denotes the normalized θ-axis vibration velocity, of which the distributions as well as the values of $\partial\xi/\partial z$ at $z = -12$ mm of the tested transducers are shown in Fig. 5. Since the distributions are bilaterally symmetrical, the values of $\partial\xi/\partial z$ equal zero at $z = 0$ [21]. Since alumina has a higher Young's modulus [24], the alumina/PZT transducer exhibits higher strain (represented by $\partial\xi/\partial z$) in its PZT disks and consequently a higher force factor than the metal/PZT ones [see Fig. 4(c)].

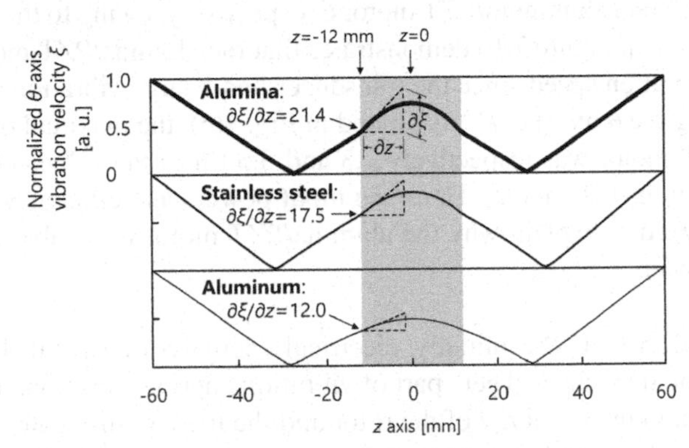

Fig. 5. Horizontal distributions of normalized θ-axis vibration velocities of alumina/, stainless-steel/, and aluminum/PZT transducers with 48-mm-long vibrating bodies.

4 Motor Performance

4.1 Load Characteristics of Alumina/PZT Motor

First, the load characteristics of the alumina/PZT motor was measured. Figure 6(a), (b), and (c) respectively plot how the rotation speed, output power, and efficiency depend on the output torque at different preloads. The applied voltage had a zero-to-peak value of 150 V. At a low preload of 110 N, the rotation speed reached 25 rad/s when no load was applied, and exhibited a sharp reduction with increasing torque. When a moderate preload of 260 N was applied, the maximal output power and efficiency reached 8.0 W and 34.3%, respectively. When the preload was enhanced to 390 N, the maximal torque reached 1.5 Nm, but the maximal power and efficiency respectively decreased to 3.6 W and 13.1% because of the high frictional loss at a large preload.

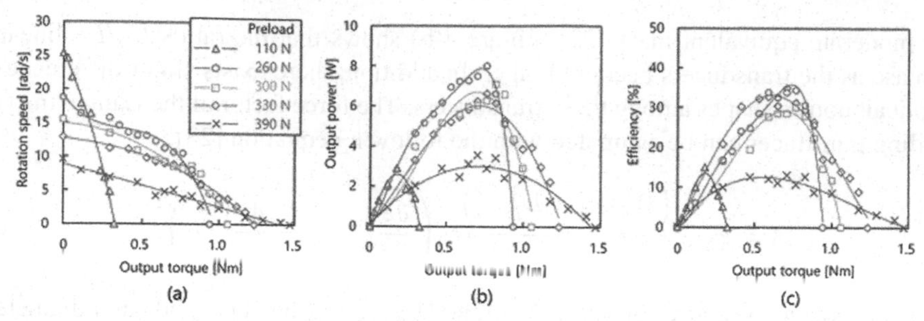

Fig. 6. Load characteristics of the alumina/PZT motor: (a) Rotation speed, (b) output power, and (c) efficiency as functions of output torque.

4.2 Comparison with Metal/PZT Motors with Identical Structures

Subsequently, the aforementioned load characteristics were compared to those of the stainless-steel/ and aluminum/PZT motors at the same voltage. As shown in Fig. 7(a), the maximal output torque of the alumina/PZT motor was 1.3 and 1.7 times the values of the stainless-steel/ and aluminum/PZT motors, respectively, owing to the relatively high force factor [2, 7, 17]. Figure 7(b) demonstrates that the alumina/PZT motor provided a higher no-load rotation speed since the transducer had a large vibration velocity on the end surface along the θ axis [2, 7]. As plotted in Fig. 7(c), the maximal output power of the alumina/PZT motor was respectively 2.3 and 3.5 times the values of the stainless-steel/ and aluminum/PZT motors. Here, the input powers and efficiencies of the tested motors are analyzed to explain why the alumina/PZT motor is capable of exhibiting a higher output power.

(1) Input power. In an USM, initially, electrical energy is transformed into vibration energy by the transducer. Then, part of vibration energy is frictionally transformed into rotational kinetic energy of the rotor and the load, or dissipates on the contact surface or in the transducer. Thus, vibration energy is assumed to the limit input power of an USM. The vibration energy of a transducer, considered as the elastic energy, can be estimated with the following equation [25]:

$$E_R = \frac{1}{2}k_m\left(\frac{v_{\theta-l}}{\omega}\right)^2 = \frac{A^2Q_H^2V_0^2}{2k_m}, \tag{2}$$

where k_m denotes the equivalent stiffness, Q_H represents the mechanical quality factor measured at an operating vibration velocity $v_{\theta\text{-}l}$. The values of Q_H can be found in Fig. 8, indicating the measured dependences of Q_H on $v_{\theta\text{-}l}$ of the tested transducers. Figure 7(c) demonstrates that the maximal input powers of the motors are proportional to the vibration energy stored in their transducers. These results validate our assumption. Besides, according to Eq. 2, the alumina/PZT transducer stores higher vibration energy dominantly due to its higher force factor.

(2) Efficiency Table 2 lists the mechanical and frictional losses when the tested motors exhibit maximal output powers. The mechanical loss is expressed as [15]

$$P_{ml} = \frac{\sqrt{k_m m_m}}{2Q_{PM}} \cdot v_{PM-\theta-l}^2, \tag{3}$$

where $v_{PM\text{-}\theta\text{-}l}$ is the vibration velocity when the maximal output power is obtained, and Q_{PM} corresponds to $v_{PM\text{-}\theta\text{-}l}$. The frictional loss equals [15]

$$P_{fl} = P_{in} - P_{ml} - P_{out}. \tag{4}$$

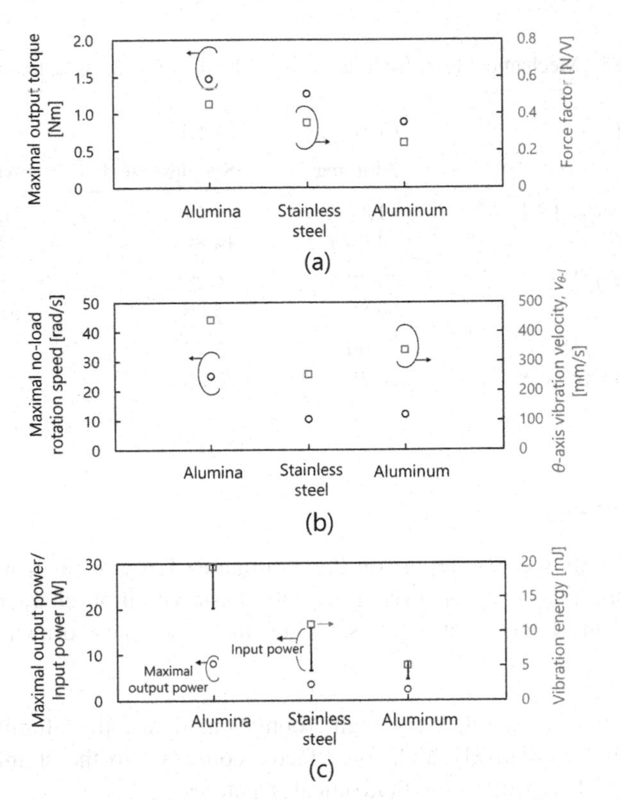

Fig. 7. Comparison in (a) maximal torques, (b) maximal no-load rotation speeds, and (c) input powers and maximal output powers among the tested motors. The bars shown in (c) represent the ranges of input powers obtained in our experiments.

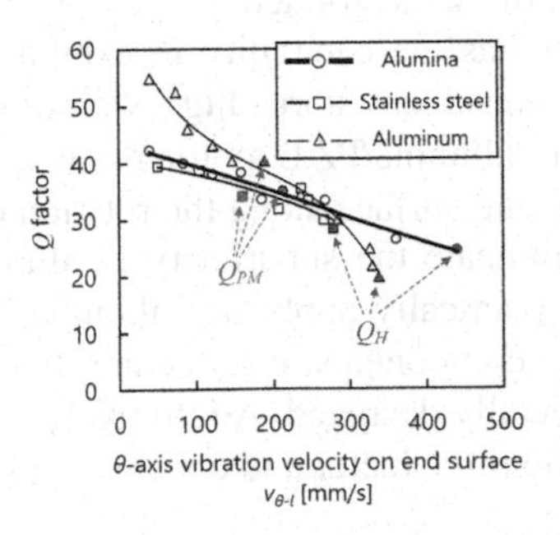

Fig. 8. Q factors as functions of vibration velocities on end surfaces of transducers.

In general, the efficiency of an USM is dominantly limited by its frictional loss [1, 2]. Table 2 demonstrates that the stainless-steel/ and aluminum/PZT motors do not show marked differences in ratios P_{fl}/P_{in} (and also efficiencies) compared to the alumina/PZT motor because they employ, as mentioned above, alumina plates as frictional materials. These results allow us to infer that the relatively high output power of our alumina/PZT motor is dominantly caused by its high vibration energy rather than efficiency.

Table 2. Mechanical loss, frictional loss, and output power of tested motors

Indicator [unit]	Ceramic	Metal	
	Alumina	Stainless steel	Aluminum
Mechanical loss P_{ml} [W] $P_{ml}/P_{in} \times 100\%$	1.14 4.90%	1.69 14.8%	0.288 3.63%
Frictional loss P_{fl} [W] $P_{fl}/P_{in} \times 100\%$	14.20 60.8%	6.28 55.0%	5.38 67.8%
Output power P_{out} [W] $P_{out}/P_{in} \times 100\%$ (efficiency)	8.00 34.3%	3.45 30.2%	2.26 28.5%

5 Conclusions

This study presents the first report on the alumina/PZT motor utilizing a Langevin-type bending transducer. After investigating the basic vibration characteristics of the transducer and the load characteristics of the motor, we have drawn the following conclusions:

1. Since alumina has a relatively high Young's modulus, the alumina/PZT transducer exhibits a relatively high force factor compared to the stainless-steel/ and aluminum/PZT transducers with identical structures.
2. The alumina/PZT motor yields maximal output torque and power of 1.5 Nm and 8 W, respectively, at the applied voltage of 150 V. Owing to the relatively high force factor, the alumina/PZT motor exhibits relatively high output torque compared to the metal/PZT motors at the same voltage.
3. The alumina/PZT motor has the capability to exhibit considerable output power dominantly because its transducer stores high vibration energy. A higher rotation speed is obtained on the alumina/PZT motor because its transducer has a larger vibration velocity on the end surface along the rotation direction.

   These conclusions indicate the superiority of alumina as vibrating bodies of high-power USMs. To practically apply our alumina/PZT motor to the chemical industry, its surrounding components, e.g. feeding lines and bearings, need to be carefully selected or specially designed. Additionally, the transducer length should be reduced to enhance the force factor and enable the motor to work at an inaudible frequency.

References

1. Ueha, S., Tomikawa, Y., Kurosawa, M., Nakamura, K.: Ultrasonic Motors—Theory and Applications. Oxford University Press, New York (1993)
2. Nakamura, K., Kurosawa, M., Ueha, S.: Characteristics of a hybrid transducer-type ultrasonic motor. IEEE Trans. Ultrason. Ferroelect. Freq. Control **38**(3), 188–193 (1991)
3. Zheng, L., Liu, S., Wang, S.: Current situation and future of Chinese industrial robot development. Int. J. Mech. Eng. Rob. Res. **5**(4), 295–300 (2016)
4. Nishioka, H., Takeuchi, A.: The development of high technology industry in Japan. In: The Development of High Technology Industries—An International Survey, 1st edn. Routledge, Taylor & Francis Group, New York (2018)
5. Waard, C.D., Milliams, D.E.: Carbonic acid corrosion of steel. Corrosion **31**(5), 177–181 (1975)
6. Ghali, E.: Corrosion Resistance of Aluminum and Magnesium Alloys: Understanding, Performance, and Testing. Wiley, Hoboken (2010)
7. Wu, J., Mizuno, Y., Nakamura, K.: Piezoelectric motor utilizing an alumina/PZT transducer. IEEE Trans. Industr. Electron. **67**(8), 6762–6772 (2020)
8. Qiu, W., Mizuno, Y., Nakamura, K.: Tribological performance of ceramics in lubricated ultrasonic motors. Wear **352**(353), 188–195 (2016)
9. Rehbein, P., Wallaschek, J.: Friction and wear behavior of polymer/steel and alumina/alumina under high-frequency fretting conditions. Wear **216**(2), 97–105 (1998)
10. Aoyagi, M., Beeby, S.P., White, N.M.: A novel multi-degree-of-freedom thick-film ultrasonic motor. IEEE Trans. Ultrason. Ferroelect. Freq. Control **49**(2), 151–158 (2002)
11. Koyama, O., Koyama, D., Nakamura, K., Ueha, S.: Ultrasonic linear motor using traveling vibration on fine ceramic twin ridge. Acoust. Sci. Techno. **29**(1), 95–98 (2008)
12. Wu, J., Mizuno, Y., Nakamura, K.: A rotary ultrasonic motor operating in torsional/bending modes with high torque density and high power density. IEEE Trans. Industr. Electron. **68**(7), 6109–6120 (2021)
13. Kurosawa, M., Nakamura, K., Okamoto, T., Ueha, S.: An ultrasonic motor using bending vibrations of a short cylinder. IEEE Trans. Ultrason. Ferroelect. Freq. Control **36**(5), 517–521 (1989)
14. Liu, D., Turner, J.A.: Numerical analysis of longitudinal ultrasonic attenuation in sintered materials using a simplified two-phase model. J. Acoust. Soc. Am. **141**(2), 1226–1237 (2017)
15. Wang, H., Ritter, T., Cao, W., Shung, K.K.: High frequency properties of passive materials for ultrasonic transducers. IEEE Trans. Ultrason. Ferroelect. Freq. Control **48**(1), 78–84 (2001)
16. Nakamura, K., Kurosawa, M., Kurebayashi, H., Ueha, S.: An estimation of load characteristics of an ultrasonic motor by measuring transient responses. IEEE Trans. Ultrason. Ferroelect. Freq. Control **38**(5), 481–485 (1991)
17. Wu, J., Mizuno, Y., Nakamura, K.: Polymer-based ultrasonic motors utilizing high-order vibration modes. IEEE/ASME Trans. Mechatron. **23**(2), 788–799 (2018)
18. Kuribayashi, M., Ueha, S., Mori, E.: Excitation conditions of flexural traveling waves for a reversible ultrasonic linear motor. J. Acoust. Soc. Am. **77**(4), 1431–1435 (1985)
19. Wu, J., Mizuno, Y., Tabaru, M., Nakamura, K.: Ultrasonic motors with polymer-based vibrators. IEEE Trans. Ultrason. Ferroelect. Freq. Control **62**(12), 2169–2178 (2015)
20. Shi, W., Zhao, H., Ma, J., Yao, Y.: An optimum-frequency tracking scheme for ultrasonic motor. IEEE Trans. Ind. Electron. **64**(6), 4413–4422 (2017)
21. Nakamura, K., Kakihara, K., Kawakami, M., Ueha, S.: Measuring vibration characteristics at large amplitude region of materials for high power ultrasonic vibration system. Ultrasonics **38**(1/2), 122–126 (2000)
22. Graff, K.F.: Wave Motion in Elastic Solids. Dover, New York (1991)

23. Koike, Y., Tamura, T., Ueha, S.: Derivation of a force factor equation for a Langevin-type flexural mode transducer. Jpn. J. Appl. Phys. **35**(5B), 3274–3280 (1996)
24. Morita, T.: Piezoelectric Phenomena. Morikita Publication, Tokyo (2017), Ch. 2
25. Timoshenko, S.P.: Vibration Problems in Engineering, 2nd edn. D. Van Nostrand Company, Inc., New York (1956). Ch. 4
26. Kurosawa, M., Kodaira, O., Tsuchitoi, Y., Higuchi, T.: Transducer for high speed and large thrust ultrasonic linear motor using two sandwich-type vibrators. IEEE Trans. Ultrason. Ferroelect. Freq. Control **45**(5), 1185–1195 (1998)
27. Yun, C.-H., Ishii, T., Nakamura, K., Ueha, S., Akashi, K.: A high power ultrasonic linear motor using a longitudinal and bending hybrid bolt-clamped Langevin type transducer. Jpn. J. Appl. Phys. **40**(5B), 3773–3776 (2001)
28. Kim, W.-S., Yun, C.-H., Lee, S.-K.: Nano positioning of a high power ultrasonic linear motor. Jpn. J. Appl. Phys. **47**(7R), 5687–5692 (2008)
29. Satonobu, J., Torii, N., Nakamura, K., Ueha, S.: Construction of mega-torque hybrid transducer type ultrasonic motor. Jpn. J. Appl. Phys. **35**(9B), 5038–5041 (1996)

MIMO Data-Driven Feedforward Tuning for the Ultra-precision Wafer Stage Using Rational Basis Functions

Weicai Huang[1,2], Kaiming Yang[1,2](✉), and Yu Zhu[1,2]

[1] State Key Lab of Tribology, Department of Mechanical Engineering, Tsinghua University, Beijing, People's Republic of China
yangkm@tsinghua.edu.cn

[2] Beijing Key Lab of Precision/Ultra-precision Manufacturing Equipments and Control, Tsinghua University, Beijing, People's Republic of China

Abstract. In this paper, a novel multi-input-multi-output (MIMO) data-driven feedforward tuning approach associated with rational basis functions (RBFs) is proposed for the ultra-precision wafer stage. Specifically, the feedforward controllers for all degrees of freedom (DOFs) of the wafer stage are all parameterized using RBFs and the global optimum is pursued. First, a novel parameter updating law for each DOF is proposed to obtain the global optimum iteratively. Then, the variables required during each iteration are estimated through closed-loop experiments. Moreover, the parameter estimation accuracy is improved through the minimization of the covariance matrixes. Finally, experiments are conducted to validate the effectiveness of the proposed approach.

Keywords: Feedforward control · MIMO systems · Data-driven control · Wafer stage · Rational basis functions

1 Introduction

Feedforward control is widely-used in precision motion systems to compensate for the error induced by the known disturbances or references [1,2]. Traditionally, feedforward control is designed based on the parametric model of the plant, and the achievable performance is limited by the modelling error [1]. By contrast, the feedforward control is designed with the measured data without modelling the plant in data-driven control [2]. As the limitation due to the modelling process is eliminated, the achievable performance of data-driven control is generally better than traditional methods.

Most pre-existing MIMO feedforward tuning approaches are developed based on their related approaches for single-input-single-output (SISO) systems [3]. However, two challenges are faced during the extension from SISO to MIMO systems: 1) the higher complexity brought in by the higher dimension; 2) the lack of commutativity between MIMO transfer function matrixes. Only a few types of data-driven methods are successfully extended towards MIMO systems,

X.-J. Liu et al. (Eds.): ICIRA 2021, LNAI 13014, pp. 353–363, 2021.
https://doi.org/10.1007/978-3-030-89098-8_34

examples include iterative learning control (ILC) and polynomial basis functions (PBFs). In ILC, the feedforward signal is updated based on the feedforward signals and tracking errors of the previous iterations [4,5]. Superior performance can be achieved with ILC for repeating tasks at the cost of extrapolation properties. By contrast, the feedforward controller is linearly parameterized and the optimal feedforward controller is pursued when PBFs are applied [3,6]. As the corresponding parameter optimization problem is convex, it is not hard to achieve global optimization with gradient-based algorithms. However, the feedforward controller parameterized using PBFs is polynomial transfer function in nature, the approximation ability to the plant inversion of general rational systems is limited, especially for MIMO systems.

In order to enhance the achievable performance based on PBFs, rational basis functions (RBFs) are proposed [7]. The numerator and denominator of the feedforward controller are both linearly parameterized while RBFs are applied. The achievable performance is improved as the ability to approximate the inverse plant of general rational systems is greatly promoted. However, nonconvex optimization is suffered when the reference-induced error is optimized directly and global optimization cannot be guaranteed [7]. In [8], the nonconvex optimization problem is converted into convex optimization problem and the global optimization is achieved using least-square method. Moreover, instrumental variable is exploited to improve the parameter estimation accuracy of the global optimum in [9]. However, RBFs haven't been applied to MIMO systems, as the commutativity is inevitable in the pre-existing approaches associated with RBFs.

In this paper, a novel MIMO data-driven feedforward tuning approach is proposed for the ultraprecision wafer stage. In order to achieve high performance and extrapolation properties simultaneously, the feedforward controllers are all parameterized using RBFs. Then, the MIMO feedforward tuning problem for the wafer stage is divided into a list of convex optimization problems for each degree of freedom (DOF), and least-square method is exploited to solve these problems in each iteration. Closed-loop experiments are designed to estimate the intermediate variables required during each iteration. A filter is introduced into the least-square method which provides extra freedom for the optimization of parameter estimation accuracy. The optimal filter is determined such that the covariance matrixes are minimized in each iteration. Finally, the reference tracking performance of the proposed approach is tested through experiments.

This paper is organized as follows. In Sect. 2, the problem to be solved in this paper is defined. In Sect. 3, the proposed approach is introduced in detail. In Sect. 4, experiments are conducted to test the performances of the proposed approach. In Sect. 5, the overall conclusions of this paper are given.

2 Problem Fomulation

The wafer stage is the critical component of the lithography machine and is required to achieve nanoscale motion accuracy. Generally, the wafer stage is controlled in all 6 DOFs, including three translations (x, y, z) and three rotations

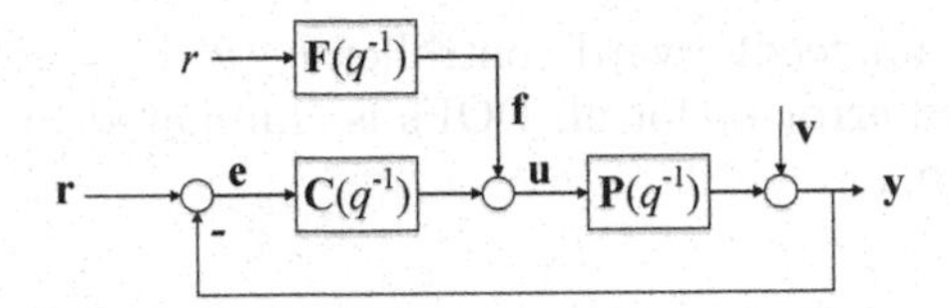

Fig. 1. The MIMO control scheme of the precision wafer stage.

(t_x, t_y, t_z). For the ease of notation, the numbers 1–6 are exploited to replace the symbols x-t_z. The control scheme of the wafer stage is shown in Fig. 1. The plant $\mathbf{P}(q^{-1})$ and feedback controller $\mathbf{C}(q^{-1})$ are both 6×6 transfer function matrixes. During the exposure of the lithography machine, the wafer stage moves in the scanning direction (y direction) and keeps positioning in all the non-scanning directions. Therefore, the reference $\mathbf{r}$ in Fig. 1 is given by $\mathbf{r} = [0, r, 0, 0, 0, 0]^T$, where r is the reference to be tracked in the scanning direction. A feedforward controller with input r is designed for each DOF, as shown in Fig. 1. Therefore, the feedforward signal $\mathbf{f}$ is given by $\mathbf{f} = \left[F^1(\theta^1)r, F^2(\theta^2)r, \cdots, F^6(\theta^6)r\right]^T$, where $F^i(\theta^i)(i=1,2,\cdots,6)$ refers to the feedforward controller of the ith DOF, and θ^i is the parameter to be optimized in $F^i(\theta^i)$. In this paper, the feedforward controllers for all DOFs are all parameterized using RBFs, i.e., $F^i(\theta^i) = A^i(\theta^{Ai})/B^i(\theta^{Bi})$ and

$$\begin{cases} A^i\left(\theta^{Ai}\right) = \sum_{j=1}^{n_{ai}} \Psi_j^{Ai}(q^{-1})\theta^{Ai}(j) \\ B^i\left(\theta^{Bi}\right) = 1 + \sum_{j=1}^{n_{bi}} \Psi_j^{Bi}(q^{-1})\theta^{Bi}(j) \end{cases} \tag{1}$$

where the parameter is given by $\theta^i = [(\theta^{Ai})^T, (\theta^{Bi})^T]^T$ with $\theta^{Ai} = [\theta^{Ai}(1), \theta^{Ai}(2), \cdots, \theta^{Ai}(n_{ai})]^T$ and $\theta^{Bi} = [\theta^{Bi}(1), \theta^{Bi}(2), \cdots, \theta^{Bi}(n_{bi})]^T$. And the basis function is given by $\Psi^i = [\Psi^{Ai}, \Psi^{Bi}]$ with $\Psi^{Ai} = [\Psi_1^{Ai}, \Psi_2^{Ai}, \cdots, \Psi_{n_{ai}}^{Ai}]^T$ and $\Psi^{Bi} = [\Psi_1^{Bi}, \Psi_2^{Bi}, \cdots, \Psi_{n_{bi}}^{Bi}]^T$.

In Fig. 1, the reference tracking error vector $\mathbf{e}$ is given by

$$\mathbf{e} = \mathbf{Sr} - \mathbf{Tf} - \mathbf{Sv} \tag{2}$$

where $\mathbf{S} = (\mathbf{I} + \mathbf{PC})^{-1}$ is the sensitivity function and $\mathbf{T} = (\mathbf{I} + \mathbf{PC})^{-1}\mathbf{P}$ is the process sensitivity function, and $\mathbf{v}$ is the measurement noise. Obviously, the reference tracking error is composed of the reference-induced component $\mathbf{e}_r = \mathbf{Sr} - \mathbf{Tf}$ and the noise-induced component $\mathbf{e}_v = \mathbf{Sv}$. Similar to [9], it is assumed that the feedback controller $\mathbf{C}(q^{-1})$ is designed such that the $\mathbf{e}_v$ is normally distributed white noise with zero mean and variance $\sigma_i^2(i = 1, 2, \cdots, 6)$ in each DOF. Therefore, it is defined that $\mathbf{e}_v = [w^1, w^2, \cdots, w^6]^T$ with $w^i(i = 1, 2, \cdots, 6)$ a normally distributed white noise with zero mean and variance $\sigma_i^2(i = 1, 2, \cdots, 6)$.

With the above control scheme and feedforward parameterization of the wafer stage, the aim of this paper is to accurately find the global optimal parameter

$\theta^i_{opt}(i = 1, 2, \cdots, 6)$ for feedforward controller $F^i(\theta^i)(i = 1, 2, \cdots, 6)$ such that the reference-induced error $\mathbf{e}_r$ for all DOFs is eliminated in the presence of the noise-induced error $\mathbf{e}_v$.

3 The Proposed Parameter Tuning Algorithm

3.1 Global Optimization Algorithm for the Feedforward Controllers

It can be derived from (2) that the global optimum $\boldsymbol{\theta}_{opt}$ which makes $\mathbf{e}_r = 0$ needs to satisfy

$$e^i_r(\boldsymbol{\theta}_{opt}) = S_i r - \sum_{j=1}^{6} T_{ij} F^j(\theta^j_{opt}) r = S_i r - F^i\left(\theta^i_{opt}\right) T_{ii} r - \kappa^i\left(\boldsymbol{\theta}_{opt}\right) = 0 \quad (3)$$

where e^i_r is the reference-induced error in the ith DOF, $T_{ij} = \mathbf{T}(i, j)$ is the element of $\mathbf{T}$ in the ith rank and jth column. Similarly, $S_i = \mathbf{S}(i, 2)$ is the element of $\mathbf{S}$ in the ith rank and 2nd column. Besides, the optimal parameter matrix $\boldsymbol{\theta}_{opt} = [\theta^1_{opt}, \theta^2_{opt}, \cdots, \theta^6_{opt}]$ and the coupling term $\kappa^i\left(\boldsymbol{\theta}_{opt}\right) = \sum_{j=1, j\neq i}^{6} F^j(\theta^j_{opt}) T_{ij} r$.

View (3) as the equation of θ^i_{opt} and solve equation (3) directly using least-square method, yields

$$\theta^i_{opt} = \left[\left(\phi^i\left(\boldsymbol{\theta}_{opt}\right)\right)^T \phi^i\left(\boldsymbol{\theta}_{opt}\right)\right]^{-1} \left(\phi^i\left(\boldsymbol{\theta}_{opt}\right)\right)^T \left(S_i r - \kappa^i\left(\boldsymbol{\theta}_{opt}\right)\right) \quad (4)$$

where $\phi^i\left(\boldsymbol{\theta}_{opt}\right) = \left[\phi^{Ai}\left(\boldsymbol{\theta}_{opt}\right), \phi^{Bi}\left(\boldsymbol{\theta}_{opt}\right)\right]$ and

$$\begin{cases} \phi^{Ai}\left(\boldsymbol{\theta}_{opt}\right) = \left[\Psi^{Ai}_1(q^{-1}), \Psi^{Ai}_2(q^{-1}), \cdots, \Psi^{Ai}_{n_{ai}}(q^{-1})\right] T_{ii} r \\ \phi^{Bi}\left(\boldsymbol{\theta}_{opt}\right) = -\left[\Psi^{Bi}_1(q^{-1}), \Psi^{Bi}_2(q^{-1}), \cdots, \Psi^{Bi}_{n_{bi}}(q^{-1})\right] \left(S_i r - \kappa^i\left(\boldsymbol{\theta}_{opt}\right)\right) \end{cases} \quad (5)$$

It is observed from (4) and (5) that θ^i_{opt} is unavailable when $\theta^j_{opt}(j \neq i)$ is unknown. Hence, Eq. (4) cannot be used directly. In response to the above problem, the following parameter updating law is proposed for the ith DOF of the wafer stage

$$\theta^i_{k+1} = \theta^i_k + \left[\left(\phi^i\left(\boldsymbol{\theta}_k\right)\right)^T \phi^i\left(\boldsymbol{\theta}_k\right)\right]^{-1} \left(\phi^i\left(\boldsymbol{\theta}_k\right)\right)^T \eta^i\left(\theta^i_k\right) \quad (6)$$

where the subscript "k" denotes the iteration number, $\boldsymbol{\theta}_k = [\theta^1_k, \theta^2_k, \cdots, \theta^6_k]$, and

$$\eta^i\left(\theta^i_k\right) = B^i\left(\theta^{Bi}_k\right) e^i_r\left(\boldsymbol{\theta}_k\right) = B^i\left(\theta^{Bi}_k\right)\left(S_i r - \kappa^i\left(\boldsymbol{\theta}_{opt}\right)\right) - A^i\left(\theta^{Ai}_k\right) T_{ii} r \quad (7)$$

It can be deduced from (6) and (7) that θ^i_{k+1} is the solution to

$$S_i r - F^i\left(\theta^i_{k+1}\right) T_{ii} r - \kappa^i\left(\boldsymbol{\theta}_k\right) = 0 \quad (8)$$

After the convergence is achieved with (6), $\boldsymbol{\theta}_{k+1} = \boldsymbol{\theta}_k = \boldsymbol{\theta}_\infty$ holds, which means equations (3) and (8) become the same and $\mathbf{e}_r(\boldsymbol{\theta}_\infty) = 0$ holds. Therefore, global optimization can be achieved after convergence.

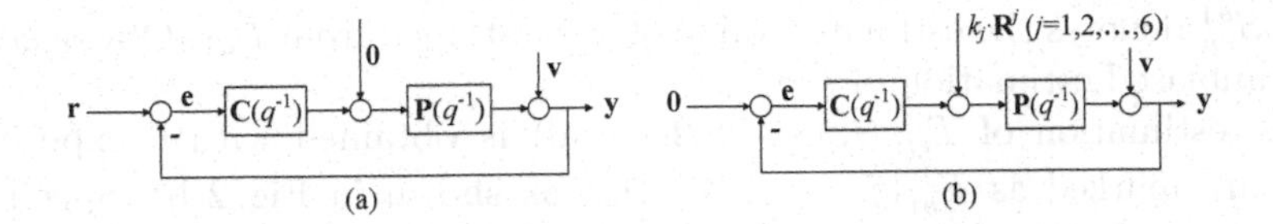

Fig. 2. The control schemes of experiment E_{a1} (a) and E_{b1}^{j} (b).

In order to improve the parameter estimation accuracy of $\boldsymbol{\theta}_{k+1}$, A filter $L_k^i(q^{-1})$ is introduced to (6) as follows

$$\theta_{k+1}^i = \theta_k^i + \left[\left(L_k^i\phi^i\left(\boldsymbol{\theta}_k\right)\right)^T\left(L_k^i\phi^i\left(\boldsymbol{\theta}_k\right)\right)\right]^{-1}\left(L_k^i\phi^i\left(\boldsymbol{\theta}_k\right)\right)^T\left(L_k^i\eta^i\left(\theta_k^i\right)\right) \quad (9)$$

Similar to (6), θ_{k+1}^i obtained with (9) is the solution to

$$L_k^i(q^{-1})\left(S_ir - F^i\left(\theta_{k+1}^i\right)T_{ii}r - \kappa^i\left(\boldsymbol{\theta}_k\right)\right) = 0 \quad (10)$$

Obviously, the solution to (8) is also the solution to (10). Therefore, the addition of $L_k^i(q^{-1})$ does not change the value of θ_{k+1}^i in noiseless condition, and global optimization is available using (9). Moreover, in the following description, it is shown that the introduction of $L_k^i(q^{-1})$ is helpful to improve the parameter estimation accuracy in terms of variance in noisy condition.

3.2 Parameter Updating in Noisy Condition

In order to make (9) applicable, some intermediate variables need to be estimated, including S_ir, $T_{ij}r$ and $\eta^i(\theta_k^i)$. Of all these variables, $\eta^i(\theta_k^i)$ can be estimated from the tracking error in the kth iteration as follows

$$\hat{\eta}^i(\theta_k^i) = B^i\left(\theta_k^{Bi}\right)e_k^i \quad (11)$$

where e_k^i is the reference tracking error in the ith DOF and the kth iteration. Combining (11) with (7), the relation between the estimated value and the real value of $\eta^i(\theta_k^i)$ is as follows

$$\hat{\eta}^i(\theta_k^i) = \eta^i(\theta_k^i) - B^i\left(\theta_k^{Bi}\right)w_k^i \quad (12)$$

where w_k^i is the measurement noise in the ith DOF and the kth iteration. As w_k^i is with zero mean, $\hat{\eta}^i(\theta_k^i)$ is the unbiased estimation of $\eta^i(\theta_k^i)$.

The estimation of S_ir can be realized with a reference tracking trail when no feedforward is applied, which is marked as E_{a1}, as shown in Fig. 2(a). The reference tracking error of E_{a1} in the ith DOF is as follows

$$e_{a1}^i = S_ir - w_{a1}^i \quad (13)$$

where w_{a1}^i is the measurement noise in the ith DOF in E_{a1}. Ignore the noise term in (13) and the estimation of S_ir can be given by

$$\hat{S}_{ri}^{a1} = e_{a1}^i = S_ir - w_{a1}^i \quad (14)$$

where $\hat{S}_{ri}^{a1}$ denotes the estimated value of $S_i r$ obtained from E_{a1}. Obviously, $\hat{S}_{ri}^{a1}$ is the unbiased estimation of $S_i r$.

The estimation of $T_{ij}r(i,j = 1,2,\cdots,6)$ is obtained with 6 experiments which are marked as $E_{b1}^{j}(j = 1,2,\cdots,6)$, as shown in Fig. 2(b). Specifically, the $\mathbf{R}^j$ in Fig. 2(b) is given by $\mathbf{R}^j = \underbrace{[0,\cdots,0,\overbrace{r}^{j\text{th}},0,\cdots,0]}_{6}$. Therefore, the measured output of E_{b1}^{j} in the ith DOF is given by

$$y_{b1}^{i} = k_j T_{ij} r + w_{b1}^{ij} \tag{15}$$

where w_{b1}^{ij} is the measurement noise in the ith DOF in E_{b1}^{j}. based on (15), the estimation of $T_{ij}r$ is as follows

$$\hat{T}_{rij}^{b1} = \frac{1}{k_j} y_{b1}^{i} = T_{ij} r + \frac{1}{k_j} w_{b1}^{ij} \tag{16}$$

Similar to $\hat{\eta}^i(\theta_k^i)$ and $\hat{S}_{ri}^{a1}$, $\hat{T}_{rij}^{b1}$ is also the unbiased estimation of $T_{ij}r$.

In order to avoid the biasedness caused by the autocorrelation terms of the noises while calculating θ_{k+1}^{i} with (9), E_{a1} and $E_{b1}^{j}(j = 1,2,\cdots,6)$ are all repeated once. The repeated experiments are marked as E_{a2} and $E_{b2}^{j}(j = 1,2,\cdots,6)$ respectively, and subscript or superscript "$a2$" and "$b2$" are employed to mark the variables related to E_{a2} and $E_{b2}^{j}(j = 1,2,\cdots,6)$ respectively.

The signal to noise ratio(SNR) of all the $T_{ij}r(i,j = 1,2,\cdots,6)$ estimated from E_{b1}^{j} is not always satisfying. Actually, the SNR of an arbitrary $\hat{T}_{rij}^{b1}$ is the measure of the coupling effect from the jth DOF to the ith DOF of the wafer stage. Low SNR indicates that the coupling effect is weak, and the corresponding $\hat{T}_{rij}^{b1}$ can be neglected while parameter updating. Hence, based on the $\hat{T}_{rij}^{b1}$ obtained from E_{b1}^{j}, the coupling term $\kappa^i(\boldsymbol{\theta}_k)$ can be estimated as follows

$$\hat{\kappa}_{b1}^{i}(\boldsymbol{\theta}_k) = \sum_{j=1,j\neq i}^{6} \tau_{ij} F^j(\theta_k^j) \hat{T}_{rij}^{b1} \tag{17}$$

where $\tau_{ij} = 0$ holds when the SNR of $\hat{T}_{rij}^{b1}$ is low, or τ_{ij} is equal to 1. Then, the estimation of $\phi^i(\boldsymbol{\theta}_k)$ is given by $\hat{\phi}_1^i(\boldsymbol{\theta}_k) = [\hat{\phi}_1^{Ai}(\boldsymbol{\theta}_k), \hat{\phi}_1^{Bi}(\boldsymbol{\theta}_k)]$ and

$$\begin{cases} \hat{\phi}_1^{Ai}(\boldsymbol{\theta}_{opt}) = \left[\Psi_1^{Ai}(q^{-1}), \Psi_2^{Ai}(q^{-1}), \cdots, \Psi_{n_{ai}}^{Ai}(q^{-1})\right] \hat{T}_{rii}^{b1} \\ \hat{\phi}_1^{Bi}(\boldsymbol{\theta}_{opt}) = -\left[\Psi_1^{Bi}(q^{-1}), \Psi_2^{Bi}(q^{-1}), \cdots, \Psi_{n_{bi}}^{Bi}(q^{-1})\right] \left(\hat{S}_{ri}^{a1} - \hat{\kappa}_{b1}^{i}(\boldsymbol{\theta}_k)\right) \end{cases} \tag{18}$$

Here the subscript "1" means that $\hat{\phi}_1^i(\boldsymbol{\theta}_k)$ is calculated based on the data from E_{a1} and $E_{b1}^{j}(j = 1,2,\cdots,6)$. $\hat{\phi}_2^i(\boldsymbol{\theta}_k)$ can be obtained in a similar way with the data from E_{a2} and $E_{b2}^{j}(j = 1,2,\cdots,6)$. With the above estimated values, the estimation of θ_{k+1}^{i} in (9) can be calculated as follows

$$\hat{\theta}_{k+1}^{i} = \theta_k^i + \left[\left(L_k^i \hat{\phi}_1^i(\boldsymbol{\theta}_k)\right)^T \left(L_k^i \hat{\phi}_2^i(\boldsymbol{\theta}_k)\right)\right]^{-1} \left(L_k^i \hat{\phi}_1^i(\boldsymbol{\theta}_k)\right)^T \left(L_k^i \hat{\eta}^i(\theta_k^i)\right) \tag{19}$$

Similar to [8], it can be proved that the $\hat{\theta}_{k+1}^i$ in (19) obtained in noisy condition is the unbiased estimation of the ideal value θ_{k+1}^i in (9).

3.3 Optimization of the Parameter Estimation Accuracy in Terms of Variance

From (19), the difference between the estimated value and ideal value of θ_{k+1}^i is given by

$$\hat{\theta}_{k+1}^i - \theta_{k+1}^i = \left[\left(L_k^i \hat{\phi}_1^i(\boldsymbol{\theta}_k)\right)^T \left(L_k^i \hat{\phi}_2^i(\boldsymbol{\theta}_k)\right)\right]^{-1} \left(L_k^i \hat{\phi}_1^i(\boldsymbol{\theta}_k)\right)^T L_k^i \left(\hat{\eta}^i(\theta_k^i) - \hat{\phi}_2^i(\boldsymbol{\theta}_k)(\theta_{k+1}^i - \theta_k^i)\right) \tag{20}$$

With the above estimations in Subsect. 3.2, we have that

$$\begin{aligned}\hat{\eta}^i\left(\theta_k^i\right) - \hat{\phi}_2^i\left(\boldsymbol{\theta}_k\right)\left(\theta_{k+1}^i - \theta_k^i\right) = -B^i\left(\theta_k^{Bi}\right) w_k^i - \frac{A^i(\theta_{k+1}^{Ai} - \theta_k^{Ai})}{k_i} w_{b2}^{ii} - \\ \left[B^i(\theta_{k+1}^{Bi} - \theta_k^{Bi}) - 1\right] \left[w_{a2}^i + \sum_{j=1, j\neq i}^{6} \frac{\tau_{ij} F^j(\theta_k^j) w_{b2}^{ij}}{k_j}\right]\end{aligned} \tag{21}$$

Considering that w_k^i, w_{b2}^{ii}, w_{a2}^i and w_{b2}^{ij} are all noise signals generated from the ith DOF of the wafer stage and share the same statistic characteristics, it is assumed that $w_k^i \approx w_{b2}^{ii} \approx w_{a2}^i \approx w_{b2}^{ij}$. Therefore, we have that

$$\hat{\eta}^i(\theta_k^i) - \hat{\phi}_2^i(\boldsymbol{\theta}_k)(\theta_{k+1}^i - \theta_k^i) \approx -\left[B^i(\theta_k^{Bi}) + \frac{A^i(\theta_{k+1}^{Ai} - \theta_k^{Ai})}{k_i} + \left[B^i(\theta_{k+1}^{Bi} - \theta_k^{Bi}) - 1\right]\left[1 + \sum_{j=1, j\neq i}^{6} \frac{\tau_{ij} F^j(\theta_k^j)}{k_j}\right]\right] w_k^i \tag{22}$$

Based on (20) and (22), the covariance matrix of $\hat{\theta}_{k+1}^i$ is given by

$$\begin{aligned}P_{k+1}^i = \mathbb{E}\left[\left(\hat{\theta}_{k+1}^i - \theta_{k+1}^i\right)\left(\hat{\theta}_{k+1}^i - \theta_{k+1}^i\right)^T\right] = \left[\left(L_k^i \phi_1^i(\boldsymbol{\theta}_k)\right)^T \left(L_k^i \phi_2^i(\boldsymbol{\theta}_k)\right)\right]^{-1} \times \\ \left(L_k^i \phi_1^i(\boldsymbol{\theta}_k)\right)^T \bar{L}_k^i \Omega_k \left(\bar{L}_k^i\right)^T \left(L_k^i \phi_1^i(\boldsymbol{\theta}_k)\right) \left[\left(L_k^i \phi_1^i(\boldsymbol{\theta}_k)\right)^T \left(L_k^i \phi_2^i(\boldsymbol{\theta}_k)\right)\right]^{-T}\end{aligned} \tag{23}$$

where $\bar{L}_k^i$ is the Toeplitz matrix of the filter $L_k^i(q^{-1})$, please refer to [3,9] for the definition of Toeplitz matrix. And Ω_k is as follows

$$\Omega_k = \mathbb{E}\left[\left(\hat{\eta}^i(\theta_k^i) - \hat{\phi}_2^i(\boldsymbol{\theta}_k)(\theta_{k+1}^i - \theta_k^i)\right)\left(\hat{\eta}^i(\theta_k^i) - \hat{\phi}_2^i(\boldsymbol{\theta}_k)(\theta_{k+1}^i - \theta_k^i)\right)^T\right] = \sigma_i^2 \bar{X}_k^i \left(\bar{X}_k^i\right)^T \tag{24}$$

where $\bar{X}_k^i$ is the Toeplitz matrix of $X_k^i(q^{-1})$, which is given by

$$X_k^i(q^{-1}) = -B^i\left(\theta_k^{Bi}\right) - \frac{A^i(\theta_{k+1}^{Ai} - \theta_k^{Ai})}{k_i} - \left[B^i(\theta_{k+1}^{Bi} - \theta_k^{Bi}) - 1\right]\left[1 + \sum_{j=1, j\neq i}^{6} \frac{\tau_{ij} F^j(\theta_k^j)}{k_j}\right] \tag{25}$$

Similar to [9], it can be proved that the optimal filter $L^i_{k-opt}(q^{-1})$ such that P^i_{k+1} is minimized is given by

$$L^i_{k-opt}(q^{-1}) = \left(X^i_k(q^{-1})\right)^{-1} \tag{26}$$

However, it can be observed from (25) and (26) that $L^i_{k-opt}(q^{-1})$ cannot be obtained directly, as the ideal value θ^i_{k+1} is generally unknown in practical implementations. Therefore, a novel estimation method is proposed to estimate $L^i_{k-opt}(q^{-1})$ iteratively. In the following description, the superscript "m" is employed as the iteration number while estimating $L^i_{k-opt}(q^{-1})$ and θ^i_{k+1}. Specifically, with the initial parameter $\hat{\theta}^{i<0>}_{k+1} = \theta^i_k$ given, the detail procedure in the mth iteration is as follows

1 Estimate $X^i_k(q^{-1})$ as follows

$$X^{i<m>}_k = -B^i(\theta^{Bi}_k) - \frac{A^i(\theta^{Ai<m-1>}_{k+1} - \theta^{Ai}_k)}{k_i} - [B^i(\theta^{Bi<m-1>}_{k+1} - \theta^{Bi}_k) - 1]\left[1 + \sum_{j=1, j\neq i}^{6} \frac{\tau_{ij} F^j(\theta^j_k)}{k_j}\right] \tag{27}$$

2 Estimate $L^i_{k-opt}(q^{-1})$ as follows

$$L^{i<m>}_k(q^{-1}) = \left(X^{i<m>}_k(q^{-1})\right)^{-1} \tag{28}$$

3 Estimate θ^i_{k+1} as follows

$$\theta^{i<m>}_{k+1} = \theta^i_k + \left[\left(L^{i<m>}_k \hat{\phi}^i_1(\boldsymbol{\theta}_k)\right)^T \left(L^{i<m>}_k \hat{\phi}^i_2(\boldsymbol{\theta}_k)\right)\right]^{-1} \left(L^{i<m>}_k \hat{\phi}^i_1(\boldsymbol{\theta}_k)\right)^T \left(L^{i<m>}_k \hat{\eta}^i(\theta^i_k)\right) \tag{29}$$

4 Check if the parameter $\theta^{i<m>}_{k+1}$ is convergent. If convergence is achieved, let $L^i_{k-opt}(q^{-1}) = L^{i<m>}_k(q^{-1})$ and $\hat{\theta}^i_{k+1} = \theta^{i<m>}_{k+1}$ and end this procedure; if not, let $m = m + 1$, return to step 1) and start new iteration.

3.4 Procedures of the Proposed Approach

In conclusion, the overall procedures of the proposed approach is as follows:

1 Select the initial parameter $\theta^i_0 (i = 1, 2, \cdots, 6)$ for all DOFs, set the iteration number $k = 0$, and conduct the initial reference tracking trail;
2 Conduct the trails E_{a1}, E_{a2}, $E^j_{b1}(j = 1, 2, \cdots, 6)$ and $E^j_{b2}(j = 1, 2, \cdots, 6)$ and calculate estimation of $S_i r$ and $T_{ij} r$;
3 Estimate $\eta^i(\theta^i_k)(i = 1, 2, \cdots, 6)$ with (11);
4 Calculate $\hat{\phi}^i_1(\boldsymbol{\theta}_k)$ and $\hat{\phi}^i_2(\boldsymbol{\theta}_k)$ for all DOFs with (17) and (18);
5 Update the parameter $\theta^i_{k+1}(i = 1, 2, \cdots, 6)$ and filters $L^i_{k-opt}(q^{-1})(i = 1, 2, \cdots, 6)$ for all DOFs with (27)–(29) iteratively;
6 Conduct the reference tracking trail with the updated parameter $\boldsymbol{\theta}_{k+1}$, if the reference tracking performance is convergent, end this procedure; if not, let $k = k + 1$, go back to step 3 and start new iteration.

Remark 1. The potential instability of the feedforward controller $F^i(\theta^i_k)$ is dealt with by the stable inversion approach in the Appendix of [8].

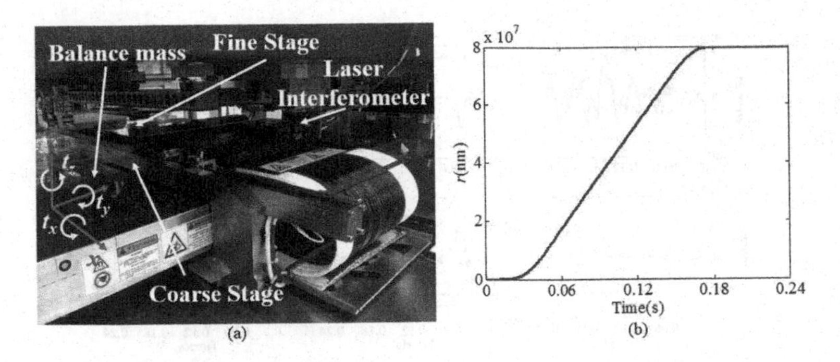

Fig. 3. (a) The ultraprecision wafer stage; (b) The reference to be tracked in the experiments.

4 Experiments

4.1 Setups

Experiments are conducted on an ultraprecision wafer stage, as shown in Fig. 3(a). The wafer stage is a typical dual-stage system, with a coarse stage achieving microscale accuracy in the long range and a fine stage achieving nanoscale positioning in a millimeter scale range. A balance mass is utilized to absorb the reaction force of the wafer stage. The coarse stage is driven by four planar motors and the fine stage is driven by eight voice coil motors. The 6-DOF displacement of the wafer stage is measured by laser interferometer with subnanometer-scale resolution. In this paper, experiments are conducted on the fine stage and the performances of the following two methods are compared

M_1: MIMO data-driven feedforward tuning method associated with PBFs [3];
M_2: The proposed MIMO data-driven feedforward tuning method associated with RBFs

The feedforward controllers are parameterized as follows in M_2.

$$\begin{cases} A^i\left(\theta^{Ai}\right) = \sum_{j=1}^{n_{ai}} \left(\frac{1-q^{-1}}{T_s}\right)^j \theta^{Ai}(j) \\ B^i\left(\theta^{Bi}\right) = 1 + \sum_{j=1}^{n_{bi}} \left(\frac{1-q^{-1}}{T_s}\right)^j \theta^{Bi}(j) \end{cases} \tag{30}$$

where the sampling time $T_s = 0.0002$s, the dimensions $n_{ai} = [4,4,4,4,4,4]$ and $n_{bi} = [1,2,2,1,1,1]$. In M_1, the feedforward controllers are given by $F^i\left(\theta^i\right) = A^i\left(\theta^{Ai}\right)$ $(i = 1,2,\cdots,6)$. The initial parameters for both methods are set as zero, and ten iterations are performed. The reference to be tracked is a fourth-order multi-segment polynomial trajectory, as shown in Fig. 3(b).

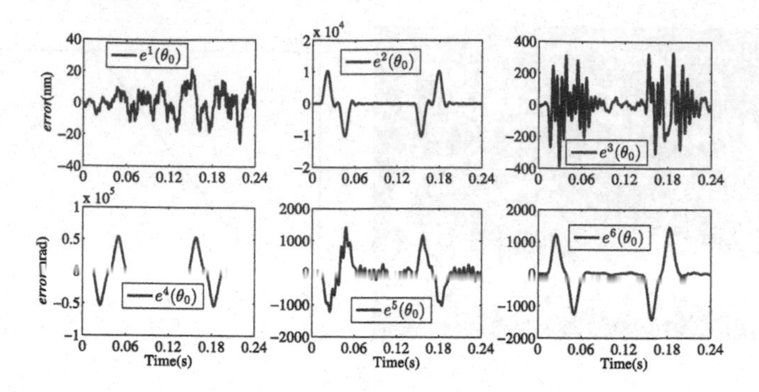

Fig. 4. The reference tracking errors of all DOFs with initial parameter.

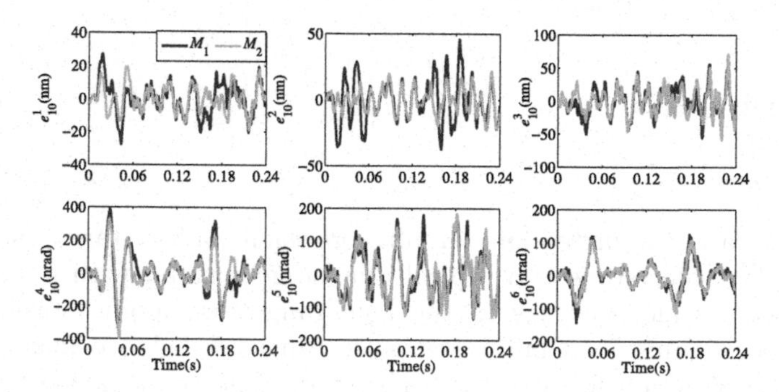

Fig. 5. The reference tracking errors of M_1 and M_2 after 10 iterations.

4.2 Results

With M_1 and M_2, the achieved performances after 10 iterations are shown in Figs. 5 and 6. It can be observed that the reference tracking errors are greatly attenuated with both methods compared to the initial condition in Fig. 4. By contrast, the performance of M_2 is better than M_1 in most directions, especially in x and y directions. This results demonstrates the superior performance of the proposed approach.

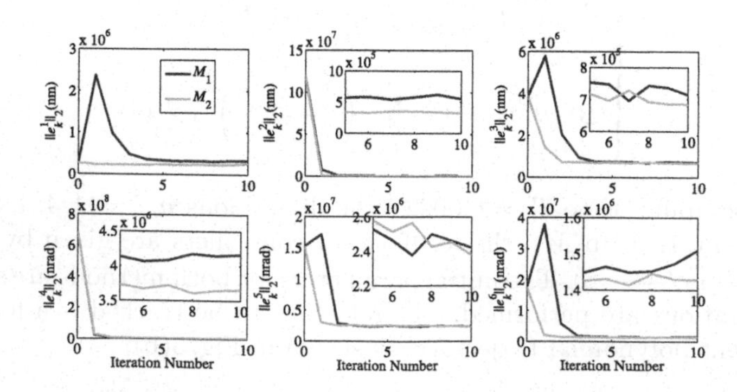

Fig. 6. The evolutions of the 2-norm of reference tracking errors of all DOFs.

5 Conclusions

This paper proposed a novel MIMO data-driven feedforward tuning approach associated with rational basis functions for the ultraprecision wafer stage. In the proposed approach, the optimal parameters of all DOFs were determined by sequentially solving a series of equations using least square method. The estimation method of the intermediate variables was given so that the proposed parameter updating law is able to work in noisy condition. Moreover, the parameter estimation accuracy of the updated parameter in terms of variance was optimized, an iterative procedure was proposed to estimate the optimal filter and parameter in the absense of plant model. The performance of the proposed approach was tested on an ultraprecision wafer stage, and the results validated the excellent reference tracking performance and convergence performance of the proposed approach.

References

1. van Zundert, J., Oomen, T.: On inversion-based approaches for feedforward and ILC. Mechatronics **50**, 282–291 (2018)
2. Li, M., Zhu, Y., Yang, K., Yang, L., Hu, C.: Data-based switching feedforward control for repeating and varying tasks: with application to an ultraprecision wafer stage. IEEE Trans. Ind. Electron. **66**(11), 8670–8680 (2019)
3. Jiang, Y., Zhu, Y., Yang, K., Hu, C., Yu, D.: A data-driven iterative decoupling feedforward control strategy with application to an ultraprecision motion stage. IEEE Trans. Ind. Electron. **62**(1), 620–627 (2015)
4. Bristow, D.-A., Tharayil, M., Alleyne, A.G.: A survey of iterative learning control. IEEE Control Syst. Mag. **26**(3), 94–114 (2006)
5. Blanken, L., Oomen, T.: Multivariable iterative learning control design procedures: from decentralized to centralized, illustrated on an industrial printer. IEEE Trans. Control Syst. Technol. **28**(4), 1534–1541 (2020)
6. Heertjes, M., Hennekens, D., Steinbuch, M.: MIMO feed-forward design in wafer scanners using a gradient approximation-based algorithm. Control Engineering Practice **18**, 495–506 (2010)
7. Bolder, J., Oomen, T.: Rational basis functions in iterative learning control-with experimental verification on a motion system. IEEE Trans. Control Syst. Technol. **23**(2), 722–729 (2020)
8. Huang, W., Yang, K., Zhu, Y., Li, X., Mu, H., Li, M.: Data-driven rational feedforward tuning: with application to an ultraprecision wafer stage. Proc. Inst. Mech. Eng. Part I J. Syst. Control Eng. **234**(6), 748–758 (2020)
9. Huang, W., Yang, K., Zhu, Y., Lu, S.: Data-driven parameter tuning for rational feedforward controller: achieving optimal estimation via instrumental variable. IET Control Theor. Appl. **15**(7), 937–948 (2021)

A High-Performance Normal-Stressed Electromagnetic Fast Tool Servo

Zi-Hui Zhu[1], Dongran Shen[1], Peng Huang[1], LiMin Zhu[2], and Zhiwei Zhu[1(✉)]

[1] School of Mechanical Engineering, Nanjing University of Science and Technology, Nanjing, China
zw.zhu@njust.edu.cn

[2] School of Mechanical Engineering, Shanghai Jiao Tong University, Shanghai, China

Abstract. This paper reports on the design and control of a high-performance fast tool servo (FTS) having a moderate stroke and a high bandwidth, which is actuated by the normal-stressed electromagnetic force. An analytical model of both the mechanical mechanism and electromagnetic driving circuit is established. Assisted by this model, the structural parameters are optimized through a typical genetic algorithm. For the control system, a damping controller combining a lead-lag compensator is firstly adopted to modify the system dynamics, and a PID controller with a feed-forward compensator is further employed for the motion control. Both the open-loop and closed-loop tests are conducted to demonstrate the performance of the FTS.

Keywords: Fast tool servo · Flexure mechanism · Electromagnetic

1 Introduction

Fast tool servo (FTS) is demonstrated to be a very promising technology in ultra-precision machining for the efficient fabricating capacity of complex optical surfaces with nanometric accuracy [1]. With the increasing demands on more complicated surfaces, especially on the micro/nanostructured functional surfaces featuring highly intensive features in microscale, many efforts have been devoted to improve the performance of the FTS during last three decades.

Nowadays, many types of FTS have been developed, which can be categorized according to the driving principle into the piezoelectrically actuated, the voice coil motor (VCM) actuated and Normal-stressed Electromagnetic actuated system. Although the piezoelectric actuator (PEA) has many superior advantages, it is strongly limited by the small travel and the contact nature at the connection end. In general, to achieve a higher stroke, complicated flexure mechanisms are required to amplify the travel, which may significantly increase the inertia force, resulting a low resonant frequency. Overall, the stroke of the PEA based

Supported by National Natural Science Foundation of China (U2013211), Outstanding Youth Foundation of Jiangsu Province of China (BK20211572) and Fundamental Research Funds for the Central Universities (30921013102).

X.-J. Liu et al. (Eds.): ICIRA 2021, LNAI 13014, pp. 364–373, 2021.
https://doi.org/10.1007/978-3-030-89098-8_35

FTS usually is limited in dozens of micrometers. When the stroke is over hundred microns, the resonant frequency will rapidly decrease to several hundred Hertz or less. With respect to the VCM based on the shear-stressed Lorentz electromagnetic force, it can reach large stroke, however, the inherent low force density may lead to low bandwidth even using the large-sized VCM [2]. Generally, VCM based FTS can achieve long stroke up to millimeter, but the resonant frequency is limited in dozens of Hertz. Compared with the PEA and VCM, the linearized reluctance actuator (LRA) using the normal-stressed electromagnetic force has relatively high force density and large reachable stroke. It was successfully applied to generate the uniaxial ultra-fine motion, for example, the ultra-fast motion for FTS [3], the axial motion for the rotary spindle [4], and short stroke with high bandwidth for FTS [5]. In general, the current research of FTS is mainly focused on the small stroke with ultra-high frequency or long stroke with low frequency.

In this paper, taking advantage of the normal-stressed electromagnetic force, a medium stroke FTS with high-frequency response is introduced to satisfy the requirement for high-performance servo motion in diamond turning, and the corresponding optimal design and trajectory tracking strategy are detailed with further verification through experimental tests.

2 System Configuration of the FTS

The mechanical structure of the FTS is illustrated in Fig. 1(a) and (c), which mainly consists of a flexure mechanism shown in Fig. 1(b) and a linearized reluctance actuator shown in Fig. 1(d). When an excitation current is applied to the coil winding, different magnetic flux density will be generated on the two working gaps of the armature, which will lead to an actuation force F on the armature. With the guidance of the flexure beams, the tool holder will be pushed to move along the axial direction by the actuation force.

The linearized reluctance actuator consists of six stators, four permanents, two coil wingdings, and an armature. The stators are made of soft magnetic material to conduct the magnetic field, and they are divided into three parts including a front stator, a rear stator, and four middle stators for the simplification of manufacturing and fitting. The permanents are arranged around the armature to provide a biasing direct current (DC) flux $\bar{B}$, and the coil wingdings provide an alternating current(AC) flux $\tilde{B}$. The armature is initially suspended in the center of the front and rear stator by the core frame with an initial working gap d_0 shown in Fig. 1(d).

The flexure mechanism is constructed by two sets of crossed parallel leaf-spring flexure hinges which are connected to the core frame and integrally manufactured with the front and rear frame respectively. Moreover, the structure can provide very high lateral and rotary stiffness, which makes it have strong resistance to the disturbance of cutting force.

Assuming that the armature has a displacement of z, the actuation force F generated by the difference of the fluxes can be expressed as [3]:

$$F = \frac{A\bar{B}_a}{d_0} NI + \frac{2A\bar{B}_a{}^2}{d_0 \mu_0} z \tag{1}$$

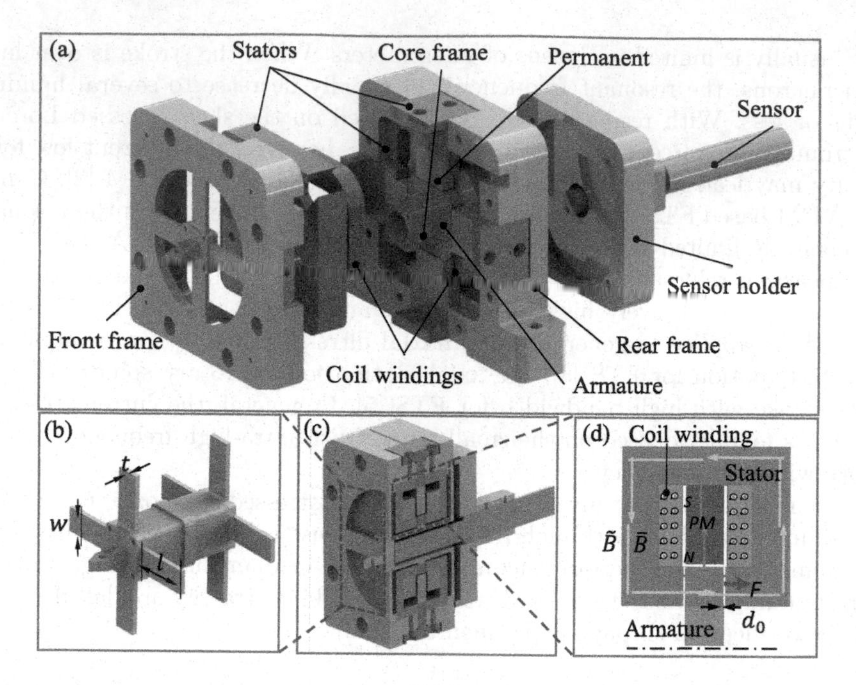

Fig. 1. Structure of the FTS, (a) 3D explosion view, (b) The flexure mechanism, (c) The crossed-section view of the assembled structure, and (d) Magnetic circuit model.

where A is the pole area of the armature, $\bar{B}_a$ is the average biasing DC flux density in the two working gaps with $z = 0$, μ_0 is the vacuum permeability, NI is the excitation current in the coil wingdings.

Meanwhile, considering the saturation constraint of the armature, the maximum excitation current NI_{max} should satisfy the following inequality

$$\frac{d_0 + |z_{max}|}{d_0}\bar{B}_a + \frac{\mu_0 NI_{max}}{4d_0} \leq B_{sat} \tag{2}$$

According to the Castigliano's second theorem, the axial motion stiffness of flexure mechanism can be expressed as [6]:

$$k = \frac{40Ewt^3}{5l^3 + (12 + 11\mu)\, lt^2} \tag{3}$$

where l, t and w are the length, thick and width of the flexure beam, respectively, shown in Fig. 1(c), μ and E are the Poisson's ratio and elastic modulus of the adopted material respectively.

By applying a maximum allowable current I_{max}, the stroke z_{max} of the FTS can be derived through solving the following equation

$$F\left(I_{max}, z_{max}\right) = kz_{max} \tag{4}$$

Following the Lagrange's equation, the resonant frequency of the mechanical system can be obtained as

$$f = \frac{1}{2\pi}\sqrt{\frac{k}{M}} \tag{5}$$

where M is the equivalent mass of the moving part.

3 Parameters Determination and Simulation

To satisfy the desired stroke (over ± 125 μm) and maximize the first order resonant frequency, the structure parameters are optimally designed according to the model established in Sect. 2. Moreover, the finite element analysis (FEA) is conducted to verify the effectiveness of the design.

3.1 Parameters Determination

To gain a high force density as much as possible, the strong magnetic material $Nd_2Fe_{14}B$ (B_r = 1.18T) is employed for the permanent sizing 18×5×14 mm^3 with the pole face area S = 18×5 mm^2 and length l_{pm} = 14 mm. Meanwhile, the perm-alloy 1j22 with a high saturation flux density of B_{sat} = 2.3T is selected for the construction of stators and armatures. The air gap is set as d_0 = 0.2 mm to ensure the desired stroke of $z_{max} = \pm 125$ μm.

Considering the maximum allowable current of I_{max} = 3A for the adopted wire, the number of coil turns is calculated as N = 160 according to Eq. 2. From Eq. 1, the maximum actuation force is calculated as F_{max} = 350 N. Accordingly, the required axial motion stiffness is calculated as k_0 = 2.80 N/μm from Eq. 3.

The high strength aluminum alloy 7075 is employed as the material of the flexure mechanism with Young's modulus of $E = 7.1\times10^9$Pa, Poisson's ratio of $\mu = 0.33$, density of $\rho = 2.7\times10^3$ kg/m^3. Taking advantage of these properties, the structure parameters can be optimized by solving the following minimization problem through genetic algorithm

$$\min \quad obj = \frac{|k - k_0|}{k_0} + \frac{1}{f} \tag{6}$$

According to the objective function in Eq. 6, the parameters and performances are optimized as l = 22 mm, w = 9.2 mm, t = 1.8 mm, k = 2.80 N/μm and f = 1287 Hz by using genetic algorithm in the Optimization Tool module of MATLAB. The significant geometric parameters are summarized in Table 1.

Table 1. The significant geometric parameters.

Symbol	l	w	t	d_0	S	l_{pm}
Values	22 mm	9.2 mm	1.8 mm	0.2 mm	18×5 mm^2	14 mm

3.2 Finite Element Verification

To verify the effectiveness of the established analytical model, both the electromagnetic and mechanical simulation are conducted by FEA in this section. The significant structure dimensions are set as given in Table 1.

The electromagnetic analysis of the developed linearized reluctance actuator is conducted through the software of Maxwell. The magnetization curve of perm-alloy 1j22 material used for stators and armature is set as Fig. 2(b). To simulate the magnetic field distribution and actuating force under maximum excitation condition, a current of I_{max} =3A is applied to the coil wingdings. By checking the magnetic flux distribution shown in Fig. 2(a), the maximum flux density is observed as 1.98T which is smaller than the saturation flux density. Meanwhile, the corresponding maximum actuation force is observed as 326 N, showing a good agreement with the analytical result. Moreover, different driving currents with NI varying from 0A to 720A are applied to the coil wingdings, and the actuation force is recorded as shown in Fig. 2(c), which suggests a good linearity between the current and actuation force with $NI < 480$A

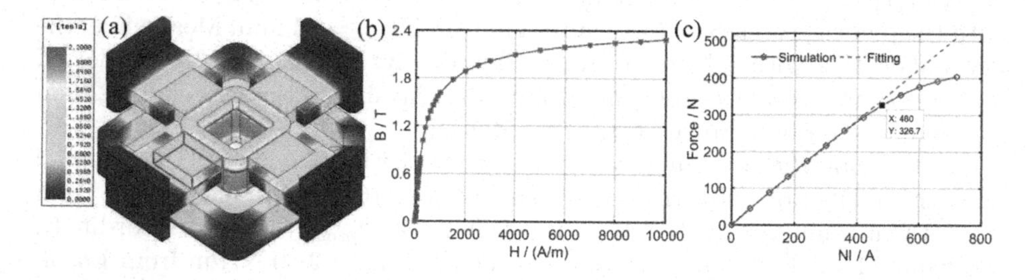

Fig. 2. Electromagnetic simulation result, (a) The flux density distribution, (b) The magnetization curve of the perm-alloy 1j22, and (c) The actuation force related to the excitation current

For the flexure mechanism, both the static and dynamic behavior are investigated. The material of flexure mechanism is set as high strength aluminum alloy 7075. By applying a maximum actuation force 326 N on the armature, the axial deformation is simulated to be 133.5 μm shown in Fig. 3(a). Accordingly, the stiffness is derived as 2.44 N/μm, suggesting a deviation of 14.7% between analytical and FEA results. Meanwhile, the maximum equivalent stress is simulated to be 101 Mpa, as shown in Fig. 3(b), which is much smaller than the yield strength (455 Mpa) of the adopted material. As shown in Fig. 3(c) and (d), the first and second resonant frequencies are observed to 1201 Hz 1837 Hz respectively. Meanwhile, the corresponding resonant mode of the first resonant frequency is consistent with the desired axial motion.

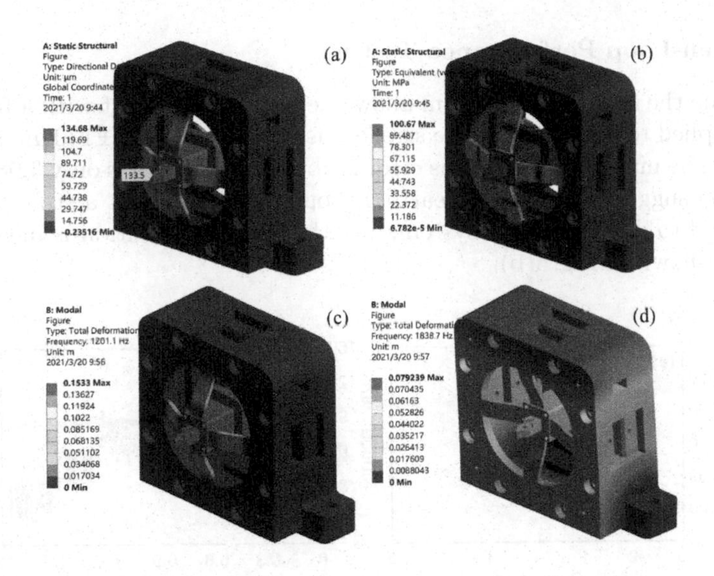

Fig. 3. The FEA results, (a) The resulting deformation with maximum actuation force, (b) Equivalent stress with maximum actuation force, (c) The first, and (d) The second resonant mode.

4 Performance Testing and Discussion

4.1 Experiment Setup

The assembled prototype is photographically illustrated in Fig. 4, and the overall size is about $124\times98\times97\,\mathrm{mm}^3$. For the experiment setup, one servo amplifier (SMA5005-1, Glenteck Corporation, USA) with fixed gain of 0.8 A/V was adopted to amplify the command for the coil. Meanwhile, a capacitive displacement sensor (5503) with a charge amplifier (5810) from the Micro-sense Corporation, USA, was adopted to measure the axial motion. Through the Matlab/Simulink based desk-top real-time modulus, a data acquisition board (PCI-6259, NI Corporation, USA) was used for collecting and sending out signal. The sampling frequency was 20 kHz for the testing (Fig. 4).

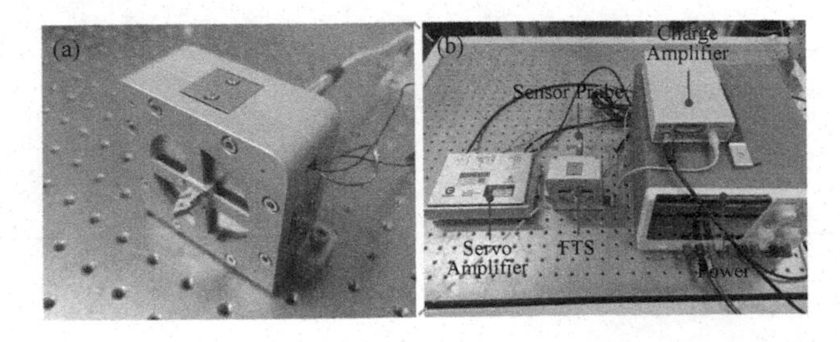

Fig. 4. The prototype of the designed FTS and experiment setup.

4.2 Open-Loop Performance Testing

To examine the stroke, a maximum allowable harmonic current with a frequency 1 Hz is applied to the coil, and the stroke was measured to be $\pm 130\,\mu$m shown in Fig. 5(a). The maximum hysteresis error is about $8\,\mu$m which is only 3.08% of the full motion suggesting a good linearized property. In addition, a stair command is employed to drive the FTS system, and the motion resolution is measured to be 30 nm shown in Fig. 5(b).

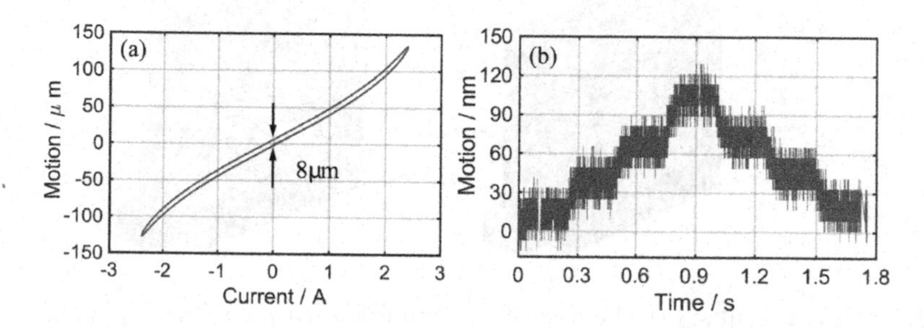

Fig. 5. Static performance of the FTS, (a) The measured stroke related to the applied current, (b) The resolution test result

To examine the dynamic performance, a small amplitude sweep excitation with frequency linearly varying from 0.1 to 2000 Hz is conducted for the coil to excite the FTS system. Based on the input excitation command and the measured output displacement data, the transfer function in s-domain is identified as following by using System Identification module of MATLAB

$$P(s) = \frac{-1.5598 \times 10^7 (s - 3.82 \times 10^4)(s + 7934)(s + 431.5)}{(s + 316.5)(s^2 + 1651\,s + 3.576 \times 10^6)(s^2 + 89.26s + 5.038 \times 10^7)} \quad (7)$$

Accordingly, both the practical and estimated bode diagram can be obtained and illustrated in Fig. 6. The resonant frequency is observed to 1121 Hz, showing a good agreement with the analytical result.

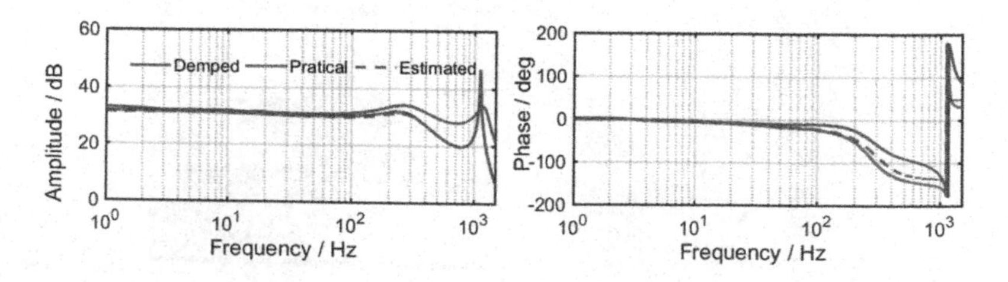

Fig. 6. Frequency response of the FTS.

4.3 Closed-Loop Performance Testing

As presented in Fig. 7, a control system is designed to verify the tracking performance of the FTS. Since the amplitude of the transfer function decays rapidly before the first resonant frequency, a second-order lead-lag compensator $L(s)$ is adopted to revise the amplitude and phase response, which is

$$L(s) = \frac{(1 + b_1 T_1 s)(1 + b_2 T_2 s)}{(1 + T_1 s)(1 + T_2 s)} \tag{8}$$

where b_1, b_2, T_1, and T_2 are five adjustable parameters. By checking the amplitude and phase of the revised system, the parameters can be tuned. Meanwhile, the damping controller $C_{dp}(s)$ constructed by the positive acceleration, velocity and position feedback (PAVPF) [7] is employed to suppress the resonant vibration to achieve high-bandwidth tracking.

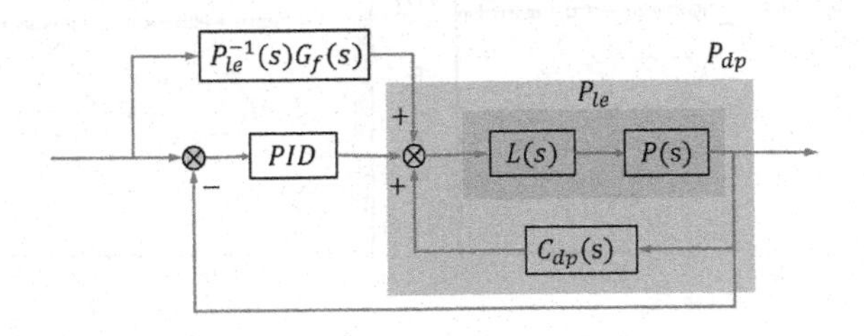

Fig. 7. Schematic of the control system.

After applying a lead-lag compensator, the controlled system can be simplified to a third-order system denoted by $P_{le}(s)$. Hence, a second-order PAVPF controller $C_{dp}(s)$ can be adopted to damp the system, which is [7]

$$C_{dp}(s) = \frac{\alpha_2 s^2 + \alpha_1 s + \alpha_0}{s^2 + 2\xi_d \omega_d + \omega_d^2} \tag{9}$$

where α_0, α_1, α_2, ξ_d and ω_d are five adjustable parameters.

From Fig. 7, the transfer function of the damped control system can be expressed by

$$P_{dp}(s) = \frac{P_{le}(s)}{1 - P(s)C_{dp}(s)} = \frac{\prod_{i=1}^{2}(s - z_i)}{\prod_{i=1}^{5}(s - p_i)} \tag{10}$$

where z_l and p_k are the zeros and poles of the damped system. By deliberately setting the desired damping ratio of the dominant poles, the adjustable parameters can be derived from Eq. 10 [7].

Accordingly, the lead-lag compensator $L(s)$, the simplified transfer function $P_{le}(s)$ and damping controller $C_{dp}(s)$ are identified as follows

$$\begin{cases} L(s) = \frac{7.036\times10^{-8}s^2+6.631\times10^{-4}s+1}{2.814\times10^{-9}s^2+1.203\times10^{-4}s+1} \\ P_{le}(s) = \frac{18985s^2-8.442\times10^{7}s+4.073\times10^{12}}{s^3+1929s^2+5.041\times10^{7}s+9.208\times10^{10}} \\ C_{dp}(s) = \frac{-0.0111s^2+248.3s+1.980\times10^{5}}{s^2+8385s+8.381\times10^{7}} \end{cases} \tag{11}$$

After implementing the PAVPF, the plant to be controlled is changed to be $P_{dp}(s)$, and the frequency response function of the damped system is shown in Fig. 6. By adopting a low pass filter $G_f(s)$, a classical PID controller combining a feed-forward compensator $P_{le}^{-1}(s)G_f(s)$ is adopted for the out-loop controller. The parameters for the PID controller are tuned through the PID tuner toolbox in MATLAB.

To investigate the closed-loop performance of the FTS, a harmonic trajectory with an amplitude of 60 μm and a frequency 20 Hz was employed as a desired trajectory. The resulting motions are illustrated in Fig. 8(a) with the errors being in Fig. 8(b). Compared to the PID and PID+damp controller, the tracking performance of the PID+Lead+damp controller is improved significantly, and the maximum tracking error is about ±0.8 μm (±0.67% of the full motion span).

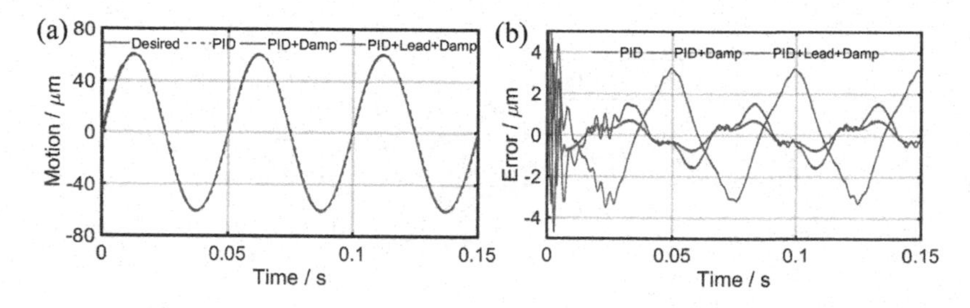

Fig. 8. Tracking performance comparison, (a) The actual and desired motion, and (b) The tracking error.

4.4 Performance Comparison and Discussion

To better demonstrate the advancement of this design, the main performance of some typical moderate stroke FTS, including a commercial one from the PI Corporation (Germany P-621.1), are summarized in Table 2. Compared to those FTS, a much higher nature frequency is achieved in this study. Moreover, the high lateral stiffness and linearity make it more suitable for diamond turning.

Table 2. Performance comparison.

Ref.	This study	[8]	[9]	[10]	[11]	P-621.1
Stroke (μm)	260	200	214	119	410	120
Freq. (Hz)	1121	130	250	220	355	800

5 Conclusion

A high-performance fast tool servo (FTS) is developed herein for the diamond turning by using the normal-stressed electromagnetic actuation. The analytical model is established for the optimal design of the FTS, which is then verified through finite element analysis. Experimental test on the prototype demonstrates a motion resolution better than 30 nm and a stroke larger than $\pm 130\,\mu$m. The resonant frequency is identified to 1121 Hz by conducting the sweep excitation. By constructing a dual-loop control system, the motion error is about $\pm 0.67\%$ for tracking a harmonic trajectory with a frequency 20 Hz.

References

1. Fang, F., Zhang, X., Weckenmann, A., Zhang, G., Evans, C.: Manufacturing and measurement of freeform optics. CIRP Ann. **62**(2), 823–846 (2013)
2. Liu, Q., Pan, S., Yan, H., Zhou, X., Wang, R.: In situ measurement and error compensation of optical freeform surfaces based on a two dof fast tool servo. Int. J. Adv. Manuf. Technol. **86**, 793–798 (2016)
3. Lu, X.-D., Trumper, D.: Ultrafast tool servos for diamond turning. CIRP Ann. **54**(1), 383–388 (2005)
4. Lu, X.D., et al.: Rotary-axial spindles for ultra-precision machining. CIRP Ann. **58**(1), 323–326 (2009)
5. Nie, Y.H., Fang, F.Z., Zhang, X.D.: System design of Maxwell force driving fast tool servos based on model analysis. Int. J. Adv. Manuf. Technol. **72**, 25–32 (2014)
6. Zhao, D., Zhu, Z., Huang, P., Guo, P., Zhu, L., Zhu, Z.: Development of a piezoelectrically actuated dual-stage fast tool servo. Mech. Syst. Sig. Process. **144**, 106873 (2020)
7. Li, L., Li, C.X., Gu, G., Zhu, L.M.: Positive acceleration, velocity and position feedback based damping control approach for piezo-actuated nanopositioning stages. Mechatronics **47**, 97–104 (2017)
8. Yang, F., Yang, H.K., Chen, Z.H., Wang, G.L.: Analysis and design of voice-coil actuator used in fast tool servo. J. Nat. Univ. Defense Technol. **31**(4), 42–47 (2009)
9. Zhu, X., Xu, X., Wen, Z., Ren, J., Liu, P.: A novel flexure-based vertical nanopositioning stage with large travel range. Rev. Sci. Instrum. **86**(10), 105112 (2015)
10. Shiou, F.J., Chen, C.J., Chiang, C.J., Liou, K.J., Liao, S.C., Liou, H.C.: Development of a real-time closed-loop micro-/nano-positioning system embedded with a capacitive sensor. Meas. Sci. Technol. **21**(5), 54007–54016 (2010)
11. Muraoka, M., Sanada, S.: Displacement amplifier for piezoelectric actuator based on honeycomb link mechanism. Sens. Actuators A Phys. **157**(1), 84–90 (2010)

Vibration Test and Analysis of a Six-Degree-of-Freedom Cooperative Robot Manipulator End

Lianzhe Guan[1], Guohua Cui[1], Zhenshan Zhang[1](✉), Ying Pan[1], and ChunLin Li[2]

[1] Institute of Intelligent Cooperative Robot Application Technology, Shanghai University of Engineering and Technology, Shanghai 201600, China
[2] School of Foreign Languages, Shanghai University of Engineering and Technology, Shanghai 201600, China

Abstract. In order to study the vibration characteristics of the end of a six-degree-of-freedom manipulator, based on the vibration transmission principle of "excitation source-vibration transfer system-receiver", a solution for end-of-freedom vibration analysis and testing of a six-degree-of-freedom robot is proposed. By combining the robot body vibration experiment and the single-axis working vibration experiment test, this scheme can effectively analyze the influencing factors of the end vibration during the robot movement, so as to trace the source of the fault. The test results show that as the speed of the robot increases, the end-of-end vibration performance will become more obvious; the change of the robot's motion trajectory will also affect the severity and duration of the end-of-robot vibration; single-axis working vibration test and vibration test. Combined with analysis, it can be inferred that the second-axis is the main cause of the vibration at the end of the robotic arm in this test. The study obtained the dynamic characteristics of the robot under different working conditions, which provided a basis for the optimization of the performance of the manipulator.

Keywords: Robot · Vibration test · Natural frequency · Vibration characteristics

1 Introduction

Robots are widely used in the fields of machinery, automobiles, electronics, aerospace, medical equipment, and housekeeping services. At the same time, the requirements for mechanical precision are also increasing [1, 2]. However, during welding, assembling, deburring and other operations, the end is prone to vibration [3], which leads to a large gap between the actual motion performance of the robot and the expected high-precision motion performance [4]. To meet the needs of many light manufacturing tasks [5], it is necessary to study the vibration performance of the robot.

G. Lianzhe—Robot fault diagnosis and health assessment.
Z. Zhenshan—Tribology, fluid lubrication, mechanism dynamics and reliability.

X.-J. Liu et al. (Eds.): ICIRA 2021, LNAI 13014, pp. 374–383, 2021.
https://doi.org/10.1007/978-3-030-89098-8_36

In order to meet the requirements for the reliability and efficiency of the robot, it is necessary to study the vibration performance of the robot. Commonly used vibration performance analysis methods include simulation vibration analysis and vibration test methods. Gao Weiya et al. [6] conducted vibration analysis on dangerous working conditions of the robotic arm and optimized the structure. After optimization, the amplitude of the end of the robotic arm was significantly reduced. Chen Jiwen et al. [7] took the palletizing robot arm as the research object, and studied its dynamic characteristics under dangerous conditions, so as to optimize the structure topology and realize the lightweight design of the arm structure. Zhang Tie et al. [8] obtained the vibration shape information of the robot joint based on the flexible body dynamics model, and studied the method of robot joint vibration suppression. Shen Xiaodong et al. [9] established a system model of a hole-making robot, and studied the relationship between joint deformation and changes in drilling force and drilling speed. Mousavi et al. [10] proposed the nonlinear strong coupling relationship between robot vibration and cutting parameters, robot posture and stiffness, and studied the vibration of the robot on a predetermined trajectory. Gao Weiya et al. [6] conducted a vibration simulation analysis on the dangerous working conditions of the robotic arm, and established a parameterized model of the robot based on this, so as to design the lightweight robot. The result showed that the overall mass of the optimized robot was reduced. And the end vibration performance is significantly reduced. Zhang Yuling et al. [11] designed a single-degree-of-freedom test bench based on the active control method of arm stiffness, and analyzed the vibration characteristics of the end of the manipulator under different preloads. Tang Shuiyuan et al. [12] took the JZF-1 explosion-proof assembly robot as the research object and used the vibration test method to obtain the structural characteristics of the robot in different postures. The results show that the middle part of the robot arm is the key vibration weak link. Xiong Xiong et al. [13] proposed a dynamic characteristic evaluation index based on vibration test data to quantify the structural dynamic performance of industrial robots through the frequency response function obtained by the vibration test of the hammer excitation. Mejri [14] verified that the cutting speed of the robot tool is the root cause of the robot's cutting vibration through the vibration test, and pointed out that the cutting position and feed direction of the robot have a great influence on the vibration of the cutting head position of the robot. Bisu [15] took the KUKA KR240-2 six-axis industrial robot as the research object, and studied the vibration of the robot tool at different positions in the rotating cutting and rotating non-cutting state through vibration experiments, and concluded that the robot posture stiffness affects its end vibration The main factor. Tang Zhixian [16] took SCARA robot as the research object, obtained the vibration characteristics of the robot through modal analysis and dynamic analysis, and used experiments to verify the validity of the simulation results.

Because the transmission system of the robot is very complicated, and the boundaries between different movable parts are difficult to determine, it is difficult to accurately establish a model to describe the dynamic characteristics of the joints, and its dynamic characteristics such as motion accuracy and stiffness are affected by many factors in practical applications influences. In this paper, based on the vibration test data, the robot body experiment and the single-axis motion experiment are combined to analyze the end vibration factors of the robot motion, and then evaluate the dynamic characteristics of the robot. This research can be used for the diagnosis and prediction of robot vibration and structural dynamics. The optimized design of the characteristics provides the basis.

2 Principle and Method of Vibration Test of Mechanical Arm

The vibration transmission path refers to the process of vibration from the vibration source to the receiver through different paths through the structure. According to the operating characteristics and working status of the robot manipulator, the principle diagram of the vibration test of the manipulator structure is shown in Fig. 1.

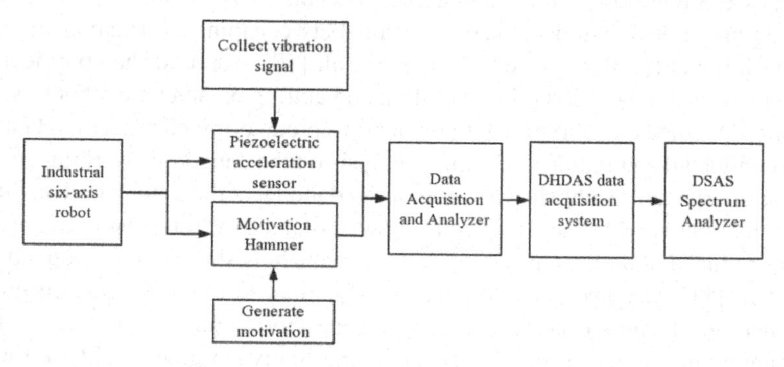

Fig. 1. Principle diagram of vibration test

In order to measure the dynamic characteristics of the robot body components, this paper carried out a vibration test on the robot prototype. Add simple harmonic excitation in the time domain to the input channel of the system $f(t) = f_0 \sin(\omega t)$, In the forced vibration of a single degree of freedom system, the differential equation of motion of the vibration system is

$$M\ddot{x} + c\dot{x} + kx = f_0 \sin(\omega t) \tag{1}$$

Where f_0 is the excitation force amplitude; ω is the excitation frequency; M is the mass;c is the damping coefficient; k is the stiffness coefficient.

If the excitation is expressed as a complex number $f(t) = f_0 \exp(j\omega t)$, the vibration differential equation is rewritten as

$$M\ddot{x} + c\dot{x} + kx = f_0 \exp(j\omega t) \tag{2}$$

The complex frequency response function of Eq. (2) is

$$H(jw) = \frac{1}{M(jw)^2 + c(jw) + k} \tag{3}$$

The complex frequency response function $H(jw)$ contains not only the amplitude information of the response, but also the phase information between the excitation and the response, which fully expresses the dynamic characteristics of the system, and is an important tool for studying the dynamic characteristics of the system. According to the transmission principle of "excitation source-vibration transmission system-receiver" of vibration transmission, the vibration analysis of the robot end mainly considers the influence of the excitation source and the structural characteristics of vibration transmission.

3 The Construction of the Vibration Test Platform of the Robotic Arm

Table 1. The main movement of each axis of the robot

Shaft number	Main movement
First axis	Rotational movement of the robotic arm
Second axis	Large-range pitch motion of the robotic arm
Third axis	Pitching motion of the big arm
Fourth axis	Rotational movement of the forearm
Fifth axis	Pitching movement of the wrist
Sixth axis	Rotational movement of the end effector

The selected research object is an industrial six-degree-of-freedom robot. The six-axis of the robot is composed of the base, waist, big arm, elbow, forearm, and wrist. Table 1 shows the main movements of each axis of the robot. Among them, the movement of the first three axes basically determines the spatial position of the end effector, and the completed action requires a large amount of kinetic energy, which is mainly driven by a high-power motor. The movement of the rear three axes is mainly used to control the spatial attitude of the end effector. The required kinetic energy is relatively small, and low-power motors are used.

This experimental test is carried out in an environment without vibration and noise and at normal room temperature. The vibration test system consists of two parts: a measurement system and an analysis system. The measurement system part includes a hammer exciter, an acceleration sensor (one-way acceleration sensor, three-way acceleration sensor), and the analysis system uses a 24-channel vibration measurement instrument for industrial robot vibration analysis. A three-way acceleration sensor is arranged at the end effector of the robot, and a one-way acceleration sensor is arranged at each joint of the robot. The layout position of the robot prototype and the sensor is shown in Fig. 2, and the structure model of the robot arm is shown in Fig. 3.

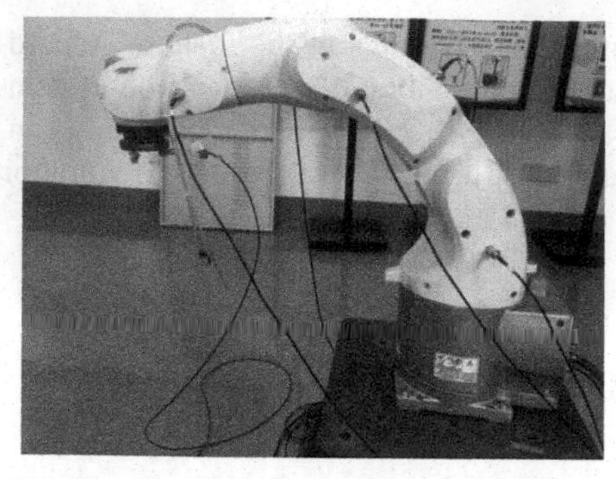

Fig. 2. The layout of the robot arm prototype and the sensor

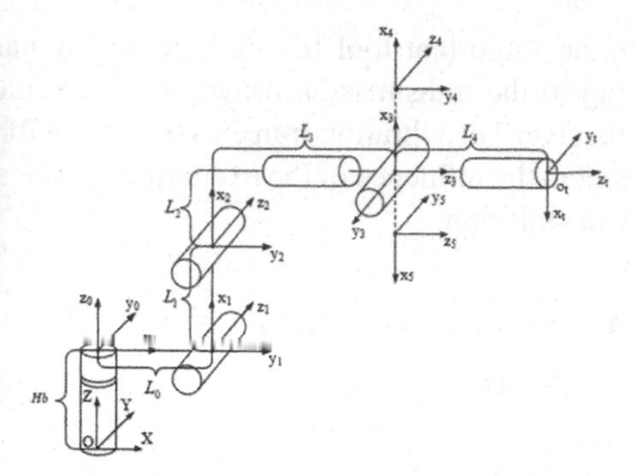

Fig. 3. Mechanical arm structure model

4 Test Plan and Process

Set the robot's motion trajectory and speed, and keep the load on the end of the robot unchanged. The one-way acceleration sensor and the three-way acceleration sensor are used to collect the vibration signal of the robot when it is moving, so as to test the vibration data of the robot.

The test procedure is divided into: (1) Change the movement trajectory of the robot. The robot moves according to the set trajectory, and the linear motion and circular motion of the robot are tested respectively, and the vibration signal of each axis is analyzed. (2)Change the movement speed of the robot. The robot moves in a circular motion, keeping the load added at the end of the robot unchanged, and adjusting the motion speed to 10%, 30%, 50%, 70%, and 90%, respectively, to analyze the effective value of the vibration signal at the end of the robot. (3)Single-axis motion of the robot arm. Keep the load added to the end of the robot unchanged, and test the vibration data when each axis is moving in a single axis.

5 Test and Result Analysis

5.1 The Effect of Motion Trajectory on the Vibration of the End of the Robotic Arm

According to the above test plan, the robot manipulator is made to move on the set trajectory, and the test results are shown in Figs. 4 and 5.

The test results show that: (1) The oscillation time of the robot is longer when it makes a circular motion. (2) When the robot is in circular motion, the vibration of the third, fourth and sixth axes without linear motion is large, which may be caused by the large displacement of the third, fourth and sixth axes during circular motion. (3) On the whole, the vibration magnitude of the robot in circular motion is smaller than that in linear motion.

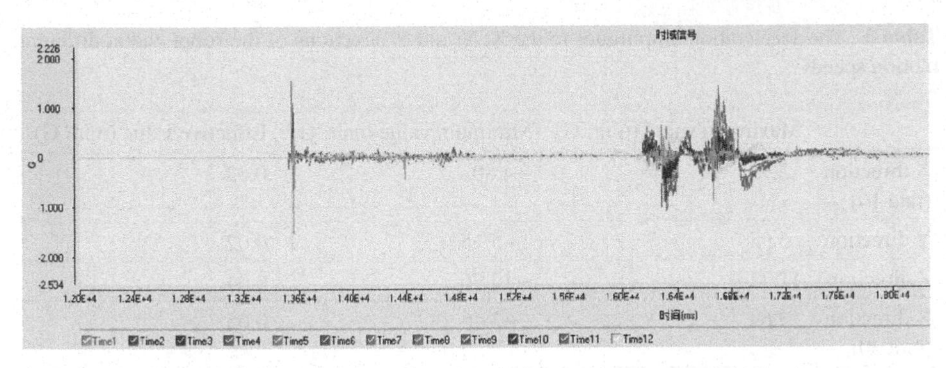

Fig. 4. Time domain signal of the robot moving in a straight line

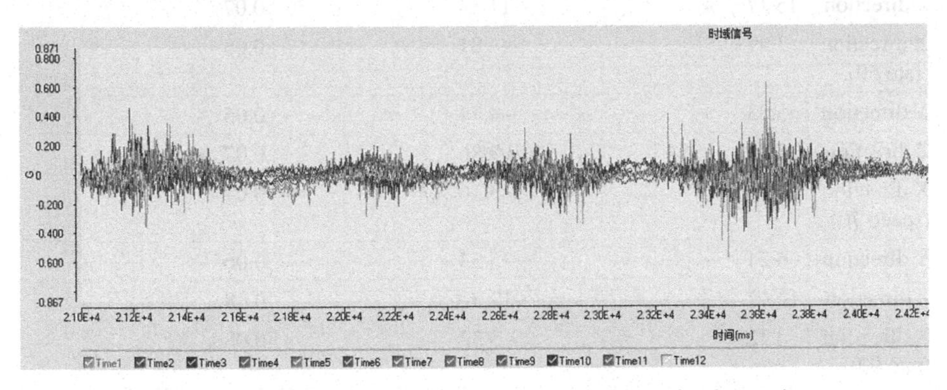

Fig. 5. The time domain signal of the robot making a circular motion

5.2 The Effect of Movement Speed on the Vibration of the End of the Robotic Arm

Control the robot to make a circular motion and change the speed of the robot. The test results are shown in Table 2 and Fig. 6.

The test results show that: (1) The test results in Table 2 show that the overall vibration of the six axes still increases with the increase of the robot's movement speed, and the end tends to be stable. (2) The test results in Table 2 show that during the robot movement along the set trajectory, the Z-direction vibration performance is more obvious than the X-direction, and the X-direction vibration performance is more obvious than the Y-direction. And the vibration magnitude of the robot end increases with the increase of its moving speed. (3) It can be seen from Fig. 6 that the main frequency of vibration during the movement of the robot body is around 19.69 Hz and 24.69 Hz, and the vibration performance of the robot when the robot is at 24.69 Hz is significantly greater. Both frequencies are lower than the natural frequency of the robot body, so it can be inferred that the vibration during its motion is not caused by the resonance of the body, but caused by the single-joint rotational vibration of each joint. Therefore, the single-axis vibration test is required.

Table 2. The acceleration amplitudes in the X, Y, and Z directions of the robot end at different motion speeds

	Maximum value (unit: G)	Minimum value (unit: G)	Effective value (unit: G)
X direction (rate 10)	3.93	−4.80	0.03
Y direction	5.98	−5.15	0.02
Z direction	11.72	−12.96	0.04
X direction (rate 30)	3.69	−3.46	0.05
Y direction	4.31	−5.91	0.04
Z direction	15.17	−11.24	0.07
X direction (rate 50)	3.47	−4.31	0.05
Y direction	5.83	−4.74	0.05
Z direction	13.70	−9.86	0.07
X direction (speed 70)	4.57	−5.16	0.07
Y direction	6.24	−5.33	0.06
Z direction	12.12	−12..61	0.08
X direction (rate 90)	4.22	−5.32	0.08
Y direction	6.46	−4.82	0.06
Z direction	10.58	−12.24	0.08

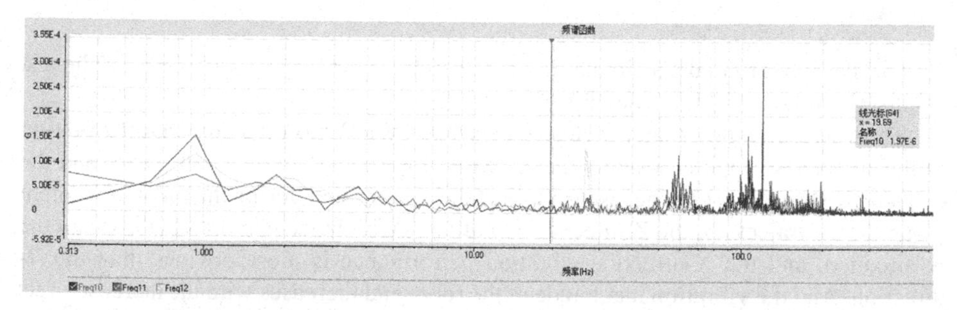

Fig. 6. Spectrum function diagram of the robot movement process

5.3 The Effect of Single-Axis Motion on the End Vibration of the Manipulator

According to the structure and use characteristics of the mechanical arm, the power transmission and energy consumption are mainly the first three axes, and the vibration source mainly comes from these three axes. From the vibration test data, it can be seen that the first axis has a small effect on the end vibration of the robot. Therefore, the end vibration is tested under the excitation of the second and third axis operation, and the vibration of the end piece under the excitation of the second and third axis independent operation is obtained, as shown in the Figs. 7, 8, 9 and 10 shows:

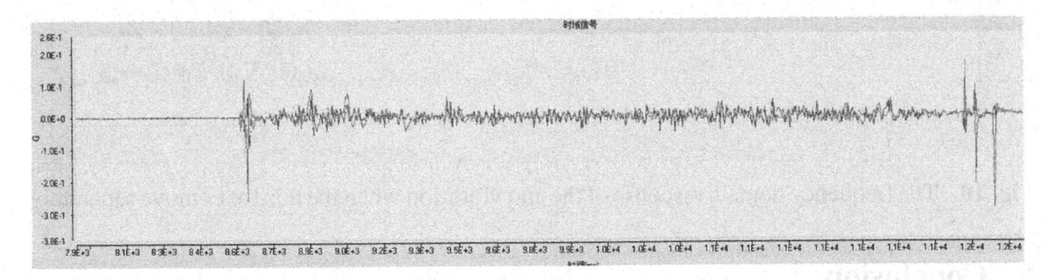

Fig. 7. Time domain response of end vibration when the second axes move separately

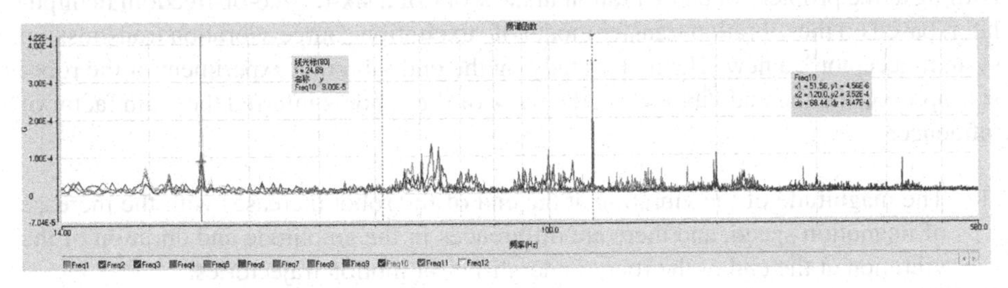

Fig. 8. Frequency domain response of end vibration when the second axes move separately

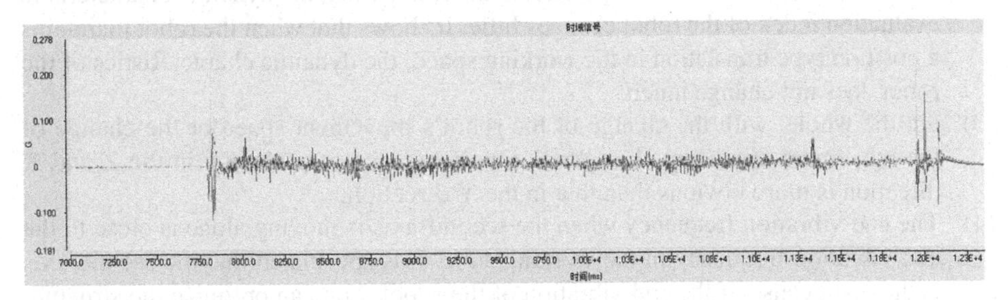

Fig. 9. Time-domain response of end vibration when third axes are moving separately

Through the analysis of the test data, it can be seen that the vibration range of the second-axis start is −0.26G–+0.12G, the vibration range at the moment of stop is −0.30G– +0.17G, the vibration range is ±0.1G during stable operation, and the vibration frequency is mainly Around 24.69 Hz. The vibration range of third-axis start is −0.116G– +0.208G, the vibration range at the moment of stop is −0.1G –+0.213G, the vibration range is ±0.05G when running smoothly, and the vibration frequency is mainly distributed around 27.19 Hz. The test results show that the main vibration frequency when the second-axis is moving alone is close to the end vibration frequency when moving in the working state. It is speculated that the main influencing factor of the end vibration in the robot's work is the vibration generated during the second-axis movement.

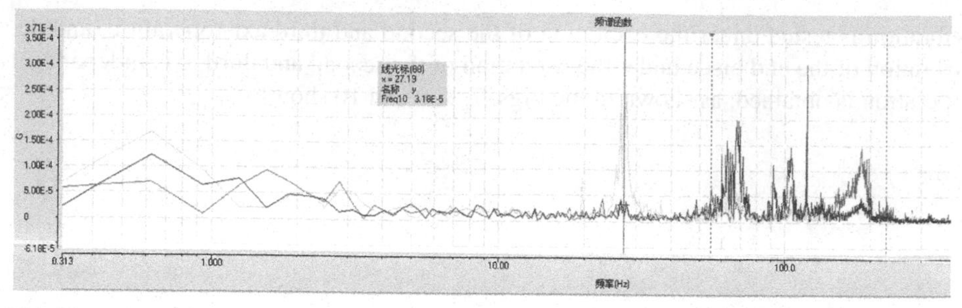

Fig. 10. The frequency domain response of the end vibration when the third axes move separately

6 Conclusion

Aiming at the problem of end vibration in the work of a six-degree-of-freedom manipulator, based on the vibration transfer model of "excitation source-vibration transmission system-acceptor", a new scheme for studying the end vibration experiment of the robot is proposed, and the end vibration in the work of the robot is inferred the main factor of influence.

(1) The magnitude of the vibration at the end of the robot increases with the increase of its motion speed, and there are differences in the amplitude and duration of the vibration at the end of the robot under different motion trajectories.
(2) When each axis of the robot moves independently, the movement is mainly manifested as a movement in a single direction. At this time, the dynamic characteristic evaluation index of the robot changes little. It shows that when the robot maintains a position type translation in the working space, the dynamic characteristics of the robot does not change much.
(3) On the whole, with the change of the robot's movement speed or the change of the movement trajectory, the vibration performance of each axis in the Z and X direction is more obvious than that in the Y direction.
(4) The end vibration frequency when the second-axis is moving alone is close to the end vibration frequency in the working state. It is speculated that the second-axis is the main cause of the end vibration of the robot. You can optimize the structure of the second-axis, replace the second-axis motor or replace the resonance. Wave gears to improve the end vibration of the robot.

Acknowledgement. This paper was funded by 2020 Collaborative Education Program of Higher Education Department of the Ministry of Education, No. 202002110006; 2021 National Foreign Languages Teaching and Research Program for Universities, No. 202110003; National Natural Science Foundation of China, No.51775165.

References

1. Wenqiang, C.: Research status and development trend of industrial robots. Equipment Manage. Maintenance **24**, 118–120 (2020)

2. Wei, L., Xue, X.: Research status and development trend of industrial robots. Inf. Record. Mater. **20**(07), 48–49 (2019)
3. Tobias, S.A., Fishwick, W.: The chatter of lathe tools under orthogonal cutting conditions. ASME Trans. **80**(5), 1079–1088 (1958)
4. Petit, F., Lakatos, D., Friedl, W., et al.: Dynamic trajectory generation for serial elastic actuated robots. IFAC Proc. **45**(22), 636–643 (2012)
5. Summer, M.: Robot capability test and development of industrial robot positioning system for the aerospace industry. In: SAE 2005 AeroTech Congress & Exhibition, Grapevine, TX, SAE Technical Papers 2005-01-3336, (2005)
6. Weiya, G., Junjun, Z., Leishu, Y., et al.: Vibration analysis and structure optimization of the robotic arm of a detection robot. J. Southwest Univ. Sci. Technol. **33**(3), 82–87 (2018). https://doi.org/10.3969/j.issn.1671-8755.2018.03.016
7. Chen, J., Chen, Q., Hu, X., et al.: Lightweight design of palletizing robot arm structure. Modular Mach. Tool Autom. Process. Technol. **5**, 19–22 (2019). https://doi.org/10.13462/j.cnki.mmtamt.2019.05.005. 26
8. Zhang, T., Qin, B., Liu, X.: Analysis and suppression of robot joint vibration based on flexible body dynamic model. Vibr. Test. Diagn. **39**(02), 242-248+438 (2019)
9. Xiaodong, S., Changyi, L., Baishou, Z.: Dynamic simulation of robot hole making process considering joint flexibility. Mach. Design Manuf. **03**, 196–200 (2015)
10. Mousavi, S., Gagnol, V., Bouzgarrou, B.C., Ray, P.: Dynamic modeling and stability prediction in robotic machining. Int. J. Adv. Manuf. Technol. **88**(9–12), 3053–3065 (2016). https://doi.org/10.1007/s00170-016-8938-0
11. Zhang Yuling, G., Yongxia, Z.J., et al.: Research on end vibration characteristics of manipulator arm under active control of stiffness. Chinese J. Theoret. Appl. Mech. **52**(4), 985–995 (2020). https://doi.org/10.6052/0459-1879-20-075
12. Shuiyuan, T., Jianmin, Z., Xinyi, W., Xianfeng, W.: Research on the design method and application of robot mechatronics—modal analysis of vibration test. J. Beijing Instit. Technol **02**, 206–210 (1994)
13. Xiong, X., Qin, H.: Quantitative evaluation of the dynamic performance of industrial robots based on vibration testing. Metrol. Test. Technol. **46**(08), 19-22+27 (2019)
14. Mejri, S., Gagnol, V., Le, T.-P., Sabourin, L., Ray, P., Paultre, P.: Dynamic characterization of machining robot and stability analysis. Int. J. Adv. Manuf. Technol. **82**(1–4), 351–359 (2015). https://doi.org/10.1007/s00170-015-7336-3
15. Bisu, C.F., Cherif, M., Gérard, A., et al.: Dynamic behavior analysis for a six axis industrial machining robot. Adv. Mater. Res. **42**(3), 65–76 (2012)
16. Zhixian, T., Ying, P.: Simulation vibration analysis and vibration test of SCARA manipulator. Manuf. Autom. **42**(06), 21–26 (2020)

Design and Analysis of a New Symmetrical Fast Steering Mirror

Lianhui Jia[1], Lichao Ji[2], Wule Zhu[3], Xiaolei Zhou[1], and Xu Yang[2](✉)

[1] China Railway Engineering Equipment Group Co., Ltd., Zhengzhou 450016, China
[2] Institute of Marine Science and Technology, Shandong University, Qingdao 266237, China
[3] State Key Laboratory of Fluid Power and Mechatronic Systems, Zhejiang University, Hangzhou 310027, China

Abstract. In this paper, the development of a symmetrical fast steering mirror (SFSM) actuated by a piezoelectric actuator is conducted, aiming at the high-speed and high-accuracy angle control of light beam. First, the compliant mechanism of SFSM is specially designed and modelled. Because of the symmetrical design, high decoupling output motion can be achieved. The compact compliant structure also leads to fast response and frictionless motion. For investigating the performance of proposed SFSM, finite element analysis is carried out in terms of kinematic, stiffness and dynamic. Results confirms that the proposed SFSM is featured with high bandwidth, good output decoupling and good disturbance rejection, and could be a promising solution for the future practical applications.

Keywords: Fast steering mirror · Compliant mechanism · Symmetrical design

1 Introduction

Fast steering mirror could regulate the reflection angle of light beam in a fast and accurate manner, which has found broad application in measurement, manufacture, and communication. For example, the single-axis fast steering mirror can be used to compensate for the flatness error of the polygon facets in prepress photography [1]. With an additional fast steering mirror, a high-speed scanning of laser beam could be realized for the laser machine, which help to minimize the thermal cracks of brittle materials [2]. Fast steering mirror can also be used to correct the angular error of laser beam caused by spacecraft jitter, resulting in a stable optical path for sending and receiving communication signals from Mars [3]. Even the variety of applications put forward different requirements, high bandwidth and high resolution are the basic requirements for the fast steering mirrors.

In the past few years, various fast steering mirrors have been designed which can be generally classified into electromagnetic actuated and piezoelectric actuated types. The steering mirror driven by voice coil can realize a precision positioning over a large range. Kluk *et al.* employed two flux-steering electromagnetic actuators to configure a two-axis mirror which has a motion range as large as ±3.5 mrad [4]. Csencsics *et al.* investigated the eddy current effects and proposed to adopt a layered yoke structure to

X.-J. Liu et al. (Eds.): ICIRA 2021, LNAI 13014, pp. 384–392, 2021.
https://doi.org/10.1007/978-3-030-89098-8_37

improve the system dynamics [5]. Furthermore, Csencsics *et al.* integrated a hybrid-reluctance-force actuator and a flexure mover into a fast steering mirror [6]. Long *et al.* designed a two-axis rotary electromagnetic actuator with symmetric structure and cross topology armature for the fast steering mirror [7]. However, the voice coil based steering mirrors exhibit small output stiffness and low operating bandwidth, and also are easily influenced by the structural vibration. Hence, some researchers devoted to design the fast steering mirrors with piezoelectric actuators. Chen *et al.* designed a 2-DOF fast steering mirror with three parallel branch chains which are actuated by three piezoelectric actuators [8]. Zhou *et al.* proposed to control each axis of the mirror with two parallel connected piezoelectric actuators [9]. Through integrating four symmetrical and parallel piezoelectric actuators, Zhu *et al.* designed a two-axis fast steering mirror which can move around two perpendicular directions [10]. Even much research work has been carried out, the fast steering mirror still needs to be optimized in terms of working bandwidth, output decoupling and disturbance rejection.

In this paper, a new symmetrical fast steering mirror (SFSM) is designed, which mainly consists of a piezoelectric actuator and a deflection compliant mechanism. The deflection mechanism can transform the output motion of piezoelectric actuator into the rotation of reflection panel by means of self-deformation. For being free of traditional mechanical joints, high motion resolution and high dynamic can be easily achieved. Moreover, the symmetrical design of deflection mechanism helps to restrain the undesired parasite motions.

The organization of this paper is as follows. Section 2 presents the mechanical layout of proposed SFSM. The kinematic and static characteristics are detailly analyzed in Sect. 3. Simulations are carried out to investigate the performance of proposed SFSM in Sect. 4. Finally, conclusions are drawn in Sect. 5.

2 Mechanical Layout of SFSM

Figure 1 displays the mechanical layout of proposed symmetrical fast steering mirror (SFSM). As can be seen from the figure, a deflection compliant mechanism driven by a piezoelectric actuator is designed to realize the fast and accurate rotation of mirror. As parts of the deflection compliant mechanism, two input decoupling mechanisms are designed to protect the piezoelectric actuator from undesired lateral force. The output motion of piezoelectric actuator is transformed into the reverse motion of two translational beams (termed as up translation beam and down translation beams respectively) by means of two compliant levers. Four flexure beams are designed to guide the motions of the two translational beams. The two translational beams are also linked by two reflection panels (termed as up reflection and down reflection panels respectively), configuring a parallelogram. Single-notch circular flexure hinges are employed to be the rotation joints of the parallelogram. In this way, the reverse motion of two translational beams can be transformed into the rotational motion of two reflection panels. The reflection center is close to the rotation center of the up reflection panel, which can attenuate the undesired parasite motions of the reflection center. With the proposed SFSM, the reflection angle of light beam can be regulated in a fast and accurate manner, as shown in Fig. 2.

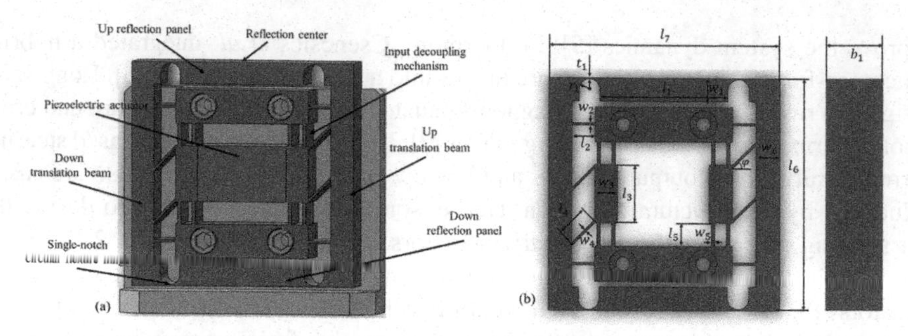

Fig. 1. (a) Three-dimensional structure of SFSM, (b) Front side view of SFSM.

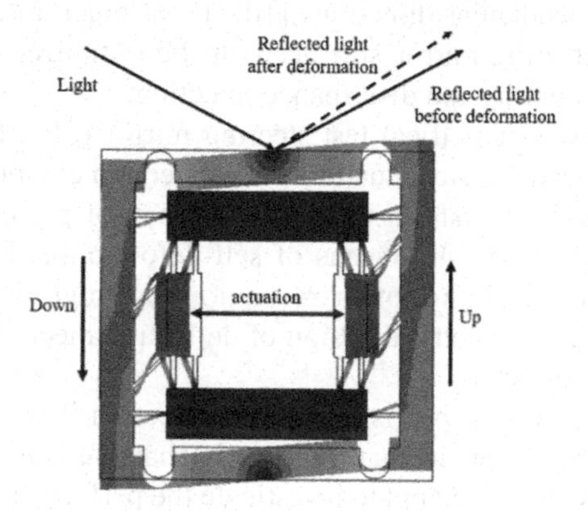

Fig. 2. Working principle of SFSM.

3 Analytical Model of SFSM

According to the motion characteristics of proposed SFSM, a kinematic model is specially established, as shown in Fig. 3. The input decoupling mechanism, the compliant levers, and the guiding beams are modeled with the rigid links. The single-notch circular flexure hinges are modelled as rotatory joints.

According to the transient velocity theory, the angle velocity of the compliant lever ω_4 is solved by

$$\omega_4 = V_P / l_{PO}, \ l_{PO} = l_4 \times sin\varphi \tag{1}$$

where l_4 is the length of the compliant lever, V_P is the instantaneous speed of point P, l_{PO} is the distance from point P to point O (instantaneous center of speed).

From (1), the instantaneous speed of point Q (V_Q) can be obtained as

$$V_Q = \omega_4 \times l_{QO}, \ l_{QO} = l_4 \times cos\varphi \tag{2}$$

where l_{QO} is the distance from point Q to point O.

Considering the initial angle parameter of the compliant lever, the speed relationship between the two end-points of the compliant lever can be solved.

$$V_Q = V_P \times cot\varphi \tag{3}$$

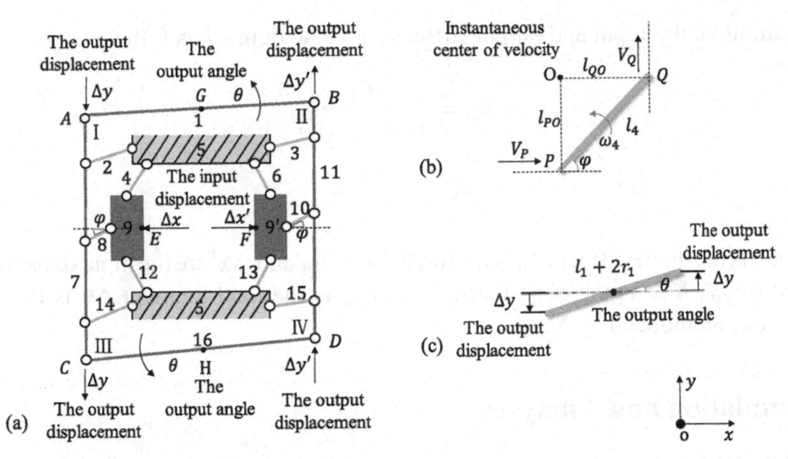

Fig. 3. (a) Kinematic model of SFSM, (b) Model of the compliant lever, (c) Model of the reflection panel.

In the case of small deformation, we can further deduce the following equation

$$\Delta x = \Delta y \times tan\varphi \tag{4}$$

where Δx is the input displacement of point P and Δy is the output displacement of point Q.

According to model shown in Fig. 2 (c), the output angle of reflection panel can be obtained as

$$\Delta\theta = \frac{2 \times \Delta y}{l},\ l = l_1 + 2 \times r_1 \tag{5}$$

Combining Eqs. (4) and (5), we can obtain the relationship between the input displacement Δx and the output angle $\Delta\theta$.

$$\Delta\theta = \frac{2 \times \Delta x \times cot\varphi}{l},\ l = l_1 + 2 \times r_1 \tag{6}$$

While the structure beam rotates around the instantaneous center, the reflection center will also have parasitic motions in x and y directions. For investigating the parasitic motions, an x-directional parasitic ratio is defined as

$$R_x = \frac{d_x}{\Delta\theta} \tag{7}$$

where d_x is the x-directional displacement of reflection center.

Besides, an y-directional parasitic ratio is defined as

$$R_y = \frac{d_y}{\Delta\theta} \tag{8}$$

where d_y is the y-directional displacement of reflection center.

In addition, the input and output stiffness are also defined as follows

$$K_{in} = \frac{f_{in}}{(\Delta x + \Delta x')} \tag{9}$$

$$K_{out} = \frac{M_{out}}{\Delta \theta} \tag{10}$$

where K_{in} is the input stiffness, f_{in} is the input force, Δx and $\Delta x'$ are the input deformations induced by f_{in}, K_{out} is the output stiffness, M_{out} is the load moment, $\Delta\theta$ is the output angle variation induced by M_{out}.

4 Simulation and Analysis

A three-dimensional model of the proposed SFSM with the dimensions shown in Table 1 is established for performance analysis, with the SolidWork Software. A piezoelectric actuator (type: NAC2003-H08, size: 5 × 5 × 8 mm, maximum stroke: 9 μm, output stiffness: 117 N/μm, resonant frequency: 120 kHz) is adopted to drive the SFSM virtual prototype. The model is defined with the material of titanium alloy (young's modulus: 9.6×10^{10} Pa, poisson's ratio: 0.36, density: 4620 kg/m^3). Based on this model, finite element analysis is carried out in the ANSYS Workbench Software to investigate its performance in terms of kinematic, stiffness and dynamic.

Table 1. Dimensions of the SFSM virtual prototype.

Par.	t_1	r_1	l_1	l_2	l_3	l_4	l_5	l_6	l_7
Val. (mm)	0.1	0.9	11.2	2	5	2.83	2	20	20
Par.	b_1	w_1	w_2	w_3	w_4	w_5			
Val. (mm)	5	2	0.2	2	0.2	0.3			

4.1 Mesh Generation

In the ANSYS Workbench Software, a meshing model of the proposed SFSM is firstly built as shown in Fig. 4. For calculation accuracy as well as calculation speed, the mesh size function is set to be proximity and curvature, and the relevance is configured to be −50. According to the Fig. 4, each flexible beam is divided into two or more layers of elements which results in better accuracy. As a result, the resulted mesh model has about 1.2×10^5 elements.

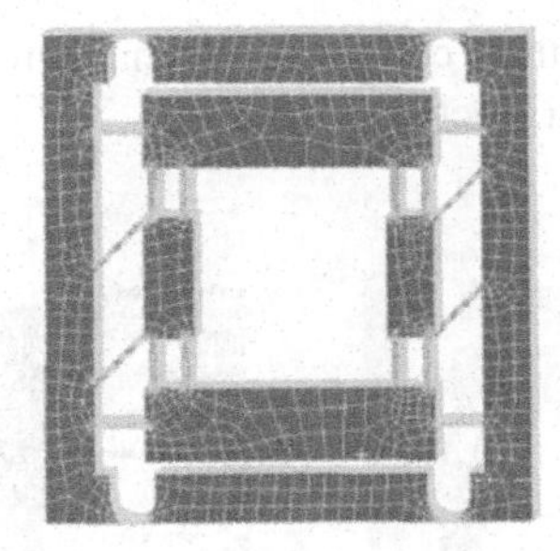

Fig. 4. Finite-element model of SFSM virtual prototype.

4.2 Kinematic Analysis

Applying two input displacements (1×10^{-6} m) to the input points E and F respectively, the induced stress and displacement distribution of SFSM prototype can be calculated in the ANSYS Workbench, as shown in Fig. 5(a) and Fig. 5(b). Two symmetrical points M and N are selected on the up reflection panel of SSFM, and point M and N are about 4 mm apart from the reflection center. According to Eq. (6), the kinematic ratio between out angle and input displacement is figured out to be 67.38 rad/m. In addition, the resulted maximum stress of SFSM prototype is also figured out, as shown in Fig. 5(a). According to the maximum stress shown in the figure, the ratio of maximum stress over out angle is obtained to be 1.17×10^{11} Pa/rad.

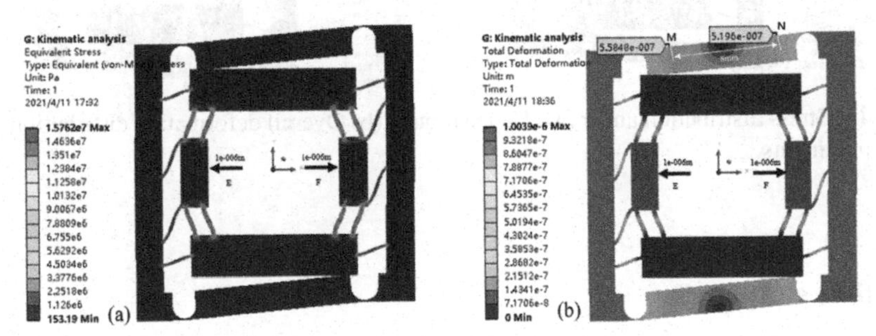

Fig. 5. (a) Stress distribution under two 1 μm inputs, (b) Overall deformation distribution under two 1 μminputs.

Corresponding to the overall deformation shown in Fig. 5(b), the deformation in x and y directions are further revealed in Fig. 6(a) and Fig. 6(b). According to Eqs. (7) and (8), the x-directional parasitic ratio is revealed to be 1.89×10^{-4} m/rad. In addition, the y-directional parasitic ratio is obtained to be 1.67×10^{-4} m/rad. Therefore, the output motion of proposed SFSM prototype is featured good decoupling capability.

Increasing the two input displacements to 5.9 5.9×10^{-5} m, the stress of SFSM reaches the tensile yield strength (9.3×10^{8} Pa), as shown in Fig. 7(a). In the meantime, the compliant mechanism of SFSM reaches the maximum output angle, which is revealed to be 7.95×10^{-3} rad (according to the displacements of the points M and N), as shown in Fig. 7(b). In practice, the maximum output angle of SFSM is not only limited by the maximum tensile yield strength of compliant mechanism, but also is restricted by the maximum stroke of piezoelectric actuator.

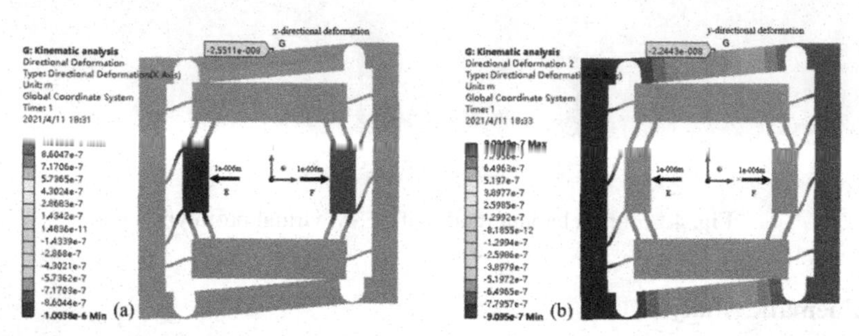

Fig. 6. (a) x-directional deformation under two 1 μm inputs, (b) y-directional deformation under two 1 μm inputs.

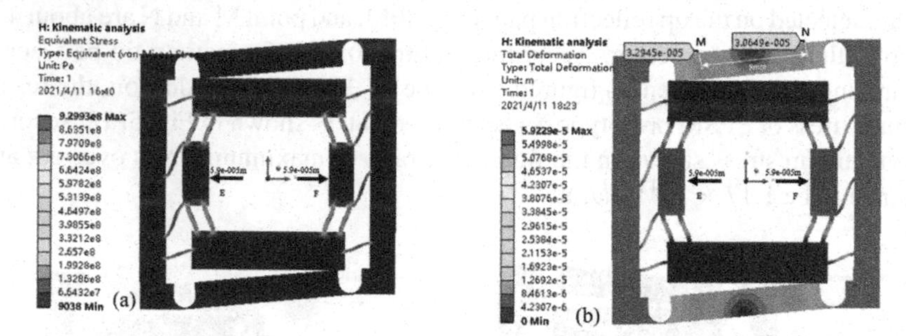

Fig. 7. (a) Stress distribution under two 59 μm inputs, (b) Overall deformation distribution under two 59 μm inputs.

4.3 Stiffness Analysis

Applying two unit forces (1 N) to the input points E and F respectively, the resulted deformation of SFSM prototype is obtained as Fig. 8 (a). Based on the simulation result, the input stiffness of SFSM prototype is calculated as 2.99×10^6 N/m according to Eq. (9). Creating two planes which are perpendicular to the reflection panel and 0.5 mm away from the reflection center on the SFSM prototype in the ANSYS Workbench. The two planes divide the up reflection panel into three parts. Applying a moment (1 Nm) around the z-axis on the middle part of the up reflection panel of the SFSM prototype, the resulted deformation can be obtained as Fig. 8 (b). In particular, this mesh model has about 1.0×10^5 elements. The angle deformation of reflective center is calculated according to displacement difference between point M and N which are both 4 mm away from the reflection center. By relating the applied load moment and induced output deformation, the output stiffness is figured out to be 75.16 Nm/rad.

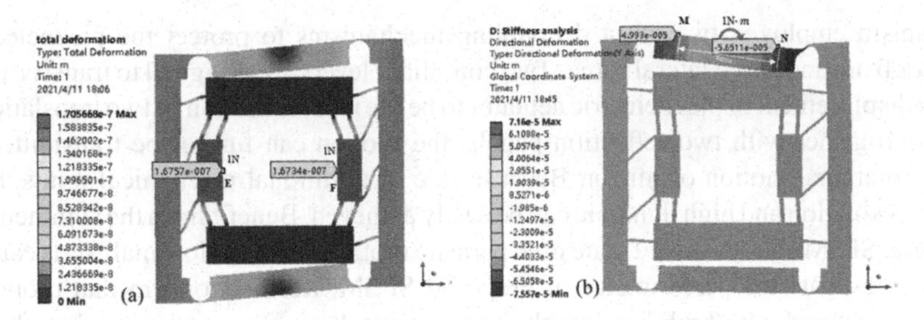

Fig. 8. (a) Input stiffness analysis result, (b) Output stiffness analysis result.

4.4 Modal Analysis

In addition, modal analysis is carried out to investigate the open-loop bandwidth and the disturbance resistance capability of SFSM prototype. The first to four orders modal shapes of the SFSM prototype are revealed and analyzed, as shown in Fig. 9. As can be seen from the figure, the first-mode natural frequency is as large as 7276.6 Hz, which means a high open-loop bandwidth.

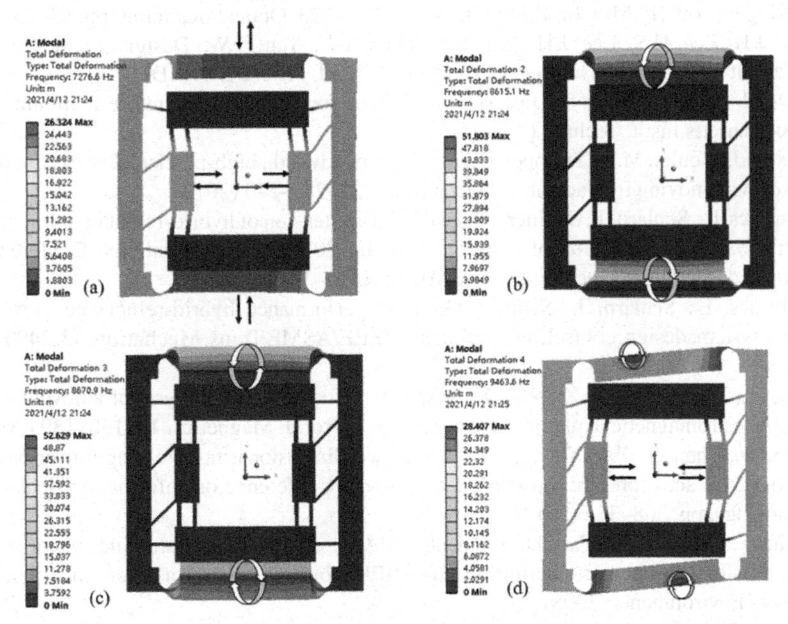

Fig. 9. (a) 1st order modal shape, (b) 2nd order modal shape, (c)3rd order modal shape, (d) 4th order modal shape.

5 Conclusion

This paper proposes a new symmetrical fast steering mirror (SFSM), which consists of a deflection compliant mechanism and a piezoelectric actuator. The deflection compliant mechanism employed two input decoupling mechanisms to protect the piezoelectric actuator from undesired lateral force. Two compliant levers are designed to transform the output displacement of piezoelectric actuator to be the reverse motion of two translational beams. Together with two reflection panels, the motion can further be transmitted to be the rotational motion of mirror. Because free of traditional mechanical joints, high motion resolution and high dynamic can be easily achieved. Benefit from the symmetrical structure, SFSM can also avoid undesired parasite motions. Simulation analysis is carried out to investigate the performance of proposed SFSM. Results confirm that proposed SFSM is featured with high bandwidth, good output decoupling and good disturbance resistance.

Acknowledgments. This work is supported by the National Key R&D Program of China under Grant 2019YFB2005303, National Natural Science Foundation of China under Grant 51805363.

References

1. Sweeney, M., Rynkowski, G., Ketabchi, M., Crowley, R.: Design considerations for fast-steering mirrors (FSMs). In: Proceedings of SPIE 4773, Optical Scanning, pp. 63–73 (2002)
2. Park, J.H., Lee, H.S., Lee, J.H., Yun, S.N., Ham, Y.B., Yun, D.W.: Design of a piezoelectric-driven tilt mirror for a fast laser scanner. Jpn. J. Appl. Phys. **51**, 09MD14 (2012)
3. Hawe II, L.E.: Control of a fast steering mirror for laser-based satellite communication. Massachusetts Instit. Technol (2006)
4. Kluk, D.J., Boulet, M.T., Trumper, D.L.: A high-bandwidth, high-precision, two-axis steering mirror with moving iron actuator. Mechatronics **22**, 257–270 (2012)
5. Csencsics, E., Schlarp, J., Schitter, G.: Bandwidth extension of hybrid-reluctance-force-based tip/tilt system by reduction of eddy currents. In: 2017 IEEE International Conference on Advanced Intelligent Mechatronics (AIM), pp. 1167–1172 (2017)
6. Csencsics, E., Schlarp, J., Schitter, G.: High-performance hybrid-reluctance-force-based tip/tilt system: design, control, and evaluation. IEEE/ASME Trans. Mechatron. **23**, 2494–2502 (2018)
7. Long, Y., Wang, C., Dai, X., Wei, X., Wang, S.: Modeling and analysis of a novel two-axis rotary electromagnetic actuator for fast steering mirror. J. Magnetics. **19**, 130–139 (2014)
8. Chen, W., Chen, S., Wu, X., Luo, D.: A new two-dimensional fast steering mirror based on piezoelectric actuators. In: 4th IEEE International Conference on Information Science and Technology, pp. 308–311 (2014)
9. Zhou, Q., Ben-Tzvi, P., Fan, D., Goldenberg, A.A.: Design of fast steering mirror systems for precision laser beams steering. In: 2008 IEEE International Workshop on Robotic and Sensors Environments (2008)
10. Zhu, W., Bian, L., An, Y., Chen, G., Rui, X.: Modeling and control of a two-axis fast steering mirror with piezoelectric stack actuators for laser beam tracking. Smart Mater. Struct. **24**, 075014 (2015)

Sample-Efficiency, Stability and Generalization Analysis for Deep Reinforcement Learning on Robotic Peg-in-Hole Assembly

Yuelin Deng, Zhimin Hou, Wenhao Yang, and Jing Xu(✉)

Department of Mechanical Engineering, Tsinghua University, Beijing 100084, China
jingxu@tsinghua.edu.cn

Abstract. In the field of robotic assembly, deep reinforcement learning (DRL) has made a great stride in the simulated performance and holds high promise to solve complex robotic manipulation tasks. However, a huge number of efforts are still needed before RL algorithms could be implemented in the real-world tasks directly due to the risky but insufficient interactions. Additionally, there is still a lack of analyzation in the sample-efficiency, stability and generalization ability of RL algorithms. As a result, Sim2Real, analyzing RL algorithms in simulation and then implementing in real-world tasks, has become a promising solution. Peg-in-hole assembly is one of the fundamental forms of the robotic assembly in industrial manufacturing. In the paper, we set up a simulation platform with physical contact models of both single and multiple peg assembly configurations; we then provide the commonly used RL algorithms with an empirical study of the sample-efficiency, stability and generalization, ability; we further propose a new algorithm framework of Actor-Average-Critic (AAC) for better stability and sample-efficiency performance. Besides, we also analyze the existing reinforcement learning with hierarchical structure (HRL) and demonstrate its better generalization ability into new assembly tasks.

Keywords: Deep reinforcement learning · Peg-in-hole assembly

1 Introduction

Peg-in-hole assembly is a fundamental form of industrial assembly tasks. The existing research has achieved a certain level through the conventional controllers e.g., PD and impedance controllers, however, such acquired performance largely depends on contact state recognition. It is still difficult to address the multi-peg assembly problem via conventional controllers due to the complicated contact model [7]. Additionally, modern manufacturing requires not only quantity, but also quality and efficiency which introduce the need for greater stability and generalization ability against environmental uncertainties.

Y. Deng, Z. Hou and W. Yang—Joint first author.

X.-J. Liu et al. (Eds.): ICIRA 2021, LNAI 13014, pp. 393–403, 2021.
https://doi.org/10.1007/978-3-030-89098-8_38

Reinforcement learning (RL) holds the promise to learn assembly skills without contact state recognition [8]. Particularly, deep reinforcement learning (DRL) combined with learning representation skills, has solved a large range of complex robotic manipulation tasks [9]. DRL techniques have been extensively explored to learn the competitive assembly skills [5], however, the sample efficiency, stability, and generalization ability are less discussed. The real-world peg-in-hole assembly with small clearance requires both vision information and accurate force/torque feedback, which guides the assembly movement and avoid failure caused by large contact force. To our best knowledge, there are less existing work on the simulation platform for peg-in-hole assembly tasks basing on force feedback. The first motivation of this paper is to set up a peg-in-hole assembly simulator with physical contact prediction ability. With the contact force detected and read from the force sensor in the simulation platform, the sample-efficiency, stability and generalization ability of the current RL algorithms can then be analyzed.

The sample-efficiency is an essential issue for real-world applications as interactions with real-world environments are at great cost. Compared with the on-policy RL that requires new samples for every training step, off-policy RL can reuse the past experience and reduce the sample complexity [10]. Deep Deterministic Policy Gradient (DDPG) is one of the popular off-policy RL algorithms to learn a high-dimensional deterministic policy. Therefore, many DDPG-like algorithms have been put forward and researched to handle the peg-in-hole assembly tasks [2,15]. Exploration is another essential topic for RL algorithms to reduce the sample complexity and search for the optimal skills, therefore, some approaches were proposed to guide the efficient exploration basing on conventional controllers or human demonstration [15]. Recently, model-based RL assisted with an estimated dynamic environment model is showing a great promise to reduce the need for interactions [9]. Guided policy search (GPS) has shown a competitive performance to learn manipulation skills with the estimated local dynamic model [9].

DRL implemented basing on deep neural networks has achieved great success to solve complex control tasks [9]. However, most of the RL algorithms are sensitive to the selection of hyper parameters [10]. Compared with supervised learning methods, RL training is often unstable without the fixed targets, which is not allowed for real-world tasks. Some practical training techniques have been studied to improve the stability e.g., mini-batch training [10], delayed target network to address the over-estimation issues in function approximation [14]. To address the bad actions brought by random explorations, the residual RL learns advanced skills basing on the pre-trained policy or conventional controllers [6]. Another attractive direction is to reformulate the action space [13], meaning that the action will be constrained by a safe and efficient scheme. Three action spaces(PD control, inverse dynamic, and impedance control) to solve the commonly used robotic tasks are compared and discussed [13].

DRL holds the promise to learn the skills with better robustness and generalization. Random initial positions, peg-hole clearance, and sizes of pegs are often considered as the environmental variables to demonstrate the robustness and generalization ability of DRL methods [2,15]. The adaptability of learned

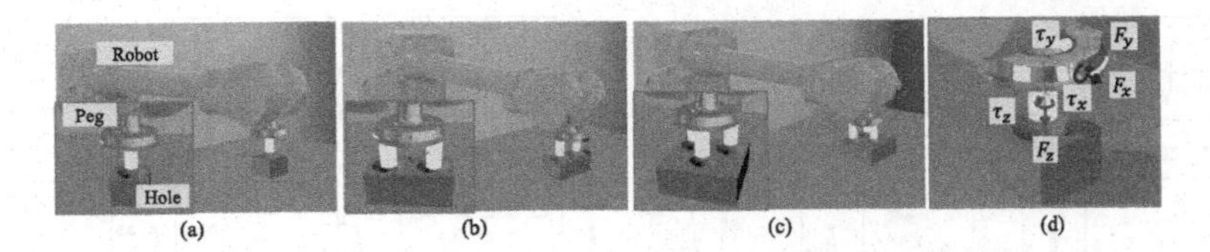

Fig. 1. Structure of peg-in-hole assembly. (a) Single (b) Dual (c) Triple (d) F/T sensor.

assembly skills to the new tasks with different shapes of pegs have been discussed [2]. To improve the task generalization and environment adaptation ability, a deep model fusion framework is proposed to fuse the previously learned knowledge from different assembly environments. Currently, the complex control tasks are sometimes structured hierarchically, hierarchical reinforcement learning (HRL) aims to learn a hierarchical policy with some basic skills, which can be easily transferred to new tasks [4]. However, up to now, the generalization for peg-in-hole assembly has not been discussed sufficiently.

The motivation of this paper is to promote the research for solving peg-in-hole assembly tasks by DRL and provide a guidance to develop the practical RL algorithm. The contributions are concluded as:

1. We set up a force-based simulation platform on Webots to train and test the RL agents for peg-in-hole assembly tasks in different task configurations.
2. We propose a data-efficient actor-average-critic (AAC) RL framework with both better training stability and higher sample-efficiency. We then verify the performance of the proposed algorithm framework and compare it with the methods from existing literature in different peg-in-hole assembly situations (with different clearance, friction, and the number of pegs), through our simulation platform.
3. We evaluate the transfer performance of the assembly skills learned on unseen assembly tasks with different numbers of pegs. We demonstrate that the assembly skills learned by HRL could be transferred to different assembly situations (with different number of pegs) easier than the existing methods.

2 Peg-in-Hole Assembly Platform

The peg-in-hole assembly tasks (including single, dual and triple peg situations) are set up in Webots simulator [1], as shown in Fig. 1(a)–(c).

2.1 Hardware

Robot: There is a variety of manipulators to choose from in Webots e.g., UR, PUMA, ABB IRB, etc. The robots can also be built given the Unified Robot Description Format (URDF). In this paper, ABB IRB 4600, a 6-DOF manipulator is selected to implement the peg-in-hole assembly tasks. All the motors of the robot in Webots are implemented with an underlying PID controller, the robot is controlled by giving the velocity and target angle of each motor.

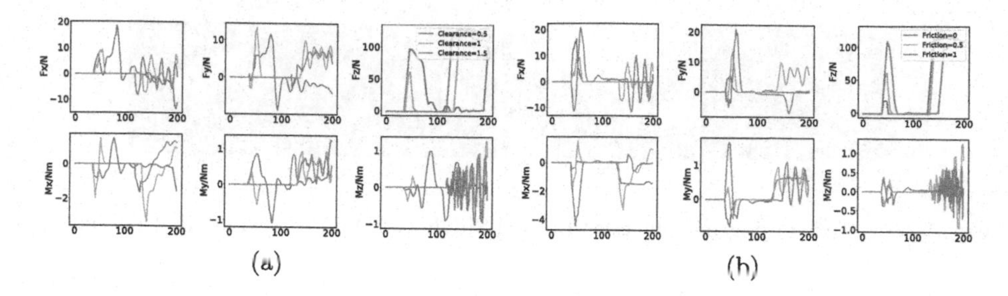

Fig. 2. F/T sensor signals (a) Different peg-hole clearance. (b) Different friction.

F/T Sensor: There is no available multi-axis force sensor in Webots, however, a motor can output force feedback to indicate the required force to hold its position. We built a chain of joints at a single point of the robot end-effector and apply motor behaviour, as shown in Fig. 1(d), which includes three rotational motors and three linear motors.

This chain of joints will serve as a 6-axis F/T sensor to detect the contact forces. However, due to the properties of the simulation engine, this error could be integrated to a noticeable level, which we consider as a uncontrollable random noise. To test the performance of the F/T sensor, a PD controller is applied to control the single peg-in-hole assembly in different situations. The force signals from F/T sensor are plotted as shown in Fig. 2.

Components: Three different numbers of cylindrical peg-hole components (single, dual, and triple) are set up. We assume that each peg-hole pair has the same configuration for dual and triple peg-in-hole assembly tasks. We added chamfers to the hole entrance so that we can focus on the insertion process instead of worrying about an extra search stage to align the peg. Additionally, the Coulomb friction coefficients and the clearance of peg-hole pairs can be changed to simulate different peg-in-hole assembly tasks. The parameters and their adjustable range are listed in Table 1.

Table 1. Adjustable parameters and properties

Name	Adjustable region
Number of peg-hole pairs	1,2,3
Coulomb friction coefficients	0 - inf
Hole size	30 mm
Peg-hole clearance	0.5 mm–1.5 mm
Hole position noise range	[−1, 1]mm

2.2 Software

The Webots platform is compatible with the system including Windows, Linux, and Mac OS. The connection between the simulator and the RL algorithms is implemented by the Python API, which reads the sensor feedback from the simulator as RL input, and outputs the control commands to the robot. Different assembly tasks can be set up via the Webots-Python interface given the parameters in Table 1.

3 RL Algorithms Analysis

3.1 Problem Statement

RL can solve all phases integrally based on the force feedback similar as "feeling" of human beings, which formulates the assembly process as the *Markov Decision Process(MDPs)*. At each time step t, according to the observed state $\boldsymbol{s}_t$, the robot select the "optimal" action $\boldsymbol{a}_t$ based on the policy $\pi(\boldsymbol{a}_t|\boldsymbol{s}_t)$. The robot will receive the reward signals r_t from environment to evaluate the state-action pair, and then the environment transit to the next state $\boldsymbol{s}_{t+1} = p(\boldsymbol{s}_{t+1}|\boldsymbol{a}_t, \boldsymbol{s}_t)$. The policy is updated through maximizing the expected cumulative reward, as:

$$\mathbb{E}_{a_t \sim \pi, s_{t+1} \sim p(\cdot)} \left[\sum_{t=0}^{T-1} \gamma^t r_{t+1}(\boldsymbol{s}_t, \boldsymbol{a}_t) \right] \tag{1}$$

where γ denotes the discount factor. The action value function $Q^\pi(\boldsymbol{s}, \boldsymbol{a}) = \mathbb{E}_\Omega[G_t|\boldsymbol{s}_t = \boldsymbol{s}, \boldsymbol{a}_t = \boldsymbol{a}]$ is defined to estimate the expected cumulative reward given the trajectory Ω, $G_t = \sum_{i=t}^{T-1} \gamma^{i-t} r_{i+1}(\boldsymbol{s}_i, \boldsymbol{a}_i)$ denotes the cumulative reward of one episode from $t = 0$ to T. The objective can be rewritten as:

$$\mathcal{J}(\pi) = \mathbb{E}_{(\boldsymbol{s},\boldsymbol{a}) \sim \mathcal{P}_\pi} [Q^\pi(\boldsymbol{s}, \boldsymbol{a})] \tag{2}$$

where $\mathcal{P}_\pi$ denotes the state-action marginals of the trajectory distribution induced by a policy π.

As shown in Fig. 1, the state for all the peg-in-hole assembly tasks often consists of the pose(position and orientation) of pegs and the force(force and moment) read from the F/T sensor, therefore, the state denotes as:

$$\boldsymbol{s}_t = [\underbrace{P_t^x, P_t^y, P_t^z}_{\text{position(mm)}}, \underbrace{O_t^x, O_t^y, O_t^z}_{\text{orientation}}(\circ), \underbrace{F_t^x, F_t^y, F_t^z}_{\text{force(N)}}, \underbrace{T_t^x, T_t^y, T_t^z}_{\text{torque(Nm)}}] \tag{3}$$

where the subscripts x, y and z denote the axes of the robot base coordinate. The basic action space $\boldsymbol{a}_t$ is denoted as follows:

$$\boldsymbol{a}_t = [\Delta P_t^x, \Delta P_t^y, \Delta P_t^z, \Delta O_t^x, \Delta O_t^y, \Delta O_t^z] \in \mathbb{R}^6 \tag{4}$$

which are utilized to control the pose of the pegs along the corresponding axis in the robot base coordinate.

A reward function is defined to decrease the assembly steps and avoid the unsafe contact force, as

$$r_t = \begin{cases} -0.1, \ done = False \ and \ safe = True \\ 1 - \dfrac{N}{N_{max}}, \ done = True \ and \ safe = True \\ -1 + \dfrac{D_t}{D_e}, \ safe = False \end{cases} \tag{5}$$

where N_{max} denotes the maximum number of steps in one episode; $N \in [0, N_{max})$ denotes the final number of steps. To encourage fewer number of assembly steps, the robot will receive -0.1 every step. Additionally, the penalty is designed according to the force signals, the assembly will be interrupted with reward -1 once the value of forces exceeding the setting boundary. D_e denotes the desired insertion depth, D_t denotes the final insertion depth when the assembly is not completed due to the forces exceeds the safe boundary.

3.2 Sample-Efficiency

In practice, two directions has achieved success in improving the sample-efficiency.

Residual RL for Control: The robotic assembly tasks can be divided into two parts. The first part is robot-related, which can be solved through the typical controller. And the another part is environment-related, involving the dynamics that can be addressed by RL algorithms [6]. In residual RL, the reward function is often formulated as two parts:

$$r_t = f(\boldsymbol{s}_{r,t}) + g(\boldsymbol{s}_{e,t}) \tag{6}$$

where $\boldsymbol{s}_{r,t}$ is the robot-related state and $f(\boldsymbol{s}_{r,t})$ is the reward to encourage the robot to approach the target position. $g(\boldsymbol{s}_{e,t})$ is the environment-related state in terms of different friction and contact model due to different geometries of assembly components.

Accordingly, the control command also consists of two parts:

$$\boldsymbol{u}_t = \pi_C(\boldsymbol{s}_{r,t}) + \pi_\theta(\boldsymbol{s}_{e,t}) \tag{7}$$

where $\pi_C(\boldsymbol{s}_{r,t})$ is the hand-designed conventional controller without considering the dynamics of environments. $\pi_\theta(\boldsymbol{s}_{e,t})$ is the residual DRL policy to compensate the dynamics of environment and supplement the flexibility of the conventional controller. For peg-in-hole assembly tasks, any RL algorithms (DDPG, TD3, SAC) can be implemented to learn the residual RL policy.

Action Space Reformulation: Action space reformulation can be considered as introducing the prior of policy structure. For different hardware and tasks requirements. PD controller given the parameters $\boldsymbol{K}_p$ and $\boldsymbol{K}_d$, is often applied to feed the control errors without any assumption about the system, as:

$$\begin{aligned} \boldsymbol{u}_t &= \boldsymbol{K}_p \circ \boldsymbol{e}_t + \boldsymbol{K}_d \circ \dot{\boldsymbol{e}}_t \\ \boldsymbol{e}_t &= \boldsymbol{q}_d - \boldsymbol{q}_t, \quad \dot{\boldsymbol{e}}_t = \dot{\boldsymbol{q}}_d - \dot{\boldsymbol{q}}_t \end{aligned} \tag{8}$$

where $\circ$ denotes the Hadamard product; $\boldsymbol{u}_t$ denotes the control torques in joint-space, therefore, the new action space includes the desired joint angle $\boldsymbol{q}_d$ and the desired joint velocity $\dot{\boldsymbol{q}}_d$.

The contact model of peg-in-hole assembly tasks for impedance controller is often formulated in the task-space, as:

$$\mathcal{M}_d(\ddot{\boldsymbol{x}}_t - \ddot{\boldsymbol{x}}_e) + \mathcal{B}_d(\dot{\boldsymbol{x}}_t - \dot{\boldsymbol{x}}_e) + \mathcal{K}_d(\boldsymbol{x}_t - \boldsymbol{x}_e) = \mathcal{F} \tag{9}$$

where $\mathcal{M}_d$, $\mathcal{B}_d$ and $\mathcal{K}_d$ denotes the desired inertia, damping and stiffness matrix in task-space, respectively. For position-control robot, the control command is position offset in task-space. For torque-control robot, the control command $\boldsymbol{u}$ is actuated torque on joints to compensate the dynamics from robot and feedforward external force.

Consequently, instead of controlling the $\{\boldsymbol{q}_t, \dot{\boldsymbol{q}}_t, \boldsymbol{x}_t, \dot{\boldsymbol{x}}_t, \tau\}$ directly, the reformulated action space often consist of the desired reference value $\{\boldsymbol{q}_d, \dot{\boldsymbol{q}}_d, \boldsymbol{x}_d, \dot{\boldsymbol{x}}_d, \mathcal{F}_d\}$ and the corresponding parameters $\{\boldsymbol{K}_p, \boldsymbol{K}_d\}$.

3.3 Stability

The RL stability can be achieved through reducing the over estimation and providing the stable target value [3,14].

Over Estimation: Over estimation in function approximation is a main factor that introduces instability, which has been addressed in all actor-critic algorithms (e.g., TD3, SAC) [3] to reduce the variance and improve the final performance. For actor-critic architecture, TD3 and DDPG have the same objective via minimizing the loss:

$$1/2\left(r_{t+1} + \gamma Q(\boldsymbol{s}_{t+1}, \mu(\boldsymbol{s}_{t+1})|\boldsymbol{\omega}) - Q(\boldsymbol{s}_t, \boldsymbol{a}_t|\boldsymbol{\omega})\right)^2 \tag{10}$$

where $Q(\cdot|\boldsymbol{\omega})$ is the action value function estimated by a neural network given the state-action pair. $\boldsymbol{a}_{t+1} \sim \mu(\boldsymbol{s}_{t+1}|\boldsymbol{\theta}) \; : \; \mathcal{S} \to \mathcal{A}$ denotes the deterministic policy estimated by a neural network. The past experience $(\boldsymbol{s}_t, \boldsymbol{a}_t, r_{t+1}, \boldsymbol{s}_{t+1})$ is sampled from replay buffer $\mathcal{M}$. TD3 implements two independent network (Q_1, Q_2) to estimate the action value function, the target value in (10) is corrected as $r_{t+1} + \gamma \min_{i=1,2} Q_i(\boldsymbol{s}_{t+1}, \mu(\boldsymbol{s}_{t+1}|\boldsymbol{\theta})|\boldsymbol{\omega}_i)$. Afterwards, the deterministic policy is updated based on the gradient of action value function with respect to the parameters of μ.

Averaged-Critic: The target network is commonly used trick to provide a stable objective and speed the convergence [10]. The parameters $\bar{\boldsymbol{\theta}}$ target network is slow-updating from the estimated network parameters $\boldsymbol{\theta}$ through the scheme $\bar{\boldsymbol{\theta}} \leftarrow \zeta_\theta \boldsymbol{\theta} + (1 - \zeta_\theta)\bar{\boldsymbol{\theta}}$. Another approach to reduce the variance of target approximation error is to replace the target network with the average over the previous learned action value estimation. In this paper, we extend this idea to the continuous control RL algorithms to improve the stability and final performance. The target value in (10) is replaced by the average over the past K estimated Q as:

$$r_{t+1} + \gamma \frac{1}{K} \sum_{i=0}^{K} Q_i(\boldsymbol{s}_{t+1}, \mu(\boldsymbol{s}_{t+1}|\bar{\boldsymbol{\theta}})|\boldsymbol{\omega}_i) \tag{11}$$

where $Q_i(\boldsymbol{s}_{t+1}, \mu(\boldsymbol{s}_{t+1}|\bar{\boldsymbol{\theta}})|\boldsymbol{\omega}_i)$ represents the past ith estimated action-value.

3.4 Generalization Improvement

Two important directions are often adopted to improve the generalization of RL algorithms. **Dynamic randomized:** To improve the robustness to environmental noise, randomized dynamic was utilized to train the RL algorithms [12], the dynamics parameters β are sampled from ρ_β for each episode as $s_{t+1} = p(s_{t+1}|a_t, s_t, \beta)$. The dynamics parameters are as a part of the input to estimate the action value function, which can reduce the variance of the policy gradient via compensating the environmental dynamics.

HRL: Recently, HRL is researched to solve the complex manipulation tasks through learning sub-policies. These learned sub-policies may represents the basic skills to solve the complex tasks, therefore, some HRL work has achieved better performance through transferring the sub-policies [4,11]. Compared to the objective (2), the hierarchical policy is derived with a discrete variable $o \in \mathcal{O}$ as $\pi_h(a_t|s_t) = \sum_{o \in \mathcal{O}} \pi_{\mathcal{O}}(o|s_t)\pi_o(a_t|s_t)$, $\pi_o(a_t|s_t)$ denotes the lower-level policy to generate the action, $\pi_{\mathcal{O}}(o|s_t)$ denotes the upper-level policy to select the lower-level policy given the state. The hierarchical policy $\pi_h(a_t|s_t)$ can be learned given the objective (2) based on the policy gradient theorem and additional constraints.

4 Experiment

We implement the proposed actor-average-critic (AAC) with different average length ($K = 2, 3, 5, 10$), the DDPG [10], TD3 [3] and a HRL algorithm (adInfoHRL) [11] on three different assembly tasks with different clearance and friction setting. For the equal comparison, the hyper-parameters of AAC are selected as TD3, and the hyper-parameters of other algorithms are selected as the original papers, all the algorithms are implemented with action space reformulation in (8).

4.1 Performance: Sample-Efficiency and Stability

As shown in Figs. 3–5, all the algorithms are trained for 10^5 steps with five seeds and cumulative reward is tested every 10^3 steps. Darker lines represents average reward over five seeds. Shaded region shows the standard deviation of average reward. The assembly task becomes more complex from single to triple, DDPG and TD3 cannot perform the assembly tasks effectively, obviously, the proposed AAC with different average length can achieve both the better sample efficiency and stability performance in most of the above assembly tasks.

4.2 Performance: Generalization

To demonstrate the generalization ability, we implement the HRL with different options ($\mathcal{O} = 2, 4, 6$) in dual and triple peg-in-hole assembly tasks. As shown in Fig. 6(a) and 7(a), the HRL with four options can achieve largely better sample efficiency and stability performance than TD3 and DDPG. To demonstrate there kinds of transfer performance, the learned assembly policy is loaded, then tested or re-trained again in a new assembly tasks (Figs. 3–5).

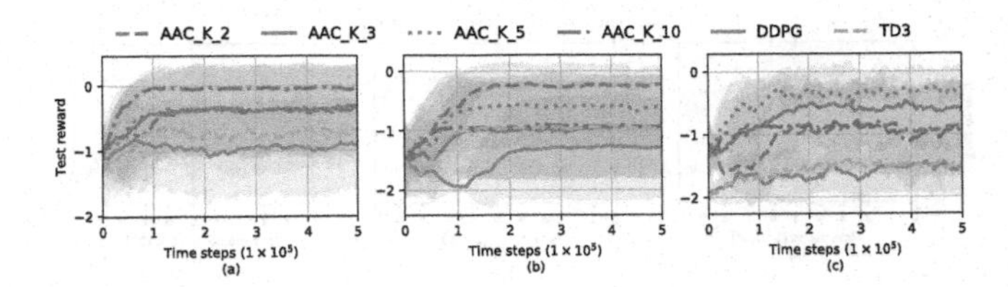

Fig. 3. Single peg-in-hole assembly. (a) Friction-0.5-clearance-1 mm. (b) Friction-0.5-clearance-0.5 mm. (c) Friction-1-clearance-1 mm.

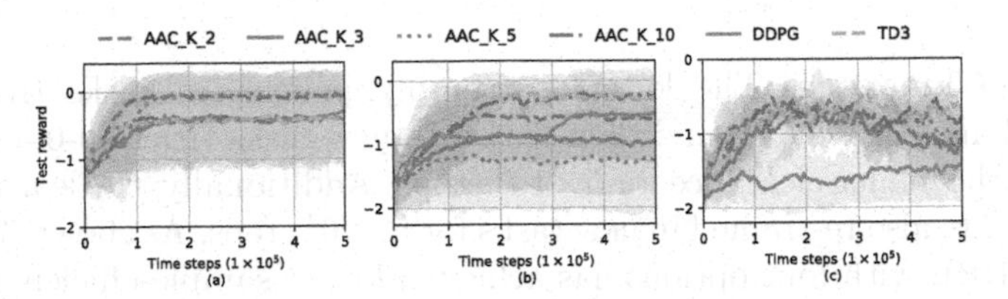

Fig. 4. Dual peg-in-hole assembly. (a) Friction-0.5-clearance-1 mm. (b) Friction-0.5-clearance-0.5 mm. (c) Friction-1-clearance-1 mm.

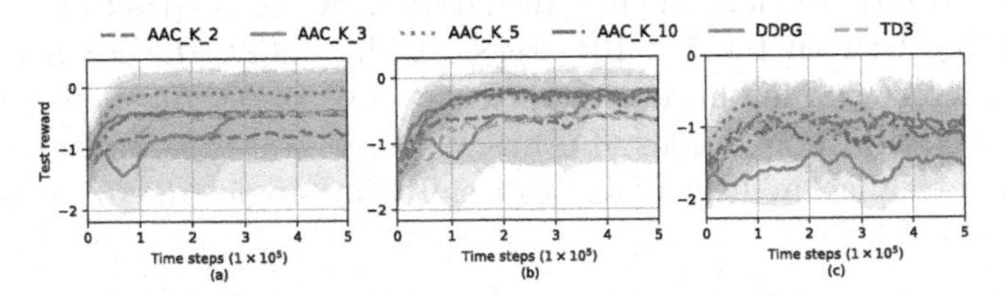

Fig. 5. Triple peg-in-hole assembly. (a) Friction-0.5-clearance-1 mm. (b) Friction-0.5-clearance-0.5 mm. (c) Friction-1-clearance-1 mm.

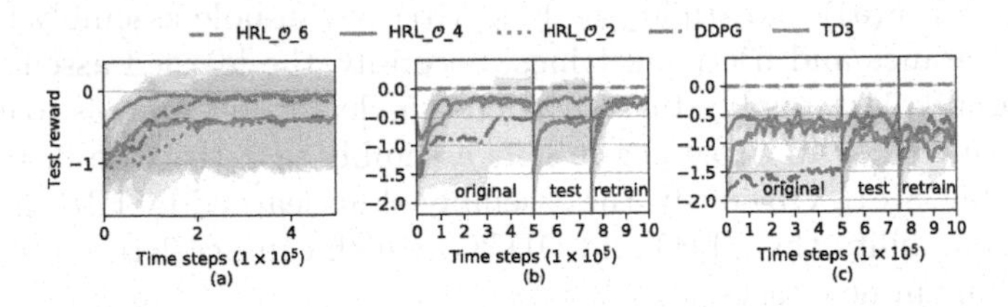

Fig. 6. Dual peg-in-hole assembly. (a) Friction-0.5-clearance-1 mm. (b) Friction-0.5-clearance-0.5 mm. (c) Friction-1-clearance-1 mm.

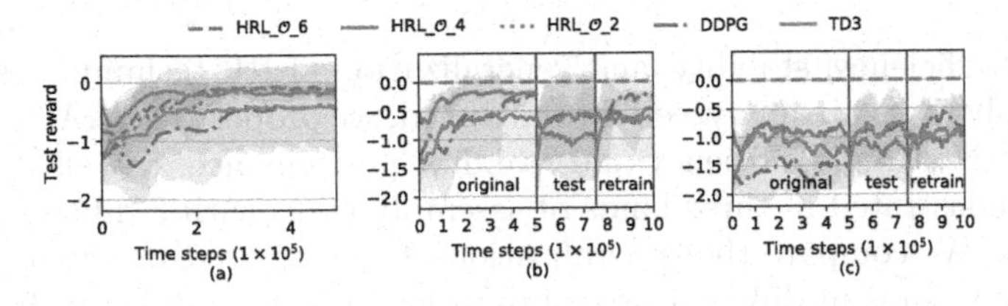

Fig. 7. Triple peg-in-hole assembly. (a) Friction-0.5-clearance-1 mm. (b) Friction-0.5-clearance-0.5 mm. (c) Friction-1-clearance-1 mm.

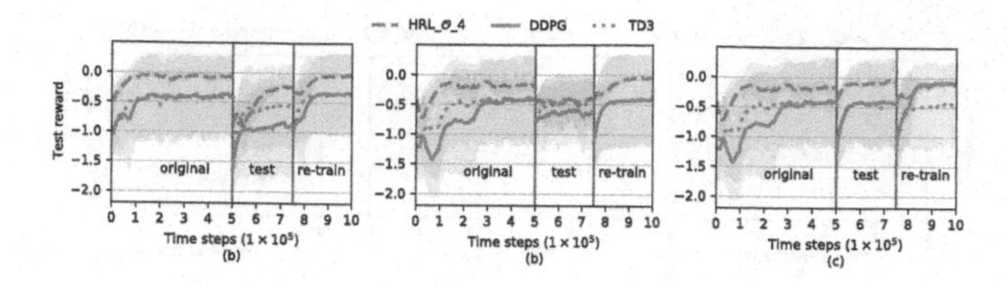

Fig. 8. Different assembly tasks transfer performance. (a) Single-to-dual. (b) Dual-to-triple. (c) Single-to-triple.

Different Clearance: The learned assembly policies in tasks (friction-0.5-clearance-1 mm) are loaded and then tested in new tasks (friction-0.5-clearance-0.5 mm) with smaller clearance for 5×10^4 steps. Additionally, the learned assembly policies are also re-trained in new tasks for 5×10^4 steps. As shown in Fig. 6(b) and 7(b), HRL with four options has achieved better sample-efficiency and stability performance in new assembly tasks.

Different Friction: Likewise, the learned assembly policies in tasks (friction-0.5-clearance-1mm) are loaded and tested in new tasks (friction-1-clearance-1mm) with big friction for 5×10^4 steps. Additionally, the learned assembly policies are also re-trained in new tasks for 5×10^4 steps. As shown in Fig. 6(c) and 7(c), HRL with four options also has achieved better sample-efficiency and stability performance in new tasks, especially in single and dual peg-in-hole assembly tasks.

Different Number of Pegs: The assembly tasks with different number of pegs have different contact model. The learned assembly policy in single peg-in-hole assembly task (friction-0.5-clearance-1 mm) is loaded, then tested and re-trained in dual (see Fig. 8(a)) and triple (see Fig. 8(b)) peg-in-hole assembly tasks with same the clearance and friction settings. Secondly, the learned assembly policy in dual peg-in-hole assembly task (friction-0.5-clearance-1 mm) is loaded, then tested and re-trained in triple peg-in-hole assembly task (friction-0.5-clearance-1 mm) (see Fig. 8(c)). Obviously, the assembly skills learned by HRL have better generalization ability than TD3 and DDPG, which can accelerate the assembly policy learning in new tasks.

5 Conclusion and Discussion

The sample-efficiency, stability, and generalization of RL techniques on peg-in-hole assembly tasks are analyzed and concluded, we propose the AAC framework to improve the sample-efficiency and stability performance, the effectiveness of AAC is demonstrated in three kinds of assembly tasks with 3 different settings, respectively. We compare the generalization of our method and two commonly used RL algorithms in different assembly tasks with three different friction and clearance settings. Consequently, the assembly skills learned by HRL can achieve better generalization performance than commonly used TD3 and DDPG. This lead us to believe that our algorithm have high potential to transfer a policy learned in simulation to real-world efficiently, which will be our future work.

References

1. Webots r2019b (2019). https://www.cyberbotics.com/#webots
2. Fan, Y., Luo, J., Tomizuka, M.: A learning framework for high precision industrial assembly. In: 2019 International Conference on Robotics and Automation (ICRA), pp. 811–817. IEEE (2019)
3. Fujimoto, S., Hoof, H., Meger, D.: Addressing function approximation error in actor-critic methods. In: International Conference on Machine Learning, pp. 1582–1591 (2018)
4. Hou, Z., Fei, J., Deng, Y., Xu, J.: Data-efficient hierarchical reinforcement learning for robotic assembly control applications. IEEE Trans. Ind. Electron. **68**(11), 11565–11575 (2020)
5. Inoue, T., De Magistris, G., Munawar, A., Yokoya, T., Tachibana, R.: Deep reinforcement learning for high precision assembly tasks. In: 2017 IEEE/RSJ International Conference on Intelligent Robots and Systems (IROS), pp. 819–825. IEEE (2017)
6. Johannink, T., et al.: Residual reinforcement learning for robot control. In: 2019 International Conference on Robotics and Automation (ICRA), pp. 6023–6029. IEEE (2019)
7. Johannsmeier, L., Gerchow, M., Haddadin, S.: A framework for robot manipulation: skill formalism, meta learning and adaptive control. In: 2019 International Conference on Robotics and Automation (ICRA), pp. 5844–5850. IEEE (2019)
8. Kober, J., Bagnell, J.A., Peters, J.: Reinforcement learning in robotics: a survey. Int. J. Robot. Res. **32**(11), 1238–1274 (2013)
9. Levine, S., Wagener, N., Abbeel, P.: Learning contact-rich manipulation skills with guided policy search. In: 2015 IEEE International Conference on Robotics and Automation (ICRA), pp. 156–163, May 2015
10. Lillicrap, T.P., et al.: Continuous control with deep reinforcement learning. arXiv preprint arXiv:1509.02971 (2015)
11. Osa, T., Tangkaratt, V., Sugiyama, M.: Hierarchical reinforcement learning via advantage-weighted information maximization. In: International Conference on Learning Representations (2019)
12. Peng, X.B., Andrychowicz, M., Zaremba, W., Abbeel, P.: Sim-to-real transfer of robotic control with dynamics randomization. In: 2018 IEEE International Conference on Robotics and Automation (ICRA), pp. 1–8. IEEE (2018)
13. Varin, P., Grossman, L., Kuindersma, S.: A comparison of action spaces for learning manipulation tasks. In: 2019 IEEE/RSJ International Conference on Intelligent Robots and Systems (IROS), pp. 6015–6021 (2019)
14. Wu, D., Dong, X., Shen, J., Hoi, S.C.: Reducing estimation bias via triplet-average deep deterministic policy gradient. IEEE Trans. Neural Netw. Learn. Syst. **31**(11), 4933–4945 (2020)
15. Xu, J., Hou, Z., Wang, W., Xu, B., Zhang, K., Chen, K.: Feedback deep deterministic policy gradient with fuzzy reward for robotic multiple peg-in-hole assembly tasks. IEEE Trans. Ind. Inform. **15**(3), 1658–1667 (2018)

Kinetostatic Modeling of Piezoelectric Displacement Amplifiers Based on Matrix Displacement Method

Dezhi Song, Benliang Zhu, Hai Li, and Xianmin Zhang(✉)

Guangdong Province Key Laboratory of Precision Equipment and Manufacturing Technology, South China University of Technology, Guangzhou 510640, Guangdong, People's Republic of China
zhangxm@scut.edu.cn

Abstract. Displacement amplification ratio and input stiffness modeling of three types of piezoelectric displacement amplifiers (PDAs) were investigated in this paper. The main feature is that we further improved the accuracy of matrix displacement model (MDM) by setting a node on both input and output points, and simplified modeling process of the MDM by utilizing axial symmetry of PDAs. Firstly, we deduced the stiffness matrix of the flexure element based on the general flexure hinge's compliance formulas. Then, we established the MDM of a PDA by discretizing a displacement amplifier into flexure elements and assembling element stiffness matrices. The finite element method was employed to verify the superiority of presented model and to compare with typical analytical models and conventional MDMs. The results show that the presented model of both the displacement amplification ratio and input stiffness has better accuracy on the whole, thanks to consider deformation of both the input and output points during modeling and abandons considering input and output ends as a lumped mass.

Keywords: Piezoelectric actuator · Displacement amplifier · Flexure element · Matrix displacement method · Stiffness matrix

1 Introduction

Thanks to excellent characteristics of nanoscale resolution, large output force and fast frequency response, piezoelectric actuators have been increasingly used in many precision engineering fields such as precision positioning [1], micro/nano manipulation [2], ultra-precision machining [3], and so on. However, a major drawback of the piezoelectric actuator is its quite small output displacement range. To expand the piezoelectric actuator's motion range, PDA consisting of flexure hinge mechanisms is widely utilized.

Displacement amplifiers can be categorized into a lever type and a triangular amplification type [4]. The latter type mainly includes the rhombus-type amplifier, the bridge-type amplifier and the Scott-Russell amplifier. Owing to the characteristics of large amplification ratio and compact dimensions, the rhombus-type and the bridge-type amplifier

X.-J. Liu et al. (Eds.): ICIRA 2021, LNAI 13014, pp. 404–414, 2021.
https://doi.org/10.1007/978-3-030-89098-8_39

have been universally used in various piezo-driven devices. Displace-ment amplification ratio and input stiffness are significant kinematic and static properties, respectively. In order to predict the performance and optimize design of the PDA, how to establish accurate mechanical model has become the research concentration of the displacement amplification mechanisms.

During the past decades, a variety of analytical methods have been developed for kinetostatic modeling of the amplifiers, such as the Castigliano's second theorem [5], the Euler-Bernoulli beam theory [6–8], the Timoshenko beam theory [9] and the combination of the Castigliano's second theorem and elastic beam theory [10]. These methods are widely used in simple displacement amplifiers. However, the formulations for the complex mechanisms are usually too complicated to use. In addition to the analytical methods, recent years have witnessed the development of the matrix displacement model (MDM) [11], which is a more general method for modeling of compliant mechanisms. Ling et al. [12] developed a kinetostatic model of a lever amplification-based XY positioning stage utilizing the MDM. However, he simplified the input ends by considering it as a lumped mass. Ling [13] subsequently proposed a general two-port model of the bridge-type amplifier on a basis of the MDM and transfer matrix method, while the output port was still assumed to be the lumped mass.

In the above matrix displacement models, input and output ends of the amplifiers are not given thorough consideration. The predecessors universally established the model without considering its symmetry, and therefore, increased the modeling workload and made the model complex. Hence, the improved MDMs of three types of amplifiers were presented with consideration of both input and out ports. Furthermore, we decreased the modeling workload by modeling a quarter of the amplifiers due to the symmetry.

2 Element Stiffness Matrix

2.1 Compliance Matrix of the Flexure Hinge

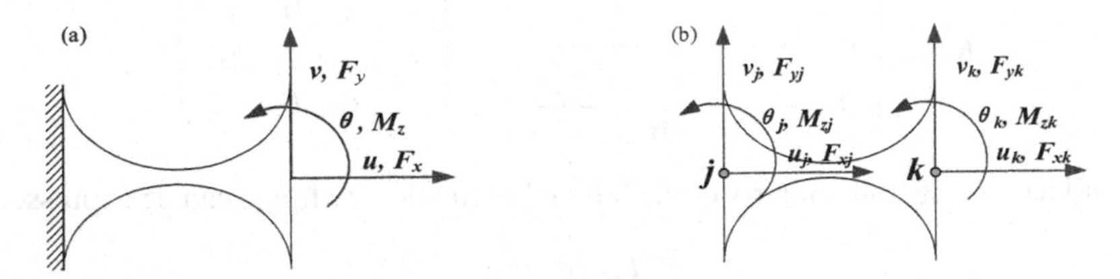

Fig. 1. Schematic of a flexure hinge (a) and a flexure element (b).

Schematic of the generalized flexure hinge is shown in Fig. 1(a). With one end of the hinge fixed, another end is loaded by axial force F_x, lateral force F_y and bending moment M_z. Rewrite the loads into vector form as

$$\mathbf{F} = [F_x\ F_y\ M_z]^T \tag{1}$$

corresponding deformation vector is

$$\mathbf{U} = [u \;\; v \;\; \theta]^T \tag{2}$$

The relation between the load vector and the deformation vector at free end is

$$\mathbf{U} = \mathbf{CF} \tag{3}$$

where **C** is compliance matrix

$$\mathbf{C} = \begin{bmatrix} c_{11} & 0 & 0 \\ 0 & c_{22} & c_{23} \\ 0 & c_{32} & c_{33} \end{bmatrix} \tag{4}$$

where, c_{11}–c_{33} can be calculated by the Castigliano's second theorem

$$\begin{cases} c_{11} = \int_l \frac{1}{EA(x)}dx, \; c_{22} = \int_l \frac{x^2}{EI_z(x)}dx + \int_l \frac{\zeta}{GA(x)}dx \\ c_{23} = \int_l \frac{x}{EI_z(x)}dx, \; c_{32} = c_{23}, \; c_{33} = \int_l \frac{1}{EI_z(x)}dx \end{cases} \tag{5}$$

where l is the axial dimension of the flexure hinge. E and G are the Young's Modulus and shear modulus relating the material, respectively. ζ is the shear effect parameter, and it will be 1.2 if the cross-section is rectangle. $A(x)$ and $I_z(x)$ mean the cross-sectional area and z-axis moment of inertia x unit length from the loading end. It is worth mentioning that scholars have formulated various flexure hinge according to Eq. (5), which means one can utilize existed conclusions directly.

2.2 Stiffness Matrix of the Flexure Hinge/Element

The stiffness matrix of the right end of the flexure hinge can be obtained by getting inverse of Eq. (4)

$$\mathbf{K}_r = \begin{bmatrix} \frac{1}{c_{11}} & 0 & 0 \\ 0 & \frac{c_{33}}{c_{22}c_{33}-c_{23}^2} & -\frac{c_{23}}{c_{22}c_{23}-c_{23}^2} \\ 0 & -\frac{c_{23}}{c_{22}c_{23}-c_{23}^2} & \frac{c_{22}}{c_{22}c_{23}-c_{23}^2} \end{bmatrix} = \begin{bmatrix} k_{11} & 0 & 0 \\ 0 & k_{22} & k_{23} \\ 0 & k_{23} & k_{33} \end{bmatrix} \tag{6}$$

Similarly, the stiffness matrix of the left end of the flexure hinge can be expressed as

$$\mathbf{K}_l = \begin{bmatrix} k_{11} & 0 & 0 \\ 0 & k_{22} & -k_{23} \\ 0 & -k_{23} & k_{33} \end{bmatrix} \tag{7}$$

Accordingly, Eq. (3) correspondingly becomes

$$\mathbf{F} = \mathbf{K}_r\mathbf{U} \; or \; \mathbf{F} = \mathbf{K}_l\mathbf{U} \tag{8}$$

No. i of the flexure element is shown in Fig. 1(b). Here, the flexure hinge shown in Fig. 1(a) is regarded as a flexure element and its fixed end is simulated by loads.

Firstly, considering the load equilibrium relationship of the i th flexure element and Eq. (6) and (8), one can acquire the relation between the element load and the kth nodal displacement by assuming that the kth node is fixed.

$$\begin{bmatrix} F_{xj} \\ F_{yj} \\ M_{zj} \\ F_{xk} \\ F_{yk} \\ M_{zk} \end{bmatrix} = \begin{bmatrix} -1 & 0 & 0 \\ 0 & -1 & 0 \\ 0 & l & -1 \\ 1 & 0 & 0 \\ 0 & 1 & 0 \\ 0 & 0 & 1 \end{bmatrix} \begin{bmatrix} F_{xk} \\ F_{yk} \\ M_{zk} \end{bmatrix} = \mathbf{K}_k \cdot \begin{bmatrix} u_k \\ v_k \\ \theta_k \end{bmatrix} \tag{9}$$

there is $c_{23} = l_i c_{33}/2$ for axial and transversely symmetric flexure hinges [14], and one knows $k_{22} = -2k_{23}/l_i$ according to Eq. (6), and therefore

$$\mathbf{K}_k = \begin{bmatrix} -k_{11} & 0 & 0 \\ 0 & -k_{22} & -k_{23} \\ 0 & k_{23} & -l_i k_{23} - k_{33} \\ k_{11} & 0 & 0 \\ 0 & k_{22} & k_{23} \\ 0 & k_{23} & k_{33} \end{bmatrix} \tag{10}$$

Similarly, one can acquire the relation between the element load and the jth nodal displacement

$$\mathbf{K}_j = \begin{bmatrix} k_{11} & 0 & 0 \\ 0 & k_{22} & -k_{23} \\ 0 & -k_{23} & k_{33} \\ -k_{11} & 0 & 0 \\ 0 & -k_{22} & k_{23} \\ 0 & -k_{23} & -l_i k_{23} - k_{33} \end{bmatrix} \tag{11}$$

Actually, the kth and j^{th} nodes are free. Therefore, one can obtain the element stiffness of the flexure hinge by combining $\mathbf{K}_j$ and $\mathbf{K}_k$

$$\mathbf{K}^e = [\mathbf{K}_j \ \mathbf{K}_k] \tag{12}$$

One can verify that Eq. (12) becomes beam element stiffness if $A(x)$ and $I_z(x)$ are constant [15]. It is noteworthy that one need transfer $\mathbf{K}^e$ from the local coordinate to the global coordinate.

3 Matrix Displacement Model of Amplifiers

3.1 Rhombus-Type PDA

Rhombus-type displacement amplifier is shown in Fig. 2(a). Piezoelectric actuator connects with the amplifier by preload mechanism. When a certain voltage is input to the actuator, the actuator will generate driving force to act on the input ends A and C of the amplifier. The driving force further leads to the deformation of the input ends, linking parts and output ends, and finally resulting in an output displacement along vertical direction at the output ends B and D. One can get an amplified output displacement comparing with input displacement by reasonable design of the dimension of the amplifier. The modeling procedure consists of following steps.

Step 1: Discrete the amplifier into individual flexure elements and number them.

The MDM of the rhombus-type amplifier is illustrated in Fig. 2(b). Since the symmetrical structure of the amplification mechanism, only a quarter of the structures is needed to established the model. ①–③ are modeled input ends, linking parts and output ends, respectively, which are regarded as beam elements. 1–4 are four nodes from input ends to output ends.

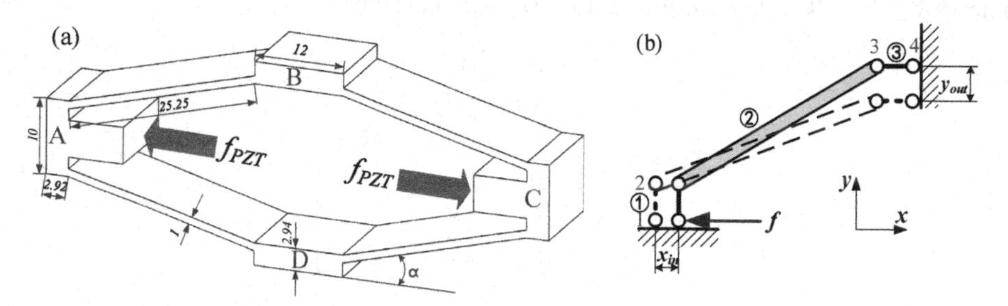

Fig. 2. Diagram of a rhombus displacement amplifier (a) and its MDM (b).

Step 2: Establish the element stiffness Matrix.

One can deduced compliance of the beam from Eq. (4)

$$\begin{cases} c_{11} = \dfrac{l}{EA}, \quad c_{22} = \dfrac{4l^3}{EI_z} + \dfrac{1.2}{GA} \\ c_{23} = \dfrac{6l^2}{EI_z}, \quad c_{32} = c_{23}, \quad c_{33} = \dfrac{12l}{EI_z} \end{cases} \tag{13}$$

where $I_z = d_i h_i^3/12$, and $A = d_i h_i (i = 1, 2, 3)$. Then, submitting Eq. (13) into Eq. (6)–(11) finally results Eq. (12), the element matrix in the local coordinate. Taking the example of the element ②, one knows the angle between the linking beam and the horizontal direction is $\boldsymbol{\alpha}$. Hence, the coordinate transfer matrix is

$$\mathbf{T} = \begin{bmatrix} \cos\alpha & \sin\alpha & 0 \\ -\sin\alpha & \cos\alpha & 0 \\ 0 & 0 & 1 \end{bmatrix} \tag{14}$$

One can further obtain the global stiffness matrix by substituting Eq. (12) into

$$\overline{\mathbf{K}}^e = \mathbf{T}'\mathbf{K}^e\mathbf{T} \tag{15}$$

Step 3: Establish MDM based on equilibrium equations of the nodal load

From Fig. 2(b), one can set the jth node of ith element as $\mathbf{F}_{i,j}$, and similarly, $\mathbf{F}_{i,k}$:

$$\begin{cases} \mathbf{F}_j = \left[F_{xj} \; F_{yj} \; M_{zj} \right]^T \\ \mathbf{F}_k = \left[F_{xk} \; F_{yk} \; M_{zk} \right]^T \end{cases}, \quad i = 1, 2, 3 \tag{16}$$

In Fig. 2(b), the equilibrium equations of the nodal load are

$$\begin{cases} node\ 1 : \mathbf{F}_{1,1} = \mathbf{F}_{in} = [f\ 0\ 0]^T = [f_{PZT}/2\ 0\ 0]^T \\ node\ 2 : \mathbf{F}_{1,2} + \mathbf{F}_{2,1} = 0 \\ node\ 3 : \mathbf{F}_{2,2} + \mathbf{F}_{3,1} = 0 \\ node\ 4 : \mathbf{F}_{3,2} = \mathbf{F}_{out} = [F_{x4}\ 0\ M_{z4}]^T \end{cases} \tag{17}$$

One can turn the element stiffness matrix in the element stiffness equation into block matrix as follows

$$\mathbf{F}_i = \begin{bmatrix} \mathbf{F}_{i,j} \\ \mathbf{F}_{i,k} \end{bmatrix} = \overline{\mathbf{K}}_i^e \begin{bmatrix} \mathbf{U}_j \\ \mathbf{U}_k \end{bmatrix} = \begin{bmatrix} \mathbf{K}_{i,1} & \mathbf{K}_{i,2} \\ \mathbf{K}_{i,3} & \mathbf{K}_{i,4} \end{bmatrix} \begin{bmatrix} \mathbf{U}_j \\ \mathbf{U}_k \end{bmatrix} \tag{18}$$

where $\mathbf{U}_j$, $\mathbf{U}_k$ are j^{th} and k^{th} node of ith element, respectively.

$$\begin{cases} \mathbf{U}_j = \left[u_j \; v_j \; \theta_j \right]^T \\ \mathbf{U}_k = \left[u_k \; v_k \; \theta_k \right]^T \end{cases}, \quad i = 1, 2, 3 \tag{19}$$

Inserting Eq. (18) into Eq. (17), yields

$$\begin{cases} \mathbf{K}_{1,1}\mathbf{U}_1 + \mathbf{K}_{1,2}\mathbf{U}_2 = \mathbf{F}_{in} \\ \mathbf{K}_{1,3}\mathbf{U}_1 + \mathbf{K}_{1,4}\mathbf{U}_2 + \mathbf{K}_{2,1}\mathbf{U}_2 + \mathbf{K}_{2,2}\mathbf{U}_3 = 0 \\ \mathbf{K}_{2,3}\mathbf{U}_2 + \mathbf{K}_{2,4}\mathbf{U}_3 + \mathbf{K}_{3,1}\mathbf{U}_3 + \mathbf{K}_{3,2}\mathbf{U}_4 = 0 \\ \mathbf{K}_{3,3}\mathbf{U}_3 + \mathbf{K}_{3,4}\mathbf{U}_4 = \mathbf{F}_{out} \end{cases} \tag{20}$$

Further rewrite Eq. (20) as follows

$$\mathbf{F} = \begin{bmatrix} \mathbf{F}_{in} \\ \mathbf{0} \\ \mathbf{0} \\ \mathbf{F}_{out} \end{bmatrix} = \begin{bmatrix} \mathbf{K}_{1,1} & \mathbf{K}_{1,2} & 0 & 0 \\ \mathbf{K}_{1,3} & \mathbf{K}_{1,4} + \mathbf{K}_{2,1} & \mathbf{K}_{2,2} & 0 \\ 0 & \mathbf{K}_{2,3} & \mathbf{K}_{2,4} + \mathbf{K}_{3,1} & \mathbf{K}_{3,2} \\ 0 & 0 & \mathbf{K}_{3,3} & \mathbf{K}_{3,4} \end{bmatrix} \begin{bmatrix} \mathbf{U}_1 \\ \mathbf{U}_2 \\ \mathbf{U}_3 \\ \mathbf{U}_4 \end{bmatrix} = \mathbf{KU} \tag{21}$$

where $\mathbf{K}$ is the global stiffness matrix of the structure shown in Fig. 2(b). Here, the MDM of the rhombus-type amplifier is obtained. Therefore, the input and output displacements can be obtained by substituting displacement and load boundary conditions into Eq. (21), which further gives displacement amplification ratio and input stiffness

$$\begin{cases} R_{amp} = \dfrac{y_{out}}{x_{in}} \\ K_{in} = \dfrac{f}{x_{in}} \end{cases} \tag{22}$$

3.2 Bridge-Type PDA and Compound Bridge-Type PDA

Here, a bridge-type displacement amplifier is designed as shown in Fig. 3(a). It shares the same working principle with the rhombus-type one. To predict its performance, the MDM of the amplifier is established as illustrated in Fig. 3(b). The modeling procedure is the same as Sect. 3.1.

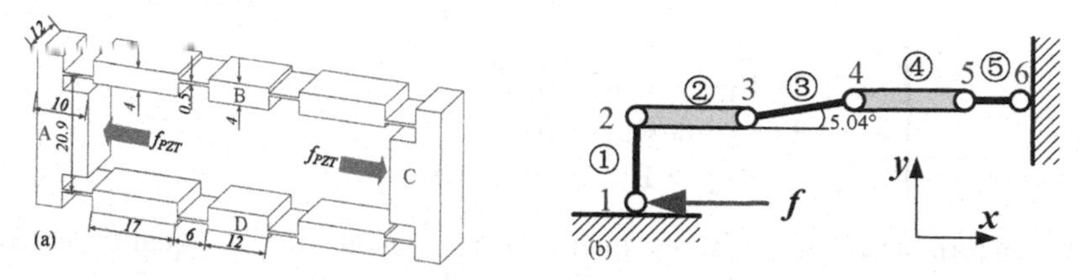

Fig. 3. Diagram of a bridge-type amplifier (a) and its MDM (b).

A compound bridge-type amplifier is further designed based on the bridge-type as shown in Fig. 4(a). Compared with bridge-type amplifier, the compound bridge-type add a bridge in every quarter corner and is considered having greater stiffness. MDM of the compound bridge-type amplifier is established in order to predict its amplification ratio and input stiffness, as is demonstrated in Fig. 4(b). The rest modeling procedure is similar to Sect. 3.1. It is noticeable that the right circular hinge is adopted in the compound bridge-type amplifier. The compliance equations like Eq. (5) have been calculated by Wu et al. [16].

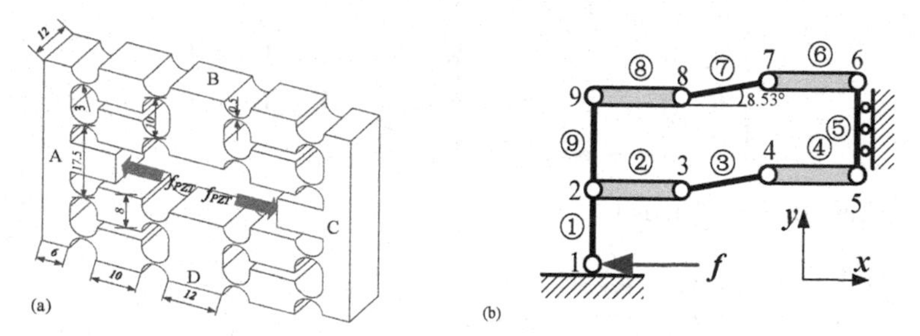

Fig. 4. Diagram of a compound bridge-type amplifier (a) and its MDM (b).

4 Simulation and Verification

Three examples were studied to verify the superiority of the presented model in this part. In the view of universally acknowledged accuracy, the finite element method (FEM) software ANSYS Workbench is employed as the benchmark. Every example compares the accuracy of the presented model with the typical analytical model and the conventional MDMs, which include the MDM without considering both input and output ends (MDM_{WIO}), the MDM considering only the output end (MDM_O) and the MDM considering only the input end (MDM_I).

4.1 Rhombus-Type

The angle between the linking beam and the horizontal direction α is considered as a variable. The Young's modulus and Shear modulus for the rhombus-type amplifier are 71.9 GPa and 27 GPa. The rest of the parameters relating to dimensions are shown in Fig. 3. The finite element analysis is carried out with 50N loading on both right and left input ends. One can probe the deformation of the input point and output point as the input displacement and output displacement in the ANSYS.

The typical analytical model [8] is demonstrated as follows.

$$\begin{cases} x_{in} = \left(\dfrac{l_2 \cos^2 \alpha}{EA} + \dfrac{l_2^3 \sin^2 \alpha}{12EI} \right) \cdot \dfrac{f_{PZT}}{2} \\ y_{out} = \dfrac{l_2^3 \sin \alpha \cos \alpha}{12EI} \cdot \dfrac{f_{PZT}}{2} \end{cases} \tag{23}$$

Substituting the input displacement x_{in} and output displacement y_{out} into Eq. (22), one can obtain the amplification ratio and input stiffness.

The MDMs of the rhombus-type amplifier without considering both input and output ends, without considering input ends and without considering output ends are demonstrated in Fig. 5.

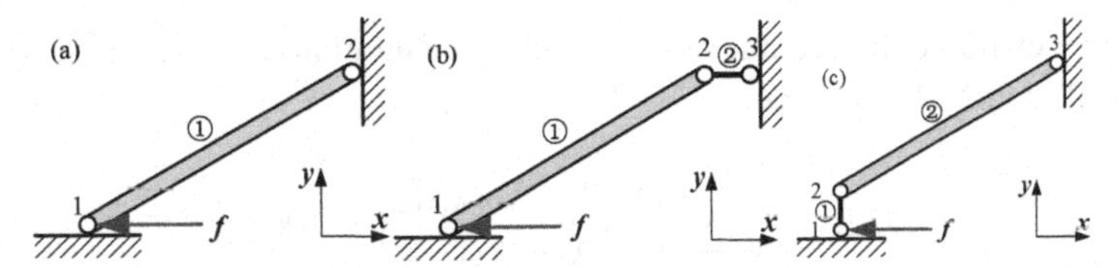

Fig. 5. The models of the rhombus-type amplifier: (a) MDM_{wIO}, (b) MDM_O, (c) MDM_I.

The results calculated using the analytical model, the conventional MDM, the presented model and the FEM are shown in Fig. 6. Conclusions can be drawn as follows.

(1) All models share the same shape of amplification ratio and input stiffness curve with the results of the FEM. The amplification ratio of the rhombus-type rapidly increases in a quite small angle range and reaches the peak at about 2.5°. After the peak, the amplification ratio decreases slowly as shown in Fig. 6(a). From Fig. 6(b), one can observe that input stiffness of the rhombus-type amplifier drops rapidly within 5° and decreases slowly after 5°.

(2) One can also find the accuracy difference among different models. From Fig. 6(a), one can see that there are large errors in amplification ratio with the analytical model, the MDM_{wIO} and the MDM_O in small angle range, while the presented model (MDM_{IO}) and the MDM_I match well with the results of the FEM. Particularly, the presented model matches the best with the results of the FEM over 7.5°. Misled prediction and assessment of amplification ratio will be made if one uses the model without considering input end in small angle range. From Fig. 6(a), one can see that the input stiffness curve of the presented model matches the best with the results of

the FEM in 0–30° range. The input stiffness curve of the MDM_I also fits well but not as well as the presented model.

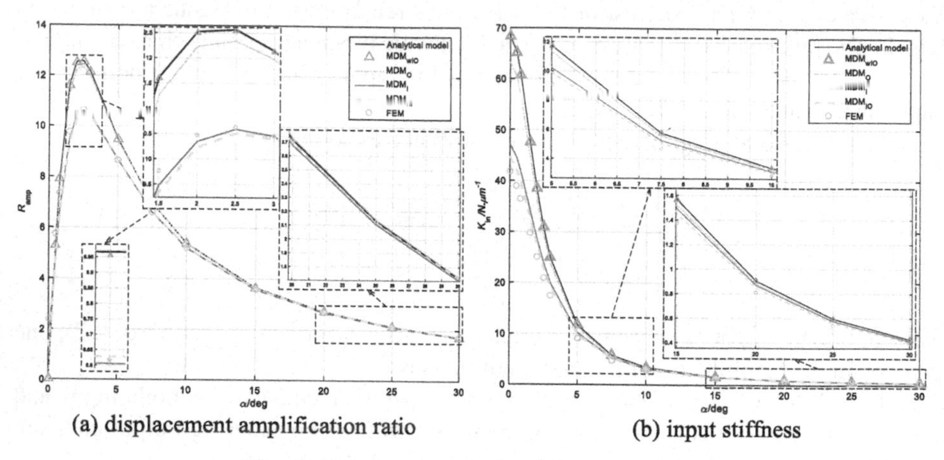

(a) displacement amplification ratio (b) input stiffness

Fig. 6. Calculated results by different methods.

To further evaluate the accuracy of the model, mean relative errors in comparison to the results of the FEM are adopted

$$MRE = \frac{1}{N}\sum_{i=1}^{N}\left|\frac{X(i) - X_{FEM}(i)}{X_{FEM}(i)}\right| \times 100\% \tag{24}$$

where $X(i)$ and $X_{FEM}(i)$ mean the calculated results of the investigated model and the FEM, respectively, when different angles $\boldsymbol{\alpha}$ are employed. The MRE of the different models are listed in Table 1.

From Table 1, one can find that the presented model has nearly the lowest MRE of the kinetostatic performance on the whole. MREs of amplification ratio and input stiffness of the presented model are lower 11% and 29% than the MDM_{wIO}, respectively. Although the MRE of the amplification ratio of the presented model is slightly higher than the MDM_I, the MRE of the input stiffness significantly lower than the MDM_I. Moreover, the amplification ratio of the presented model matches better than the MDM_I with the results of the FEM over 7.5° as shown in Fig. 6(a). From the above analysis, we can attribute the better accuracy of the proposed rhombus-type model to considering both input and output ends.

4.2 Bridge-Type and Compound Bridge-Type

Error analysis of different models are carried out for the designed bridge-type and compound bridge-type amplifiers in this section. The referred analytical models are Ling's model [8] for the bridge-type amplifier and Xu's static model [6] for the compound one. The Young's modulus and Poisson's ratio for the bridge-type amplifier are 206 GPa

Table 1. The MRE of different models

	MRE of the amplification ratio	MRE of the input stiffness
Analytical model [8]	14.89%	36.57%
MDM_{WIO}	14.78%	36.55%
MDM_O	14.21%	14.26%
MDM_I	5.44%	11.41%
MDM_{IO}	5.95%	7.29%

and 0.3, respectively. And the Young's modulus and Poisson's ratio for the compound bridge-type are 71 GPa and 0.33, respectively. The finite element analysis is carried out with 100N loading on the input ends of the bridge-type and the compound bridge-type amplifiers. Relative errors (REs) to the FEM of different models of the bridge-type and the compound bridge-type are tabulated in Table 2.

Table 2. Relative errors of different models.

Models \ REs	Bridge-type		Compound Bridge-type	
	Amplification ratio	Input stiffness	Amplification ratio	Input stiffness
Analytical model	2.8%	4.5%	32.9%	36.4%
MDM_{WIO}	2.2%	4.7%	29.8%	65.2%
MDM_O	2.1%	4.3%	22.5%	23.8%
MDM_I	1.5%	3.7%	18.1%	43.9%
MDM_{IO}	1.3%	3.4%	2.7%	2.0%

One can obviously find that the presented models of the two amplifiers have the lowest relative errors of both amplification ratio and input stiffness. Particularly, the presented model of the compound bridge-type amplifier improves the accuracy of amplification ratio and input stiffness significantly compared with the other models. The presented models considering deformation of both input and output ends keep the relative errors within 3.5% and therefore are more accurate than the MDM without considering input or output ends.

5 Conclusions

This work proposed a kinetostatic method for the piezoelectric displacement amplifiers based on the matrix displacement model (MDM). In the example of three types of amplifiers, we minimized modeling workload by utilizing symmetry of the amplifiers and compared the accuracy of the proposed models with the typical analytical and the MDMs without considering input/output ends. The results demonstrate that mean relative errors of amplification ratio and input stiffness of the rhombus-type amplifier are lower 11% and 29% than the MDM without considering input and output ends, respectively. Particularly, the presented model considerably improved the accuracy of the compound bridge-type amplifier's amplification ratio and input stiffness. Therefore, it is necessary to consider the deformation of both the input and output ends instead of without considering neither ends or just one of the ends.

Acknowledgements. This research was supported by the National Natural Science Foundation of China (Grant No. 52035013), the Guangdong Basic and Applied Basic Research Foundation (Grant No. 2021B1515020053 & 2021A1515012418).

References

1. Wang, R.Z., Zhang, X.M.: Parameters optimization and experiment of a planar parallel 3-DOF nanopositioning system. IEEE Trans. Industr. Electron. **65**(3), 2388–2397 (2016)
2. Chen, W.L., Lu, Q.H., Zhang, X.M., et al.: A novel compression-based compliant orthogonal displacement amplification mechanism for the typical actuators used in micro-grasping. Sens. Actuators A **297**, 111463 (2019)
3. Wang, Z.W., Zhang, X.M., Yin, Z.Q.: Design and stiffness modeling of a compact 3-dof compliant parallel nanopositioner for the tool servo of the ultra precision machining. In: 2018 IEEE International Conference on Robotics and Biomimetics, pp. 964–971. IEEE (2018)
4. Chen, F.X., Zhang, Q.J., Gao, Y.Z., et al.: A Review on the flexure-based displacement amplification mechanisms. IEEE Access **8**, 205919–205937 (2020)
5. Lobontiu, N., Garcia, E.: Analytical model of displacement amplification and stiffness optimization for a class of flexure-based compliant mechanisms. Comput. Struct. **81**(32), 2797–2810 (2003)
6. Xu, Q.S., Li, Y.M.: Analytical modeling, optimization and testing of a compound bridge-type compliant displacement amplifier. Mech. Mach. Theory **46**(2), 183–200 (2011)
7. Qi, K.Q., Xiang, Y., Fang, C., et al.: Analysis of the displacement amplification ratio of bridge-type mechanism. Mech. Mach. Theory **87**, 45–56 (2015)
8. Ling, M., Cao, J., Zeng, M., et al.: Enhanced mathematical modeling of the displacement amplification ratio for piezoelectric compliant mechanisms. Smart Mater. Struct. **25**(7), 075022 (2016)
9. Liu, P., Peng, Y.: Kinetostatic modeling of bridge-type amplifiers based on Timoshenko beam constraint model. Int. J. Precis. Eng. Manuf. **19**(9), 1339–1345 (2018)
10. Zhang, Q., Zhao, J., Peng, Y., et al.: A novel amplification ratio model of a decoupled XY precision positioning stage combined with elastic beam theory and Castigliano's second theorem considering the exact loading force. Mech. Syst. Signal Process. **136**, 106473 (2020)
11. Wang, H., Zhang, X.M.: Input coupling analysis and optimal design of a 3-DOF compliant micro-positioning stage. Mech. Mach. Theory **43**(4), 400–410 (2008)
12. Ling, M.X., Cao, J.Y., Howell, L.L., et al.: Kinetostatic modeling of complex compliant mechanisms with serial-parallel substructures: a semi-analytical matrix displacement method. Mech. Mach. Theory **125**, 169–184 (2018)
13. Ling, M.X.: A general two-port dynamic stiffness model and static/dynamic comparison for three bridge-type flexure displacement amplifiers. Mech. Syst. Signal Process. **119**, 486–500 (2019)
14. Lobontiu, N.: Note: bending compliances of generalized symmetric notch flexure hinges. Rev. Sci. Instrum. **83**(1), 249–253 (2012)
15. Bathe, K.J.: Finite Element Procedures. 2nd edn. Prentice Hall, New Jersey (2014)
16. Wu, Y.F., Zhou, Z.Y.: Design calculations for flexure hinges. Rev. Sci. Instrum. **73**(8), 3101–3106 (2002)

Design and Development of a Portable Electrical Impedance Tomography System

Jiahao Xu[1,2], Jiewei Lu[1,2], Song Zhang[1,2], Ningbo Yu[1,2], and Jianda Han[1,2](✉)

[1] College of Artificial Intelligence, Nankai University, Tianjin, China
hanjianda@nankai.edu.cn

[2] Tianjin Key Laboratory of Intelligent Robotics, Nankai University, Tianjin, China

Abstract. Electrical impedance tomography (EIT) is a non-invasive detection technique for human tissue imaging. However, most of the current EIT systems are complex and expensive in design. In this paper, we have developed a portable and economic EIT system, which just consists of five parts, including a data acquisition card (DAQ), a power module, a signal generator, a voltage-controlled current source, and a multiplexer module. The fast Fourier transform is used to extract the amplitude of the voltage signals collected by the DAQ. The Tikhonov regularization algorithm is adopted to reconstruct the image of conductivity distribution. Experiments on a practical phantom were designed and conducted to validate the performance of the system. The results showed that the developed system has an average signal-to-noise ratio of 65 dB for measurement channels and a high imaging signal accuracy of 0.99. The conductivity distribution in the phantom was successfully reconstructed by the developed EIT system.

Keywords: Electrical impedance tomography · Voltage-controlled current source · Multiplexer · Image reconstruction · Tikhonov regularization

1 Introduction

Electrical impedance tomography (EIT) is a non-invasive functional imaging technique [4]. By measuring impedance information from the body surface, EIT can reconstruct the conductivity distribution of tissues in the body [7]. Experiments have proved that the conductivity of various tissues in the human body changes with physiological activities or pathological changes [1]. Compared with other tomographic methods, such as CT and MRI, EIT has the advantages of no radiation, fast detection, and low cost. It has a broad application prospect in clinical real-time monitoring and early diagnosis of disease [5].

This work was supported by the National Natural Science Foundation of China (U1913208, 61873135) and the Chinese fundamental research funds for the central universities.

X.-J. Liu et al. (Eds.): ICIRA 2021, LNAI 13014, pp. 415–427, 2021.
https://doi.org/10.1007/978-3-030-89098-8_40

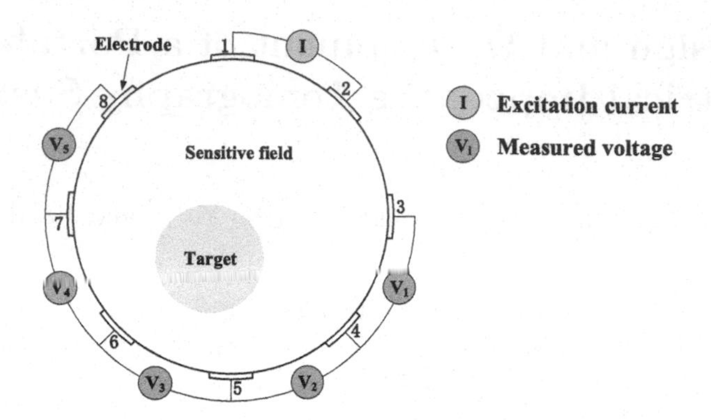

Fig. 1. Adjacent drive and measurement pattern based on the four-electrode method.

In EIT measurement, the excitation current is injected into a sensitive field to obtain voltage signals at the field boundary [9]. This paper takes the 8-electrode EIT system as an example to illustrate the principle of EIT measurement. Figure 1 shows the adjacent drive and measurement pattern based on the four-electrode method. Firstly, the electrode pair 1/2 is selected as the first excitation electrode pair, and the measurement electrode pair is 3/4, 4/5... 7/8. Then, the electrode pair 2/3 is selected to inject current, and the voltages are measured from the remaining electrodes. Until the eighth pair of electrode 8/1 is used as the excitation electrode for measurement. Finally, 40 boundary voltages are collected as one frame of imaging data. To achieve the above measurement, the EIT system is required to generate current excitation and collect voltage signals.

In the past few decades, many research groups have studied the design of EIT system. Early EIT systems used data acquisition cards (DAQ) for stimulation and measurement [15]. For example, Sheffiled MK3.5 system contains 8 DAQs, leading to a very large system structure [13]. Then, the combination of microprocessor and peripheral circuit is used to make the EIT system more integrated and improve the signal-to-noise ratio (SNR) and bandwidth of the system. KHU Mark 2.5 system based on DSP can maintain a SNR of 80 dB within a frequency of 250 kHz [12]. Another wideband EIT based on FPGA has a frequency range of 1 kHz to 1 MHz, which can be used for frequency-difference imaging [10]. With advances in electronics, several groups made wearable EIT devices using ASIC technology for lung and brain imaging [8,14]. Although the above EIT systems have good performance, most of them are too complex and expensive for portable applications.

In this paper, we designed a portable and economic EIT system. Firstly, the hardware structure of the EIT system with 8 electrodes is built by a DAQ, a power module, a signal generator, a voltage-controlled current source, and a multiplexer module. Secondly, FFT is used to extract the amplitude of voltage signals for imaging. Next, EIT image reconstruction is calculated by using the

Tikhonov regularization algorithm. Finally, the performance of the designed EIT system was verified using a practical phantom.

2 System Design

2.1 System Architecture

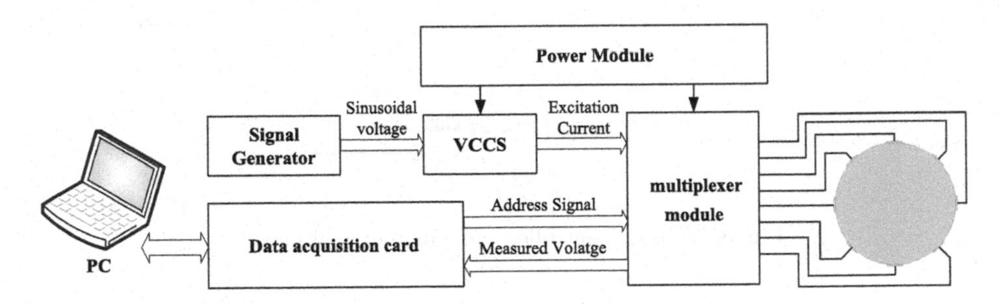

Fig. 2. The overall structure of the EIT system.

Figure 2 shows the overall structure of the EIT system. The system mainly includes five parts: a power module, a signal generator, a voltage-controlled current source (VCCS), a multiplexer module and a data acquisition card (DAQ). The power module provides a voltage of ± 12 V for the VCCS and the multiplexer module. The signal generator uses AD9834 device to generate a sinusoidal voltage with adjustable frequency. The VCCS converts the sinusoidal voltage into an excitation current with the same frequency. The multiplexer module is used for the selection of excitation and measurement electrodes. The DAQ, as the lower computer, controls the work of the multiplexer module and communicates with PC.

ART USB3103A is chosen as the data acquisition card, which is cost-effective for this system. It has 16 analog input channels and 16 programmable digital I/O channels. The analog channel has a maximum conversion rate of 500 Ksps and a high resolution of 0.03 mV, which satisfies the measurement requirements. USB3103A provides an integrated C++ interface, allowing users to write data acquisition programs in Microsoft Visual Studio on PC. During the measurement, the PC sends acquisition task commands to DAQ. Then, the DAQ sends address signals to the multiplexer module by 16 digital output channels, while collecting voltage signals through 1 analog input channel. The measured voltage signals are transmitted to the PC for processing and imaging.

2.2 Voltage-Controlled Current Source

In EIT, the human body is injected with a current excitation at a certain frequency, which is provided by a current source. To generate a high-precision sinusoidal current, a voltage-controlled current source was built based on two operational amplifiers (OP27GS), as shown in Fig. 3. In the circuit, R_L represents

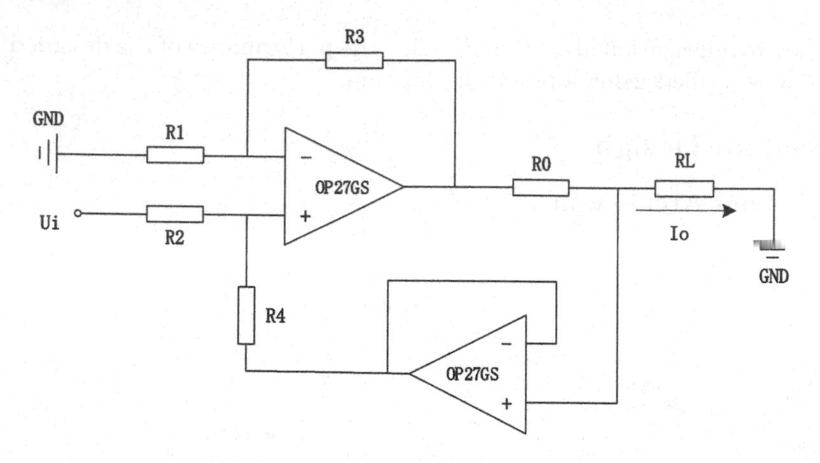

Fig. 3. Voltage-controlled current source circuit.

the load on the output. When R1 = R2 = R3 = R4 is satisfied, the relationship between the output current I_o and the input voltage U_i can be expressed as:

$$I_o = \frac{U_i}{R_o} \tag{1}$$

where R_0 is set 500 Ω, denoting the gain of the VCCS. The voltage signal with a peak-to-peak value of 500 mV generated by the signal generator is used as the input voltage of VCCS. Theoretically, 1 mA peak-to-peak (also equal to 0.5 mA amplitude) current can be obtained at the output of VCCS. In general, the amplitude of the current injected into the human body cannot exceed 5 mA [2]. So the designed VCCS is safe for measuring the human body.

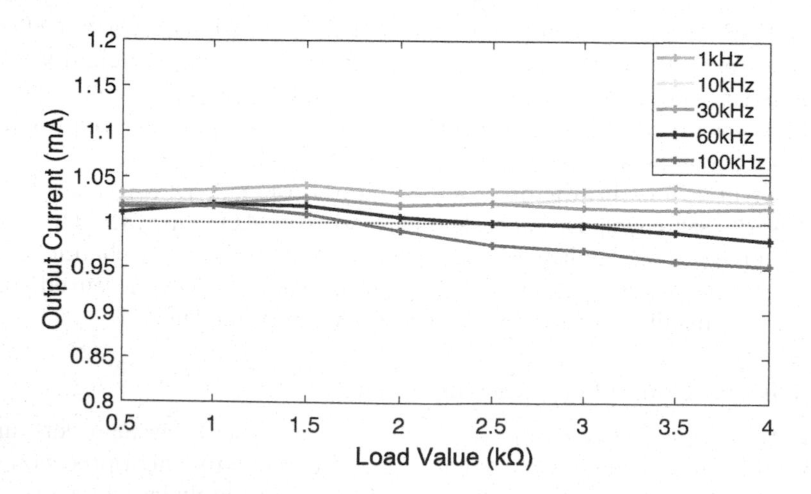

Fig. 4. VCCS output performance test results.

The high-quality VCCS is required to drive different loads and has a stable output current. Generally, the resistance of the human body surface ranges from 0.8 kΩ to 2 kΩ. Hence, resistors with a range of 0.5 kΩ to 4 kΩ are used as the loads for the VCCS accuracy test. Output current test under different frequencies and loads was carried out, as presented in Fig. 4. With the increase of frequency and load, the output current can be maintained around 1 mA. The peak-to-peak value 1 mA is used as the standard value to calculate the relative error (RE) of the output current. The RE of the output current is not more than 4.7% within the frequency of 100 kHz and the load of 4 kΩ. In conclusion, the designed VCCS has a stable output current within a certain range.

2.3 Multiplexer Module

The multiplexer module in the EIT system is used to select the excitation electrodes and the measurement electrodes. Figure 5 illustrates the multiplexer module structure. The multiplexer module consists of four ADG1606 devices, representing the current positive I_+, the current negative I_-, the voltage positive V_+ and the voltage negative V_-, respectively. ADG1606 is a high-speed analog multiplexer with low on-resistance. The ADG1606 switches one of 16 independent inputs/outputs (S1–S16) to a common input/output (D), as determined by the 4-bit binary address lines (A_0,A_1,A_2 and A_3). Because the designed EIT system is an 8-electrode array, each ADG1606 only uses channels S1–S8.

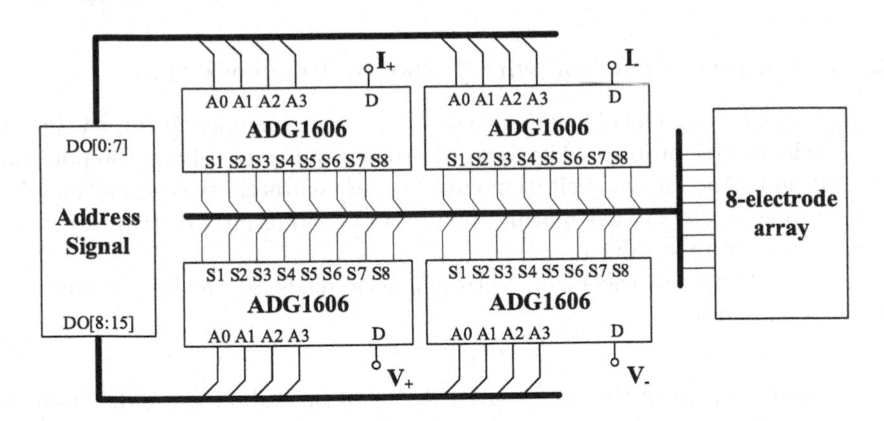

Fig. 5. Multiplexer module structure.

When the multiplexer module receives the 16 address signals sent by DAQ, each ADG1606 can open the corresponding channels. The I_+ and I_- ports are connected with the excitation electrode pair to inject excitation current into sensitive field. The V_+ and V_- ports are connected with the measurement electrode pair to collect voltage signals. The transition time of ADG1606 is no more than 0.22 us when switching from one address state to another. Compared with the measurement time, the time consumed by channel switching is negligible, which ensures the efficiency of the system during measurement.

3 Image Reconstruction

3.1 Amplitude Extraction Based on FFT

In EIT measurement, the voltage signal X on the measurement electrodes is collected by an analog input channel of DAQ at the frequency of 500 kHz. Obviously, X contains a component with the same frequency as the excitation current, which represents the impedance information in the sensitive field. FFT is used to extract the amplitude of the corresponding frequency in the voltage signal. The amplitude of the component with frequency k is calculated as follows:

$$R = \sum_{n=0}^{N-1} X(n) \cos(2\pi k \cdot \frac{n}{N}) \tag{2}$$

$$I = \sum_{n=0}^{N-1} -X(n) \sin(2\pi k \cdot \frac{n}{N}) \tag{3}$$

$$A = \frac{2}{N}\sqrt{R^2 + I^2} \tag{4}$$

where n and N denote the nth and total number of sampling points, respectively. R and I are the real and imaginary parts, respectively. A is the extracted amplitude of the component with frequency k, which is used for imaging.

3.2 Image Reconstruction with Tikhonov Regularization

The mathematical model of EIT image reconstruction includes the forward problem and the inverse problem. The forward problem refers to solving the potential distribution by the known excitation current and conductivity distribution. The inverse problem refers to calculating the internal conductivity distribution using the boundary voltage values.

Based on Maxwell's theory of electromagnetic fields [9], the forward problem can be expressed as:

$$\nabla \cdot (\sigma \cdot \nabla \varphi) = 0 \tag{5}$$

where σ and φ represent the conductivity distribution and potential distribution, respectively. Equation (5) is a second-order partial differential equation, which is usually solved by the finite element method (FEM). FEM can divide the sensitive field into a finite number of small elements so that the problem can be calculated more efficiently. Then Eq. (5) can be linearized near the baseline conductivity σ_0 as:

$$J\delta\sigma = \delta U \tag{6}$$

where J is the Jacobian matrix, denoting the sensitivity of the boundary voltage to the change of conductivity. $\delta\sigma$ is the difference of conductivity distribution in the sensitive field. δU is the change of the boundary voltages, also called a frame of imaging signal. The Jacobian matrix is calculated by using the method in literature [3]. In follow-up experiments, we choose saline with a conductivity of 0.1 S/m as the baseline for measurement.

The inverse problem is seriously ill-posed, which is generally regarded as an optimization problem to seek approximate solutions. By setting an objective function, the estimated value of the forward problem is constantly corrected with the measured value. When the objective function is iterated to the minimum value, the approximate solution of the equation is found. In this paper, the Tikhonov (TK) regularization algorithm is used to calculate the inverse problem. The objective function based on TK regularization is given by:

$$\arg\min_{\delta\sigma}\left\{\frac{1}{2}\|J\delta\sigma - \delta U\|^2 + \lambda\|\delta\sigma\|_2\right\} \tag{7}$$

where λ is a hyperparameter, which controls the weight between the error term and the regularization term. The solution of the objective function (7) can be obtained by one iteration of the Gauss-Newton method, which greatly saves the calculation time [11].

4 Experiments and Results

To verify the performance of the developed system, an EIT experiment platform was built, as shown in Fig. 6. A circular tank with a diameter of 150 mm was used as the practical phantom for experiments. Eight copper electrodes with a width of 20 mm are evenly arranged around the tank. The imaging targets are polyurethane rods with a diameter of 30 mm and conductivity of 0 S/m, which are used to be identified from the saline with a conductivity of 0.1 S/m.

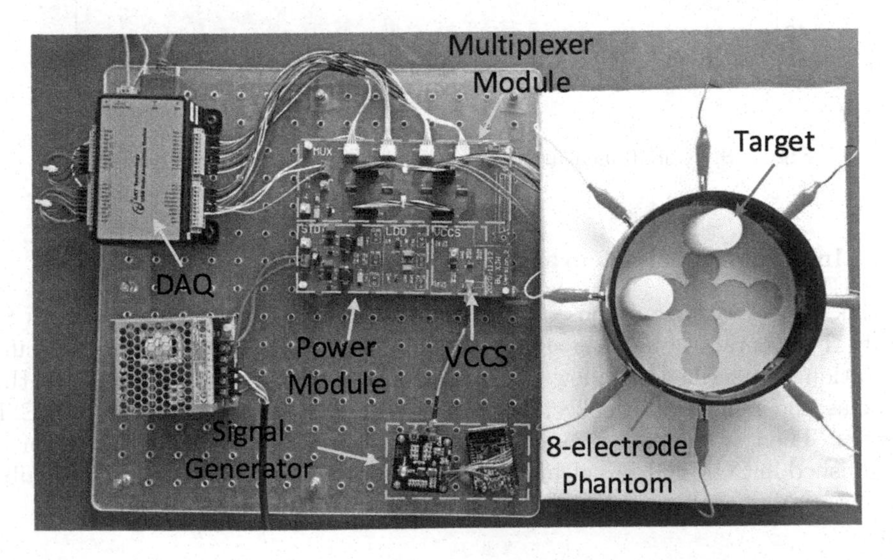

Fig. 6. The EIT experimental platform.

4.1 Measurement Channel SNR

Under homogeneous background solution (saline), the boundary voltage was collected through 40 measurement channels at 1, 10, 30, 60, 100 kHz. Then the SNR in each measurement channel was calculated by the following equation:

$$\mathrm{SNR} = -20\log_{10}\frac{\sqrt{\frac{1}{N}\sum_{i=1}^{N}\left(v_i^2 - \bar{v}^2\right)}}{\bar{v}} \tag{8}$$

where N is the number of collection at a certain frequency, and $N = 100$. v_i and $\bar{v}$ are the ith measurement value and the mean of all measurements, respectively. The results of SNR are shown in Fig. 7, in which the maximum, minimum and average values of 40 measurement channels were calculated at the above frequencies. It can be seen that the SNR of all the measurement channels is mainly distributed between 60 dB and 70 dB. The average SNR of the system measurement channels is 65 dB, which is satisfactory performance.

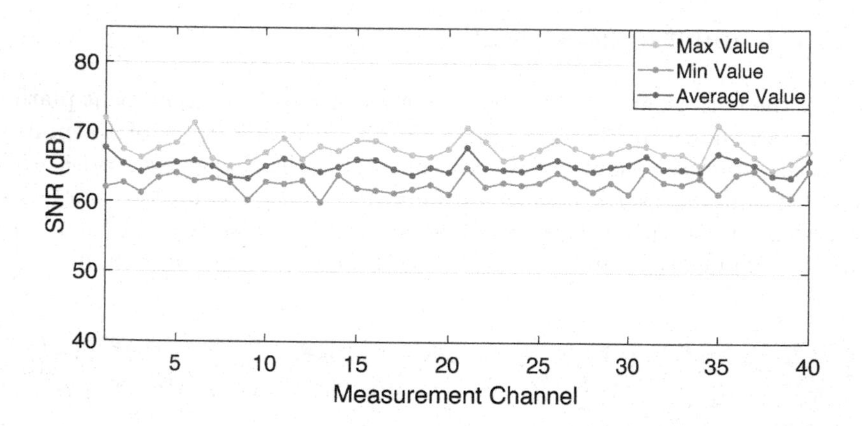

Fig. 7. SNR of 40 measurement channels at the tested frequencies.

4.2 Imaging Signal Accuracy

In this paper, we first proposed a method of using simulation experiments to verify the accuracy of imaging signals measured by the EIT system. The specific operation is described as follows. Firstly, the polyurethane rod is placed in the tank as the phantom to collect the actual imaging signal (AIS) by the EIT system. Then, a simulation model completely consistent with the phantom is established in COMSOL to calculate the simulation imaging signal (SIS). Finally, the correlation coefficient (Cor) between AIS and SIS is calculated by:

$$\mathrm{Cor} = \frac{\sum_{i=1}^{n}\left(\Delta U_{a_i} - \Delta\bar{U}_a\right)\left(\Delta U_{s_i} - \Delta\bar{U}_s\right)}{\sqrt{\sum_{i=1}^{n}\left(\Delta U_{a_i} - \Delta\bar{U}_a\right)^2\left(\Delta U_{s_i} - \Delta\bar{U}_s\right)^2}} \tag{9}$$

where ΔU_a and ΔU_s represent the actual imaging signal and the simulation imaging signal, respectively. $\Delta\bar{U}_a$,$\Delta\bar{U}_s$ are the mean value of ΔU_a,ΔU_s. The value of Cor represents the accuracy of the imaging signal collected by the designed EIT system. Twenty-one sets of tests were conducted at a frequency of 30 kHz, in which the polyurethane rod were placed in different positions in the tank.

The Cor of all test groups are above 0.99 as shown in Fig. 8. These results indicate that there exists a strong correlation between AIS and SIS. Figure 9 shows three groups of comparisons between AIS and SIS. It can be seen that when the phantom is consistent with the simulation model, the AIS and SIS also tend to be consistent. The correlation coefficient between the actual measurement results and the simulation results is greater than 0.99, which proves that the designed EIT system has high measurement accuracy.

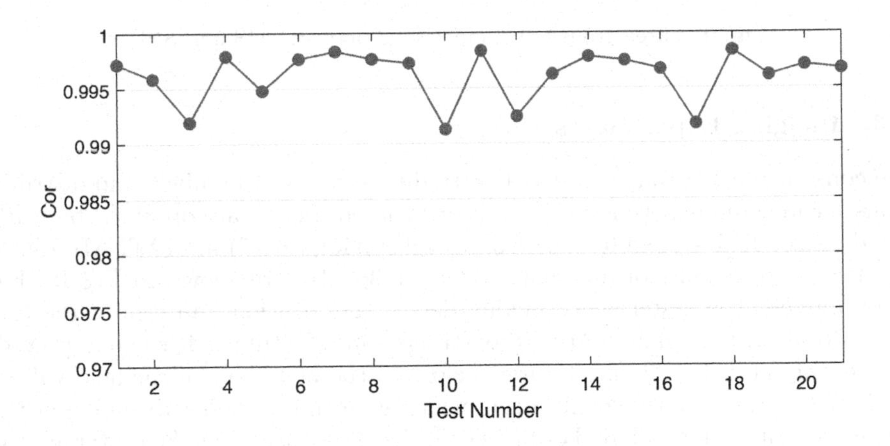

Fig. 8. The correlation between AIS and SIS.

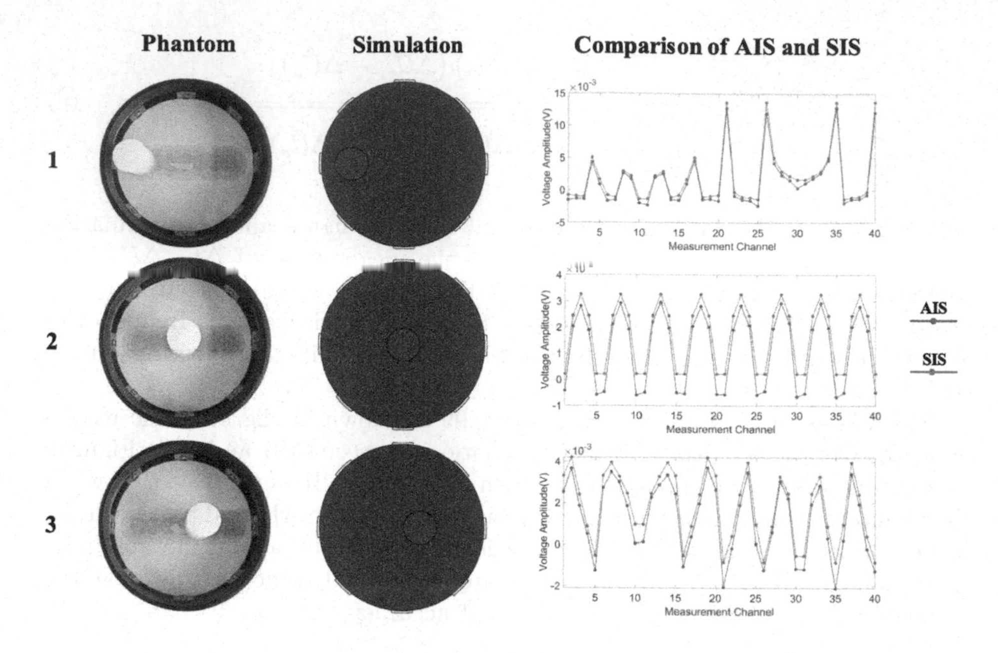

Fig. 9. Three groups of comparisons between AIS and SIS.

4.3 Imaging Experiments

We constructed 12 groups of models with different target numbers and distributions for imaging experiments. The reconstructed images are obtained by using the designed EIT system to collect data and solving Eq. (7) in MATLAB. Image reconstruction results of different models at 30 kHz are shown in Fig. 10. For most models, the system can roughly detect the location and size of the targets. However, the quality of the reconstructed image still needs to be improved. Comparing models 1, 2 and 3, targets close to the boundary of the field will be distorted in the reconstructed image, which is an inherent disadvantage of the Jacobian matrix method [6]. For model 12, the three targets at the center are too close to distinguish them accurately. These problems require improvements in image reconstruction algorithms. Figure 11 shows image reconstruction results at different frequencies. As can be seen from Fig. 11, the EIT system can accurately identify targets at frequencies f = 1 kHz, 30 kHz, 60 kHz and 100 kHz. There is no significant difference in the imaging results of the phantom at different frequencies, which demonstrates that the system has good stability within f = 100 kHz. In summary, the imaging experiments indicate that the designed EIT system can realize impedance acquisition and imaging in the sensitive field within 100 kHz.

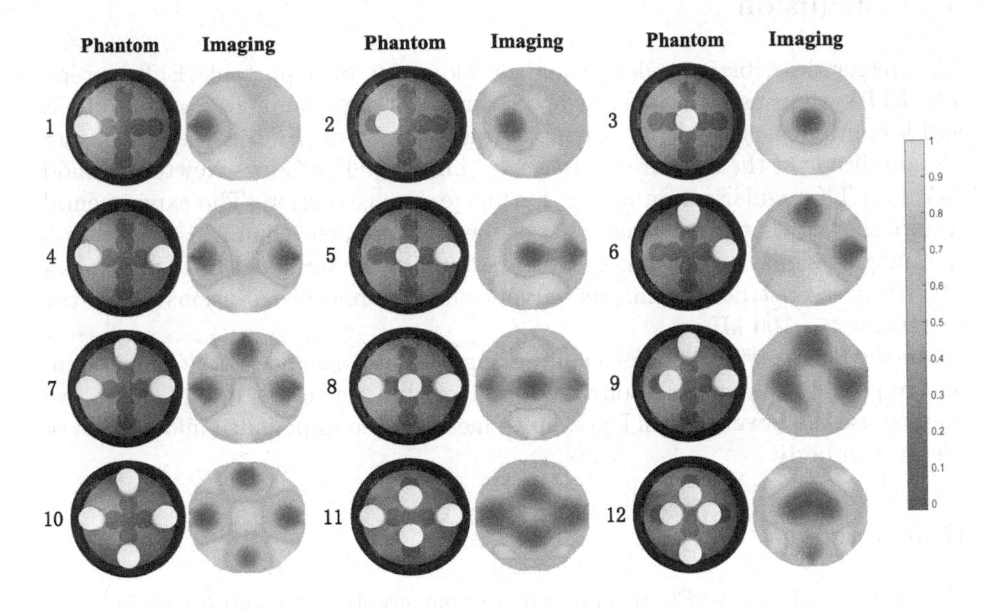

Fig. 10. Image reconstruction results of different models at 30 kHz.

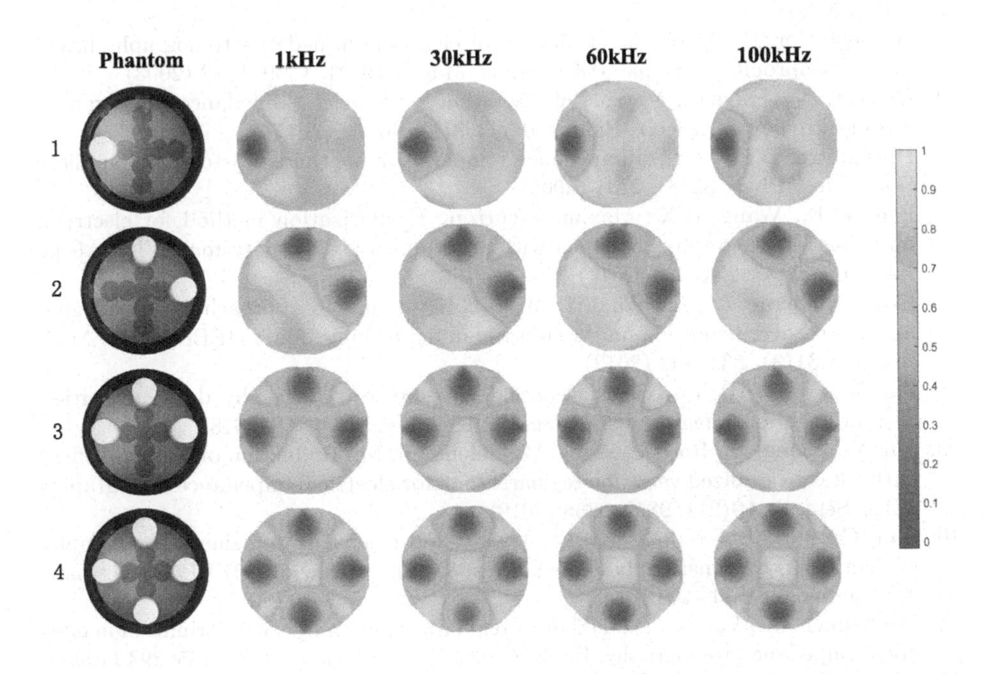

Fig. 11. Image reconstruction results at different frequencies.

5 Conclusion

This paper described the design and development of a portable EIT system. The EIT system uses DAQ as the lower computer and combines a stable VCCS and a multiplexer module for driving and measurement. FFT is used to extract the amplitude of the measured voltages for imaging. The Gauss-Newton method based on TK regularization is used for image reconstruction. The experimental results showed that the average SNR of the measurement channel is 65 dB, and the imaging signal accuracy is greater than 0.99. The designed EIT system can effectively reconstruct the imaging targets with different distributions within the frequency $f = 100$ kHz.

In the future, we will focus on optimizing the performance of the EIT system and improving the image reconstruction algorithm. In terms of application, we plan to use the developed EIT system to measure the impedance information of the human body.

References

1. Adler, A., Boyle, A.: Electrical impedance tomography: tissue properties to image measures. IEEE Trans. Biomed. Eng. **64**(11), 2494–2504 (2017)
2. Association, J.S.: Medical electrical equipment - part 1: general requirements for basic safety and essential performance. Japanese J. Radiol. Technol. **56**(11), 1332–1333 (2005)
3. Brandstatter, B.: Jacobian calculation for electrical impedance tomography based on the reciprocity principle. IEEE Trans. Magn. **39**(3), 1309–1312 (2003)
4. Chitturi, V., Farrukh, N.: Spatial resolution in electrical impedance tomography: a topical review. J. Electr. Bioimpedance **8**(1), 66–78 (2017)
5. Djajaputra, D.: Electrical impedance tomography: methods, history and applications. Med. Phys. **32**(8), 2731 (2005)
6. Fan, W.R., Wang, H.X.: Maximum entropy regularization method for electrical impedance tomography combined with a normalized sensitivity map. Flow Meas. Instrum. **21**(3), 277–283 (2010)
7. Gomez-Laberge, C., Arnold, J.H., Wolf, G.K.: A unified approach for eit imaging of regional overdistension and atelectasis in acute lung injury. IEEE Trans. Med. Imaging **31**(3), 834–842 (2012)
8. Shi, X., et al.: High-precision electrical impedance tomography data acquisition system for brain imaging. IEEE Sens. J. **18**(14), 5974–5984 (2018)
9. Shi, Y., Zhang, X., Rao, Z., Wang, M., Soleimani, M.: Reduction of staircase effect with total generalized variation regularization for electrical impedance tomography. IEEE Sens. J. **19**(21), 9850–9858 (2019)
10. Tan, C., Liu, S., Jia, J., Dong, F.: A wideband electrical impedance tomography system based on sensitive bioimpedance spectrum bandwidth. IEEE Trans. Instr. Measurement, 1–11 (2019)
11. Vauhkonen, M., Vadasz, D.: Tikhonov regularization and prior information in electrical impedance tomography. IEEE Trans. Med. Imaging **17**(2), 285–293 (1998)
12. Wi, H., Sohal, H., Mcewan, A.L., Woo, E.J., Oh, T.I.: Multi-frequency electrical impedance tomography system with automatic self-calibration for long-term monitoring. IEEE Trans. Biomed. Circuits Syst. **8**(1), 119–128 (2014)

13. Wilson, A.J., Milnes, P., Waterworth, A.R., Smallwood, R.H., Brown, B.H.: Mk3.5: a modular, multi-frequency successor to the mk3a eis/eit system. Physiol. Measurement **22**(1), 49–54 (2001)
14. Wu, Y., Jiang, D., Bardill, A., de Gelidi, S., Bayford, R., Demosthenous, A.: A high frame rate wearable eit system using active electrode asics for lung respiration and heart rate monitoring. IEEE Trans. Circuits Syst. I Regul. Pap. **65**(11), 3810–3820 (2018)
15. Yerworth, R.J., Bayford, R.H., Cusick, G., Conway, M., Holder, D.S.: Design and performance of the uclh mark 1b 64 channel electrical impedance tomography (eit) system, optimized for imaging brain function. Physiol. Meas. **23**(1), 149–158 (2002)

Systematic Analyses of Passive Vibration Isolation System for Atomic Force Microscopes

Shenghang Zhai[1,2,3], Peng Yu[1,2], Jialin Shi[1,2], Tie Yang[1,2], and Lianqing Liu[1,2(✉)]

[1] State Key Laboratory of Robotics, Shenyang Institute of Automation, Chinese Academy of Sciences, Shenyang, China
lqliu@sia.cn

[2] Institutes for Robotics and Intelligent Manufacturing, Chinese Academy of Sciences, Shenyang 110169, China

[3] University of Chinese Academy of Sciences, Beijing, China

Abstract. Atomic force microscope (AFM) is a powerful tool for imaging a wide range of materials with nanometer resolution, which is sensitive to mechanical vibration from the ground or buildings. For improving imaging performance, vibration isolation systems are often employed. Air tables are often used to attenuate the vibration transmission from ground to precision instruments, but it performs poorly for low-frequency vibration isolation. Suspending AFM using long common bungee cords is a simple and effective vibration isolation method which perform well in both low-frequency and high-frequency domain. However, it requires a lot of space and the bungee cords will failure when using a long time. Here we developed a vibration isolation system that uses linear steel springs to suspend the AFM for high precision imaging. The simplified model is used to explain how the spring constant and the damper affects the vibration isolation performance, through experiment the relationship between the spring constant and the damper was verified. The accelerometer and the AFM were employed to show the isolation performance, it shows the noise level of the AFM can be reduced from 71 pm to 21 pm.

Keywords: Atomic force microscope · Passive vibration isolation · High precision imaging

1 Introduction

Atomic force microscope (AFM) is a powerful tool for imaging a wide range of materials with nanometer resolution [1–4], which is sensitive to mechanical vibration from ground or buildings. Therefore, vibration isolation systems are needed to reduce the vibration transmitted to the AFM [5]. Commonly used vibration isolation techniques can be classified into active vibration isolation system and passive vibration isolation system. Passive vibration isolation system have air tables [6], Folded continuous beams [7], suspension using bungee cords [8], etc. Active vibration isolation system usually uses external sensors to sense the vibration and control the actuator to generate proper motion or force to compensate the sensed vibration [9]. Active damping techniques [10,

X.-J. Liu et al. (Eds.): ICIRA 2021, LNAI 13014, pp. 428–439, 2021.
https://doi.org/10.1007/978-3-030-89098-8_41

11] and active anti-phase control techniques [12] are commonly employed to control the active vibration isolation system.

Air tables are often used to attenuate the vibration transmission from ground to precision instrument, but it performs poorly for low-frequency vibration isolation [13]. In addition, the air table needs air to support the platform as a spring, it can't be used in the space or vacuum environment. For the active vibration isolation system, it can isolate not only the high-frequency vibration but the low-frequency vibration. But it needs extra energy for the sensors and actuators, and suffers the unstable problem [14].

Suspending AFM using long common bungee cords is a simple and effective vibration isolation method which perform well in both low-frequency and high-frequency domain. It is a proper vibration isolation method for high precision AFM imaging and other special occasions need high vibration isolation performance, and it doesn't need an extra power supply. However, low stiffness bungee cords are needed so as to get high vibration isolation performance. It means the system needs a lot of space and the bungee cords will face failure problems when using a long time. Gero [13] used steel spring replaced the bungee coeds suspend the whole air table, the vibration isolation performance is better than the commercial passive vibration isolation system. However, this system didn't take the parameters into consideration and didn't evaluate the vibration isolation performance by the precision instruments.

In this paper, we proposed a passive vibration isolation system for high precision AFM imaging, set the mechanical model of the system, analyzed how the parameters (the spring stiffness, the damping, and the total mass) of the system affect the vibration isolation performance through the mechanical model, and then the vibration isolation performance was measured experimentally, verified the effect of the spring stiffness and the damping on the vibration isolation performance. Finally, the noise of the AFM with vibration isolation and without vibration isolation was compared, the result shows that the passive vibration system in this paper can reduce the noise from 71 pm to 20 pm.

2 Analyses of Passive Vibration Isolation

Figure 1 shows the schematic diagram of the entire vibration isolation system, the AFM is placed on the vibration isolation platform, one end of the four linear steel springs are

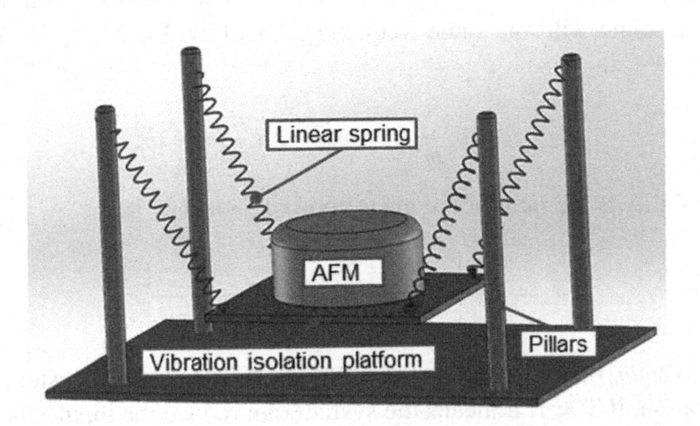

Fig. 1. Schematic diagram of the entire passive vibration isolation system.

connected to the vibration isolation platform, the other end of the linear steel springs are connected to the four pillars. As a result, the AFM is suspended by the system. A small rod is fixed at the center of the vibration isolation platform and put into a container filled with damper, through changing the depth into the container, the damping can be changed.

The passive vibration isolation system can be equivalent to one simplified system that is composed of a mass [6, 15], a resilient member and viscous damper as shown in Fig. 2, where the x_1 is the vertical position of the fixed base, x_2 is the vertical position of the moving platform. The equation of motion result as

$$m\ddot{x}_2 + c(\dot{x}_2 - \dot{x}_1) + k(x_2 - x_1) = 0 \tag{1}$$

Where m is the total mass of vibration isolation platform and the AFM, k and c are the total stiffness of the four linear steel springs in the vertical direction and the total damping, respectively.

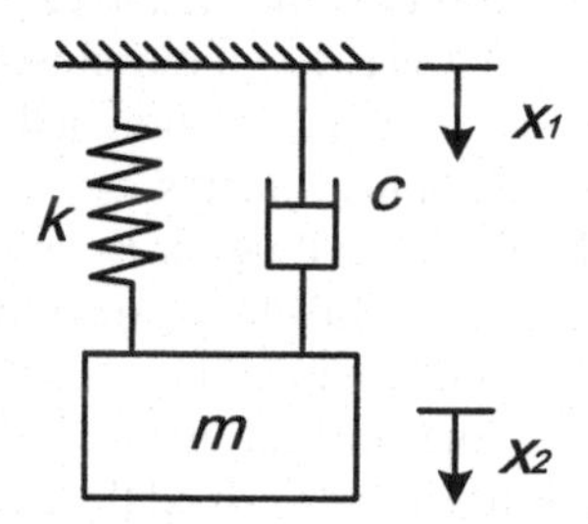

Fig. 2. The equivalent model of the passive vibration isolation system

Let $\omega_n^2 = k/m$,ω_n is the natural frequency,

$$\ddot{x}_2 + \frac{c}{m}\left(\dot{x}_2 - \dot{x}_1\right) + \omega_n^2(x_2 - x_1) = 0 \tag{2}$$

The performance of the vibration isolator can be evaluated through transmissibility, which can be defined as the ratio of the magnitude of transmitted vibration to harmonic excitation. If the vibration of the fixed base is a harmonic motion, then the motion of the mass is harmonic with the same frequency, let $x_1 = X_1 e^{i\omega t}$, $x_2 = X_2 e^{i(\omega t + \emptyset)} = X_2 e^{i\emptyset} e^{i\omega t} = \widehat{X_2} e^{i\omega t}$, then

$$\widehat{X_2} - \omega^2 + \frac{c}{m} i\omega + \omega_n^2 = X_1 \frac{c}{m} i\omega + \omega_n^2 \tag{3}$$

Transmissibility is the ratio of the magnitude of x_2 and x_1 as

$$T = \left|\frac{\widehat{X_2}}{X_1}\right| = \left|\frac{2\xi\omega_n\omega i + \omega_n^2}{-\omega^2 + 2\xi\omega_n\omega i + \omega_n^2}\right| \tag{4}$$

Where $\xi = c/(2m\omega_n)$ is the damping ratio, If $T > 1$, it means the system can amplify the input vibration, if $T < 1$, it means the system can reduce the input vibration.

As shown in Fig. 3, when the mass doesn't change, the natural frequency increase with the stiffness k. The vibration which the frequency below $\sqrt{2}\omega_n$ will directly couple to the AFM without attenuation. Above $\sqrt{2}\omega_n$, vibrations will be attenuated, and with the increase of the frequency, the transmissibility will be reduced quickly.

The damping doesn't affect the natural frequency of the system, but it has an effect on the transmissibility, as shown in the Fig. 4. When the stiffness and the mass stay constant, increase of the damping ratio will reduce the amplification around the natural frequency, but it will also reduce the vibration isolation ability above $\sqrt{2}\omega_n$.

The mass has also an effect on the performance of the vibration isolation as shown in the Fig. 5. With the increase of the total mass the natural frequency decreased, for the same frequency harmonic vibration, the heavier the mass is, the smaller the transmissibility is.

From the analyses above, we know that a high-performance passive vibration isolator should use low stiffness spring and add proper damping. If the damping was too big, the vibration isolation ability will be reduced. If the damping was too small, the system will amplify the excitation vibration around the natural frequency, and the whole system should take a long time from unstable state to stable state.

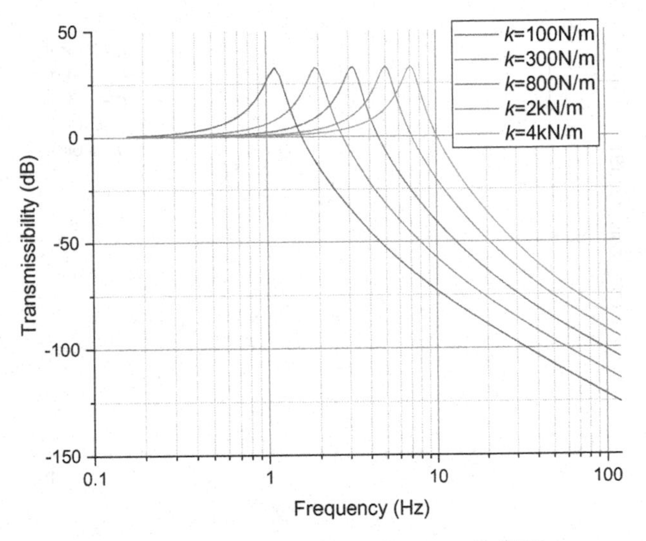

Fig. 3. Effect of stiffness on the transmissibility

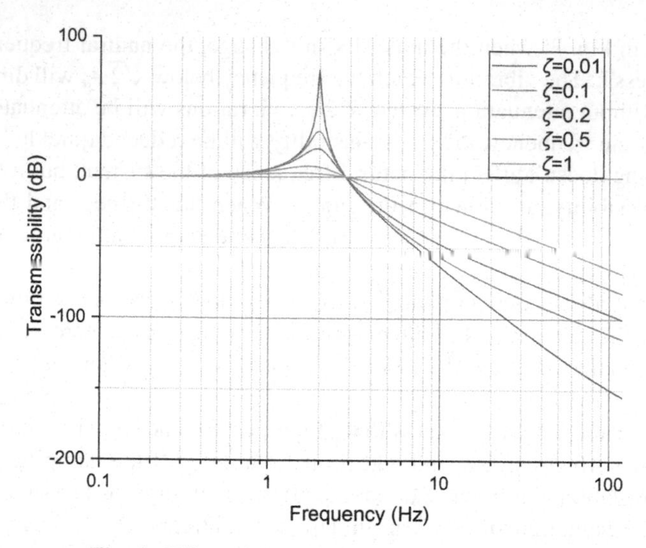

Fig. 4. Effect of damping on the transmissibility

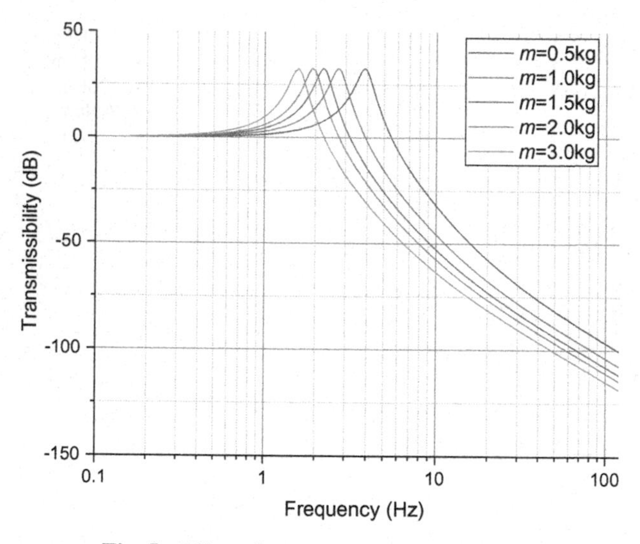

Fig. 5. Effect of mass on the transmissibility

3 Experiment and Result

3.1 Experiment Platform

Figure 6 shows the setup of the passive vibration isolation system, the vibration isolation platform is suspended by the four pillars, the four linear steel springs join the isolation platform and the pillars. The isolation platform has two layers which are connected by another four pillars. An AFM (Multimode III, Bruker Nano Inc) is placed on the upper lawyer of the vibration isolation platform. In order to measure the vibration of the platform when the AFM is imaging, a high precision accelerometer (Fortis, Güralp Systems Ltd., United Kingdom) is placed on the lower platform. In order to add adjustable damping into the system, a screw (GB70.1 M6*40) is mounted on the lower lawyer of the platform, the screw is put 20 mm into a container filled with damping fluid, through changing the materials of the damping fluid, the damping can be adjusted.

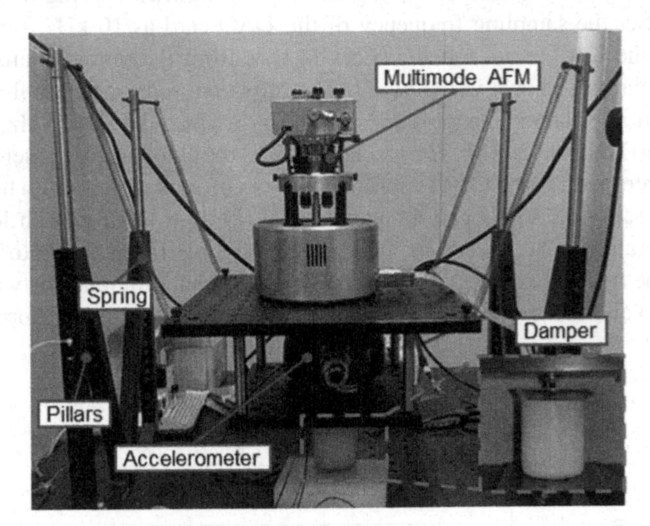

Fig. 6. The passive vibration isolation system setup for AFM

The total mass of the blackboard, the AFM and the accelerometer is 5.7 kg. The vibration isolation performance is evaluated by the high precision accelerometer, which has bandwidth 0–200 Hz, dynamic range exceed 160 dB, full range from 0.5 g to 4 g. In this experiment the 0.5 g full range was chose. The accelerometer can output 3 axis acceleration analog signal, the signal was measured by the DAQ card (M2p.59xx, Spectrum Instrumentation GMBH, Germany), set the sample frequency as 10 kHz, set the sample size as 512 k. The excitation of the isolation system is the laboratory floor and acoustic vibration.

3.2 Effect of Stiffness on the Vibration Isolation Performance

From the analyses of the Sect. 2, we know that the stiffness is a key factor about the vibration isolation performance. When the total mass doesn't change, the smaller the stiffness of the spring is, the smaller the natural frequency of the system is, and the system have wider bandwidth that can isolate vibration, the performance of the vibration isolation is much better. However, it is impossible to reduce the stiffness to infinite small, because the reduction of stiffness will increase the elongation of the spring, which means more space needed for the whole system.

If the scan rate of the AFM is above 10 Hz, the vibration below 10 Hz have little impact on the imaging, so we hope the passive vibration isolation system can isolate vibration above 10 Hz.

We firstly explored how the stiffness affect the vibration isolation performance, theoretically the system using low stiffness spring can perform better. Two kinds of springs were used for comparison, the spring constant of the low stiffness spring and the high stiffness spring were 76.2 N/m and 60.4 N/m, respectively. The acceleration of the accelerometer without vibration isolation (the accelerometer was placed on a common table), suspend by the low stiffness spring and the high stiffness spring were measured, respectively. Set the sampling frequency of the DAQ card as 10 kHz, sample size as 512 k, using the fast Fourier transform (FFT) transform the time domain data to the frequency domain, as shown in the Fig. 7. The red curve represents the floor vibration without vibration isolation, the vibration is almost distributed 0 Hz–50 Hz, and it has a maximum at 30 Hz about 200 ug. The green curve and the blue curve represent the passive vibration isolator with high stiffness spring and low stiffness spring. With high stiffness spring the vibration in 30 Hz can be attenuated 31.3 ug to 1.5 ug, with low stiffness spring the vibration in 30 Hz can be attenuated to 0.2 ug. From 50 Hz to 200 Hz, the vibration of the floor relatively small, there is no significant difference between the high stiffness group and the low stiffness group in the vibration isolation performance.

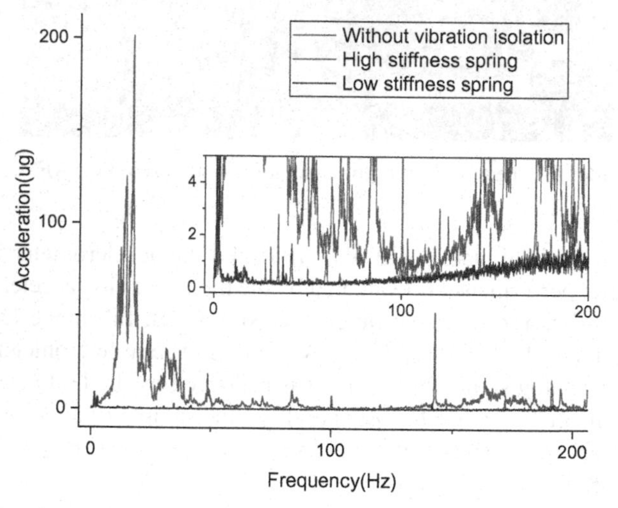

Fig. 7. Effect of spring stiffness on the performance of the passive vibration isolation system, the insert figure shows the zoomed-in-view of the image. (Color figure online)

3.3 Effect of Damping on the Vibration Isolation Performance

From the analyses of the Sect. 2, we know that the damping is another key factor for the vibration isolation performance. If the system is lightly damped, and is disturbed somehow, the subsequent transient vibration will take a long time to die out. If the damping is too big, the vibration isolation performance will be reduced. Therefore, it is critical to select a proper damping for the passive vibration isolation system. Two kinds of damping fluid (DP266, resistance torque:20 gf.cm@25 °C, DP322, resistance torque:

3 gf.cm@25 °C, Shenzhen Ecco Lubrication Co., Ltd, China) was selected for evaluating the effect of damping on the vibration isolation performance. The acceleration of the accelerometer without extra damping, with little damping and with rich damping was measured respectively. Transform the time domain data to the frequency domain through FFT as shown in the Fig. 8. Without damping, as the red curve shown, there are several high peaks among 1 Hz–3 Hz, with damping, the peaks can be reduced significantly. The high damping group can reduce more than the little damping group. From 3 Hz to 200 Hz, there is no significant difference between the without damping group and little damping group, they perform better than the rich damping group.

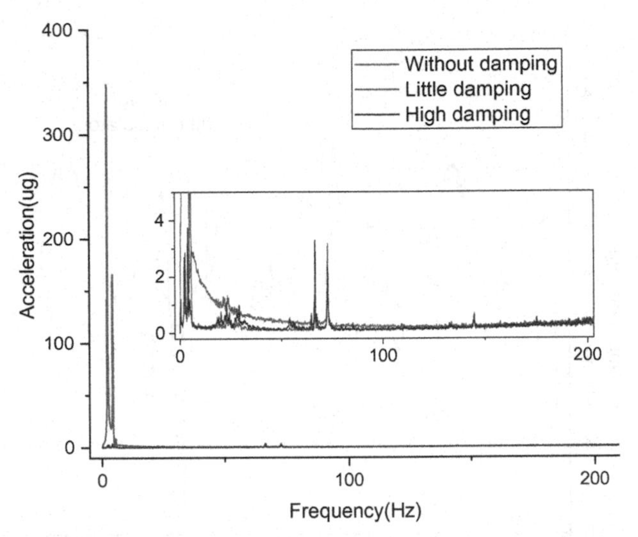

Fig. 8. Effect of damping on the performance of the passive vibration isolation system, the insert figure shows the zoomed-in-view of the image.

3.4 Effect on the AFM Mechanical Noise Reduction

The passive vibration isolation platform was designed for high precision AFM imaging, its effectiveness should be evaluated by the final imaging quality. The imaging quality is related to the AFM z axis mechanical noise. Therefore, we measured the z axis mechanical noise with vibration isolation and without vibration isolation.

Set the AFM worked on the contact mode. If the stiffness between prob and sample is infinite high, the height image will keep constant. But the stiffness can't be infinite high, so the mechanical noise will be coupled to the image. Set the scan size was 0 × 0 nm^2 with 128 × 128 pixels, continue to save 8 images of trace channel and retrace channel. Splicing the image data to a serial time domain data as shown in the Fig. 9, the red curve and the blue curve represent the mechanical vibration noise with vibration isolation and without vibration isolation. With vibration isolation, the peak to peak of the mechanical noise is much lower than that without vibration isolation. Without vibration isolation, the root mean square (RMS) of noise is 1.6 nm, and it is 0.8 nm with the vibration isolation. The noise is reduced one-half. Transform the time domain data to the frequency domain as shown in the Fig. 10, the red curve and the blue curve represent the frequency spectrum of the vibration with vibration isolation and without vibration isolation respectively. There is a high peak among 0 Hz ~ 2.5 Hz, without vibration isolation the height of the peak is 0.57 nm, when using the vibration isolation, the height of the peak was reduced to 0.27 nm, from 2.5 Hz to 30 Hz. The vibration can be reduced by five times.

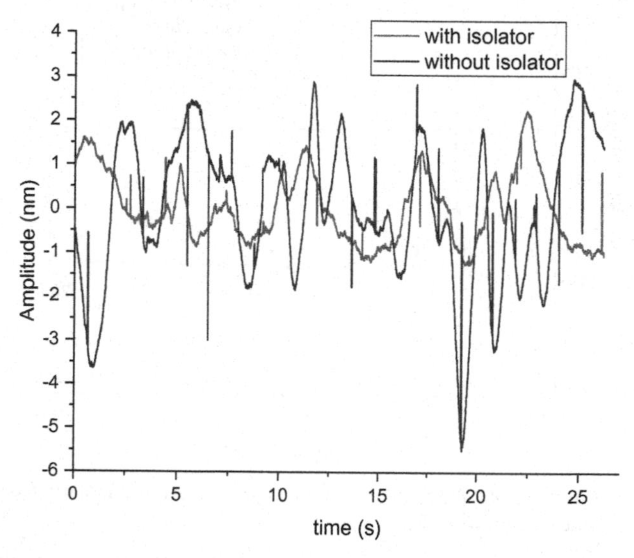

Fig. 9. The mechanical noise of the AFM in the time domain (Color figure online)

Finally, we conducted the AFM imaging test to demonstrate the performance of the AFM. As the Fig. 11. shows, the image on the left hand was conducted with vibration isolation, the image on the right hand was conducted without vibration isolation. Without vibration isolation the RMS of the whole image is 71 pm, with vibration isolation the RMS is 20 pm.

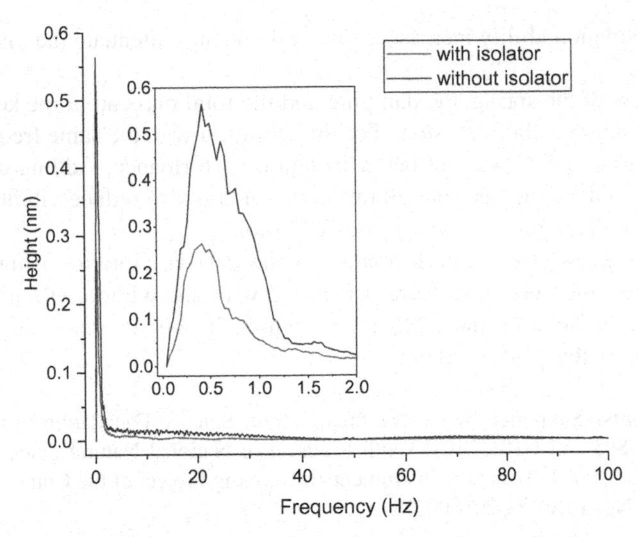

Fig. 10. The mechanical noise of the AFM in the frequency domain, the insert figure shows the zoomed-in-view of the image. (Color figure online)

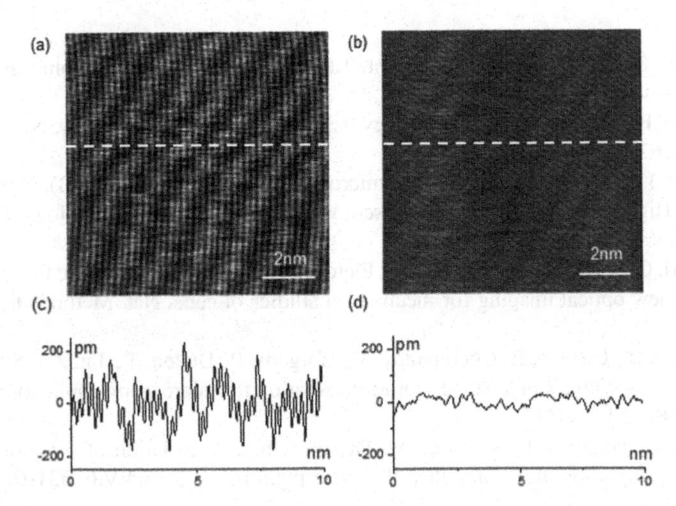

Fig. 11. Comparison of the AFM image with vibration isolation and without vibration isolation, (a) is the AFM image without vibration isolation, (b) is the AFM image with vibration isolation, (c) is a line section of the image (a), (d) is a line section of the image (b).

4 Conclusion

In this paper, the math model of the passive vibration isolation system was established firstly, the vibration isolation ability was represented by the transmissibility, the vibration isolation system can reduce the transmits of the vibrations whose frequency are higher than $\sqrt{2}\omega_n$ from the floor to the AFM. With the increase of the frequency of the vibration, the vibration isolation ability increased. The system can't attenuate the vibration lower than the $\sqrt{2}\omega_n$.

The stiffness of the spring, the damping and the total mass are three key factors for the passive vibration isolation system. For the vibration with the same frequency, using the lower stiffness spring will get better isolation performance. Adding damping into the system can reduce the resonate vibration, but it can also reduce vibration isolation performance for the higher frequency floor vibration.

In order to investigate the performance of the passive vibration isolation system, mechanical noise measurements were performed with and without vibration isolation. Without vibration isolation the RMS of the whole image roughness is 71 pm, with vibration isolation the RMS is 20 pm.

Acknowledgments. Supported by the National Natural Science Foundation of China (Grants 61925307,61927805 and U1813210), Youth Program of National Natural Science Foundation of China (Grants 61903359) and the Instrument Developing Project of the Chinese Academy of Sciences (Grant No. YJKYYQ20180027).

References

1. Binnig, G., Gerber, C., Stoll, E., Albrecht, T.R., Quate, C.F.: Atomic resolution with atomic force microscope. Epl. **3**(12) (1987)
2. Dufrêne, Y.F.: Towards nanomicrobiology using atomic force microscopy. Nat. Rev. Microbiol. **6**(9), 674–680 (2008)
3. Giessibl, F.J.: Advances in atomic force microscopy. Rev. Mod. Phys. **75**(3), 949–983 (2003)
4. Ando, T.: High-speed atomic force microscopy coming of age. Nanotechnology **23**(6), 062001 (2012)
5. Chaudhuri, O., Parekh, S.H., Lam, W.A., Fletcher, D.A.: Combined atomic force microscopy and side-view optical imaging for mechanical studies of cells. Nat. Methods **6**(5), 383–387 (2009)
6. Voigtländer, B., Coenen, P., Cherepanov, V., Borgens, P., Duden, T., Tautz, F.S.: Low vibration laboratory with a single-stage vibration isolation for microscopy applications. Rev. Sci. Instrum. **88**(2), 1–8 (2017)
7. Kamesh, D., Pandiyan, R., Ghosal, A.: Passive vibration isolation of reaction wheel disturbances using a low frequency flexible space platform. J. Sound Vib. **331**(6), 1310–1330 (2012)
8. Hansma, H.G., Laney, D.E.: Applications for atomic force microscopy of DNA. Biophys. J. **68**(5), 1672–1677 (1995)
9. Liu, C., Jing, X., Daley, S., Li, F.: Recent advances in micro-vibration isolation. Mech. Syst. Signal Process. **56**, 55–80 (2015)
10. Wang, C., Xie, X., Chen, Y., Zhang, Z.: Investigation on active vibration isolation of a Stewart platform with piezoelectric actuators. J. Sound Vib. **383**, 1–19 (2016)
11. Tjepkema, D., Van, D.J.: Sensor fusion for active vibration isolation in precision equipment. J. Sound Vib. **331**(4), 735–749 (2012)
12. Gu, F., Rehab, I., Tian, X., Ball, A.D.: The influence of rolling bearing clearances on diagnostic signatures based on a numerical simulation and experimental evaluation. Int. J. Hydromechatronics **1**(1), 1–16 (2018)

13. Hermsdorf, G.L., Szilagyi, S.A., Rösch, S., Schäffer, E.: High performance passive vibration isolation system for optical tables using six-degree-of-freedom viscous damping combined with steel springs. Rev. Sci. Instrum. **90**(1), 1–7 (2019)
14. Chen, D., Yin, B., Liu, J., Li, W., Wu, L., Li, H.: Active disturbance rejection control design for fast AFM. Key Eng. Mater. **645**, 670–674 (2015)
15. Li, Q., Zhu, Y., Xu, D., Hu, J., Min, W., Pang, L.: A negative stiffness vibration isolator using magnetic spring combined with rubber membrane. J. Mech. Sci. Technol. **27**(3), 813–824 (2013)

Loop Compensation Method to Improve Stability of Power Amplifier for Inertial Stick-Slip Driving

Yuhan Zhang, Daqian Zhang, Bowen Zhong(✉), Zhenhua Wang, and Lining Sun

The College of Mechanical and Electrical Engineering, Soochow University, Suzhou, China
zhbw@suda.edu.cn

Abstract. Inertial stick-slip driving is widely applied in precision positioning and needs a high stable power amplifier with fast step response to drive it. However, the inertial stick-slip driving is a capacitive load, which easily leads to the output voltage instability of power amplifier, and its ripple voltage also affects the positioning accuracy of inertial stick-slip driving. Therefore, this paper presents a loop compensation method to improve stability of power amplifier for inertial stick-slip driving. In this paper, a high dynamic amplification circuit is designed to realize fast step response, and the loop compensation method is proposed to eliminate self-excited oscillation and improve stability of the high dynamic amplification circuit. Then the experimental system is built to test the stability of the power amplifier through monitoring output voltage. Experimental results show that when the power amplifier drives inertial stick-slip platform of 0.5 μF, its phase margin is up to 80° and its static ripple to less than 3 mV. The results show that the method can improve stability by 400 times and the power amplifier can achieve high stability.

Keywords: Inertial stick-slip driving · High dynamic amplification circuit · Loop compensation

1 Introduction

Inertial stick–slip driving [1–3], as a novel trans-scale precision positioning techniques [4–7], can achieve big scope movement, high displacement resolution and high positioning accuracy. So inertial stick–slip driving has extensive applied foreground in precision position [8]. The inertial stick-slip usually needs a power amplifier to drive it. Figure 1 presents a simple inertial stick-slip driving system and the power amplifier amplifies driving signal generated by the signal generator, which is used to drive the inertial stick-slip to achieve precision positioning.

The driving principle [9–11] of inertial stick-slip driving is shown as Fig. 2 when driven by saw tooth waveform. During the fast deformation, the piezoelectric actuator elongate rapidly under the excitation of step response signal. During the slow deformation, the piezoelectric actuator shrink slowly under the excitation of slow signal. The inertial block follows piezoelectric actuator, and the slider block stays stationary because

X.-J. Liu et al. (Eds.): ICIRA 2021, LNAI 13014, pp. 440–453, 2021.
https://doi.org/10.1007/978-3-030-89098-8_42

of friction force (stick). Therefore, the slider moves forward one step in one cycle of the driving signal. When the frequency and amplitude of the driving signal are fixed, the faster the step response signal is during the fast deformation, the farther the slider block will slide.

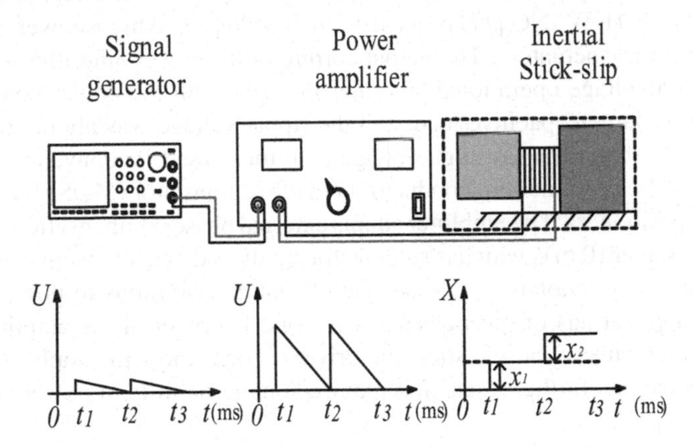

Fig. 1. Inertial stick-slip driving system

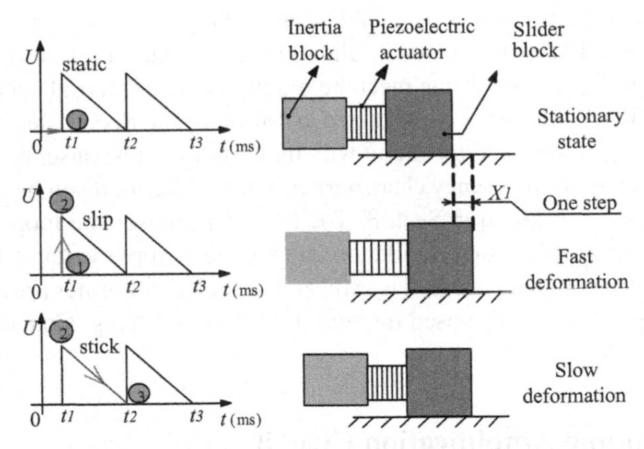

Fig. 2. Principle diagram of inertial stick-slip driving

The inertial stick-slip driving has two modes of motion. In the stepping-mode, it can achieve the continuous step movement when driven by periodic signal. The power amplifier with fast step response [12–14] can help improve the movement its velocity. In the scanning-mode, when the distance from the target is within the one-step displacement, the inertial block moves forward slowly with the piezoelectric actuator, and inertial stick-slip slowly approaches to the target and completes precise positioning. The ripple voltage of power amplifier will affect displacement accuracy of piezoelectric actuator and the positioning accuracy of inertial stick-slip driving [15, 16]. In addition, when driving capacitive load, the amplifier with fast response is easy to produce self-excited oscillation due to the additional phase shift. It is difficult for power amplifiers to have both fast step response and high stability. Therefore, designing a high stable amplifier with fast step response is greatly significant for inertial stick–slip driving.

Several power amplifiers had been proposed for driving piezoelectric actuators in some literature. D. H. WANG [17] proposed a high-voltage and high-power amplifier for driving piezoelectric actuators. The output current of the power amplifier was enlarged by using a high-voltage operational amplifier in series with the power booster section in order to drive large capacitive load, and the ripple voltage was about 20 mV. M. S. Colclough [18] proposed a fast high-voltage amplifier based on conventional complementary pair of metal-oxide semiconductor field effect transistor (MOSFET) for driving piezoelectric positioners. The amplifier adopts external phase compensation to make ripple voltage less than 10 mV, which is suitable for small load and small current occasions. In conclusion, many scholars have made significant contributions to power amplifiers for different applications of piezoelectric actuators. However, these amplifiers cannot meet the requirements of inertial stick-slip driving. To promote the application of inertial stick–slip driving, further research is required on the stability of power amplifier for inertial stick-slip driving.

Therefore, this paper presents a loop compensation method to improve stability of power amplifier with fast response for inertial stick-slip driving. The current study is to design a high dynamic amplification circuit based on discrete components. Then the loop compensation method is proposed to improve stability by analyzing self-excited oscillation. By setting up an experimental system, the performance parameters of the amplifier are tested. The experimental results show that the stability and ripple voltage of the amplifier with fast response can meet the requirements of inertial stick-slip driving.

The remainder of this paper is organized as follows: The Sect. 2 describes the principle of a multistage amplification circuit with high dynamic response. By analyzing the high-frequency and low-frequency characteristics of the circuit, it can be concluded that the circuit has self-oscillation in Sect. 3. The Sect. 4 proposes the loop compensation method to eliminate self-excited oscillation and suppress ripple voltage. In the Sect. 5, the experiments show that the method is valid and the power amplifier present low-ripple voltage and high stability when used for inertial stick-slip driving. Concluding remarks are given in Sect. 6.

2 High Dynamic Amplification Circuit

In order to achieve fast step response of power amplifier, this paper designs an amplification circuit composed of a front amplification circuit and a deep negative feedback amplification circuit as shown in Fig. 3. The front amplification circuit composed of an operational amplifier U1 amplifies input signals. Its voltage gain is depended on the resistance R1, R2 and DR2, about 1–6. As the core circuit of the amplifier, the deep negative feedback amplification circuit is composed of four amplification stages and a feedback circuit. The resistance R3 and R17 determine its voltage gain, about 22. Therefore, the total voltage gain of the amplification circuit can be set to 60 by adjusting DR2.

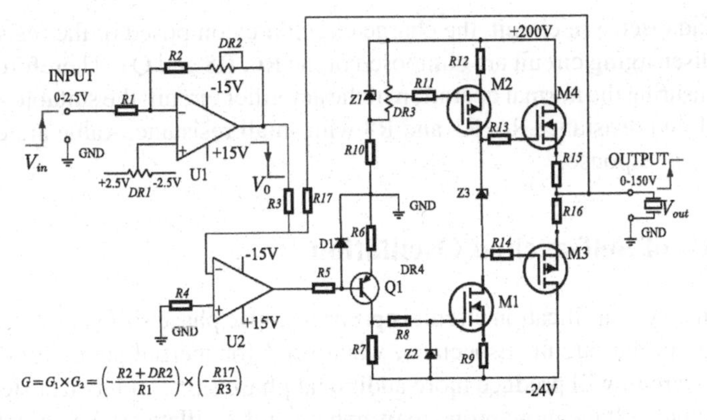

Fig. 3. High dynamic amplification circuit

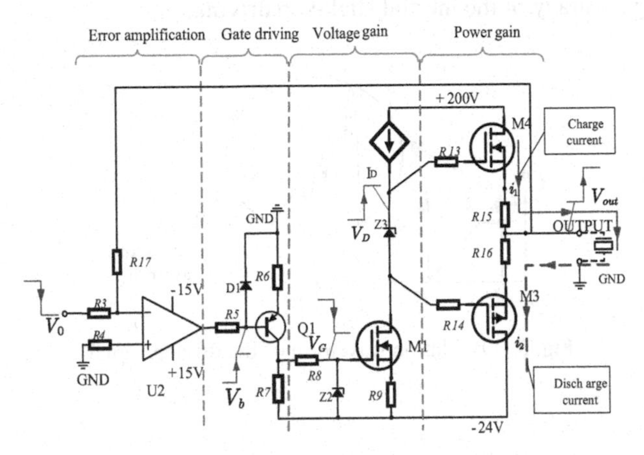

Fig. 4. The deep negative feedback amplification circuit

The deep negative feedback amplification circuit includes four stages amplification circuit to achieve high output voltage as shown in Fig. 4. The error amplification stage consists of an operational amplifier U2 in inverting configuration. The gate driving stage is a common emitter amplification circuit consisted of a bipolar transistor T1 and its associated components to drive the following MOSFET M1. The voltage amplification stage is a common source amplification circuit comprised by a high-voltage n-MOSFET M1 and a high-voltage p-MOSFET M2 configured as a current source ID. The power amplification stage is a complementary push-pull circuit constituted two opposite polarity MOSFETs M3 and M4 and the regulator Z3 can keep M3 and M4 in a micro-conducting state to avoid crossover distortion. They also can constitute charge or discharge circuits for the load when the output voltage goes up or down.

In order to achieve fast step response of the amplification circuit, the gate driving stage is further optimized. The inter-electrode capacitance of high power MOSFET will slow down the response speed of output voltage. Therefore, the gate driving circuit is required to provide a large charge and discharge current for MOSFET M1. According to the charge and discharge circuit, the charge circuit are composed of the resistor R7 and R8 and the discharging circuit are composed of the R6, R8 and Q1. Therefore, under the premise of ensuring the normal operation of the amplifier circuit, the suitable high-power transistor Q1 and resistance R6, R7 and R8 with small resistance value are designed to realize fast step response.

3 Analysis of Self-excited Oscillation

The high dynamic amplification circuit is prone to cause phase shifts, because of capacitive elements in the circuit. Especially, when used for inertial stick-slip driving, the amplification circuit will produce more additional phase shift, and its fast step response makes the output voltage more prone to overshoot and oscillate as shown in Fig. 5. The oscillation waveform makes the inertial stick-slip motion extremely unstable and affects the positioning accuracy of the inertial stick-slip driving.

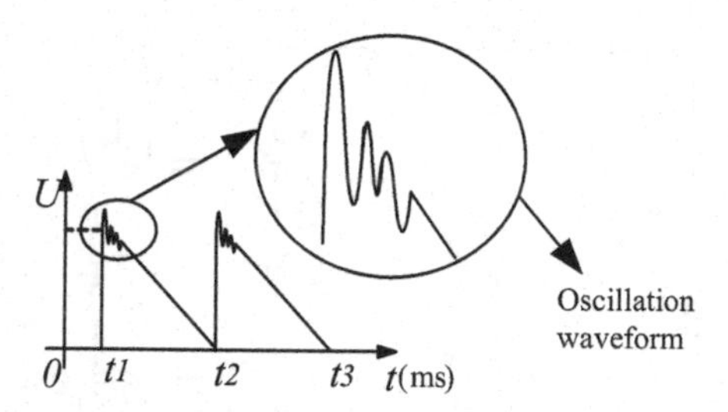

Fig. 5. Oscillation waveform of the driving signal

3.1 Low Frequency Analysis

The deep negative amplification circuit has a very high open-loop voltage gain because the voltage amplification stage sets up a current source that can be equivalent to one megohm resistance Rd. The low-frequency equivalent circuit of small signal is obtained according to the static equivalent circuit as shown in Fig. 6. The voltage gain of the multistage amplification circuit is equal to the product of the voltage gain of each stage amplification circuit in the open loop state.

Therefore, the total voltage gain of the amplification circuit can be calculated by formula (1).

$$A_L = A_{L0}A_{L1}A_{L2}A_{L3} = A_0 * \frac{\beta * R_C}{R_b + r_{be} + (1+\beta)R_e} \frac{g_{m1} * R_d}{1 + g_{m1} * R_s} \frac{g_{m2} * R_{ds}}{1 + R_{ds} * g_{m2}} \quad (1)$$

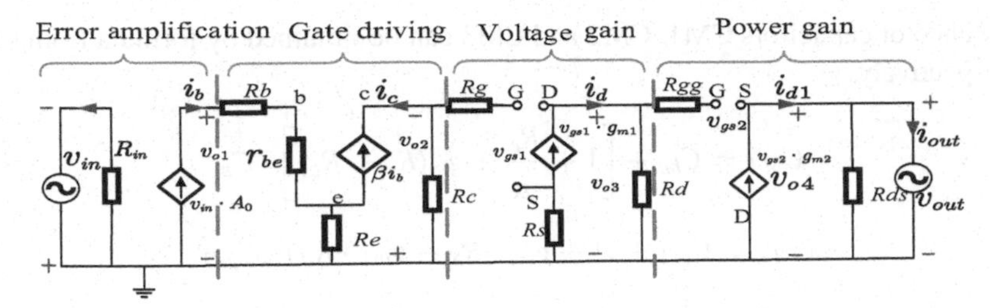

Fig. 6. Low frequency equivalent circuit diagram

3.2 High Frequency Analysis

In order to study the high frequency characteristics of amplification circuits, the effects of inter-electrode capacitance of semiconductor devices need to be considered. The internal capacitance of the integrated operational amplifier is small. However, due to the high input resistance, its main pole frequency f0 is very low and it usually behaves as a first-order inertial loop. The high frequency equivalent circuit of triode and MOSFET is shown in the Fig. 7.

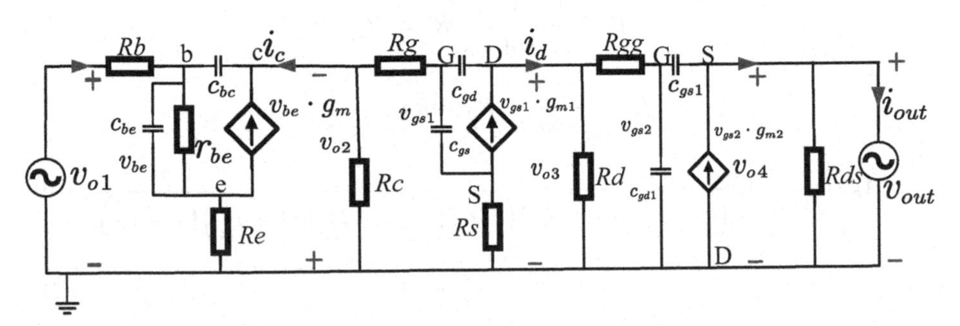

Fig. 7. High frequency equivalent circuit diagram

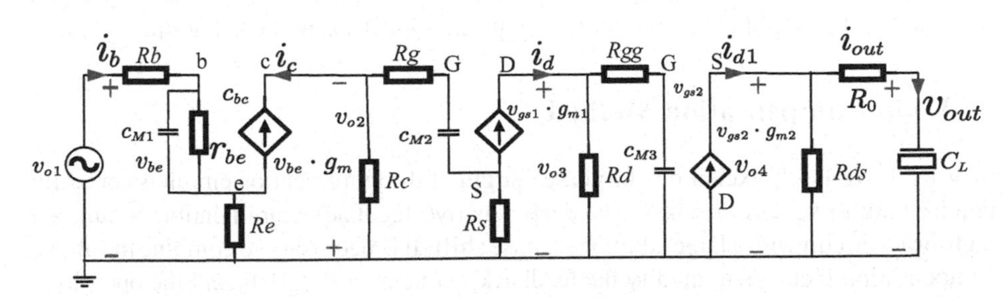

Fig. 8. High frequency equivalent circuit

In order to calculate other pole frequency of amplification circuit, it is necessary to simplify the high frequency equivalent circuit. Considering the influence of the capacitors of each stage amplification circuit on the high-frequency characteristics, the equivalent high-frequency circuit in the Fig. 7 is simplified as shown in the Fig. 8. The equivalent capacitance of capacitors CM1, CM2 and CM3 can be obtained by formula from (2) to (4), respectively.

$$C_{M1} = C_{be} + \left[1 + \frac{R_e}{r_{be}} + g_m(R_c + R_e)\right] \cdot C_{bc} \quad (2)$$

$$C_{M2} = C_{gs1} + (1 + g_{m1} \cdot R_S + g_{m1} \cdot R_d) \cdot C_{gd1} \quad (3)$$

$$C_{M3} \approx C_{gd2} \quad (4)$$

According to the formula from (2) to (4), three pole frequencies of the circuit can be obtained by formula from (5) to (7), and one pole frequency is added by formula (8) under the load 0.5 μF.

$$f_1 = \frac{1}{2\pi R_1 C_{M1}} \quad R_1 = r_{be}//[R_b + (1 + \beta) \cdot R_e] \quad (5)$$

$$f_2 = \frac{1}{2\pi R_2 C_{M2}} \quad R_2 = \frac{R_g}{1 + g_{m1} \cdot R_S} \quad (6)$$

$$f_3 = \frac{1}{2\pi R_3 C_{M3}} \quad R_3 = R_{gg} \quad (7)$$

$$f_4 = \frac{1}{2\pi R_0 R_L} \quad R_0 = R_{ds}//\frac{1}{g_{m2}} \quad (8)$$

$$A_H(jw) = A_L \frac{1}{\left(1 + J\frac{f}{f_0}\right)\left(1 + j\frac{f}{f_1}\right)\left(1 + j\frac{f}{f_2}\right)\left(1 + j\frac{f}{f_3}\right)\left(1 + j\frac{f}{f_4}\right)\left(1 + j\frac{f}{f_5}\right)} \quad (9)$$

$$\mathrm{F} = -\frac{R_3}{R_{17}} \quad (10)$$

The open-loop transfer function of amplification circuit can be approximately expressed as the formula (9), and the gain of the feedback loop is expressed in formula (10). It can be seen that the amplification circuit is very prone to self-excited oscillation.

4 Loop Compensation Method

In order to avoid self-excitation, the phase margin of the amplification circuit is generally required not to be less than 45°. The deep negative feedback amplification circuit is a high-order circuit with a large additional phase shift. It is necessary to combine the above compensation methods to modify the feedback gain curve 20 lg|1/F| and the open-loop gain curve 20 lg|AH| to ensure that their closed slope is 20 dB/dec.

Firstly, an isolation resistance compensation is used to eliminate the effect of the pole at frequency f4 of the inertial stick-slip load. The resulting zero fz4 is used to compensate the pole fp4.

Then, to compensate for the open loop path, a small capacitance CT and a current limiting resistor RT between the gate and drain of M1 can greatly increase the input capacitance at this level, changing the pole frequency f2 into f'2, as shown in the Fig. 11. The formula (3) and (6) are rewritten respectively as (11) and (12). At the same time, according to the formula (13), new zero frequency fz2 are added to eliminate the effects of pole frequency f1.

$$f_{p4} = \frac{1}{2\pi (R_0 + R_L) \cdot C_L} \tag{11}$$

$$f_{z4} = \frac{1}{2\pi R_L \cdot C_L} \tag{12}$$

$$C'_{M2} = C_{M2} + (1 + g_{m1} \cdot R_S + g_{m1} \cdot R_{d2}) \cdot C_T \tag{13}$$

$$f'_2 = \frac{1}{2\pi R_2 C'_{M2}} \tag{14}$$

$$f_{z2} = \frac{1}{2\pi R_T \cdot C_T} \tag{15}$$

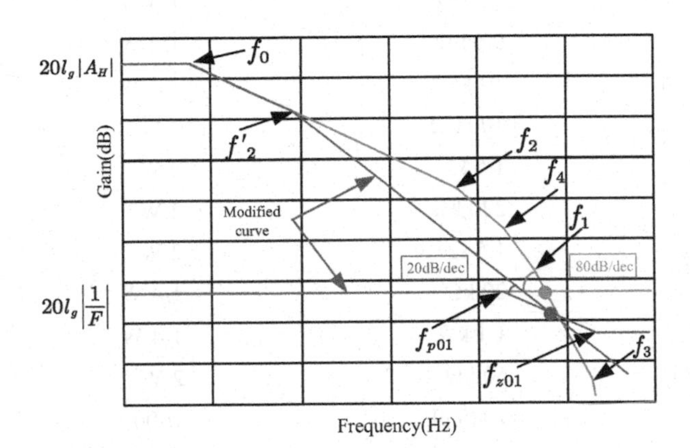

Fig. 9. Modified bode diagram for amplification circuit

Finally, to compensate for feedback path, the resistor Rf and capacitor Cf are connected to the resistor R17. According to the formula (16) and (17), the compensation method adds a pole fp01 and a zero fz01 to the feedback gain curve. The value of Rf1Cf1 can be determined by formula (20). As shown in the Fig. 9, the closed slope of the modified open-loop gain curve and the modified feedback gain curve is 20 dB/dec, and their gain expressions can be expressed as Eq. (18) and (19) respectively.

$$f_{z01} = \frac{1}{2\pi R_f \cdot C_f} \tag{16}$$

$$f_{p01} = \frac{1}{2\pi \left(R_{17} + R_f\right) \cdot C_f} \tag{17}$$

$$A'_{H1} = \frac{A_L}{\left(1 + j\frac{f}{f_0}\right)\left(1 + j\frac{f}{f_2'}\right)} \tag{18}$$

$$F_{(j\omega)} = F \cdot \frac{1 + j\omega(R_f + R_{f1}) \cdot C_f}{1 + j\omega R_{f1} C_{f1}} \tag{19}$$

$$\begin{cases} \left| \frac{A_L \cdot F}{\left(1+j\frac{f_s}{f_0}\right)\left(1+j\frac{f_s}{f_2'}\right)} \right| = 1 \\ f_{p01} < f_s < f_{z01} \end{cases} \tag{20}$$

Table 1. The components and values of amplification circuit

Components	Values	Note
R1, R2, R3	10 kΩ	1/4 W
R4, R5	2.4 kΩ	1/4 W
R6, R7	0.6 kΩ	1/4 W
R8, R11, R13, 14	10 Ω	1/4 W
R9, R12	100 kΩ	2 W
R15, R16	1.5 Ω	2 W
R17	220 kΩ	2 W
RL	1.5 Ω	2 W
Rf	20 kΩ	1/4 W
RT	1.1 kΩ	1/4 W
DR1, DR2, DR3	50 kΩ	2 W
CT	1000 pF	1000 V
Cf	20 pF	1000 V
Z1, Z2, Z3	10 V	1 W
D1	1N4148	1 W
U1, U2	OPA277(model)	SO8 (packaging)
Q1	S8550	TO92
M1	IRFP450	TO247
M2, M3	IXTH11P50	TO247
M4	IXFH26N50	TO247

In general, open-loop compensation changes the distance between the main pole and the other poles, and feedback compensation reduces the phase shift. This loop compensation method enables the amplification circuit to have at least 45° phase margin, so as to improve the stability of the amplification circuit and reduce ripple voltage. The optimized amplification circuit is shown in the Fig. 10, and the associated components and their values for the optimized amplification circuit are given in Table 1.

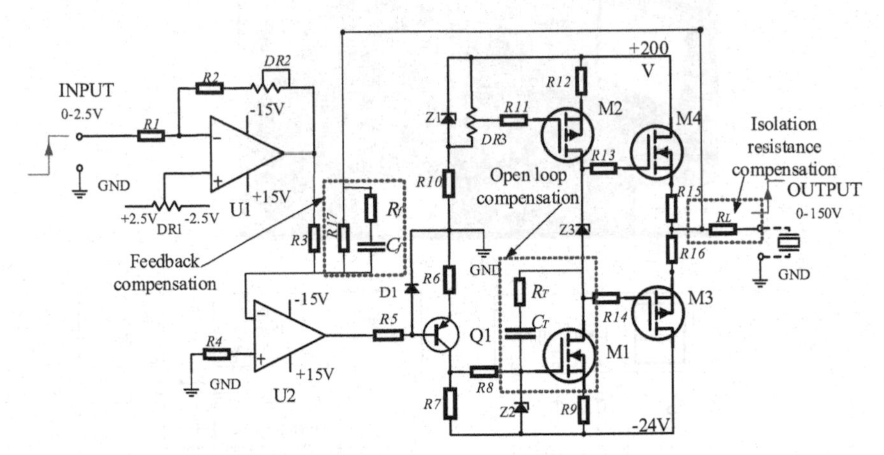

Fig. 10. Optimized amplification circuit

Therefore, according to the formula, the gain 20 lg|1/F| can be calculated as 27 dB. The gain 20 lg|AL| can be calculated as 215 dB. The pole f0 can be calculated as 10 Hz. The pole f '2 can be calculated as 507 Hz. The pole fp01 can be calculated as 40 kHz. The zero fz01 can be calculated as 398 kHz.

5 Experiments

The experiment system was built as shown in Fig. 11(a) to test loop response, stability and ripple voltage of the power amplifier, including a prototype amplifier, a signal generator, an oscilloscope, a digital multimeter, a loop analyzer and a inertial stick–slip piezoelectric actuator.

When the system is operating, the signal generator provides a voltage signal to the amplifier. The output voltage is applied to the stick–slip piezoelectric actuator. The loop response (20 lg|AF|) of the amplifier can be acquired according to Fig. 11(b). The experiment data are captured by the oscilloscope, and the digital multimeter is used to detect the output voltage, according to Fig. 11(c).

5.1 Loop Response Testing

In order to verify whether the loop response of the amplifier are consistent with the theory, as shown in Fig. 12, the open-loop gain curve and feedback gain curve of the driving power supply are tested when driving an inertial stick-slip platform of 0.5 μF. According to Fig. 12(a), the open-loop gain curve has two pole frequencies f0 and f '2, and the trend of the curve with frequency is consistent with the theory.

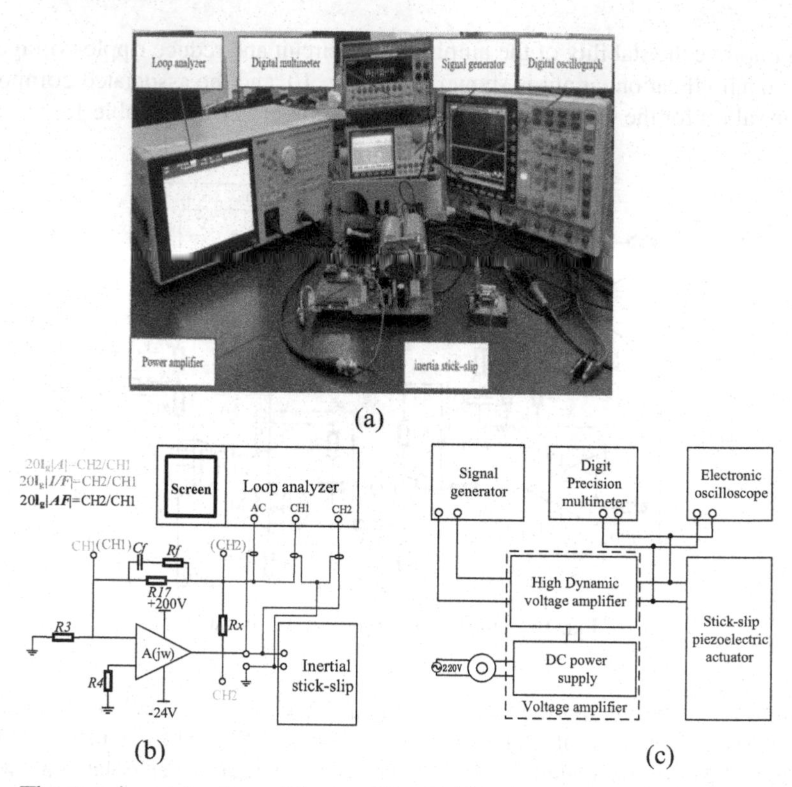

(a)

(b) (c)

Fig. 11. The experimental system of the amplifier (a) The photograph (b) Schematic diagram of loop response testing (c) Schematic diagram of stability and ripple voltage testing

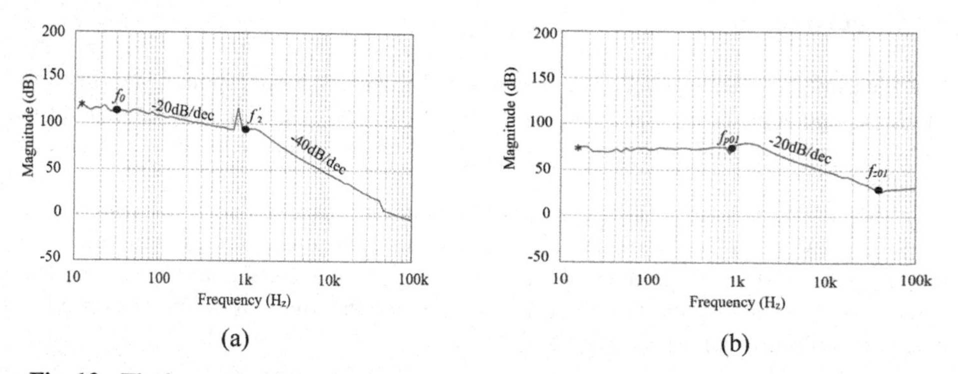

(a) (b)

Fig. 12. The loop gain of the circuit (a) The gain of open loop (b) The gain of feedback loop

The experimental results show that the open-loop gain curve and the feedback gain curve of the power amplifier can intersect with the closed slope of 20 dB/dec, which satisfies the stability condition of the amplifying circuit. Figure 13 shows the loop response (20 lg|AF|) of the power amplifier after being compensated. The phase margin of the amplification circuit is 80°, which is expected. When the compensation capacitance value is increased within a certain range, the value of phase margin can be increased, which is consistent with the theoretical analysis.

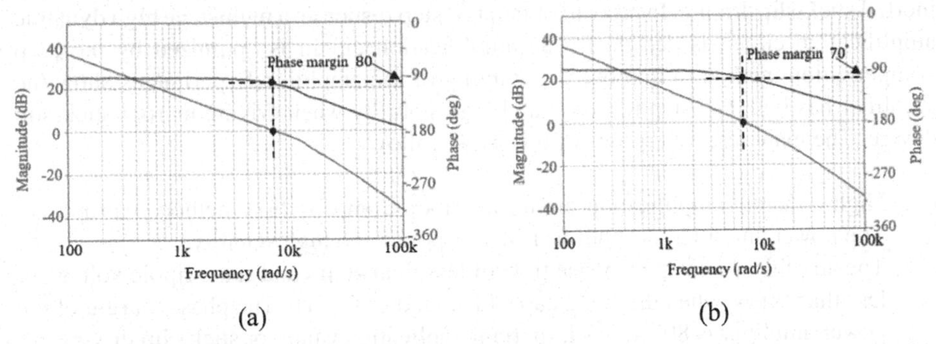

Fig. 13. The loop response of the amplifier after being compensated (a) The loop response with Cf = 20pF (b) The loop response with Cf = 50pF

This reason is the larger the compensation capacitor Cf is, the closer the frequency fs of the closed intersection of the open-loop gain curve and the feedback gain curve is to the zero frequency fz01, which will reduce the phase margin of the amplifier, but the drive power supply is still in a highly stable state.

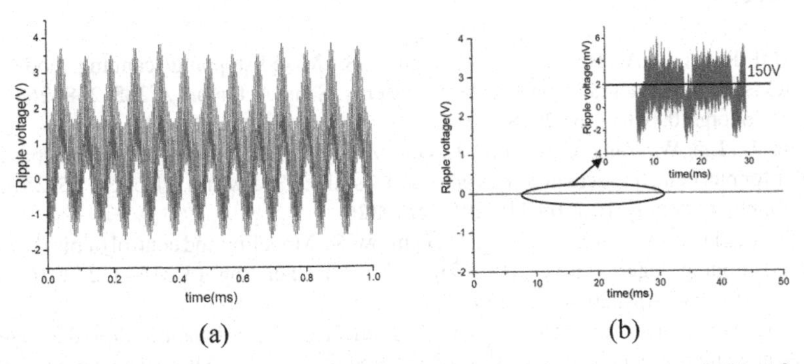

Fig. 14. Ripple voltage diagram of amplifier (a) Before being compensated (b) After being compensated

Figure 14 presents the voltage fluctuation of the power amplifier before and after being compensated, when the power amplifier outputs 150 V and is loaded into the stick-slip piezoelectric actuator. Before being compensated, the output voltage oscillates significantly with a peak-to-peak value exceeding 5 V. After being compensated, the voltage fluctuation is significantly reduced to 10 mV.

The experimental results show that the compensation circuit can eliminate the self-excited oscillation and the power amplifier can output the driving signal stably.

The experimental results show that, when the power amplifier drives the inertial stick-slip of 0.5 μF, its the static ripple voltage is less than 3 mV, which is conducive to precise positioning of the inertial stick-slip driving.

6 Conclusions

In this paper, a high stable power amplifier with fast step response is proposed for inertial stick-slip driving. In order to obtain fast step response, a multistage high dynamic amplification circuit is designed. The amplification circuit is optimized by the loop compensation method to improve the stability of the amplifier. Experiments show the amplifier has both fast step response and high stability when driving inertial stick-slip device. The main research results of this paper follow:

(1) High dynamic amplification circuit and loop compensation method is proposed. The power amplifier has both fast step response and high stability.
(2) The amplifier has step response time of less than 40 μs and static ripple voltage of less than 3 mV when driving a capacitive load of 0.5 μF. The phase margin of the power amplifier is 80°. It has important application value for stick-slip driving.

Acknowledgments. The work was supported by the National Natural Science Foundation of China (No.52175541), the National key research and development program (No. 2018YFB1304900) and the Jiangsu province Natural Science Foundation (No. BK20181439).

References

1. Oubellil, R., Voda, A., Boudaoud, M., Régnier, S.: Mixed stepping/scanning mode control of stick-slip SEM-integrated nano-robotic systems. Sens. Actuators A **285**, 258–268 (2019). https://doi.org/10.1016/j.sna.2018.08.042
2. Cheng, L., Liu, W., Yang, C., Huang, T., Hou, Z.-G., Tan, M.: A neural-network-based controller for piezoelectric-actuated stick-slip devices. IEEE Trans. Ind. Electron. **65**, 2598–2607 (2018). https://doi.org/10.1109/TIE.2017.2740826
3. Gu, G.-Y., Zhu, L.-M., Su, C.-Y., Ding, H., Fatikow, S.: Modeling and control of piezo-actuated nanopositioning stages: a survey. IEEE Trans. Automat. Sci. Eng. **13**, 313–332 (2016). https://doi.org/10.1109/TASE.2014.2352364
4. Jae, Y.L., Won, C.L., Young-Ho, C.: Serially connected weight-balanced digital actuators for wide-range high-precision low-voltage-displacement control. J. Microelectromech. Syst. **18**, 792–798 (2009). https://doi.org/10.1109/JMEMS.2009.2023843
5. Rios, S., Fleming, A.: Design of a charge drive for reducing hysteresis in a piezoelectric bimorph actuator. IEEE/ASME Trans. Mechatron. **21**(1), 51–54 (2015). https://doi.org/10.1109/TMECH.2015.2483739
6. Ronkanen, P., Kallio, P., Vilkko, M., Koivo, H.N.: Displacement control of piezoelectric actuators using current and voltage. IEEE/ASME Trans. Mechatron. **16**, 160–166 (2011). https://doi.org/10.1109/TMECH.2009.2037914
7. Fleming, A.J., Moheimani, S.O.R.: Improved current and charge amplifiers for driving piezoelectric loads. In: Agnes, G.S., Wang, K.-W. (Eds.), pp. 242–252. San Diego, CA (2003). https://doi.org/10.1117/12.483971
8. Saadabad, N.A., Moradi, H., Vossoughi, G.: Dynamic modeling, optimized design, and fabrication of a 2DOF piezo-actuated stick-slip mobile microrobot. Mech. Mach. Theory **133**, 514–530 (2019). https://doi.org/10.1016/j.mechmachtheory.2018.11.025

9. Zhong, B., Zhu, J., Jin, Z., He, H., Sun, L., Wang, Z.: Improved inertial stick-slip movement performance via driving waveform optimization. Precis. Eng. **55**, 260–267 (2019). https://doi.org/10.1016/j.precisioneng.2018.09.016
10. Nguyen, X.-H., Mau, T.-H., Meyer, I., Dang, B.-L., Pham, H.-P.: Improvements of piezo-actuated stick-slip micro-drives: modeling and driving waveform. Coatings **8**, 62 (2018). https://doi.org/10.3390/coatings8020062
11. Zhong, B., Wang, Z., Chen, T., Chen, L., Sun, L.: The study on the effect of step time on the movement of the cross-scale stage based on the stick-slip driving. In: 2013 International Conference on Manipulation, Manufacturing and Measurement on the Nanoscale, pp. 357–361. IEEE, Suzhou, China (2013). https://doi.org/10.1109/3M-NANO.2013.6737450
12. Liu, P., Chen, T., Hsu, S.: Area-efficient error amplifier with current-boosting module for fast-transient buck converters. IET Power Electronics. **9**, 2147–2153 (2016). https://doi.org/10.1049/iet-pel.2015.0322
13. Tu, Y.J., Jong, T.L., Liaw, C.M.: Development of a class-D audio amplifier with switch-mode rectifier front-end and its waveform control. IET Pwr. Electr. **4**, 1002 (2011). https://doi.org/10.1049/iet-pel.2010.0344
14. Wu, C.-H., Chang-Chien, L.-R.: Design of the output-capacitorless low-dropout regulator for nano-second transient response. IET Pwr. Electr. **5**, 1551 (2012). https://doi.org/10.1049/iet-pel.2011.0286
15. Marano, D., Palumbo, G., Pennisi, S.: Step-response optimization techniques for low-power three-stage operational amplifiers for large capacitive load applications. In: 2009 IEEE International Symposium on Circuits and Systems, pp. 1949–1952. IEEE, Taipei, Taiwan (2009). https://doi.org/10.1109/ISCAS.2009.5118171
16. Milecki, A., Regulski, R.: Investigations of electronic amplifiers supplying a piezobimorph actuator. Mech. Syst. Signal Process. **78**, 43–54 (2016). https://doi.org/10.1016/j.ymssp.2016.01.011
17. Wang, D.H., Zhu, W., Yang, Q., Ding, W.M.: A high-voltage and high-power amplifier for driving piezoelectric stack actuators. J. Intell. Mater. Syst. Struct. **20**, 1987–2001 (2009). https://doi.org/10.1177/1045389X09345559
18. Colclough, M.S.: A fast high-voltage amplifier for driving piezoelectric positioners. Rev. Sci. Instrum. **71**, 4323 (2000). https://doi.org/10.1063/1.1319984

Human-Robot Collaboration

IoT-Enabled Robot Skin System for Enhancement of Safe Human-Robot Interaction

Gbouna Vincent Zakka[1], Zeyang Hou[1], Gaoyang Pang[2], Huayong Yang[1], and Geng Yang[1](✉)

[1] State Key Laboratory of Fluid Power and Mechatronic Systems, School of Mechanical Engineering, Zhejiang University, Hangzhou 310027, China
yanggeng@zju.edu.cn

[2] School of Electrical and Information Engineering, The University of Sydney, Sydney, NSW 2006, Australia

Abstract. Collaborative robots along with skilled manpower offer flexible and intelligent services by combining the cognitive power of humans with robot's flexibility in small-batch production. To realize this, collaborative robots are required to collaborate with humans in close range sharing the same workspace. In such a case, ensuring safe Human-Robot Collaboration (HRC) is a big challenge. Safety solutions are expected to be smart enough to monitor the robot environment, analyse the collected data, and alert the user in case of any anomalies for timely intervention. This work, therefore, proposes an Internet of Things (IoT)-enabled smart robot skin system as a safety solution to enhance safety in the collaborative environment. It is flexible and modular, based on self-capacitive technology for proximity-touch sensing. Leveraging the IoT technology, wireless communication within the skin system and remote programming of the skin network were implemented. More so, real-time data monitoring, analysis, and visualization are implemented leveraging the flexible ThingSpeak cloud service data processing capabilities. Based on the processed data, cloud-based event handling, which updates the user of the robot skin performance, was also implemented. This approach will enable remote maintenance, and enhance the safety function of the skin system for effective HRC.

Keywords: Internet of Things · Human-robot collaboration · Robot skin

1 Introduction

New areas for the application of robot-engineered systems have emerged due to the rapid advancement in design, manufacturing, electronics, materials, computing, communication, and system integration [1]. Therefore, the focus is now shifting gradually towards collaborative robots handling real-world objects under arbitrary circumstances, working alongside humans. This recent development shall continue to grow as we enter the age of smart factories, Industry 4.0, social robots, Internet of Things (IoT), telehealth, where

X.-J. Liu et al. (Eds.): ICIRA 2021, LNAI 13014, pp. 457–468, 2021.
https://doi.org/10.1007/978-3-030-89098-8_43

robots are expected to work with humans in close quarters. An important consideration to the full realization of these advances is the deployment of artificial skin on the robot body to enable the robot's understanding of the various dimensions of physical interactions needed to deal with unconstrained environments. Therefore, the development of low-cost, reliable, and easy-to-use artificial skin would have major technical, economic, and social impacts [2].

Therefore, to meet the growing demand for artificial skin systems for Safe Human-Robot Collaboration (HRC), various artificial skin alternatives have been explored, with off-the-shelf electronic components on Flexible Printed Circuit Boards (FPCBs). A capacitive sensing system solution based on off-the-shelf components was presented in [3] to address the issue of collision avoidance in a partially modeled environment. In [4], a proximity sensor module made up of capacitive and time-of-flight (ToF) on an FPCB for collision avoidance was proposed. A tactile proximity sensor made of a multi-layered FPCB for close human-robot interaction (HRI) was proposed in [1]. Other safety solutions based on flexible materials that can conform to the arbitrary curve surface of robots have been explored recently [5–7]. These solutions have illustrated how artificial skin can enhance safe HRC.

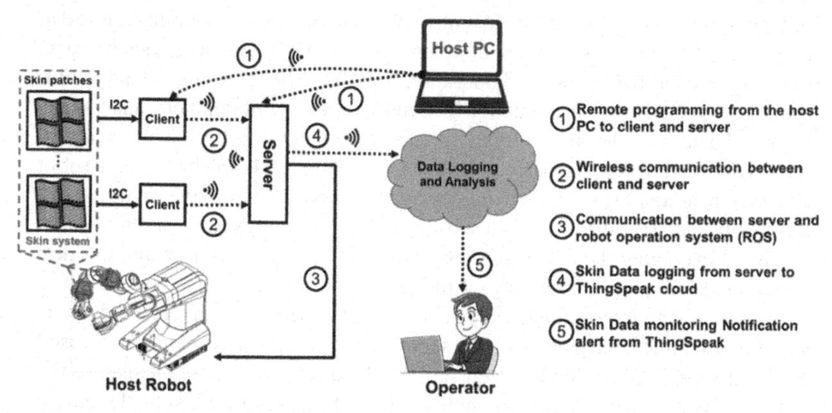

Fig. 1. Conceptual diagram of the IoT-enabled smart robot skin system. PC denotes personal computer. I2C denotes inter-integrated circuit, which is a wired communication protocol.

However, the efficient implementation of the robot skin on a robot for safe HRC is still a technological challenge. Traditionally, management solutions are needed to manage networked equipment, devices, and services. The robot skin similarly deployed on the robot body creates management requirements. Thus, traditional management functionalities such as remote control, monitoring, and maintenance are considered of paramount significance for robot skin systems. IoT technology, on the other hand, provides the platform for the implementation of such management systems and has found application in many areas [8]. It aims to add smartness to conventional things, making things easily accessible in less time and effort anywhere, which will boost efficiency, productivity, and flexibility. This, therefore, motivates the application of IoT technology in robot skin systems as proposed in this work. The aim is to introduce smartness to the robot skin system to enhance safety performance.

In this work, a flexible approach to enhance the safety performance of robot skin systems leveraging IoT technology is proposed (see Fig. 1). The robot skin module in our prior research work [9], which has the advantage of an embedded platform for the development of an intelligent system, was used in this work. The contribution of this work is threefold: 1) In situ programming which will enhance flexibility and adaptability by reconfiguring the skin system; 2) Cloud-based monitoring which provides a user-friendly interface to monitor the performance of the skin system efficiently, and 3) Cloud-based event handling which provides users with automatic notifications of potential issues with the robot skin for timely resolution.

2 The IoT-Enabled Robot Skin System Architecture

2.1 Perception Layer

In this layer, a capacitive proximity-touch sensor based on self-capacitive technology is deployed on the robot to sense the physical environment in real-time. It is flexible and modular, able to scale up to form a large-area sensing system. The basis for a single module of the sensor is a custom-designed FPCB (see Fig. 2) [9]. When an obstacle is detected by the robot skin, the color of the LEDs changes from green to red. The MCU used in this work is NodeMCU, a low-cost IoT hardware platform. It is a complete and self-contained Wi-Fi networking solution for developing IoT applications. NodeMCU is used in this work to provide wireless communication within the robot skin system and the ThingSpeak cloud. In this work, a server is a NodeMCU. A client is a skin patch connected to a NodeMCU.

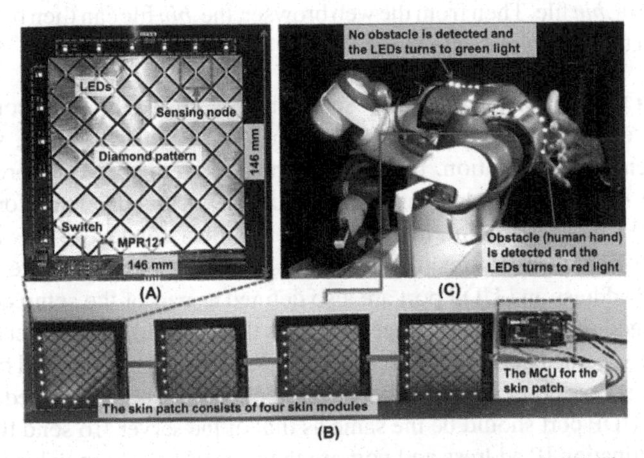

Fig. 2. The robot skin system. (A) The sensor module. (B) The skin patch. (C) The skin is deployed on a collaborative robot.

2.2 Network Layer

Remote Programming. In situ programming will enhance flexibility and adaptability by reconfiguring the skin system. In this work, we, therefore, propose a remote programming system for the skin network (see Fig. 3).

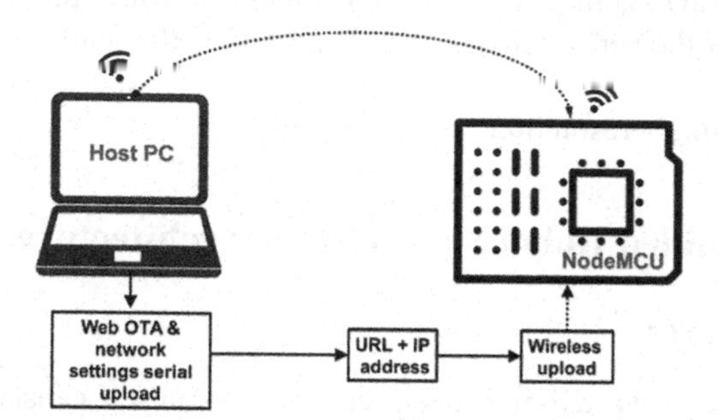

Fig. 3. The overall process of remote programming from the host personal computer (PC) to the microcontroller unit (MCU)

The approach was realized through Over the Air (OTA) programming to upload/update new sketches using a web browser. OTA web configuration with network settings was uploaded first serially. This will then create a Uniform Resource Locator (URL) with the IP address where new sketches can be uploaded without serial connection. To therefore upload the sketch via a web browser, open the generated URL, which has the IP address in a browser, and then a page to upload a sketch will open. The sketch to be uploaded should be converted to a *.bin* file. Then from the web browser, the *.bin* file can then be selected, and the sketch will then be uploaded to the MCU.

Server-Client Communication. The User Datagram Protocol (UDP) communication protocol, one of the core members of the Internet Protocol (IP) suite, was used for the server-client communication. In this implementation, the server is programmed to operate in the Access Point (AP) mode and station (STA) mode. Therefore, the server will act as a Wi-Fi access point, and the client will then connect to this access point. Firstly, the Wi-Fi credentials for the server to act as an access point are defined. The server local IP address and UDP port are also defined as part of the setup configuration. Using the defined setup details, the access point is then created, and the server can begin listening to the specified port for any incoming packet. Since the client will be connecting directly to the access point set up in the server application, the Wi-Fi credentials of the client and the UDP port should be the same as that of the server. To send the data to the server, the destination IP address and port are then specified. These values are obtained using the UDP *remoteIP* and UDP *remotePort* method. On the server's side, once the packet is received, the UDP *parsePacket* method is used, which processes the received packet and returns the size of the packet. Finally, the UDP read method is then used to read the received packet and returns the packet size, the IP address of the sender, the port number, and the received packet.

2.3 Data Processing and Application Layer

Cloud-Based Robot Skin Monitoring. Log data of robot skin systems are invaluable for managing and maintenance of the skin system. The user can monitor in real-time for some anomalies or inactivity and can therefore gauge the working condition of the robot skin system. The ThingSpeak cloud platform was used to implement this function. Channels and Messages are the two key concepts used by the ThingSpeak platform to exchange and store data. A channel is a data structure that is identified by a unique identifier (ID) and Application Programming Interface (API) keys for reading from and writing to it. Each channel contains eight fields of data. Messages, on the other hand, are considered as the base information unit. This can be defined as the information transferred when an entry is written to a ThingSpeak channel using the channel's API writing key by a device. Therefore, to send data to the ThingSpeak cloud, an account, a channel, and a field were created. With the writing API keys from the created channel, an application was designed to configure the NodeMCU to write data to the ThingSpeak cloud data acquired by the robot skin system was then be logged to the cloud. Utilizing the data processing capability of the ThingSpeak platform, the Matlab algorithm was then designed within the platform to analyze the received robot skin data.

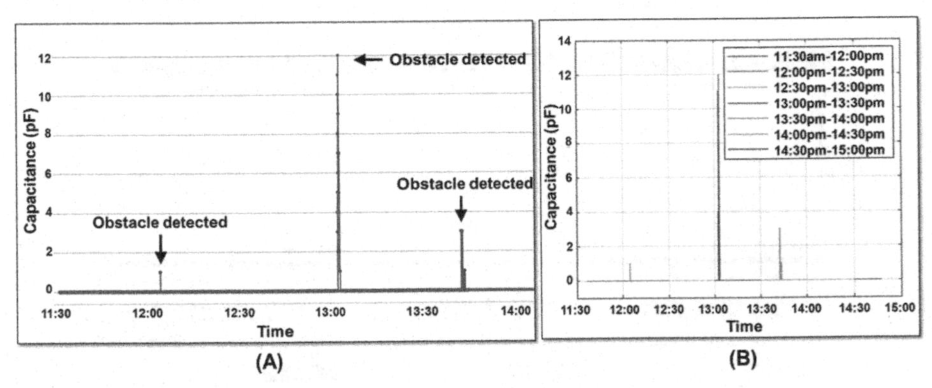

Fig. 4. Cloud-based robot skin monitoring. (A) Data entries for three hours thirty minutes. (B) Analysis and visualization of the received data at thirty-minute intervals showing more clearly the time obstacles were detected.

To demonstrate the significance of robot skin data logging to cloud which allows intuitive monitoring of the robot skin performance, in this implementation, the robot skin data was monitored for three hours thirty minutes. During this period, the robot skin data were continually sent to the cloud and visualized (see Fig. 4A). To extract and visualize specific information for easy monitoring of the performance of the robot skin, a Matlab analytic algorithm was designed to extract and display the robot skin data every thirty minutes. To implement this, two channels were created, and the robot skin data was feed to the field one of the first channels. The analytical Matlab script was designed to run automatically every thirty minutes, and then extract the data entries for thirty minutes and writes them to the second channel. The subsequent data entries in channel one were all acted on in a similar pattern until the end of the three-hour-and-thirty-minute period.

To visualize the data written to the second channel, a Matlab visualization algorithm that collects the data and visualizes them at run time was designed. The final visualization result shows the robot skin performance at every thirty minutes interval (see Fig. 4B). As seen from the result, it can be observed that an obstacle has entered the detectable range of the robot skin between 12:00 pm–12:30 pm, 1:00 pm–1:30 pm, and 1:30 pm–2:00 pm with the capacitance change of about 1 pF, 12 pF, and 3 pF respectively. Other specific features from the logged data can be extracted, analyzed, and visualized according to the application's needs. This advantage of developing an algorithm to monitor different situations from the robot skin data and the operator having access to the results regardless of time and place is appealing in the design of the robot skin system.

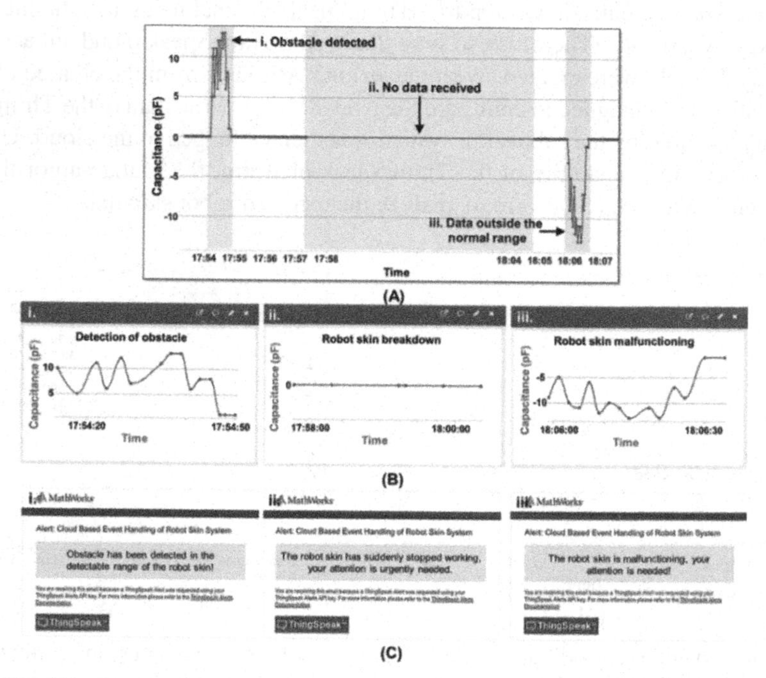

Fig. 5. Cloud-based event handling of robot skin system. (A) Continuous data was received on channel one showing different detected events. (B) Detected events visualization on channel two: (i.) When an obstacle was detected in the detectable range of the robot skin, (ii.) when no data was received, (iii.) when the received data was out of range of normal values. (C) Email alert notification sent to the user: (i.) Email alert when an object was detected; (ii.) Email alert when no data was received, (iii.) Email alert when data was out of range.

Cloud-Based Event Handling of Robot Skin System. Being able to resolve an incident or any anomalies as quickly as possible is very paramount in the robot skin system implementation. Such timely resolution of any potential problem with the robot skin system will enhance greater safety in HRC. Therefore, in this work, an event detection algorithm was designed within the ThingSpeak platform to analyze the received robot skin data for some events and to alert the user via email when those events occur. The events to analyze and check for were: when an obstacle is detected, when no data is received from the robot skin, and when the data from the robot skin is out of the normal range of values. When any of these events is detected, an email notification will be sent to the user. The user is, therefore, able to provide a timely solution to the issues, for example, by using web-based remote programming as described in the earlier section to resolve failures due to software-related issues.

In the ThingSpeak cloud server, two channels were created, one to receive the real-time robot skin data and the second for event detection management. The continuous data received by the first channel was then visualized (see Fig. 5A). The event detection algorithm then acts on every data that is written to the first channel to check for any of the aforementioned events. As events are detected, the detected events are visualized on three different fields on channel two for easy monitoring (see Fig. 5B), and the email alert notification was sent to the user (see Fig. 5C). To detect the first event which is the detection of an approaching object, the algorithm is to check from the received data if the data is greater than 1. Once this is detected, the detected event data will be extracted and written to the second channel and then visualized [see Fig. 5B(i.)]. At the same time, an email alert will be sent to the user to notify that an obstacle has been [see Fig. 5C(i.)]. For second event detection, the algorithm checks for when no data entry is received on channel one. If no data is received, 0 will be written to the second channel [see Fig. 5B(ii.)]. An email alert to notify the user that the robot skin has suddenly stopped working will be sent as well to the user [see Fig. 5C(ii.)]. For the detection of the third event, the received robot skin data was analyzed to check whether it is outside the normal range of values which is 0 pF to 14 pF. When this was detected, the event data was visualized on the second channel [see Fig. 5B(iii.)], and an email alert to notify the user that the robot skin is malfunctioning was sent [see Fig. 5C(iii.)]. This timely notification about the status of the performance of the robot skin will help to keep the user updated about the working condition of the robot skin.

Proposed System Performance Evaluation. The Transaction Processing Performance Council (TPC), a body that maintains benchmarks to measure performance in a standardized, objective, and verifiable manner, published a TPC-IoT benchmark which serves as a guide for analyzing IoT systems performance [10]. Therefore, two metrics based on the TPC-IoT criterion namely ingestion capacity and time delay were used to evaluate the system performance of this work.

Ingestion Capacity. The term injection is used to describe the acquisition of sensor measurement, and ingestion is used to refer to the whole process of loading data to the IoT system through sets of components This process is coordinated by the gateway (NodeMCU) and the IoT platform (ThingSpeak). For this evaluation, two factors, namely data transmission and receiving capacity was considered to assess the performance of the system. These two factors were chosen as they are fundamental operations in any IoT application. In this evaluation, we aim to study the impact of different injection rates on the performance of the system. To evaluate the data transmission capacity, the sensor measurements were acquired at different injection rates (0.1 s to 1 s) and transmitted to the ThingSpeak platform. Figure 6A gives the relationship in the form of percentage of the data acquired and transmitted at different rates. As shown in the result, it can be observed that at faster injection rates (0.1 s, 0.2 s 0.3 s, and 0.4 s), the percentage of data transmitted (15.89, 28.34, 41.03, and 52.92) was relatively low. This indicates that much of the data acquired were not successfully sent to the platform. This is due to the limitation of the data update interval of the ThingSpeak platform. However, as the injection rate lowers, the percentage of data transmitted increases significantly. From the rate of 0.8 s, 0.9 s, and 1.0 s, it can be overserved that all the data acquired were successfully sent to the platform. From the result, maximum data ingestion can be achieved between the rate of 0.7 s and 1 s. Although a faster rate would be more desirable, however achieving maximum ingestion from 0.7 s is still sufficient to realize a good system performance. For the evaluation of the data receiving capacity, the data transmitted to the platform and the data received by the platform at different rates were recorded. Since maximum ingestion could be achieved from 0.7 s, the injection rate was set from 0.7 s to 10 s to study the data receiving capacity of the system. Figure 6B shows the relationship between the data sent and received by the platform at different rates. As seen from the result, all the data sent were received successfully by the platform. As expected, at a faster acquisition rate, more data were transmitted, and at a slower rate, fewer data were sent.

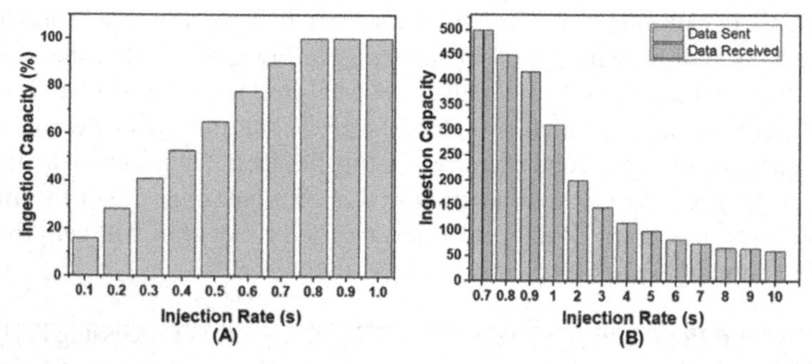

Fig. 6. Ingestion capacity evaluation of the proposed system. (A) Data transmission capacity of the system at different injection rates. (B) Data receiving capacity of the system at different injection rates.

Time Delay. The IoT application designed in this work initiates rule-based triggers which are automatically raised once an event occurs. It is therefore paramount to evaluate the time it takes for the system to handle the events and triggers a response. For this evaluation, firstly, the delay was measured as the time difference between when data was received by the platform and when the data was transmitted and secondly between when an event occurred and when the response was received. For the data transmission and receiving delay, the data sent and received were recorded with the timestamp at different data injection rates for 10 min each. At the end of 10 min for each rate, the time difference for all the data was calculated. The mean time difference was then calculated to estimate the time delay for each rate, as shown in Fig. 7A. From the result, the time delay was mostly between 0.083 s and 0.873 s. Some anomalies can be seen from the

1 s and 2 s rates with a time delay of about 1.898 s and 1.715 s. This could be because of network fluctuation during this period. In general, the time delay is induced by the network connectivity and the ThingSpeak platform computation time. For the delay in response to an occurred event, the data transmitted to the platform was recorded with a timestamp for 2 h. At the interval of 15 min, a human hand approaches the sensor surface to trigger the obstacle detection event. As described in the earlier section, when an event is detected, an email alert is sent to the user. The email alert was configured to be sent with a timestamp to keep track of the exact time the email was received. The time difference between when the event was triggered, and the email alert received was then calculated, with the result shown in Fig. 7B. From the result, the maximum delay (3 s) was observed in the 45th minute, and the minimum delay (1 s) was seen at the 120th minute, and for the remaining time intervals, the time delay was 2 s. This delay accounts for the time it takes for the data to arrive on the ThingSpeak platform, data analysis time in the platform, and the time it takes for the email alert to be received. This is usually affected by the Wi-Fi network being used.

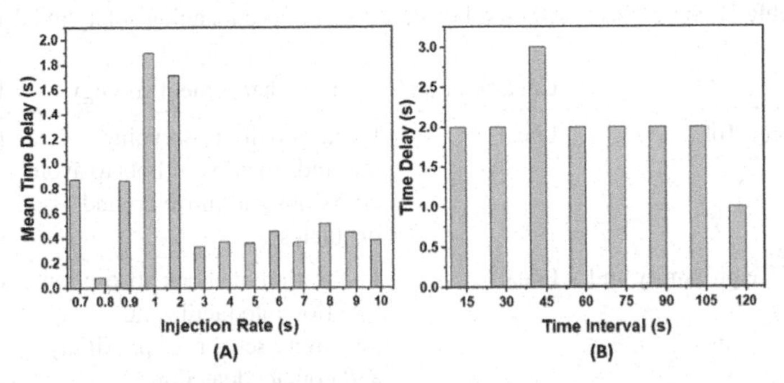

Fig. 7. Time delay evaluation of the proposed system. (A) Mean time delay between data transmission and receiving at different injection rates. (B) The time delay between when an event occurs and when notification alert is received at different time intervals.

3 Discussion

This work, through IoT technology, has addressed some of the issues that can enhance the safety performance of robot skin systems where less effort has been made to address in the literature. It has demonstrated how remote configuration, cloud-based monitoring, and cloud-based event handling of robot skin systems can help enhance the safety performance of the robot skin system. In the literature, as shown in Table 1, several strategies based on robot skin systems have been developed over the years to enhance safety in HRC. Although these strategies have produced excellent results, however not much attention has been given to management functionalities such as remote control, monitoring, and maintenance of the robot skin systems. In this regard, this work has proposed an approach through IoT technology that holds promising potential in enhancing the safety performance of robot skin systems. ThingSpeak IoT platform, which offers the flexibility required for the development and testing of new algorithms for IoT applications, was used to implement the cloud-based functions in this work. It is therefore paramount to ascertain how fast and reliable the platform is as it is used in real-time applications. In this regard, the performance of the ThingSpeak platform was evaluated in real application scenarios. From the ingestion performance evaluation, the platform was found to be reliable as all the data transmitted were successfully received, which is also consistent with the performance evaluation conducted in [11]. From the time delay analysis, a time delay between 0.083 s and 0.873 s was observed between when the data was sent and received by the platform except for some anomalies, which occurred only twice. Also, the delay in response to an occurred event was between 1 s and 3 s. Compared with the latency analysis of the ThingSpeak platform in [12] which found a round trip delay time to be between 1 s and 3 s, it can be seen that the performance of our proposed system is consistent. Although the performance of the proposed system is comparable to existing studies, a faster performance will be desirable. In the future, we hope to investigate latency minimization methods as proposed in [13].

Table 1. Strategies based on robot skin system for enhancing safety in HRC.

Sensor type	Configuration	Safety enhancement strategy	Ref
Capacitive and ToF	Local	Using proximity servoing methods to allow robots to avoid collisions and move around obstacles	[14]
Self-capacitive proximity and tactile sensor	Local	Integration of shock absorption function into tactile and proximity sensor for proximity and contact detection	[15]
Hall effect and ToF	Local	Use of multi-modal sensor array for contact detection, proximity sensing, and mapping	[16]
Capacitive imaging-based sensor	Local	A capacitive imaging array for obstacle detection and obstacle trajectory maps for collision avoidance	[17]
Self-capacitive proximity sensor	Local/remote	Real-time remote monitoring and analysis of the working condition of the robot skin system and alert notification for anomalies or inactivity for timely resolution	This work

4 Conclusion

In this work, an approach to enhance the safety performance of robot skin systems leveraging IoT technology was proposed. Through IoT technology, wireless communication within the robot skin system was implemented. Remote programing, which allows in situ programming of the robot skin system wirelessly, was also implemented. Cloud-based monitoring of the robot skin system was implemented, which has shown the benefit of allowing the operator to monitor the performance of the robot skin regardless of time and place. In addition, through cloud-based event handling, as implemented in this work, the operator can receive timely notification about any abnormal situation that occurs with the robot skin. This has the benefit of timely resolution of the abnormal situation with the robot skin, which can guarantee greater safety and enhance HRC. The application of IoT technology in robot skin systems has demonstrated promising potential in enhancing safety in HRC.

Acknowledgments. This work was supported in part by the National Natural Science Foundation of China under Grant 51975513 and Grant 51890884, and in part by the Natural Science Foundation of Zhejiang Province under Grant LR20E050003.

References

1. Göger, D., Alagi, H., Wörn, H.: Tactile proximity sensors for robotic applications. In: 2013 IEEE International Conference on Industrial Technology, pp. 978–983. IEEE Press, Cape Town, South Africa (2013)
2. Dahiya, R., Yogeswaran, N., Liu, F., et al.: Large-area soft e-skin: the challenges beyond sensor designs. Proc. IEEE **107**(10), 2016–2033 (2019)
3. Xia, F., Campi, F., Bahreyni, B.: Tri-mode capacitive proximity detection towards improved safety in industrial robotics. IEEE Sens. J. **18**(12), 5058–5066 (2018)
4. Ding, Y., Wilhelm, F., Faulhammer, L., et al.: With proximity servoing towards safe human-robot-interaction. In: 2019 IEEE/RSJ International Conference on Intelligent Robots and Systems, pp. 4907–4912. IEEE Press, Macau, China (2019)
5. Heng, W., Yang, G., Pang, G., et al.: Fluid-driven soft coboskin for safer human–robot collaboration: Fabrication and adaptation. Adv. Intell. Syst. **3**(3), 202000038 (2020)
6. Ye, Z., Yang, G., Pang, G., et al.: Design and implementation of robot skin using highly sensitive sponge sensor. IEEE Trans. Med. Robot. Bio. **2**(4), 670–680 (2020)
7. Pang, G., Yang, G., Heng, W., et al.: CoboSkin: soft robot skin with variable stiffness for safer human–robot collaboration. IEEE Trans. Ind. Electron. **68**(4), 3303–3314 (2020)
8. Yang, G., Deng, J., Pang, G., et al.: An IoT-enabled stroke rehabilitation system based on smart wearable armband and machine learning. IEEE J. Transl. Eng. Health Med. **6**, 1–10 (2018)
9. Gbouna, Z.V., Pang, G., Yang, G., et al.: User-interactive robot skin with large-area scalability for safer and natural human-robot collaboration in future telehealthcare. IEEE J. Biomed. Health Inform. (2021). https://doi.org/10.1109/JBHI.2021.3082563
10. Nambiar, R., Poess, M.: Industry standards for the analytics era: TPC roadmap. Lect. Notes Comput. Sci. **10661**, 1–6 (2018)
11. Viegas, V., Dias Pereira, J.M., Girao, P., et al.: A preliminary study of loop-time delays in IoT platforms: the thingspeak case. In: 2019 International Conference on Sensing and Instrumentation in IoT Era, pp. 1–4. IEEE Press, Lisbon, Portugal (2019)

12. Viegas, V., Pereira, J.D., Girão, P., et al.: Study of latencies in Thingspeak. Adv. Sci. Technol. Eng. Syst. J. **6**(1), 342–348 (2021)
13. Bhandari, S., Sharma, S.K., Wang, X.B.: Latency minimization in wireless IoT using prioritized channel access and data aggregation. In: 2017 IEEE Global Communications Conference, pp. 1–6. IEEE Press, Singapore (2017)
14. Ding, Y.T., Wilhelm, F., Faulhammer, L., et al.: With proximity servoing towards safe human-robot-interaction. In: 2019 IEEE/RSJ International Conference on Intelligent Robots and Systems, pp. 4007–4012. IEEE Press, Macau, China (2019)
15. Tsuji, S., Kohama, T.: Self-capacitance proximity and tactile skin sensor with shock-absorbing structure for a collaborative robot. IEEE Sens. J. **20**(24), 15075–15084 (2020)
16. Abah, C., Orekhov, A.L., Johnston, G.L.H., et al.: A multi-modal sensor array for safe human-robot interaction and mapping. In: 2019 International Conference on Robotics and Automation, pp. 3768–3774. IEEE Press, Montreal, QC, Canada (2019)
17. Ma, G.G., Soleimani, M.: A versatile 4D capacitive imaging array: a touchless skin and an obstacle-avoidance sensor for robotic applications. Sci. Rep. **10**(1), 11525 (2020)

Compliance Auxiliary Assembly of Large Aircraft Components Based on Variable Admittance Control

Chao Wu, Ye Shen, Gen Li, Pengcheng Li, and Wei Tian(✉)

Nanjing University of Aeronautics and Astronautics, Nanjing 210016, China
tw_nj@nuaa.edu.cn

Abstract. With the increasing demand for automation in aviation manufacturing, it is significant to realize intelligent manufacturing by using industrial robots for the compliant auxiliary assembly of large aircraft components. Based on the requirements of human-machine collaboration, the characteristics of variable admittance control are analyzed by taking the large aircraft components as the assembly object, and the specific implementation methods of variable admittance control based on operation intention recognition for different phases of the auxiliary assembly process are studied. Later, the force signal processing is simulated and an auxiliary assembly test prototype is built to verify the conclusion. The results show that using the proposed method can effectively improve the phenomenon of reverse acceleration mutation and contact bounce of robot, then the robot's compliance and assembly efficiency in the process of auxiliary assembly are enhanced.

Keywords: Variable admittance control · Compliance control · Human-machine collaboration · Aircraft assembly

1 Introduction

In recent years, industrial robots have been applied to aircraft automatic assembly in major manufacturing companies due to the upgrading of its hardware, aiming to improve assembly accuracy and assembly safety. For the variety of the overall components of aircraft products, as well as the complex coordination between components, the support of multiple sophisticated technologies is needed in the whole assembly process [1]. While there still contains many full manual steps in the domestic aviation assembly, such as the hoisting and maintenance during the product transportation and fabrication, which not only requires the cooperation of multiple workers, but also has low efficiency. Besides, the assembly precision, period, and reliability of the whole process cannot be guaranteed [2]. What's more, as for parts in large size and weight, a high knock risk may exist for the assembly environment is hard to maintain manually [3]. The human-machine collaboration with flexibility provides a well comprehensive solution for such a complex assembly environment [4, 5].

A reasonable compliance control can not only ensure the safety and efficiency of cooperative tasks, but also can greatly reduce the burden of human beings in the task

X.-J. Liu et al. (Eds.): ICIRA 2021, LNAI 13014, pp. 469–479, 2021.
https://doi.org/10.1007/978-3-030-89098-8_44

[6]. Passive compliant control is limited for dissatisfying the assembly load and accuracy in the compliance assembly tasks. While the position force hybrid control in active compliance control has higher requirements on the environment and low sensitivity to the collision. Therefore, impedance control [7] and admittance control [8] are generally adopted. Gan et al. [9] proposed a force-tracking control strategy based on adaptive variable admittance in an unstructured environment, where the admittance model parameters were adaptively adjusted online according to the change of contact force. However, the selection law of admittance parameters was not mentioned. Akgun et al. [10] put forward an adaptive compliance control by estimating the force applied to the four-bar linkage through parameter identification, though avoiding the use of force sensor, the dynamic model for six degrees of freedom robot is difficult. Tsumugiwa et al. [11] studied a variable impedance control of the robot at low speed ($\leq$0.02 m/s), whose stability and efficiency performed well compared with invariable control. However, the comparison with other variable control methods was not experimented and the method was limited to low-speed conditions. Lecours et al. [12] proposed a variable admittance control method based on human intention, which enabled the robot to respond safely to human actions, but ignored the case of reverse intention. At present, researches on human-machine collaboration compliance control are mostly focused on cooperative robots, namely light-load and light-weight robots [13–15], and relatively absent on heavy-load industrial robots. For the heavy-load robots have more stringent safety requirements than the light-load robot, to ensure the safety of human beings and the compliance in the assembly process, the acceleration mutation in reverse intention, the speed switch and the contact bounce ignored in the above researches should be considered.

Therefore, to realize the compliance-assisted assembly of large aircraft components with heavy-load industrial robot, a variable admittance control based on human intention recognition is proposed in this paper, which can switch the assembly speed for different requirements of different assembly stages and change the admittance coefficients in real-time to adapt to the physical interaction behavior of operators. Thus, the acceleration mutation of the robot in reverse intention and the contact bounce phenomenon can be improved. Finally, the effectiveness of the proposed method is compared to the other variable admittance control method by simulation and experiment.

2 Variable Admittance Control Model

The core idea of admittance control is to establish a dynamic relationship between position deviation and force deviation based on the second-order differential equation of "mass-damp-spring", which is also known as position-based impedance control [16]. In the Cartesian coordinate system, the corresponding one-dimensional admittance equation is:

$$f_H = m(\ddot{x} - \ddot{x}_d) + c(\dot{x} - \dot{x}_d) + k(x - x_d) \tag{1}$$

where f_H represents the mutual force applied by the operator, m represents the virtual mass, c represents the virtual damping, k represents the virtual stiffness. $\ddot{x}_d$, $\dot{x}_d$, x_d relatively represent the real acceleration, velocity, and position of the robot end; $\ddot{x}_d$, $\dot{x}_d$, x_d relatively represent the expected acceleration, velocity, and position of the robot end.

Since the system in this paper simulates free motion in most cases, k, $\ddot{x}_d$, $\dot{x}_d$ *and* x_d are set to zero, then Eq. (1) can be written as

$$f_H = m_v\ddot{x} + c_v\dot{x} \tag{2}$$

When m_v and c_v change, different dynamic characteristics of robots will be shown. Therefore, the admittance coefficients can be adjusted to meet the needs of different environments.

For speed control, the Laplace transform of robot speed can be written as:

$$\dot{X}(s) = \frac{F_H(s)}{m_v s + c_v} = \frac{F_H(s)/c_v}{\frac{m_v}{c_v}s + 1} = F_H(s)H(s) \tag{3}$$

where $\dot{X}(s)$ is the Laplace transform of $\dot{x}$, and $F_H(s)$ is the Laplace transform of f_H. Then the transfer equation is

$$H(s) = \frac{\frac{1}{c_v}}{\frac{m_v}{c_v}s + 1} \tag{4}$$

which shows that the ratio of virtual mass to virtual damping will affect the poles of the system. Therefore, to stabilize the system, the ratio should be kept within a reasonable range while changing the admittance parameters. Besides, the reasonable range varies with the movement of different loads and different axes, which needs to be calibrated through experiments.

The Cartesian coordinate system is divided into position space (XYZ direction) and attitude space (ABC direction) in this paper. In the following, the position space is taken as an example to illustrate, and the control model of attitude space can be obtained by analogy. The admittance coefficient control rules are set below:

2.1 Variable Admittance Coefficient Setting

$$c_v = \begin{cases} c_d - \lambda_1|\ddot{x}_b| & \text{for acceleration} \\ c_d & \text{for stop} \\ c_d + \lambda_2|\ddot{x}_b| & \text{for deceleration} \end{cases} \tag{5}$$

where c_d is the default value of virtual damping, c_v is the effective virtual damping, λ_1 and λ_2 are the adjustable coefficients used to control the change rate of virtual damping. To roughly estimate λ_1 and λ_2, an ideal variation region denoted as $\left[c'_{\min}, c'_{\max}\right]$ is set, and the maximum acceleration denoted as $|\ddot{x}|_{\max}$ is given.

$$\begin{cases} \lambda_1 \approx \frac{c_d - c'_{\min}}{|\ddot{x}|_{\max}} \\ \lambda_2 \approx \frac{c'_{\max} - c_d}{|\ddot{x}|_{\max}} \end{cases} \tag{6}$$

Therefore, if the operator intends to accelerate the robot, the damping should be decreased; if the operator intends to slow down the robot, the damping should be increased; as the robot is at rest, the admittance coefficients should equal the default values. Additionally, at the beginning of the movement, the speed of the robot shouldn't be too high, but the default damping requires high with a not too low default mass.

Some researchers [17] have found that if the virtual mass remains constant, some adverse effects will occur. According to Eq. (4), when the system needs to be accelerated, the ratio of virtual mass to virtual damping will increase significantly, resulting in a slower response; when the system needs to be decelerated, the ratio will decrease, resulting in the system instability. Therefore, the virtual mass should vary with the virtual damping Denote $k = c_v / c_d$, and the virtual mass is controlled as follows:

$$m_v = \begin{cases} \max(km_d, \Omega) & \text{for acceleration} \\ m_d & \text{for stop} \\ km_d \cdot \Lambda & \text{for deceleration} \end{cases} \tag{7}$$

where m_v is the effective virtual quality, Ω is a parameter to smoothly transit in reverse motion without compromising comfort, expressed as

$$\Omega = \varepsilon m_{vb} \tag{8}$$

where ε $(0.98 < \varepsilon < 1)$ is to adjust the smoothness of the speed change, m_{vb} is the effective virtual quality of the previous moment.

Λ is set to guarantee the robot gaining a larger negative acceleration through a lower virtual mass in the deceleration stage, under the premise of maintaining the continuity of the function.

$$\Lambda = 1 - a\left(1 - e^{-b(c_v - c_d)}\right) \tag{9}$$

where $a(0 < a < 1)$ is used to adjust the steady-state ratio of virtual mass and virtual damping, b is used to adjust the smoothness of the ratio change.

2.2 Human Intention Recognition

The acceleration, deceleration, and stop processes in Eq. (5) and (7) can be judged logically as shown in Fig. 1.

1) If $\ddot{x}_b \neq 0$ and $\ddot{x}_b \cdot \dot{x}_b \geq 0$, the human intention is to accelerate.
2) If $\ddot{x}_b \neq 0$ and $\ddot{x}_b \cdot \dot{x}_b < 0$, the human intention is to decelerate.
3) If $\ddot{x}_b = 0$ and $\dot{x}_b \neq 0$, the human intention is to decelerate.
4) If $\ddot{x}_b = \dot{x}_b = 0$, the human intention is to stop.

where $\dot{x}_b$ is the velocity of the robot end of the previous moment, $\ddot{x}_b$ is the acceleration of the robot end of the previous moment. As a human's operational intention can be divided into three situations [12]: acceleration, deceleration and stop, the inference strategy in Fig. 1 videlicet the process of human intention recognition.

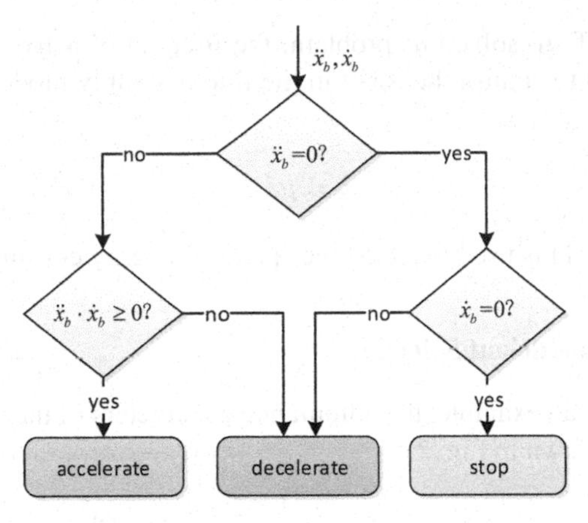

Fig. 1. Logical judgment flow chart

2.3 Proportionality Coefficient

Some scholars [17, 18] have pointed out that when the admittance parameters (virtual mass and virtual damping) are set to high, are easier to perform, but greater forces are required to move the robot at a given speed or acceleration. Conversely, when the parameters are set to lower, the operator can move the robot quickly with less forces, but the fine motions are more difficult to perform. The former working mode is named fine assembly mode, and the latter mode is named fast-moving mode in this paper. In application, the working mode of the robot needs to be changed quickly, so a coefficient p is introduced to divide the process into the two modes.

To facilitate the implementation of the actual industrial robot controller, Eq. (2) is discretized [9], then the following equation can be obtained

$$x = x_b + \dot{x}t = x_b + \Delta x \tag{10}$$

where, the subscript b represents the previous moment, and Δx is the displacement of the robot. To separate the working mode, let

$$\Delta x_2 = \Delta x/p \tag{11}$$

where Δx_2 is the displacement that robot receives. After preparation [19], a speedy conversion between two modes can be realized by setting different p values.

The fine assembly mode is also different from the fast-moving mode in that the assembly part needs to contact the assembly body. Assume the kinetic energy of the assembly part in the X direction is W, for

$$W = \frac{1}{2}m\dot{x}^2 \tag{12}$$

W is related to $\dot{x}$. When the assembly part touches the assembly body, the kinetic energy and elastic potential energy of the system will constantly transform, thus making the assembly bounce off. To solve this problem, the method of active kinetic energy consumption is adopted to control the robot in the fine assembly mode, which is shown as follow

$$\dot{x} = \mu \dot{x}_b \tag{13}$$

where $\mu(0.9 < \mu < 1)$ is used to adjust the speed of energy consumption.

2.4 Admittance Coefficient Selection

Taking the X-axis as an example, the admittance parameters of the robot are calibrated, and the results are shown in Fig. 2.

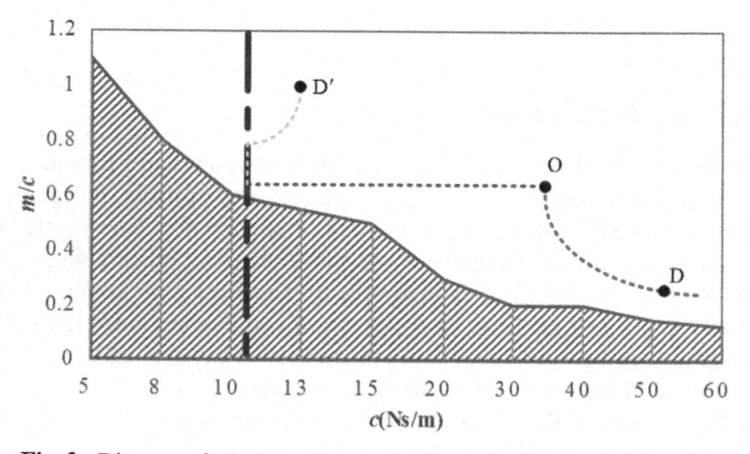

Fig. 2. Diagram of admittance parameters variation (Color figure online)

The shaded region represents the unstable region. When the virtual damping and virtual mass are set in the shaded region, the operator will feel a vibration. When the operator applies the same force, the closer the selected admittance coefficient is to the longitudinal axis, the faster the robot moves; the closer the robot is to the horizontal axis, the more flexible the robot's movement will be. As can be seen from Fig. 2, when the virtual damping is less than a certain value, the shaded area will increase rapidly, so a virtual damping minimum value should be set, denoted as $c_{\min}$ (the black dotted line in Fig. 2).

According to Fig. 2, the admittance parameters can be expressed as point O in the stop state, which is called the default admittance parameter. And in the acceleration stage, the parameters will vary within the dashed blue line, while in the deceleration stage, the parameters will vary within the red dotted line. In the reverse phase, assuming the parameters start at the point D, where the acceleration and velocity happen to go from the opposite direction to the same direction, then the parameters at the next moment will change from the point D′ along the yellow dotted line until coinciding with the blue one.

3 Simulation of the Admittance Controls

To better illustrate the effectiveness of the method proposed, implementations of fixed admittance control, variable admittance control in reference [12] (named reference control) and the variable admittance control proposed in this paper are simulated and compared in the stage of fast-moving.

To comprise the cases for the entire human-machine collaboration, a typical actual force signal was collected by the six-dimensional force sensor used in the subsequent experiment, which is shown in Fig. 3.

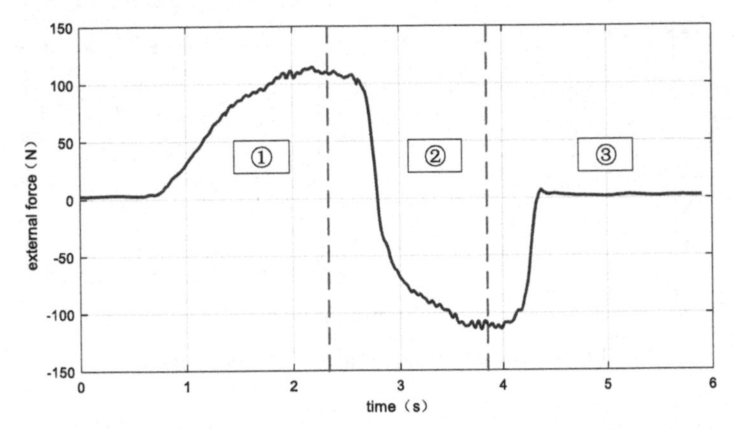

Fig. 3. A typical actual force signal (Color figure online)

where the force applied by the operator recorded in ① increases gradually, indicating that the operator wants the robot to accelerate; in ②, the force suddenly decreases,indicating that the operator wants the robot to reverse; and the operator stops touching the sensor in ③, indicating that the operator wants the robot to stop.

Select parameters according to Fig. 2, four control models were established by using simulation software:

1) Fixed admittance control: condition 1 with high admittance parameters ($m = 36, c = 60$) and condition 2 with low admittance parameters ($m = 6, c = 10$).
2) Reference control: the default admittance parameters are $m_d = 18\,\mathrm{N} \cdot \mathrm{s}^2/\mathrm{m}$ and $c_d = 30\,\mathrm{N} \cdot \mathrm{s/m}$, set $c_{\min} = 10\,\mathrm{N} \cdot \mathrm{s/m}$, $\lambda_1 = \lambda_2 = 5$, $a = 0.6$, $b = 20.\lambda_1 = \lambda_2 = 5$, $a = 0.6$, $b = 20$.
3) Variable admittance control in this paper: Set $\varepsilon = 0.99$, the other parameters are consistent with the reference control.

Then the processing results of the force signal described in Fig. 3 are shown in Fig. 4, using four control models.Which shows that in the fixed admittance control method, the robot's moving speed decreases with the increase of the admittance parameters. Although the robot's moving speed is faster in the case of low admittance parameters, the response delay is larger. Compared with the fixed admittance control method, the variable admittance control proposed in this paper can maintain the characteristics of rapid movement while reducing the response delay. As shown in the enlarged figure, there exists an obvious acceleration abrupt change in the reverse stage when using reference control, however, the robot speed can be smoothly transitioned by using the variable admittance control method in this paper, finally the compliance was improved without affecting the overall performance.

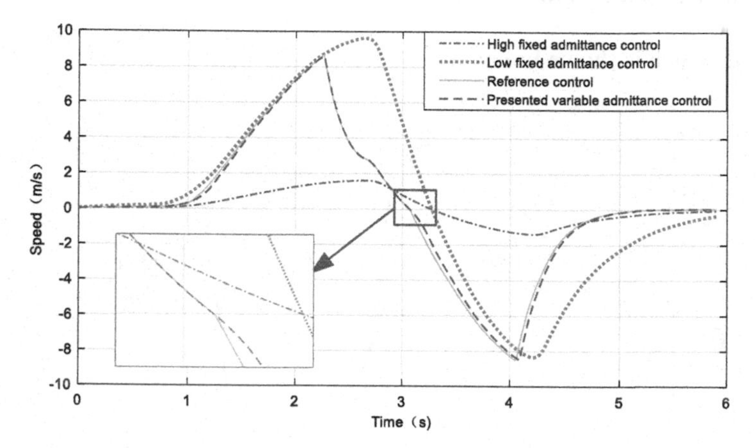

Fig. 4. The processing results of the force signal

4 Experiment

KR210–2700 robot of KUKA company was used in the experiment, and Axia80-M50 sensor of ATI company was selected as the six-dimensional force sensor. To simulate the large aircraft components, the aluminum alloy frame is designed to be loaded. The weight of the frame is about 40 kg, and the counterweight block can be placed anywhere in the frame to simulate different types of components. The communication mode between the upper computer and the robot is RSI, while between the upper computer and the force sensor is UDP, and the communication cycle is 4 ms.

4.1 Reverse Intention Experiment

A reverse intention experiment in the fast-moving mode was carried out under light load to ensure the safety of the experiment. Before the compliance control, a calibration to determine the "shadow area" of the robot should be down. Similar to Fig. 2, the x-axis of the robot is calibrated, the specific selections are shown in Table 1

Table 1. Variable admittance control parameters of reverse intention experiment

Parameters	c_d	m_d	$c_{\min}$	λ_1	λ_2	a	b	ε	p	μ
Value	30	18	10	5	5	0.4	2	0.99	2	0.99

Several experimenters were selected for reverse operation, the specific process is shown in Fig. 5.

Fig. 5. The process of reverse intention experiment

The experimenters reported that if the reverse force changed greatly during the reverse process, the acceleration of the robot would increase instantly when using the reference control model, while there was no acceleration mutation when using the model proposed, which was consistent with the simulation results in Sect. 3.1.

4.2 Bolting Experiment

Since the difference between the reference control model and the variable admittance control model in the fast-moving stage has been demonstrated by the reverse intention experiment, the difference between the two control methods in the fine assembly stage is focused on in this section. Therefore, a bolting test is designed to compare the feasibility of the two methods, as shown in Fig. 6.

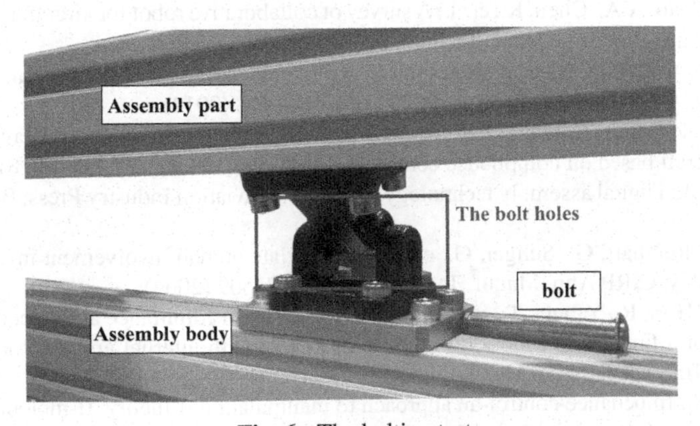

Fig. 6. The bolting test

In theory, since the reference control does not set the mode switching function, its fine assembly stage is essentially the same as its fast-moving stage, which means fine movement is hard to be carried out. Besides, the reference control method only considered controlling the robot in free space, when the assembly part contact with the assembly body, the assembly body will be regarded as a relatively high stiffness arm, which will cause the assembly part to "bounce off", making it harder to reach the designated assembly position.

Video [20] recorded the human-machine collaboration process in the fine assembly stage, where several tests were carried out for each admittance control model, and the theoretical analysis was proved to be consistent with the experimental result. In the video, the bounce-off and repeated collision indeed appeared in the bolt connection process while using the reference control, which brings about the failure of the assembly. However, the method presented in this paper can complete the assembly due to the consideration of the constraint space.

5 Conclusion

A variable admittance control method was proposed in view of the situation in compliant auxiliary assembly technology for large industrial robots in this paper. The relationship between the coefficient of admittance was analyzed, then the concept of human intention recognition was introduced, subsequently a method to determine various parameters was given, and the variable admittance control for different assembly stage was realized by considering the acceleration mutation in the stage of the robot inverse kinematics and improving the control model to solve bounce phenomenon. Finally, the method proposed was verified for effectively improving the reverse acceleration mutation problem and the contact bounce phenomenon through the comparative simulation results of force signal processing, as well as the reverse intention experiment and bolting experiment, the flexibility of robot-assisted assembly is advanced.

References

1. Wu, D., Zhao, A.A., Chen, K., et al.: A survey of collaborative robot for aircraft manufacturing application. Aeronaut. Manuf. Technol. **62**(10), 24–34 (2019)
2. Li, T., Hu, X.X., Yao, W., et al.: Research on application of robot in space equipment automatic assembly. Aeron. Manuf. Technol. **0**(21), 102–104+108 (2014)
3. Hu, R.Q., Zhang, L.J., Meng, S.H., et al.: Robotic assembly technology for heavy component of spacecraft based on compliance control. J. Mech. Eng. **54**(11), 85–93 (2018)
4. Feng, Z.M.: Digital assembly technology for aircraft. Aviation Industry Press, Beijing, China (2015)
5. Bley, H., Reinhart, G., Seliger, G., et al.: Appropriate human involvement in assembly and disassembly. CIRP Ann. Manuf. Technol. **53**(2), 487–509 (2004)
6. Sano, Y., Hori, R., Yabuta, T., et al.: Comparison between admittance and impedance control method of a finger-arm robot during grasping object with internal and external impedance control. Trans. Japan Soc. Mech. Engineers **79**(807), 4330–4334 (2013)
7. Hogan, N.: Impedance control-an approach to manipulation. I-theory. II-mplementation. III-applications. J. Dynamic Syst. Meas. Control, **107**, 1–24 (1985)

8. Seraji, H.: Adaptive admittance control: an approach to explicit force control in compliant motion. In: IEEE International Conference on Robotics and Automation, pp. 2705–2712. IEEE, Washington, DC, USA (2002)
9. Gan, Y.H., Duan, J.J., Dai, X.Z.: Adaptive variable impedance control for robot force tracking in unstructured environment. Control Decis. **34**(10), 2134–2142 (2019)
10. Akgun, G., Cetin, A.E., Kaplanoglu, E.: Exoskeleton design and adaptive compliance control for hand rehabilitation. Trans. Inst. Meas. Control. **42**(3), 493–502 (2020)
11. Tsumugiwa, T., Yokogawa, R., Hara, K.: Variable impedance control based on estimation of human arm stiffness for human-robot cooperative calligraphic task. In: IEEE International Conference on Robotics and Automation, IEEE, Marina Bay Sands, Singapore (2017)
12. Lecours, A., Mayer-St-Onge, B., Gosselin, C.: Variable admittance control of a four-degree-of-freedom intelligent assist device. In: IEEE International Conference on Robotics & Automation, pp. 3903–3908. IEEE, Paul, Minnesota, USA (2012)
13. Kormushev, P., Calinon, S., Caldwell, D.G.: Imitation learning of positional and force skills demonstrated via kinesthetic teaching and haptic input. Adv. Robot. **25**(5), 581–603 (2011)
14. Rozo, L., Calinon, S., Caldwell, D.G., et al.: Learning physical collaborative robot behaviors from human demonstrations. IEEE Trans. Rob. **32**(3), 513–527 (2016)
15. Li, Y., Ge, S.S.: Force tracking control for motion synchronization in human-robot collaboration. Robotica **34**(6), 1260–1281 (2016)
16. Dong, J.W., Zhou, Q.Q., Xu, J.M.: Research on robot impedance control. In: 37th Chinese Control Conference, p. 7. Technical Committee on Control Theory, Chinese Association of Automation, Wuhan, Hubei, China (2018)
17. Ikeura, R., Inooka, H.: Variable impedance control of a robot for cooperation with a human. In: IEEE International Conference on Robotics and Automation, pp. 3097–3102. IEEE, Nagoya, Japan (1995)
18. Duchaine, V., Gosselin, C.M.: General model of human-robot cooperation using a novel velocity based variable impedance control. In: 2nd Joint EuroHaptics Conference and Symposium on Haptic Interfaces for Virtual Environment and Teleoperator Systems, pp. 22–24. IEEE, Tsukuba, Japan (2007)
19. Li, G., Li, P.C., Wu, C., et al.: Research on optimization algorithm of robot load gravity compensation based on genetic algorithm. Aeronaut. Manuf. Technol. **64**(5), 52–59 (2021)
20. BILI Homepage. https://www.bilibili.com/video/BV1Kf4y1p7LU. Accessed 21 Jun 2021

Variable Admittance Control by Measuring Arm Impedance

Rui Duan[1,2], Chongchong Wang[1(✉)], Chin-Yin Chen[1], Guilin Yang[1], and Long Chen[3]

[1] Zhejiang Key Laboratory of Robotics and Intelligent Manufacturing Equipment Technology, Ningbo Institute of Materials Technology and Engineering, Chinese Academy of Sciences, Beijing 315201, China
wangchongchong@nimte.ac.cn

[2] Tanyuan University of Science and Technology, Taiyuan 030024, China

[3] Faculty of Science and Technology, University of Macau, Macau 999078, China

Abstract. This paper presents a variable admittance control (VAC) method of human-robot collaboration (HRC) based on human body impedance. In HRC, when the stiffness of a person's arm increases, the control based on fixed admittance parameters become unusable. VAC is proposed, which makes the damping coefficient in the robot admittance parameters determined by the stiffness of the end of the human arm and the system's stability. The damping parameters are changed according to the speed of the end of the robot, so that the HRC is easy to complete.

Keywords: Human-robot collaboration · Admittance control · Human arm impedance model

1 Introduction

As robot technology becomes more and more widely used in industrial productions, robots increasingly need to work in close coordination with people in the same space and achieve natural interaction between humans and robots while ensuring safety. One of the many challenges facing the development of robots is the need to shift from a one-way working mode of human-operated robots to a two-way collaboration mode.

In the process of HRC, due to the complex kinematics of the human body and the excellent flexibility of limb movement, the force and movement of the robot need to match the natural impedance of the human body. Especially when performing complex tasks in dynamic environments, modelling the impedance of the human body is a prerequisite for an accurate and compliant motion of robots [1].

Ficuciello changes the admittance control damping through the end speed of the robot, but does not consider human characteristics [2]. Tsumugiwa et al. have proposed in a new variable impedance control law based on an estimation of the human stiffness [3]. Vincent Duchaine proposed a variable admittance

X.-J. Liu et al. (Eds.): ICIRA 2021, LNAI 13014, pp. 480–490, 2021.
https://doi.org/10.1007/978-3-030-89098-8_45

control for real-time measurement of human impedance [4]. But the result of identification of human impedance characteristics is not good.

In this paper, through the identification of human body impedance, the identification result is used as the reference value for determining the admittance control parameter, so as to improve the interaction ability of HRC. In the admittance control of HRC, damping plays a dominant role in the admittance control [5]. When the human performs fine movements at low speeds, high admittance parameters are required, and when performing large movements at high speeds, low admittance parameters are required. The selection of admittance parameters needs to consider the stability of HRC.

The paper is organized as follows. Section 2 describes the modeling of the human arm impedance model, which is used to study the stability of HRC. Section 3 analyzes the stability of HRC and introduces a method to adjust the admittance controller parameters. Section 4 describes the experimental ac,.tivities carried out to identify the impedance parameters of the human arm, and reports the results of actual control experiments. Conclusions are drawn in Sect. 5

2 Human Arm Impedance Model

H. Patel found that the impedance provided by the operator to the robot are mainly related to the arm [6]. Therefore, this article only considers the impedance of the human arm. S. Abiko, A. Konno and others proposed that the accurate model of the human arm should not only consider the mechanical characteristics, but also consider the neuromuscular characteristics [7]. Therefore, a simplified human arm model is used, which is based on the following assumptions:

1. The arm only performs translational movement on the horizontal plane;
2. The impedance of the human arm can be decomposed into two parts in the X and Y directions on the plane of motion;

It can be seen from the experimental results that the introduction of these assumptions does not affect the effectiveness of the model [8], but it greatly simplifies the derivation of the model, especially the parameter identification link. According to the previous assumptions, so, the impedance model of the human arm along the x and y directions, can be expressed by the following second-order differential equation:

$$F_h(q_h, t) = M_h(q_h, t)\ddot{q}_h + D_h(q_h, t)\dot{q}_h + K_h(q_h, t)q_h \tag{1}$$

where $F_h(q_h, t)$ denote the force exerted by the human arm in the X and Y directions, $M_h(q_h, t)$, $D_h(q_h, t)$ and $K_h(q_h, t)$ denote equivalent mass, damping and stiffness matrix of human upper limbs related to time position, and q_h denote the reality displacement of the human

The experimental results show that the stiffness matrix K_h for the human arm is a non-diagonal matrix. But under the assumption, the diagonal component represents the dominant effect. The transformation equation to convert the stiffness matrix into the stiffness ellipse is as follows [8]:

$$\begin{bmatrix} F'_x(t) \\ F'_y(t) \end{bmatrix} = K_h \begin{bmatrix} \cos(t) \\ \sin(t) \end{bmatrix}, t \in [0, 2\pi] \tag{2}$$

where $F'_x(t)$, $F'_y(t)$ respectively represent the magnitude of the reaction force in the x-direction and y-direction corresponding to the unit displacement at the end of the arm.

The stiffness of the human arm varies, and it depends on its mechanical properties and neuromuscular properties. Therefore, the related neuromuscular properties must be considered, namely the reaction time of voluntary and involuntary actions, and the change in stiffness caused by muscle contraction. Can be expressed as follows:

$$K_h(q_h, t) = K_{post}(q_h, t) + K_{cont}(t) \tag{3}$$

where $K_{post}(q_h, t)$ is mainly related to posture of arm. $K_{cont}(t)$ is related to the phenomenon of co-contraction of muscles, and its size depends on the strength of the contraction. The stiffness caused by muscle contraction can be modeled as:

$$K_{cont}(t) = \bar{K}_{cont}\sigma(t - \tau) \tag{4}$$

Where $\bar{K}_{cont}$ is maximum stiffness increases due to co-contraction, σ is a normalized muscular co-contraction index, Value is 0 to 1. τ is a delay accounting for involuntary reaction time.

Experiments show [9], the posture stiffness can be decomposed into two orthogonal axis, and each models is modeled as a quadratic function of the position along the corresponding axis, that is, the x-axis direction can be expressed as:

$$K_{post_x}(x, t) = b_1 x^2(t) + b_2 x(t) + b_3 \tag{5}$$

where b_1, b_2 and b_3 represent related gain coefficients, and the Y-direction is similar to the X-direction.

3 Variable Admittance Control

3.1 Admittance Control Model

The input of the admittance controller is the force and torque applied by the human, and the output is to modify the position of the robot's Cartesian space movement (see Fig. 1). In the actual operation process, the operator further closes to form a loop outside the robot position control loop. In this case, the impedance characteristics of the human and the admittance characteristics set by the robot affect each other, which may cause severe vibration or even instability of the robot.

For the convenience of explanation, one-dimensional equation is used to discuss the idea of admittance control in HRC. The general admittance control equation is:

$$F_h = M_r(\ddot{X} - \ddot{X}_d) + D_r(\dot{X} - \dot{X}_d) + K_r(X - X_d) \tag{6}$$

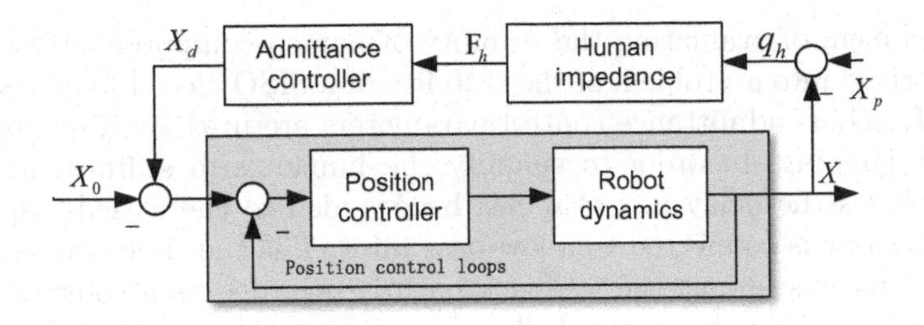

Fig. 1. Control architecture for admittance control.

where X_p denote the desired displacement of the human, X_p denote the reality displacement of the human. F_h denote the forces and torques of the human,X_d denote the desired additional displacement. The admittance control equation is the relationship between the additional motion X_d of the end effector and the external force F_h, M_r , D_r , K_r are virtual mass, damping, and stiffness matrices.

In order to enable the end effector to follow and adapt to the force exerted by the human body in human-computer interaction, the stiffness K_r is generally set to 0, that is, the end effector dynamics is set to the equation of the mass-damping system:

$$F_\mathrm{h} = M_r(\ddot{\mathrm{X}} - \ddot{X}_d) + D_r(\dot{\mathrm{X}} - \dot{X}_d) \tag{7}$$

3.2 Human-Machine Interaction Stability Analysis

From a stability analysis point of view, the proportion of the mass term in the stiffness Eq. (1) is very low. And damping always has the effect of stabilizing the system. In this case, only the stiffness component is considered, which is equivalent to analyzing the worst case. That is, the impedance of a person can be expressed as:

$$K_h(x,t) = b_1x^2(t) + b_2x(t) + b_3 + \bar{K}_{cont}\sigma(t-\tau) \tag{8}$$

In order to simplify the control system, In HRC, the value of external position control is usually 0.

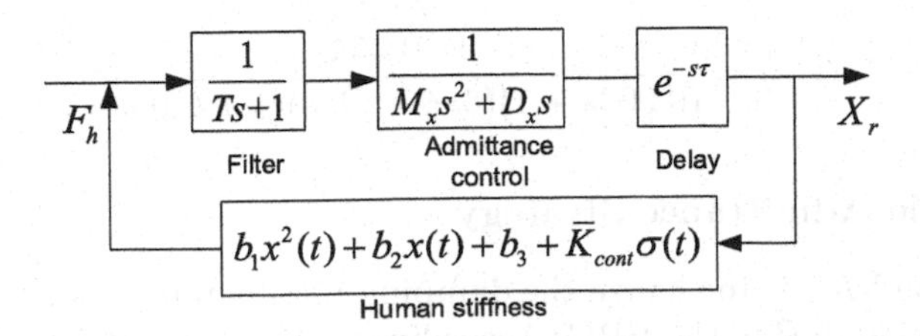

Fig. 2. The closed-loop system for absolute stability analysis.

The problem of evaluating the stability of human-computer interaction can be transformed into a problem of the stability of a SISO closed-loop system (see Fig. 2). M_x, D_x is admittance control parameters are in the x-direction virtual mass and damping. In order to simplify the human arm stiffness model, the model excludes the delay link that has been added to the closed-loop system. The force sensor is connected to a low-pass filter. T is the time constant.

For nonlinear systems, the feedback path is regarded as a constant gain by the absolute stability theorem, which requires the nonlinearity of the feedback to satisfy the sector condition [10]:

$$\alpha x^2 < \mathrm{K}_h(x,t)x^2 < \beta x^2 \qquad \forall t > 0 \tag{9}$$

where $\alpha, \beta \in R$, $\alpha < \beta$ and $x \in [x_{\min}, x_{\max}]$, with x_{min} , x_{max} is the range of x values for the absolute stability of the closed-loop system of interest. Since $K_h(x,t)$ is a quadratic function, it is easy to get the range of the function in the target area α and β.

In addition, admittance control is an asymptotically stable system. For processing convenience, using Pader approximation equation to replace the delay. consider the first-order Pader approximation [10]:

$$G(s) = \frac{1}{Ts+1}\frac{1}{M_x s^2 + D_x s}e^{-s\tau} = \frac{1}{Ts+1}\frac{1}{M_x s^2 + D_x s}\frac{1 - s\tau/2}{1 + s\tau/2} \tag{10}$$

In summary, the transfer function of the closed-loop system is approximately:

$$G_T(s) = \frac{G(s)}{1 + \alpha G(s)} \tag{11}$$

where α is one of the coefficients defining the boundary of the sector. Using absolute stability theory [10], the function (12) is strictly positive and real.

$$Z_T(s) = 1 + (\beta - \alpha)G_T(s) = \frac{1 + \beta G(s)}{1 + \alpha G(s)} \tag{12}$$

If the Nyquist diagram of $G(s)$ does not enter a circle whose center is in the complex plane on the real axis, and intersects the real axis at $-1/\alpha$ and $-1/\beta$. Under the condition of ensuring the stability of HRC, once M_x is given, iterative methods can be used to find the smallest D_x.

$$G(s) = \frac{1{-}0.05s}{(0.003s + 1)(5s^2 + D_x s)(1 + 0.05s)} \tag{13}$$

3.3 Variable Admittance Strategy

The goal of the VAC is to change the damping parameters of the robot to adapt to human motion during the HRC. According to [11] the admittance parameter needs to match the human impedance parameter. When people perform fine operations at low speeds, the stiffness is greater and high admittance parameters are required, High speed opposite. Therefore, the damping is changed according to the Cartesian speed of the end effector to improve the velocity and accuracy of the operation. The setting of admittance parameters is the control boundary deduced by stability.

The relationships used to vary the damping for each of the Cartesian principal directions is [2]:

$$D_x(\dot{x}) = \max\{D_{\max}e^{-b|\dot{x}|}, D_{\min}\} \tag{14}$$

where D_{max} , D_{min} is Damping is the upper and lower bounds of damping. b Is the adjustment factor.

4 Experimental Results

In all experiments, the aubo-i5 cobot is equipped with an ATI six-dimensional force/torque sensor and a holding tool (See Fig. 3). The robot allowed an external real-time Linux PC to connect via 5 ms Ethernet. In this specific application, PC is used to close the admittance loop on the joint position loop of the industrial controller and record the force/torque and joint information.

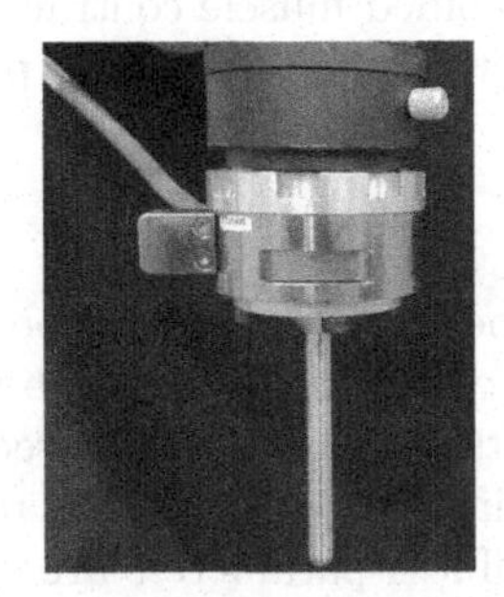

Fig. 3. The handling tool and ATI sensor.

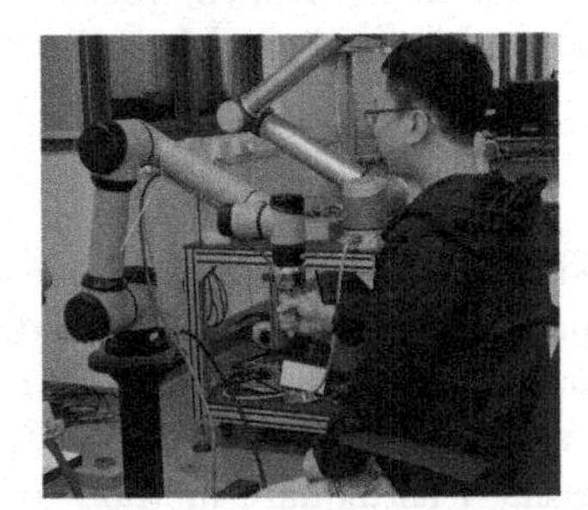

Fig. 4. Human arm impedance identification experiments.

4.1 Human Arm Impedance Identification

The experiment was organized as follows:

1. A chair is placed in front of the robot and always remains in the same position. The person sits there and holds the tool at the end. Try to ensure that only the arm is moved to keep the body still during the test (See Fig. 4).
2. Five different postures are used, corresponding to different end effector positions (see Fig. 7(a)).

Positions for human arm impedance identification, coordinates with respect to absolute robot frame. The point P_0 x,y,z coordinates are 548.87 mm, −121.50 mm and 239.69 mm, The x-axis is separated by 100 mm, and the y-axis is separated by 150 mm .

In order to identify the impedance of the human arm, the robot performs a very small movement in the X-direction and the Y-direction (See Fig. 5), recording the reaction of the human arm's forces in both directions (See Fig. 6). In

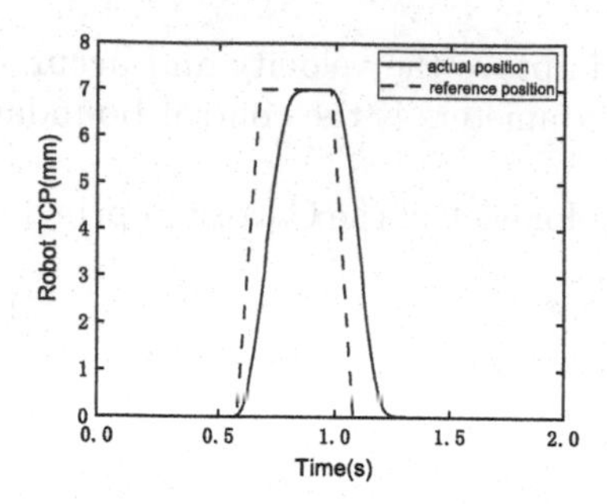

Fig. 5. Position profile in a cycle

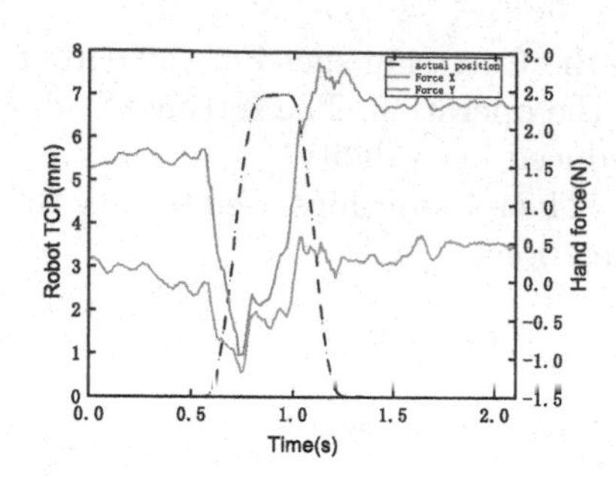

Fig. 6. Human arm reaction in a cycle

order to consider the actual performance, the trapezoidal planning is adopted, and the interval between the ascending and descending ends of the two ends is moved 7 mm after 0.1 s and kept for 0.4 s.

A complete test consists of 7 cycles, each cycle has a trapezoidal displacement in the x or y direction, and there is a 6-s rest between two cycles. The same test was performed twice, the first time the subject avoided muscle contraction, and the second time the subject was asked to contract as much muscle as possible.

4.2 Identification Results

Based on the position and force information, linear regression can be used to estimate the stiffness matrix of the arm. It can be seen from the Fig. 6 that the response of the human arm to the stimulus lasts for about half a second. In the experiments, as reported in [12], a 200-ms window was used. The results of using the least square method to identify the stiffness parameters are shown in the Table 1.

Table 1. Results of human arm stiffness identification.

Relaxation:				
Position	K_{xx}(N/m)	K_{xy}(N/m)	K_{yx}(N/m)	K_{yy}(N/m)
P_0	245.21	81.89	58.23	148.1
P_1	204.04	122.02	177.68	313.94
P_2	329.15	37.01	50.26	107.18
P_3	282.88	135.18	58.41	126.52
P_4	227.47	102.58	143.7	222.75
Contraction:				
Position	K_{xx}(N/m)	K_{xy}(N/m)	K_{yx}(N/m)	K_{yy}(N/m)
P_0	449.65	144.6	70.14	152.58
P_1	311.54	205.04	242.6	364.81
P_2	552.75	69.2	116.36	120.45
P_3	476.35	178.75	61.87	148.53
P_4	401.87	159.68	175.3	295.3

It can be seen from the result that moving from point P_1, P_0, P_2 along the Y direction, the K_{xy}, K_{yx} and K_{yy} components gradually decrease, and the K_{xx} component gradually increases. Moving from point P_3, P_0, P_4 along the X direction, the K_{xx} gradually decrease, the K_{yx} , K_{yy} component gradually increases, and the K_{xy} decrease first and then increase. Regardless of the trend of muscle tension and relaxation, the result is the same.

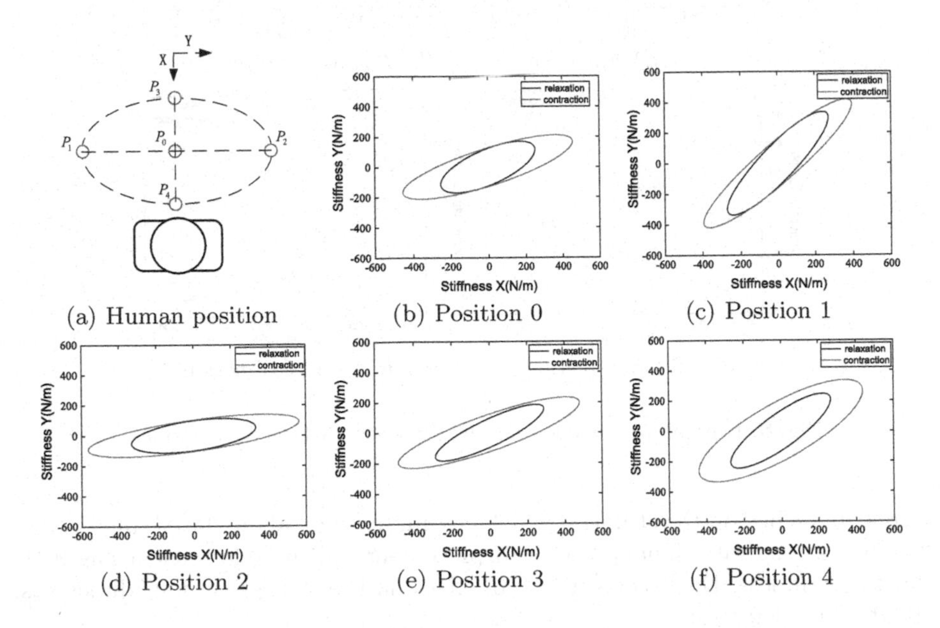

(a) Human position (b) Position 0 (c) Position 1

(d) Position 2 (e) Position 3 (f) Position 4

Fig. 7. Human position and stiffness ellipsoids.

Using the data in Table 1, the symmetric part of the stiffness matrix can be determined, and the Eq. (2) can be used to obtain the stiffness ellipsoid associated with each position (See Fig. 7(a)). Form Fig. 7(b)–(f) show that the direction and size of the stiffness ellipse in different postures of the arm are different. The trend of the stiffness ellipse is similar to that reported [13], However, due to individual differences, the results obtained are slightly different. From the result of the stiffness ellipse, the tension of the muscle is mainly affected in the direction of the long axis of the ellipse, that is, the direction where the arm resists the most. For the y direction, the major axis of the stiffness ellipse rotates clockwise, and the direction of the minor axis is always along the direction of the human arm. For the x direction, the shape of the stiffness ellipse is almost unchanged (Fig. 8).

Exploiting again data in Table 1 and the model introduced in Sect. 2, the impedance relationship of the human arm along a linear path in the X or Y direction can be derived. Figure 9 show the stiffness-position relation for a linear path along positions P_3, P_0, P_4 and P_1, P_0, P_2. It can be seen from the results that K_{xx} is almost a linear function in any direction, and the muscle tension has

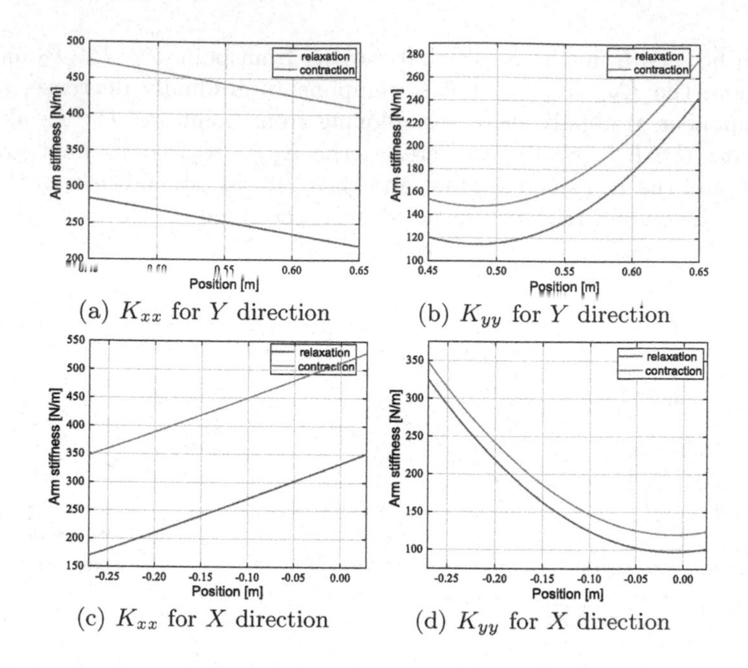

(a) K_{xx} for Y direction (b) K_{yy} for Y direction

(c) K_{xx} for X direction (d) K_{yy} for X direction

Fig. 8. Human arm stiffness model for liner path along positions

a great influence on the stiffness, which can increase by about 180 N/m at most. K_{yy} has a quadratic function with displacement and muscle tension has little effect on stiffness. Displacement has an effect on the stiffness in four directions, about 100 – 200 N/m.

It can be seen from the results that the stiffness of the arm in the X-direction changes significantly in both diastole and tension, while there is almost no change in the Y-direction. In the Y-direction, the human arm stiffness changes significantly with position, while the X-direction arm stiffness is not sensitive to position changes.

4.3 Variable Admittance Control Experiment

From the Sect. 3 stability conditions and the data in Fig. 7, Available admittance parameters, X direction $D_{\max} = 70$, $D_{\min} = 40$, Y direction $D_{\max} = 40$, $D_{\min} = 30$. The VRC strategy is:

$$\begin{aligned} D_x(\dot{x}) &= \max\{70e^{-7|\dot{x}|}, 40\} \\ D_y(\dot{y}) &= \max\{40e^{-7|\dot{y}|}, 30\} \end{aligned} \tag{15}$$

In this scheme, the process of driving the end effector through force F is shown in Fig. 2. Where $K = 200\,\text{N/m}$ corresponds to the middle of the arm stiffness during a task value. By comparing the force F applied by the human to the tip of the robot, the performance difference obtained by the selection of different parameters can be compared.

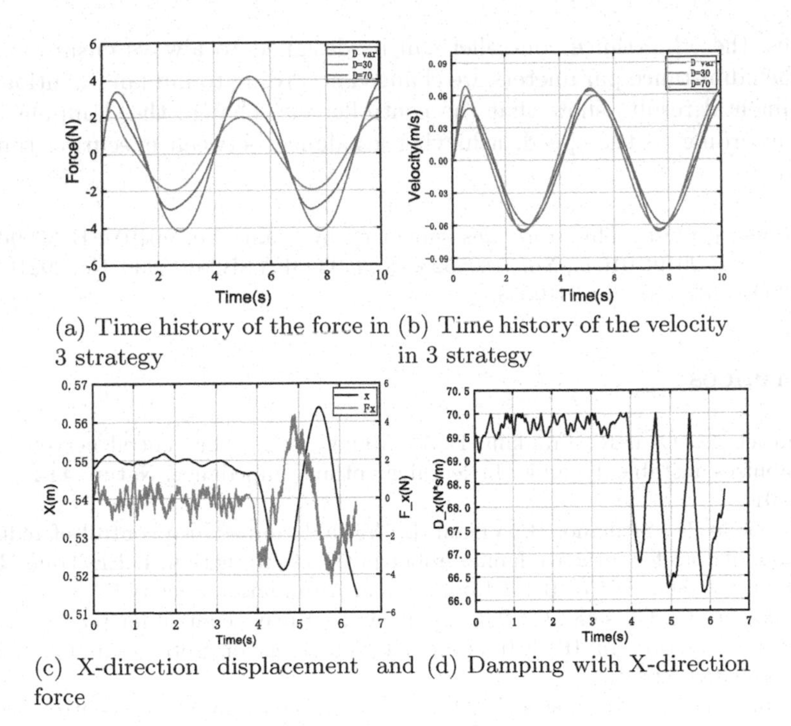

(a) Time history of the force in 3 strategy (b) Time history of the velocity in 3 strategy

(c) X-direction displacement and force (d) Damping with X-direction

Fig. 9. VRC experiment results.

In Fig. 9(a), the variable damping scheme is compared with the constant damping case set to the minimum value (D = 30 Ns/m) and maximum value (D = 70 Ns/m). It can be observed that the force required to move the end effector under the variable damping condition reaches an intermediate value compared with the force required under the minimum and maximum constant damping conditions. On the other hand in Fig. 9(b) when the speed is high, the velocity profile under variable damping is very close to the velocity profile under damping. For low speeds, the profile at maximum damping is closer.

Since the arm stiffness is mainly sensitive to the upward position of the X-direction, the experiment mainly carried out the movement in the X-direction. The relationship between force and displacement with respect to time is shown in the Fig. 9(c) and the virtual damping parameters can change accordingly with the speed (see Fig. 9(d)). Thereby reducing the power consumption during high-speed exercise and improving the execution time and efficiency of HRC.

5 Conclusions

This paper proposes a human-machine collaboration method for variable admittance control based on human body impedance. Due to the nonlinear simplified model of human arm impedance, its parameters can be easily identified by least squares, the admittance controller can be designed to always ensure stability, and the admittance parameters are changed according to human intentions. The experimental results show that the controller can change the damping in real time according to the speed, achieving a balance between execution time and accuracy.

Acknowledgment. This work was supported by grant: No. 2019YFB1309900, No. 92048201, No. 2019C01043, No. 51805523, No. 174433KYSB20190036, No. 2021C01067, No. U20A20282, No. 2018B10058.

References

1. Burdet, E., Osu, R., Franklin, D.W., Milner, T.E.: The central nervous system stabilizes unstable dynamics by learning optimal impedance. Nature **414**, 446–449 (2001)
2. Ficuciello, F., Siciliano, B., Villani, L.: Variable impedance control of redundant manipulators for intuitive human-robot physical interaction. IEEE Trans. Robot. **31**(4), 850–863 (2015). IEEE Robotics and Automation Society
3. Duchaine, V., Gosselin, C.: Safe, stable and intuitive control for physical human-robot interaction. In: IEEE International Conference on Robotics and Automation 2009. ICRA (2009)
4. Tsumugiwa, T., Yokogawa, R., Hara, K.: Variable impedance control based on estimation of human arm stiffness for human-robot cooperative calligraphic task. In: IEEE International Conference on Robotics & Automation (2017)
5. Bascetta, L., Ferretti, G.: Ensuring safety in hands-on control through stability analysis of the human-robot interaction. Robot. Comput. Integrated Manufact. **57**(JUN), 197–212 (2019)
6. Patel, H., Oneill, G., Artemiadis, P.: On the effect of muscular cocontraction on the 3-d human arm impedance. IEEE Trans. Biomed. Eng. **61**(10), 2602–2608 (2014)
7. Tahara, K., Kino, H.: Iterative learning control for a redundant musculoskeletal arm: acquisition of adequate internal force. In: IEEE/RSJ International Conference on Intelligent Robots and Systems (2010)
8. Wang, S., Zuo, G., Xu, J., Zheng, H.: Human hand impedance characteristics during reaching movements. In: International Conference on Biomedical Engineering and Informatics (2011)
9. Woo, H.S., Lee, D.Y.: Exploitation of the impedance and characteristics of the human arm in the design of haptic interfaces. IEEE Trans. Ind. Electron. **58**(8), 3221–3233 (2011)
10. Slotine, J.-J.E: Applied nonlinear control (2004)
11. Lecours, A., Mayer-St-Onge, B., Gosselin, C.: Variable admittance control of a four-degree-of-freedom intelligent assist device. In: IEEE International Conference on Robotics and Automation (2012)
12. Erden, M.S., Billard, A.: End-point impedance measurements at human hand during interactive manual welding with robot. In: IEEE International Conference on Robotics and Automation (2014)
13. Perreault, E.J., Kirsch, R.F., Crago, P.E.: Effects of voluntary force generation on the elastic components of endpoint stiffness. Exper. Brain Res. **141**(3), 312–323 (2001)

Event-Driven Collision-Free Path Planning for Cooperative Robots in Dynamic Environment

Zhiqiang Wang, Jinzhu Peng(✉), Shuai Ding, Mengchao Dong, and Bo Chen

School of Electrical Engineering, Zhengzhou University, Zhengzhou 450001, China
jzpeng@zzu.edu.cn

Abstract. This paper presents an event-driven safe collision-free path planning method for robotic manipulator in human-robot cooperation. To meet the rapidity requirement of real-time robotic systems, the event-driven is introduced, and the collision prediction based on kinematics is used to trigger the rapid-exploring random tree (RRT) planner, while the quintic polynomial path planner is used when the event is not triggered. The fast planning and dynamical obstacle avoidance can be then achieved by the combination of quintic polynomial and RRT planner. By introducing the event-driven method, the safe collision-free path planning can be abstracted and standardized in human-robot cooperation. Finally, the simulation results show that the effectiveness of the proposed event-driven quintic-RRT path planning method.

Keywords: Event-driven · Path planning · Cooperative robots

1 Introduction

Human-robot cooperation is a promising field, since it makes the best use of the human intelligence and the robot labor. In the process of human and robotic manipulator cooperating with each other, dynamic obstacle avoidance is a very important topic when human and robots perform different tasks in sharing workspace. In general, the safe human-robot cooperation can be ensured provided that the dynamic obstacle avoidance can be achieved.

The event-driven method is an ideal solution to achieve dynamic obstacle avoidance. The event-driven method was presented in [1] and has been applied to various fields. It generally has the benefits of reducing calculation or communication burden [2,3]. Inspired by [1], Dimarogonas *et al.* [4] discussed both centralized and distributed event-driven methods, and concluded that event-driven methods require fewer state updates in control systems. Other than that, the periodic event-driven method was presented in [5,6]. In [7], event-driven

Supported by the National Natural Science Foundation of China (Grant No. 61773351), and the Program for Science & Technology Innovation Talents in Universities of Henan Province (Grant No. 20HASTIT031).

X.-J. Liu et al. (Eds.): ICIRA 2021, LNAI 13014, pp. 491–499, 2021.
https://doi.org/10.1007/978-3-030-89098-8_46

method was applied to multi-threads computing, resulting in reducing the system complexity and a more robust system. The event-driven control was also utilized to reduce the communication among the controller, actuators and sensors [8]. Zhang *et al.* [9] proposed an event-driven guaranteed cost control method combined with reinforcement learning, to deal with actuator faults of nonlinear systems. Similarly, Mu *et al.* [10] proposed an event-driven integral reinforcement learning algorithm by using a novel dynamic event triggering scheme, and partially unknown nonlinear systems can be then controlled. In [11], the event generator for robust control problems was proposed and the uncertain systems can be then asymptotically stabilized.

Taken together, after introducing the event-driven method, the system will only respond to the events, and not execute corresponding actions when events are not triggered, so that the computing cost of the system will be reduced. In this paper, the event-driven method is applied to obtain the safe collision-free path for robotic manipulator, so as to create a safe workspace in the process of human-robot cooperation. The event trigger is defined as the predicted collision. If the event is triggered, the advanced dynamic obstacle avoidance planner is invoked; if the event is not triggered, the trivial planner is invoked. The methodology has been used in some similar works [12,13], but is not formalized as an event-driven method. This kind of path planning problem can be unified by adopting the formalized event-driven methodology.

For applying the event-driven method, the trajectory of human arm needs to be predicted, so that more information and time are reserved for the planning module, and the accuracy and rapidity of the system are improved. The prediction methods of human arm generally takes the input as the position of human arm, and some methods also need to input the speed and acceleration. There exists lots of human arm prediction algorithms such as kinematics-based modeling [14], hidden Markov model [15], Gaussian probability model [16], neural network [17], etc. By utilizing the human arm trajectory prediction algorithms, the possible collision points can be predicted.

After obtaining the collision prediction, the advanced path planning method can be used to avoid possible collision. At present, the path planning algorithms that can cope with dynamic environment including artificial potential field [18], probabilistic roadmap [19], rapid-exploring random tree [20] (RRT), optimization algorithm [21], learning-based algorithm [22], dynamic movement primitives [23], and so on. The RRT ensures rapidity, so it is usually used on real-time systems. However, using the RRT alone is not efficient, since there is no necessity to use it in non-collision areas. Therefore, in non-collision areas, polynomial interpolation, spline interpolation and other algorithms can be used to generate the path, which can further reduce computation cost and generate a smooth path. The quintic polynomial interpolation algorithm is used with RRT in this paper.

Finally, the simulation experiment scheme in this paper demonstrates that the proposed event-driven quintic-RRT path planning method can be used to plan the dynamic collision avoidance path in human-robot cooperation. By comparing the paths planned before and after using the proposed method, the advantages of proposed method are shown. The proposed method can be easily improved by using better prediction or path planning methods.

2 Event-Driven Quintic-RRT Path Planning

2.1 Quintic Polynomial

Among the aforementioned path planning algorithms in Sect. 1, quintic polynomial is one of the most commonly used path planning algorithms, which has the advantages of less computation, more flexibility, and smooth path. After specifying the position, velocity and acceleration of the starting point and the target point, a smooth path can be then derived. Path planning for the robotic manipulator is usually implemented in joint space, so the quintic polynomial algorithm can be denoted as,

$$\theta(t) = a_0 + a_1 t + a_2 t^2 + a_3 t^3 + a_4 t^4 + a_5 t^5 \tag{1}$$

where θ is the joint angle, t is time or the number of waypoints, and $a_{0,1,\ldots,5}$ are the coefficients.

After giving the position, velocity and acceleration information of the starting point and the target point, one can substitute these information into the equation and get the general solution of each coefficient, which is,

$$\begin{cases} a_0 = \theta_0 \\ a_1 = \dot{\theta}_0 \\ a_2 = \frac{\ddot{\theta}_0}{2} \\ a_3 = \frac{20\theta_f - 20\theta_0 - (8\dot{\theta}_f + 12\dot{\theta}_0)t_f - (3\ddot{\theta}_0 - \ddot{\theta}_f)t_f^2}{2t_f^3} \\ a_4 = \frac{30\theta_0 - 30\theta_f + (14\dot{\theta}_f + 16\dot{\theta}_0)t_f + (3\ddot{\theta}_0 - \ddot{\theta}_f)t_f^2}{2t_f^4} \\ a_5 = \frac{12\theta_f - 12\theta_0 - (6\dot{\theta}_f + 6\dot{\theta}_0)t_f - (\ddot{\theta}_0 - \ddot{\theta}_f)t_f^2}{2t_f^5} \end{cases} \tag{2}$$

where subscript 0 and f stand for starting and target point.

The disadvantage of quintic polynomial algorithm is that it is necessary to manually set the position, velocity and acceleration of the starting point and the target point, which can not be changed during execution, resulting in the inability to respond to environmental changes in time. Therefore, in human-robot cooperation tasks, an algorithm that can quickly respond to changes in environment and human operators is needed.

2.2 RRT

RRT is suitable for quick path searching in high-dimensional space, since it conducts sampling in all dimensions, and does not need to do other calculations on the environment except collision detection. The features of RRT make it usually be used to solve the path planning problem of multi-degree-of-freedom manipulator. The routine of RRT is shown in Algorithm 1.

Algorithm 1: RRT

```
1  Initialize the initial node n_init, the maximum iteration number N, the step
   length Δn, and the goal position G;
2  Q.init(n_init);
3  for k = 1 : N do
4      n_rand=rand();                          // Determin a random position
5      n_near=NEAREST(n_rand,G);               // Find the nearest node
6      n_new=NEW(n_near,Δn); /* Determin the new collision-free node that
       along the random position */
7      Q.add(n_new);                           // Add the new node
8      if || Q(end) - G|| < Δn then
9          Q.add(G);
10         break;
11     end
12 end
13 return Q
```

2.3 Event-Driven Method

The core idea of the event-driven method is to dispose the events only when events are triggered. For example, the event-driven control method is to conduct communication, calculation and action execution when the control error can not be tolerated, while the traditional control method conducts the executions from the beginning to the end, which increases system overhead.

Similarly, the event-driven path planning method can be described as: when events are not triggered, trivial planners are used; when events are triggered, advanced planners are invoked. Noted that there may be multiple triggers and planners to deal with different scenes. In this paper, the event is defined as the predicted collision. It is beneficial to introduce the event-driven ideology into path planning, since the planning problem can be then abstracted and studied systematically in this way.

2.4 Collision Prediction Based on Kinematics

In the process of human-robot cooperation, human actions bring great uncertainty, so it is necessary to predict the trajectory of human arm, giving more information and reserving more time for the planning module, and improving the accuracy and rapidity of the system. Multiple human arm prediction methods can be utilized such as kinematics-based modeling, hidden Markov model, Gaussian probability model, neural network, etc. In this paper, the kinematics-based prediction method is adopted: assume that the human arm moves at a constant speed along a straight line, then calculate the predicted arm position by using the current position and position at the last timestep. Theoretically, the proposed method is applicable to any other prediction algorithm and any kind of arm motion.

After combining quintic polynomial and RRT algorithms by introducing event-driven method to predict the collision, one can use the fast and simple quintic polynomial in the area where no collision predicted, and use the RRT in the area where the collision is predicted. This method satisfies the obstacle avoidance requirements and the real-time requirements of the system. The detailed system block diagram and pseudocode for the proposed method is shown in the Fig. 1 and Algorithm 2, respectively. Noted that dotted line represents that the event will not be always triggered.

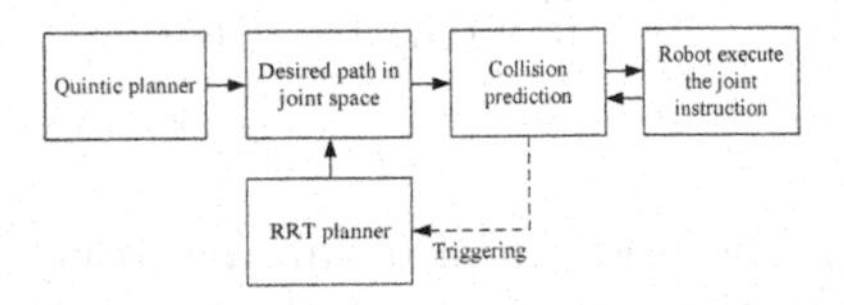

Fig. 1. The system block diagram of event-driven collision-free path planning method for cooperative robots in dynamic environment.

Algorithm 2: Event-driven quintic-RRT path planning

---

1 Initialize the initial and final joint angles θ_0 and θ_f;
2 Plan the joint angles Q by using the quintic planner;
3 $\theta = null$;
4 **while** *not* $\theta == \theta_f$ **do**
5   Detect event by using kinematics-based method;
6   **if** *Event is triggered* **then**
7   Change the corresponding joint angles Q by using the RRT planner;
8   **end**
9   θ = next joint angles in Q;
10   Robotic manipulator move to θ;
11 **end**

---

3 Simulation

Redundant manipulator has good flexibility and can avoid singularity to some extent, so it is more suitable for human-robot cooperation tasks. This paper adopts a 7-DOF redundant robotic manipulator for simulation, and its DH parameters are shown in Table 1.

The physical collision model is also desired. Each link of the manipulator is defined as a cylinder with a radius of 0.2 m. This is because, except the radius of the manipulator links, an additional expansion distance is needed to eliminate the modeling error and keep a sufficient safe distance from obstacles. To simplify the calculation, the hand and forearm of human body are modeled as a cylinder

Table 1. The DH parameters of the 7-DOF redundant robotic manipulator.

Link	θ	d	a	α
1	0	0.360	0	$-\frac{\pi}{2}$
2	0	0	0	$\frac{\pi}{2}$
3	0	0.420	0	$-\frac{\pi}{2}$
4	0	0	0	$\frac{\pi}{2}$
5	0	0	0.400	$-\frac{\pi}{2}$
6	0	0	0	$\frac{\pi}{2}$
7	0	0.126	0	0

with a radius of 0.04 m and a length of 0.4 m, and other parts of human body are omitted. By defining these physical collision models, the proposed method can be then verified in the simulation environment.

The procedure of the proposed method can be described as follows. First, the starting point and target point of the robotic manipulator are given in the joint space, with zero velocity and zero acceleration. Then the initial path is planned with trivial quintic polynomial algorithm, and the robotic manipulator executes the planned path. During the execution of the robotic manipulator, the collision prediction module is invoked to determine when the collision will occur. If no measures are taken, collisions may occur at several consecutive waypoints. If there is at least one collision point, the collision event will be triggered, and then the advanced RRT planner is invoked which is used to plan the obstacle avoidance path that from the first collision point to the last collision point. In other words, a segment of quintic polynomial path will be replaced by a series of new waypoints given by RRT. The end point of the path planned by RRT is on the path planned by quintic polynomial, and the collision event will not be triggered at that moment. After that, the robotic manipulator will return to the path planned by the quintic polynomial, and continue to predict collision until it reaches the target point. Finally, the robotic manipulator will stop moving, and the online operation stage ends.

Based on the above definition, the simulation results are shown. In Fig. 2, the desired target position can be reached by using quintic polynomial, RRT and quintic-RRT algorithms. A collision path with 60 waypoints is planned by the quintic polynomial algorithm, and a collision-free path consists of 104 waypoints is planned by RRT without event-driven method. The collision-free path planned by the proposed event-driven quintic-RRT method consists of 90 waypoints. It can be seen that the robotic manipulator changed the original path at the 31st waypoint. At the 77th waypoints, the path planned by RRT ends, and then the path planned by quintic polynomial is adopted. The event trigger status is shown in Fig. 3. The quintic polynomial path is smooth and the shortest. However, according to Fig. 3, the quintic polynomial path is not collision-free. In contrast, the RRT path is the longest but collision-free. The event-driven quintic-RRT path can achieve to combine the merits of the quintic polynomial path and the

RRT path, which are avoiding the dynamic obstacle and minimizing the path length.

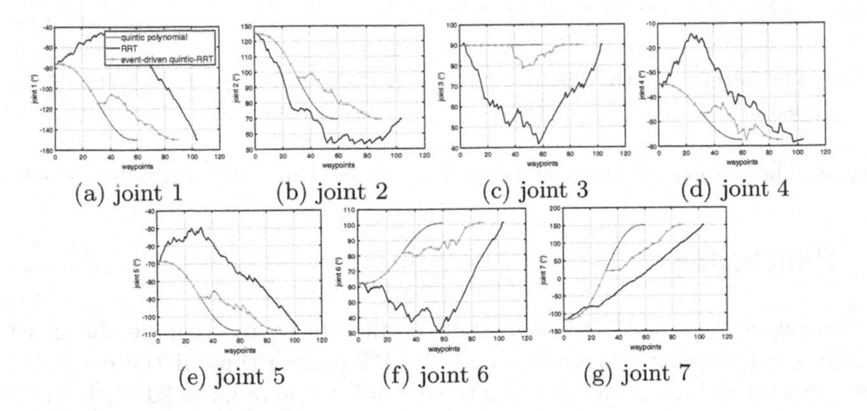

(a) joint 1 (b) joint 2 (c) joint 3 (d) joint 4

(e) joint 5 (f) joint 6 (g) joint 7

Fig. 2. Waypoints planned by the three algorithms.

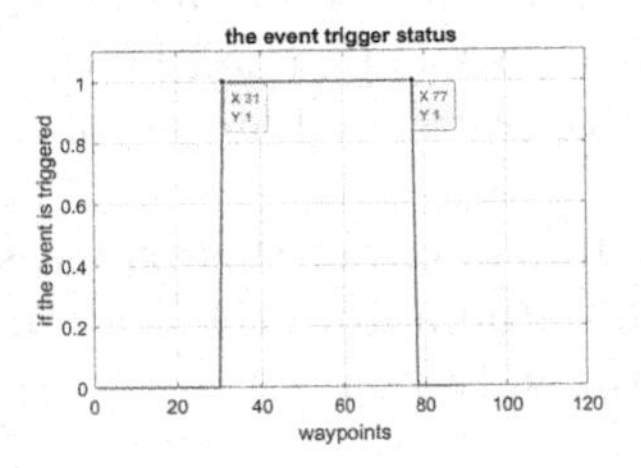

Fig. 3. The status of event trigger.

The initial positions of human arm (light yellow) and robotic manipulator (red and blue) are shown in the Fig. 4(a). The human arm will move downward, and if the robotic manipulator does not respond to the arm, collision will occur, see Fig. 4(b). As the proposed event-driven quintic-RRT path planning method is used, the event is triggered after the collision is predicted, the RRT planner is invoked, and a path above the human arm is planned (Fig. 4(c)). Finally, the robotic manipulator arrive the target point (Fig. 4(d)). The quintic polynomial algorithm is efficient, but the dynamic obstacles can not be avoided, which may be harmful to human operator. The RRT without event-driven can achieve to avoid obstacle, but the planned path is unnecessarily long, which reduces the efficiency of the robotic manipulator. The proposed event-driven quintic-RRT path planning method can achieve to avoid dynamic obstacles by utilizing RRT, and adopt the path planned by quintic polynomial algorithm where the event is not triggered, which guaranteeing human safety and efficiency.

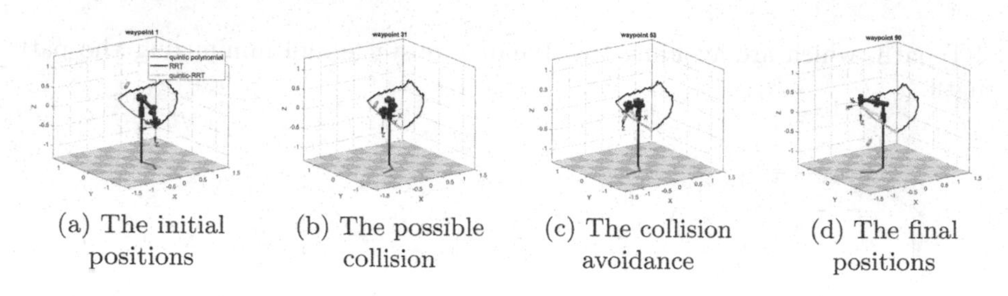

(a) The initial positions (b) The possible collision (c) The collision avoidance (d) The final positions

Fig. 4. The process of human-robot cooperation in simulation. (Color figure online)

4 Conclusion

In this paper, the event-driven method is introduced to combine the merit of the quintic polynomial algorithm and the RRT algorithm. If the event is not triggered, the quintic path is used; if the event is triggered, the RRT planner will be invoked to generate a feasible path. The proposed event-driven quintic-RRT path planning method can achieve to solve the problem of dynamic obstacle avoidance of robotic manipulator in the process of human-robot cooperation. In the simulation experiment of a 7-DOF robotic manipulator, it can be seen that an effective collision-free path is planned by the proposed method, and the potential threat to human operator in human-robot cooperation can be then avoided.

It is easy to extend the framework of this paper, such as applying advanced collision prediction algorithms to improve the performance, implementing RRT-connect to find shorter path, and so on.

References

1. Tabuada, P.: Event-triggered real-time scheduling of stabilizing control tasks. IEEE Trans. Autom. Control **52**(9), 1680–1685 (2007)
2. Zhang, X.M., Han, Q.L.: Event-based H_∞ filtering for sampled-data systems. Automatica **51**, 55–69 (2015)
3. Heemels, W.P.M.H., Johansson, K.H., Tabuada, P.: An introduction to event-triggered and self-triggered control. In: 2012 IEEE 51st IEEE Conference on Decision and Control (CDC), pp. 3270–3285 (2012)
4. Dimarogonas, D.V., Frazzoli, E., Johansson, K.H.: Distributed event-triggered control for multi-agent systems. IEEE Trans. Autom. Control **57**(5), 1291–1297 (2012)
5. Heemels, W.P.M.H., Donkers, M.C.F., Teel, A.R.: Periodic event-triggered control for linear systems. IEEE Trans. Autom. Control **58**(4), 847–861 (2013)
6. Zhang, X., Han, Q., Zhang, B.: An overview and deep investigation on sampled-data-based event-triggered control and filtering for networked systems. IEEE Trans. Ind. Inform. **13**(1), 4–16 (2017)
7. Dabek, F., Zeldovich, N., Kaashoek, F., Mazières, D., Morris, R.: Event-driven programming for robust software. In: Proceedings of the 10th Workshop on ACM SIGOPS European Workshop: Beyond the PC - EW10, p. 186. ACM Press, Saint-Emilion, France (2002)

8. Lunze, J., Lehmann, D.: A state-feedback approach to event-based control. Automatica **46**(1), 211–215 (2010)
9. Zhang, H., Liang, Y., Su, H., Liu, C.: Event-driven guaranteed cost control design for nonlinear systems with actuator faults via reinforcement learning algorithm. IEEE Trans. Syst. Man Cybern. Syst. **50**(11), 4135–4150 (2020)
10. Mu, C., Wang, K., Qiu, T.: Dynamic event-triggering neural learning control for partially unknown nonlinear systems. IEEE Trans. Cybern. 1–14 (2020)
11. Zhang, Q., Zhao, D., Wang, D.: Event-based robust control for uncertain nonlinear systems using adaptive dynamic programming. IEEE Trans. Neural Netw. Learn. Syst. **29**(1), 37–50 (2018)
12. Duguleana, M., Barbuceanu, F.G., Teirelbar, A., Mogan, G.: Obstacle avoidance of redundant manipulators using neural networks based reinforcement learning. Robot. Comput. Integrated Manufact. **28**(2), 132–146 (2011)
13. Wang, Q., Wang, Z., Shuai, M.: Trajectory planning for a 6-DoF manipulator used for orthopaedic surgery. Int. J. Intell. Robot. Appl. **4**(1), 82–94 (2020)
14. Weitschat, R., Ehrensperger, J., Maier, M., Aschemann, H.: Safe and efficient human-robot collaboration part I: estimation of human arm motions. In: 2018 IEEE International Conference on Robotics and Automation (ICRA), pp. 1993–1999. IEEE, Brisbane, QLD (2018)
15. Liu, H., Wang, L.: Human motion prediction for human-robot collaboration. J. Manufact. Syst. **44**, 287–294 (2017)
16. Mainprice, J., Berenson, D.: Human-robot collaborative manipulation planning using early prediction of human motion. In: 2013 IEEE/RSJ International Conference on Intelligent Robots and Systems, pp. 299–306. IEEE, Tokyo (2013)
17. Cheng, Y., Zhao, W., Liu, C., Tomizuka, M.: Human motion prediction using semi-adaptable neural networks. arXiv:1810.00781 (2019)
18. Khatib, O.: Real-time obstacle avoidance for manipulators and mobile robots. In: Proceedings of1985 IEEE International Conference on Robotics and Automation, vol. 2, pp. 500–505. Institute of Electrical and Electronics Engineers, St. Louis, MO, USA (1985)
19. Rodrigues, R.T., Basiri, M., Aguiar, A.P., Miraldo, P.: Low-level active visual navigation: increasing robustness of vision-based localization using potential fields. IEEE Robot. Autom. Lett. **3**(3), 2079–2086 (2018)
20. Lavalle, S.M.: Rapidly-exploring random trees: A new tool for path planning. Annual Res. Rep. **1**(1), 1–4 (1998). Department of Computer Science
21. López, J., Sanchez-Vilariño, P., Cacho, M.D., Guillén, E.L.: Obstacle avoidance in dynamic environments based on velocity space optimization. Robot. Auton. Syst. **131**, 103569 (2020)
22. Hoppe, S., Lou, Z., Hennes, D., Toussaint, M.: Planning approximate exploration trajectories for model-free reinforcement learning in contact-rich manipulation. IEEE Robot. Autom. Lett. **4**(4), 4042–4047 (2019)
23. Ijspeert, A.J., Nakanishi, J., Hoffmann, H., Pastor, P., Schaal, S.: Dynamical movement primitives: learning attractor models for motor behaviors. Neural Comput. **25**(2), 328–373 (2013)

The Analysis of Concurrent-Task Operation Ability: Peripheral-Visual-Guided Grasp Performance Under the Gaze

Yuan Liu[1], Wenxuan Zhang[1], Bo Zeng[2,3], Kuo Zhang[1](✉), Qian Cheng[1], and Dong Ming[1]

[1] Tianjin University, Tianjin 300072, China
ryanliu@tju.edu.cn
[2] Beijing Institute of Precision Mechatronics and Controls, Beijing 100076, China
[3] Laboratory of Aerospace Servo Actuation and Transmission, Beijing 100076, China

Abstract. Gazing at the target is a necessary part of human-machine interaction, providing the basic scene information, such as watching videos, handling documents, manufacturing and assembling workpieces. In these situations, for accomplishing a concurrent-task (such as getting a drink, taking a file, picking up a tool), a peripheral-visual-guided grasp is needed. Compared with the natural grasp without visual restrictions, the peripheral-visual-guiding will limit the grasp performance. However, few studies focus on the quantitative investigation of human concurrent-task operation ability. In this paper, we built a general concurrent-task scene: peripheral-visual-guided grasp under gazing at a target. Ten voluntary participants were required to keep their eyes gazing at the screen target and accomplish the concurrent grasp task in three restrictions: (1) keep both head and trunk still, (2) keep trunk still while the head can turn, (3) both trunk and head can turn. The maximum range of object placement able to accomplish accurate grasp, is defined as the workspace in each restriction. The rotation angle of head and trunk, and grasped object position, were collected using IMU and VICON. Three precision grasp workspace were obtained within the range of 44.631 ± 1.348°, 67.315 ± 2.075°, and 80.835 ± 4.360°, corresponding to each restriction. Besides, the workspace division results were further verified by the head and trunk peak rotation angles. The results can be applied to provide a design guideline of assistive devices to enhance human concurrent-task operation ability.

Keywords: Workspace · Eye-hand coordination · Peripheral vision · Human enhancement

1 Introduction

"Man owes a good part of his intelligence to his hands" Jean Piaget [1]. In daily life, the application of machinery and complex tools has become ubiquitous, and most activities of daily living (ADL) require the control of both hands in collaboration with various devices to accomplish. In addition to haptic control, vision plays a very important role in

X.-J. Liu et al. (Eds.): ICIRA 2021, LNAI 13014, pp. 500–509, 2021.
https://doi.org/10.1007/978-3-030-89098-8_47

the motion control of the hand. Thus, eye-hand coordination has been widely investigated for a variety of visuomotor tasks. Eye motions usually start about 150 ms before hand movements and can guide the limb to move [2, 3]. This behavioral pattern has also been proved in more complex movements. Such as reading [4], driving [5, 6], etc. Eye motions not only guide the hand in time but also provide direct and useful key information to control hand movements.

The definition of concurrent-task is to complete multiple tasks at the same time, vision and manual tasks are two important parts of concurrent-task. In human-machine interaction, gazing at the target is a necessary part of the visual task and providing the basic scene information. Such as watching videos, playing computer games, handling documents, manufacturing and assembling workpieces. In these situations, for accomplishing a concurrent-task (such as getting a drink, taking a file, picking up a tool), a peripheral-visual-guided grasp is needed to provide motion information. In this case, central vision is fixed on a specific spot, compared with the natural grasp without visual restrictions, the motion performance of the limbs will be greatly impaired and eye-hand coordination will be reduced [7–9], it has been proven in vehicle steering tasks. When subjects were denied eye movements and looked straight ahead (center gaze), completion time increased by 30.3 s, from 249.7 s to 280 s, the wheel was turned further away from center less of the time, driving trajectory deviates from the normal significantly [10, 11]. In Addition to visual restriction, the torso could also be restricted during concurrent-tasks, such as assembling large aircraft interior workpieces, where maintenance workers have difficulty turning around [12]. Thus, some investigations analyze three torso restriction conditions in the peripheral-visual-guided grasp: (1) head movement, (2) arm movement, keep the head still, (3) head and arm movement, verify that torso restrictions have an impact on the movement times, and grip aperture [13]. To the best of our knowledge, although efforts have been made to analyze the motion performance under the peripheral-visual-guided grasp, few have deterministically division of the workspace.

The contribution of this paper is to divide the workspace into three precision areas with the restrict of the torso under gazing. Compared to previous research, we visualized the workspace of its capabilities. Adding assistive devices (for example wearable robots) in the non-precision area could enhance the concurrent-task operation ability. In addition, it could be applied to find the workspace of ability deficit for the patient with upper-limb impairment by comparing with the healthy people. In the first section, we describe the methodology of the experiment, introduces the participants, equipment, procedure, and analyzing the data. The second section acquires the quantitative division results of the workspace. The third section analyzes the results specifically, and the last section evaluates the research design, procedure, results, and limitations.

2 Method

2.1 Participants

Ten participants completed the experiment (6 male and 4 female, mean age 21.4 years), recruited at Tianjin University. Participants had the normal or corrected-to-normal vision and did not have any known neuromuscular disorders. As confirmed by a handedness questionnaire (Oldfield 1971). Their arms ranged in length from 65.3 to 76.7 cm, with an average length of 70.9 cm, measuring the distance from the acromion to the stylion radiale. All participants were right-handed. They were naive to the purpose of the experiment and were provided with a subsistence allowance.

2.2 Apparatus

Participants sit on an adjustable chair with a height range of 48–62 cm. In front of them is a table with a height of 75 cm. This is according to previous research, in such case the upper limbs were located in the most dexterous area [14]. The object is a rectangular block made of PVC (1.5 × 3 × 1 cm). The table is labeled to facilitate the placement of objects, the moving angle of the object for each grasp is the angle between each of the two labeled lines. When the moving angle of the object is 5° or more, there will be the obvious rotation of the trunk, to dividing each area more precisely, we set the moving angle of the object at 2°. Visual stimulate was presented on a 24 cm flat-screen monitor, which was placed 80 cm in front of the object. Prime stimulate was a picture with a fixation cross (see Fig. 1).

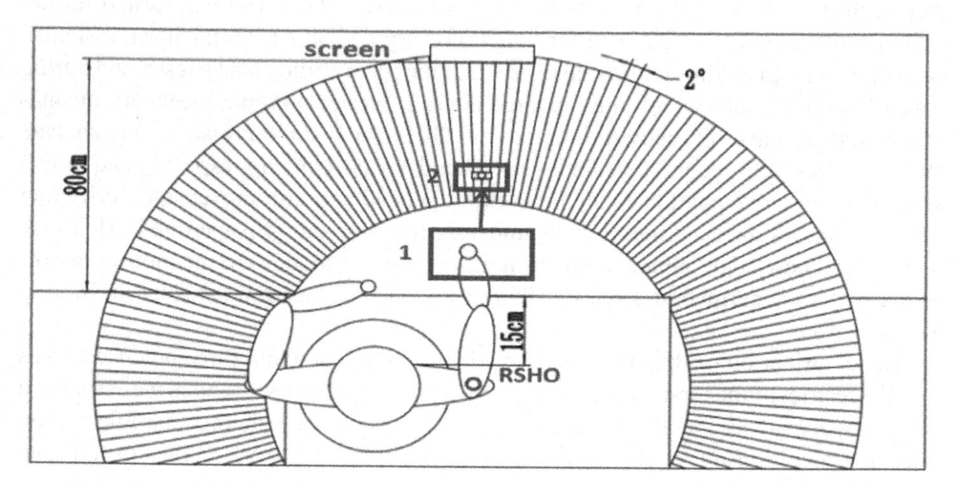

Fig. 1. Table layout for the experiment, indicating the hand start and object locations.

An optical motion capture system (VICON Motion Systems, Oxford, UK), consisting of sixteen Bonita cameras with a temporal and spatial resolution of 200 Hz and 1 mm, respectively, was used to record kinematic data. Eight retro-reflective markers (10 mm in diameter) were placed on the right shoulder (RSHO), the right elbow (RELB), the right medial epicondyle (RMEP), the styloid process of the radius (WRT), the styloid process of the ulna (WRP), the third right finger (RFIN), the thumbnail (TB) and the index fingernail (IDX) of the upper limb and hand. One retro-reflective marker was placed on the object center (OC). Other markers are intended to create a more accurate upper-limb model. Five inertial measurement units (IMUs) on the subject's head, upper spine, right upper arm, forearm, and pelvis (see Fig. 2). Establish the upper limb position and record the head and trunk rotation angles.

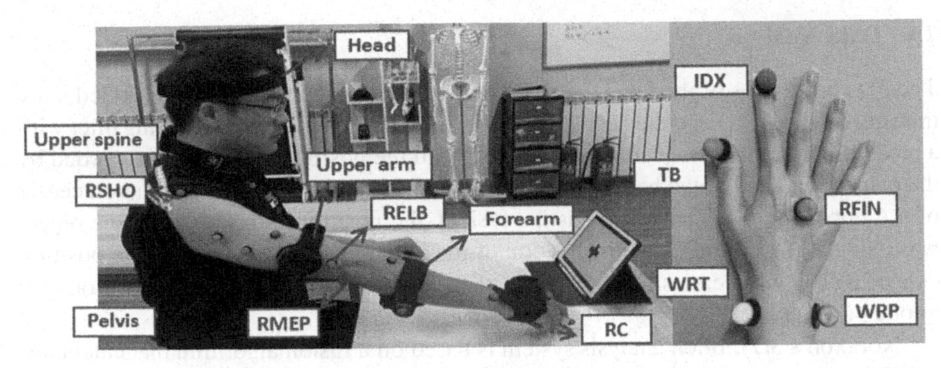

Fig. 2. Illustration of marker and IMUs placement.

2.3 Procedure

The participants grasped the object from the initial position to the maximum reachable range at the height of the chair of 48 cm, 55 cm, and 62 cm. The initial position is on the vertical midline of the table, the distance is based on the maximum grasp distance for different participants without elbow flexion. They sat with the waist straight, the position of the RSHO coincided with the table midline, and the vertical distance of the chest from the edge of the table was 15 cm. Participants kept still for 5 s to calibrate the system. When starting, the participant kept the head and trunk square to the table with the right hand in the position of box 1 in Fig. 1. The participant should grasp the object with their thumb and index finger after hearing the sound of "Started", then lift the object, put it back on the table, and withdraw hand to the initial position. Kept the eye gazed on the fixation cross in the center of the screen and keep the elbow no flexion. At the end of each grasp, the position of the object moves on a circle, axis by the center of the arm's rotation, with the maximum grasp distance being the radius, with no requirement for the speed of the grasp or the position where the object is put back. To prevent the effect of habituation, after the experiment at the same chair height was completed, the participants took a 10-min break before adjusting the chair height for the next experiment. The participants performed the task in three types: (1) keep both head and trunk still, (2) keep trunk still while the head can turn, (3) both trunk and head can turn. These three types of tasks cover most of the concurrent-tasks types under the gaze in daily life and are designated as the three areas, and when the first area cannot complete the grasping task, it turns to the second area, and likewise, turns to the third area. Successful completion of the grasping task was defined as a precise grasp of the object by the participants whose TB, IDX, and OC should on the same level and reach that position in one go. The experiment was considered to be finished until the participant could not complete a precise grasp in the case of the trunk and head can turn. The experiment took approximately 90 min to complete.

2.4 Data Analysis

The 3D coordinates of the retro-reflective markers were reconstructed and labeled. Any missing data were interpolated using a cubic spline and filtered using a Woltring filter with a mean predicted mean square error of 5 mm^2 and the position information recorded by the camera was calibrated before the experiment (Vicon Nexus 2.6). The rotation center of the upper limb was calculated as the RSHO, the position information of the object was calculated as the OC. Normalization of the location of RSHO, record the position data of RSHO in the cartesian coordinates of each participant and unify the position data to one point.

Noraxon's 3D motion analysis system is based on a fusion algorithm that calculates the 3D rotation angle of each sensor in the IMU in absolute space (Noraxon MR3 16.3). The movement angle is recorded by the head sensor, the pelvis sensor, the upper arm sensor, and the forearm sensor, normalization of angle information. A synchronization box is used to connect Noraxon to Vicon, and when Vicon starts recording, Noraxon generates step signals synchronously to facilitate the extraction of angular information about head and trunk rotation during each grasp.

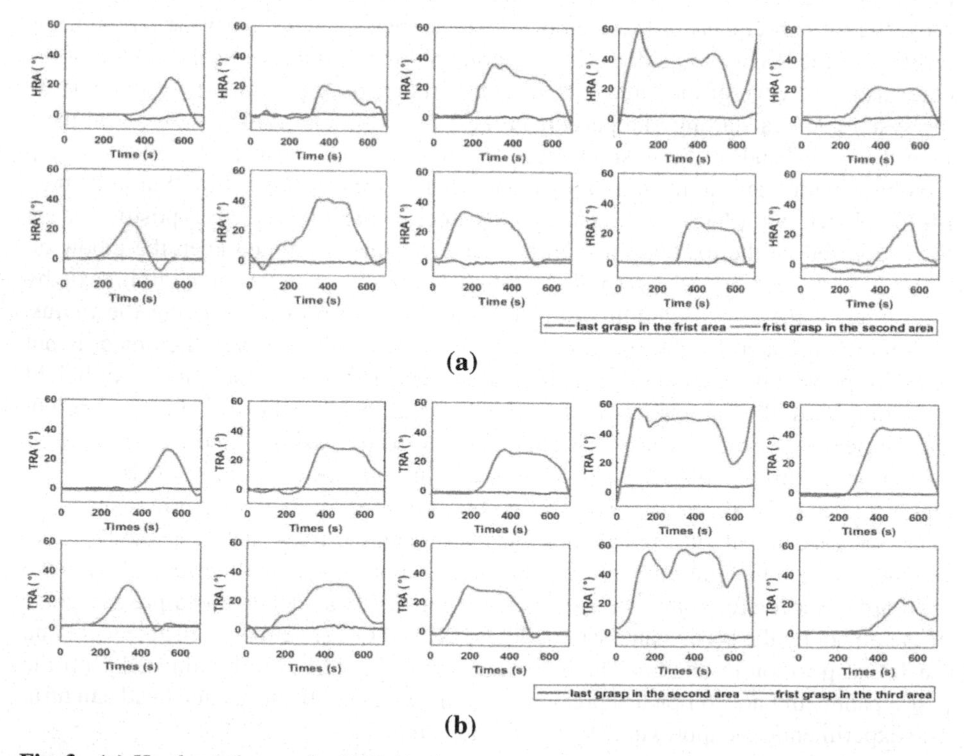

Fig. 3. (**a**) Head rotation angle (HPRA) change curves of ten participants. Red lines represent the last grasp in the first area, Blue lines represent the first grasp in the second area. X-axis is the number of frames recorded by IMU. (**b**) Trunk rotation angle (TPRA) change curves of ten participants. Red lines represent the last grasp in the second area, Blue lines represent the first grasp in the third area. X-axis is the number of frames recorded by IMU. (Color figure online)

The blue and red lines in Fig. 3a show the head rotation angle during the last grasp in the first area and the first grasp in the second area for each participant, The blue and red lines in Fig. 3b show the trunk rotation angle during the last grasp in the second area and the first grasp in the third area for each participant, the X-axis is the number of frames recorded by IMU and the Y-axis is the rotation angle. When entering the next area of grasp, there is a significant change in head and trunk rotation angles, both greater than 20°. Therefore, the change in the rotation angle of the head and trunk during each grasp can be used as the foundation for the workspace division. When the data fluctuated significantly from the previous one, the angle of the object from the midline during that grasp was the angle of each area.

Data analysis is performed using Graphpad Prism 7.0. Calculated mean and ±1 SD of the angle of three areas. Using one-way ANOVA with the angle of each area and the height of the chair (48 cm, 55 cm, 62 cm), To observe the difference in workspace division when the relative height differs. Extracted the peak head, and trunk rotation angles during each grasp, related peak head and trunk rotation angles, and object placement with four-parameter nonlinear regression analysis, to investigate the differences in head and trunk rotation angles between participants in different areas.

3 Result

3.1 Workspace Division When Chair-Height Differs

Figure 4 shows the mean and ±1 SD of each area angle when chair height differs. When chair-height is 62 cm, 55 cm and 48 cm, the angle of the first area is 44.143 ± 1.537°, 45.363 ± 2.241°, 44.779 ± 2.094° (mean ± SD, $F = 0.9489$; $p = 0.3997$). The angle of the second area is 66.904 ± 4.005°, 68.466 ± 2.057°, 67.364 ± 3.030° ($F = 0.6561$; $p = 0.5269$). The angle of the third area is 78.453 ± 3.552°, 82.317 ± 6.419°, 82.554 ± 4.372° ($F = 0.6561$; $p = 0.5269$). There was no significant between the angle and chair height in each area (all $p > 0.05$).

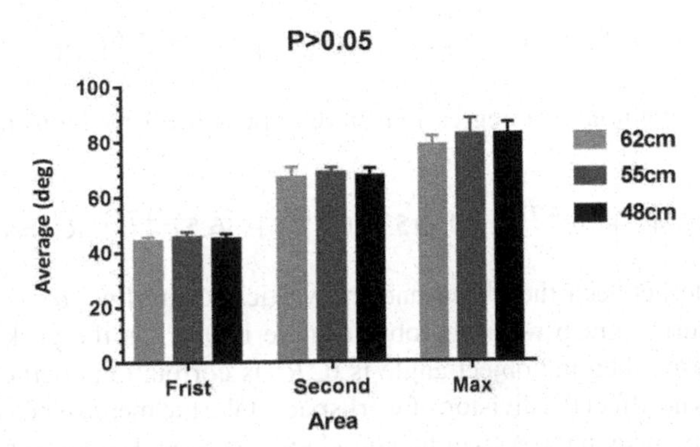

Fig. 4. Mean angle of each area when chair-height is 62 cm, 55 cm and 48 cm. Error bars represent ±1 SD.

3.2 Workspace Division Angle

Figure 5 shows the mean and ±1 SD of the angle of each area, division the workspace into three areas, ignore the height factor. The angle of the first area is 44.631 ± 1.348°, the angle of the second area is 67.315 ± 2.075°, the angle of the third area is 80.835 ± 4.360°. It is significant between the angle of each area (all $p < 0.0001$). The ±1 SD of three areas increased.

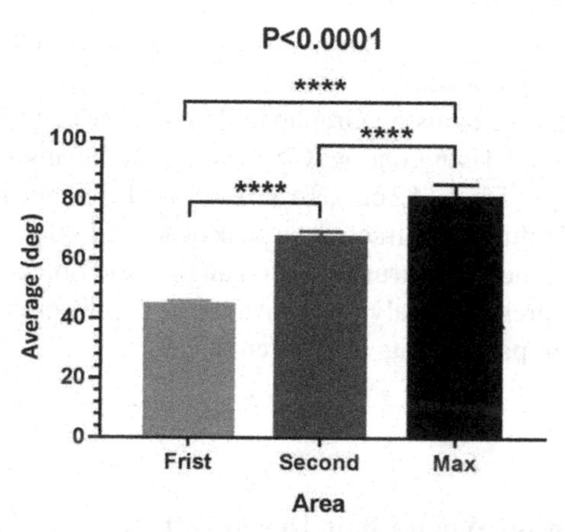

Fig. 5. Mean angle of each area. Error bars represent ±1 SD. Asterisks indicate significant differences, $* * **p < 0.0001$.

3.3 Head and Trunk Peak Rotation Angle of Each Area

A four-parameter nonlinear regression curve of peak head rotation angles and object position is:

$$\alpha_1 = 0.3717 + (\theta^{8.728} \times 40.8383/(\theta^{8.728} + 26.17^{8.728}), R^2 = 0.9048 \quad (1)$$

A four-parameter nonlinear regression curve of peak trunk rotation angles and object position is:

$$\alpha_2 = 0.8047 + (\theta^{24.77} \times 40.2153/(\theta^{24.77} + 36.57^{24.77}), R^2 = 0.9615 \quad (2)$$

θ is the angle between the object and the vertical centerline, α_1 is the peak head rotation angle during grasp when the object angle is θ. α_2 is the peak trunk rotation angle during grasp when the object angle is θ. R^2 is correlation coefficient. Since the chair height does no affect the division of workspace, take the mean of peak rotation angle α_1 and α_2 at each grasp in three chair height when θ is same. Figure 6a and b shows the hand and trunk regression curve, when $\theta > 44^\circ$, α_1 gradually increasing, when $\theta > 67^\circ$, α_2 gradually increasing, and then stabilization. The four-parameter regression curve can achieve a better fitting effect.

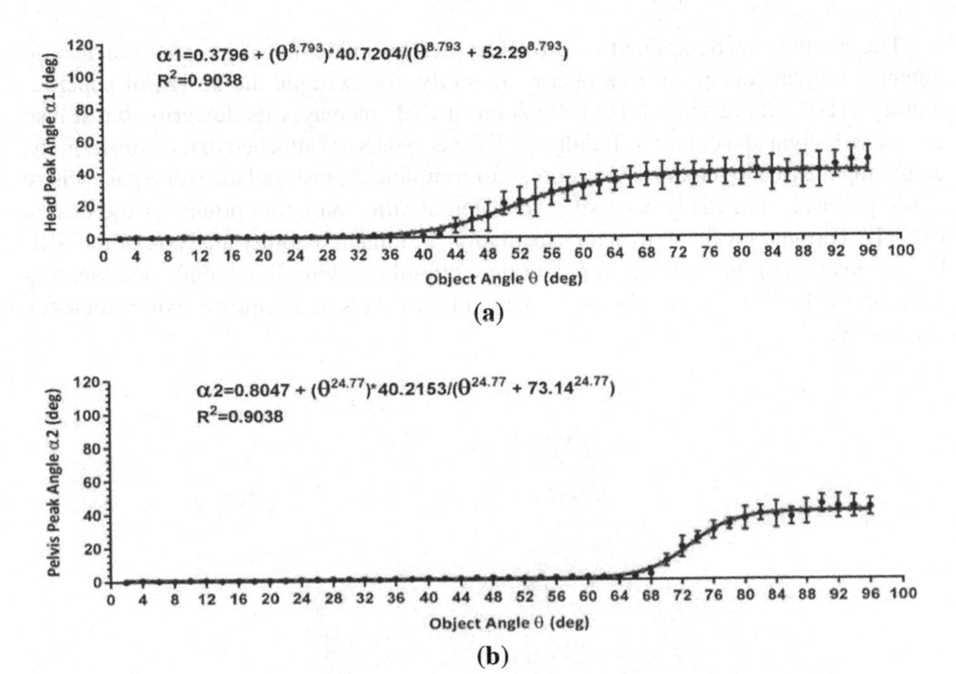

Fig. 6. (**a**) A four-parameter nonlinear regression curve of peak head rotation angles and object position. Error bars represent ±1 SD. Red dashed line indicates the 95% confidence interval. (**b**) A four-parameter nonlinear regression curve of peak trunk rotation angles and object position. Error bars represent ±1 SD. Red dashed line indicates the 95% confidence interval. (Color figure online)

4 Discussion

The aim of the present study is a quantitative division of the workspace under the gaze to research peripheral-visual-guided grasp ability in concurrent-task and obtained the variation of peak rotation angles of the head and trunk in each area. Considering the restricted condition of the human torso, the workspace is divided into three areas by constraining the rotation of the trunk and head. Our results show that the angle of the three areas was the same for each participant, this experiment was effective in workspace division. Using one-way ANOVA, the angle of each area was no significant in different chair height, shows the relative height do not affect on the workspace division. Figure 7 reveals the visualization model for workspace divisions.

A four-parameter nonlinear regression curve of peak head (Fig. 6a) or trunk (Fig. 6b) rotation angles and object position is acquired. The peak rotation angle of the head or trunk is gradually increased at the beginning of different areas. After increasing to a fixed angle, the peak rotation angle remains stable at each grasp. There was some variation in the peak rotation angle between participants, but the growth trend was essentially the same, indicates that the research on the division of human upper limb workspace is reliable.

The results can be applied to provide a design guideline of assistive devices to enhance human concurrent-task operation ability, for example the design of supernumerary robotic limbs (SRL). High freedom of SRL increases its dexterity, but it also increases weight and control difficulty [15]. Since SRLs are attached to the human body, analyzing the freedom required by humans to complete the task and the workspace where is not precisely and easily accessible in different situation is important for the design of SRL. Through inverse kinematic calculation and manipulability analysis of the SRL by the position of this area could obtain the optimal freedom, link length, and wearing position of the SRL, reduce the body burden of workers and improve work efficiency and safety.

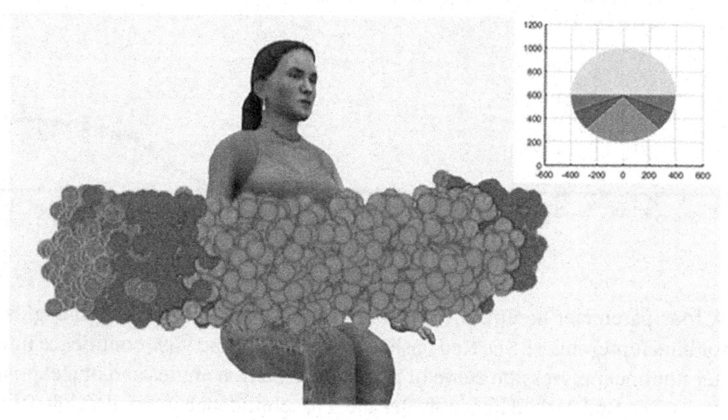

Fig. 7. The visualization model for workspace divisions, the points are the data for different participants, and the hemispheres are the results of the workspace division. Blue is the first area, red is the second area, green is the third area. (Color figure online)

5 Conclusion and Limitation

This research was investigated peripheral-visual-guided grasp performance under the gaze in concurrent-task. Could be applied to provide a design guideline of assistive devices to enhance human concurrent-task operation ability. However, this experiment only considered the maximum grasp range of the human body, i.e., the working area when grasping without flexion the elbow. And the relative position of the object was below the level of the human body when the arm was flatly extended, instead of dividing all the accessible workspace of the human upper limb, it could be improved by further research.

Acknowledgement. This work was supported in part by the National Natural Science Foundation of China (51905375), the China Post-doctoral Science Foundation Funded Project (2019M651033), Foundation of State Key Laboratory of Robotics and System (HIT) (SKLRS-2019-KF-06), and Peiyang Elite Scholar Program of Tianjin University (2020XRG-0023).

References

1. Navarro, J., Hernout, E., Osiurak, F., et al.: On the nature of eye-hand coordination in natural steering behavior. PLoS ONE **15**(11), 0242818 (2020)
2. Rossetti, Y., Desmurget, M., Prablanc, C.: Vectorial coding of movement: vision, proprioception, or both? J. Neurophysiol. **74**(1), 457–463 (1995)
3. Smirnov, A.S., Alikovskaia, T.A., Ermakov, P.N., et al.: Dart throwing with the open and closed eyes: kinematic analysis. Comput. Math. Methods Med. **2019**(2), 1–10 (2019)
4. Maturi, K.S., Sheridan, H.: Expertise effects on attention and eye-movement control during visual search: evidence from the domain of music reading. Atten. Percept. Psychophys. **82**(5), 2201–2208 (2020)
5. Wang, Z., Zheng, R., Kaizuka, T., Nakano, K.: Relationship between gaze behavior and steering performance for driver–automation shared control: a driving simulator study. IEEE Trans. Intell. Vehicles **4**(1), 154–166 (2019)
6. Kober, H., Wager, T.D.: Meta-analysis of neuroimaging data. WIREs Cogn. Sci. **1**(2), 293–300 (2010)
7. Gene-Sampedro, A., Alonso, F., Sánchez-Ramos, C., et al.: Comparing oculomotor efficiency and visual attention between drivers and non-drivers through the Adult Developmental Eye Movement (ADEM) test: a visual-verbal test. PLoS ONE **16**(2), 0246606 (2021)
8. Roux-Sibilon, A., Trouilloud, A., Kauffmann, L., et al.: Influence of peripheral vision on object categorization in central vision. J. Vis. **19**(14), 7 (2019)
9. Vater, C., Williams, A.M., Hossner, E.-J.: What do we see out of the corner of our eye? The role of visual pivots and gaze anchors in sport. Int. Revi. Sport Exerc. Psychol. **13**(1), 81–103 (2019)
10. Wolfe, B., Dobres, J., Rosenholtz, R., et al.: More than the useful field: considering peripheral vision in driving. Appl. Ergon. **65**, 316–325 (2017)
11. Marple-Horvat, D.E., Chattington, M., Anglesea, M., et al.: Prevention of coordinated eye movements and steering impairs driving performance. Exp. Brain Res. **163**(4), 411–420 (2005)
12. Bright, L.: Supernumerary robotic limbs for human augmentation in overhead assembly tasks. Robot. Sci. Syst. (2017)
13. Suzuki, M., Izawa, A., Takahashi, K., et al.: The coordination of eye, head, and arm movements during rapid gaze orienting and arm pointing. Exp. Brain Res. **184**(4), 579–585 (2008)
14. Nakabayashi, K., Iwasaki, Y., Iwata, H.: Development of evaluation indexes for human-centered design of a wearable robot arm. In: International Conference, pp. 305–310 (2017)
15. Guggenheim, J., Hoffman, R., Song, H., et al.: Leveraging the human operator in the design and control of supernumerary robotic limbs. IEEE Robot. Autom. Lett. **5**(2), 2177–2184 (2020)

Research on Collision Detection and Collision Reaction of Collaborative Robots

Zhihong Zhu[1], Zhihao Gong[1], Yuexiang Zhang[1], Simin Chen[2], Zhiqiang Zhao[1], Xing Zhou[1,2], Meng Gao[2], and Shifeng Huang[1,2](✉)

[1] The National CNC Engineering Technology Research Center, Huazhong University of Science and Technology, Wuhan 430074, China
d201677154@hust.edu.cn

[2] Foshan Institute of Intelligent Equipment Technology, Foshan 528000, China

Abstract. The first-order model-based generalized momentum observer (GMO) is a low pass filter with high-frequency attenuation characteristics and detection delay. Therefore, it cannot detect short impacts (extremely short time interval) sensitively. According to the frequency distribution of modeling errors, collision signal, and measurement noise, we design three different second-order model-based GMOs, which improve the performance of collision detection effectively. Specifically, these GMOs include a second-order damped system (abbreviated as Damp), a second-order damped system with PD regulation (abbreviated as PD), and a band-pass filter (abbreviated as BPF). Typically, the collision reaction function can be realized by switching to the torque control mode with gravity compensation. However, the robot cannot continue to move along its original trajectories after switching from position to torque control, and cannot continue to perform subsequent tasks. To address this problem, we design a collision reaction strategy in position control mode by mapping the output of the observer to the increment of joint velocity, realizing a safe collision reaction.

Keywords: Collision detection · Collision reaction · Collaborative robots · Physical human-robot interaction (pHRI)

1 Introduction

With the development of manufacturing from automation to intelligence, the application of collaborative robots is more and more extensive. It has potential possibilities of a collision between human and the environment due to the incorrect operation or machine failure of robots. Therefore, in recent years, the importance of collision safety and contact force estimation has become increasingly prominent.

It is an intuitive idea to realize the collision detection by monitoring the external wrench acting on the robot links, but it might need to use external sensors. For instance, the robot link is covered with sensitive skin [1], making the collision position and force can be detected directly. However, it is costly and the sensitive skin is easy to be worn, which increases the maintenance costs. The dynamic model-based method for monitoring signal of the external torque is the typical technology. In [2], it proposes for the first

X.-J. Liu et al. (Eds.): ICIRA 2021, LNAI 13014, pp. 510–520, 2021.
https://doi.org/10.1007/978-3-030-89098-8_48

time to use the internal sensors of robot to detect collisions, and the joint motion components and driving current sampled in actual operation are substituted into the dynamic model to directly calculate the external joint torque. However, this method needs joint angular acceleration, and the double differential of joint position will introduce large measurement noise. Considering the defects of the above methods, external moment observers are proposed [3–5]. First, an energy-based observer is proposed, which reflects the change of external torque through the change of system energy. This method cannot detect the external force at rest, nor the external force perpendicular to the moving direction of the robot, and the monitoring signal directly reflects the energy rather than the information of external torque. Second, a speed-based external force moment observer is proposed, which solves the problem that the specific external force of the energy observer cannot be detected. However, due to the inverse operation of the inertia matrix, the coupling relationship between the monitoring signal and the external torque is nonlinear. Third, an external moment observer based on generalized momentum (GMO) is proposed. It can directly reflect the change of external torque of each joint through the monitoring signal. There is no calculation demands for the second-order differential of joint positions and inverse inertial matrix. The calculation is relatively simple. However, due to the low-pass filtering relationship between the monitoring signal and the external torque signal, there are detection delay and high-frequency attenuation.

In order to improve the detection accuracy of GMO, reference [6] uses GMO to estimate the friction torque, then uses the least square method (LS) to solve the friction model parameters. Reference [7] combines a Kalman filter with GMO to reduce the measurement noise of joint parameters. In reference [8], it is analyzed that monitoring value of GMO is easy to diverge with the adjustment of gain matrix and put forward a state observer to improve the detection accuracy.

When the collision is detected, in order to ensure the personal safety or the safety of the robot itself, the robot needs to implement the reaction function quickly. The most simple and intuitive reaction strategy is to stop the robot, but it cannot solve the safety problem of quasi-static contact. Reference [9, 10] propose a zero-gravity torque reaction method by switching to torque control with gravity compensation. It leaves the robot floating in space in response to the collision force. On this basis, reference [11] proposes a reflex torque reaction strategy, which uses the motor torque so as to react to the external collision force along the same resulting direction.

In this paper, we construct different second-order observers and compare their detection performance with GMO in terms of detection delay, sensitivity and accuracy. Moreover, we design a collision reaction strategy in position control mode. By mapping the output of the observer to the increment of joint velocity, the evacuation point is calculated and sent to the controller, realizing the safe evacuation of the robots. Our contributions are:

- According to the frequency distribution of error, collision signal and measurement noise, we put forward the improvement strategy of GMO.
- We propose a collision reaction strategy in position control mode.

2 Preliminaries

The generalized momentum $\boldsymbol{p}$ is defined:

$$p = M\dot{q} \tag{1}$$

where $\boldsymbol{M}$ is the symmetric and positive-definite inertia matrix, and $\dot{\boldsymbol{q}}$ is the joint velocity vector.

The relation between the generalized momentum $\boldsymbol{p}$ and the external joint torque $\boldsymbol{\tau}_{\text{ext}}$ [12] can be expressed as:

$$\begin{cases} \dot{p} = \tau_m + \tau_{ext} - \tau_f + C^T\dot{q} - G = \tau_m + \tau_{ext} - \tau_f + \beta(q, \dot{q}) \\ \beta(q, \dot{q}) = C^T\dot{q} - G \end{cases} \tag{2}$$

where $\boldsymbol{q}$ is the joint position vector, $\boldsymbol{C}$ is the matrix of centrifugal force and Coriolis force, $\boldsymbol{G}$ is the gravity vector, $\boldsymbol{\tau}_f$ is the friction torque vector, and $\boldsymbol{\tau}_m$ is the active motor torque vector. $\boldsymbol{M}$, $\boldsymbol{C}$, $\boldsymbol{G}$ can be obtained by dynamic identification experiments, which are not covered in this paper.

Set the $\boldsymbol{r}_1$ as the external joint torque monitoring value. Considering the output signal of the first-order inertia system changes with the step signal inputting, the system model can be expressed as:

$$r_1 = \frac{K_1}{K_1 + s}\tau_{ext} \Rightarrow \dot{r}_1 = K_1(\tau_{ext} - r_1) \tag{3}$$

where $\boldsymbol{K}_1$ is the coefficient matrix, and $\mathbf{r}_1$ is the GMO output vector.

In the observer model, considering the errors of dynamics and friction model, we can define $\mathbf{r}_1 = \hat{\tau}_{ext}$. The Eq. (2) is rewritten as:

$$\hat{\dot{p}} = \tau_m + \hat{\tau}_{ext} - \hat{\tau}_f + \hat{\beta}(q, \dot{q}) = \tau_m + r_1 - \hat{\tau}_f + \hat{\beta}(q, \dot{q}) \tag{4}$$

where $\hat{\dot{\boldsymbol{p}}}$, $\hat{\boldsymbol{\tau}}_{ext}$, $\hat{\boldsymbol{\tau}}_f$, $\hat{\boldsymbol{\beta}}(\boldsymbol{q}, \dot{\boldsymbol{q}})$ are the estimated values of observer, and $\boldsymbol{p}$, $\boldsymbol{\tau}_{ext}$, $\boldsymbol{\tau}_f$, $\boldsymbol{\beta}(\boldsymbol{q}, \dot{\boldsymbol{q}})$ are the theoretical values.

Assuming $\hat{\boldsymbol{\tau}}_f \approx \boldsymbol{\tau}_f$, $\hat{\boldsymbol{\beta}}(\boldsymbol{q}, \dot{\boldsymbol{q}}) \approx \boldsymbol{\beta}(\boldsymbol{q}, \dot{\boldsymbol{q}})$, and combining (2)–(4), we can get:

$$\begin{gathered} \tau_{ext} - r_1 = \dot{p} - \hat{\dot{p}} \\ \dot{r}_1 = K_1(\dot{p} - \hat{\dot{p}}) \end{gathered} \tag{5}$$

Integrate the above equations, and the observer form is constructed as follows:

$$r_1(t) = K_1 \int [\dot{p}(t) - \hat{\dot{p}}(t)]dt = K_1[p(t) - \int (\tau_m - \hat{\tau}_f + r_1 + \hat{\beta})dt] \tag{6}$$

The performance of the observer is affected by the transfer function. Based on the first-order model, the system output has a delay to the input. Increasing the gain can improve the system response, but it also increases the error. The first-order model also has low-pass filtering characteristics, which will reduce the detection performance of high-frequency signals.

3 Second-Order Observer

3.1 The Frequency Distribution of the Signal

The frequency distribution of the signal caused by collision events on the basis of the GMO model is shown in Fig. 1 as

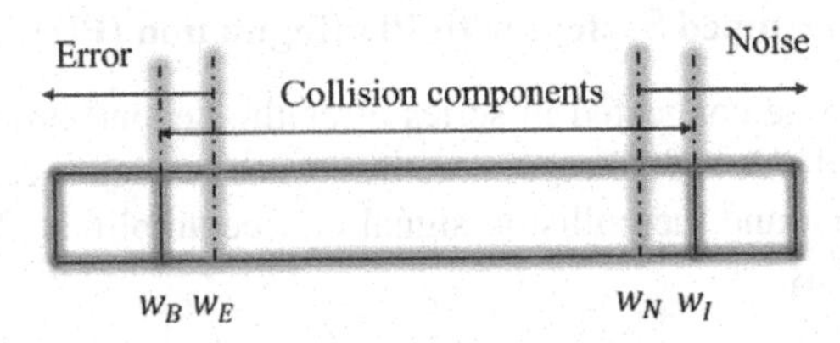

Fig. 1. The frequency distribution of the monitoring signal

where w_E is the upper limit of the frequency of the robot inertial and friction identification error, w_B is the frequency of soft collision detection, w_I is the upper limit of the frequency of hard collisions, the frequency range $w_B \sim w_I$ constitutes the frequency bandwidth of collision components, and w_N is the lower limit of measurement noise frequency. The frequency of measurement noise is generally high, and the high-frequency disturbance caused by hard collisions may be mixed in the measurement noise.

The above frequency distribution is only used to define the signal frequency region caused by normal collision. In view of different robots, the distribution of measurement noise is different, and the model error also changes with identification accuracy. Significantly, qualitative analysis of the specific frequency component of GMO improvement strategy is proposed.

3.2 Second-Order Damped System (Damp)

Replacing the GMO transfer function with the Second-order Damped System:

$$\frac{r_D}{\tau_{ext}} = \frac{w_n^2}{s^2 + 2\xi w_n s + w_n^2} \tag{7}$$

$$\frac{r_D}{\tau_{ext}} = \frac{K_1}{s^2 + K_1 K_2 s + K_1} \Rightarrow \ddot{r}_D = K_1(\tau_{ext} - r) - K_1 K_2 \dot{r}_D \tag{8}$$

where $K_1 = w_n^2$, $K_1 K_2 = 2\xi w_n$, K_1, K_2 are coefficient matrixes, r_D is the output vector of Damp Observer (The second-order observer, whose transfer function is Damp).

Integrating the formulation (8) to get:

$$r_D(t) = K_1 \int [\int (\tau_{ext} - r_D)dt - K_2 r_D]dt \tag{9}$$

Assuming that $\hat{\tau}_f \approx \tau_f$, $\hat{\beta}(q, \dot{q}) \approx \beta(q, \dot{q})$, $r_D = \hat{\tau}_{ext}$, the formulation of time domain relation can be calculated as:

$$r_D(t) = K_1 \int [p(t) - \int (\tau_m - \hat{\tau}_f + \hat{\beta} + r_D)dt - K_2 r_D]dt \tag{10}$$

According to experience, when the damping is 0.707, the amplitude-frequency characteristic curve is close to a straight line in the low frequency range, and the system response sensitivity and oscillation stability can be balanced optimally. Compared with GMO, the system is easy to obtain lower detection delay. However, the model error still exists, and the high frequency signal detection performance has not been improved.

3.3 Second-Order Damped System with PD Regulation (PD)

The PD regulator can be connected in series after the Second-order Damped System. Using the slope of 20 dB/dec of PD regulator, the corner frequency can be selected within the range of $w_B \sim w_I$, and the collision signal can be amplified. The system transfer function is expressed as:

$$\frac{r_P}{\tau_{ext}} = \frac{K_1(K_3 s + 1)}{s^2 + K_1 K_2 s + K_1} \tag{11}$$

where K_1, K_2, K_3 are coefficient matrixes and r_P is the output vector of PD Observer (The second-order observer, whose transfer function is PD).

Similar to the above derivation process, the formulation of time domain relation is given:

$$\begin{cases} r_P(t) = K_1\{K_3 p(t) + \int [-K_3\hat{p} + (K_3 - K_2)r_P + p(t) - \int \hat{p} dt]dt\} \\ \hat{p} = \tau_m - \hat{\tau}_f + \hat{\beta} + r_P \end{cases} \tag{12}$$

The PD improves the collision detection performance by increasing the amplitude of the collision signal in the frequency domain. This method fails to achieve any improvement in the detection accuracy.

3.4 Band-Pass Filter (BPF)

If the amplitude of the model error can be reduced without changing the amplitude of the collision signal, the detection accuracy of the system will be inevitably improved. Since the GMO is a low-pass filter model, a feasible method is to obtain a band-pass filter by connecting a high-pass filter in series on the basis of GMO, as shown in Fig. 2, which attenuates low-frequency errors and high-frequency noise. The amplitude of the collision signal remains unchanged.

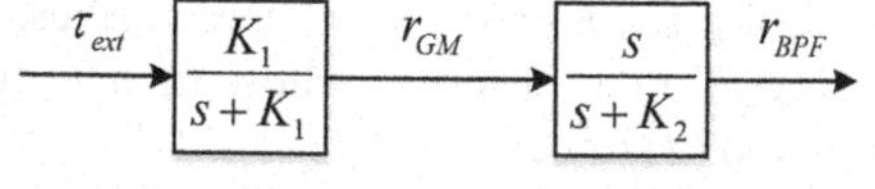

Fig. 2. Schematic diagram of band-pass filter model.

The system transfer function is as follows:

$$\frac{r_B}{\tau_{ext}} = \frac{K_1 s}{s^2 + (K_1 + K_2)s + K_1 K_2} \Rightarrow \ddot{r}_B = K_1(\tau_{ext} - r_B)s - K_2 r_B s - K_1 K_2 r_B \tag{13}$$

where K_1, K_2 are coefficient matrixes and r_B is the output vector of BPF Observer (The second-order observer, whose transfer function is BPF).

Similar to the above derivation process, the formulation of time domain relation is given:

$$\begin{cases} r_B(t) = K_1 p(t) - \int (K_1 \hat{\dot{p}} + K_2 r_B + \int K_1 K_2 r_B dt)dt \\ \hat{\dot{p}} = \tau_m - \hat{\tau}_f + \hat{\beta} + r_B \end{cases} \tag{14}$$

4 Collision Reaction Strategy in Position Control Mode

When the collision occurs, the magnitude and direction information of $\boldsymbol{r}^*$ is converted into the increment of joint angular velocity:

$$\Delta \dot{q} = K_r r^* \tag{15}$$

where $\boldsymbol{K}_r$ is the damping coefficient matrix, $\boldsymbol{r}^*$ is the output of the observer when the collision is detected, which is a constant vector. Then, by using the angular velocity when the collision is detected, the evacuation point can be calculated:

$$\begin{cases} \dot{q}(t+1) = \dot{q}(t) + K_r r^* \\ q_d(t+1) = q(t) + \dot{q}(t+1) \cdot T \end{cases} \Rightarrow q_d = T(K_r r^* + \dot{q}) + q \tag{16}$$

where $\boldsymbol{q}(t)$ is the current position vector, $\boldsymbol{q}_d(t+1)$ is the position vector of the next time step, and T is the sampling time period.

By adjusting the value of $\boldsymbol{K}_r$, the evacuation distance can be easily adjusted. The algorithm block diagram is shown in Fig. 3.

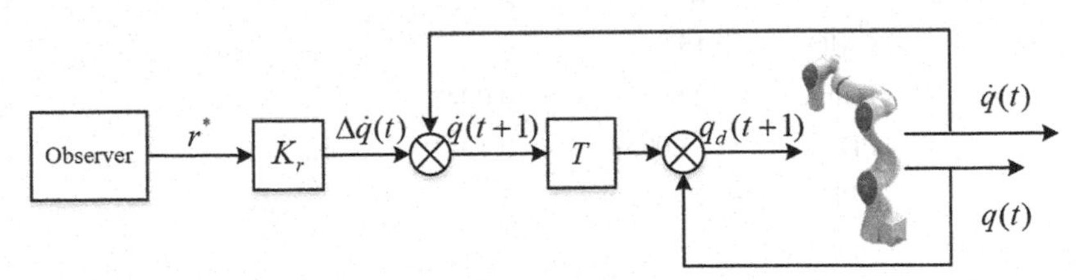

Fig. 3. Principle of the Collision reaction strategy in position control mode.

This method makes the robot safely evacuate in the direction of external force without switching the control mode. If the collision point is memorized when a collision occurs, the robot can be controlled to return to the collision point and execute the subsequent tasks when the obstacle disappears. However, it is powerless for collisions during the evacuation process.

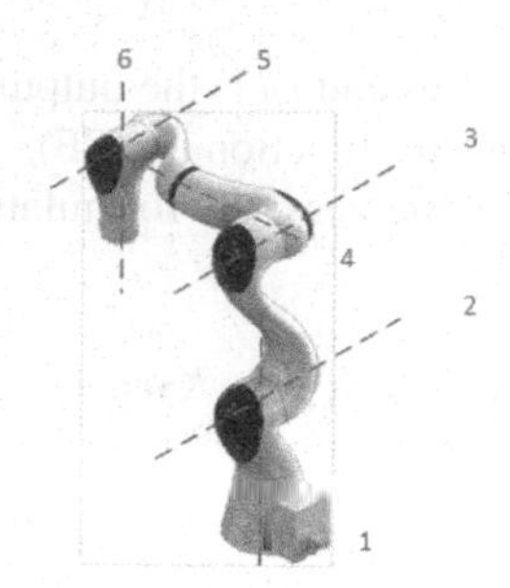

Fig. 4. An HSR-Co605 Robot manufactured by Huashu robot company.

5 Experiment

We performed several tests on collision detection with a collaborative robot HSR-Co605 (see Fig. 4), which is developed by the HSR company, and validated the designed robot reaction strategy.

The presented three second-order observers in this paper (Damp, PD, and BPF) are compared with GMO one by one to show the characteristics and effectiveness in terms of the detection delay, sensitivity and accuracy. The coefficients of each observer are shown in Table 1.

The soft collisions and hard collisions are applied to 3-rd link of the tested robot respectively, and the monitoring signals of the joint 3 are shown in Fig. 5.

Table 1. Coefficients of the second-order observers and GMO (The main diagonal elements of each coefficient matrix are the same)

	K_1	K_2	K_3
GMO	15	–	–
Damp	225	0.0943	–
PD	53.0959	0.1941	0.2
BPF	18	5	–

In Fig. 5, the blue curve is the GMO output, the red curve is the second-order observer output, the three subfigures from the top to the bottom are the Damp, PD, BPF, respectively. The dotted line is the threshold, where the dark color represents the second-order observer threshold and the light color represents GMO threshold. The results are summarized in Table 2.

Table 2. Performance evaluation of the second-order observers and GMO

	Link 3 applied soft collisions (joint 3)			Link 3 applied hard collisions (joint 3)		
	Upper threshold (NM)	Inspected sampling point	Maximum amplitude (NM)	Upper threshold (NM)	Inspected sampling point	Maximum amplitude (NM)
GMO	6.000	2610	12.82	0.9178	2362	2.382
Damp	1.814	2602	8.437	0.3818	2366	1.638
PD	6.016	2609	15.17	0.9172	2364	2.578
BPF	2.037	2589	8.341	0.4498	2357	2.414

If the ratio of the maximum amplitude of the observer output signal to the upper limit of the threshold is taken as the criterion to measure the collision detection sensitivity of the observers, it can be calculated from the above table that the detection sensitivity of the second-order observers is better than that of GMO in terms of soft collisions and hard collisions, among which Damp Observer has the highest soft collisions detection sensitivity and BPF Observer has the highest hard collisions detection sensitivity. If the sampling position where the observer output signal exceeds the threshold for the first time is taken as the collision detection point, it can be seen from the above table that the detection delay of BPF Observer is significantly reduced.

After the collision, the reaction strategy in position control mode is adopted. According to the expectation, the robot will evacuate along the direction of external force and stop after evacuating to the appropriate distance. The whole process from collision detection to reaction is shown in Fig. 6.

In Fig. 6, (a) (b) (c) the robot moves along the predetermined trajectory, (d) the collision force is applied artificially, (e) the robot detects the collision, (f) (g) the robot responds to the collision and evacuates from the collision point along the direction of the external force, (h) the robot evacuates to the appropriate distance and stops. It can be observed from the above figure that when the collision occurs, joint 3 is the main evacuation joint of the robot. Taking joint 3 as the object, the torque monitoring values and desired angular velocity before and after the collision are collected, and the results are shown in Fig. 7.

As shown in Fig. 7, after the observer output value exceeds the set threshold, the desired angular velocity drops rapidly to zero. The robot stops quickly, and then move to the calculated evacuation point. The velocity planning of the robot is completed by the controller. After the collision is detected, the observer will not monitor the change of external torque at this time, and the output keeps zero.

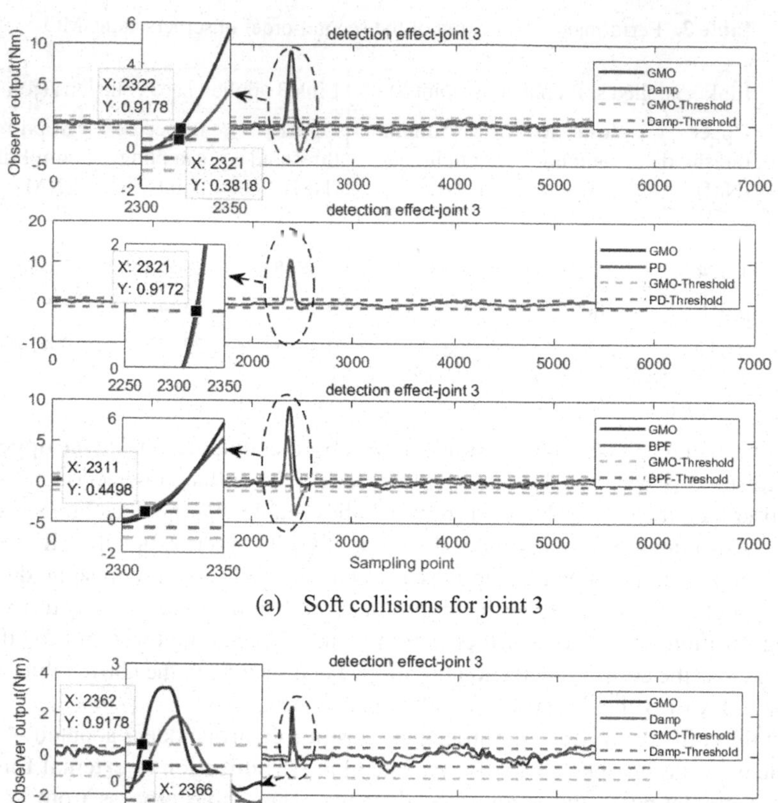

(a) Soft collisions for joint 3

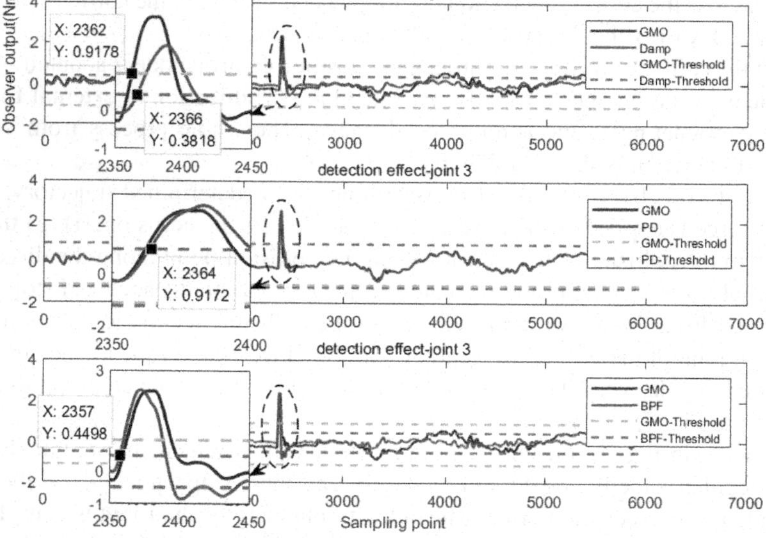

(b) Hard collisions for joint 3

Fig. 5. Performance comparison between second-order observers and GMO. (Color figure online)

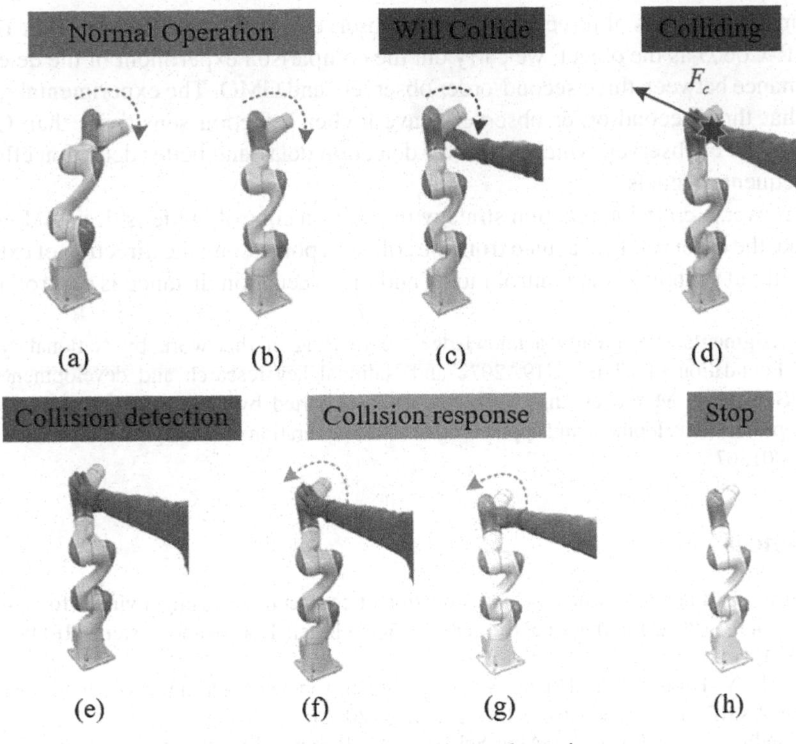

Fig. 6. Collision detection and reaction

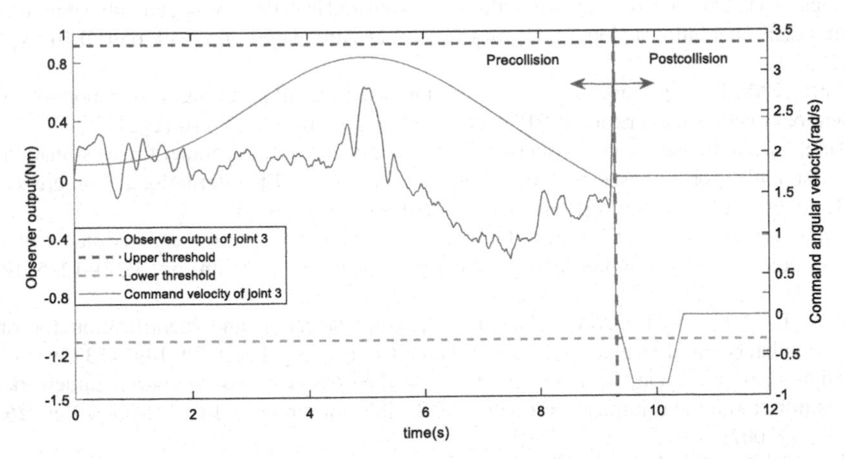

Fig. 7. Observer output and desired angular velocity before and after collision.

6 Conclusion

In this paper, according to the frequency distribution of error, collision signal and measurement noise, three kinds of second-order observers are designed on the basis of GMO.

The transfer functions of three second-order observers are Damp, PD, and BPF. Taking the HSR-Co605 as the object, we carry out the comparison experiment of the detection performance between three second-order observers and GMO. The experimental results show that three second-order observers have higher detection sensitivity than GMO, especially BPF Observer, which has lower detection delay and better detection effect of high-frequency signals.

Moreover, a collision reaction strategy in position control mode is designed, which can make the robot safely evacuate from the collision point along the direction of external force without switching the control mode, and the evacuation distance is controllable.

Acknowledgments. We greatly acknowledge the funding of this work by National Natural Science Foundation of China, U19A2072 and National key research and development plan, 2017YFB1303702. Moreover, this work was also supported by the Foshan core technology research project (development and application of a new generation of intelligent industrial robots), 1920001001367.

References

1. Magrini, E., Flacco, F., Luca, A.D.: Estimation of contact forces using a virtual force sensor. In: 2014 IEEE/RSJ International Conference on Intelligent Robots and Systems (IROS). IEEE (2014)
2. Yamada, Y., Hirasawa, Y., Huang, S., et al.: Human-robot contact in the safeguarding space. IEEE/ASME Trans. Mechatron. **2**(4), 230–236 (2002)
3. Haddadin, S., De Luca, A., Albu-Schäffer, A.: Robot collisions: a survey on detection, isolation, and identification. IEEE Trans. Robot. **33**(6), 1292–1312 (2017)
4. Luca, A.D., Mattone, R.: Actuator failure detection and isolation using generalized momenta. In: IEEE International Conference on Robotics & Automation, pp. 634–639. IEEE Xplore (2003)
5. Park, K.M., Kim, J., Park, J., et al.: Learning-based real-time detection of robot collisions without joint torque sensors. IEEE Robot. Autom. Lett. **6**(1), 103–110 (2021)
6. Tian, Y., Zhi, C., Jia, T., et al.: Sensorless collision detection and contact force estimation for collaborative robots based on torque observer. In: 2016 IEEE International Conference on Robotics and Biomimetics (ROBIO), pp. 946–950. IEEE (2017)
7. Wahrburg, A., Morara, E., Cesari, G., et al.: Cartesian contact force estimation for robotic manipulators using Kalman filters and the generalized momentum, pp. 1230–1235. IEEE (2015)
8. Ren, T., Dong, Y., Dan, W., Chen, K.: Collision detection and identification for robot manipulators based on extended state observer. Control Eng. Pract. **79**, 144–153 (2018)
9. Albu-Schaffer, A., Ott, C., Hirzinger, G.: A unified passivity based control framework for position, torque and impedance control of flexible joint robots. Int. J. Robot. Res. **26**(1), 23–39 (2007)
10. Albu-Schffer, A., Hirzinger, G.: Cartesian impedance control techniques for torque controlled light-weight robots. In: Proceedings - IEEE International Conference on Robotics and Automation, vol. 1, pp. 657–663 (2002)
11. Luca, A.D., Albu-Schaffer, A., Haddadin, S., et al.: Collision detection and safe reaction with the DLR-III lightweight manipulator arm. In: 2006 IEEE/RSJ International Conference on Intelligent Robots and Systems. IEEE (2006)
12. Haddadin, S.: Towards Safe Robots: Approaching Asimov's 1st Law, pp. 31–50. RWTH Aachen (2011)

An Outdoor Navigation System for Mobile Robots Based on Human Detection

Haixiang Zhou[1], Minghui Hua[1], Haiping Zhou[2,3], Wanlei Li[1], and Yunjiang Lou[1(✉)]

[1] Harbin Institute of Technology Shenzhen, Shenzhen 518000, China
louyj@hit.edu.cn

[2] Beijing Institute of Precision Mechatronics and Controls, Beijing, China

[3] Laboratory of Aerospace Servo Actuation and Transmission, Beijing, China

Abstract. The reliable navigation of mobile robots outdoors depends on high-precision maps, accurate localization and efficient path planning. The complexity and variability of outdoor scenes bring great challenges to the navigation of mobile robots. The safety of mobile robot operation is the premise of application. In this paper, a navigation system including localization, human detection and path planning is proposed for the safety of mobile robot in outdoor tasks. We use Support Vector Machine (SVM) to carry out human detection for 3D point cloud before path planning. And special avoidance processing is carried out for clusters labeled as human. The system can enhance the safety of robot and human in the operation of mobile robot.

Keywords: Localization · Safety · Human detection · Path planning

1 Introduction

Mapping, localization and path planning are the three essential abilities of an intelligent mobile robot. At present, the application scenarios of mobile robots are still mainly indoor scenes. Outdoor scenes are more complex than indoor scenes. The complexity of outdoor scenes is mainly reflected in the large scene, the uneven flatness of the ground, and the great changes of the environment over time. The algorithm of building map by lidar is very mature, and the main factor that hinders mobile robot to work outdoors is navigation.

There are many localization algorithms based on point cloud registration, such as ICP [1], NDT [2] and SALO [3]. The purpose of registration is to realize that a pair of point clouds can be aligned in the same coordinate system, so that the transformation of point clouds between two scans can be calculated. Due to the unstructured and high complexity of outdoor scene, the error and

This work was supported partially by the National Key Research and Development Program of China (2020YFB1313900), and partially by the Shezhen Science and Technology Program (No. JCYJ20180508152226630).

X.-J. Liu et al. (Eds.): ICIRA 2021, LNAI 13014, pp. 521–531, 2021.
https://doi.org/10.1007/978-3-030-89098-8_49

calculation difficulty of ICP algorithm will be increased. In this paper, we will use IMU as an auxiliary sensor, so we will choose NDT algorithm to locate the mobile robot.

Another difficulty with navigation is path planning. The quality of path planning is related to the efficiency and safety of mobile robot operation, which requires comprehensive consideration of multiple factors such as time, path length, obstacle avoidance and so on. Safety in mobile robot task has become the biggest problem whether mobile robot can be used in practical production.

In this paper, we add human detection to before path planning. And we carry out special treatment for pedestrians in local path planning. There are two kinds of target detection methods based on 3D point cloud, including traditional detection algorithm and deep learning detection algorithm. Due to the limited computing power of the mobile robot processor, the deep learning method cannot meet the real-time requirements, so we use the traditional detection algorithm for human detection.

The traditional detection algorithm mainly completes the target detection task through ground recognition, target segmentation and target classification. There are many traditional point cloud segmentation algorithms, which can segment the point cloud according to the distance between points [4], or judge whether the adjacent points are from the same object based on the angle measurement method of adjacent points [5]. After the segmentation of the point cloud is completed, classifiers are used to classify the segmented point cloud. Common classifiers include SVM [6] and Adaboost [7].

Especially, at the complex scene with many pedestrians outdoors, this paper propose a navigation system including localization, human detection, global path planning and local path planning. Figure 1 shows the flow chart of the system.

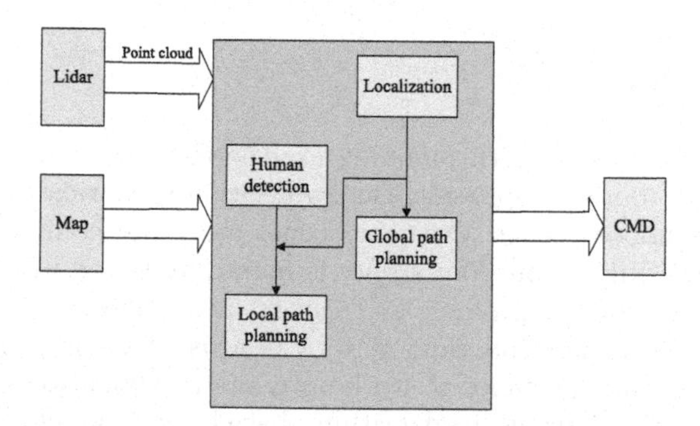

Fig. 1. Flow chart of this system

In the rest of this paper, we first introduce localization, human detection and local path planning algorithm in our system. Then we introduce the experiment we have done to verify the feasibility and practicability of the system. Finally, we discuss the results and summarize the system proposed in this paper.

2 Method

2.1 Localization

In this paper, we use 3-D Normal Distributions Transform algorithm for localization [8]. NDT is a process to transform point data in a voxel into normal distribution. The voxel is a cubic lattice in a 3-D space, where the whole space is divided into voxels. Reference scan point vectors $x_i = (x_i, y_i, z_i)^T$ are voted in the appropriate voxels by rounding the coordinate value of x_i. An average p_k and covariance matrix $\sum_k$ of the voxel k are calculated as follows:

$$p_k = \frac{1}{M_k} \sum_{i=0}^{M_k-1} x_{ki} \tag{1}$$

$$\Sigma_k = \frac{1}{M_k} \sum_{i=0}^{M_k-1} (x_{ki} - p_k)(x_{ki} - p_k)^T \tag{2}$$

Where M_k is the number of poingts in voxel k and $x_k i$ is a coordinate vector of the voted point in voxel k. Estimation value $e(x)$ of ND voxel k is defined as follows:

$$e(x) \sim \exp\left(-\frac{(x - p_k)^T \sum_k^{-1} (x - p_k)}{2}\right) \tag{3}$$

The purpose of localization is to find the rotation matrix R and translation vector t. 3-D coordinate transform equations for input scan points x_i to reference scan points $x_i{}'$ are given as follows:

$$x_i{}' = Rx_i + t \tag{4}$$

where R is rotation matrix of euler angle α, β, γ to rotate z, y, x and $t = (t_x, t_y, t_z{}^T)$.

Scan matching is a parameter search problem that finds the best rotation matrix R and translation vector t for the input scan to match the existing reference scan. The measure function of NDT is defined as follows:

$$E(X, R, t) = \sum_{i}^{N-1} \exp \frac{-(x_i{}' - p_i)^T \Sigma_i^{-1} (x_i{}' - p_i)}{2} \tag{5}$$

Where p_i and $\sum_i$ are mean vector $p_{k'}$ and covariance matrix $\sum_{k'}$ of voxel k' corresponding to the vector x_i' respectively. If the value for the measure function E is high, then input scan and reference scan are well-corresponding. Newton's nonlinear function optimizer is utilized to find a rotation matrix R and a translation vector t such that $E(X, R, t)$ is maximized on the NDT scan matching.

2.2 Human Detection

Human detection consists of three main components: a cluster detector for 3D LiDAR point clouds, a multi-target tracker and a human classifier. Figure 2 shows the process of human detection.

At each new iteration, 3D point cloud is segmented by the cluster detector. Positions and velocities of the clusters are estimated in real-time by a multi-target tracking system. At the same time, the classifier labels each cluster as 'human' or 'non-human'.

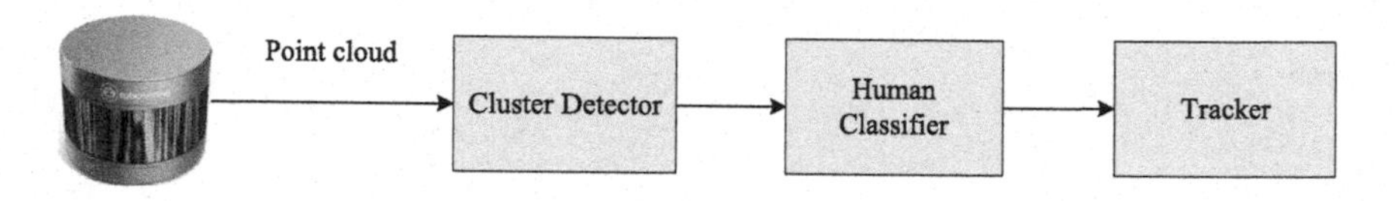

Fig. 2. Process of human detection

Point Cloud Cluster Detector. The input of point cloud cluster detector is point cloud $P = \{p_i | p_i = (x_i, y_i, z_i) \in R^3, i = 1, 2, ..., I\}$, where I is the total number of points in a single scan. In this process, we divide the point cloud into non-overlapping clusters based on 3D Euclidean distance. As in Rusu (2009) [9], if $C_i, C_j \subset P$ are two non-overlapping clusters, the non-overlapping condition can be written as follows:

$$C_i \cup C_j = \emptyset, fori \neq j \Rightarrow \min \|p_i - p_j\| >= d \tag{6}$$

where $p_i \subset C_i$, $p_j \subset C_j$, and d is a distance threshold.

The size of d has a huge effect on the results. If d is too small, single objects could be split into multiple clusters. If too large, multiple objects could be merged into a single cluster. In our case, due to lidar's vertical angular resolution, the vertical distance between two consecutive points can be very large, as shown in Fig. 3. We therefore adapt the threshold d to the actual scan range r:

$$d = 2 \cdot r \cdot \tan \frac{\theta}{2} 1 \tag{7}$$

where θ is the vertical angular resolution of lidar.

In addition, we divide the horizontal space into nested circular regions around the sensor, and fix a constant distance threshold within each of them. We consider a set of value d_i at fixed intervals Δd, where $d_{i+1} = d_i + \Delta d$. For each of them, we compute the maximum cluster detection range r_i using the inverse of Eq. (2), and determine the corresponding radius $R_i = \lfloor r_i \rfloor$, where R_0 is centre of the sensor. The width of a region with constant threshold d is $l_i = R_i - R_{i-1}$. In our application, we define circular regions 2–3 wide using $\Delta d = 0.1m$, which is a good resolution to detect potential human clusters [10].

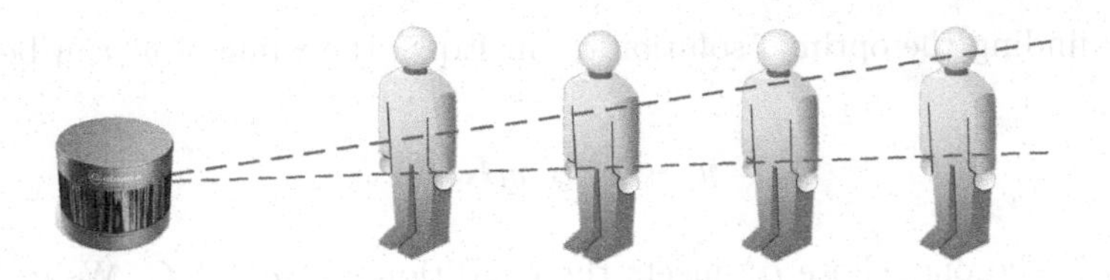

Fig. 3. The further the person is from the sensor, the sparser is the relative point cloud

Human Classifier. A nonlinear Support Vector Machine (SVM) is used for human classification [11]. SVM method has a solid theoretical foundation in mathematics, which is good at dealing with small data samples.

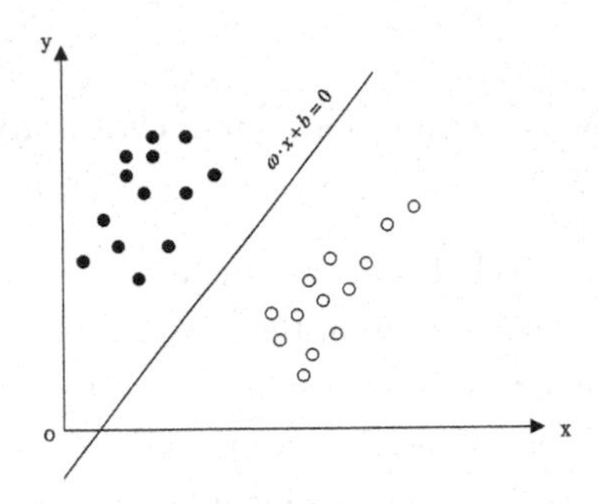

Fig. 4. SVM

Considering a training data set in feature space $T = \{(x_1, y_1), ..., (x_N, y_N)\}$, where $x_i \in R^n$, $y_i = \{+1, -1\}$, x_i is the ith feature vecotr and y_i is the class tag. The purpose of SVM is to find hyperplanes $x^T\omega + b = 0$ to segment the feature space. As shown in Fig. 4.

Nonlinear support vector machines can be described by the following equation:

$$\begin{aligned} &\min_{\alpha} \frac{1}{2} \sum_{i=1}^{N} \sum_{j=1}^{N} \alpha_i \alpha_j y_i y_j K(x_i, x_j) - \sum_{i=1}^{N} \alpha_i \\ &s.t. \sum_{i=1}^{N} \alpha_i y_i = 0, 0 <= \alpha_i <= C \end{aligned} \tag{8}$$

Where $K(x_i, x_j)$ is a kernel function and C is a penalty parameters. In this paper, we choose Gaussian kernel as follows:

$$K(x, z) = \exp\left(-\frac{\|x - z\|^2}{2}\right) \tag{9}$$

After finding the optimal solution α^* in Eq. 8, the value of b^* can be calculated:

$$b^* = y_j - \sum_{i=1}^{N} \alpha_i^* y_i K(x_i, x_j) \tag{10}$$

where y_j is the one whose α_j^* meets the condition $0 < \alpha_j^* < C$. We can make classification decisions for new feature vectors through the following equation:

$$f(x) = sign\left(\sum_{i=1}^{N} \alpha_i^* y_i K(x, x_i) + b^*\right) \tag{11}$$

In order to train the SVM, we extract seven different features from each point cloud cluster, which are shown in Table 1. Features were proposed by Navarro Serment et al. [12].

Table 1. Features for human classification

Feature	Description	Dimension
f_1	Number of points included in the cluster	1
f_2	Minimum cluster distance from the sensor	1
f_3	3D covariance matrix of the cluster	6
f_4	Normalized moment of inertia tensor	6
f_5	2D covariance matrix in 3 zones including the upper half, the left and right lower halves	9
f_6	The normalized 2D histogram for the main PC plane	98
f_7	The normalized 2D histogram for the secondary PC plane	45

The full set of features from cluster C_j forms a vector $(f_1, f_2, ..., f_7)$. At each iteration of the learning process, a binary SVM is trained to label human and non-human clusters based on these features.

2.3 Local Path Planning

In this paper, we use Dynamic Window Approach (DWA) [13] for local path planning. Before simulating the moving trajectory of the robot, it is necessary to establish the Kinematics of the robot. Figure 5 shows the Kinematics model of mobile robot.

In each sampling period, the movement path of the robot is approximated, and the movement path within each sampling period is regarded as a straight line. Then, the position of the robot at time $t + 1$ is $(x(t + 1), y(t + 1))$:

$$\begin{aligned} x(t+1) &= x(t) + v(t)\Delta t \cos(\theta(t)) \\ y(t+1) &= y(t) + v(t)\Delta t \sin(\theta(t)) \\ \theta(t+1) &= \theta(t) + \omega(t)\Delta t \end{aligned} \tag{12}$$

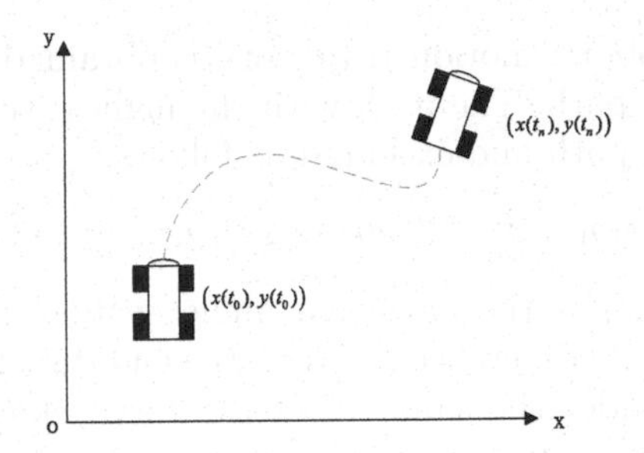

Fig. 5. Typical motion model of mobile robot

where $\omega(t)$ and $v(t)$ are respectively the linear velocity and angular velocity of the mobile robot in the world coordinate at time t.

Therefore, the two cores of the dynamic window method are as follows: (1) According to the obstacle environment and the mechanical characteristics of the robot, the speed constraint is formed, and the dynamic window is generated to conduct speed sampling. (2) According to the evaluation function, the corresponding predicted trajectory of the sampled velocity is scored, so as to obtain the optimal path and execute it.

During velocity sampling, the speed of mobile robot is mainly subject to the following constraints:

(1) The mobile robot is limited by its maximum and minimum speed, which is also the maximum range V_i of DWA algorithm to solve the speed:

$$V_i = \{v \in [v_{\min}, v_{\max}] \cap \omega \in [\omega_{\min}, \omega_{\max}]\} \tag{13}$$

(2) Due to the influence of the mobile robot's own motor, the torque provided by the deceleration of its growth rate is limited. Therefore, in the period of the forward movement of the simulated robot, there is a dynamic window, that is, the speed in the window is the actual speed V_j that the robot can achieve under the influence of its own mechanical characteristics:

$$V_j = \{(v, \omega)|v \in [v_c - \dot{v}_b \Delta t, v_c - \dot{v}_a \Delta t] \cap \omega \in [\omega_c - \dot{\omega}_b \Delta t, \omega_c + \dot{\omega}_a \Delta t]\} \tag{14}$$

Where v_c is the current linear velocity of the robot, $\dot{v}_a$ and $\dot{v}_b$ are respectively the maximum acceleration and maximum deceleration, ω_c is the current angular velocity, $\dot{\omega}_a$ and $\dot{\omega}_b$ are respectively the maximum angular acceleration and maximum angular deceleration.

(3) In order to achieve safe obstacle avoidance and avoid collision with obstacles occupying a certain space, the range V_k can be obtained under the condition of deceleration and maximum acceleration to further reduce the dynamic window range:

$$V_k = \{(v, \omega)|v \leqslant \sqrt{2 \cdot dist(v, \omega) \cdot \dot{v}_b} \cap \omega \leqslant \sqrt{2 \cdot dist(v, \omega) \cdot \dot{\omega}_b}\} \tag{15}$$

In conclusion, according to the mechanical characteristics of the robot itself and the obstacle environment, the dynamic window can be defined as

$$V_r = V_i \cap V_j \cap V_k \tag{16}$$

After the mobile robot's motion trajectory is obtained, evaluation function is needed to score each path. The path with the highest score is selected as the comprehensive optimal path and executed as follows:

$$G(v,\omega) = \alpha \cdot heading(v,\omega) + \beta \cdot dis\tan ce(v,\omega) + \gamma \cdot velocity(v,\omega) \qquad (17)$$

Where, $heading(v,\omega)$ is the deviation angle evaluation function, which is used to evaluate the angle difference between the end direction of the trajectory and the target point under the simulated trajectory velocity. $distance(v,\omega)$ is the safety factor evaluation function. If an obstacle is detected as a pedestrian, the coefficient will be enlarged according to the distance between the mobile robot and the pedestrian; if it is other obstacles, the coefficient will be reduced. $velocity(v,\omega)$ is the speed evaluation subfunction, whose function is to select the fastest path among the sampling tracks that can achieve safe obstacle avoidance, so as to reach the target point as soon as possible [14].

3 Experiments

In order to verify the feasibility of the system proposed in this paper, we conduct a series of experiments. Our experiment was carried out on the Scout mobile robot, which is an UGV with differential wheel. The sensors used in the experiment include a 16-wire lidar and IMU. In addition, we fixed a MINI PC on the mobile robot as the upper computer to issue instructions to the robot.

3.1 3D Point Cloud Map

The following figure shows the 3D point cloud map established by the point cloud information obtained by the 3D lidar (Fig. 6).

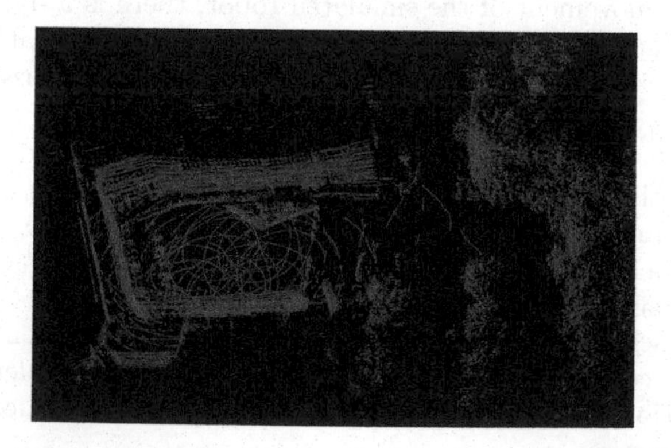

Fig. 6. 3D point cloud map

The accuracy of the 3D point cloud map is very high, which can restore most of the details in the environment and meet the positioning requirements in the localization process.

3.2 Human Detection

Figure 7 shows the visualization of human detection:

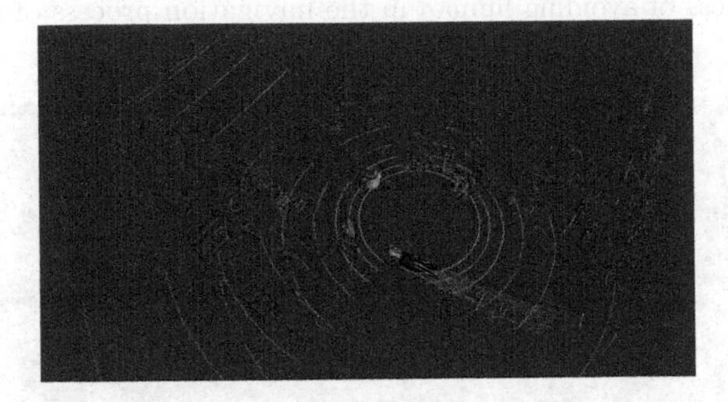

Fig. 7. Visualization of human detection

In Fig. 7, point cloud clusters judged as human by the classifier are represented by external cuboids. The figure shows that almost all pedestrians are detected, but some non-pedestrian obstacles are judged as pedestrians. The classification results tend to be conservative, which will affect the efficiency of path planning, but it is beneficial to the operation safety of mobile robots.

3.3 Path Planning

Path planning was carried out by integrating point cloud data obtained by lidar, maps, human detection results and other information. Figure 8 shows the result of path planning.

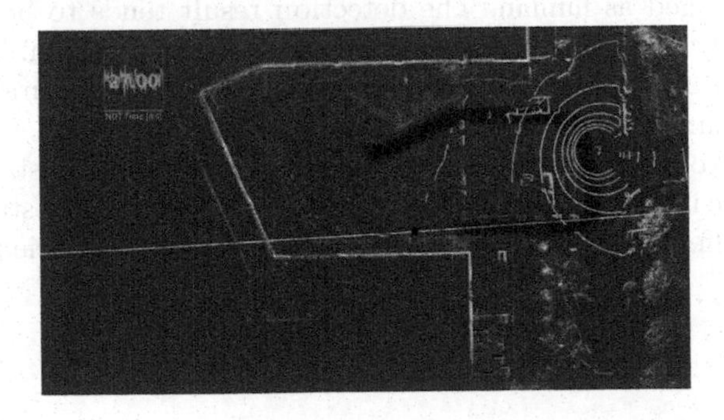

Fig. 8. Path planning

3.4 Mobile Robot Navigation

With the map and target points known, the mobile robot can complete the tasks of autonomous navigation and obstacle avoidance through this system. Figure 9 shows the process of avoiding human in the navigation process of mobile robot.

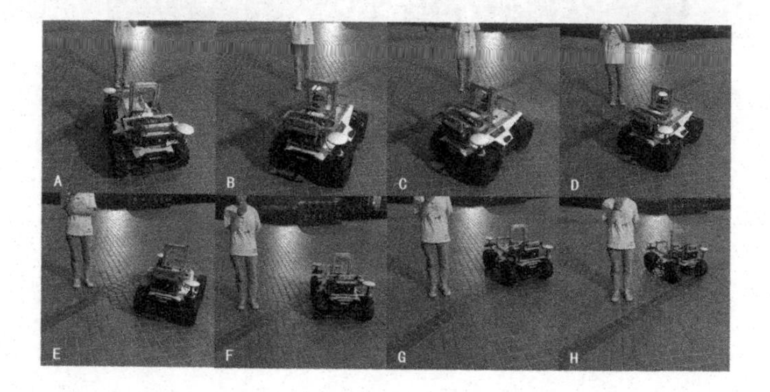

Fig. 9. The process of avoiding human

4 Conclusion

In this paper, we propose a system for outdoor navigation. In view of the complexity of outdoor scenes and the safety of mobile robot tasks, human detection was carried out before the local path planning of the robot. In the local path planning process, we adopt a special obstacle avoidance strategy for the obstacle labeled as "human".

In the experiment, we verified the feasibility and practicability of this system. In human detection, pedestrians are basically detected, but some non-pedestrian obstacles are judged as human. The detection result tends to be conservative, but it is helpful to ensure the safety of mobile robot operation. The results of mobile robot outdoor navigation show that the system can accurately carry out local path planning to avoid human in the path.

For future work, we plan to integrate visual sensor for obstacle detection, extract semantic information of obstacles, implement different obstacle avoidance strategies for different obstacles, and increase the safety and efficiency of outdoor tasks.

References

1. Segal, A., Haehnel, D., Thrun, S.: Generalized-ICP. In: Robotics: Science and Systems, Seattle, WA, vol. 2, p. 435 (2009)
2. Magnusson, M., Lilienthal, A., Duckett, T.: Scan registration for autonomous mining vehicles using 3D-NDT. J. Field Robot. **24**(10), 803–827 (2007)

3. Cho, Y., Kim, G., Kim, A.: Deeplo: geometry-aware deep lidar odometry. arXiv preprint arXiv:1902.10562 (2019)
4. Douillard, B., et al.: On the segmentation of 3D lidar point clouds. In: 2011 IEEE International Conference on Robotics and Automation, pp. 2798–2805. IEEE (2011)
5. Bogoslavskyi, I., Stachniss, C.: Efficient online segmentation for sparse 3D laser scans. PFG-J. Photogramm. Remote Sens. Geoinf. Sci. **85**(1), 41–52 (2017)
6. Vapnik, V.N.: An overview of statistical learning theory. IEEE Trans. Neural Netw. **10**(5), 988–999 (1999)
7. Freund, Y., Schapire, R.E.: A decision-theoretic generalization of on-line learning and an application to boosting. J. Comput. Syst. Sci. **55**(1), 119–139 (1997)
8. Takeuchi, E., Tsubouchi, T.: A 3-D scan matching using improved 3-D normal distributions transform for mobile robotic mapping. In: 2006 IEEE/RSJ International Conference on Intelligent Robots and Systems, pp. 3068–3073. IEEE (2006)
9. Rusu, R.B.: Semantic 3D object maps for everyday manipulation in human living environments. KI-Künstliche Intelligenz **24**(4), 345–348 (2010)
10. Yan, Z., Duckett, T., Bellotto, N.: Online learning for 3D lidar-based human detection: Experimental analysis of point cloud clustering and classification methods. Auton. Robot. **44**(2), 147–164 (2020)
11. Cortes, C., Vapnik, V.: Support-vector networks. Mach. Learn. **20**(3), 273–297 (1995)
12. Navarro-Serment, L.E., Mertz, C., Hebert, M.: Pedestrian detection and tracking using three-dimensional ladar data. Int. J. Robot. Res. **29**(12), 1516–1528 (2010)
13. Seder, M., Petrovic, I.: Dynamic window based approach to mobile robot motion control in the presence of moving obstacles. In: Proceedings 2007 IEEE International Conference on Robotics and Automation, pp. 1986–1991. IEEE (2007)
14. Fox, D., Burgard, W., Thrun, S.: The dynamic window approach to collision avoidance. IEEE Robot. Autom. Mag. **4**(1), 23–33 (1997)

Experiments of Composite Learning Admittance Control on 7-DoF Collaborative Robots

Xin Liu, Zhiwen Li, and Yongping Pan(✉)

School of Computer Science and Engineering, Sun Yat-sen University, Guangzhou 510006, China
{liux389,lizh63}@mail2.sysu.edu.cn, panyongp@mail.sysu.edu.cn

Abstract. With the increasing demand for collaborative robots in various industries, the implementation of interaction control algorithms with superior performances for collaborative robots becomes significant. This paper develops a composite learning admittance control (CLAC) approach and implements it to an industrial collaborative robot with 7 degrees of freedom named Franka Emika Panda. The interaction performance of the CLAC in the task space is verified by both force tracking and compliance control experiments. Experimental results show that the CLAC enables the Panda robot with favorable force tracking capability under a specific admittance model, high tracking accuracy in the free state, and satisfied compliance under motion constraints.

Keywords: Adaptive control · Collaborative robot · Composite learning · Compliance control · Human-robot interaction

1 Introduction

Unlike traditional industrial robots, collaborative robots can interact directly with people or environments while ensuring safety [1]. Robot interaction control can be broadly classified into two types: Hybrid force/position control [2] and impedance control [3]. Hybrid force/position control requires a motion controller along the tangent of kinematic constraints and a force controller along the normal, which is not practical for unstructured environments. In contrast, impedance control achieves compliance by controlling the dynamic relationship between motions and interaction forces, which means it does not require frequent switching of control modes for different environments and offers the possibility of controlling motion and forces simultaneously. Many researchers have improved impedance control from different aspects. For examples, to improve flexibility, hybrid impedance control that combines impedance control and hybrid force/position control was proposed in [4]; to enhance force tracking, impedance control with improved impedance models was proposed in [5–9].

Solving uncertain factors of collaborative robots to improve the impedance control performance is an important topic. Adaptive control is commonly combined with basic impedance control as it has the ability to handle parametric

X.-J. Liu et al. (Eds.): ICIRA 2021, LNAI 13014, pp. 532–541, 2021.
https://doi.org/10.1007/978-3-030-89098-8_50

uncertainty and to overcome the influence of nonparametric uncertainty through robust modifications [10]. For example, two adaptive control approaches were injected into position-based impedance control (i.e., admittance control) for unconstrained motion control in [11]. More adaptive impedance control schemes can be found in [12–20]. Nevertheless, only simulations are provided in these studies except [17–20], and experimental platforms are mostly robotic arms with no more than 3 degrees of freedom (DoFs). Recently, an innovative composite learning robot control (CLRC) approach was proposed in [21], where fast and accurate parameter estimation is achieved under a weak excitation condition named interval excitation. Further studies on CLRC can be founded in [22–27]. However, experiments are based on robotic arms with no more than 3 DoFs in the tracking control results of [22–25], and only simulations are provided in the impedance control results of [26, 27].

In this paper, we develop a composite learning admittance control (CLAC) approach for a 7-DoF collaborative robot named Franka Emika Panda using an experimental platform built with Robot Operating System (ROS) and Franka Control Interface (FCI). The aims of this study are to experimentally verify the force tracking and compliance control abilities of the CLAC in the task space and to obtain preliminary experimental results of the CLAC on a 7-DoF robot after solving the issues about highly coupled dynamics, imprecise friction dynamics, and rapid computation for 1 kHz real-time torque control.

In the rest of this paper, Sect. 2 shows the dynamics modeling of the Panda robot; Sect. 3 illustrates the CLAC design; Sect. 4 provides experimental results; Sect. 5 draws conclusive remarks. Throughout this article, $\mathbb{R}$, $\mathbb{R}^+$, $\mathbb{R}^n$ and $\mathbb{R}^{m\times n}$ denote the spaces of real numbers, positive real numbers, real n-vectors and real $m \times n$-matrices, respectively, diag $(x_1, \cdots, x_n)$ denotes a diagonal matrix with elements x_1 to x_n, $\|\boldsymbol{x}\|$ denotes the Euclidean norm value of $\boldsymbol{x}$, and L_∞ denotes the space of bounded signals, where $x_i \in \mathbb{R}$, $\boldsymbol{x} \in \mathbb{R}^n$, and $i = 1$ to n.

2 Panda Robot Dynamics

The Panda robot is a joint torque-controlled robotic arm usually described by a flexible-joint robot model [28]. Nevertheless, as the joint torque control loop is not opened for programming, this study only considers the rigid link dynamics described in the following Euler-Lagrangian form:

$$M(\boldsymbol{q})\ddot{\boldsymbol{q}} + C(\boldsymbol{q}, \dot{\boldsymbol{q}})\dot{\boldsymbol{q}} + G(\boldsymbol{q}) + F(\dot{\boldsymbol{q}}) = \boldsymbol{\tau} + \boldsymbol{\tau}_{\text{ext}} \quad (1)$$

where $\boldsymbol{q}(t) = [q_1(t), q_2(t), \ldots, q_7(t)]^T \in \mathbb{R}^7$ is a joint position, $M(\boldsymbol{q}) \in \mathbb{R}^{7\times 7}$ is a symmetric positive-definite inertia matrix, $C(\boldsymbol{q}, \dot{\boldsymbol{q}}) \in \mathbb{R}^{7\times 7}$ is a centripetal and Coriolis matrix, $G(\boldsymbol{q}) \in \mathbb{R}^7$ is a gravity torque, $\boldsymbol{\tau}(t) \in \mathbb{R}^7$ is a control torque, $\boldsymbol{\tau}_{\text{ext}}(t) \in \mathbb{R}^7$ is an external torque, and $F(\dot{\boldsymbol{q}}) \in \mathbb{R}^7$ is a friction torque given by

$$F(\dot{\boldsymbol{q}}) = F_v\dot{\boldsymbol{q}} + F_c\text{sgn}(\dot{\boldsymbol{q}}) \quad (2)$$

with $F_v = \text{diag}(f_{v1}, f_{v2}, \cdots, f_{v7})$ and $F_c = \text{diag}(f_{c1}, f_{c2}, \cdots, f_{c7})$, in which f_{vi}, $f_{ci} \in \mathbb{R}^+$ ($i = 1$ to 7) are coefficients of viscous friction and Coulomb friction, respectively. The left side of the equality (1) can be linearly parameterized by

$$M(\boldsymbol{q})\dot{\boldsymbol{v}} + C(\boldsymbol{q}, \dot{\boldsymbol{q}})\boldsymbol{v} + G(\boldsymbol{q}) + F(\dot{\boldsymbol{q}}) = \Phi^T(\boldsymbol{q}, \dot{\boldsymbol{q}}, \boldsymbol{v}, \dot{\boldsymbol{v}})W \tag{3}$$

where $\Phi \in \mathbb{R}^{N\times 7}$ is a regression matrix, $W \in \mathbb{R}^N$ is a base parameter vector, $\boldsymbol{v} \in \mathbb{R}^7$ is an auxiliary variable, and N is the number of the base parameters. Note that Φ can be derived from the Euler-Lagrangian formulation (1), and the base parameter vector W is unknown but constant.

Let $\boldsymbol{q}_d(t) = [q_{d1}(t), q_{d2}(t), \ldots, q_{d7}(t)]^T \in \mathbb{R}^7$ be a desired joint position satisfying $\boldsymbol{q}_d$, $\dot{\boldsymbol{q}}_d$, $\ddot{\boldsymbol{q}}_d \in L_\infty$, $\boldsymbol{e}(t) := \boldsymbol{q}_d(t) - \boldsymbol{q}(t)$ be a tracking error, $\boldsymbol{e}_f(t) := \dot{\boldsymbol{e}}(t) + \Lambda\boldsymbol{e}(t)$ be a filtered tracking error, $\hat{W}(t) \in \mathbb{R}^N$ be an estimate of W, $\tilde{W}(t) := W - \hat{W}(t)$ be a parameter estimation error, and $\tilde{W}_{\text{noF}}$ be a parameter estimation error without considering fiction, where $\Lambda \in \mathbb{R}^{7\times 7}$ is a positive-definite diagonal matrix.

3 Composite Learning Admittance Control

The CLAC consists of the CLRC in the inner position control loop to guarantee high tracking accuracy, and the admittance control in the outer control loop to achieve compliance behavior. Unlike the general adaptive control, the CLRC constructs a generalized prediction error using both instantaneous data and online data memory, and combines the generalized prediction error and the tracking error to construct the parameter update law, which has the advantages of obtaining accurate and smooth parameter estimation under a weakened excitation condition termed interval excitation [21]. The form of CLRC under human-robot interaction is briefly introduced below, and more details can be referred to [21].

The control torque $\boldsymbol{\tau}$ is designed as follows:

$$\boldsymbol{\tau} = K_c\boldsymbol{e}_f + \Phi^T(\boldsymbol{q}, \dot{\boldsymbol{q}}, \dot{\boldsymbol{q}}_r, \ddot{\boldsymbol{q}}_r)\hat{\boldsymbol{W}} - \boldsymbol{\tau}_{\text{ext}} \tag{4}$$

with $\dot{\boldsymbol{q}}_r := \dot{\boldsymbol{q}}_d + \Lambda\boldsymbol{e}$, where $K_c \in \mathbb{R}^{n\times n}$ is a positive-definite diagonal matrix of control gains. Then, one gets the closed-loop tracking error dynamics

$$M(\boldsymbol{q})\dot{\boldsymbol{e}}_f = -K_c\boldsymbol{e}_f - C(\boldsymbol{q}, \dot{\boldsymbol{q}})\boldsymbol{e}_f + \Phi^T(\boldsymbol{q}, \dot{\boldsymbol{q}}, \dot{\boldsymbol{q}}_r, \ddot{\boldsymbol{q}}_r)\tilde{W}. \tag{5}$$

Applying (3) into (1) lead to a linearly parameterized robotic model

$$\boldsymbol{\tau} + \boldsymbol{\tau}_{\text{ext}} = \Phi^T(\boldsymbol{q}, \dot{\boldsymbol{q}}, \dot{\boldsymbol{q}}, \ddot{\boldsymbol{q}})W. \tag{6}$$

To obviate using the acceleration $\ddot{\boldsymbol{q}}$, a linear filtering operator $L(s) = \frac{\alpha}{\alpha+s}\{*\}$ is applied on each side of (6), which gives

$$\boldsymbol{\tau}_f(t) = \Phi_f^T(\boldsymbol{q}, \dot{\boldsymbol{q}})W \tag{7}$$

with $\Phi_f(\boldsymbol{q}, \dot{\boldsymbol{q}}) := L(s)\{\Phi(\boldsymbol{q}, \dot{\boldsymbol{q}}, \dot{\boldsymbol{q}}, \ddot{\boldsymbol{q}})\}$ and $\boldsymbol{\tau}_f := L(s)\{\boldsymbol{\tau} + \boldsymbol{\tau}_{\text{ext}}\}$, where $\alpha \in \mathbb{R}^+$ is a filtering parameter, and s is a Laplace transform operator. Let $\hat{\boldsymbol{\tau}}_f(t) \in \mathbb{R}^7$ be a predicted counterpart of $\boldsymbol{\tau}_f$. A torque predicted model is given by

$$\hat{\boldsymbol{\tau}}_f(t) = \Phi_f^T(\boldsymbol{q}, \dot{\boldsymbol{q}})\hat{W}(t). \tag{8}$$

The generalized prediction error $\varepsilon(t) \in \mathbb{R}^N$ is is defined as

$$\varepsilon(t) := \Theta(t)W - \Theta(t)\hat{W} \tag{9}$$

where $\Theta(t) \in \mathbb{R}^{N \times N}$ is an excitation matrix given by

$$\Theta(t) := H(s)\{\Phi_f(\boldsymbol{q}, \dot{\boldsymbol{q}})\Phi_f^T(\boldsymbol{q}, \dot{\boldsymbol{q}})\} \tag{10}$$

and the term $\Theta(t)W$ is calculable by

$$\Theta(t)W = H(s)\{\Phi_f(\boldsymbol{q}, \dot{\boldsymbol{q}})\boldsymbol{\tau}_f\} \tag{11}$$

in which $H(s) = \frac{\beta}{s+\beta}\{*\}$ is a linear filtering operator, and $\beta \in \mathbb{R}^+$ is a filtering parameter. The composite learning law is given by

$$\dot{\hat{W}} = \Gamma(\Phi(\boldsymbol{q}, \dot{\boldsymbol{q}}, \dot{\boldsymbol{q}}_r, \ddot{\boldsymbol{q}}_r)\boldsymbol{e}_f + \kappa\boldsymbol{\varepsilon}) \tag{12}$$

where $\Gamma \in \mathbb{R}^{N \times N}$ is a positive-definite diagonal matrix of learning rates, and $\kappa \in \mathbb{R}^+$ is a weighting parameter. The superior tracking capability of the above CLRC design has been shown in [21].

For simplicity, this study only considers the robot end-effector position but omits the orientation. Figure 1 shows the blcok diagram of the CLAC for force tracking control, where $\boldsymbol{x}_r \in \mathbb{R}^3$, $\boldsymbol{x}_d \in \mathbb{R}^3$, and $\boldsymbol{x} \in \mathbb{R}^3$ represent a Cartesian reference position, a Cartesian desired position sent to the inner-loop position controller, and a Cartesian robot position, respectively, $\boldsymbol{x}_e \in \mathbb{R}^3$ denotes a position of the environment, $K_e \in \mathbb{R}^3$ denotes an unknown stiffness of the environment considered as a linear spring, $F_d \in \mathbb{R}^3$ is a desired contact force, $F_e \in \mathbb{R}^3$ is an interaction force between the robot and the environment, and $\tilde{F} = F_e - F_d$ is a force tracking error. The admittance model for force tracking is chosen as a second-order linear system with a transfer function $G(s) = 1/(M_d s^2 + B_d s + K_d)$, which is expressed as follows [7,9]:

$$F_e - F_d = M_d\ddot{\tilde{\boldsymbol{x}}} + B_d\dot{\tilde{\boldsymbol{x}}} + K_d\tilde{\boldsymbol{x}} \tag{13}$$

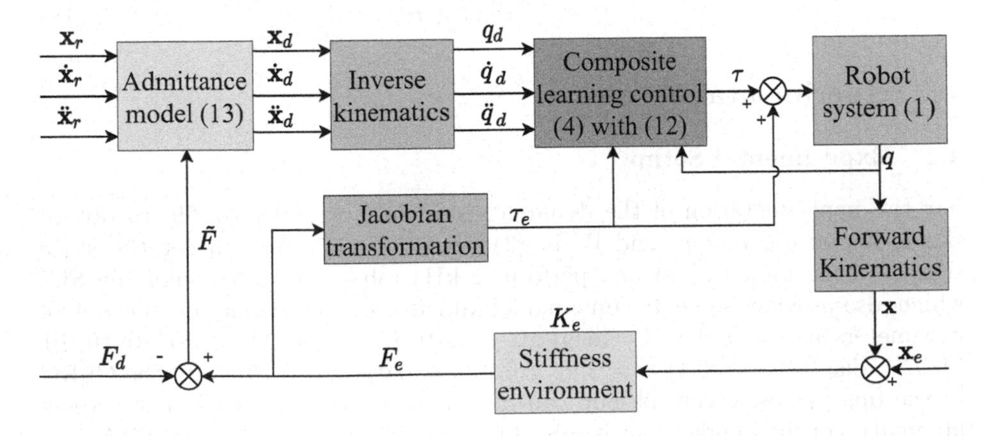

Fig. 1. A block diagram of Composite learning robot admittance control

where $M_d \in \mathbb{R}^{6\times6}$, $B_d \in \mathbb{R}^{6\times6}$, and $K_d \in \mathbb{R}^{6\times6}$ are the desired inertia, damping, and stiffness matrices, respectively, which are all diagonal and positive-definite. It is assumed in the inner control loop that $\boldsymbol{q}_d$ and $\boldsymbol{x}_d$ can be tracked accurately by the CLRC, so that the interaction force $F_e = K_e(\boldsymbol{x}_e - \boldsymbol{x}) = K_e(\boldsymbol{x}_e - \boldsymbol{x}_d)$ and the Cartesian tracking error $\tilde{\boldsymbol{x}} = \boldsymbol{x} - \boldsymbol{x}_r = \boldsymbol{x}_d - \boldsymbol{x}_r$. An offset from the admittance model (13) is applied to correct the Cartesian reference position $\boldsymbol{x}_r$ to generate the Cartesian desired position as $\boldsymbol{x}_d = \boldsymbol{x}_r + \tilde{\boldsymbol{x}} = \boldsymbol{x}_r + G(s)\{\tilde{F}\}$.

Let $f_d \in \mathbb{R}^+, f_e \in \mathbb{R}^+, \tilde{f} \in \mathbb{R}^+, m_d \in \mathbb{R}^+, b_d \in \mathbb{R}^+, k_d \in \mathbb{R}^+, k_e \in \mathbb{R}^+, x_d \in \mathbb{R}^+, x_r \in \mathbb{R}^+$ and $\tilde{x} \in \mathbb{R}^+$ be an element of $F_d, F_e, \tilde{F}, M_d, B_d, K_d, K_e, \boldsymbol{x}_d, \boldsymbol{x}_r$ and $\tilde{\boldsymbol{x}}$ respectively. Since each Cartesian dimension is independent, one dimension is chosen during illustration for simplicity, such that (13) becomes

$$f_e - f_d = m_d\ddot{\tilde{x}} + b_d\dot{\tilde{x}} + k_d\tilde{x}. \tag{14}$$

Applying $f_e = k_e(x_e - x_d)$ and $x_d = x_r + G(s)\{\tilde{f}\}$ to $\tilde{f} = f_e - f_d$ yields

$$\tilde{f} = k_e(x_e - x_r) - k_e G(s)\{\tilde{f}\} - f_d. \tag{15}$$

Substituting $G(s) = 1/(m_d s^2 + b_d s + k_d)$ to (15), one obtains

$$\left(1 + \frac{k_e}{m_d s^2 + b_d s + k_d}\right)\{\tilde{f}\} = k_e(x_e - x_r) - f_d. \tag{16}$$

Letting $s = 0$ in (16), one gets a steady-state force tracking error

$$\tilde{f}_{ss} = \frac{k_d}{k_d + k_e}[k_e(x_e - x_r) - f_d]. \tag{17}$$

The environment position x_e and the environment stiffness k_e cannot be obtained precisely. According to (14), the desired stiffness k_d can be chosen as 0, which results in $\tilde{f}_{ss} = 0$ independent of the environment model with any stiffness k_e. Thus, one gets the force tracking impedance model as follows [7,9]:

$$f_e - f_d = m\ddot{\tilde{x}} + b\dot{\tilde{x}}. \tag{18}$$

4 Experimental Studies

4.1 Experimental Setup

For the implementation of the dynamics modeling, we refer to [29] to obtain the forms of $\Phi(\boldsymbol{q}, \dot{\boldsymbol{q}}, \boldsymbol{v}, \dot{\boldsymbol{v}})$ and W in (3) with $N = 57$. We can get the state signals of the Panda robot and perform 1 kHz robot control through the FCI which also provides some dynamic model information. The values of the control parameters are selected as $\Lambda = \text{diag}(20, \cdots, 20)$, $K_c = \text{diag}(20, 20, 20, 20, 10, 10, 5)$, $\Gamma = \text{diag}(0.1, \cdots, 0.1)$, $\kappa = 2$, $\alpha = 0.2$, $\beta = 0.05$, and $\hat{W}(0) = 0$. The CLRC in the inner position control loop can be calculated within 0.1 ms. The inverse kinematics of the Panda robot is solved by using the numerical solver TRAC-IK [31] which can complete the calculation within 0.37 ms. The Panda robot equips with 7 joint torque sensors to measure resultant forces acting on the link side. We use the estimated interaction forces of the 7 joints provided by the FCI (based on the joint torque sensors) to obtain the required interaction forces of the robot end-effector by Jacobian pseudo-inverse transformation. The estimated interaction forces include both actual interaction forces and some unmodeled efforts such as friction and sensing noise, resulting in some side effects on the subsequent experimental results.

4.2 Force Tracking Experiments

We consider only the z-axis admittance model in this subsection and utilize the robot's gripper to squeeze a horizontal rigid table to test the force tracking ability. The impedance parameters are set as $m_d = 20$, $b_d = 3000$ and $k_d = 0$. The end-effector initially moves to the position with $x = 0.1\,\mathrm{m}$, $y = -0.45\,\mathrm{m}$ and $z = 0.65\,\mathrm{m}$. The z-axis is then given the desired force and the end-effector presses down continuously until it reaches the table. Subsequently, the end-effector attempts to stabilize on the contact surface until 110 s, performs a sinusoidal motion on the y-axis with period 8 s and amplitude 0.05 m between 110 and 190 s, and again attempts to stabilize on the contact surface after 190 s.

When the desired interaction forces f_d of the z-axis admittance model are set to 10 N and 25 N, the actual interaction forces f_e are shown in Fig. 2(a) and 2(b), respectively, and the norm of the parameter estimation error $\tilde{W}_{\mathrm{noF}}$ in the

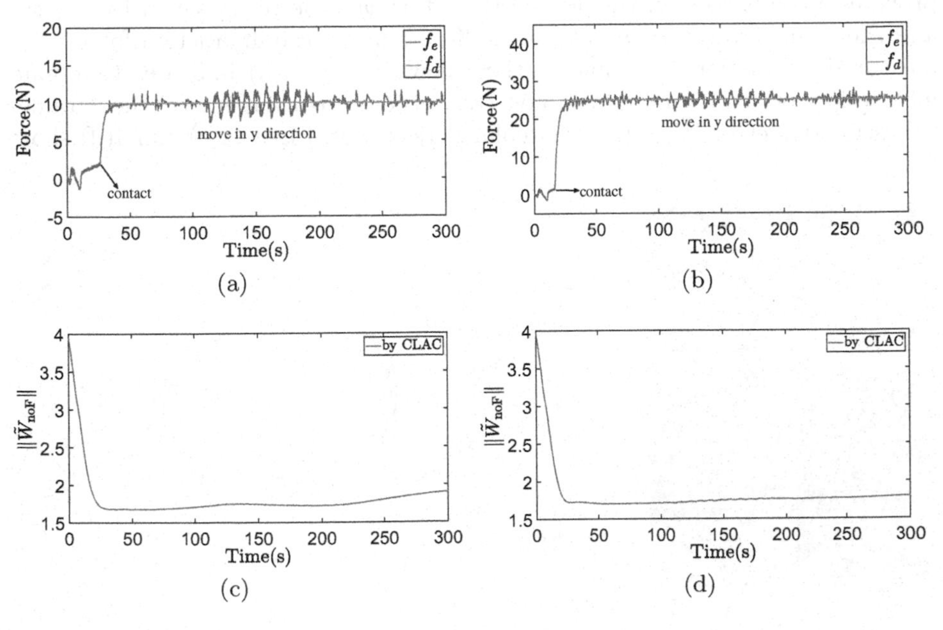

Fig. 2. Force tracking experiments of CLAC in task space. (a) Variation of f_e with $f_d = 10\,\mathrm{N}$. (b) Variation of f_e with $f_d = 25\,\mathrm{N}$. (c) Variation of $\|\tilde{W}_{\mathrm{noF}}\|$ with $f_d = 10\,\mathrm{N}$. (d) Variation of $\|\tilde{W}_{\mathrm{noF}}\|$ with $f_d = 25\,\mathrm{N}$.

inner-loop CLRC are shown in Fig. 2(c) and 2(d), respectively. It is observed that the interaction force f_e keeps at the desired value with accuracy about 1 N when the end-effector is static, but the accuracy decreases to be about 3 N when the end-effector is moving along the table surface. With the consideration of unmodeled efforts, the above result still implies favorable force tracking performance. In addition, the parameter estimation error $\tilde{W}_{\mathrm{noF}}$ converges quickly regardless of the value of the desired force, and the slight fluctuations result from the construction of the generalized prediction error in the CLRC which is disturbed by inaccurate measurement of interaction forces.

4.3 Compliance Control Experiments

We consider the admittance model (14) with x-, y- and z-axis directions of the task space in this subsection, and maintain the parameter settings as before, in which $f_d = 0\,\mathrm{N}$, $m_d = 10$, $b_d = 200$, and $k_d = 100$. The desired position for this experiment is a counterclockwise circular trajectory in the y-z plane with radius 0.2 m and period 16 s. To avoid frequently generating trajectory correction terms for measured interaction torques with noise, we add a dead zone with 2 N to the measured interaction torque of each joint. To demonstrate the compliance of the CLAC, we randomly apply an interaction force to deflect the end-effector towards the negative direction of the y-axis.

It is shown in Fig. 3(a) that the desired trajectory $\boldsymbol{x}_d$ changes accordingly under the interaction force, which means that the CLAC is compliant. Figure 3(b) provides a comparison of the desired trajectory $\boldsymbol{x}_d$, the reference trajectory $\boldsymbol{x}_r$ and the actual trajectory $\boldsymbol{x}$ of the end-effector under the interaction force. The measured interaction force information shown in Fig. 4(a) indicates that four interactions occur. Figure 4(b) demonstrates the fast convergence of the parameter estimation error $\tilde{W}_{\mathrm{noF}}$ under circular trajectories. Figure 4(c) and 4(d) show

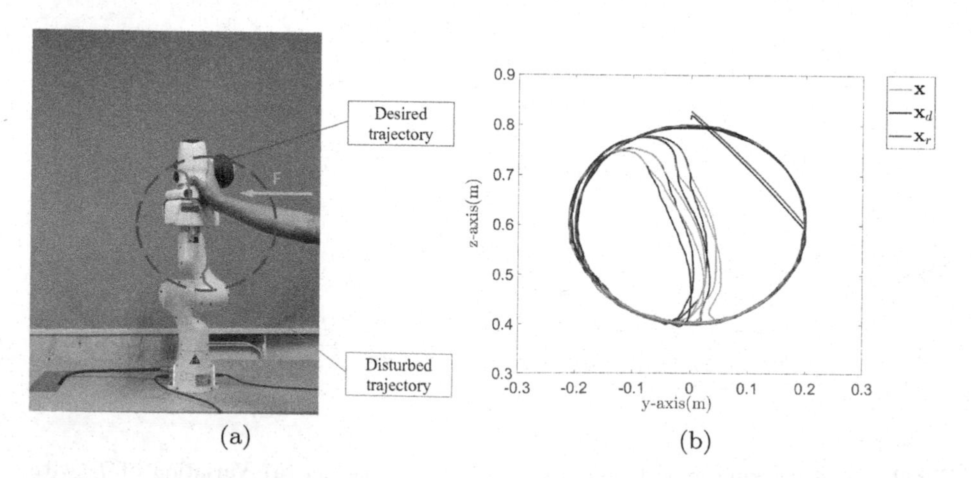

Fig. 3. Compliance control experiments of CLAC in task space. (a) Schematic diagram of the experiments. (b) Variation of $\boldsymbol{x}_d$, $\boldsymbol{x}_r$, and $\boldsymbol{x}$ during the experiments.

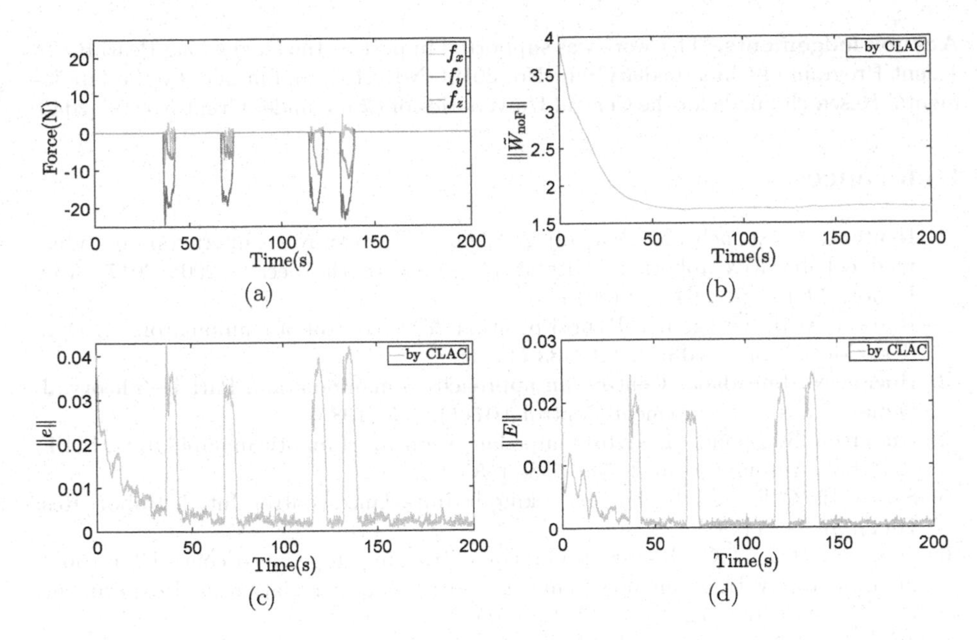

Fig. 4. Compliance control experiments of CLAC when the robot end-effector performs circular motion in the y-z plane. (a) Variation of interaction forces f_x, f_y and f_z. (b) Variation of $\|\tilde{W}_{\text{noF}}\|$ (c) Variation of $\|\boldsymbol{e}\|$.(d) Variation of $\|\boldsymbol{E}\|$.

the norms of the joint tracking error $\boldsymbol{e}$ and the Cartesian position tracking error $\boldsymbol{E} = \boldsymbol{x}_d - \boldsymbol{x}$ without considering the rotational pose deviation, respectively. It is observed that $\|\boldsymbol{e}\|$ can be reduced to about 0.003 and $\|\boldsymbol{E}\|$ can be reduced to about 0.001 after the interaction force is withdrawn, i.e., the average tracking error is less than 0.001 rad per joint in the joint space, and is less than 0.4 mm per direction in the Cartesian space. These results fully demonstrate that the CLAC can maintain both compliance during interactive motion and high tracking accuracy during unconstrained motion. Experimental results of pure position tracking control on the Panda robot can be referred to [32].

5 Conclusions

In this paper, a CLAC approach has been developed for a 7-DoF collaborative robot named Franka Emika Panda, where its effectiveness has been illustrated through both task-space force tracking and interaction control experiments. In our further research, the Panda robot would be equipped with a more accurate 6-axis force/torque sensor at the wrist to increase the accuracy of external force measurement, and variable impedance algorithms would be incorporated into the control framework of this study to enable richer applications.

Acknowledgements. This work was supported in part by the Guangdong Pearl River Talent Program of China under Grant No. 2019QN01X154, and in part by the Fundamental Research Funds for the Central Universities of China under Grant No. 19lgzd40.

References

1. Hentout, A., Aouache, M., Maoudj, A., Akli, I.: Human-robot interaction in industrial collaborative robotics: a literature review of the decade 2008–2017. Adv. Robot. **33**(15–16), 764–799 (2019)
2. Raibert, M.H., Craig, J.J.: Hybrid position/force control of manipulators. J. Dyn. Syst. Meas. Contr. **103**(2), 126–133 (1981)
3. Hogan, N.: Impedance Control: an approach to manipulation: Part I—Theory. J. Dynamic Syst. Measurement Control **107**(1), 1–7 (1985)
4. Anderson, R., Spong, M.: Hybrid impedance control of robotic manipulators. IEEE J. Robot. Automation **4**(5), 549–556 (1988)
5. Seraji, H., Colbaugh, R.: Force tracking in impedance control. Int. J. Robot. Res. **16**(1), 97–117 (1997)
6. Jung, S., Hsia, T.C., Bonitz, R.G.: Force tracking impedance control for robot manipulators with an unknown environment: theory, simulation, and experiment. Int. J. Robot. Res. **20**(9), 765–774 (2001)
7. Jung, S., Hsia, T., Bonitz, R.: Force tracking impedance control of robot manipulators under unknown environment. IEEE Trans. Control Syst. Technol. **12**(3), 474–483 (2004)
8. Roveda, L., Vicentini, F., Tosatti, L.M.: Deformation-tracking impedance control in interaction with uncertain environments. In: Proceeding of IEEE/RSJ International Conference on Intelligent Robots and Systems, pp. 1992–1997, Tokyo, Japan (2013)
9. Duan, J., Gan, Y., Chen, M., Dai, X.: Adaptive variable impedance control for dynamic contact force tracking in uncertain environment. Robot. Auton. Syst. **102**, 54–65 (2018)
10. Peng, L., Woo, P.Y.: Neural-fuzzy control system for robotic manipulators. IEEE Control Syst. Mag. **22**(1), 53–63 (2002)
11. Lu, W.S., Meng, Q.H.: Impedance control with adaptation for robotic manipulations. IEEE Trans. Robot. Autom. **7**(3), 408–415 (1991)
12. Kelly, R., Carelli, R., Amestegui, M., Ortega, R.: On adaptive impedance control of robot manipulators. In: Proceeding of International Conference on Robotics and Automation, Scottsdale, AZ, USA, pp. 572–577 (1989)
13. Carelli, R., Kelly, R.: An adaptive impedance/force controller for robot manipulators. IEEE Trans. Autom. Control **36**(8), 967–971 (1991)
14. Wedeward, K., Colbaugh, R.: New stability results for direct adaptive impedance control. In: Proceeding of International Symposium on Intelligent Control, Monterey, CA, USA, pp. 281–287 (1995)
15. Chien, M.C., Huang, A.C.: Adaptive impedance control of robot manipulators based on function approximation technique. Robotica **22**(4), 395–403 (2004)
16. Jiang, Z.H.: Impedance control of flexible robot arms with parametric uncertainties. J. Intell. Rob. Syst. **42**(2), 113–133 (2005)
17. Sharifi, M., Behzadipour, S., Vossoughi, G.R.: Model reference adaptive impedance control of rehabilitation robots in operational space. In: Proceeding of IEEE RAS/ EMBS International Conference on Biomedical Robotics and Biomechatronics, Rome, Italy, pp. 1698–1703 (2012)

18. Sharifi, M., Behzadipour, S., Vossoughi, G.: Model reference adaptive impedance control in cartesian coordinates for physical human–robot interaction. Adv. Robot. **28**(19), 1277–1290 (2014)
19. Sharifi, M., Behzadipour, S., Vossoughi, G.: Nonlinear model reference adaptive impedance control for human-robot interactions. Control. Eng. Pract. **32**, 9–27 (2014)
20. Li, X., Pan, Y., Chen, G., Yu, H.: Adaptive human-robot interaction control for robots driven by series elastic actuators. IEEE Trans. Rob. **33**(1), 169–182 (2017)
21. Pan, Y., Yu, H.: Composite learning robot control with guaranteed parameter convergence. Automatica **89**, 398–406 (2018)
22. Guo, K., Pan, Y., Yu, H.: Composite learning robot control with friction compensation: a neural network-based approach. IEEE Trans. Industr. Electron. **66**(10), 7841–7851 (2019)
23. Guo, K., Pan, Y., Zheng, D., Yu, H.: Composite learning control of robotic systems: a least squares modulated approach. Automatica **111**, Art. No. 108612 (2020)
24. Huang, D., Yang, C., Pan, Y., Cheng, L.: Composite learning enhanced neural control for robot manipulator with output error constraints. IEEE Trans. Industr. Inf. **17**(1), 209–218 (2021)
25. Jiang, Y., Wang, Y., Miao, Z., Na, J., Zhao, Z., Yang, C.: Composite-learning-based adaptive neural control for dual-arm robots with relative motion. IEEE Trans. Neural Networks Learn. Syst., to be published (2021). https://doi.org/10.1109/TNNLS.2020.3037795
26. Sun, T., Peng, L., Cheng, L., Hou, Z.G., Pan, Y.: Composite learning enhanced robot impedance control. IEEE Trans. Neural Networks Learn. Syst. **31**(3), 1052–1059 (2020)
27. Sun, T., Cheng, L., Peng, L., Hou, Z., Pan, Y.: Learning impedance control of robots with enhanced transient and steady-state control performances. Sci. China: Inf. Sci. **63**(9), 1–13 (2020)
28. Albu-Schaffer, A., Ott, C., Hirzinger, G.: A unified passivity-based control framework for position, torque and impedance control of flexible joint robots. Int. J. Robot. Res. **26**(1), 23–39 (2007)
29. Kawasaki, H., Bito, T., Kanzaki, K.: An efficient algorithm for the model-based adaptive control of robotic manipulators. IEEE Trans. Robot. Autom. **12**(3), 496–501 (1996)
30. Gaz, C., Cognetti, M., Oliva, A., Giordano, P.R., De Luca, A.: Dynamic identification of the franka emika panda robot with retrieval of feasible parameters using penalty-based optimization. IEEE Robot. Automation Lett. **4**(4), 4147–4154 (2019)
31. Beeson, P., Ames, B.: TRAC-IK: An open-source library for improved solving of generic inverse kinematics. In: Proceeding of IEEE/RAS International Conference on Humanoid Robots, Seoul, South Korea, pp. 928–935 (2015)
32. Liu, X., Li, Z., Pan Y.: Preliminary evaluation of composite learning tracking control on 7-DoF collaborative robots. In: IFAC Conference on Modelling, Identification and Control of Nonlinear Systems, Tokyo, Japan, pp. 1–6 (2021)

Analysis and Optimization of Interior Noise Based on Transfer Path Analysis

Wenqiang Liu[1,2(✉)], Junfeng Hu[2], Fengxin Jiang[2], Bing Gong[2], Xiaolong Deng[2], and Yongjiang Xu[2]

[1] Tsinghua University, Beijing 100084, China
liuwenqiang@geely.com
[2] Ningbo Geely Royal Engine Components Co., Ltd., Ningbo 315336, China

Abstract. The identification and solution to noise and vibration problems are important ways to continuously improve the comfort in the mechanical field. The car is a more complex mechanical system, and the transfer path of noise and vibration into the car is very complex. It is difficult to identify the path problem by conventional noise and vibration test and analysis measures. Using the transfer path analysis method, the key transfer path of noise and vibration can be identified more quickly. Aiming at the problem of 340–440 Hz and 460–560 Hz acceleration roughness noise in a newly developed car, the OTPA model is built, and the main contribution paths of the problematic frequency band of the car are the right suspension active side tie rod and the right suspension passive side bracket. The scheme of adding vibration absorber is verified, and the acceleration roughness amplitude in this frequency band is reduced by about 3–5 dB (A). The NVH problem can also be solved by increasing the dynamic stiffness of the active side mount bracket with CAE. It is proved that by further improving the dynamic stiffness of the active end mount bracket, the resonance amplitude of the passive end can be reduced, which provides a new idea for similar problems.

Keywords: Noise · OTPA · Transmission path · Contribution analysis

1 Introduction

With the development of science and technology, consumers have higher and higher requirements for the comfort of automotive products, especially the NVH (Noise, Vibration, Harshness) performance of the vehicle, which directly affects the customer's evaluation of the vehicle quality. There are many factors that affect the NVH problem of vehicles. The engine and transmission are the main noise sources. Transfer paths such as mounting system and connecting pipeline will transmit the vibration and noise of the powertrain to the vehicle, thus reducing the quality of the vehicle [1–3].

In the research and development process of a GEELY vehicle, it is found that there are obvious 340–440 Hz and 460–560 Hz acceleration knocking noises in the vehicle, which seriously affect the NVH quality of the vehicle. Since the powertrain is equipped on other vehicles without this problem, it is considered that the biggest cause of the

X.-J. Liu et al. (Eds.): ICIRA 2021, LNAI 13014, pp. 542–551, 2021.
https://doi.org/10.1007/978-3-030-89098-8_51

problem is the transfer path resonance, which amplifies the noise inside the vehicle, thus leads to this problem. Therefore, the key to solve this problem is to identify the transfer path. In this paper, through the OTPA analysis method, the transfer path of the vehicle is analyzed, and the main contribution path of noise is found. The accuracy of this method is verified by adding vibration absorber, and an effective engineering solution is further found by CAE method.

2 Principle of OTPA Analysis Method

In 1997, P. J.G. Vander Linden et al. and Wim Hendricx et al. introduced the basic principle of this method of quantifying the airborne noise, and elaborated and analyzed the contribution of the body parts which affect the interior noise; In 2016, Guo Shihui and others introduced the transfer path analysis method under working condition load, and introduced the basic principle of TPA (Transfer Path Analysis); In 2019, Jiang Shaowei conducted the research on vehicle interior noise identification based on working condition transfer path analysis method. Through the comparison of research methods in domestic and abroad, the faster test method is operational transfer path analysis (OTPA), which can quickly analyze the contribution of each transfer path to the sound pressure at the measured position, and whether the contribution is caused by large excitation or sensitive path [4–6].

The principle of TPA is to decompose the noise of the response point into the superposition of the different excitation onto each path. The formula is

$$X(\omega) = \sum_{i=1}^{n} H_i(\omega)F_i(\omega) \tag{1}$$

Where $X(\omega)$ is the response and $H_i(\omega)$ is the transfer function of each path, $F_i(\omega)$ for each excitation. For noise response

$$\mathrm{P}(\omega) = \sum_{i=1}^{n} NTF_i(\omega)F_i(\omega) + \sum_{j=1}^{m} NTF_j(\omega)Q_j(\omega) \tag{2}$$

Where $F_i(\omega)$ is the load of each vibration source of the system, $Q_i(\omega)$ is the sound load of each sound source acting on the system, and $NTF_i(\omega)$, $NTF_j(\omega)$ is the transfer function from the excitation point of vibration source and sound source to the response point respectively.

OTPA is based on the response and the transitivity matrix of the response [7–14]. The main theoretical formula is as follows

$$\begin{bmatrix} H_{11} & \cdots & H_{1n} \\ H_{21} & \cdots & H_{2n} \\ \vdots & \vdots & \vdots \\ H_{(m-1)1} & \cdots & H_{(m-1)(n-1)} \\ H_{m1} & \cdots & H_{mn} \end{bmatrix} = \begin{bmatrix} X_{11} & \cdots & X_{1m} \\ X_{21} & \cdots & X_{2m} \\ \vdots & \vdots & \vdots \\ X_{(j-1)1} & \cdots & X_{(j-1)(m-1)} \\ X_{j1} & \cdots & X_{jm} \end{bmatrix}^{-1} \begin{bmatrix} Y_{11} & \cdots & Y_{1n} \\ Y_{21} & \cdots & Y_{2n} \\ \vdots & \vdots & \vdots \\ Y_{(j-1)1} & \cdots & Y_{(j-1)(n-1)} \\ Y_{j1} & \cdots & Y_{jn} \end{bmatrix} \tag{3}$$

Where H is the transfer function matrix, X is the input signal matrix (including vibration signal and noise signal), Y is the target point response matrix, m is the number of channels at the excitation point, j is the number of test steps (it is recommended that j is greater than or equal to 3), and n is the number of channels at the response point. After the singular value decomposition of the X matrix, the

$$\begin{bmatrix} X_{11} \\ X_{12} \\ \vdots \\ X_{1(m-1)1} \\ X_{1m} \end{bmatrix} \begin{bmatrix} H_{11} \\ H_{21} \\ \vdots \\ H_{(m-1)1} \\ H_{m1} \end{bmatrix} = \begin{bmatrix} Y_1 \\ Y_2 \\ \vdots \\ Y_{(m-1)} \\ Y_m \end{bmatrix} \sum\nolimits_{\rightarrow} \ldots Y_{1,syn} \begin{bmatrix} Y_{1,syn} \\ Y_{2,syn} \\ \vdots \\ Y_{(m-1),syn} \\ Y_{m,syn} \end{bmatrix} \sum\nolimits_{\rightarrow} \ldots Y_{syn} \quad (4)$$

The singular value decomposition of the X matrix also realizes the main component decomposition. The smaller part of the eigenvalues need to be discarded in the calculation.

3 OTPA Transmission Path Identification

The main parameters of the vehicle studied in this paper are shown in Table 1. The main problem frequency bands of the car in the partial load acceleration condition are 340–440 Hz and 460–560 Hz. The spectrum diagram of interior noise is shown in Fig. 1. As can be seen from Fig. 1, the energy of these two bands is very prominent during the whole operation condition, with which can be judged as two resonance bands. Because the masking effect of road noise, wind noise and other noises are relatively small at low vehicle speed which resulting the noise of these two bands are very obvious in the vehicle carbine, and consequently sound quality of the vehicle is seriously affected. In the total cost of power, the vibration signals of the two resonance bands are not obvious, so it can be judged that there is resonance of parts in the transmission path, which causes problems in the car.

Table 1. Vehicle and engine parameters

Vehicle	Engine					
Weight	Displacement	Air influence	Injection method	Rated power	Rated speed	Maximum torque
1670 kg	1.969 L	Turbocharged	Direct injection	175 kW	5500 r/min	350 Nm

The structure transfer path of the car mainly includes the suspension system, heating water pipe, air conditioning pipe, transmission half shaft, shift cable, exhaust hook, etc. the radiated noise is mainly transmitted into the car through the front wall of the car body. Therefore, OTPA method can be used to analyze and sort the possible structure

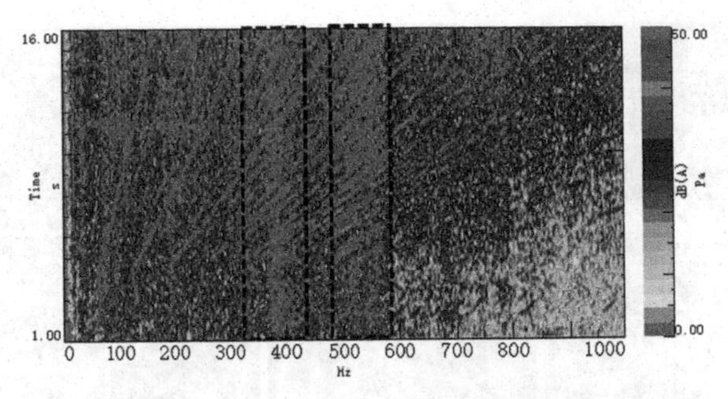

Fig. 1. Interior noise spectrum under acceleration condition

contribution path and air contribution path, and quickly identify the main contribution path. In OTPA analysis, it is necessary to fully understand all the transmission paths of the whole vehicle, and the layout of the measurement points of the required path is highly required. If the selection is not reasonable, the transfer function error will be relatively large, which will affect the accuracy of the noise analysis of the powertrain in the vehicle. The transfer function is calculated under different working conditions, so there are high requirements for the selection of test conditions. Therefore, the sensors should be arranged carefully and the operating conditions should not be similar to each other [14, 15].

This paper mainly considers four mounting points, left and right axle, air conditioning pipeline and heating pipeline in the structural path analysis by totally dampening the intake and decoupling the exhaust hook. The acceleration sensor is used to obtain the vibration data of each structural path as the excitation input of the structural path. The air path mainly considers the influence of cabin on the interior noise. The microphone measurement points are in the near field of engine, transmission and generator transformer. The data of near field noise is used as the excitation input of air path. Therefore, the specific digital analysis model and sensor test points are shown in Fig. 2 and Fig. 3.

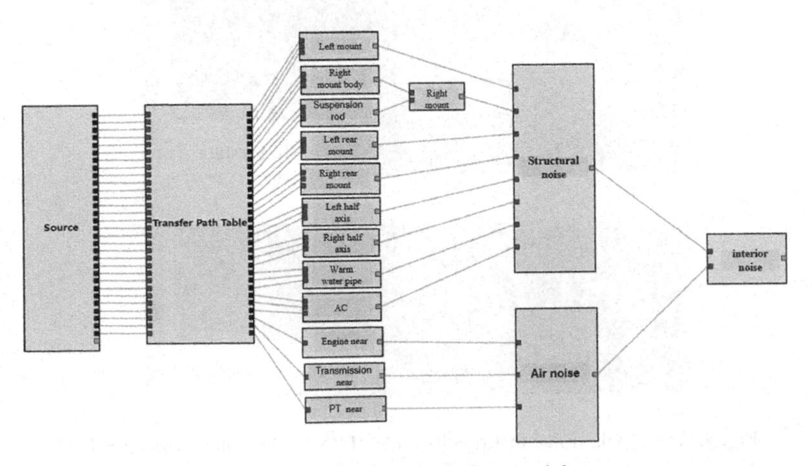

Fig. 2. OTPA calculation model

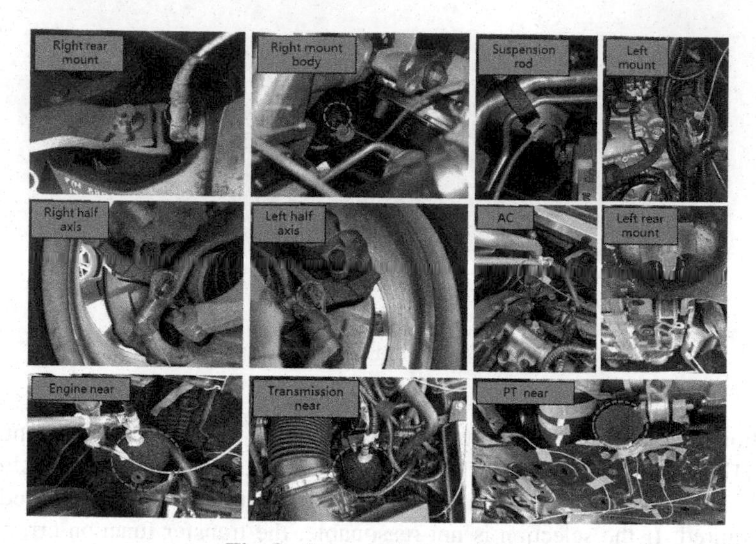

Fig. 3. Sensor pointing photos

In order to get a more accurate transfer function, not only the main path measurement points should not be lost and repeated, but also the number of working steps should not be less than 3 times of the number of channels. In this paper, the working condition data of 30% throttle opening is used to analyze the problem, and the rest of the working condition data is used to solve the transfer function. As shown in Fig. 4, the spectrum comparison between the test results of 30% throttle opening and the calculation results of vehicle interior noise shows that the calculation results of the problem frequency band are close to the measured results, which indicates that the model has included the main contribution path of the problem and ensures the accuracy of path contribution analysis.

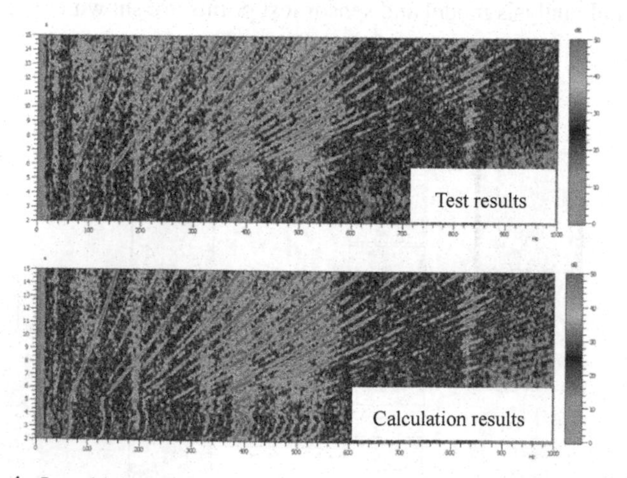

Fig. 4. In-vehicle noise test results and OTPA model calculation results

Figure 5 shows the analysis results of each path contribution of vehicle interior roughness sound. It can be found that the paths that have the greatest impact on vehicle interior roughness sound are the right suspension rod and the right suspension passive side bracket. In order to reduce the rough sound and improve the interior sound quality, these two main paths need to be optimized and verified.

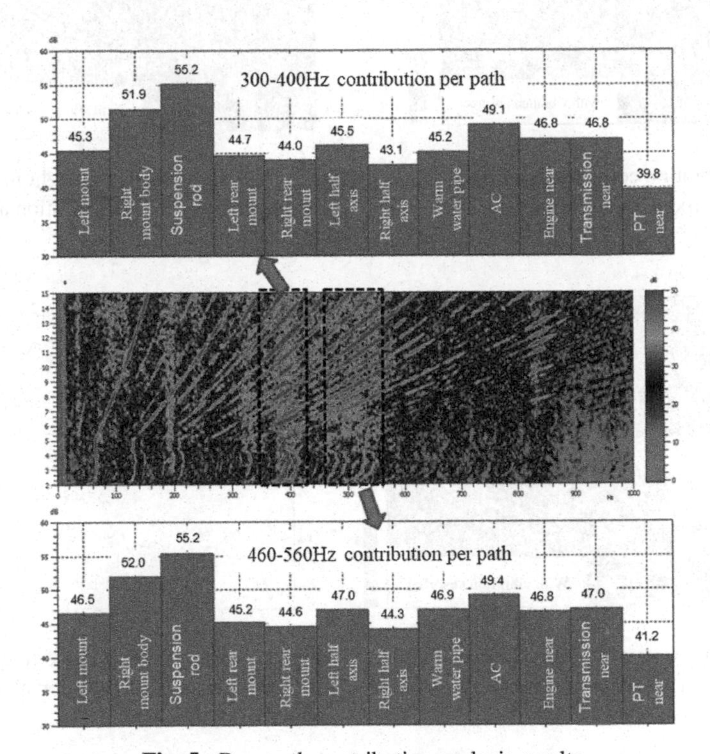

Fig. 5. Pass path contribution analysis results

Figure 6 and Fig. 7 show the vibration spectrum changes of the right suspension rod and the passive side bracket with vibration absorber added respectively. It can be concluded from the comparative effect of the frequency color charts in Fig. 6 and Fig. 7 that after adding the vibration absorber, the vibration of the right suspension rod and the passive side bracket decreases significantly, and the vibration peaks at the characteristic frequency bands of 340–440 Hz and 460–560 Hz basically disappear. It can be seen from Fig. 8 and Fig. 9 that after adding vibration absorbers to the right suspension rod and the right suspension passive side bracket, the amplitude of 340–440 Hz and 460–560 Hz acceleration roughness sound in the vehicle decreases by about 3–5 dB (A), the subjective evaluation of the interior roughness sound basically disappears, and the sound quality of the interior acceleration condition is significantly improved. It also proves that the path which has the greatest impact on the noise in this frequency.

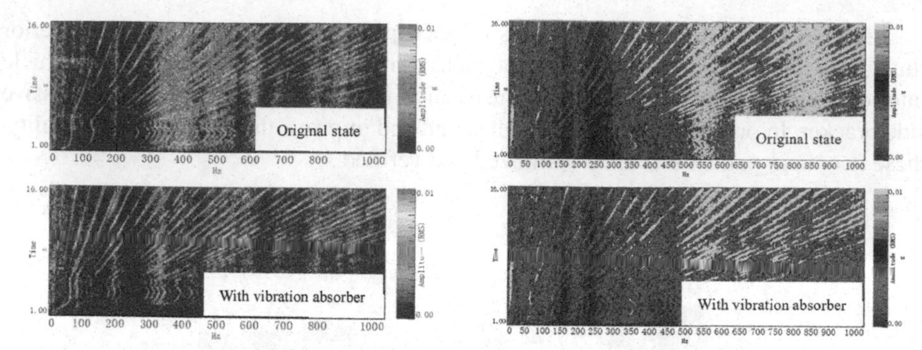

Fig. 6. Vibration comparison of right suspension rod with vibration absorber

Fig. 7. Vibration change of right mount passive side bracket with vibration absorber added

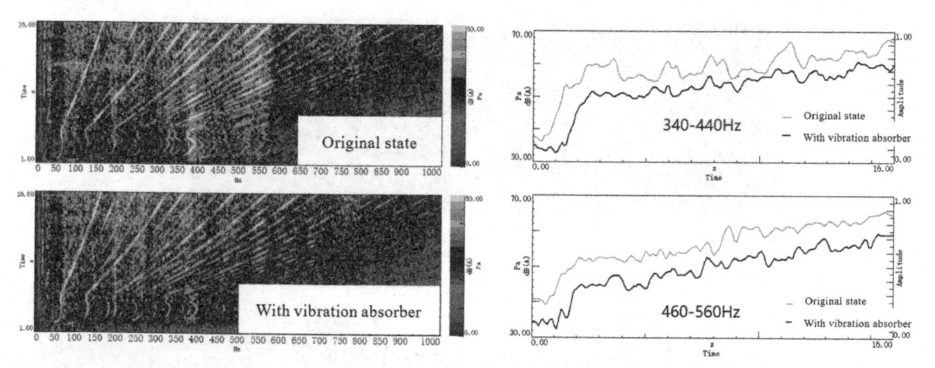

Fig. 8. Comparison of interior noise of vehicle with vibration absorber on passive side of right mount

Fig. 9. Comparison of noise frequency slice in vehicle with vibration absorber added to right mount

4 CAE Optimization Analysis of the Delivery Path

It can be seen from the previous text that the right suspension rod and the right suspension passive side bracket are the main paths for the noise transmission of the vehicle. Obviously, the resonance of the two brackets enlarges the excitation of the powertrain, resulting in the noise problem in the vehicle. Although this problem can be solved by adding vibration absorbers, it involves the layout space and will increase the cost. It is obviously very difficult to directly change the suspension passive side bracket or even the body sub frame.

According to the formula of transmission rate of mount, the influence of transmission path can be reduced by optimizing the active side and passive side of mount. When the structural characteristics of the mount remain unchanged, the transmissibility is only related to the frequency ratio and damping ratio. When the mount attenuates the excitation (without resonance), the amplitude of the force transmitted to the foundation is proportional to the amplitude of the excitation force. Reducing the excitation force of the active side mount can also reduce the excitation transmitted to the passive side of the car body. The formula of transmissibility is as follows:

$$T = \left|\frac{F_a}{F}\right| = \sqrt{\frac{1 + (2\xi r)^2}{(1 - r^2)^2 + (2\xi r)^2}} \tag{5}$$

Where T is the transmissivity, F_α is the amplitude of the force transmitted to the foundation, F is the amplitude of the exciting force, r is the ratio of the exciting frequency to the natural frequency, and ξ is the viscous damping ratio [15].

Therefore, the author considers to optimize and improve the active side of the right mount. By reducing the dynamic flexibility of the active side of the right mount, the vibration excitation force transmitted from the powertrain to the vehicle body is reduced, and the risk of noise transmission to the vehicle is reduced. Therefore, the CAE simulation of powertrain is established, as shown in Fig. 10. The simulation model of CAE powertrain system includes cylinder head, cylinder block, crankcase, cylinder head cover, mounting bracket, oil pan, right mounting bracket, engine cooling bracket and other parts system. The engine moving parts system is simulated by concentrated mass and divided by two-order tetrahedron, with a total of 700,000 units and 1.2 million nodes. The boundary condition is: in the free state of the whole movable assembly, unit load is applied to the center point of the right suspension active end bracket for calculation, and the origin dynamic stiffness of the suspension active end bracket in the assembly state of the movable assembly is finally obtained.

Fig. 10. Build the powertrain simulation model

The analysis results are shown in Fig. 11. The blue line is the original dynamic flexibility of the active end bracket, and the green line is the optimized dynamic flexibility of the active end bracket. It can be seen from the figure that the Z direction is much larger than the empirical value of 1e−4 mm/N, so it is necessary to reduce the dynamic flexibility and improve the dynamic stiffness of the active side bracket. The dynamic stiffness of the active side bracket is improved by adding a bolt connection. The results of dynamic flexibility analysis after adding bolts are shown in the green line of Fig. 11. The dynamic flexibility of the z-direction active side bracket is reduced by about 72% (the peak value is reduced from 6.7e−04 mm/N to 1.85e−04 mm/N), and the first-order modal frequency of the active side bracket is increased from 820 Hz to 993 Hz. At the same time, the dynamic flexibility decreases about 60% at 340–440 Hz and 64% at 460–560 Hz, which reduces the vibration response of the active side of the mount and controls the vibration transmission to the passive side of the mount. After optimization, the interior noise is reduced by about 2 dB(A) in the range of 340–560 Hz, which basically solves the problem. Therefore, although the problem of the vehicle is caused by the resonance of the right suspension rod and the right suspension passive side bracket, in the actual project, when various restrictions cannot be changed, the problem can be solved by greatly improving the dynamic stiffness of the active end bracket, reducing the engine transmission excitation, and reducing the vibration transmission rate. This idea has important reference significance for the optimization of transfer path of practical engineering problems.

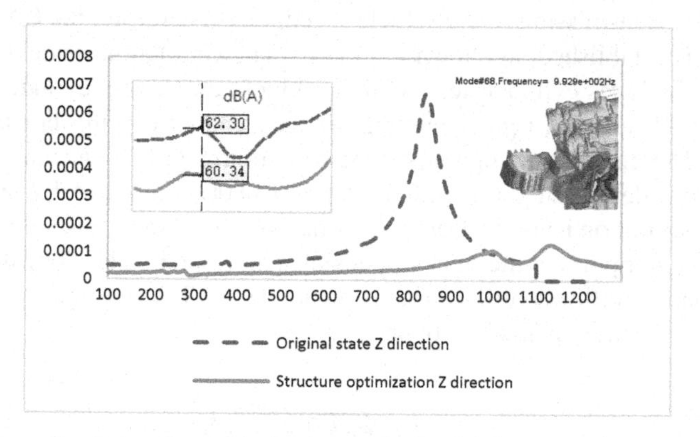

Fig. 11. Test results of structure optimization analysis of right suspension active side bracket

5 Conclusion

In this paper, based on the main noise frequency band of a car powertrain, the digital analysis of transmission path is carried out. Using the test and analysis method of working condition transmission path, the contribution order of the noise of the car in 340–440 Hz and 460–560 Hz is quickly found, and the main contribution path of each noise area is obtained. The main paths are the right suspension rod and the right suspension passive side bracket. The main conclusions are as follows.

(1) OTPA transmission path analysis method plays an important role in the vibration and noise analysis of the whole vehicle. Through this method, the main noise contribution paths of the powertrain can be quickly identified, and the contribution sizes can be sorted, so as to provide the basis for solving the problem quickly;
(2) For the design of the mounting system, not only the vibration isolation rate of the system should be considered, but also the design of the mounting bracket is very important. It is necessary to control the resonance frequency and dynamic stiffness of the active and passive end bracket of the mounting bracket, so as to reduce the transmission of vibration and the noise inside the vehicle as much as possible;

(3) In practical engineering, the resonance of the passive end of the mounting bracket can be controlled by improving the dynamic stiffness of the active end, so as to reduce the impact on the interior noise, which also has important reference significance for the optimization of the transmission of other paths.

References

1. Hu, S., Zuo, S., Lou, Y.: Vibration reduction and noise reduction of vehicle battery pack based on improved transfer path. Vibr. Test Diagn. **39**(06), 1348–1354 (2019)
2. Xue, L., Huang, S.: Research status and prospect of automotive NVH technology. Automot. Parts (05), 78–81 (2013)
3. Chang, H., Liu, W., Wu, D.: Application of transmission path analysis technology in NVH development. In: 2012 LMS China User Conference, pp. 1–8 (2013)
4. Hu, C., Gao, Y., Liu, S., Feng, H.: Noise identification and control of excavator based on spectrum and coherence analysis. Vibr. Test Diagn. **33**(06), 1032–1038 (2013)
5. Vander Linden, P.J.G., Fun, J.K.: Using mechanical-acoustic reciprocity for diagnosis of structure borne sound in vehicles. SAE Technical Paper Series Paper Number 931340, Michigan, USA (1993)
6. Xiao, Z., Jin, C., Wei, N., Liang, X., Zhao, Y., Tian, S.: Application of OTPA combined with sound field analysis in road noise development. Noise Vibr. Control **40**(02), 140–145+151 (2020)
7. Waisanen, A.S., Blough, J.R.: Road noise TPA simplification for improving vehicle sensitivity to tire cavity resonance using helium gas. SAE 2009-01-2092
8. Yang, N.: Analysis and Research on Transmission Path of Vehicle Interior Noise. Hebei University of Technology (2004)
9. Xu, M., Zhang, J., Kong, C., He, W., Zhai, N.: Structural path contribution analysis of interior noise based on structural force identification. Automot. Technol. **12**, 28–31 (2013)
10. Zhao, T.: Identification and Control of Vehicle Interior Noise Based on Transfer Path Analysis. Jilin University (2008)
11. Guo, S., Liu, Z., Zang, X., Fan, Y., Zhou, D.: Analysis method of transmission path under working load. Noise Vibr. Control **36**(02), 104–107 (2016)
12. Jiang, S.: Research on Vehicle Interior Noise Recognition Based on Condition Transfer Path Analysis. Shenyang University of technology (2019)
13. Long, Y.: Optimization Design of Powertrain Mounting System Based on Improved Transfer Path Analysis Method. Jilin University (2010)
14. Lu, Y., Cheng, W., Lu, J., Zhang, Z.: Research progress of transmission path analysis method for operating conditions. J. Hebei Univ. Sci. Technol. **36**(04), 359–367 (2015)
15. Wang, J.: Optimization Design and Dynamic Characteristic Analysis of Automobile Powertrain Mounting System. Hefei University of Technology (2010)

Enhanced Admittance Control for Time-Varying Force Tracking of Robots in Unknown Environment

Chengguo Liu[1], Ye He[1,2](✉), Kuan Li[1], and Xue Zhao[1]

[1] College of Mechanical Engineering, Chongqing University, Chongqing, China
[2] Key Laboratory of Mechanical Transmission of Chongqing, Chongqing, China

Abstract. In this paper, we introduce admittance control as an approach to control the physical interaction between robot and environment, and propose an enhanced admittance controller (EAC) framework with a well-designed control scheme that improves the system response while possessing the ability to suppress transient force overshoot and maintain steady-state force tracking. Within this framework, we analyze the pre-fuzzy PID, environmental parameter estimation, computed torque control, and propose a time-varying force control theory analysis based on the traditional target admittance model, and introduce an adaptive algorithm to compensate the environmental uncertainty, and verify the stability of the system based on the Routh criterion and Lyapunov equation. Finally, simulations are performed to verify the proposed control scheme in terms of system response, transient and steady-state force overshoot, and steady-state force tracking. Finally, simulations are performed to verify the effectiveness of the proposed control scheme in terms of system response, transient and steady-state force control performance.

Keywords: Pre-fuzzy PID · Environmental estimate · Transient force overshoot · Steady-state force tracking · Time-varying force tracking

1 Introduction

The role of force control in the field of robotic contact operations is becoming more and more prominent, distinguishing industrial robots with traditional position control, especially in unknown environments where it cannot achieve the goals we require. Examples include: precision assembly [2, 3], finishing processes such as grinding and polishing deburring [4, 5], medical surgery [6], and human-robot interaction [7–9]. The implementation of these tasks not only requires high requirements for robot trajectory tracking, but more importantly, its superior regulation strategy for contact forces, i.e., it needs to have suitable flexibility for force variations due to environmental uncertainties in order to avoid adverse effects that may lead to task failure or worse. In-depth research has revealed two main research approaches to be discussed: hybrid force/position control [10] and impedance control [11–13].

In order to make the robot have a wider application scenario as well as to better assist the workers to complete the tedious labor, therefore, the flexible control algorithms that combine position control with force control were soon proposed. Raibert and

X.-J. Liu et al. (Eds.): ICIRA 2021, LNAI 13014, pp. 552–562, 2021.
https://doi.org/10.1007/978-3-030-89098-8_52

Craig proposed hybrid force/position control [10], which was refined by Mason [14]. Hybrid force/position control is to place the force and position control in two orthogonal subspaces to control the force and position in different directions. This is also considered as its biggest limitation, which may lead to vibration in the system due to incorrect switching of states, thus affecting the control system performance, for which adaptive Jacobi position/force control [15] was proposed to solve the problem, but still system instability may occur. Stability.

Further is the impedance control proposed by Hogan [11–13] based on the study of stiffness control and damping control. It not only presents the dynamic relationship between the manipulator and the environment first, but also gives a unified framework for considering both unconstrained and constrained motion control problems. The core idea is to control the dynamic interactions between motion and contact forces through impedance equations, controlling both force and position in one direction.

We generally classify them into two categories: power-based impedance control and position-based impedance control (The following are collectively referred to as admittance control), as shown in Fig. 1. The former focuses on motion tracking, while the latter focuses on force tracking [16, 17], and usually impedance is applied to rigid environments, and admittance control is applied to flexible environments, and the goal of both approaches is to improve control performance through indirect force control while satisfying interactive force requirements [18]. Admittance control has better control robustness and easy to implement, and the control effect mainly depends on the accuracy of position control, and the current position control of the robot can achieve high control accuracy, so this paper is based on admittance control in the robot in industrial finishing and medical physical therapy rehabilitation field for force control.

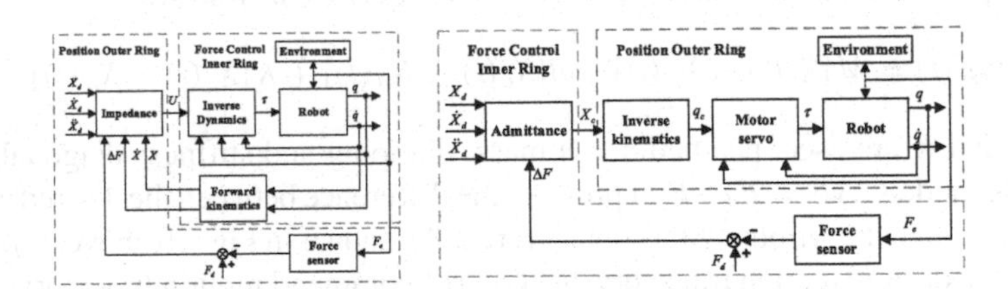

Fig. 1. a. Impedance control b. Admittance control

Due to the highly coupled nature of the robotic system and its complex nonlinear dynamics, together with the unknown nature of the environmental parameters, time-varying nonlinear external disturbances, the parameter selection for impedance control has high requirements [19]. One is the combination of intelligent control algorithms to modify impedance parameters in real time to achieve better force tracking, sliding mode impedance control [20]. Second, In search of a more accurate understanding of the environment, such as adaptive environment estimation [21].

Initial research was mainly devoted to non-deforming or slowly deforming more structured environments. However, in certain interaction operations of robots, the position and stiffness of the interaction target (ultra-thin aircraft wings) are fragile in practice as their position and stiffness are unstructured and time-varying. Many of the above intelligent impedance control algorithms are good at maintaining steady-state force tracking performance, but suppressing transient force overshoot is very problematic. Therefore, an excellent impedance interactive controller should have superior performance in transient, steady state and response speed.

Based on the above discussion, we highlight the contributions of this paper as follows:

i. Environmental dynamics is considered in the optimal interaction analysis between robot and environment, which is described as a linear system with unknown dynamics. The corresponding environment identification algorithm is proposed to improve the steady-state performance of force control.
ii. The traditional pre-PID admittance control is modified, and the corresponding fuzzy control strategy is proposed to ensure the steady-state tracking accuracy and response speed, and reduce the transient overshoot of the system.
iii. Based on the task-oriented algorithm, the adaptive parameters are updated in real time without the need of training data, which ensures the optimal force interaction between the robot and any dynamic environmental forces.

2 Principle of Admittance Control Algorithm

2.1 Overview of Admittance Control

The purpose of admittance control is to establish the dynamic relationship between the user-specified contact force F_d and the difference between the desired position X_d and the actual position X at the end of the robot. Its models are as follows:

$$\Delta F(t) = M[\ddot{X}_c(t) - \ddot{X}_d(t)] + B[\dot{X}_c(t) - \dot{X}_d(t)] + K[X_c(t) - X_d(t)] \tag{1}$$

M, B and K are user-specified $n \times n$ mass, damping and stiffness diagonal matrix admittance models. $\Delta F = F_d - F_e$ represents the difference between the desired tracking force and the actual environmental contact force. As shown in Fig. 1b above, X_d, X_c and X denote the desired motion trajectory, the control command motion trajectory, and the actual motion trajectory, respectively. At present, the robot position control technology is quite mature, so for the sake of simplicity we consider $X = X_c$. F_d and F_e denote the desired tracking force and the actual robot-environment contact force, respectively. k_e denotes the environmental stiffness and X_e denotes the environmental position. In this case we consider the environment to be rigid, so we have:

$$F_e = K_e(X - X_e) \tag{2}$$

For the convenience of expression, without loss of generality, take one-dimensional space as an example, replace ΔF, F_d, F_e,X_d, X_e,X with Δf, f_d, f_e,x_d, x_e,x. Under the above assumption $x = x_c$, we consider $f_d(t)$ and $x_e(t)$ as time-varying and k_e as time-invariant, thus:

$$\Delta f(t) = m\left\{\frac{\ddot{f}_d(t) - \Delta\ddot{f}(t)}{k_e} + \ddot{x}_e(t) - \ddot{x}_d(t)\right\}$$
$$+b\left\{\frac{\dot{f}_d(t) - \Delta\dot{f}(t)}{k_e} + \dot{x}_e(t) - \dot{x}_d(t)\right\} + k\left\{\frac{f_d(t) - \Delta f(t)}{k_e} + x_e(t) - x_d(t)\right\} \tag{3}$$

From the **Laplace** transform and the final value theorem, we know that:

$$e_{ss} = \lim_{s\to 0} s\Delta f(s) = \frac{s\left(ms^2 + bs + k\right)\{f_d(s) + k_e \bullet x_e(s) + k_e x_d(s)\}}{ms^2 + bs + k + k_e} \tag{4}$$

It follows that to converge the force tracking error to zero at steady state reference trajectory $x_d = x_e + \frac{f_d}{k_e}$ or stiffness parameter $k = 0$. However, it is not desirable to set the stiffness parameter to 0. In layman's terms, the stiffness k represents the active mind of the robot, and by setting it to zero the robot cannot express its skills and "it will be at the mercy of others". When the robot is in a non-contact environment and $\Delta f(t)$ is not zero, the robot will not have the "active rebound" feature and will move to the robot singularity under the force until the alarm.

3 Environmental Parameter Estimation

Based on the above, the environment estimation is proposed to cope with the case of unknown time-varying environmental parameters that cannot be measured precisely, in order to make $f_e(t) \to f_d(t)$ when $t \to \infty$.

Assuming that the environmental position and stiffness are $\hat{x}_e$ and $\hat{k}_e$, respectively, the external forces are estimated as:

$$\hat{f}_e = \hat{k}_e(x - \hat{x}_e) = \hat{k}_e x - \hat{k}_e\hat{x}_e = \hat{k}_e x - \hat{k}_x \tag{5}$$

Definition $\varphi = \begin{bmatrix}\varphi_1 \\ \varphi_2\end{bmatrix} = \begin{bmatrix}\hat{k}_e - k_e \\ \hat{k}_x - k_x\end{bmatrix}$, Combining with (5), we get:

$$\hat{f}_e - f_e = \begin{bmatrix} x & -1 \end{bmatrix}\varphi \tag{6}$$

For system (6), establish the Lyapunov function:

$$V = \varphi^T P\varphi \tag{7}$$

P is a positive definite matrix defined in advance, if we define:

$$\dot{\varphi} = -P^{-1}\begin{bmatrix} x \\ -1 \end{bmatrix}\left(\hat{f}_e - f_e\right) \tag{8}$$

Derivation of (7) and substituting (8) yields:

$$\dot{V} = 2\varphi^T P\dot{\varphi} = -2(\hat{f}_e - f_e)^2 \leq 0 \tag{9}$$

It follows that the system is stable. It can be calculated that:

$$\begin{cases} \dot{\hat{k}}_e = -\alpha x(\hat{f}_e - f_e) \\ \dot{\hat{k}}_x = \beta(\hat{f}_e - f_e) \end{cases} \tag{10}$$

Then:

$$\hat{k}_e(t) = \hat{k}_e(0) - \alpha \int_0^t x(\hat{f}_e - f_e)dt - \varepsilon_1 \int_0^t \ddot{k}_e(t)dt \tag{11}$$

$$\hat{x}_e(t) = \hat{x}_e(0) + \beta \int_0^t \frac{\hat{f}_e - f_e}{k_e}(\frac{\alpha}{\beta}x\hat{x}_e + 1)dt - \varepsilon_2 \int_0^t \hat{x}_e(t)dt \tag{12}$$

where ε- modification to enhance the robustness of the system. Due to the uncertainty of the environmental model and the error of the identification algorithm, the steady-state force error still exists. Therefore, the pre-fuzzy PID adaptive admittance control in the next section is proposed to improve the force tracking accuracy to the greatest extent.

4 Enhanced Admittance Control (EAC)

4.1 Theoretical Analysis

A closer look at (1) shows that the admittance function is used as a second-order low-pass filter. The force feedback signal is converted to a position correction signal within the framework of the admittance control. This greatly tests the trajectory tracking capability of the internal position controller, and a high gain PD controller is usually used to improve the fast response capability. However, too high gain can also bring adverse effects to the system, such as affecting system stability and robustness. In order to improve the force fast convergence capability, also to mitigate the effect of high gain of the internal position controller, plus the great benefit to improve the force tracking capability, the pre-PID controller is used to improve the overall system performance:

$$k_p e_f + k_i \int_0^\tau e_f + k_d \bullet \dot{e}_f = m\ddot{e} + b\dot{e} + ke \tag{13}$$

e_f and e are used instead of Δf and Δx. k_p, k_i, k_d are the PID controller gains.

$$e_{ss} = \lim_{s \to 0} s\Delta f(s) = \frac{s(ms^2 + bs + k)\{f_d(s) + k_e \bullet x_e(s) + k_e x_d(s)\}}{ms^2 + (b + k_e k_d)s + k + k_e + k_e k_p + \frac{k_e k_i}{s}} \tag{14}$$

Obviously, after adding the pre-PID controller, the steady-state force error only changes the denominator, so an appropriate pre-PID controller is desirable. It does not change the reference trajectory x_d, and reduces the steady-state error. However, the parameters must be carefully selected, and excessive gains will also cause system oscillations. The robot-environment interaction model can be represented by the Fig. 2.

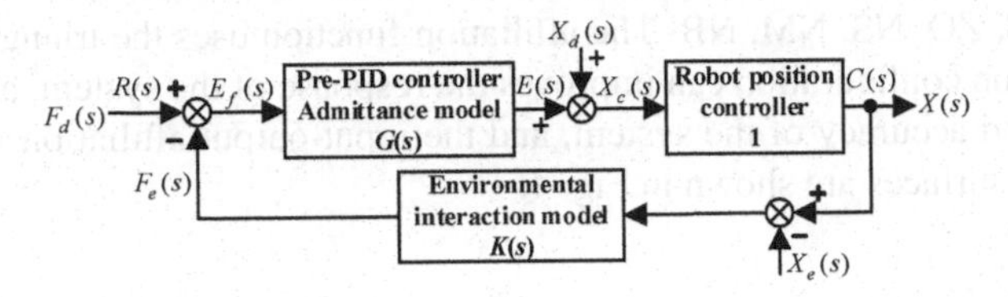

Fig. 2. Robot-environment interaction model.

$$\begin{cases} G(s) = \frac{k_p + \frac{k_i}{s} + k_d s}{ms^2 + bs + k} \\ K(s) = k_e \end{cases} \tag{15}$$

By superposition principle, the output of the closed-loop system $C(s)$ is:

$$C(s) = \frac{\left(k_d s^2 + k_p s + k_i\right) F_d(s) + \left(ms^3 + bs^2 + ks\right) X_d(s) + k_e\left(k_d s^2 + k_p s + k_i\right) X_e(s)}{ms^3 + (b + k_d k_e)s^2 + (k + k_p k_e)s + k_i k_e} \tag{16}$$

It can be clearly seen from Eq. (16) that, after increasing the pre-PID controller, the system has closed loop zero, which has the advantage that the system-rise time and peak time will be shortened, but the overshoot will increase. But with the increase of k_p(k_i, k_d is usually a very small constant), the system stability and overshoot will increase significantly. Based on this analysis, keeping the value of k_p constant is not an excellent choice. Therefore, under the desire to both avoid transient force overshoot and maintain steady-state precision force tracking. Enhanced admittance control is proposed to correct k_p online by formulating suitable fuzzy rules. It is also combined with the above-mentioned environmental parameter identification algorithm in order to further improve the system performance. The control framework is shown in Fig. 3.

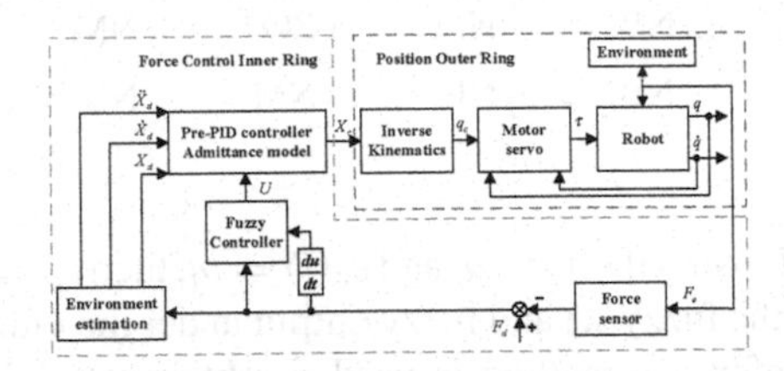

Fig. 3. Enhanced admittance control block diagram.

4.2 Design of Fuzzy Controller

Assume that $[-X_e, X_e], [-X_{ec}, X_{ec}]$ and $[Y_1, Y_2]$ are the fundamental domains represented by $e_f, \dot{e}_f, k_p$. They are usually normalized to seven labels: positive big, positive medium, positive small, zero, negative small, negative medium, and negative big. Corresponds to PB, PM, PS, ZO, NS, NM, NB. The affiliation function uses the triangle function, a suitable function configuration can improves the response of the system, also maintains the stability and accuracy of the system, and the input-output affiliation functions and control output surfaces are shown in Fig. 4:

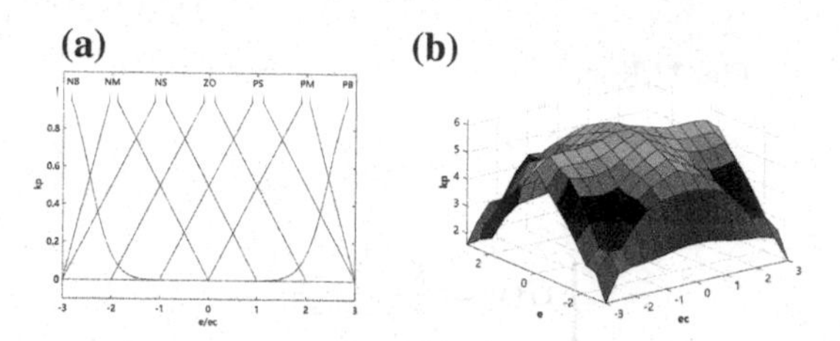

Fig. 4. a. Fuzzy affiliation function. b. Fuzzy control output surface

The fuzzy rules reflect the correspondence between input and output, and they are established based on the experience and intuition of experts and skilled workers. Based on the simulation results, the system fuzzy rules are established as shown in Table 1.

Table 1. Fuzzy rules.

		e						
		NB	NM	NS	ZO	PS	PM	PB
e_c	NB	NB	NB	NM	NM	NM	NB	NB
	NM	NB	NB	NM	ZO	NM	NB	NB
	NS	NM	NM	NS	PM	NS	NM	NM
	ZO	PB	PM	PS	PB	PS	PM	PB
	PS	NM	NM	NS	PM	NS	NM	NM
	PM	NB	NB	NM	ZO	NM	NB	NB
	PB	NB	NB	NB	NM	NB	NB	NB

As shown in Table 1, using if $e(i) = a_j$ and $e_c(i) = b_j$, then $k_p = c_j$ ($j = 1, ..., 49$), a_j, b_j and c_j correspond to the fuzzy set and fuzzy output under the j fuzzy rule, respectively. The Mamdani fuzzy inference method is used to obtain the values of the affiliation functions corresponding to different fuzzy sets from the fuzzy rule table, and the fuzzy system output is obtained from this. The fuzzy output set is defuzzified using the single-point fuzzification, center-of-gravity defuzzification method.

$$k_p = \frac{\sum_{j=1}^{49} a_j(e)b_j(e_c)\mu_j}{\sum_{j=1}^{49} a_j(e)b_j(e_c)} \tag{17}$$

Thus, a fuzzy controller is used to adjust the PID gain online in order to compensate for the errors caused by external disturbances and environmental uncertainties.

4.3 Stability and Boundary Analysis

To derive the stability condition for the proposed enhanced admittance control algorithm, the system characteristic equation can be obtained from Eq. (16) as:

$$ms^3 + (b + k_e k_d)s^2 + (k + k_e k_p)s + k_e k_i = 0 \tag{18}$$

This gives the Routh-Hurwitz array as (In this case, the mass parameter m is taken as 1):

$$\begin{array}{ccc} s^3 & 1 & k + k_e k_p \\ s^2 & b + k_e k_d & k_e k_i \\ s^1 & \frac{(b+k_e k_d)(k+k_e k_p)-k_e k_i}{b+k_e k_d} & 0 \\ s^0 & k_e k_i & 0 \end{array} \tag{19}$$

Then the system stability needs to meet:

$$(b + k_e k_d)(k + k_e k_p) - k_e k_i > k_e k_d \bullet (k + k_e k_p) - k_e k_i = k_e^2 \left(\frac{k k_d}{k_e} + k_d k_p - \frac{k_i}{k_e} \right) > 0 \tag{20}$$

The environmental stiffness k_e is considered large enough to allow easy selection of the control gain to satisfy (20). So, a broad range of parameters is specified:

$$\begin{cases} 0 \le k_i, k_d \le 0.05 \\ 0 \le k_p \le 10 \end{cases} \tag{21}$$

5 Simulation

5.1 Simulation Parameter Setting

The simulation environment simulates the actual operating conditions, we use white noise to simulate sensor noise, simultaneous addition of friction on the system. Assuming that Coefficient of viscous friction: $F_c = 4$ N, coulomb coefficient of friction: $c_v = 1.2$ Ns /m. The admittance model of quality m, damping b and stiffness k are 1 Ns^2/m, 200 Ns /m and 1200 N /m (Fig. 5).

$$F_f = -\text{sign}(\dot{x})(c_v|\dot{x}| + F_c) \tag{22}$$

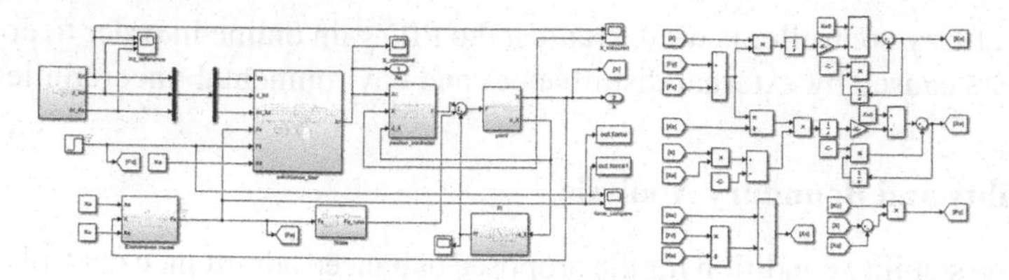

Fig. 5. The diagram of enhance admittance control simulation process

5.2 Simulation Results

Simulation was performed to simulate the tracking performance of the controller for time-varying forces during continuous changes in the stiffness of the environment during rehabilitation physiotherapy, assuming: $K_e = 4000 + 800\sin(\frac{\pi}{2}t)$ N /m. At the same time, we also want to verify the robustness of the algorithm.

As shown in Fig. 6, as K_p increases, the transient force overshoot also increases, which generates greater vibration and force overshoot, at the same time, the steady-state force error also decreases and the system response becomes faster. Therefore: we use a fuzzy controller to adjust the proportional gain K_p online to reduce the system vibration and steady-state error, while speeding up the system response.

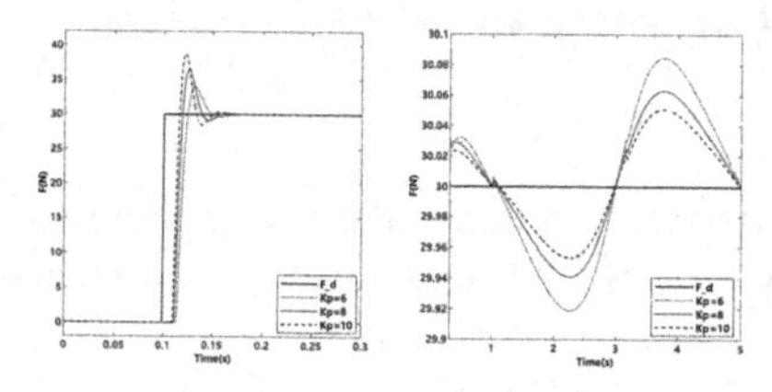

Fig. 6. The effect of different K_p on the system

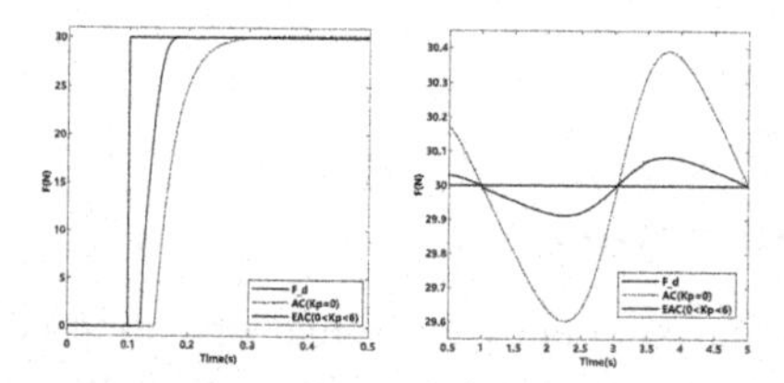

Fig. 7. Comparison of EAC and AC controllers

The force overshoot of the interaction process transient can be avoided. As in Fig. 7, the force tracking performance of EAC (environment estimation + Pre-Fuzzy PID admittance controller) and AC (environment estimation + admittance controller) is shown, where we restrict the update range of K_p is between [0, 6]. During the continuous change of stiffness, we can clearly see that the EAC has better tracking accuracy for time-varying forces than the AC controller, and there is almost no transient force overshoot in the EAC. Importantly, the system response of the EAC is much better than the AC controller.

6 Conclusion

In order to solve the human-robot interaction problem of time-varying force tracking in the case of non-structural dynamic environmental parameters unknown (surface treatment, medical rehabilitation), this paper first analyzes the impedance and admittance control algorithms and proposes the necessity of environmental parameter estimation (EE), therefore proposes the corresponding control algorithm based on **Lyapunov** function, and also analyzes the advantages of enhanced admittance control (EAC) and the need to update the proportional gain K_p online. Therefore, the corresponding control algorithm is proposed, the corresponding fuzzy rule and affiliation function are established and its stability is verified based on the Routh criterion, and the control principle is clarified. The EAC controller is verified to have superior system response, superior transient force overshoot suppression capability, excellent steady-state force performance through controlled simulation experiments. In summary, the EAC controller meets all the specifications of an excellent admittance interaction controller.

Acknowledgments. This research is supported by the key R&D project "Key Technology and Application Research of High-Power Direct Drive Spindle Unit for High-end CNC Lathe" of Chongqing Technology Innovation and Application Development Special Project (project number cstc2019jscx-fxydX0022).

References

1. Chen, F., Zhao, H., Li, D.: Contact force control and vibration suppression in robotic polishing with a smart end effector. Robot. Comput. Integr. Manuf. **57**, 391–403 (2019). https://doi.org/10.1016/j.rcim.2018.12.019
2. Gracia, L., Solanes, J.E., Muñoz-Benavent, P.: Human-robot collaboration for surface treatment tasks. Interact. Stud. Soc. Behav. Commun. Biol. Artif. Syst. **20**(1), 148–184 (2019). https://doi.org/10.1075/is.18010.gra
3. Yao, B., Zhou, Z., Wang, L.: Sensorless and adaptive admittance control of industrial robot in physical human−robot interaction. Robot. Comput. Integrat. Manuf. **51**, 158–168 (2018). https://doi.org/10.1016/j.rcim.2017.12.004
4. Ji, W., Wang, L.: Industrial robotic machining: a review. Int. J. Adv. Manuf. Technol. **103**(1–4), 1239–1255 (2019). https://doi.org/10.1007/s00170-019-03403-z
5. Park, H., Park, J., Lee, D.-H.: Compliance-based robotic peg-in-hole assembly strategy without force feedback. IEEE Trans. Indust. Electron. **64**(8), 6299–6309 (2017). https://doi.org/10.1109/tie.2017.2682002
6. Yuen, S.G., Perrin, D.P., Vasilyev, N.V.: Force tracking with feed-forward motion estimation for beating heart surgery. IEEE Trans. Rob. **26**(5), 888–896 (2010). https://doi.org/10.1109/TRO.2010.2053734
7. Ferraguti, F., Talignani Landi, C., Sabattini, L.: A variable admittance control strategy for stable physical human–robot interaction. Int. J. Robot. Res. **38**(6), 747–765 (2019). doi: https://doi.org/10.1177/0278364919840415
8. Keemink, A.Q.L., van der Kooij, H., Stienen, A.H.A.: Admittance control for physical human–robot interaction. Int. J. Robot. Res. **37**(11), 1421–1444 (2018). https://doi.org/10.1177/0278364918768950

9. Solanes, J.E., Gracia, L., Muñoz-Benavent, P.: Adaptive robust control and admittance control for contact-driven robotic surface conditioning. Robot. Comput.-Integrat. Manuf. **54**, 115–132 (2018). https://doi.org/10.1016/j.rcim.2018.05.003
10. Raibert, M.H., Craig, J.J.: Hybrid position/force control of manipulators. J. Dyn. Syst. Meas. Contr. **103**(2), 126–133 (1981). https://doi.org/10.1115/1.3139652
11. Hogan, N.: Impedance control: an approach to manipulation: Part I—Theory. J. Dyn. Syst. Meas. Contr. **107**(1), 1–7 (1985). https://doi.org/10.1115/1.3140701
12. Hogan, N.: Impedance control: an approach to manipulation: Part II—Implementation. J. Dyn. Syst. Meas. Contr. **107**(1), 8–16 (1985). https://doi.org/10.1115/1.3140702
13. Hogan, N.: Impedance control: an approach to manipulation: Part III—Applications. J. Dyn. Syst. Meas. Contr. **107**(1), 17–24 (1985). https://doi.org/10.1115/1.3140713
14. Mason, M.T.: Compliance and force control for computer controlled manipulators. IEEE Trans. Syst. Man Cybern. **11**(6), 418–432 (1981). https://doi.org/10.1109/tsmc.1981.4308708
15. Wang, H., Xie, Y.: Adaptive Jacobian force/position tracking control of robotic manipulators in compliant contact with an uncertain surface. Adv. Robot. **23**(1–2), 165–183 (2009). https://doi.org/10.1163/156855308X392726
16. Seraji, H., Colbaugh, R.: Force tracking in impedance control. Int. J. Robot. Res. **16**(1), 97–117 (1997). https://doi.org/10.1177/027836499701600107
17. Ueberle, M., Mock, N., Buss, M.: VISHARD10, a novel hyper-redundant haptic interface. In: Proceedings - 12th International Symposium on Haptic Interfaces for Virtual Environment and Teleoperator Systems, HAPTICS, 2004, pp. 58–65. doi: https://doi.org/10.1109/HAPTIC.2004.1287178. doi: https://doi.org/10.1017/s026357479700057x.
18. Cao, H., He, Y., Chen, X., Liu, Z.: Control of adaptive switching in the sensing-executing mode used to mitigate collision in robot force control. J. Dyn. Syst. Measur. Control **141**(11) (2019). doi: https://doi.org/10.1115/1.4043917
19. Sciavicco, B.S.L.: Modelling and Control of Robot Manipulators (2012)
20. Ketelhut, M., Kolditz, M., Göll, F., Braunstein, B.: Admittance control of an industrial robot during resistance training. IFAC-PapersOnLine **52**(19), 223–228 (2019). https://doi.org/10.1109/ACCESS.2019.2924696
21. Xu, W., Cai, C., Yin, M.: Time-varying force tracking in impedance control. In: 2012 IEEE 51st IEEE Conference on Decision and Control (CDC), pp. 344–349 (2012). doi: https://doi.org/10.1109/TCST.2017.2739109

Variable Impedance Force Tracking with a Collaborative Joint Based on Second-Order Momentum Observer Under Uncertain Environment

Jie Wang[1], Yisheng Guan[1(✉)], Haifei Zhu[1], and Ning Xi[2]

[1] Biomimetic and Intelligent Robotics Lab (BIRL), Guangdong University of Technology, Guangzhou 510006, China
ysguan@gdut.edu.cn

[2] Department of Industrial and Manufacturing Systems Engineering, The University of Hong Kong, Hong Kong SAR, China

Abstract. Collaborative robots are the focus of the development of next-generation industrial robot. In order to improve the flexibility and force control performance of collaborative robot joint, this paper proposes a variable impedance force tracking algorithm based on a second-order momentum observer for the collaborative joints developed in our laboratory. The algorithm first designs a second-order feedforward generalized momentum external force observation method for force perception, which is used as the torque outer loop of the force tracking algorithm. Secondly, a simplified robot-environment variable impedance system is constructed through the dynamic transfer function, which is used as a position inner loop to optimize the gain adjustment process and dynamic response speed by establish a differential error model. The simulation shows the robustness of the scheme under unknown environment stiffness and moving environment. The experimental results support the claim that this method can track the desired force in real time under uncertain environment.

Keywords: Collaborative joint · External torque observer · Variable impedance control · Force tracking

1 Introduction

In recent years, collaborative robots have begun to enter people's field of vision, and are mainly used in various human-robot interaction scenarios [1]. collaborative robots usually come into force contact with unknown environment (human

The work in this paper is in part supported by the Frontier and Key Technology Innovation Special Funds of Guangdong (Grant No. 2017B050506008, 2017B090910008), and the Key Research and Development Program of Guangdong Province (Grant No. 2019B090915001).

X.-J. Liu et al. (Eds.): ICIRA 2021, LNAI 13014, pp. 563–574, 2021.
https://doi.org/10.1007/978-3-030-89098-8_53

body), such as polishing, precision assembly and other tasks. The error caused by pure position control will cause excessive contact force between the contact surfaces [2]. Therefore, the sensitivity of the robot/compliance control problem is very important to improve the performance of the robot.

Force perception is the prerequisite for achieving compliance control of collaborative robots. Olsson proposed a terminal load force sensing method applied to industrial robots, which realized force sensing by solving the parameters of the zero point and installation angle of the six-dimensional torque sensor [3]. Park proposed a sensor zero compensation method, which realizes external force observation by adjusting the robot's posture [4]. Kebria first applied the joint sensor to the UR robot, and adopted a double-encoding structure, with high force sensing accuracy [5]. In addition, there are traditional external force sensing methods based on electric current, piezoelectric effect, and inverse dynamics.

Impedance control method is widely used at home and abroad, which Often used as a strategy for force tracking. Impedance control was proposed by Hogan [6]. Jeon built a position control inner loop based on the traditional impedance idea, and adjusted the trajectory of the robot's end position by using the acquired torque information [7]. Reference [8] designed an impedance controller, which not only considers the impedance relationship between the end effector and the object, but also considers the impedance relationship between the end and the environment. Reference [9] proposed an adaptive master/slave control strategy for hybrid force/position tracking to compensate for trajectory tracking errors. Erhart proposed an impedance-based multi-robot collaborative dual-arm mobile operation control architecture [10]. Another effective method of collaborative joint impedance control is to use a cascade structure of internal torque control loop and external impedance control loop [11,12].

The rest of this paper is organized as follows. In Sect. 2, the dynamic model is given and proposes an external force detection algorithm based on the second-order momentum observer. In Sect. 3, a variable impedance force tracking algorithm based on momentum observation is proposed, which can realize variable impedance control according to the observation threshold and impedance gain under uncertain environments. Simulation experiments are presented in Sect. 4 to verify the robustness of the algorithm. In Sect. 5, we conduct verification experiments on the control algorithm proposed in this paper based on self-developed collaborative joint system. Finally, we give a conclusion in Sect. 6.

2 External Force Detection Based on Second-Order Momentum Observer

2.1 Integrated Collaborative Joint Dynamic Model

The physical and transmission structure of the collaborative joint developed in this research is shown in Fig. 1. According to the flexible joint dual-mass spring system proposed by Spong [13], The dynamic model designed in this paper is as follows:

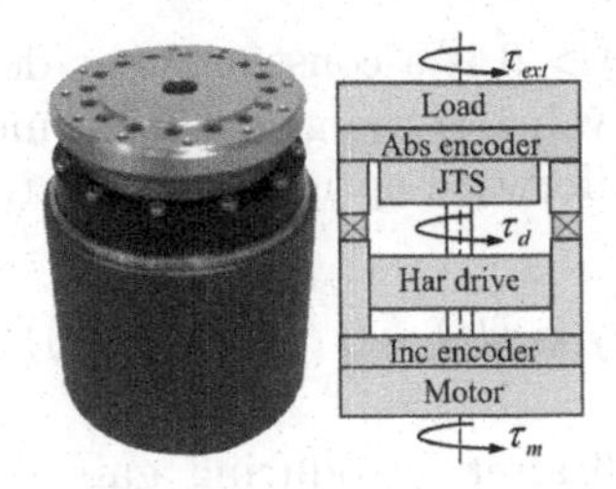

Fig. 1. Collaborative joint and transfer model

$$M(q)\ddot{q} + C(q,\dot{q})\dot{q} + G(q) = J^{-1}\tau_d - \tau_{ext} - \tau_f \tag{1}$$

where q, $\dot{q}$, $\ddot{q} \in R^{n\times 1}$ are the joint angle, joint angular velocity and joint angular acceleration of the manipulator respectively, $M(q) \in R^{n\times n}$ is the symmetrical inertia matrix of the manipulator, $C(q,\dot{q}) \in R^{n\times n}$ is the centrifugal inertia matrix, which including coriolis force and centrifugal force, $G(q) \in R^{n\times 1}$ is the gravity term, τ_d is the joint output torque, J^{-1} is the inverse matrix of the joint, and τ_{ext} is the external disturbance torque.

2.2 Second-Order Momentum External Force Observer

In order to sensitively and accurately determine the direction and magnitude of the external force at the joint level. this section proposes a second-order momentum observer's external force detection algorithm. Consider the concentrated observation of disturbance moments, observation torque can be expressed as $\tau_c = \tau_{ext} - \tau_f$, Eq. 2 can be further transformed into the external force disturbance equation as follows:

$$\tau_c = J^{-1}\tau_d - M(q)\ddot{q} + C(q,\dot{q})\dot{q} + G(q) \tag{2}$$

The generalized momentum equation that defines the collaborative joint is as follows:

$$p = M(q)\dot{q} \tag{3}$$

where $M(\dot{q})$ is a positive definite symmetric matrix, and $M(\dot{q}) - 2C(q,\dot{q})$ is an antisymmetric matrix [14], and the derivative can be obtained:

$$\dot{M}(q) = C(q,\dot{q}) + C^T(q,\dot{q}) \tag{4}$$

According to the momentum decoupling characteristics of flexible collaborative joints, the disturbance vector is defined as $r = \tau_c$, and the disturbance external torque observation algorithm is designed as follows:

$$\dot{r}(t) = K_a\left[-K_b r(t) + [p(t) - \hat{p}(t)]\right] \tag{5}$$

where $K_a, K_b = \text{dia}(k_{a,b,i}) > 0$ is a constant that depends on the system, $\hat{p}(t)$ is the real-time estimated value of the generalized momentum. Based on Eq. 6, construct the dynamic feedforward optimization factor O_f as follows:

$$O_f = \hat{p}(t) - \int_0^t (\dot{p}(t) - \hat{r}(t))dt \tag{6}$$

Combining Eq. 7, the Eq. 6 after introducing gain constant optimization is as follows:

$$\dot{r}(t) = -K_a K_b \hat{r} + K_a p(t) - K_a K_c \tau_c - K_a \int_0^t \left[\tau_d + \hat{C}^T(q,\dot{q})\dot{q} - G(q) + \hat{r}\right] dt \tag{7}$$

The block diagram of the external force observation algorithm is shown in Fig. 2, compared with the classic first-order momentum observer $\dot{\hat{r}} = K(\tau_c - r)$, the feedforward optimization factor O_f in the algorithm proposed in this paper can eliminate the suppression system overshoot and high-frequency noise.

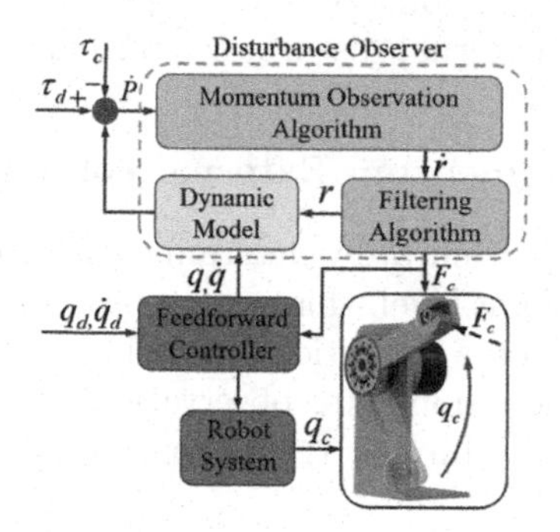

Fig. 2. Framework of external force observation algorithm

3 Variable Impedance Force Tracking Algorithm Based on External Force Observer

3.1 Model of Variable Impedance Force/position Tracking

Impedance control does not directly control position or force, but adjusts the relationship between force and position. The classic second-order mass-damping-stiffness constant impedance model is:

$$M_e\ddot{e} + B_e\dot{e} + K_e e = f_e \tag{8}$$

where f_e is the force error, e is the position error, M_e, K_e, B_e are the environmental quality, environmental stiffness and damping respectively. This study takes the two-axis variable impedance model as the research object, as shown in

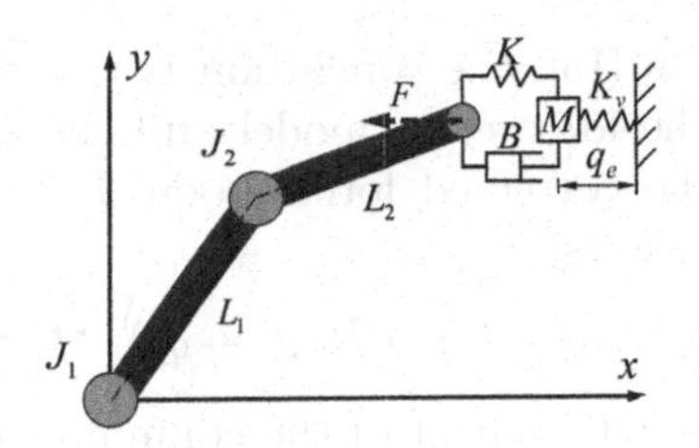

Fig. 3. Robot-environment impedance contact model

Fig. 3, suppose the variable stiffness of the environment is K_v, the expected contact force between the end of the manipulator and the environment is $F_d = K_v(q_e - q_c)$, M, K, B are the variable inertia matrix, the variable stiffness matrix and the variable damping matrix of the end of the manipulator.

According to [15], it is impossible to directly observe and compensate the force tracking error in the interactive environment in real time. Therefore, it is necessary to construct the variable impedance function by constructing the force error model based on K_v, and then integrating it into the impedance model. Suppose q_e, q_d, q_a, q_c respectively as the environment position, the reference position of the end of the robot arm, the command (corrected) position and the actual position, F_c is the real-time external disturbance force. When the end of the robotic arm is in contact with the environment, a force error model and a position error model are generated as follows:

$$e_f = F_d - F_c = K_v(q_e - q_d - Ke_f) - F_c \tag{9}$$

$$e_q = Ke_f \tag{10}$$

Considering that it is difficult to directly compensate the position error in practice, so choose to indirectly compensate the torque observation error to construct the target variable impedance model:

$$M(\ddot{q}_a - \ddot{q}_d) + B(\dot{q}_a - \dot{q}_d) + K_v(q_a - q_d) = \mu\dot{e}_f + \eta e_f \tag{11}$$

where $\ddot{q}_a$ is the actual end acceleration value, $\dot{q}_a$ is the end actual speed value, $\ddot{q}_d$ is the end reference acceleration, $\dot{q}_d$ is the end reference speed, and μ, η are system constant.

3.2 Force Tracking Control Method

When the robot moves freely in space, the expected contact force is $F_d = 0$, the actual external disturbance force is $F_c = 0$, and the force error model is $e_f = F_d - F_c = 0$, Eq. 13 can be expressed as follows:

$$M(\ddot{q}_a - \ddot{q}_d) + B(\dot{q}_a - \dot{q}_d) + K_v(q_a - q_d) = 0 \tag{12}$$

When the robot is in contact with the environment, it is expected that $F_C \Rightarrow F_d$. We assume $K = 0$, the amount of impedance change depends on K_v. Similar methods are mentioned in [15], the conclusion is that the steady state contact force is constant. And the force error model and the steady-state position are $e_f \Rightarrow 0$. Substituting the corrected force model into Eq. 13, and solving the corrected trajectory as follows:

$$\ddot{q}_c = \ddot{q}_d + \left[e_f - B\left(q_a - q_d\right)\right] M^{-1} \tag{13}$$

Considering the real-time adjustment of the communication cycle T of the controller, the corrected trajectory is substituted into Eq. 11, the real-time tracking torque after position compensation is inversely solved:

$$F_c = K_v\left(q_e - q_d - Ke_f\right) - e_f \tag{14}$$

The above-mentioned position loop derivation process is used as the inner loop of variable impedance control to modify the trajectory, and the external force detection algorithm based on the second-order momentum observer designed in the previous section is used as the outer moment of the moment loop. The implementation process of the variable impedance force tracking algorithm based on external force detection designed in this paper is shown in Fig. 4.

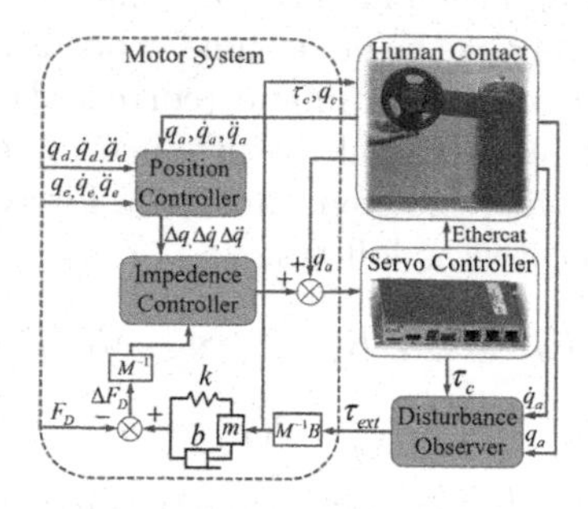

Fig. 4. Framework of variable impedance force tracking algorithm

4 Simulation Verification

In this section, the proposed control algorithm is tested by simulating the tracking performance under different environment conditions in MATLAB, the sampling period of the controller is selected as $T = 2\,\text{ms}$.

4.1 External Force Observation with Variable Force Environment

The observation result of the external torque r will directly affect the accuracy of force tracking. In this section, we apply a constant force and a time-varying force to simulate the actual system being disturbed by constant force and sudden collision, see Fig. 5. External disturbance force is defined as

$$F = \begin{cases} 3 & t \in [1, 2] \\ 5t - 15 & t \in [3, 4] \end{cases} \tag{15}$$

4.2 Force Tracking with Time Varying Desired Force

the decoupled external torque r is the input of the variable impedance inner loop, that is $F_c = r$. K_v is set as a constant in this section, we apply a sinusoidal curve to simulate a time-varying desired force, see Fig. 6. Simultaneously compared with the force tracking algorithm [16].

4.3 Force Tracking with Time Varying Impedence Environment

In order to verify the performance of the algorithm in a variable impedance environment, we set different impedance coefficients K_v at different periods of time, see Fig. 7. F_c is set as a constant in this section.

$$K_v = \begin{cases} 10000 \; t \in [1,2] \\ 8000 \quad t \in [2,3] \\ 2000 \quad t \in [3,4] \end{cases} \tag{16}$$

Figure 5 shows that the external force observation algorithm can effectively track the external disturbance force, and there will be no overshoot and oscillation under the time-varying force. The external disturbance observation can be completed by setting a smaller threshold $F_t = 0.3$ Nm. The algorithm can realize the collision stop function at the same time. Figure 6 shows that under time varying desired force, the initial overshoot of the algorithm proposed in this paper is obviously smaller than that [16], and the force tracking error is also smaller. It can be seen that the error [16] reached $F_e = -2$ Nm. Figure 7 shows that under the time-varying impedance environment, compared with [16], the tracking performance of this algorithm at the time of impedance change is more stable, and the overshoot [16] is reached $F_e = -0.7$ Nm.

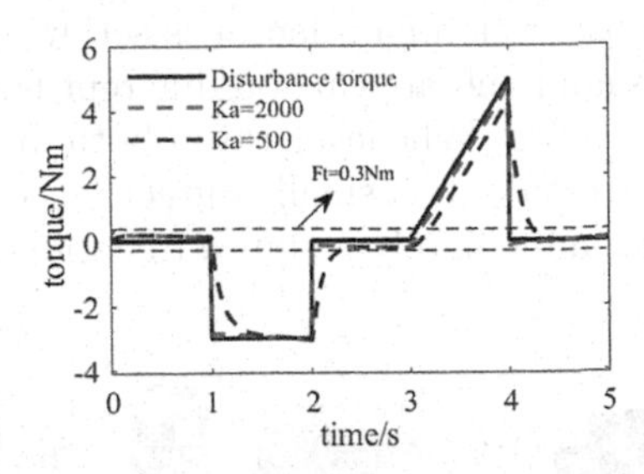

Fig. 5. External force observation with variable force environment

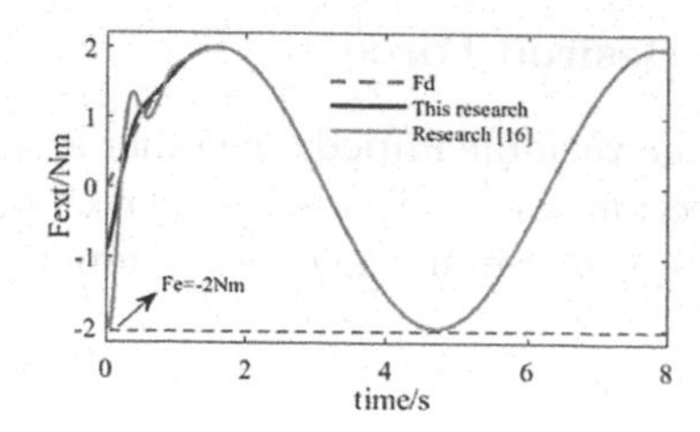

Fig. 6. Force tracking with time varying desired force

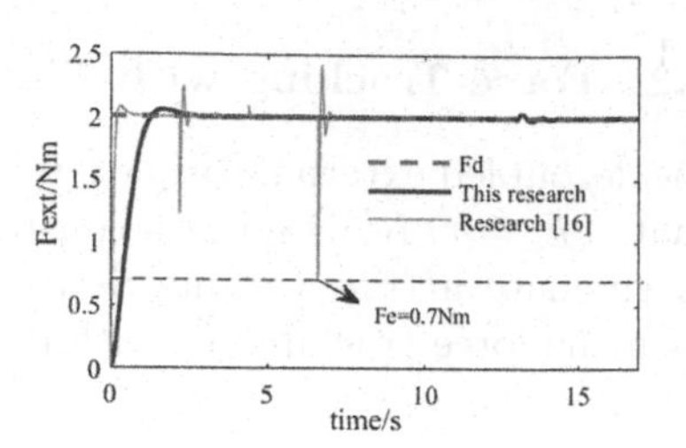

Fig. 7. Force Tracking with time varying impedence environment

5 Experiments with a Collaborative Joint

In order to verify the effectiveness of the variable impedance force tracking controller proposed in this paper, a collaborative joint development system is built based on the integrated joint developed by our laboratory, as shown in Fig. 8. The control system uses Ethercat bus to communicate with the host computer, the communication period is 2 ms. This experiment is divided into three parts. The first is external force disturbance experiment under different environments (human, soft sponge, balloon), the second is force tracking experiment in fixed unknown environment (foam, cartons, metal part), and the last is force tracking experiment in moving environment (human).

5.1 Joint External Force Observation (sensitive Collision Detection) Experiment

As show in Fig. 9, the robot starts to move freely from the initial point $q_0 = 150°$, and the external force observation gain factor is set $K_a = 2000$ (the same as in the simulation). The human hand appears on different trajectories of the robotic arm multiple times at random. Whenever the robotic arm touches the palm, it can be stopped suddenly with only a small contact force. When the palm leaves the end, the robotic arm immediately resumes free movement.

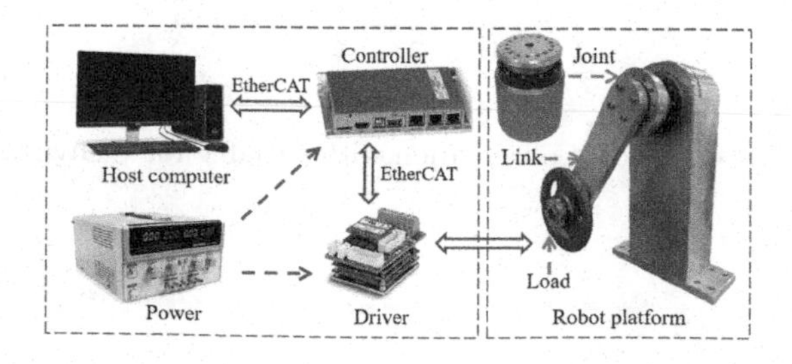

Fig. 8. Experiment platform

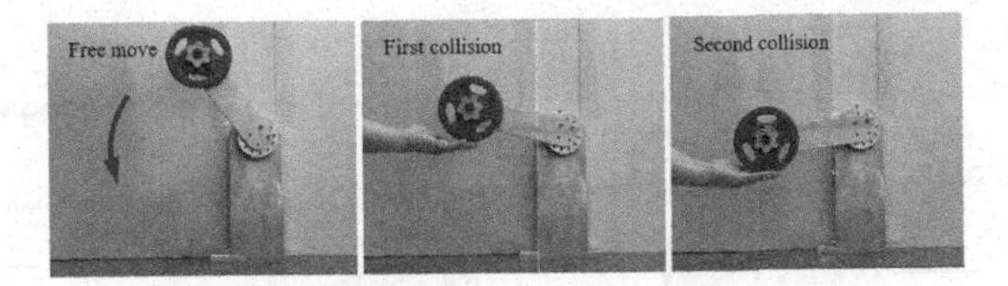

Fig. 9. Robot-human variable force observation

Figure 10 shows the low-density sponge and balloon collision experiment, represents objects of different stiffness and unknown environment respectively.

As shown in Fig. 12, when contact occurs, the observer responds immediately. The threshold of the human hand collision experiment is 0.2 Nm, and the threshold of the variable stiffness environment is 0.3 Nm. Verified the sensitivity and accuracy of the external force observation algorithm.

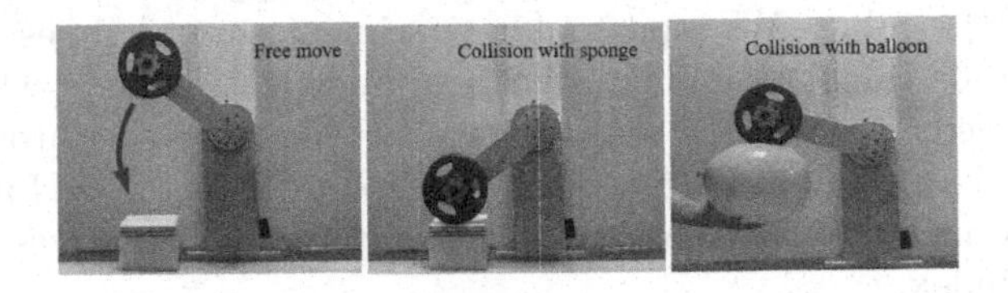

Fig. 10. Robot-variable stiffness environment force observation

5.2 Joint Force Tracking with Stationary Environment

Figure 11 is a tracking comparison with the desired force under objects with different stiffness (foam, cartons, metal part) and different desired force. Figure 13 shows that the Collaborative joint has a shorter adaptation time when it comes into contact with objects with greater stiffness. It is obviously that force tracking can be achieved under different expected torques. When $F_d = 0.2$ Nm, the error is the smallest, but the adaptive period is longer, when $F_d = 0.8$ Nm, the adaptive period is the shortest, which is 2.3 s.

Figure 14 shows the force tracking error of three different stiffness objects in the same expected force tracking experiment. It can be seen that the force errors $e_f < 0.04$ Nm.

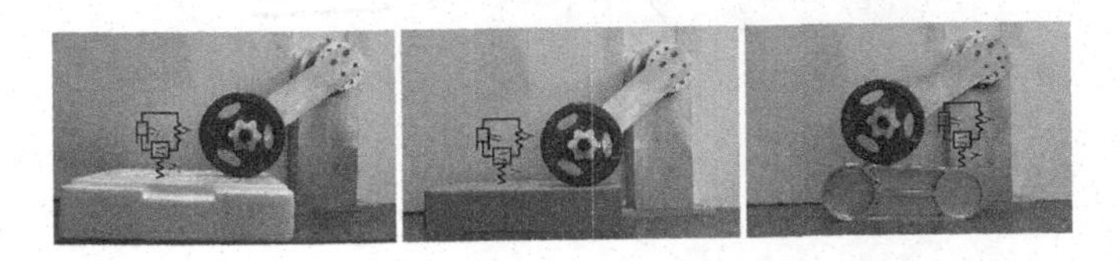

Fig. 11. Robot-variable stiffness objects force tracking

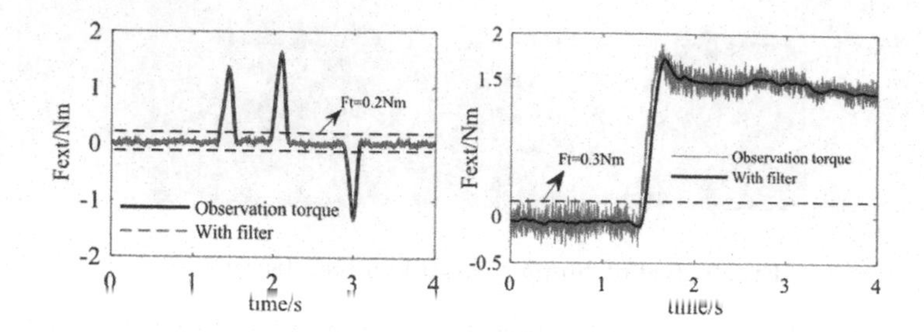

Fig. 12. Performance analysis of external force observation

5.3 Joint Force Tracking with Moving Environment

Figure 15 shows the Variable impedance force tracking in a moving environment. Human hand represent the moving environment. The robot starts to move from the initial position, and the human hand randomly appear in the trajectory and interact with the robot. We repeat the experiment several times and take the average value for calculation. The robot responds immediately when a human hand is observed, it reaches a steady state (F_d = 2.9 Nm) through adaptive force tracking in a small time ($t < 0.5s$), and $e_f < 0.15$ Nm, see Fig. 16. Table 1 shows the parameter settings in this experiment and Table 2 gives the values of performance index.

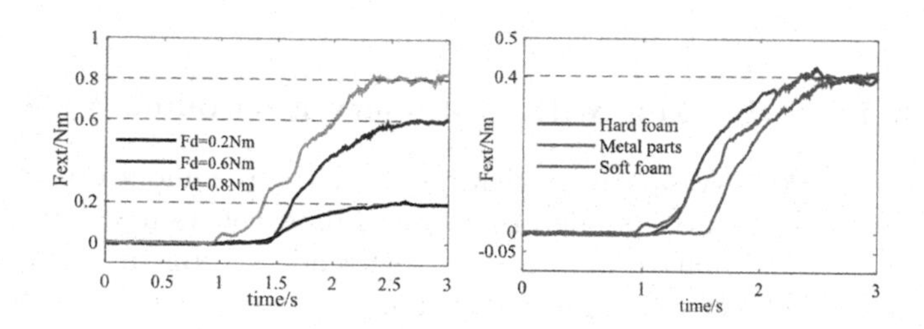

Fig. 13. Force tracking performance analysis

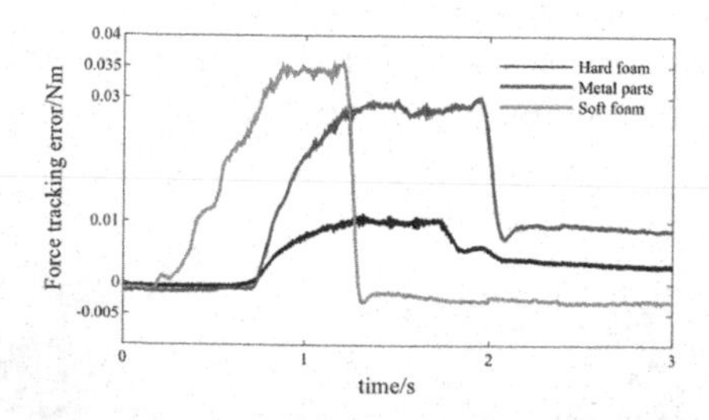

Fig. 14. Force tracking error analysis

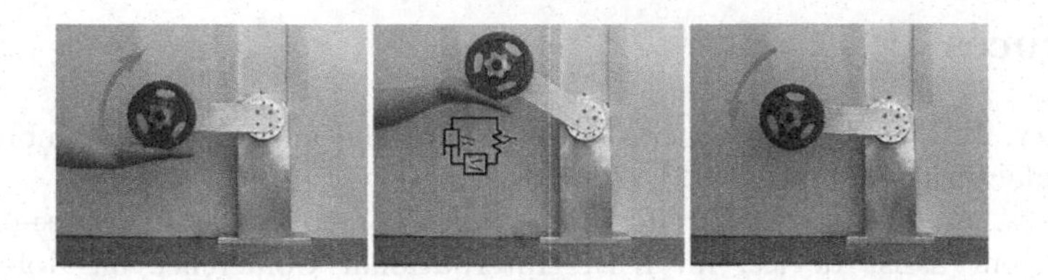

Fig. 15. Robot-palm variable impedance force tracking

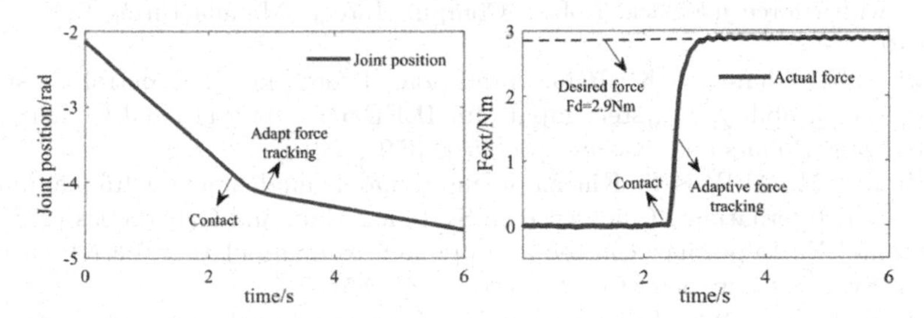

Fig. 16. performance analysis

Table 1. Parameters setting.

Parameters	Values
Load m	1.5 kg
Link length L	250 mm
Communication cycle T	2 ms
Variable impedance coefficient M	1.2
Variable impedance coefficient B	45

Table 2. Experimental results.

Performance index	Values
Desired force F_d	2.9 Nm
Average error e_f	0.15 Nm
Average response time t	0.5 s

6 Conclusion

In this paper, a collaborative joint variable impedance force tracking algorithm based on the second-order momentum observer is presented. The stability and robustness of the algorithm are verified through analysis and simulation, The experimental results support the claim that this method can track the desired force in real time under variable impedance and unknown environment. The future, we will implement the algorithm proposed in this paper on a 7-DOF collaborative robot, and improve the performance of the algorithm.

References

1. Krüger, J., Lien, T.K.: Cooperation of human and machines in assembly lines. J. Manufacturing Technol. **58**(2), 628–646 (2009)
2. Lecours, A., Mayer, B.: Variable admittance control of a four-degree-of-freedom intelligent assist device. In: IEEE International Conference on Robotics and Automation (2012)
3. Olsson, T., Haage, M.: Cost efficient drilling using industrial robots with high-bandwidth force feedback. Robot. Comput.-Integr. Manufacturing **26**(1), 24–38 (2010)
4. Park, Y.J., Chung, W.K.: Reinterpretation of force integral control considering the control ability of system input. In: IEEE/RSJ International Conference on Intelligent Robots and Systems, pp. 3586–3591 (2010)
5. Kebria, P.M., Al-Wais, S.: Kinematic and dynamic modelling of UR5 manipulator. In: IEEE International Conference on Systems, Man, and Cybernetics (2017)
6. Hogan, 1.N.: Impedance control: an approach to manipulation: Part I-Theory. J. Dyn. Syst. Measurement. Control, pp. 1–7 (1985)
7. Hyo, W.J., Jeong S.K.: Embedded design of position based impedance force control for implementing interaction between a human and a ROBOKER. In: IEEE International Conference on Control, Automation and Systems, pp. 1946–1950 (2008)
8. Yong, C., Junku, Y.: A unified adaptive force control of underwater vehicle-manipulator systems. In: IEEE/RSJ International Conference on Intelligent Robots and Systems, pp. 553–558 (2003)
9. Aghili, F.: Self-tuning cooperative control of manipulators with position/orientation uncertainties in the closed-kinematic loop. In: IEEE/RSJ International Conference on Intelligent Robots and Systems, pp. 4187–4193 (2011)
10. Erhart, S., Sieber, S.: An impedance-based control architecture for multi-robot cooperative dual-arm mobile manipulation. In: IEEE/RSJ International Conference on Intelligent Robots and Systems, pp. 315–322 (2013)
11. Shah, S.S., Raheja, U.: Input impedance analyses of charge controlled and frequency controlled LLC resonant converter. In: IEEE Energy Conversion Congress and Exposition, pp. 1–5 (2018)
12. Yu, G., Donglai, Z.: Adaptive impedance matching optimal control method for cascaded DC-DC power supply system. In: IEEE Industrial Electronics Society, pp. 751–755 (2017)
13. Marino, R., Spong, M.: Nonlinear control techniques for flexible joint manipulators: a single link case study. In: IEEE International Conference on Robotics and Automation, pp. 1030–1036 (1986)
14. Lee, S., Kim, M.: Sensorless collision detection for safe human-robot collaboration. In: IEEE/RSJ International Conference on Intelligent Robots and Systems, pp. 2392–2397 (2015)
15. Iqbal, K., Zheng, Y.F.: Predictive control application in arm manipulator coordination. In: IEEE International Symposium on Intelligent Control, pp. 409–414 (1997)
16. Liu, H., Lu, W.: Force tracking impedance control with moving target. In: IEEE International Conference on Robotics and Biomimetics, pp. 1369–1374 (2017)

An Surface Defect Detection Framework for Glass Bottle Body Based on the Stripe Light Source

Junxu Lu[1], Xu Zhang[1](✉), and Chen Li[2]

[1] School of Mechatronic Engineering and Automation, Shanghai University, Shanghai, China
xuzhang@shu.edu.cn

[2] School of Mechanical Science and Engineering, Huazhong University of Science and Technology, Wuhan, China

Abstract. Quality inspection is an essential technology in the glass product industry. Machine vision has shown more significant potential than manual inspection at present. However, the visual inspection of the bottle for defects remains a challenging task in a quality-controlled due to the difficulty in detecting some notable defects. To overcome the problem, we propose a surface defect detection framework based on the stripe light source. First, a novel method, DTST, determines the stripe type in the background with traditional image processing methods. Then, according to the result of DTST, the stripe type is divided into vertical stripes and horizontal stripes. For the former, a defect detection method based on DDVS that uses machine learning technology is proposed to detect cold mold defects precisely. A defect detection method named DDHS that uses deep learning technology is proposed to precisely detect wrinkled skin defects for the latter. The proposed framework is tested for data sets obtained by our designed vision system. The experimental results demonstrate that our framework achieves good performance.

Keywords: Defect detection · Stripe light source · Machine learning · Deep learning

1 Introduction

Glass bottles are easy to mold, relatively inexpensive, highly resistant to internal pressure, and easy to recycle. Thus, glass bottles are widely used in the food and beverage packaging industry, especially for packaging beer and carbonated beverages [1]. However, glass bottles are also susceptible to be damaged during transportation. Once a bottle that contains any defects is filled, many potential hazards, such as product deterioration and bottle explosion, may occur during product storage and transportation. Hence, the quality of each glass bottle could be thoroughly checked [2].

Supported by organization x.

X.-J. Liu et al. (Eds.): ICIRA 2021, LNAI 13014, pp. 575–585, 2021.
https://doi.org/10.1007/978-3-030-89098-8_54

Manual inspection methods suffer from inherently inconsistent and unreliable detection results because of the subjectiveness and alertness of the inspection worker. As a non-contact sensing technology, machine vision has shown great potential for the quality inspection of industrial products [3]. This paper focuses on the defect detection of the bottle body with machine vision technique.

Approaches directly related to the bottle body are briefly introduced. Most of these methods are composed of two processes: localization and defect detection. Zheng et al. [4] first employed a gray projection image to extract the central axis of the bottle body. Then, it proposed the whole bottle body detection algorithm based on BP neural network and PCA dimension reduction. It also proposed the detection algorithm on the different regions aimed at glass bottles with complex surfaces. Tian et al. [5] proposed a defect detection method that is based on edges, threshold, local fast OTSU segmentation, and wavelet transform. Ju et al. [6] first employed the edge pixel scanning method to calculate the bottle size. Then, proposed a defect detection method, which is based on threshold segmentation.

Although many surface defect detection methods have been reported to date, there are still many problems for bottle body inspection, e.g., defect detection for cold mold defects and wrinkle defects, which is no noticeable difference with the background. For these particular defects, we propose a novel defect detection framework for the glass bottle body. The main contribution of this paper is twofold:

1) Cold mold defects may affect the entire bottle area. This paper proposes a bottle body defect detection method named DDVS, which is based on vertical stripes.
2) Wrinkle defects may affect the partial bottle area. This paper proposes a bottle body defect detection method named DDHS, which is based on horizontal stripes.

2 An Surface Defect Detection Framework for Glass Bottle Body

The framework's content is mainly divided into two parts, and the first one is system structure design, the second one is the proposed method.

2.1 System Structure Design

This paper research and develop an automatic inspecting apparatus to detect the transparent defects on the Glass Bottle body.

To acquire a clear image of the bottle, we designed a machine vision apparatus, which is composed of three main parts: 1) electromechanical device, 2) imaging system and 3) processing module. The designed imaging system can get a good shot with the aid of an optoelectronic switch installed at the central line position of the camera and light source, as shown in Fig. 2. The striped light source and the camera are on both sides of the conveyor belt, and the striped light source continuously switches between horizontal and vertical stripes. Moreover, the camera focuses on the glass bottle (Fig. 1).

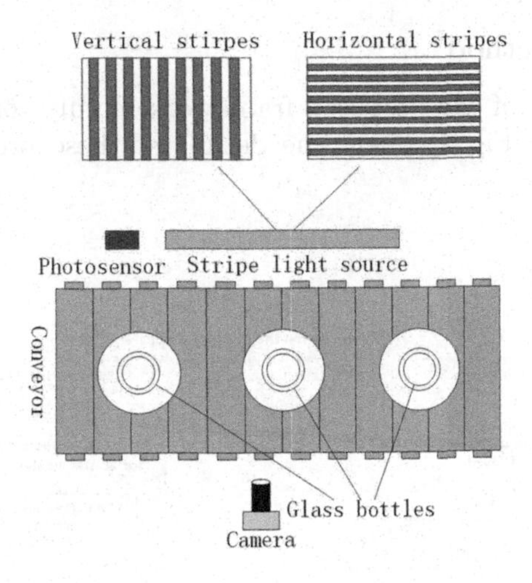

Fig. 1. Structure of the vision system

When a test bottle is transmitted to the position of the optoelectronic switch, the optoelectronic sensor generates a valid signal. Programmable logic controller (PLC) receives the signal and immediately generates two trigger signals for triggering the camera to capture two bottle images, which are under the horizontal and vertical stripes, as shown in Fig. 2. There is no apparent difference between the defects and the background under the backlight for the cold mold defects and wrinkle defects. However, the stripe width of the glass bottle with cold-molded defects, which is under the vertical stripes, fluctuates wildly; the stripe area of the glass bottle with wrinkle defects, which is under the horizontal stripes, has apparent changes.

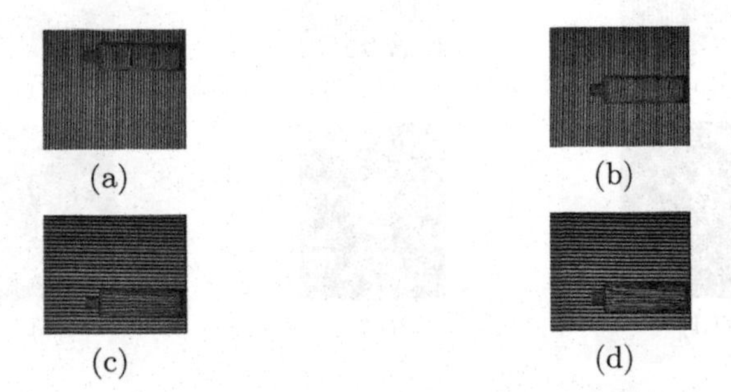

Fig. 2. Acquired images. (a) Standard glass bottle under vertical stripes (b) Bad glass bottle under vertical stripes (c) Standard glass bottle under horizontal stripes (d) Standard glass bottle under horizontal stripes

2.2 Proposed Method

The entire flowchart of the proposed framework mainly consists of VSDD and HSDD, as shown in Fig. 3. Then, the details of these proposed methods are described.

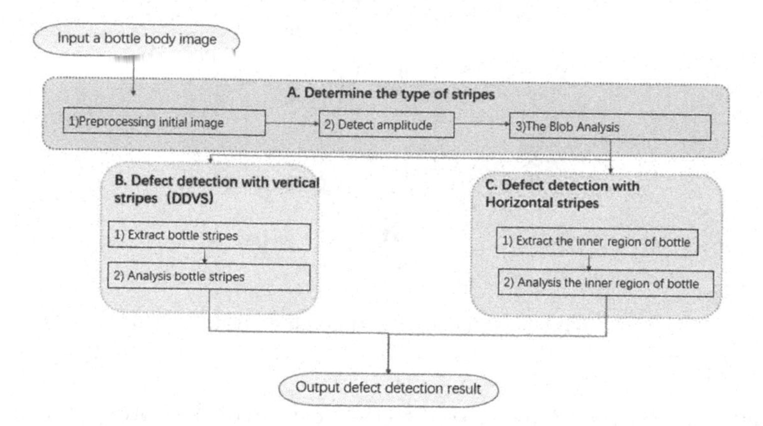

Fig. 3. Proposed defect detection framework for the bottle body image

2.3 Determine the Type of Stripes (DTST)

The algorithm to determine the stripes type is as shown as in Fig. 4

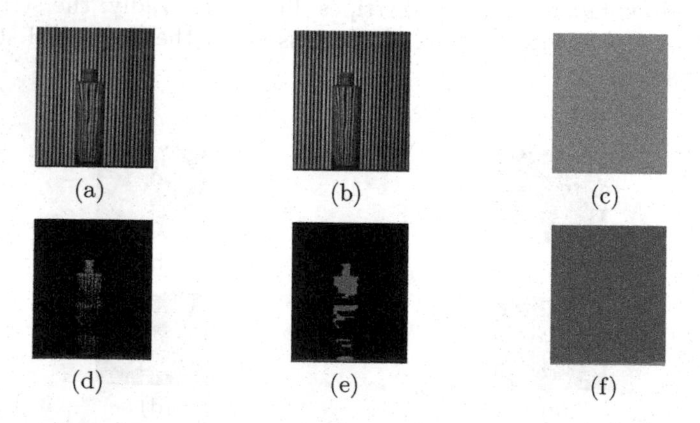

Fig. 4. The algorithm of determining the stripes type (a) smoothed image (b) emphasized image (c) EdgeAmpiltude image (d) Segmented region (e) Region in the vertical stripes (f) Region in the horizontal stripes

In order to effectively suppress noise, Gaussian filtering is used to preprocess the image. The original image is convolved with the two-dimensional Gaussian kernel, which is as follows:

$$G(x,y) = \frac{1}{2\pi\delta^2} e^{-x^2+y^2/2\delta^2} \tag{1}$$

Where δ is the standard deviation.

In order to enhance the difference between the defect and the background, the low-pass filter first is used to calculate the smooth gray level, and then calculate the sharpened image:

$$\text{res} = round(orig - mean) * factor + orig \tag{2}$$

Where res is the gray value after sharpening, mean is the smooth gray value, factor is the contrast enhancement coefficient, orig is the initial gray value.

The Sobel algorithm is employed to calculate the grayscale gradient map along the x-direction, and convolve the preprocessed image with the mask, which is

$$mask = \begin{pmatrix} \frac{1}{4} & 0 & -\frac{1}{4} \\ \frac{1}{2} & 0 & -\frac{1}{2} \\ \frac{1}{4} & 0 & -\frac{1}{4} \end{pmatrix} \tag{3}$$

Then, the global binarization is employed, which selects the pixels from the input image whose gray values fulfill the following condition: [−80, −10] or [10, 128]. Finally, the determination area is obtained through morphological operations. If the pixel area of the determination area is greater than 100 w, the background is vertical stripes; otherwise, it is horizontal stripes.

2.4 Defect Detection with Vertical Stripes (DDVS)

This paper mainly uses the volatility of stripes to detect cold mold defects. The algorithm is mainly divided into two parts: extract the bottle body stripes and analyze the stripes' volatility to determine whether the bottle contains defects. The algorithm of extract stripes is shown in Fig. 5.

First, the glass bottle region, which is based on the characteristics of vertical stripes, is accurately located. The shape of the bottle in the region was transformed to the most oversized axis-parallel rectangle fitting into the bottle, as shown in Fig. 5(a). The largest axis-parallel rectangle region with a rectangular structuring element which is 150 × 100 pixels, was eroded. Binarization and Blob analysis is used to extract the stripes shown in Fig. 5(c).

Then the volatility of the stripes is analyzed. The width array of each stripe, and its deviation array, used to get feature variables, such as average deviation, maximum deviation, minimum width, average width, maximum width, are calculated. Moreover, a support vector machine can be employed to classify whether the bottle contains defects.

(a)

(b)

(c)

Fig. 5. The algorithm of extract stripes (a) bottle region (b) inner region (c) extracted stripes

In order to avoid the difference between the data from affecting the experimental results, all the original data are normalized, and the normalization formula is

$$x = \frac{x_i - x_{\min}}{x_{\max} - x_{\min}} \tag{4}$$

Where x is processed data, x_1 is actual value, $x_{\min}$ is the minimum value, and $x_{\max}$ is the maximum value [7].

In order to solve the non-linear classification problem, the feature variables are mapped to the high-dimensional space, which realizes accurate classification [7]. In this paper, a mixed kernel function is constructed to improve the accuracy of classification, which is as follow:

$$K(x, y) = aK_1(x, y) + bK_2(x, y) \tag{5}$$

where $K_1(x, y)$ is polynomial kernel function, $K_2(x, y)$ is radial basis kernel function, a and b are weight factor.

SVM's classification accuracy rate is used as the objective function, which is used to optimize the SVM iteratively. The classification accuracy rate of each iteration is recorded, and the optimal mode is selected.

2.5 Defect Detection with Horizontal Stripes (DDHS)

The algorithm is mainly divided into two parts: extract the stripe's inner region and analyzed the inner region, which is used to determine whether the bottle contains wrinkled skin defects. The algorithm of extract stripes is shown in Fig. 6.

The glass bottle region, which is based on the characteristics of horizonal stripes, is accurately located. The shape of the bottle region is transformed to the largest axis-parallel rectangle fitting into the bottle. The largest axis-parallel rectangle region is eroded with a rectangular structuring element which is 150 × 100 pixels.

Then deep learning technology is employed to detect glass bottles to determine whether the glass bottle contains defects. This paper uses the classic

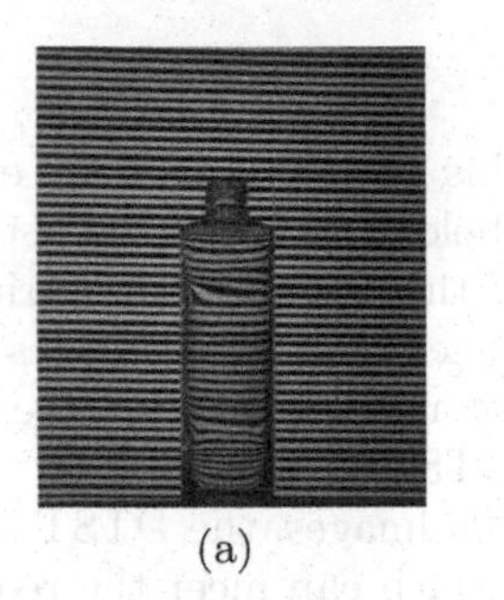

(a)

(b)

Fig. 6. The algorithm of extract stripe inner region (a) bottle region. (b) inner region

AlexNet network [8], which is divided into a total of eight layers, five convolutional layers, and three fully connected layers. The convolutional layer contains the excitation function ReLU, local response normalization, and downsampling. The basic structure of the network is shown in Fig. 7.

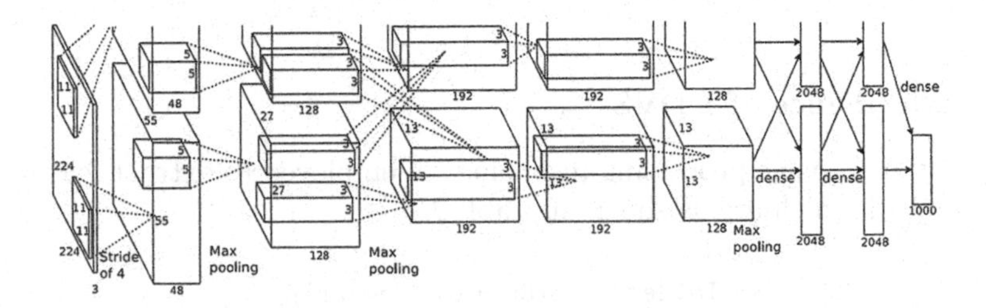

Fig. 7. An illustration of the architecture of AlexNet CNN

Firstly, the data set is preprocessed. The bottle inner region is converted into a three-channel image which is 224 × 224 pixels. Then the data set is divided into training set (70%), validation set (15%), test set (15%). And training parameters are set, such as batch size, initial learning rate and learning rate step ratio etc. And the accuracy of the classification in the training is set as the objective function, and continuously iteratively optimize to find the optimal network. Finally, the optimal network is evaluated with test sets.

3 Experiment

In this section, the performances of the proposed DTST, DDVS, DDHS, and the entire framework are individually evaluated on the calibrated data sets.

3.1 Performance of DTST

In order to verify the reliability of DTST, this paper designs an experiment that takes 300 images. In there, 150 images belong to horizontal stripes, and 150 images belong to vertical stripes. Firstly, the image of the stripe type is calibrated, and DTST is employed to decide the experimental images. Then the judgment result is compared with the calibration result to acquire the error rate, which is used to evaluate the performance of DTST.

The test results are shown in Table 1. For 300 images, the DTST accuracy is 100%, and the average test time is 14.387ms, which can meet the requirements of the actual application.

Table 1. The confusion matrix of DTST.

Class	Predicted class		
Actual class		Vertical	Horizontal
	Vertical	150	0
	Horizontal	0	150

3.2 Performance of DDVS

The traditional image processing technology is employed to extract feature variables, some of which are as shown in Table 2.

Table 2. Feature variables table.

No.	Max deviation	Average deviation	Min width	Average width	Max width	Classification
1	12.92	6.96265	16.201	24.7687	35.9981	0
2	14.6729	7.7491	18.3072	31.023	44.2626	0
3	7.78322	5.86844	11.2551	23.4678	34.0365	0
4	21.2204	7.84197	13.2927	27.0117	44.2044	0
5	8.51773	5.83497	15.1949	25.3302	34.961	0
6	4.56197	3.01544	15.5027	22.6391	24.8976	1
7	4.37018	3.47605	23.2773	27.4486	30.6519	1
8	5.29801	3.70247	18.4781	23.3912	27.3395	1
9	5.43173	4.09966	18.538	26.7987	32.3676	1
10	6.06601	4.37687	11.9483	25.9269	35.5059	1

After obtaining the data features, use the data set to train the SVM classifier. The training process is as shown in Fig. 8(a). After 28 iterations, the optimal classifier is obtained. The minimum classification error is 4.3%, and the kernel constraint level is 0.45308.

Accuracy rate, recall rate, area under the curve are used to evaluate the performance of the training model. The test results are shown in Fig. 8(b) (c) for the new data set. The accuracy rate is 95.9%, the recall rate is 97.9%, and the AUC value is 0.99. It shows good algorithm performance and can meet the requirements of the actual application.

3.3 Performance of DDHS

The traditional image processing technology is employed to extract the internal area of the glass bottle and calibrates its category, which obtains training data set. Some of the data set are shown in Fig. 9.

The data set is divided into a training set (70%), validation set (15%), and test set (15%). Training set and test set are used to train networks, and test set is used to evaluate the detection effect of the optimal network. The training error and validation error of DDHS were 99.3% and 98.9%, respectively.

3.4 Performance of the Entire Framework

To further evaluate the performance of the framework on the bottle body, the data set including 3000 bottle images are used for testing. The test results are shown in the Table 3.

Table 3. The test result of 3 data set

Data set	Good	Bad	Consumed time/ms
Cold mold bottles	28	976	257.631
Wrinkled skin bottles	78	922	248.588
Standard bottles	992	8	267.336

The confusion matrix of the proposed framework is shown in the Table 4. The precision and recall of the proposed framework were 95.00% and 92.10%, respectively. Moreover, the average consumed time of our framework is about 257.687ms. The experimental results show that the proposed methods can achieve good performances.

Table 4. Confusion matrix of the proposed framework

	Predicted class		
Actual class		Good	Bad
	Good	992	8
	Bad	106	1898

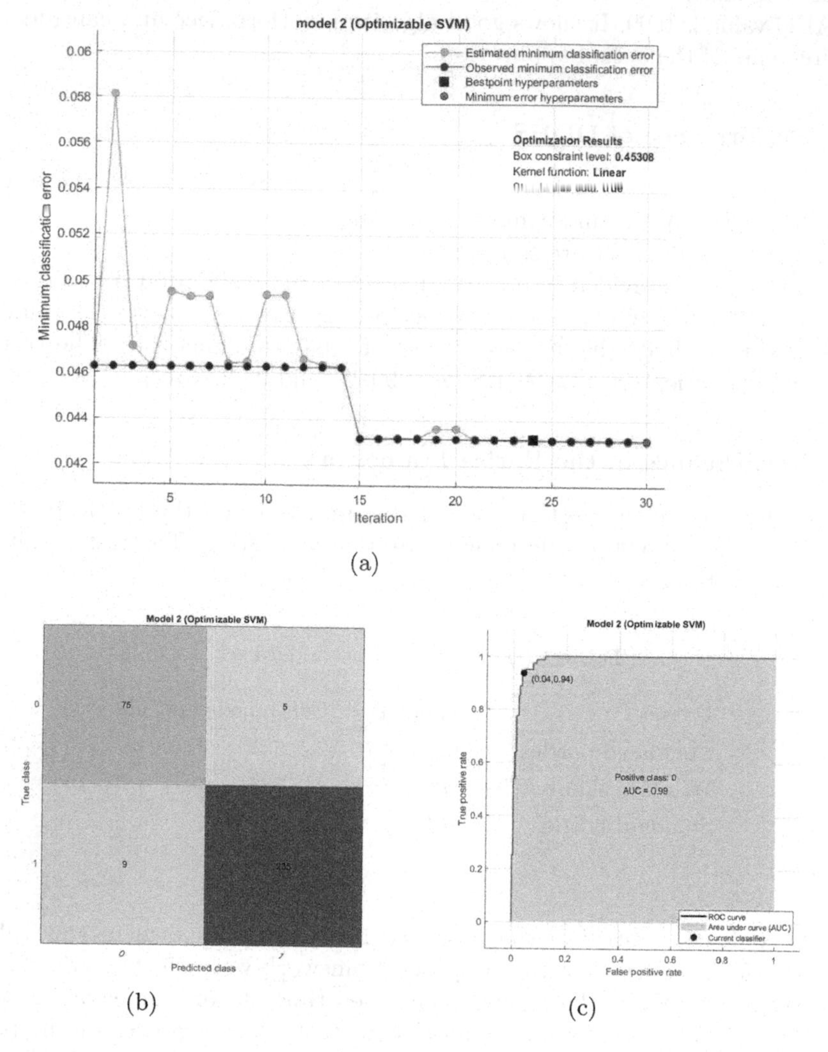

Fig. 8. Model performance evaluation (a) Training process (b) Confusion matrix (c) The graph of AUC

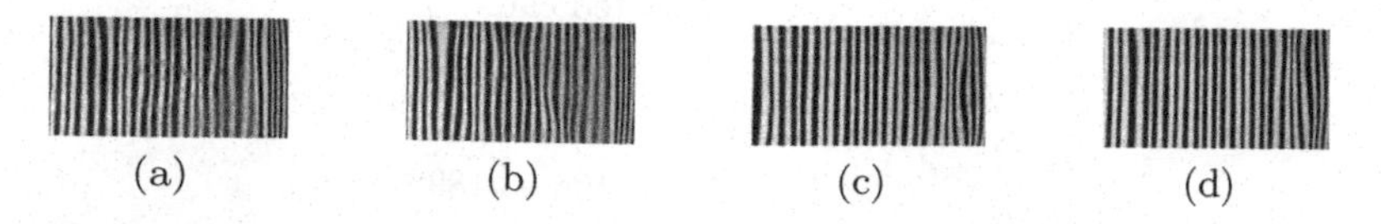

Fig. 9. Training data set (a) (b) defective sample (c) (d) standard sample

4 Conclusion

This paper proposed a surface defect detection framework based on the stripe light source to overcome the difficulty in detecting cold mold defects and wrinkle defects. It is evaluated on the data set acquired by our designed vision system. The precision and recall of the proposed framework were 99.5% and 94.7%, respectively. The average consumed time of our framework is about 257.687 ms. The experimental results show that the proposed framework can achieve good performances.

Acknowledgement. This study is supported by the key research project of the Ministry of Science and Technology (Grant No. 2018YFB1306802) the National Natural Science Foundation of China (Grant No. 51975344) and China Postdoctoral Science Foundation (Grant No. 2019M662591).

References

1. Rahman, A.: Shape and level bottles detection using local standard deviation and hough transform. Int. J. Electr. Comput. Eng. (IJECE) **8**(6) (2018)
2. Xin, Z.: An improve feature selection algorithm for defect detection of glass bottles. Appl. Acoust. **174** (2021). (in Chinese)
3. Fu, L.: Medicine glass bottle defect detection based on machine vision. In: Chinese Control And Decision Conference. Nanchang, China 2019, pp. 5681–5685 (2019). (in Chinese)
4. Yexin, Z.: Bottle body detection algorithm and application in high speed high accuracy empty bottle detection robot. Hunan University (2017). (in Chinese)
5. Tian, B.: Glassware defect detection system based on machine vision. Shandong University (2018). (in Chinese)
6. Wei, J.: Inspection of medicinal glass bottle's dimension and defect based on machine vision. Chongqing University of Technology (2012). (in Chinese)
7. We, Z.: Classification and recognition method for bearing fault based on IFOA-SVM. Mech. Transm. **45**(02), 148–156 (2021). (in Chinese)
8. Krizhevsky, A.: ImageNet classification with deep convolutional neural networks. In: International Conference on Neural Information Processing Systems, pp. 1097–1105. Curran Associates Inc. (2012)

Transfer Learning - Based Intention Recognition of Human Upper Limb in Human - Robot Collaboration

Mengchao Dong, Jinzhu Peng(✉), Shuai Ding, and Zhiqiang Wang

School of Electrical Engineering, Zhengzhou University, Zhengzhou 45001, China
jzpeng@zzu.edu.cn

Abstract. Under the wave of rapid industrial development, automated production is gradually shifting to intelligent and customized production. The human-robot collaboration (HRC) system, as an effective way to improve the intelligence and flexibility of automated production, has received great attention from people. Recognizing human intentions quickly and accurately is the foundation for a safe and efficient HRC. In this work, we propose a novel approach of intention recognition, which transforms intention recognition into the recognition of feature images. The trajectory of human movement is projected and reconstructed into feature images, and transfer learning is implemented on Alexnet to complete the recognition of feature images to indirectly realize the recognition of the intention. We evaluate the proposed approach on a self-made dataset. The experimental results show our method can accurately recognize the intention in the early stage of human motion.

Keywords: Human-robot collaboration · Intention recognition · Transfer learning

1 Introduction

As the focus of manufacturing shifts from mass production to proprietary customization, the demand for flexibility in automated production continues to increase. HRC, which provides flexibility and intelligence to automated production, has received extensive attention in recent years [1,2]. The concept of HRC is that humans and robots can complete collaborative tasks safely and efficiently in a shared space [3,4]. In order to achieve safe and efficient HRC, robots need not only current skills, but also some understanding of human movements, such as the recognition of human actions and the inference of intentions [5,6].

The recognition of human actions and intentions has received continuous attention from researchers. Ding et al. [7] proposed a hierarchical spatiotemporal pattern for human activities, which predicts ongoing activities from videos

This work is partially supported by the National Natural Science Foundation of China (61773351), the Program for Science & Technology Innovation Talents in Universities of Henan Province (20HASTIT031).

X.-J. Liu et al. (Eds.): ICIRA 2021, LNAI 13014, pp. 586–595, 2021.
https://doi.org/10.1007/978-3-030-89098-8_55

containing the beginning of activities. Spatiotemporal patterns were modelled by a Hierarchical Self-Organising Map (HSOM). Yang et al. [8] approximated the traditional 3D convolution with an efficient asymmetric one-directional 3D convolutions. By incorporating multi-scale 3D convolution branches, an asymmetric 3D-convolution neural network (CNN) deep model is constructed to complete the task of human behavior recognition. Martinez et al. [9] proposed three changes to the standard recurrent neural network (RNN) model used for human motion in terms of architecture, loss function and training process, which constructed a simple and scalable RNN architecture for human motion prediction. Liu et al. [10] introduced a novel human motion prediction framework, which combines RNN and inverse kinematics (IK) to predict the motion of the human arm. A modified Kalman filter (MKF) was applied to adapt the model online.

In the field of HRC, researchers attempted to achieve the recognition of human intention to improve the safety and efficiency of HRC systems. Pérez-D'Arpino et al. [11] proposed a data-driven method which combines the synthesizes anticipatory knowledge of human actions and subsequent action steps to predict the intention of human stretched out a hand in real time. Liu et al. [12] utilised back propagation(BP) neural network to complete the prediction of human "grasping" behavior intention. The Euclidean distance, azimuth angle, and relative speed of the human upper limb "grabbing" behavior intention in color images and depth images were used as expression factors, and D-S evidence theory was used to fuse the predicted probabilities of different expression factors. Yan et al. [13] proposed a typical multi-stacked long-term short-term memory (LSTM) neural network for human intention recognition, which combines the advantages of a single LSTM layer and overcomes the shortcoming of the long-term time dependence of RNN learning. Liu et al. [14] introduced a deep learning system that combines CNN and LSTM with visual signals to accurately predict human motion. Wang et al. [15] developed a teaching-learning-collaboration (TLC) model for the collaborative robot to learn from human demonstrations and assist its human partner in shared working situations. The maximum entropy inverse reinforcement learning algorithm was used by robots to learn from the human assembly demonstration and matches the optimal assembly strategy according to human intentions.

In the HRC system, it is excellent that the robot recognizes the human's intentions even though the completion of the human action is low. The accuracy of the above method in identifying intentions at an early stage of human action is unsatisfactory. In this paper, we propose a method that the trajectory information of human motion is transformed into feature image and the purpose of human intention recognition is indirectly realized by transfer learning. Vicon motion capture apparatus are used to collect the motion trajectory information of the human upper limb movement, and transform it into feature images through operations such as projection and reorganization. In this way, a one-to-one correspondence between the movement of human upper limb and the feature image is formed. Alexnet is used for transfer learning to realize the classification of feature images, thereby indirectly completing the intention recognition of the corresponding actions. The developed method is verified experimentally on the self-made data set.

The remainder of this paper is organized as follows. Section 2 describes the environment of data collection, the generation process of feature images, and transfer learning. Section 3 introduces the validation setup, the implementation details, and discusses the outcome of the experiments. Section 4 offers some concluding remarks.

2 Methodology

2.1 Data Collection

In the process of human-robot collaboration, as a result of human behavior habits prompted people normally complete most of the operation process with the right arm. Therefore, the research on human upper limb intention recognition in this paper is mainly reflected in the human right arm. The data collection environment is shown in Fig. 1. Vicon motion capture apparatus are used to capture the marks on the shoulder, elbow and wrist of the human right arm to realize motion capture and experimental data recording.

(a) (b)

Fig. 1. Environment for data collection. (a) Human right arm with marked points. (b) Vicon motion capture apparatus.

In order to reflect the action process of human upper limb in the process of human-robot collaboration, this paper defines three kinds of upper limb movements with the right arm as the main body. This paper uses the abbreviation Act 1, Act 2, and Act 3 to distinguish three kinds of actions and their representative intentions. The duration of each action is about 3 s. The detailed execution process of the three upper limb movements is shown in Fig. 2.

2.2 Generate Feature Image

The powerful performance of neural network in computer vision is obvious, especially in the application of image recognition and object classification. CNN was also used to do data processing and feature extraction on video sequences. In

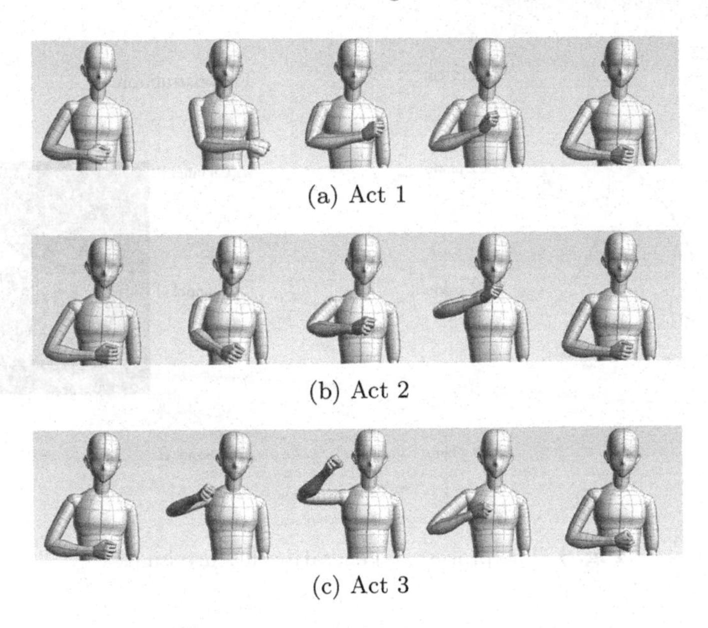

(a) Act 1

(b) Act 2

(c) Act 3

Fig. 2. Three types of actions

order to make full use of the advantages of CNN in image, the intention recognition problem based on human upper limb motion trajectory is transformed into the classification of feature images. By projecting and reconstructing the motion trajectory of the human upper limb in the 3D space, the feature images is encoded, which implies the temporal and spatial relationship of the human upper limb movement. This method is used to realize the one-to-one transformation and correspondence of motion trajectory to feature images.

During the movement of the human arm, the movement space of the wrist joint is significantly larger than the shoulder joint and the elbow joint. For different arm movements, the variation of wrist movement trajectory is significant, which can better express the intention of the current movement. In this paper, only the wrist movement trajectory is selected as the input for the intention recognition of the current movement, which reduced the amount of calculation and shortened the time for the intention recognition to a certain extent. Take the movement trajectory of a wrist joint in action 1 as an example, project it to the XY, XZ, and YZ planes in a 3D space, and the projected trajectory lines are respectively represented by three colors of red, green, and blue. The three projection images occupy the three corresponding channels of the black RGB image to obtain the feature image corresponding to the motion trajectory. This feature image, which implies the motion trajectory and intention of the human upper limb right arm, is fed into the CNN as input. The generation process of feature images is shown in Fig. 3.

In order to make the presentation clear and easy to understand, the complete movement trajectory of the wrist joint of the movement is used in the projection

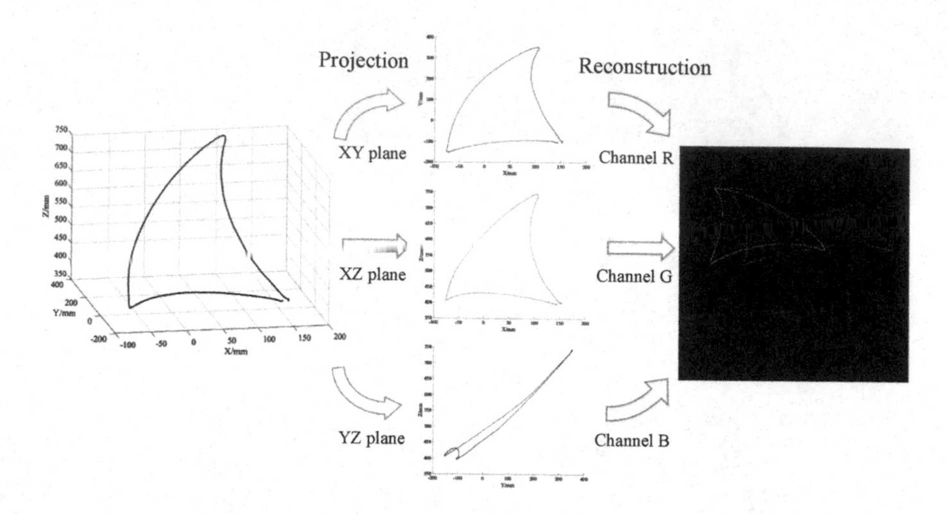

Fig. 3. The process of generating feature images

example. The data used in the actual experiment is only the first half of the action, which will be introduced in detail in the subsequent experiment part. When creating the scale of an RGB image, make sure that it can fully bear the projected image of the current action wrist trajectory in the XY, XZ, and YZ planes. Since the pixel coordinates of the image are all integers, the coordinate value of the projected trajectory must be rounded.

2.3 Transfer Learning

The Alexnet network model is a CNN widely used in computer vision. It was first applied by Hinton G and Krizhevsky A in the 2012 ImageNet Vision Challenge. AlexNet network includes five layers of convolutional layers and three fully connected layers. This structure can ensure its strong learning and analysis capabilities, accelerate the processing of data sets, shorten the network training process and research cycle, and reduce the occurrence of convolutional neural network overfitting.

For deep learning models, a large amount of labeled data is required for training, but collecting and labeling data requires a lot of time and resources, and even if there is a certain amount of training data, training an excellent convolutional neural network model takes a long time. In order to solve the shortcomings of complex annotation data and long training time, the CNN model is pre-trained by using the idea of transfer learning, and then the adaptive training is carried out on the self-made data set. For transfer learning, the first few layers of the neural network trained with other big data sets are retained; then the final fully connected layers and output layer of the network are adaptively adjusted according to their own requirements; and finally the self-made data set is put into the network for fine-tuning training. It can significantly reduce the training time and provide a more effective way for small data set training. The CNN migration model in this paper is Alexnet, which trained on the ImageNet 2012 classification data set. The first five pre-trained convolutional layer parameters are transferred to the intention recognition model. Pooling layer does not need to keep training parameters, and other training parameters are initialized in a random manner. The parameters of the Softmax layer determined by the categories of intention recognition. After transfer training on the self-made data set, a CNN model for recognizing the movement intention of human upper limbs is obtained. The overall scheme of AlexNet transfer learning is shown in Fig. 4.

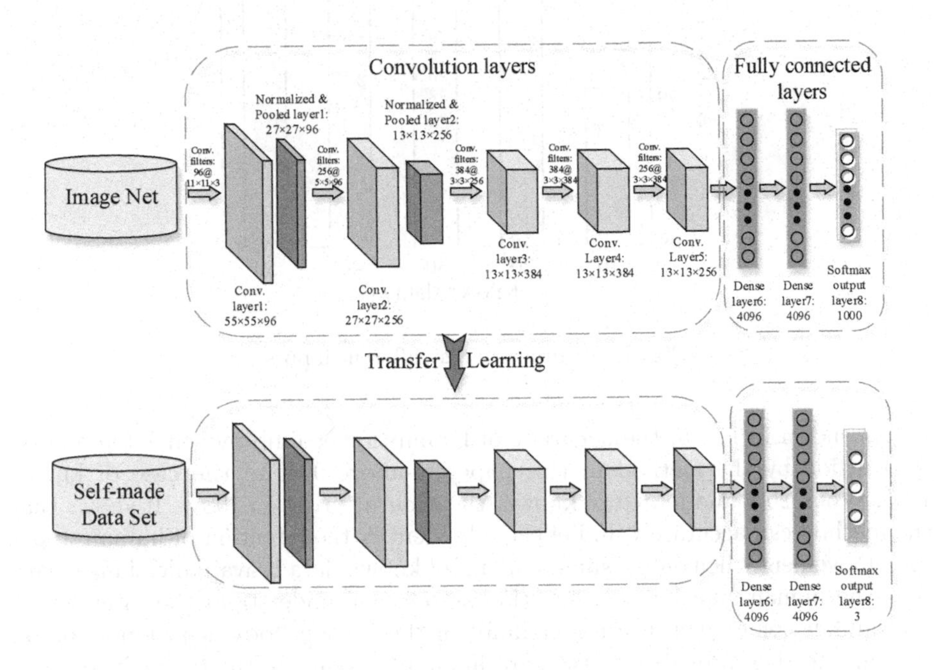

Fig. 4. The transfer learning of Alexnet network

3 Experimental

Vicon motion capture apparatus is used to capture and record the motion information of the three joints of human upper limb shoulder, elbow and wrist at the frequency 300 Hz to form a self-made data set. These included Act 1 103, Act 2 120 and Act 3 131. The ratio of training set to test set is 7:3.

In 2.2, it is mentioned that the complete movement trajectory of the wrist is used in the example of feature image generation to make the expression clear and easy to understand. The significance of studying human upper limb movement intention recognition in the context of human-robot collaboration lies in judging the current intention of the human before the movement is completed. It is meaningless for the human-robot collaboration system to recognize the intention of the current action when the action is about to end. Therefore, we divide the completion of the action as the known data, which includes known 10%, known 20%, known 30%, known 40% and known 50%. The accuracy of intention recognition under the five different input situations is shown in Fig. 5.

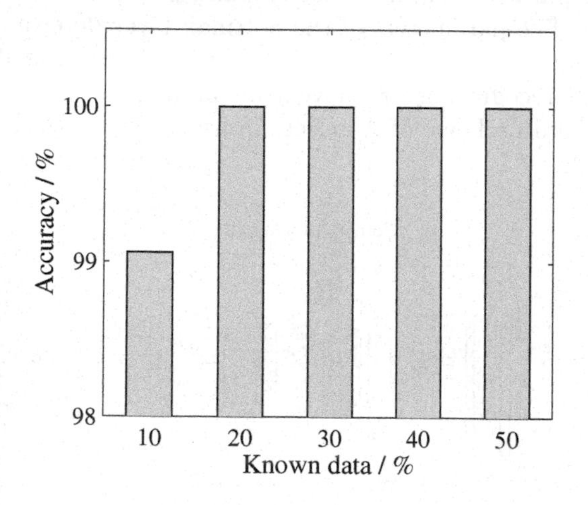

Fig. 5. Accuracy under different inputs

As shown in Fig. 5, the accuracy of human upper limb movement intention recognition by the method in this paper is above 99% in the case of known 10%. When 20% and more are known, the accuracy rises to 100%. It shows that the method can accurately and effectively identify the intention of human upper limb movement when only a small amount of known data is available. This means that the human-machine collaboration system can understand the intention of the human being when the upper limbs of the human body is just making an action, and then provide assistance to human to complete the task efficiently.

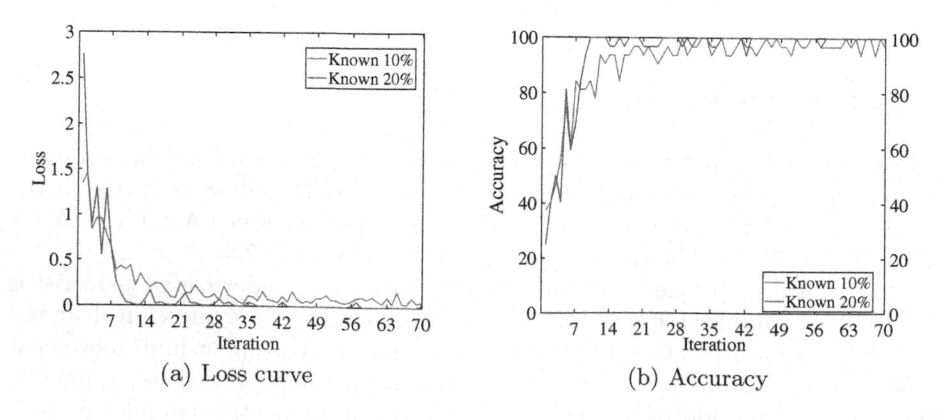

(a) Loss curve (b) Accuracy

Fig. 6. The accuracy and loss curve of the training process

In Fig. 6, the loss curve and accuracy change of the model training process are given. Since the accuracy has reached 100% when 20% and above are known, only known 10% and known 20% are shown.

In the case of known 10%, loss decreased significantly and the training accuracy increased rapidly in the first two epochs, and the model gradually improved in the subsequent epochs. When the known data is increased to 20%, the loss value decreases significantly in the first epochs, and the training accuracy rate reaches close to the maximum. The more known data about the human upper limb movements, the faster the intention recognition network will converge, which reflects that as the human upper limb movements progress, the differences between individual movements are gradually increasing. The model test results in the case of known 10% and known 20% are shown in Fig. 7.

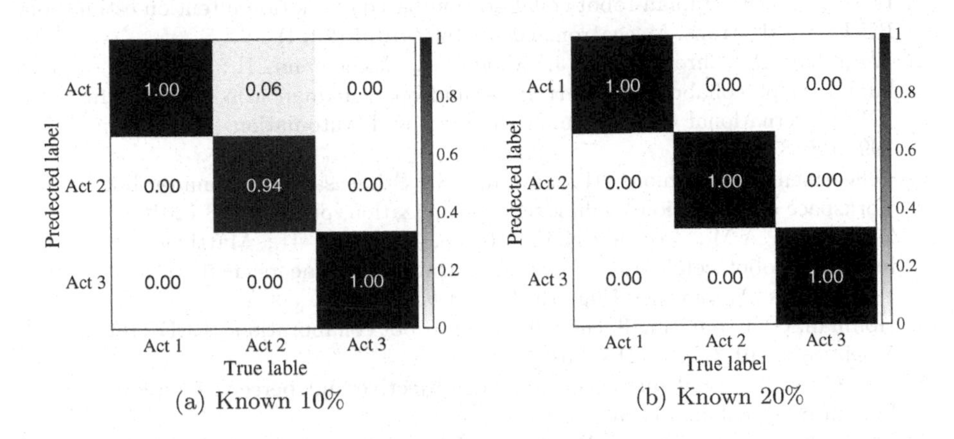

Fig. 7. Confusion matrix

In the case of known 10%, the model trained by the proposed method only produces errors in the intention recognition of Act 2. It has a 6% error rate for Act 2, all of which focuses on misidentifying Act 2 as Act 1. At the beginning of the action, the differences between different actions are not so obvious that it leads to false recognition. With the increase of known data, the accuracy of intention recognition is also improving. As shown in Fig. 7b, when the known data reaches 20%, the model corrects the previous wrong judgment of Act 2 so that the recognition accuracy reaches 100%.

4 Conclusion

This paper proposed a novel indirect recognition method for human upper limb movement intention in HRC system. In this method, the trajectory of human upper limb motion is projected and reconstructed to generate a feature image that implies the current motion information. AlexNet is used for transfer learning to complete the recognition of feature images, consequently the intention recognition of the action corresponding to the feature image can be realized indirectly. The experimental results indicate the accuracy of intention recognition is 99.1% when only 10% of the action is completed. When the first 20% action information is input, the accuracy of intention recognition reaches 100%. This means for an action that lasts 3 s, the robot can understand the intention of the human upper limb motion in 0.6 s. This intention recognition method provides a long enough time for the HRC system to determine whether to provide cooperation for humans and how to collaborate safely and efficiently.

References

1. Li, Y., Ge, S.S.: Human-robot collaboration based on motion intention estimation. IEEE/ASME Trans. Mechatron. **19**(3), 1007–1014 (2014)
2. Weitschat, R., Ehrensperger, J., Maier, M., Aschemann, H.: Safe and efficient human-robot collaboration Part I: estimation of human arm motions. In: 2018 IEEE International Conference on Robotics and Automation (ICRA), pp. 1993–1999 (2018)
3. Pellegrinelli, S., Admoni, H., Javdani, S., Srinivasa, S.: Human-robot shared workspace collaboration via hindsight optimization, pp. 831–838 (2016)
4. Zanchettin, A.M., Ceriani, N.M., Rocco, P., Ding, H., Matthias, B.: Safety in human-robot collaborative manufacturing environments: metrics and control. IEEE Trans. Autom. Sci. Eng. **13**(2), 882–893 (2016)
5. Hoffman, G.: Evaluating fluency in human-robot collaboration. IEEE Trans. Hum. Mach. Syst. **49**(3), 209–218 (2019)
6. Liu, C., et al.: Goal inference improves objective and perceived performance in human-robot collaboration (2018)
7. Ding, W., Liu, K., Cheng, F., Zhang, J.: Learning hierarchical spatio-temporal pattern for human activity prediction. J. Vis. Commun. Image Represent. **35**, 103–111 (2016)
8. Wu, C., Han, J., Li, X.: Time-asymmetric 3d convolutional neural networks for action recognition. In: 2019 IEEE International Conference on Image Processing (ICIP), pp. 21–25 (2019)
9. Martinez, J., Black, M.J., Romero, J.: On human motion prediction using recurrent neural networks. In: 2017 IEEE Conference on Computer Vision and Pattern Recognition (CVPR), pp. 4674–4683 (2017)
10. Liu, R., Liu, C.: Human motion prediction using adaptable recurrent neural networks and inverse kinematics. IEEE Control Syst. Lett. **5**(5), 1651–1656 (2021)
11. Pérez-D'Arpino, C., Shah, J.A.: Fast target prediction of human reaching motion for cooperative human-robot manipulation tasks using time series classification. In: 2015 IEEE International Conference on Robotics and Automation (ICRA), pp. 6175–6182 (2015)
12. Liu, H., Wang, L.: Human motion prediction for human-robot collaboration. J. Manuf. Syst. **44**, 287–294 (2017)
13. Yan, L., Gao, X., Zhang, X., Chang, S.: Human-robot collaboration by intention recognition using deep LSTM neural network. In: 2019 IEEE 8th International Conference on Fluid Power and Mechatronics (FPM), pp. 1390–1396 (2019)

14. Liu, Z., Liu, Q., Xu, W., Liu, Z., Zhou, Z., Chen, J.: Deep learning-based human motion prediction considering context awareness for human-robot collaboration in manufacturing. Procedia CIRP **83**, 272–278 (2019)
15. Wang, W., Li, R., Chen, Y., Diekel, Z.M., Jia, Y.: Facilitating human-robot collaborative tasks by teaching-learning-collaboration from human demonstrations. IEEE Trans. Autom. Sci. Eng. **16**(2), 640–653 (2019)

Real-Time Design Based on PREEMPT_RT and Timing Analysis of Collaborative Robot Control System

Yanlei Ye[1], Peng Li[1], Zihao Li[1], Fugui Xie[1,2], Xin-Jun Liu[1,2](✉), and Jinnhui Liu[3]

[1] The State Key Laboratory of Tribology, Department of Mechanical Engineering, Tsinghua University, Beijing 100084, China
xinjunliu@mail.tsinghua.edu.cn

[2] Beijing Key Lab of Precision/Ultra-Precision Manufacturing Equipments and Control, Tsinghua University, Beijing 100084, China

[3] Yantai Tsingke+ Robot Joint Research Institute Co., Ltd., Yantai 264006, China

Abstract. The design of the collaborative robot control system is restricted by the following key factors: real-time system, field bus, robot frame and human-computer interaction module. Academia and industry have an urgent need for high-quality, high-stability, distributed and easy-to-operate control systems with the development of the robotics industry. Robot systems architecture based on ROS and ROS2 have attracted more and more attention due to the flexibility, openness, modularity, scalability and friendliness. The main contribution of this article is to partially construct the real-time architecture of the collaborative robot control system. In addition, the timing jitter of the system, EtherCAT master station and ROS under different frequencies and loads is analyzed. First, a Linux kernel based on PREEMPT_RT patch is constructed, the basic architecture is described, and the timing performance of the robot system is improved through the application of high-precision timers, priority assignment, scheduling mode, and energy consumption management. Furthermore, the timing jitter performance of the operating system is analyzed. The maximum jitter of the system and the EtherCAT master station is about 10 us, and the average jitter is 1 us. Finally, the timing jitter of the built-in timing callbacks in ROS and ROS2 is evaluated. The design of the real-time system and performance analysis provide critical support for the design of the robot controller.

Keywords: PREEMPT_RT · Timing Jitter · ROS · ROS2 · EtherCAT Master

1 Introduction

A real-time system means that it must respond to events in the environment within precise timing [1, 2]. Therefore, for the system, not only is the logic correct, but also the data interaction needs to be performed within a given time or termination time. Delayed or advanced trigger timing will lead to instability and unpredictable damage of the system. For example, the control of nuclear plants, railway switching systems, flight control

X.-J. Liu et al. (Eds.): ICIRA 2021, LNAI 13014, pp. 596–606, 2021.
https://doi.org/10.1007/978-3-030-89098-8_56

systems, space missions, autonomous driving, robots, etc. all require precise real-time control. The real-time system does not mean that the faster the better, but to maintain a small timing jitter, to meet the receiving, calculating and sending of data in the timing period. In a typical case, for the control of the robot, the untimely data reception and transmission will make the motion planning unsmooth, resulting in vibration and noise in the system. Real-time systems are usually divided into soft real-time, firm real-time and hard real-time. The classification of a real-time system depends on the processing of the system after the deadline is lost [3, 4].

The Linux system is generally used in server and desktop environments. Native Linux usually has a delay of several hundred milliseconds. ROS and ROS2 are also developed based on the Ubuntu version of Linux [5]. In order to use the powerful Linux ecological environment, including drivers, desktops, human-computer interaction interfaces, especially the use of ROS architecture, the Linux kernel system needs to be modified to make it have real-time performance. Usually, the alternative methods are based on dual-kernel methods [3] (also known as Pico-kernel, Nano-kernel, Dual kernel) such as Xenomai [6, 7] and RTAI [2, 8]. The basic idea is that the microkernel lies between the computer hardware and the Native Linux kernel. Real-time tasks run on the microkernel. The microkernel takes over the interrupt and manages it directly from the bottom. When there is no real-time program running on the microkernel, the Linux kernel can obtain the running time. In 2016, the Linux Foundation started the real-time Linux collaboration project. This project aims to coordinate the development of the real-time environment kernel, especially the development of the PREEMPT_RT patch [9]. The development of the real-time PREEMPT_RT patch began in 2005 [10, 11], in order to reduce the delay of the kernel and improve predictability. Figure 1 shows the architecture of Xenomai and PREEMPT_RT.

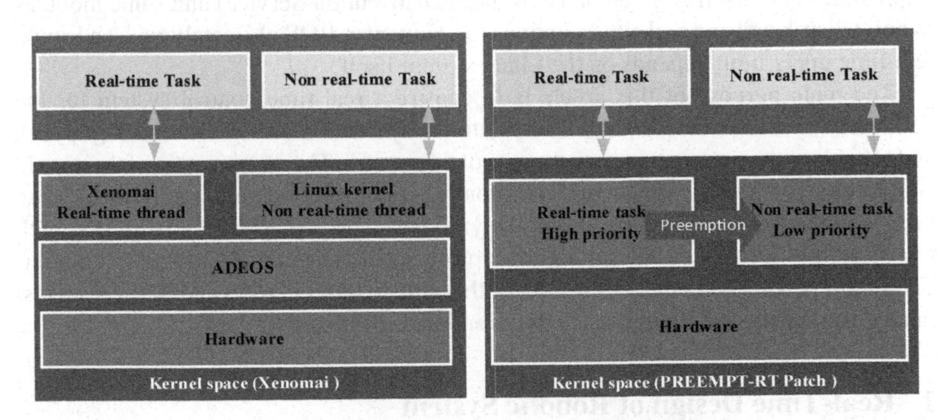

Fig. 1. Architecture of Xenomai and PREEMPT_RT

Compared with Xenomai's dual-core solution, PREEMPT_RT's preemption solution is to interrupt threading and set a priority for each process. The priority of the real-time thread is higher than that of the interrupt processing thread, which will avoid the interruption of the real-time process by the hardware interrupt. With the high-precision clock event layer, the real-time performance of the system can be improved. The core of the PREEMPT_RT solution is to minimize the amount of non-preemptible kernel code, so that preemption can be triggered at any level, and to reduce the amount of code that must be changed. Compared with Native Linux kernel threads, preemption can be interrupted. Of course, in user space, the timing jitter is weaker than Xenomai. This is a trade-off process. Although dual cores have advantages in terms of latency, the problem is that the maintained cores need to be supported on new hardware, and a large number of interface configurations need to be performed on the running program, which greatly increases the developer's workload and maintenance costs. PREEMPT_RT does not manage real-time applications in a dual-core manner. This will make the use of real-time programs and non-real-time programs in the user space similar, and only need to add scheduling type and priority configuration in the real-time program. The purpose of this article is not to use Linux to construct a new RTOS system, but to combine PREEMPT_RT to make the Linux kernel have real-time capabilities, and combine the EtherCAT master and ROS architecture to analyze timing performance.

In order to realize real-time communication of distributed equipment, EtherCAT fieldbus is used. Famous robotics system lab in ETH Zurich designed ANYmal quadruped robot using EtherCAT protocol [12]. PAL Robot Company designed the TALOS humanoid robot [13], the whole robot is based on EtherCAT, and each joint is controlled by torque controller to complete industrial tasks. Some open source EtherCAT masters can be used such as SOME [14] and IGH [15, 16] and some commercial master stations such as Acontis and CODESYS can also be used. The article will combine the Acontis master station to analyze the timing jitter performance of the EtherCAT master.

The construction of the real-time system is intended for the development of ROS-based robotic systems. Since the initial establishment of ROS did not fully consider the real-time problem. Compared with ROS, the real-time performance of ROS2 has been improved. ROS2 itself is based on DDS (Data Distribution Service) and some modules to build distributed and real-time solutions [5]. However, ROS ultimately runs on Linux, real-time upper limit depends on the Linux system itself.

The main purpose of this article is to construct real-time control system for the collaborative robot (the designed manipulator body has 7 motor shafts and 1 gripper), and to analyze the timing jitter performance of the system. The organization of this article is as follows: in the second part, real-time control system architecture is described; in the third part, the real-time performance is analyzed based on PREEMPT_RT Linux kernel, which mainly focuses on the timing jitter of operating system and EtherCAT master; The fourth part analyzes the timing jitter of the built-in timer in ROS and ROS2 systems; finally, the conclusion and prospect are given in the fifth part.

2 Real-Time Design of Robotic System

This part mainly introduces the design architecture of the real-time robot control system. The Linux kernel with PREEMPT_RT patch carries out real-time thread processing. The motion library and EtherCAT master run in high-priority and adopt the FIFO scheduling mode. Trajectory planning, collision detection, simulation, human-computer interaction interface and visual inspection run in non-real-time threads and do not require separate processing. In addition, some other tools and drivers also run in non-real-time threads.

The communication and interaction between programs use ROS messages, services and actions. The Motion library is built in the way of ROS_Controller. The Motion module sends and receives EtherCAT master data in real-time. The method of Cyclic Synchronous Position Mode (CSP) is adopted to control the servo movement. The master station and the slave station as well as the slave station are respectively connected with an Ethernet network cable. The complex and computationally intensive nodes is transplanted to another PC as ROS slave nodes, such as deep learning and reinforcement learning calculations. To achieve distributed data processing, and to decouple the impact of large loads on real-time control. At the same time, some peripheral components, such as grippers, cameras and IO peripherals, are connected to the controller. The overall architecture diagram is shown in Fig. 2.

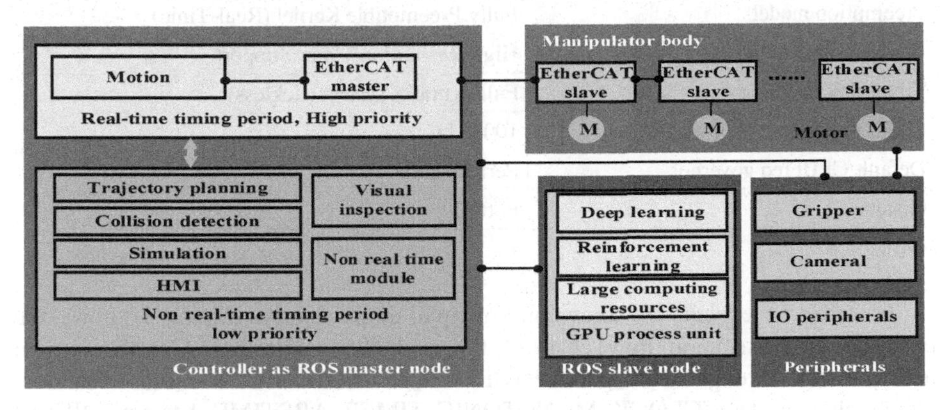

Fig. 2. Block diagram of robot control system architecture

The configuration list of the established system is given in Table 1. The Linux system is Ubuntu 20.04, the Linux kernel is 5.10.25, and the patch is PREEMPT_RT.

Table 1. System setup

Configuration	Description
Ubuntu	20.04
PREEMPT_RT	5.10.25-rt35
Linux kernel	5.10.25
CPU	Intel i7 9700K 8 cores 3.6 GHz
RAM	DDR4-3200 16 GB
ROS	Noetic
ROS2	Foxy
EtherCAT master	Acontis
Network interface controller	Intel I210(1 Gbit/s)

3 Timing Analysis of Real-Time System Based on PREEMPT_RT Patch

In this part, the timing jitter of the real-time system under different timing frequencies and different loads will be evaluated, including the analysis of real-time jitter and maximum jitter. The PREEMPT_RT patch is built in Linux kernel, and some configurations of the kernel are shown in Table 2:

Table 2. PREEMPT_RT Kernel Configuration

Configuration	Description
Preemption model	Fully Preemptible Kernel (Real-Time)
Timers' subsystem	High Resolution Timer Support
Timer tick handling	Full dynticks system (tickless)
Timer frequency (1000 Hz)	1000 Hz
Default CPUFreq governor	Performance
C-state	Forbid

In order to explain the drawing principle more clearly. Figure 3 shows the corresponding execution flow chart of the evaluation algorithm. Get the current accurate time through clock_gettime (CLOCK_MONOTONIC, &ts_now) function. Use clock_nanosleep (CLOCK_MONOTONIC, TIMER_ABSTIME, &ts_nest, NULL) function to realize timing delay.

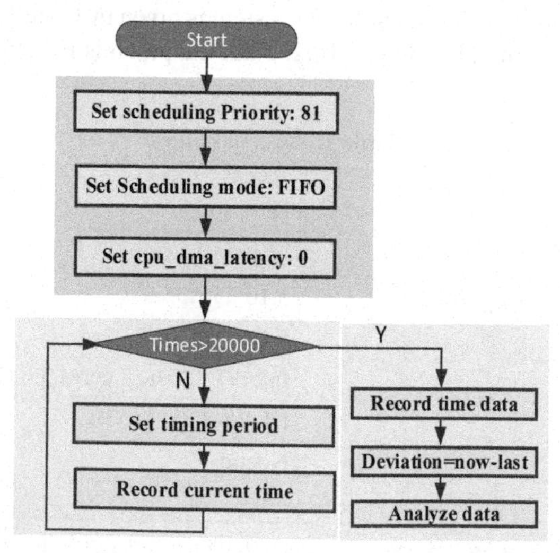

Fig. 3. Real-time performance evaluation algorithm based on PREEMPT_RT

Figure 4 shows the jitter of the Native Linux system, with a timing cycle of 1000 us. 20000 cycles were tested in the experiment. The horizontal axis is the number of recordings, and the vertical axis is the time deviation between the two counts. The maximum jitter is 400 us, with large fluctuations. And the system only runs the test program, and does not set a heavy load. Obviously, the Native system is not real-time. It cannot be used for multi-axis high-precision motion control.

On the built real-time kernel, the timing jitter evaluation is carried out for different timing periods, and the corresponding curve is drawn and shown in Fig. 5. The maximum jitter of the system is less than 5 us. The jitter fluctuation is small, and the real-time jitter and average jitter are maintained at 1 us. Obviously, the real-time system based on PREEMPT_RT patch shows good real-time performance, especially with small average jitter. In order to test the stability of the system, we analyzed the data of 100 us, 500 us, and 2000 us control cycles. As shown in Figs. 5, 6, and 7. The maximum period jitter is below 10 us. It should be noted that browsers and other large-scale software are not executed during the experiment.

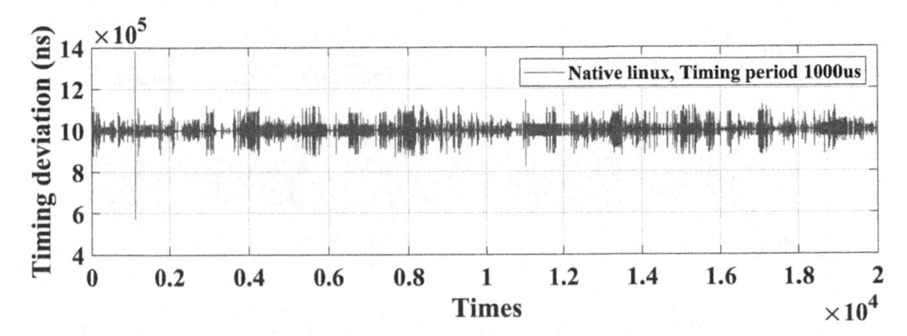

Fig. 4. Native Linux, Timing period 1000 us

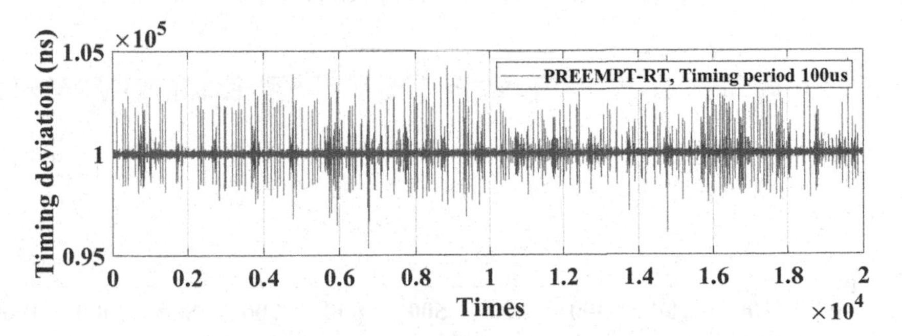

Fig. 5. PREEMPT-RT, Timing period 100 us

The jitter under different timing periods is analyzed above. In order to better explore the system and evaluate the real-time performance, a loading test is carried out in the configured system environment. As shown in the Fig. 8, the 8 cores of the CPU are running close to full load. In this case, the timing cycle is 1000 us, and the curve is

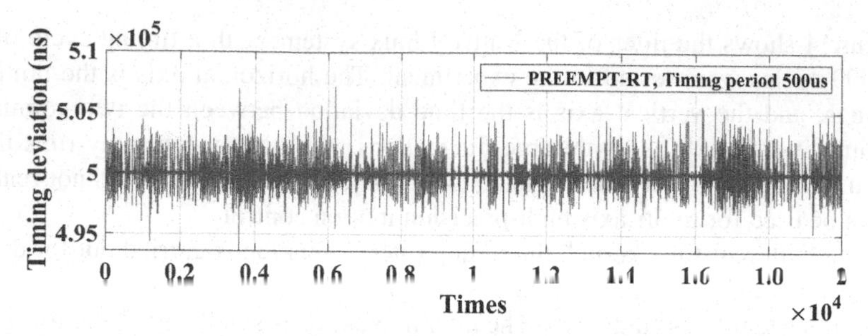

Fig. 6. PREEMPT-RT, Timing period 500 us

shown in Fig. 9. Further, the different jitter amplitudes are counted, and the corresponding histogram is drawn in Fig. 10. It can be seen that the maximum jitter is less than 5 us, and the average jitter is 1 us. The jitter generally exhibits a normal distribution. Similarly, in the experiment, a running cycle of 20000 times was set.

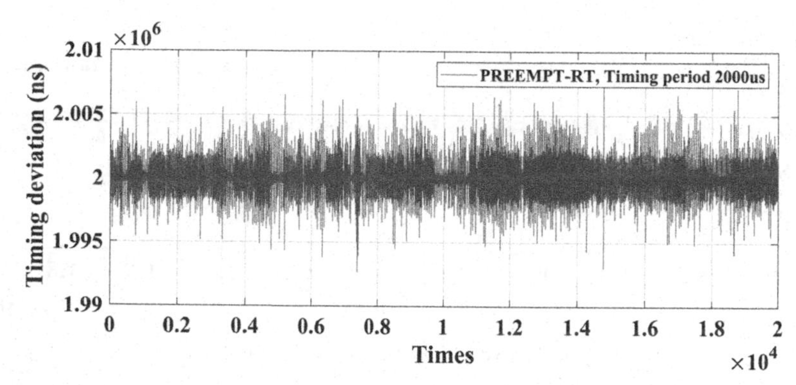

Fig. 7. PREEMPT-RT, Timing period 2000 us

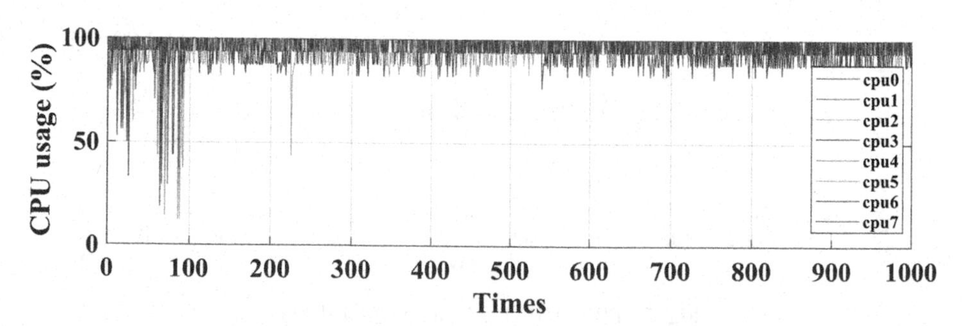

Fig. 8. PREEMPT-RT, Timing period 1000 us, CPU100%

The EtherCAT master station needs to run on a real-time system to strictly guarantee real-time performance. This system has an EtherCAT master station and 8 EtherCAT slave stations (Including robotic arm and gripper). In the experiment, the data of 1661 s

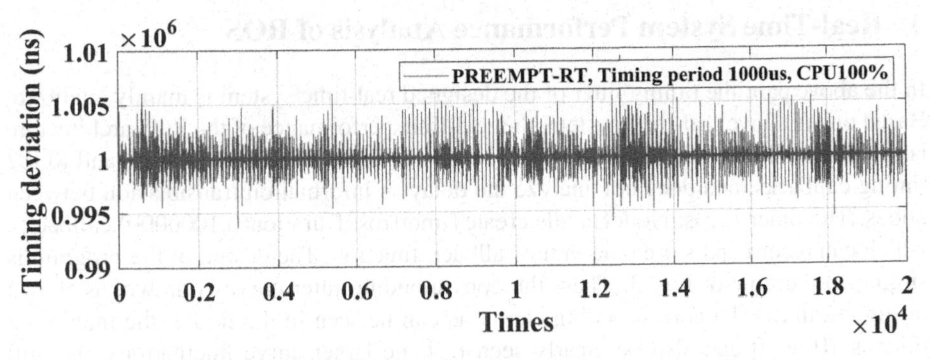

Fig. 9. PREEMPT-RT, Timing period 1000 us, CPU100%

duration is obtained, and get the delay of the EtherCAT master station, and the maximum jitter is 12 us (Fig. 11).

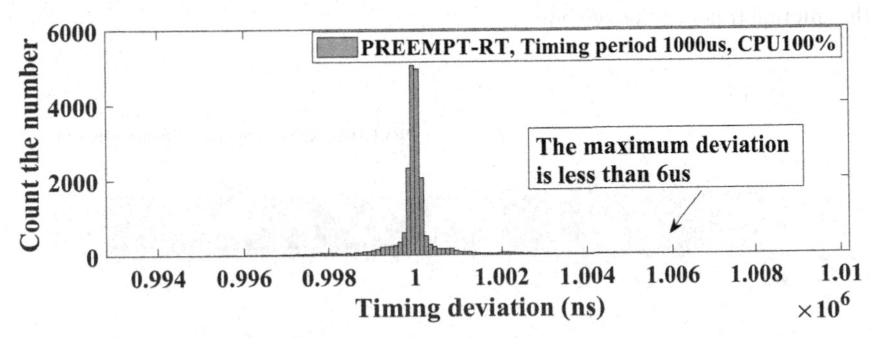

Fig. 10. PREEMPT-RT, Timing period 1000 us, CPU100%, Histogram

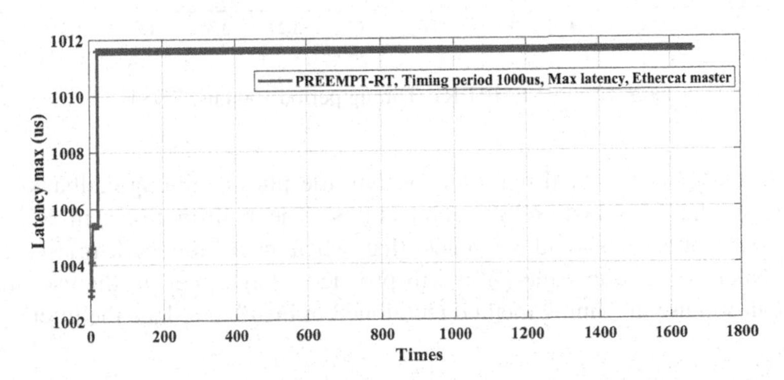

Fig. 11. PREEMPT-RT, Timing period 1000 us, EtherCAT master

It can be seen that the real-time system based on PREEMPT_RT design has greatly improved the real-time performance compared to the native Linux system. The stress test also showed stable real-time performance. This part of the research provides a basic guarantee for our subsequent control system design.

4 Real-Time System Performance Analysis of ROS

In the above part, the timing jitter of the designed real-time system is mainly analyzed. Based on the above real-time system, the real-time performance of the ROS architecture is evaluated. This article mainly focuses on the real-time discussion of ROS and ROS2 timing callbacks, and does not analyze the delay of information transmission between nodes. The timer is ros::NodeHandle.createTimer(ros::Duration(0,1000000),callback), which can record and store data in the callback function. The design of the program is similar to the flow of Fig. 3. Then, the corresponding jitter curve is drawn, as shown in Fig. 12 draws the corresponding curve, as can be seen in the figure, the maximum jitter is 10 us. It can also be clearly seen that the larger curve fluctuations, but still kept in a small range. In the same way, we analyze the ROS2 system, and the real-time configuration is the same as above. Relying on the ROS2 architecture, the system's timing callback function is adopted (rcl::Node::create_wall_timer(1ms, timer_callback, this)). The corresponding result is shown in Fig. 13. The maximum jitter is close to 80 us. The overall jitter is relatively large. Of course, this also depends on the specific settings of the internal timer of the system.

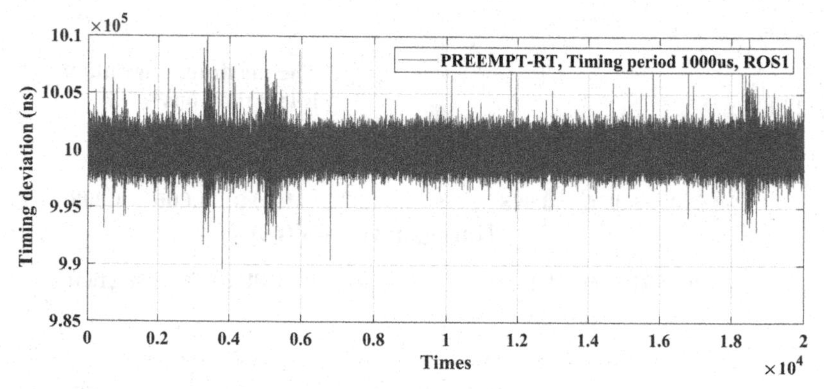

Fig. 12. PREEMPT-RT, Timing period 1000 us, ROS1

Based on the constructed real-time system, the jitter of timing callbacks is evaluated under the ROS and ROS2 architectures. The built-in timer in ROS has a small timing jitter. It should be noted that when using the built-in ROS2 timer (rclcpp::Node.create_wall_timer()), it will produce a large jitter. In the use of ROS2, People can still use the timing method mentioned in Fig. 3 to reduce the jitter.

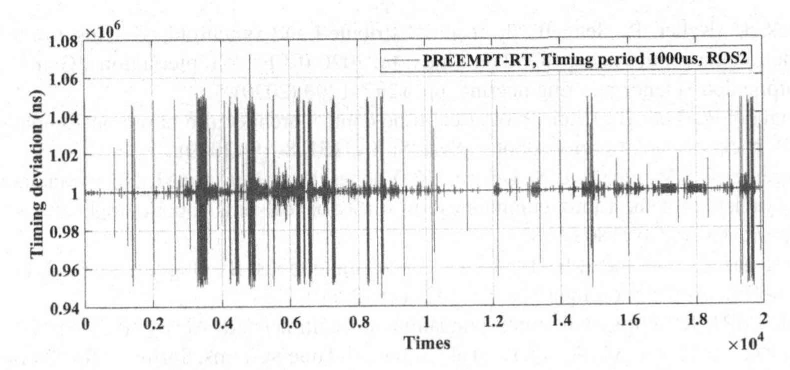

Fig. 13. PREEMPT-RT, Timing period 1000 us, ROS2

5 Conclusions and Further Work

The complex controller needs to meet the real-time requirements to achieve high-precision and stable trajectory control. At the same time, it also needs to run some non-real-time tasks, so that the system has complete functions. The article is based on the PREEMPT_RT kernel scheme to realize the real-time guarantee of the system. The contribution of this article is to analyze the timing jitter of the system, EtherCAT master station and ROS under different frequencies and loads. The PREEMPT_RT Linux kernel and timer are constructed, and the maximum timing jitter the system and the EtherCAT master is about 10 us. Using the built-in timers of ROS and ROS2, the maximum jitter is 10 us and 80 us respectively. In addition, EtherCAT slaves have good cycle synchronization (hardware clock calibration). For the 1 kHz control cycle, it can meet the good control effect. In the subsequent research, the influence of other factors such as communication delay and throughput between nodes on real-time communication will be further analyzed to better improve real-time performance.

Acknowledgement. This work is supported by the National Key Research and Development Program of China (Grant No. 2019YFB130185), the Key Research and Development Program of Shandong Province (Grant No. 2019JZZY010432), and a grant from the Institute for Guo Qiang, Tsinghua University (Grant No. 2019GQG0007).

References

1. Yoon, H., Song, J., Lee, J.: Real-time performance analysis in Linux-based robotic systems. In: Proceedings of the 11th Linux Symposium, pp. 331–340 (2009)
2. Reghenzani, F., Massari, G., Fornaciari, W.: The real-time linux kernel: a survey on PREEMPT_RT. ACM Comput. Surv. **52**(1), 1–36 (2019)
3. Karamousadakis, M.A.: Real-time programming of EtherCAT master in ROS for a quadruped robot (2019)
4. Kopetz, H.: Real-Time Systems: Design Principles for Distributed Embedded Applications. Springer, New York (2011). https://doi.org/10.1007/978-1-4419-8237-7

5. Puck, L., Keller, P., Schnell, T., et al.: Distributed and synchronized setup towards real-time robotic control using ROS2 on Linux. In: 2020 IEEE 16th International Conference on Automation Science and Engineering, pp. 1287–1293 (2020)
6. Delgado, R., You, B., Choi, B.W.: Real-time control architecture based on Xenomai using ROS packages for a service robot. J. Syst. Softw. **151**, 8–19 (2019)
7. Gutiérrez, C.S.V., Juan, L.U.S., Ugarte, I.Z., Vilches, V.M.: Real-time Linux communications: an evaluation of the Linux communication stack for real-time robotic applications (2018). https://arxiv.org/abs/1808.10821
8. Mantegazza, P., Bianchi, E., Dozio, L., Papacharalambous, S., Hughes, S., Beal, D.: RTAI: Real-Time Application Interface, pp. 142–148 (2000)
9. PREEMPT_RT. https://wiki.linuxfoundation.org/realtime/start. Accessed 29 Apr 2021
10. Buttazzo, G., Lipari, G., Abeni, L., et al.: Soft Real-Time Systems. Springer, New York (2005)
11. Oliveira, D.B., Oliveira, R.S.: Timing analysis of the PREEMPT RT Linux kernel. Softw. Pract. Exp. **46**(6), 789–819 (2016)
12. Hutter, M., Gehring, C., Lauber, A., et al.: Anymal-toward legged robots for harsh environments. Adv. Robot. **31**(17), 918–931 (2017)
13. TALOS Humanoid Robot. https://pal-robotics.com. Accessed 29 Apr 2021
14. SOME. https://github.com/OpenEtherCATsociety/SOEM. Accessed 29 Apr 2021
15. IGH. https://www.etherlab.org/en/ethercat/index.php. Accessed 29 Apr 2021
16. Zurawski, Z.: Industrial Communication Technology Handbook, 2nd edn. CRC Press, New York (2017)

Robotic Machining

Modeling of Chatter Stability in Robot Milling

Daxian Hao[1](✉), Gang Zhang[2], Huan Zhao[1], and Han Ding[1]

[1] State Key Laboratory of Digital Manufacturing Equipment and Technology, Huazhong University of Science and Technology, Wuhan 430074, China

[2] Wuxi Research Institute of Huazhong University of Science and Technology, Wuxi 214100, China

Abstract. Compared with CNC machines, articulated robots show significant pose-dependent dynamic characteristics. Thus, the chatter mechanism of robot milling is murky. Furthermore, trajectory accuracy is also dependent on robot pose, resulting in complexity of the robot milling stability. In this paper, robot trajectory tracking error, independent from tool speed, can be seen as an external disturbance to the classical regenerative chatter model. With the consideration of robot trajectory tracking error, a new robot milling dynamic model was proposed to study the chatter stability in robot milling process. Robot milling tests have been carried out to verify the model.

Keywords: Robotic milling · Regenerative chatter · Trajectory tracking error

1 Introduction

The wide adoption of robotic milling processes has been hampered by chatter [1]. Robotic machining chatter is a complex elasto-dynamic phenomenon, which is excited by absorbing energy from the periodic cutting force. Chatter-related problems in robotic machining can result in poor surface quality and workpiece accuracy, tool abrasion or damage, and even in damaged spindles or robots [2]. A lot of research in the field of machine tools is currently being conducted on the machining chatter and vibration suppression mechanisms [3]. However, the use of robots in machining is relatively new, as compared to conventional machine tools, and a lot of research remains to be done, including the investigation of the chatter mechanism and developing solutions.

The chatter in robotic machining generally breaks down into mode coupling chatter and regenerative chatter. Modal coupling chatter in robot milling was first reported by Pan et al. in 2006 [2]. Guo et al. [4] investigated the vibration mechanism in the robotic boring process and pointed out that the vibration occurring during robotic boring is a forced vibration with displacement feedback. Cordes et al. [5] adopted semi-discrete time and frequency domain methods to analyze the stability of the robot milling process. Tunc [6] studied the dynamics of a hexapod platform for robotic milling and found that the dynamic characteristics varied under differing robot positions.

This paper examines the influence of the robot trajectory tracking error (TTE) on milling stability. The purpose is to more broadly explain the robot milling chatter mechanism, propose solutions to suppress milling chatter, and improve the efficiency and

X.-J. Liu et al. (Eds.): ICIRA 2021, LNAI 13014, pp. 609–619, 2021.
https://doi.org/10.1007/978-3-030-89098-8_57

quality of robot milling. The effect of the robot trajectory on the dynamic chip thickness and multiple time delay effects following an error are analyzed. The relationship between the trajectory tracking error and the process damping coefficient during machining is discussed. A modified semi-discretization method is used to predict the robot milling stability lobe diagram (SLD). Then, the robot milling system parameters are obtained. The conclusion is given in the last section.

2 Modeling of Robot TTE in Robot Milling

2.1 Multiple Delay Modeling Under the Robot TTE Effect

To analyze this influence, a kinematics model of the milling process considering the robot TTE is established, as shown in Fig. 1, where XOY is the coordinate system of the workpiece, and the feed direction is along the X-axis. $X_1O_1Y_1$ and $X_2O_2Y_2$ are respectively defined as the tool coordinate system of the previous cutting position and that of the current cutting position, originating at the center of the tool.

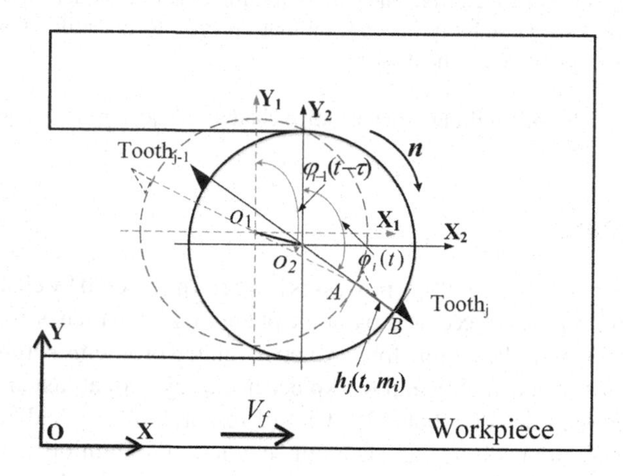

Fig. 1. Kinematics model considering robotic TTE.

In order to simplify the trajectory error fluctuation function, the TTE is approximated as a simple harmonic function, and the system trajectory error period is T_{er}. The general equation of the tool TCP position, considering the robot TTE is:

$$\begin{cases} x_{0\text{er}} = V_f t + e_x \sin(\omega_{ex} t + \theta_x) \\ y_{0\text{er}} = e_y \sin(\omega_{ey} t + \theta_y) \end{cases} \tag{1}$$

where $x_{0\text{er}}$ and $y_{0\text{er}}$ are the positions of TCP in the XOY coordinate system; V_f is the feed speed of the robot milling system; e_x and e_y are the amplitudes of the robot trajectory error in the X-axis and Y-axis, respectively; ω_{ex} and ω_{ey} are the angular frequencies, which are decided by the TTE fluctuation period, and θ_x, θ_y are the phase angles of the robot in the X-axis and Y-axis directions. The formula for the instantaneous chip thickness in the milling process considering the influence of the robot trajectory error can be given by:

$$h(\phi_j) = f_z \sin\phi_j(t) + (x(t-\tau) - x(t)) \sin\phi_j(t) + (e_x \sin(\omega_{ex}(t-\tau) + \theta_x) - e_x \sin(\omega_{ex}t + \theta_x)) \sin\phi_j(t) + (y(t-\tau) - y(t)) \cos\phi_j(t) + (e_y \sin(\omega_{ey}(t-\tau) + \theta_y) - e_y \sin(\omega_{ey}t + \theta_y)) \tag{2}$$

In the experiments in this article, the TTE is much larger in the X-axis direction than in the Y-axis direction, and can thus be ignored. The dynamic thickness formula can be further simplified as follows:

$$\begin{aligned} h(\phi_j) \approx & (x(t-\tau) - x(t)) \sin\phi_j(t) + (e_x \sin(\omega_{ex}(t-\tau) + \theta_x) \\ & - e_x \sin(\omega_{ex}t + \theta_x)) \sin\phi_j(t) + (y(t-\tau) - y(t)) \cos\phi_j(t) \end{aligned} \tag{3}$$

The system is not only affected by the single time delay of the tool passing frequency in the milling process, but also by the time delay effect caused by the periodic fluctuation of the TTE. The stability of the system is affected by multiple time delays. The maximum delay period T_m of the system is controlled by the least common multiple of the robotic TTE period and by the tool passing period.

$$T_m = \mathrm{lcm}[T_{\mathrm{er}}, \tau] \tag{4}$$

Where, τ is cutting period. Hence, the maximum time delay period T_m can be approximated as T_{er}. Then, the multiple delay term m of the system is the rounding ratio of the robot trajectory error period to the milling tool passing period.

$$N_m = \mathrm{fix}[T_{er}/\tau] \tag{5}$$

The milling axial force is not considered. The cutting force in the X and Y directions is given by:

$$\begin{cases} F_{x,j} = -F_{tj} \cos\phi_{\mathrm{j}} - F_{rj} \sin\phi_{\mathrm{j}} \\ F_{y,j} = +F_{tj} \sin\phi_{\mathrm{j}} - F_{rj} \cos\phi_{\mathrm{j}} \end{cases} \tag{6}$$

where $F_{tj} = K_t a h(\phi_j)$, $F_{rj} = K_r F_{tj}$, a is the axis depth of cut, respectively, K_t and K_r are the tangential and radial cutting-force coefficients.

Then, the dynamic milling force model, considering the effects of multiple delays can be written as [7]:

$$F(t) = \begin{pmatrix} F_x(t) \\ F_y(t) \end{pmatrix} = \frac{1}{2} a K_t \sum_{l=1}^{N_m} [\mathbf{H}_l(t)][\Delta_{\mathrm{sum}}(t)] \tag{7}$$

$$[\Delta_{\mathrm{sum}}(t)] = \begin{bmatrix} \Delta x \\ \Delta y \\ \Delta er \end{bmatrix}, \quad [\mathbf{H}_l(t)] = \begin{bmatrix} h_{l,xx} & h_{l,xy} & h_{l,\mathrm{xer}} \\ h_{l,yx} & h_{l,yy} & h_{l,\mathrm{yer}} \end{bmatrix}$$

Where, N_m is the maximum number of delays, $[\Delta_{\mathrm{sum}}(t)]$ denotes the system dynamic displacement, and $[\mathbf{H}_l(t)]$ are time-varying directional dynamic milling force coefficients.

The displacement $\{\Delta_{\text{sum}}(t)\}$ can be regarded as a combined dynamic displacement. It is not only related to the milling period, but also linked with the robot TTE. Therefore, the dynamic displacement caused by the robot trajectory error should be taken into account in the regenerative chatter model.

2.2 A Modified Dynamic Model Considering the Robotic TTE

The dynamic cutting force Eq. (7) of the system can be converted into Eq. (8) by considering the influence of process damping [7]:

$$\{F(t)\} = \frac{1}{2}aK_t\sum_{l=1}^{N_m}[[\boldsymbol{H}_l(t)][\Delta_{\text{sum}}(t)]] - [\boldsymbol{c}_p]\dot{q}(t) \tag{8}$$

where $[\boldsymbol{c}_p]$ is the process damping matrix, $[\boldsymbol{c}_p] = \begin{bmatrix} c_x^p & \\ & c_y^p \end{bmatrix}$, and c_x^p, c_y^p are the average process damping coefficients in two directions on the milling plane.

The relationship between the indentation coefficient and the average process damping is obtained as:

$$c_p = \frac{K_d b}{A\pi}\int_{\pi/2\omega_{ez}}^{3\pi/2\omega_{ez}} U(t)\cos(\omega_{ez}t)dt = \frac{K_d I}{A\pi} \tag{9}$$

where c_p is the average process damping coefficient, ω_{ez} is the angular frequency of TTE in the Z-axis direction, b is the radial depth of cut, K_d is the indentation coefficient, I is the indentation volume, and $U(t)$ represents the cross-sectional area of the contact between the tool and the workpiece. The indentation volume I is proportional to the square of the error fluctuation amplitude A.

Then, the dynamic differential equation of the robot milling system is established as follows:

$$[\boldsymbol{M}]\ddot{q}(t) + [\boldsymbol{C}+\boldsymbol{c}_p]\dot{q}(t) + [\boldsymbol{K}]q(t) = \frac{1}{2}aK_t\sum_{l=1}^{N_m}[[\boldsymbol{H}_l(t)][\Delta_{\text{sum}}(t)]] \tag{10}$$

where $[\boldsymbol{M}]$ is the mass matrix of the dynamic system, $[\boldsymbol{C} + \boldsymbol{c}_p]$ is the damping matrix, $[\boldsymbol{K}]$ is the stiffness matrix, $q(t)$ is the displacement vector, and $F(t)$ is the force vector. For the robot milling task, it is simplified into a 2-DOF system.

Equation (7) is substituted into Eq. (10), then the state space expression of Eq. (11) can be expressed as:

$$\dot{\mathbf{U}}(t) = \mathbf{A}(t)\mathbf{U}(t) + \sum_{l=1}^{m}\mathbf{B}_l(t)\mathbf{U}(t-\tau_l) \tag{11}$$

where $A(t)$ and $B(t)$ are periodic matrices determined by considering the dynamic cutting force of the regeneration effect, and satisfy the requirements of $A(t+T) = A(t)$, $B(t+T) = B(t)$, and U is the state space vector.

In the present paper, a modified semi-discretization method is designed to solve the multiple delay system. First, the system is discretized to determine the step length of discrete time Δt as [8]:

$$\Delta t = \frac{T_m}{k} \tag{12}$$

where k is the discrete number of the periodic T_m.

Then, Eq. (11) is approximately discrete as:

$$\dot{\mathbf{U}}(t) = \mathrm{A}_j\mathbf{U}(t) + \sum_{l=1}^{m} \mathbf{B}_{l,j}\mathbf{U}_{\tau_l,j} \tag{13}$$

where $\mathrm{A}_j = \frac{1}{\Delta t}\int_{t_j}^{t_{j+1}} \mathrm{A}(t)dt$, $\mathbf{B}_{l,j} = \frac{1}{\Delta t}\int_{t_j}^{t_{j+1}} \mathbf{B}_l(t)dt$.

The solution of Eq. (13) is:

$$\mathbf{U}(t) = e^{A_j(t-tj)}[\mathbf{U}_j + \sum_{l=1}^{m} \mathbf{A}_j^{-1}\mathbf{B}_{l,j}\mathbf{U}_{\tau_l,j} - \sum_{l=1}^{m} \mathbf{A}_j^{-1}\mathbf{B}_{l,j}\mathbf{U}_{j-m_l+1}] \tag{14}$$

The discretized solution can be written as:

$$\mathbf{U}(t) = \mathbf{Q}_j\mathbf{U}_j + \sum_{l=1}^{m} (w_{l,b}\mathbf{R}_{l,j}\mathbf{U}_{j-m_l} + w_{l,a}\mathbf{R}_{l,j}\mathbf{U}_{j-m_l+1}) \tag{15}$$

Where, $\mathbf{Q}_j = e^{A_j(t-tj)}$, $\mathbf{R}_{l,j} = (\mathbf{Q}_j - \mathbf{I})\mathbf{A}_j^{-1}\mathbf{B}_{l,j}$.

Equation (15) can be recast into a discrete map form as follows:

$$\mathbf{V}_{j+1} = \mathbf{Z}_j\mathbf{V}_j \tag{16}$$

where $\mathbf{V}_j$ is the state vector, $\mathbf{Z}_j$ is the coefficient matrix:

According to Eq. (16), the formula can be established by coupling the solutions of the k successive time intervals in the least period T_m:

$$\mathbf{V}_k = \Psi\mathbf{V}_0 \tag{17}$$

where $\mathbf{\Psi} = \mathbf{Z}_{k-1}\mathbf{Z}_{k-2}, \ldots, \mathbf{Z}_1\mathbf{Z}_0$, and $\mathbf{\Psi}$ is the Floquet transition matrix. According to Floquet theory, the system is asymptotically stable if the norms of all eigenvalues of $\mathbf{\Psi}$ are less than 1. Finally, the critical stable axial depth is obtained by solving the eigenvalues of the Floquet transition matrix that are equal to 1 at different rotational speeds through a cycle method.

3 Robotic Milling System Experiments

3.1 Experiment and Analysis of Robot TTE and Modal Tests

In accordance with the modified semi-discretization method, to plot the stability lobe of the system, a set of trajectory accuracy experiments were conducted to obtain the characteristics of the TTE of the robot. ABB IRB 6660, a pre-machining robot with high accuracy and rigidity, was used in the experiments. A Leica laser tracker was used for non-contact measurement, with a reflector fixed on the tool end, as shown in Fig. 2.

Similarly, the error of TCP in the X-axis and Y-axis during the operation can be collected, Fig. 3(a) and Fig. 3 (b). The error in the X-axis is much larger than that in the Y-axis. Therefore, the system can be simplified by ignoring the error in the Y-axis, and only considering that in the X-axis. FFT (Fast Fourier Transform) is then applied to the

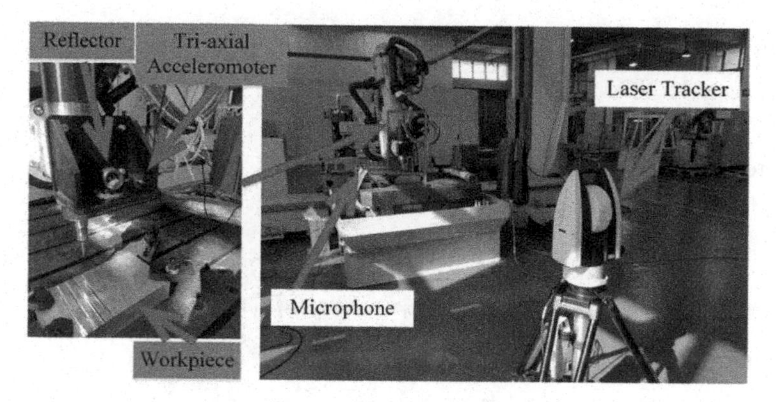

Fig. 2. Robot end milling system.

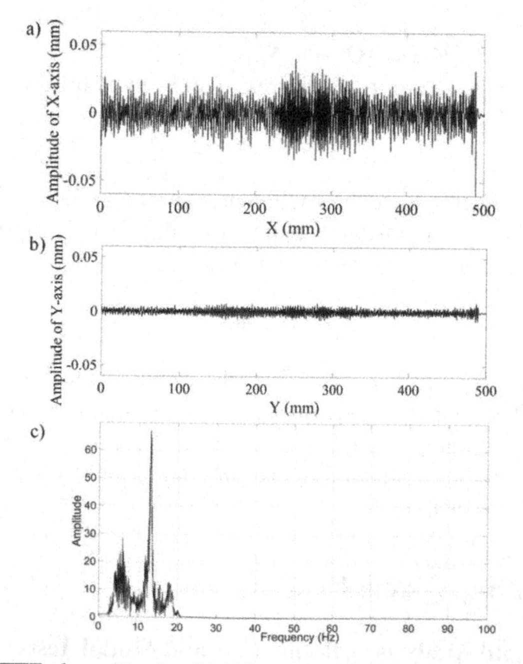

Fig. 3. (a) The TCP TTE along X-axis; (b) TCP TTE along Y-axis. (c) Spectrum of the filtered signal along X-axis.

collected signal to obtain the spectrum in the X-axis, as shown in Fig. 3 (c). When the robot moves along the X-axis, there is an obvious peak, and the frequency is about 13 Hz, showing that the TTE of the robot is fluctuating. However, no distinct peak appears when the robot is moving along the Y-axis.

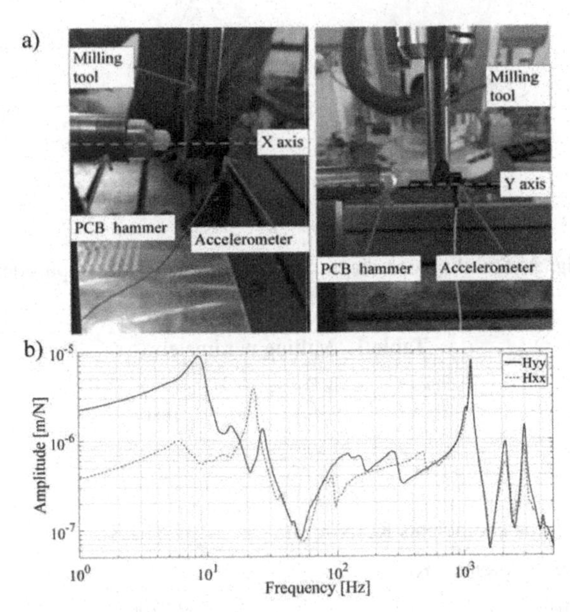

Fig. 4. (a) Accelerometer arrangement for hammer tests; (b) Measured FRF of robot milling system.

In order to analyze the stability of the robotic milling system, it is necessary to obtain the frequency response function (FRF) of the milling system, which is obtained by modal tests, as shown in Fig. 4(a). The milling tool tip is hammered in the X-axis and Y-axis. The experimental data is smoothed and fitted to obtain the FRF of the system (Fig. 4(b)). The modal parameters of the robot milling system are identified by FRF, after which they are imported into the modified model to analyze the stability of the system.

3.2 Generation of Stability Lobe Diagrams

A semi-discretization method was designed to analyze the stability. Three different amplitudes, A = 0 mm, A = 0.02 mm and A = 0.04 mm, were selected to plot the SLDs (stability lobe diagrams). The SLDs are shown in Fig. 5. It can be seen that the larger the error amplitude of the robot, the higher the stability region of the system will be.

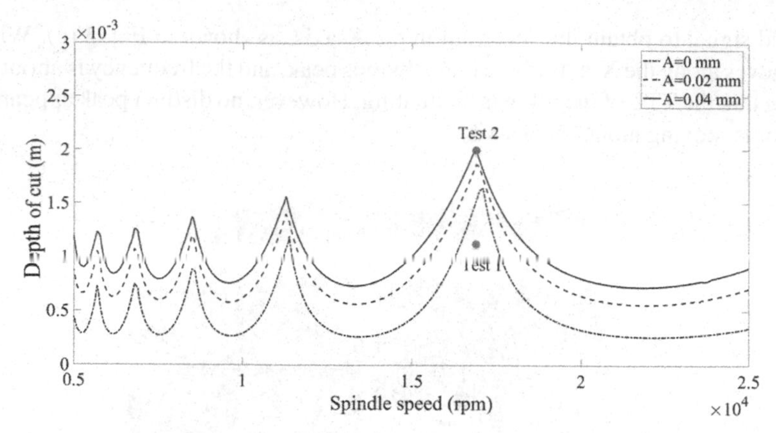

Fig. 5. Stability lobe diagrams with different TTE amplitudes.

Table 1. Milling parameters

Parameters	Test 1	Test 2
Depth of cut (mm)	1	2
Number of flutes	2	
Tangential cutting-force coefficients Kt (N/m^2)	7.81 E8	
Radial cutting-force coefficients Kr	0.3	
Spindle speed (rpm)	16700	
Workpiece material	Aluminum alloy 6061	
Feed speed (mm/s)	10	
Prediction results	Stability	Chatter

According to the obtained SLDs, two groups of experimental cutting parameters, Test 1 and Test 2, were planned (Table 1). The cutting parameter of Test 2 is in the instability domain when A = 0 mm, and is at the critical stability boundary when the amplitude A = 0.04 mm. Test 1 is a comparative experiment, in which the error fluctuation amplitude parameters are all in the stable domain.

3.3 Robot Milling Tests and Verification

After the robot milling tests, the vibration and displacement signals collected in the machining process are analyzed. Firstly, the time domain signals of the accelerometer in the milling process and its corresponding processing effect in Test 1 and Test 2 are shown in Fig. 6 (a) and (b). When the robot milling is along the X direction, the amplitude of the accelerometer signal fluctuates even if the milling parameters are set in the stable region in Test 1. This phenomenon may be mainly caused by the robot trajectory error, which expands the combined dynamic displacement in the robot milling process.

FFT is performed on the time domain signal. The peak frequency is 1117 Hz in Test 1 as shown in Fig. 6 (c), which is twice the tooth passing frequency. The spectrum

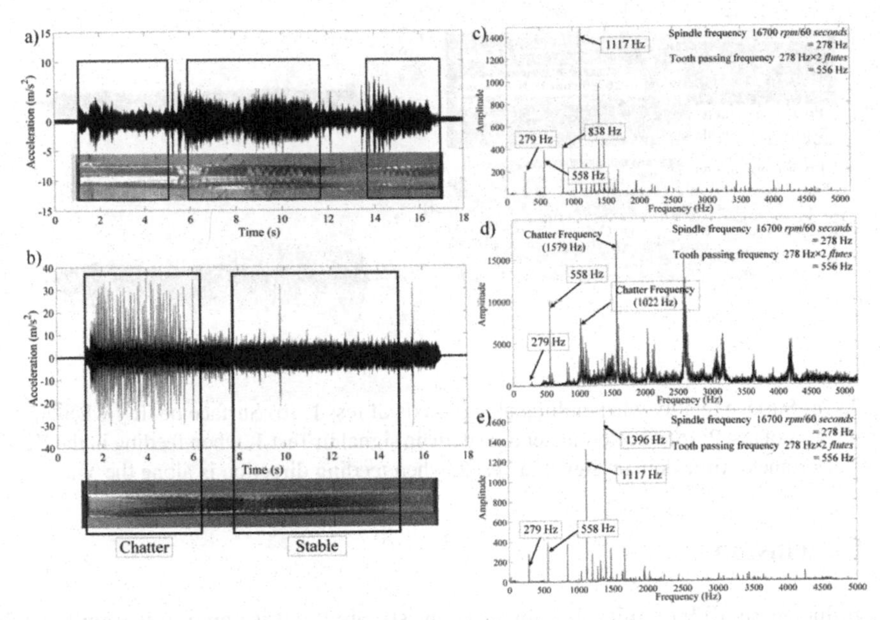

Fig. 6. (a) The accelerometer time domain signal with the corresponding milling result of test 1; (b) The accelerometer signal in time domain with the corresponding milling result of test 2; (c) Spectrum of acceleration signal in Test 1 is similar along X directions; (d) Chatter part spectrum of Test 2 along the X direction; (e) Stable part spectrum of Test 2

indicates that there is no chatter vibration along this path, and the vibration is dominated by the tool passing frequency. In Test 2, the chatter part, as shown in Fig. 6 (d), the peak appears at 1579 Hz, which is an asynchronous frequency of the tool passing. The spectrum diagram of the stable part in Test 2, as shown in Fig. 6 (e), shows that the slotting process is stable, and that no chatter frequency occurs.

A set of comparative experiments were also conducted by changing the robotic feeding direction, but selecting approximate robotic poses, thus allowing the robot to feed along the Y-axis direction of the BASE coordinate. This makes the surface quality better when slotting along the Y-axis, while the surface quality is greatly improved, as can be seen in Fig. 7(a) and (b). This coupling motion is directional, and is caused by a high TTE of the robot, which is one of the typical differences between robots and machine tools.

According to Fig. 3(a) and (b), when the robot moves along the Y-axis, the accuracy is significantly higher than when it is on the X-axis. Meanwhile, the prediction result of the stability region is close to the traditional regenerative chatter theory, in which the robot TTE is not considered. Time domain signals during the robot processing are shown in Fig. 7(c) and (d). Related experiments were analyzed in previous research.

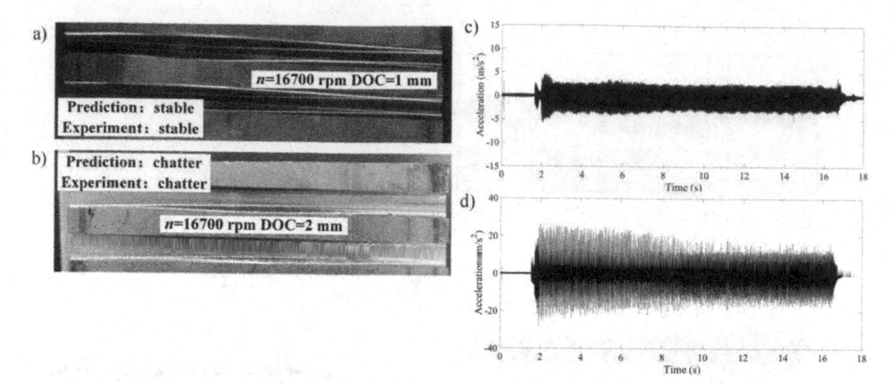

Fig. 7. (a) Surface quality when milling along Y-axis of test 1; (b) Surface quality when milling along Y-axis of test 2; (c) Accelerometer time domain signal in Test 1, when feeding in the Y-axis; (d) Accelerometer time domain signal in Test 2, when feeding direction is along the Y-axis

4 Conclusion

The influence of TTE on milling stability is investigated in this paper. It is found that the robot TTE mechanism increases process damping in robot milling. A modified semi-discretization method is used to construct the milling SLD for robot milling and the results are verified by milling experiments.

A dynamic cutting force model considering the trajectory accuracy error is established. The robotic TTE mainly caused by the motion coupling and positioning accuracy of the robot is shown to be related to processing stability in high-speed robotic milling. It affects the dynamic chip thickness and time delay, further disturbs the feedback modulation of the classical regenerative chatter model, and thus inhibits the occurrence of regenerative chatter in the milling process. The robot TTE is proven by milling experiments to affect the stability domain.

Acknowledgements. This work was supported by the National Key R&D Program of China No. 2018YFB1308900, the Natural Science Foundation of Hubei Province, China under Grant No. 2020CFA077 and the Basic national Defense Research of China No. JCKY2018205C004.

References

1. Lei, Y., Zengxi, P., Donghong, D., Shuai, S., Mcps, L.W.: A review on chatter in robotic machining process regarding both regenerative and mode coupling mechanism. IEEE/ASME Trans. Mech. 1–11 (2018)
2. Pan, Z., Zhang, H., Zhu, Z., Wang, J.: Chatter analysis of robotic machining process. J. Mater. Process. Technol. **173**(3), 301 (2006)
3. Alexander, V., Valente, A., Shreyes, M., Christian, B., Erdem, O., Tunc, L.T.: Robots in machining. CIRP Ann. Manuf. Tech. **68**, 799–822 (2019)
4. Guo, Y., Dong, H., Wang, G., Ke, Y.: Vibration analysis and suppression in robotic boring process. Int. J. Mach. Tool Manuf. **101**, 102–110 (2006)

5. Cordes, M., Hintze, W., Altintas, Y.: Chatter stability in robotic milling. Robot. Comput. Integr. Manuf. **55**, 11–18 (2006)
6. Tunc, L.T., Stoddart, D.: Tool path pattern and feed direction selection in robotic milling for increased chatter-free material removal rate. Int. J. Adv. Manuf. Technol. **89**(9–12), 2907–2918 (2017). https://doi.org/10.1007/s00170-016-9896-2
7. Wan, M., Zhang, W.H., Dang, J.W., et al.: A unified stability prediction method for milling process with multiple delays. Int. J. Mach. Tools Manuf. **50**(1), 29–41 (2010)
8. Insperger, T., Stepan, G.: Updated semi-discretization method for periodic delay-differential equations with discrete delay. Int. J. Numer. Meth. Eng. **61**, 117–141 (2017)

Teleoperation Robot Machining for Large Casting Components

Meng Wang[1], Panfeng Wang[1](✉), Tao Sun[1], Yuecheng Chen[2], Binbin Lian[1], Baomin Hou[2], Xueman Zhao[1], and Yimin Song[1]

[1] Key Laboratory of Mechanism Theory and Equipment Design, Ministry of Education, Tianjin University, Tianjin 300350, China
panfengwang@tju.edu.cn
[2] Tianjin Zhongyiming Technology Co., Ltd., Tianjin 300400, China

Abstract. Casting is an important means of components manufacturing. However, the machining of residual characteristics (RC) is a long-standing challenge, especially for large casting components. The reason is that the RC of casting components are different in size, complex in shape and random in distribution. It is difficult to deal with that by manual, serial robots, or machine tools. To address this problem, this paper proposes a robot-based teleoperation machining approach. A novel five degrees of freedom (5-DoF) hybrid machining robot developed by us is utilized as the slave robot. Then the teleoperation machining experiment of a large casting planet carrier was carried out. The results show that the machining efficiency is improved by about 10 times and the completion rate is above 95%. This provides a new idea for the machining of large components.

Keywords: Large casting components · Teleoperation machining · 5-DoF hybrid robot

1 Introduction

With the development of aerospace, wind power, ship, and other industries, the size of the equipment used is increasing. In order to meet the performance requirements of these equipment, the size of the core components is also gradually increasing. It is worth noting that more than 40% of them are casting components. Unfortunately, many RC will be brought inevitably by casting process, especially for large casting components. The RC refer to the casting heads, flashes, burrs, and other materials that need to be removed. Their removal quality and efficiency are greatly related to the quality and cost of subsequent machining and assembly. Therefore, the machining of RC is an indispensable process to produce casting components [1].

At present, there are two ways for the machining of RC of casting components: manual machining and automatic machining. And the manual machining is the most frequent operation method adopted by most enterprises [2]. Because it has the advantages such as good flexibility, low cost, and poor level of technology demanded, compared to the other methods. The RC of the casting components are removed layer by layer by the

X.-J. Liu et al. (Eds.): ICIRA 2021, LNAI 13014, pp. 620–627, 2021.
https://doi.org/10.1007/978-3-030-89098-8_58

human operator using hand-held grinding machines or cutting machines. Nevertheless, this method is labor-intensive, inefficient, and inconsistent. What's more, the harsh working environment will bring workers pneumoconiosis, cancer, and other health problems [3].

The automatic machining is mainly divided into two categories: serial robot and machine tool. The former is represented by the Koyama robot [4]. It has been widely used in grinding or polishing casting components such as wheel hub and wind turbine blade due to its good flexibility and low cost. Although the machining efficiency is improved dramatically, it can only complete the machining of the casting components with a small allowance. That is mainly limited to the low stiffness determined by its serial topology structure [5, 6]. On the contrary, the latter is represented by the Maus machine tool [7]. It is generally reformed from the traditional milling machine tool with high stiffness and high accuracy. Hence, it can grind or cut casting components with hard material. However, its expense is very high and the flexibility is poor, only suitable for machining the small or medium-sized casting components.

Furthermore, the two types of automatic machining need to be programmed. It generally employs the teaching programming method or off-line programming method based on CAD/CAM. Whereas, the two programming approaches are cumbersome and the preparation period is very long [8, 9]. So, they are usually used in the production of large quantities of components. Besides, the RC size of large casting components is huge and the distribution is stochastic. It is hard to decide the global machining trajectory in advance. If the above two methods are adopted, it can only be machined case by case. This will greatly decrease the machining efficiency. Therefore, the machining of large casting components has been a huge challenge for a long time.

To settle the machining problem of large casting components, this paper proposes a teleoperation machining approach based on robot. A 5-DoF hybrid robot with good flexibility and high stiffness is developed by us, which is used as the slave robot. This method integrates master-slave control and automatic control. They are used for global position switching and local precision machining respectively. It not only promotes the machining efficiency of large casting components, but ensures the safety and health of human operators.

This paper is organized as follows: Sect. 2 introduces the 5-DoF hybrid machining robot and the teleoperation machining approach for large casting components. The machining experiment of a large casting planet carrier is exampled to validate the effectiveness of the machining method proposed in Sect. 3. Section 4 draws the conclusions.

2 Teleoperation Machining Approach Based on Robot

2.1 Machining Robot

A novel machining robot with hybrid (serial-parallel-serial) topology structure is proposed in this section. It is utilized as the slave robot. As shown in Fig. 1, it consists of a vertical guide rail, a planar 5R (R denotes revolute joint) parallel mechanism, and two revolute heads. The planar 5R parallel mechanism is composed of five revolute joints, and it is symmetric about the y-axis in the initial pose. These two revolute heads connect with the end of the parallel mechanism one by one. The end-effector attaches to revolute head 2. They are all mounted together on the vertical guide rail.

A 2-DoF translational motion can be carried out by the 5R parallel mechanism in the plane *x*-*O*-*y*. The vertical guide rail moves along the *z*-axis. The two revolute heads rotate around the *y*-axis and *z*-axis, respectively. They strengthen the flexibility of the robot. Therefore, the three translational and two rotational motion in three-dimensional space could be realized by the hybrid robot.

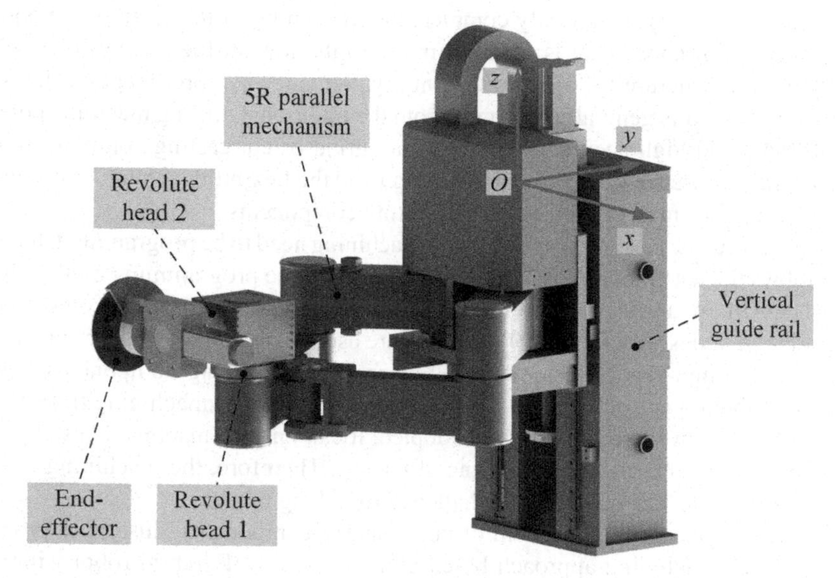

Fig. 1. The virtual prototype of the 5-DoF hybrid robot.

It's worth mentioning that the 5-DoF hybrid machining robot has the high stiffness of the parallel mechanism and the good flexibility of the serial mechanism. The large RC can be removed at one time. Hence, using the hybrid robot to machine large casting components will greatly improve production efficiency and save expenses.

2.2 Teleoperation Machining Approach

The teleoperation machining method based on master-slave control is proposed in this section. However, the hand's shaking, communication delay between the master device and slave robot will bring non-ideal operations [10, 11]. That is likely to destroy the casting components body and cause great economic loss. Hence, the automatic control is integrated into the machining method to overcome this problem. The schematic diagram of the teleoperation machining system is shown in Fig. 2. The operation procedures are divided into the following two stages.

***Stage*1**: global position switching.

1) The machining scene is transmitted to the control platform through the imaging system in the form of video. The human operator will see the large casting component and the hybrid machining robot.
2) Depending on the relative position between the local RC to be machined and the end-effector, the global motion trajectory is planned by the human operator. Then the human operator drags the hand shank of the master device to move towards the local RC direction.
3) The kinematic information of the master device is received by the control system. And it is converted into motion command based on the heterogeneous mapping algorithm. Then the actuators make the end-effector of the slave robot move towards the local RC and stop near it.

This teleoperation steps are shown in the left half of Fig. 3.

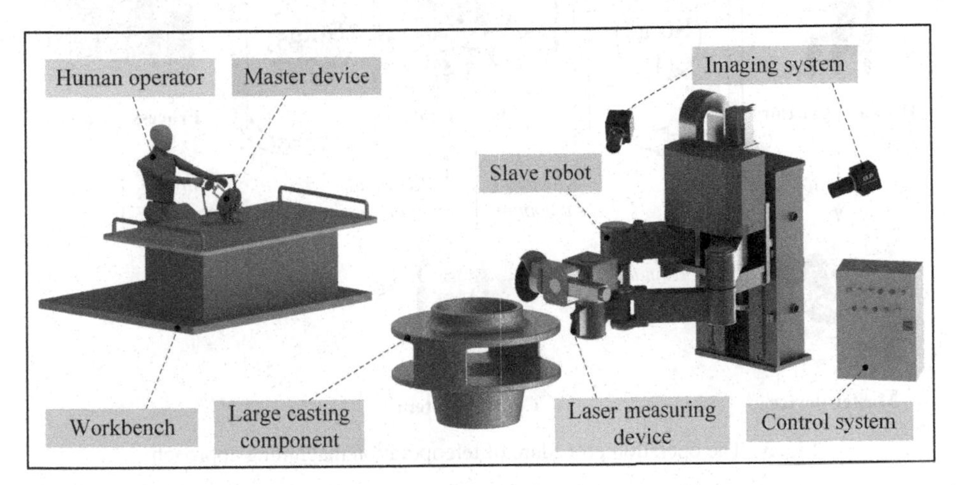

Fig. 2. The schematic diagram of the teleoperation machining system.

***Stage*2**: local precision machining.

1) After finishing *Stage* 1, the morphology features of the local RC are extracted by the laser measuring device. The machining trajectory is generated based on motion planning algorithm.
2) Then the size information of the local RC is input into the process database library. The machining parameters such as spindle speed rate, feed speed, and cutting depth are matched and delivered to the control system.
3) Later, the local machining will be carried out by the slave robot automatically based on the previous two steps.

The local machining steps are shown in the right half of Fig. 3.

It is important to emphasize that *Stage* 1 only needs to be carried out once for the same local RC. However, *Stage* 2 needs to be implemented for many times probably until the local RC is removed to an acceptable level. Then, a similar procedure is followed for another local RC.

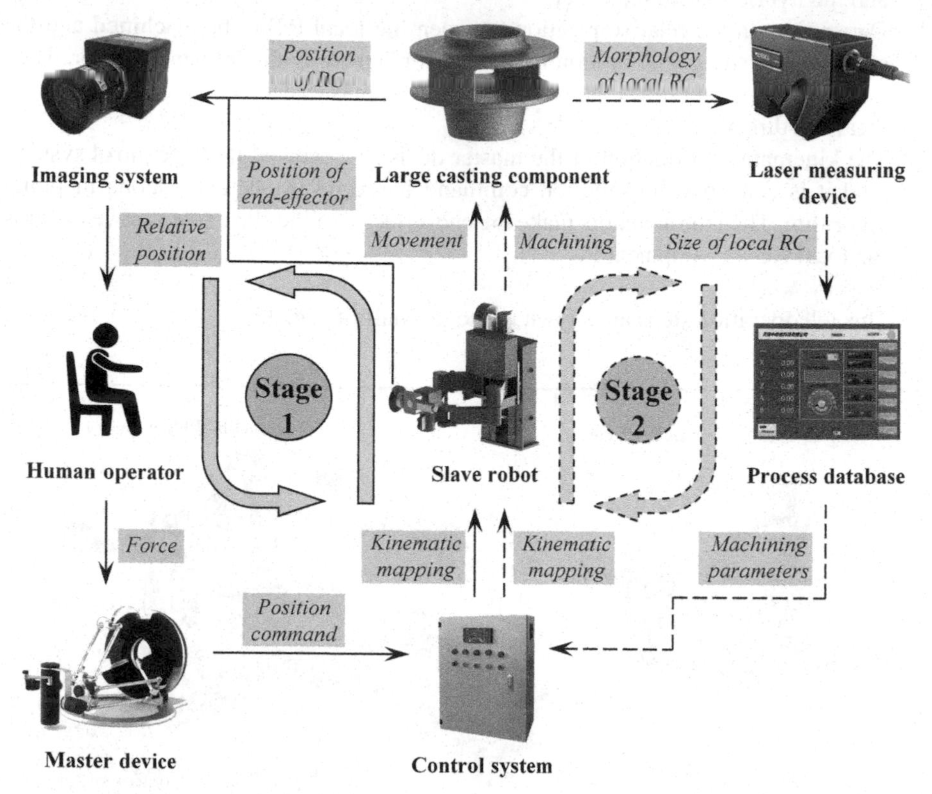

Fig. 3. The operation procedure of teleoperation machining approach.

3 Experiment and Discussion

In this section, the machining experiment of a large casting planet carrier with a diameter of 1000 mm was carried out, as shown in Fig. 4. This component is one of the most important parts in a wind turbine's gearbox. It is usually made of ductile iron (QT700-2 GB1348-2009) by centrifugal casting. After casting, the maximum width and length of RC on the casting component are two centimeters and tens of centimeters, respectively.

The machining scene is displayed as Fig. 5. The end-effector is an electroplated diamond wheel with mesh sizes 40, a radius of 190 mm, and a width of 12 mm.

The RC_A and RC_B of the large casting planet carrier after machining are shown in Fig. 6. The comparison of the effects before and after machining by manual and teleoperation methods is displayed in Fig. 7. From that, we can learn the following results.

Fig. 4. The RC of large casting planet carrier.

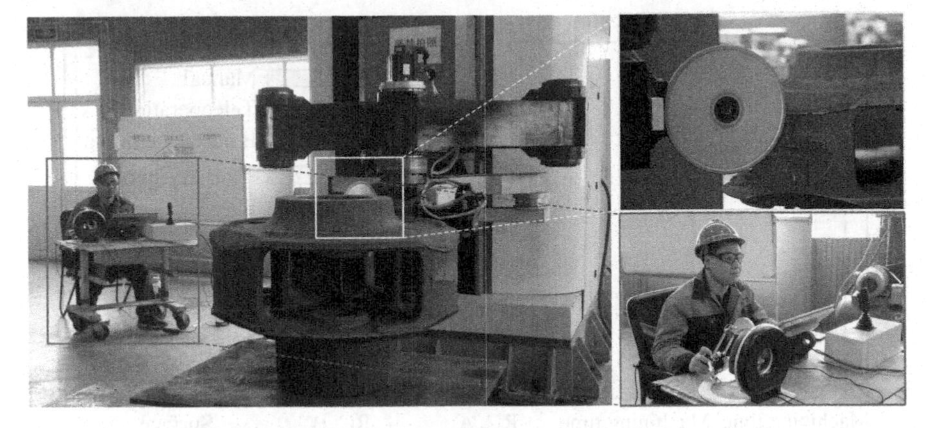

Fig. 5. The machining scene of large casting planet carrier.

1) Compared to the 396 s and 438 s for machining RC_A and RC_B by adopting manual method, teleoperation machining only takes 37 s and 45 s. The machining efficiency is improved by 10 times. The main reason is that the teleoperation method based on master-slave control does not need motion modeling. Also, the movement of the machining robot can be adjusted by the human operator in real-time.
2) The introduction of the imaging system allows the human operator and the workpiece to be further away from each other. That frees the human operators from the hostile working environment and enhances their safety and health. Furthermore, the width of RC_A was reduced from 17.0 mm to 0.40 mm and the width of RC_B was decreased from 20.0 mm to 0.36 mm. The machining completion rate is greater than 95%.
3) Due to the use of automatic control, the non-ideal operations are avoided. In another word, the manufacturing cost is decreased and the machining quality is improved. The surface roughness of RC_A and RC_B after machining is 3.759 μm and 3.286 μm, respectively.

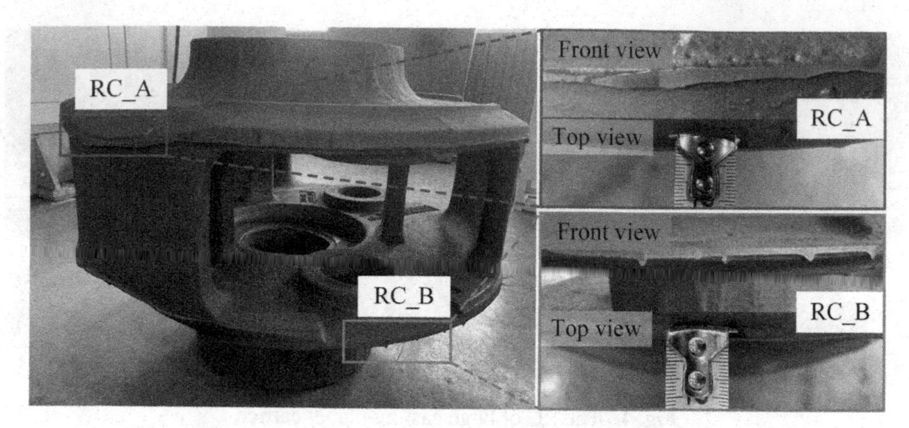

Fig. 6. The large casting planet carrier after machining.

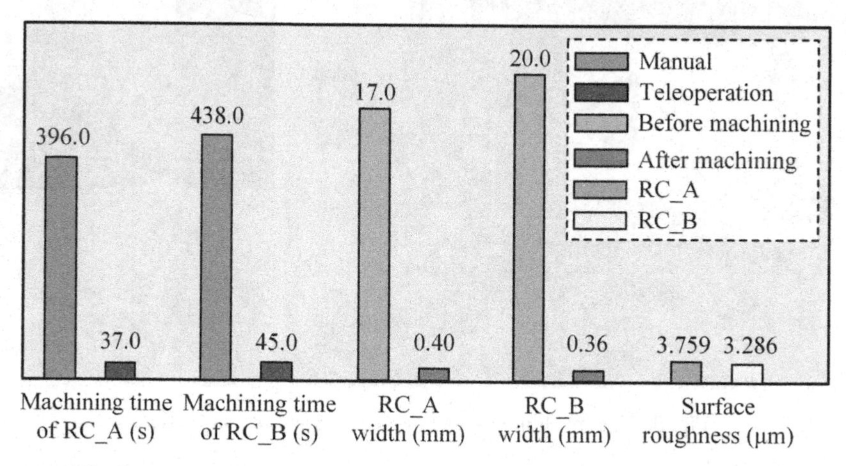

Fig. 7. The comparison of RC_A and RC_B before and after machining.

4 Conclusions

The requirements of large casting components are increasing with the development of equipment in aerospace, ship and other industries. Because the RC of casting components will appear after casting. So, the machining of RC is a necessary process. It determines the manufacturing efficiency and quality of components. However, the existing manual, serial robot and machine tool methods are not suitable for the machining of large casting components.

Focus on solving the problem, the contributions in this paper are summarized as follows.

1) A novel 5-DoF hybrid machining robot, utilized as the slave robot, is invented. It has the good flexibility and high stiffness performances.

2) A teleoperation machining approach based on robot is proposed. This method contains two stages: global position switching and local precision machining. It improvs the machining efficiency, guarantees the human operator safety and health and decreases the manufacturing cost.
3) The machining experiment of a large casting planet carrier was implemented. The results show that the machining time, compared to the manual method, is reduced by about 90%. The completion rate is above 95% and the surface roughness is about 3.50 μm.

In the future, the teleoperation machining method integrating force feedback will be studied to enhance the controllability of the operation. Furthermore, more experiments with different casting components will be implemented to make the approach perfect.

Acknowledgements. This research has been supported by National Natural Science Foundation of China (No. 51875391, No. 51875392); Tianjin Science and Technology Plan Project of China (No. 20YDLZGX00290, No. 18PTLCSY00080).

References

1. Sabourin, L., Robin, V., Gogu, G.: Robotized cell dedicated to finishing operations, by machining and polishing, on large cast parts. Mec. Ind. **12**(6), 495–502 (2011)
2. Cho, C.-H., Kim, J.-H., Soowon, C., et al.: Improvement of a deburring tool for intersecting holes with reduced irregular cutting of burr edge. Proc. Inst. Mech. Eng. Part B J. Eng. Manuf. **227**(11), 1693–1703 (2013)
3. Miao, C., Ye, H., Hu, X.Q.: Occupational health risk assessment of a foundry enterprise based on ICMM method. Ind. Hyg. Occup. Dis. **46**(2), 129–133+136 (2020)
4. http://www.barinder.jp/english/products.php.
5. Dumas, C., Caro, S., Cherif, M.: Joint stiffness identification of industrial serial robots. Robotica **30**(4), 649–659 (2012)
6. Bogue, R.: Finishing robots: a review of technologies and applications. Ind. Robot. **36**(1), 6–12 (2009)
7. http://www.maus.it/
8. Lee, H.M., Kim, J.B.: A survey on robot teaching: categorization and brief review. Appl. Mech. Mater. **330**, 648–656 (2013)
9. Zhou, B., Zhang, X., Meng, Z., et al.: Off-line programming system of industrial robot for spraying manufacturing optimization. In: Proceedings of the 33rd Chinese Control Conference, pp. 8495–8500. IEEE, Nanjing (2014)
10. Siciliano, B., Khatib, O. (eds.): Springer Handbook of Robotics. Springer, Cham (2016). https://doi.org/10.1007/978-3-319-32552-1
11. García, C.E., Postigo, J.F., Castro, A., et al.: On-line estimation of communication time delay in a robotic teleoperation system. Lat. Am. Appl. Res. **33**(4), 371–377 (2003)

Research on Short Term Power Load Forecasting Combining CNN and LSTM Networks

Yineng Zhuang[1], Min Chen[1], Fanfeng Pan[2](✉), Lei Feng[1], and Qinghua Liang[1]

[1] School of Mechanical Engineering, Shanghai Jiao Tong University, Shanghai 200240, China
[2] China New Energy (Shanghai) Limited Company, Shanghai, China

Abstract. Accurate prediction of power load plays an important role in the optimal scheduling of resources. However, the lack of power data in the traditional automatic acquisition system inevitably affects the subsequent data analysis. With the help of on-site real-time monitoring, the integrity of data collection can be ensured. In this paper, a load forecasting model based on the fusion of Convolutional Neural Networks (CNN) and Long Short-Term Memory (LSTM) is proposed. Through training the historical data collected by the on-duty robot, a complete network model is constructed. The network extracts the effective sequence features of the input data through CNN network, and gets the load prediction results through LSTM network. The experimental results show that the fusion network of CNN and LSTM obtains higher prediction accuracy than present algorithms.

Keywords: On-duty robot · Load forecasting · Convolutional neural network · Long Short-Term Memory

1 Introduction

An accurate prediction of power load provides preferable load scheduling and resource optimization [1] and ensures the common supply of communal power. However, the traditional data monitoring method of distribution stations is confronted with difficulty in solving the lack of power data due to malfunction in the collection, communication and other steps, further affects the result of subsequent data analysis. On-duty robots [2–5] of the distribution station site can further assist and improve data collection in monitoring method, ensuring the construction of subsequent load forecasting models.

Though further identification and collection of the on-site data by on-duty robot, both historical data and load forecasting models are constructed. The existing load forecasting method is divided into non-time series input and time series input depending on whether the input feature is time-related. The algorithms with non-time series input include Random Forest (RF) [6], Support Vector Regression (SVR) [7] and Ridge Regression (RR) [8], etc. Algorithms with no-time series input lack time relevance, thus are prone to get large fluctuations in the results of prediction and the accuracy is not high enough. Another type of algorithms with time series input predicts future load based on historical data, including autoregressive integrated moving average model [9], Kalman filter model

X.-J. Liu et al. (Eds.): ICIRA 2021, LNAI 13014, pp. 628–638, 2021.
https://doi.org/10.1007/978-3-030-89098-8_59

[10], exponential smoothing model [11], Markov model [12], etc. These models are not effective in nonlinear data.

With the application of deep learning in data mining, the Long Short-Term Memory (LSTM) [13] network overcomes the gradient disappearance and gradient explosion problems of the Recurrent Neural Network (RNN) [14], and becomes more popular in the prediction problem.

Zhang Yufan [15] first applied deep LSTM network on power load forecasting and made certain achievement. Chen Zhengyu [16] then combined LSTM with XGBoost model, and used the error reciprocal method to improve prediction accuracy. Furthermore, Hu Bo [17] proposed a medium long-term load curve fitting method combining improved variational mode decomposition and LSTM to reduce the fitting error. The above algorithms are derived from mathematical methods, but are complicated and inefficient in certain applications.

With the auxiliary data acquisition function of on-duty robot in distribution station, this paper proposes a model forecasting the load trend of distribution box by fusion of Convolutional Neural Network (CNN) and LSTM network (CNN-LSTM). CNN model is applied to extract the effective features from input data, and input them into LSTM network to get effective prediction results. The result of CNN-LSTM network is compared with RF, SVR, RR, LSTM and other algorithms to verify the prediction ability and the result shows that this CNN-LSTM model has better performance in forecasting power load.

2 Process of Power Distribution Load Forecasting

2.1 Data Collection and Short-Term Load Forecasting

Data acquisition system in distribution station is composed of background host computer and slave computer at on-duty robot. Figure 1 shows the on-duty robot in distribution

Fig. 1. On-duty robot on distribution station site

station. Database in host computer stores historical data of distribution device. On-duty robot identifies the digital numbers in distribution cabinet through its binocular camera in real time to collect operation data of certain device. Host computer then compares these real time data with historical one. When error is within a certain range, real time data will be considered to be normal and be saved in database.

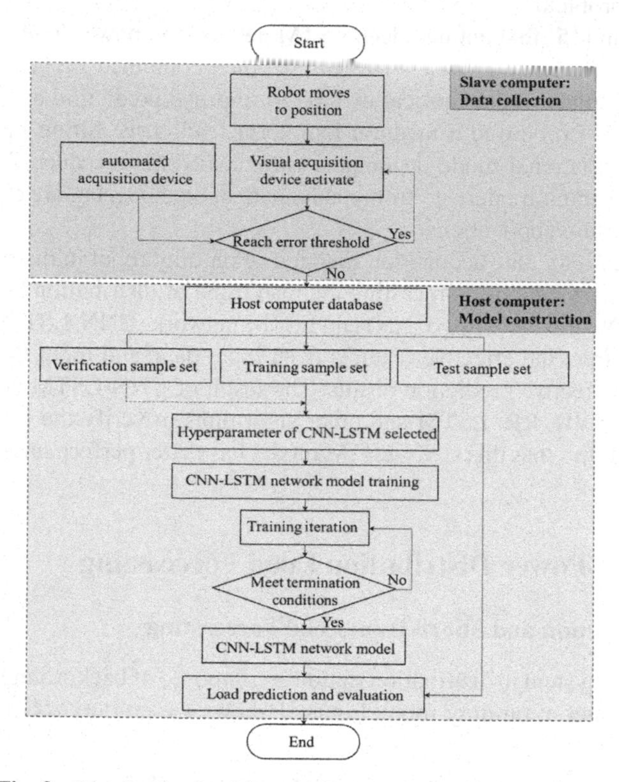

Fig. 2. Distribution load forecasting process based on duty robot

Figure 2 shows the process of power distribution load forecasting based on the on-duty robot. To collect operation data from distribution cabinet, on-duty robot first reaches the designated position in distribution station through its navigation system. By comparing the data collected from vision system with historical data, results will be saved on to the host computer database.

In order to predict the load of power distribution cabinet, the host computer data divides the historical data into training set, verification set and test set. The trained CNN-LSTM network is applied for short-term load predication.

2.2 Construction of the CNN-LSTM Model by Host Computer

Collected data is stored in the host computer database by on-duty robot. The model is built based on the historical data. The CNN-LSTM model proposed in this paper is composed of CNN and LSTM network.

Applying convolution transformation and down-sampling process and using methods of local connections and weights sharing, CNN model extracts the local features of the data. So that the number of neuron parameters is reduced and the feature vectors are dense and complete.

RNN model is against the problem of gradient disappearance and gradient explosion dealing with long-distance dependence. LSTM network overcomes it and can well maintain the long-short distance dependence of time series data. The basic units of LSTM model include forget gate, input gate and output gate. The network structure of LSTM units is shown in Fig. 3.

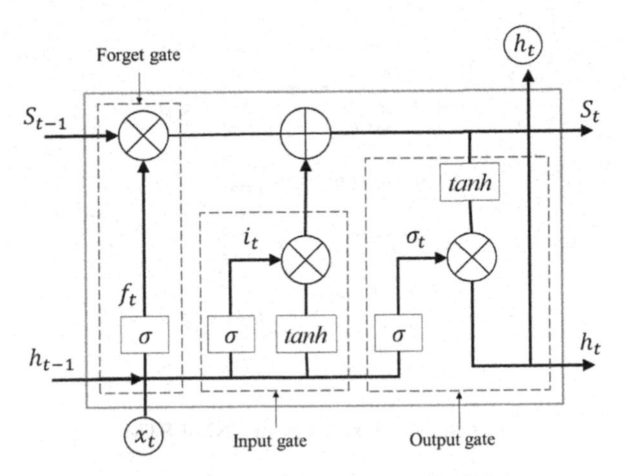

Fig. 3. Unit network structure of LSTM

$$f_t = \sigma\left(W_{fx}x_t + W_{fh}h_{t-1} + b_f\right) \tag{1}$$

$$i_t = \sigma\left(W_{ix}x_t + W_{ih}h_{t-1} + b_i\right) \tag{2}$$

$$g_t = tanh\left(W_{gx}x_t + W_{gh}h_{t-1} + b_g\right) \tag{3}$$

$$o_t = \sigma\left(W_{ox}x_t + W_{oh}h_{t-1} + b_o\right) \tag{4}$$

$$S_t = g_t \odot i_t + S_{t-1} \odot f_t \tag{5}$$

$$h_t = tanh(S_t) \odot o_t \tag{6}$$

where W_{ix}, W_{ih}, W_{gx}, W_{gh}, W_{ox}, W_{oh} is the neural network weight parameter matrix. b_i, b_g, b_o is the corresponding bias. f_t, i_t, o_t is the forget gate, input gate and output gate at time t. g_t is the input node. h_t is the intermediate output state. S_t is the unit state.

σ is the sigmoid activation function, controlling the state of gate. $\odot$ is the Hadamard operation.

Figure 4 shows the CNN-LSTM model. It contains 3 layers of convolutional layers (Conv1D). The convolutional kernel processes the input data of upper layer to get feature as an output, and the output then is used as the input of the lower layer. There is a polling layer (MaxPooling1D) between 2 convolutional layers, which eliminates non-key features to reduce the number of parameters, further improving training efficiency and estimation accuracy. CNN model extracts certain data that responses time features from the input layer of time series. The number of convolution kernel of convolution layer is determined by experiments so that parameters with best effect are selected.

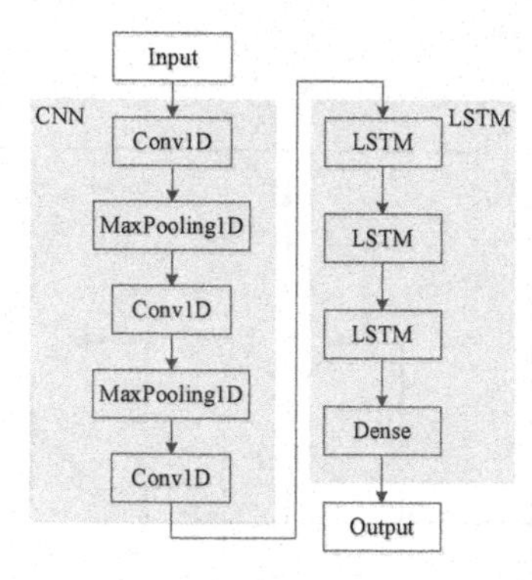

Fig. 4. Model structure of CNN-LSTM

In the short-term load forecasting of distribution cabinet, data is collected once per minute. The variation trend of load in the next day is predicted through the historical data.

3 Data Process and Model Training with On-Duty Robot

3.1 Construction of Time Series Feature

According to the data collected by host computer, the input features related to time series are constructed. Data points are taken out of the time interval t in data processing. The output value at time T is $y = x(T)$, then the input features following time interval are arranged as $\{x = x(T - t), x(T - 2t), \ldots, x(T - nt)\}$. Training sample $\{x, y\}$ is constructed of the above input and output data. The time series feature construction is shown in Fig. 5.

According to the division of time series feature data, let the data set matrix be X

$$X = \begin{bmatrix} x_{11} & x_{12} & \cdots & x_{1n} \\ x_{21} & x_{22} & \cdots & x_{2n} \\ \vdots & & \ddots & \vdots \\ x_{m1} & x_{m2} & \cdots & x_{m1} \end{bmatrix} \quad (7)$$

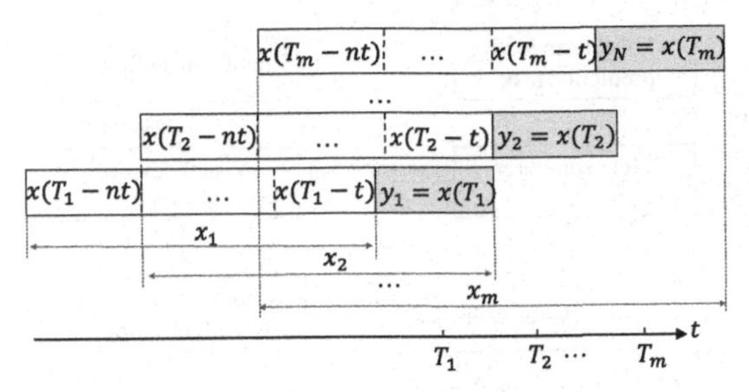

Fig. 5. Time slider

where n represents the number of input features, m represents the number of training samples. When the model predicts output at time T, then $x_{k1}, x_{k2}, \ldots, x_{kn}$ respectively represent the input feature of $\{x(T_k - nt), x(T_k - nt + t), \ldots, x(T_k - t)\}$.

Normalize each column of the original data set matrix to get the matrix of $\tilde{X}$:

$$\tilde{X} = \begin{bmatrix} \tilde{x}_{11} & \tilde{x}_{12} & \ldots & \tilde{x}_{1n} \\ \tilde{x}_{21} & \tilde{x}_{22} & \ldots & \tilde{x}_{2n} \\ \vdots & & \ddots & \vdots \\ \tilde{x}_{m1} & \tilde{x}_{m2} & \ldots & \tilde{x}_{m1} \end{bmatrix} \quad (8)$$

For feature in column k and row l, $\tilde{x}_{lk} = \frac{x_{lk} - min\{X_k\}}{max\{X_k\} - min\{X_k\}}$, $1 \le l \le m$, $min\{X_k\}$ is the minimum value in column k and $max\{X_k\}$ is the maximum value in column k.

3.2 Training of Model

The historical data is divided into training set, verification set and test set. CNN-LSTM model is trained by training set and verification set. The convolution kernel of each layer in CNN model, and W_{ix}, W_{ih}, W_{gx}, W_{gh}, W_{ox}, W_{oh} and b_i, b_g, b_o of each layer in LSTM network are all part of machine learning. The model is trained by adjusting the hyperparameters of the network including the size of convolution kernel, the number of batches, the selection of pooling method, the number of LSTM hidden neurons, the number of iterations, etc. The result is validated through verification set to achieve the best effect (Fig. 6).

3.3 Criteria of Experimental Evaluation

In experimental evaluation, root mean squared error (RMSE) and mean absolute percentage error (MAPE) are selected as evaluation criteria of the prediction effect:

$$l_{RMSE} = \sqrt{\frac{\sum_{i=1}^{n} (\hat{y}_i - y_i)}{n}} \quad (9)$$

$$l_{MAPE} = \frac{1}{n} \sum\nolimits_{i=1}^{n} \left| \frac{\hat{y}_i - y_i}{y_i} \right| \quad (10)$$

where n represents the total number of samples, $\hat{y}_i$ is the value of forecasting load, y is the true value of load, l_{MAPE} and l_{RMSE} represents the effects of the prediction.

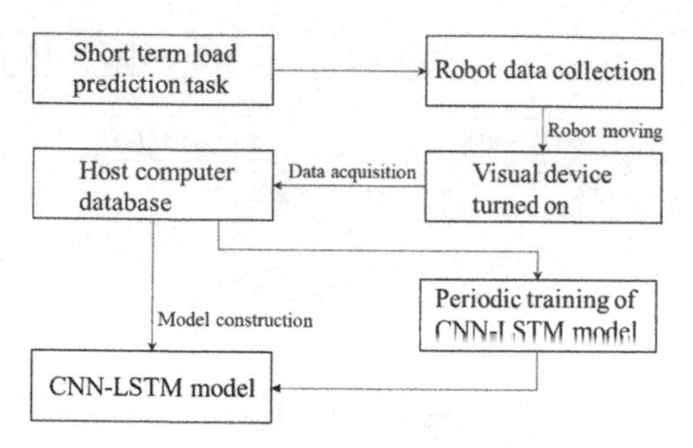

Fig. 6. Model training process

4 Results Analysis

The experimental condition in this paper is Intel(R) Core™ i7-5500U CPU, 8GB RAM. CNN-LSTM model is trained under the deep learning framework of Tensorflow and Keras.

4.1 Description of Samples

The data sample in this paper is based on 58962 continuous monitoring records of on-duty robot in a power distribution room of Jiangsu Province from August to September in 2019. Sampling interval is one data per minute, including the active power and reactive power of the distribution cabinet. Active power is used for training and prediction without loss of generality. The results of prediction randomly select more than 1000 pieces of data sample in test set.

4.2 Selection of Hyperparameter

To improve the accuracy of prediction, moderate hyperparameters need to be selected by experiments. In experiment, the number of layers of LSTM network, the number of nodes of LSTM network and the number of iterations arethe hyperparameters. The number of iterations is 50 and batch size for each training is 100. By gradually increasing the number of layers and then changing the number of hidden nodes in each layer of LSTM network, different results of prediction are obtained, as shown in Table 1.

Table 1 shows that when the number of LSTM network layers is 3, and the numbers of nodes of each layer are {16, 32, 16}, MAPE of the result achieves minimum Thus the result determine the primary structural parameters of LSTM network.

In this paper, 60% of experimental data is divided into training set, 20% into verification set and 20% into test set. Mean Squared Error (MSE) is used as the objective loss function:

$$L = \sqrt{\frac{1}{S}\sum\nolimits_{i=1}^{S}\left(\hat{y}_i - y_i\right)^2} \tag{11}$$

where y is the true value, $\hat{y}_i$ is the predicted value from models and L is the objective loss function.

Structural parameters of CNN-SLTM network in experiments are shown in Table 2.

Table 1. Prediction results under different network layers of CNN-LSTM

CNN Layer	LSTM network	Iteration	RMSE	MAPE
2	{16, 16}	50	0.0181	2.74%
2	{32, 32}	50	0.0192	3.29%
2	{64, 64}	50	0.0177	2.95%
3	{16, 32, 16}	50	0.0183	2.65%
3	{32, 64, 32}	50	0.0187	3.55%
3	{64, 100, 64}	50	0.0178	2.73%

Table 2. Network structure parameters of CNN-LSTM

Network Structure	Parameters
Sample	58962
Input series length	20
Output series length	1
CNN layer	3
CNN filter	{16, 16, 16}
Filter convolution kernel	{5, 5, 2}
LSTM layer	3
LSTM hidden node	{16, 32, 16}
Iteration	50
Cost function	MSE
Training optimizer	RMSProp

4.3 Experimental Results

After the network structure and hyperparameters are determined, prediction results by CNN-LSTM model are compared with other present algorithms (RF model, SVR model, LSTM model). Table 3 shows MAPE and RMSE of the results from test set and training set. SVR model and RF model are implemented in Sklearn library, LSTM model and CNN-LSTM model are implemented in Keras framework.

Results show that MAPE and RMSE of CNN-LSTM model in test set are respectively 2.73% and 0.0178, indicating that CNN-LSTM model performs well among the other 3 algorithms. Compared with SVR and RF algorithms, accuracy of CNN-LSTM is improved by 55% and 54% on MAPE, and 36% and 44% on RMSE. LSTM model performs well in prediction of time series, while CNN-LSTM model is further improved based on LSTM and has a better prediction effect. From test set results, CNN-LSTM improves MAPE by 3.9% compared with LSTM.

Table 3. Prediction results of different algorithms

Algorithms	MAPE		RMSE	
	Training set	Test set	Training set	Test set
SVR	6.09%	5.30%	0.0265	0.0257
RF	5.89%	5.83%	0.0312	0.0295
LSTM	2.99%	2.54%	0.0179	0.0174
CNN-LSTM	2.73%	2.44%	0.0165	0.0178

Data is collected by on-duty robot, and then trained to generate CNN-LSTM model. Load forecast of the distribution cabinet lasts for more than 1 day. Forecasting effect of the 4 algorithms is shown in Fig. 7. Prediction value from RF and SCR fluctuate greatly, and CNN-LSTM model has a relatively higher accuracy. The sub-figure shows that when load is at its bottom, RF and SVR have relatively large errors while CNN-LSTM model can still match the changing trend and follows the curve of real load better than LSTM model.

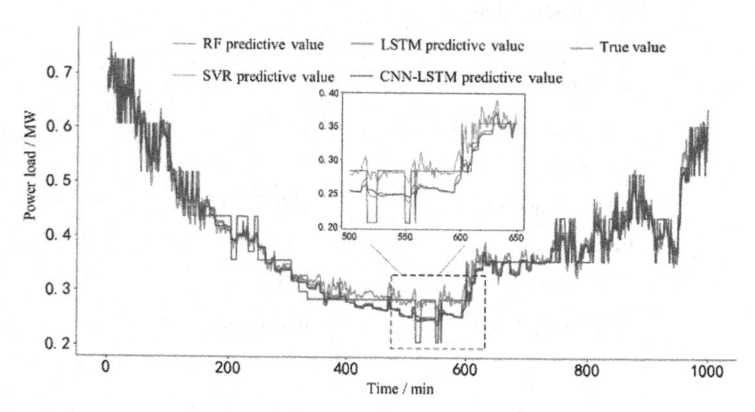

Fig. 7. Comparison of prediction results between CNN-LSTM and other algorithm

5 Conclusions

This paper proposes a short-term load forecasting method. First, on-duty robot combined with automated acquisition device collects data into database, then these historical data is trained through CNN-LSTM model. It processes time series by extracting effective features through CNN layers first, then LSTM model takes these features as inputs to analyze and predict the changing trend of load for more than 1 day.

CNN-LSTM model analyzes data collected by on-duty robot from power distribution cabinet. By selecting appropriate hyperparameters, training is carried out through a three-layer CNN network and a three-layer LSTM network using training set of historical data.

CNN-LSTM model constructed by host computer uses CNN model to process the input data and maintains long-short distance dependence of time series by LSTM network.

In comparison with other algorithms, CNN-LSTM model has higher accuracy in prediction. Thus, CNN-LSTM have advantages on processing time series data and can be applied in prediction tasks of other scenarios.

The short-term load forecasting method towards distribution cabinet proposed in this paper is based on fusion of CNN and LSTM network to improve the accuracy of prediction. Failure conditions are not considered in this paper so the future work needs to be carried out towards load forecasting under failure conditions.

Acknowledgement. This work was supported by Science and Technology Project of State Grid Corporation Headquarters: "Research and Application Verification on Intelligent Cloud Robot for Distribution Station" (5700-202018266A-0-0-00).

References

1. Yanhua, Z., Jinquan, W., Youhuai, T., et al.: Digital technology of low-voltage distribution cabinet. Electr. Power Autom. Equipment **33**(03), 158–161 (2013)
2. Tao Zhiyuan, X., Weiqing, W.S., Qinghua, L.: Path planning applied on a field robot for hot-line working in 110 kV intelligent substation. Mach. Des. Res. **34**(01), 17–25 (2018)
3. Shibao, X., Shicheng, L., Qinghua, L.: System development and analysis of the live-line maintenance tracked mobile robot navigation system. Mach. Des. Res. **33**(03), 26–34 (2017)
4. Lixia, Z., Zhicheng, Y., Jingsheng, Z.: No Counterweight underactuated cable inspection robot study and test quasi-zero stiffness isolator. Mach. Des. Res. **33**(03), 30–34 (2017)
5. Qing, Z., Xinping, L., Xiaolong, Z., et al.: Design and research of the mechanical structure for power transmission lines inspection robot. Mach. Des. Res. **32**(04), 46–49 (2016)
6. Wanhua, L., Hong, C., Kun, G., et al.: Research on electrical load prediction based on random forest algo-rithm. Comput. Eng. Appl. **52**(23), 236–243 (2016)
7. Pai, P.-F., Hong, W.-C.: Support vector machines with simulated annealing algorithms in electricity load forecasting. Pergamon **46**(17), 2669–2688 (2005)
8. Xiaofei, Z., Juncheng, G., Yubao, S.: Abnormal electricity behavior recognition of graph regularization nonlinear ridge regression model. Comput. Eng. **44**(06), 8–12 (2018)
9. Kaytez, F.: A hybrid approach based on autoregressive integrated moving average and least-square support vector machine for long-term forecasting of net electricity consumption. Energy, 197 (2020)
10. Khodaparast, J., Khederzadeh, M.: Least square and Kalman based methods for dynamic phasor estimation: a review. Prot. Control Mod. Power Syst. **2**(2), 1–8 (2017)
11. Ming, Z., Shulei, L., Liang, W., et al.: Wind power prediction model based on the combined optimization algorithm of ARMA model and BP neural networks. East China Electr. Power **41**(02), 347–352 (2013)
12. Haitao, C., Jun, Y., Yingchun, S., et al.: Life prediction method of relay protection device based on could model and Markov Chain. Power Syst. Prot. Control **47**(16), 94–100 (2019)
13. Zhang, W., Qin, J., Mei, F., et al.: Short-term power load forecasting using integrated methods based on long short-term memory. Sci. China (Technol. Sci.) **63**(04), 614–624 (2020)
14. Sherstinsky, A.: Fundamentals of recurrent neural network (RNN) and long short-term memory (LSTM) network. Phys. D, 404 (2020)

15. Zhang, Y., Ai, Q., Lin, L., et al.: A very short-term load forecasting method based on deep LSTM RNN at zone level. Power Syst. Technol. **43**(06), 1884–1892 (2019)
16. Zhenyu, C., Jinbo, L., Chen, L., et al.: Ultra short-term power load forecasting based on combined LSTM-XGBoost model. Power Syst. Technol. **44**(02), 614–620 (2020)
17. Bo, H., Xiao, P., Weichun, G., et al.: Typical period load curve fitting method based on IVMD-LSTM for improving the gridding of wind power. Renew. Energy Resour. **38**(03), 366–372 (2020)

Effect of Postures and Cutting Process on Robot Milling Performance and Surface Quality

Lvchen Feng[1], Wenxiang Zhao[2], Zhibo Zhang[1], Pei Yan[2(✉)], Minghui Cheng[1], Li Jiao[2], and Tianyang Qiu[2]

[1] School of Mechanical Engineering, Beijing Institute of Technology, Beijing 100081, China
[2] Key Laboratory of Fundamental Science for Advanced Machining (Beijing Institute of Technology), Beijing 100081, China
pyan@bit.edu.cn

Abstract. The application of industrial robots in machining fields is of great significance to the integral machining of large complex structures. However, subject to the structural rigidity, it can not become the mainstream means to replace the traditional machine tool processing. In order to study the machining performance of the robot milling system and improve the machining quality, the influence of different robot postures on the machining quality was studied, and the better machining posture was selected. The influence of different process parameters on the machining performance of the robot milling system was analyzed, which provides the basis for selecting machining parameters for the robot milling system. The results show small cutting depth and high spindle speed are suitable for robot milling. Finally, by comparing the effects of milling mode and cutting direction on the robot machining system, the anisotropic characteristics of the end stiffness of the robot were revealed. The experiment results show that the machining performance of the robot milling is better when the cutting direction is along *Y*.

Keywords: Robot milling · Posture · Cutting process · Milling performance · Surface quality

1 Introduction

With the continuous development and progress of aerospace industry, large complex structures are widely used in the rockets arrows components, aircraft envelope, satellite cabin and other high-precision equipment. One of the main technical factors to realize the high function of large equipment is the integration of the main structure [1].

At present, there have been many researches on the application of robot machining. EI (Electroimpact Inc) [2] designed and developed a high precision robotic drilling and milling system for Boeing 737 wing skins in 2011, with milling accuracy up to ±0.13 mm and surface roughness up to 1.6 μm. The team of Melkote S. N. in Georgia Institute of Technology [3–6] proposed an industrial robotic milling system for the multi-axis machining of large aerospace structures, and carried out a series of studies to improve the machining accuracy based on this system.

X.-J. Liu et al. (Eds.): ICIRA 2021, LNAI 13014, pp. 639–647, 2021.
https://doi.org/10.1007/978-3-030-89098-8_60

As for the process parameters of robot machining, Liang et al. [7] studied the robot drilling experiments of Ti-6Al-4V and 7075-T6 laminated materials, and found that the spindle speed and feed speed had significant effects on the machining quality of holes. Slamani et al. [8, 9] quantified the influence of cutting parameters (feed rate, cutting speed) on the precision of parts under specific conditions, and studied the comprehensive influence of cutting process parameters on the cutting force components of robotic high-speed cutting composites.

The application of industrial robots in the field of processing is of great significance for the integrated design and manufacture of large and complex structural parts. However, the machining quality of industrial robots is difficult to be effectively guaranteed. Cutting force and vibration are important factors affecting the robot machining performance. In order to explore the influence of machining technology on the robot performance, this study firstly studies the effect of different postures on the robot milling quality. Then, the effect of different milling processes on the cutting signal and machining quality in robot milling were studied under the better machining posture.

2 Experiment Design

The robot milling experimental platform is shown in Fig. 1. During the experiment, the Kistler 9527B three-way plate dynamometer was used to measure the milling force of the robot in real time. The DASP acceleration sensor was used to collect the cutting vibration signals in the process of robot milling. A portable profilometer was used to collect the linear roughness Ra of the processed surface along the feed direction, and the sampling length was set as 2.5 mm.

To study the effect of the postures on the machining performance of the robot, this paper designed the robot milling experiment with different postures. The process parameters were presented in Table 1, and the robot posture setting was shown in Fig. 2.

To study the effect of process parameters on the cutting force, cutting vibration and machining quality in robot milling, a single factor experiment was designed. The rotational speed of the end spindle of the robot, the feed and the axial cutting depth were selected as the process parameters. Four levels were taken for each factor, and the other factors were kept at the level of the second in each group of tests. The milling mode was down milling with the feed direction of *X*, and the cutting width was 5 mm. Test factors and levels are presented in Table 2.

Table 1. Cutting parameters of different postures experiments

Types	n (r/min)	f_z (mm/z)	a_p (mm)	a_e (mm)	Milling mode	Cutting direction
Value	4000	0.05	1	5	Down	X

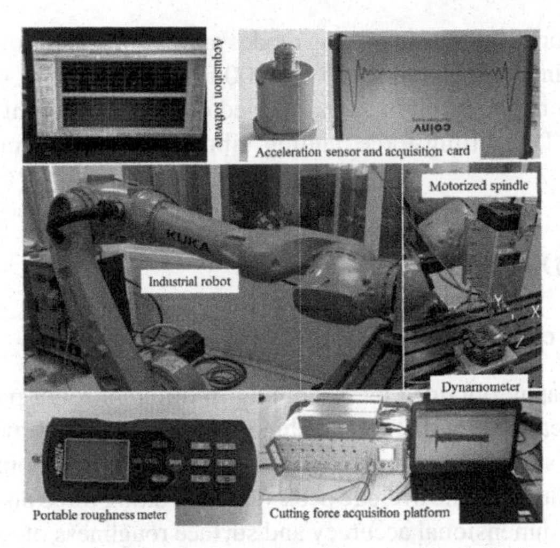

Fig. 1. Industrial robot milling experimental platform

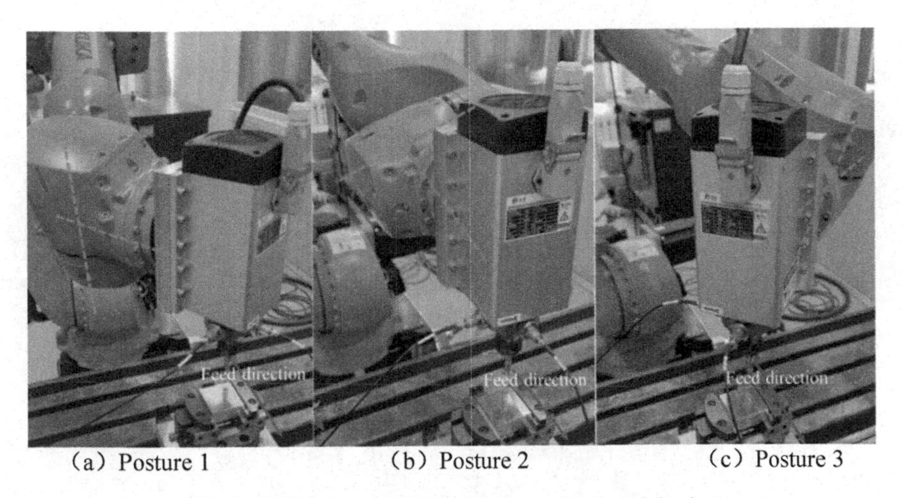

(a) Posture 1 (b) Posture 2 (c) Posture 3

Fig. 2. Different processing postures of industrial robots

Table 2. Single factor experiment and its level

Factors	Level
n (r/min)	2000, 4000, 6000, 8000
f_z (mm/z)	0.025, 0.05, 0.075, 0.1
a_p (mm)	0.8, 1.2, 1.6, 2.0

Further, to explore the effect of milling modes and cutting directions on the machining performance, keeping the spindle speed of 4000 r/min, the feed of 0.05 mm/z, and the cutting depth of 1 mm, the machining was carried out under different milling modes and cutting directions. The cutting force, cutting vibration and machining quality process were analyzed.

3 Result and Discussion

3.1 Robot Postures

The surface topography obtained by machining in different robot postures is shown in Fig. 3. It can be clearly seen that the machined surface obtained by posture 2 is more uniform, while the surface obtained by posture 1 and posture 3 both show color depth changes in the height map. The average peak value of cutting force and vibration acceleration as well as the dimensional accuracy and surface roughness of the machined workpiece were selected as the evaluation indexes for machining. The results of cutting force, vibration acceleration, dimensional accuracy, line roughness and surface roughness in different postures are shown in Fig. 4.

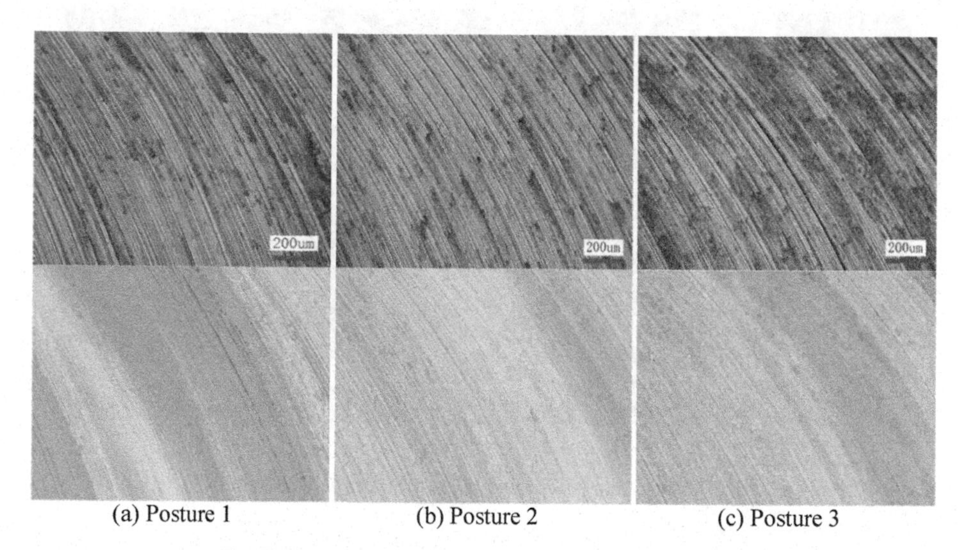

(a) Posture 1 (b) Posture 2 (c) Posture 3

Fig. 3. Surface topography for different robot postures

The experimental results show that the cutting force is not affected by the change of robot postures. It can be concluded that the combined vibration accelerations corresponding to the three postures are 131.65 m/s^2, 116.92 m/s^2, and 129.72 m/s^2, respectively. Among them, posture 2 is significantly lower than the other two, but there is almost no difference in cutting force, which indicates that posture 2 has better robot stiffness performance and is conducive to the milling process. The dimensional errors of the workpiece corresponding to the three postures are basically the same, which indicates

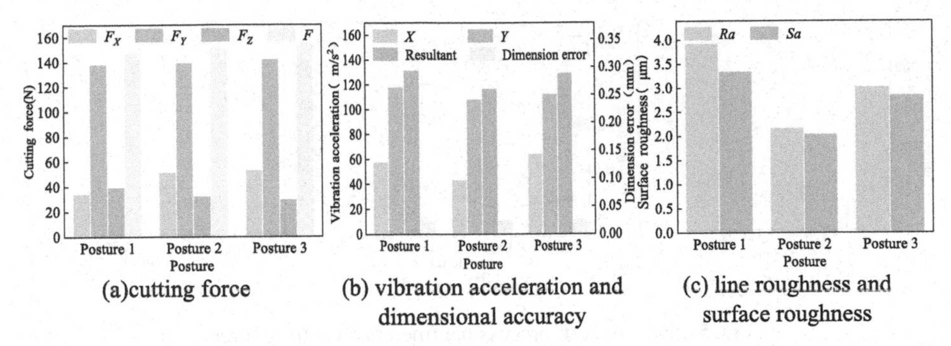

(a)cutting force (b) vibration acceleration and dimensional accuracy (c) line roughness and surface roughness

Fig. 4. Experimental results of different robot postures

that the deformation of the robot end is small when the cutting force is within the load range. The surface roughness obtained by different postures is shown in Fig. 4(c). The line roughness Ra and the surface roughness Sa corresponding to posture 2 are smaller than the other two, so its surface quality is the best, which is the same as the result of cutting vibration.

The above analysis shows that different robot postures have an important effect on its machining performance. In the application of robot milling, the appropriate robot postures should be selected firstly to improve its machining quality.

3.2 Process Parameters

Cutting Force

The effect of process parameters on cutting force is shown in Fig. 5. The changing of the resultant cutting force is basically the same as that of main cutting force. The effect of cutting depth and feed rate on the cutting force of robot milling is significant, while the effect of spindle speed is relatively small, which is basically consistent with the changing trend of traditional machine tool machining, indicating that the cutting force of robot milling process is mainly determined by the cutting process parameters. Due to the structural rigidity and load requirements of the industrial robot, the cutting data can not be too large, so that the cutting force is controlled in a certain range to meet the machining performance requirements of the robot.

Cutting Vibration

The effect of process parameters on cutting vibration is shown in Fig. 6. It can be seen that the variation curves of vibration acceleration with cutting depth and feed are basically consistent with the trend of cutting force, which indicates that the vibration amplitude at the end of the robot milling system is mainly determined by cutting force at a certain spindle speed. The cutting vibration in robot machining can be controlled by changing the cutting data. With the increase of the spindle speed, the vibration of the spindle will become more intense, which makes the cutting vibration of the robot end change greatly, showing a different trend from the cutting force curve.

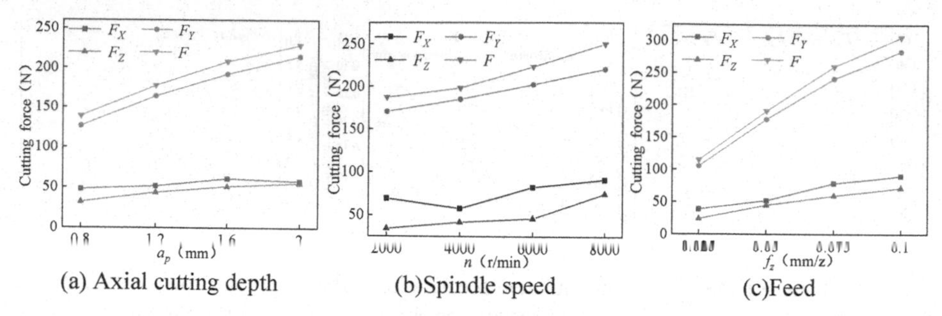

(a) Axial cutting depth (b)Spindle speed (c)Feed

Fig. 5. The effect of process parameters on cutting force

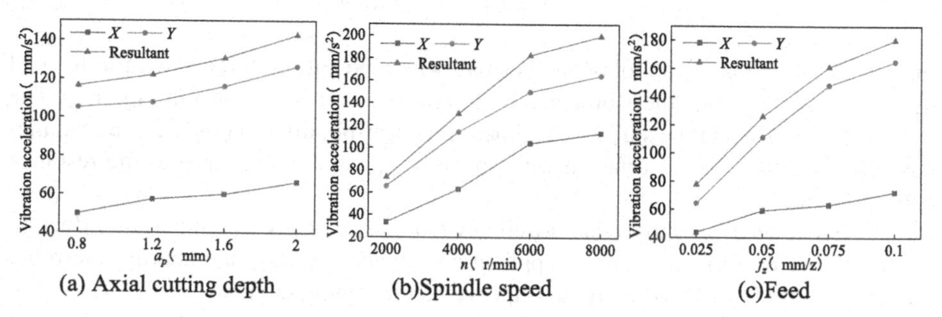

(a) Axial cutting depth (b)Spindle speed (c)Feed

Fig. 6. The effect of process parameters on cutting vibration

Dimension Accuracy

The effect of process parameters on machining dimension error is shown in Fig. 7. When the cutting depth is small (less than 1.2 mm), the dimension error of the workpiece fluctuates within a certain range. When the cutting depth is large, the size error will be positively correlated with the cutting depth, which is obviously different from the traditional machine tool. It is because of the low robot structure stiffness. When the cutting depth is large, the cutting force will make the end of the robot produce a certain deformation, resulting in the increase of the machining error. To meet the robot machining accuracy requirements, the cutting depth should not be too large.

With the increase of feed, the dimension error increases first and then fluctuates, which does not increase with the cutting force. It also indicates that the effect of cutting depth on the machining error is not only affected by the cutting force, and the larger cutting depth may lead to the increase of the random error of the robot milling system. When the feed is very small (0.025 mm/z), the dimension error is almost negligible. This is because the cutting force and vibration are very small at this time, and the machining is relatively stable, but the machining efficiency is low. With the increase of spindle speed, the dimension error gradually decreases, which is contrary to the changing trend of cutting force and vibration. This indicates that high spindle speed can enhance the dynamic stiffness of robot milling process and improve the machining accuracy. Therefore, the robot machining system is more suitable for high speed milling process.

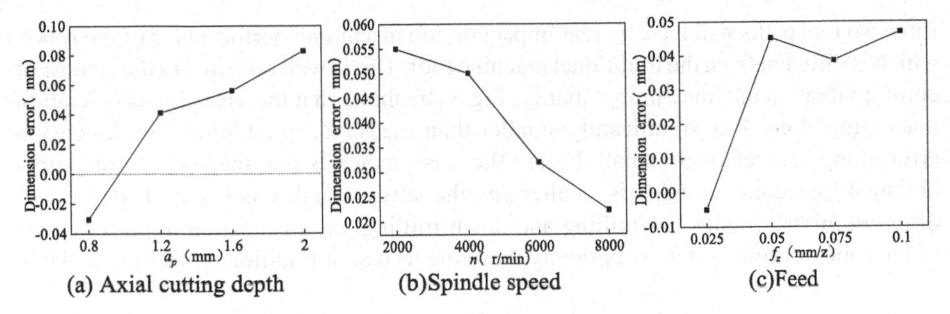

Fig. 7. The effect of process parameters on dimension error

Surface Roughness

The effect of process parameters on surface roughness is shown in Fig. 8. With the increase of cutting depth, the linear roughness Ra gradually increases, which is due to the combined action of cutting force and cutting vibration, resulting in the reduction of machining surface quality. Surface roughness Sa fluctuates to a certain extent, and the value is higher at large cutting depth. The effect of spindle speed further proves that the robot milling system is more suitable for high speed milling. The effect of feed on Ra is basically the same as that of cutting force and cutting vibration. The surface quality of machining is determined by the vibration characteristics of robot milling system. When the feed rate is within a certain range (less than 0.075 mm/z), the Sa does not change significantly, and the surface uniformity is good. When the feed is large, there will be a significant increase in Sa.

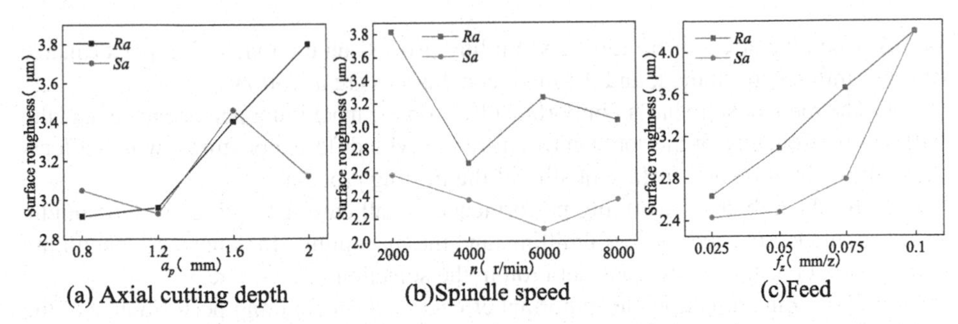

Fig. 8. The effect of process parameters on surface roughness

3.3 Milling Modes and Cutting Directions

Figure 9 shows the effect of different milling modes and cutting directions on the cutting performance of the robot. The resultant cutting force obtained by the four tool paths are basically the same, indicating that the milling mode and path have little effect on resultant cutting forces. However, the end stiffness of the robot will be different in different directions, and different milling paths will change the direction of the cutting force, so tool paths will have a great impact on the machining performance of the robot, which is different from the traditional machine tool. This is well reflected in the change of cutting vibration and machining quality: Fig. 9(b) shows that the vibration acceleration machining along *Y* is significantly smaller than that in *X*. In addition, the dimension error along *Y* is relatively small. Figure 9(c) also indicates that the surface roughness obtained by feeding along *Y* is smaller and the surface quality is better. The effect of different milling modes (up-milling and down-milling) on the machining performance of the robot milling system is basically the same as that of traditional machine tools.

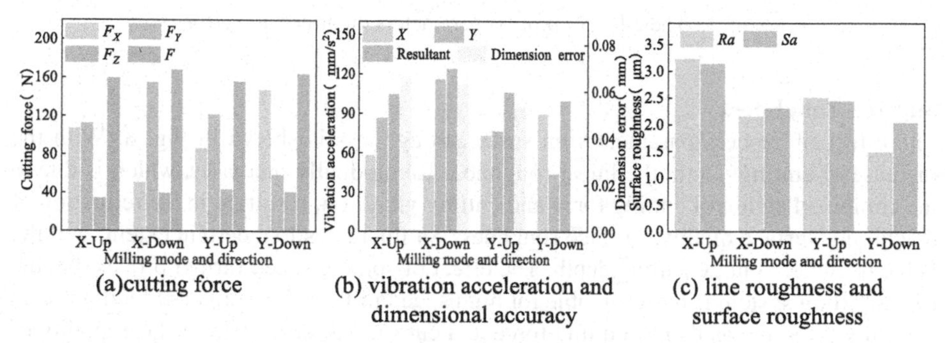

Fig. 9. Experimental results of different milling modes and cutting directions

4 Conclusion

In this paper, the effect of postures and milling process on the machining performance of robot milling are studied, and the main conclusions are as follows:

(1) The robot posture has an important effect on the machining performance, and the stiffness performance of the robot end is different with different postures, which affects the cutting vibration and surface quality of the milling process.

(2) To obtain better machining performance, small cutting depth and high spindle speed should be selected for robot milling, and the machining efficiency and machining quality should be taken into consideration in the selection of feed rate.

(3) The cutting direction has important effects on the machining performance of the robot milling system. The dimensional accuracy and surface quality obtained by feeding along *Y* are better than that along *X*.

Funding. This work was supported by the National Key Research and Development Program of China, China [Grant number 2020YFB2010600] and National Natural Science Foundation of China, China [Grant number 52075040].

References

1. Uriarte, L., et al.: Machine tools for large parts. CIRP Ann-Manuf Technol **62**(2), 731–750 (2013). https://doi.org/10.1016/j.cirp.2013.05.009

2. Russell, D.: High-accuracy robotic drilling/milling of 737 inboard flaps. SAE Int. J. Aerospace **4**(2), 1373–1379 (2011)
3. Cen, L.J., Melkote, S.N.: Effect of robot dynamics on the machining forces in robotic milling. In: Wang, L., Fratini, L., Shih, A.J. (eds.) 45th SME North American Manufacturing Research Conference, vol. 10. Procedia Manufacturing. Elsevier Science Bv, Amsterdam, pp. 486–496 (2017). https://doi.org/10.1016/j.promfg.2017.07.034
4. Cen, L.J., Melkote, S.N.: CCT-based mode coupling chatter avoidance in robotic milling. J. Manuf. Process. **29**, 50–61 (2017). https://doi.org/10.1016/j.jmapro.2017.06.010
5. Cen, L.J., Melkote, S.N., Castle, J., Appelman, H.: A wireless force-sensing and model-based approach for enhancement of machining accuracy in robotic milling. IEEE-ASME Trans. Mechatron. **21**(5), 2227–2235 (2016). https://doi.org/10.1109/tmech.2016.2567319
6. Cen, L.J., Melkote, S.N., Castle, J., Appelman, H.: A method for mode coupling chatter detection and suppression in robotic milling. J. Manuf. Sci. Eng.-Trans. ASME **140**(8), 9 (2018). https://doi.org/10.1115/1.4040161
7. Liang, J., Bi, S.S.: Design and experimental study of an end effector for robotic drilling. Int. J. Adv. Manuf. Technol. **50**(1–4), 399–407 (2010). https://doi.org/10.1007/s00170-009-2494-9
8. Slamani, M., Gauthier, S., Chatelain, J.F.: Analysis of trajectory deviation during high speed robotic trimming of carbon-fiber reinforced polymers. Robot. Comput. Integrated Manuf. **30**(5), 546–555 (2014). https://doi.org/10.1016/j.rcim.2014.03.007
9. Slamani, M., Gauthier, S., Chatelain, J.F.: A study of the combined effects of machining parameters on cutting force components during high speed robotic trimming of CFRPs. Measurement **59**, 268–283 (2015). https://doi.org/10.1016/j.measurement.2014.09.052

Development of Assembly Process Information Management System Using Model-Based Definition Technology

Pengfei Zeng[1,2,3](✉), Hao Wang[1,2,3], Chunjing Shi[1,2,3], Weiping Shao[1,2,3], and Yongping Hao[2,3]

[1] School of Mechanical Engineering, Shenyang Ligong University, Shenyang 110159, China
[2] Key Laboratory of Advanced Manufacturing Technology and Equipment of Liaoning Province, Shenyang Ligong University, Shenyang 110159, China
[3] R&D Center of CAD/CAM Technology, Shenyang Ligong University, Shenyang 110159, China

Abstract. Aim at the actual requirements and product assembly characteristics from a certain manufacturing enterprise, an assembly process information management system is designed and developed by the model-based definition (MBD) technology. Product assembly process is analyzed, and MBD information model of assembly process is constructed. The UG Open API and Visual C++ are chosen as the base platform and programming language environment for the system development. The design work of the system is completed, and recognition and extraction method of MBD information of assembly process is also established. Concept model design of database is implemented, and data storage tables are built. Using the Microsoft SQL Server as the database management system, the integration with a workshop information management system is realized on the basis of completing the system implementation. The quality of process design of product assembly and the efficiency of workshop assembly production are effectively improved.

Keywords: Model-based definition · Assembly process · Information recognition · Management system development · System integration

1 Introductions

With the increasing of market competition and the diversification of customer demands, the competition of enterprise product development is becoming increasingly fierce. Design and manufacturing activities are the most important link of the product life cycle [1–3]. The manufacturing process and assembly process are one of the key factors affecting product costs [4, 5]. The enterprise pays attention to the optimization and improvement of product assembly design and assembly process to improve product quality and competitiveness all the time [6, 7]. Assembly automation is the goal of the enterprise. Three dimension assembly and digital assembly based on MBD will

X.-J. Liu et al. (Eds.): ICIRA 2021, LNAI 13014, pp. 648–658, 2021.
https://doi.org/10.1007/978-3-030-89098-8_61

significantly improve the quality and efficiency of product assembly production [8–11]. Assembly process of MBD and information system integration is the basis for full three-dimensional digital assembly. It is also hot issues related to technology in this field.

At present, rich results have been achieved in the field of 3D assembly and process planning. Reference [12] built an MBD-based platform of modeling system of 3D assembly technology. Reference [13] combined product process information with 3D solid modeling to establish MBD model of vane pump assembly design for realizing the integration UG and workshop assembly management system. Reference [14] researched the application of digital assembly technology based on MBD in spacecraft field to transfer process design result to workshop assembly site in the form of 3D assembly instruction. Reference [15] proposed a parallel change process of MBD aircraft assembly tooling development to simulate change process so as to shorten duration and save resource consumption.

Aiming at the actual requirements of assembly process design and integration from a manufacturing enterprise, based on UG 3D software and secondary development environment, an assembly process information management system is established using MBD. Using 3D design software UG and its Open API technology, the information management system is designed and developed including some key characteristics, such as assembly process information recognition, assembly component constraint relationship and assembly MBD information acquisition et al. By the automatically identifying and standard extracting assembly process information, the storage of structured assembly process information and attributes and the information oriented to assembly workshop management are implemented, which can effectively improve the assembly process quality and assembly production efficiency.

2 Requirement Analysis of System Development

2.1 Description of Product Assembly Technological Process

Product assembly process design begins with product design modeling and always provides process specifications and technical preparation for assembly activities of product life cycle. By investigating the needs of enterprise, we can learn about the activity process of enterprise product assembly process design and the required information set of assembly process, as shown in Fig. 1. According to actual production needs and technical conditions, enterprise divides the product structure and its configuration mode into assembly unit, so as to express assembly sequence through the form of assembly tree. Then assembly baseline and positioning information are determined, and the best assembly planning and process organization schema is select. At the same time, the assembly process information and process information are clearly defined, including positioning, clamping, connection of between parts, procedure inspection, etc., the required tools, process equipment, tooling resources are determined. Finally, the material requirements and supply is confirmed, process data models and assembly process instructions are issued, the layout of work sites and the facilities of assembly station are improved.

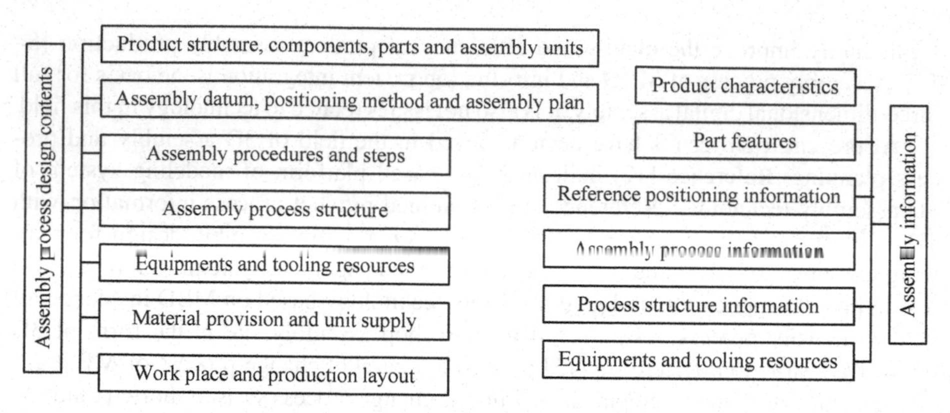

Fig. 1. Product assembly process design information relationship

2.2 MBD Information Model of Assembly Process

Assembly process information is an important part in the product manufacturing process, is the core element of assembly process planning design. It plays the role of media and bridge from product modeling, process design, to workshop manufacturing, information management and production control, and determines final product quality directly. The MBD model issued by design department is a three-dimensional geometric model parts required and process information set for product assembly process, explains intermediate manufacturing process so as to directly guide production practice. Considering the process data and information involved in the part processing and assembly process, an MBD information model for assembly process is established, as shown in Fig. 2.

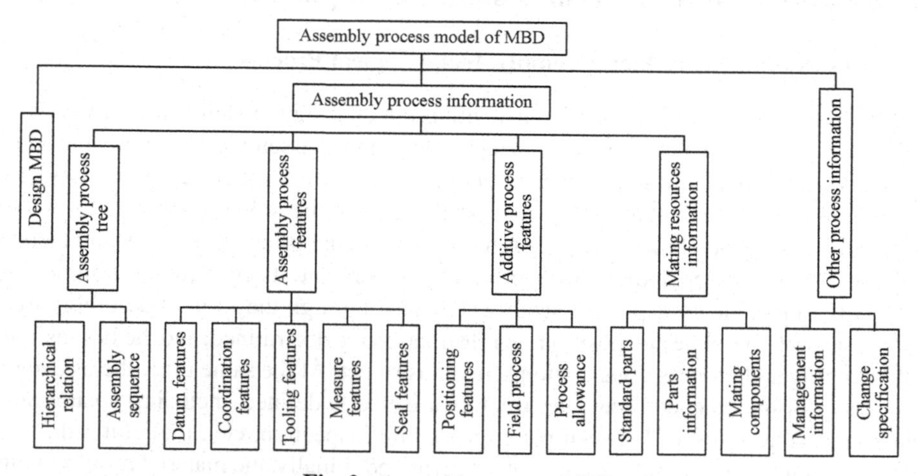

Fig. 2. Assembly process MBD

2.3 UG Open API and Secondary Development Technology

The UG is an interactive computer-aided design, analysis and manufacturing software developed by Siemens, which is widely used in aerospace, machinery, automotive, electronics and shipbuilding industries. In addition, UG software provides users with a variety of development tools, interfaces and solutions. The open design of UG can support a variety of programming languages, such as C++, Java, C# and .Net. User can choose the familiar programming language according to the specific application requirements, and realize the rapid development of functional modules meet their own needs through the secondary development tool set UG Open provided by UG.

Development tools related to UG Open include classic application programming interface, generic application programming interface, logging, knowledge-driven automation, and other UG toolkits. The classic application programming interface is used in earlier versions, including UG Open C, UG Open C++, and UG Open GRIP. The UG Open C++ is an object-oriented application interface that takes full advantage of object-oriented features and is inherited, encapsulated, and polymorphic. In the Visual C++ 6.0 integrated development environment, using the MFC application wizard, an application framework is created based on the MBD information structure. Through menu invocation, direct activation, callback function and user exit, the link interactive application with the UG software platform are complete, and the system function modules are developed.

3 System Design and Information Extraction Key Technology

3.1 UG NX Function Menu Design

The user interface of UG software provides several types of menus. User can activate the corresponding function through select the menu item. According to the MBD information model of the assembly process and combing with the actual enterprise needs, the secondary development is implemented in the UG environment. Through adding function menus, the MBD function modules are integrated into the UG software. UG supports dynamic link library, we can use ActiveX to set related controls in the DLL engineering files created by Visual C++ programs, and then load the UG Open API function set into the application under the controls. Through the menu creation and environment variable configuration, the DLL engineering file is automatically loaded when the UG software is started, and the corresponding secondary development function under the menu is executed.

3.2 Function Design of MBD Information Management

Through the above analysis and combined with the actual characteristics of enterprise product assembly, the assembly process information management system based on MBD is designed. The functional structure of the MBD system is been shown in Fig. 3. System design and the setting of function modules include assembly model retrieval, product assembly structure tree, MBD model information extraction, assembly parts property

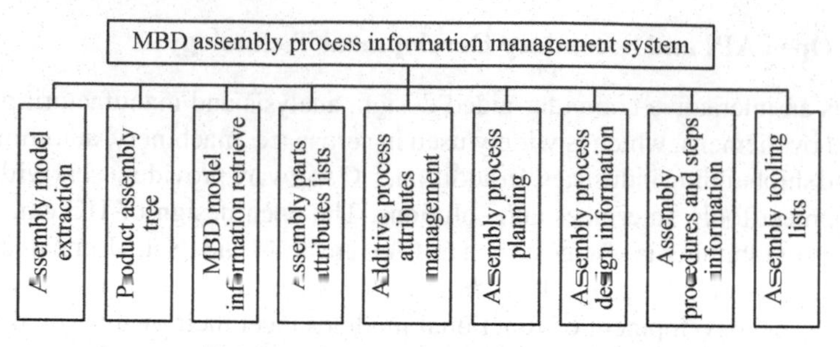

Fig. 3. MBD system function structure

table, subsidiary process property management, assembly process specification, assembly process design information, assembly procedure step information and assembly tooling list.

The identification information of the product, assembly and assembly part compiled by the assembly process is directly extracted from the MBD design model. Using the relationship of assembly units between the upper and lower layers (parent item and child item), the assembly of each layer is identified by the assembly serial number according to the assembly sequence, and the corresponding components and parts information is identified and supplemented within the assembly unit, so as to obtain three-dimensional model of each part and attached information under the assembly unit. The process information retrieved and defined by the MBD model information can be organized into an assembly process specification, and the work content, technical requirement, and operation method of the corresponding assembly process are formed. The generation of digital procedure card need invocate procedure step information and identify equipment, tools, tooling and auxiliary materials information used in each step. The assembly tooling list collects some information such as tools, toolings, molds, gauges, and other auxiliary tools used in the assembly process.

3.3 Information Traverse of MBD Assembly Tree

Assembly usually consists of many parts and components. There are also components and parts underneath the components. In order to display data items with a certain hierarchical structure, and to reflect the assembly affiliation between components and assembly constraints. An assembly tree function module is created in the MBD system, and the corresponding assembly information extraction method is designed. The node information traversal process for the product assembly tree is shown in Fig. 4. Firstly, we need get the event identifier of the assembly tree root node. Secondly, all the child nodes are gotten under the root node of the assembly tree. Third, we repeat the previous step for each child node until all parts of child nodes are obtained. At the same time, corresponding measures are taken to extract the assembly information contained in the MBD model of each part node.

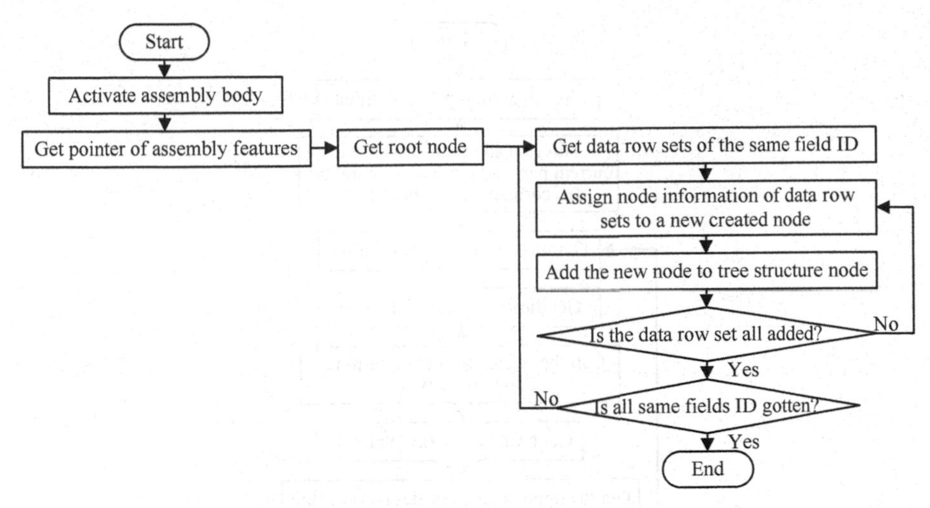

Fig. 4. MBD assembly model traverse flow

3.4 MBD Model Information Extraction Method

Using UF objects provided by UG/Open API function set, some identification information can been identified and obtained, such as part dimension tolerance and geometric tolerance. The extraction process of part dimension information is shown in Fig. 5. Firstly, the related functions are used to query the current display part identification, traverse obtained all parts identification, and name and assembly feature, and get the quantity of identification features. Then, the correlation function is called to obtain the dimension symbol name, determine the dimension type, and get the basic dimension value and the upper and lower deviation values. The extracted information of MBD model includes diameter, angle, radius, inclination angle, cylinder diameter, hole dimension, arc length, thickness, vertical length, and horizontal length, etc.

In addition, for geometric tolerances, we need extract tolerance types, tolerance values, and baseline information. Firstly, the current part ID is queried, the part tolerance object and the number of tolerances are obtained by traversal method. Then, the tolerance type and the tolerance value are obtained by the corresponding function. Finally, the geometric tolerance information of all parts is obtained in assembly. When the geometrical tolerance of the MBD model is retrieved and extracted, the labeled PMI information is identified and extracted by traversing the components and parts identification. The corresponding dimensional information and tolerance information are both automatically extracted through the developed application programs, and are automatically stored in the corresponding data table of database so as to provide information and data support for subsequent assembly process formulation.

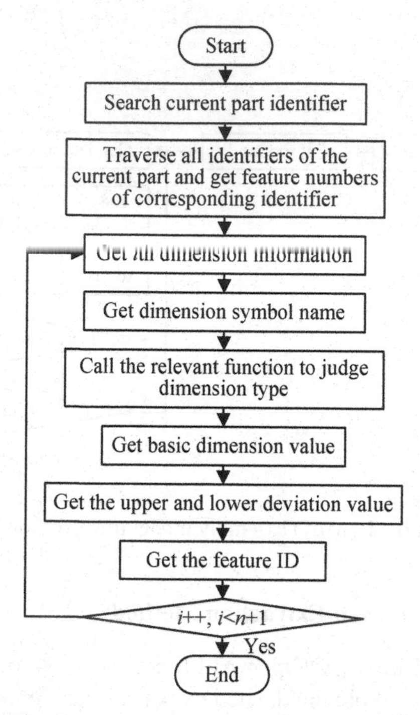

Fig. 5. Parts dimension extraction process

4 System Development and Implementation

4.1 Database Storage Design

Combing with the conceptual data modeling method of database design, the E-R model of the assembly process management process is constructed, as shown in Fig. 6. The E-R model is used to determine the relationship between entities in the assembly process integration management, which provides the basis for database management and physical storage design of information. In the process of assembly process management, the corresponding assembly process line is formulated according to user requirements, production plan, basic parts information and assembly design model. The process line creates related process file, the process file corresponds to the process information, and the personnel information corresponds to the personnel file. The tooling information corresponds to the tooling specification. The device information corresponds to device specification. These

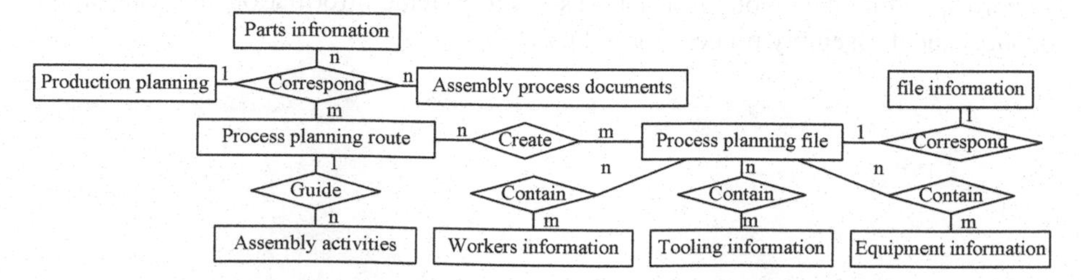

Fig. 6. Assembly process information management E-R model

are all one-to-one relationships. The process file corresponds to personnel information, tooling information and equipment information. This is a one-to-many relationship.

4.2 Data Relationship Design of Process Information Management

The SQL Server is used as the database management platform of system development. Combined the E-R concept relationship mode and the related database design theory, related database tables are established to improve the management of assembly process MBD model information, as shown in Table 1. These database tables include product basic information table, product assembly process planning information table, product assembly parts detail information table, etc. In the database, the base table is the product basic information table. The structuralization, standardization and ordering management of the assembly process information are realized through the connection between other tables and the product basic information table.

Table 1. MBD assembly system database tables.

No.	Table name	Primary key	Description
1	Parts inspection records	Inspection record No	Parts and components verify and inspect information
2	Assembled parts information	Assembled part ID	Assembled parts basic information
3	Product basic information	Product ID	Assembled product basic information
4	Assembly process plan mating information	Plan ID	Related plan mating information in assembly process
5	Assembly process scheduling mating information	Scheduling ID	Related scheduling mating information in assembly process
6	Process type information	Process type No	Manufacturing process corresponding type
7	Assembly process planning information	Process planning ID	Product assembly process related information
8	Product assembly lists	Product assembly No	Parts and components lists from assembled product
9	Assembly procedure information	Procedure No	Product assembly procedure contents and related information
10	Tooling information	Tooling ID	Tooling related information used in assembly process

4.3 System Implementation

According to the actual needs of the enterprise production, under the UG NX 8.0 software system environment, the MBD information system is implemented through the UG Open

API, using Visual C++ 6.0 as programming language and the MFC class library provided by the operating system. The Microsoft SQL Server 2008 is used as a back-end database management system to complete storage and manage of information and data. Through the control interface of the ADO (ActiveX Data Object), the manipulation of data is completed. A use case interface of the assembly process MBD information management system is shown in Fig. 7.

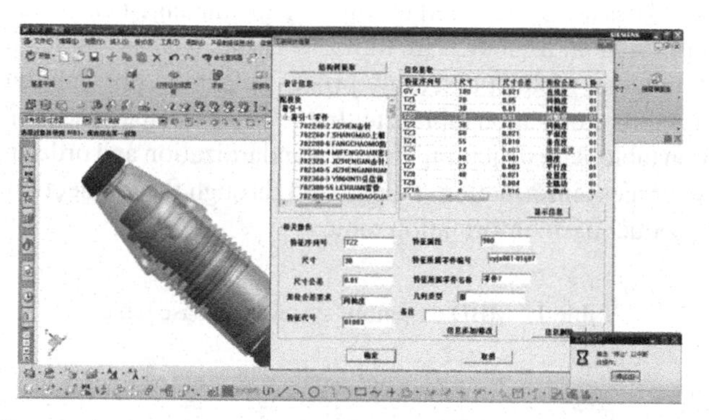

Fig. 7. MBD assembly process information management system interface

4.4 Cross System Information Integration

When the MBD assembly process information system is developed, the integration with a workshop information management system previously developed is realized [16]. The corresponding data and information are obtained by the assembly MBD system, and related assembly components and parts information are identified and assembly constraints relationships are stored in the database. These provide basis data for the production preparation of assembly workshop and the formation of assembly planning. The guidance and model data are provided for the assembly workshop operation through the MBD assembly information model. The data interaction and integration process between two different information systems is shown in Fig. 8.

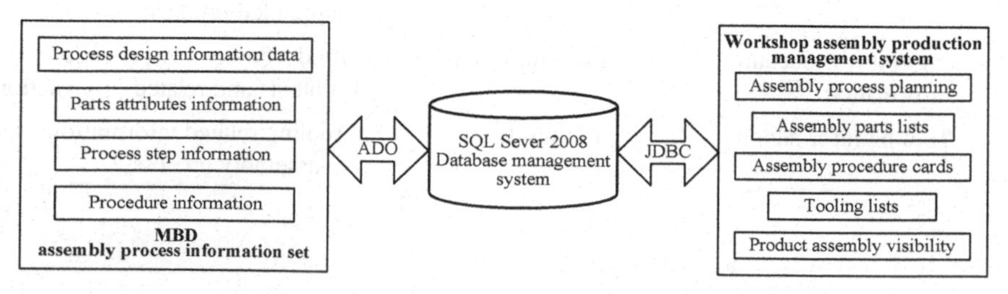

Fig. 8. System integration process data flow

5 Conclusions

The development of assembly process information system is completed by the MBD technology. Information set of design process of product assembly process is described. The MBD information model of assembly process is established. The UG Open API and secondary development technology are chosen as the environment of implementing application programs. The function menus are designed within UG NX software environment. The function design of MBD information management system for assembly process is completed. Information traversal method of assembly structure tree of MBD and related information extraction method of MBD model are both realized. The database concept design and setup of data tables are achieved. Using Visual C++ as programming language and SQL Server 2008 as database management system, the implementation of system is completed. Through the ADO control interface to achieve the application program's operation on the underlying data. Finally, the integration of the MBD system with a workshop information management system is implemented. The MBD model information, features and constraint relationships extracted and identified are used to guide production operations of assembly workshop and provide model data. The system effectively improves the level of enterprise assembly design and operation production, and improves the digital management level and core competitiveness of the enterprise.

Acknowledgements. This work was supported by the State Administration of Science, Technology and Industry for National Defense under Grant No. JSZL2020208A001, and the project is sponsored by "Liaoning BaiQianWan Talents Program".

References

1. Xu, Z.-J., Wang, P., Wang, Q.-H., Li, J.-R.: Integrating part modeling and assembly modeling from the perspective of process. J. Intell. Manuf. **30**(2), 855–878 (2016). https://doi.org/10.1007/s10845-016-1288-9
2. Shi, Y., Qiao, L.: Integration of assembly process and product design information based on three-dimensional lightweight model. Manuf. Autom. **36**(7), 10–15 (2014)
3. Huang, B., Zhang, S., Huang, R., Li, X., Zhang, Y.: An effective retrieval approach of 3D CAD models for macro process reuse. Int. J. Adv. Manuf. Technol. **102**(5–8), 1067–1089 (2018). https://doi.org/10.1007/s00170-018-2968-8
4. Bahubalendruni, M., Deepak, B., Biswal, B.: An advanced immune based strategy to obtain an optimal feasible assembly sequence. Assem. Autom. **36**(2), 127–137 (2016)
5. Hadj, R., Belhadj, I., Trigui, M., Aifaoui, N.: Assembly sequences plan generation using features simplification. Adv. Eng. Softw. **119**, 1–11 (2018)
6. Ruan, S., Liu, J., Tang, C., Zhuang, C., Zhou, Z.: Multi-dimensional assembly process data management system for complex product. Comput. Integr. Manuf. Syst. **21**(3), 656–668 (2015)
7. Ahmad, M., Ferrer, B., Ahmad, B., Vera, D., Lastra, J., Harrison, R.: Knowledge-based PPR modelling for assembly automation. CIRP J. Manuf. Sci. Technol. **21**, 33–46 (2018)
8. Fan, Y.: Model based definition technology and its practices. Aeronaut. Manuf. Technol. **6**, 42–47 (2012)
9. Alemanni, M., Destefanis, F., Vezzetti, E.: Model-based definition design in the product lifecycle management scenario. Int. J. Adv. Manuf. Technol. **52**(1–4), 1–14 (2011)

10. Ruemler, S., Zimmerman, K., Hartman, N., Hedberg, T., Feeny, A.: Promoting model-based definition to establish a complete product definition. J. Manuf. Sci. Eng. **139**(5), 051008 (2017)
11. Fang, Z., Li, Z., Arokiam, A., Gorman, T.: Closed loop PMI driven dimensional quality lifecycle management approach for smart manufacturing system. Procedia CIRP **56**, 614–619 (2016)
12. Ye, S., Tang, J., Bao, J., Huang, W.: MBD-based three dimensional assembly process modeling and application. J. Drainage Irrigation Mach. Eng. **33**(2), 179–184 (2015)
13. Guo, M., Hao, Y., Zeng, P., Xu, S.: MBD-based vane pump modeling and process information for integration studies. Group Technol. Prod. Modern. **34**(2), 38–42 (2017)
14. Song, X., Lu, Y., Liu, Z., Yang, C., Wang, P., Xue, F.: Research on application of digital assembly technology based on MBD in spacecraft field. IOP Conf. Ser. Mater. Sci. Eng. **408**, 0122026 (2018)
15. Yin, L., Xing, K.: MBD-based parallel change management for aircraft assembly tooling. Adv. Eng. Res. **162**, 450–453 (2018)
16. Zeng, P., Ren, K., Zhang, X., Shi, C., Hao, Y.: Development of integrated workshop production management system for lean production. J. Shenyang Ligong Univ. **37**(1), 51–57 (2018)

Dimensional Optimization of the 3RRR Parallel Manipulator with Specified Workspace

Hua Shao[1,2], Canyu Shi[1,2](✉), Jun Deng[1,2], and Hua Zhang[1,2]

[1] Key Laboratory of Metallurgical Equipment and Control Technology (Wuhan University of Science and Technology), Ministry of Education, Wuhan 430081, People's Republic of China

[2] Hubei Key Laboratory of Mechanical Transmission and Manufacturing Engineering, Wuhan, China

Abstract. 3RRR planar parallel manipulator is an important module for the high accuracy mobile manufacturing of large planar components, and the dimensional parameters optimized design with a specified workspace is also important for the parallel manipulator engineering design. This paper proposed a new global performance index - task workspace transmission index (TWTI) for solving the design problem of parallel manipulators with specified workspace. The inverse kinematic model of the 3RRR manipulator is presented, and the optimum dimensional parameters are achieved. The result shows that good kinematic performance with specified task workspace can be obtained. This index is also applicable for the dimensional parameters optimization of other parallel manipulator with specified workspace.

Keywords: Mobile manufacturing · Parallel manipulator · Dimensional optimization · Task workspace transmission index

1 Introduction

Mobile manufacturing is an effective method for large complex components machining [1–3], such as wind turbine blade, train carriage, and aircraft wing [4]. Usually the mobile manufacturing equipment is composed with a mobile vehicle and a machining manipulator, the manipulator could be a 6-DOF serial industry robot [5], sometimes maybe a 6-DOF Stewart-Gough parallel manipulator [6], or a novel 5-DOF hybrid manipulator, but they are both difficult to get high accuracy and large workspace. A new type of mobile manufacturing equipment, which consists of a mobile vehicle, a 6-DOF serial industry robot and a compact 5-DOF parallel manipulator is presented for wind turbine blades polishing [7].

Due to the high positioning accuracy, compact structure and load capacity, parallel manipulators have great potential in mobile manufacturing, and the parallel manipulators should designed with good kinematic performance, short dimensional parameter to reduce the load of the serial or mobile robot, there are many research on the performance evaluation and optimized design of parallel manipulators. Many kinematic performance indices such as global conditioning index [8], global velocity index [9] and

X.-J. Liu et al. (Eds.): ICIRA 2021, LNAI 13014, pp. 659–668, 2021.
https://doi.org/10.1007/978-3-030-89098-8_62

global force/motion transmission index [10] have been proposed and applied widely in many parallel manipulators. These global indices are based on the average local performance value among the reachable workspace of the parallel manipulators, but usually the task workspace is specified in the design process, and the task workspace must be less than the reachable workspace.

3RRR is an important planar manipulator used for planar translation and rotation. Many research regarding kinematic optimized design [11], reactionless design [12], dynamic analysis [13] and actuation mode [14] have been obtained, but few work about parameters design within specified workspace is taken out. Parameters optimization within specified workspace is an important subject for parallel manipulator engineering design.

This paper presents a dimensional parameters optimal index of a 3RRR manipulator with specified workspace, when the manipulator is used for large area mobile manufacturing of aircraft wings. First the description of the 3RRR parallel manipulator is made, then the inverse kinematic model is presented, respectively. A new task workspace transmission index, is proposed to evaluate the performance among the specified task workspace. Then the optimum dimensional parameters are achieved based on the non-dimensional region and limited maximum radius dimension. Finally the designer can obtain the respected kinematic performance with specified task workspace.

2 Inverse Kinematics of the 3-RRR Parallel Manipulator

2.1 Description of the 3-RRR Parallel Manipulator

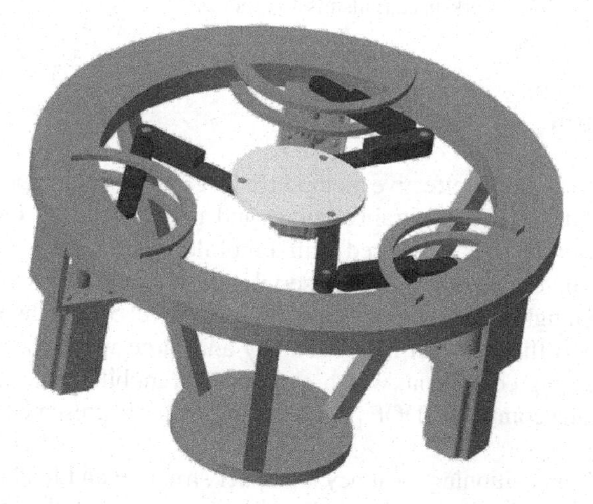

Fig. 1. CAD model of the 3-RRR planar parallel manipulator.

The 3RRR planar parallel manipulator is shown in Fig. 1 and its kinematic scheme is illustrated in Fig. 2. It is comprised of end-effector $C_1C_2C_3$, base $A_1A_2A_3$ and three identical RRR legs. The revolute joints A_1, A_2, and A_3 are active, the end-effector can achieve two DOFs of translation and one DOF of rotation. The radius of the base $A_1A_2A_3$ is denoted as r_1 and r_2 is the radius of the $C_1C_2C_3$, l_1 and l_2 are the length of links A_iB_i and B_iC_i, respectively. The global coordinate system $O - XY$ and the local coordinate system $O' - X'Y'$ are defined as shown in Fig. 2. Therefore, the position of the end-effector $C_1C_2C_3$ can be denoted as $\left[x_{O'}\ y_{O'}\ 0 \right]^T$, and the rotation angle θ is the angle between OX and $O'X'$.

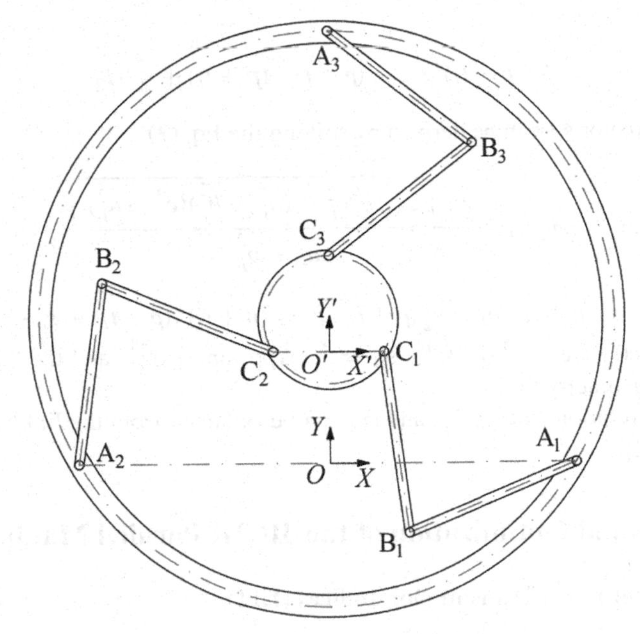

Fig. 2. Kinematic scheme of the 3RRR manipulator.

2.2 Inverse Kinematic Model

From the kinematic scheme, we can get that the vector equations of each leg.

$$\overrightarrow{OA_i} + \overrightarrow{A_iB_i} + \overrightarrow{B_iC_i} = \overrightarrow{OO'} + \boldsymbol{E}\overrightarrow{{}_{'}O'C_i} \quad (i = 1, 2, 3) \tag{1}$$

where $\overrightarrow{OA_i}$, $\overrightarrow{A_iB_i}$, $\overrightarrow{B_iC_i}$ and $\overrightarrow{OO'}$ are the vectors belong to coordinate $O - XY$, and, $\overrightarrow{{}_{'}O'C_i}$ is the vector belongs to coordinate $O' - X'Y'$.

These vectors could be expressed as $\overrightarrow{OA_1} = [r_1 \cos 30^\circ \quad 0 \quad 0]^T$, $\overrightarrow{OA_2} = [-r_1 \cos 30^\circ \quad 0 \quad 0]^T$, $\overrightarrow{OA_3} = [0 \quad r_1 + r_1 \sin 30^\circ \quad 0]^T$, $\overrightarrow{{}_{'}O'C_1} = [r_2 \cos 30^\circ \quad 0 \quad 0]^T$, $\overrightarrow{{}_{'}O'C_2} = [-r_2 \cos 30^\circ \quad 0 \quad 0]^T$, $\overrightarrow{{}_{'}O'C_3} = [0 \quad r_2 + r_2 \sin 30^\circ \quad 0]^T$, and $\boldsymbol{E}$ is the rotational matrix between coordinate $O - XY$ and $O' - X'Y'$:

$$\boldsymbol{E} = \begin{bmatrix} \cos\theta & -\sin\theta & 0 \\ \sin\theta & \cos\theta & 0 \\ 0 & 0 & 0 \end{bmatrix}.$$

By using unit vectors s_{i1} and s_{i2} to represent the direction of links A_iB_i and B_iC_i, respectively, the following equation can be derived from Eq. (1)

$$l_1 s_{i1} + l_2 s_{i2} = p + t_i \tag{2}$$

where $t_i = E,\overrightarrow{O'\ C_i} - \overrightarrow{OA_i} p = \left[x_{O'}\ y_{O'}\ 0\right]^T$.

For the reason of that s_{i2} is a unit vector, the following equation can be derived from Eq. (2).

$$2l_1(p + t_i)^T s_{i1} = (p + t_i)^T(p + t_i) + l_1^2 - l_2^2 \tag{3}$$

Then the vector s_{i1} can be derived by solving the Eq. (3)

$$\sin\varphi_{i1} = \frac{b_i c_i \pm \sqrt{b_i^2 c_i^2 - (a_i^2 + b_i^2)(c_i^2 - a_i^2)}}{a_i^2 + b_i^2} \tag{4}$$

where $a_i = e_1^T(p + t_i)$, $b_i = e_2^T(p + t_i)$, $c_i = (p + t_i)^T(p + t_i) + l_1^2 - l_2^2$, $e_1 = \left[1\,0\,0\right]^T$, $e_2 = \left[0\,1\,0\right]^T$, $s_{i1} = \left[\cos\varphi_{i1}\ \sin\varphi_{i1}\ 0\right]^T$, and the "±" should be "+" from the geometry limit.

The input rotational angle φ_{i1} and φ_{i2} can be obtained from the link vector s_{i1} and s_{i2}, respectively.

3 Dimensional Optimization of the 3RRR Parallel Manipulator

3.1 Solution of Local Transmission Indices (LTI)

The LTI is a good index to evaluate the local performance of motion/force transmissibility for the planar 3-RRR parallel manipulator. Chen [11] has presented the input transmission index (ITI) and the output transmission index (OTI) of the 3RRR parallel manipulator LTI.

The ITI of leg i presents the power coefficient between its TWS and its input twist screw [10], it can be derived as

$$\lambda_i = |\sin\gamma_i| \tag{5}$$

where $\gamma_i = \pi - \varphi_{i1} + \varphi_{i2}$.

The OTI of leg i presents the power coefficient between its TWS and its output twist screw, it can be derived as

$$\eta_i = |\sin\alpha_i| \tag{6}$$

where α_i is the small angle between line B_iC_i and line C_iP_i, as shown in Fig. 3.

Therefore, the LTI of the planar 3RRR parallel manipulator can be presented as

$$\psi = min\{\lambda_i\ \eta_i\}(i = 1, 2, 3) \tag{7}$$

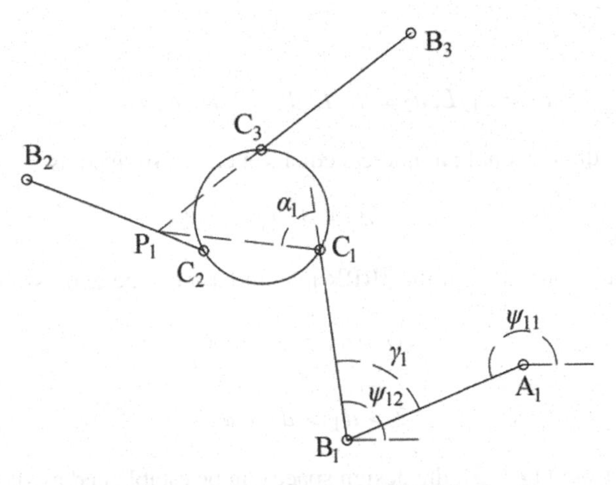

Fig. 3. The solution process of OTI for leg 1

3.2 Task Workspace Design Indices

The global transmission index (GTI) is widely used to evaluate the transmission performance among the global reachable workspace, but usually a limited workspace is specified in the equipment engineering design, so that the performance outside the task workspace is not important for the design, so the GTI will not be suitable at this case. In this paper, a task workspace transmission index (TWTI) is proposed for the performance evaluation of specified workspace.

The TWTI is defined as

$$\Gamma_{TW} = \frac{\int_{TW} \psi dTW}{\int_{TW} dTW} \tag{8}$$

where TW is the area of specified task workspace. The physical meaning of TWTI is the mean of LTI in TW. A larger Γ_{TW} demonstrates better motion/force transmissibility in TW, it will be applied to the dimensional optimal design of the 3RRR manipulator.

As an example the task workspace of the manipulator is defined as $-w_0 < x_{O'} < w_0$, $-w_0 < y_{O'} < w_0$, $-\pi/6 < \theta < \pi/6$, and $w_0 = 100$ mm.

3.3 Dimensional Parameters Design Space

Liu [15] showed an normalization method to get a finite non-dimensional parameters design space, the dimensional parameters r_1, r_2, l_1, l_2 and w_0 will be normalized first, and it is showed in [16] that the areas of reachable workspace of the 3-RRR parallel manipulator can reach the maximum value if $l_1 = l_2$, so in this paper the length of links A_iB_i and B_iC_i are assumed to be equal for convenience.

Let

$$L = (\mathrm{r}_1 + \mathrm{r}_2 + l_1 + w_0)/4 \tag{9}$$

and

$$d_1 = r_1/L, d_2 = r_2/L, d_3 = l_1/L, d_4 = w_0/L \tag{10}$$

Thus, a non-dimensional parameters equation can be derived as

$$d_1 + d_2 + d_3 + d_4 = 4 \tag{11}$$

The geometry constrains of the 3RRR manipulator can be expressed as

$$d_2 + d_3 > d_1 + d_4 \tag{12}$$

$$2 > d_1 > d_2 + d_4 \tag{13}$$

On the bases of (11)–(13), the design space can be established as shown in Fig. 4, it could presented by the parameters d_1, d_2 and d_4.

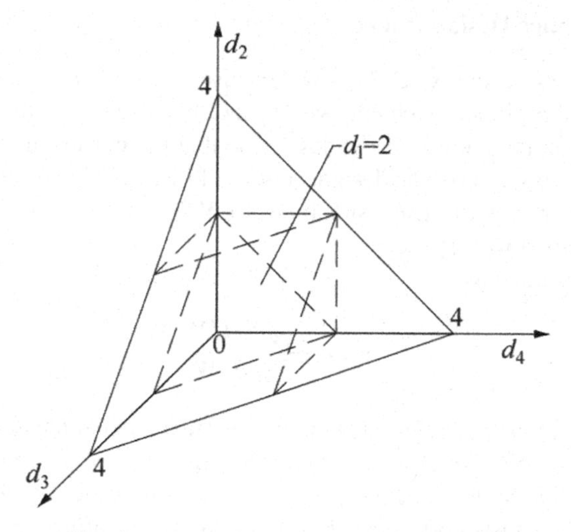

Fig. 4. The design space of the manipulator

3.4 Dimensional Optimization

Based on the task workspace transmission index Γ_{TW}, the optimization of the manipulator geometry parameters could be implemented. The proportionality coefficients d_1, d_2 and d_4 could be used to define the parameter design space.

Figure 5 gives the distribution of the task workspace transmission index Γ in the design space. It shows that Γ_{TW} decreases when d_4 increases, and Γ increases as d_1, d_2 increase. Γ gets its maximum value 0.9216 when

$$d_1 = 1.5067, d_2 = 0.05, d_3 = 2.3933, d_4 = 0.05$$

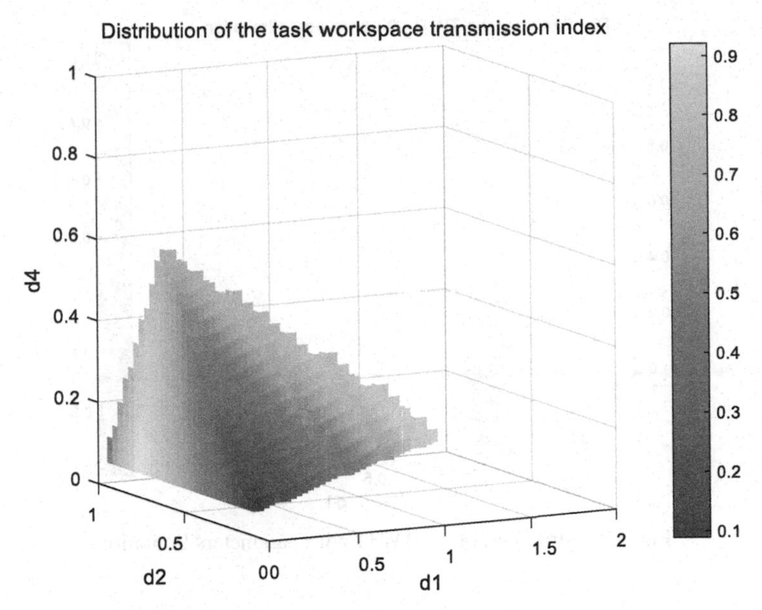

Fig. 5. Distribution of the Task Workspace Transmission Index

Then the dimensional parameters can be obtained as

$$r_1 = 3013\,\text{mm},\ r_2 = 100\,\text{mm},\ l_1 = l_2 = 2393\,\text{mm}$$

This result is too big for the application of mobile manufacturing, so additional parameters limitation will be carried out to reduce the maximum dimensional parameters.

$$d_1 < 6d_4,\ d_2 > d_4$$

Figure 6 gives the distribution of the TWTI Γ in the design space with parameters limitation. It shows that the design space is much smaller than the space without parameters limitation shown in Fig. 5. Γ gets a different maximum value 0.8367 when

$$d_1 = 1.5067,\ d_2 = 0.2717,\ d_3 = 1.95,\ d_4 = 0.2717$$

Then the dimensional parameters can be obtained as

$$r_1 = 554.6\,\text{mm},\ r_2 = 100\,\text{mm},\ l_1 = l_2 = 358.9\,\text{mm}$$

4 LTI Distribution of the Optimized 3RRR Manipulator

The performance of the designed manipulator is evaluated by the distributions of local transmission index, as shown in Fig. 7(a). For the difficulty to find detailed information from this 4D distribution map, the contour map of section $\theta = 30°$, $y = 0.235$m, $x = 0.024$m are presented as Fig. 7(b), (c) and (d), respectively. It shows that the manipulator has good motion/force transmissibility in the task workspace.

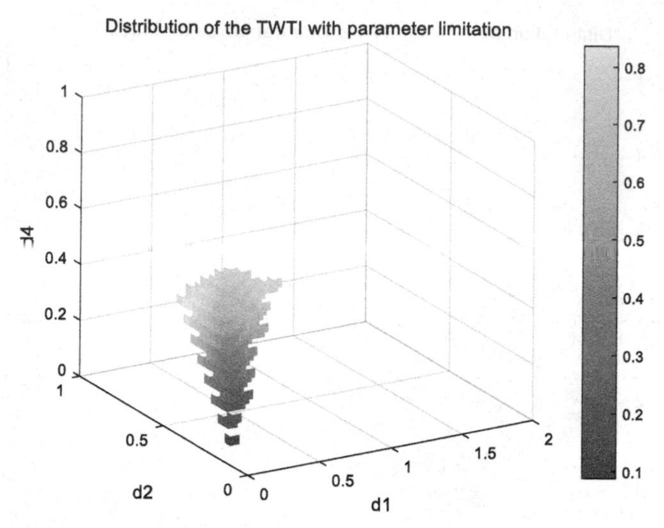

Fig. 6. Distribution of the TWTI with parameters limitation

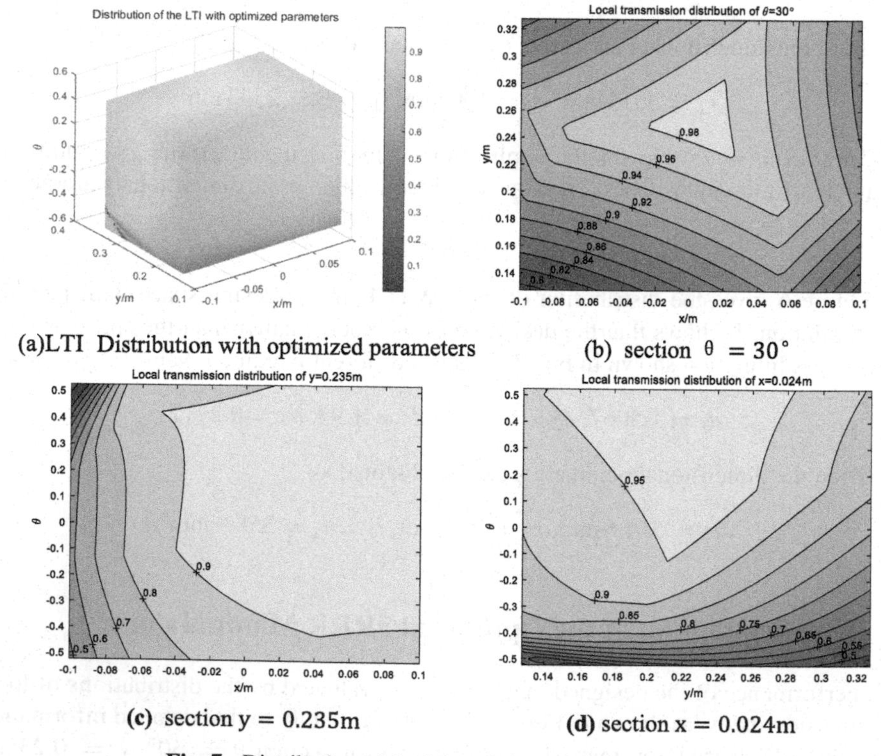

(a)LTI Distribution with optimized parameters (b) section θ = 30°

(c) section y = 0.235m **(d)** section x = 0.024m

Fig. 7. Distribution of the LTI with optimized parameters

5 Conclusions

In this paper, an mobile manufacturing manipulator based on 3RRR parallel mechanism is built, and a task workspace transmission index TWTI is proposed for performance evaluation of the manipulator with specified workspace limitation. The parameter design space is defined by normalizing both the geometry parameters of the manipulator and the task workspace. Based on the performance charts, the TWTI can be represented graphically, then the optimum dimensional parameters are achieved, then the improved dimensional parameters result with manipulator size limitation arc carricd out. Thc designed manipulator would have desirable performance.

Acknowledgment. This project is supported by National Natural Science Foundation of China (Grant No. 52075395).

References

1. Tao, B., Zhao, X., Ding, H.: Mobile-robotic machining for large complex components: a review study. Sci. China Technol. Sci. **62**(8), 1388–1400 (2019). https://doi.org/10.1007/s11431-019-9510-1
2. Xie, Z., Xie, F., Liu, X., et al.: Parameter optimization for the driving system of a 5 degrees-of-freedom parallel machining robot with planar kinematic chains. J. Mech. Robot. **11**(4), 041003 (2019)
3. Xie, F., Liu, X., Wang, J., et al.: Kinematic optimization of a five degrees-of-freedom spatial parallel mechanism with large orientational workspace. J. Mech. Robot. **9**(5), 051005 (2017)
4. Susemihl, H., Moeller, C., Kothe, S., et al.: High accuracy mobile robotic system for machining of large aircraft components. SAE Int. J. Aerosp. **9**(2), 231–238 (2016)
5. Fan, Q., Gong, Z., Tao, B., et al.: Base position optimization of mobile manipulators for machining large. Robot. Comput.-Integr. Manufact. **70**, 102138 (2021)
6. Li, C., Wu, H., Eskelinen, H., et al.: Design and implementation of a mobile parallel robot for assembling and machining the fusion reactor vacuum vessel. Fusion Eng. Des. **161**, 111966 (2020)
7. Chong, Z., Xie, F., Liu, X., et al.: Design of the parallel mechanism for a hybrid mobile robot in wind turbine blades polishing. Robot. Comput.-Integr. Manufact. **61**, 101857 (2020)
8. Huang, T., Li, M., Li, Z., et al.: Optimal kinematic design of 2-DOF parallel manipulators with well-shaped workspace bounded by a specified conditioning index. IEEE Trans. Robot. Autom. **20**(3), 538–543 (2004)
9. Wu, J., Wang, L., You, Z.: A new method for optimum design of parallel manipulator based on kinematics and dynamics. Nonlinear Dyn. **61**(4), 717–727 (2010)
10. Wang, J., Wu, C., Liu, X.: Performance evaluation of parallel manipulators: motion/force transmissibility and its index. Mech. Mach. Theory **45**(10), 1462–1476 (2010)
11. Chen, Y., Liu, X., Chen, X.: Dimension optimization of a planar 3-RRR parallel manipulator considering motion and force transmissibility. In: IEEE International Conference on Mechatronics & Automation, pp. 670–675 (2013)
12. Arakelian, V.H., Smith, M.R.: Design of planar 3-DOF 3-RRR reactionless parallel manipulators. Mechatronics **18**, 601–606 (2008)
13. Zhang, X., Zhang, X., Chen, Z.: Dynamic analysis of a 3-RRR parallel mechanism with multiple clearance joints. Mech. Mach. Theory **78**, 105–115 (2014)

14. Zhang, Z., Wang, L., Shao, Z.: Improving the kinematic performance of a planar 3-RRR parallel manipulator through actuation mode conversion. Mech. Mach. Theory **130**, 86–108 (2018)
15. Liu, X., Wang, J.: A new methodology for optimal kinematic design of parallel mechanisms. Mech. Mach. Theory **42**(9), 1210–1224 (2007)
16. Liu, X., Wang, J., Gao, F.: Performance atlases of the workspace for planar 3-DOF parallel manipulators. Robotica **18**(5), 563–568 (2000)

Positioning Accuracy Compensation Method of Mobile Robot Based on Inverse Distance Weight

Yinghao Zhou(✉), Ke Wen, Lei Mao, Dongliang Ren, and Runmiao Fan

China Academy of Space Technology, Beijing, China
jacesherry@sina.com

Abstract. This paper analyzes the non-geometric factors affecting the positioning accuracy of mobile robots, as well as the easy-to-operate and high-precision requirements of mobile robot positioning accuracy compensation. A non-model calibration method based on inverse distance weight is proposed. The workspace is meshed, the D-H kinematics model of the robot is constructed, and the theoretical calculation method of the grid nodes positions is established. The measured values of the grid nodes are obtained by laser tracker measurement, the grid node error database is established, and the inverse distance weight error estimation of the target point is given. The precision compensation test system is constructed. By dividing the same working space and selecting the 50 mm and 70 mm step spacing, the compensation result of the positioning accuracy of the mobile robot is obtained. After the compensation, the positioning accuracy of the robot is increased by an order of magnitude, and the smaller the spacing, the better the compensation effect, and finally the positioning accuracy of the mobile robot is compensated to 0.11 mm.

Keywords: Grid node · Inverse distance weight · Non-model calibration · Positioning accuracy · Mobile robot

1 Introduction

In recent years, modern manufacturing technology based on robot intelligent flexible manufacturing system has developed rapidly in the world, especially in the heavy and high-end equipment manufacturing industry [1, 2]. A robotic drilling and milling system has been developed by the American Walt aircraft company, which has completed the drilling and milling of the inner flap of Boeing 737 [3]. The robot automatic drilling and riveting power race system, with powerful functions, newly developed by BROETJE company in Germany has been widely used in the aircraft automatic assembly lines of Airbus and Boeing [4]. The mobile milling robot system developed by Fraunhofer Association of Germany includes robot, 840dsl numerical control system, mobile platform, binocular vision measurement system, etc. Zhejiang University has developed

X.-J. Liu et al. (Eds.): ICIRA 2021, LNAI 13014, pp. 669–677, 2021.
https://doi.org/10.1007/978-3-030-89098-8_63

an automatic drilling system integrating laser measurement technology, computer control technology, off-line programming technology and robot technology, which greatly improves the drilling efficiency of aircraft components [5]. Tianjin University has developed a KUKA industrial robot pose measurement and online error compensation system, which is applied to the manufacturing of large spherical workpiece with complex hole system [6, 7]. The positioning accuracy of industrial robots directly affects the manufacturing quality and efficiency of the above systems.

Due to the decline of robot positioning accuracy caused by structural parameters, parts manufacturing, assembly clearance, etc., Ren Yongjie et al. [8] used the distance accuracy between two points in space to measure the absolute position accuracy of the robot, established the robot distance error model, avoided the error caused by coordinate conversion, and reduced the accuracy requirements of the measurement system. Lightcap et al. used Levenberg Marquardt algorithm to identify robot geometric parameters and joint flexibility parameters, and applied it to the kinematic calibration of Mitsubishi pa10-6ce robot [9]. Wang Gang et al. [10] Proposed a method to calibrate the robot base coordinate system accurately by using dual quaternion. Non geometric error is another important factor that affects the pose accuracy of robot, and it is complex and unpredictable in the process of robot movement, which has uncertainty on the robot positioning accuracy. Therefore, it is difficult to accurately describe the mechanism of non-geometric error factors by establishing mathematical model. Based on fuzzy interpolation, Bai et al. Proposed a non-model error calibration method [11], which divided the robot workspace into several three-dimensional mesh units according to the same step size. And by establishing the positioning error database in the orthogonal space of the robot, the positioning error at any position in the robot workspace was obtained by fuzzy interpolation. Through the grid segmentation of robot orthogonal space, Tian Wei et al. [12] of Nanjing University of Aeronautics and Astronautics also proposed the interpolation weight measurement method of error similarity to estimate the spatial interpolation of robot positioning error. Without building a complex kinematic model, the robot pose error calibration method based on non-model parameters can take into account many error factors, and directly link the pose error with the robot joint configuration.

To sum up, for the practical engineering application of mobile robot and the requirements of easy-to-operate and high-precision, through the establishment of robot motion model, the relationship between robot geometry and its motion is described, and then a positioning accuracy compensation method of mobile robot based on inverse distance weight is proposed to achieve the high positioning accuracy of mobile robot in the grid work space.

2 Robot Kinematics Model

Taking the KUKA kr500 robot as the research object, the D-H model (Denavit-Hartenberg) of the robot is established to describe the relationship between the robot geometry and its motion. For a 6-DOF serial robot, the coordinate system $O_iX_iY_iZ_i$ connected to the link i is set as O_i. The position and orientation of the coordinate system O_i relative to O_{i-1} are described by four parameters (as shown in Fig. 1), and meet the following requirements: ① the angle between X_{i-1} axis and X_i axis is joint angle θ_i, positive direction around Z_{i-1} axis; ② the distance between X_{i-1} axis and X_i axis is joint distance d_i, positive direction along Z_{i-1} axis; ③ the distance between Z_{i-1} axis and Z_i axis is connecting rod length a_i, positive direction along X_i axis; ④ the angle between Z_{i-1} axis and Z_i axis is the twist angle of the connecting rod α_i, positive direction around the X_i axis; ⑤ the X_i axis is perpendicular to the Z_{i-1} axis, and the X_i axis intersects with the Z_{i-1} axis.

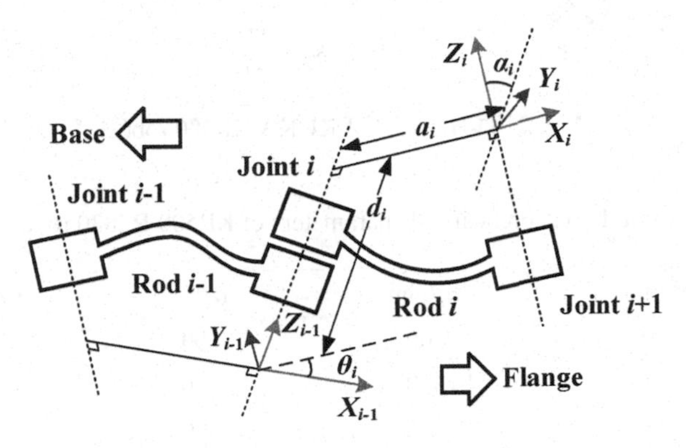

Fig. 1. Schematic diagram of D-H model

The coordinate system O_{i-1} is transferred to the coordinate system O_i through the following coordinate transformation: rotating θ_i around the axis of Z_{i-1}; moving d_i along the axis of Z_{i-1}; moving a_i along the axis of X_i; rotating α_i around the axis of X_i. So the homogeneous transformation matrix described by D-H is T_i, $i = 1–6$.

$$T_i = Rot(z, \theta_i)Trans(0, 0, d_i)Trans(a_i, 0, 0)Rot(x, \alpha_i) = \begin{bmatrix} \cos\theta_i & -\sin\theta_i\cos\theta_i & \sin\theta_i\sin\alpha_i & a_i\cos\theta_i \\ \sin\theta_i & \cos\theta_i\cos\alpha_i & -\cos\theta_i\sin\alpha_i & a_i\sin\theta_i \\ 0 & \sin\alpha_i & \cos\alpha_i & d_i \\ 0 & 0 & 0 & 1 \end{bmatrix} \tag{1}$$

According to the above D-H model parameter definition, the relevant parameters of the KUKA KR500 robot model are extracted in Table 1, and the corresponding D-H model is established, as shown in Fig. 2. And according to the forward kinematics, the coordinate transformation matrix ${}^0T_6 = {}^0T_1{}^1T_2{}^2T_3{}^3T_4{}^4T_5{}^5T_6$ from the end flange coordinate system to the robot base coordinate system is obtained. Using the Link object in MATLAB toolbox, the established robot is described, and link robot as shown in Fig. 3 is created. By modifying the different pose parameters, the correctness of the established D-H model can be verified with theoretical calculation.

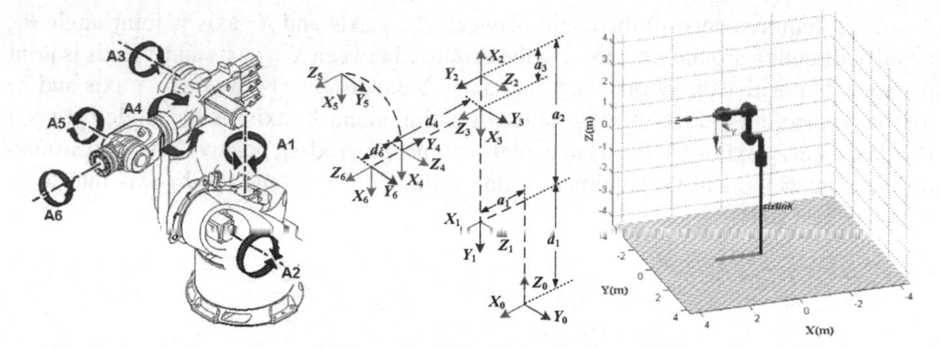

Fig. 2. D-H model of KUKA KR500 robot

Table 1. Theoretical D-H parameters of KR500-R2830 robot

	d_i/mm	a_i/mm	α_i/°	θ_i/°
1	1045	500	−90	0
2	0	1300	0	−90
3	0	55	90	180
4	1025	0	−90	0
5	0	0	90	0
6	290	0	0	0

3 Inverse Distance Weight Based Accuracy Compensation Method

The core idea of the error compensation method is to divide the robot workspace into several three-dimensional grid cells according to the same step size, and then use the laser tracker to measure the positioning error of the robot moving to each grid node, so as to establish the positioning error database in the orthogonal space of the robot. Finally, the positioning error at any position in the robot workspace is obtained by the

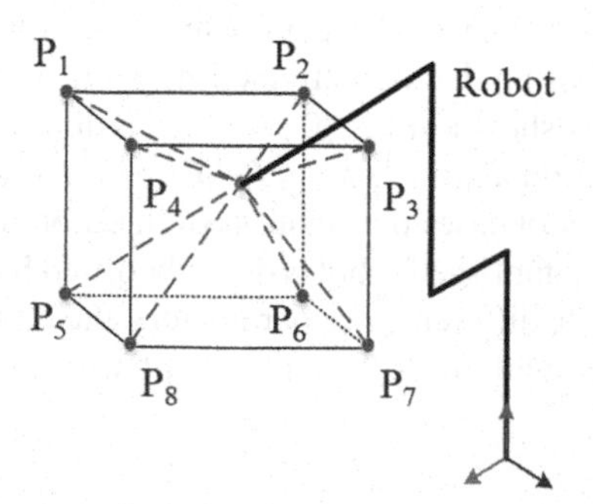

Fig. 3. Cube grid element of robot

inverse distance weight method. The workspace of the robot is segmented with the same step size, and the smallest cell of the mesh segmentation is cube structure, as shown in Fig. 4.

Each node of the mesh corresponds to a group of robot joint configurations. Due to the high precision of repeated positioning of the robot, if the robot is driven by this group of joint angles, the error between the actual end position and the theoretical position will be basically fixed, that is, each node of the mesh shown in Fig. 4 ($\boldsymbol{P}_i(x_i, y_i, z_i)$, $i = 1\ldots8$) corresponds to a group of robot positioning errors. The position of each grid node relative to the robot base coordinate system can be obtained through the robot kincmatics model. When the robot moves to each grid point in turn, the real end position of the robot at each grid point is measured by the laser tracker, and the positioning error of the robot in each direction at each grid point can be calculated.

$$\begin{aligned} e_{xi} &= x_i - x_i' \\ e_{yi} &= y_i - y_i' \\ e_{zi} &= z_i - z_i' \end{aligned} \tag{2}$$

(x_i, y_i, z_i) is the theoretical kinematics end position of the robot, (x_i', y_i', z_i') is the measured end position, (e_{xi}, e_{yi}, e_{zi}) is the end position error at the grid point.

When the robot operates, its end position must fall into a grid cell in the joint space as shown in Fig. 4. There is a correlation between the end positioning error and the position error at each node of the grid cell. Therefore, the inverse distance weight space interpolation algorithm can be used to estimate the end positioning error. Based on the distance between the robot target pose and its grid nodes, the inverse distance weight interpolation method models the positioning error. The smaller the distance between the target point and the grid point is, the more the error correlation between them is, and vice versa. The positioning error at the target point is the distance weighted average of the errors measured at each grid point. The calculation formula is as follows:

$$\begin{aligned} e_x &= \sum_{i=1}^{8} e_{xi} q_i \\ e_y &= \sum_{i=1}^{8} e_{yi} q_i \\ e_z &= \sum_{i=1}^{8} e_{zi} q_i \end{aligned} \tag{3}$$

q_i is the inverse distance weight. Since joint mesh is divided according to step distance, there are:

$$d_i = \sqrt{(x - x_i)^2 + (y - y_i)^2 + (z - z_i)^2} \tag{4}$$

$$q_i = \frac{1}{d_i} \Big/ \sum_{i=1}^{8} \frac{1}{d_i} \tag{5}$$

$\sum_{i=1}^{8} q_i = 1$. Therefore, the coordinates of the target point after correction are:

$$\begin{aligned} x &= x_0 + e_x \\ y &= y_0 + e_y \\ z &= z_0 + e_z \end{aligned} \tag{6}$$

4 Experiment and Analysis

The experiment system of the accuracy compensation is shown in Fig. 5, which mainly includes KUKA KR500 robot, API radian laser tracker and spatial analyzer data acquisition and processing software. The relative position between laser tracker and machine is fixed, and the reflecting target ball seat is fixed on the end flange of robot. The laser tracker coordinate system and the robot base coordinate system are aligned to determine the precise matrix conversion relationship between them.

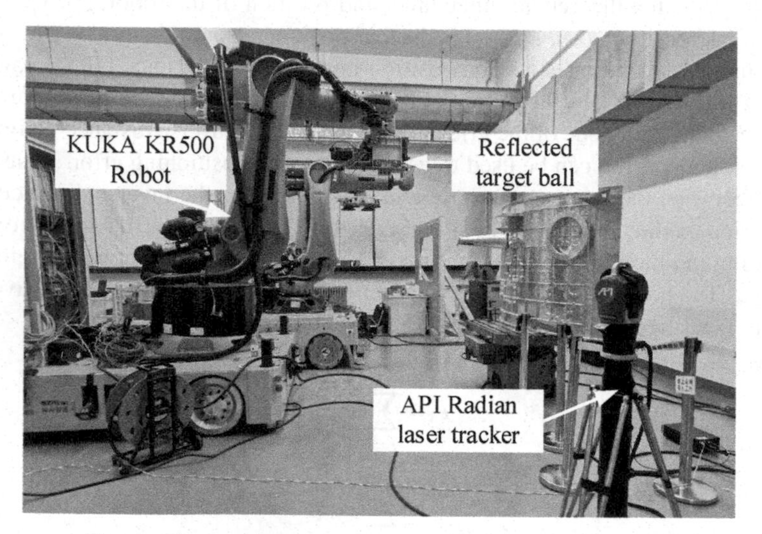

Fig. 4. The experiment system of the accuracy compensation

In the working space of 300 mm × 300 mm × 300 mm, different steps are designed to segment the space. When 70 mm step size is selected, there are 125 grid nodes; when 50 mm step size is selected, there are 343 grid nodes, Each node is measured by laser tracker, and the distribution diagram of theoretical node value and measured node value is shown in Fig. 6. Then the error database in the working space of 300 mm × 300 mm × 300 mm is established.

Then in the working space of 300 mm × 300 mm × 300 mm, another 10 different robot pose parameters are selected, corresponding to 10 different target points. Using the inverse distance weight error interpolation method, the error compensation of the target

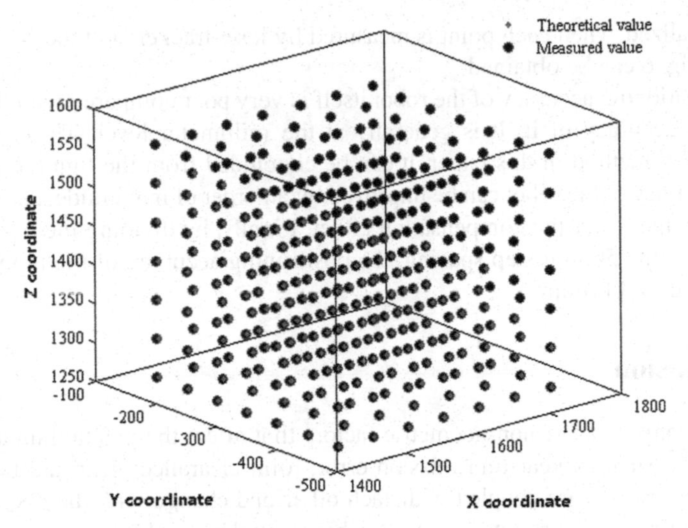

Fig. 5. Distribution diagram of theoretical node value and measured node value

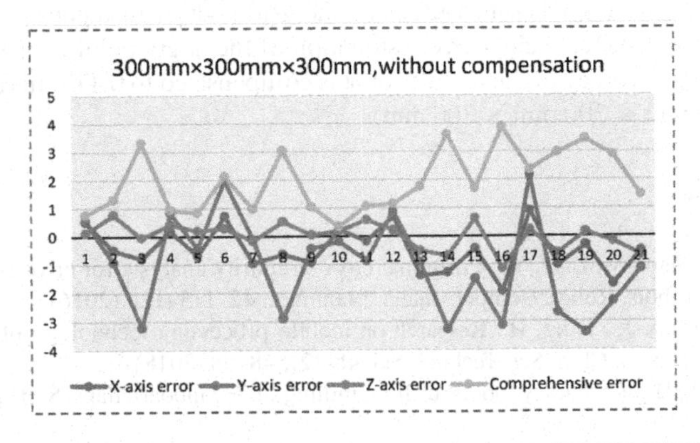

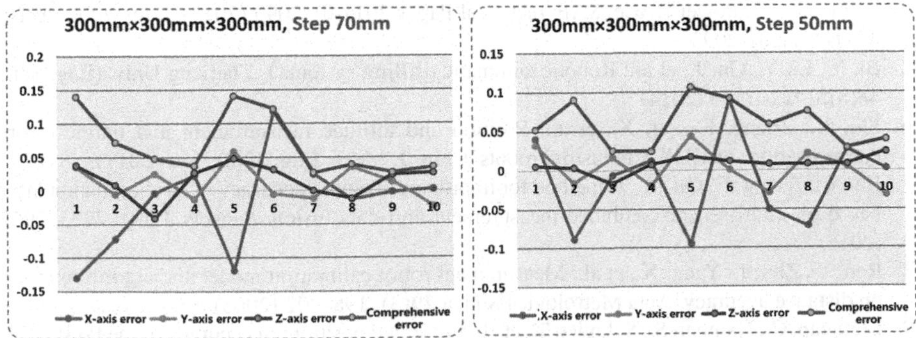

Fig. 6. Error compensation accuracy of 10 target nodes

points is realized. Then each point is measured by laser tracker, and the error results as shown in Fig. 6 can be obtained.

The positioning accuracy of the robot itself is very poor compared with the repeated positioning accuracy of it. It is generally at the millimeter level. Through the error compensation method in this paper, it can be concluded from the figure that the positioning accuracy of the robot can be improved by an order of magnitude, and the smaller the spacing, the better the compensation effect. Finally, by dividing the working space and selecting the 50 mm step spacing, the positioning accuracy of the mobile robot is compensated to 0.11 mm.

5 Conclusion

There are many kinds of non-geometric factors that affect the positioning accuracy of mobile robot, such as gear transmission error, joint clearance, gear and bearing friction, etc. The factors are coupled with each other and change with the operation cycle and environment of the robot, it is impossible to analyze and compensate each factor independently, so a non-model calibration method based on inverse distance weight is proposed. The working space of the mobile robot is divided, the theoretical calculation method of the grid nodes positions is established. The measured values of the grid nodes are obtained by laser tracker measurement, the grid node error database is established, and the inverse distance weight error estimation of the target point is given, and finally the positioning accuracy of the mobile robot is compensated to 0.11 mm (in the working space of 300 mm × 300 mm × 300 mm).

References

1. Zeng, Y., Tian, W., Liao, W.: Positional error similarity analysis for error compensation of industrial robots. Robot. Comput.-Integr. Manufact. **42**, 113–120 (2016)
2. Tao, B., Zhao, X., Ding, H.: Research on mobile processing technology of large complex component robot. Chin. Sci. Technol. Sci. **48**(12), 48–58 (2018)
3. Russell, D.: High-accuracy robotic drilling/milling of 737 inboard flaps. SAE Technical Paper 2011-01-2733
4. Dillhoefer, T.: Power RACe. SAE Technical Paper 2017-01-2093 (2017). https://doi.org/10.4271/2017-01-2093
5. Bi, Y., Li, Y., Gu, J., et al.: Robotic automatic drilling system. J. Zhejiang Univ. (Eng. Sci.) **48**(8), 1427–1433 (2014)
6. Shi, X., Zhang, F., Qu, X., et al.: Position and attitude measurement and online errors compensation for KUKA industrial robots. Chin. J. Mech. Eng. **53**(8), 1–7 (2017)
7. Liu, B., Zhang, F., Qu, X.: A method for improving the pose accuracy of a robot manipulator based on multi-sensor combined measurement and data fusion. Sensors **15**(4), 7933–7952 (2015)
8. Ren, Y., Zhu, J., Yang, X., et al.: Measurement robot calibration model and algorithm based on distance accuracy. Acta Metrologica Sinica **29**(3), 198–202 (2008)
9. Lightcap, C., Hamner, S., Schmitz, T., et al.: Improved positioning accuracy of the PA10-6CE robot with geometric and flexibility calibration. IEEE Trans. Robot. **24**(2), 452–456 (2008)
10. Wang, G., Liu, X., Gao, Y., et al.: A dual quaternion method for precise calibration of robot's base frame. J. Beijing Univ. Posts Telecommun. **40**(1), 18–22 (2017)

11. Bai, Y., Dali, W.: On the comparison of trilinear, cubic spline, and fuzzy interpolation methods in the high-accuracy measurements. IEEE Trans. Fuzzy Syst. **18**(5), 1016–1022 (2010)
12. Zeng, Y.F., Tian, W., Li, D.W., et al.: An error-similarity-based robot positional accuracy improvement method for a robotic drilling and riveting system. Int. J. Adv. Manufact. Technol. **88**(9–12), 2745–2755 (2017)

An Improved Monocular-Vision-Based Method for the Pose Measurement of the Disc Cutter Holder of Shield Machine

Dandan Peng, Guoli Zhu, Zhe Xie, Rui Liu, and Dailin Zhang(✉)

Huazhong University of Science and Technology (HUST), Wuhan 430070, China
mnizhang@mail.hust.edu.cn

Abstract. The manual disc cutter changing of the shield machine has greater security hidden danger, which leads to the urgent need to develop a disc cutter changing robot to replace the over-worn disc cutters. The rotation positioning error of the large diameter cutter head, the motion control error of the disc cutter changing robot, and the complexity of the disc cutter changing bin environment lead to the inaccuracy of the actual pose of the disc cutter. Therefore, it is necessary to increase the machine vision measurement system to accurately measure the pose of the disc cutter or its holder. This paper proposes a cutter holder positioning method, which uses the parallel line features and actual contour area of the cutter holder to determine its initial pose, and then uses the initial pose to calculates its accurate pose based on the distance matching method. The innovation of the method is that the pose measurement is carried out through the unique characteristics of the cutter holder, and it combines the initial pose estimation method based on the geometric characteristics of the cutter holder with the distance matching method which depends on the initial pose to solve the cutter holder pose. Simulation analysis shows that the accuracy of the proposed method is close to that of the most advanced PnP algorithms, and its robustness is better than others. Experimental results also verify the high accuracy of the proposed method.

Keywords: Shield machine · Disc cutter · Distance matching · Pose estimation

1 Introduction

Shield machine is a kind of semi-automatic basic equipment, but it still needs manual operations in many aspects. Due to the longtime tunneling operation of the shield tunneling machine, the disc cutters on the cutter head are easy to be seriously worn. Under the bad working environment of high pressure and high humidity, the disc cutters still need to be replaced manually at this stage [1]. The complex structure in the tunnel and many uncontrollable factors lead to many casualties caused by manual disc cutter changing at home and abroad, thus it is very necessary to replace the disc cutter with a robot [2]. To ensure that the robot can accurately grasp the disc cutter for change operation, the machine vision measurement system should have high accuracy and robustness.

X.-J. Liu et al. (Eds.): ICIRA 2021, LNAI 13014, pp. 678–687, 2021.
https://doi.org/10.1007/978-3-030-89098-8_64

Visual measurement systems can locate the object by extracting its features or by artificial markers fixed on it, because of the complexity of fixing artificial markers, they are not considered when the object has obvious features [3, 4]. The coplanar characteristic line segments and characteristic corners of the cutter holder are used to estimate its pose in this paper. The vision measurement system is fixed on the end actuator of the disc cutter changing robot, to ensure the low load and dexterous operation of the manipulator, the monocular measurement method with simple structure and relatively low cost is given priority. At present, there are many pose estimation algorithms based on monocular vision, the perspectives-n-point problem (PnP) algorithm is one of them, which aims to estimate the rotation and translation of a calibrated perspective camera by using n known 3D reference points and their corresponding 2D image projection. Direct linear transformation (DLT) is a classic PnP algorithm with high efficiency but poor robustness due to noise interference [5]. EPnP is a non-iterative algorithm, which uses virtual points to indirectly solve the pose parameters of the target, thus reduces the computational complexity of the algorithm, however, when the number of feature points is small, its solution accuracy and robustness are relatively low [6]. RPnP algorithm can present stable calculation results when there are fewer and more feature points, but it cannot get the unique solution when using least squares error, and its solving accuracy needs to be improved [7]. The basic idea of OPnP and ASPnP algorithm is to formulate the PnP problem into a functional minimization problem and retrieve all its stationary points by using the Gröbner basis technique, these two algorithms have high accuracy and robustness, and are the best PnP algorithms so far [8, 9]. These PnP algorithms have great differences in solving accuracy when the image noise is small and large. Because of the actual situation that the environment of disc cutter changing site is execrable and cutter holder images collected contain strong noise, a method with better robustness is needed to solve the pose of the cutter holder.

The distance matching (DM) method proposed in this paper can solve the pose of the cutter holder without the one-to-one correspondence between the 3D reference points and the 2D image points. Rough positioning is carried out according to the geometric features of the cutter holder, and then its exact pose is obtained by distance matching. In the initial positioning, the 6DOF of the cutter holder pose are solved respectively, the three rotational components are solved by the vanishing points obtained from the parallel lines on the cutter holder surface. The depth information is obtained by the ratio of the inside contour of the cutter holder in the image to its area in the real space, which is used to calculate the other two translational components. The cutter holder pose is accurately solved by the distance matching method, the distance between the collected image of the cutter holder and the template in the established template library is calculated, then the template corresponding to the minimum distance is selected, and its corresponding pose is used as the final estimated pose of the cutter holder.

This paper is organized as follows. Section 2 states the pose estimation algorithm of the cutter holder we propose. In Sect. 3, simulations and experiments are implemented to test the performance of the proposed algorithm. And Sect. 4 gives conclusions.

2 Mathematical Model and Method

2.1 Image Processing

Salient features of the cutter holder are extracted to accurately estimate its pose, such as lines, corners, circles, etc., which are the most commonly used geometric information in the field of robot visual positioning. It is difficult to extract the feature corners and lines directly because of the wear and corrosion on the surface of the cutter holder caused by the long time working of shield tunneling, thus we can extract the overall contour of the cutter holder image, and then obtain the salient features from it to measure the pose. Given the obvious and numerous linear features of the inside contour of the cutter holder, the characteristic straight lines on its surface are extracted for pose calculation.

The detailed algorithm for feature lines detection is summarized in the form of a block diagram in Fig. 1. The whole process of image processing is divided into two blocks: preprocessing block and feature lines extraction block.

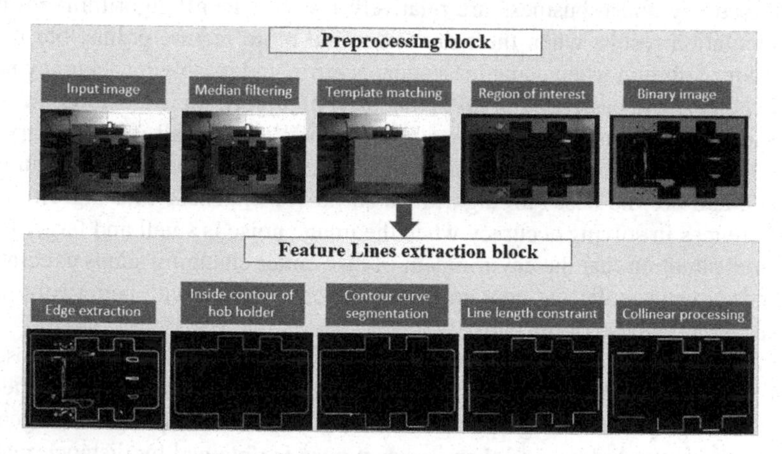

Fig. 1. Image processing flow diagram.

In the preprocessing block, the grayscale images of the cutter holder collected are processed by median filtering using a 7*7 square mask. The template matching method based on grayscale is applied to extract the region of interest (ROI) containing the whole part of the inside contour of the cutter holder from the filtered image. The global threshold segmentation method is used to segment the ROI by setting the threshold range from 70 to 255, and the binary image of the cutter holder is obtained. In the feature lines extraction block, many closed contours are obtained by edge extraction of the binary image, and the closed inside contour of the cutter holder with the longest length is selected by length sorting. Based on the curvature constraint, a straight line or arc with a length of at least 6 pixels is used to approximate the inside contour, to segment it into several line segments. The line segments whose length is smaller than the threshold, which is set as 50, are eliminated from the obtained line segments, and the remaining segments are collinearly processed to obtain the characteristic segments, which are used to solve the pose of the cutter holder.

2.2 Initial Pose Estimation

The initial pose estimation method proposed in this paper calculates the rotation and translation components of the cutter holder pose in turn. The rotation components of the cutter holder pose are obtained by using the extracted parallel characteristic line segments. The depth information is obtained by the ratio of the area of the inside contour of the cutter holder in the image to its area in space, which is helpful to calculate the remaining two translation components.

Through the above image processing method, 10 parallel lines are obtained in the horizontal and vertical directions of the cutter holder, P_1P_2 and P_4P_3, P_4P_1 and P_3P_2 in Fig. 2 are the two pairs of parallel lines arbitrarily selected from them. A pair of parallel lines in space projected onto the image plane have one and only one intersection point, which is called the vanishing point. P is the origin of the world coordinate system, the image point P_j corresponding to the space point P_{ij}, and the values of i and j are from 1 to 4. The vanishing point between parallel lines P_1P_2 and P_4P_3 is represented by P_{v1}, while the vanishing point between P_4P_1 and P_3P_2 is represented by P_{v2}. The image coordinates of P_{v1} and P_{v2} are calculated according to the characteristic line segments obtained from the above image processing. A pair of parallel lines can get a vanishing point, according to the obtained cutter holder of many pairs of parallel lines can get some vanishing points. The image point with the smallest sum of the distances to all vanishing points is used as the final vanishing point for the pose solution. The symbols of the two vanishing points obtained by the solution are also expressed by P_{v1} and P_{v2}.

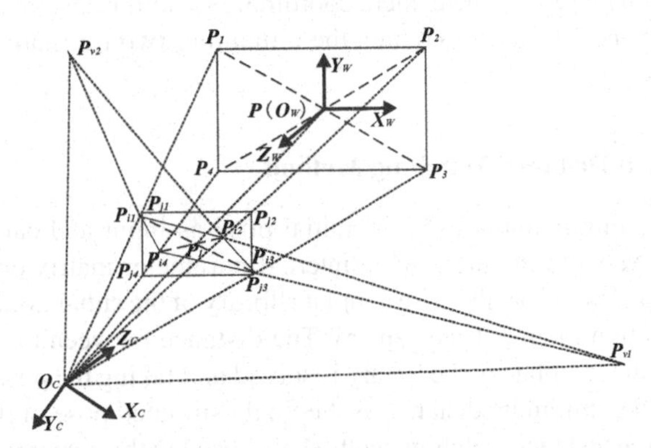

Fig. 2. Schematic diagram of pose measurement based on vanishing points.

Based on the image coordinates of P_{v1} and P_{v2}, the imaging point coordinates normalized to the image plane at the camera's focal length can be calculated as follows:

$$\begin{bmatrix} x_{1cvi} \\ y_{1cvi} \\ 1 \end{bmatrix} = \begin{bmatrix} k_x & 0 & u_0 \\ 0 & k_y & v_0 \\ 0 & 0 & 1 \end{bmatrix}^{-1} \begin{bmatrix} u_{vi} \\ v_{vi} \\ 1 \end{bmatrix} \tag{1}$$

In formula (1), (u_{vi}, v_{vi}) is the image coordinate of the vanishing point P_{v1}, (x_{1cvi}, y_{1cvi}, 1) is the coordinate of P_{v1} normalized to the image plane at the focal length, where

$i = 1, 2$. k_x and k_y are the magnification coefficients of the camera from the imaging plane coordinates to the image coordinates. (u_0, v_0) is the image coordinate of the center of the optical axis of the camera.

Geometrically, the vanishing point of a line in space is obtained by the intersection of a ray that is parallel to the line and passes through the center of the camera and the image plane, so $O_C P_{v1}$ is parallel to $P_1 P_2$ and $P_4 P_3$. Therefore, $[x_{1cv1} y_{1cv1}\ 1]^T$ is both the position vector of P_{v1} in the camera coordinate system and the direction vector of the X-axis of the world coordinate system. Similarly, $[x_{1cv2} y_{1cv2}\ 1]^T$ represents the direction of the Y-axis of the world coordinate system. By normalizing the two vectors into unit vectors, the X-axis and Y-axis components of the rotation matrix R from the world coordinate system to the camera coordinate system can be obtained. The cross-product of the two components is further solved to obtain the Z-axis rotation component of R.

The normal vector $\vec{n}$ of the plane $P_1 P_2 P_3 P_4$ is obtained according to R. The image point of P is P_1, a plane with normal vector $\vec{n}$ passes through P_1, and its intersection points with $O_C P_{i1}$, $O_C P_{i2}$, $O_C P_{i3}$, and $O_C P_{i4}$ are P_{j1}, P_{j2}, P_{j3}, and P_{j4} respectively, thus the lines $P_1 P_2$, $P_4 P_3$, $P_1 P_4$, and $P_2 P_3$ are parallel to $P_{j1} P_{j2}$, $P_{j4} P_{j3}$, $P_{j1} P_{j4}$, and $P_{j2} P_{j3}$ respectively. According to the image coordinates of P_{j1}, P_{j2}, P_{j3}, and P_{j4}, the area of the polygon $P_{j1} P_{j2} P_{j3} P_{j4}$ and the distance d_1 from point O_C to it are obtained. Using the principle of similarity, the square of the ratio of d_1 to the distance d_2 from O_C to the plane $P_1 P_2 P_3 P_4$ is equal to the area ratio of polygon $P_{j1} P_{j2} P_{j3} P_{j4}$ to the polygon $P_1 P_2 P_3 P_4$, thus the value of d_2 can be solved, which represents the depth information of the cutter holder. The coordinate of P in the camera coordinate system is also obtained according to the similarity principle, so as to obtain the remaining two translational components of the cutter holder.

2.3 The Proposed Distance Matching Method

The distance matching method takes the initial pose as input and can solve the local optimal solution. Substituting the camera internal parameter matrix obtained from the monocular camera calibration, the pose template library of the cutter holder is established through the projection model of the camera. The distance between the collected cutter holder image and the template in the library is solved and taking the pose of the template corresponding to the minimum distance as the final estimated pose of the cutter holder. The flow chart of the distance matching method proposed in this paper is shown in Fig. 3.

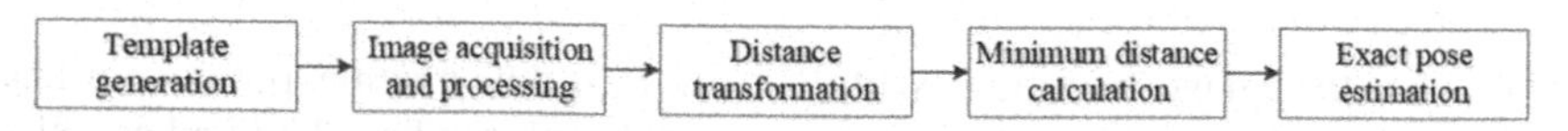

Fig. 3. Flowchart of distance matching method.

The origin of the workpiece coordinate system of the cutter holder is at the center of its surface, the Z-axis is perpendicular to its surface, and the X-axis and Y-axis are respectively parallel to the long and short sides of its outer rectangular contour. The workpiece coordinate system is set as the world coordinate system, then the relative pose between the world coordinate system and the camera coordinate system is the final required cutter holder pose. To meet the requirement that the grasping accuracy of the manipulator should reach 1 mm, the position moving step and the rotation step is set as 0.1 mm and 0.1° respectively to transform the cutter holder pose within the known relative motion range between it and the manipulator. We can get the image coordinates corresponding to the spatial feature corner points in each pose state of the cutter holder by using the camera projection model. The pose template library of the cutter holder is constructed, each template corresponds to a pose of the cutter holder in the camera coordinate system, and contains the image coordinates of 20 characteristic corner points on the surface of the cutter holder. Due to a large number of templates, compared with the way of storing images directly, the storage capacity of the template in the form of image coordinates of feature points is greatly reduced. Although it takes a long time to establish the pose template library, it is an offline operation and does not occupy the time of real-time pose estimation.

The operation converting a binary image to an approximate distance image is called a distance transformation (DT), which can produce a grayscale image, the grayscale value of each pixel in it is the distance between the pixel and the nearest background pixel. The image processing method described above is applied to the collected cutter holder image to obtain 20 feature line segments, and 20 feature corners are obtained by intersecting the extension lines of the segments. The distance transformation diagram of the binary image containing only 20 feature corners is obtained through distance transformation, as shown in Fig. 4, in which the gray value of each pixel is expressed as the distance between the pixel point and the nearest feature corner point. We can substitute the image coordinates of the characteristic corners of any template in the pose template library into the distance transformation graph to get 20 distance values, and solve their root mean square value as the distance between the actual cutter holder and the template. The pose of the template corresponding to the minimum distance in the library is selected as the estimated pose of the actual cutter holder.

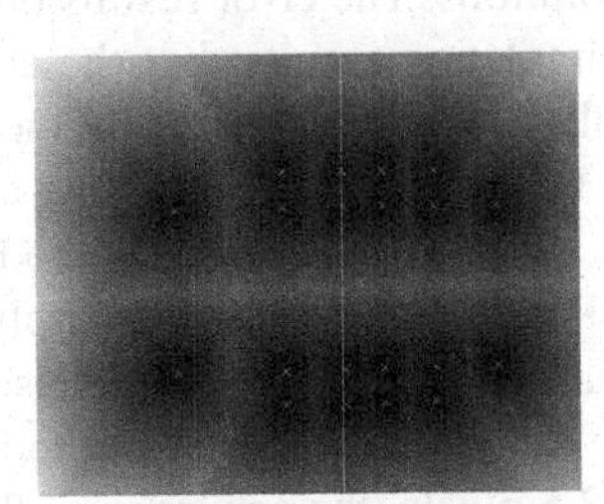

Fig. 4. DT image. The larger the distance the lighter the tone.

By changing the step size from large to small, the efficiency of searching the template corresponding to the minimum distance in the template library is improved. The initial step sizes of translation and rotation directions are set as 5 mm and 5° respectively, and the template corresponding to the minimum distance is searched from the library. The pose corresponding to the obtained template is taken as the initial pose, and the step size of translation and rotation direction is reduced to 1 mm and 1° for local search to obtain the estimated pose. The above process is repeated with the step sizes of 0.5 mm and 0.5°, 0.2 mm and 0.2°, 0.1 mm and 0.1° in the direction of translation and rotation in turn, and the template pose corresponding to the final minimum distance is taken as the estimated pose of the cutter holder.

3 Experimental Results

3.1 Experiments with Synthetic Data

The pose estimation method proposed in this paper is tested by simulation experiments. In order to accurately evaluate the accuracy and robustness of the algorithm, the experiment will be compared with various mainstream PnP algorithms, including ASPnP, DLT, EPnP, OPnP and RPnP. Given a virtual perspective camera with image size 4112×3008 pixels, pixel size 3.45 μm, and focal length 8.5 mm, the cutter holder pose is generated randomly, and the rotation and translation distribution ranges are [−60°, 60°] and [−90, 90] mm respectively. The feature points on the surface of the cutter holder are projected onto the image plane based on the camera imaging model. Different levels of Gaussian noise are added to the projected image points, and for each noise level, 1,000 test data sets are generated. The pose solving algorithms are used to solve the cutter holder pose, and the error between the calculated pose and the randomly generated pose is obtained.

Due to the harsh environment of the measurement site, the actual cutter holder images collected contain strong noise, this paper mainly studies the influence of noise intensity on the measurement accuracy. Since the features to be extracted are 20 corner points on the surface of the cutter holder, the number of input reference points is fixed as 20, and the variance of input noise is changed from 0 to 5 pixels (the step is 0.5 pixels) to verify the solving accuracy of each algorithm. The DM algorithm proposed in this paper and the other PnP algorithms are implemented in MATLAB and 500 test runs are performed. The error values of each degree of freedom are expressed by the standard deviation of the error values of 500 pose solutions. The error results obtained are shown in Fig. 5.

It can be seen from the simulation results that the accuracy of the RPnP method is lower than that of other PnP methods, followed by the accuracy of EPnP and DLT methods. The accuracy of the OPnP method is the same as that of the ASPnP method, which has the best performance among these PnP methods. With the increase of input noise, the accuracy of the PnP methods decreases sharply. Compared with these PnP methods, when the noise variance is less than 2.5 pixels, the solving accuracy of the DM method is slightly lower than that of the PnP methods. On the contrary, when the noise variance is higher than 2.5 pixels, the accuracy of the DM method is better than that of the PnP methods with the increase of noise. In the case of noise enhancement, the accuracy of the PnP method drops sharply, while the accuracy of the DM method changes little and remains relatively stable, especially in the direction of depth, which indicates that the robustness of the DM method proposed in this paper is better than the PnP method. The high precision and good robustness of the DM method under high noise indicate that it is more suitable for the construction site of disc cutter changing operation with higher noise.

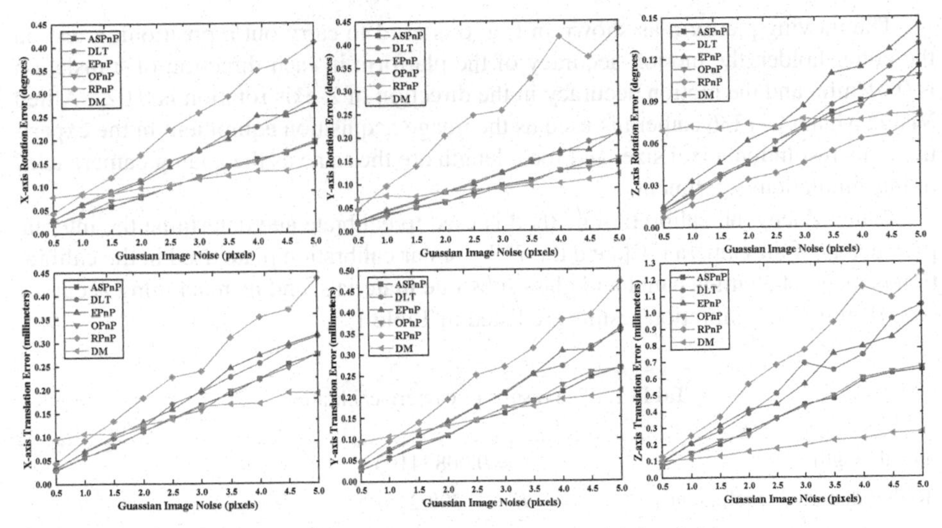

Fig. 5. Accuracy of all algorithms when noise level changes.

3.2 Experiments with Real Images

To make the method can be applied in practice, the industrial field environment is simulated in the laboratory. The experiment is carried out in a dark environment, the space in the tool changing bin is narrow, and the larger robot has a smaller movement range in the disc cutter changing bin. The motion control precision of the disc cutter changing robot is 5 mm, so the cutter holder can be accurately positioned within the range of [−10, 10] degrees and [−20, 20] millimeters in the direction of rotation and movement respectively.

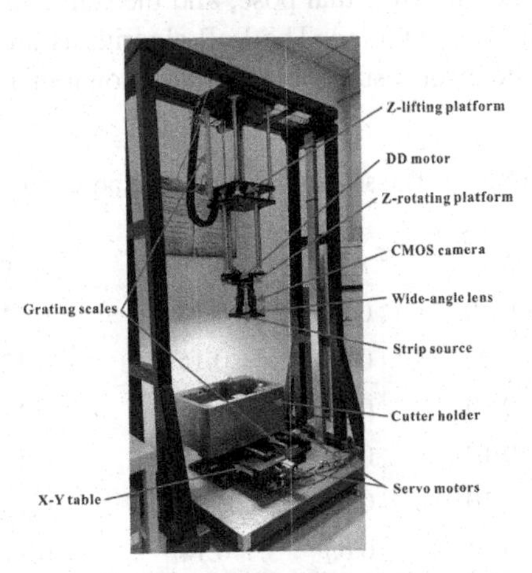

Fig. 6. The monocular vision positioning system of the cutter holder.

The moving platform as shown in Fig. 6 is used to carry out a positioning test on the cutter holder, the motion accuracy of the platform in each direction of translation is 0.01 mm, and the motion accuracy in the direction of *Z*-axis rotation is 0.02°. Allied Vision Manta G-1236 camera is used as the image acquisition equipment in the experiment, its resolution, pixel size, and focal length are the same as the analog camera used in the simulation experiment.

Zhang Zhengyou calibration method is used to calibrate and determine the internal parameters of the camera [10], and the 7*7 circular calibration plate used in the calibration is made of alumina panel and glass substrate materials, and its machining accuracy is 0.01 mm. The calibration results are listed in Table 1.

Table 1. The results of camera calibration.

Focal length	$f = 0.00854103$
Radial distortion coefficient	$K = -308.215$
Pixel size	$(Sx, Sy) = (3.45168e{-}006, 3.45e{-}006)$
Main point coordinate	$(Cx, Cy) = (2084.86, 1477.95)$

The cutter holder images are collected with the calibration monocular camera, and the cutter holder pose is changed by the moving platform within the range set above. Every time the motion platform changes the pose of the cutter holder, an image of it is collected, and the pose at that time is recorded. There are 100 cutter holder images that are randomly collected, and the pose of 100 groups of the cutter holder is recorded at the same time. The image processing, initial pose estimation, and distance matching method described above are used to solve the pose of the actual cutter holder. The calculated pose is compared with the known actual pose, and the standard deviation of the errors of each degree of freedom is obtained. The PnP algorithms are also used to solve the cutter holder pose, and the error results obtained are shown in Table 2.

Table 2. The standard deviation of pose error.

	α (°)	β (°)	γ (°)	*X* (mm)	*Y* (mm)	*Z* (mm)
ASPnP	0.04	0.10	0.03	0.15	0.11	0.18
DLT	0.07	0.09	0.03	0.15	0.12	0.19
EPnP	0.07	0.09	0.03	0.15	0.11	0.17
OPnP	0.04	0.10	0.03	0.15	0.11	0.18
RPnP	0.11	0.10	0.03	0.16	0.13	0.41
DM	0.10	0.14	0.03	0.14	0.11	0.19

The cutter holder used in this paper is a new type of cutter holder with high machining accuracy, which has not been put into use in the construction site, so the image noise of the cutter holder collected is small. The experimental results are similar to the simulation results in the case of low noise.

4 Conclusion

In this paper, a pose estimation algorithm is proposed based on the characteristics of the disc cutter holder of the shield machine, which makes use of vanishing point property and similarity principle for initial positioning, and then carries out precise positioning according to the distance matching method. The simulation results show that the proposed algorithm has high accuracy and robustness. The actual experiment proves that the algorithm can achieve high accuracy under low noise conditions. In the future, wear and corrosion treatment will be carried out on the new type of cutter holder to simulate its working state under actual working conditions.

Acknowledgments. This work was supported by the National Key Research and Development Program of China (Grant No. 2018YFB1306700).

References

1. Frenzel, C., Käsling, H., Thuro, K.: F: factors influencing disc cutter wear. Geomechan. Tunnelb. **1**(1), 55–60 (2008)
2. Yuan, J., Guan, R., Du, J.F.: Design and implementation of disc cutter changing robot for tunnel boring machine (TBM). In: 2019 IEEE International Conference on Robotics and Biomimetics (ROBIO), pp. 2402–2407. IEEE, Dali (2019)
3. Liu, Z., Liu, X., Duan, G., et al.: F: Precise pose and radius estimation of circular target based on binocular vision. Measur. Sci. Technol. **30**(2), 025006 (2019). (14pp). IOP, Kissimmee
4. Gadwe, A., Ren, H.: F: real-time 6DOF pose estimation of endoscopic instruments using printable markers. IEEE Sens. J. **19**(6), 2338–2346 (2018)
5. Abdel-Aziz, Y., Karara, H.F.: Direct linear transformation from comparator coordinates into object space in close-range photogrammetry. Am. Soc. Photogram. (1971)
6. Lepetit, V., Moreno-Noguer, F., Fua, P.: F: EPnP: an accurate O(n) solution to the PnP problem. Int. J. Comput. Vision **81**(2), 155–166 (2009)
7. Li, S., Xu, C., Xie, M.: F: a robust O(n) solution to the perspective-n-point problem. IEEE Trans. Pattern Anal. Mach. Intell. **34**(7), 1444–1450 (2012)
8. Zheng, Y., Kuang, Y., Sugimoto, S., et al.: F: revisiting the PnP problem: a fast, general and optimal solution. In: 2013 IEEE International Conference on Computer Vision, pp. 2344–2351. IEEE, Sydney (2014)
9. Zheng, Y., Sugimoto, S., Okutomi, M.: F: ASPnP: an accurate and scalable solution to the perspective-n-point problem. IEICE Trans. Inf. Syst. **E96D**(7), 1525–1535 (2013)
10. Zhang, Z.: F: a flexible new technique for camera calibration. IEEE Trans. Pattern Anal. Mach. Intell. **22**(11), 1330–1334 (2000)

Research on Escape Strategy of Local Optimal Solution for Underwater Hexapod Robot Based on Energy Consumption Optimization

Yingzhe Sun[1,2,3,4], Qifeng Zhang[1,2,3](✉), Aiqun Zhang[1,2,3], and Xiufeng Ma[1,2,3,5]

[1] State Key Laboratory of Robotics, Shenyang Institute of Automation, Chinese Academy of Sciences, Shenyang 110016, China
{sunyingzhe,zqf,zaq,maxiufeng}@sia.cn

[2] Institutes for Robotics and Intelligent Manufacturing, Chinese Academy of Sciences, Shenyang 110169, China

[3] Key Laboratory of Marine Robotics, Shenyang 1572000, Liaoning, China

[4] University of Chinese Academy of Sciences, Beijing 100049, China

[5] College of Information Science and Engineering, Northeastern University, Shenyang 110819, Liaoning, China

Abstract. Aiming at the local optimal solution problem of artificial potential field method, this paper takes energy consumption as the optimization objective to plan the escape strategy of underwater hexapod robot. The total energy consumption of underwater hexapod robot in three gait and turning gait is solved by establishing the dynamic model of the underwater hexapod robot, and the energy consumption coefficient of the whole robot is planned according to the different steps and different rotation angles of the robot. Simulate annealing method based on energy consumption difference as Metropolis criterion is used to plan the next crawling point. Finally, the correctness of the theory is verified by comparing the total energy consumption of different escape paths through simulation experiments. The simulation results show that the simulate annealing method based on energy consumption optimization can plan the local escape path of the underwater hexapod robot under the constraint of energy consumption, and has good practical application value.

Keywords: Simulate annealing · Underwater hexapod robot · Energy consumption · Gait · Path planning

1 Introduction

The path planning of underwater hexapod robot is an important index to measure the safety and feasibility of its underwater work. The essence of path planning is that the robot searches an optimal collision free path from the starting position to the target position [1–3]. At present, the common path planning algorithms are artificial potential field method [4], genetic algorithm [5], neural network [6], fuzzy logic [7] and ant colony algorithm [8]. Among them, the artificial potential field method is often used

X.-J. Liu et al. (Eds.): ICIRA 2021, LNAI 13014, pp. 688–698, 2021.
https://doi.org/10.1007/978-3-030-89098-8_65

in mobile robot motion planning system because of its simple mathematical principle, small amount of calculation, short running time and good for robot bottom control.

The principle of the artificial potential field method is to plan the next path point by calculating the resultant force of the robot in the potential field, but in the complex environment, the path will produce the point where the resultant force is 0, that is, the local optimal solution. Therefore, it is necessary to propose an improved algorithm to make the robot escape from the local optimal solution. At present, scholars at home and abroad have carried out relevant researches on the escape strategy of underwater vehicle in the local optimal solution. Li Peilun of Shanghai Jiaotong University introduced the velocity potential field function to change the static potential field into the dynamic potential field to modify the glider planning path [9]. Wu Zhengping of Three Gorges University modified the potential field function by adding current [10]. According to the ocean environment and depth data, RAO D combined RRT algorithm and A* algorithm to plan the optimal path of glider based on energy consumption optimization [11]. Pereira [12] takes minimizing collision risk as the goal to plan the path of underwater glider. The study on the local optimal solution escape strategy of underwater hexapod robot based on energy consumption optimization is relatively few. In this paper, by planning the typical gaits of the underwater hexapod robot and combining with the underwater dynamic model, the energy consumption coefficient of the robot in different crawling States is given. Finally, the simulated annealing method is planned according to the energy consumption coefficient to make the underwater hexapod escape from the local optimal solution in the artificial potential field method with low energy consumption.

2 Dynamic Modeling and Energy Consumption Calculation of Underwater Hexapod Robot

The crawling process of the underwater hexapod robot mainly relies on the joints of the crawling legs to drive the body to adjust the posture, so the total energy consumption of the whole underwater crawling process is the total power consumption of all the joints of the crawling legs(Ignore the energy consumption of lighting, sonar and other sensors). Total energy consumptionis obtained as Eq. (1):

$$\mathrm{P}_{\mathrm{total}} = \sum_{i=1}^{18} \tau_i \cdot \omega_i \cdot \eta_i \tag{1}$$

Where τ_i is joint moment, ω_i is joint angular velocity, η_i is joint drive efficiency. On the premise of known gait planning, the joint angular velocity can be solved by inverse kinematics of single leg. The driving efficiency of the motor is the inherent property of the motor, so it only needs to calculate the joint torque of the underwater hexapod robot in different gait.

2.1 Dynamic Modeling of Underwater Hexapod Robot

The dynamic modeling of the underwater hexapod robot needs to combine the particularity of the seabed environment to establish the whole machine dynamic model including the hydrodynamic force and the single leg dynamic model including the contact force. The hydrodynamic term is obtained by CFD simulation with reference to Morison formulation [13]. The contact force is obtained by combining the ground mechanics theory with the seabed parameters. The force on the whole machine is shown in Fig. 1.

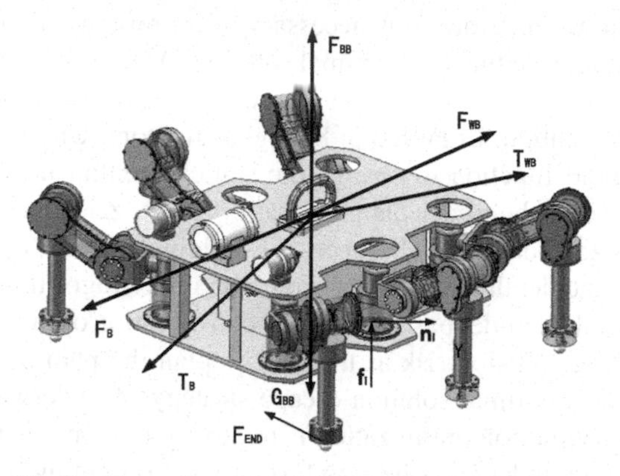

Fig. 1. Force diagram of underwater hexapod robot.

Compared with the land environment, the dynamic characteristics of the underwater hexapod robot are different from those in the air because of the buoyancy and fluid resistance. In this paper, buoyancy and hydrodynamic terms are added to the ground dynamic model. The underwater dynamic model of the modified underwater hexapod robot is obtained as Eq. (2):

$$\begin{cases} \sum_{i=1}^{6}\left({}^{B}f_i\right) + F_B + F_{WB} + G_B + F_{BB} = 0 \\ \sum_{i=1}^{6}\left({}^{B}n_i\right) + \sum_{i=1}^{6}\left[{}^{B}f_i \times \left(-{}^{B}P_i\right)\right] + T_{WB} + T_B = 0 \end{cases} \tag{2}$$

Where G_B, F_{BB} is gravity and buoyancy, $H_B = G_B\text{-}F_{BB}$ is the net buoyancy under water. f_i, n_i is the force and moment of the i-leg acting on the body. F_{WB}, T_{WB} is hydrodynamic force and hydrodynamic moment received by the body. F_B, T_B is inertial force and moment of inertia produced by body motion. In Eq. (2), the buoyancy, gravity and inertia forces of the underwater hexapod robot are related to the parameters of the whole machine, which can be calculated directly. The hydrodynamic force is obtained by CFD simulation with reference to Morison formulation. The force and moment of the crawling leg acting on the body are unknown, so it is necessary to establish a single leg dynamic model of the interaction between the crawling leg and the ground.

The crawling legs of the underwater hexapod robot connect the body and the seabed ground, and feedback the force of the seabed to the body. At the same time, the driving joint of the crawling leg is used as the power driving part of the underwater hexapod robot to realize the stable crawling of the whole robot. Therefore, it is necessary to establish the dynamic model of the crawling leg to calculate the force and torque acting on the body by the heel joint of the crawling leg, as shown in Fig. 2.

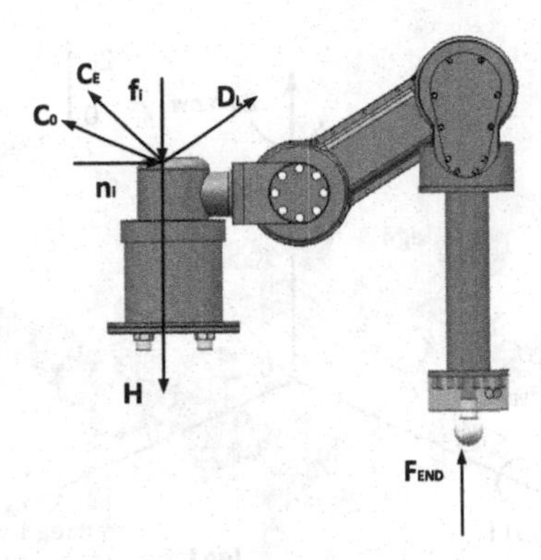

Fig. 2. Single leg dynamic model of underwater hexapod robot.

Dynamic model of one leg is obtained as Eq. (3):

$$\tau = \mathrm{M}\ddot{\theta} + \mathrm{C_E} + \mathrm{C_O} + \mathrm{H} + \mathrm{J^T F_{END}} + \mathrm{D_L} \tag{3}$$

Where τ is joint torque, $\ddot{\theta}$ is joint angular acceleration, M is inertia matrix, $\mathrm{C_E}$ is centrifugal force matrix, $\mathrm{H_B}$ is the net buoyancy under water, $\mathrm{F_{END}}$ is contact force, $\mathrm{D_L}$ is hydrodynamic force. Set:

$$\tau_k = \mathrm{M}\ddot{\theta} + \mathrm{C_E} + \mathrm{C_O} + \mathrm{H} = \left[\tau_{k1}\ \tau_{k2}\ \tau_{k3}\right]^T$$

$$\tau_s = \mathrm{J^T F_{END}} = \left[\tau_{s1}\ \tau_{s2}\ \tau_{s3}\right]^T$$

$$\tau_d = \mathrm{D_L} = \left[\tau_{d1}\ \tau_{d2}\ \tau_{d3}\right]^T \tag{4}$$

Equation (4) is changed as Eq. (5):

$$\tau = \tau_k + \tau_s + \tau_d \tag{5}$$

Where τ_k is dynamic model of general manipulator, τ_d is joint torques used to overcome hydrodynamic forces, τ_s is joint moment used to overcome contact forces. τ_d is obtained by hydrodynamic simulation. τ_s is obtained as Eq. (6):

$$\tau_s = \mathrm{J^T F_{END}} = \mathrm{J^T} \cdot \left[\mathrm{F_x}\ \mathrm{F_y}\ \mathrm{F_z}\right]^T \tag{6}$$

Where $\mathrm{J^T}$ is transpose of Jacobian matrix, $\mathrm{F_{END}}$ is contact force which can be calculated from the foot force distribution.

2.2 Distribution Method of Foot Contact Forces

When the underwater hexapod robot crawls, there are at most six feet touching the ground and at least three feet touching the ground, and the force of each foot will affect the state of the body. In order to maintain the dynamic balance of the body, it is necessary

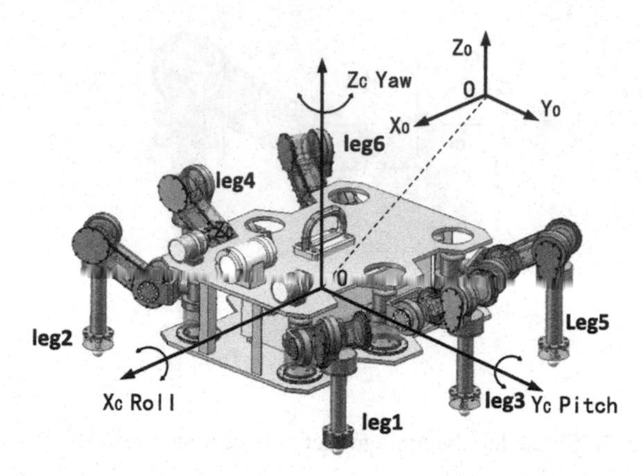

Fig. 3. Schematic diagram of coordinate system setting for underwater hexapod robot.

to establish the method of sufficient force distribution to maintain the dynamic balance of the body. In this paper, the method of moment balance is used to distribute the foot forces. Set the world coordinate system O-xyz, the body coordinate system C-xyz and the calibration of the crawling legs as shown in Fig. 3.

The main way of underwater hexapod robot is to support and advance. Therefore, in the distribution of foot force, we should first consider the distribution of Z-direction force. Suppose that the z-direction resultant force of the supporting foot on the left side (legs 1, 2 and 3) of the robot's forward direction is F_{zl}, and that of the supporting foot on the right side (legs 4, 5 and 6) of the robot's forward direction is F_{zr}. These two resultant forces should be balanced with the gravity of the whole robot. The point of gravity is regarded as the supporting point, and the forces on the left and right sides are equivalent to the fuselage, which constitutes a lever. The principle is shown in Fig. 4.

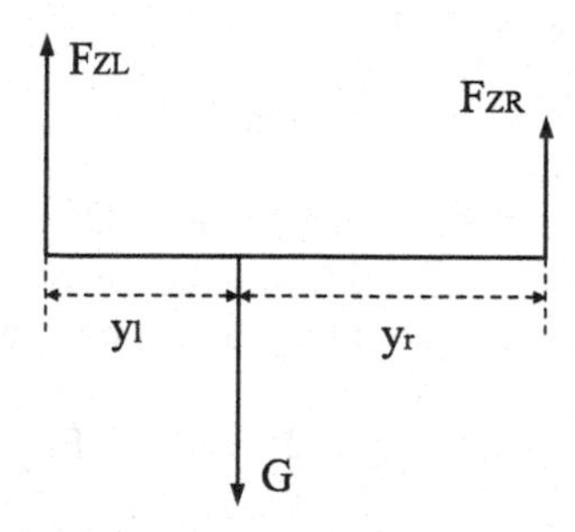

Fig. 4. Schematic diagram of Z-direction force distribution of creeper leg.

2.3 Energy Consumption Coefficient Based on Typical Gaits

The path target points of the underwater hexapod robot are discrete planning points. The whole robot needs to track the above path points through the combination of three gait and turning gait. In this paper, the total energy consumption of three gait and turning gait is calculated to fit the energy consumption coefficient of corresponding gait. Turning gait energy consumption coefficient KR and three gait energy consumption coefficient KF is obtained as Eq. (7):

$$\begin{cases} \sum_{i=1}^{2} K_{Ri} = \frac{\partial P}{\partial \theta} \\ \sum_{i=1}^{2} K_{Fi} = \frac{\partial P}{\partial l} \end{cases} \tag{7}$$

Where $\partial\theta$ is the angular element of the body rotating around the z-axis, in degrees, ∂l is the displacement of the body along the x-axis, in meter. ∂P is the total energy consumption of the crawling leg joint to complete the corresponding gait.

The energy consumption of the underwater hexapod robot in three gait is directly proportional to the water speed of the environment, and the periodic crawling time and the periodic crawling distance. Combined with the parameters of the underwater hexapod robot and the static stability criterion, the maximum crawling distance of the whole robot in a single cycle is 0.2 m [14]. Considering the underwater crawling efficiency, the minimum crawling distance is 0.1 M; According to Eqs. (1) and (4), the energy consumption of the whole machine at different crawling distances under the condition of water flow static and water flow along the x-axis negative direction of 0.1 m/s is calculated, as shown in Table 1.

Table 1. Total energy consumption of three gait.

Step (m)	***0.1***	***0.11***	***0.12***	***0.13***	***0.14***
Energy without water (KJ)	78.296	82.367	86.866	91.794	97.196
Energy with water (KJ)	78.703	82.740	87.204	92.098	97.450
Step (m)	***0.15***	***0.16***	***0.17***	***0.18***	***0.19***
Energy without water (KJ)	102.927	109.073	115.640	122.658	130.153
Energy with water (KJ)	103.168	109.270	115.781	122.726	130.229

It can be seen from Table 1 that in a single crawling cycle, the time to overcome the water resistance and do work is short, and the influence of water flow on the energy consumption of the body in a single cycle is small. Ignoring the water flow factor, the least square method is used to fit the curve of three gait coefficient K_F.

$$P = 2120\partial l^2 - 3.8\partial l + 61(\text{kJ}) \tag{8}$$

When the underwater hexapod robot completes a cycle of rotation, it not only overcomes the resistance of the water flow to the body, but also needs to overcome the resistance moment of the external water flow to the body. According to the underwater static stability criterion, the maximum rotation angle of the body in a single cycle is 10°. According to formula (1) and formula (4), the total energy consumption of single cycle with different rotation angles is calculated respectively under the environment of no water flow and water flow impacting 0.1 m/s along the negative x-axis, as shown in Table 2. Here, the calculation is based on the positive rotation of the body around the z-axis (the body itself is approximately symmetrical, assuming that the energy consumption of the negative rotation around the z-axis is the same).

Table 2 Total energy consumption of turning gait.

Angle (°)	*1*	*2*	*3*	*4*	*5*
Energy without water (KJ)	61.826	62.748	63.709	64.707	65.742
Energy with water (KJ)	61.841	62.765	63.728	64.729	65.768
Angle (°)	*6*	*7*	*8*	*9*	*10*
Energy without water (KJ)	66.818	67.933	69.087	70.280	71.509
Energy with water (KJ)	66.847	67.967	69.126	70.324	71.558

It can be seen from Table 2 that the water flow has little influence on the total energy consumption during the fixed-point rotation of the underwater hexapod robot. Because the rotation angle of the body is small, the energy consumption is mainly used to overcome the swing of the crawling leg in the water. Therefore, the rotational gait coefficient KF ignores the water flow factor, and the curve fitted by the least square method is obtained as Eq. (9):

$$P = 0.02\partial\theta^2 + 0.86\partial\theta + 60.94(\mathrm{kJ}) \tag{9}$$

3 Energy Consumption Based Simulated Annealing Local Minimum Escape Algorithm

The global path planning of the underwater hexapod robot is based on the combination of artificial potential field method and simulated annealing method. The initial robot selects the target points according to the potential field function of artificial potential field method; When the robot is trapped in a local minimum, it exits the artificial potential field planning strategy and uses simulated annealing method for escape planning.

Because the typical phenomenon of the local minimum is the repeated oscillation of the path, the distance between the planning point in step i and the planning point in step i + 2 is selected as the judgment standard. When $\left|\sqrt[2]{x_i^2 + y_i^2} - \sqrt[2]{x_{i+2}^2 + y_{i+2}^2}\right| \leq 0.1L_{BC}$, the underwater hexapod robot is considered to be trapped in a local minimum.

The idea of annealing method is applied to the local minimum escape strategy of artificial potential field method.

1) At the current point X, a virtual point X1 is added along the flow direction, and the distance between X and X1 is the step of a crawling cycle;

2) Take x as the center, step size as the radius, select 5° Select the next virtual point x2 for the deviation angle, as shown in Fig. 5.
3) The energy consumption P_{X1} and P_{X2} of the robot from x to X_1 and X_2 points are calculated respectively. If $P_{x1} \leq P_{x2}$ is satisfied and x_1 point does not coincide with the obstacle, X1 point is accepted as the current best point;
4) If $P_{x1} > P_{x2}$, calculate $P = e^{-\frac{\Delta}{T(i)}}$, $\Delta = P_{x1} - P_{x2}$, The probability P is used to choose whether to accept point X_1. If this point is less than P, and point X_2 does not coincide with the obstacle, then point x_2 is the current best choice, and T is reduced in a certain way. Set $T(i) = T \cdot \frac{36-i}{36}$;
5) Select point x_3 according to step 2, and judge the current best point according to step 3 and step 4;
6) Through a cycle comparison, the next planning coordinates of the underwater hexapod robot are selected;
7) The above process is repeated until the distance between the robot and the local minimum is more than 5 times the step size.

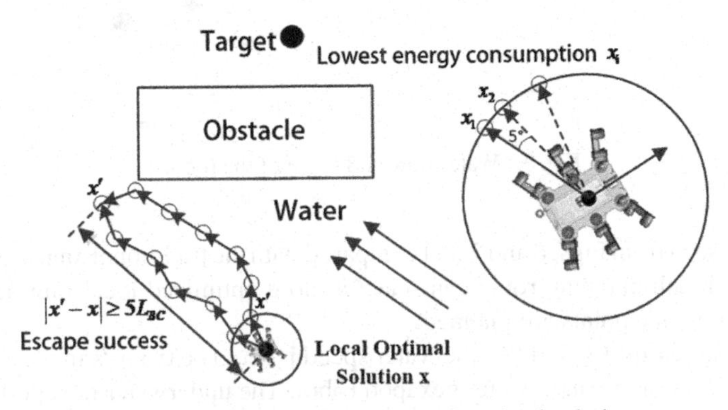

Fig. 5. The escape process of local optimal solution.

4 Modeling and Simulation Results

Through the above theoretical analysis, this paper uses MATLAB to build the simulation environment, set the water velocity along the x-axis, the flow rate is 0.5 m/s, the underwater hexapod robot starts from (0,0) point, and the target point coordinate is set to (10,10). Set the coordinates of environmental obstacles as (1,1; 3 3; 4 4; 2 6; 6 2; 5 5;) The influence distance of the obstacle is 2 m, the size radius of the obstacle is 0.5 m, the gravitational gain coefficient is 5, the repulsive gain coefficient is 15, and the distance between the adjacent points of the path is set as 0.1 m (single crawling cycle step of the underwater hexapod robot). The water flow direction is positive along the X axis. The traditional annealing method and the annealing method based on energy consumption optimization are used respectively, and the simulation results are shown in Figs. 6 and 7.

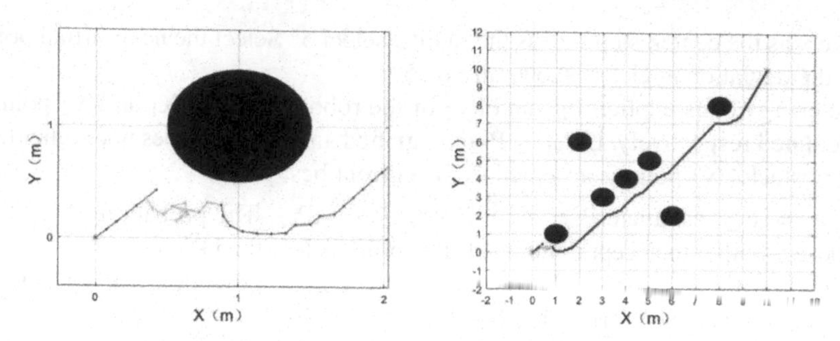

Fig. 6. Water flow and force of the robot.

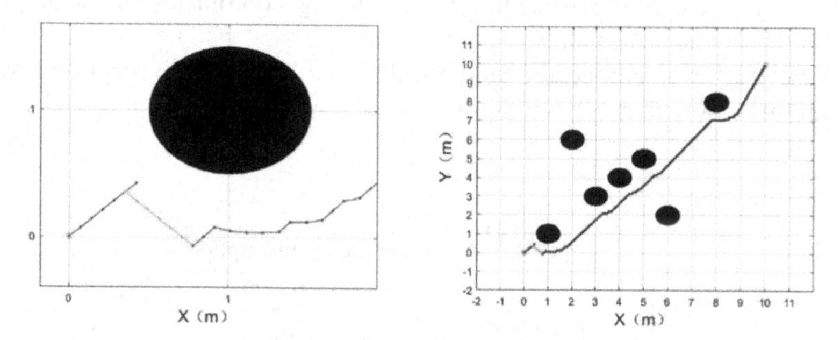

Fig. 7. Water flow and force of the robot.

It can be seen from Figs. 6 and 7 that compared with the traditional annealing method; the improved path planning route can select a more optimized local minimum escape route. Fewer escape points are planned.

According to Eqs. (8) and (9), the water speed is set to be 0.5 m/s in the straight-line crawling process of the underwater hexapod robot. The underwater hexapod robot uses three gait to crawl in a straight-line manner. The crawling cycle is 8 s and the crawling step is 0.1 m. Underwater hexapod robot completes bow rotation by using turning gait. The rotation angle is 10°. The rotation period is 8 s. The total energy consumption and the corresponding joint torque and joint angular velocity of the two gaits of leg 5 in one cycle are shown in Figs. 8 and 9.

According to Eqs. (8) and (9), the energy consumption coefficients of three gait and turning gait are brought into the path planning roadmap. The escape energy consumption of the traditional annealing method is 7157.6 kJ, and that of the simulated annealing method is 1835.2 kJ. Compared with the original algorithm, the improved algorithm can save 74% energy consumption and has good application value.

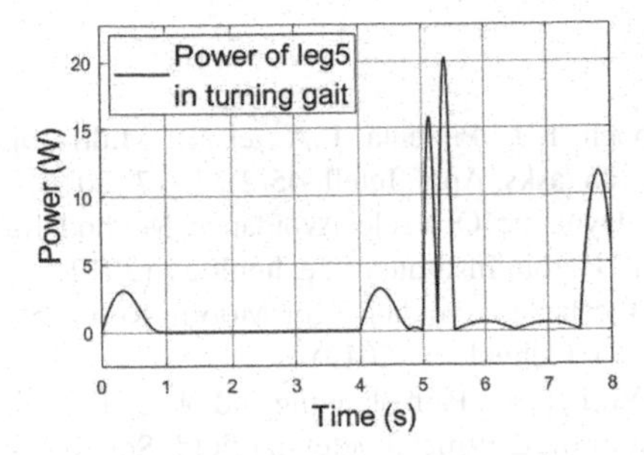

Fig. 8. Energy consumption analysis chart of turning gait.

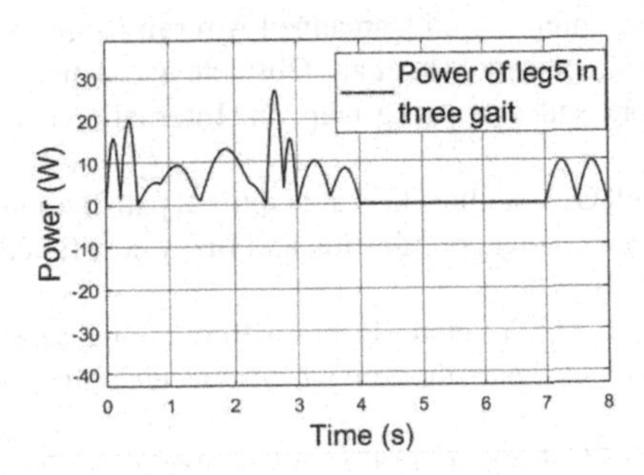

Fig. 9. Energy consumption analysis chart of three gait.

5 Conclusion

In this paper, the local minimum escape planning of the underwater hexapod robot in the artificial potential field method is taken as the research goal, and the energy consumption is used as the optimization index of the escape process. The local minimum escape strategy based on energy consumption optimization is planned. By comparing the escape path of the traditional simulated annealing method and the improved simulated annealing method, it is proved that the improved path can escape from the local minimum more efficiently. By establishing the dynamic model and typical crawling gaits of underwater hexapod robot, the energy consumption of underwater hexapod robot completing three gait and turning gait is given. The coefficient of underwater energy consumption is given. The simulation results show that the improved planning algorithm reduces turning gait in the escape process, and completes the escape with as few three gait as possible. The energy consumption of the improved annealing process is 1835.2 kJ, which is 74% less than that of the traditional annealing process. The improved path planning algorithm can plan the operation path of the underwater hexapod robot under the constraint of energy consumption, and has good planning effect.

Acknowledgment. This work was supported by the Liaoning Province youth top talent project [Grant No. XLYC1807174] and the Independent projects of the State Key Laboratory [Grant No. 2019-Z08].

References

1. Eddisaravi, K., Alitappeh, R.J., Pimenta, L.A., et al.: Multi-objective approach for robot motion planning in search tasks. Appl. Intell. **45**(2), 1–17 (2016)
2. Wang, P.P.: Study on Dynamic Obstacle Avoidance Method for Mobile Robot in Partly Unknown Environment. Harbin Institute of Technology (2012)
3. Li, Y.M. Research of Obstacles Avoidance for Mobile Robot based on Artificial Potential Field. Hefei University of Technology (2013)
4. Liu, Z.Q., Zhu, W.D., Ni, J., et al.: Path planning and obstacle avoidance method for automobile vehicle based on improved artificial potential field. Sci. Technol. Eng. **17**(16), 310–315 (2017)
5. Sonmez, A., Kocyigit, E., Kugu, E.: Optimal path planning for UAVs using genetic algorithm. In: 2015 International Conference on Unmanned Aircraft Systems (ICUAS). IEEE (2015)
6. Maningo, J., Faelden, G., Nakano, R., et al.: Obstacle avoidance for quadrotor swarm using artificial neural network self-organizing map. In: International Conference on Humanoid. IEEE (2016)
7. Garcia, M.A.P., Montiel, O., Castillo, O., et al.: Optimal path planning for autonomous mobile robot navigation with ant colony optimization and fuzzy cost function evaluation. Appl. Soft Comput. **9**(3), 1102–1110 (2009)
8. Wang, X.Y., Yang, L., Zhang, Y., et al.: Robot path planning based on improved ant colony algorithm with potential field heuristic. Kongzhi yu Juece/Contr. Decision **33**(10), 1775–1781 (2018)
9. Pei-Lun, L.I., Yang, Q.: Path planning for underwater glider based on improved artificial potential field method. Ship Sci. Technol. (2019)
10. Wu, Z.P., Tang, N., Chen, Y.L., et al.: AUV path planning based on improved artificial potential field method. Control Instrum. Chem. Indust. (2014)
11. Rao, D., Williams, S.B.: Large-scale path planning for Underwater Gliders in ocean currents (2009)
12. Pereira, A.A., Binney, J., Hollinger, G.A., et al.: Risk-aware path planning for autonomous underwater vehicles using predictive ocean models. J. Field Robot. **30**(5), 741–762 (2013)
13. Wang, K.P.: Mechanical modeling and simulation of mobile system for underwater hexapod robot. Harbin Institute of Technology (2018)
14. Xuan, Q.B.: A study on multi-legged walking robot gait planning and control system. Hangzhou Dianzi University (2013)

Trajectory-Smooth Optimization and Simulation of Dual-Robot Collaborative Welding

Jiahao Xiong[1], Zhongtao Fu[1(✉)], Miao Li[2], Zhicheng Gao[1], Xiaozhi Zhang[1], and Xubing Chen[1(✉)]

[1] School of Mechanical and Electrical Engineering, Wuhan Institute of Technology, Wuhan 430205, China

[2] School of Power and Mechanical Engineering, Wuhan University, Wuhan 430072, China

Abstract. Dual-robot collaborative welding has been widely used in the field of automation due to its flexibility. However, the jiggle in the welding process will inevitably affect the welding quality to a certain extent. For this reason, a trajectory-smooth optimization methodology based on the stationarity function is proposed in this paper. Based on the existing welding platform, the trajectory-smooth optimization of complex space weld path for dual-robot collaborative welding is studied. The motion stability function of the robot is established, and the dual-robot workspace is achieved by means of Monte Carlo method. On this basis, the optimal space point is searched to ensure the stability of the whole welding process. Finally, a saddle type space weld is taken as an example for simulation, and the results before and after optimization are compared and analyzed to verify the effectiveness and feasibility of the proposed methodology.

Keywords: Dual robots · Collaborative welding · Trajectory-smooth optimization

1 Introduction

Compared with single robot, dual robots have higher degree of freedom and flexibility [1], and the collaborative welding system of dual robots can provide a better solution for the welding of complex parts. In the welding process, the two robots move synchronously according to their respective off-line trajectories to complete the welding task [2]. However, the final welding quality of the parts is difficult to be guaranteed. In the trajectory planning, joint smoothness, velocity uniformity, impact of the joint and so on are all important factors that directly

Supported by the Natural Science Foundation of China (51805380, 51875415), the Innovation Group Foundation of Hubei (2019CFA026), and Graduate Education Innovation Foundation of Wuhan Institute of Technology (CX2020045).

X.-J. Liu et al. (Eds.): ICIRA 2021, LNAI 13014, pp. 699–709, 2021.
https://doi.org/10.1007/978-3-030-89098-8_66

affect the welding quality. Trajectory planning is an important part of the welding robot. Whether the welding robot can complete the welding task with high quality, trajectory planning occupies a very large part of the reason. Therefore, more and more researchers are actively studying the method of motion trajectory optimization to ensure the stability of the welding process and improve the ability of the collaborative welding quality of the two robots.

In general, trajectory planning methods include joint space and Cartesian space trajectory planning [3]. The joint space trajectory planning mainly uses polynomial interpolation algorithm to solve the relationship between joint variables and time in the process of motion. The end trajectory of the robot can not be determined and it is easy to cause collision. Cartesian space trajectory planning mainly uses linear interpolation method and circular interpolation method to plan the continuous path points in the workspace. Although the end trajectory is intuitive, it is possible that singularities may cause angle mutation. The two planning methods have different characteristics and are suitable for different scenarios. Wang et al. [4] proposed a smooth point-to-point trajectory planning method for industrial robots, which used high-order polynomials for trajectory planning in joint space. Chettibi et al. [5] proposed a method to generate smooth joint trajectories of robot using radial basis function interpolation method, and joint impact and running time were optimized on the resulting trajectories.

Many scholars have carried out a series of research work on optimal trajectory planning on the basis of the two kinds of trajectory planning, using intelligent algorithms such as particle swarm optimization [6] and genetic algorithm [7] to achieve trajectory planning with time optimization [8], energy optimization [9] and stationarity optimization [10]. Xidias et al. [11] proposed a time-optimal trajectory planning method using multi-population genetic algorithm. The trajectory was fitted by cubic B-spline curve, and the motion constraints of the manipulator's velocity and acceleration were taken into account. Das et al. [12] adopted an improved particle swarm optimization algorithm with evolutionary operator (IPSO) to achieve optimal collision-free trajectory planning in complex environments, and verified the robustness and effectiveness of the algorithm. Wang et al. [13] used Bezier curve to fit the joint trajectory. The kinematics equation combined with constrained particle swarm optimization algorithm is used to coordinate trajectory planning of a dual-arm robot.

Most of the studies focus on the trajectory planning and optimization methods of joint space, and there are few studies on the trajectory optimization of Cartesian space. However, for complex space weld welding tasks, since the dual-robot welding requires the welding torch to move in strict accordance with the weld trajectory, it is necessary to intuitively plan the welding torch end trajectory, so the Cartesian space trajectory planning method naturally becomes a better choice. However, one of the disadvantages of Cartesian space trajectory planning is that the inverse kinematics solution of robot is not unique and the uncertainty of robot trajectory also increases. Although the trajectory of Cartesian space coordinate frame is intuitive, singularities may occur so that the joint angle value will mutate when the robot moves from one point to another. In summary, in order to ensure the welding quality of dual robots, a motion optimization methodology based on the stationarity function is proposed. Based on the existing welding platform, the trajectory-smooth optimization of complex space weld path of dual robots collaborative welding is studied.

The remainder of the paper is structured as follows: a dual-robot welding platform is established in Sect. 2. The Monte Carlo method is used to determine the robot working space in Sect. 3. In Sect. 4, the stability function is proposed and the trajectory-smooth optimization algorithm of welding process is designed. In Sect. 5, simulation is carried out to verify the effectiveness of the proposed methodology. Finally, the conclusions of this study are given in Sect. 6.

2 Dual-Robot Welding System

As shown in Fig. 1, the dual-robot collaborative welding system, which consists of welding robot with welding gun, collaborative robot with end fixture, and unwelded workpiece, can be divided into the system establishment, welding seam discretization and optimization methodology of trajectory-smooth generation. For the dual-robot welding system, the relevant coordinate frames are established to describe the motion constraint of the two robots, in which $\{R_1\}$ and $\{R_2\}$ represent the base frames of welding robot and collaborative robot respectively, $\{E_1\}$ and $\{E_2\}$ signify the end-effector frames of welding robot and collaborative robot respectively, $\{G\}$ is the global frame, $\{M\}$ and $\{S\}$ are the main-pipe frame and welding seam frame respectively, and $\{W_g\}$ is the welding torch frame.

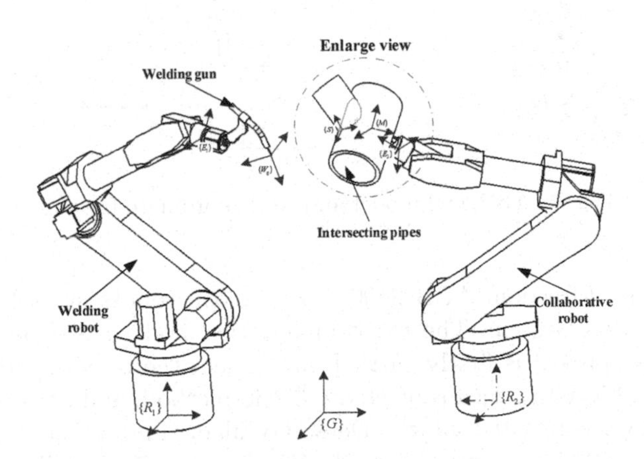

Fig. 1. Dual-robot welding system [2].

The whole welding process of dual robots constitutes a complete closed kinematic chain, so its kinematics constraint relationship can be described in form of homogeneous transformation matrix:

$$^{G}\mathbf{T}_{R_1} \cdot ^{R_1}\mathbf{T}_{E_1} \cdot ^{E_1}\mathbf{T}_{W_g} \cdot ^{W_g}\mathbf{T}_{S} = ^{G}\mathbf{T}_{R_2} \cdot ^{R_2}\mathbf{T}_{E_2} \cdot ^{E_2}\mathbf{T}_{M} \cdot ^{M}\mathbf{T}_{S} \tag{1}$$

In the working scene, the base frame $\{R_1\}$ of the welding robot is set as the global frame $\{G\}$. ${}^{G}\mathbf{T}_{R_2}$ is the transformation matrix from the base frame $\{R_1\}$ of the welding robot to the base frame $\{R_2\}$ of the collaborative robot. The calibration method can be referred to our previous dual-robot calibration work [1]. ${}^{R_1}\mathbf{T}_{E_1}$ and ${}^{R_2}\mathbf{T}_{E_2}$ are respectively the transformation matrices of the base frame $\{R_1\}$ and $\{R_2\}$ of the dual robots to their respective end-effector frames $\{E_1\}$ and $\{E_2\}$, which can be calculated by Eq. (3). ${}^{E_1}\mathbf{T}_{W_g}$ is the transformation matrix from the end frame $\{E_1\}$ of the welding robot to the welding torch frame $\{W_g\}$, which is determined according to the size of the selected welding torch, and is a constant matrix. ${}^{W_g}\mathbf{T}_S$, ${}^{E_2}\mathbf{T}_M$ and ${}^{M}\mathbf{T}_S$ are respectively the transformation matrices from the welding torch frame $\{W_g\}$ to welding seam frame $\{S\}$, from end-effector frame $\{E_2\}$ of the collaborative robot to main-pipe frame $\{M\}$ and from main-pipe frame $\{M\}$ to welding seam frame $\{S\}$. The calibration algorithm can be referred to our previous research work [2].

3 Determination of Dual-Robot Workspace

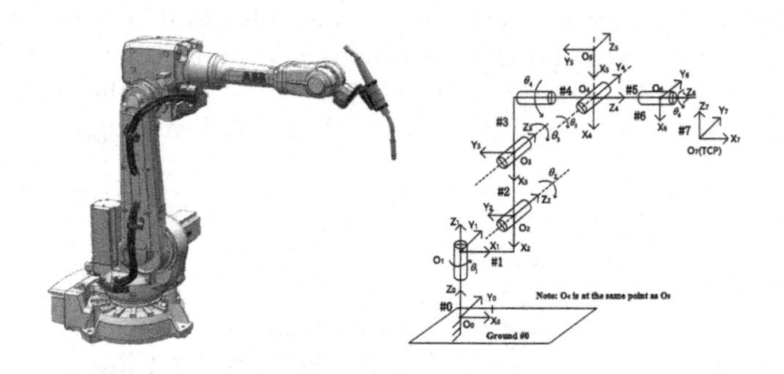

Fig. 2. ABB2600 model and link coordinate frame.

Two six degree-of-freedom ABB 2600-12 kg/1.65 m robots are selected for the dual-robot welding system. The robot kinematics model based on modified D-H parameter approach is firstly carried out. Then, the point-cloud map of the workspace is achieved by means of Monte Carlo method, and the involved work-piece boundary is extracted via the Delaunay algorithm. The robot model and the associated coordinate frames are depicted in Fig. 2, and Table 1 gives the corresponding D-H parameters and joint motion ranges.

Table 1. DH parameter table and joint angle range.

Joint i	α_{i-1} (°)	a_{i-1} (mm)	d_i (mm)	θ_i (°)	θ_{imin} (°)	θ_{imax} (°)
1	0	0	445	θ_1	−180	180
2	−90	150	0	θ_2+90	−95	155
3	0	−700	0	θ_3	−180	75
4	90	−115	795	θ_4	−400	400
5	−90	0	0	θ_5	−120	120
6	90	0	85	θ_6	−400	400

The transformation relation between $\{O_{i-1}\}$ and $\{O_i\}$ coordinate frame of two adjacent links can be obtained by the homogeneous transformation matrix as:

$$^{i-1}\mathbf{T}_i = \begin{bmatrix} \cos\theta_i & -\sin\theta_i & 0 & a_{i-1} \\ \sin\theta_i\cos\alpha_{i-1} & \cos\theta_i\cos\alpha_{i-1} & -\sin\alpha_{i-1} & -d_i\sin\alpha_{i-1} \\ \sin\theta_i\sin\alpha_{i-1} & \cos\theta_i\sin\alpha_{i-1} & \cos\alpha_{i-1} & d_i\cos\alpha_{i-1} \\ 0 & 0 & 0 & 1 \end{bmatrix} \tag{2}$$

According to the robot kinematic chain in Fig. 2, the kinematics model can be expressed via Eq. (2) as:

$$^0\mathbf{T}_6 = {^0\mathbf{T}_1} \cdot {^1\mathbf{T}_2} \cdot {^2\mathbf{T}_3} \cdot {^3\mathbf{T}_4} \cdot {^4\mathbf{T}_5} \cdot {^5\mathbf{T}_6} \tag{3}$$

For the established kinematic model, the position values of the end-effector's configuration $^0\mathbf{T}_6$ can be calculated by the chosen random values of all the joint variables in joint space. The random values can be generated via Monte Carlo method and written as:

$$\theta_i = \theta_{imin} + (\theta_{imax} - \theta_{imin}) \cdot \mathrm{rand}(N, 1) \tag{4}$$

where i represents the i-th joint, i =1, 2... 6. θ_{imin} and θ_{imax} represent the minimal and maximal angles of each joint respectively, and data sizes are shown in Table 1. N is the random points number, and θ_i is the random angle value of each joint.

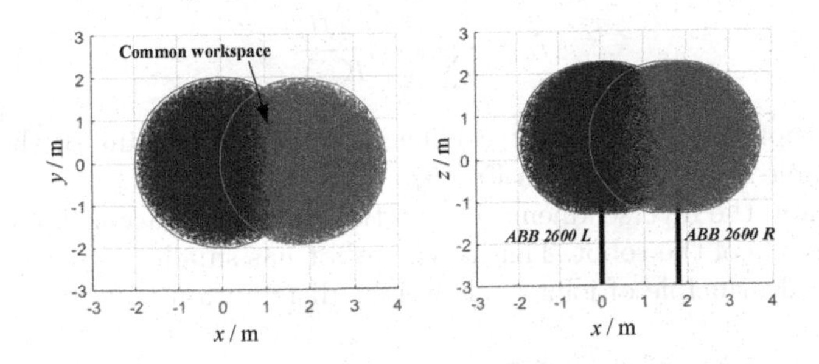

Fig. 3. Dual-robot workspace.

Setting number of random points $N = 100000$. The workspace diagram of the dual robots generated by MATLAB and the extracted boundary via Delaunay algorithm are depicted in Fig. 3. The workspace ranges of dual robots are in x direction (0 m, 2 m), y direction (0 m, 1.7 m), and z direction (0 m, 2.1 m).

4 Optimization Methodology of Trajectory-Smooth Generation

Since the dual robots workspace is irregular in Fig. 3, the rectangular space from points (0, 0, 0) to points (2.00 m, 1.77 m, 2.10 m) as the diagonal is used as the search space for optimized points, and the position of points is determined by the step size of 333 mm, 188 mm and 350 mm in the x, y and z directions of the dual robots workspace, respectively. So the total number of searched space points is 490.

Three aspects should be considered in the optimization methodology of trajectory-smooth generation: 1) The offset between the initial position of the robot and the mechanical origin position of the robot should be as small as possible, so as not to reach the joint limit during the movement; 2) The angle mutation of the robot joint should be avoided during the movement, as in Eq. (5); 3) The movement of the previous joint has an implicated influence on the subsequent joint movement. Distribute weight on each joint to ensure smaller changes in the front joint during the entire movement, as in Eq. (7). The weight coefficient is shown in the Table 2.

The stability of motion is usually described by the average joint change of each joint of the robot through the joint position sequence of two adjacent points. The function expression is as follows:

$$f_{tj} = \frac{1}{n} \cdot \sum_{i=1}^{n} (1 - \frac{|\theta_{ij} - \theta_{(i-1)j}|}{\theta_{jmax} - \theta_{jmin}}) \tag{5}$$

where n represents the number of discrete points of the weld, $i = 1, 2 \ldots n$. j represents the j-th joint, $j = 1, 2 \ldots 6$.

Usually, the following equation is used to represent the smoothness function of the motion of a welding seam welded by means of one robot system:

$$f_t = \frac{\sum_{j=1}^{n} (K_{tj} f_{tj})}{\sum_{j=1}^{n} (K_{ij})} \tag{6}$$

where K_{tj} represents the weight coefficient of the j-th joint of the robot, as shown in Table 2. f_{tj} is shown in Eq. (5).

In addition, the inverse kinematics solution is selected according to the minimal pose change of the robot. That is, the robot has smaller changes in adjacent moments, and principle of joint angle weight distribution is used:

$$J_m = \sum_{i=1}^{n} (|\theta_{ij} - \theta_{(i-1)j}| \cdot K_{mj}) \tag{7}$$

where n represents the number of discrete points of the weld, $i = 1, 2 \ldots n$. j represents the j-th joint, $j = 1, 2 \ldots 6$. K_{mj} is expressed as the weight coefficient of the j-th joint of the robot, as shown in Table 2.

Table 2 is the distribution table of the weight coefficients of each joint. The purpose is to express the influence of the weight of each joint on the stationarity through the size of the data, that is, the larger the value, the higher the weight, the greater the impact on the stability, and the 10-fold proportional number sequence is used. 100 10 1 0.1 0.01, etc. are specific weight coefficients. Of course, other values with different sizes can be used to express the weight difference, and it is not limited to the weight coefficient shown in Table 2.

Table 2. Weight coefficient.

j-th joint	Weight					
	1	2	3	4	5	6
K_{tj}	10	1	0.1	0.1	0.01	0.01
K_{mj}	100	10	1	1	0.1	0.1

The trajectory-smooth optimization algorithm is synopsized based on the established stability function, which mainly includes the following procedures as:

Step 1: Search the space position lattice with a certain step length in the working space;

Step 2: Obtain the offline trajectory of the dual robots based on the spatial position point;

Step 3: Use inverse kinematics to obtain the joint angle of the two robots according to the offline trajectory;

Step 4: Obtain the value of the stationarity function of the spatial location point;

Step 5: Obtain the location of the optimal spatial location point.

The results of all spatial points optimized by the algorithm are shown in Fig. 4. From the results, the second group spatial position (0.666 m, 0.376 m, 0 m) is the local minimum of all optimization results. The optimal point is searched precisely again near the local minimum, and the position of the point is determined with small steps of 20 mm, 20 mm, and 20 mm in the x, y and z directions in the dual-robot workspace. There are 343 spatial points accurately searched.

The results of the local accurate optimization are shown in Fig. 5. Figure 5a and Fig. 5b are respectively the evaluation values of the stationarity function of welding and collaborative robots. The weighting result is shown in Fig. 5c. The optimization results show that the stationarity function value of the 25-th group spatial positions (0.666 m, 0.376 m, 0 m) is the smallest, which is the optimal spatial point for the stability of the dual robots collaborative motion.

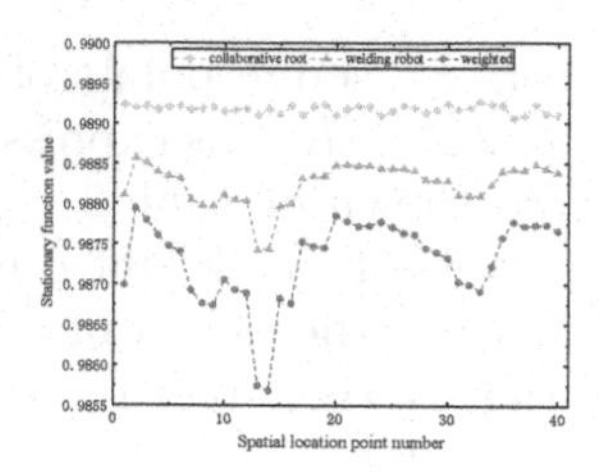

Fig. 4. Optimized result.

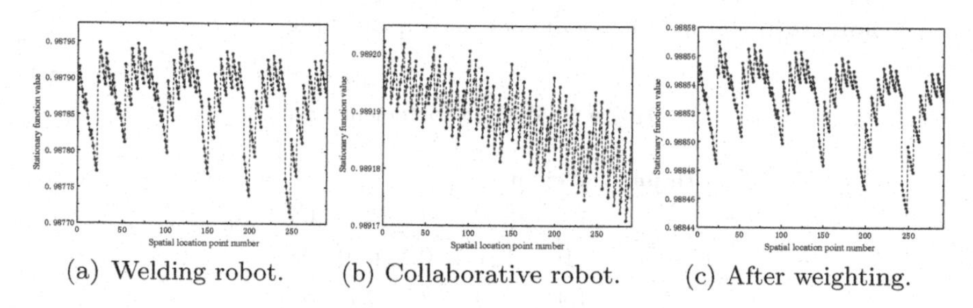

(a) Welding robot. (b) Collaborative robot. (c) After weighting.

Fig. 5. Accurate optimization results.

5 Simulation and Discussions

On the basis that the optimal spatial point position has been obtained, in order to verify the effectiveness of the method in this paper, as shown in Fig. 6, we build a simulation scene based on the spatial point. An offline welding simulation is carried out, and the results are evaluated and analyzed.

As shown in Fig. 7, the optimized angular displacement of the dual-robot at the optimal spatial point position is compared with the angular displacement of the dual-robot at any point in the space before optimization, such as (0.95 m,

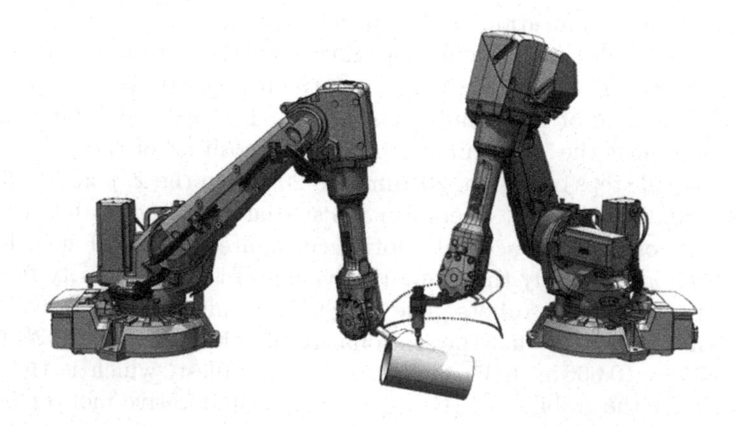

Fig. 6. Dual-robot welding setup.

0 m, 0.45 m). Figure 7a and Fig. 7c respectively show the angular displacement of the welding robot and the collaborative robot at the optimal spatial point position. Figure 7b and Fig. 7d respectively show the angular displacement of the welding robot and the collaborative robot at the spatial point (0.95 m, 0m , 0.45 m).

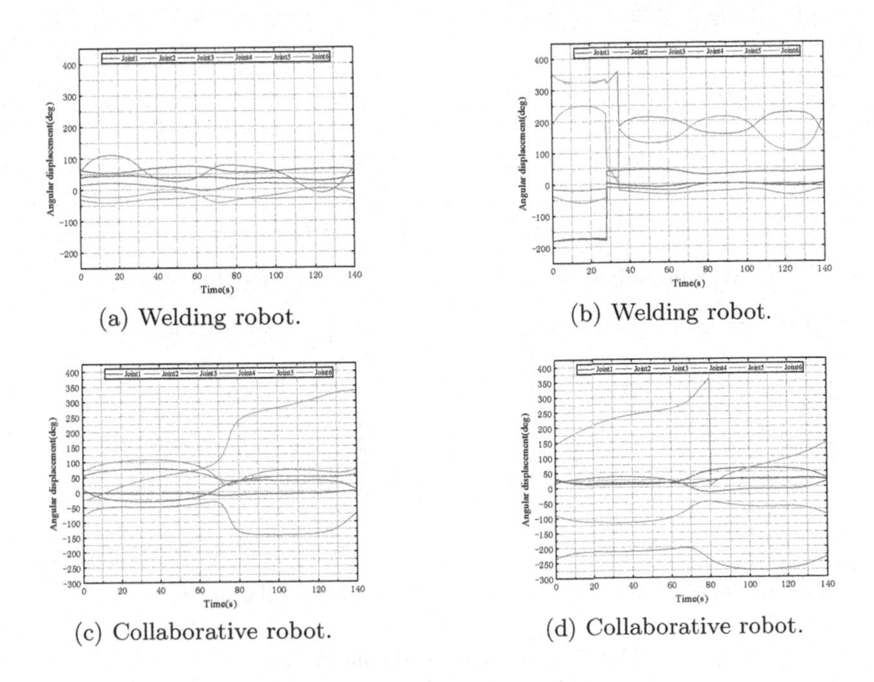

(a) Welding robot. (b) Welding robot.

(c) Collaborative robot. (d) Collaborative robot.

Fig. 7. Joint angular displacement.

From the comparison of the results in Fig. 7, it can be seen that the optimized results of the algorithm effectively avoid the sudden change of angle, and the joint angle changes smoothly, and the range of change is concentrated near the joint zero point. It can reduce the possibility of reaching the limit position of the joint. Angle optimization using the weight distribution principle also has obvious effects. The angle changes of joints 1 and 2 are small, which reduces the large-scale posture change of the entire robot and enhances the smoothness of the dual-robot collaborative welding movement.

As shown in Fig. 8, the optimized dual-robot angular displacement of the optimal spatial point position is compared with the dual-robot angular displacement of our previous research work [2]. Figure 8a and Fig. 8c show the angular displacement of the welding robot and the collaborative robot at the optimal spatial point position respectively. Figure 8b and Fig. 8d are the previous research work [2] the angular displacement of the welding robot and the collaborative robot. It can be seen from the comparison of the results in Fig. 8 that the stability of the welding robot increases after optimization, and the collaborative robot does not change much. The angular displacement of the six-th joint changed from 160° before optimization to 50° after optimization, and the angular displacement of other joints changed smoothly. On the whole, the smoothness of the optimized dual-robot movement has been improved.

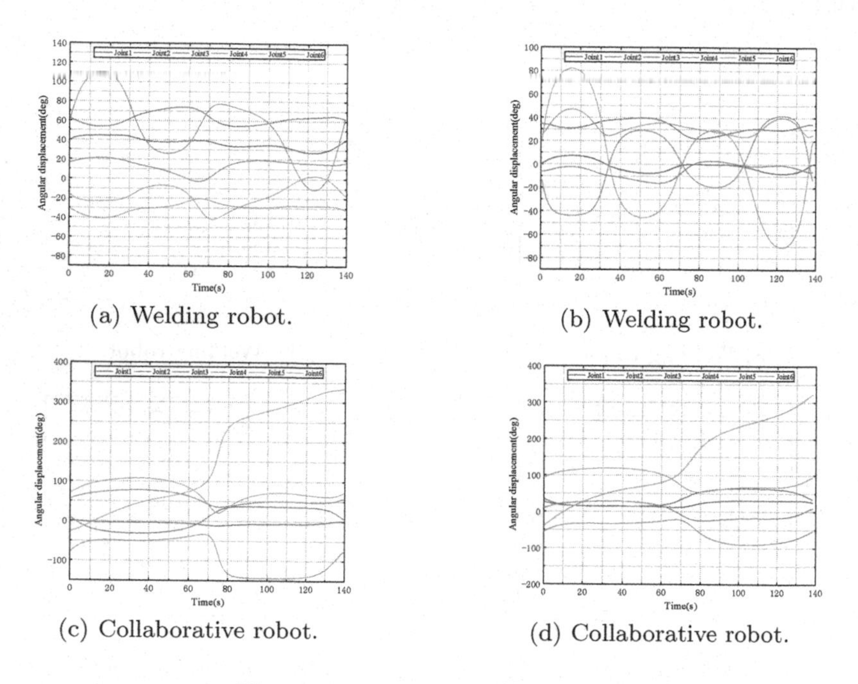

(a) Welding robot.

(b) Welding robot.

(c) Collaborative robot.

(d) Collaborative robot.

Fig. 8. Joint angular displacement.

6 Conclusions

In this paper, we propose a trajectory-smooth generation optimization algorithm based on stationarity function. Based on the established dual-robot welding platform, the trajectory-smooth optimization of complex space weld path of collaborative welding is studied. First, the Monte Carlo method is used to determine the dual robots workspace. Subsequently, taking the stability of the welding process as the optimization goal, the robot motion stability function is proposed, and the stability optimization algorithm is designed to search for the best spatial point in the workspace. Finally, the offline trajectory generated based on the optimal spatial point verifies the stability of the collaborative movement in the welding simulation, and also proves the effectiveness and feasibility of the method. In our next work, we will perform relevant actual welding experiments on the optimized offline trajectory to investigate the actual welding effect.

References

1. Fu, Z., Pan, J., Spyrakos-Papastavridis, E., Chen, X., Li, M.: A dual quaternion-based approach for coordinate calibration of dual robots in collaborative motion. IEEE Rob. Autom. Lett. **5**(3), 4086–4093 (2020)
2. Xiong, J., Fu, Z., Chen, H., Pan, J., Gao, X., Chen, X.: Simulation and trajectory generation of dual-robot collaborative welding for intersecting pipes. Int. J. Adv. Manuf. Technol., 2231–2241 (2020). https://doi.org/10.1007/s00170-020-06124-w
3. Li, Z., Li, G., Sun, Y., Jiang, G., Kong, J., Liu, H.: Development of articulated robot trajectory planning. Int. J. Comput. Sci. Math. **8**(1), 52–60 (2017)
4. Wang, H., Wang, H., Huang, J., Zhao, B., Quan, L.: Smooth point-to-point trajectory planning for industrial robots with kinematical constraints based on high-order polynomial curve. Mech. Mach. Theory **139**, 284–293 (2019)
5. Chettibi, T.: Smooth point-to-point trajectory planning for robot manipulators by using radial basis functions. Robotica **37**(3), 539–559 (2019)
6. Kim, J.J., Lee, J.J.: Trajectory optimization with particle swarm optimization for manipulator motion planning. IEEE Trans. Industr. Inf. **11**(3), 620–631 (2015)
7. Liu, Y., Guo, C., Weng, Y.: Online time-optimal trajectory planning for robotic manipulators using adaptive elite genetic algorithm with singularity avoidance. IEEE Access **7**, 146301–146308 (2019)
8. Kim, J., Croft, E.A.: Online near time-optimal trajectory planning for industrial robots. Rob. Comput. Integr. Manuf. **58**, 158–171 (2019)
9. Liu, S., Wang, Y., Wang, X.V., Wang, L.: Energy-efficient trajectory planning for an industrial robot using a multi-objective optimisation approach. Procedia Manuf. **25**, 517–525 (2018)
10. Fang, Y., Hu, J., Liu, W., Shao, Q., Qi, J., Peng, Y.: Smooth and time-optimal S-curve trajectory planning for automated robots and machines. Mech. Mach. Theory **137**, 127–153 (2019)
11. Xidias, E.K.: Time-optimal trajectory planning for hyper-redundant manipulators in 3D workspaces. Rob. Comput. Integr. Manuf. **50**, 286–298 (2018)
12. Das, P., Jena, P.K.: Multi-robot path planning using improved particle swarm optimization algorithm through novel evolutionary operators. Appl. Soft Comput. **92**, 106312 (2020)
13. Wang, M., Luo, J., Yuan, J., Walter, U.: Coordinated trajectory planning of dual-arm space robot using constrained particle swarm optimization. Acta Astronaut. **146**, 259–272 (2018)

Investigation of Robotic Belt Grinding Methods Used for Dimension Restore of Repaired Blades

Xifan Liu, Chong Lv, and Lai Zou(✉)

Chongqing University, No. 174 Shazhengjie, Shapingba, Chongqing 400044, China

Abstract. The profile accuracy and surface quality of repaired blade are directly determined by the dimension restore, which has a far-reaching impact on the service performance and fatigue life of engine after maintenance. In this paper, the robot belt grinding method for repairing blade is studied. A flexible grinding method with variable stiffness is proposed to realize the rapid and accurate material removal and the improvement of the surface quality of the repaired blade. The method of model detection and processing of the repair sample is studied, and the machining accuracy of the grinding device is improved through robot operation calibration. Finally, the feasibility of the above method is verified by the grinding experiments of the samples of the profile repair, edge repair and blade tip repair. The grinding accuracy can reach 0.07 mm, the surface roughness less than Ra0.04.

Keywords: Repaired blade · Robotic belt grinding · Model processing · Dimension restore

1 Introduction

As aero-engine blade works in the high-temperature, high-pressure and alternating load environment for a long time, the damages such as distortion, ablation, corrosion, excessive wear and cracks usually appeared. It is bad for the performance and service life of engine. Hence, it is necessary to inspect and repair engine blades regularly. Researchers have carried out a lot of research on blade remanufacturing from damage detection [1], reverse modeling [2], additive repair technology and equipment [3, 4], dimension restore and heat treatment [5]. Among them, the profile accuracy and surface quality of remanufactured blades are directly determined by dimension restore, which is very important for the service performance and life of aero-engine after maintenance.

Benoit Rosa *et al.* [6] carried out laser polishing on the surface of additive manufacturing parts, and optimized the polishing parameters according to the different surface morphology, material and thickness of additive manufacturing parts. Kim *et al.* [7] carried out electrochemical polishing on the surface of additive manufacturing parts, and found that the surface micro pits, roughness, smoothness and corrosion resistance were significantly improved. Lyczkowska *et al.* [8] carried out chemical polishing on SLM titanium alloy parts with complex spatial structure, and the surface roughness decreased significantly after polishing. Liu *et al.* [9] proposed an analytical tool correction based on

X.-J. Liu et al. (Eds.): ICIRA 2021, LNAI 13014, pp. 710–720, 2021.
https://doi.org/10.1007/978-3-030-89098-8_67

the allowance distribution to prevent the overcut phenomenon during the ECM. Simulation and experiment show that corrected tool is helpful in removing the current density peaks and thus eliminating the over cut.

Zhang *et al.* [10] carried out surface milling machining of blade formed by Plasma Deposition Manufacturing (PDM), which effectively improved the surface quality of metal remanufacturing parts, and the surface roughness reached Ra0.46 μm. Eyup [11] studied the model reconstruction and residual extraction for the repaired turbine blade, and used milling to remove the cladding layer. Yilmaz *et al.* [12] proposed a free-form surface modeling method to repair complex-structure blades. The ball end milling cutter was used to remove the cladding layer at the blade tip. Xiong *et al.* [10] realized the blade direct manufacturing by using the PDM and milling, they studied the influence law of feed rate and cutting speed on the surface accuracy, and obtained blades with surface roughness less than Ra0.46. Wu *et al.* [13] proposed an adaptive positioning method for model reconstruction and milling of repaired blade and realized the recognition and positioning of repaired area and adaptive trajectory planning of remanufactured blade. Huang *et al.* [14, 15] developed a robotic grinding system for aeroengine blade automatic repair. Zenon company of Greece [16] has developed a robotic belt grinding system for the blade dimension restore of the leading edge and exhaust edge, finally improves the surface quality of the repair blade effectively.

To sum up, researches have studied the dimension restore of repaired blade by using laser polishing, electrochemical polishing, chemical polishing, adaptive milling and robotic belt grinding and so on. Because of its flexible grinding characteristics, robotic belt grinding can ensure good surface quality while removing the cladding layer accurately. Moreover, because of its high flexibility and wide versatility, robot has unique advantages in solving the problems such as randomness of position and shape of the repair area, large local allowance and large difference in surface performance. Therefore, this paper studied a series of robotic abrasive belt grinding method on dimension restore of repaired blade.

2 Method Research

At present, the repair methods of blade damage include energy beam additive manufacturing (laser, electron beam, plasma beam), welding and brazing, etc. These repair methods will form obvious cladding layer in the damaged area, which brings three major problems to the subsequent blade dimension restore.

1. Position problem: the location, shape and size of the damaged part of the blade are very random, and there is no identical damage in theory. This puts forward high requirements for the identification and positioning of cladding layer, the precision of model processing and the flexibility of grinding equipment.
2. Local allowance: the blade is formed alternately by cold and hot during the repairing process, which makes the material easy to form shrinkage cavity, porosity and other defects in the subsurface. Therefore, a large allowance is usually reserved to ensure the material compactness of the damaged parts, while the shape and position accuracy of other undamaged parts of the blade has met the service requirements. This requires

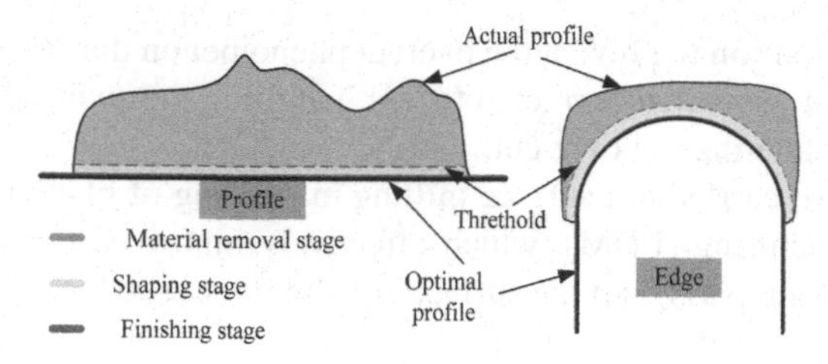

Fig. 1. Layered grinding scheme of cladding

the ability to accurately remove the residual cladding layer without damaging the surface of other areas.

3. Poor surface quality: affected by the surface tension of liquid metal, the surface roughness of additive manufacturing is usually higher than Ra5.0. At the same time, affected by the cyclic thermal stress, there is a large residual stress on the surface of cladding layer. It is urgent to improve the surface quality by surface polishing to meet the service requirements.

Moreover, the accuracy of robot belt grinding is affected by the deformation error of contact wheel, robotic positioning error and blade deformation error, which makes it difficult to control the processing quality. Therefore, the research on robotic belt grinding method for repairing blade is carried out.

2.1 Grinding Scheme and Device

According to the residual distribution of repaired blade, a layered grinding method of cladding layer is proposed, which divides the dimension restore into three steps as shown in Fig. 1.

1) In the material removal stage, most of the excess material of cladding layer was quickly removed, and the contact position of grinding tool is precisely controlled to ensure the consistency of residual allowance distribution.
2) In the shaping stage, the grinding pressure is stable, and the residual allowance in the repair area is removed accurately and evenly to ensure the contour accuracy of the blade can meet the working requirements.
3) In the finishing stage, the whole surface of the blade is polished to improve the consistency of the blade surface texture and eliminate the tool contact.

According to the layered grinding method of cladding layer, the robot belt grinding system is developed as shown in Fig. 2. It consists of robot and grinding device. The grinding device is composed of belt gear train and adaptive floating device. Belt gear train is used to control grinding speed, floating device is used to control grinding pressure, and robot is used to control grinding path and feed speed. The grinding device has two working modes: semi-locking and floating.

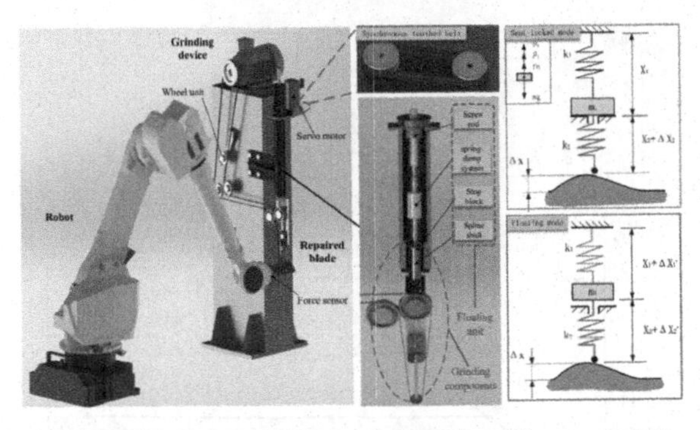

Fig. 2. Robotic belt grinding platform and variable stiffness compliance principle

1) In the semi-locking mode, the grinding head assembly is positioned by the stop block to achieve constant position grinding. At this time, the grinding depth can be accurately controlled. The initial supporting force of the stop block can be adjusted by controlling the servo motor. When the grinding pressure exceeds f_{n0} due to excessive allowance, the grinding head assembly automatically changes to the floating mode and keeps the grinding pressure about equal to f_{n0}, which is suitable for the modification stage.
2) In the floating mode, the grinding components maintains force balance under gravity and spring tension, the contact wheel moves up and down with the workpiece surface profile, and the abrasive belt can fit the surface to be machined better and ensure constant force grinding, which is suitable for the stage of shaping and polishing.

The force-position relationship of the device can be described as follows.

$$\begin{cases} mg = F_1 + F_2 + f_n + F_t \bullet \sin\theta \quad f_n \geq 0 \\ F_2 == F_n \\ f_{n0} == F_n + f_n \end{cases} \tag{1}$$

$$\begin{cases} k_2 \bullet \Delta X_2 = \Delta F \\ \Delta X_2 = \Delta X \end{cases} \tag{2}$$

$$\begin{cases} k_1 \bullet \Delta X_1' = k_2 \bullet \Delta X_2' = \Delta F \\ \Delta X_1' + \Delta X_2' = \Delta X \end{cases} \tag{3}$$

Where mg is the gravity of grinding components, F_1 is the tensile force provided by the damper, F_2 is the deformation elastic force of the contact wheel, f_{n0} is the initial supporting force of the stop, and f_n is the actual supporting force of grinding components, and F_n is the actual contact force between the belt and the workpiece. F_t is the belt tension, θ is the angle between the belt tension and the horizontal direction under semi-locked mode, about 3.6°, ΔX_1 is the deformation of the spring-damp system, ΔX_2 is the total deformation of rubber layer and sand belt, ΔF is the force change caused by residual height. k_1 is the elastic modulus of spring damper, k_2 is the equivalent elastic modulus, and $k_2 \gg k_1$.

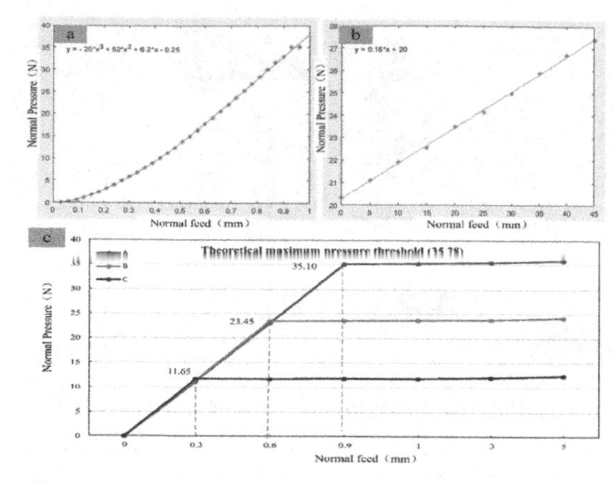

Fig. 3. Force-position relationship of grinding device

Therefore, the deformation of contact wheel can be approximated to f_{n0}/k_2, then the relation of force and position of grinding device can be described as follow.

$$F_n = \begin{cases} k_2 \bullet x & \left(\frac{f_{n0}}{k_2} \geq x \geq 0\right) \\ f_{n0} + k_1 \bullet \left(x - \frac{f_{n0}}{k_2}\right) & \end{cases} \tag{4}$$

Where k_1 is the stiffness coefficient of spring damper, which is selected as 0.16 N/mm. k_2 is affected by rubber material, rubber wheel specification, sand belt material and sand belt specification, which is difficult to be obtained by theoretical calculation directly. Therefore, the KunWei KWR75B force sensor and Renishaw RMP60 probe are used to test the relationship between force and position of the grinding device. Figure 3(a) shows the relationship in semi-locked mode, which is approximate to cubic polynomial. Figure 3(b) shows the relationship in floating mode, with good linearity.

Furthermore, the initial supporting force f_{n0} were set as 12 N, 24 N and 35 N respectively. The actual pressure thresholds were measured as 11.65 N, 23.45 N and 35.10 N correspondingly as shown in Fig. 3(c), and the error rate is less than 4.2%. It can be seen that the variable stiffness and flexibility of the grinding device are obvious, and the force position is sensitive.

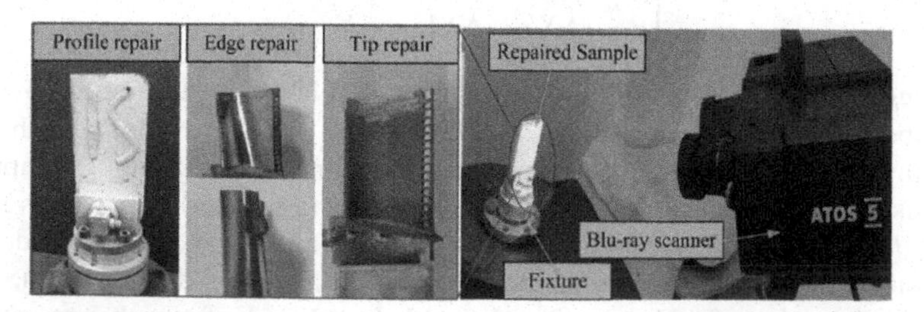

Fig. 4. Detection method of repaired blade (Color figure online)

2.2 Detection and Model Processing

The position, shape and size of the cladding layer are random, so the traditional contact measurement method can not accurately and comprehensively measure the data of the cladding layer on the repaired blade. Based on the accuracy requirements and detection indicators of the blade, this paper uses the atos blue light detector produced by GOM to obtain the actual shape and position data of the parts as shown in Fig. 4, and its measurement accuracy can reach 0.01 mm, which meets the detection requirements of the repaired blade.

Due to the influence of service process and repair process, the actual shape of remanufactured parts has changed such as torsion and offset. The design model cannot be used to the target process model. Therefore, the actual model is used to construct the processing model, as shown in Fig. 5(a).

In order to solve the problems of positioning datum failure and random position of cladding layer, the model pretreatment work was carried out. Figure 5(b) represents the establishment of model datum. The positioning coordinate system of blade is created through the integrated fixture features, and the tool coordinate system is further obtained according to the tooling size. Figure 5(b) also represents the boundary recognition and segmentation of cladding layer. PCL-features is used to calculate the normal vector of point cloud, and the edge contour of cladding layer is extracted according to the change rate of normal vector. The actual model is divided into cladding layer model and undamaged area model by PCL-segmentation.

According to the damage location, the blade can be classified into three types: surface damage, edge damage and tip damage, and the model processing is carried out respectively as shown in Fig. 6.

1) According to the characteristics of the actual model, the theoretical model is obtained by means of cavity repair, surface construction and surface extension.
2) The part of the cladding layer model mapped to the theoretical model is obtained by model registration and point cloud ratio. It is used as the machining model of cladding layer for trajectory planning.

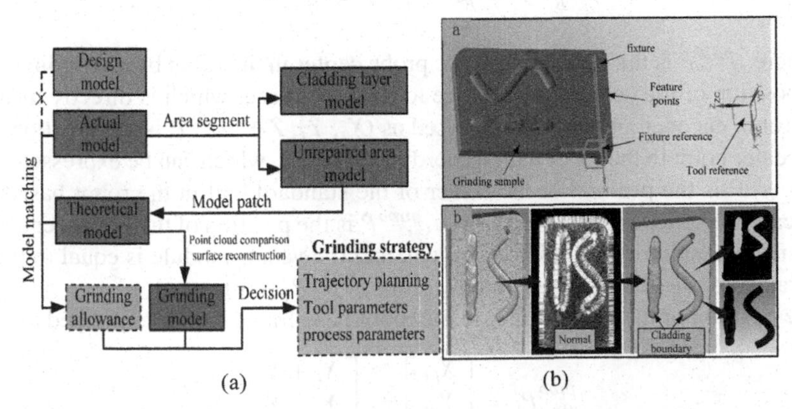

Fig. 5. Model processing. (a) Model processing of repaired blade. (b) Model pretreatment

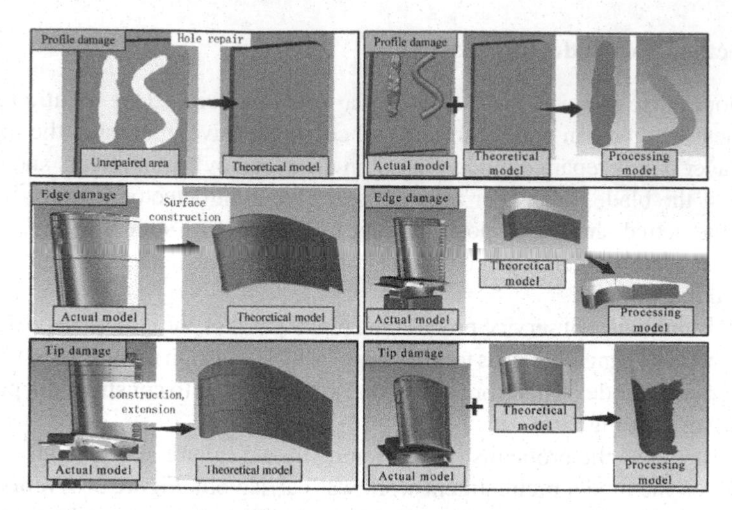

Fig. 6. Model processing

2.3 Robot Operation Calibration

In order to accurately measure the position and attitude of grinding tool to the robot base coordinate system, it is necessary to carry out robot operation calibration for the grinding platform as shown in Fig. 7(a). The whole calibration process is divided into measuring equipment calibration and tool coordinate system establishment.

1) Calibration of measuring equipment
   In order to avoid the influence of the assembly error of the measuring equipment on the calibration results, the measurement equipment is calibrated first. After the robot holding probe touches the standard ball, the I/O return signal controls the robot to stop immediately, and records the data value and coordinates of the center of the robot flange at this time. The kinematics analysis of robot belt grinding system as follows.

$$ {}_{B}^{prob}P = {}_{B}^{f}P + {}_{F}^{prob}P = {}_{B}^{C}P + {}_{C}^{prob}P \tag{5} $$

Where ${}_{B}^{prob}P$ is the coordinate of the probe center in the robot base coordinate, ${}_{B}^{f}P$ is the position of the center of the flange to the robot base, which is directly measured by the robot encoder and can be expressed as (X_i, Y_i, Z_i). ${}_{f}^{prob}P$ is the position of the probe center point in the robot flange coordinate system, which can be expressed as (X_t, Y_t, Z_t). ${}_{B}^{C}P$ is the position of the center of the standard ball in the robot base frame, which can be expressed as (X_C, Y_C, Z_C). ${}_{C}^{prob}P$ is the position of the probe center point in the standard spherical coordinate system, and the vector module is equal to the sum of the radius of the standard ball and the probe ball $(R_1 + r)$.

The position of the probe center in the robot base frame can be expressed as

$$ {}_{B}^{prob}P_i = \begin{bmatrix} X_{pi} \\ Y_{pi} \\ Z_{pi} \end{bmatrix} = \begin{bmatrix} X_i + X_t \\ Y_i + Y_t \\ Z_i + Z_t \end{bmatrix} \tag{6} $$

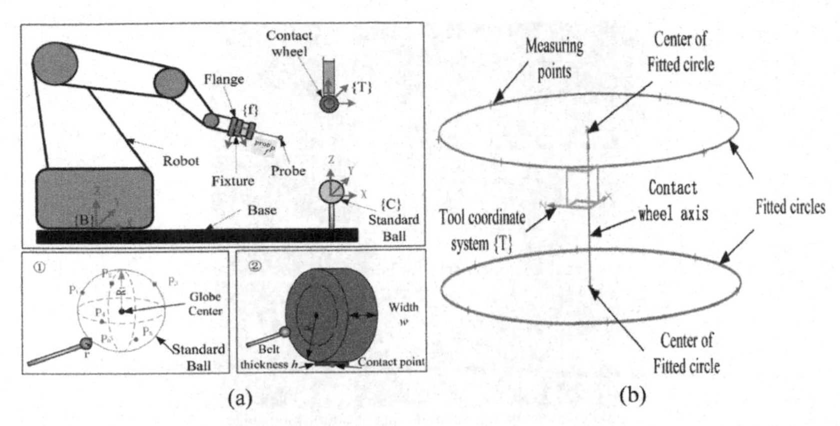

Fig. 7. Tool calibration. (a) Robot calibration process. (b) Tool coordinate establishment.

When the probe touches the standard ball, its center point should be on the same sphere, which meets the formula 7:

$$(X_i + X_t - X_C)^2 + (Y_i + Y_t - Y_C)^2 + (Z_i + Z_t - Z_C)^2 = (R_1 + r)^2 \quad (7)$$

By touching the calibration ball six times in different positions and poses, six unknowns $(X_t, Y_t, Z_t, X_C, Y_C, Z_C)$ in the foSrmula can be obtained by establishing equations.

2) Establishment of tool coordinate system
Furthermore, the calibrated measuring equipment is used to measure the grinding head, and the pose transformation matrix of the machining tool coordinate system relative to the base coordinate system is determined. In order to avoid the measurement error caused by the deformation of rubber layer, this paper chooses to measure the edge profile of contact wheel bearing.

Six measuring points are selected on the left and right sides of the bearing, and the data are imported into UG for feature fitting to obtain the center data and plane direction of the fitting circle. The tool coordinate system {T} is created with the center of the bearing axis as the origin and the axial and radial axes as the coordinate axes, as shown in Fig. 7(b). Finally, the transformation matrix ${}_B{}^tT$ can be calculated according to the coordinate system transformation. Because the actual contact point is directly below the contact point, and the distance is determined by the radius of contact wheel and the thickness of abrasive belt. It is necessary to offset the origin of workpiece coordinate system on the actual contact point in kinematic analysis.

3 Experiments

In order to verify the reliability of the above grinding methods, complex grinding experiments were carried out on the repaired samples of profile, edge and blade tip. The grinding tools and parameters for typical blade repair are shown in Table 1, and the grinding process is shown in Fig. 8.

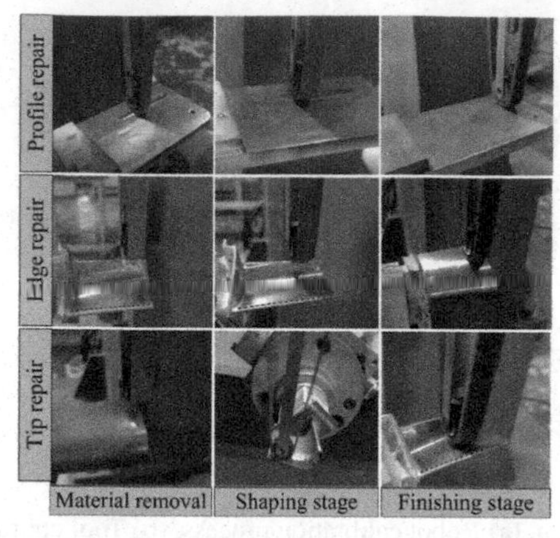

Fig. 8. Grinding process of typical remanufactured samples

The profile accuracy, roughness and residual stress of the sample after grinding were tested. The profile accuracy is characterized by the residual distribution after grinding. Figure 9(a) shows the residual comparison before and after grinding of the repaired sample. Obviously, the residual distribution of the cladding layer before grinding is extremely uneven, with the maximum residual up to 3 mm, and the residual after grinding is basically less than 0.07 mm. There is no obvious tool connection, and the edge transition is smooth.

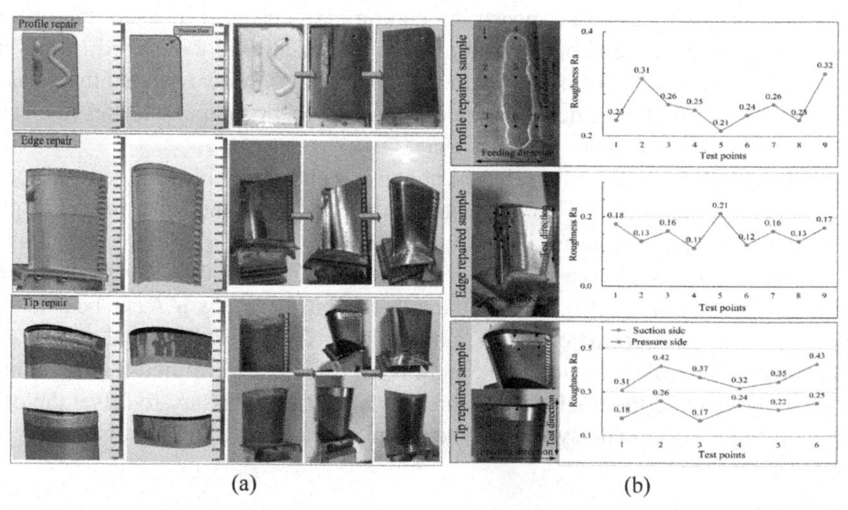

Fig. 9. Experimental results. (a) Comparison of allowance before and after grinding for typical remanufactured samples. (b) Roughness of three kinds of repaired samples after grinding.

Table 1. Grinding tools and parameters of typical remanufacturing samples

Step	Model	Allow (mm)	Belt type	Vs (m/s)	F (N)	Vw (mm/s)
Material removal	Semi-lock	×	XK870F	8	×	5
Shaping	Float	0.2	Diamond belt	18	12	20
Finishing	Float	0.02	Nylon belt	20	3	30

The surface roughness data of the repaired sample after grinding is shown in Fig. 9(b). The surface roughness of the repaired area is basically the same as that of the undamaged area after belt grinding, and the roughness is basically less than Ra0.4, which meets the service requirements of the blade.

4 Conclusion

In this paper, through the research of blade repairing robot abrasive belt grinding method, the research conclusions are as follows:

1. A robot belt grinding platform is developed, and its force position relationship is calibrated and verified. The precision control of grinding pressure within 35.2 N is realized.
2. The detection, location and model processing methods of repaired blade were investigated in this paper, and the robot operation calibration was carried out to reduce the error of robotic belt grinding system.
3. The effectiveness of the above methods is verified by grinding experiments of three kinds of typical repaired samples. The grinding accuracy is within 0.07 mm, the roughness is lower than Ra0.4.

References

1. Rinaldi, C., Antonelli, G.: ENERGY. Epitaxial repair and in situ damage assessment for turbine blades. Proc. Inst. Mech. Eng. Part A: J. Power Energy **219**(2), 93–99 (2005)
2. Tao, W., Huapeng, D., Hao, W., et al.: Virtual remanufacturing: cross-section curve reconstruction for repairing a tip-defective blade. Archiv. Proc. Inst. Mech. Eng. Part C J. Mech. Eng. Sci. **229**(17), 3141–3152 (2015)
3. Bi, G., Gasser, A.: Restoration of nickel-base turbine blade knife-edges with controlled laser aided additive manufacturing. Phys. Procedia **12**, 402–409 (2011)
4. Thukaram, S.K.: Robot Based 3D Welding for Jet Engine Blade Repair and Rapid Prototyping of Small Components. University of Manitoba (2010)
5. Sutton, B.H.E., Thodla, R.: Correction to: heat treatment of alloy 718 made by additive manufacturing for oil and gas applications. JOM **71**(6), 2137–2137 (2019)
6. Rosa, B., Mognol, P., Hascoët, J.-Y.: Modelling and optimization of laser polishing of additive laser manufacturing surfaces. Rapid Prototyp. J. **22**(6), 956–964 (2016)

7. Kim, U.S., Park, J.W.: High-quality surface finishing of industrial three-dimensional metal additive manufacturing using electrochemical polishing. Int. J. Precis. Eng. Manuf. Green Technol. **6**(1), 11–21 (2019)
8. Łyczkowska, E., Szymczyk, P., Dybała, B., et al.: Chemical polishing of scaffolds made of Ti–6Al–7Nb alloy by additive manufacturing. Archiv. Civil Mech. Eng. **14**(4), 586–594 (2014)
9. Liu, W.D., Ao, S.S., Li, Y., et al.: Elimination of the over cut from a repaired turbine blade tip post-machined by electrochemical machining. J. Mater. Process. Technol. **231**, 27–37 (2016)
10. Xiong, X., Zhang, H., Wang, G., et al.: Hybrid plasma deposition and milling for an aeroengine double helix integral impeller made of superalloy. Robot. Comput. Integr. Manuf. **26**(4), 291–295 (2010)
11. Bagci, E.: Reverse engineering applications for recovery of broken or worn parts and re-manufacturing: three case studies. Adv. Eng. Softw. **40**(6), 407–418 (2009)
12. Yilmaz, O., Gindy, N., Gao, J.: A repair and overhaul methodology for aeroengine components. Robot. Comput. Integr. Manuf. **26**(2), 190–201 (2010)
13. Wu, B., Wang, J., Zhang, Y., et al.: Adaptive location of repaired blade for multi-axis milling. J. Comput. Design Eng. **4**, 4 (2015)
14. Huang, H., Zhou, L., Chen, X.Q., et al.: SMART robotic system for 3D profile turbine vane airfoil repair. Int. J. Adv. Manuf. Technol. **21**(4), 275–283 (2003)
15. Huang, H., Gong, Z.M., Chen, X.Q., et al.: Robotic grinding and polishing for turbine-vane overhaul. J. Mater. Process. Technol. **127**(2), 140–145 (2002)
16. Whitton, S.: Adaptive robot grinding improves turbine blade repair. Indust. Robot. Int. J. Robot. Res. Appl. **30**(4), 370–372 (2003)

A Robust Blade Profile Feature Parameter Identifying Method

Zhenyou Wang[1], Xu Zhang[1,2(✉)], Zelong Zheng[2], and Jinbo Li[2]

[1] School of Mechatronic Engineering and Automation, Shanghai University, Shanghai 200444, China
Xuzhang@shu.edu.cn

[2] Huazhong University of Science and Technology Wuxi Research Institute, Wuxi 214000, China

Abstract. The blade is the key part of aero-engine. The detection of the blade profile feature parameters is an important aspect in blade manufacturing. However, the surface of the blade is free-form surface. It is difficult to extract feature parameters. In this paper, a robust feature identify method is proposed. First, the theoretical profile is segmented to concave, convex, leading edge and trailing edge by a regional identifying method. Second, the actual profile is segmented to four parts through the theoretical profile. According to the actual segmentation results, the profile feature parameter is solved. The simulation and experiment manifest that the proposed method is well robustness.

Keywords: Blade profile · Feature parameter identifying · Profile segmentation

1 Introduction

The shape of aero-engine blade is free-form surface [1]. In the measurement of the blade, non-contact measurement and contact measurement are primarily used [2]. In the Non-contact measurement, the blade is completely scanned by laser sensor. Non-contact measurement has the advantages of fast measuring speed and no damage to the workpiece. However, due to the features of the free-form surface and the laser sensor, the measuring precision of the non-contact measurement is lower than that of contact measurement [3, 4]. In the contact measurement, the general approach is the coordinate measurement machines (CMMs) with contact probe in the way of point-by-point inspection [5]. The contact measurement has the advantage of high precision. But, the efficiency of contact measurement is low. Meanwhile, the workpiece is easily to be damaged [6].

Whether non-contact measurement or contact measurement, the feature parameters of the profile are the important reference to evaluating the machining accuracy [7–9]. According to the industry standard of blade detection, blade profile errors include blade profile position error, form error and dimension error [9]. The position error includes the position error of the stacking point and the rotation error around the stacking axis. The form error is deviation from actual profile to the theoretical profile. The main dimension

X.-J. Liu et al. (Eds.): ICIRA 2021, LNAI 13014, pp. 721–731, 2021.
https://doi.org/10.1007/978-3-030-89098-8_68

error is chord length, radius of leading and trailing edges, thickness of specified position of leading and trailing edges. The position error and the form error can be calculated plain. However, affected by the undetermined feature of free-form surfaces, it is difficult to solve the dimensional error.

There are various techniques to calculate the dimensional error. In paper [10], the measuring points are filtered. Then, the feature parameters are calculated according to the fitting algorithm and geometric relations directly. Meanwhile, the influence of sampling point distribution on turbine blade feature parameter detection accuracy is analyzed. Liu [11] used minimum contain regional linear fitting and least square circle approximation to partitioning the profile. Then, the parameters are obtained by analytic method. In paper [12], a polygon convex hull method is used to extract the chord line. On this basis, based on a maximum radius error method, the edge is extracted. Then, In order to improve the precision of calculation of blade edge thickness, the upper sampling is carried out on the measured points. In summary, the present methods have several characteristics. (a) The measurement points are segmented directly. (b) The feature parameters are extracted by fitting algorithm and geometric relations.

In this paper, a robust blade profile feature parameters identify method is proposed. The method includes three steps. Step one, the theoretical profile is segmented into concave, convex, leading edge and trailing edge. Step two, according to the results of theoretical segmentation curve, the actual profile is segmented. Step three, the profile feature parameters are identified. The robustness of the proposed method is proved by simulation and experiments. The contributions from our work are listed as follows:

- The theoretical profile is segmented by an incremental search method. Actual profile is segmented by theoretical profile segment results. The proposed method avoids the segmentation failure of actual profile due to excessive error.
- We proposed a robust blade profile feature parameters identify method. The parameters can be identified accurately under the gauss noise $3\sigma = 0 \sim 0.01$.

The rest of our paper is organized as follows. In Sect. 2, the profile is segmented concave, convex, leading edge and trailing edge. In Sect. 3, the profile feature parameters are calibrated. Section 4 are simulations and experiments. Section 5 is conclusion.

2 Profile Segmentation

The complete profile is composed of concave, convex, leading edge and trailing edge. To obtain the dimensional error of blade, it is necessary to segment the whole blade shape region to the leading edge, the trailing edge, the concave and convex [9]. Because the blade is belong to free-form surface. Moreover, the profile is free curve. So, it is very difficult to identify the four parts of the profile. Therefore, the key of measuring the dimensional error is identifying the four parts. In this paper, a regional identify method based on theoretical profile is presented. For convenience of expression, the leading edge and trailing edge are collectively referred to as profile edge.

2.1 Theoretical Profile Identifying

In this section, the theoretical profile is discretized into limited points equidistantly. As shown in the Fig. 1, firstly, the minimum circumscribed circle about the profile is obtained. The point B is the intersection of the circumscribed circle and the profile. The profile point with a certain distance from ***B*** is defined as indirect edge area. Then, the points located on the indirect edge area are searched one by one by using the fitting circle algorithm. Until, the standard deviation of the fitted circle is greater than the fixed threshold. The searched points are defined as profile edge arc points. The midpoint A of the profile edge arc points is denoted as the profile edge midpoint.

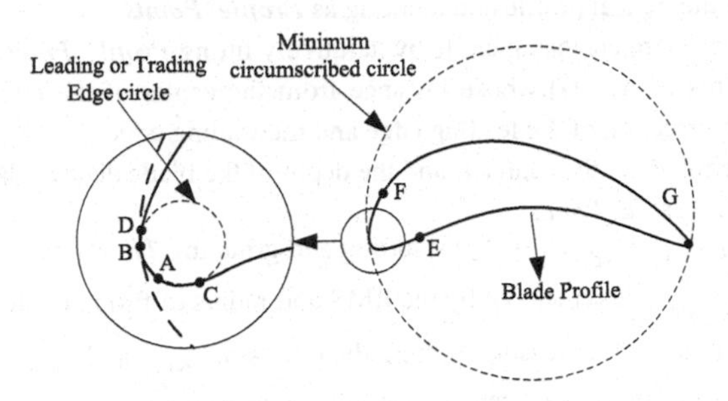

Fig. 1. Profile edge arc identify

In the Fig. 1, point ***C*** and point ***D*** are the intersection points of the profile edge arc and the theoretical profile. And the arc of CAD represents the profile edge arc. Point E and point F are intersection points of the profile edge and concave, convex. The arc of ***EAF*** represents the profile edge. The algorithm details are as follows.

Step.1 The theoretical profile is discretized into limited points equidistantly. The points are called profile points. The minimum circumscribed circle of the profile points is determined iteratively by using the three-point fixed circle method.

Step.2 The point ***B*** is the intersection of the circumscribed circle and the theoretical profile. The point ***B*** is determined as the center of the circle, and the reference depth of the blade edge is taken as the radius to make a circle, denoting as circle one. The points within circle one are taken as the points of profile edge area.

Step.3 The points in the profile edge area were grouped into a group of 10 points. And those less than 10 points were incorporated into the previous group. The groups can be expressed $group = \{g_1, \cdots, g_i, \cdots, g_m\}$. The circle is fitted to each set of points. The fitting radius is $R = \{r_1, \cdots, r_i, \cdots, r_m\}$. And the residual error is $RMS = \{rms_1, \cdots, rms_i, \cdots, rms_m\}$. The group that r_i is greater than the pre-set threshold are deleted. The groups that meet the requirements are recorded as $group\_edge = \{g_l, \cdots, g_i, \cdots, g_k\}$.

Step.4 Determine the continuity from l to k.

Step.5 If continuity. The points of g_{l-1} and g_{k+1} are putted into a new group $group\_edge\_1$ in reverse and positive order, respectively. The circle is fitted with points of $group\_edge\_1$ each time $group\_edge\_1$ is stored at a point. When the fitting residual is greater than the set threshold, the $group\_edge\_1$ stops the deposit point. The *group_edge_1* is profile edge arc points, named *EdgeDate*. The center of the fitting circle is profile edge center $O_e = (x_e, y_e)$.

Step.6 If not continuity. Updating the *group_edge* with more of the same fitting radius $R = \{r_1, \cdots, r_i, \cdots, r_m\}$. And repetition step 5.

Step.7 A circle is constructed with center of $O_e = (x_e, y_e)$ and radius of profile edge depth. The points contained in the circle are trailing edge points.

The algorithm flow is as follows.

Algorithm 1. Algorithm for theoretical profile segmentation

---

1. Discretizing the theoretical profile and marking as *Profile_Points.*
2. Obtaining the minimum enclosing circle by iteratively fitting *Profile_Points.*
3. Taking two points (***B*** and ***G***) whose distance from the center of the circle is equal to the radius as the intersection of the leading edge and the trailing edge.
4. Constructing a circle with the center ***B*** and the depth of the blade dege as the radius. The data inside the circle 1 is *TE_Points.*
5. Obtaining $group = \{g_1, \cdots, g_i, \cdots, g_m\}$, by sorting and grouping *TE_Points.*
6. The $\{g_l, \cdots, g_i, \cdots, g_k\}$ is screen out by the RMS and radius of fitting circle.
7. If $l \sim k$ is continuously increasing, putting the points of g_{l-1} and g_{k+1} into *group_TE* in reverse order and positive order. Then, re-fit the circle until $rms_i >$ threshold. And record the final *EdgeDate* and $O_e = (x_e, y_e)$.
8. If $l \sim k$ is discontinuous, eliminating discontinuous data. The new data is re-mark as $group = \{g_1, \cdots, g_i, \cdots, g_m\}$, and return step5.
9. Constructing a circle with the center $O_e = (x_e, y_e)$ and the depth of the profile dege as the radius. The data inside the circle is trailing edge points.

2.2 Identifying Actual Profile

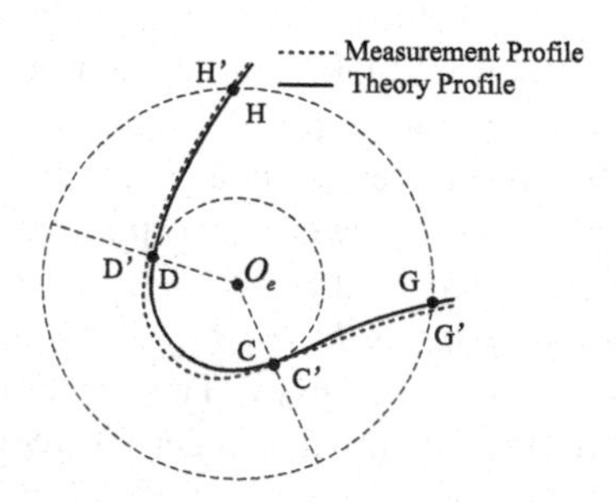

Fig. 2. Identifying the actual profile

Due to the machining error, the actual profile edge cannot completely guarantee the conformance to the CAD, so the above method can no longer be used to extract actual profile features. In Sect. 2.1, the theoretical profile is divided and the coordinate values of O_e, $\boldsymbol{C}$, $\boldsymbol{D}$, $\boldsymbol{G}$ and $\boldsymbol{H}$ are given. As shown in Fig. 2, the fan-shaped covering area consisting of acute angle $\angle CO_eD$ contains actual measurement points as actual profile edge arc point. The actual center of edge circle $O_e^{'}$ is obtained by fitting the actual profile edge arc point. Take $O_e^{'}$ as the center of the circle, the depth of the blade edge as the radius to make the circle, and the points inside the circle are the actual profile edge points.

Algorithm 2. Algorithm for identifying actual profile

1. According to the coordinate values of $\boldsymbol{O_e}$, $\boldsymbol{C}$ and $\boldsymbol{D}$ at the theoretical points, the actual profile edge arc was extracted.
2. The actual profile edge arc is fitted to obtain the circle center $O_e^{'}$.
3. Construction a circle with the center $O_e^{'}$ and the depth of the profile edge as the radius. The data inside the circle is actual profile edge points, named *Actual_PEPoints*.
4. The *Actual_PEPoints* is removed from *Actual_Profile_Points*. The residual part is concave and convex, named *Actual CCPoints* and *Actual CVPoints*.

3 Calibrating Profile Feature Parameters

According to the standard HB 20126–2012, the key parameters of blade profile are chord length, radius of leading and ending edges, and thickness of specified position of front and rear edges [9]. In Sect. 2, the radiuses of the leading and ending edge have been calculated. In this section, the chord length and the thickness of the specified blade edge position are calculated.

3.1 Calibrating Chord Length

In the standard HB5647-98, The chord length is the projection of the profile on the chord line. The chord line needs to be identified first.

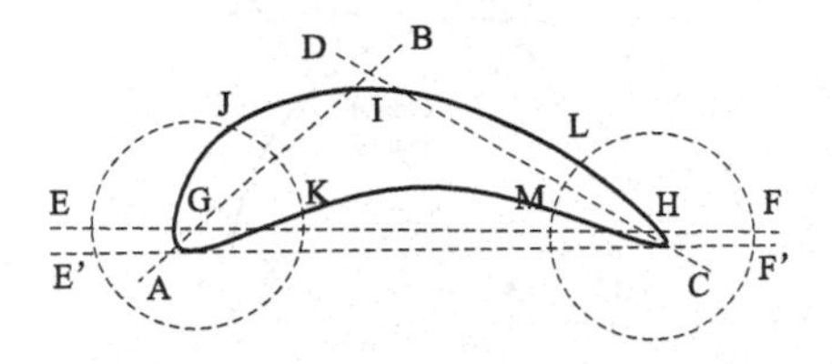

Fig. 3. Diagram of chord

According to the actual profile edge points, line $\boldsymbol{AB}$ and $\boldsymbol{CD}$ are fitted. The line $\boldsymbol{AB}$ and $\boldsymbol{CD}$ intersect at point $\boldsymbol{I}$, in the Fig. 3. The line $\boldsymbol{EF}$ goes through $\boldsymbol{G}$, $\boldsymbol{H}$ which is the center of the blade edge circle. Basis the profile edge radiuses offset the line $\boldsymbol{EF}$.

Offsetting the line ***EF*** in the opposite direction to ***I***. The offsite line ***E′F′*** is the chord. The blade point is projected onto the chord line. Then the longest distance in the chord line is chord length.

Algorithm 3. Algorithm for actual profile segmentation

1. Line ***AB*** and ***CD*** are obtained by fitting the actual profile edge points. The intersecting point is ***I***.
2. Making a straight line ***EF*** through the centers of the profile edge circles. Offsetting ***EF*** in the opposite direction to the point ***I***. The offset distance is the radiuses of the profile edge. The offsite line is marked as ***E'F'***.
3. Calculating the distance from the leading edge points to the ***E'F'***. And marking the distance as *LE_points_forward* to the point on the same side of ***I***. Otherwise, marking the distance as *LE_points_reverse*.
4. Calculating the distance from the trailing edge points to the ***E'F'***. And marking the distance as *TE_points_forward* to the point on the same side of ***I***. Otherwise, marking the distance as *TE_points_reverse*.
5. Taking the distance corresponding to the point that marked as *LE_points_reverse* and *TE_points_reverse* with the maximum distance as D_{le} and D_{te}.
6. Offsetting ***E'F'***. The offset distance is D_{le}, D_{te}. The offsite line ***E"F"*** is chord.
7. The blade point is projected onto the chord line. Then the longest distance in the chord is chord length.

3.2 Calibrating the Thickness of the Edge

As shown in the Fig. 3, the intersection points of the profile segmentation circle and the profile are four points, ***J***, ***K***, ***L*** and ***M***. The distance of ***J*** and ***K*** is trailing edge thickness. In the same way, the distance of ***L*** and ***M*** is leading edge thickness. The solution procedure of coordinates of the ***J***, ***K***, ***L*** and ***M*** are as follows (Fig. 4).

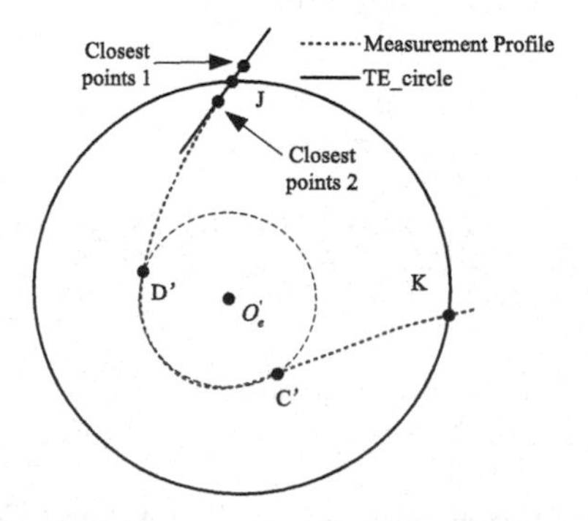

Fig. 4. Calibrating the thickness of the edge

According to the actual center of the leading and the trailing edges of the profile in Sect. 2.2, a circle, *TE_circle*, is constructed. And the radius of the circle is setting as design value. The measuring points that are closest to point J are searched. Fit a line through the two closest points. The intersection point of the line and the circle is the coordinate of point J. The rest ***K***, ***L*** and ***M*** points were calculated in turn. Then JK distance and LM distance are the required blade margin thickness.

Algorithm 4. Algorithm for actual profile segmentation

1. Constructing a circle, *TE_circle*. The center is $O_e^{'}$ in section 2.2. The radius is design value.
2. Searching two points that are closest to point ***J***. The points are fitted a line.
3. The intersection point of the line and the circle is the point ***J***.
4. Calibrating the point ***K***, ***L*** and ***M***, repetition the step 1-3.
5. Calibrating the distance from the point ***J*** to the point ***K***, and from the point ***L*** to the point ***M***.

4 Simulation and Experiments

4.1 Simulation

In order to verify the robustness of the proposed method, a theoretical blade profile point set with known parameters is selected in this section. Meanwhile, the measured value of profile is simulated by adding Gaussian noise and coordinate transformation. The position error, form error and dimension error are calculated by the proposed method.

4.1.1 Simulating Data Set

The theoretical profile is obtained by intercepting the blade CAD at Z = 53.4 mm. The theoretical profile point set is obtained by discretizing the profile. Because of machining error and measurement error, the measured profile exist fluctuation with micron level. Therefore, Gaussian noise $3\sigma = 0 \sim 0.01$ is loaded into the theoretical profile data set. According to the formula 11, the simulating measuring profile data is obtained.

$$\begin{bmatrix} x_a & y_a \end{bmatrix}^T = \begin{bmatrix} \cos(\alpha) & -\sin(\alpha) \\ \sin(\alpha) & \cos(\alpha) \end{bmatrix}^{-1} \begin{bmatrix} x_{theory} + x_{noise} - x_t \\ y_{theory} + y_{noise} - y_t \end{bmatrix} \tag{1}$$

where α is rotation error. $\begin{bmatrix} x_t & y_t \end{bmatrix}$ is displacement error. $\begin{bmatrix} x_{theory} & y_{theory} \end{bmatrix}^T$ is theoretical profile point. $\begin{bmatrix} x_a & y_a \end{bmatrix}^T$ is simulating measurement profile data. $\begin{bmatrix} x_{noise} & y_{noise} \end{bmatrix}$ is Gaussian noise. The histogram of Gaussian noise is shown below (Table 1 and Fig. 5).

Table 1. The value of the simulated profile

Descriptions	Value (mm)
Chord length	33.662
Radius of LE	0.159
Radius of TE	0.140
The thickness of LE	0.752
The thickness of TE	0.741

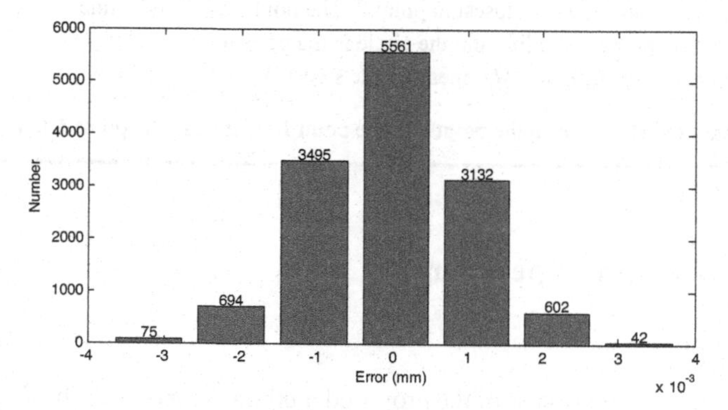

Fig. 5. The histogram of Gaussian noise

The reference dimension of the simulated profile data are shown in the table below.

4.1.2 Simulating Results

According to the method proposed in the Sect. 2, the position error, form error and dimension error of the profile are calculated in this section to verify the robustness of the proposed method. The reference data set is obtained in Sect. 4.1.1. The results are as follows.

As shown in Fig. 6, the profile is divided into four parts, concave, convex, leading edge and trailing edge. In the Table 2, the results of the proposed method are below 0.00253 mm. The simulation results offered indicate our methods are very effective to calculating the dimension error. And the proposed method has well robustness.

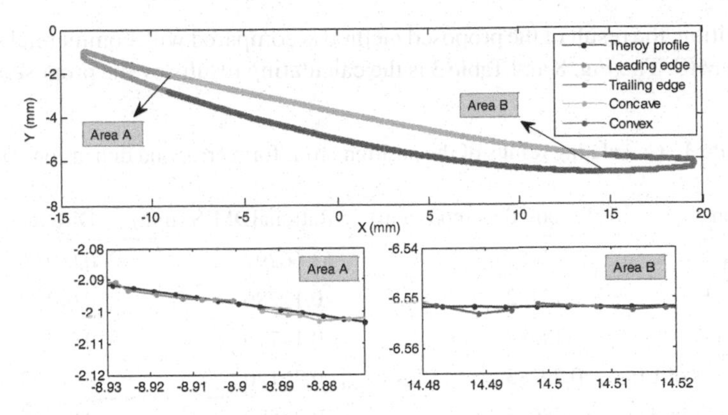

Fig. 6. The simulation results of adjusting the profile position

Table 2. Calculating results of the position error, form error and dimension error

Descriptions	Results (mm)	Error (mm)
Chord length	33.66463	0.00253
Radius of LE	0.16079	0.00179
Radius of TE	0.13201	0.00201
The thickness of LE	0.75138	0.00138
The thickness of TE	0.74295	0.00195

4.2 Experiments

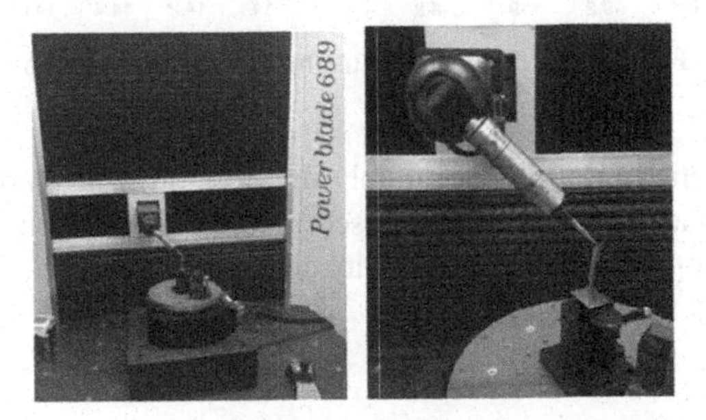

Fig. 7. Experiment

In order to verify the practicability of the proposed method, in this section, a blade is measured in the CMM. The blade is consistent with the CAD in simulation. The accuracy of the CMM used in the experiment is verified to be (2.5 + 2.5L/1000) μm where L is the measuring length. Meanwhile, the section height is 53.4 mm. The experimental picture is shown below.

In addition, the result of the proposed method is compared with commercial software, RationalDMIS. The Fig. 8 and Table 3 is the calculating results by the proposed method.

Table 3. Calculating results of the position error, form error and dimension error

Descriptions	Proposed method (mm)	RationalDMIS (mm)	Difference (mm)
Chord length	33.66463	33.66207	0.00166
Radius of LE	0.17102	0.17280	0.00178
Radius of TE	0.14588	0.14710	0.00122
The thickness of LE	0.79283	0.79120	0.00163
The thickness of TE	0.78635	0.78580	0.00055

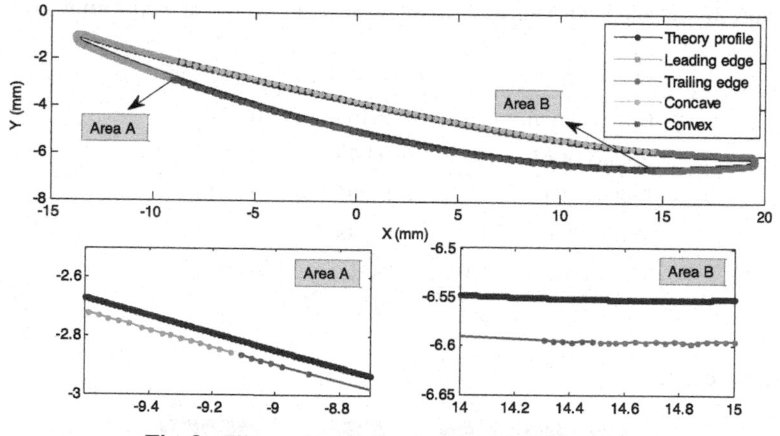

Fig. 8. The results of adjusting the profile position

In Fig. 7, the actual profile is accurately divided into four parts. In Table 3, it is obvious that difference between the proposed method and RationalDMIS software is below 2 um. The proposed method is stability and practicality.

5 Conclusion

The detection of profile feature parameters is an important step in blade manufacturing. However, the surface characteristics of blades make feature extraction difficult. In this paper, a robust blade profile feature parameter identifying method is proposed. The actual profile is segmented by theoretical profile segment results. It can improve the stability of the profile segmentation. In addition, the identify accuracy of profile feature parameters is below 2.6 um, under the gauss noise $3\sigma = 0 \sim 0.01$ in simulation. In the actual experiment, the difference between the proposed method and RationalDMIS is below 2 um. The proposed method has well robustness and practicality.

Acknowledgement. This research was partially supported by the key research project of the Ministry of Science and Technology (Grant No. 2018YFB1306802) and the National Natural Science Foundation of China (Grant No. 51975344).

References

1. Savio, E., De Chiffre, L., Schmitt, R.: Metrology of freeform shaped parts. CIRP Ann. **56**(2), 810–835 (2007)
2. Huang, J., Wang, Z., Gao, J.: Overview on the profile measurement of turbine blade and its development. Proc. SPIE Int. Soc. Opt. Eng. **7656**(1), 1–11 (2010)
3. Hageniers, O.L.: Laser coordinate measuring machine for cylindrical parts. In: Quebec Symposium. International Society for Optics and Photonics (1986)
4. Jiang, C., Lim, B., Zhang, S.: Three-dimensional shape measurement using a structured light system with dual projectors. Appl. Opt. **57**(14), 3983–3990 (2018)
5. Wang, Z.Y., Zhang, X., Shen, Y.J.: A novel 3D radius compensation method of probe stylus tip in the free-form surface profile curve scanning measurement. Meas. Sci. Technol. **31**(2020), 1–11 (2020)
6. Sato, O., Matsuzaki, K., Fujimoto, H.: Challenge for the traceable dimensional measurement using CMM in factory floor. Appl. Mech. Mater. **870**(1), 243–248 (2017)
7. Wu, D., Wang, H., Zhang, K., Zhao, B., Lin, X.: Research on adaptive CNC machining arithmetic and process for near-net-shaped jet engine blade. J. Intell. Manuf. **31**(3), 717–744 (2019). https://doi.org/10.1007/s10845-019-01474-z
8. Chen, Z.: Research of cross-sectional feature parameter extraction of blade based on measurement data
9. HB 20126–2012 Requirements of measurement for aero-engine blades and vanes using CMM
10. Bu, K., Zhang, X., Ren, S.: Research on influence of CMM sampling points on detection of feature parameters for turbine blade. J. Northwestern Polytech. Univ. **37**(4), 767–773 (2019)
11. Shu-Gui, L., Chen-Li, M., Hai-Tao, Z.: Cross-sectional feature parameters extraction of blade based on unorganized point cloud. J. Aerospace Power **5**(3), 368–378 (2016)
12. Zhi-Qiang, C.: Cross-sectional feature parameter extraction of blade based on measurement data. Sci. Technol. Eng. **7**(9), 1671–1819 (2007).

Feedrate Optimization for Pocket Machining Considering Cutting Force Constraints

Yang Lei, Jingfu Peng, and Ye Ding(✉)

State Key Laboratory of Mechanical System and Vibration, School of Mechanical Engineering, Shanghai Jiao Tong University, Shanghai 200240, China
y.ding@sjtu.edu.cn

Abstract. Pocket machining is widely used to manufacture complex structure parts in the aerospace and automotive industries. Improving productivity remains one of the core concerns for pocket machining. However, the higher feedrate also means the larger maximum cutting force during machining, which would decrease tool life and even cause tool rupture. Therefore, limiting the maximum cutting force during pocket machining is of significant importance. This paper develops the linear-programming (LP) based feedrate optimization algorithm with a novel cutting force limitation strategy. The optimized feedrates are obtained under both kinematic and cutting force constraints. First, the cubic B-spline spiral path is constructed. Then based on geometric features of the B-spline spiral toolpath for pocket machining, the cutting force is evaluated by proposing the equivalent feed per tooth and modified cutter engagement. Utilizing the linear relationship between feedrate and cutting forces in axis directions, the cutting force constraints are equivalently converted to the maximum allowable feedrate. Finally, the cutting force constraints and the kinematic constraints are integrated to formulate the LP model for feedrate optimization. The experimental results indicate that the enhanced feedrate optimization approach enables significant improvement in machining efficiency while keeping the maximum cutting forces at the same level.

Keywords: Pocket machining · Feedrate optimization · Cutting force constraint

1 Introduction

Pocket machining is widely used in the aerospace and automotive industries [1, 2]. It removes a large amount of material and is time-consuming. Improving productivity remains one of the core concerns for pocket machining.

Smooth toolpaths are essential for improving process productivity. Bieterman et al. [2] proposed a smooth toolpath generation algorithm based on the partial differential equation (PDE). The generated spiral cut pattern, represented in twice continuously differentiable parametric spline curve, is desirable for High Speed Machining (HSM). Based on the smooth spline toolpath, different feedrate optimization methods were proposed to boost productivity. The kinematic constraints, mainly containing the velocity, acceleration, and jerk limits, are commonly considered in the optimization problem.

X.-J. Liu et al. (Eds.): ICIRA 2021, LNAI 13014, pp. 732–742, 2021.
https://doi.org/10.1007/978-3-030-89098-8_69

Sencer et al. [3] used Sequential Quadratic Programming (SQP) to solve the feedrate optimization problem with kinematic constraints, but the SQP algorithm is computationally expensive. By using a linear function to approximate the nonlinear jerk constraint, the feedrate optimization problem was reduced to a Linear Programming (LP) problem [4], which can effectively get globally optimal solutions. For efficiently handling long toolpaths, Erkorkmaz et al. [5] improved the LP formulation with a windowing algorithm to process different portions of long toolpaths in parallel.

However, the higher feedrate also means the larger maximum cutting force during machining, which would decrease tool life and even cause tool rupture. To avoid tool failure, the cutting force constraint, depending on the accurate estimation of cutting forces, should be considered for feedrate optimization. In limiting cutting forces, most methods in literature utilize the conventional linear cutting force model [6] to express the cutting force as a linear function of the feedrate and get the maximum allowable feedrate [7–9]. Different from the linear toolpath, the feed direction, actual cutting feedrate and cutter engagement vary along the spline toolpath [10]. Due to the variations in the geometric features along the spline toolpath, the conventional linear cutting force model used in the linear toolpath cannot be directly applied. Thus, it is preferred to enhance the linear cutting force model with geometric features of the toolpath.

In this paper, a novel cutting force limitation strategy is proposed. The cutting force limitation strategy improves upon the linear cutting force model in Ref. [6] based on geometric features of the B-spline spiral toolpath for pocket machining and converts the cutting force constraints to the maximum allowable feedrate, which is then added to the LP formulation to get the enhanced feedrate optimization algorithm.

Henceforth, the B-spline spiral path for pocket machining is constructed in Sect. 2. The novel cutting force limitation strategy is proposed in Sect. 3. The enhanced feedrate optimization algorithm combining the LP formulation with the proposed cutting force limitation strategy is described in Sect. 4. In Sect. 5, the effectiveness of the enhanced feedrate optimization algorithm is demonstrated through experiments. In the last section, the contributions of the present work and future recommendations are presented.

2 Spiral Path Generation

The smooth toolpath is paramount to maintain higher feedrates during cornering motions and ultimately yields a shorter machining time. So the smooth toolpath generation algorithm proposed in Ref. [2] is adopted to obtain the spline spiral path.

The cubic B-spline spiral path is constructed as follows [2]: Firstly, the eigenvalue problem for an elliptic PDE is solved subject to Dirichlet's boundary conditions. Then the solution contours are generated with a specified maximum stepover. The spiral points between the solution contours are obtained from equiangular points on each contour by linear interpolation next. Finally, the cubic B-spline spiral path is generated with the obtained spiral points. The overall B-spline spiral path generation scheme is illustrated in Fig. 1.

In this work, the helical interpolation is adopted for tool entry so there is a circle island for the pocket area. The cubic B-spline spiral path, $\mathbf{r}(u)$, generated for a rounded rectangular region with a circle island is shown in Fig. 1.

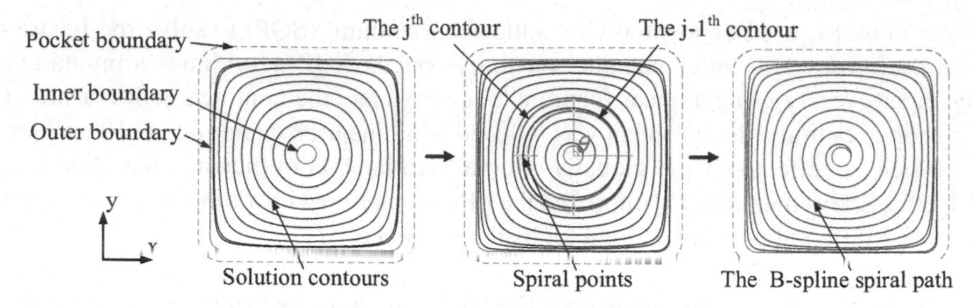

Fig. 1. Overall B-spline spiral path generation scheme

3 Cutting Force Limitation Strategy

The accurate estimation of cutting forces is crucial to establishing the cutting force constraints for the spline spiral path feedrate optimization. In this section, the linear cutting force model in Ref. [6] is improved based on geometric features of the B-spline spiral toolpath. Then the cutting force constraints are converted to the maximum allowable feedrate. As the axial cutting force is pretty small compared to the tangential and radial cutting forces, it is neglected in this work.

3.1 Cutting Force Estimation

The cutter is divided into a series of axial disk elements, as shown in Fig. 2(a). The tangential and radial cutting forces on the *k-th* disk element at the *j-th* tooth of the cutter, shown in Fig. 2(b), depend on the uncut chip thickness h and the elementary axial depth of cut (DOC) dz as [6]:

$$\begin{bmatrix} dF_{t,j,k} \\ dF_{r,j,k} \end{bmatrix} = g(\phi_{j,k}) \begin{bmatrix} K_{te}+K_{tc}h(\phi_{j,k}) \\ K_{re}+K_{rc}h(\phi_{j,k}) \end{bmatrix} dz \tag{1}$$

where K_{te}, K_{tc}, K_{re} and K_{rc} are cutting force coefficients. $\phi_{j,k}$ is the rotation angle of the differential flute element. ϕ_{st} and ϕ_{ex} are the start angle and exit angle for the cutter respectively. The switching function $g(\phi_{j,k})$ is equal to one when the flute element is engaged in the cut, i.e. $\phi_{st} \leq \phi_{j,k} \leq \phi_{ex}$, and zero otherwise.

For each disk element, the cutting forces along the feed and normal directions at the tooth j can be then expressed as:

$$\begin{bmatrix} dF_{feed,j,k} \\ dF_{normal,j,k} \end{bmatrix} = \begin{bmatrix} -dF_{t,j,k}\cos\phi_{j,k} - dF_{r,j,k}\sin\phi_{j,k} \\ dF_{t,j,k}\sin\phi_{j,k} - dF_{r,j,k}\cos\phi_{j,k} \end{bmatrix} \tag{2}$$

The total feed and normal direction forces on the cutter can be obtained by summing Eq. (2) over all the disk elements:

$$\begin{bmatrix} F_{feed} \\ F_{normal} \end{bmatrix} = \begin{bmatrix} \sum\limits_{k=1}^{N_A}\sum\limits_{j=1}^{N_t} dF_{feed,j,k} \\ \sum\limits_{k=1}^{N_A}\sum\limits_{j=1}^{N_t} dF_{normal,j,k} \end{bmatrix} \tag{3}$$

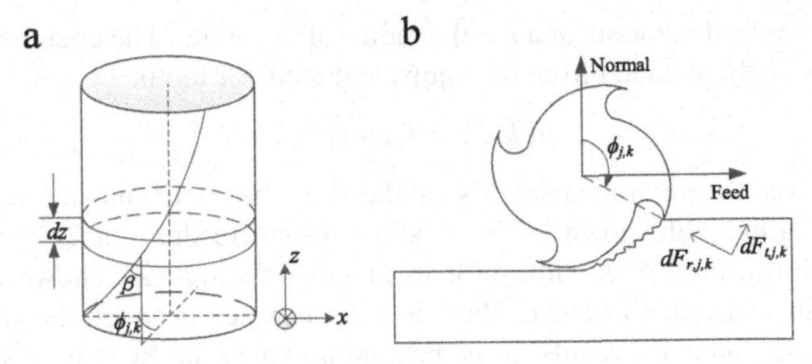

Fig. 2. Geometry of milling process. (a) Discretization of the cutter. (b) Cutting force geometry

where N_A is the number of the disk elements and N_t is the number of the cutter teeth.

Denoting θ as the angle between the feed direction and x direction of the workpiece coordinate frame, the forces can be projected into x and y directions as:

$$\begin{bmatrix} F_x \\ F_y \end{bmatrix} = \begin{bmatrix} F_{feed}\cos\theta - F_{normal}\sin\theta \\ F_{feed}\sin\theta + F_{normal}\cos\theta \end{bmatrix} \tag{4}$$

From Eq. (1), it is obvious that the simulated cutting forces depend on the uncut chip thickness calculation and engagement modeling. For the spiral spline toolpath, the variations in the radius of curvature (ROC) and the radial DOC for different cutter location (CL) points necessitate modifications in chip thickness and cutter engagement formulations to obtain a more accurate cutting force model. In this work, the equivalent feed per tooth f_e defined at the centroid of the cutting cross section is adopted to calculate the uncut chip thickness. The cutter engagement is also modified considering the curve ROC ρ and the radial DOC a. The equivalent feed per tooth and the cutter engagement, shown in Fig. 3, can be calculated by:

$$f_e = f_t \cdot (\rho + r - a/2)/\rho \tag{5}$$

$$\phi_{st} = \cos^{-1}\left(\left(\rho^2 + r^2 - (\rho + r - a)^2\right)/(2\rho r)\right),\ \phi_{ex} = \pi \tag{6}$$

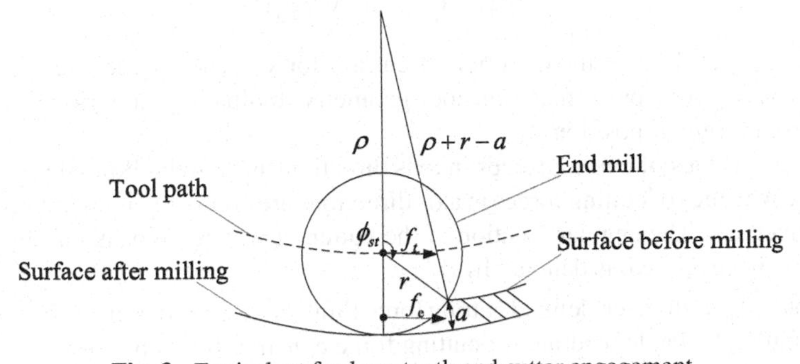

Fig. 3. Equivalent feed per tooth and cutter engagement

where f_t denotes feed per tooth and r is the radius of the cutter. The uncut chip thickness can be subsequently obtained from the equivalent feed per tooth:

$$h(\phi_{j,k}) = f_e \sin(\phi_{j,k}) \tag{7}$$

The improved cutting force model is validated by the pocket machining experiment along the spiral toolpath shown in Fig. 1 at a constant feedrate of 220 mm/min. The simulated and measured peak cutting forces in axis directions are shown in Fig. 4. As can be seen, in some time instances there is a discrepancy between the simulated and measured cutting force, especially at the beginning where the ROC is relatively small. These errors may mainly come from the estimation errors of ROC. As shown in Eq. (5), the ROC is the denominator of the equivalent feed per tooth. When the ROC is small, small errors of ROC will lead to large changes in estimation. In general, a close match between the measured and simulated cutting forces is evident.

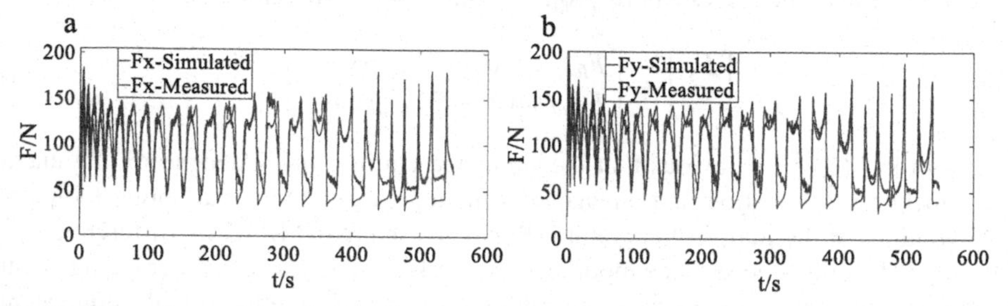

Fig. 4. Comparison of the simulated and measured peak cutting forces. (a) x direction; (b) y direction.

3.2 Cutting Force Constraints Conversion

As the proposed equivalent feed per tooth f_e has the advantage of being proportional to feed per tooth f_t according to Eq. (5), at each CL point the cutting force corresponding to each axis direction is still linearly dependent on the feedrate V in the following form, similar to [7, 8]:

$$F^{\mu}(\phi) = A_0^{\mu}(\phi) + A_1^{\mu}(\phi)V \tag{8}$$

where $\mu \in \{x, y\}$ denotes the direction of cutting forces. The coefficients, A_0^{μ} and A_1^{μ} are functions of workpiece material, tool geometry, toolpath geometric features, and instantaneous angular position.

The peak values of cutting forces in axis directions need to be limited to avoid tool failure. Peak values of cutting forces in axis directions are computed numerically at tooth passing intervals. The angular location of the cutter, at which the peak cutting force in μ direction is generated, is denoted by ϕ_{peak}^{μ}.

Denote $F_{\lim}^{\mu}$ as the user-defined cutting force limit values in μ direction. From Eq. (8), the maximum allowable feedrate for cutting force constraints can be solved as:

$$V_{\lim} = \min\left\{\frac{F_{\lim}^{x} - A_0^{x}(\phi_{peak}^{x})}{A_1^{x}(\phi_{peak}^{x})}, \frac{F_{\lim}^{y} - A_0^{y}(\phi_{peak}^{y})}{A_1^{y}(\phi_{peak}^{y})}\right\} \tag{9}$$

Above, the cutting force constraints in axis directions have been converted to the maximum allowable feedrate at each CL point using the improved model.

4 Feedrate Optimization Using Linear Programming

Feedrate optimization for the cubic B-spline spiral path $\mathbf{r}(u) = (x(u), y(u))$ in Sect. 2 is carried out in this part.

The objective of feedrate optimization for minimizing travel time is equivalent to maximize the feedrate along the toolpath [4]. The maximum allowable feedrate $V_{\lim}$ for cutting force constraints solved in Eq. (9) and the user-defined maximal feedrate V_{com} determine the feedrate bound $V_{\max}$. The acceleration and jerk constraints limit maximal acceleration $\mathbf{A}_{\max}$ and jerk $\mathbf{J}_{\max}$ in axis directions respectively. So the optimization problem can be formulated as follows [5]:

$$\begin{aligned} &\max \int_0^1 \dot{u} du \\ &s.t. \ \ \|\mathbf{v}\| = \left\|\mathbf{r}'\dot{u}\right\| \le V_{\max} = \min\{V_{\lim}, V_{com}\} \\ &\quad\quad |\mathbf{a}| = \left|\mathbf{r}''\dot{u}^2 + \mathbf{r}'\ddot{u}\right| \le \mathbf{A}_{\max} \\ &\quad\quad |\mathbf{j}| = \left|\mathbf{r}'''\dot{u}^3 + 3\mathbf{r}''\dot{u}\ddot{u} + \mathbf{r}'\dddot{u}\right| \le \mathbf{J}_{\max} \end{aligned} \tag{10}$$

where $(\ )'$ and "·" denote the derivative of the corresponding variable with respect to curve parameter$u$ and time t, respectively. $\mathbf{r}$ is the cubic B-spline toolpath. $\mathbf{v}$, $\mathbf{a}$ and $\mathbf{j}$ are the machining feedrate, acceleration and jerk respectively.

A new parameter q, which is the square of parametric velocity $\dot{u}$, is introduced to linearize the inequalities in Eq. (10) and the relationship between q and u can be inferred as:

$$\dot{u} = \sqrt{q}, \ddot{u} = \frac{1}{2}q', \dddot{u} = \frac{1}{2}q''\sqrt{q} \tag{11}$$

Substituting Eq. (11) into Eq. (10) and squaring both sides in the velocity constraint, the problem is reformulated as:

$$\begin{aligned} &\max \int_0^1 q du \\ &s.t. \ \ \left\|\mathbf{r}'\right\|^2 q \le V_{\max}^2 \\ &\quad\quad \left|\mathbf{r}''q + \frac{1}{2}\mathbf{r}'q'\right| \le \mathbf{A}_{\max} \\ &\quad\quad \left|\mathbf{r}'''q + 3\mathbf{r}''q' + \frac{1}{2}\mathbf{r}'q''\right|\sqrt{q} \le \mathbf{J}_{\max} \end{aligned} \tag{12}$$

By discretizing the toolpath, the geometric derivatives q' and q'' can be linearized and the objective of feedrate optimization is transformed into a linear form [4]:

$$\max \sum_{i=1}^{K} q(u_i) \tag{13}$$

where K is the number of discretized CL points for the toolpath.

However, the $\sqrt{q}$ term in the jerk inequality still reflects a nonlinearity. To solve this problem, a precomputed upper bound q^* obtained by considering only the velocity and acceleration constraints in Eq. (12) is utilized. By multiplying $\sqrt{q^*/q}$ on both sides and relaxing the right-hand side, the jerk constraint can be transformed to the linear form [4]:

$$\left|\mathbf{r}'''q + 3\mathbf{r}''q' + \frac{1}{2}\mathbf{r}'q''\right|\sqrt{q^*} \preceq \mathbf{J}_{\max}\left(\frac{3}{2} \quad \frac{q}{2q^*}\right) \tag{14}$$

The final linear programming problem to be solved has the objective function Eq. (13) with velocity and accelerate constraints in Eq. (12) and linear-form jerk constraint in Eq. (14). The optimal feedrate sequence can be solved in two steps: Firstly, the upper bound q^* is obtained by considering only the velocity and acceleration constraints. Then the jerk constraint of Eq. (14) with obtained upper bound q^* is incorporated to get the final results.

5 Experimental Verification

The effectiveness of the enhanced feedrate optimization algorithm proposed in Sect. 4 is validated by pocket machining experiments. All experiments are performed on a DMG 5-axis CNC machine tool, as shown in Fig. 5. The 6061 aluminum alloy is used throughout the experiments. The two-flute end mill with a diameter of 10 mm and a helix angle of 25° is adopted. The spiral toolpath in Sect. 2 is executed. The cutting forces are measured by the Kistler 9255C dynamometer, which is a mature commercial product and has been widely used in industry. The cutting conditions are fixed as: the spindle speed of 1000 rpm, the axial DOC of 1 mm. The chord error, feedrate, acceleration, and jerk limits considered are 0.5 μm, 480 mm/min, 200 mm/s^2, 5000 mm/s^3 respectively.

Fig. 5. The experimental set-up for pocket machining

Vericut software is used to estimate the radial DOC for the CL points on the spiral toolpath in this paper. The simulated radial DOC results for the CL points are presented in Fig. 6. It can be seen from Fig. 6 that the radial DOC increases when the tool enters the corner and decreases when the tool leaves the corner, which is consistent with previous scholars' analysis [11, 12].

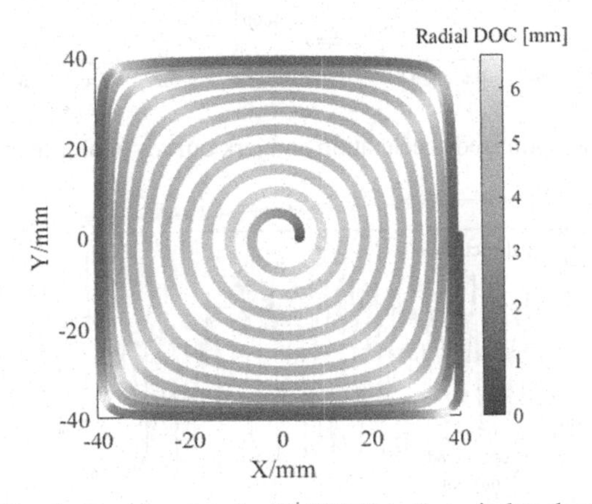

Fig. 6. The simulated radial DOC in the spiral toolpath

The simulated radial DOC results for the CL points and the corresponding feedrates optimized with a force limit of 130N in axis directions are simultaneously shown in Fig. 7. The acceleration and jerk curves with constraints are shown in Fig. 8, from which we can see that the acceleration and jerk bounds are satisfied. To analyze the optimization result, the machining process can be divided into two segments, presented with different colors (red and blue). The measured peak cutting forces in axis directions with optimized feedrates are shown in Fig. 9. In Fig. 7 and Fig. 9, the machining process under the optimized feedrate is compared with the case under constant feedrate 200 mm/min.

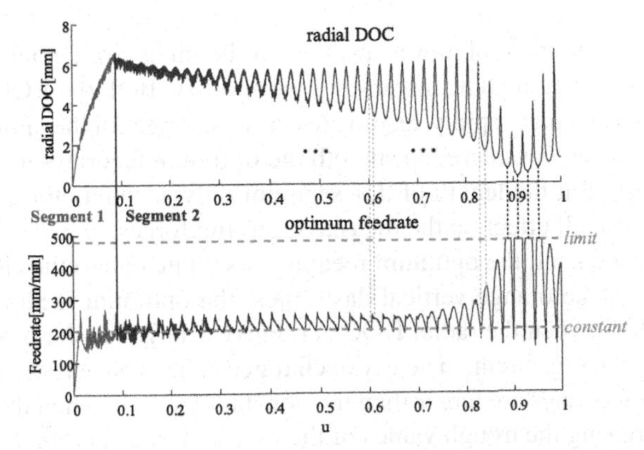

Fig. 7. The optimum feedrates and corresponding radial DOC

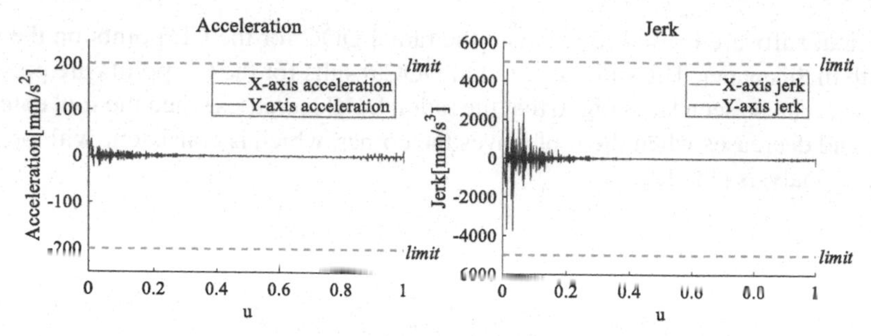

Fig. 8. The optimized acceleration and jerk curves (Color figure online)

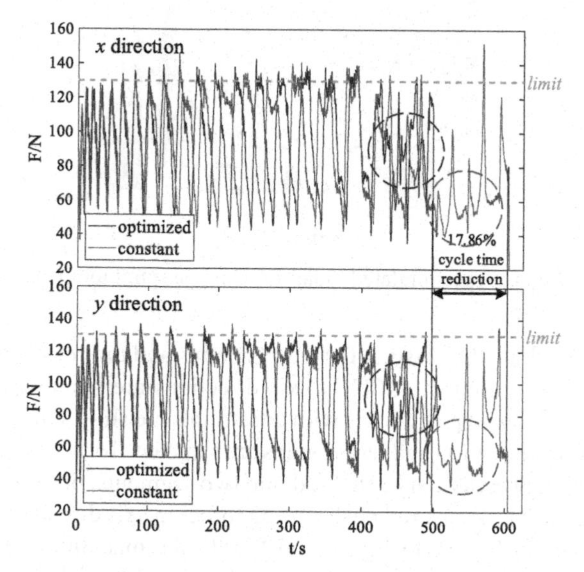

Fig. 9. Cutting forces and cycle time comparison (Color figure online)

In the first segment, the tool moves away from the inner circle boundary and enters the workpiece, so the radial DOC increases continuously. Both the ROC and the radial DOC synchronously change rapidly, leading to drastic changes of the optimum feedrates. A low-pass filter can be considered to smooth the optimum feedrates to avoid chatter in the milling process. But the length of this segment only accounts for 2% of the whole toolpath and has a small effect on the measured cutting forces.

In the second segment, the optimum feedrates oscillate due to the changes in radial DOC. As shown by the orange vertical dash lines, the optimum feedrates are close to the constant feedrate when the radial DOC is relatively large, so the peak values of the cutting force curve in Fig. 9 remain nearly unchanged. The cyan vertical dash lines show that the optimum feedrates are larger than the constant feedrate when the radial DOC is relatively small, raising the trough values of the cutting force curve in Fig. 9. When the radial DOC is pretty small, the optimum feedrates can reach the user-defined maximal value, as shown by the black vertical dash lines. As presented in Fig. 9, the range of the corresponding cutting force trough values of the proposed method within the blue circle is larger than that of the constant feedrate case within the red circle.

It can be seen from the above analysis that the proposed enhanced feedrate optimization algorithm with the cutting force limitation strategy works well. By applying the enhanced feedrate optimization scheme, the overall cycle time is reduced by 17.86%.

To further validate the effectiveness of the enhanced feedrate optimization algorithm, two more groups of experiments for the optimized feedrates under different force limits in axis directions and corresponding constant feedrate are conducted. The maximum cutting forces and the cycle time are compared, as shown in Table 1. The results indicate that the enhanced feedrate optimization approach enables significant improvement in machining efficiency while keeping the maximum cutting forces at the same level. The experimental maximum forces are slightly higher than the limited forces due to cutting force estimation errors, but the cutting force constraints overall performed reasonably well on limiting the maximum forces.

Table 1. Comparison of cutting time, maximum forces in x and y directions

Group	Constant feedrate (mm/min)	Limited force (N)	Max F_x (N)	Max F_y (N)	Cycle time (s)	Time saved
1	220	–	157.52	152.98	550.10	16.54%
	–	140	154.76	144.05	459.11	
2	240	–	164.38	162.08	504.55	15.46%
	–	150	161.15	152.91	426.55	

6 Conclusion

This paper develops the linear-programming based feedrate optimization algorithm with a novel cutting force limitation strategy. The cutting force limitation strategy improves upon the linear cutting force model based on geometric features of the B-spline spiral toolpath for pocket machining and converts cutting force constraints to the maximum allowable feedrate, which is then added to the LP formulation to get the enhanced feedrate optimization algorithm. The effectiveness of the enhanced feedrate optimization approach is verified through pocket machining experiments.

The proposed feedrate planning approach can efficiently decrease the machining time while keeping the maximum cutting forces at a certain level to prolong tool life and avoid tool rupture. More efforts are favourable to improve the accuracy of the modified cutting force model so as to enhance the effectiveness of cutting force constraints.

Acknowledgements. This work was supported by the National Natural Science Foundation of China [grant numbers: 51822506, 51935010].

References

1. Held, M.: On the Computational Geometry of Pocket Machining. Springer, Heidelberg (1991)
2. Bieterman, M.B., Sandstrom, D.R.: A curvilinear tool-path method for pocket machining. J. Manuf. Sci. Eng. Trans. ASME **125**, 709–715 (2003). https://doi.org/10.1115/1.1596579
3. Sencer, B., Altintas, Y., Croft, E.: Feed optimization for five-axis CNC machine tools with drive constraints. Int. J. Mach. Tools Manuf. (2008). https://doi.org/10.1016/j.ijmachtools.2008.01.002
4. Fan, W., Gao, X.S., Lee, C.H., Zhang, K., Zhang, Q.: Time-optimal interpolation for five-axis CNC machining along parametric tool path based on linear programming. Int. J. Adv. Manuf. Technol. (2013). https://doi.org/10.1007/s00170-013-5083-x
5. Erkorkmaz, K., Chen, C.Q.G., Zhao, M.Y., Beudaert, X., Gao, X.S.: Linear programming and windowing based feedrate optimization for spline toolpaths, CIRP Ann. Manuf. Technol. **66** (2017). https://doi.org/10.1016/j.cirp.2017.04.058
6. Altintas, Y.: Manufacturing Automation. Cambridge University Press, Cambridge (2011)
7. Altintas, Y, Merdol, S.D.: Virtual high performance milling. CIRP Ann. Manuf. Technol. **56** (2007). https://doi.org/10.1016/j.cirp.2007.05.022
8. Merdol, S.D., Altintas, Y.: Virtual cutting and optimization of three-axis milling processes. Int. J. Mach. Tools Manuf. **48** (2008). https://doi.org/10.1016/j.ijmachtools.2008.03.004
9. Erkorkmaz, K., Layegh, S.E., Lazoglu, I., Erdim, H.: Feedrate optimization for freeform milling considering constraints from the feed drive system and process mechanics. CIRP Ann. Manuf. Technol. **62** (2013). https://doi.org/10.1016/j.cirp.2013.03.084
10. Wei, Z.C., Wang, M.J., Ma, R.G., Wang, L.: Modeling of process geometry in peripheral milling of curved surfaces. J. Mater. Process Technol. (2010). https://doi.org/10.1016/j.jmatprotec.2010.01.011
11. Zhang, L., Zheng, L.: Prediction of cutting forces in milling of circular corner profiles. Int. J. Mach. Tools Manuf. (2004). https://doi.org/10.1016/j.ijmachtools.2003.10.007
12. Han, X., Tang, L.: Precise prediction of forces in milling circular corners. Int. J. Mach. Tools Manuf. **88** (2015). https://doi.org/10.1016/j.ijmachtools.2014.09.004

Conceptual Design and Kinematic Optimization of a Gantry Hybrid Machining Robot

Jie Wen[1], Fugui Xie[1,2](✉), Weiyao Bi[1], and Xin-Jun Liu[1,2]

[1] The State Key Laboratory of Tribology, Department of Mechanical Engineering (DME), Tsinghua University, Beijing 100084, China
xiefg@mail.tsinghua.edu.cn

[2] Beijing Key Lab of Precision/Ultra-Precision Manufacturing Equipments and Control, Tsinghua University, Beijing 100084, China

Abstract. High-efficiency and high-precision machining of large-scale structural parts is one of the difficult problems in the field of machining. Such parts usually have the characteristics of large size, large aspect ratio, and complex geometric shapes. This paper attempts to propose a gantry machining equipment with five-axis parallel processing module. It can realize large stroke positioning and small range precise operation. Large aspect ratio parts can be processed once without repeatedly clamping workpieces and setting tools, which improves machining efficiency and precision. According to this design concept, the tilt and torsion angles are used to describe the end posture. On this basis, the motion/force transmission performance and mechanism geometric restraint conditions are considered. This paper comprehensively evaluates the performance of the mechanism by integrating the local transmission index (LTI) and the angle between the screw and the spindle, and then optimizes the design of the parallel processing module through parameter optimization algorithms. The optimization results show that the parallel processing module has excellent kinematic transmission performance in the allowable workspace. It means that parallel processing module with gantry structure can realize high-efficiency and high-precision machining of structural parts with large aspect ratio.

Keywords: Large aspect ratio workpiece · Large stroke positioning · Motion/Force transmissibility · Parameter optimization

1 Introduction

With the prosperous development of the aerospace industry, high-efficiency and precision manufacturing process of structural parts [1] with large aspect ratio (such as aircraft ribs) has become a prominent problem in industry. These kinds of parts usually have the characteristics of complex geometry [2] and large material removal rate. The machining equipments are required to have the ability of compound angle machining, large-stroke machining and high-efficiency precision machining [3].

It is well-known that the serial mechanism usually has large workspace and simple kinematics, but suffers from large inertia. In contrast, the parallel mechanism has low

X.-J. Liu et al. (Eds.): ICIRA 2021, LNAI 13014, pp. 743–753, 2021.
https://doi.org/10.1007/978-3-030-89098-8_70

inertia [4] and compact structure [5], but suffers from limited workspace [6] and complex kinematics [7, 8]. By the means of rational design and optimization [9], the hybrid mechanism can combine both the advantages of serial and parallel mechanisms. Based on this idea, many companies have developed prototypes with good performance, such as Tricept [10], Exechon [11] and Ecospeed [12]. It can well solve the processing problems of complex parts. But there is still a problem encountered, i.e., how to process complex structural parts with large aspect ratios at one setup, to eliminate the influence of repeated clamping and tool setting on the final machining accuracy. In response to this problem, this paper intends to use the hybrid configuration of large stroke gantry and portable five-axis parallel machining module [13, 14] to establish a gantry truss hybrid machining robot to achieve one-time processing of structural parts with large aspect ratios and to improve the processing accuracy.

The gantry truss hybrid machining robot is based on the 5-DoF PKM designed by Xie et al. in 2019 [15]. On basis, this paper will carry out its dimension synthesis, and LTI [16] is used to evaluate the motion/force transmission characteristics. Additionally, in order to avoid mechanism interference, the geometric constraints of the mechanism itself are taken as one of the evaluation indicators. The motion/force transmission index and the geometric constraint characteristics of the mechanism are integrated as the evaluation criteria for the optimization of the parallel processing module discussed in this paper. The parameter optimization process needs to optimize multiple parameters at the same time, in order to shorten the optimization time, this paper uses a combination of local optimization and global optimization, i.e., reducing the parameter value range through the local optimization, then optimizing multiple parameters at the same time through genetic algorithm to obtain the optimal parameters.

The reminder of this paper is organized as follows. In Sect. 2, the gantry truss hybrid processing robot is presented to meet the processing requirements of complex structural parts with large aspect ratio. Section 3 establishes parameter optimization indices based on motion/force transmission index and mechanism characteristics. Taking a certain LTI value as the constraints, the parameters are optimized by meeting the required of workspace. Section 4 evaluates the optimized parameters. Section 5 concludes the paper.

2 Conceptual Design

Many structural parts in the aerospace industry have large aspect ratios (at least 2 m in length) and complex geometry. In order to realize the processing of such workpieces, processing equipment is required to have a large stroke positioning space and local five-axis high-precision and flexible processing capabilities. In this paper, the processing requirements of the workpieces to be machined are summarized in Table 1.

For the processing of complex structural parts analyzed above, the workspace of the five-axis machine tool is not enough to cover all the areas to be processed, and the repeated clamping of the workpiece and the tool will affect the final processing accuracy.

Therefore, this paper intends to establish a gantry truss-type hybrid processing robot.with a hybrid configuration of large stroke gantry and portable five-axis processing module. The gantry structure is adopted to realize the large-stroke precise positioning of the parallel processing module, and the advantages of the parallel robot in terms of

Table 1. Processing requirements of some large aspect ratio workpieces

Parameters	Value
x-axis workspace	600 mm
y-axis workspace	600 mm
z-axis workspace	500 mm
A/B-axis rotational angles	20°
Stroke of the gantry	3000 mm

structural compactness, movement flexibility, and dynamic response rate are fully utilized to perform local fine processing of structural parts. The one-time forming of the large aspect ratios structural parts reduces the downtime of clamping and improves the processing efficiency. There is no need to repeat the tool setting during the processing process, which reduces the error generation links and improves the processing accuracy.

The CAD model of gantry truss machining equipment is shown in Fig. 1. The movable gantry structure with 3000 mm stroke is convenient for loading and unloading large complex parts. Single clamping can meet the processing requirements of the whole parts, without repeated clamping of the workpiece and the tool. The five-axis parallel machining module of the gantry is based on DiaRoM-II [17], and the parameters are optimized to match the machining requirements and realize the high-precision machining of complex structural parts in its workspace.

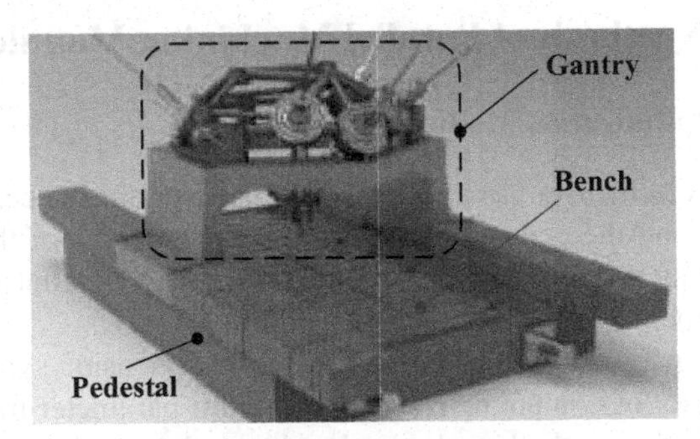

Fig. 1. The CAD model of gantry truss machining equipment

To facilitate the design and analysis of the parallel module in the following sections, the kinematic scheme of the parallel mechanism is shown in Fig. 2, the coordinate system and structural parameters are as follows.

The global coordinate system $O - xyz$ is fixed with the moving gantry. The origin O coincides with the center of the circle $B_1B_2B_3$. Coordinate $O' - x'y'z'$ is fixed to the spindle. The origin O' is the tool center point.

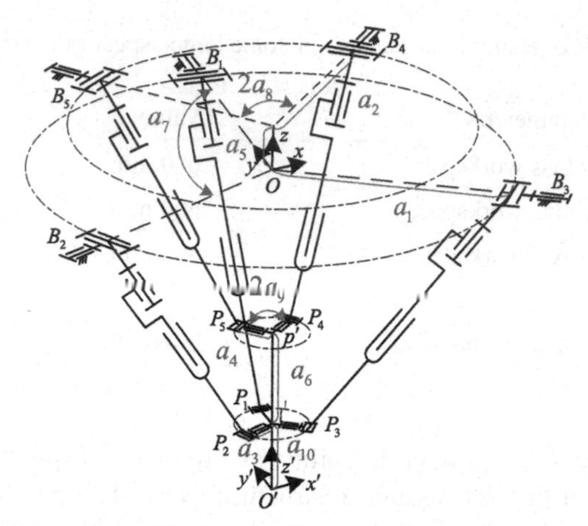

Fig. 2. Kinematic scheme of five-axis parallel machining module

For the five-axis parallel machining module, there are two kinds of parameters: the motor hinge point position (B_i , $i = 1, 2, ...5$) and the spindle hinge point position (P_i , $i = 1, 2, ...5$). To simplify the expression, the parameters of the parallel mechanism compose a set ϕ.

$$\phi = \{ a_1, a_2, a_3, a_4, a_5, a_6, a_7, a_8, a_9, a_{10} \} \quad (1)$$

3 Dimension Synthesis of Parallel Machining Module

3.1 Parameter Optimization Method Introduction

To ensure that the mechanism has excellent kinematic performance, the motion/force transmission index and the geometric constraint characteristics of the mechanism are integrated as the evaluation criteria for the optimization of the parallel processing module discussed in this paper.

There are many parameters to be optimized in the dimension synthesis process. To avoid consuming a lot of computing resources in multi-parameter traversal, this paper proposes an optimization method combining local optimization and global optimization: grouping the most relevant parameter in pairs, and narrowing the value space of each parameter through local optimization. Then use genetic algorithm for global optimization, and finally the optimal parameter in the parameter space is obtained. Dimension synthesis index.

According to the description in Ref. [17], the input motion and the corresponding force transmission driven by the five branches of the parallel module are expressed as the input twist screw $\mathbf{S}^j_{\text{ITS}}$and the transmission wrench $\mathbf{S}^j_{\text{TWS}}$($j = 1, 2, 3, 4, 5$), respectively.

$$\$_{\text{ITS}} = \begin{cases} \$_{\text{ITS-11}} = \left(\mathbf{0}; \frac{\mathbf{B}_1\mathbf{P}_1}{|\mathbf{B}_1\mathbf{P}_1|}\right) \\ \$_{\text{ITS-22}} = \left(\mathbf{0}; \frac{\mathbf{B}_2\mathbf{P}_2}{|\mathbf{B}_2\mathbf{P}_2|}\right) \\ \$_{\text{ITS-33}} = \left(\mathbf{0}; \frac{\mathbf{B}_3\mathbf{P}_3}{|\mathbf{B}_3\mathbf{P}_3|}\right) \\ \$_{\text{ITS-44}} = \left(\mathbf{0}; \frac{\mathbf{B}_4\mathbf{P}_4}{|\mathbf{B}_4\mathbf{P}_4|}\right) \\ \$_{\text{ITS-55}} = \left(\mathbf{0}; \frac{\mathbf{B}_5\mathbf{P}_5}{|\mathbf{B}_5\mathbf{P}_5|}\right) \end{cases}, \quad \$_{\text{TWS}} = \begin{cases} \$_{\text{TWS-11}} = \left(\frac{\mathbf{B}_1\mathbf{P}_1}{|\mathbf{B}_1\mathbf{P}_1|}; \frac{\boldsymbol{O'P}_1 \times \mathbf{B}_1\mathbf{P}_1}{|\mathbf{B}_1\mathbf{P}_1|}\right) \\ \$_{\text{TWS-22}} = \left(\frac{\mathbf{B}_2\mathbf{P}_2}{|\mathbf{B}_2\mathbf{P}_2|}; \frac{\boldsymbol{O'}\mathbf{P}_2 \times \mathbf{B}_2\mathbf{P}_2}{|\mathbf{B}_2\mathbf{P}_2|}\right) \\ \$_{\text{TWS-33}} = \left(\frac{\mathbf{B}_3\mathbf{P}_3}{|\mathbf{B}_3\mathbf{P}_3|}; \frac{\boldsymbol{O'}\mathbf{P}_3 \times \mathbf{B}_3\mathbf{P}_3}{|\mathbf{B}_3\mathbf{P}_3|}\right) \\ \$_{\text{TWS-44}} = \left(\frac{\mathbf{B}_4\mathbf{P}_4}{|\mathbf{B}_4\mathbf{P}_4|}; \frac{\boldsymbol{O'}\mathbf{P}_4 \times \mathbf{B}_4\mathbf{P}_4}{|\mathbf{B}_4\mathbf{P}_4|}\right) \\ \$_{\text{TWS-55}} = \left(\frac{\mathbf{B}_5\mathbf{P}_5}{|\mathbf{B}_5\mathbf{P}_5|}; \frac{\boldsymbol{O'}\mathbf{P}_5 \times \mathbf{B}_5\mathbf{P}_5}{|\mathbf{B}_5\mathbf{P}_5|}\right) \end{cases} \tag{2}$$

The unit output twist screw of the mechanism can be expressed as

$$\$_{\text{OTS}}^{j} = \left(\mathbf{s}_j\,;\mathbf{r}_j \times \mathbf{s}_j\right),\ j = 1, 2, 3, 4, 5 \tag{3}$$

where $|\mathbf{s}_j| = 1$, $\$_{\text{OTS}}^{j}$ can be determined by

$$\begin{cases} \$_{\text{OTS}}^{m} \circ \$_{\text{TWS}}^{n} = 0,\ m \neq n \\ \$_{\text{OTS}}^{m} \circ \$_{\text{CWS}} = 0 \end{cases},\quad m/n = 1, 2, 3, 4, 5 \tag{4}$$

The input transmission index (ITI) of the j-th kinematic chain is

$$\gamma_{I-j} = \frac{\left|\$_{\text{ITS}}^{j} \circ \$_{\text{TWS}}^{j}\right|}{\left|\$_{\text{ITS}}^{j} \circ \$_{\text{TWS}}^{j}\right|_{\max}} \tag{5}$$

According to the Eq. (5), $\gamma_{I-j} = 1$. The output transmission index (OTI) of the j-th kinematic chain is

$$\gamma_{O-j} = \frac{\left|\$_{\text{OTS}}^{j} \circ \$_{\text{TWS}}^{j}\right|}{\left|\$_{\text{OTS}}^{j} \circ \$_{\text{TWS}}^{j}\right|_{\max}} \tag{6}$$

where $\gamma_{O-j} \in [0, 1][0, 1]$. Therefore, LTI can be expressed as

$$\gamma = \min\{\gamma_{I-j},\ \gamma_{O-j}\} = \min\{\gamma_{O-j}\},\ j = 1, 2, 3, 4,\ 5 \tag{7}$$

In addition to the motion/force transmission performance, it is necessary to establish the motion constraint equation of the mechanism to ensure that there is no interference among the motion branches and between each branch and the spindle.

The frame and the bottom circle of the spindle are connected by kinematic chains $B_iP_i,\ (i = 1, 2, 3)$. To ensure that kinematic chains $B_iP_i,\ (i = 1, 2, 3)$ do not interfere with kinematic chains $B_jP_j,\ (j = 4, 5)$. The geometric constraint equation can be established by

$$\arctan(\frac{a_4 - a_3}{a_6}) < \arccos(\frac{\overrightarrow{s'p'} \cdot \overrightarrow{P_iB_i}}{\left|\overrightarrow{s'p'}\right|\left|\overrightarrow{P_iB_i}\right|}),\ (i = 1, 2, 3) \tag{8}$$

The frame and spindle bottom circle are connected by kinematic chains $B_jP_j,\ (j = 4, 5)$. To ensure that $B_jP_j,\ (j = 4, 5)$ kinematic chains do not interfere with the spindle. The geometric constraint equation can be established by

$$0 < \arccos(\frac{\overrightarrow{s'p'} \cdot \overrightarrow{P_jB_j}}{\left|\overrightarrow{s'p'}\right|\left|\overrightarrow{P_jB_j}\right|}),\ (j = 4, 5) \tag{9}$$

Through the above LTI index and motion constraint equations, the motion/force transmission performance and motion constraint performance of the mechanism can be well described. To increase the z-direction workspace as much as possible, the minimum value of LTI is set as 0.4.

3.2 Local Parameter Optimization

Local optimization narrows the parameter value space through the control variable method. To avoid obtaining the local optimal value, select any value from the completed local optimization parameter to replace the initial value. Considering the stroke of the gantry truss, the diameter of the maximum circle a_1 is 2000 mm. Considering the symmetry of the structure, set $a_7 = 120°$.The tool length parameter a_{10} needs to be selected according to the actual requirements, and the tool length is set to 200 mm during the optimization process. The remaining seven parameters need to be optimized. According to the previous design [15], with the largest circle diameter as the numerical scaling reference, the size of each structure is scaled proportionally. The initial parameters (unlabeled unit mm) are set as

$$\phi_{\mathrm{Init}} = \{2000, 1680, 280, 280, 195, 265, 120°, 45°, 120°, 200\} \tag{10}$$

The algorithm of local optimization is shown in Table 2.

a_2, a_5 Parameter Optimization. Considering the installation position of hook hinge on the truss, when the hierarchical height a_5 of the motor hinge point takes different values, the circular diameter a_5 of the upper hinge point will also change in the position arrangement. Therefore, a_2 and a_5 are regarded as the same set of parameters to study the influence of a_2 and a_5 on the motion/force transmission characteristics. a_2 and a_5 are taken as variables at the same time to solve the minimum OTI(ITI = 1, so LTI = OTI) of all points when the cutting point of the machining module meets the linkage of 20° A/B and

Table 2. Local optimization algorithm

Algorithm 1 Parametric optimization algorithm
Input: The range of optimization parameters
Output: Minimum OTI of all workspace points
For each parameter in the parameter range
Calculate the inverse kinematics solution of all workspace points
Calculate the minimum OTI of all workspace points through Eqs. (2)-(7)
Check motion constraints through Eqs. (8)-(9)
Record current optimized parameters
Record the minimum OTI of all workspace points which is segmented by z value
End for
Return matrix of minimum OTI of all workspace points

z takes each value in the optimization design workspace($z \in [-1200\text{ mm}, -200\text{ mm}]$, interval 100 mm). The results are shown in Fig. 3. The parameter (unlabeled unit mm) setting is

$$\phi_1 = \{2000, a_2, 280, 280, a_5, 265, 120°, 45°, 120°, 200\} \tag{11}$$

where $a_2 \in [1500\text{ mm}, 2000\text{ mm}]$, $a_5 \in [0, 200\text{ mm}]$.

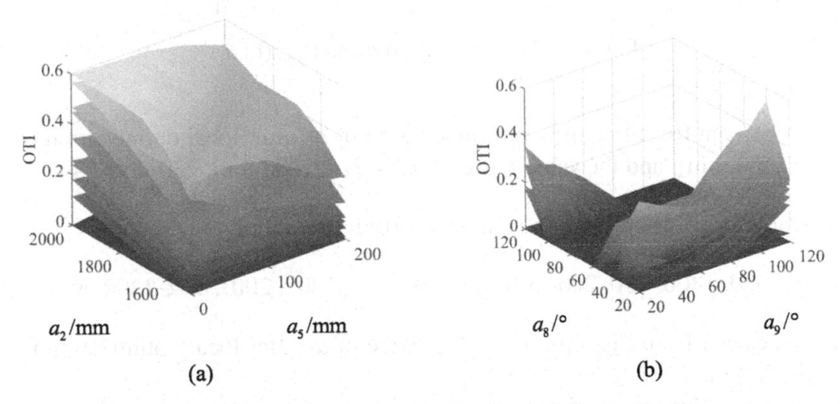

Fig. 3. Effects of OTI: (a) a_2 and a_5; (b) a_8 and a_9.

From Fig. 3(a), it can be seen that the greater the circular diameter a_2, and the smaller the a_5 value, the better the motion/force transmission performance, i.e., the greater the OTI value. When $a_2 > 1900$ mm and $a_5 < 150$ mm, the trend of OTI values slows down. Considering the working stroke of the gantry module is 3 m and the maximum circular diameter of the mechanism a_1=2000 mm. After local optimization, the value space of a_2 is $a_2 \in [1700\text{ mm}, 1900\text{ mm}]$ and the value space of a_5 is $a_5 \in [50\text{ mm}, 100\text{ mm}]$.

Similarly, the parameters of three groups of (a_8, a_9), ($a_3 a_4$) and a_6 are optimized. The parameter setting and optimization results are as follows.

a_8, a_9 Parameter Optimization. The parameter (unlabeled unit mm) setting is

$$\phi_2 = \{2000, 1800, 280, 280, 100, 265, 120^\circ, a_8, a_9, 200\} \quad (12)$$

where $a_8 \in [30^\circ, 110^\circ], a_9 \in [20^\circ, 120^\circ]$.

It can be seen from Fig. 3(b), the value space of a_8 after local optimization is $a_8 \in [30^\circ, 50^\circ]$, and the value space of a_9 is $a_9 \in [100^\circ, 120^\circ]$.

a_3, a_4 Parameter Optimization. The parameter (unlabeled unit mm) setting is

$$\phi_3 = \{2000, 1800, a_3, a_4, 100, 265, 120^\circ, 30^\circ, 120^\circ, 200\} \quad (13)$$

where $a_3 \in [300\text{ mm}, 400\text{ mm}],\ a_4 \in [200\text{ mm}, 300\text{ mm}]$.

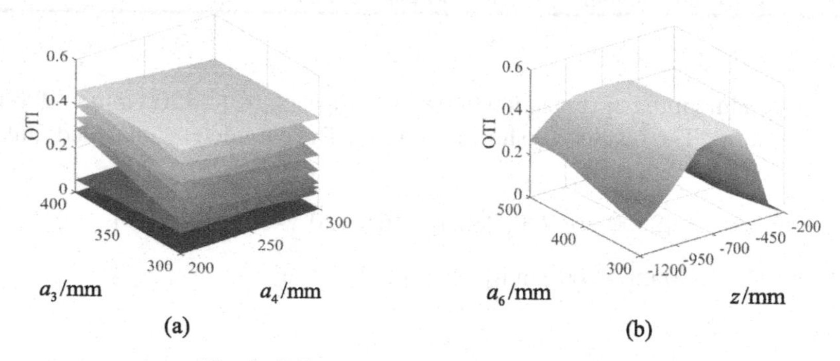

Fig. 4. Effects on OTI: (a) a_3 and a_4; (b) a_6.

It can be seen from Fig. 4(a), the value space of a_3 after local optimization is $a_3 \in [350\text{ mm}, 400\text{ mm}]$, and the value space of a_4 is $a_4 \in [200\text{ mm}, 250\text{ mm}]$.

a_6 Parameter Optimization. The parameter (unlabeled unit mm) setting is

$$\phi_4 = \{2000, 1800, 350, 200, 100, a_6, 120^\circ, 30^\circ, 120^\circ, 200\}, a_6 \in [200, 350] \quad (14)$$

It can be seen from Fig. 4(b), the value space of a_6 after local optimization is $a_6 \in [300\text{ mm}, 350\text{ mm}]$.

So far the optimization of local parameters is completed, the workspace and local parameter value space are as follows:

Workspace: $x, y \in [-300\text{ mm}, 300\text{ mm}]$, $z \in [-1000\text{ mm}, -500\text{ mm}]$, $\theta \in [0, 20^\circ]$, $\varphi \in [0, 360^\circ]$.

Value space of parameters after local optimization: $a_2 \in [1700\text{ mm}, 1900\text{ mm}]$, $a_3 \in [350\text{ mm}, 400\text{ mm}]$, $a_4 \in [200\text{ mm}, 250\text{ mm}]$, $a_5 \in [50\text{ mm}, 100\text{ mm}]$, $a_6 \in [300\text{ mm}, 350\text{ mm}]$, $a_8 \in [30^\circ, 50^\circ]$, $a_9 \in [100^\circ, 120^\circ]$.

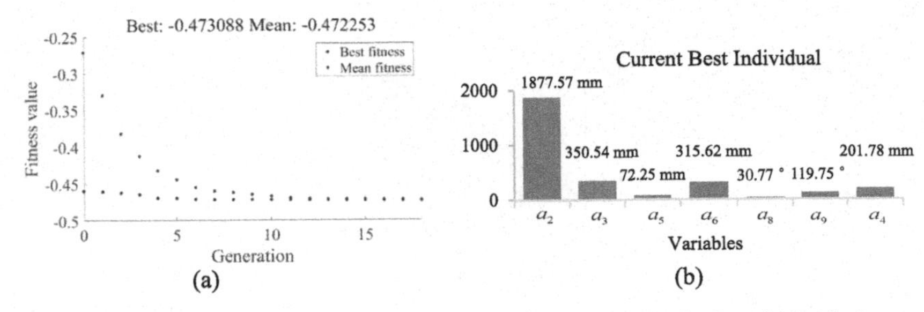

(a) (b)

Fig. 5. Genetic optimization results: (a) Iterative process; (b) Best individual.

3.3 Global Parameter Optimization

The global parameter optimization is accomplished by genetic algorithm toolbox of MATLAB. Taking the value space of each parameter after local optimization as the boundary, the number of samples per generation is 200, and iterates to convergence. The genetic optimization results are shown in Fig. 5.

According to the above optimization results of genetic algorithm, the best individual parameter is: $a_2 = 1877.57$ mm, $a_3 = 350.54$ mm, $a_4 = 201.78$ mm, $a_5 = 72.25$ mm, $a_6 = 315.62$ mm, $a_8 = 30.77°$, $a_9 = 119.75°$. The minimum OTI in the workspace is about 0.473. The parameters are rounded to obtain the optimized parameter structure of the final parallel module (unmarked unit is mm)

$$\phi_{fin} = \{\ 2000, 1880, 350, 200, 70, 320, 120°, 30°, 120°, 200\} \tag{15}$$

Assuming time-consuming T can traverse the entire workspace once. All parameters to be optimized are divided into six layers. From Fig. 5(a) we can see that the global optimization converges in the 15th generation. The optimization algorithm proposed in this paper takes time 3114T, while full traversal method takes time 279936T. It can be seen that the proposed optimization algorithm significantly improves the optimization efficiency.

4 Evaluation of Optimization Results

The simulation results record the minimum OTI values of all points in the entire workspace to evaluate the motion/force transmission performance of the mechanism after parameter optimization. Figure 6 shows the OTI distribution map and contour map of the layer where the OTI minimum value is located.

After rounding the parameters, Fig. 6 shows that the minimum value of OTI in the full workspace is about 0.47, which compliance with design expectations in Sect. 3.1. It shows that the mechanism has good motion/force transmission performance in its workspace.

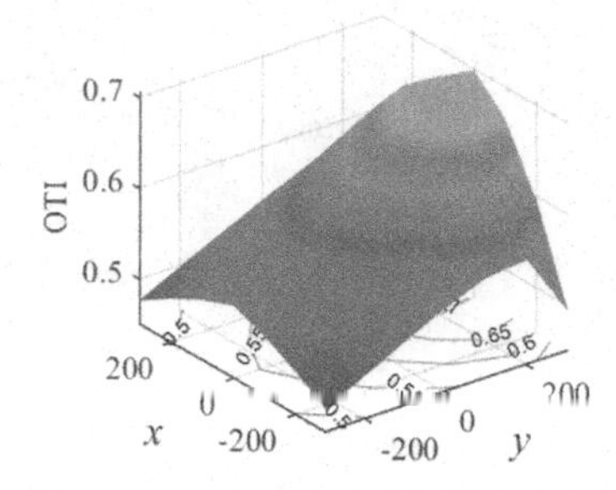

Fig. 6. OTI_{min} contour map of optimization results

5 Conclusion

To solve the processing problem of complex structural parts with large aspect ratio, this paper proposes a hybrid configuration of large stroke gantry and portable five-axis parallel machining module to establish a gantry truss hybrid machining robot which can realize the processing space of 3000 mm × 600 mm × 500 mm. The gantry structure can realize the large-stroke precise positioning of the parallel processing module, and the parallel robot can realize local fine processing of structural parts. In this way, the complex structural parts with large aspect ratio can be processed at one setup, without repeated tool setting, which can reduce the error generation links and improves the processing accuracy.

The parameter optimization process has a large amount of calculation, resulting in high time cost. To reduce the amount of calculation in the optimization process. This paper proposes a method which combines local optimization and global optimization. The parameter value space is reduced by local optimization, and the final optimal structural parameters are obtained by global optimization. The local optimization algorithm reduces the parameter value space, and the genetic algorithm obtains the final optimization result. The optimization result shows that the mechanism has good motion/force transmission performance in workspace.

The design presented in this paper is helpful to the development of the robot capable of the processing complex structural parts with large aspect ratio, and a prototype will be developed in the near future.

Acknowledgements. This work is supported by National Key R&D Program of China under Grant 2019YFA0706701, and the National Natural Science Foundation of China under Grants 51922057 and 91948301.

References

1. Alatorre, D., et al.: Teleoperated, in situ repair of an aeroengine: overcoming the internet latency hurdle. IEEE Rob. Autom. Mag. **26**(1), 10–20 (2018)
2. Tang, T., Zhang, J.: Conceptual design and kinetostatic analysis of a modular parallel kinematic machine-based hybrid machine tool for large aeronautic components. Rob. Comput. Integr. Manuf. **57**, 1–16 (2019)

3. Xie, H., Li, W.L., Zhu, D., Yin, Z., Ding, H.: A systematic model of machining error reduction in robotic grinding. IEEE/ASME Trans. Mechatron. **25**(6), 2961–2972 (2020)
4. Ma, N., Yu, J.J., Dong, X., Axinte, D.: Design and stiffness analysis of a class of 2-DoF tendon driven parallel kinematic mechanism. Mech. Mach. Theory **129**, 202–217 (2018)
5. Ma, N., et al.: Parametric vibration analysis and validation for a novel portable hexapod machine tool attached to surfaces with unequal stiffness. J. Manuf. Process. **47**, 192–201 (2019)
6. Wang, G., Li, W.L., Jiang, C., Zhu, D.H.: Simultaneous calibration of multicoordinates for a dual-robot system by solving the AXB= YCZ problem. IEEE Trans. Rob. **37**, 1172–1185 (2021)
7. Dong, X., Palmer, D., Axinte, D., Kell, J.: In-situ repair/maintenance with a continuum robotic machine tool in confined space. J. Manuf. Process. **38**, 313–318 (2019)
8. Wang, M., Palmer, D., Dong, X., Alatorre, D., Axinte, D., Norton, A.: Design and development of a slender dual-structure continuum robot for in-situ aeroengine repair. In: 2018 IEEE/RSJ International Conference on Intelligent Robots and Systems (IROS), pp. 5648–5653. IEEE. (2018)
9. Liu, X.J., Bi, W.Y., Xie, F.G.: An energy efficiency evaluation method for parallel robots based on the kinetic energy change rate. Sci. Chin. Technol. Sci. **62**(6), 1035–1044 (2019). https://doi.org/10.1007/s11431-019-9487-7
10. Wang, Y., Liu, H., Huang, T., Chetwynd, D.G.: Stiffness modeling of the tricept robot using the overall Jacobian matrix. J. Mech. Rob **1**(2), 021002 (2009)
11. Fernandez, A.J.S., Jimenez, V.C. Olazabal, M.G.: Kinematical system for a movable platform of a machine. Patent EP1245349 A1 (2002)
12. Wahl J.: Articulated tool head. US Patent No.: US6431802 B1 (2000)
13. Xie, F.G., Liu, X.J., Luo, X., Wabner, M.: Mobility, singularity, and kinematics analyses of a novel spatial parallel mechanism. J. Mech. Rob. **8**(6), 061022 (2016)
14. Shen, X., Xie, F.G., Liu, X.J., Xie, Z.H.: A smooth and undistorted toolpath interpolation method for 5-DoF parallel kinematic machines. Rob. Comput. Integr. Manuf. **57**, 347–356 (2019)
15. Liu, X.J., Xie, Z.H., Xie, F.G., Wang, J.S.: Design and development of a portable machining robot with parallel kinematics. In: 16th International Conference on Ubiquitous Robots (UR), pp. 133–136. IEEE. (2019)
16. Wang, J., Wu, C., Liu, X.J.: Performance evaluation of parallel manipulators: motion/force transmissibility and its index. Mech. Mach. Theory **45**(10), 1462–1476 (2010)
17. Xie, Z.H., Xie, F.G., Liu, X.J., Wang, J.S.: Global G3 continuity toolpath smoothing for a 5-DoF machining robot with parallel kinematics. Rob. Comput. Integr. Manuf. **67**, 102018 (2021)

Kinematics and Dimensional Synthesis of a Novel 2RPU-PUR Parallel Manipulator

Wenxi Ji, Zhentao Xie, and Xinxue Chai(✉)

Faculty of Mechanical Engineering and Automation, Zhejiang Sci-Tech University, Hangzhou 310018, Zhejiang, People's Republic of China
chaixx@zstu.edu.com

Abstract. In this paper, we focus our attention on a novel 2RPU-PUR (P, R, U denotes prismatic, revolute and universal joint, respectively) parallel manipulator with two rotations and one translation (2R1T). The proposed manipulator has three limbs, which is composed of two identical RPU limbs and one PUR limb. Firstly, the degrees of freedom of the mechanism are obtained based on screw theory. Then, the velocity analysis and singularity analysis of the manipulator are analyzed. Three types of singularity, i.e., inverse singularity, forward singularity and combine singularity of the parallel manipulator are obtained based on the Jacobian matrix. Finally, the local of motion/force transmission and global transmission index are used to evaluate the performance of the manipulator. And the link parameters of the manipulator are optimized to obtain a better transmission performance. It is shown that the proposed 2RPU-PUR PM has a good performance for high precision application.

Keywords: Parallel manipulator · Screw theory · Kinematics analysis · Performance evaluation

1 Introduction

Five-axis machine tools are widely used in the manufacturing of complex parts such as high precision molds, impellers and aircraft structural parts [1]. Most of the existing five-axis machine tools are based on serial manipulators. There are only a few PMs (parallel manipulators) that can be successfully applied to five-axis processing.

As the counterpart to serial manipulators, PMs have advantages of high rigidity, good dynamic response and large payload capability. Compared with the six degrees-of-freedom (DOFs) PM, the lower-mobility PMs not only have the advantages as general PMs, but also have the advantages of simpler structure, easier control, and lower manufacturing cost [2, 3]. Therefore, lower-mobility PMs gradually become a research hotspot. The type of Three DOFs PMs with 2R1T motion is an important branch of the lower-mobility PMs, which are useful in five-axis machining [4], aircraft wing assembly [5], and friction stir welding [6]. For 5-axis machining applications, there are some 2R1T PMs proposed, such as Z3 head [7], A3 head [8], Tricept [9], and Exechon [10].

X.-J. Liu et al. (Eds.): ICIRA 2021, LNAI 13014, pp. 754–765, 2021.
https://doi.org/10.1007/978-3-030-89098-8_71

It is required to have the characteristics of high precision and stiffness for five-axis machining tools. The joints are the weaken part of manipulators and the main sources of manipulators error. Thus, the number of joints in PMs should be as few as possible [11]. However, the parallel part of the Tricept is a 3UPS-UP PMs with 21 1-DOF joints. The parallel part of the Tricept [9] is composed of four limbs which imply more manufacturing costs and more error sources. The famous Z3 head [7] is a 3PRS PMs with 15 1-DOF joints. Compared by the Z3 head, the parallel part of Exechon has only 13 1-DOF joints. But it is actuated with unfixed linear actuators, which can increase the moveable mass and reduce the performance of dynamic response. To the best of our knowledge, the fewest joints of the PMs with 2R1T motion are 12 [12–15].

Above all, 2R1T PMs have great potential to be used in five-axis machine tools. And this paper presents a novel 2R1T 2RPU-PUR PM, which belongs to the family of RPR-equivalent PMs [15]. Besides, it is an overstrained PM with higher stiffness and precision [16]. This paper is organized as follows. Section 2 introduces a 2RPU-PUR PM. Mobility and inverse kinematics are obtained in Sect. 3 and 4. The velocity and singularity analysis of the mechanism are analyzed in Sect. 5 and Sect. 6. In Sect. 7, The performance of 2RPU-PUR PM is evaluated based on the motion/force transmission index, and the link parameters of the manipulator are optimized to obtain a better transmission performance. In the end, Sect. 8 presents the conclusion.

2 Description of the 2RPU-PUR Parallel Manipulator

As shown in Fig. 1, this paper presents a PM whose moving platform is connected to the base through three limbs.

In this PM, limb 1 and limb 2 are identical RPU limbs (R: revolute joint; P: prismatic joint; U: universal joint), and the third one is PUR limb. Thus, this manipulator can be named 2RPU-PUR PM, where the prismatic joint is selected as actuation. For the limb 1 and limb 2, the axes of the R joint are parallel to each other and the P joints are perpendicular to the revolute axes of the R joints. The first revolute axes of U joint in limb 1 and limb 2 are parallel to the revolute axes of the R joints in the limb 1 and 2. Meanwhile, the second revolute axes of the U joint are the coaxial in limb 1 and 2. For the limb 3, the first revolute axis of the U joint is parallel to the revolute axis of the R joint in limb 1. And the second axis of the U joint is parallel to the revolute axis of the R joint connected to the moving platform in limb 3.

Let A_1 and A_2 denote the centers of the R joints in 1 and limb 2, respectively, and A_3 denotes the center of the U joint in limb 3. The centers of the U joints in limb 1 and limb 2 are denoted by B_1 and B_2, and the center of the R joint in limb 3 are denoted by B_3. Without loss of generality, it is assumed that A_1, A_2 and A_3 are coplanar in the initial configuration, and it is also assumed that B_1, B_2 and B_3 are coplanar. The coordinate frames are established as shown in Fig. 2. A fixed reference frame *O-xyz* is attached to the fixed base. And the origin point O is at the middle of line A_1A_2. Let x-axis always point in the direction of A_3O, y-axis point along OA_1, z-axis is defined according to the right-hand rule. A moving coordinate frame *o-uvw* is attached to the moving platform. Let the u-axis point in the direction of B_3o and the v-axis aligned with oB_1, the w-axis is defined according to the right-hand rule. The architectural parameters of 2RPU-PUR PM are defined as follows: $B_1B_2 = 2l_1$, $B_3o = l_2$, $A_1A_2 = 2l_4$, $A_3O = l_3$, $A_3B_3 = l$.

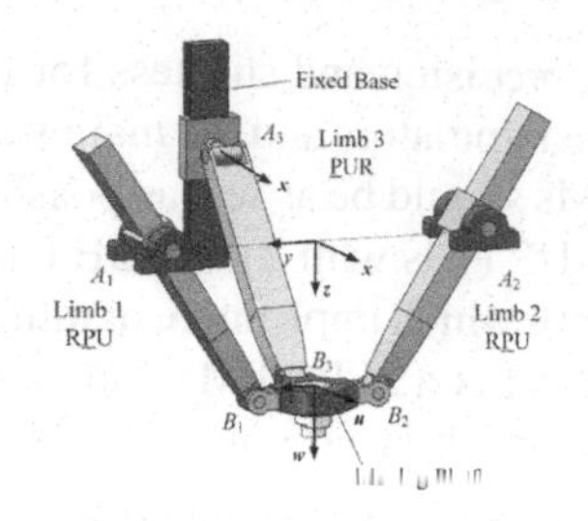

Fig. 1. 2RPU-PUR PM Model

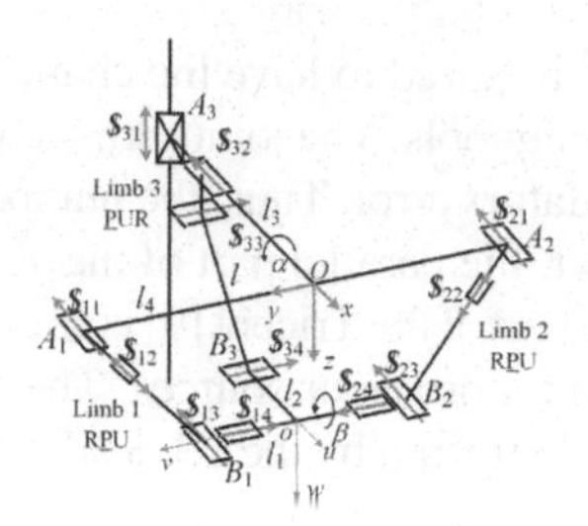

Fig. 2. 2RPU-PUR PM schematic representation

3 Mobility Analysis

Screw theory [2, 17] is widely used in the type synthesis and mobility analysis of PM. Here we use screw theory to analyze the mobility of 2RPU-PUR PM.

According to the screw theory, a screw can be called a twist when it is used to represent an instantaneous motion of a rigid body. And a screw can be called a wrench screw when it is used to represent a force or a couple acting on a rigid body. With respect to the fixed coordinate frame *O-xyz*, the position vectors of $\boldsymbol{B}_i$ ($i = 1, 2, 3$) are defined as $[0\ y_{B1}\ z_{B1}]^{\mathrm{T}}$, $[0\ y_{B2}\ z_{B2}]^{\mathrm{T}}$ and $[x_{B3}\ y_{B3}\ z_{B2}]^{\mathrm{T}}$. Similarly, the position vectors of $\boldsymbol{A}_i$ ($i = 1, 2, 3$) are defined as $[0\ l_4\ 0]^{\mathrm{T}}$, $[0\ -l_4\ 0]^{\mathrm{T}}$ and $[-l_3\ 0\ d_3]^{\mathrm{T}}$. According to the above definition, the twist system of limb 1 can be written as

$$\begin{cases} \$_{11} = (1\ 0\ 0; 0\ 0\ -l_4\ ) \\ \$_{12} = (0\ 0\ 0; 0\ \frac{y_{B1}-l_4}{\sqrt{(y_{B1}-l_4)^2+z_{B1}^2}}\ \frac{z_{B1}}{\sqrt{(y_{B1}-l_4)^2+z_{B1}^2}}\ ) \\ \$_{13} = (1\ 0\ 0; 0\ z_{B1}\ -y_{B1}\ ) \\ \$_{14} = (0\ -\cos\alpha\ -\sin\alpha; -y_{B1}\sin\alpha + z_{B1}\cos\alpha\ 0\ 0\ ) \end{cases} \tag{1}$$

where α denotes the rotation angle around the direction of x-axis, $\$_{11}$ and $\$_{12}$ denote the unit twist associated with R joint and P joint of the limb 1, $\$_{13}$ and $\$_{14}$ denote the unit twist associated with U joint of the limb 1.

The wrench system of limb 1 can be calculated according to Eq. (1) based on the screw theory

$$\begin{cases} \$_{11}^r = (0\ 0\ 0; 0\ -\sin\alpha\ \cos\alpha\ ) \\ \$_{12}^r = (1\ 0\ 0; 0\ z_{B1}\ -y_{B1}\ ) \end{cases} \tag{2}$$

where $\$_{11}^r$ denotes a constraint couple which is perpendicular to the direction of x-axis and B_1B_2, $\$_{12}^r$ denotes a constraint force in the direction of the R joint of the limb 1.

Similarly, the wrench systems of limb 2 are given by

$$\begin{cases} \$_{21}^r = (0\ 0\ 0; 0\ -\sin\alpha\ \cos\alpha\ ) \\ \$_{22}^r = (1\ 0\ 0; 0\ z_{B2}\ -y_{B2}\ ) \end{cases} \tag{3}$$

where $\$_{21}^r$ and $\$_{22}^r$ are the same as $\$_{11}^r$ and $\$_{12}^r$.

And the twist system of limb 3 can be given by

$$\begin{cases} \$_{31} = (0\,0\,0; 0\,0\,1) \\ \$_{32} = (1\,0\,0; 0\,d_3\,0) \\ \$_{33} = (0\ -\cos\alpha\ -\sin\alpha;\ d_3\cos\alpha\ -l_3\sin\alpha\ l_3\cos\alpha) \\ \$_{34} = (0\ -\cos\alpha\ -\sin\alpha;\ -y_{B3}\sin\alpha + z_{B3}\cos\alpha\ x_{B3}\sin\alpha\ -x_{B3}\cos\alpha) \end{cases} \tag{4}$$

where $\$_{3i}$ is the same as $\$_{1j}$ ($i = 1, 2, \ldots, 5; j = 1, 2, 3$).

The wrench system of limb 3 can be calculated according to Eq. (4) based on the screw theory

$$\begin{cases} \$_{31}^r = (0\,0\,0; 0\ -\sin\alpha\ \cos\alpha) \\ \$_{32}^r = (x_{B3}\sin\alpha + l_3\sin\alpha\ m\ 0; -md_3\ d_3\ -l_3m)/\sqrt{(x_{B3}\sin\alpha + l_3\sin\alpha)^2 + m^2} \end{cases} \tag{5}$$

where $m = d_3 + y_{B3}\sin\alpha - z_{B3}\cos\alpha$, and $\$_{31}^r$ is the same as $\$_{11}^r$, $\$_{32}^r$ denotes a constraint force pass the point A_3 in the direction of $\mathbf{s}_{32}^r$, where $\mathbf{s}_{32}^r = \left[x_{B3}\sin\alpha + l_3\sin\alpha\ m\ 0\right]^{\mathrm{T}}/\sqrt{(x_{B3}\sin\alpha + l_3\sin\alpha)^2 + m^2}$.

Combining Eq. (2), Eq. (3) and Eq. (5) with screw theory, the wrench system of the 2PRU-PUR PM is given by

$$\begin{cases} \$_1^r = (0\,0\,0; 0 - \sin\alpha\ \cos\alpha) \\ \$_2^r = (1\,0\,0; 0\ z_{B1}\ -y_{B1}) \\ \$_3^r = (x_{B3}\sin\alpha + l_3\sin\alpha\ m\ 0; -md_3\ d_3\ -l_3m)/\sqrt{(x_{B3}\sin\alpha + l_3\sin\alpha)^2 + m^2} \end{cases} \tag{6}$$

The moving platform twist system of the 2RPU-PUR PM can be obtained as

$$\begin{cases} \$_1^m = (0\,0\,0;\ 0\,0\,1) \\ \$_2^m = (1\,0\,0;\ 0\,d_3\,0) \\ \$_3^m = (0\ \cos\alpha\ \sin\alpha;\ p\ q\ 0) \end{cases} \tag{7}$$

where p and q guarantee that the reciprocal product $\$_3^m$ and the constraint screws equal to zero. p and q are omitted here for they won't be used in the subsequent analysis. Equation (7) shows that moving platform has three DOFs, including two rotations whose revolute axes parallel to x-axis and B_1B_2, respectively, and one translation along z-axis.

4 Inverse Kinematics

Kinematic analysis is the basis of the subsequent motion analysis and control of the manipulator. The purpose of the kinematics analysis is to establish the relationship between the output parameters of the moving platform and the input parameters of driving joint. The P joint of each limb is selected as the driving joint of the manipulator.

The inverse kinematics involves obtaining the input parameters (d_1, d_2, d_3) when given the output parameters (α, β, z_o), where d_i $(i = 1, 2)$ denotes the distance between A_iB_i, and β denotes the rotational angel around the B_1B_2.

The transformation from the coordinates system *o-uvw* to the coordinate system *O-xyz* can be described by a vector $\boldsymbol{p} = [0 \; -(z_o - d_3)\tan\alpha \; z_o]^{\mathrm{T}}$, and a rotation matrix ${}^{O}\boldsymbol{R}_o$, which can be expressed as

$$
{}^{O}\boldsymbol{R}_o = \begin{bmatrix} \cos\beta & 0 & \sin\beta \\ \sin\alpha\sin\beta & \cos\alpha & -\sin\alpha\cos\beta \\ -\cos\alpha\sin\beta & \sin\alpha & \cos\alpha\cos\beta \end{bmatrix} \tag{8}
$$

The position vectors of point A_i, B_i $(i = 1, 2, 3)$ expressed in the fixed coordinate frame *O-xyz* are given as

$$
\begin{cases} \boldsymbol{A}_1 = [0 \; l_4 \; 0]^{\mathrm{T}} \\ \boldsymbol{A}_2 = [0 \; -l_4 \; 0]^{\mathrm{T}} \\ \boldsymbol{A}_3 = [-l_3 \; 0 \; d_3]^{\mathrm{T}} \end{cases}, \quad \begin{cases} \boldsymbol{B}_1 = {}^{O}\boldsymbol{R}_o[0 \; l_1 \; 0]^{\mathrm{T}} + \boldsymbol{p} \\ \boldsymbol{B}_2 = {}^{O}\boldsymbol{R}_o[0 \; -l_1 \; 0]^{\mathrm{T}} + \boldsymbol{p} \\ \boldsymbol{B}_3 = {}^{O}\boldsymbol{R}_o[-l_2 \; 0 \; 0]^{\mathrm{T}} + \boldsymbol{p} \end{cases} \tag{9}
$$

A loop-closure equation can be written for limb i as

$$
\boldsymbol{A}_i - \boldsymbol{B}_i = \boldsymbol{A}_i\boldsymbol{B}_i (i = 1, 2, 3) \tag{10}
$$

Substituting the position vectors $\boldsymbol{A}_i$ and $\boldsymbol{B}_i$ $(i = 1, 2, 3)$ into Eq. (10), three constraint equations are obtained as

$$
\begin{cases} (-l_1\cos\alpha + l_4 - (d_3 - z_0)\tan\alpha)^2 + (l_1\sin\alpha + z_o)^2 = d_1^2 \\ (-l_4 + l_1\cos\alpha - (d_3 - z_0)\tan\alpha)^2 + (l_1\sin\alpha - z_o)^2 = d_2^2 \\ (-l_3 + l_2\cos\beta)^2 + (l_2\sin\alpha\sin\beta + (z_0 - d_3)\tan\alpha)^2 + (z_o - d_3 + l_2\cos\alpha\sin\beta)^2 = l^2 \end{cases} \tag{11}
$$

According to Eq. (11), the inverse kinematics of the 2RPU-PUR PM combined with the Fig. 1 are given as

$$
\begin{cases} d_1 = \sqrt{(-l_1\cos\alpha + l_4 - (d_3 - z_0)\tan\alpha)^2 + (l_1\sin\alpha + z_o)^2} \\ d_2 = \sqrt{(l_1\cos\alpha - l_4 - (d_3 - z_0)\tan\alpha)^2 + (l_1\sin\alpha - z_o)^2} \\ d_3 = z_0 + l_2\cos\alpha\sin\beta - \sqrt{l^2\cos^2\alpha - (-l_3 + l_2\cos\beta)^2\cos^2\alpha} \end{cases} \tag{12}
$$

The forward kinematics are omitted here in that it won't be used in the followed analysis.

5 Velocity Analysis

The velocity analysis is to obtain the mapping i.e. Jacobian matrix between the actuated velocities $\left[\dot{d}_1 \; \dot{d}_2 \; \dot{d}_3\right]^{\mathrm{T}}$ and the output velocities $\left[\dot{\alpha} \; \dot{\beta} \; \dot{z}_o\right]^{\mathrm{T}}$.

The first derivative of the Eq. (12) with respect to time is given as

$$\boldsymbol{J}_q \cdot \begin{bmatrix} \dot{d}_1 \\ \dot{d}_2 \\ \dot{d}_3 \end{bmatrix} = \boldsymbol{J}_x \cdot \begin{bmatrix} \dot{\alpha} \\ \dot{\beta} \\ \dot{z}_o \end{bmatrix} \tag{13}$$

where

$$\boldsymbol{J}_q = \begin{bmatrix} J_{q11} & 0 & 0 \\ 0 & J_{q22} & 0 \\ 0 & 0 & J_{q33} \end{bmatrix}, \quad \boldsymbol{J}_X = \begin{bmatrix} J_{X11} & J_{X12} & J_{X13} \\ J_{X21} & J_{X22} & J_{X23} \\ J_{X31} & J_{X32} & J_{X33} \end{bmatrix},$$

$J_{q11}=d_1, J_{q22}=d_2, J_{q33}=z_o - d_3 + l_2 \cos\alpha \sin\beta, J_{X11} = d_3 l_1 \cos\alpha + l_1 l_4 \sin\alpha + (z_0 - d_3)(l_4 + l_1 \tan\alpha \sin\alpha) + (z_0 - d_3)^2 \tan\alpha, J_{X12} = 0, J_{X13} = l_1 \sin\alpha + z_o + \sec\alpha (d_3 - z_0)(l_4 \sec\alpha - l_1) - (d_3 - z_0)^2 \tan\alpha \sec^2\alpha, J_{X21} = l_1 l_4 \sin\alpha - d_3 l_1 \cos\alpha + (d_3 - z_o)(l_4 + l_1 \sin\alpha \tan\alpha) + (z_o - d_3)^2 \tan\alpha, J_{X22} = 0, J_{X23} = z_o - l_1 \sin\alpha + (z_0 - d_3)(l_1 - l_4 \sec\alpha) \sec\alpha + (z_o - d_3)^2 \tan\alpha \sec^2\alpha, J_{X31} = (z_o - d_3)(z_o - d_3 + l_2 \cos\alpha \sin\beta) \tan\alpha,$ $J_{X32} = (z_o - d_3) l_2 \cos\alpha \cos\beta + l_3 l_2 \sin\beta \cos^2\alpha, J_{X33} = z_o - d_3 + l_2 \cos\alpha \sin\beta.$

Multiplying $\boldsymbol{J}_q^{-1}$ in both sides, the velocity equation can be given as

$$\begin{bmatrix} \dot{d}_1 \\ \dot{d}_2 \\ \dot{d}_3 \end{bmatrix} = \boldsymbol{J}_q^{-1} \boldsymbol{J}_X \cdot \begin{bmatrix} \dot{\alpha} \\ \dot{\beta} \\ \dot{z}_o \end{bmatrix} = \boldsymbol{J} \cdot \begin{bmatrix} \dot{\alpha} \\ \dot{\beta} \\ \dot{z}_o \end{bmatrix} \tag{14}$$

where $\boldsymbol{J}$ is a 3×3 Jacobian matrix.

6 Singularity Analysis

Singularity analysis can be solved based on the previous velocity analysis. Generally, the singularities can be divided into three types: forward kinematic singularities, inverse kinematic singularities, and combined singularities [18].

6.1 Inverse Singularity

The inverse singularity, also called boundary singularity, often appears at the boundary of workspace. The inverse singularity occurs when Jacobian matrix satisfies

$$|\boldsymbol{J}_q| = 0 \text{ and } |\boldsymbol{J}_X| \neq 0 \tag{15}$$

According to Eq. (15), the inverse singularity occurs when one of J_{q11}, J_{q22} or J_{q11} is equal to zero. The inverse singularity is satisfied only when J_{q33} is equal to zero. From Eq. (15), one can obtain that A_3B_3 is perpendicular to the axis of the prismatic pair of the limb 3. The forward singularity of the 2R$\underline{P}$U-$\underline{P}$UR PM is shown in Fig. 3.

6.2 Forward Singularity

A forward singularity occurs when Jacobian matrix satisfies

$$|\boldsymbol{J}_q| \neq 0 \text{ and } |\boldsymbol{J}_X| = 0 \tag{16}$$

From Eq. (16), the forward singularities of 2RPU-PUR PM can be divided into two cases. One is that limb 1 or 2 is colinear with the B_1B_2, the other is that limb 2 is colinear with the B_3o. Thus, there are two forward singular types as shown in Fig. 4.

6.3 Combined Singularity

Under the combined singular configuration, the configuration of the inverse and the forward singularities appear at the same time. Jacobian matrix satisfies

$$|\boldsymbol{J}_q| = 0 \text{ and } |\boldsymbol{J}_x| = 0 \tag{17}$$

Based on the result above, the combined singular configurations are shown in Fig. 5.

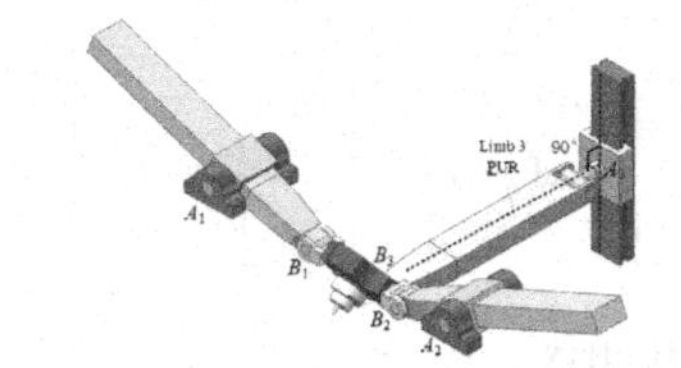

Fig. 3. Inverse kinematic singularity

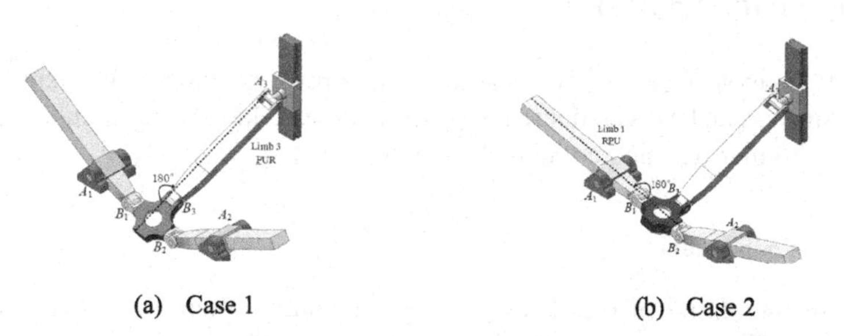

(a) Case 1 (b) Case 2

Fig. 4. Forward kinematic singularity

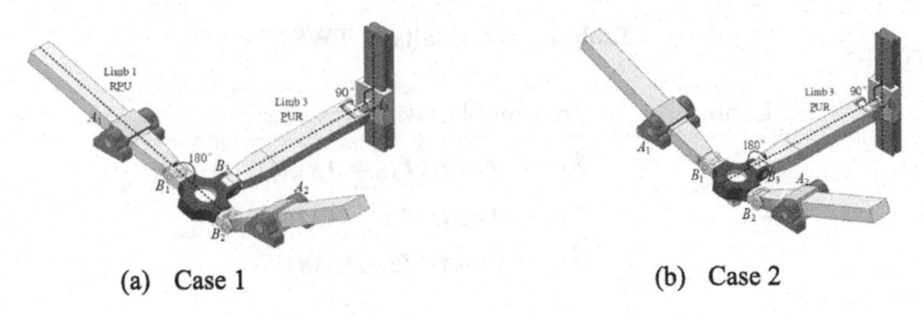

(a) Case 1 (b) Case 2

Fig. 5. Combined singularity

7 Performance Analysis

Performance evaluation is the basis for optimal design of the manipulator. Some indices have been proposed and are commonly used in evaluating PMs; e.g., the manipulability [19] and condition number [20–22]. But the above indices may not be fit when they applied to PMs with combined mobility [23]. Here we use the motion/force transmission index [24, 25] to evaluate the performance of the 2RPU-PUR PM.

7.1 Local Transmission Index (LTI) for PMs

According to motion/force transmission index, the input transmission index (ITI), output transmission index (OTI) and local transmission index (LTI) can be defined as

$$\lambda_i = \frac{|\$_{Ai} \circ \$_{Ti}|}{|\$_{Ai} \circ \$_{Ti}|_{\max}} \tag{18a}$$

$$\eta_i = \frac{|\$_{Oi} \circ \$_{Ti}|}{|\$_{Oi} \circ \$_{Ti}|_{\max}} \tag{18b}$$

$$\Gamma = \min\{\lambda_i, \eta_i\}(i = 1, 2 \cdots, n) \tag{18c}$$

where λ_i, η_i and Γ denote the ITI, OTI, and LTI of the ith limb, respectively. $\$_{Ai}$ denotes the input twist screw of the ith limb; $\$_{Oi}$ denotes the output twist screw (OTS) of the ith limb; $\$_{Ti}$ denotes the transmission wrench screw (TWS) of the ith limb. The range of Γ is from zero to one. And, the larger Γ is, the better motion/force transmissibility is.

Without loss of generality, let us take the limb 1 as an example. Equation (1) shows the twist system and Eq. (2) shows the wrench system.

The input twist screw of limb 1 can be given as $\$_{A1} = \$_{11}$, according to the reciprocal product of the TWS $\$_{Ti}$ and all the passive twist screw in the limb is equal to 0, we can get the following conclusion.

$$\$_{Ti} \circ \$_{1j} = 0\,(j = 2, 3, 4) \tag{19}$$

According to Eq. (18), the TWS of limb 1 is given by

$$\$_{T1} = (\boldsymbol{r}_{A1B1}; \boldsymbol{r}_{1a} \times \boldsymbol{r}_{A1B1}) \tag{20}$$

Table 1. The results of TWS

Limb	Transmission wrench screw
1	$\$_{T1} = (r_{A1B1}; r_{1a} \times r_{A1B1})$
2	$\$_{T2} = (r_{A2B2}; r_{2a} \times r_{A2B2})$
3	$\$_{T3} = (r_{A3B3}; r_{4a} \times r_{A3B3})$

where $\boldsymbol{r}_{A1B1}$ denotes the unit vector along A_1B_1, and $\boldsymbol{r}_{1a}$ denotes the position vectors of A_1 point.

Similarly, all the TWS of the three limbs can be obtained as summarized in Table 1.

When limb i^{th} is actuated and the other limbs are locked, and using reciprocal screw theory, we can get the OTS of the 2RPU-PUR PM, and the wrench screw system are shown in Table 2.

Table 2. The Wrench Screw System

Limb	Locked Limb	The Wrench Screw
1	2, 3	$[\$_{C1}; \$_{C2}; \$_{C3}; \$_{T2}; \$_{T3}]$
2	1, 3	$[\$_{C1}; \$_{C2}; \$_{C3}; \$_{T1}; \$_{T3}]$
3	1, 2	$[\$_{C1}; \$_{C2}; \$_{C3}; \$_{T1}; \$_{T2}]$

Substituting $l_1 = l_2 = 100$ mm, $l_3 = l_4 = 300$ mm, $l = 440$ mm into Eq. (18), we can get the LTI.As shown in Fig. 6, the LTI of the area surrounded by the white line is greater than 0.7. It can be found that the performance of the 2RPU-PUR PM has good transmission performance in the workspace, and it is symmetrically distributed, which is in line with the symmetrical mechanical structure. Moreover, with the appropriate increase of z, the overall performance is gradually improved. But the length of z is limited by the structure design, so it can't be too long.

7.2 Global Transmission Index (GTI) for PMs

Considering LTI only represents the performance of the motion/force transmission in a single configuration. Thus, the global transmission index (GTI) is used here to evaluate the global transmission performance of the robot. According to the concept of the transmission angle, it is assumed that the PM has good motion/force transmissibility when $\Gamma \geq 0.7$ in this paper. The workspace where $\Gamma \geq 0.7$ is defined to form the good transmission workspace (GTW). The GTI can be defined as

$$\sigma = \frac{\int_{S_G} dW}{\int_{S} dW} \tag{21}$$

where W is the reachable orientation workspace. And S_G and S denote the areas of the GTW and overall reachable workspace, respectively. Obviously, σ ranges from 0 to 1. The closer σ is to 1, the better transmission performance of 2RPU-PUR PM has.

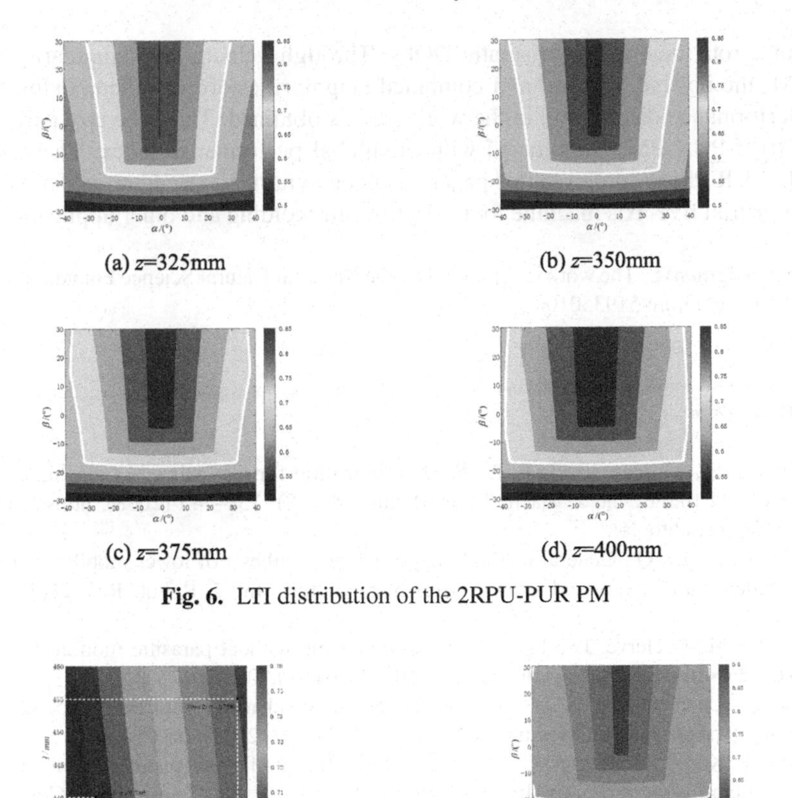

(a) z=325mm (b) z=350mm

(c) z=375mm (d) z=400mm

Fig. 6. LTI distribution of the 2RPU-PUR PM

Fig. 7. Variation of global transmission index

Fig. 8. Distribution diagram of local transmission index

In order to optimize the performance, taking the length of limb 3 l and the size of the fixed platform l_1 as the design variables and letting $l_1 = l_2$, the global transmission index σ corresponding to different variables is calculated as shown in Fig. 7 and the optimization paraments can be chosen. We take the LTI of Fig. 6(d) as the optimization object and select $l_1 = l_2 = 128$ mm and $l = 455$ mm as the optimization paraments. The LTI of the optimized PM can be obtained, as shown in Fig. 8. Compared with Fig. 6(d) and Fig. 8, it can be seen that the area surrounded by the white line is obviously larger, which shows that the GTW is increased significantly.

It should be noted that the above optimization is aimed at the performance in the attitude workspace when $z = 400$ mm. In the design and application, we should consider the selection of suitable z for better optimization.

8 Conclusion

A novel 2R1T manipulator with 12 single DOF joints is proposed. The DOF of the PM is analyzed by using Screw theory. It is concluded that the 2R$\underline{\text{P}}$U-$\underline{\text{P}}$UR PM can output 2 rotational and 1 prismatic DOFs. Through velocity and singularity analysis of PM, the inverse, forward and combined singularities are obtained. Using the LTI, the performance distribution in the workspace is obtained. Then, the size paraments of the 2RPU-PUR PM is optimized with the global performance index. The new 2R1T 2RPU-PUR PM proposed in this paper has good symmetry structure, which is suitable for industrial five-axis machine tools, friction stir welding and other applications.

Acknowledgments. The work is supported by the National Natural Science Foundation of China (NSFC) under Grant 51935010.

References

1. Sencer, B., Altintas, Y., Croft, E.: Feed optimization for five-axis CNC machine tools with drive constraints. Int. J. Mach. Tools Manuf. **48**(7–8), 733–745 (2008). https://doi.org/10.1016/j.ijmachtools
2. Huang, Z., Li, Q.: General methodology for type synthesis of lower mobility symmetrical parallel manipulators and several novel manipulators. Int. J. Robot. Res. **21**(2), 131–146 (2002)
3. Li, Q., Marie Hervé, J.: 1T2R parallel mechanisms without parasitic motion. IEEE Trans. Rob. **26**, 401–410 (2010). https://doi.org/10.1109/TRO.2010.2047528
4. Li, Q., Chen, Q., Yu, X., et al.: 3-DOF parallel mechanism with two perpendicular and non-intersecting axes of rotation: CN, 201110357878.7 (2011). (in Chinese)
5. Yan, J., Kong, X., Higgins, C., et al.: Kinematic design of a new parallel kinematic machine for aircraft wing assembly. In: Proceedings of the IEEE 10th International Conference on Industrial Informatics, Beijing, China (2012)
6. Li, Q., Wu, W., Xiang, J., et al.: A hybrid robot for friction stir welding. ARCHIVE Proc. Inst. Mech. Eng. Part C J. Mech. Eng. Sci. 1989–1996 **203–210**, 2639–2650 (2015). https://doi.org/10.1177/0954406214562848
7. Wahl, J.: Articulated tool head: US, 6431802, 13 August 2002
8. Yanbing, N.I., Zhang, B., Sun, Y., et al.: Accuracy analysis and design of A3 parallel spindle head. Chin. J. Mech. Eng. (2016). CNKI:SUN:YJXB.0.2016-02-003
9. Siciliano, B.: The Tricept robot: Inverse kinematics, manipulability analysis and closed-loop direct kinematics algorithm. Robotica **17**(04), 437–445 (1999). https://doi.org/10.1017/s0263574799001678
10. Zhao, Y., Jin, Y., Zhang, J.: Kinetostatic modeling and analysis of an Exechon parallel kinematic machine (PKM) module. Chin. J. Mech. Eng. **29**(1), 33–44 (2016). CNKI:SUN:YJXB.0.2016-01-004
11. http://www.exechonworld.com/document/200804/article34.htm
12. Xu, L., Zhu, X., Wei, Y., Li, Q., Chen, Q.: Kinematic analysis and dimensional synthesis of a new 2R1T parallel kinematic machine. In: ASME 2018 International Design Engineering Technical Conferences and Computers and Information in Engineering Conference (2018)
13. Li, Q., Xu, L., Chen, Q., et al.: New family of RPR-equivalent parallel mechanisms: design and application. Chin. J. Mech. Eng. **02**(v.30), 11–15 (2017)
14. Xu, L., Li, Q., Tong, J., Chen, Q.: Tex3: an 2R1T parallel manipulator with minimum DOF of joints and fixed linear actuators. Int. J. Precis. Eng. Manuf. **19**(2), 227–238 (2018). https://doi.org/10.1007/s12541-018-0026-y
15. Li, Q., Herve, J.M.: Type synthesis of 3-DOF RPR-equivalent parallel mechanisms. IEEE Trans. Rob. **30**(6), 1333–1343 (2017). https://doi.org/10.1109/TRO.2014.2344450

16. Chao, Y., Li, Q., Chen, Q., Xu, L.: Elastostatic stiffness modeling of overconstrained parallel manipulators. Mech. Mach. Theory **122**, 58–74 (2018). https://doi.org/10.1016/j.mechmachtheory.2017.12.011
17. Dai, J.: Screw algebra and Lie group, Lie algebra. Mach. Des. Res. (2014). (in Chinese)
18. Gosselin, C.M., Angeles, J.: Singularity analysis of closed-loop kinematic chains. IEEE Trans. Rob. Autom. **6**(3), 281–290 (1990). https://doi.org/10.1109/70.56660
19. Stoughton, R.S, Arai, T.: Modified stewart platform manipulator with improved dexterity. IEEE Trans. Rob. Autom. **9**(2), 166–173 (1993). https://doi.org/10.1109/70.238280
20. Angeles, J., López-Cajún, C.S.: Kinematic isotropy and the conditioning index of serial robotic manipulators. Int. J. Rob. Res. **11**(6), 560–571 (1992). https://doi.org/10.1177/027836499201100605
21. Altuzarra, O., Hernandez, A., Salgado, O., Angeles, J.: Multiobjective optimum design of a symmetric parallel schnflies-motion generator. J. Mech. Des. **131**(3) (2009). https://doi.org/10.1115/1.3066659. Paper No. 031002
22. Altuzarra, O., Pinto, C., Sandru, B., Hernandez, A.: Optimal dimensioning for parallel manipulators: workspace, dexterity, and energy. J. Mech. Des. **133**(4) (2011). https://doi.org/10.1115/1.4003879. Paper No. 041007
23. Merlet, J.: Jacobian, manipulability, condition number, and accuracy of parallel robots. J. Mech. Des. 128(1), 199–206 (2006). https://doi.org/10.1007/978-3-540-48113-3_16
24. Wang, J., Chao, W., Liu, X.J.: Performance evaluation of parallel manipulators: motion/force transmissibility and its index. Mech. Mach. Theory **45**(10), 1462–1476 (2010). https://doi.org/10.1016/j.mechmachtheory.2010.05.001
25. Liu, X.J., Wu, C., Wang, J.: A new approach for singularity analysis and closeness measurement to singularities of parallel manipulators. J. Mech. Rob. **4**(4), 041001 (2012). https://doi.org/10.1115/1.400

Dynamic Model and Vibration Characteristic of the Cartesian Robotic Arm with Deploying Motion

Yufei Liu[1,2,3(✉)], Yang Nie[1], and Jinyong Ju[1,2]

[1] School of Mechanical Engineering, Anhui Polytechnic University, Wuhu 241000, China
liuyufeiahpu@126.com
[2] School of Artificial Intelligence, Anhui Polytechnic University, Wuhu 241000, China
[3] School of Mechanical Engineering, Hefei University of Technology, Hefei 230009, China

Abstract. This paper investigates the dynamic and vibration characteristic of the Cartesian robotic arm with deploying motion (CRMdm), which is a common application scenario in engineering. During the investigations, the dynamic model and vibration equation of the CRMdm are derived based on Hamilton variational principle. According to the established dynamic equation, the transverse vibration response of the manipulator during the operation motion were studied. The results show that the vibration responses of the CRMdm gradually enhance, with the increase of motion velocity, the trend becomes more obvious while the vibration response frequency decreases, which are different from the conventional dynamic characteristics of manipulator. To further investigate the vibrations characteristic of the CRMdm, the first three mode shapes of the CRMdm are investigated.

Keywords: Robotic arm · Dynamic model · Vibration characteristic · Mode shapes · deploying motion

1 Introduction

Robotic arms are widely used in industrial fields, for such operations as materials handling, assembly, painting, and mechanical processing [1–3]. One of the typical industrial robot units is the Cartesian robot arm which can be modeled as a driving base, a robotic arm and an end-effector [4, 5]. The dynamic responses especially vibrations encountered by the flexibility of the robotic arm and its inertia significantly influence the operating performance and productivity, therefore, to ensure the operation stability and precision for effective applications, the dynamic especially the vibrations of the robotic arm remain a significant issue needed to be investigated [6, 7].

Numerous researches have focused on the dynamics of the robotic arm [8]. Based on the dynamic modeling, the vibration characteristics as well as the vibration suppression categories can be conducted. In these precious studies, there are mainly two basic vibration suppression categories are proposed, namely, the active control strategy and the motion planning. The active control strategy is usually conducted with piezoelectric intelligent materials that have the property of offering corresponding control voltages

X.-J. Liu et al. (Eds.): ICIRA 2021, LNAI 13014, pp. 766–773, 2021.
https://doi.org/10.1007/978-3-030-89098-8_72

against the residual vibrations of the robotic arm [9, 10]. With the piezoelectric actuator, many control algorithms have been proposed to suppress the vibrations, which can be viewed in the researches of Shin [11], Qiu [12] and Lin [13]. Syed [14] further conducted a comparative between positive position feedback and negative derivative feedback for vibration control of the robotic arm. Furthermore, robotic arms can be designed as smart structures with magnetorheological or electrorheological materials, and in this case, the aroused vibrations can be controlled with the appropriate control of the smart structures [15]. Eshaghi [16, 17] presented a comprehensive review of reported studies on the applications of magnetorheological or electrorheological fluids for realizing active and semi-active vibration suppression in sandwich structures and investigated the vibration characteristics of a sandwich plate treated with magnetorheological fluid. Compared with active control strategies, the method of motion planning to reduce the residual vibrations and obtain the desired trajectory is another attractive approach, which is conducted without changing the intrinsic properties of the structure. Using singular perturbation theory, Mirzaee [18] proposed a composite control scheme that makes the orientation of the arm track a desired trajectory while suppressing its vibration. Reis [19] proposed an approach to trajectory planning based on LQR theory and applied to a single flexible link robot to improve the performance under parameter uncertainty. Xin [20] presents an optimization method of trajectory planning with minimum residual vibration for space manipulator system, and the particle swarm optimization algorithm is employed to achieve minimum residual vibration by optimizing redundant coefficients of movement in the investigation.

As presented in these studies of the robotic arms, the dynamic models and vibrations analysis are the basic of effective vibration suppression. In these studies, the robotic arms are usually modeled as a beam or plate model with specialized rotational motions or translational motions. And the vibrations of the robotic arms with translational or rotational motions have attracted numerous attentions as indicated in Ref. [5, 12, 15, 21, 22]. Moreover, considering the influence of the motion disturbances, the beam or plate model of the robotic arm is similar to a parametrically excited system [23]. Chen et al. [24–26] investigated the nonlinear vibrations of the axial moving beam structures. Yabuno [27] investigated the nonlinear normal modes of a vertical cantilever beam excited by a principal parametric resonance. Feng [28] and Li [29] studied the principal parametric and internal resonances of flexible beams and plates. Pratiher and Dwivedy [30–32] investigated the nonlinear vibrations of lateral moving robotic arms for the driving base suffering harmonic excitations. These indepth nonlinear dynamics and vibrations investigations of the beam or plate model as well as the robotic arms are significant for obtaining actual dynamic characteristics and further vibration control. However, these investigations of dynamics and vibrations of the beam or plate model as well as the robotic arms are obviously different with the robotic arm studied in this paper. As the takeout robot, which executes the plastic molding process, the robotic arm conveys a deploying, and in this case, the dynamic and vibration characteristics of the referred robotic arm will be more complex. The dynamics of the CRMdm should be studied which are obviously different with these existent investigations.

Dynamic model and vibration characteristics of the CRMdm are investigated in this paper. For these investigations, a coupled dynamic model of the CRMdm is established in Sect. 2, and based on the proposed dynamic model, the vibration equation of the

CRMdm is derived. In Sect. 3, the results of the vibration responses characteristics of the CRMdm are presented. In Sect. 4, the paper is concluded with a brief summary.

2 Dynamic Model and Vibration Equation of the CRMdm

The dynamic model of the CRMdm is constructed as shown in Fig. 1. In order to investigate the vibration characteristics of the CRMdm, the dynamic equation of motion is established based on the Hamilton variational principle in this section. It is assumed that the robotic arm satisfies the Bernoulli-Euler beam theory and the transverse vibrations in the *x*-*y* plane are the primary motions considered.

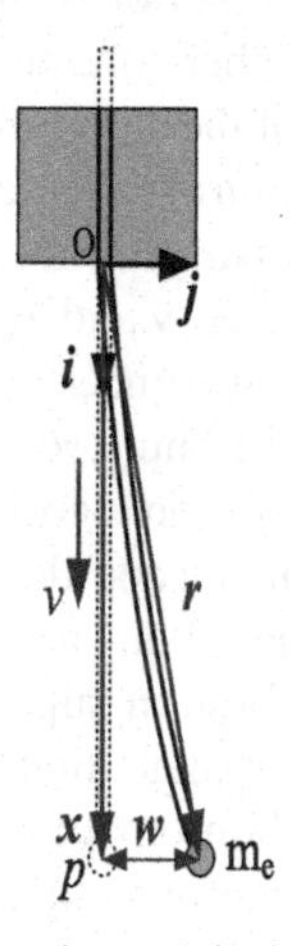

Fig. 1. Dynamic model of the CRMdm

During the investigation, we use the dynamic responses of the end-effector to illustrate the dynamic characteristics of the CRMdm. As shown in Fig. 1, the motion displacement of the end-effector, denoted as point *p*, can be expressed as

$$\mathbf{r} = x\mathbf{i} + w\mathbf{j} \tag{1}$$

where $x(t)$ represents the distance of the end-effector from the hub, $w(x,t)$ represents the transverse vibration displacement of the end-effector.

According to Eq. (1), the motion velocity of the end-effector can be obtained as

$$\mathbf{v} = v\mathbf{i} + (\frac{\partial w}{\partial t} + v\frac{\partial w}{\partial x})\mathbf{j} \tag{2}$$

here v denotes the velocity of deploying motion.

The dynamic equation of motion for the CRMdm is obtained from the Hamilton variational principle [33], which can be expressed as

$$\int_{t_0}^{t_1} \delta(E_T - E_V)\mathrm{d}t = 0 \tag{3}$$

where E_T is the kinetic energy of the CRMdm and can be expressed as

$$E_T = \frac{1}{2}\rho A \int_0^L \mathbf{v} \cdot \mathbf{v}\mathrm{d}x + \frac{1}{2}m_e\left[(\mathbf{v} \cdot \mathbf{v})|_{x=L}\right]^2 \tag{4}$$

here the first part represents the kinetic energy of the robotic arm, the second part represents the kinetic energy of the end-effector, ρ is the mass density, A is the cross-sectional area, A = b × h, b and h are the width and the thickness, respectively, m_e is the mass of the end-effector.

According to Eq. (2), it further obtains that

$$E_T = \frac{1}{2}\rho A \int_0^L \left[v^2 + \left(\frac{\partial w}{\partial t} + v\frac{\partial w}{\partial x}\right)^2\right]\mathrm{d}x + \frac{1}{2}m_e\left[v^2 + \left(\frac{\partial w}{\partial t} + v\frac{\partial w}{\partial x}\right)^2\right]\Bigg|_{x=L} \tag{5}$$

E_V is the potential energy of the CRMdm which mainly considers the bending elastic potential energy of the manipulator and can be given by

$$E_V = \frac{1}{2}EI \int_0^L \left(\frac{\partial^2 w}{\partial x^2}\right)^2 \mathrm{d}x \tag{6}$$

where E is Young's modulus, I is the cross-sectional moment of inertia about the neural axis, $I = bh^3/12$.

Based on Eqs. (5) and (6), it obtains that

$$\begin{aligned}
\delta E_T &= \delta\left\{\frac{1}{2}\rho A \int_0^L \left[v^2 + \left(\frac{\partial w}{\partial t} + v\frac{\partial w}{\partial x}\right)^2\right]\mathrm{d}x + \frac{1}{2}m_e\left[v^2 + \left(\frac{\partial w}{\partial t} + v\frac{\partial w}{\partial x}\right)^2\right]\Bigg|_{x=L}\right\} \\
&= \rho A \int_0^L \left(\frac{\partial w}{\partial t} + v\frac{\partial w}{\partial x}\right)\frac{\partial}{\partial t}\delta w\mathrm{d}x + m_e\left(\frac{\partial w}{\partial t} + v\frac{\partial w}{\partial x}\right)\Bigg|_{x=L}\frac{\partial}{\partial t}\delta w \\
&= -\rho A \int_0^L \left[\frac{\partial^2 w}{\partial t^2} + v\frac{\partial^2 w}{\partial x \partial t} + \dot{v}\frac{\partial w}{\partial x} + v\left(\frac{\partial^2 w}{\partial x \partial t} + v\frac{\partial^2 w}{\partial x^2}\right)\right]\delta w\mathrm{d}x \\
&\quad - m_e\left[\frac{\partial^2 w}{\partial t^2} + v\frac{\partial^2 w}{\partial x \partial t} + \dot{v}\frac{\partial w}{\partial x} + v\left(\frac{\partial^2 w}{\partial x \partial t} + v\frac{\partial^2 w}{\partial x^2}\right)\right]\Bigg|_{x=L}\delta w
\end{aligned} \tag{7}$$

$$\delta E_V = EI \int_0^L \frac{\partial^4 w}{\partial x^4}\delta w\mathrm{d}x \tag{8}$$

Substituting Eqs. (7) and (8) into Eq. (3), it obtains that

$$\begin{aligned}&\int_0^L \left\{ \rho A\left[\frac{\partial^2 w}{\partial t^2} + 2v\frac{\partial^2 w}{\partial x \partial t} + \dot{v}\frac{\partial w}{\partial x} + v^2\frac{\partial^2 w}{\partial x^2}\right] + EI\frac{\partial^4 w}{\partial x^4}\right\}dx \\ &+ m_e\left[\frac{\partial^2 w}{\partial t^2} + 2v\frac{\partial^2 w}{\partial x \partial t} + \dot{v}\frac{\partial w}{\partial x} + v^2\frac{\partial^2 w}{\partial x^2}\right]\Bigg|_{x=L} = 0\end{aligned} \tag{9}$$

According to the assumed modes method [34], $w(x, t)$ can be expressed as

$$w(x,t) = \sum_{i=1}^{n\to\infty} \phi_i(x)q_i(t) \tag{10}$$

where $q_i\ (t)$ denotes the ith generalized co-ordinate, $\phi_i\ (x)$ denotes the ith orthogonal mode shape that can be described as

$$\phi(x) = \sin\beta x - \sinh\beta x - \frac{\sin\beta L + \sinh\beta L}{\cos\beta L + \cosh\beta L}(\cos\beta x - \cosh\beta x) \tag{11}$$

According to the orthogonality of mode shapes [35, 36], it obtains that

$$\begin{aligned}&\left[m_e\phi_i^2(L) + \rho A\right]\ddot{q}_i + \left[2v\rho A\int_0^L \phi_i\phi_i' dx + 2v m_e\phi_i(L)\phi_i'(L)\right]\dot{q}_i \\ &+ \begin{bmatrix}\dot{v}\rho A\int_0^L \phi_i\phi_i' dx + v^2\rho A\int_0^L \phi_i\phi_i'' dx \\ +\dot{v} m_e\phi_i(L)\phi_i'(L) + v^2 m_e\phi_i(L)\phi_i''(L) + \rho A\omega_i^2\end{bmatrix} q_i = 0\end{aligned} \tag{12}$$

here $m_i = \int_0^L \rho A\phi_i dx$.

3 Results and Discussion

In this section, numerical simulation is used to analyze the dynamic and vibration characteristics of the of the CRMdm. In the numerical simulations, the structural parameters of the robotic arm are set as follows: width b = 45 mm, thickness h = 3 mm, Young's modulus E = 25.24 GPa, volumetric density ρ = 2030 kg/m^3 and Poisson's ratio μ = 0.30, the mass of the end-effector m_e = 100 g which is used to describe the payload during the operations. The initial displacement and initial velocity are defied as 0.0001 m and 0.02 m/s, the transverse vibration displacement of the end-effector is observed to evaluate the vibration of the manipulator during the operation.

In order to facilitate the comparative analysis, we define the manipulator to complete a same distance handling operations with different motion velocities. The vibration responses of the CRMdm during the deploying movement are shown in Fig. 2. It can be observed that, the transverse vibrations of the CRMdm convey an obvious enhancement trend during the operation motion and the trend becomes more obvious with the increase of the velocity, while the vibration response frequency decreases. It can be concluded that the vibration amplitude and frequency characteristics of the manipulator change during the deploying movement, which is different from the conventional dynamic characteristics of the manipulator, and will also increase the difficulty of control.

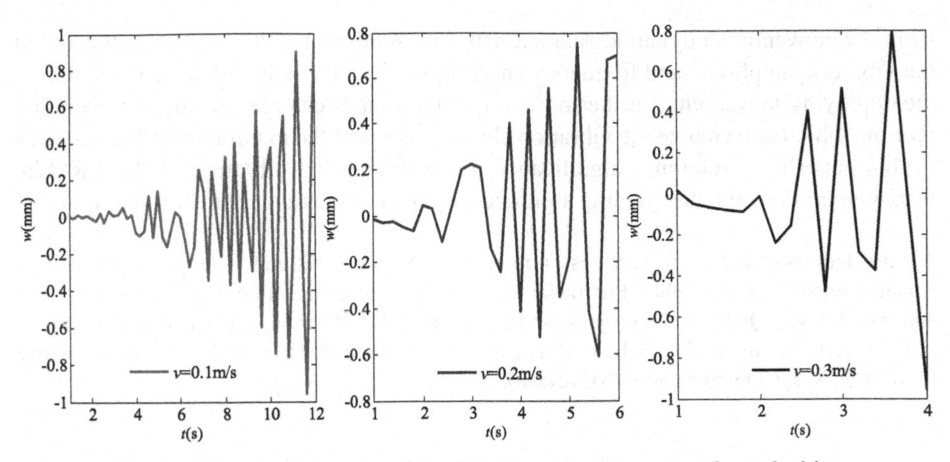

Fig. 2. Vibration responses of the CRMdm with different motion velocities

To further investigate the vibrations characteristic of the CRMdm, the first three mode shapes of the CRMdm are investigated as shown in Fig. 3. It can be obtained that the mode shapes changes during the operation motion, which will change the vibration characteristics of the manipulator, and this has important guiding significance for the vibration control of the CRMdm.

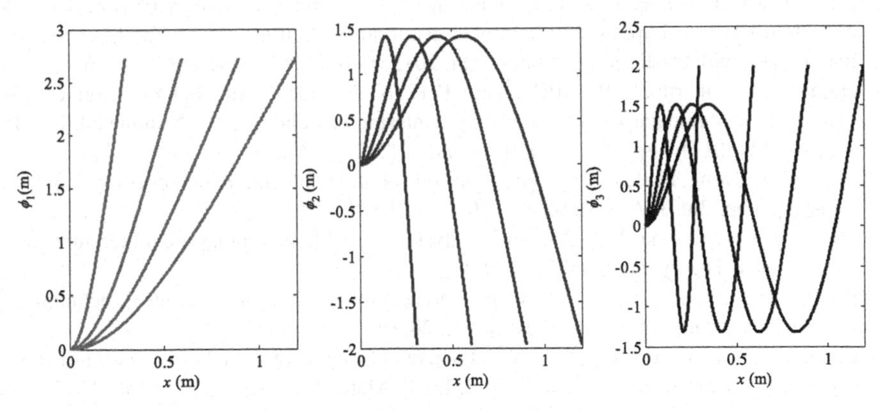

Fig. 3. Mode shapes of the CRMdm

4 Conclusions

In this paper, the dynamic model and vibration equation of the CRMdm are derived based on Hamilton variational principle, and the transverse vibration responses and mode shapes of the manipulator during the operation motion were studied. The results indicate that transverse vibrations of the CRMdm convey an obvious enhancement trend during the operation motion and the trend becomes more obvious with the increase of the velocity, while the vibration response frequency decreases, which are different from the conventional dynamic characteristics of manipulator. It can be concluded that the vibration amplitude and frequency characteristics of the manipulator change during the deploying movement. Furthermore, the mode shapes changes during the operation motion, which will change the vibration characteristics of the manipulator. The research in this paper is of reference significance for analyzing the vibration of the CRMdm during operation, and has guiding significance for the vibration control of manipulator.

Acknowledgments. This research is supported by National Natural Science Foundation of China (No.51805001), Anhui Provincial Natural Science Foundation (No. 1808085QE137 and No. 1908085QE193), the project funded by China Postdoctoral Science Foundation (No. 2017M612060), Anhui Polytechnic University and Jiujiang District Industrial collaborative Innovation Special Fund Project (2021cyxtb1).

References

1. Neto, P., Moreira, A.P. (eds.): WRSM 2013. CCIS, vol. 371. Springer, Heidelberg (2013). https://doi.org/10.1007/978-3-642-39223-8
2. Zi, B., Sun, H., Zhang, D.: Design, analysis and control of a winding hybrid-driven cable parallel manipulator. Rob. Comput. Integr. Manuf. **48**, 196–208 (2017)
3. Kim, B., Chung, J.: Residual vibration reduction of a flexible beam deploying from a translating hub. J. Sound Vib. **333**, 3759–3775 (2014)
4. Liu, Y.F., Li, W., Wang, Y.Q., Yang, X.F., Ju, J.Y.: Dynamic model and vibration power flow of a rigid-flexible coupling and harmonic-disturbance exciting system for flexible robotic manipulator with elastic joints. Shock Vibr. **2015**, 1–10 (2015). Article ID 541057
5. Hamed, Y.S., Alharthi, M.R., AlKhathami, H.K.: Nonlinear vibration behavior and resonance of a Cartesian manipulator system carrying an intermediate end effector. Nonlinear Dyn. **91**(3), 1429–1442 (2017). https://doi.org/10.1007/s11071-017-3955-6
6. Cheol, H.P., Dong, I.I.P., Joo, H.: Vibration control of flexible mode for a beam-type substrate transport robot. Int. J. Adv. Rob. Syst. **10**, 1–7 (2013)
7. Dwivedy, S.K., Eberhard, P.: Dynamic analysis of flexible manipulators, a literature review. Mech. Mach. Theory **41**, 749–777 (2006)
8. Rahimi, H.N., Nazemizadeh, M.: Dynamic analysis and intelligent control techniques for flexible manipulators: a review. Adv. Robot. **28**, 63–76 (2014)
9. Chen, Y., Hu, J., Huang, G.: A design of active elastic metamaterials for control of flexural waves using the transformation method. J. Intell. Mater. Syst. Struct. **27**, 1337–1347 (2016)
10. Aridogan, U., Basdogan, I.: A review of active vibration and noise suppression of plate-like structures with piezoelectric transducers. J. Intell. Mater. Syst. Struct. **26**, 1455–1476 (2015)
11. Shin, C., Hong, C., Jeong, W.B.: Active vibration control of beam structures using acceleration feedback control with piezoceramic actuators. J. Sound Vib. **331**, 1257–1269 (2012)
12. Qiu, Z.C.: Adaptive nonlinear vibration control of a Cartesian flexible manipulator driven by a ball screw mechanism. Mech. Syst. Signal Process. **30**, 248–266 (2012)
13. Lin, C.Y., Jheng, H.W.: Active vibration suppression of a motor-driven piezoelectric smart structure using adaptive fuzzy sliding mode control and repetitive control. Appl. Sci. Basel **7**, 1–17 (2017)
14. Syed, H.H.: Comparative study between positive position feedback and negative derivative feedback for vibration control of a flexible arm featuring piezoelectric actuator. Int. J. Adv. Rob. Syst. **14**, 1–9 (2017)

15. Wei, K.X., Bai, Q., Meng, G., Ye, L.: Vibration characteristics of electrorheological elastomer sandwich beams. Smart Mater. Struct. **20**, 1–8 (2011)
16. Mehdi, E., Ramin, S., Subhash, R.: Dynamic characteristics and control of magnetorheological/electrorheological sandwich structures: a state-of-the-art review. J. Intell. Mater. Syst. Struct. **27**, 2003–2037 (2016)
17. Mehdi, E., Ramin, S., Subhash, R.: Vibration analysis and optimal design of multi-layer plates partially treated with the MR fluid. Mech. Syst. Signal Process. **82**, 80–102 (2017)
18. Mirzaee, E., Eghtesad, M., Fazelzadeh, S.A.: Trajectory tracking and active vibration suppression of a smart Single-Link flexible arm using a composite control design. Smart Struct. Syst. **7**, 103–116 (2011)
19. Reis, J.C.P., da Costa, J.S.: Motion planning and actuator specialization in the control of active-flexible link robots. J. Sound Vib. **331**, 3255–3270 (2012)
20. Xin, P., Rong, J., Yang, Y., Xiang, D., Xiang, Y.: Trajectory planning with residual vibration suppression for space manipulator based on particle swarm optimization algorithm. Adv. Mech. Eng. **9**, 1–16 (2017)
21. Liu, Y.F., Li, W., Yang, X.F., Wang, Y.Q., Fan, M.B., Ye, G.: Coupled dynamic model and vibration responses characteristic of a motor-driven flexible manipulator system. Mech. Sci. **6**, 235–244 (2015)
22. Lu, E., Li, W., Yang, X.F., Fan, M.B., Liu, Y.F.: Modelling and composite control of single flexible manipulators with piezoelectric actuators. Shock Vibr. **2016**, 1–14 (2016). Article ID 2689178
23. Fossen, T.I., Nijmeijer, H. (eds.): Parametric Resonance in Dynamical Systems. Springer, New York (2012). https://doi.org/10.1007/978-1-4614-1043-0
24. Chen, L.Q., Yang, X.D.: Stability in parametric resonance of axially moving viscoelastic beams with time-dependent speed. J. Sound Vib. **284**(3–5), 879–891 (2005)
25. Yang, X.D., Chen, L.Q.: Stability in parametric resonance of axially accelerating beams constituted by Boltzmann's superposition principle. J. Sound Vib. **289**(1), 54–65 (2006)
26. Ding, H., Chen, L.: Nonlinear dynamics of axially accelerating viscoelastic beams based on differential quadrature. Acta Mech. Solida Sin. **22**(3), 267–275 (2009). https://doi.org/10.1016/S0894-9166(09)60274-3
27. Yabuno, H., Nayfeh, A.H.: Nonlinear normal modes of a parametrically excited cantilever beam. Nonlinear Dyn. **25**, 65–77 (2001)
28. Feng, Z.H., Hu, H.Y.: Principal parametric and three-to-one internal resonances of flexible beams undergoing a large linear motion. Acta. Mech. Sin. **19**(4), 355–364 (2003)
29. Li, S.B., Zhang, W.: Global bifurcations and multi-pulse chaotic dynamics of rectangular thin plate with one-to-one internal resonance. Appl. Math. Mech. **33**(9), 1115–1128 (2012)
30. Pratiher, B.: Non-linear response of a magneto-elastic translating beam with prismatic joint for higher resonance conditions. Int. J. Non-Linear Mech. **46**(9), 685–692 (2011)
31. Pratiher, B., Dwivedy, S.K.: Nonlinear response of a flexible Cartesian manipulator with payload and pulsating axial force. Nonlinear Dyn. **57**(12), 177–195 (2009)
32. Pratiher, B., Dwivedy, S.K.: Non-linear dynamics of a flexible single link Cartesian manipulator. Int. J. Non-Linear Mech. **42**(9), 1062–1073 (2007)
33. Mei, F.X.: Analytical mechanics. Beijing Institute of Technology Press, Beijing (2013)
34. Singiresu, S.R.: Mechanical Vibrations, 5th Edn. Pearson Education Inc. (2010)
35. Ge, S.S., Lee, T.H., Gong, J.Q.: A robust distributed controller of a single-link Scara/Cartesian smart materials robot. Mechatronics **9**, 65–93 (1999)
36. Reza Moheimani, S.O., Fleming, A.J.: Fundamentals of Piezoelectricity. Piezoelectric Transducers for Vibration Control and Damping. Springer, London (2006)

An Adsorption Machining Robot and Its Force Design

Zijian Ma[1], Fugui Xie[1,2](✉), Jiakai Chen[1], Xin-Jun Liu[1,2], and Jindou Wang[3]

[1] The State Key Laboratory of Tribology, Department of Mechanical Engineering (DME), Tsinghua University, Beijing 100084, China
xiefg@mail.tsinghua.edu.cn

[2] Beijing Key Lab of Precision/Ultra-Precision Manufacturing Equipments and Control, Tsinghua University, Beijing 100084, China

[3] Yantai Tsingke+ Robot Joint Research Institute Co., Ltd., Yantai 264006, China

Abstract. Machining robots have great advantages in large structure components manufacture and maintenance due to their flexibility and low cost. Among them, adsorption robots have the potentials to reach the blind areas of the large-scale structure. But the complex force interaction problem and high requirement of processing quality during machining make the robot design become a very challenging problem. This study proposes an adsorption machining robot consisted of end-effector and omnidirectional adsorption mobile platform. The friction adsorption locking device and elevating omnidirectional wheel are designed to satisfy the high machining quality. The driving force and adsorption force matching design is presented to make the robot capable of climbing on low friction coefficient surface. This design can increase twice robot's climbing ability from 5° to 10° by providing extra adsorption force. The results in this paper are helpful to the development of the robot.

Keywords: Machining robot · Adsorption robot · Omnidirectional mobile · Force matching design · Force balance design

1 Introduction

Robots have great potentials in large complex structures components machining due to their flexibility and low cost. In order to satisfy different demands of processing, many kinds of machining robots have been proposed. For example, AGV mobile robot for aircraft wing drilling [1–3], continuum robotic machine tool for aero-engine maintenance [4–6], walking machine tool [7] and etc. And methods such as error compensation [8, 9], motion control based on tracking error prediction [10] and calibration of multirobot systems [11] are presented to promote the robots processing quality. However, the existing robots have blind areas during machining, such as the top area of large workpiece. Mobile machining on workpiece surface through adsorption robot is a feasible way to solve this problem. Wall-climbing robot is one kind of adsorption robots, they can execute some low-load tasks, such as large metal tank inspection [12], ship surface rust

X.-J. Liu et al. (Eds.): ICIRA 2021, LNAI 13014, pp. 774–784, 2021.
https://doi.org/10.1007/978-3-030-89098-8_73

removal [13], high building windows cleaning [14]. However, low load ability and low stiffness limit their application in machining. As for adsorption machining robot, the complex force interaction and high processing quality requirement during machining makes its design become a challenging problem.

Wheel locomotion is one of the most commonly used adsorption mobile mechanism. Many researches have been done in its design to ensure the robot can adhere stably. Some researchers design the adsorption force through no slipping and no overturn conditions, then driving force is determined through adsorption force [15, 16]. This method is simple and effective in most cases, but driving wheels slipping may occur when robot moving on low friction coefficient surface. The wheel-slippage problem has been studied in some researches [17], no wheel-slippage is taken as a limitation condition to get the proper driving force. However, wall climbing ability is still be limited without additional measures. Besides, cutting force is rarely be considered in the existing design.

This paper proposes an adsorption machining robot which can realize omnidirectional movement on workpiece surface. A friction adsorption locking device is designed to ensure that the robot has well machining quality under complex load environment during machining. To make the robot have better climbing performance and movement stability, a matching design is carried out to get the proper driving force and adsorption force within the load range.

The remainder of this paper is organized as follows: in Sect. 2, the concept design and kinematic modeling is introduced. In Sect. 3, matching design is presented to optimize the driving force and adsorption force. The conclusion is presented in Sect. 4.

2 Concept Design and Kinematic Modeling

2.1 Concept Design

Inspired by ants gnawing bones in nature, making robot adsorb on workpiece surface can realize large components in-suit processing, and this method can also realize the machining of blind area. This paper proposed an adsorption machining robot which is composed of a parallel end-effector and an omnidirectional adsorption mobile platform, as shown in Fig. 1. The parallel part has 5 degrees of freedom (DoFs) motion ability and can realize complex surface machining. The kinematic analysis [18], optimization [19] and energy efficient design [20] of the parallel part have been studied in our previous works.

The omnidirectional adsorption platform is consisted of omnidirectional mobile platform and friction adsorption locking device. To ensure the robot will not scratch the workpiece during moving, the wheel locomotion is chosen as the motion mechanism and the driven wheel of omnidirectional wheel material is made of polyurethane. The mobile platform has three omnidirectional wheels which are installed at interval of 120°. Through the relative motion of three omnidirectional wheels, the robot can realize flexible 3-DoF movement among different working stations. Because of the complex force interaction during machining, weak contact stiffness between wheel and workpiece surface will directly influence the machining quality. To solve this problem, an elevating device is installed on omnidirectional wheel. When robot reaches the target position, the elevating device will raise the wheel away from the surface and lay down the adsorption locking device before machining. The adsorption locking device is composed of electromagnet and friction material (Fig. 1). After wheels are raised, the electromagnetic adsorption force can press the high friction materials (i.e. brake material) on the workpiece surface to achieve the lock function. Friction material has higher stiffness than polyurethane, so it could guarantee machining quality well. To decrease the load of the elevating device, the electromagnet has the function of adjusting adsorption force.

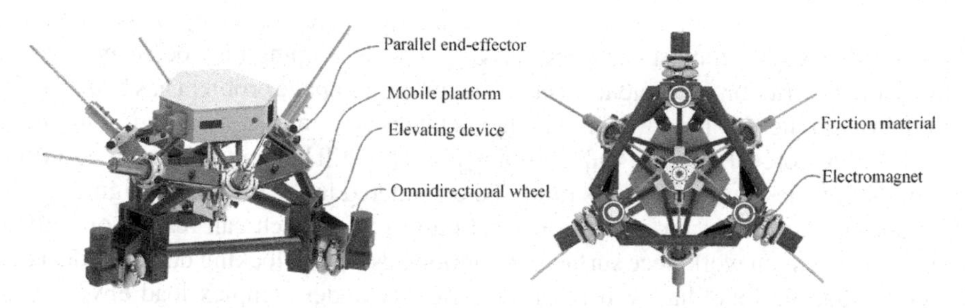

Fig. 1. The structure of the omnidirectional adsorption mobile platform.

2.2 Kinematic Analysis of Omnidirectional Mobile Platform

Omnidirectional mobile platform can realize 3-DoF movement through relative motion of three wheels. In order to get the wheel speed in different motion to realize control, kinematic analysis of the mobile platform should be carried out.

Transformation matrix $\mathbf{R}_1(\theta)$ of the wheel velocity under robot coordinate $x_a o_a y_a$ and $\mathbf{R}_2(\theta)$ of the coordinate $x_a o_a y_a$ under the universal coordinate XOY(Fig. 2) can be derived as

$$\mathbf{R}_1(\theta) = \begin{bmatrix} 1 & 0 & L \\ -\cos\theta_1 & -\sin\theta_1 & L \\ -\sin\theta_2 & \cos\theta_2 & L \end{bmatrix} \mathbf{R}_2(\alpha) = \begin{bmatrix} \cos\alpha & \sin\alpha & 0 \\ -\sin\alpha & \cos\alpha & 0 \\ 0 & 0 & 1 \end{bmatrix} \tag{1}$$

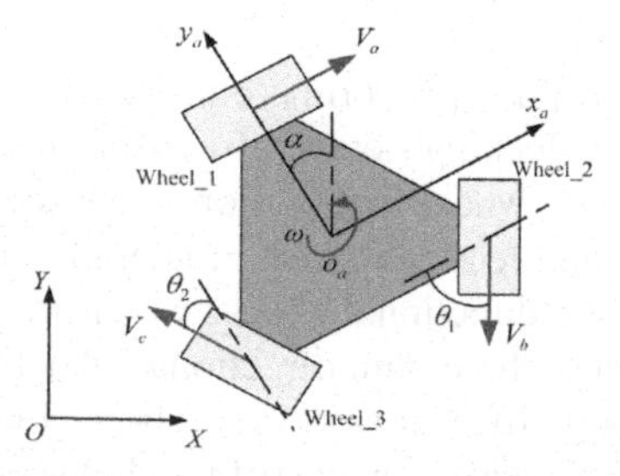

Fig. 2. Kinematic model of omnidirectional mobile platform

By using the transformation matrix above, the relationship between wheel speed and universal coordinate can be expressed as following

$$\begin{bmatrix} V_a \\ V_b \\ V_c \end{bmatrix} = \mathbf{R}_1(\theta) \begin{bmatrix} V_{X'} \\ V_{Y'} \\ \omega \end{bmatrix} = \mathbf{R}_1(\theta)\mathbf{R}_2(\alpha) \begin{bmatrix} V_X \\ V_Y \\ W \end{bmatrix} \tag{2}$$

Finally, kinematic equation can be derived as Eq. (3). This equation indicates the mobile platform can change its direction by adjusting the velocity of the three wheels.

$$\begin{bmatrix} V_a \\ V_b \\ V_c \end{bmatrix} = \begin{bmatrix} \cos\alpha & \sin\alpha & L \\ -\cos\theta_1\cos\alpha + \sin\theta_1\sin\alpha & -\cos\theta_1\sin\alpha - \sin\theta_1\cos\alpha & L \\ -\sin\theta_2\cos\alpha - \cos\theta_2\sin\alpha & -\sin\theta_2\sin\alpha + \cos\theta_2\cos\alpha & L \end{bmatrix} \begin{bmatrix} V_X \\ V_Y \\ W \end{bmatrix} \tag{3}$$

3 Adsorption Force and Driving Force Design

To ensure the robot can work steadily, design of adsorption force and driving force in different stages are discussed respectively.

3.1 Machining Stage

In this stage, the omnidirectional wheels have been lifted, and the friction material contacts directly with the workpiece surface. To make sure the robot will not slip and overturn when machining, the minimum adsorption force should guarantee total force along X and total overturning moment equal to zero with considering the safety factor

$$\begin{cases} n\sum \mathbf{M}_{AY+}(A) + \sum \mathbf{M}_{AY-}(A) = 0 \\ \sum \mathbf{F}_{X+} + n\sum \mathbf{F}_{X-} = 0 \end{cases} \tag{4}$$

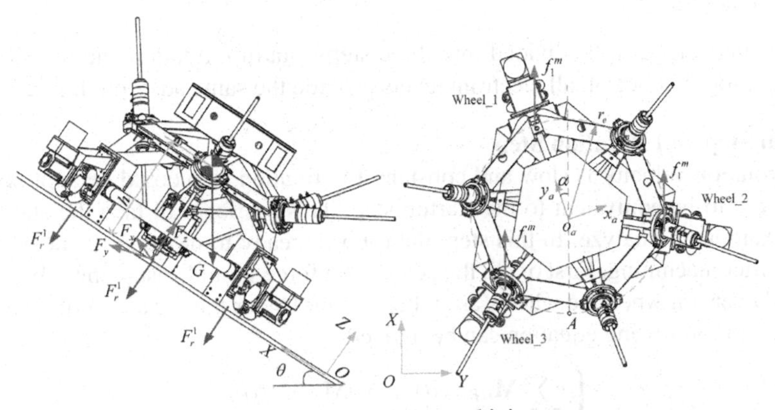

Fig. 3. Force analysis in machining stage

According to the force analysis shown in Fig. 3, following equation can be derived

$$\begin{cases} (G\sin\theta \cdot h + F_z \cdot r_e)n = \\ G\cos\theta \cdot r_e + F_r^1 \cdot r_e(1+\cos\alpha) + F_r^1 \cdot r_e\left(1-\cos(\frac{\pi}{3}+\alpha)\right) + F_r^1 \cdot r_e\left(1-\cos(\frac{\pi}{3}-\alpha)\right) \\ \left(G\sin\theta + F_y\right)n = 3f_1^m\, 3f_1^m = \left(3F_r^1 + G\cos\theta - F_z\right) \cdot \mu_1 \end{cases} \tag{5}$$

In the above formulas, G and h are the weight of robot and height of gravity center; F_x, F_y, F_z are the cutting force along the x_a, y_a, z_a axis; f_1^m is the static friction force during machining; F_r^1 is the minimum adsorption force during machining; r_e is the electromagnet placement radius; n is the safety factor; μ_1 is the static friction coefficient between workpiece surface and friction material; θ is the incline angle of surface; α is the robot pose angle; $O-XYZ$ is the global coordinate; $o_a - x_a y_a z_a$ is the robot coordinate.

According to the given parameters, adsorption force distribution surface can be illustrated in Fig. 4. According to the result, adsorption force only changes with the angle θ. The maximum adsorption force is 3567 N when the angle θ equals to 90°. Considering that the robot should be capable of machining on the 90° incline surface, the adsorption force F_r^1 should larger than 3567 N.

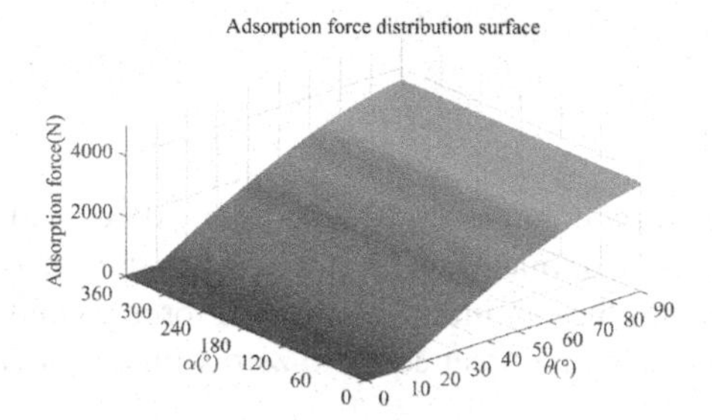

Fig. 4. Adsorption force distribution surface in machining stage

3.2 Moving Stage

The moving stage can be divided into three steps: startup, rotation and translation. In order to simplify control, all electromagnets provide the same adsorption force.

Startup Step and Rotation Step

When rotation velocity is slow and constant, the rotation step could be approximately considered to be equivalent to the startup step. This section will take the startup step as an example to analyze. In this step, the robot is ready to move to the next working station after machining. First of all, the adsorption force should ensure the robot will not slip and overturn when the wheels are fallen down. Under the condition of considering safety factor, following equation can be derived

$$\begin{cases} n\sum \mathbf{M}_{BY+}(B) + \sum \mathbf{M}_{BY-}(B) = 0 \\ \sum \mathbf{F}_X + n\sum \mathbf{F}_{X-} = 0 \\ F_{Ni} \le F_{limit} \end{cases} \tag{6}$$

From the force analysis shown in Fig. 5, following equation can be derived

$$\begin{cases} (G\sin\theta \cdot h)n = \\ G\cos\theta \cdot r_w + F_r^2 \cdot r_w(1+\cos\alpha) + F_r^2 \cdot r_w\left(1-\cos\left(\frac{\pi}{3}+\alpha\right)\right) + F_r^2 \cdot r_w\left(1-\cos\left(\frac{\pi}{3}-\alpha\right)\right) \\ (G\sin\theta)n = f_1^s + f_2^s + f_3^s \ \sum f_i^s = \left(3F_r^2 + G\cos\theta\right)\cdot \mu_f \\ F_{Ni} \le F_{wheel\_limit} \ F_{Ni} \le F_{up\_limit} \end{cases} \tag{7}$$

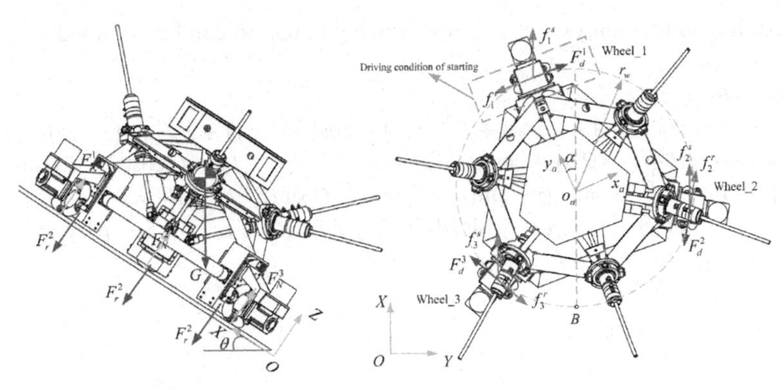

Fig. 5. Force analysis in startup and rotation step

In Eq. (7), μ_f is static friction coefficient between workpiece surface and omnidirectional wheel; r_w is the omnidirectional wheel placement radius; f_i^s is the static friction force; F_{wheel_limit} and F_{up_limit} are the limit loads of wheel and elevating device. And the support force F_{Ni} could be calculated through equilibrium condition

$$\sum_{i=1}^{3}\mathbf{M}_{Ni} = 0; \sum_{i=1}^{3} F_{Ni} = 3F_r^2 + G\cos\theta \tag{8}$$

This step should also satisfy the driving force constraints (the gray dotted line in Fig. 5) that the omnidirectional wheel could overcome the rolling friction force and move without slipping. The formula is shown as follows

$$f_i^r \le F_{di} \le f_i^s \Rightarrow \mu_r F_{Ni} \le F_{di} \le \mu_f F_{Ni} \tag{9}$$

In this formula, μ_r is rolling friction coefficient between workpiece surface and omnidirectional wheel; F_{di} is the driving force; f_i^r is the rolling friction force. Further analysis of F_r^i and F_{di} will be discussed in translation step to determine the final parameters of adsorption device.

Translation Step
In this step, it is assumed that the robot moves at a constant speed. The robot should satisfy no overturning, force balance in the direction of motion, no side slide condition, no rotation, load limitation and driving force constraint. And no overturning condition should consider the safety factor. Following equation can be derived

$$\begin{cases} n\sum \mathbf{M}_{BY+}(B) + \sum \mathbf{M}_{BY-}(B) = 0 & \sum \mathbf{M}_{z_a}(o_a) = 0 \\ \sum \mathbf{F}_v = 0 & F_{Ni} \le F_{limit} \\ \sum \mathbf{F}_{\perp v} = 0 & f_i^r \le F_{di} \le f_i^s \end{cases} \tag{10}$$

According to the analysis in Fig. 6, following equation can be obtained

$$\begin{cases} (G\sin\theta \cdot h)n = \\ G\cos\theta \cdot r_w + F_r^3 \cdot r_w(1+\cos\alpha) + F_r^3 \cdot r_w\left(1-\cos\left(\frac{\pi}{3}+\alpha\right)\right) + F_r^3 \cdot r_w\left(1-\cos\left(\frac{\pi}{3}-\alpha\right)\right) \\ F_{d1}\sin\beta - \cos\left(\frac{\pi}{6}-\beta\right)F_{d2} + \cos\left(\beta+\frac{\pi}{6}\right)F_{d3} - f_1^r - f_2^r - f_3^r - G\cos(\beta-\alpha)\sin\theta = 0 \\ -F_{d1}\cos\beta + \sin\left(\frac{\pi}{6}-\beta\right)F_{d2} + \sin\left(\beta+\frac{\pi}{6}\right)F_{d3} - G\sin(\beta-\alpha)\sin\theta = 0 \\ r_w(-F_{d1}) - r_wF_{d2} - r_wF_{d3} + r_w\sin(\beta)f_1^r + r_w\sin(\beta)f_3^r - r_w\cos\left(\frac{\pi}{6}-\beta\right)f_2^r = 0 \\ F_{Ni} \le F_{wheel\_limit};\ F_{Ni} \le F_{up\_limit} \\ \mu_r F_{Ni} \le F_{di} \le \mu_f F_{Ni} \end{cases} \tag{11}$$

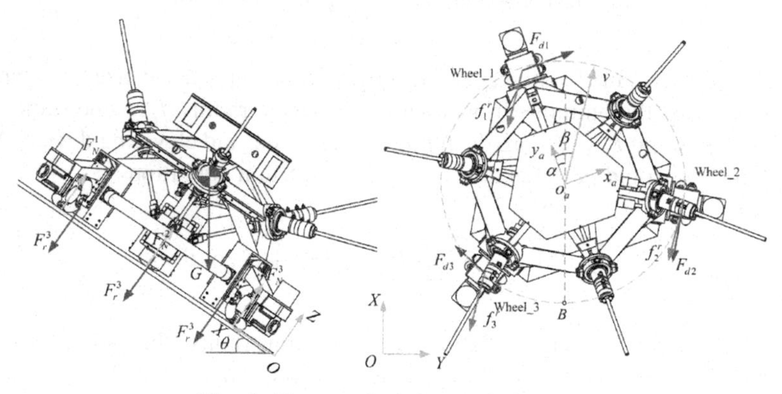

Fig. 6. Force analysis in translation step

For simplicity of design, angle α is set to zero, and only the angle θ and β are changed. As presented in Figs. 7 (a–c), the robot can achieve omnidirectional motion on incline that no more than 5°. And the detail of driving force distribution when θ= 5° is shown in Fig. 7 (d). The result indicates that F_{d1} and F_{d2} are the main limited driving force. The low static friction coefficient μ_f between polyurethane and metal surface results in a low static friction force, and the insufficient static friction force limits the promotion of driving force and causes low climbing performance. To solve this problem, a matching design will be carried out in Sect. 3.3.

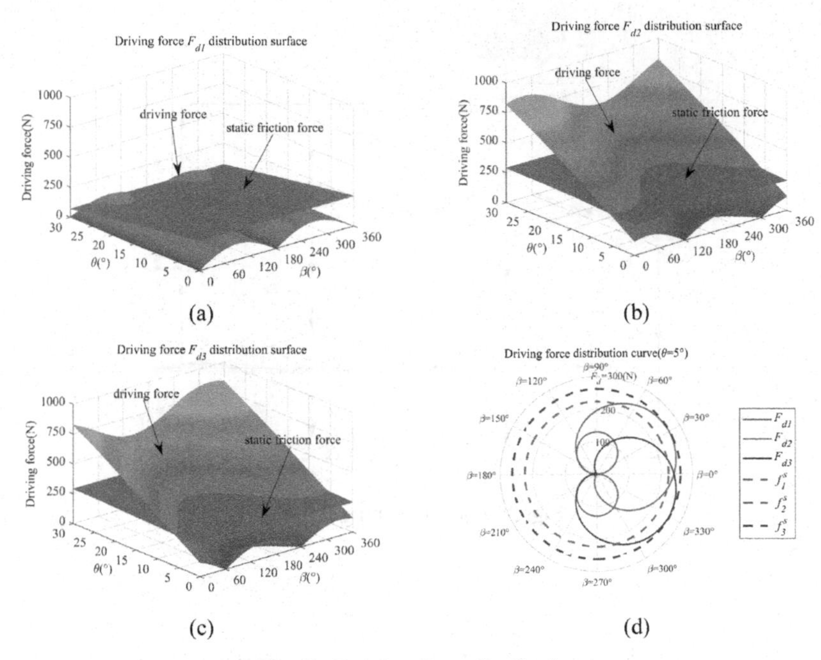

Fig. 7. Driving force distribution

3.3 Force Matching Design

Inspired by the experience that vehicles with some cargoes have better climbing ability. Adding additional adsorption force to make robot realize better climbing performance seems to be a viable solution. So, a driving force and adsorption force matching design is proposed and presented in Fig. 8. This design mainly increases the static friction force by providing extra adsorption force within the load range, so as to make the robot climb incline with larger angle.

According to the matching design, driving force distribution surface can be meshed as shown in Fig. 9 (a). Considering the lightweight design of robot and installation space, driving force is chosen as 530 N which corresponds to angle θ equals to 10°. Compared with the results shown in Fig. 7, the maximum value of θ could be enhanced two times. Relative adsorption force distribution surface is presented in Fig. 9 (b).

The air gap distance between workpiece surface and electromagnet will directly influence the value of adsorption force. In order to increase the air gap distance as much as possible to make the robot have a better passing ability, the minimum value plane in Fig. 9 (b) is chosen as the adsorption force distribution. And the adsorption force can be adjusted through PWM signal. The driving force curves that θ equal to 0° and 10° are shown in Fig. 10 (a) and Fig. 10 (b). No slipping occurs during the omnidirectional motion. In conclusion, the adsorption mobile platform has better climbing ability and better motion stability by using the presented matching method.

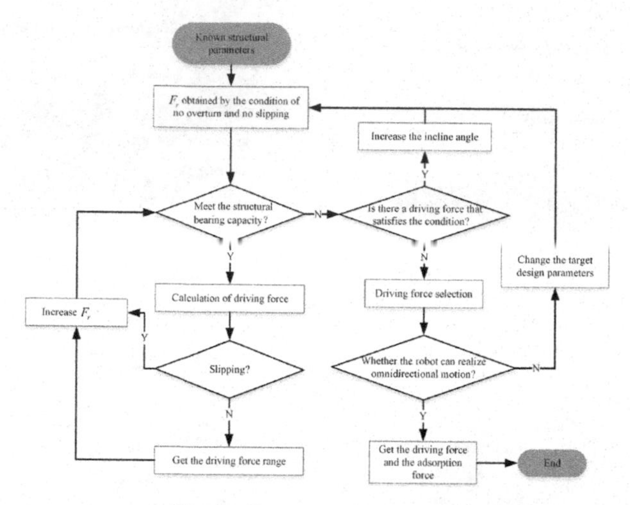

Fig. 8. Force matching design

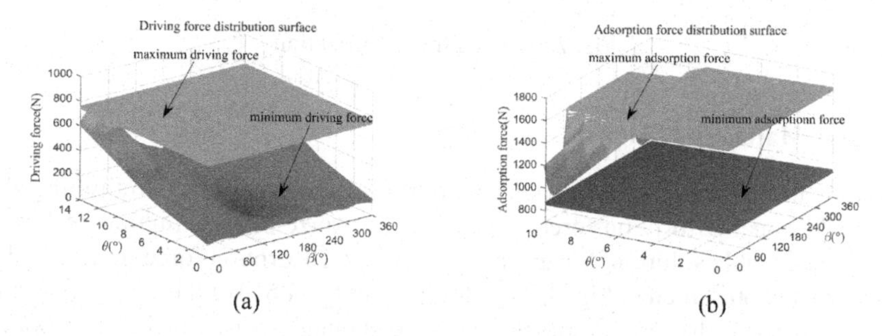

(a) (b)

Fig. 9. Matching design results

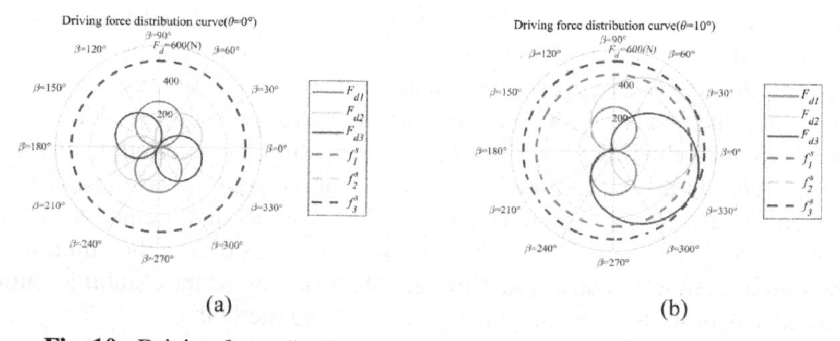

(a) (b)

Fig. 10. Driving force distribution curve after using force matching design

4 Conclusion

This paper proposes an adsorption machining robot which can process or maintain large structure in-situ. Omnidirectional mobile platform with friction adsorption locking device and elevating omnidirectional wheels are designed to guarantee high-performance processing quality. Then kinematic problem of omnidirectional mobile platform is analyzed. Adsorption force during machining stage should larger than 3567 N to guarantee the robot can process on 90° incline surface. Due to the low friction coefficient between workpiece surface and wheel in moving stage, the robot only has 5° climbing ability without extra adsorption force. By carrying out matching design of adsorption force and driving force, robot's climbing performance has been improved by two times. The maximum climbing ability can reach 10° and corresponding driving force is designed as 530 N. And the adsorption force in moving stage should vary between 862 N and 934 N for different incline angles. This research can provide guidance for prototype design. Future work will mainly focus on the development of the robot and its positioning and cutting stability.

Acknowledgements. This work is supported by National Key R&D Program of China under Grant 2019YFA0706701 and the National Natural Science Foundation of China under Grants 51922057 and 91948301.

References

1. Susemihl, H., et al.: High accuracy mobile robotic system for machining of large aircraft components. SAE Int. J. Aerosp. **9**, 231 (2016)
2. Moeller, C., et al.: Real time pose control of an industrial robotic system for machining of large scale components in aerospace industry using laser tracker system. SAE Int. J. Aerosp. **10**, 100 (2017)
3. Xie, F.G., Mei, B., Liu, X.J., Zhang, J.B., Yue, Y.: Novel mode and equipment for machining large complex components. J. Mech. Eng. **56**(19), 70–78 (2020)
4. Dong, X., Palmer, D., Axinte, D., Kell, J.: In-situ repair/maintenance with a continuum robotic machine tool in confined space. J. Manufact. Process. **38**, 313 (2019)
5. Ma, N., Yu, J.J., Dong, X., Axinte, D.: Design and stiffness analysis of a class of 2-DoF tendon driven parallel kinematics mechanism. Mech. Mach. Theory **129**, 202 (2018)
6. Wang M.F., David P., Dong X., Alatorre D., Axinte D., Norton A.: Design and development of a slender dual-structure continuum robot for in-situ aeroengine repair. In: 2018 IEEE/RSJ International Conference on Intelligent Robots and Systems (IROS), vol. 5648 (2018)
7. Liu, J.M., Tian, Y., Gao, F.: A novel six-legged walking machine tool for in-situ operations. Front. Mech. Eng. **15**, 351 (2020)
8. Xie, H., Li, W.L., Zhu, D.H., Yin, Z.P., Ding, H.: A systematic model of machining error reduction in robotic grinding. IEEE/ASME Trans. Mechatron. **25**, 2961 (2020)
9. Mei, B., Xie, F.G., Liu, X.J., Yang, C.B.: Elasto-geometrical error modeling and compensation of a five-axis parallel machining robot. Precis. Eng. **69**, 48 (2021)
10. Xie, Z.H., Xie, F.G., Liu, X.J., Wang, J.S., Mei, B.: Tracking error prediction informed motion control of a parallel machine tool for high-performance machining. Int. J. Mach. Tools Manufact. **164**, 103714 (2021)

11. Wang, G., et al.: Simultaneous calibration of multicoordinates for a dual-robot system by solving the AXB = YCZ problem. IEEE Trans. Robot. **37**(4), 1172–1185 (2021). https://doi.org/10.1109/TRO.2020.3043688
12. Zhang, Y., Dai, Z., Xu, Y., Qian, R.: Design and adsorption force optimization analysis of TOFD-based weld inspection robot. In: Journal of Physics. Conference Series, vol. 1303, p. 12022 (2019)
13. Yi, Z.Y., Gong, Y.J., Wang, Z.W., Wang, X.R.: Development of a wall climbing robot for ship rust removal. In: 2009 International Conference on Mechatronics and Automation, vol. 4610 (2009)
14. Nazim, M.-N., Siswoyo, H., Ali, M.H.: Portable autonomous window cleaning robot. Procedia Comput. Sci. **133**, 197 (2018)
15. Chen, Y., Wang, C.M.: The design of permanent magnetic adhesion system for wall-climbing robot. Appl. Mech. Mater. **300–301**, 531 (2013)
16. Wu, M.H., Pan, G., Zhang, T., Chen, S.B., Zhuang, F., Zheng, Y.: Design and optimal research of a non-contact adjustable magnetic adhesion mechanism for a wall-climbing welding robot. Int. J. Adv. Robot. Syst. **10**, 63 (2013)
17. Hongwei, T., Chaoquan, T., Gongbo, Z., Xin, S., Qiao, G.: Design and magnetic force analysis of patrol robot for deep shaft rigid cage guide. In: Yu, H., Liu, J., Liu, L., Ju, Z., Liu, Y., Zhou, D. (eds.) ICIRA 2019. LNCS (LNAI), vol. 11744, pp. 362–374. Springer, Cham (2019). https://doi.org/10.1007/978-3-030-27541-9_30
18. Xie, F.G., Liu, X.J., Luo, X., Wabner, M.: Mobility, singularity, and kinematics analyses of a novel spatial parallel mechanism. J. Mech. Robot. **8**, 61022 (2016)
19. Xie, F.G., Liu, X.J., Wang, J.S., Wabner, M.: Kinematic optimization of a five degrees-of-freedom spatial parallel mechanism with large orientational workspace. J. Mech. Robot. **9**, 051005 (2017)
20. Liu, X., Bi, W., Xie, F.: An energy efficiency evaluation method for parallel robots based on the kinetic energy change rate. Sci. China Technol. Sci. **62**(6), 1035–1044 (2019). https://doi.org/10.1007/s11431-019-9487-7

Research on Mobile Robot Processing System Based on Shape and Position Measurement

Zhanxi Wang(✉), Yongnian Han, Gang Wang, Yiming Zhang, and Hang Chen

Northwestern Polytechnical University, Xi'an 710072, China
zxwang@nwpu.edu.cn

Abstract. Aiming at the problem of on-line inspection and processing due to the alignment errors in the assembly process of large aeronautical components, an adaptive digital processing and assembly method is proposed, and a mobile robot processing system for milling mating surfaces is designed. The preprocessing algorithm of a large number of scattered point cloud data based on the laser profile scanner is studied, and the LSPIA method based on the selection of key points is used to reconstruct the NURBS parametric surface, and the fitting error is analyzed. According to the assembly requirements of multiple sets of mating surfaces at the same time, a calculation method for the minimum machining allowance of each area to be processed is proposed, and the processing track is optimized. Finally, the proposed method is verified by building a prototype of the equipment. The experimental results show that the method meets the technical indicators required by the system, and it can be applied to engineering practice, effectively reducing labor intensity.

Keywords: Aeronautical components · Point cloud data processing · Surface reconstruction

1 Introduction

The main research object of this paper is the detection and processing method of the space contour of the aerospace component mating surface, which is mainly used in the detection of the excess tolerance of the mating surface and the secondary online adjustment of the mating surface of the mating component in the aviation industry. In the traditional docking method, the processing and correction of the mating surfaces between the components are carried out by manual work such as grinding and scraping by the workers, and the alignment accuracy of the mating surface is detected by the workers using traditional measurement methods such as red tan powder and stopper. This approach is time-consuming, laborious, and the assembly quality is unstable. In contrast, the mobile robot milling processing system has high efficiency, good space accessibility, and stable accuracy, and can well adapt to the assembly operations of large and complex components.

X.-J. Liu et al. (Eds.): ICIRA 2021, LNAI 13014, pp. 785–796, 2021.
https://doi.org/10.1007/978-3-030-89098-8_74

Domestic and international aviation industry in terms of assembly technology a lot of research, including digital measurement technology, virtual assembly technology, automatic control technology, so that aircraft assembly toward automation direction. Since the advent of industrial robots, robots have been widely used in the field of aviation manufacturing because of their high flexibility and low cost [1, 2]. Such as EI and Boeing jointly developed for F/A-18E/F rear flaps of the ONCE robot automatic hole-making system [3]. In recent years, 3D measurement technology has been relatively mature, in industrial testing has been widely used [4]. For example, Ahn uses the detection light emphasis algorithm, optically slices the structure, reconstructs the 3D point cloud, and overlays the sliced image to reconstruct the 3D model of the dental cast [5]. Gashongore uses on-board laser contour scanners and CCD cameras for 3D reconstruction of indoor space [6]. In terms of surface reconstruction techniques, progressive and Iterative Approximation for Least Squares B-spline Curve and Surface Fitting (LSPIA) proposed by Deng, is an effective algorithm suitable for large-scale data, and its limit curve surface is the result of the least square fit for a given data point [7]. Subsequently, Lin proves that the LSPIA algorithm can solve the singular least-squares fitted equation [8].

According to the requirements of the over-tolerance detection task of multiple sets of mating surfaces in the docking assembly process of large aerospace components at the same time, the design of a space contour detection and processing system based on a laser scanner and a mobile robot is carried out. According to the point cloud data collected by the inspection system, the research on the three-dimensional surface reconstruction of the mating surface is carried out, and the process optimization of the mating surface is completed. Finally, it is verified by experiment.

2 Detection and Processing System for Mating Surfaces of Large Aeronautical Components

The detection and processing system for mating surfaces of large aeronautical components mainly includes four subsystems: attitude adjustment system of component A, attitude adjustment system of component B, the measurement system and the milling system of mobile robot, and Fig. 1 shows the overall layout of the system. At the assembly site, component A and component B are symmetrically arranged on both sides of the fuselage axis, and two sets of component A and four sets of component B need to be assembled. In one station, after the robot completes the mating surface machining of component A, the AGV mobile platform in the robot milling system carries the robot to the next station to complete the processing tasks of the other stations.

In accordance with the processing process, the position of component A and B under the spatial 3D coordinate system is first established by means of a laser tracker through the ground target measurement point. Then, through attitude adjustment system of component A and B respectively to complete its space multi-degree of freedom adjustment, and finally move the robot processing equipment to a fixed station to complete the milling of the mating surface. Finally, the precise butt assembly of part A and part B in the three-dimensional coordinate system of space is realized.

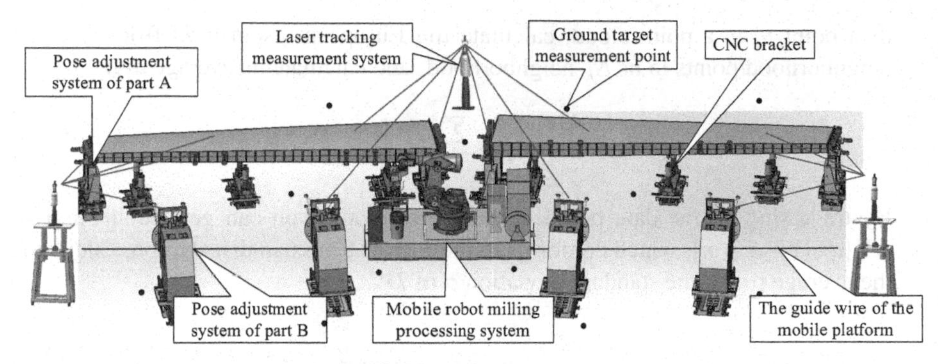

Fig. 1. System layout

3 Research on Detection and Processing Methods

3.1 Processing of Point Cloud Data

Pre-processing of Point Cloud Data. The original point cloud data obtained by the robot end effector carrying the laser contour scanner for multiple movement scans belongs to the scattered point cloud. Therefore, the geometric topological relationship between the point cloud data needs to be established first. An improved scanning line method is proposed to build a point cloud data topology relationship. The specific method is described as follows:

1. By setting the Z value of all point coordinates to 0, the 3D point cloud data is projected onto the xOy plane and the projected data is a 2D point cloud.
2. The point cloud data is then grouped according to the size of the Y value, and the resulting two-dimensional point cloud data is stored in a list, sorted by Y-value size, the same way, and then a point-cloud list sorted by X size.
3. To obtain the topological neighborhood of a key point, first select the point set of the scan line where it is located as the search range, select the points in its neighborhood according to the search radius, and calculate the Euclidean distance from the key point respectively, keep the points whose Euclidean distance is less than the search radius, and sort them by size.

   The original point cloud data obtained by laser scanning will contain a large number of noise point clouds, and different denoise methods should be adopted according to their spatial properties for different noises [9, 10].

- For the drift points contained in the obtained point cloud data, filter the point cloud by setting a threshold in the X, Y, and Z directions, and if the coordinates of a point in the point cloud exceed the set threshold, you can determine that it is a useless background point and eliminate it.
- For sparse point clouds that deviate from the main point cloud in outliers, sparse point clouds suspended around the principal point cloud can be filtered out according to the distance distribution from the data point to its neighborhood point. For each

data point p_i in a point cloud, calculate the European distance d_{ij} from p_i to all neighborhood points in its N_k neighborhood, and calculate its average dv_i:

$$dv_i = \frac{\sum_{j=1}^{k} d_{ij}}{k} \tag{1}$$

By traversing all the data points in the point cloud, you can get a collection of $DV\{dv_i|1 \leq i \leq n\}$, which conforms to the law of Gaussian distribution, calculates the average μ and the standard deviation σ of DV

$$\mu = \frac{1}{n}\sum\nolimits_{i=1}^{n} dv_i \tag{2}$$

$$\sigma = \sqrt{\frac{1}{n}\sum\nolimits_{i=1}^{n} (dv_i - \mu)} \tag{3}$$

Set a threshold Ω to exclude points whose average distance is outside the threshold range from the point cloud data.

$$dv_i > \mu + \Omega \cdot \sigma \tag{4}$$

- There are some noise burrs in the data that interfere with the accuracy of subsequent surface reconstructions, so the cloud data of the junction point is smoothed using a threshold-bound neighborhood median method. Firstly, we need to build a new data store S and copy the point cloud data that has been processed to that cache. Then, reading the data point p_i in turn, and the filter's template center coincides with the read data point p_i, and sort by the data point Z value from small to large, the data point in the middle position is recorded as p_{ci}; Finally, the Z-coordinate difference Δz between the p_i and the p_{ci} is calculated, and the threshold evaluation of the p_i of the original data point is made by that difference. Determine the size of the difference Δz and threshold T, if the Δz is greater than the threshold T, p_i coordinates change to (x_i, y_i, z_{ci}), and if Δz is less than threshold T, the original data point is retained.

The NURBS Surface Fitting of the Space Profile of the Docking Surface. The LSPIA method based on the selection of key points is used to reconstruct the non-uniform rational B-spline (NURBS) parametric surface, and a model reflecting the actual intact surface is obtained. In this method, the point with the largest local curvature in the point cloud data is selected as the initial control vertex, and combined with uniformly distributed selected control points, the initial control vertex mesh is formed together to construct the initial surface. Through continuous iteration and adjustment of the control vertices, a series of fitted surfaces are obtained, the limit surface of which is the surface fitted by the least square method, and the number of surface iterations can be determined by the surface fitting error.

3.2 The Optimized Design of the Docking Process

Calculate the Offset Machining Margin. In the same coordinate system, calculate the distance between the corresponding points on the mating surface of part B and part A. Because there are 4 pairs that need to be paired at the same time, in order to ensure that all pairs fit together, the maximum values in ΔZ_{max1}, ΔZ_{max2}, ΔZ_{max3}, ΔZ_{max4} are used as virtual aligning distances for parts A and B to achieve full aligning. Move component B along the alignment direction by a distance of ΔZ_{max}, and the overlap between components A and B is the minimum machining allowance required for the aligning surface of component A, as shown in Fig. 2.

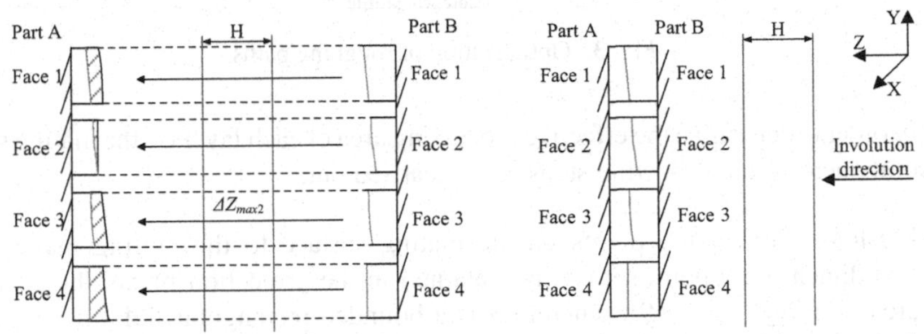

(a) The calculation of the processing margin (b) The aligning state after processing

Fig. 2. The calculation of the processing margin of the docking surface

Surface Bias Layering Algorithm. In order to improve processing efficiency, different surface offset algorithms are used in different processing stages:

- Rough machining uses a flat-end milling cutter to perform layered multiple cuts. Define a set of cutting planes perpendicular to the tool rotation axis, the distance between each cutting plane is the machining depth of each layer, and then plan a two-dimensional tool path on the cutting plane.
- Use a ball-end knife to semi-finish the blank to remove the stepped cutting residue. For the mating surface with a relatively flat surface, the theoretical profile can be offset along the Z-direction equidistantly, and the tool path can be generated. For surface $S(u, v) = (x(u, v), y(u, v), z(u, v))$, the surface equation for offset distance d along Z is

$$S^{'}(u, v) = (x(u, v), y(u, v), z(u, v) + d) \tag{5}$$

- In surface finishing, the tool location surface is an equidistant surface obtained by offsetting a tool radius along the normal direction of the processed surface. For the same point on each layer, its x, y, z coordinates change at the same time. The equidistant surface of surface $S(u, v)$ after the offset distance d along the method is

$$S^{'}(u, v) = S(u, v) + d \cdot N(u, v) \tag{6}$$

Toolpath Planning

The optimization of the processing path of flat end milling cutter. In order to improve the efficiency of rough machining of the mating surface, the effective machining area on the cutting plane can be obtained first. Project the mating surface onto the cutting plane, and divide a number of grids evenly within the projection range, as shown in Fig. 3.

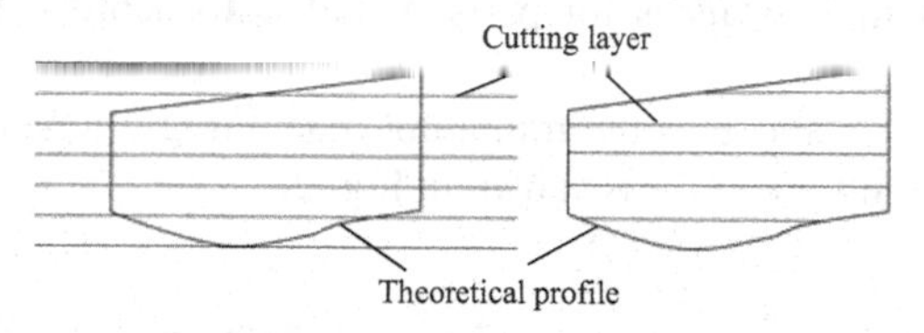

Fig. 3. Optimization of roughing paths

In order to determine the effective processing area of each layer on the mating surface of component A, the following steps can be carried out:

1. Meshing: Project the points on the mating surface to the cutting plane, and a two-dimensional point set can be obtained in the projection plane, the projection area is $A(x, y) \in C^2(S)$, wherein S is a bounded rectangular field, it needs to be divided equally. The determination of the grid spacing needs to be determined by the processing effect of the matching surface.
2. Identification and identification of mesh machinability: The four corners of the grid are perpendicular to the plane, and the perpendicular has 4 points of intersection with the rough surface, of which the point with the largest Z value is recorded as recorded as Z_h. In the same way, the vertical lines of the four corners of the grid and the theoretical surface also have four intersection points, and the highest point is recorded as Z_l, as shown in Fig. 4. At the same time, it is necessary to leave a margin for finishing, and set the offset distance as Z_m, obviously, for the cutting plane $Z = Z_c$, only when the $Z_h \geq Z_c \geq Z_l + Z_m$ is satisfied, this mesh is a machinable mesh, otherwise it is an unprocessable mesh.

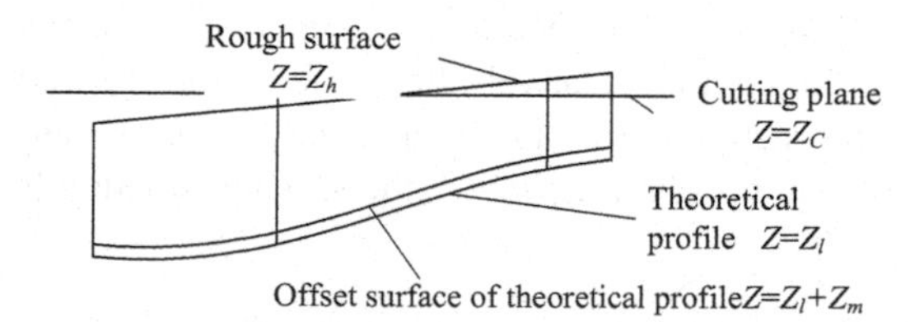

Fig. 4. The determination of the machinable mesh

3. Boundary identification of machinable areas: Judge and mark the boundary grids, all boundary grid search ends, the boundary grid loop of this machinable area is formed, and the link list on the cutting plane is established. Store all the boundary mesh loops in the linked list. According to the topological relationship between the loops, all the effective processing areas of the cutting plane can be determined.

4. Obtaining multi-layer effective processing area: For cutting plane $Z_c > \max(Z_l + Z_m)$, the search can be extended to the periphery based on the effective processing area boundary calculated by the previous layer. For the cutting plane of $Z_c < \max(Z_l + Z_m)$, the entire profile needs to be solved once to obtain the set of machinable areas of the cutting plane of this layer.
5. Tool path generation based on boundary mesh: Use the direction-parallel to plan the tool path on the cutting plane. First, plan the initial tool path on the cutting plane according to the row spacing, and then intersect the tool location point grid obtained by the calculation to obtain the tool path of each processing area.

The Optimization of the Processing Path of the Ball End Mill

After rough machining, there will be step-like residues on the mating surface. For this reason, a ball-end knife is used for semi-finishing to make the machining allowance of each point on the mating surface uniform to ensure the quality of finishing. Ball-end tool trajectory planning is mainly to determine the appropriate step length control algorithm and raw spacing control algorithm.

1. Determination of raw spacing: Take the raw spacing L as a fixed value. To ensure that the residual height h at all processing points is within the required range, take the value of L is:

$$L = \sqrt{\frac{8Rh}{1 \pm \rho_{max} R}} \tag{7}$$

Among them, tool radius R and residual height h can be determined according to machining needs, ρ is the curvature at the processing point.

2. Determination of step size: The processing step is planned by equal chord-error method, assuming that the current tool path curve is $C(t)$ and the tool contact position is $P(t_i)$, the radius of curvature at this time is:

$$R_i = \frac{\left|C'(t_i)\right|^3}{\left|C'(t_i) \times C''(t_i)\right|} \tag{8}$$

Assuming that the curvatures between the two tool contacts are equal, and the arc is used instead of the curve, the processing step size can be obtained by the radius of curvature and the allowable bow height error e:

$$L_i = 2\sqrt{2eR_i - e^2} \tag{9}$$

Then, using the second-order Taylor expansion, the processing step is converted to the parameter increment Δt_i, and then the next contact $P(t_{i+1})$ is struck.

$$t_{i+1} = t_i + \Delta t_i = t_i + \frac{L_i}{\left|C'(t_i)\right|} \tag{10}$$

The initial step is calculated by traditional method, and then the core is checked to compare the actual bow height error ε the size of the bow high tolerance e:

- $e - \varepsilon > \xi$: The step size is too conservative, so the parameter increment can be taken as $\Delta t = (1 + \alpha)\Delta t$ until$0 \le e - \varepsilon \le \xi$, then the iteration ends;
- $0 \le e - \varepsilon \le \xi$: Prove that the step size is just right and the parameter increment is unchanged;
- $e - \varepsilon < 0$: Prove that the step size is out of the allowed value, so the parameter increment is taken as $\Delta t = (1 - \alpha)\Delta t$ until $0 \le e - \varepsilon \le \xi$, then the iteration ends.

When $t_{i+1} \ge 1$, t_{i+1} is the last tool contact on the path.

4 Experimental Validation

4.1 Extracts the Characteristics of the Docking Face

Taking the mating surface 1 in component A as a representative, the original point cloud data distribution obtained by the scanner after completing the measurement of the mating surface is shown in Fig. 5(a). It can be seen that there are a lot of interference points in the point cloud. Taking the Z coordinate of the point cloud data as the statistical object, draw a histogram of the frequency distribution of the point cloud Z value, as shown in Fig. 5(b). Among them, the coordinate point with the Z coordinate value in the range of [89.7, 101] is the effective point cloud data of the required mating surface. The distribution of point cloud data through pre-processing is shown in Fig. 6. Obviously, after processing, the outliers have been eliminated, and at the same time, the noise on the mating surface has been filtered out.

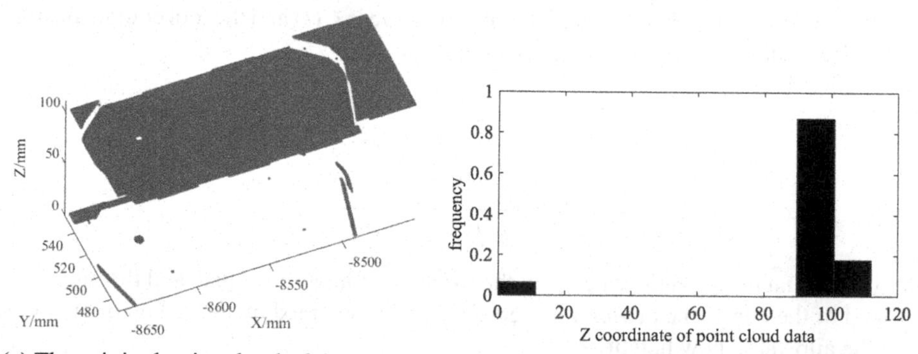

(a) The original point cloud of the A1 (b) Z coordinate frequency distribution histogram

Fig. 5. The original point cloud of the A1

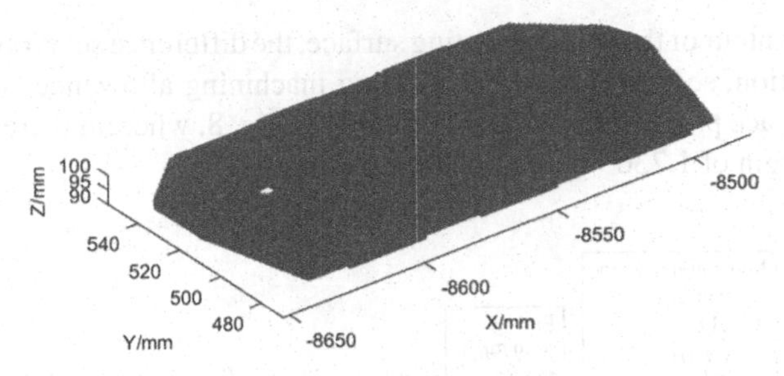

Fig. 6. The distributed point cloud data after processing

4.2 The Fitting of the Spatial Profile of the Closed Surface and the Calculation of the Machining Margin

Subsequent corrections to the machining allowance of the mating surface of part A are subtractive machining. In order to avoid new matching problems when the processed profile of part A is aligned with the mating surface of part B, the mating surface is reasonably expanded into a corresponding regular quadrilateral of equal length and width according to the original design size. According to the above method, extract the spatial contour features of all mating surfaces, and use the zero point of the machining coordinate system as the reference to unify the scanning coordinate system of the different positions of the part A and the part B. The fitting effect is shown in Fig. 7.

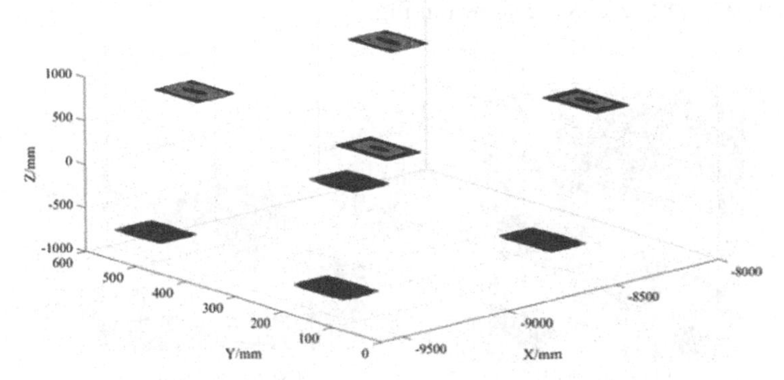

Fig. 7. The contour fit effect of the docking face in space

The spatial contour data of the mating surface is the data in the processing coordinate system, so first transform the point coordinates on the mating surface to the world coordinate system. Perform a coordinate rotation transformation on all the data, and use the alignment direction n of the part B as the new Z axis to obtain the alignment distance ΔZ_{max}. In the same space coordinate system, using the space contour of the mating surface B as the reference, calculate the minimum machining allowance required for the original mating surface of the mating component A. For the original mating surface and the corrected mating surface of component A, the machining allowance of the component A at each point (x, y) in the xoy plane is obtained through coordinate calculation. Based on the spatial contour of the original mating surface, the difference set is obtained through Boolean operation, so as to obtain the required machining allowance. The calculation results of the space processing margin are shown in Fig. 8, wherein there is a maximum processing margin of 1.7303 mm at the face 4 of part A.

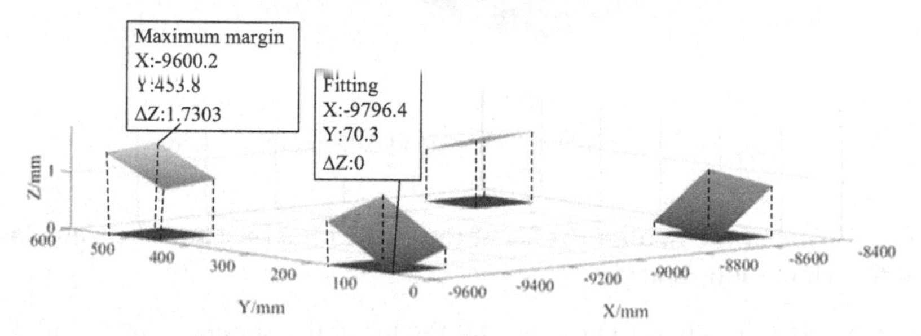

Fig. 8. The calculation of the processing margin

4.3 Testing of Processing Results

After scanning all the contours of the 8 mating surfaces with a laser scanner, the machining allowance of each mating surface is calculated, and then the electric spindle clamps the milling cutter to sequentially mill each mating surface according to the optimized processing path. The results are shown in Fig. 9.

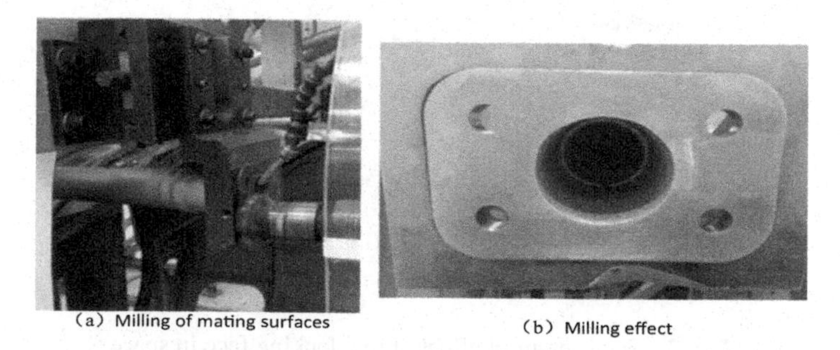

Fig. 9. The processing experiment site for component B docking surface

After the processing is completed, in order to accurately detect the matching accuracy of the mating surface after processing, respectively use the laser inspection verification method and the enterprise inspection standard: "feeler gauge + manual application of red lead powder" for verification, the results are as follows:

1. Laser detection and verification: A laser profile scanner is used to scan the mating surface of part A after the machining allowance is removed. The data preprocessing and spatial contour fitting are also performed according to the above method, and

the mating surface of the part B spatial contour reference is virtual assembled and aligned.

In the calculation results as shown in Fig. 10, the mating surface 4 is aligned, and there is the maximum fit gap at the mating surface 1. It can be seen that after the mating surface of component A is corrected, the maximum gap of the mating surface is 0.021 mm, that is, the preliminary judgment meets the requirements of the alignment accuracy: the gap is less than or equal to 0.10 mm.

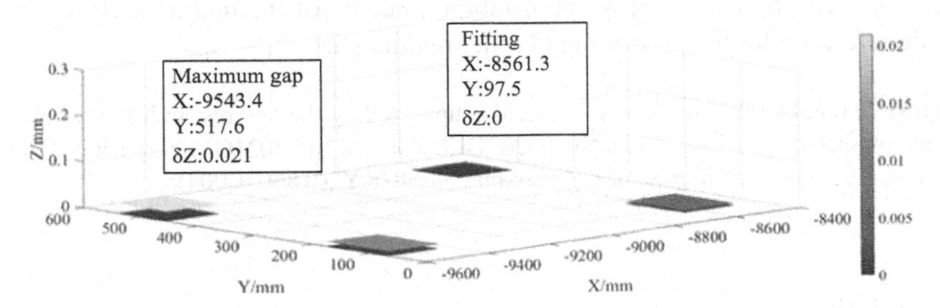

Fig. 10. The fitting gap that has been corrected for the closing surface

2. Feeler gauge + manual application of red lead powder: Use feeler gauge and manual application of red lead powder for testing to verify the fit accuracy of the two mating surfaces. The results are shown in Fig. 11.

Fig. 11. Detects the fit of the docking surface

It can be seen that the red lead powder on the mating surface of component B is evenly distributed, and after field testing, the mating degree of the contact surface is greater than 80%, that is, the mating accuracy after the modified processing meets the technical requirements. Through many product experiments, and using the above two methods to test, the results obtained are basically the same. This shows that the aforementioned point cloud data collection method, mating surface feature extraction and mating surface machining allowance calculation algorithm are correct and can be applied to engineering practice.

5 Conclusion

In the aviation industry, there are a large number of secondary on-line processing tasks with out-of-tolerance in the assembly process. The content of this paper can meet the engineering needs of the docking processing site. At the same time, the point cloud data preprocessing method and the extraction method of mating surface space contour features in this article are universal, and they also have good reference significance when processing other point data and detecting other feature targets of different shapes. The adaptive processing and process optimization methods of the mating surface can also provide reference for the processing of other mating parts.

Acknowledgments. This work was financially supported by the key research and development program of Shaanxi province (Grant No. 2020ZDLGY06-10, No. 2021GY-302), national defense basic scientific research program of China (Grant No. JCKY2018607C004).

References

1. Butunoi, P.A., Stan, G., Ungureanu, A.L.: Research regarding improvement of torsional stiffness for planetary speed reducers used in the actuation of industrial robots. Appl. Mech. Mater. **809–810**(s5–6), 718–723 (2015)
2. Wang, G., Wu, D., Chen, K.: Current status and development trend of aviation manufacturing robot. Aeronaut. Manufact. Technol. **479**(10), 26–30 (2015)
3. Yan, Q., Chen, W.: Automatic modification of local drilling holes via double pre-assembly holes. World J. Eng. Technol. **03**(3), 191–196 (2015)
4. Sheng, H., et al.: Research on object recognition and manipulator grasping strategy based on binocular vision. Int. J. Front. Med. **2**(2) (2020)
5. Ahn, J.S., et al.: A three-dimensional scanning apparatus based on structured illumination method and its application in dental scanning. In: Applied Optical Metrology II (2017)
6. Gashongore, P.D., et al.: Indoor space 3D visual reconstruction using mobile cart with laser scanner and cameras. In: Eighth International Conference on Graphic and Image Processing (ICGIP 2016) (2017)
7. Deng, C., Lin, H.: Progressive and iterative approximation for least squares B-spline curve and surface fitting. Comput. Aided Des. **47**, 32–44 (2014)
8. Lin, H., Cao, Q., Zhang, X.: The convergence of least-squares progressive iterative approximation for singular least-squares fitting system. J. Syst. Sci. Complex. **31**(6), 1618–1632 (2018). https://doi.org/10.1007/s11424-018-7443-y
9. Lee, S.H., Kim, H.U., Kima, C.S.: ELF-Nets: deep learning on point clouds using extended laplacian filter. IEEE Access **7**, 156569–156581 (2019)
10. Moorfield, B., Haeusler, R., Klette, R.: Bilateral filtering of 3D point clouds for refined 3D roadside reconstructions. In: Azzopardi, G., Petkov, N. (eds.) CAIP 2015. LNCS, vol. 9257, pp. 394–402. Springer, Cham (2015). https://doi.org/10.1007/978-3-319-23117-4_34

Load Balance Optimization Based Multi-robot Cooperative Task Planning for Large-Scale Aerospace Structures

Jiamei Lin, Wei Tian(✉), Pengcheng Li, and Shaorui Lu

Nanjing University of Aeronautics and Astronautics, Nanjing, China
tw_nj@nuaa.edu.cn

Abstract. Regarding the manufacturing and assembling requirements of large-scale structures in the aerospace industry, the existing large-scale machine tool integral manufacturing and offline split manufacturing methods have multiple limitations in cost, efficiency and quality. Taking advantages of flexibility, multi-robot system can be used for in-situ machining of large cabin, which has received extensive attention in the field of aerospace manufacturing. However, the research on multi-robot cooperative task planning is still insufficient, and there are problems such as uneven task allocation and difficult to eliminate interference between robots. This paper proposes a multi-robot cooperative task planning method to solve the problem of unbalanced load of multi-robot systems in the processing of large structures. The proposed algorithm utilizes the K-Means clustering principle to determine the base position of robots according to the density of task. The auction algorithm combined with the load distribution is studied to coordinate the assignment of tasks. The collision avoidance algorithm based on the unilateral coordination strategy is proposed to optimize the processing path of robots. The effectiveness of the algorithm in terms of time efficiency under different working conditions is demonstrated by the simulation of the large wing structure assembly. The results proved that the proposed algorithm effectively reduces the completion time and achieves the load balance of the multi-robot system. The proposed algorithm plans a conflict-free path and improves the efficiency of the multi-robot system.

Keywords: Multi-robot system · Base position planning · Auction algorithm · Path planning · Collision avoidance

1 Introduction

In the manufacturing process of core equipment in major engineering fields such as aerospace field, shipbuilding field and other major engineering fields, there is a kind of large structures with large space size and high precision requirements. In this context, the study of using multiple robots to perform in-situ processing of large structures is a new way to break through the limit of product size and solve the problems of high-precision, high-efficiency and flexible manufacturing. Because of the large size of structures and

X.-J. Liu et al. (Eds.): ICIRA 2021, LNAI 13014, pp. 797–809, 2021.
https://doi.org/10.1007/978-3-030-89098-8_75

the different types of tasks, the robot cannot process all the tasks at a fixed position. It seems very important that quickly and effectively lay out multiple robots according to the allocation of tasks. At the same time, the uneven allocation of tasks on structures lead to different loads among robots. In order to improve processing efficiency of multi-robots, it is necessary to reduce the idle consumption and balance the load of the robots. Therefore, the task allocation and scheduling algorithm should be studied. Moreover, multiple robots share the operating space during task execution, and their processing ranges overlap. In order to ensure the safety and reliability of the task execution, the space contention and interference avoidance problems among multiple robots cannot be ignored.

Aiming at the problem of robot base position planning, the current research methods mainly include grid method [1], hypothetical base method [4–6] and so on. In Literature [2], the workspace and joint angle limits were taken as constraint conditions, and the robot's base position and joint angle were optimized based on the improved genetic algorithm and quasi-Newton algorithm. Literature [3] introduced genetic algorithm on the basis of the above method, which overcomes the shortcomings of traditional nonlinear programming algorithm and improves the optimization quality. According to the base position planning problem of spraying robot in literature [4–6], a reverse motion chain of the robot was established with wrist center as an imaginary base. The feasible area of the position was obtained by solving the spatial intersection of each target point, and the position was optimized based on the operability and stability indexes. However, the above methods only concentrate on the station planning of single robot, without considering the space contention in the position planning of multi-robot system. There is no position planning strategy for multi-robot systems in major engineering fields such as aerospace at present.

Aiming at the task allocation problem of multi-robot scheduling plan coupling, currently there are mainly hierarchical auctions [7, 8], time hierarchical trees [9], local centralized assignment [10] and other algorithms. In addition to the above methods, literature [11] proposed a multi-robot coordinated learning scheduling strategy based on network. Literature [12] proposed a task allocation algorithm for a multi-robot system. The system divides tasks into disjoint groups with priority constraints, which improves the execution ability of the robot group. In reference to the problem of multi-robot task allocation in aircraft structure assembly, literature [13] proposed a scheduling algorithm that uses cooperative robots to balance the workload between robots, and achieves a conflict-free scheduling. Literature [14] proposed a centralized algorithm and developed a multi-agent task sequencer similar to real-time processor scheduling technology, which can handle tightly coupled time and space constraints. However, the above methods have a strong pertinence to solve the problem. They only focus on specific types of inter-task constraints, and have poor performance when dealing with time-sensitive multi-robot cooperative tasks.

In view of the space contention problem in the motion coordination of multi-robot systems, some research methods have been produced. For example, literature [15, 16] proposed to design a resource contention penalty function in the assignment problem. Literature [17] proposed to use the probability of conflict events to analyze specific resource contention. Literature [18] proposed the deployment of shared resources based on mobile agent. However, these methods only assign a single task to the robot, or the robot only performs cyclic and reciprocating tasks, which cannot optimize the time efficiency of long-term cooperative planning.

Based on the research of the above-mentioned literatures, this paper adopted the processing region division strategy based on the clustering principle according to the clustering density of the task, and the task completion rate of the multi-robot system was improved. On this basis, the auction algorithm combined with the load distribution is proposed to coordinate the task allocation and balance the load of the multi-robot system, the distance-based task priority assignment strategy is proposed to reduce the interleaving between robot paths. Aiming at the problem of interference between robots, a collision avoidance algorithm based on unilateral coordination strategy is proposed, which can effectively avoid the collision between robots and improve the processing efficiency of robot system.

The structure of this paper is as follows: Sect. 2 describes the task planning problem in large structure assembly. Section 3 introduces the division method of machining area and the planning of robot base position. Section 4 describes the task auction algorithm based on the different load distribution of robot teams, and introduces the path planning method and collision avoidance strategy. The simulation results are given and discussed in Sect. 5. Section 6 summarizes the whole paper.

2 Problem Statement

This section describes the task allocation and load balance of the multi-robot system in the assembly process of large-scale aerospace structures, and gives the objective function of the problem. As shown in Fig. 1, the uneven task allocation leads to the unbalanced load of the multi-robot system. At the same time, the processing paths of robot 1 and robot 2 are staggered, which will greatly increase the probability of collisions between the two robotic arms. Therefore, it is necessary to carry out research on task allocation strategy and path planning.

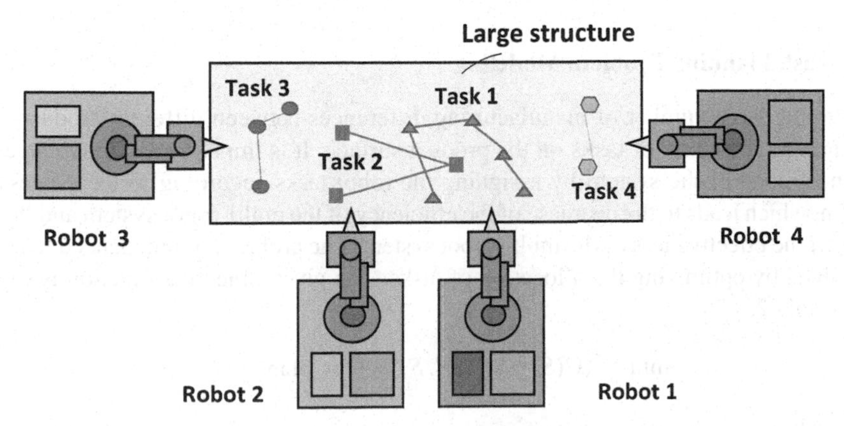

Fig. 1. The large-scale aerospace structure assembled by multi-robot system.

2.1 Processing Area Division

Given a set of n tasks to be processed $\boldsymbol{X}$ and a robot team $\boldsymbol{R}$ with m robots. Due to the large size of structures and uneven allocation of tasks, the base position planning of robot will directly affect the completion rate of tasks, and the base position of robot is related to the division of processing area. The traditional processing area is divided evenly according to the area of the structure, without considering the density of tasks. To solve the problem of low completion rate of task, it can be solved by dividing the processing area in combination with task density. The given objective function is shown in formula 1:

$$\max \sum_{i=1}^{m} \text{reach}(r_i, \boldsymbol{C}_i),\ r_i \in \boldsymbol{R},\ \boldsymbol{C}_i \subseteq \boldsymbol{C} \tag{1}$$

subject to

$$\bigcup_{i=1}^{m} \boldsymbol{C}_i = \boldsymbol{X}$$

$$\boldsymbol{C}_i \neq \emptyset (i = 1, 2, \cdots, m)$$

$$\forall \boldsymbol{C}_i, \boldsymbol{C}_j \subseteq \boldsymbol{C}, i \neq j, \boldsymbol{C}_i \cap \boldsymbol{C}_j = \emptyset$$

where $\boldsymbol{C} = \{\boldsymbol{C}_1, \boldsymbol{C}_2, \cdots, \boldsymbol{C}_{\text{m}}\}$ is m divisions of n tasks, $\boldsymbol{C}_i = \left\{x_1^i, x_2^i, \cdots, x_{n_i}^i\right\}$ is the set of tasks in the i-th division, including the machinable and non-machinable tasks of the robot r_i, and each set is not empty, and there can only be one division for a task in. In formula 1, reach$(r_i, \boldsymbol{C}_i)$ is the statistics of the number of machinable tasks in the division $\boldsymbol{C}_i$ of the robot r_i. In order to achieve the highest completion rate of the total tasks, a suitable task division should be found and the base position of each robot should be determined according to the division of the processing area.

2.2 Task Planning Problem Modeling

Due to the large number of manufacturing differences between different products and the uneven allocation of tasks on the product surface. It is impossible to guarantee the balanced load of the system by assigning the robot tasks according to the processing region, which leads to the decrease of the efficiency of the multi-robot system and failure to exert the effectiveness of the multi-robot system. The problem of unbalanced load can be solved by optimizing the allocation of tasks. The given objective function is shown in formula 2:

$$\min \text{S}^2(\text{C}(\boldsymbol{S})), \boldsymbol{S}_j \subseteq \boldsymbol{S}, \boldsymbol{S}_j = \text{Opt\_plan}\left(r_j, \boldsymbol{T}_j\right) \tag{2}$$

subject to

$$\boldsymbol{T}_j = \left\{x_1^j, x_2^j, \cdots, x_{j-1}^j, x_{j+2}^j, \cdots, x_{n_j}^j, x_k^i\right\}, \boldsymbol{T}_j \subseteq T$$

$$\boldsymbol{K}_j = \left\{x_j^j, x_{j+1}^j\right\}, \boldsymbol{K}_j \subseteq \boldsymbol{K}$$

$$\boldsymbol{T} \cup \boldsymbol{K} = \boldsymbol{X}$$

where $\boldsymbol{T}_j$ is the task set of the robot, which includes all the machinable tasks of the robot r_j in the division $\boldsymbol{C}_j$, and the tasks x_k^i that the robot r_i cannot be machined but the robot r_j can be machined in the division $\boldsymbol{C}_i$, and $\boldsymbol{K}_j$ is the unmachinable task set of the robot r_j and other robots. All the set of machinable tasks $\boldsymbol{T}$ and the set of unmachinable tasks $\boldsymbol{K}$ constitute all the tasks set $\boldsymbol{X}$. In formula 2, $\boldsymbol{S}_j = \text{Opt\_}plan(r_j, \boldsymbol{T}_j)$ is the schedule of robot r_j for its task set $\boldsymbol{T}_j$, $C(\boldsymbol{S})$ is the function of robot scheduling cost, and $\mathrm{S}^2(C(\boldsymbol{S}))$ is the variance function of multi-robot system scheduling cost. The task planning modeling should find an appropriate task allocation method, so that the load of the robots can achieve the maximum balance and improve the processing efficiency of the system.

3 Base Position Planning Strategy of Multi-robot System

Suppose the task set is $\boldsymbol{X} = \{x_1, x_2, \cdots, x_n\}$, where x_i is the task to be processed, and m is the number of initial clusters. By calculating the distance between each task and the clustering center, the division region of the task is determined, so as to minimize the total distance J_C of each task to the clustering center:

$$J_C = \sum_{k=1}^{m} \sum_{x_i \in \boldsymbol{C}_k} distance(x_i, z_k)$$

where z_k is the center of the k-th cluster, and its calculation formula is

$$z_k = \frac{x_1^k + x_2^k + \cdots + x_{n_k}^k}{n_k}$$

distance(x_i, z_k) is the distance from the task x_i to the clustering center

$$\text{distance}(x_i, z_k) = \|x_i - z_k\|$$

The processing area is divided according to the task density, and the independent task clustering division $\boldsymbol{C} = \{\boldsymbol{C}_1, \boldsymbol{C}_2, \cdots, \boldsymbol{C}_\mathrm{m}\}$ is obtained. The coordinates of each clustering center z_i are projected to the structure frame direction to obtain the fuzzy position area Position 1–4 of the robot r_i (the number of fuzzy position area of robot is related to the shape of large structures). By judging the number of tasks that can be processed by the robot r_i in each fuzzy position area, a preferred position area Position a is determined to satisfy formula 3:

$$a = \arg\max_j(\text{reach}(r_i^j, X)) \tag{3}$$

where reach(r_i^j, X) is the number of tasks that can be processed by the robot r_i in the fuzzy position area Position j. The fuzzy position area Position a with the largest number of tasks is selected as the preferred position area, so that the number of tasks that can be processed by the robot at this position is better than other fuzzy position areas. Finally, the optimal position of the robot is determined by combining the position and pose of the robot during the processing task in the preferred position area.

4 Load Distribution Optimization Based Task Allocation Method of Multi-robot System

4.1 Task Reallocation Algorithm

Through the clustering of tasks, the robots obtain the tasks in their respective divisions, but this division may include tasks that the robot cannot process. In order to improve the task completion rate of the robot system, we redistribute the tasks that cannot be processed in the clustering division, and divide these tasks to other robots to achieve the goal of completing the total processing tasks to the greatest extent. The above task allocation method is shown in Algorithm 1.

Algorithm 1: Reallocation process of unable to process tasks

1. **for each** $x_j^i \in \boldsymbol{C}_i$, $\boldsymbol{C}_i \subseteq \boldsymbol{C}$ **do**
2.   **if** reachability$\left(x_j^i, r_i_base\right) = 1$ **then**
3.   add x_j^i to $\boldsymbol{T}_i$, $\boldsymbol{T}_i = \boldsymbol{T}_i \cup x_j^i$
4.   **else**
5.   add x_j^i to $\boldsymbol{K}_i$, $\boldsymbol{K}_i = \boldsymbol{K}_i \cup x_j^i$
6.   **end if**
7. **end for**
8. **for each** $x_j^i \in \boldsymbol{K}_i$, $\boldsymbol{K}_i \subseteq \boldsymbol{K}$ **do**
9.   **for each** $r_k \in \boldsymbol{R} \setminus r_i$ **do**
10.   **if** reachability$\left(x_j^i, r_k_base\right) = 1$ **then**
11.   add r_k to *candidate*
12.   **end if**
13.   **end for**
14. $a = \arg\min_k \mathrm{C}(\boldsymbol{S}_k), r_k \in candidate$
15.   add x_j^i to $\boldsymbol{T}_a$, $\boldsymbol{T}_a = \boldsymbol{T}_a \cup x_j^i$
16.   delete x_j^i from $\boldsymbol{K}_i$, $\boldsymbol{K}_i = \boldsymbol{K}_i \setminus x_j^i$
17. **end for**

In line 1 and 2, the robot judges the machinability of the tasks in their respective clusters. In line 3, if the task can be processed, add it to the machinable task set. In line 4–7, otherwise add it to the unmachinable task set. In line 8–9, take out task from the set of unmachinable tasks in sequence. In line 10, the remaining robots judge the machinability of the task. In line 11–13, if the task can be processed, the robot is added to the ranks of candidate robots. In line 14, select the robot with the least cost from the candidate robots to obtain this task. If the candidate robot set is empty, the new task is auctioned. In line 15–17, add this task to the new task set, and delete this task from the unmachinable task set.

4.2 Task Auction Algorithm Combined with Load Distribution

In order to solve the problem of unbalanced load of multi-robot system and increase the effectiveness of multi-robot system, this paper adopts auction algorithm based on market method to balance the load between multi-robots. By optimizing the assignment order of tasks, the interference between robot paths is reduced. In view of the unbalanced load distribution of multi-robot system, a priority bidding strategy combined with bidding levels is proposed. **Method 1** is an auction algorithm where robots have different levels of bidding rights. **Method 2** is an auction algorithm where robots have the same level of bidding rights. In the task auction process, the specific applicable method is selected according to the load distribution of the robot system.

Method 1: Single-Robot Bidding. The task is auctioned by the robot with high load, and the other robots bid and have different levels of bidding rights. The lower the load, the higher the bidding rights of the robot. At the beginning of each round of auction, the robots bid in order according to the bidding level. If the high-level robot trades successfully, the low-level robot will not bid. After each auction is successful, the ownership of the auction rights will be determined again. The single-robot bidding auction algorithm is shown in Algorithm 2.

Algorithm 2: Single-robot bidding auction algorithm

1. **for each** $r_i \in \boldsymbol{R}$ **do**
2. $\boldsymbol{T}_i = \boldsymbol{T}_i^{sequence}$ //sort $\boldsymbol{T}_i$ according to distance(x_j^i, r_i_base)
3. $\boldsymbol{S}_i = \text{Opt\_plan}(r_i, \boldsymbol{T}_i)$
4. **end for**
5. $a = \arg\max_i \mathrm{C}(\boldsymbol{S}_i)$, $r_a \to auctioneer$
6. $b = \arg\min_i \mathrm{C}(\boldsymbol{S}_i)$, $r_b \to bidder$
7. r_a take task x_i^a from out $\boldsymbol{T}_a$
8. **for** each task x_i^a from $\boldsymbol{T}_a$ **do**
9. **if** reachability(x_i^a, r_b_base) $== 1$ and $\mathrm{C}(\boldsymbol{S}_b \cup x_i^a) < \mathrm{C}(\boldsymbol{S}_a)$ **then**
10. add x_i^a to $\boldsymbol{T}_b$, $\boldsymbol{T}_b = \boldsymbol{T}_b \cup x_i^a$
11. $\boldsymbol{T}_b = \boldsymbol{T}_b^{sequence}, \boldsymbol{S}_b = \text{Opt\_plan}(r_b, \boldsymbol{T}_b)$
12. delete x_i^a from $\boldsymbol{T}_a, \boldsymbol{T}_a = \boldsymbol{T}_a \setminus x_i^a$
13. $\boldsymbol{T}_a = \boldsymbol{T}_a^{sequence}, \boldsymbol{S}_a = \text{Opt\_plan}(r_a, \boldsymbol{T}_a)$
14. $a = \arg\max_i \mathrm{C}(\boldsymbol{S}_i)$, $r_a \to auctioneer$
15. $b = \arg\min_i \mathrm{C}(\boldsymbol{S}_i)$, $r_b \to bidder$
16. **end if**
17. **end for**

In line 1–4, sort the tasks of each robot according to the distance from its corresponding robot base position. In line 5, the robot with the highest cost is selected as the auctioneer. In line 6, the robot with the lowest cost is selected as the bidder. In line 7–8, the robot with the highest cost takes out the task with the farthest distance from the base position for auction. In line 9, the robot with the lowest cost determines whether the task is within its reach, and whether the cost after adding the task exceeds the current maximum cost. In line 10–13, if the task is reachable and the cost does not exceed the current maximum cost, the robot with the lowest cost will add this task to its task set, delete this task from the original task set, reorder the processing task sequence of the robot, and update the cost of the multi-robot system. In lines 14–16, select the current highest-cost and lowest-cost robots as the auctioneer and bidder for the new round of auction. If the task of the highest-cost robot is traversed once, and none of the robots with the lowest cost has performed a task auction, the robot with the second lowest cost has the priority to bid, and the auction process in line 7–16 is repeated until no auction task occurs.

This method is suitable for situations where the lowest load value in a multi-robot system is far away from other load values. It can quickly achieve the load balance of the multi-robot system. However, the task assignment result under this method will change with the change of the number of iterations, and the task processing path will be staggered. Method 1 is suitable for the situation that the task size is small.

Method 2: Multi-robot Bidding. The task is auctioned by the robot with high load, and the remaining robots bid and have the same level of bidding rights. At the beginning of each round of auction, the remaining robots bid at the same time, and the robot with the lowest value of load will get this task. After each auction is successful, the ownership of the auction rights will be determined again. The multi-robot bidding auction algorithm is shown in Algorithm 3.

Algorithm 3: Multi-robot bidding auction algorithm

1. **for each** $r_i \in \boldsymbol{R}$ **do**
2. $\boldsymbol{T}_i = \boldsymbol{T}_i^{sequence}$ // sort $\boldsymbol{T}_i$ according to distance(x_j^i, r_i_base)
3. $\boldsymbol{S}_i =$ Opt_plan($r_i, \boldsymbol{T}_i$)
4. **end for**
5. $a = \arg\max_i \mathrm{C}(\boldsymbol{S}_i)$, $r_a \to auctioneer$
6. r_a take task x_i^a from out $\boldsymbol{T}_a$
7. **for each task** x_i^a **from** $\boldsymbol{T}_a$ **do**
8. **for each** $r_j \in \boldsymbol{R} \setminus r_a$ **do**
9. **if** reachability(x_i^a, r_j_base) $== 1$ **and** $\mathrm{C}(\boldsymbol{S}_j \cup x_i^a) < \mathrm{C}(\boldsymbol{S}_a)$ **then**
10. add r_j to *candidate*
11. **end if**
12. **end for**
13. $b = \arg\min_i \mathrm{C}(\boldsymbol{S}_i \cup x_i^a), \boldsymbol{S}_i =$ Opt_plan$(r_i, \boldsymbol{T}_i), r_i \in candidate, r_b \to bidder$
14. add x_i^a to $\boldsymbol{T}_b$, $\boldsymbol{T}_b = \boldsymbol{T}_b \cup x_i^a$
15. $\boldsymbol{T}_b = \boldsymbol{T}_b^{sequence}, \boldsymbol{S}_b =$ Opt_plan$(r_b, \boldsymbol{T}_b)$
16. delete x_i^a from $\boldsymbol{T}_a, \boldsymbol{T}_a = \boldsymbol{T}_a \setminus x_i^a$
17. $\boldsymbol{T}_a = \boldsymbol{T}_a^{sequence}, \boldsymbol{S}_a =$ Opt_plan$(r_a, \boldsymbol{T}_a)$
18. $a = \arg\max_i \mathrm{C}(\boldsymbol{S}_i)$, $r_a \to auctioneer$
19. **end for**

In line 1–4, sort the tasks of each robot according to the distance from its corresponding robot base position. In line 5, the robot with the highest cost is selected as the auctioneer. In line 6–7, the robot with the highest cost takes out the task with the farthest distance from the base position for auction. In line 8–12, the remaining robots judge the reachability and cost of this task. If it is reachable and the cost after adding this task is less than the current maximum cost, the robot will be added to the ranks of candidate bidders. Otherwise, the robot will exit this time task bidding. In line 13, select the robot with the least cost from the candidate bidders to become the bidder and obtain this task. In line 14–17, add this task to the task set of the bidder, delete this task from the original task set. The task sequence of the robot is reordered, and the cost of the robot is updated. In line 18, the robot with the highest current cost is selected as the new auctioneer, and the auction process in line 6–18 is repeated until no auction task occurs.

This method is suitable for situations where the highest load value in the multi-robot system is far away from other load values. During the bidding process, the multi-robot system has achieved the multicomputer communication. Each robot selects the most suitable task, although increase the complexity of the bidding process, but reduces interleaving between paths. The results of the task allocation will not change with the change of the number of iterations.

5 Simulation and Analysis

The proposed algorithm has been tested in the large wing structure assembly mission. The multi-robot system needs to handle the drilling and fastening tasks distributed on the wing. The robot operates on the wing and all the tasks are distributed in the area of 6 m × 3 m on the upper surface of the wing. The goal of this task is to minimize the time consumption of completing all tasks, balance the load of the multi-robot system, and improve the processing efficiency of the system so that the product can be delivered faster. A homogeneous robot team is considered in the simulation and all tasks are identical, so the time takes to complete any task for any robot is equal.

This algorithm is compared with the clustering algorithm and the task allocation algorithm based on uniform partition:

Clustering Algorithm: The task is divided into areas according to the task aggregation density, the robot selects the base position according to the clustering center. And the task clustering result is the task allocation results, there is no task allocation optimization process. When there is interference between robots, it is reflected in the load of the robots by increasing the time to deal with the failure.

Task Allocation Algorithm Based on Uniform Partition: According to the size of the structure, the processing area is evenly divided and the robots are evenly placed. The tasks of each robot are the tasks in each uniform partition, there is no task allocation optimization process. When there is interference between robots, it is reflected in the load of the robots by increasing the time to deal with the failure.

5.1 Simulation 1

The performance of three algorithms under a variant number of tasks is tested in the first place. In each simulation, tasks are randomly allocated in a rectangular area, and the four robots determine their base positions according to the corresponding algorithm. The moving speed of the robot actuators is 1 m/s, and the processing time for each task is 15 s. Three examples of the number of tasks were used which is 100 in **Simulation 1.1**, 200 in **Simulation 1.2**, and 300 in **Simulation 1.3**. Different task numbers represent different levels of spatial competition, and the larger number of tasks means the more interference may occur. 100 random allocation instances were generated and used in each set of simulations so that we can statistically compare the performance of different algorithms. Boxplot is a statistical graph used to show the allocation of a set of data, in which the center mark of the box represents the median value of thc data, and the bottom and upper edges represent 25% and 75% of the data values, respectively.

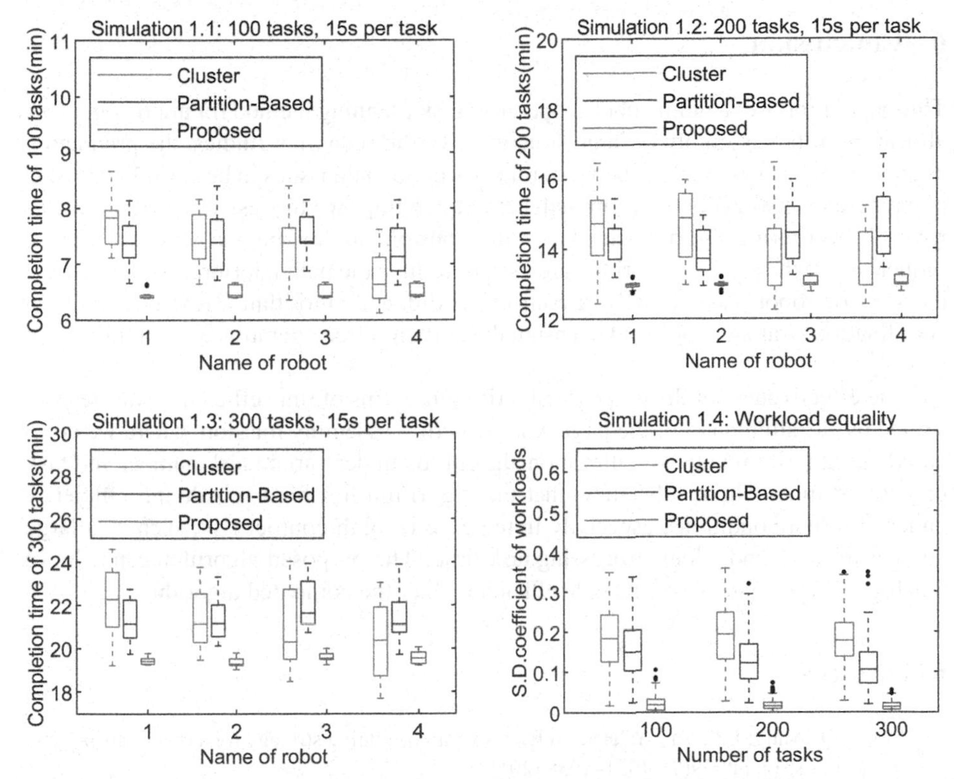

Fig. 2. Comparison of the completion time of the three algorithms under different number of tasks.

The result is shown in Fig. 2. The **Simulation 1.1** to **1.3** are the task completion time of four robots under different task numbers, and the **Simulation 1.4** is the standard deviation coefficient of workloads of the robot team under the three algorithms. According to **Simulation 1.1** to **1.3**, it can be observed that as the number of tasks increases, the space contention problem of the multi-robot system becomes more and more obvious, and the task completion time continues to increase. When using the clustering algorithm, the load of the robot is related to the allocation of tasks and does not change with the increase of the number of tasks. When using the task allocation algorithm based on uniform partitions, as the number of tasks increases, the allocation of tasks in each partition tends to be uniform, and the load of the multi-robot system tends to be balanced. But the dynamically changing load cannot be effectively balanced between the robots when interference occurs, resulting in great fluctuations in the actual load of the robot system. The algorithm proposed in this paper shows good load uniformity in the task allocation stage, and when there is interference between robots, the actual task completion time is much shorter than the other two algorithms. In **Simulation 1.4**, we can observe the allocation of the standard deviation coefficients of the workload between the robots under the three algorithms. The method proposed in this paper is superior to the other two algorithms in terms of load uniformity and scheduling efficiency.

6 Conclusion

This paper proposes a multi-robot cooperative task planning method for multi-robot task allocation in the assembly of large structures. In the scenarios studies, the production efficiency, as well as conflicts between robots are the main issues to be considered when planning a teamwork. In our proposed method, we improve the task completion rate of robots by clustering algorithm and task redistribution. Optimizing the allocation of task sequence and the auction algorithm are used to reduce the path interference and balance the load of robot team. Finally, a collision avoidance algorithm based on unilateral coordination strategy is adopted to ensure the collision free operation of the multi-robot team.

The effectiveness of the proposed algorithm in terms of time efficiency was demonstrated by the simulation of the large wing structure assembly mission, where we compared our algorithm with two allocation algorithms under variant task number and task processing time. The results show that the algorithm has high scheduling efficiency under different conditions, especially in the case of high conflict risk, such as a large number of tasks and a long processing task time. The proposed algorithm can achieve much greater advantage on schedule efficiency than the compared algorithms.

References

1. Lin, X.Q., Yang, J.Z., Yue, Y., et al.: A base position planning strategy for a mobile inspection robot. J. Astronaut. **39**(9), 1031–1038 (2018)
2. Mitsi, S., Bouzakis, K.D., Sagris, D., et al.: Determination of optimum robot base location considering discrete end-effector positions by means of hybrid genetic algorithm. Robot. Comput. Integr. Manuf. **24**(1), 50–59 (2008)
3. Wang, J., Cao, C., Ding, W., et al.: Layout optimization on dual-robot machining center based on genetic algorithm. China Mech. Eng. **27**(2), 173–178 (2016)
4. Ren, S., Yang, X., Xu, J., et al.: Determination of the base position and working area for mobile manipulators. Assem. Autom. **36**(1), 80–88 (2015)
5. Ren, S.N., Yang, X.D., Wang, G.L., et al.: Base position planning of mobile manipulator for large parts painting. J. Jilin Univ. (Eng. Technol. Edn.) **46**(6), 1995–2002 (2016)
6. Shunan, R.Y., et al.: A method for optimizing the base position of mobile painting manipulators. IEEE Trans. Autom. Sci. Eng **14**(1), 370–375 (2017)

7. Jones, E.G., Dias, M.B., Stentz, A.: Tiered auctions for multi-agent coordination in domains with precedence constraints. In: Proceedings of the 26th Army Science Conference (2008)
8. Jones, E.G., Dias, M.B., Stentz, A.: Time-extended multi-robot coordination for domains with intra-path constraints. Auton. Robot. **30**(1), 41–56 (2011)
9. Lemaire, T., Alami, R., Lacroix, S.: A distributed tasks allocation scheme in Multi-UAV context. In: IEEE International Conference on Robotics and Automation, New Orleans, LA, USA, 2004. Proceedings ICRA, vol. 4, pp. 3622–3627 (2004)
10. Mackenzie, D.C.: Collaborative tasking of tightly constrained multi-robot missions.In: Multi-Robot Systems: From Swarms to Intelligent Automata, Washington D.C. Proceedings Second International Workshop on Multi Robot Systems, vol. 2, pp. 39–50 (2003)
11. Wang, Z., Gombolay, M.: Learning scheduling policies for multi-robot coordination with graph attention networks. IEEE Robot. Autom. Lett. **5**(3), 4509–4516 (2020)
12. Luo, L., Chakraborty, N., Sycara, K.: Multi-robot assignment algorithms for tasks with set precedence constraints. In: Proceedings of IEEE International Conference on Robotics and Automation, pp. 2526–2533 (2011)
13. Tereshchuk, V., Stewart, J., Bykov, N., Pedigo, S., Devasia, S., Banerjee, A.G.: An efficient scheduling algorithm for multi-robot task allocation in assembling aircraft structures. In: IEEE Robot. Autom. Lett. **4**, 3844–3851 (2019)
14. Gombolay, M.C., Wilcox, R.J., Shah, J.A.: Fast scheduling of robot teams performing tasks with temporospatial constraints. IEEE Trans. Robot. **34**(1) 220–239 (2018)
15. Nam, C., Shell, D.A.: Assignment algorithms for modeling resource contention in multirobot task allocation. IFFF. Trans. Autom. Eng. **12**(3), 889–900 (2014)
16. Nam, C., Shell, D.A.: Assignment algorithms for modeling resource contention and interference in multi-robot task allocation. In: IEEE International Conference on Robotics and Automation (ICRA), Hong Kong, pp. 2158–2163. IEEE (2014)
17. Palmer, A.W., Hill, A.J., Scheduling, S.J.: Modelling resource contention in multi -robot task allocation problems with uncertain timing. In: 2018 IEEE International Conference on Robotics and Automation (ICRA), Brisbane, pp. 3693–3700 (2018)
18. Semwal, T., JhaS, S., Nair, S.B.: On ordering multi-robot task executions within a cyber physical system. ACM Trans. Auton. Adapt. Syst. **12**(4), 1–27 (2017)

Surface Defect Detection of High Precision Cylindrical Metal Parts Based on Machine Vision

YuJie Jiang[1], Chen Li[2], Xu Zhang[1(✉)], JingWen Wang[1,2], and ChuZhuang Liu[1,2]

[1] Shanghai University, School of Mechatronic Engineering and Automation, Shanghai, China
xuzhang@shu.edu.cn

[2] Haiphong University of Science and Technology, School of Mechanical Science and Engineering, Wuhan, China

Abstract. The surface quality of high precision cylindrical metal parts is an important index to measure its quality. Most of the existing detection methods still use manual visual inspection. Manual detection is inefficient and difficult to ensure the standard of detection. It is difficult to make an effective judgment for the defects in the critical index, and it is more prone to miss detection and misjudgment. In this paper, the seamless steel pipe used for the shock absorber of bike is taken as the main research object, and machine vision is used for its surface defecting. Combined with the characteristics of arc and high reflection on the surface of steel pipe, an image acquisition and processing system composed of linear light source, linear array camera, encoder and rotation system is proposed. Refer to the national standard GB/T9797-2005, the defects mainly include pit, spalling, pitting, speckle, which is determined by Fourier transform, gradient threshold, and line detection by their four different characteristics. Finally, a complete experimental platform with clamping, blowing, detection, and classification functions is built to test. The experimental results show that the stability, accuracy and detection efficiency of the steel pipe detection system based on machine vision is high, which can meet the needs of daily production detection.

Keywords: Cylindrical metal parts · Machine vision · Fourier transform · Gradient threshold · Line detection · Edge detection

1 Introduction

With the rapid development of industry, the use of cylindrical metal parts has become increasingly broad, and the surface quality of cylindrical metal parts is an important index to measure the overall quality of cylindrical metal parts. At the same time, property mutation, fatigue damage and corrosion are often concentrated in this area, which greatly reduces the working performance of

X.-J. Liu et al. (Eds.): ICIRA 2021, LNAI 13014, pp. 810–820, 2021.
https://doi.org/10.1007/978-3-030-89098-8_76

cylindrical metal parts in the complex and harsh environment [1]. At present, the vast majority of cylindrical metal parts plants in the industry use the traditional manual visual inspection method or traditional non-destructive testing method in the detection of surface defects [2]. The manual visual inspection method is not only low speed and efficiency, but also difficult to ensure the standard of detection. It is difficult to make an effective judgment for the defects in the critical index, and it is more prone to miss detection and misjudgment. The speed of traditional detection methods and the kinds of defects that can be detected is limited, so it is impossible to evaluate the surface quality of products comprehensively. The above two traditional detection methods can not meet the requirements of height, high precision and high accuracy in the industry. With the development of computer and lens hardware in recent years, machine vision detection technology is growing rapidly. Machine vision system works for a long time, which can ensure the stability of detection and labeling, and is suitable for long-time observation, analysis and recognition tasks.

Using machine vision to detect defects is widely used in various fields, and there are many references in the aspect of algorithm analysis. Zhou F et al. [3] proposed an automatic surface defect detection method based on bilinear model to solve the problems of complex texture and changeable interference factors in metal surface defect detection. Renukalatha S et al. [4] Based on the principle of support vector machine to detect and diagnose glaucoma diseases. Li K et al. [5] proposed a method for detecting the surface defects and size of graphite sealing ring based on machine vision, and achieved good results through the combination of Canny algorithm and template matching. Du-MingTsai et al. [6] proposed a morphological method to detect surface defects with circular tool marks.

For the object with curved and highly reflective surface, the imaging difficulty is greater than that of plane. The imaging effect of incident light on the convex surface is worse, which further increases the difficulty of image algorithm. The existing detection theory based on computer vision can be divided into three categories: laser scanning method [7], area array CCD imaging method [8,9], and linear array CCD imaging scanning method [10]. Therefore, this paper proposes to use the combination of linear array camera and linear light source to obtain the image. As long as the incident light angle is reasonable and the intensity distribution is uniform, the ideal surface defect image can be obtained, and then it can be recognized.

2 Image Acquisition

2.1 Design of Image Acquisition System

The diameter of the tested seamless steel pipe is 2.5 cm and the length is 30 cm to 40 cm. It is mainly used as the supporting steel pipe in the shock absorption of bicycles. The surface of steel pipe is a curved surface, which is more difficult to image than plane. When the area array light source is used, the arc-shaped outer surface makes the distance between the center of the light source and all parts of the irradiation area too large. At this time, if the area array camera is used directly, there will be too much redundant information, so the steel pipe image needs to use multiple images for splicing. At the same time, the light intensity is inconsistent on the curved steel pipe surface, and the light compensation is also needed for the steel pipe image. It greatly increases the difficulty of image processing and defect location. The combination of linear array camera and linear light source can not only greatly improve the image quality, but also speed up the image acquisition and processing (Fig. 1).

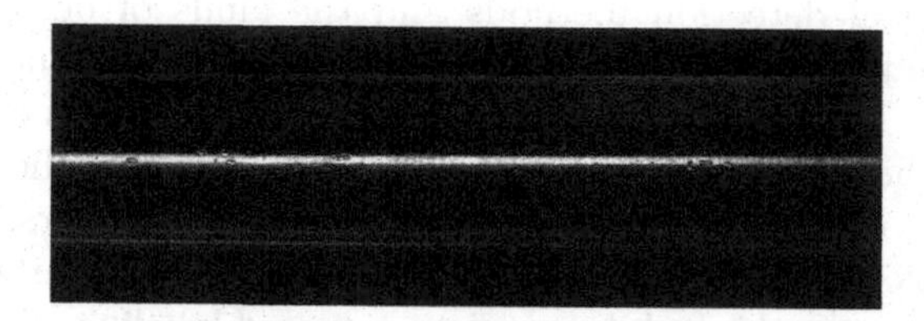

Fig. 1. Steel tube shot by area array camera

The design detection accuracy of steel pipe defect detection is required to be 1 mm. The GigE linear array camera produced by DALSA is selected, and its resolution is 4096 × 1 pixels. When the working distance of camera is 75 cm, the field of view H = 4000 mm. According to the formula, the minimum detection accuracy can be calculated (Fig. 2)

$$Rs = \frac{H}{N} = \frac{4000}{4096} = 0.97\,\text{mm} \tag{1}$$

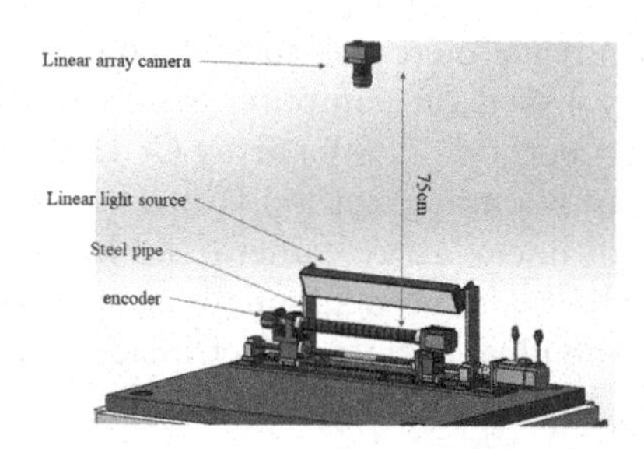

Fig. 2. Schematic diagram of detection station

Under the premise of ensuring the field of view width to meet the whole length of steel pipe, the minimum defect size Rs can meet the minimum detection accuracy, so the camera and lens meet the requirements. The encoder uses Omron's three wire DC encoder, which acquires the image in the mode of hard trigger. When the camera receives the rising edge of the signal from the encoder, it will trigger a shot. The circumference of the steel pipe is about 5 cm, and 1000 images are shot 1000 times in a rotation cycle, that is, one image is shot every

$50\,\mu$ of the steel pipe, and the image is automatically synthesized and processed by the linear array camera, thus the curved surface of the steel pipe is expanded into a plane image.

2.2 Types and Causes of Defects

According to the national standard GB/T9797-2005, the surface coating of steel pipe shall be bright, and there shall be no spots, pits, speckle, pitting and other defects. Therefore, the surface of seamless steel pipe should be smooth with uniform reflectivity. In the dark field, the smooth plane produces diffuse reflection and makes the image dark, while the defect image shows different light intensity (Fig. 3).

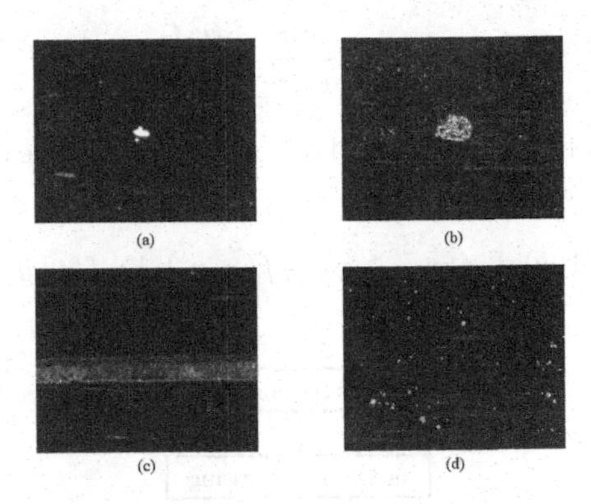

Fig. 3. Pictures of defects (a) Pit (b) Speckle (c) Spalling (d) Pitting

Pits are local depressions distributed on the surface of steel pipe with different areas, which are high brightness dots in the image; The spot is the rust spot formed by corrosive substances such as water during storage or processing. The image shows a round block bright spot, which is less bright than the reflection of pits; Spalling is the continuous surface spalling caused by the scratch of steel pipe, which is shown as a strip-shaped bright stripe on the image; Pockmarks are small pits on the surface, which have many bright spots in an area.

3 Image Processing

The surface defect detection of steel pipe is mainly realized by using complex image processing technology. There are mainly four kinds of defects on the surface of steel pipe, such as spots, pits, spalling and pitting. Firstly, preprocessed is used on the acquired surface image. According to the different characteristics of each defect, Fourier transform, gradient threshold and line detection are used to determine the defect. The algorithm processing is shown in the block diagram (Fig. 4).

3.1 Image Enhancement

In order to make the details of the image more prominent, image enhancement is used to enhance the contrast of the details of the image. Image enhancement algorithms [11] include: image enhancement based on histogram equalization, which is suitable for low contrast images; The image enhancement based on Laplace operator increases the sharpness of the whole image; Image enhancement based on logarithmic transformation is used to emphasize the low gray part of the image; Image enhancement based on gamma transform is mainly used for image correction. From the analysis of the original image, the more appropriate enhancement method is Laplacian. Enhancement Laplacian is a kind of second-order differential linear operator. For two-dimensional image f (x, y), the simplest Laplacian of second-order differential is defined as

$$\nabla^2 f = \frac{\partial^2 f}{\partial x^2} + \frac{\partial^2 f}{\partial y^2} \tag{2}$$

Combined with the definition of second order differential, we can get the following results (Fig. 5)

$$\nabla^2 f(x, y) = f(x+1, y) + f(x-1, y) + f(x, y+1), f(x, y-1) - 4f(x, y) \tag{3}$$

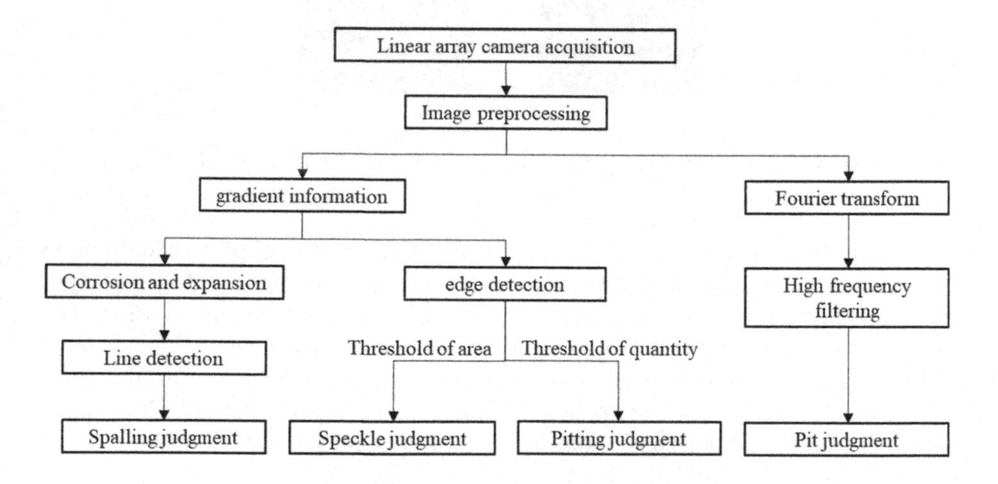

Fig. 4. Algorithm flow chart

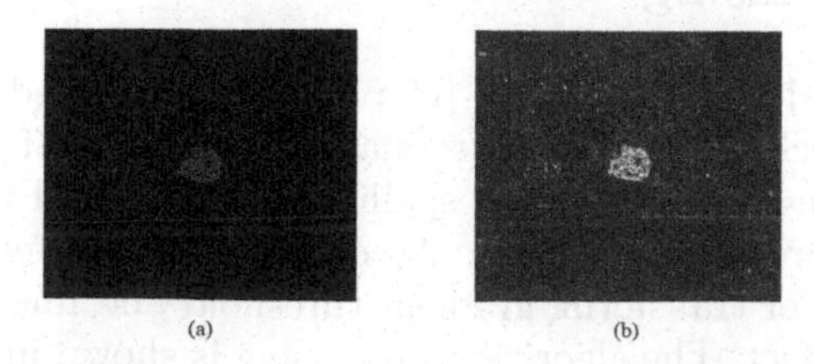

Fig. 5. (a) Before enhancement (b) After enhancement

3.2 Line Detection Based on Gradient Image

Acquisition of Gradient Image. The gradient value of image is the change rate of gray value. The gradient of image function f (x, y) at point (x, y) is a quantity with size and direction. The gradient of the image in position (x, y) can be expressed as a vector

$$\nabla f(x,y) = [Gx, Gy]^T = [\frac{\partial f}{\partial x}\frac{\partial f}{\partial y}] \tag{4}$$

$Gy = \frac{\partial f}{\partial y}$ can be called the first derivative in the y-axis direction, and it can be expressed by

$$dy(i,j) = I(i,j+1) - I(i,j) \tag{5}$$

The x-axis direction is the same. In the spalling image, the gray value along the x-axis direction changes more consistently, while in the y-axis direction, the gray value has obvious mutation edge. Therefore, the edge information of defects can be obtained by calculating the gradient image along the y-axis direction (Fig. 6).

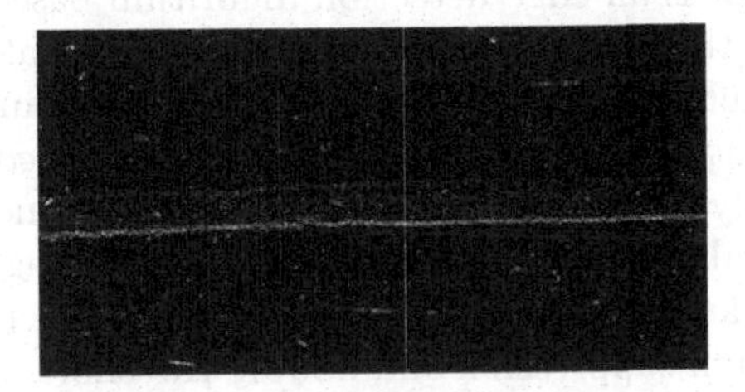

Fig. 6. Y-axis gradient image

Line Detection. The principle of Hough transform [12] is to transform the pixels in the image space into the lines in the parameter space by using the point line duality. The intersection of the middle lines in the parameter space corresponds to the lines in the image space. The problem of line detection in image space is transformed into the problem of cumulative statistics in parameter space. Hough detection sets a threshold T1 for the accumulation of line length. Only the line greater than T1 is regarded as a line, and the lines with similar distance and smaller angle are combined to avoid detecting the same line as multiple lines with smaller angle. After the above processing, an obvious straight line can be observed in the image. Here, we set a higher threshold to detect the straight line in the processed image. If the straight line is detected, it indicates that there is speckle defect in the image of steel pipe (Fig. 7).

Fig. 7. Results of linear detection

3.3 Image Segmentation Algorithm Based on Edge

The segmentation method based on edge detection tries to solve the segmentation problem by detecting the edges of different regions [13], which is one of the most commonly used methods. In the image, we mainly use the intensity of gray value change of pixels on the edge between regions to make judgment, which is one of the main assumptions of edge detection. First order or second order differential operators are often used for edge detection.

The Soble operator is an edge detection algorithm based on the first derivative, which combines the difference operation with the local average. Each point of the image is convoluted by two templates. Take the maximum value of two convolutions as the output value of the point. Laplacian is a second-order differential operator, which mainly uses the characteristic that the second-order differential of image presents zero value at the edge due to the step of image at the edge. Canny edge detection is also a first-order differential operator detection algorithm, based on which non maximum suppression and double threshold are added. Using non maximum suppression can not only effectively suppress multi response edges, but also improve the positioning accuracy of edges. Using double threshold can effectively reduce the missing rate of edges. From the edge results of the three algorithms in the figure below, Canny edge detection has the best effect (Fig. 8).

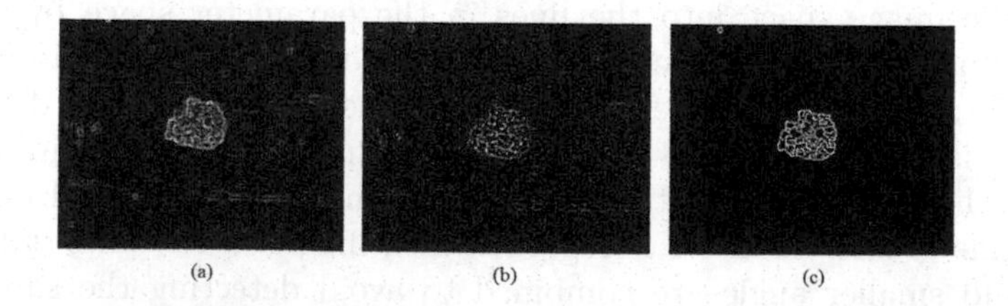

Fig. 8. Edge detection image (a) Soble (b) Laplacian (c) Canny

The minimum bounding rectangle is found for canny processed image. For the calculated rectangle, set the threshold of area size, and convert to the actual distance, that is, the area greater than 0.1 cm^2 specified in the national standard is regarded as the defect of spot. At the same time, the threshold value is set for the number of rectangles, that is, the defects with more than 200 defects are regarded as pockmarks (Fig. 9).

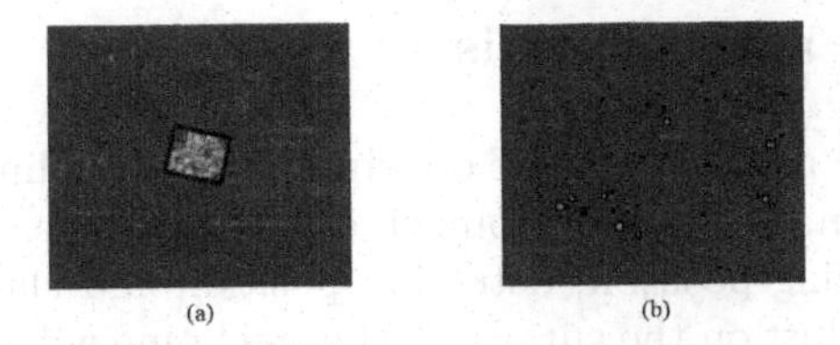

Fig. 9. Pitting and speckle

3.4 Frequency Domain Method

For the discrete signal, the frequency value indicates the intensity of signal transformation or the speed of signal change [14]. The higher the frequency, the more dramatic the transformation, the smaller the frequency, the smoother the signal. In the corresponding image, high frequency signal is often edge signal and noise signal in image, while low frequency signal includes image contour and background light signal with frequent image change. Firstly, the time domain information of the image is converted to the frequency domain by Fourier transform. For the two-dimensional discrete Fourier transform, the definition is

$$X(k,l) = \sum_{m=0}^{M-1}\sum_{n=0}^{N-1} x(n,m)e^{-j\frac{2\pi}{N}kn}e^{-j\frac{2\pi}{M}lm} = \sum_{m=0}^{M-1}\sum_{n=0}^{N-1} x(n,m)w_N^{kn}w_M^{lm} \quad (6)$$

$$w_N = e^{-j\frac{2\pi}{N}}, w_M = e^{-j\frac{2\pi}{M}}$$

The main feature of pits is the highlight area in the image. The image is analyzed in frequency domain. High frequency information is filtered out by using high frequency filter, that is, the position of pits in the image. Then the inverse Fourier transform is used to change the image back to the time domain. At this time, the part left after high-frequency filtering is the image information of pits. At this time, whether there are pits or not is transformed into the problem of whether there is residual image information in the inverse transformed image. The existence of defects can be determined by traversing the image (Fig. 10).

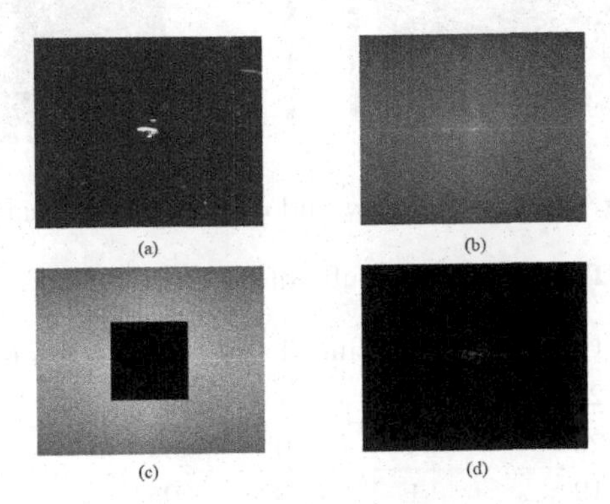

Fig. 10. (a) Original image (b) After fourier transform (c) High frequency filter (d) Inverse Fourier transform

4 Experiment and Analysis

In order to meet the requirements of docking production line to achieve a high degree of automation, the test equipment consists of four stations: lifting and drying position, feeding position, detection position and classification position. Because the oil and dust on the surface of the steel pipe will have a great impact on the image quality of the steel pipe, the rod less cylinder is used to drive the air knife to blow the surface of the steel pipe first. The feeding part uses the cylinder claw to grab the steel pipe from the jacking position to the detection station. In the detection station, the steel pipe is clamped and fixed by the bidirectional ball screw and the thimble at both ends, and a pressure sensor is installed at the forward motor of the screw to control the clamping degree, so as to prevent the scratch of the steel pipe head due to excessive clamping. The classification position is automatically classified according to the results given by the detection algorithm (Fig. 11).

Take 100 seamless steel pipe samples, of which 20 are defect free, 20 are pit defects, 20 are pitting defects, 20 are spot defects, 20 are spalling defects. The test is divided into two times. The first time, each defect type is tested separately to test the recognition accuracy of the algorithm. The second time, all 100 steel pipes are disturbed to test the stability of the algorithm (Table 1).

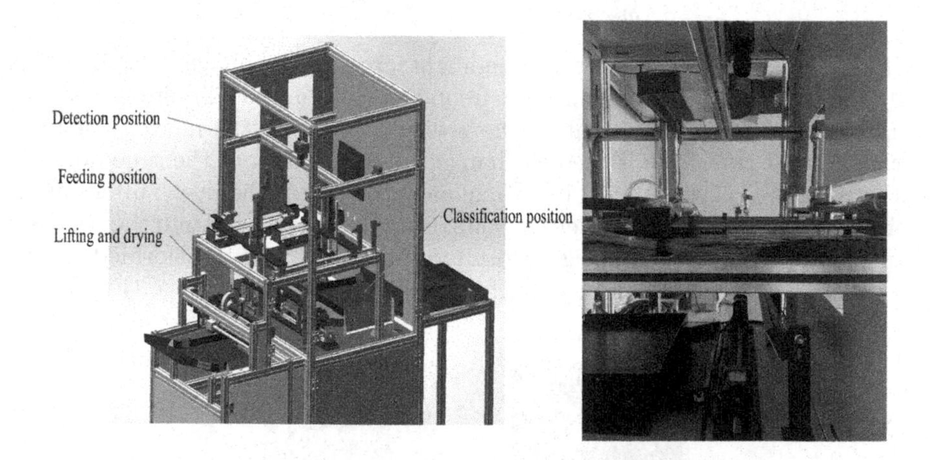

Fig. 11. General view and camera station map

Table 1. Test results of different defects

Defect type	Total	Unqualified	Missing	False	Detection rate%	Noise factor%
Pit	20	20	0	1	100	5
Spalling	20	20	0	0	100	0
Pitting	20	19	1	3	95	15
speckle	20	19	1	0	95	0

Table 2. Multiple test results

Number	Total	Qualified	Unqualified	Pass rate %	Average test time
First time	100	21	79	21	1.51
Second time	100	20	80	20	1.43
Third time	100	22	78	22	1.52

According to the above table, the detection efficiency and recognition accuracy are relatively high, the average detection time is about 1.5 s, and the repeated detection effect is stable, which basically meets the requirements of surface detection. For the most obvious spalling feature, the algorithm has the highest recognition accuracy. The misjudgment rate and missed detection rate of pitting are both high, because pitting is easy to be judged as pits or speckle when the corrosion area or depth is large, which leads to high misjudgment rate. In the continuous test, the main reason for the fluctuation of the pass rate is the accidental collision and surface fingerprints when taking the steel pipe, which indicates that the cleanliness of the surface inspection of the steel pipe to be inspected has a certain impact on the accuracy of the judgment of the steel pipe, and also reflects that the surface of the steel pipe will inevitably have defects in the production process (Table 2).

5 Summary

In this paper, the machine vision method is used to detect the defects on the surface of steel pipe. The seamless steel pipe used for shock absorber is taken as the research object. Combined with the bending high reflection surface of steel pipe, the traditional area array camera image acquisition will cause uneven illumination and information redundancy. An experimental platform composed of linear array camera, linear light source, encoder and rotation system is built to obtain the image of steel pipe. According to the different characteristics of four kinds of defects of steel pipe: pit, pitting, spot and spalling, a defect detection algorithm combining frequency domain transformation, gradient threshold and line detection is proposed and tested. The test results show that the average detection time of the algorithm is about 1.5 s and the average recognition rate is 93%. It can effectively distinguish four kinds of defects and avoid the disadvantages of traditional manual visual inspection.

Acknowledgements. This research was partially supported by the key research project of the Ministry of Science and Technology (Grant No. 2018YFB1306802), the National Natural Science Foundation of China (Grant No. 51975344) and China Postdoctoral Science Foundation (Grant No. 2019M662591).

References

1. Feng, L., et al.: Welding of porosity defects in large diameter thick wall seamless steel pipe. J. Plastic Eng. **000**(004), 102–108 (2014)
2. Jianguo, Z.: Combined nondestructive testing technology and its application in online automatic inspection of seamless steel pipe. Iron Steel **34**(6), 60–64 (1999)
3. Zhou, F., Liu, G., Xu, F., et al.: A generic automated surface defect detection based on a bilinear model. Appl. Sci. **9**(15), 3159 (2019)
4. Renukalatha, S., Suresh, K.V.: Classification of glaucoma usings implified-multiclass support vector machine. Biomed. Eng. Appl. Basis Commun. **31**(05), 1950039 (2019)
5. Kui, L., Manlong, C., Lizhi, Y., et al.: Surface quality inspection method of graphite sealing ring based on machine vision. J. Shanxi Univ. Technol. (Natural Science Edition) **37**(2), 29–34 (2021)
6. Tsai, D.-M., Molina, D.: Morphology-based defect detection in machined surfaces with circular tool-mark patterns. Measurement (2019)
7. Kopineck, H.J., et al.: Automatic surface inspection of continuously and batch annealed cold rolled steel strip. MPT **5**(66), 69 (1987)
8. Xu, K., Xu, J., Ban, X.: Research on pattern recognition method for automatic surface quality monitoring system of cold rolled strip. Iron Steel **37**(006), 28–31 (2002)
9. Li, C.: Image processing technology and its application in strip surface defect detection. Haiphong University of science and technology (2002)
10. Shi, J.: Research on steel plate surface defect detection system based on linear CCD. Yanshan University
11. Wang, H., Zhang, Y., Shen, H., et al.: Overview of image enhancement algorithms. Optics of China, 2017 (4)
12. Huairen, Y., Musheng, Y.: Line extraction algorithm based on improved Hough transform. Infrared Technol. **37**(11), 970–975 (2015)
13. Jiang, L., Han, R., Yuan, Y., Zheng, Y.: Application of Canny operator in fabric defect detection. J. Beijing Inst. Fashion (Natural Science Edition) (4), 57–63
14. Li, X., Li, F., Zhao, R., et al.: Thresholdless window Fourier transform filtering method. Acta PHOTONICA Sinica (2014)

Research on Robot Grinding Force Control Method

MingJian Sun[1,2], Kai Guo[1,2](✉), and Jie Sun[1,2]

[1] Key Laboratory of High-Efficiency and Clean Mechanical Manufacture, National Demonstration Center for Experimental Mechanical Engineering Education, School of Mechanical Engineering, Shandong University, Jinan 250061, China
kaiguo@sdu.edu.cn

[2] Research Center for Aeronautical Component Manufacturing Technology and Equipment, Shandong University, Jinan 250061, China

Abstract. The contact force control between the tool and the workpiece in the industrial robot grinding process is essential to improve the surface quality and processing efficiency of the workpiece. This paper focus on the control of grinding force in the robot grinding process. Firstly, the grinding force control device and method are developed. Then, the force control mathematical model of the system is established and the fuzzy PID strategy of constant force control is especially designed. Finally, the force control tracking experiments has been conducted to verify the system performance. The experimental results show that the contact force control can be controlled smoothly, which can ensure the constant contact force between the workpiece and the tool of the grinding process.

Keywords: Robot grinding · Contact force control · Force control tracking

1 Induction

The degree of industrial automation usually represents the level of manufacturing industry. At present, the grinding and polishing processing of high-precision complex curved workpieces such as aero-engine blades, aircraft composite parts surfaces, and high-precision molds in most regional manufacturing industries are still mainly operated by skilled technicians manually [1, 2], which results in low processing efficiency, poor working environment for technicians, and difficult to guarantee quality consistency. According to the statistics of manufacturing process time distribution, more than 52% of the total manufacturing time is occupied by the grinding and polishing process [3].

In order to improve the efficiency of polishing, ensure the consistency of the processing performance and provide a better working environment, it is very necessary to apply the high degree of freedom and flexible operation of the robot to the processing process [4–7], as early as in the 1980s, Japanese scholars conducted experiments on robot polishing and concluded through experiments that reasonable contact force control can effectively ensure the surface quality of the processed workpiece [8–10]. Most of the current research mainly focuses on the robot's own joint motor adjustment to

X.-J. Liu et al. (Eds.): ICIRA 2021, LNAI 13014, pp. 821–829, 2021.
https://doi.org/10.1007/978-3-030-89098-8_77

achieve force control, using different types of force/position hybrid control to achieve automation of the machining process [11–13], which is difficult and more complicated to control, with lower accuracy and control bandwidth [14–16].

To overcome the shortage of force/position hybrid control, Yongjie Shi [17] proposed a new decoupled polishing system, which can keep the polishing force constant based on the curvature variation of the part surface, however, the magnetorheological torque servo device adopted makes the ratio of system weight to power relatively high and does not guarantee the compliance of the softness of the contact force. Ping W [18] proposed an open polishing robot control system that uses the deformation of the spring to protect the tool and workpiece from damage, which increases the flexibility of the polishing system and can protect the mold surface from damage when the polishing force exceeds a preset value, but the demerit is that due to the passive nature of the spring, it can only function at the contact limit state of the tool and workpiece, and it is also difficult to guarantee a constant contact force control. Pusan University, Korea [19, 20] proposed a flexible polishing tool head, mainly composed of a spring, a universal joint, and a ball spline shaft, which basically realized flexible control of axial contact force, but the control accuracy was relatively low. Zhou [21] developed a force-position decoupled pneumatic servo-polishing force control system, which obtains good real-time polishing force control performance using a neural proportional-integral-derivative controller. Likewise, Fraunhofer IPT [22] designed a fully automated polishing unit based on a pneumatic servo system to achieve a fully automated and repeatable polishing process, solving the problem of time-consuming polishing work and high labor costs.

Given the excellent flexibility of pneumatic actuators [23, 24], this paper proposes an active compliant flange device with a flexible telescopic airbag as the core component, based on a fuzzy PID control method, which can guarantee the constant control of the contact force during polishing. Firstly, the overall structure design of the active compliant flange is carried out, secondly, the force control mathematical model of the system is established, then the fuzzy PID constant force control strategy is studied. Finally, the performance of the system is verified by using continuous step response constant force tracking experiment.

2 Device Structure Design and Kinetics Analysis

2.1 Structure Design

In actual work, the active compliant flange needs to be installed at the end of the robot flange to perform grinding and polishing operations on the workpiece. Considering the limitation of the robot payload and the vibration problem caused by the high-speed rotation of the pneumatic grinding head, the overall design of the device should meet the guidelines of low in weight, compact structure and easy disassembly [25, 26]. The overall structure design of the active compliant flange proposed in this paper is shown in Fig. 1.

The active compliant flange unit is connected to the end of the robot via a flange connector on the left side, and the embedded controller as well as transition plate are integrated into the left flange. The transition plate is used to mount components such as solenoid directional valve, proportional pressure valve, and linear guides. The device

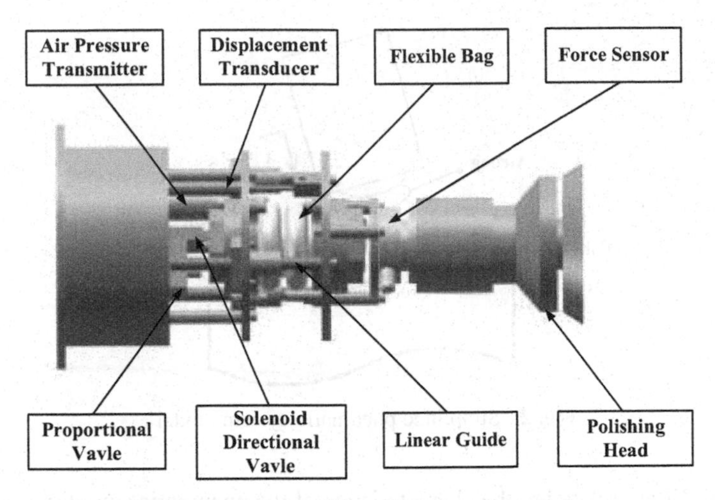

Fig. 1. Active compliant flange structure design

itself has one floating degree of freedom using three sets of linear guides as guiding components and airbags as power actuating components. The rebound LVDT linear displacement transducer shell is fixed on the flange site, and the linear displacement transducer tie rod is fixed on the active side of the airbag, which can read the displacement value of the airbag in real-time, while the air pressure transmitter can read the internal pressure in the airbag with the controller to realize the pre-pressure alarm function. The grinding force is transmitted to the grinding head through the airbag, and the pressure sensor feeds back the grinding force to the controller, which regulates the proportional pressure valve and the solenoid directional valve to control the amount of gas entering the airbag according to the working requirements, thus controlling the pressure inside the airbag and realizing the compliant feedback control of the grinding force.

2.2 Kinetic Modeling

The purpose of mathematical modeling is to abstract the connection between complex objects into a concrete mathematical form and explore the universal law of problem-solving. The mathematical model can accurately express the mathematical change relationship between input and output. In the active compliant flange grinding force control loop, the control input is the excitation voltage of the proportional pressure valve, the system output is the grinding force, A simplified pneumatic system model is shown in Fig. 2.

As depicted in Fig. 2, the main power output component of the system is the flexible telescopic airbag, which relies on its round-trip telescopic motion to provide a grinding force perpendicular to the contact surface. In the output force control, the pressure output from the airbag is mainly controlled, for which the pressure inside the airbag depends on the control of the proportional pressure valve. for the proportional pressure valve, the dynamic relationship between the input voltage u and output pressure P_1 can be assumed as the first-order which is given by [27].

$$\frac{P_1(s)}{U(s)} = \frac{K_1}{\tau s + 1} \tag{1}$$

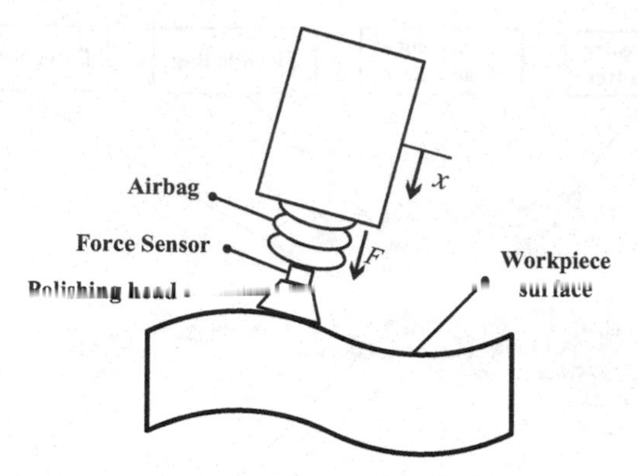

Fig. 2. Simplified pneumatic system model

where τ and K_1 are related to the characteristics of the pneumatic actuator. Considering the motion of the airbag, the force balance equation of the cylinder motion can be derived according to Newton's second law.

$$M\ddot{x} + C\dot{x} + Kx + F_L - G_\alpha = F - F_f \tag{2}$$

where M is the mass of the moving part, X is the displacement of the tool along the normal direction of the working surface, C is the air damping constant, K is the spring coefficient, F_L is the actual contact force between the tool and the workpiece, G_α is the gravitational component of the moving part along the axial direction, F is the airbag output force, and F_f is the sum of the static friction and Coulomb friction. Take Laplace transforms of Eq. (2) yields.

$$Ms^2X(s) + CsX(s) + KX(s) + F_L(s) - G_\alpha(s) = F(s) - F_f(s) \tag{3}$$

where the relationship between the active compliant flange displacement $X(s)$ and the grinding force $F(s)$ can be expressed as follows [28].

$$\frac{F(s)}{Y(s)} = \frac{Km}{G_m(s)} = K_m\left(\frac{s^2}{\omega_n^2} + 2\frac{\zeta}{\omega_n}s + 1\right) \tag{4}$$

where ω_n and ζ are the nature frequency and damping ratio of the model. The values of K_m, ω_n and ζ can be derived from the frequency response of the lightly damped experiments. Due to the nonlinearity nature of the frictional force, F_f can be regarded as an external disturbance [29]. Hence, the transfer function of the system from the input voltage u to the output polishing force F can be derived as

$$\frac{F(s)}{U(s)} = \frac{K_m A_1 K_1\left(s^2 + 2\zeta\omega_n s + \omega_n^2\right)}{\left[\left(K_m + M\omega_n^2\right)s^2 + \left(2K_m\omega_n\zeta + C\omega_n^2\right)s + \omega_n^2(K_m + K)\right](\tau s + 1)} \tag{5}$$

From Eq. (5), it can be concluded that the grinding force control system with the flexible telescopic airbag as the core can be assumed as a third-order system. The modeling of the force control system is thus completed, and the relationship between the control input and the output grinding force of the force control system is established.

3 Control Strategy

The fuzzy PID takes the error e and the rate of change of the error e_c as input and utilizes the preset fuzzy rules to identify the fuzzy relationship between the three parameters of the PID and the controller input. In the mode of operation, the three parameters are modified online to meet the different requirements of the control parameters according to the fuzzy control principle, thus providing an excellent dynamic and static performance of the object under control [30, 31]. In this section, based on the force control mathematical model of pneumatic actuator, the constant force control process of fuzzy PID algorithm is studied and simulation experiments are conducted to verify the effectiveness of the designed algorithm. The fuzzy PID control block diagram for the active compliant flange is given in Fig. 3. The inputs to the fuzzy controller are the force tracking error e and the rate of change of the error e_c, defined as follows.

$$e = F_r - F_L \tag{6}$$

$$e_c = \frac{de}{dt} \tag{7}$$

where F_r is the preset target contact force, and F_L is the actual output grinding force of the active compliant flange detected by the force sensor. The fuzzy controller makes online adaptive adjustment of the controller parameters and outputs the control quantity to the control model of the active compliant flange system, the output grinding force is fed back through the force sensor to achieve the closed-loop accurate control of grinding force.

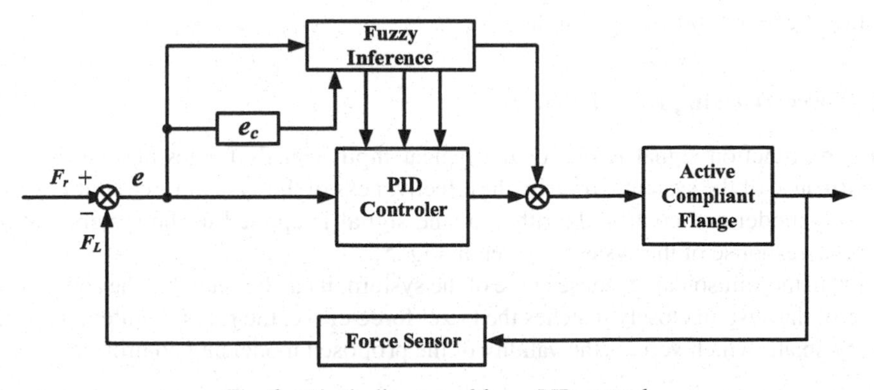

Fig. 3. Block diagram of fuzzy PID control

The system control model is established by Simulink as depicted in Fig. 4, which mainly composed of the internal pressure calculation of the flexible telescopic airbag, online calculation of fuzzy PID control parameters, data transformation, and waveform recording.

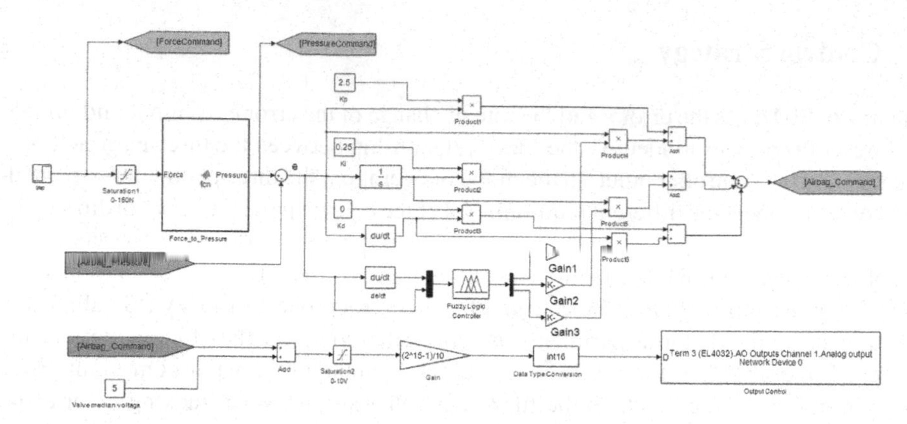

Fig. 4. System control model

4 Experimental Application

4.1 Experimental Platform Description

In the active compliant flange control system, the host computer is PC and the hypogyny-machine is XPC real-time control system, which mainly collects and converts the analog signals from sensors and proportional regulators, distributes the processed position information and contact force information to the host computer, the host computer executes control algorithm and sends the result back to the hypogyny-machine, The hypogyny-machine completes the control of the proportional pressure valve and solenoid directional valve according to the calculation result and then controls the airbag pressure to achieve the target contact force output of the system.

4.2 Force Tracking Experiment

The sine function signal is one of the typical input signals for testing the dynamic performance of the system. To verify the effectiveness of the system force control mathematical model and control algorithm, a sine signal is applied to the system and the dynamic response of the system is given in Fig. 5.

From the sinusoidal response curve of the system, it can be seen that the actual force curve of the system closely matches the target force curve, the pressure jitter amplitude is very small, which verifies the validity of the proposed model and control strategy.

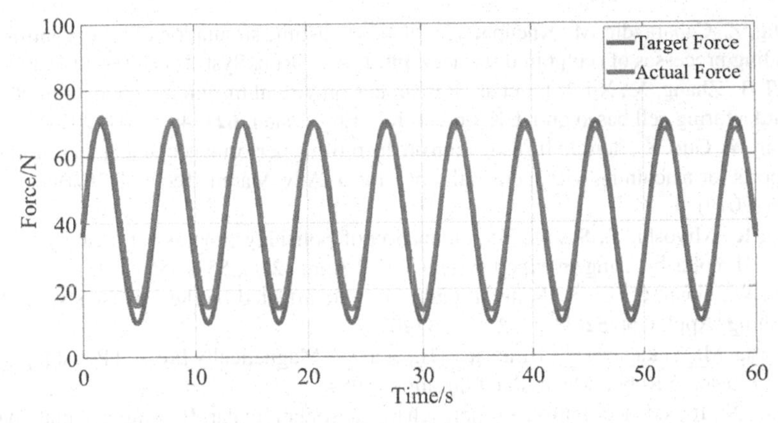

Fig. 5. System sinusoidal input response

5 Conclusion

A robot grinding force control device and method are introduced, and the performance of the system is verified by using variable target force tracking experiments, and the results show that.

1. The active compliant flange device in this paper solves the problem of direct contact between the grinding tool and the workpiece in the traditional grinding process, which improves the grinding quality.
2. In contrast of traditional air cylinders, this paper uses the flexible telescopic airbag as the power component, which is friction-free and can also absorb grinding vibration during the floating and telescoping process. By controlling the internal pressure of the airbag, the active flexible flange can ensure the output of stable grinding force.
3. The experimental results show that the designed fuzzy PID control algorithm has small overshoot, high response speed, and smoother contact force control, indicating that the proposed active flexible flange device and method has excellent constant force tracking performance and can meet the requirements of rapid changes in contact force requirements in actual industrial production.

References

1. Ma, Z., Peng, L., Wang, J.: Ultra-smooth polishing of high-precision optical surface. Optik Int. J. Light Electron Optics **124**(24), 6586–6589 (2013)
2. Birkgan, S.E., Bachurin, V.I.: Computer simulation of the two-stage ion polishing of a silicon surface. J. Surf. Invest. X-ray, Synchrotron Neutron Tech. **8**(3), 524–529 (2014)
3. Saito, K.: Finishing and polishing of free-form surface. Bull. Jpn Soc. Precis. Eng. **18**(2), 104–109 (1984)
4. Dieste, J.A., et al.: Automatic grinding and polishing using spherical robot. Procedia Eng. **63**, 938–946 (2013)

5. Xiang, Z., Cabaravdic, M., Kneupner, K., et al.: Real-time simulation of robot controlled belt grinding processes of sculptured surfaces. Int. J. Adv. Rob. Syst. **1**(1), 109–114 (2008)
6. Yu, T.B., Zhang, X., Xu, X.L., et al.: Researches on virtual machining simulation of flexible manufacturing cell based on KUKA robot. Key Eng. Mater. **621**, 499–504 (2014)
7. Zhang, Y., Guo, K., Sun, J.: Investigation on the milling performance of amputating clamping supports for machining with industrial robot. Int. J. Adv. Manuf. Technol. **102**(9–12), 3573–3586 (2019)
8. Saito, K., Miyoshi, T., Sasaki, T.: Automation of polishing process for a cavity surface on dies and molds by using an expert system. CIRP Ann. **42**(1), 553–556 (1993)
9. Peng, W., Guan, C., Li, S.: Material removal mode affected by the particle size in fluid jet polishing. Appl. Opt. **52**(33), 7927–7933 (2013)
10. Kunieda, M., Nakagawa, T., Hiramatsu, H., et al.: A Magnetically Pressed Polishing Tool for a Die Finishing Robot. Macmillan Education (1984)
11. Hogan, N.: Impedance control: an approach to manipulation: part II—implementation (1985)
12. Zhang, Y., Guo, K., Sun, J., Sun, Y.: Method of postures selection for industrial robot joint stiffness identification. IEEE Access **9**, 62583–62592 (2021)
13. Mason, M.T.: Compliance and force control for computer controlled manipulators. IEEE Trans. Syst. Man Cybern. **11**(6), 418–432 (1981)
14. Guo, K., Pan, Y., Zheng, D., et al.: Composite learning control of robotic systems: a least squares modulated approach. Automatica **111**, 108612 (2020)
15. Tian, F., et al.: Modeling and control of robotic automatic polishing for curved surfaces. CIRP J. Manuf. Sci. Technol. **14**, 55–64 (2016)
16. Guo, K., Pan, Y., Yu, H.: Composite learning robot control with friction compensation: a neural network-based approach. IEEE Trans. Industr. Electron. **66**(10), 7841–7851 (2019)
17. Shi, Y., et al.: NC polishing of aspheric surfaces under control of constant pressure using a magnetorheological torque servo. Int. J. Adv. Manuf. Technol. **58**(9–12), 1061–1073 (2012)
18. Haibo, Z., et al.: A hybrid control strategy for grinding and polishing robot based on adaptive impedance control. Adv. Mech. Eng. **13**(3), 16878140211004034 (2021)
19. Ahn, J.H., Lee, M.C., Jeong, H.D., Kim, S.R., Cho, K.K.: Intelligently automated polishing for high quality surface formation of sculptured die. J. Mater. Process. Technol. **130–131**, 339–344 (2002)
20. Ahn, J.H., Shen, Y.F., Kim, H.Y., Jeong, H.D., Cho, K.K.: Development of a sensor information integrated expert system for optimizing die polishing. Robot. Comput. Integr. Manuf. **17**(4), 269–276 (2001)
21. Zhou, W., et al.: Development of a real-time force-controlled compliant polishing tool system with online tuning neural proportional–integral–derivative controller. Proc. Inst. Mech. Eng. Part I J. Syst. Control Eng. **229**(5), 440–454 (2015)
22. Brecher, C., et al.: Development of a force controlled orbital polishing head for free form surface finishing. Prod. Eng. Res. Devel. **4**(2–3), 269–277 (2010)
23. Tuell, M.T.: Aspheric optics: smoothing the ripples with semi-flexible tools. Optical Eng. **41**(7), 1473 (2002)
24. Ahn, K.K., Anh, H.: Design and implementation of an adaptive recurrent neural networks (ARNN) controller of the pneumatic artificial muscle (PAM) manipulator. Mechatronics **19**(6), 816–828 (2009)
25. Xu, Y., Guo, K., Sun, J., et al.: Design, modeling and control of a reconfigurable variable stiffness actuator. Mech. Syst. Signal Process. **160**, 107883 (2021)
26. Xu, Y., Guo, K., Li, J., et al.: A novel rotational actuator with variable stiffness using S-shaped springs. IEEE/ASME Trans. Mechatron. **1**, 2249–2260 (2020)
27. Liao, L., Xi, F.J., Liu, K.: Adaptive control of pressure tracking for polishing process. J. Manuf. Sci. Eng. **132**(1), 011015 (2010)

28. Güvenç, L., Srinivasan, K.: Force controller design and evaluation for robot-assisted die and mould polishing. Mech. Syst. Signal PR **9**(1), 31–49 (1995)
29. Fan, C., Hong, G.S., Zhao, J., Zhang, L., Zhao, J., Sun, L.: The integral sliding mode control of a pneumatic force servo for the polishing process. Precision Eng. **55**, 154–170 (2019)
30. Lu, J., Chen, G., Hao, Y.: Predictive fuzzy PID control: theory, design and simulation. Inf. Sci. **137**(1–4), 157–187 (2001)
31. Guo, K., Li, M., Shi, W., et al.: Adaptive tracking control of hydraulic systems with improved parameter convergence. IEEE Trans. Ind. Electron. 1 (2021)

Research on Recognition Algorithm of Scattered Rivets Based on Feature in Robot Automatic Nail Feeding System

Lin Liu, Zhiwei Zhuang, Pengcheng Li, Wei Tian(✉), and Chao Wu

Nanjing University of Aeronautics and Astronautics, Nanjing, China
tw_nj@nuaa.edu.cn

Abstract. Robot vision-based automatic nail feeding system is a research hotspot in the field of automatic drilling and riveting for aircraft assembly. As for traditional robot vision-based automatic nail feeding system, it is difficult to recognize scattered rivets. This paper proposes the feature-based algorithm for scattered rivets recognition for monocular vision-based nail feeding system. Firstly, the image of scattered rivets is pre-processed and the division method of maximum entropy threshold is employed to divide the reflective band of rivet. Then, through morphological filtering and other operations to get the size of the reflective band. Finally, compare the size of the reflective tape with the size characteristics of the rivet itself, and finally achieve the purpose of rivet identification.in this paper has a strong robustness for the blocking state and brightness variation of scattered rivets. Besides, it has scale invariance and rotational invariance. This algorithm has a quick speed and high recognition rate and accuracy, and meets the requirement for online rivet recognition of robot vision-based automatic nail feeding system.

Keywords: Robot vision · Automatic nail feeding system · Morphological filtering · Feature recognition

1 Introduction

In order to ensure the automation degree and work efficiency of the automatic drilling and riveting, the automatic transmission of aviation rivets is naturally involved in the automatic drilling and riveting technology. With the continuous improvement of the riveting rate of aircraft assembly machines and the continuous development of automatic drilling and riveting technology, higher requirements have been correspondingly put forward for automatic nail feeding technology. A study by the French AHG Company shows that 90% of the failures of the automatic drilling and riveting system are caused by the nail feeding system [1]. It can be seen that the automatic nail feeding technology plays a very critical role in the automatic assembly of aircraft.

The AHG company has developed a storage box nail feeding system. The rivets are placed in the coil of the storage box in an orderly manner. The code plate of the storage box records the information of the stored rivets and stores the information of different types of rivets. Multiple storage boxes are installed in the storage box rack [2]. American

X.-J. Liu et al. (Eds.): ICIRA 2021, LNAI 13014, pp. 830–841, 2021.
https://doi.org/10.1007/978-3-030-89098-8_78

EI company has developed a box type nail feeding system, a drawer type nail feeding system and other nail feeding devices. The nail feeding system occupies a large space and the types of rivets conveyed are small. In recent years, domestic research has also been conducted on automatic nail feeding technology. Zhejiang University has developed a vertical pipe nail feeding system [3] and a drawer type nail feeding system [4]. They need to pre-arrange and load the rivets into the nail storage tube through a vibrating hopper or manual methods, this method leads to low system flexibility. Nanjing University of Aeronautics and Astronautics has developed a vertical array pipeline nail delivery system [5], which can adapt to the delivery of various types of rivets, but due to the limited length of the nail storage pipeline, the number of rivets that can be stored is small. In order to improve the flexibility of the nail feeding system, Nanjing University of Aeronautics and Astronautics took the lead in developing an automatic nail feeding system based on a vision robot [6], using a small industrial robot equipped with a binocular vision system and a nail grabbing end effector. Because the scattered rivets block each other, the two images collected by binocular vision are quite different. The solution cannot identify the rivets that are stacked interlaced with each other. In order to solve this problem, Nanjing University of Aeronautics and Astronautics has also developed an automatic nail feeding system based on monocular vision [7], which uses a monocular camera and a laser displacement sensor to realize the identification and positioning of scattered rivets.

Aiming at the difficulty in identifying scattered rivets in the automatic nail feeding system based on monocular vision, this paper proposes a feature-based scattered rivets recognition algorithm, which divides the reflective band area of the rivet from the image and calculates the size of the rivet reflective tape which is compared with the size characteristics of the rivet itself, so as to achieve the purpose of rivet identification.

2 Preprocessing of Scattered Rivets Images

Figure 1 shows the final effect of processing rivets images with different filtering methods. Bilateral filtering used in this paper considers the spatial distance of the pixel and the gray value of the pixel at the same time, which can achieve the purpose of edge preservation and denoising.

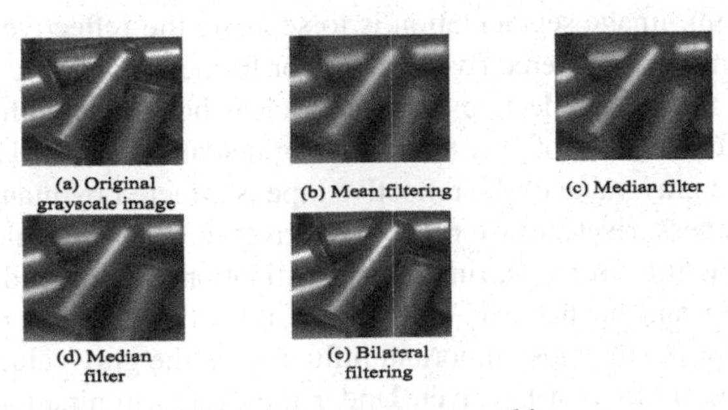

Fig. 1. Rivet image smoothing

3 Random Rivet Recognition Algorithm Based on Feature

The traditional template matching algorithm is one of the most basic image recognition methods. Because of its simplicity, practicality and robustness, it has been widely used in the field of industrial inspection. However, the template matching algorithm needs to search the entire image. In this search area, most of the image backgrounds are meaningless, which causes a lot of waste of computing resources. In addition, because the rivets in the storage box are staggered and stacked with different postures, it is difficult for the template matching algorithm to be suitable for the recognition of scattered rivets. To this end, this paper proposes a feature-based scattered rivets recognition algorithm. By segmenting the reflective band area of the rivet from the image as the recognized feature, the feature is compared with the size feature of the rivet itself, so as to achieve the purpose of identifying rivets.

3.1 Segmentation of Rivet Images

Figure 2 is an image of scattered rivets in the storage box. It can be seen from the figure that there is an obvious bright reflective tape on the central axis of the rivet rod. The size of the reflective tape is related to the size of the rivet, and there is a clear difference between the gray value and the background, so it can be used as a feature for rivet recognition.

Fig. 2. The image of scattered rivets

In this algorithm, image segmentation is to separate the reflective tape of the rivet from the background in a scattered rivets image for further processing. In this example, the gray value of the rivet reflective tape has a clear boundary with the background gray value. Therefore, the use of gray threshold segmentation can well separate the rivet reflective tape. The gray value of the reflective tape is affected by many factors such as illumination brightness, rivet surface reflectivity, rivet inclination angle, etc.

When taking pictures of rivets, ring light illumination is used, and the illumination brightness is stable and unchanged. The surface reflectivity of the rivet can also be regarded as the same, so the most important influence on the gray value of the reflective tape is the inclination angle of the rivet. Under the same illumination conditions, the rivets are tilted at different angles, and the grayscale changes of the rivets are shown in Fig. 3. In order to distinguish the grayscale value distribution more clearly, the rivet image is pseudo-colorized by grayscale layering method, the degree value is divided into 6 Gy-scale intervals, and each gray-scale interval is mapped into six colors of black, red, green, blue, yellow, and white according to the gray value from small to large. It can be seen from the figure that with the tilt angle increases, the brightness of the reflective tape gradually decreases. When the tilt angle is greater than 15°, as the tilt angle increases, the shape of the reflective tape gradually degenerates from a rectangle to a slender triangle. Therefore, only when the rivet tilt at 15°, the ideal reflective tape area can be segmented.

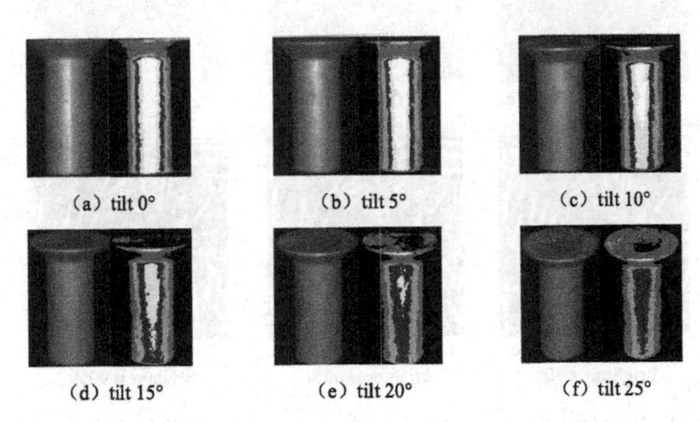

Fig. 3. The gray changes of rivets under different tilt angles

The most critical step of gray-scale threshold segmentation is to determine the segmentation threshold t, and use the threshold t as a dividing line to segment the target area and the background area. There are many methods to determine the segmentation threshold t. The first is the histogram threshold bimodal method. When the grayscale image is relatively simple and the grayscale distribution of the target object is relatively regular, the background and the target object respectively form two crests on the grayscale histogram of the image, and the gray value of the trough between the two crests is used as the segmentation threshold t, which can separate the background and the target object. This method is very sensitive to changes in illumination. When the brightness of the light changes, the trough's position also shifts accordingly. At this time, the size of the segmentation threshold t needs to be adjusted artificially to adapt to the change of illumination. The second method is to use maximum entropy threshold segmentation [13].

The maximum entropy threshold segmentation method uses the concept of entropy in information theory, so that the selected segmentation threshold t can maximize the amount of information in the grayscale statistics of the target area and the background area. The maximum entropy threshold segmentation method is not sensitive to changes in lighting brightness. Figure 4 shows the images collected under different lighting brightness and the results of maximum entropy threshold segmentation on these images. It can be seen that although the lighting brightness changes greatly, the binary images obtained by the maximum entropy threshold segmentation are almost exactly the same. Therefore, the maximum entropy threshold segmentation method is robust to changes in lighting brightness. The maximum entropy threshold segmentation method can automatically calculate the segmentation threshold t, eliminates the tedious operation of manually setting the segmentation threshold. It can also be seen from Fig. 4 that the maximum entropy threshold segmentation method successfully completes the separation of the rivet reflective band from the background, and there are many short rod-shaped connected domains in the binary image, which creates good conditions for later feature extraction.

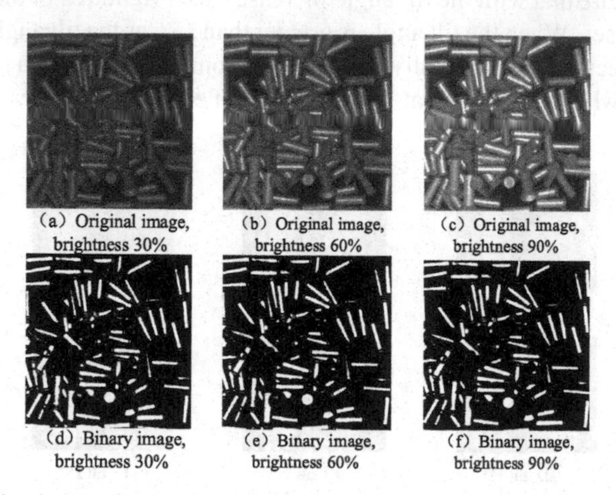

Fig. 4. The results of maximum entropy threshold segmentation

3.2 Morphologic Filter

Although the binary image after threshold segmentation has been able to separate the rivet reflective tape from the background, however, it can be seen from Fig. 4 that there are still many white spots in the binary image. In addition, the rivets that overlap each other have a small distance between their reflective tapes, these are not conducive to later feature extraction and recognition, Therefore, it is necessary to remove white spots and separate the reflective tapes of adjacent rivets from each other, and morphological processing can solve this problem well.

Open operation is a commonly used morphological processing method, it can remove white spots through the operation of first corrosion and then expansion and make the remaining highlight areas are better separated, and can keep its size unchanged. The open operation can handle the white spots in the rivet image and separate the adjacent rivet reflective tapes. Using a circular kernel with a radius of 5 pixels, the scattered rivets image is open operated. The images before and after the processing are shown in Fig. 5. It can be seen that the white spots are basically eliminated, and the separation effect of adjacent rivets is obvious.

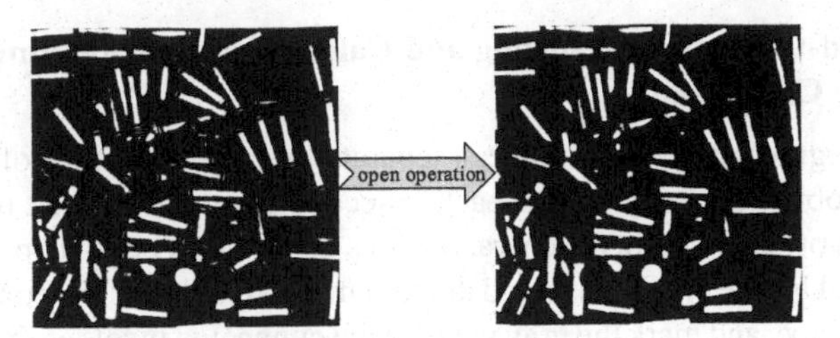

Fig. 5. Open operation

3.3 Removal of Small Blocks

Although the open operation can eliminate the white spots, there are still many small bright blocks in the image. These small blocks are formed by the reflection of the rivet rods that are blocked. Although they are also reflective tapes for nail rods, they obviously do not meet the requirements for rivet grabbing, so it is necessary to remove these small bright blocks. The area of the small block is much smaller than that of the complete nail rod reflective tape, so the area of the small block can be used as the dividing condition to remove the small block. First, we mark the connected domains, and count the number of pixels of each connected domain as the area, and remove the connected domains whose area is less than a certain threshold. The threshold can be estimated by the following formula:

$$t_s = \frac{\eta^2 d(L - H)}{5} \tag{1}$$

Among them: d is the diameter of the nail rod, η is the scale factor, which represents the number of pixels in the pixel coordinate system corresponding to the length of 1mm in the world coordinate system, L is the total length of the rivet and H is the height of the countersunk head, L-H is the length of the shank.

Figure 6 shows the effect before and after the treatment. It can be seen that the complete nail rod reflective tape is well preserved.

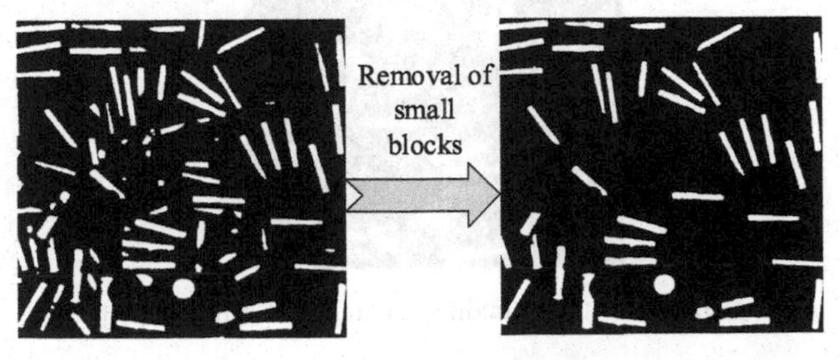

Fig. 6. Removal of small blocks

3.4 Connected-Component Labeling and Calculation of Minimum Bounding Rectangle Calculation

The threshold segmentation algorithm can separate the reflective strips of rivets from the background to obtain a binary image, each reflective strip is separated from each other on the image. In order to obtain the characteristics of each reflective strip, they need to be separated mark. Use the eight connected domain method to define the connected domain of the reflective tape, and mark the gray value of the connected pixels as the same number, in the subsequent processing, each rivet reflective tape can be distinguished according to the gray value of the pixel, so that each reflective tape is individually calculated. Figure 7 shows the connected domain labeling result of the binary image of scattered rivets. A total of 39 regions are marked.

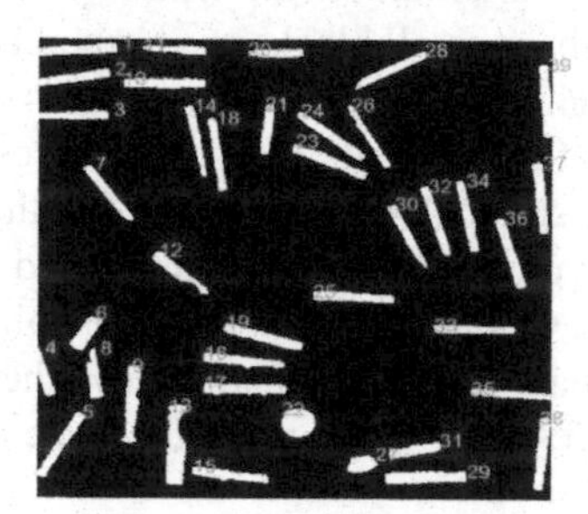

Fig. 7. Connected domain labeling result

According to the calculation method of the minimum bounding rectangle of convex polygon studied by Toussaint, the minimum bounding rectangle of each connected domain is calculated according to the evaluation index of the smallest area, and the positions of the four vertices of the rectangle can be obtained. The red rectangular box in Fig. 8 is the calculated minimum bounding rectangle.

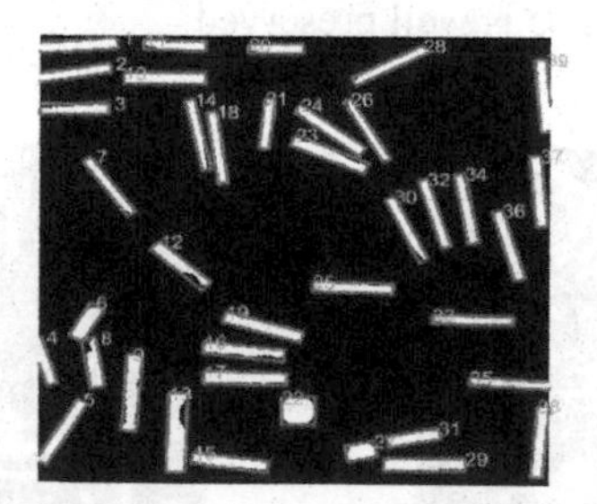

Fig. 8. Minimum bounding rectangle of reflective tape

3.5 Extraction and Recognition of Rivet Features

It can be seen from Fig. 8 that the slender reflective belt is the expected rivet position, but there are still some short and thick areas in the figure., which are formed by the rivet head or the covered rivet rod. These parts do not meet the rivet grasping requirements, so it is necessary to remove them. The length L_k and width W_k of the rectangle can be calculated from the 4 vertices of the minimum circumscribed rectangle. The length $\mathrm{L_k} L_k$ and width W_k of the minimum circumscribed rectangle are used as the characteristics of rivet recognition. These thick and short areas can be removed, and finally only the slender reflective tape is retained. The judgment conditions for the long reflective tape are:

$$\begin{cases} \dfrac{L_\mathrm{k} \cdot d}{W_\mathrm{k} \cdot (L - H)} \geq 1.2 \\ 0.5 \leq \dfrac{L_\mathrm{k}}{\eta (L - H)} \leq 1.5 \end{cases} \tag{2}$$

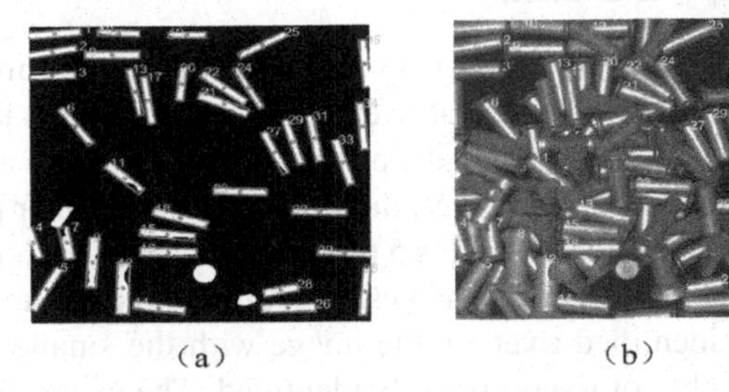

Fig. 9. The result of feature recognition

Figure 9 (a) shows the rivet image after removing the thick and short areas. Rivets that are partially covered do not affect the grasping, because we use a vacuum suction cup with a small diameter to suck the rivets. After removal, there are still 36 reflective tape areas. Figure 9 (b) is the final rivet recognition result. The identified position and the rivet in the original image Comparing the positions, it is obvious that the positions of the two are completely matched, indicating that the feature-based rivet recognition algorithm is accurate and has a high recognition rate.

In application scenarios with many types of rivets, if the traditional template matching algorithm based on the teaching model is used, a lot of effort will be spent in the early stage to teach the standard template for each rivet. The quality of the template taught is also heavily dependent on the operator's cognitive level of machine vision, and this algorithm is sensitive to lighting changes, which means that once the lighting environment changes, the template taught in the previous period is no longer applicable. Using the feature-based rivet recognition algorithm proposed in this article, only the size of the rivet and camera parameters are required. The gray-scale segmentation threshold, area segmentation threshold, and feature judgment conditions involved in the algorithm are all automatically calculated. No matter how many types of rivets, the camera parameters are the same. You only need to set it once. The dimensions can be obtained directly from the rivet database. In addition, the algorithm is robust to changes in illumination. Therefore, the recognition of scattered rivets can be realized by one-click operation without the need to set those tedious image processing parameters, so that even ordinary operators who do not understand machine vision, it is also easy to use the automatic nail feeding system based on robot vision.

4 Conclusion Verification of Feature-Based for Scattered Rivet Recognition Algorithm

The verification of the feature-based scattered rivet recognition algorithm is divided into three parts: algorithm validity verification, robustness verification of lighting brightness changes and algorithm real-time verification. In this paper, five kinds of rivets with large differences in shape and size are selected for experiments to verify the performance of the algorithm.

4.1 Algorithm Validity Verification

The validity verification of the algorithm is to verify that the algorithm can accurately identify scattered rivets. First put 5 kinds of rivets in 5 boxes, and then put 20% of rivets in each box for the first time. After taking the picture, continue to add about 20% of the rivets to the rivet box, and then stir the rivets thoroughly. Take another picture to obtain the second set of rivet images. By doing this 5 times, 5 images of each type of rivet in a different stacked state can be obtained. Then use recognition algorithm to identify each picture. And frame the identified rivets in the image with the smallest circumscribed rectangle. Count the number of rivets correctly identified. The experimental results are shown in Fig. 10. By comparing the data of 5 groups of the same type of rivets, it can be seen that the collection heights of the 5 groups of images are different. The size of each rivet image is also different. However, the number of marked rivets is not much different. This shows that the algorithm has a certain scale invariance. The rivets in the box are scattered randomly. The posture of the rivet has a lot of randomness. But the algorithm can still recognize rivets in various poses. This shows that the algorithm

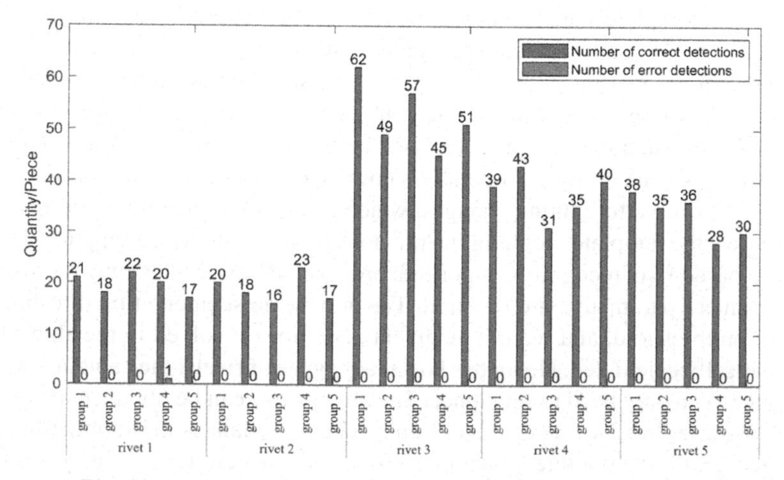

Fig. 10. Recognition results of scattered rivets based on features

is suitable for rivet images in different stacking states. Therefore, the feature-based scattered rivet recognition algorithm is effective. The algorithm has scale invariance and rotation invariance, high stability and strong practicability.

4.2 Robustness Verification of Lighting Brightness Changes

Choose a type of rivet to verify the robustness of the algorithm in changing the brightness of thc lighting, fill the storage box with rivets, and keep the rivets static. Place a sheet of monochrome paper in the camera's field of view to ensure that the monochrome paper does not block the rivets. The brightness of the ring light source is calculated from the gray value of the monochrome paper area and converted to a percentage. Adjust the brightness of the light source to gradually increase from 0% to 100%. And collect 20 rivet images at the same time. Then use the feature recognition algorithm to identify each rivet picture, and use the smallest circumscribed rectangle to compose the rivet identified in the original image. The number of correctly identified rivets and incorrectly identified rivets are counted. The experimental results are shown in Fig. 11. It can be seen that as long as the illumination brightness is greater than 10%, the feature recognition algorithm can identify a sufficient number of rivets. Experiments prove that the feature recognition algorithm proposed in this paper is robust to changes in lighting brightness.

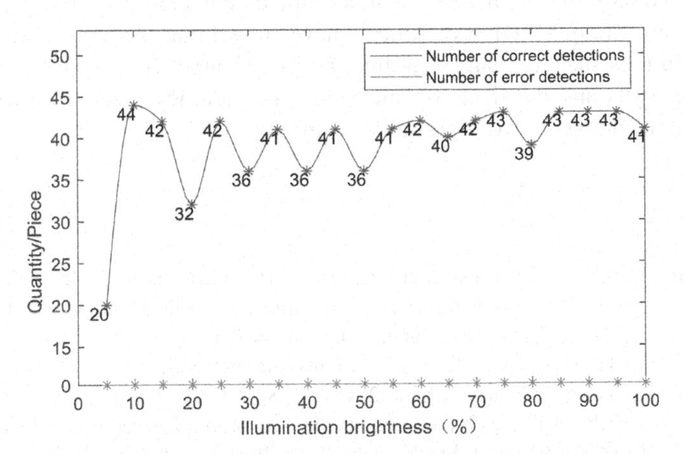

Fig. 11. Recognition results of the algorithm under different illumination brightness

4.3 Algorithm Real-Time Verification

Choose a type of rivet for testing. Before taking each photo, stir the rivets thoroughly and take photos to obtain 50 rivet images. Apply the algorithm to process these 50 images. The result is shown in Fig. 12. The algorithm takes up to 75 ms and the average time is 52.6 ms. Automatic nail feeding systems usually require less than 500 ms of image processing time. It can be seen that the algorithm can fully meet the high real-time requirements of system operation.

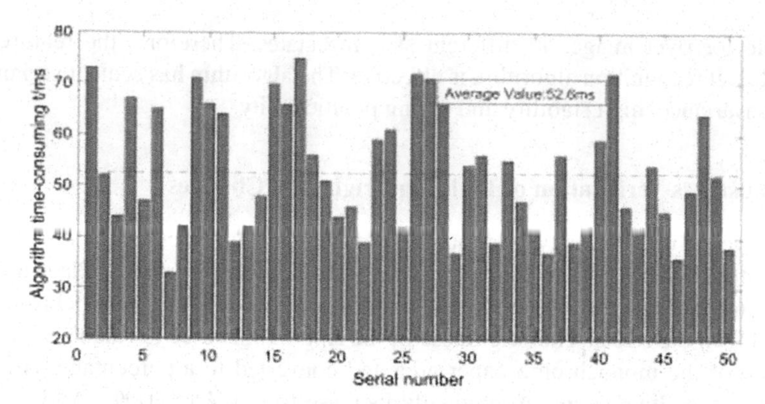

Fig. 12. Time consumption of feature recognition algorithm

5 Conclusion

Since the scattered rivets is difficult to be identified by the monocular vision-based robot feeding system, a feature-based random rivet recognition algorithm is proposed. The preprocessing method of the rivet image is studied. After the preprocessing, the contrast of the light and dark of the image is clear, and the characteristics of the key parts have been effectively strengthened; Experiments have shown that the algorithm is robust to adverse conditions such as changes in illumination, changes in rivet poses, and mutual occlusion of rivets and the rivet size measurement accuracy meets the requirements, ensuring that the delivered rivets are fully qualified.

References

1. Li, H.: Design and research of automatic drilling and riveting terminal device and its automatic nail supply device based on dual-robot collaboration. M.S. thesis, Department Mechanical Engineering, Zhejiang University, Hangzhou, China (2010)
2. Jiang, T., Fang, H., Dong, X.L.: Construction of visual inspection system for riveting quality of aircraft skin. Aeron. Manuf. Technol. **2017**(10), 88–91 (2017)
3. Long, H., Zhu, B.R., Fang, Q.: Experimental study of automatic rivet feeding system of robot automatic drilling & riveting. J. Mech. Electr. Eng. **29**(04), 404–408 (2012)
4. Zhang, J.X., Zhu, W.D.: Pneumatic rivet feeding technology for automatic aircraft panel drilling and riveting machine. Acta Aeronautica ET Astronautica Sinica. **39**(01), 304–313 (2018)
5. Zhou, Z.F., Tian, W., Liao, W.H.: Research on rivets auto-feeding system for aircraft components assembly line. Aviat. Precis. Manuf. Technol. **50**(1), 34–37+30 (2014)
6. Li, Y.F., Tian, W., Li, B.: A robot rivet feeding system for automatic drilling and riveting. Aeronaut. Manuf. Technol. **62**(10), 44–50 (2019)
7. Tian, W., Zheng, Y., Zhuang, Z.W., Liao, W.H., Zhang, L., Li, B., Hu, J.S.: An automatic nail feeding system and method based on robot vision. CN Patent 110523909A, 03 Dec 2019
8. Yu, T.W., Peng, X., Du, L.Q., Chen, T.H.: Real-time detection of parts by assembly robot based on deep learning framework. Acta Armamentarii. **41**(10), 2122–2130 (2020)

9. Boularias, A., Kroemer, O., Peters, J.: Learning robot graspingfrom 3-D images with markov random fields. In: Proceedings of the 2011 IEEE/RSJ International Conference on Intelligent Robots and Systems, San Francisco, USA, pp. 1548–1553. IEEE (2013)
10. Schmitt, R., Cai, Y.: Recognition of dynamic environments for robotic assembly on moving workpieces. Int. J. Adv. Manuf. Technol. **71**(5–8), 1359–1369 (2014)
11. Papazov, C., Haddadin, S., Parusel, S.: Rigid 3D geometry matching for grasping of known objects in cluttered scenes. Int. J. Robot. Res. **31**(4), 538–553 (2012)
12. Schmidt, B., Wang, L.: Automatic work objects calibration via a global–local camera system. Robot. Comput. Integr. Manuf. **30**(6), 678–683 (2014)
13. Kapur, J.N., Sahoo, P.K., Wong, A.K.C.: A new method for gray-level picture thresholding using the entropy of the histogram. Comput. Vis. Graph. Image Process. **29**(3), 273–285 (1985)

Author Index

Printed in the United States
by Baker & Taylor Publisher Services

Printed in the United States
by Baker & Taylor Publisher Services